ISBN 978-0-282-90365-7
PIBN 10872591

FB &c Ltd, Dalton House, 60 Windsor Avenue, London, SW19 2RR.
Company number 08720141. Registered in England and Wales.

For support please visit www.forgottenbooks.com

UNITED STATES COMMISSION OF FISH AND FISHERIES.

PART VI.

REPORT

OF

THE COMMISSIONER

FOR

1878.

A.—INQUIRY INTO THE DECREASE OF FOOD-FISHES.

B.—THE PROPAGATION OF FOOD-FISHES IN THE WATERS OF THE UNITED STATES.

WASHINGTON:
GOVERNMENT PRINTING OFFICE.
1880.

LETTER

FROM THE

COMMISSIONER OF FISH AND FISHERIES,

TRANSMITTING

His report for the year 1878.

JANUARY 13, 1879.—Ordered to lie on the table and be printed.

UNITED STATES COMMISSION, FISH AND FISHERIES,
Washington, D. C., January 9, 1879.

GENTLEMEN: I have the honor to transmit herewith my report for the year 1878, as United States Commissioner of Fish and Fisheries, embracing, first, the result of inquiries into the condition of the fisheries of the sea-coast and lakes of the United States; and, second, the history of the measures taken for the introduction of useful food-fishes into its waters.

Very respectfully, your obedient servant,

SPENCER F. BAIRD,
Commissioner.

Hon. WM. A. WHEELER,
President United States Senate, and

Hon. S. J. RANDALL,
Speaker of the House of Representatives.

CONTENTS.

I.—REPORT OF THE COMMISSIONER.

A—GENERAL CONSIDERATIONS.

B.—INQUIRY INTO THE HISTORY AND STATISTICS OF FOOD-FISHES.

C.—THE PROPAGATION OF FOOD-FISHES.

D.—THE RELATION OF FISH CULTURE TO THE AMERICAN FISHERIES.

E.—TABLES OF DISTRIBUTION OF FISH FOR 1878.

II.—APPENDIX TO REPORT OF COMMISSIONER.

APPENDIX A.—PATENTS; LEGISLATION; PROTECTION.

APPENDIX B.—FISHERY EXPOSITIONS.

* The eggs of the California salmon were hatched out in 1878 but not distributed, for the most part, until 1879. The hatching and distribution were made by the State fishery commissions, except when otherwise stated. The imperfections of the returns will be remedied in the next report.

† Not hatched until 1879. The hatching and distribution made by the New Jersey Fish Commission. The eggs were collected by F. N. Clark, of Northville, Mich.

APPENDIX C.—THE SEA FISHERIES.

APPENDIX D.—DEEP-SEA RESEARCH.

APPENDIX E.—THE NATURAL HISTORY OF MARINE ANIMALS.

* The species involved is the Quinnat salmon (*Salmo quinnat*).

---

* Published in 1876.

APPENDIX H—MISCELLANEOUS.

REPORT OF THE COMMISSIONER.

A.—GENERAL CONSIDERATIONS.

1.—INTRODUCTORY REMARKS.

The present report is intended to include an account of the operations of the United States Fish Commission for the calendar year 1878, although the history of a portion of its work, especially that connected with the propagation of salmon, is continued to the date of the actual planting and disposition of the young fish in 1879. It constitutes the sixth volume of the series, although relating to the eighth year of the existence of the Commission.

As in previous years, the history of the work of 1878 shows a continued increase in the scale of operations, commensurate with the increased appropriations made by Congress. This, however, has involved no material addition to the expense of the management, the clerical force remaining the same, notwithstanding an enormous increase of correspondence, especially with the fish commissioners of States, fish culturists, and generally persons interested in having private or public waters supplied with the fish covered by the work of the Commission.

In the accompanying reports, the operations of the Commission will be treated, as heretofore, under two heads—Inquiry and Propagation.

Under the former is included the history, condition, and statistics of the great fisheries and the proper methods of prosecuting them; and with this are closely connected questions as to the natural or adventitious causes influencing the abundance of fish and the methods by which such abundance may be increased.

Under the second head is given the history of measures taken to actually increase the supply of desirable fishes in particular waters, either by artificial propagation, or by transfer from other localities, or both combined.

The first-mentioned division of the work, including research into the character of the fishes belonging to the North American fauna, has been in charge of Mr. G. Brown Goode, assisted by Dr. T. H. Bean. The collection and investigation of marine invertebrates has been conducted by Prof. A. E. Verrill, assisted by Mr. Richard Rathbun, Mr. Sanderson Smith, and Mr. Warren Upham. The work of propagation of food-fishes was under the superintendence of Mr. James W. Milner, assistant commissioner, aided by Mr. Frank N. Clark, and with the very valuable co-operation of Mr. T. B. Ferguson, Fish commissioner of Maryland. To all these gentlemen I am under great obligations for efficient assistance in carrying out the objects of the Commission. To Dr. Farlow I am indebted for an important research upon the peculiar reddening of salted codfish, to which further reference will be made.

2.—OBJECTS OF THE UNITED STATES FISH COMMISSION.

It had been my intention in this report to go into very minute details in explanation of the plan of research adopted by the Commission for carrying out its objects and the actual results that have been accomplished. This, however, will be more conveniently deferred until the next report, which will chronicle some important changes; and I will here present only a brief synopsis of the subject, in illustration of the extent of the general programme and the amount of labor involved in carrying it out; as also shadowing forth the benefit to American fisheries to be hoped for as the result of such action. Not the least important feature in the research is the securing of statistics for the proper treatment of international questions connected with the common use, by the United States and the British Provinces, of the waters of the North Atlantic.

The results hoped for by the inquiries initiated or contemplated may be summarized as follows:

1. The preparation of a series of reports upon the various groups of aquatic animals and plants of North America, especially those that have a direct relation to the wants or luxuries of mankind; these to be published as monographs in successive volumes of the Commission, to be illustrated by wood-cuts and otherwise, as may be necessary for the proper comprehension of the subject. The aim, of course, will be to present the descriptions of the various species in intelligible phraseology, and to add accounts of the habits and peculiarities of the species, with their relation to each other and the physical conditions of their surroundings. This will include, among others, an illustrated history of the various food-fishes of the United States, and towards which great progress has been made, especially in the preparation of a large number of admirable illustrations, executed by Mr. H. L. Todd.

2. The utilization of the very extensive facilities at the command of the commission in the interest of educational and scientific establishments in the United States, by securing large numbers of specimens of aquatic animals and plants which, after reserving the first series for the National Museum, will be distributed, properly labelled, to colleges and academies and scientific societies. A vast amount of material of this kind has already been gathered, and is now in the hands of specialists, who are engaged in preparing it for the treatment referred to. It is hoped the coming year to distribute many hundreds of thousands of specimens.

3. A complete account of the physical character and conditions of

the waters of the United States, as to chemical composition, temperature, &c., with special reference to their availability in nurturing the proper species of food-fishes.

4. A history and description of the various methods employed in North America, in the pursuit, capture, and utilization of fishes and other aquatic animals, with suggestions as to imperfections of existing methods and the presentation of devices and processes not hitherto adopted in the United States. A careful study of all the circumstances connected with this division of the proposed work of the Fish Commission has shown that an exchange of experiences may be of very great importance in improving the old fisheries and developing new ones. Several methods of fishing employed in Europe and unknown in the United States can be introduced to very great advantage; but so far no special effort has been made to bring this about. Among the noteworthy of these is the system of beam trawling, so universal and so productive in Europe, and by means of which the flat fishes, especially the turbot and sole, are obtained in immense quantities, in otherwise unproductive localities and at moderate cost. It may almost be said that there is a larger investment in this fishery than in any other in Great Britain; and yet it is practically entirely unknown in the United States, its use having been confined to the operations of the United States Fish Commission, of Professor Agassiz, and perhaps the Chicago Academy of Sciences. There is no doubt that beam trawling will add enormously to the facilities for procuring wholesome food at a very cheap rate. The sandy coast of the United States, especially south of Cape Cod, is pre-eminently adapted to the use of this apparatus, and there are thousands of square miles over which it can be carried with no possibility of exhausting the supply.

Another method of fishing, in great part unknown, or at least unpracticed in the United States, is that of taking codfish by means of gill-nets. Could this be introduced on our shores, especially in connection with the vast schools of cod that come in winter on our coast to spawn, it would relieve fishermen of their great embarrassment, namely, that of procuring bait. During the winter season it is frequently almost impossible to obtain bait of the proper kind, and without which fish cannot be taken. In the Loffoden Islands there is a fishery very similar to that referred to, in which, during the winter, large numbers of fish are taken, one-half of which, and these the finest and fattest, are caught in gill-nets without any bait whatever. The American methods of the treatment of fish in preparation for market can also be greatly improved by adopting foreign experience.

5. Statistics of the various branches of the American fisheries from the earliest procurable dates to the present time, so as to show the development of this important industry and its actual condition. There is no nation so badly provided with such statistics as the United States; and in the absence of appointed methods of gathering them the task

will be a very onerous one, but the later it is deferred the more and more difficult will it be, with but little on record. Old men, still living, alone possess the traditions in regard to the existence and progress of many of our most important elements of the fisheries, and it has been a special object of the Commission, at its several stations, to find such depositories and to collect, by the help of a phonographer, all the facts they can furnish, as also to overhaul old account-books and other memoranda more or less fugitive in their character. A great amount of such history has already been secured, especially in regard to the mackerel-, cod-, and halibut-fisheries.

6. The establishment either by the general government or in connection with the States of a thoroughly reliable and exhaustive system of recording fishery statistics for the future, to be combined annually and published by some of the public departments of the government. Something of this kind is done by the Treasury Department for a few branches of the fisheries, but the result is necessarily inadequate and incomplete.

7. The bringing together in the National Museum not only of a complete collection of the aquatic animals and plants referred to, but of illustrations of all apparatus or devices used in the prosecution of fisheries at home and abroad, together with specimens of the results.

In the winter of 1874–'75 Congress made an appropriation to enable the Departments of the Government to present at the International Exhibition at Philadelphia a complete display of the resources of the United States. A portion of this fund having been assigned to the Fish Commission, the occasion was embraced to commence such a collection as that referred to. This was exhibited at Philadelphia and was highly appreciated. Since that time every opportunity has been made use of to secure additional objects of the same kind, showing the earlier and perhaps obsolete methods and applications, as well as those that are now in current use. To these have been added illustrations of the methods and apparatus of artificial propagation of fishes, or of technical fish culture.

8. An investigation of the movements and habits of the various kinds of fish, to serve as a basis of legislation, either by the general government or by the States.

It is very difficult to establish data of this kind upon facts furnished by any one State; it is only by considering the subject in its relations to the whole country that an equitable system of legislation can be suggested. Dates and conditions that answer admirably for one part of the country will be entirely unsuited to another, especially so far as relates to the periods during which fish should not be taken. The question, too, of keeping open the natural channels of the water, so that fish may ascend to their source, is one that will generally require the action of the general government.

Other inquiries involved are the introduction into the water of substances injurious to fish, either of a mechanical or chemical nature, &c.

A corollary to the above is the determination of the best form of fishways for the different conditions of American rivers, of methods of chemical or mechanical purification of the waters, &c.

9. By means of the information to be thus obtained, it will be possible to determine what regulations shall be made by the general government or by the States in respect to close seasons or intermissions of capture, the size of the fish to be caught, the enforced use of fishways, regulations as to introduction of refuse, &c. All this will require careful consideration, so as to avoid infringing upon natural or vested rights, while doing everything to the best interest of the community.

10. The stocking the various waters of the United States with the fish most suited to them, either by artificial propagation or transfer, and the best methods and apparatus for accomplishing this object.

3. ASSISTANCE RENDERED THE COMMISSION.

The act of Congress authorizing the prosecution of the labors of the United States Fish Commission instructs the Heads of the various departments of the Government to render to it all necessary and possible aid; and, as in previous years, the most generous and liberal interpretation of the law has been given by them.

To the Secretary of the Navy, Hon. R. W. Thompson, obligations are especially due for aid, without which the success of the Commission would have been much diminished, both in the branch of Inquiry and Propagation.

The most notable favor rendered by the Secretary has been the fitting out of the iron steamer Speedwell, and placing it at the disposal of the Commission for the summer-work, this being the third year of its detail for such service. Only second in importance to this was the furnishing of two steam launches, with two firemen each, to be used in connection with the propogation of shad in Albermarle Sound and in the Susquehanna River.

In accordance with the instructions of the Secretary of the Navy to the commandant, the facilities of the navy-yard at Washington have been freely extended in the fitting out or repairs of the scows and other vessels belonging to the Fish Commission.

The Treasury Department, through the Bureau of Revenue Marine, has also rendered a hearty co-operation by transporting the hatching barges of the Commission to and from their various stations. The revenue-cutter Ewing, under Captain Fengar, towed these boats from Washington to Norfolk *en route* for the scene of operations on Albemarle Sound, and, at its conclusion, from Norfolk to Havre de Grace, for service there, and finally back to Washington when the hatching season was completed for the year.

The revenue-cutter E. A. Stevens, under command of Capt. J. G. Baker, and stationed at Newbern, was also instructed to render similar aid, whenever necessary, in the Albemarle Sound waters.

To the Light-House Board of the Treasury Department is due the means of initiating and prosecuting important observations upon water temperatures in the vicinity of various light-houses and light-ships along the coast the necessary blanks being furnished by the Commission. These, when filled, were delivered to the light-house inspectors and by them forwarded through the Light-House Board to the Commission.

The War Department has furnished eight Springfield rifles and eight hundred cartridges for the purpose of the protection of the United States salmon-hatching station on the McCloud River against lawless depredators, white and Indian. General McDowell also supplied a detail of men for special service during the critical period of operations.

The Signal Office of the War Department, under General Myer, has also extended important co-operation, by continuing the series of observations of water temperatures initiated several years ago at the request of the Commission. It has been possible by this means to get a general idea of the variations of temperature in the principal streams of the country, and thus to supply, incidentally, information necessary to judicious action in connection with the introduction of the different kinds of food fish.

The observations taken at the sea-coast stations of the Signal-Office are also of great importance in determining the conditions of the movements of the pelagic fish, such as the mackerel, menhaden, blue-fish, &c.; and the extension of this system promised by the Chief Signal-Officer, by which all the coast telegraph and life-saving stations and light-houses and light-ships are to be included in the series of observations and furnished with the best kind of instruments, is also of very great importance.

Partly for the service of the Commission, and also to assist in the commerce and fisheries of the coast, the Chief Signal-Officer made Gloucester a storm-warning station during the summer of 1878, thus adding greatly to the facilities of the work. The forecasts of weather were also sent daily, arriving some time before the receipt of the Boston papers.

To the Patent Office of the Interior Department is due, through Dr. Dyrenforth, chief examiner, a list of all the patents relating to fish and fish culture issued in Great Britain and some other countries, as well as in the United States.

For the purpose of better facilitating the operations at the McCloud River salmon station the Post-Office Department authorized the establishment of the post-office of Baird, in Shasta County, by means of which the station and its vicinity generally are provided with the necessary postal facilities. Previously, the nearest convenient post-office had been at Redding, a number of miles distant, and for the receipt of the mail therefrom the station was dependent upon the courtesy of the stage-drivers.

To Colonel Casey, Superintendent of Public Buildings and Grounds in Washington, the Commission is indebted for the construction and improvement of the carp ponds on Monument Lot, the work being executed with great economy and with satisfactory results.

The public and official acknowledgments of the Commission for important services rendered are also equally due to many private establishments and individuals. The most important of these is the Maryland Fish Commission, under the direction of Mr. T. B. Ferguson. By combining operations at various times with this organization, the United States Commission has been enabled not only to secure the valuable superintendence and aid of Mr. Ferguson in its work, but the free use of important apparatus, and a consequent reduction of the absolute expense.

The Druid Hill Park Commission, of the city of Baltimore, is also entitled to mention in this connection for authorizing the use of the park for the cultivation by the Commission of the German carp, golden ide, and other fishes. For this purpose it constructed several ponds at a large expense to itself, for the continued culture of these fish, thus serving as an auxiliary station to the establishment at Washington. This is a matter of very great importance, as the ponds in Washington are very low, and the locality has been overflowed by the Potomac River several times within the last thirty years; and as this may at any time occur again, involving the loss of all the fish, the Baltimore station will furnish the means of renewing the supply at Washington.

A large number of railroads throughout the country, a list of which will be furnished hereafter (see p. xxxvii–xxxviii), have also co-operated with the Commission. The special favor conferred is that of receiving quartermaster's orders for the transportation of messengers, and in permitting the cans containing the young fish to be carried in the baggage cars of express trains without extra charge, and allowing at the same time the attendance of one or more messengers.

The extent of this favor can be better appreciated by the fact that not unfrequently there are two messengers, with twelve to eighteen 50-quart milk cans filled with water, to be transported on a passenger train.

Acknowledgments due to other co-operating bodies and to individuals, will be made in their proper place.

B.—INQUIRY INTO THE HISTORY AND STATISTICS OF FOOD-FISHES.

4.—FIELD OPERATIONS DURING THE SUMMER OF 1878.

The ability to carry on the researches along the coast of the United States, for the purpose of solving the problems referred to in a preceding page, has been dependent in a great measure upon the facilities furnished by the Navy Department for the purpose; and I have already mentioned that the liberal interpretation of the law of Congress made by the predecessor of the present Secretary of the Navy and carried out by the latter in the earlier years of his administration, has been continued during the year 1878.

The United States steamer Speedwell, assigned to the United States Fish Commission in 1877, was also placed at its disposal in 1878. Commander L. A. Beardslee, who had been in charge of the steamer Blue-Light during the field-work of 1873, 1874, and 1875, was placed in command of the Speedwell; Commander Kellogg, who was in charge of the vessel in 1877, having been assigned to other duty.

All the necessary repairs to the Speedwell were made at the Portsmouth navy-yard, at which place she had been laid up during the preceding winter.

After a careful inquiry into of different points on the sea-coast from which a critical scientific research might profitably be made in the interest of the fisheries, Gloucester, Mass., was selected, and on the 9th of July I established my headquarters there for the season, accompanied by the entire clerical force of the Commission.

After due inquiry, a suitable wharf and buildings were rented on Fort Hill, at the mouth of Gloucester Harbor. Rooms for laboratories, offices, storage, &c., as also a large apartment, used afterwards for the hatching of codfish, were included in the accommodations supplied. The wharf, directly on which the buildings were situated, fronted about 150 feet on two sides.

The Speedwell arrived on the 18th of July, and from that time until her departure, on the 30th of September, the work was carried on without serious interruption other than that caused by the weather, excepting for one period, from the 4th of August until the 14th, when she was at Portsmouth undergoing certain necessary repairs.

The personnel of the Speedwell consisted, in addition to her commander, Captain Beardslee, of Dr. J. F. Bransford, surgeon; H. E. Drury, paymaster; R. W. Galt, engineer; James H. Smith, executive officer; James. H. Kuhl, mate. The wharf, buildings, and apparatus were in charge of Capt. H. C. Chester, under whose superintendence also the work of dredging and trawling was usually conducted. The total force of the steamer, including petty officers and men, amounted to about 40.

The laboratory work was, as usual, under the special charge of Mr. G. Brown Goode and Prof. A. E. Verrill; Mr. Goode, aided by Dr. Bean, taking charge more particularly of the fishes, while Professor Verrill, assisted by Mr. Richard Rathbun, and for a portion of the time by Mr. Warren Upham, superintended the dredging and trawling work and the collection of marine invertebrates.

For a portion of the season Dr. W. G. Farlow was engaged in carrying on some researches into the peculiar condition to which salted codfish is liable during the moist summer weather. Small red specks show themselves upon the fish and rapidly spread, in time covering it completely. This is accompanied by a tendency to decomposition, which spoils the fish for market. As a very important subject, I invited Dr. Farlow's attention to it, and his report will be found in the appendix herewith. He considers the affection to be due to the presence of a minute red alga, possibly derived from the salt used in curing the fish. The Cadiz salt, examined by him, was found to contain the spores of this alga in large quantity, being tinted of a pink color thereby. These were doubtless derived from the vats or evaporating places of the salt. The Trapani salt, also used by the fishermen of Gloucester, was found to be free from this admixture, and its use is therefore recommended. The attempt to eradicate the affection will require that the holds of the vessels and the salt-houses be kept perfectly free from the introduction of this plant.

During a visit by Prof. W. O. Atwater to Gloucester, during the summer, an arrangement was made with him to prosecute a series of investigations upon the food-qualities of various species of fishes and their availability for the manufacture of fertilizers, involving many chemical analyses.

The various researches prosecuted during the summer's campaign will be presented hereafter in the form of special reports.

The usual collections were made, especially by means of the dredge and trawl, and the specimens secured are held for the National Museum and for distribution to educational establishments throughout the United States.

An extremely valuable mass of information was obtained during the summer, by Mr. Goode, in connection with the early history of the Gloucester fisheries, and by means of questioning some old fishermen and sailors he secured full details as to the inception and early history of the mackerel, halibut, cod, and other fisheries. This will be embodied with the series of investigations undertaken for the purpose of securing statistical information on the American fisheries, the importance of which was referred to in the previous report. Many specimens of fishery apparatus were also secured, some of them obsolete or displaced by modern apparatus, others illustrating the present condition of operations; all, however, of interest.

In addition to the collections made, many soundings and temperatures of the water were taken, the condition of the bottom ascertained, &c.; and an important generalization was made by Professor Verrill, based upon certain collections of fossil remains brought in by fishermen from various parts of the fishing banks. These were evidently of Tertiary age, but of a formation and distribution differing remarkably from anything known on the mainland, and suggesting to Professor Verrill the existence of a Tertiary deposit off the coast, hitherto unknown. While some of the species are the same as those found on the mainland, others are entirely different and appear to be new to science.

The work of the Commission was greatly facilitated during the period of its stay by the establishment, by order of General Myer, of a storm-warning station at Gloucester. This was erected on the top of the custom-house, one of the highest edifices in the city and visible for a great distance. Apart from its aid to the work of the Commission it enabled the large fleet of Gloucester fishermen to regulate their departure to sea with great advantage.

As usual, the Commission had many visitors during the summer; some for the purpose of taking a special part in its work, and others to familiarize themselves with its general operations.

Among the visitors were a number of gentlemen belonging to the Boston Fish Bureau, and familiar with the coast fisheries, to whom I had the pleasure of exhibiting the Pole flounder, seen by them for the first time. A similar experience was had with fish merchants and skippers of Gloucester. Reporters from the principal Boston papers, and some from New York, were also included in the number of those receiving the attention of the Commission.

A special incident of the season was a call from the Secretary of the Navy, Hon. R. W. Thompson, on the Tallapoosa, on the 25th of July. The Secretary was accompanied by several of the officers of the department, as the chief naval constructor, the chief engineer, the Paymaster-General, the attending surgeon, &c. The vessel remained in port for two days, and the occasion was taken to show the Secretary and party the operations of the Commission on the Speedwell, in the way of trawling and dredging.

For the purpose of determining more particularly the character of the animal life on the Grand Banks, especially of the ocean birds, which are used in great numbers by the fishermen for bait; I made arrangements with Captain Collins to carry Mr. R. L. Newcomb on a halibut trip to the banks. He was absent from the 28th of August to the 18th of September, and brought back many interesting specimens of birds as well as of marine invertebrates. It was found that the birds serving as bait were for the most part a species of petrel, of which many hundreds are often taken on a single trip by means of the hook and line.

The most active field-work of the Commission closed for the season on the departure of Professor Verrill on the 12th of September; but other branches were continued until the departure of the Speedwell for Washington on the 30th of that month. She reached her station in good season and was laid up in the Washington navy-yard for the winter.

I remained in Gloucester until the 15th of October for the purpose of finishing up certain statistical inquiries and of making the necessary arrangements for the propagation of codfish, to which reference will be made in a succeeding section of the report. Leaving on the 15th, I reached Washington with my party on the 24th. Mr. Milner, with Mr. R. E. Earll, Frank N. Clark, and Capt. H. C. Chester remained behind in connection with the last-mentioned interest.

C.—THE PROPAGATION OF FOOD FISHES.

5.—WORK ACCOMPLISHED IN 1878.

The Quinnat or California Salmon (*Salmo quinnat*).

The McCloud River Station.—The heavy rains of the winter of 1877–1878, and during the spring of the latter year, caused great damage in the valley of the McCloud River, and especially to the works on the United States Salmon Reservation. Many of the buildings were swept away, and the dam and works for raising water to the hatching-house were entirely ruined. A special allowance of $2,500 was made to Mr. Stone for restoring the station to the proper condition; and, reaching the ground on the 9th of May, he immediately went to work to reduce the disorder and render the works satisfactory for future operations.

The establishment by the Postmaster-General of the post-office of Baird, on the reservation, on the 3d of May, 1878, was of very great service to the Commission in keeping up its communication with the outside world. Previously the nearest convenient post-office was that of Redding, 22 miles distant, and the party at the works was dependent upon the courtesy of the stage-drivers for bringing along the mail. This act proved of service, not only to the reservation itself, but to the settlers scattered around, who appreciated the advantage to them in diminishing their travel.

As in previous years, there were various alarms in regard to lawless whites and Indians who threatened to raid the establishment and burn the buildings, as also to take possession of the penned-up fish, and thus nullify the work of the Commission. An application made to the War Department for arms was met by the issue of eight Springfield rifles and eight hundred cartridges. This equipment, supplemented by the detail by General McDowell of some soldiers, placed the establishment in a satisfactory condition of defense, and no violence was attempted.

The season of 1878 proved to be the most productive in the history of the establishment, and the number of eggs obtained, fourteen millions, was truly enormous, far exceeding those taken in any one season by all the salmon establishments in the world put together. According to Mr. Stone's estimate 18,000,000 could easily have been secured if desired, but the take was limited to the number applied for by the State commissioners and those needed to maintain the supply in the Sacramento River.

The first eggs of the season were taken on the 20th of August, and from that time until the 5th of October, when the last car was loaded with salmon eggs, the time of Mr. Stone and his assistants was employed without intermission. The fish were unusually abundant, thousands being often taken at a haul.

A notable feature in the season was the small size of the parent fish, these averaging less than nine pounds, some of the mother fish, full of ripe eggs, weighing only from six to eight pounds. This, Mr. Stone thinks, is due to the stoppage of the large salmon by the fishing at the canneries on the Lower Sacramento, allowing only the smaller fishes to pass.

It will doubtless somewhat astonish many persons at the East who are not familiar with the scale of operations on which the business is conducted at the West, to be told that from seven to nine thousand salmon were several times taken at the station in a single day.

The number of eggs actually secured and embryonized was so large that two cars were required to transport them to the East. Of these, the first left Redding with 4,000,000 on the 29th of September, arriving at Chicago on the 3d of October. The second, with 3,250,000, arrived at Chicago on the 7th of October. They were met by Mr. Fred. Mather, and the distribution was immediately made from that city by express. The details of distribution will be found in the schedule attached to Mr. Stone's report.

As usual, a large number were hatched out and planted in the McCloud, for the purpose of keeping up the supply in the Sacramento, 2,500,000 being thus treated; 500,000 eggs were presented to Canada, 100,000 to England, 100,000 to France, 100,000 to Holland, 250,000 to Germany, and 200,000 were sent to New Zealand.

In the report for 1877 mention is made of the shipment to various foreign countries of California salmon. The half million of eggs sent to New Zealand arrived in perfect condition, and were distributed by the government, to the several provinces. The latest advices speak of the young fish being seen in every direction, and promising to be the ancestors of a numerous progeny. Owing to various causes, however, the consignments to Germany, France, England, and the Netherlands in 1877, were failures, only about 25,000 eggs of the German lot surviving. These had been packed in a special manner by Mr. Mather, and escaped the fate of the rest.

Owing to the very high opinion entertained by European fish culturists of the California salmon as a food fish, both on account of the ease of its cultivation and the fact that it resists higher temperatures of water than the Atlantic salmon, it was determined to renew the experiment, by a transmission in 1878; and Mr. Mather was authorized to repack the eggs in his own way and accompany them to their destination. In accordance with instructions he, therefore, met the car containing the eggs from California at Chicago and received 250,000 for Germany, 100,000 for France, 100,000 for Great Britain, and 100,000 for the Netherlands. These he carried to his residence at Newark, and after repacking them by his own method, he took passage by the Bremen steamer Oder, and arrived at Bremerhaven on the 23d of October. The consignment for France was shipped from Southampton, on the way, and that for England was sent, for the most part, to the Southport Aquarium. The eggs for the Netherlands were met by an agent of the government at Bremen and transported to Amsterdam, where they were hatched out by the Zoological Garden.

The new venture proved to be a perfect success, a very small percentage of the eggs failing to be hatched out. Of the eggs of the German consignment 45,000 were sent directly to Mr. Haack at Hüningen, for introduction into the Rhine. One hundred and fifteen thousand were sent to Mr. Schuster, at Freiburg, for the Danube and the Rhine; and 30,000 to Hameln, for the river Weser. Various smaller lots were distributed to other places; and all were successfully hatched out and placed in their destined waters.

The 100,000 sent to France were also hatched out with comparatively small loss and introduced into various rivers of the republic. Those for the Netherlands were equally successful. The number actually received in Amsterdam was estimated at 85,000, and of these over 60,000 produced healthy fish, and were planted in various streams.

In the general table of distribution of California salmon will be found the indications of the various streams in which the fish were placed respectively.

Later in the season a consignment of the land-locked salmon was sent to the Société d'Acclimatation in Paris. These, however, owing to some unexplained casualty, arrived in poor condition, and comparatively few were saved.

Full details in regard to the work at the McCloud River station will be found in Mr. Stone's article, given in the Appendix.

Clackamas Station.—In the report of 1877 reference is made to the fears of the salmon-canners on the Columbia River as to a threatened diminution of the fish, and to the arrangement made through Mr. Stone for the establishment of a station for artificial propagation. This, after considerable delay, was established in the Clackamas River, but owing to the lateness of the season when the work was completed only a small number of eggs were obtained. These were supplemented by a transmission from the McCloud River, and a successful result accomplished.

The work was continued in 1878; but the funds available for the purpose being very limited, I agreed to assign a portion, not to exceed $5,000 of the appropriation, to the work, believing that in so doing I was properly carrying out the intention of Congress.

The first eggs were taken on the 5th of September; and up to the 30th 2,081,000 had been taken from 478 females. Some casualties were experienced in the course of the season by the heavy rains, which caused the dam to break; but a reasonable percentage of eggs was satisfactorily hatched out and introduced into the river. The principal part of the hatching and depositing in the river was done between the 24th of December, 1878, and the 2d of January, 1879, the number of young turned in being estimated at 1,203,000. The percentage of loss would have been much less but for the necessity at one time of moving the eggs from the hatching house to the river and back on account of a flood.

The details of the work will be found in the appended report of Mr. William F. Hubbard, the assistant superintendent.

At the close of the season it was found that the bills, in regard to which proper vouchers could be rendered, and applicable to the actual work, and not simply to the construction of permanent improvements, amounted to about $3,600, which was duly paid to Mr. J. G. Megler, the secretary of the Oregon and Washington Fish Propagation Company, to which the works belonged.

Proposed salmon-hatching station for the Southern States.—It is well known to all fish culturists that the expense of moving impregnated eggs of fish is very much less than that of transporting the same number of the young fish, as the former, with proper precautions, can be forwarded by express to any part of the United States, while the latter require the constant care and attendance of a messenger, and a much larger space, in proportion, for their accommodation.

The demand for the California salmon on the part of the southern and middle tier of Mississippi Valley States has suggested the propriety of a station where the eggs can be received and hatched, and from which the fry can be distributed at much less expense than from Baltimore, Maryland; Northville, Michigan, and other stations, where the hatching in question has been carried on.

An extensive correspondence was entered into with parties in Tennessee, Northern Alabama, Mississippi, &c., and several points were visited by Mr. Clark to ascertain their adaptation for the purpose. The especial requisites are, an ample supply of pure spring water of a temperature as much under 60° as possible; a proper fall of water; and convenient relation to a railroad center from which the fish can be distributed to assigned depositaries. Of course the place must be healthy, and one where the desired facilities will be freely granted by the owners of the ground.

Several localities were found possessing more or less of the necessary requisites. Among these were Huntsville, Ala.; Vicksburg, Miss.; Bon Aqua Springs, Tenn.; Birmingham, Ala., &c. The highest temperature found, of 63½°, was at Vicksburg; the lowest, about 59°. The outbreak of the yellow fever in Tennessee during the summer of 1878 prevented any action on the subject. This, however, is only deferred for the present, and it is hoped that another season, when a selection will be arrived at, it will be possible to arrange a temporary establishment where the eggs of California salmon and possibly of California trout may be successfully hatched. It will not be necessary to keep the works in operation for more than a month for either of these occasions; so that the expense will be comparatively trifling.

Atlantic salmon (*Salmo salar.*)

In view of the uncertainty as to the results of earlier efforts connected with securing the eggs of the Atlantic salmon, operations were suspended in 1877, and this intermission continued in 1878, it being thought desirable to wait for evidence that the work had been successful.

I am happy to say that during the present year the indications of success have been so unquestionable as to warrant the re-establishment of the Bucksport station, and it is hoped that the result for 1879 will show a good progress in this connection. It may be stated in general terms that nearly every stream on the Atlantic coast as far south as the Susquehanna in which young salmon were introduced as far back as 1874 and 1875 has proved to contain adult spawning fish in 1878.

An exact statement of the catch of salmon in the rivers along the coast is impossible, but the daily newspapers have been filled with the records of capture from Denny's River, in Eastern Maine, to the Susquehanna, in Maryland. In addition to this the correspondence of the Commission contains numerous references to captures of salmon, some of which I will proceed to present.

The increase in the rivers of Maine, although decided, has not been much a matter of specific statement, as the salmon have never been entirely absent from its waters, and consequently their occurrence in the rivers excited less remark.* The case was quite different, however, in the Merrimack, where salmon of late years have only been seen at very rare intervals. As the result of the action of the commissioners of Maine and New Hampshire, large numbers of salmon were observed while ascending the fishway in the dam at Lowell for the purpose of performing the function of spawning in the headwaters of the rivers, especially in the Pemigewasset, where many young were afterward seen.

For the details of these runs of salmon I refer to the extracts from the reports of the commissioners of Massachusetts and New Hampshire given in the Appendix.

In October, 1878, a salmon weighing 11 pounds was caught in Narragansett Bay, between Narragansett and Wickford. Other instances of captures in the same waters are recorded.

The weirs in Martha's Vineyard Sound, especially at Menemsha, secured a considerable number of salmon, most of which were sent to the

* In connection with the subject of salmon in Maine, it should be borne in mind that the fish from which the eggs are taken at the Penobscot or Bucksport station are not destroyed by the operation, but are returned to the water uninjured. Mr. Atkins has been in the habit of affixing a platinum tag to each fish before returning it to the water, bearing a number corresponding to a record of the date of capture, the weight before spawning, the weight of the eggs taken, and the weight of the fish when restored to the water. Mr. Atkins found several instances of a second capture of the same fish. Thus he records No. 768 as having been stripped on the 1st of November, 1875. It then weighed 21 pounds 7 ounces, and yielded 5 pounds 7 ounces of eggs. When turned back into the river it weighed 15 pounds. The same fish was recaptured at Lincolnville, Me., on the 14th of June, 1877, weighing 26 pounds.

Another fish, No. 1010, which on the 9th of November, 1875, weighed 18 pounds 2 ounces, had 4 pounds 10 ounces of eggs, and when dismissed weighed 13½ pounds, was retaken, also near Lincolnville, on the 13th of June, 1877, weighing 30½ pounds, thus showing an increase of 12½ pounds in two years.

New York market with the other captures. The greatest success was, however, experienced in the Connecticut, where the catch from the beginning to the end of the season is considered as amounting to not less than 600 individuals, varying in size from 9 to 20 pounds, most of them finding a market in New York. A great deal of enthusiasm was excited in the early part of the season by these captures, and the fish first taken were sold readily for a dollar per pound, and even more.

One of the earliest catches in the Connecticut was on the 4th of May, when a fish weighing 11 pounds was sent to Benjamin West, of New York, from Saybrook. On the 10th of May Mr. S. B. Miller reported a salmon taken in a seine near the west end of Long Island. Two were taken eight miles from the mouth of the Connecticut on the previous day, and on the same day 12 other salmon were received in New York from the Connecticut, one weighing 19 pounds, and all selling for from 85 cents to $1 per pound. Mr. James A. Bill, fish commissioner of Connecticut, on the 14th day of May informed me that within his knowledge 80 fish had been taken up to that time in the Connecticut River between its mouth and Windsor, these varying from 8½ to 18 pounds in weight. From 6 to 12 were captured daily.

It is known that in addition to what were caught by the fishermen in the Connecticut many others entered it, as shown by the holes made in the gill-nets. These holes were at first supposed to be caused by sturgeons, but it was subsequently ascertained that they were due to large salmon that could not be held by the thin twine.

There were no authentic cases of the occurrence of salmon in the Hudson during the year. This is easily explicable from the fact that no young were introduced by the commissioners of the State, they being unwilling to take any steps in this direction until the proper means for their protection, as well as that of the shad, against the gill and stake nets should be passed by the legislature. A very few planted by private enterprise yielded no positive result, although several rumors of captures were given in the newspapers.

The case was very different in regard to the Delaware River, in which quite a number of deposits were made, partly by the fish commissions of the State and of the United States, and partly by individuals. The earliest introduction of salmon in this river was made in 1871 by Mr. Thaddeus Norris at the expense of some public-spirited citizens of Philadelphia, the eggs having been hatched out on the Hudson River, and the young transported to the Delaware. Only about 2,000 survived the journey. In 1872 12,000 eggs were purchased of Mr. Wilmot, at Newcastle, Ontario, and hatched out near Easton, Pa., with a loss of only ten per cent. The young were placed in the Bushkill, a tributary of the Delaware, near Easton.

The next lot of salmon planted in the Delaware consisted of 5,000 fry, the sole product of 750,000 eggs received from Germany by the United States Fish Commission, in the winter of 1872–'73.

These were hatched out by Dr. Slack, at Bloomsburg, and planted in the Muscanetnong in the spring of 1873. Subsequently the commissioners of New Jersey and Pennsylvania introduced other lots, and it is difficult to say how many of these deposits contributed to the results of 1878.

The first show of salmon in the Delaware was in the autumn of 1877, when a large fish was seen directly engaged in the act of spawning at the mouth of the Bushkill River, this quite probably being one of Mr. Norris's fish. It was killed by a rifle-ball in ignorance of its true character, and sent to me for identification. It is now preserved in the collections of the National Museum.

On the 19th of January Dr. Abbott, of Trenton, reported the capture of a salmon 16 inches in length at Trenton, this being probably a grilse.

On the 6th of April Mr. E. J. Anderson, fish commissioner of New Jersey, announced the taking of two salmon, one weighing 18 and the other 23½ pounds, in the Delaware. One of these was also sent to the United States commission at Washington, where it is preserved, together with an excellent cast.

Later in the year the catches in the Delaware were quite numerous, the total number, according to the fish commissioner of the State, amounting to some hundreds.

The southernmost locality in which salmon have been taken is the Susquehanna, a fine one of 19 pounds having been caught in a gill-net in the vicinity of Spesutie Island, just below Havre de Grace, and obtained by Mr. James W. Milner, in charge of the United States shad-hatching operations there, and sent to Washington. This was a fresh-run fish, in perfect condition, and formed the subject of an admirable drawing and plaster cast. It probably was derived from a lot of salmon planted by Mr. Ferguson, fish commissioner of Maryland, in one of the tributaries of the Susquehanna.

For fuller details of the occurrence of salmon in the Eastern and Middle States and in Maryland I refer to the appendix, where a condensed statement, as prepared by Mr. C. W. Smiley from reports of State commissioners, will be found.

While these facts show conclusively that the experiment of introducing the *Salmo salar* into the more northern rivers of the Atlantic States by the United States has been a success, it will be readily understood that the great object will be to establish a continued run to be kept up by naturally spawned fish, a result which should be continually aimed at. It is not to be expected that the general government or the States will continue indefinitely their effort to obtain eggs and plant the young fish, especially as the time may come when this resource will not be at their command.

Where rivers are entirely destitute of salmon, either from an exhaustion of the supply or from never having existed there, artificial propagation must begin the work. But unless this is supplemented by the enactment and enforcement of laws forbidding absolutely the capture of the fish for a period of four to six years, and then establishing a close time of several days in each week up to a certain period, after which no fish at all shall be taken, the efforts now being made might as well be intermitted first as last. There is no object in going to the expense for the purpose of furnishing a few fishermen with a supply of fish to be sold for their benefit, and not administered for the good of the community. The magnitude of the results will be in direct proportion to the enactment and enforcement of the proper legislation.

Schoodic salmon (*Salmo salar*, subsp. *Sebago*.)

Grand Lake Stream Station.—Of the various species of *Salmonidæ*, treated by the United States Fish Commission, the fish variously known as landlocked salmon, Schoodic salmon, Sebago salmon, Glover's salmon, Win-ni-nish, &c., is one that is most eagerly sought after by State fish commissioners, fishing clubs, and fish culturists generally. An exact miniature of the sea salmon or *Salmo salar* in appearance, flavor, game qualities, &c., the difference in size was for a long time considered sufficient to establish it as a distinct species. Late researches, however, prosecuted by Professor Gill and Professor Jordan, among the large collections at Washington, have satisfied these gentlemen that it must be regarded as a dwarfed form, hardly even a variety, of the *Salmo salar*, owing its reduced proportions to its abode in lakes or ponds, and consequently more limited range than it would have in the ocean, although its continual sojourn in fresh water may have had something to do with it. The westernmost locality where it is found on the New England coast appears to be Sebago Pond, a large body of water which discharges into Casco Bay, north of Portland. Here it is called Sebago salmon or Sebago trout, and attains a considerably larger size than in most other waters, as in the Sebec Lakes, northwest of Bangor, in certain ponds in the Mount Desert region, and the Schoodic Lakes of Maine and New Brunswick, which are perhaps its best-known localities. It is also seen in the Saint John's River and certain ponds of New Brunswick and Nova Scotia, as well as in the tributaries of the Saint Lawrence. To what extent it is taken on the south shore of the Gulf of Saint Lawrence or on the coast of Labrador, I am unable to say.

In various parts of the British provinces it is known as the Win-ninish, which would perhaps be a much more appropriate appellation than the term landlocked salmon, since other species of the Salmonidæ present themselves under similar circumstances. A similar variety occurs in Sweden, and possibly elsewhere in Europe, and relating to the same species, *Salmo salar*.

In the opinion of many persons, and especially of Mr. Samuel Wilmot, the salmon of Lake Ontario belongs to the same division, although in size it more nearly corresponds with the sea-going salmon. Formerly immense numbers of these fish existed in Lake Erie and ascended its tributaries on both the Canadian and American sides to spawn. They have, however, for the most part, been exterminated on the American side, and but for the efforts of Mr. Wilmot would probably have experienced the same fate on the north shore. A number of years ago, however, that gentleman, finding a few pairs in a small tributary of Lake Ontario, near Newcastle, undertook their artificial propagation, and so successful were his efforts that he increased the number enormously, although no great increase in the number of captures has resulted. This is probably due to the fact that they cannot be taken at the time when they are fresh run and in good condition for food. Their present spawning-grounds are very near the lake, and, as in the short rivers of California, they come into the streams only when they are nearly ripe, and remain a very short time, returning at once to the lake. It would seem that, to have a satisfactory river salmon fishery, the stream must be long enough for the fish to remain a considerable time in it, so that they may enter it before they are ripe and give an opportunity for their capture by suitable devices.

The advantages of this landlocked form, which, so far as the United States Fish Commission is concerned, it is proposed hereafter to term the Schoodic salmon, unless the name Win-ni-nish be considered preferable, are the readiness with which the eggs can be obtained, the hardiness of the fish, and their perfect adaptability to a great variety of circumstances and temperature. They are said to resist warmth of water better than even the brook trout and to be an available fish not only for lakes and ponds, but also for long reaches of deep water in rivers through which there is comparatively little current, such as are found in the Saint John's River in New Brunswick and elsewhere.

The Schoodic salmon has, for several years, occupied the attention of the United States Fish Commission, and the successive reports will show what has been done in this connection. A trial made several years ago in Sebec Lake and this year at Sebago Pond, have led to the conclusion that the Schoodic lakes of Maine, and perhaps New Brunswick, will furnish the best stations for the collection and distribution of eggs. The locality controlled by the United States Fish Commission is situated not far from the tannery of the Messrs. Shaw Brothers, on Grand Lake Stream, the outlet of Grand Lake, one of the Schoodic chain of eastern Maine, and at no great distance to the west of Calais. This, for many years, had been the resort of fishermen in the proper season, the fish occurring in immense numbers and furnishing admirable sport. By arrangement with the Messrs. Shaw, certain privileges of water and fishing were obtained by payment of an annual rental, on the usual condition as established by the laws of Maine, that one-fourth of the eggs obtained should be hatched out and the young returned to the waters. In addition to this, a considerable per-cent. of the remainder goes into the waters of the State in other localities. Here, the United States, in conjunction with the States of Maine, Massachusetts, New Hampshire, and Connecticut (two or more of them), has carried on operations under the superintendence of Mr. Charles G. Atkins and with varying success, for which reference may be made to the detailed report of Mr. Atkins. There have been some difficulties from time to time in getting a proper head of water for developing the eggs to a suitable stage for shipping, and numerous obstacles have been found in the securing of the fish. These, however, have now all been palliated or overcome.

The taking of eggs in 1878 was begun on the 7th of November, and closed on the 4th of December, at which time the return of the parent fish to the lake ended. The total number of eggs for the season amounted to 1,723,000. One great advantage connected with the taking of eggs from salmon as well as trout is that the parents are not injured, but by careful handling may be returned to the water in good condition, so that another year they may yield an additional supply. Great care is exercised in this respect, so that neither at Bucksport or Grand Lake Stream are many fish absolutely lost.

While, by actual experiment, about 90.1 per-cent. of all the eggs taken were impregnated and embryonized by the artificial process, scarcely more than 10 per cent. would have been by natural propagation. If we consider the immense number of even impregnated eggs consumed by the white perch and other vermin of the lake, and compare the remainder with the absolute propagation artificially, the vast disproportion of results can be readily appreciated.

Mr. Atkins, in referring to the impregnation of the Schoodic salmon states that at Bucksport the successful impregnation of 96 to 98 per cent. of the sea salmon was accomplished. This difference from the experience with the former he considers to be due to possible circumstances affecting the fish in their somewhat artificial detention in fresh water, from which the sea-run individuals escape.

Owing to various circumstances beyond the control of Mr. Atkins, such as an abrupt change to colder weather, a certain portion of the eggs collected were destroyed. But, of the 1,723,000, there were 1,470,000 embryonized, or carried to that point where the eyes of the young fish could be seen through the envelope. Of these, 370,000 were retained for Grand Lake Stream, and of the remainder 1,110,000 were shipped by the United States Fish Commission, and distributed among a number of States. The rest went to Massachusetts, Connecticut, and New Hampshire. Of the 370,000 retained for Grand Lake Stream, 350,000 healthy young fish were hatched out and turned into the water.

The details of the distribution of these fish will be found in the tables of Mr. Atkins's report, to which I refer for much interesting information.

Sebago Station.—An earnest appeal by Mr. E. M. Stilwell, fish commissioner of Maine, determined the United States Fish Commission to make an experiment in regard to securing a supply of eggs of the Sebago Pond variety of landlocked salmon, in view of its much greater size than that found at Grand Lake Stream, and of the greater accessibility of the locality.

It will be remembered that the Sebago is a large stream, situated in Southwestern Maine, which discharges through the Presumpscat River into Casco Bay to the north of Portland. The landlocked salmon found in it have always been celebrated for their beauty and weight, a size of six or eight and ten pounds, and even more being not unfrequent.

Unsuccessful efforts were made some years ago to obtain spawning fish from Sebago Pond, for the purpose of securing their eggs. It was imagined that, owing to the protection afforded by recent legislation and the removal of certain obstructions in the water, a new effort might be more satisfactory. Acting upon this impression, Mr. Atkins was directed to establish a station, for the purpose of an experiment, which he accordingly did, leaving Mr. Buck, one of his assistants of long experience, in charge. After giving the matter a fair trial, the enterprise was abandoned, as, with all the devices in the way of nets, &c., only ten males and six females were captured, and the entire number of fish entering the river for the purpose of spawning was estimated at scarcely more than fifty. The largest fish taken was a female, weighing 8 pounds 10 ounces after spawning, and the average was about three pounds.

Whitefish.—(*Coregonus clupeiformis.*)

The great amount of attention paid to the artificial propagation of the whitefish by the commissioners of the lake States, especially of New York, Ohio, Michigan, and Wisconsin, has rendered it unnecessary for the United States Commission to take up the subject to any great extent, although Mr. Frank N. Clark usually collects several hundreds of thousands of eggs, and develops them at his fish-culture establishment at Northville, Mich., for any desired assignment. These, for the most part, have been sent to the commissioners of California, and also to various parties in Pennsylvania, New Jersey, Wisconsin, &c.

The actual distribution made will be found in the appropriate page of the tables.

Shad.—(*Alosa sapidissima.*)

As in previous years, the propagation and distribution of shad was conducted under the able and efficient superintendence of Mr. James W. Milner, co-operating for a portion of the time with Mr. T. B. Ferguson, the fish commissioner of Maryland, whose help, as in previous years, is gratefully acknowledged.

To Mr. Milner's report, in the appendix to the present volume, I refer for details of the work accomplished, confining myself here to a mere abstract.

Albemarle Sound Station.—In previous reports reference has been made to the advantages of substituting Mr. Ferguson's cone and bucket apparatus for the floating hatching-boxes, so unsatisfactory in tidal waters. Desirous of testing the experiment with this apparatus on a large scale, operations were commenced much earlier in the season than usual, and at a southern station, in Albemarle Sound. The barges used by the Maryland commission in its work in 1877 were purchased and thoroughly equipped by the United States Commission, and towed by the revenue-cutter Ewing, in command of Captain Fengar, to Norfolk, whence a private tug carried them to Avoca, a plantation and fishing landing of Dr. W. R. Capehart, situated near the mouth of the Chowan River. The Maryland steamer Lookout was also employed in the service by an arrangement with Mr. Ferguson.

In addition to the Lookout, a steam-launch, furnished by the Navy Department, rendered essential aid in visiting distant landings for the collection of spawn and in transporting young fish from the station to the steamer for shipment via Franklin to various portions of the Southern States.

The work commenced about the 1st of April, with the benefit of every possible aid from Dr. Capehart, and up to the 1st of May about 10,000,000 eggs had been secured; the largest number taken in any one night being 1,605,000, on the 15th of April. The shipments of fish to remote points began on the 11th of April, amounting in all to about 5,000,000. These were distributed in part by the United States Commission, and in part by the fish commissioners of Maryland, North Carolina, and Virginia, who were furnished with what they could well transport to waters within these States.

A remarkable feature of the fishery season on the North Carolina coast consisted in the unprecedented number of alewives, or fresh-water herring, captured at various landings, as many as 400,000 having been taken at one haul. The glut of these fish was so great, that at one time they were sold at 50 cents per thousand; indeed, it became necessary to stop using the seines ten days earlier than usual on account of the difficulty of handling so many fish.

Mr. Ferguson, having been appointed one of the Commissioners to the Paris Exhibition, was obliged to leave Avoca before the close of the season, and the work was then continued by Mr. Milner and his assistants. On the 2d of May the station was closed, and the barges and launches were towed to Norfolk by the revenue-cutter E. A. Stevens. At Norfolk the Ewing again took charge of the tow and reached Havre de Grace with her charge on the 11th of May, where the hatching work was resumed under direction of Mr. Frank N. Clark—Mr. Milner, however, having general supervision.

The station selected this year at Havre de Grace was the same as that used in 1877, namely, a sheltered harbor between Spesutia Island and the western shore. The work was prosecuted on four barges and aided by two navy launches, a second one having been furnished for the purpose by the Navy Department. The steamer Lookout was dispatched to the Potomac for the purpose of collecting eggs of shad and hatching them on that river.

The entire take of eggs at Havre de Grace amounted to 12,730,000; the largest number secured at one time being 1,940,000 from 97 spawners, on the 29th of May. The total shipments and distribution of fish from this point amounted to over 9,000,000.

In the absence of Mr. Ferguson, the interests of Maryland were cared for by Mr. Thomas Hughlett, another member of the State fish commission.

The total production of the season at the three stations of Avoca, the Potomac, and at Havre de Grace amounted to 15,500,000 fish. The shipments extended to all parts of the United States, as far even as California, a fourth transmission having been made to the Sacramento River—a stream in which the success of the work in the past has been notably manifest.

Special acknowledgments are due on the part of the United States Fish Commission to Col. Marshall Parks, the president of the Albemarle and Chesapeake Canal, who not only tendered the use of the canal, passing all of our vessels to and fro free of toll charges, but having learned that toll had been collected from the steamer Lookout on her first voyage of reconnaissance, made in December, 1876, generously refunded the amount collected.

Col. Parks has, throughout all of our operations on Albemarle Sound, given us every aid, and by his cordial co-operation has evidenced his interest in the development of the resources and the future prosperity of that region.

A pleasant feature of the shad hatching operations at Havre de Grace consisted in the visits made by various persons to the station. Thus, on the 5th of June, I accompanied the President and the Secretary of the Navy, with a party of other invited guests, in a special car, returning the same night, and at a later date, a number of members of Congress. Many reporters from New York, Philadelphia, and Baltimore also embraced the opportunity to become familiar with the aims and results of the Commission, and to publish an account of the same.

To complete the history of the operations of the year 1878, connected with the propagation of shad on the Atlantic coast, I may remark that Colonel McDonald, fish commissioner of Virginia, made a station at Tobago Bay, near the mouth of the Roanoke River, and hatched out about 1,960,000 fish between the 3d of May and the 1st of June. All of these were placed directly in the Roanoke, and cannot fail to make their presence known within the next three or four years.

I am gratified in being able to state that the labors of the United States Commission in introducing shad into new or depleted waters have commenced to show results during the year 1878. Some of the earliest efforts in regard to stocking the rivers with shad were prosecuted in connection with the Sacramento River, a shipment of 12,000 fish having been made June 19, 1871, by Seth Green at the expense of the

California commission, followed in subsequent years by transmissions by the U. S. Fish Commission. The Sacramento River may now be considered as fairly supplied with fish, numerous adults having been taken during the year, although they have been sold surreptitiously, in consequence of a prohibitory law. It is to be hoped that with a few additional shipments the stock will soon be self-sustaining, and possibly that the adjacent rivers north and south will receive an ample supply.

For the Mississippi Valley, we have a very satisfactory result of the operations of the Fish Commission in the Ohio River at Louisville, where several hundreds of fish were captured in 1878 and exposed for sale in the Louisville market. The citizens are naturally jubilant at this great addition to their food resources, and stoutly maintain that, compared with the shad of the Connecticut, the Delaware, and the Susquehanna, those of the Ohio are by far the finest. Should this run continue, I hope to give further information in regard to it in a future report.

As nearly as we can ascertain, these fish have all been derived from a deposit of 30,000 made in the Allegheny River by Seth Green, and 200,000 by Mr. Wm. Clift in the year 1872, at Salamanca, in Western New York, in both cases in behalf and at the expense of the U. S. Commission.

For the purpose of ascertaining the facts in regard to the occurrence of shad at points in the Mississippi Valley other than Louisville, the commissioners of fisheries of Kentucky caused a circular of inquiry to be published in the principal newspapers, asking to be informed on this subject. Several responses were received, and among them one from Mr. John F. Oliver, of Steubenville, Ohio, who on the 25th of September wrote to say that a number of shad were caught in the Ohio at that place early in the season, on their way up, very many having been brought into market. He urges the importance of legislative measures for the protection of these fish, at least for a time, stating that two fishermen at Wing and Wing Rock, three miles above Steubenville, on the West Virginia side of the river, caught with hoop or set nets six or seven bushels of shad.

Dr. Paul Sears, of Mount Carmel, Ill., also writes to say that parties fishing with set nets in March, April, May, and June, caught what they supposed to be a new species of hickory shad (*Pomolobus mediocris*), but which he found on examination a different variety, in not having the lower jaw protruding as in the hickory shad, and in being thicker below the dorsal fin. These are points in which the true shad differs from its ally, and render the fact of its occurrence at Mount Carmel unquestionable.

In addition to these statements, Mr. George Spangler announced on the 3d of May the capture of about a dozen shad near Madison, Ind. The first sold for a trifle, but the price rose considerably when the fish were identified.

Mr. George F. Akers, of Nashville, Tenn., wrote on the 21st of May that many shad were taken during the year near Nashville.

On the 20th of March a four-pound shad was caught at Wetumpka, Ala., in the Coosa River, and on the 18th of April several shad were taken at Rome, in George Creek, according to the report of Dr. George A. Hampton.

Specimens of the Ohio River and Alabama shad were sent to the National Museum for identification.

A very decided increase in the catch of shad in the Roanoke River in 1878 is ascribed to the fact of the introduction of so many young fish in previous years, as the result of the operations of the United States Fish Commission on that river.

It may here be remarked that the fishermen, at least on the Potomac and in Albemarle Sound, distinguish what they call a May shad, a fish coming in later than the ordinary shad, in Albemarle Sound appearing from the middle to the end of May. These are said to be very fat, with short, thick tails, and with the back more golden than blue. Whether we are to establish two species of shad, as has been done with the herring, one composing an earlier run and another a later, has not been shown for want of sufficient material.

Herring.—(*Clupea harengus.*)

Experiments prosecuted at Gloucester before the eggs of the cod were ripe showed satisfactorily that the sea herring could be multiplied artificially on a sufficiently large scale for economical purposes. A large run of the spawning fish came on the coast in October, and, for a few days at least, ripe eggs could be had in any desired abundance.

Mr. Clarke fitted up an extempore apparatus by placing slides of glass vertically in a long box, somewhat in the style of the Williamson apparatus, so that the same water was made to flow through a series of compartments. The glass plate was laid flat in shallow pans, and the eggs dropped upon them, adhering tenaciously wherever they touched. A portion of the milt being added, a small quantity of water was introduced so as to dilute it, and by coming in contact with the eggs, produce the desired impregnation in the current of water. The eggs hatched out rapidly, and a very considerable number of young were produced and placed in Gloucester Harbor.

Partial experiments, indicating the same general result, were made in 1877 at Noman's Land, by Mr. Vinal Edwards, of the United States Fish Commission, and mentioned in detail in the report for the year 1877. About the same time Dr. Meyer, of Kiel, made a very elaborate investigation upon the development of the herring and the means of retarding it. He suggests that the result as published may be applied to lengthening the hatching period of the egg of the American shad and alewife, in connection with the effort to transmit these fish to Germany.

It is proposed to test this question more fully during the coming year;

but the conditions are quite different in the case of these species. The shad and alewife have non-adhesive eggs (those of the latter slightly so at first), and are hatched out in warm water, or with a rising temperature. The egg of the herring, on the other hand, is adhesive, and is hatched at a low and descending temperature, the difference in physical conditions demanding different treatment, the nature of which the proposed experiments will no doubt settle satisfactorily. In Dr. Meyer's apparatus glass plates are arranged horizontally with the eggs on the under side, a condition impossible in the case of the alewife.

The Carp.—(*Cyprinus carpio.*)

The Druid Hill Park Station.—I have already referred in previous reports to the various experiments of the Commission looking to the introduction into the United States of the best varieties of the German carp, a species considered to be of very great utility, especially to the South, for food purposes, and bidding fair to stand in the same relation to the farmer among fishes that domestic fowl do among birds. The fish brought in by Mr. Hessel in 1877 were cared for at the Druid Hill Park, under the direction of Mr. T. B. Ferguson, commissioner of fisheries of the State of Maryland, and by permission of the park authorities. It was considered advisable; however, to have a portion of the supply in Washington, where the fish could be more immediately under supervision. Inquiry was therefore made as to a suitable location for the fish, either in ponds already built, or to be constructed. It was, of course, thought best that they should be placed on government ground, where there would be no question as to rental. Several small lakes on the Soldiers' Home property were at first thought of, but the governors were unwilling to allow the changes necessary to fit them for the purpose, and it was with difficulty that suitable ground could be found for the construction of new receptacles. The work was under the direction of Mr. Hessel, who had a survey made, and laid out the contour. Soon after the work was begun, it was ascertained that the supply of water that had been relied upon for this purpose was inadequate, and the enterprise was abandoned. In this emergency Mr. Hessel had his attention called to the so-called "Babcock Lakes," two in number, which have a surface of about 6 acres each, situated on the Monument Lot, and separated by a driveway. These were found suitable in every way as to size, supply of water, &c. Application was made to Congress, at its special session, for the privilege of using these ponds, and for an appropriation to adapt them to the required service. The application was granted, and $5,000 allowed to put them in order. The work was conducted by Colonel Casey, with all due economy, the plans being furnished by Mr. Hessel. The ponds were drawn off and graded, so that a series of ditches, radiating from one point, would concentrate therein the contents of the pond. A basin or collector was built at this outlet, walled with brick and armed with heavy plank, and a suitable gateway and overflow was established.

This portion of the work was completed in the spring, and after the water was let in, two-thirds of the fish in Druid Hill Park were brought over and placed in their new abode. For the purpose of having a suitable series of hatching ponds, the surface of an island in the west pond was elevated, and the area subdivided so as to form two basins of suitable dimensions. These were fitted up properly with reservoirs and ditches, so that they could be drawn off on the same general principle as that adopted for the larger pond. Into these were placed several of the breeding carp, and quite a number of the young fish resulting therefrom. In the mean time work was also prosecuted on the westernmost of the two ponds; but owing to the adverse weather and incessant rains of the spring it could not be completed, as the warm weather suggested the necessity of restoring the water to its place to prevent malarious exhalations. This was accordingly done, and further action deferred for the time. A second appropriation for $2,400 was used in completing the work, and especially in paving the bank to prevent the washing of the wind and waves. As an additional means of putting the ponds in proper order for the discharge of the necessary functions, a series of brick tanks were planned (six in number), in which the fish could be classified when the ponds were drawn off, and those taken out that served for shipment, and the others returned. These were to be 20 feet long, and respectively 2, 5, and 9 feet in width, with a uniform depth of 5½ feet. The work on the ponds was postponed, owing to the fear of endangering the health of the city by making the necessary excavations on the island for the walls of the bank, and the completion was delayed by various vexatious causes, so that it was not till the early part of the winter that they were completed, and to disturb the fish in their winter quarters was not considered desirable. The construction, however, is available for service, and it is hoped that in 1879 an extensive distribution of fish may be made.

Of the fish brought from Germany by Mr. Hessel in 1877, the following were found alive and in good condition in the Druid Hill Park Pond when drained in the spring of 1878: 10 mirror carp, 90 leather carp, 80 scale carp, 40 gold orfe, 50 King or Hungarian tench, 20 common tench, 2 golden tench. As already explained, the three varieties of carp all belong to the species *Cyprinus carpio*. The gold orfe is a variety of the *Idus melanotus*, a large, fine Cyprinoid fish of Europe, somewhat resembling in size and shape the fall fish (*Semotilus shotheus*) of American waters, and of a brilliant red something like that of the common gold fish. The tench (*Tinca vulgaris*) like the carp, occurs in several varieties, the best being the king tench. The gold tench is a red form of the species just mentioned.

Of the fish above enumerated, there were retained in the Baltimore ponds the ten mirror carp, one-fourth the stock of the leather and scale carp, the hungarian tench, and gold orfes, respectively; all the common tench and the two golden tench. There were brought to Washington 65 leather carp, 48 scale carp, 10 golden ides, and 14 tench, which were distributed as follows:

	Babcock Lake, or East Pond.	Island Pond No. 1.	Island Pond No. 2.	Arsenal Pond.	Total.
Leather carp	39	13	13	....	65
Scale carp	....	....	....	48	48
Golden ides	10	....	....	....	10
King tench	14	....	....	....	14
Total	63	13	13	48	137

All the mirror carp and the golden tench, about half of the scale carp, three-quarters of the stock of ides, and most of the tench remained in Baltimore.

In order to diminish the danger of loss of the carp in the Monument Park by disease, inundation, or theft, the offer of Major McKee, commandant of the United States Arsenal, to accommodate a portion of them in the ice-pond of the arsenal grounds, was gladly accepted, and all of the scale carp, 48 in number, were placed therein on the 23d of May. So far as known, these fish continued in excellent condition throughout the year and without loss.

The very severe weather of the end of December, 1878, and beginning of January, 1879, caused the two carp ponds to freeze over sufficiently thick to bear skaters, and the opportunity was eagerly embraced by large crowds of both sexes. As any disturbance overhead was likely to seriously injure the carp in the east pond, a notification was placarded around it forbidding entrance on the ice on any pretense whatever. No restriction, however, was made in regard to the western pond, and while the deprivation was cheerfully borne, the community enjoyed the facilities allowed to their fullest extent. The superintendent, Dr. Hessel, was directed to prevent the crowd from coming upon the island in the west pond, on which tanks and hatching apparatus were located, but was authorized to allow ladies and children to enter the house, a privilege gladly embraced, and to such an extent that sometimes as many as sixty persons were in the building at one time. A few days of incessant skating cut up the ice so that several applications were made by the public to have the surface of the pond flooded, and thereby make a new skating surface. It was found impossible, however, to meet the request of the petitioners, as there was no plug of sufficient size in the west pond to produce any effect.

The Cod (*Gadus morrhua*).

The Gloucester Station.—A most important increase in the range of the work of the United States Fish Commission, in the way of the propagation of food fishes, was made during the year in connection with the various species of the cod family, especially of the true codfish. While engaged in the prosecution of researches into the condition of the fisheries at Gloucester, my attention was called to the fact, in the early autumn, of the approaching ripeness of the cod, haddock, &c., and it was determined, after conference with Mr. Milner, to institute experiments looking towards the artificial propagation of the cod, it being known from the researches of Sars that the eggs of that fish are non-adhesive and that they are discharged in the open sea, and float freely at the surface. With this information as a basis, preparations were made to utilize a portion of the wharves and buildings leased by the Commission at Gloucester for the erection of the necessary cod-hatching apparatus. A steam-engine, pumps, and other appliances were ordered on from the shad-hatching barges at Baltimore, and the work of fitting up was vigorously prosecuted under the direction of Capt. H. C. Chester and Mr. Sauerhoff, the whole work being under the charge of Mr. Milner.

The cones, so serviceable in the hatching of shad, were first tried; but did not work satisfactorily, in consequence of the changed conditions, the eggs being lighter and floating at the surface instead of sinking to the bottom as with the shad. After numerous trials to overcome this principal difficulty, a device was hit upon by Captain Chester, which, in a great measure, answered the desired object; and as it became possible to secure an ample supply of eggs, the experiment was prosecuted vigorously and ultimately crowned with success. Several millions of cod were hatched out and turned into the harbor of Gloucester, where, in the ensuing summer, they could be readily observed around the wharves, and even taken with a hook, the unwonted sight attracting the greatest interest of the fishermen and residents.

Mr. Frank N. Clark, who had had charge of the shad-hatching work at Havre de Grace, also supervised the hatching of cod at Gloucester, and introduced some important improvements in the apparatus.

Mr. Milner was obliged to return to Washington by illness, and Captain Chester having also been incapacitated from a similar cause, the establishment was broken up in the early part of January, 1879, and the apparatus dismantled and boxed, ready to be returned to the southern stations.

Other species of *Gadidæ*, as the haddock, etc., were experimented with upon a small scale, and the feasibility of artificial propagation of the species of the cod family fully established.

The only very serious difficulty experienced during these experiments was that from the turbidity of the water, this being necessarily taken from the harbor, and more or less polluted, especially in stormy weather, by the dock mud.

It is confidently believed that if a vessel can be constructed and anchored in the proper quality of water an enormous propagation of fish can be accomplished. There is apparently no limit to the number of eggs that can be secured, in view of the fact that a mature cod will furnish from two to nine millions, and the number of spawners taken in the vicinity of Cape Ann almost every day being very great. Of course it requires special conveniences to do this work, particularly during the inclement season of winter. The season during which the eggs can be obtained, however, is a very long one, extending from November to March and April.

For a detailed account of the whole experiment and of the observations made during its progress, I refer to the article by Mr. R. E. Earll, in the appendix. This may justly be claimed as perhaps the most important contribution ever made to our stock of information respecting the natural history of our principal food-fish.

In connection with the work upon the codfish, satisfactory experiments were also made in regard to hatching the sea herring, as detailed elsewhere.

The Sole (*Solea vulgaris*).

Reference has been made heretofore to a wish to meet the oft-expressed desire of citizens of the United States that the European sole might be introduced into American waters; and not daunted by the essential non-success of the work of last year (by which only two were successfully transported to our shores), arrangements were made with Mr. C. L. Jackson and Mr. Long, of the aquarium at Southport, in England, to secure a supply of young fish and hold them in readiness for further action. In accordance with this, over eleven hundred were brought in during the season. Many deaths occurred in this number in a few days after being captured, but 165 surviving were kept alive in the tanks for a considerable period of time.

Mr. Mather, of whose visit to Europe in connection with the transportation of the eggs of the California salmon mention has been made on another page, went to Southport, on his return to the United States, and took charge of the fish. Unfortunately, however, the necessarily crowded quarters, and, possibly, the fact that they were brought in tin cans, which rusted very rapidly, proved adverse to a successful experiment, and the entire lot died, one after another, before the return voyage was completed.

A portion of this ill success was thought by Mr. Mather to be due to a pump in use for aërating the water, the packing of which had been saturated with some chemical substance which exerted a deleterious influence. In this, as well as in the previous experiment, the United States Fish Commission is indebted to the courtesy of the Cunard Steamship Company for important facilities.

The Sponge of commerce.

Among the more recent enterprises in the way of artificial propagation of aquatic animals is that relating to the artificial propagation of the sponge of commerce. Prof. Oscar Schmidt, of the University of Gratz, has been so successful in his preliminary efforts in this direction, that the Austrian Government has authorized him to attempt the development of this industry on the coast of Dalmatia. The process is very simple, consisting in selecting the proper season in the spring, and dividing a living, marketable sponge into numerous pieces, and fastening them to stakes, which are driven into the sea bottom so as to submerge them. These fragments at once begin to grow out, and at the end of a certain time each one becomes an entire sponge.

According to Dr. Schmidt, three years is a sufficient length of time to obtain from very small fragments sponges worth several cents apiece. In one experiment the cost of raising 4,000 sponges amounted to about $45, including the interest for three years on the capital employed. The sales amounted to $80, leaving a profit of $35.

It was my intention to give a detailed account of the practical results of the work prosecuted by the Commission from the beginning, showing the aggregate of work done and the promise of future success, by the reappearance as adults of the young fish which had been planted in their localities. Owing, however, to necessary delay in the preparation and the publication of this report, it has been thought expedient to keep this history for the report of 1879, when it is hoped that sufficient evidence will be given to show that all reasonable anticipations of a successful outcome have been realized, and that the future holds in store great possibilities of ever-increasing food resources, which, so far as the United States is concerned, is to have a very important economical bearing.

It must be borne in mind, too, that the United States Fish Commission is only one of many in operation in the same direction in the country, very many states now having commissioners devoted to their work, and all more or less successful either in the artificial propagation of fishes in extending the distribution of species already occurring in the waters, or in the introduction and enforcement of protection of fishes during the critical periods, without which the most extensive efforts in fish culture will fail of their object.

D.—HUMAN AGENCIES AS AFFECTING THE FISH SUPPLY, AND THE RELATION OF FISH CULTURE TO THE AMERICAN FISHERIES.*

6.—INFLUENCE OF CIVILIZED MAN ON THE ABUNDANCE OF ANIMAL LIFE.

It may safely be said that wherever the white man plants his foot and the so-called civilization of a country is begun the inhabitants of the air, the land, and the water, begin to disappear. The bird seeks a new

* This article, exactly in its present form, was written for presentation elsewhere, but not published. It was intended to constitute a popular exposition of the subject to the end of 1878, and consequently includes to a considerable extent data contained in the previous pages.

abiding place under the changed conditions of the old, but the return of the season brings him again within the dangerous influence, until taught by several years of experience that his only safety is in a new home. The quadruped is less fortunate in this respect, environed as he is by more or less impassable restrictions, such as lofty mountains, deep rivers and lakes, and abrupt precipices, and sooner or later reaches the point of comparative extinction, or reduction to such limited numbers as not to invoke any continuance of special attack.

The fish, overwhelmingly numerous at first, began to feel the fatal influence in even less time than the classes already mentioned, especially such species as belong to the fresh waters and have a comparatively limited range.

The cause of this rapid deterioration is not to be found in a natural and reasonable destruction for purposes of food, of material for clothing, or other needs. The savage tribes, although more dependent for support upon the animals of the field and forest than the white man, will continue for centuries in their neighborhood without seriously diminishing their numbers. It is only as the result of wanton destruction for purposes of sport or for the acquisition of some limited portion only of the animal that a notable reduction is produced and the ultimate tendency to extinction initiated.

Of the abundance of animal life in North America, in the primitive days of its occupation by the European immigrant, we have an ample history in the accounts of the earlier travelers. Buffaloes in enormous hordes reached almost to the Atlantic coast, wherever extensive plains existed. The antelopes rivaled in numbers those of Central and South Africa. The deer of various species were distributed over the entire continent from the Arctic regions southward, and from the Atlantic to the Pacific. The moose existed far south of its present limit. The elk was a familiar inhabitant of Pennsylvania and Virginia. Wild fowl, such as ducks, geese, swans, &c., of many species, were found during the winter in countless myriads in the Chesapeake and other Southern bays and sounds.

Now what remains of this multitude? The buffalo has long since disappeared from the vicinity of the Mississippi River, the deer is nearly exterminated in many localities, though still holding its own under favorable circumstances, and the antelope is restricted to limited areas. The wild fowl, congregated at one time in bodies miles in extent, are now scarcely to be seen, although still proportionably more abundant in the winter season on the coast of California and towards the mouth of the Rio Grande in Texas than anywhere else.

Perhaps a still more striking illustration is seen in the fishes. It is still within the recollection of many old people (showing how plentiful the fish must have been) that the apprentice and pauper, in the vicinity of the Connecticut River, protested against eating salmon more than twice a week. This noble fish abounded in all the waters of New England as far west as the Connecticut and even to the Housatonic, though we have no evidence that they ever occurred in the Hudson River or farther to the south. The shad was found in every stream of the coast from Georgia to the Gulf of Saint Lawrence, and, although still ascending most of these waters during the spring, has been sadly reduced in abundance. Within even fifty years no waters of the same extent in the world could show such numbers of shad and herring as the Potomac River below the Great Falls. Martin's Gazeteer of Virginia, published in 1834, at Alexandria, states that the preceding year twenty-five and a half millions of shad were taken by the various Potomac fisheries, as well as seven hundred and fifty millions of fresh-water herring. This, by a moderate estimate, would amount to six hundred millions pounds of fish secured in six weeks in this single system of waters. This Gazeteer also states that during the same year nearly one million barrels of fish were packed on the Potomac, requiring as many bushels of salt. These were consumed in the United States or shipped to the West Indies and elsewhere. What is the condition of things at the present time? In 1866 the catch of shad on the Potomac had dwindled to 1,326,000, in 1878 to 224,000, the latter not 1 per cent. of the yield of 1833. The catch of herring in 1833, estimated, as stated, at 750,000,000, had been reduced in 1866 to 21,000,000, in 1876 to 12,000,000, and in 1878 to 5,000,000; again less than 1 per cent. of the yield of the first-mentioned period.

A similar reduction has taken place in the abundance of the striped bass or rock-fish, a species inferior to none in its excellence and economical value for food. John Josselyn, gent., in 1660, says that three thousand bass were taken at one haul of the net in New England. Thomas Morton, in 1632, says, of the Merrimac, that he has seen stopped in the river at one time as many fish as would load a ship of a hundred tons, and that at the going out of the tide the river was sometimes so full of them that it seemed if one might go over on their backs dry-shod.

Mr. Higginson, in 1630, says that the nets usually took more bass than they were able to land. Even so recently as 1846, one hundred and forty-eight tons are said to have been taken on Martha's Vineyard at two hauls of the seine. *Per contra*, the catch in the Potomac in 1866 amounted to 316,000 pounds; in 1876, to 100,000; in 1878, to 50,000.

Many more instances of the enormous abundance of the anadromous fishes (marine species running up from the ocean into fresh waters for the purpose of spawning) in different parts of the country in former times could easily be adduced. Similar illustrations of the former abundance of fishes exclusively inhabitants of the salt water can be brought forward to any extent. In the early days of the Republic the entire Atlantic shore of the United States abounded in fish of all kinds. Where cod, mackerel, and other species are now found in moderate quantities, they occurred in incredible masses.

The halibut, one of the best of our fishes, was so common along the New

England coast as not to be considered worthy of capture, and was considered a positive nuisance when taken. It is only within a few years that our people have come to learn their excellence and value, but they have already disappeared almost entirely from the inshores of New England, and have even gradually become exterminated in nearly all waters of less than five hundred feet in depth.

The inquiry now arises as to the causes of the terrible depletion of the inhabitants of the water, and one so detrimental to human interests. The question relates in part to an actual extermination, and in part to a disappearance from accessible fishing-grounds. The practical result to the fishing interest is about the same in either case.

It is quite safe to assume that most species of the ocean fishes, in their abundance and ability to escape the pursuit of man, are less amenable to destructive influences than those of the interior waters, the halibut being perhaps one of the few exceptions of a species that may be considered actually exterminated over a certain area. That the supply of nearly all other kinds in the inshore fisheries of America everywhere has diminished in enormous ratio is unquestioned. What were and are the causes, and what the remedy?

One most plausible solution of the problem is to be found in the very close relationships between the so-called anadromous fishes and those permanently resident in the ocean. The anadromous species are represented by the salmon, the shad, the fresh-water herring or alewife, and some other kinds, which, although spending the greater part of their life in the ocean, periodically enter the fresh waters, in greater or less numbers, and ascend as high as they can for the purpose of finding suitable places wherein to deposit their spawn. This done, the parent fish soon returns, leaving the young to follow. The young shad or herring remain in the rivers three or four months and then go down to the ocean. The salmon is more persistent, the young remaining from one to two years, after which they too descend to the sea, and, like the shad and herring, for the most part there attain their entire growth. It is not thought that either the parent fish or the young go to any great distance from the mouths of the rivers, and it is believed that the fish born in one stream never think of entering any other than that in which they first made their appearance.

Bearing in mind the countless myriads of these fishes formerly entering our rivers—the shad and herring along the entire coast of the United States to the Bay of Fundy, the salmon from the Connecticut eastward—and noting the extent to which they are preyed upon by the more rapacious inhabitants of the sea, we may understand why such multitudes of the larger fish formerly approached the shores in pursuit until deterred by the increasing shoalness or freshness of the water. Even then, however, they would remain near the shore, lying in wait for the parents and their young returning in such vast quantities during the later months of the year. In all probability these constituted a chief inducement to the movement of the predacious fish to the coast in such numbers during the spring and summer. In autumn and winter the sea-herring and the fish of the cod family visit the shores for quite another purpose, namely, to deposit their eggs. But from whatever motive, the fact remains that years ago throughout the twelve months an ample supply of the finest fishes was within the reach of everyone, so that a fisherman with a small hand-line and an open boat was able to support his family without any difficulty.

Now, with the continued reduction in abundance of the salmon, shad, and fresh-water herring, the summer fisheries have dwindled and nearly disappeared, leaving only those of winter with its inclement weather to furnish occupation to the fishermen, and compelling him in the most dangerous season of the year to betake himself to the Georges, La Have, Quereau, and other banks, especially to the Grand Banks of Newfoundland, to prosecute his work in expensive vessels, and exposed to perils and privations of a terrible character.

Assuming, then, that the chief agency in the decrease of the ocean-shore fisheries has been the reduction in the number of the anadromous fish, *i. e.*, those passing up from the ocean into the fresh waters to spawn, let us inquire into the causes of the diminution of the latter. They certainly were very plentiful in the early days of European colonization in America, but at that period all the rivers were open to the sea, without dams or other artificial obstructions. Few or no saw-mills cast into them sawdust and other refuse; no gas-works polluted them with coal-tar, creosote, &c., and paper-mills, factories, &c., running off poisonous compounds, were unknown. The fishing apparatus was confined to lines and nets of no great extent, not sufficient to barricade the streams and impede the upward movement of the fish.

After the settlement of the country began, these possible dangers came to have an actual existence. It is probably to the erection of dams, however, that the first great diminution was due. The salmon, the shad, and the herring proceeding from the ocean to the headwaters of their native stream, were met by an impassable barrier, which they were unable to surmount, cutting them off from their favorite spawning-ground, and, indeed, in many cases, from the only localities where the operations of reproduction could be properly performed. They wore themselves out in fruitless attempts to overcome these obstructions, and were compelled finally to return to the ocean without depositing, or at least utilizing, their spawn. A second year, a third, and even a fourth would probably make but little difference in the number making the attempt to ascend, this being due to the fact that four years is the average period from birth at which most fish are mature and able to exercise the reproductive act. By the end of the fourth year, the last crop of young fish hatched in the upper waters of the river will have made its appearance as mature males and females. After this the diminution takes place with increasing rapidity until, five or six years afterwards, the fish are found to have disappeared entirely from the stream. So much for the dams. As for the other causes, sawdust and other refuse matter get into the gills of fish and produces irritation and subsequent death. Coal-tar refuse is known to be a very great detriment to the healthful condition of water so far as fish are concerned, and it is probable that a part, at least, of the decrease of shad and herring in the Potomac is due to the discharge from the gas-works of Washington and Alexandria.

The rapid increase in the size and number of the nets, whether pounds, seines, drift or gill nets, that has manifested itself within the last twenty years has doubtless had a similar effect with the dams in producing a decrease. The fish are harassed and worried by them, and hindered in an equal degree from reaching their spawning-ground, and thus another drain on the supply is added to the many already in operation.

What, now, are the remedies to be applied to recover from this lamentable condition of the American fisheries (a condition which, we may remark, has existed in all countries of Europe, but which in some of them has already been greatly lessened by the proper measures)? These are twofold. One consists in the enactment and enforcement of legislation protecting what we have, and allowing natural agencies to play their part in the recovery; the other consists in the application of the art of artificial propagation of the fish. Either, alone, in some circumstances, will answer a very good purpose. The two combined constitute an alliance which places at our command the means of recovering our lost ground to a degree which, but for the experience of the last ten years, would hardly be credible.

7.—POLITICAL AND SOCIAL IMPORTANCE OF INCREASE OF FISH SUPPLY.

Now let us glance at the importance in the political economy of the United States of an increase in the supply of fish for food. We are at present a people of 49,000,000 souls, which, by the end of the present century, will probably amount to double the number. The production of animal food on land depends in large part upon the amount of soil available for grazing; but, with the increase of population, the necessity of a more lucrative yield makes it imperative to prosecute the cultivation of the cereals or other articles of direct food to man, thus restricting the area of pasture-lands. Many countries of Europe have already reached that period when they look to foreign nations for their supply of animal food. America furnishes a great part; the less populated regions of Europe the remainder. The increase in the price of what is called "butcher's meat," though gradual, is inevitable, and every year a larger and larger percentage of the population will be unable to secure it. In this emergency we must look to the water for the means of supply. In former days the inhabitants of the sea-coast and rivers obtained a very large portion of their animal food from the water; and in proportion as this state of affairs is restored will the condition of the future population be improved. The legislation required consists in the enactment of laws for the introduction of fish-ladders, by means of which the spawning-fish can reach the headwaters of the rivers; in a prohibition against discharging sawdust, gas-refuse, chemicals, &c., into the water; in a limitation as to the pounds, number, and size of mesh of nets, and especially in the establishment of close seasons during the week, during which the capture of fish by nets shall be forbidden, and an absolute prohibition of their capture after a certain date in the year. These dates will necessarily vary with the kinds of fish to be protected.

8.—MODE OF INCREASING THE SUPPLY OF FISH.

Even, however, with all these regulations, supposing them to be thoroughly enforced, there remains much to be done. Our rivers, capable of accommodating very many tons of fish, must be restocked, or there will be no result from our labors. This is not to be accomplished by the transfer of the parent fish from one point to another, especially as the shad and alewife will not survive a few moments removal from the water. It is through artificial propagation that the restoration of certain species of fish to their former place of abode, and the introduction of fish to waters where they were before unknown, is to be accomplished.

Fish-culture and fish-rearing, in a certain sense, are nothing new. The Chinese and other Oriental nations have practiced a form of the art for ages. In Europe it has been prosecuted for centuries. The transfer of fish from one sea to another was accomplished by the Romans of old. The bringing of fish into restricted waters, where they are supplied with food and allowed to grow and multiply, or even the gathering of eggs after they have been laid and impregnated by the fish, represents the so-called fish-culture of China. The young fish hatched under favorable circumstances are supplied with food and reared carefully, sometimes even in tubs or jars, and in the course of a few years furnish a remunerative return to their owners. Such nurture or maintenance of fish, under circumstances when they can multiply and attain their growth, is, however, not fish-culture proper in its modern sense. This is based upon the artificial impregnation of the fish and is practiced by stripping the eggs from mature females; by fertilizing these eggs by the milt of the male; by placing the eggs thus fertilized in a condition favorable for their development, and by the protection of the resultant fish until they are able to take care of themselves; they may be then kept indefinitely in ponds or turned out at once into suitable waters. It is this operation which has constituted the basis of recent effort, and which has been crowned with such triumphant success.

At first sight it would seem impossible for man with his limited opportunities and means to compete with or even to supplement nature in the process of maintaining or increasing the supply of a certain species by artificial impregnation, but a consideration of the subject will show what really can be claimed. The process of natural impregnation of fishes is, for the most part, external; that is to say, the eggs of the female and the milt of the male are discharged at or about the same time into the water, the two being close together, so that, as far as they come in contact with the milt, the eggs become fertilized. Observation, however, has shown that a very small proportion of the eggs are actually fertilized, possibly 10 per cent. being a liberal estimate. Again, these eggs, some of which remain three months before they are developed, some but a few days, or even hours, are exposed continually to the attacks of vast hordes of animals of all sizes, especially minnows, crabs, frogs, birds, &c., by which still another large percentage is consumed.

Still further: The young fish when hatched out is almost as helpless as the egg, being unable to defend itself from danger, and is devoured with great eagerness by the same class of enemies, as being an especial delicacy. As a general rule it is believed that a yield of five young fish, with the yolk-bag absorbed, and the fins fully formed, and able to take care of themselves, is a liberal allowance for each thousand eggs. By the artificial method of propagation, 90 per cent. of the eggs should be thoroughly fertilized, and when the fish is hatched out it is kept from its adversaries until able to look out for itself. It may safely be assumed that eight hundred and fifty fish out of a thousand eggs are produced artificially as compared with the five in a thousand produced naturally. The ratio of production may, in round numbers, be claimed to be nearly two hundred to one in favor of artificial production, possibly much more.

A few words in regard to the history of this wonderful art may not be amiss. Among the first to practice artificial hatching was a German named Jacobi, who, about the middle of the last century, announced the success of experiments with the German trout. There are vague rumors of something earlier, but they are not matters of history. In 1844, Remy and Gehin, two illiterate Frenchmen, rediscovered the art and brought it to the notice of the French Government, by which they were liberally rewarded, and steps were taken to exercise it, although with but little result, notwithstanding the efforts were directed by eminent naturalists, such as Quatrefages and Coste. At a later day, however, the practice of artificial impregnation of fish-eggs in Europe became more and more common, until at the present time there are several national and a large number of state and private establishments occupied, for the most part, in hatching and rearing the various species belonging to the trout and salmon family.

In America, the first practical action in the way of artificial production of fish was begun in 1853 with the trout. The experimenters were Drs. Garlick and Ackley, of Cleveland, Ohio; and to them we owe the initiation of actual fish-culture in this country. With the proof of success attending their efforts, as shown by these gentlemen in their fish-ponds near Cleveland, it was not long before many persons entered the same field, the trout, as before, being at first the exclusive object of attention.

In 1867, the attention of the New England commissioners of fisheries was directed to the possibility of increasing by artificial means the abundance of the shad, the enormous diminution of which was felt to be a serious evil. It is to Seth Green, of New York, that we owe the idea of the possibility of reproducing shad and the initiation of the steps necessary to carry it into effect. While the methods of stripping the parents and of impregnating the eggs were essentially the same, the devices employed for hatching out the eggs of the trout were all found to be unavailable, not only unfit in themselves, but powerless to accomplish the work on a sufficiently large scale to make it of any economical value. A floating box with a wire-cloth bottom first suggested itself to Mr. Green. This was filled with impregnated eggs, and anchored in the river, where it occupied a horizontal position, and a partially successful result obtained. It was found, however, that in a horizontal box the eggs were not sufficiently exposed to the action of the water and that they collected in the ends and corners, where the larger number perished. The idea then occurred to Mr. Green that by nailing two parallel strips of wood obliquely across the opposite sides of the box, the bottom would be maintained obliquely to the water. The experiment worked like a charm, and from that time until very recently the Green floating box has been the apparatus almost universally employed for this purpose.

It is impossible here to go into any further account of the numerous modifications of fish-hatching apparatus for special cases and particular kinds of fish, although I shall refer hereafter to certain improved devices now employed by the United States Fish Commission as superior to all others known to it.

As already stated, the successes of Drs. Garlick and Ackley induced great numbers of people to take part in the work, but it was not until about the time that Seth Green obtained a patent for his invention that any State action was brought into play on a large scale in hatching shad, although something had previously been done with salmon. It was about this time that the New England States appointed commissioners of inland fisheries to see that certain legislation was enforced, and to take measures for the improvement of the general supply. Other States followed, and now there are nearly thirty having fish commissioners appointed to attend to the subject.

9.—OPERATIONS OF THE UNITED STATES FISH COMMISSION.

A new era in the history of fish culture was entered upon in the establishment of the United States Fish Commission in 1871. Its original object, as authorized by law of Congress, was an investigation into the causes of the alleged decrease of the fishes of the sea-coast and lakes, and the recommendation of measures for their restoration. The Commission was organized by the appointment of Spencer F. Baird, then assistant secretary of the Smithsonian Institution, as Commissioner, and his first work was prosecuted during the summer of 1871, along the New England coast. Many investigations were made and a rational theory in regard to the condition and improvement of certain coast fisheries was prepared and published by him in the following year.

In 1872, a committee of the American Fish Culturists Association urged upon Congress the importance of an appropriation to supply useful food-fishes to such rivers and lakes of the country as were the common property of the nation, but which, not being under the jurisdiction of one State, had been left unattended to. An appropriation was ultimately made and put at the command of the United States Commissioner. These appropriations have been made year by year, and year by year new varieties of fish have been taken into consideration, and the field of operations extended, although still confined almost exclusively to species of national importance, and their introduction into rivers and lakes which State or private enterprise cannot cover.

Among the more important species now cared for by the United States Fish Commission, may be mentioned the shad, the fresh-water herring, or alewives, the striped bass, the salmon of Maine, the land-locked salmon of Maine, the salmon of California, and the German carp.

The importance of increasing the supply of shad already existing in a given river is easily appreciable, and the desirability of introducing them into rivers where they had been previously unknown is equally evident. As the result, partially or entirely, of the efforts of the United States Commission, the Sacramento River, and many streams of the Mississippi Valley and of the Gulf of Mexico, where this fish was previously unknown, have been largely stocked with it, and it is hoped that in a few years it will constitute a very important element of the food supply of the country. A statement of what has been done in this connection will be found in the reports of the Commission. Thus, in the year 1873, about 200,000 diminutive fish, averaging a quarter of an inch in length, were placed in the headwaters of the Alleghany River, in Wesern New York. These fish, or such of them as escaped the perils of infancy, passed down to the Gulf of Mexico and there obtained their growth. In 1877, or at the end of the four years required for their full development, they re-entered the Mississippi on their return to the place whence they had started in 1873. On their passage upward they passed Louisville at a time when the river seines were in full operation, and the fishermen were surprised to find among their hauls large numbers of fine-looking fish of a kind entirely unknown to them. It was soon shown, however, by those familiar with this famous fish, that they were the genuine white shad, of which it is estimated that no less than 600, from three to five pounds in weight, were taken during the run past Louisville. Additional captures were recorded at other points of the Ohio and its tributary rivers. Specimens of these shad are now carefully preserved in the National Museum.

The eastern salmon has for many years been unknown in the waters of the United States, except to a limited degree in the Kennebec, Penobscot, and other streams of Maine. The work of restocking the original haunts with this fish was commenced in 1866, by the State of New Hampshire, and followed subsequently by several of the New England States, and in 1872 with the very important co-operation of the United States Fish Commission, which of late years has borne the chief expenses of the outlay.

What, now, has been the result, and especially in the Connecticut River, which formerly abounded in large numbers of the salmon, and which has been the principal scene of operations? Young salmon in greater or less numbers have been introduced by the States of Massachusetts, Connecticut, Vermont, and New Hampshire, beginning in 1867, but not in any considerable quantity, until supplemented and strengthened by the United States Commission in 1873, which from that time took the lead in the production.

Great incredulity had been manifested by most persons as to any practical result from artificial propagation, and, as year after year passed without bringing the expected run of salmon into the Connecticut, sneers and jokes at the expense of the United States and State commissions multiplied. The occurrence, however, of one or two large salmon in the Connecticut in 1876, and of a dozen in 1877, interfered with this skepticism, which was changed into enthusiastic appreciation by the appearance in 1878 of large numbers of fine, fat salmon, such as have not been seen in the river for many years. No less than 500 fish, each of from ten to twenty pounds in weight, were captured at the mouth of the river, and sold in the New York market for the most part, at prices ranging from 75 cents to a dollar per pound. This, in all probability, did not represent anything like the number of fish that entered the river, but merely those that were taken in the shad-nets, apparently very imperfectly and ill-adapted to the capture of so heavy a fish. Increasingly larger and larger yields may be expected in the future, at least up to 1880; their continuance beyond that time may depend upon the legislation of the States through a part of which the Connecticut river flows.*

The California salmon has great advantage over the ordinary species in much greater hardiness and capacity for existence in waters warmer by many degrees than those to which the eastern salmon is habituated. It has been introduced by millions in the tributaries of the Mississippi, the Gulf of Mexico, and the Atlantic Ocean. It has been transported to Australia, New Zealand, and the Sandwich Islands, to Germany, France, and the Netherlands, where the eggs have been thankfully received, hatched out with perfect success, and successfully planted. Wherever taken it has been looked upon as one of the most important subjects of fish-culture. Without any exception, the distribution of the

* The experience in the Merrimac and the Delaware Rivers, and to a like degree on the Susquehanna is much the same as that mentioned for the Connecticut. Specimens of salmon from all these rivers are preserved in the National Museum.

California salmon has been made under the auspicies of the United States Fish Commission.

The German carp, one of the latest species that has occupied the attention of the United States Commission, is one of the most important. There are many varieties, three in particular being best known. Of these, one is covered with large scales, something as in the goldfish; another has lost all the scales, except along the lateral line, while the third is entirely destitute of scales. These are known respectively as the scale, mirror, and leather carp.

The carp has been domesticated in Europe from time immemorial, and represents among the finny tribe the place occupied by poultry among birds. It is a fish adapted to the farmers' ponds and to mill-dams, less so to clear gravelly rivers with a strong current. Where there is quiet water with a muddy bottom and abundant vegetation, there is the home for the carp; there it will grow with great rapidity, sometimes attaining a weight of three to four pounds in as many years. It is a vegetable-feeder and not dependent upon man for its sustenance. As an article of food the better varieties rank in Europe with the trout, and bring the same price per pound.

I have already referred to the use of the Seth Green floating box for the hatching out of shad. I now call attention to the very important improvement, in greater part the invention of Mr. Ferguson, fish commissioner of Maryland, by which the floating box has been superseded by a new apparatus worked by steam, in which a thousand shad can be produced with the same facility as a single one by the old method.

As the result of the first year's experiments with the new apparatus, in 1878, of the United States Fish Commission, 16,000,000 shad were hatched out, and in large part deposited in streams all over the country by the Commission's messengers.

During the past winter of 1878–'79, the United States Fish Commission has, however, made a step far in advance of its previous efforts, and of the most novel and striking character. While the establishment and increase of the fresh-water fisheries has been of the utmost importance, especially those of the anadromous species, the Commission has of late been considering the possibility of artificially multiplying the marine species, confident that by this measure a vastly greater sphere of usefulness will be entered upon. The first experiments have been made with the cod, a fish which is the staple of American marine industry, and which involves the investment of a large sum of money and the labor of many thousands of men. This visits the coast of New England in the winter for the purpose of spawning; Cape Ann, Mass., being an especially favorite ground for the purpose.

A temporary establishment was fitted up in the autumn of 1878, in Gloucester Harbor, Massachusetts, for the purpose of a series of preliminary experiments looking towards the artificial propagation of the cod, commencing in November. During these trials many difficulties were encountered and overcome. It was found that the principle on which the work was to be done was diametrically opposite to that used in connection with the fresh-water varieties, as the egg of the cod floats on the top of the water instead of sinking to the bottom. This obstacle was finally surmounted, and many millions of the young cod were hatched out and planted in the ocean adjacent to Gloucester Harbor. A number of these fish were sent to Washington.

It is now believed to be possible not only to greatly increase the supply of the cod where it is at present found, but, by carrying the young to new localities, to establish cod fisheries as far south as the coast of North Carolina, where the fishermen may find regular occupation during the winter—now his poorest season—in capturing these fish in large quantities and supplying the adjacent markets or even exporting them.

The same apparatus and mode of treatment can be used for hatching mackerel, halibut, sea-herring, and other species, so that we have at our command the means of so improving and increasing the American fisheries as to obviate the necessity in the future of asking a participation in the inshore fisheries of the British provinces and thus enable us to dispense with fishery treaties or fishery relations of any kind with the British or other governments.

TABLE I.—*Distribution of shad, from April 11 to June 14, 1878, by the United States and Maryland Commissions Fish and Fisheries.*

Introduction of fish.				Date of transfer.	Transfer in charge of—	Number of fish.
State.	Stream.	Tributary of—	Town or place.			
Alabama	Tombigbee River	Mobile Bay	Demopolis	Apr. 13, 1878	J. F. Ellis	116,000
	Escambia River	Pensacola Bay	Pollard	May 24, 1878	F. A. Ingalls	100,000
	Tallapoosa River	Alabama River	Salisbury	June 9, 1878	J. M. Donaldson	50,000
Arkansas	Sabine River	Washita River	Benton	June 1, 1878	H. E. Quinn	50,000
	[illegible]o River	do	[illegible]lphia	June 1, 1878	do	50,000
California	Sacramento River	Pacific Ocean	Teharra	June 11, 1878	F. N. Clark	150,000
District of Columbia	Potomac River	Chesapeake Bay	Washington	May 27, 1878	United States Fish Commission	400,000
Georgia	Ocmulgee River	Altamaha River	Macon	Apr. 25, 1878	H. E. Quinn	60,000
	Flint River	Appalachicola River	Albany	Apr. 25, 1878	do	60,000
	do	do	Montezuma	May 25, 1878	J. F. Ellis	150,000
	Etowah River	Coosa River	[illegible]lle	June 9, 1878	J. M. Donaldson	50,000
Illinois	Kaskaskia River	Mississippi River	Farlow	May 27, 1878	C. W. Scheurmann	100,000
	Rock River	do	Rickford	June 2, 1878	do	100,000
	Embarrass River	Wabash River	Charleston	June 9, 1878	W. H. Hines	50,000
	Mississinowa River	do	Marion	June 9, 1878	do	50,000
Indiana	Elkhart River	Lake Michigan	Elkhart	June 1, 1878	J. F. Ellis	100,000
	Wabash River	[illegible]o River	Terre Haute	June 9, 1878	H. E. Quinn	65,000
Iowa	Des Moines River	[illegible]pi River	[illegible]r Rapids	June 8, 1878	B. F. Shaw	24,000
	Boyer River	do	Logan	June 8, 1878	do	36,000
Kentucky	Cumberland River	Ohio River	Somerset	May 26, 1878	H. E. Quinn	60,000
	Green River	do	McKinney's Station	May 26, 1878	do	60,000
			High Bridge	June 1, 1878	F. A. Ingalls	175,000
	Green River	Ohio River	Bowling Green	June 3, 1878	William Russ	100,000
Louisiana	Amite River	Lake Ponchartrain	Tickfaw	May 27, 1878	W. M. Russ	100,000
Maryland	Potomac River	Chesapeake Bay	Various points	Apr. 21, 1878	William Hamlen	100,000
	do	do	Potomac Point	May 1, 1878	do	100,000
	Choptank River	do	Greensborough	May 17, 1878	Thomas Hughlett	100,000
	Potomac River	do	[illegible]co Creek	May 23, 1878	United States Fish Commission	50,000
	Nanticoke River	do	Federalsburg	May 24, 1878	Thomas Hughlett	100,000
	Potomac River	do	Glymont	May 24, 1878	United States Fish Commission	75,000
	Susquehanna River	do	Havre de Grace	May 25, 1878	do	150,000
	North East River	do	[illegible]ll's [illegible]t	May 26, 1878	do	350,000
	Susquehanna River	do	Havre de [illegible]ace	May 26, 1878	do	100,000
	Spesutie Narrows	do	do	May 27, 1878	do	200,000
	do	do	do	May 28, 1878	do	400,000
	Potomac River	do	Fort Washington	May 28, 1878	do	100,000
	do	do	Glymont	May 28, 1878	do	300,000
	[illegible]o River	do	Salisbury	May 29, 1878	Thomas Hughlett	175,000
	Manokin River	Tangier Sound	Princess Anne	May 29, 1878	do	75,000
	Spesutie Narrows	Chesapeake Bay	Havre de Grace	May 29, 1878	United States Fish Commission	1,500,000
	do	do	do	May 30, 1878	do	500,000
	Chester River	do	Millington	May 31, 1878	[illegible]as Hughlett	250,000

	Tread [illegible]	[illegible]	Easton	[illegible] 1, 1878	S. M. Rixey	50,000
	[illegible]	[illegible] B[illegible]y	N[illegible]ar [illegible]	[illegible] 1, 1878	do	75,000
	[illegible] Narrows	[illegible] Bay	[illegible] de [illegible]	[illegible] 1, 1878	[illegible] Fish [illegible]ommission	500,000
	do	do	do	[illegible] 3, 1878	do	500,000
	Choptank [illegible]	[illegible]	Green [illegible]	June 3, 1878	S. M. [illegible]	100,000
	Tuck[illegible] River	Ch[illegible] [illegible]	[illegible]	[illegible] 3, 1878	do	50, 0
	[illegible]	Ch[illegible] Bay	Laurel	[illegible] 5, 1878	J. M. [illegible]	150, 0
	do	do	do	[illegible] 6, 1878	David [illegible]tt	50, 0
	[illegible] N[illegible]ows	do	[illegible] de [illegible]	[illegible] 6, 1878	United S[illegible]s F[illegible]h [illegible]mission	500, 0
	[illegible]	do	S[illegible]w Mill	[illegible] 10, 1878	S M. Rixey	150,000
	Bush [illegible]	do	[illegible] c[illegible]sing	June 10, 1878	W. F Page	150,000
	Spesutie N[illegible]ws	do	[illegible] de [illegible]	[illegible] 10, 1878	United S[illegible]s [illegible]h [illegible]mission	100, 0
	do	do	do	[illegible] 11, 1878	do	185,000
	Gunpowder River	do	[illegible]	[illegible] 14, 1878	U. Simmons	80,000
	[illegible] Narrows	do	[illegible] de [illegible]	[illegible] 14, 1878	[illegible] t[illegible]es [illegible]h [illegible]mission	120,000
Mississippi	Big [illegible]	M[illegible]sippi [illegible]	Vaughn Station	Apr. 13, 1878	C. W. [illegible]	144,000
	[illegible] River	Y[illegible]oo [illegible]	Friar's [illegible]	Apr. 23, 1878	J. F. [illegible]	100,000
	[illegible]	do	H[illegible]y Springs	Apr. 23, 1878	[illegible]W. Schuermann	40, 0
	[illegible]	do	Railroad crossing	Apr. 23, [illegible]	do	40, 0
	[illegible]	do	[illegible]	Apr. 23, 1878	do	40,000
	[illegible]	[illegible] Bay	[illegible]	May 15, 1878	J. F. Ellis	90, 0
	do	do	[illegible]	[illegible] 15, 1878	do	60, 0
	[illegible]	[illegible] River	Meridian	May 23, 1878	R. E. Earle	100, 0
Missouri	Sal[illegible]	[illegible]	N[illegible]	[illegible] 20, 1878	C. W. [illegible]	120, 0
	[illegible]pi [illegible]	[illegible] of [illegible]	Saint [illegible]	[illegible] 8, 1878	F. A. Ingalls	100, 0
North Carolina	Salmon [illegible]	[illegible]	[illegible]	Apr. 11, 1878	United S[illegible] Fish Commission	2,500
	Roanoke River	Albemarle [illegible]	[illegible]	Apr. 11, 1878	W. G. [illegible]	139,000
	N[illegible]	[illegible] Sound	N[illegible]	Apr. 12, 1878	S. G. Worth	100,000
	[illegible] River	[illegible]	Railroad [illegible]	Apr. 12, 1878	W. G. [illegible]	150,000
	Salmon [illegible]	do	The Mill	Apr. 15, [illegible]	[illegible] Fish [illegible]mission	120,000
	N[illegible]	[illegible] Sound	Raleigh	Apr. 20, 1878	H. E. [illegible]	50 0
	[illegible]	do		Apr. 22, [illegible]	[illegible]	100,000
	[illegible]	[illegible] River	Avoca	Apr. 24, 1878	United States Fish Commission	800,000
	Cape Fear [illegible]	[illegible]	[illegible]ville	Apr. 24, 1878	Col L. L. Polk	100,000
	[illegible] Creek	[illegible]	[illegible]	Apr. 25, 1878	United States Fish Commission	200,000
	[illegible]	Albemarle [illegible]	[illegible]	Apr. 25, 1878	do	200,000
	[illegible] Sound	[illegible]	[illegible] Hill Fishery	Apr. 25, 1878	do	115,000
	[illegible]	[illegible] Sound	[illegible]	Apr. 26, 1878	do	250,000
	[illegible]	do	[illegible]	Apr. 26, 1878	do	25,000
	[illegible] Sound	[illegible]	[illegible]	Apr. 26, [illegible]	do	70, 0
	[illegible]	Cho[illegible] [illegible]	[illegible]	Apr. 26, 1878	Thomas Taylor	200,000
	Salmon [illegible]	do	A[illegible]	Apr. 28, 1878	United States Fish [illegible]mission	300,000
	do	do	do	Apr. 29, 1878	do	18, 0
	do	do	do	[illegible] 1, 1878	do	45,000
	do	do	do	May 2, 1878	do	200,000
	Yadkin [illegible]	[illegible] River	[illegible]	[illegible] 7, 1878	S. G. Worth	50,000
	[illegible]	[illegible]	[illegible] Station	[illegible] 7, 1878	do	50,000
Ohio	S[illegible]y [illegible]	Lake Erie	[illegible]	[illegible] 9, 1878	H. E. [illegible]	60,000
South Carolina	[illegible]	[illegible] Bay	Railroad [illegible]sing	[illegible] 1, 1878	S. G. Worth	50,000
	Broad River	Santee [illegible]	[illegible]	June 1, 1878	do	50,000

TABLE I.—*Distribution of shad from April 11 to June 14, 1878, &c.*—Continued.

Introduction of fish.				Date of transfer.	Transfer in charge of—	Number of fish.
State.	Stream.	Tributary of—	Town or place.			
Tennessee	Middle fork of Forked Deer River.	[illegible]d Deer River	Humboldt	May 18, 1878	H. E. Quinn	50,000
	South fork of Obion River	Mississippi [illegible]r	Huntingdon	May 18, 1878	do	50,000
Virginia	Nottoway River	[illegible]n [illegible]r	Nottoway Mills	Apr. 11, 1878	W. G. Williamson	111,000
	James River	[illegible]e Bay	Richmond	Apr. 13, 1878	do	115,000
	South branch Nansemond River	[illegible]s River	Seaboard and Roanoke Railroad crossing.	Apr. 22, 1878	H. B. Nicholas	40,000
	Roanoke River	[illegible]e Sound	Salem	Apr. 24, 1878	W. F. Page	100,000
	Appomattox River	James River	Petersburg	Apr. 25, 1878	H. B. Nicholas	60,000
	Mattapony River	York [illegible]r	Milford Station	Apr. 30, 1878	do	90,000
	do	do	do	May 1, 1878	H. D. Johnson	90,000
	Little River	South Anne [illegible]r	Taylorsville	May 1, 1878	W. F. Page	100,000
	Blackwater River	Chowan [illegible]r	Franklin	May 2, 1878	Thomas Taylor	150,000
	do	do	do	May 2, 1878	do	100,000
	South branch Nansemond River	James [illegible]r	Seaboard and Roanoke Railroad crossing.	May 3, 1878	Page & Johnson	50,000
	Potomac River	Chesapeake Bay	Freestone	May 23, 1878	United States Fish Commission	50,000
	Shenandoah River	Potomac River	Riverton	May 28, 1878	W. F. Page	200,000
	Rivanna River	James River	Shadwell	June 1, 1878	do	175,000
	South River	Shenandoah River	Waynesborough	June 4, 1878	do	150,000
						15,700,500

TABLE II.—*Distribution of California salmon reared from eggs collected in 1878.*

States.	Where finally hatched.	Waters stocked.	Tributaries in which fish were placed.	Locality.	Date of transfer.	Number of fish.
California	United States hatchery	Sacramento River	McCloud and Little Sacramento Rivers.	McCloud and Sacramento, Cal	Oct. —, 1878	2,000,000
Illinois	Geneva Lake hatching-house	Fox River	Geneva Lake	Geneva, Wis	1879	200,000
	do	do	Crystal Lake	Crystal Lake, Ill	1879	20,000
	do	Illinois River	Rock River	Rockford, Ill	1879	50,000
	do	do	Fox River	Cacy, Ill	1879	20,000
Iowa	Anamosa, Iowa	Missouri River	Sioux River		Jan. 27, 1879	4,000
	do	Mississippi River	Mud Lake		Jan. 31, 1879	10,000
	do	do	Skunk River		Jan. 31, 1879	10,000
	do	do	Wall Lake		Jan. 31, 1879	10,000
	do	do	Towner's Lake		Jan. 31, 1879	7,000
	do	do	Iowa River		Jan. 31, 1879	7,000
	do	do	Des Moines River		Feb. 12, 1879	10,000
	do	do	East [illegible] River		Feb. 12, 1879	10,000
	do	do	Middle River		Feb. 12, 1879	16,000
	do	do	Wall Lake		Feb. 12, 1879	20,000
	do	do	Maple River		Feb. 12, 1879	10,000
	do	do	Maguoketa River		Feb. 20, 1879	10,000
	do	do	Turkey River		Mar. 20, 1879	3,000
	do	do	Lisborn River		Mar. 24, 1879	4,000
	do	do	Wapsee River		Mar. 10, 1879	10,000
	do	do	Big Rock River		Mar. 17, 1879	3,000
	do	do	Boone River		Mar. 22, 1879	10,000
	do	do	Upper Des Moines River		Mar. 22, 1879	10,000
	do	Missouri River	Plymouth River		Mar. 22, 1879	15,000
	do	Mississippi River	Cedar River		Mar. 28, 1879	5,000
	do	do	East Skunk River		Apr. 1, 1879	2,500
	do	do	West Skunk River		Apr. 1, 1879	2,500
	do	do	Maltby Kellogg		May 14, 1879	2,500
	do	do	Streams along C. B. and Q. R. R.		May 19, 1879	12,000
	do	do	Independence River		May 20, 1879	4,000
	do	do	Volga River		May 20, 1879	5,000
	do	do	Turkey River		May 20, 1879	5,000
	do	do	Cedar River		May 30, 1879	10,000
	do	do	Spring Branch		May 30, 1879	5,000
Kansas	Cedar Rapids, Iowa	Missouri River	Stranger River	Stranger, Kans	1879	2,500
	do	do	Verdigris River	Independence, Kans	1879	2,500
	do	do	Delaware River	Delaware, Kans	1879	3,000
	do	Kansas River	Red Vermillion River	Centralia, Kans	1879	2,000
	do	do	Spring Creek	Wetmore, Kans	1879	1,000
	do	Big Blue River	Mill Creek	Washington, Kans	1879	1,000
	do	do	Black Vermillion River	Frankford, Kans	1879	2,000
	do	Kansas River	Bear Creek	Barretts, Kans	1879	1,000

TABLE II.—*Distribution of California salmon reared from eggs collected in 1878*—Continued.

States.	Where finally hatched.	Waters stocked.	Tributaries in which fish were placed.	Locality.	Date of transfer.	Number of fish.
Kansas	Cedar Rapids, Iowa	Kansas River	Big Blue River	Blue Rapids, Kans	1879	5,000
	do	Big Blue River	Little Blue River	Waterville, Kans	1879	2,000
	do	Kansas River	Republican River	Concordia, Kans	1879	5,000
	do	do	Solomon River	Beloit, Kans	1879	5,000
	do	do	Soldier River	Topeka, Kans	1879	1,000
	do	do	Silver Lake	Silver Lake, Kans	1879	500
	do	do	Vermillion River	Wamego, Kans	1879	2,500
	do	do	Big Blue River	Manhattan, Kans	1879	5,000
	do	do	Republican River	Junction City, Kans	1879	3,000
	do	do	Chapman's Creek	Chapman's Creek, Kans	1879	2,000
	do	do	Solomon River	Solomon City, Kans	1879	2,000
	do	Smoky Hill River	Saline River	Saline, Kans	1879	3,000
	do	do	Spring Creek	Brookville, Kans	1879	1,000
	do	Kansas River	Smoky Hill River	Ellsworth, Kans	1879	5,000
	do	Smoky Hill River	Big Creek	Hays City, Kans	1879	5,000
	do	do	do	Ellis, Kans	1879	5,000
	do	Kansas River	Wakasa River	Ottawa, Kans. (?)	1879	2,000
	do	[illegible]ri River	Osage River (?)	Reading, Kans	1879	2,500
	do	Arkansas River	Neosho River	Emporia, Kans	1879	5,000
	do	Neosho River	Cottonwood River	Florence, Kans	1879	2,000
	do	Arkansas River	Walnut River	Eldorado, Kans	1879	3,000
	do	do	Little Arkansas River	Halstead, Kans	1879	2,500
	do	Little Arkansas River	Lake Inman	McPherson, Kans	1879	2,000
	do	Arkansas River	Cow Creek	Hutchinson, Kans	1879	3,000
	do	do	Walnut River	Great Bend, Kans	1879	3,000
	do	do	Pawnee Creek	Larned, Kans	1879	5,000
Maine	Pembroke hatching-house	Saint Croix River	Keen's Lake Stream	Keen's Lake, Me	Feb. 6, 1879	9,700
	do	Bay of Fundy	Penmaquan River	Pembroke, Me	Feb. 13, 1879	3,000
Maryland	Druid Hill hatching-house	Bush River	Winter's Run	Wilna, Md	Feb. 1, 1879	4,000
	do	Chesapeake Bay	Patuxent River	Savage Station, Md	Feb. 1, 1879	10,000
	do	do	Chester River	Millington, Md	Feb. 13, 1879	12,000
	do	do	Choptank River	Henderson, Md	Feb. 21, 1879	8,000
	do				Feb. 24, 1879	7,000
	do	Tangier Sound	Black Water	Cambridge, Md	Feb. 26, 1879	3,000
	do	Transguaking River	Chicacomico		Feb. 26, 1879	3,000
	do	Tangier Sound	Transguaking River	Airey's Station, Md	Feb. 26, 1879	3,000
	do	Chesapeake Bay	Patapsco and Patuxent Rivers	Hood's Mill and Airey, Md	Feb. 28, 1879	12,000
	do				June 6, 1879	236
Michigan		Lake Michigan	Paw Paw River	Berrien County, Michigan	Jan. 8, 1879	25,000
		Dowagiac River	Pokagon Creek	Cass County, Michigan	Jan. 9, 1879	10,000
		do	Peceanic River	do	Jan. 10, 1879	5,000
		Pine Lake	Dowagiac River	Van Buren County, Michigan	Jan. 10, 1879	10,000
		Saint Joseph River	Walron Creek	Saint Joseph County, Michigan	Jan. 23, 1879	10,000
		do	Fates Creek	do	Jan. 23, 1879	10,000

		Lake Michigan	Grand River	[illegible], [illegible] Mich	Jan. 30, 1879	30,000
		do	Mani[illegible] [illegible] River	[illegible] City, [illegible] Mich	[illegible] 6, 1879	25,000
		[illegible]	[illegible] River	[illegible], [illegible] Mich	Feb. 13, 1879	30,000
		[illegible] River	[illegible] River	[illegible] City, [illegible] Mich	[illegible] 13, 1879	30,000
		[illegible] River	Round Lake	[illegible] City, [illegible] Mich	Apr. 18, 1879	500
		do	[illegible] Lake	do	Apr. 18, 1879	500
			[illegible]	[illegible], [illegible] Mich	July 12, 1879	20
		[illegible] River	[illegible]	[illegible] City, [illegible] Minn	1879	5,000
Minnesota	Red Wing, Minn.	[illegible] River	Silver Lake	W[illegible] City, [illegible] Minn	1879	1,000
	do	[illegible] River	Mary's [illegible]	[illegible] City, [illegible] Minn	1879	1,000
	do	[illegible] River	[illegible] Lake	[illegible] City, [illegible] Minn	1879	1,000
	do	Minn[illegible] River	Chain Lake	do	1879	1, 0
	do	do	[illegible] Lake	[illegible] City, [illegible] Minn	1879	[illegible]0
	do	do	[illegible] Lake	do	1879	2,000
	do	do	[illegible] Lake	[illegible] [illegible], [illegible] Minn	1879	5,000
	do	[illegible] River	[illegible] Lake	[illegible] City, [illegible] Minn	1879	2,500
	do	[illegible] River	[illegible] Lake	do	1879	2,500
	do	do	[illegible] Lake	W[illegible] City, [illegible] Minn	1879	[illegible]0
	do	[illegible] River	[illegible]	[illegible] City, [illegible] Minn	1879	3,000
	do	do	do	[illegible] City, [illegible] Minn	1879	3,000
	do	[illegible] River	Lake [illegible]	[illegible] [illegible], [illegible] Minnesota	1879	2,000
	do	Saint [illegible] River	Lake [illegible]		1879	3,000
	do		Lake Foss	Stevens County, Minnesota	1879	2, 0
	do	Pomme de Terre River	[illegible] River	Franklin, Mo	[illegible] —, 1878	75,000
Missouri	California	[illegible] River	Spring [illegible]	Carthage, Mo	[illegible] —, 1878	75, 0
	do	A[illegible] River	[illegible] River	Reno, Nev	[illegible] —, 1879	1 , 0
Nevada	[illegible] station	[illegible] River	[illegible]	Carson City, Nev	[illegible] —, 1879	10,000
	do	[illegible] River	[illegible] River	Campton and Plymouth, N. H.	Feb. 1, 1879	317,000
New Hampshire	[illegible]outh, N. H.	[illegible] River	[illegible] Lake	Wakefield, N. H	Mar. 14, 1879	5, 0
	do	Salmon [illegible]	[illegible] Lake	Milton, N. H	Mar. —, 1879	5, 0
	do	do	[illegible]	[illegible], N. H	Mar. —, 1879	5,000
	do	do	[illegible]	[illegible], N. H	Mar. —, 1879	5,000
	do	do	[illegible]	Wolfeborough, N. H	Mar. —, 1879	10,000
	do	Lake Winnipiseogee	[illegible]	[illegible] [illegible], N. H	Mar. —, 1879	20,000
	do	Merrimac River	[illegible]	Eighty [illegible] [illegible]	1879	156,200
New Jersey	[illegible], N. J	Delaware River	[illegible] River	Atl[illegible] City, New Jersey	1879	50, 00
	do	Great Egg Harbor	Alloway's [illegible]	Salem City, New Jersey	1879	25,000
	do	Delaware River	[illegible] River	[illegible] County, New Jersey	1879	25,000
	do	do	[illegible]	[illegible] City, New Jersey	1879	25,000
	do	do	[illegible] River	do	1879	25,000
	do	[illegible] Bay	[illegible]	Somerset [illegible], New Jersey	1879	30,000
	do	Raritan River	[illegible] River	do	1879	30,000
	do	[illegible] River	[illegible] River	[illegible] City, New Jersey	1879	30,000
	do	[illegible] River	[illegible]	[illegible] City, New Jersey	1879	5, 00
	do	[illegible]	Singum [illegible]		1879	5,000
	do	Singum [illegible]	[illegible] Lake		1879	3,000
	do	[illegible] Lake	[illegible] Lake	Passaic County, New Jersey	1879	22,000
	do	[illegible] Lake	Swartswood Lake	Sussex County, New Jersey	1879	10,000
	do	Swartswood Lake	[illegible] Lake		1879	[illegible]0
	do	V[illegible]	Cline's [illegible]		1879	10,000
	do	Cline's [illegible]				

TABLE II.—*Distribution of California salmon reared from eggs collected in* 1878—Continued.

States.	Where finally hatched.	Waters stocked.	Tributaries in which fish were placed.	Locality.	Date of transfer.	Number of fish.
New York		Hudson River	Tributaries of Hudson River	Greene County, New York	Dec. 5, 1878	20,000
		Genesee River	Spring Creek	Monroe County, New York	Dec. 31, 1878	10,000
	Caledonia, N. Y	Lake Ontario	do	Livingston County, New York	Jan. 15, 1869	10,000
	do	Hemlock Lake	Spring Brooks	Ontario County, New York	Feb. 26, 1879	36,000
	do	Summer Hill Lake	do	Cayuga County, New York	Mar. 1, 1879	9,000
	do	Lake Ontario	Spring Creek	Livingston County, New York	Mar. 11, 1879	1,000
North Carolina	Henry's, N. C	Cape Fear River	North Fork, Deep River	Friendship, N. C	Jan. 2, 1879	18,000
	do	Catawba River	Johns River and Upper Creek	Morgantown, N. C	Jan. 10, 1879	30,000
	do	do	Little River	Bridgewater, N. C	Jan. 11, 1879	30,000
	do	Roanoke River	Doe River	Danbury, N. C	Jan. 13, 1879	20,000
	do	Broad River	Broad River	Hickory-nut Gap, N. C	Jan. 15, 1879	45,000
	do	Roanoke River	Town Fork River	Germantown, N. C	Jan. 18, 1879	15,000
	do	Pee Dee River	Yadkin River	Patterson's (Caldwell Co.), N. C	Dec. 16, 1878	30,000
	do	do	do	do	Dec. 18, 1878	30,000
	do	Broad River	North Pacolet River	Near Hendersonville, N. C	Dec. 20, 1878	5,000
	do	do	Green River	do	Dec. 20, 1878	20,000
	do	Cape Fear River	Bull Run Creek	Near Jamestown, N. C	Dec. 27, 1878	20,000
	do	do	North Fork, Deep River	do	Dec. 31, 1878	24,500
Total						4,460,356
Canada	Newcastle hatchery	Lake Ontario	Wilmot's Creek	Province of Ontario	Spring, 1879	1,000
	do	Lake Huron	Saugeen River	do	Spring, 1879.	500
	do	Lake Ontario	River Trent	do	Spring, 1879.	200

TABLE III.—*Distribution of land-locked salmon reared from eggs collected in* 1878.

N. B.—These eggs were for the most part hatched out and planted by the State Commissioners of Fisheries. The returns, however, are so imperfect that it has been thought best to defer giving the record until the next report.

TABLE IV.—*Distribution of whitefish reared from eggs collected in* 1878.*

States.	Where finally hatched.	Waters stocked.	Tributaries in which fish were placed.	Locality.	Date of transfer.	Number of fish.
New Jersey	Bloomsburg, N. J	Shepherd's Pond		Morris County, New Jersey	Feb. 17, 1879	45,000
	do	Lake Hopatcong		do	Feb. 17, 1879	45,000

* The eggs were collected by Frank N. Clarke, Northville, Mich, and hatched and planted by the New Jersey Commissioners of Fisheries.

APPENDIX A.

PATENTS; LEGISLATION; PROTECTION.

I.—LIST OF PATENTS ISSUED IN THE UNITED STATES, GREAT BRITAIN, AND CANADA, UP TO THE END OF 1878, RELATING TO FISH AND THE METHODS, PRODUCTS, AND APPLICATIONS OF THE FISHERIES.

By Robert G. Dyrenforth,
Principal Examiner, United States Patent Office.

AMERICAN PATENTS.

FISHING.

Including Fish-hooks, Fish-traps, Fishways, Floats, Harpoons and Spears, Nets and Seines, Oyster Culture, Pisciculture, Reels, Rods, Sinkers.

DECOYS.

17,192.	Bogle, R	May	5, 1857.
74,458.	Wales, N	Feb.	11, 1868.
93,293.	Foster, J	Aug.	3, 1869.
102,799.	Fisher, E	May	10, 1870.
156,239.	Strater and Sohier	Oct.	27, 1874.

FISH-HOOKS.

	Engelbrecht and Skiff	July	28, 1846.
	Pendleton, S	Aug.	21, 1847.
	Johnson, J	Aug.	21, 1847.
	Ellis and Gritty	Aug.	15, 1848.
	Jenks, W	Sept.	5, 1848.
6,207.	Johnson, J	March	20, 1849.
8,853.	Buel, J. T	April	6, 1852.
10,761.	Sigler, H	April	11, 1854.
10,771.	Buel, J. T	April	11, 1854.
13,068.	De Saxe, C	June	12, 1855.
13,081.	Cook, R. F	June	19, 1855.
13,649.	Johnson, J	Oct.	9, 1855.
14,706.	Buel, J. T	April	22, 1856.
17,803.	McLean, D	July	14, 1857.
25,507.	Haskell, R	Sept.	20, 1859.
31,396.	Morris, W. L	Feb.	12, 1861.
43,694.	Lenhart, A. I	Aug.	2, 1864.
44,368.	Gardiner, jr., N. A	Sept.	20, 1864.
50,799.	Crandell, G	Nov.	7, 1865.

51,651. Davis and Johnson.........................Dec. 19, 1865.
51,951. Livermore, H. B .... Jan. 9, 1866.
54,251. Johnson and Howarth......................April 24, 1866.
54,684. Chapman, W. D ....May 15, 1866.
58,404. Goodwin, W. C ....Oct. 2, 1866.
59,844. King, Jacob ....Nov. 20, 1866.
59,893. Crosby, C. O..............................Nov. 20, 1866.
60,786. Rhodes, E. R. and J. W...................Jan. 1, 1867.
62,042. Lee, jr., Benjamin ....Feb. 12, 1867.
68,027. Angilard, F ....Aug. 27, 1867.
69,221. Kidder, D ....Sept. 24, 1867.
70,868. Lenhart, A. I.............................Nov. 12, 1867.
70,913. Sterling, E ....Nov. 12, 1867.
77,365. Fish, R. A................................April 28, 1868.
79,446. Christian, J. B...........................June 30, 1868.
80,151. Dennetts, A. A ....July 21, 1868.
86,154. Hiltz, M..................................Jan. 26, 1869.
94,893. Kemlo, Francis ....Sept. 14, 1869.
94,894. Kemlo, Francis ....Sept. 14, 1869.
94,895. Kemlo, Francis ....Sept. 14, 1869.
95,755. Angers, F. T..............................Oct. 12, 1869.
104,930. Chapman, William D ....July 5, 1870.
111,898. Arnold, L ....Feb. 21, 1871.
115,434. Chapman, William D.......................May 30, 1871.
117,719. Arnold, L ....Aug. 8, 1871.
121,182. Mann, J. H...............................Nov. 21, 1871.
123,844. Sinclair, G..............................Feb. 20, 1872.
129,053. Pitcher, E...............................July 16, 1872.
139,180. Mullaly, J...............................May 20, 1873.
141,910. Allen, B. F..............................Aug. 19, 1873.
143,146. Harper and Smith ....Sept. 23, 1873.
146,443. Fitzgerald, T. F.........................Jan. 13, 1874.
146,764. James, W. H ....Jan. 27, 1874.
148,926. Cahoon, M. V. B..........................March 24, 1874.
149,123. Hazzard, A. W ....March 31, 1874.
151,394. Huard and Dunbar ....May 26, 1874.
153,854. Skinner, G. M ....Aug. 4, 1874.
157,480. Perry, R. J ....Dec. 8, 1874.
163,980. Dunlap, E. L ....June 1, 1875.
167,784. Pierce, G. R.............................Sept. 14, 1875.
171,697. Place, J. H..............................Jan. 4, 1876.
171,768. Buel, J. T ....Jan. 4, 1876.
171,769. Buel, J. T ....Jan. 4, 1876.
177,639. Hill, L. S...............................May 23, 1876.
181,308. Brush, H. C ....Aug. 22, 1876.
184,627. Jones, F ....Nov. 21, 1876.

185,914. Gregg, W. H .... Jan. 2, 1877.
186,134. Jahne and Moors .... Jan. 9, 1877.
189,805. Smith, B. F .... April 17, 1877.
190,222. King, J. O .... May 1, 1877.
191,165. Miller, G. C .... May 22, 1877.
196,648. Edgar, B .... Oct. 30, 1877.
197,935. Holt, H. H .... Dec. 11, 1877.
199,926. Mitchell, J. A .... Feb. 5, 1878.
208,581. Falvey, J .... Oct. 1, 1878.

FISH-TRAPS.

16,217. Van Hoesen, L .... Dec. 9, 1856.
22,644. Gray, R .... Jan. 18, 1859.
23,154. Bowman, D .... March 8, 1859.
75,075. Talbot, D. C .... March 3, 1868.
76,489. McCaughans, T. B .... April 7, 1868.
77,893. Koehler, J .... May 12, 1868.
85,199. Beach, E. B .... Dec. 22, 1868.
113,292. Hammond, J. E .... April 4, 1871.
123,164. Fuller, O. M .... Jan. 30, 1872.
131,439. Harcourt and Cottingham .... Sept. 17, 1872.
132,476. Livaudais, C .... Oct. 22, 1872.
135,113. Goodman, S. A .... Jan. 21, 1873.
141,588. Pavonarius and Michtle .... Aug. 5, 1873.
156,648. Peck, L. A .... Nov. 10, 1874.
163,498. Kepner, R. B .... May 18, 1875.
178,375. McRoberts, J .... June 6, 1876.
181,844. Hitchcock, J. C .... Sept. 5, 1876.
188,503. Davis, S. M .... March 20, 1877.
194,253. Lasater, J. M .... Aug. 14, 1877.
198,894. McBryde, D. S .... Jan. 1, 1878.
201,504. Davis, G .... March 19, 1878.
202,818. Hesse, O .... April 23, 1878.
202,962. Robertson, S. N. and E. A .... April 30, 1878.
204,168. Roney, R. J .... May 28, 1878.
204,538. Clark and Roberts .... June 4, 1878.

FISHWAYS.

57,159. Livermore, A .... Aug. 14, 1866.
126,257. Brewer, J. D .... April 30, 1872.
132,349. Brackett, E. A .... Oct. 22, 1872.
154,216. Brewer, J. D .... Aug. 18, 1874.
206,715. Fisher, D .... Aug. 6, 1878.
208,408. McDonald, M .... Sept. 24, 1878.

FLOATS.

10,795. De Saxe, Charles .... April 18, 1854.
12,060. Hoards, J. W .... Dec. 12, 1854.

86,609. Terrell, J. A........................ ....Feb. 2, 1869.
99,572. Ingram, J..................................Feb. 8, 1870.
127,218. Brown and Jarvis, jr.........................May 28, 1872.
128,885. Jewell, E...................................July 9, 1872.
165,867. Quinn, W. T.................................July 20, 1875.
179,490. Sanders, L..................................July 4, 1876.
186,232. Davis, D. W. and S. H........................Jan. 16, 1877.
188,755. Redfield, P. S...............................March 27, 1877.

HARPOONS AND SPEARS.

7,709. Warner and Gaylord.......... ....Oct. 8, 1850.
16,014. Horton, E....................................Nov. 4, 1856.
20,343. Garl, J......................................May 25, 1858.
35,476. Roys, Thomas W...............................June 3, 1862.
144,110. Knapp, J. W..................................Oct. 28, 1873.
168,335. Jincks, M....................................Oct. 5, 1875.
172,312. Hedges, S. P.................................Jan. 18, 1876.
206,694. Taylor, W........................ ....Aug. 6, 1878.

NETS AND SEINES.

Evarts, Russ...March 21, 1838.
Hale, B. W...June 4, 1838.
Tracy, Cyrus...Sept. 19, 1838.
Cook, H.................. ....March 17, 1843.
Downs, John..April 25, 1843.
Carr, Shannon, and Carr..............................Sept. 14, 1844.
10,794. De Saxe, Charles...............................April 18, 1854.
20,125. Hall, Thomas...................................April 27, 1858.
20,725. Merritt, jr., B.................................June 29, 1858.
34,887. Goodwin, F.....................................April 8, 1862.
39,676. Randolph, W....................................Aug. 25, 1863.
55,635. Field, E. A....................................June 19, 1866.
56,917. Ferl and Larkins...............................Aug. 7, 1866.
59,429. Maxwell, William...............................Nov. 6, 1866.
62,481. Crossman, C. C.................................Feb. 26, 1867.
76,284. Wills, Daniel..................................March 31, 1868.
76,387. Bell, Thomas...................................April 7, 1868.
80,274. Collines, John.................................July 28, 1868.
82,490. Cartwright, T..................................Sept. 29, 1868.
82,913. Allen, George D................................Oct. 13, 1868.
83,429. Wilcox, W. S...................................Oct. 27, 1868.
83,493. Harper, S......................................Oct. 27, 1868.
87,740. Werdmüller, F. A...............................March 9, 1869.
99,713. Sabins, P. G...................................Feb. 8, 1870.
113,572. Ryder, jr., B..................................April 11, 1871.
113,817. Tiernan, P. E..................................April 18, 1871.

117,957.	Alexander, L. H	Aug. 15, 1871.
120,974.	Jeffery, R	Nov. 14, 1871.
124,635.	Smith, Henry	March 12, 1872.
137,930.	Ketcham, C. E	April 15, 1873.
144,888.	Campbell, J. O	Nov. 25, 1873.
155,140.	Brewster, C	Sept. 22, 1874.
167,189.	Nason, C. F	Aug. 31, 1875.
194,434.	Howes, S. B	Aug. 21, 1877.
197,313.	Bates, L	Nov. 20, 1877.

OYSTER CULTURE.

127,903.	Lyford, B. F	June 11, 1872.
130,631.	Frazier, E. H	Aug. 20, 1872.
149,921.	Cook, O	April 21, 1874.

PISCICULTURE.

68,871.	Green, S	Sept. 17, 1867.
72,177.	Drexler, C	Dec. 17, 1867.
78,952.	Furman, W. H	June 16, 1868.
80,775.	Smith, A. J	Aug. 4, 1868.
105,176.	Collins, A. S	July 12, 1870.
116,112.	Stone, L	June 20, 1871.
116,995.	Sabin, R. E	July 11, 1871.
136,834.	Holton, M. G	March 18, 1873.
148,035.	Clark, N. W	March 3, 1874.
149,198.	Clark, N. W	March 31, 1874.
151,080.	Bryan, O. N	May 19, 1874.
160,002.	Bond, A	Feb. 23, 1875.
166,413.	Roth, J	Aug. 3, 1875.
173,262.	Brackett, E. A	Feb. 8, 1876.
180,085.	Wilmot, S	July 18, 1876.
199,527.	Ferguson, T. B	Jan. 22, 1878.
207,333.	Wright, I. H	Aug. 20, 1878.

REELS.

	Tiffany, A	May 26, 1838.
15,466.	Bailey, J. A	Aug. 5, 1856.
16,626.	Deacon, Edward	Feb. 10, 1857.
24,987.	Billinghurst, William	Aug. 9, 1859.
27,305.	Palmer, M. S	Feb. 28, 1860.
41,494.	Dougherty, A	Feb. 9, 1864.
43,460.	Van Gieson, W. H	July 5, 1864.
43,485.	Ellis, D	July 12, 1864.
43,546.	Cummings, Thomas W	July 12, 1864.
49,663.	Stuart, W. M	Aug. 29, 1865.
55,653.	Hatch, A	June 19, 1866.
56,937.	Hartill, A. B	Aug. 7, 1866.

71,344. Von Hofe, Julius ....Nov. 26, 1867.
78,546. Stacy, E. F..................................June 2, 1868.
82,377. Bradley, W. H................................Sept. 22, 1868.
83,740. Stetson, John.................................Nov. 3, 1868.
87,188. Xavier, Francis...............................Feb. 23, 1869.
88,026. Foster, C. S. H ....March 23, 1869.
95,839. Ross, J. J ....Oct. 12, 1869.
96,652. Altmaeir, P. A ....Nov. 9, 1869.
103,668. Sheldon, G. G ....May 31, 1870.
112,326. Decker, E. L ....March 7, 1871.
121,020. Terry, S. B ....Nov. 14, 1871.
128,137. Fowler, A. H ....June 18, 1872.
134,917. Mooney, G ....Jan. 14, 1873.
135,283. Noe, C. L....................................Jan. 28, 1873.
147,414. McCord, C. W ....Feb. 10, 1874.
150,883. Orvis, C. F ....May 12, 1874.
161,314. Winans and Whistler ....March 23, 1875.
162,845. McDonald, A. L ....May 4, 1875.
166,241. Winslow, H ....Aug. 3, 1875.
175,227. Winans and Whistler ....March 21, 1876.
177,544. Noe, C. L ....May 16, 1876.
191,813. Philbrook, F. J ....June 12, 1877.
195,578. Copeland, G. T ....Sept. 25, 1877.

RODS.

10,795. De Saxe, Charles ....April 18, 1854.
20,309. Underwood and Bargis ....May 18, 1858.
25,693. Pritchard, Henry.............................Oct. 4, 1859.
35,339. Von Hofe, Julius ....May 20, 1862.
58,833. Isaacs, R. N ....Oct. 16, 1866.
72,667. Montrose, J. H...............................Dec. 24, 1867.
100,895. Hubbard, W. J ....March 15, 1870.
119,251. Tout, Thomas.................................Sept. 26, 1871.
137,015. McHarg, J. B.................................March 18, 1873.
140,656. Smith, W. M ....July 8, 1873.
142,126. Senieur, F ....Aug. 26, 1873.
154,141. Hill, B ....Aug. 18, 1874.
164,828. Graves, J. L ....June 22, 1875.
169,181. Leonard, H. L ....Oct. 26, 1875.
170,188. Perry, C ....Nov. 23, 1875.
173,534. Endicott, F ....Feb. 15, 1876.
198,879. Fisher, C. R ....Jan. 1, 1878.
206,264. J. A. Robertson..............................July 23, 1878.
207,665. Leonard, H. L ....Sept. 3, 1878.
208,500. Van Altena, H................................Oct. 1, 1878.

SINKERS.

14,587.	Smith, William	April	1, 1856.
39,192.	Woodbury, William	July	7, 1863.
46,453.	Decker, E. F	Feb.	21, 1865.
56,857.	Martin, J. R	July	31, 1866.
58,211.	Burnham, L. A	Sept.	25, 1866.
61,625.	Martin, J. R	Jan.	29, 1867.
71,879.	Hiltz, Martin	Dec.	10, 1867.
77,628.	Lothrop, L. D	May	5, 1868.
77,774.	Smith, William H	May	12, 1868.
83,681.	Albee, S	Nov.	3, 1868.
84,885.	Leach and Hutchins	Dec.	15, 1868.
86,786.	Tellgmann, F	Feb.	9, 1869.
93,220.	Osgood, R. T	Aug.	3, 1869.
118,772.	Camp, H	Sept.	5, 1871.
155,266.	Sprague, H. L	Sept.	22, 1874.
167,687.	Pitcher, E	Sept.	14, 1875.
175,949.	Dixon, G. W	April	11, 1876.
182,428.	Forbes, jr., J	Sept.	19, 1876.

FISH, PRESERVATION OF.—ICHTHYOCOLLA.

FOOD.—PRESERVATION OF FISH.

7,895.	Westacott, R. G	Jan.	7, 1851.
15,452.	Wright, J	July	29, 1856.
26,427.	Gross, M	Dec.	13, 1859.
31,736.	Piper, E	March	19, 1861.
44,340.	Reeves, J. F	Sept.	20, 1864.
45,765.	Staunton, J. G	Jan.	3, 1865.
48,723.	Robinson, B	July	11, 1865.
59,833.	Gilmore, E. T	Nov.	20, 1866.
66,616.	Noble, B. G	July	9, 1867.
66,732.	Noble, B. G	July	16, 1867.
70,435.	Heron, G. H	Nov.	5, 1867.
74,378.	Kellogg, T. D	Feb.	11, 1868.
80,775.	Smidth, A. J	Aug.	4, 1868.
81,987.	Cutler, W. D	Sept.	8, 1868.
83,533.	Nunan, P	Oct.	27, 1868.
83,836.	Cutler, W. D	Nov.	10, 1868.
84,801.	Crowell, E	Dec.	8, 1868.
84,855.	Burnham, E. E	Dec.	15, 1868.
85,913.	Davis, William	Jan.	19, 1869.
86,040.	Sim, Thomas	Jan.	19, 1869.
87,986.	Stephens, B. F	March	16, 1869.
88,064.	Nickerson, J	March	23, 1869.
90,334.	Atwood, jr., J	May	25, 1869.

90,944.	Havard, C., and Harmony, N. X	June 8, 1869.
93,183.	Dotch, J. E	Aug. 3, 1869.
95,179.	Adams, R. A	Sept. 28, 1869.
96,288.	Thorp, G. T	Oct. 26, 1869.
97,145.	Adams, R. A	Nov. 23, 1869.
108,983.	Dotch, J. E	Nov. 8, 1870.
109,820.	Howell, D. Y	Dec. 6, 1870.
112,129.	Davis and Davis	Feb. 28, 1871.
113,395.	Brown, M. W	April 4, 1871.
125,102.	Vazquez and Rosenberg	March 26, 1872.
127,115.	Stanley, I. L	May 21, 1872.
128,320.	Mosquera, B	June 25, 1872.
131,820.	Henley, T. F	Oct. 1, 1872.
132,316.	Pharo, E. A	Oct. 15, 1872.
135,113.	Goodman, jr., S. A	Jan. 21, 1873.
143,386.	Sharp, William	Sept. 30, 1873.
148,317.	Müller, G. A	March 10, 1874.
149,256.	Shriver, A. K	March 31, 1874.
150,183.	Osborn, G. K	April 28, 1874.
152,181.	Shriver, A. K	June 16, 1874.
161,596.	Davis and Davis	April 6, 1875.
162,119.	Tait, A. H	April 13, 1875.
171,662.	Goodale, S. L	Jan. 4, 1876.
178,916.	Dunbar, G. W., G. H., and F. B	June 20, 1876.
181,549.	Andrews, O	Aug. 29, 1876.
186,204.	Goodale, S. L	Jan. 16, 1877.
187,122.	Gauthier, C. W	Feb. 6, 1877.
201,834.	Rich, S	March 26, 1878.
204,647.	Woodruff, Lyman	June 4, 1878.
204,966.	Griffin, S. W	June 18, 1878.
205,830.	Bliss, Samuel	July 9, 1878.

ICHTHYOCOLLA, ETC.

	Waldron	March 4, 1812.
	Hall	March 23, 1822.
	Hastings	Aug. 14, 1822.
	Norwood *et al*	Jan. 21, 1834.
5,978.	Rowe	Dec. 19, 1848.
53,636.	Lewis and Stanwood	April 3, 1866.
78,016.	Robinson	May 19, 1868.
106,212.	Rowe	Aug. 9, 1870.
134,690.	Manning	Jan. 7, 1873.
148,317.	Müller	March 10, 1874.
149,165.	Stanwood	March 31, 1874.
177,764.	Stanwood	May 23, 1876.

ENGLISH PATENTS.

FISHING.

Including DECOYING, FLOATS, FISHWAYS, FISH-TRAPS, HARPOONS AND SPEARS, HOOKS, NETS, OYSTER CULTURE, PISCICULTURE, PARALYZING BY ELECTRICITY, REELS, RODS, TACKLING, SINKERS.

DECOYING.

Grent .... No. 59 of 1632.
Williams and Marwood .... No. 295 of 1692.
De Chabannes .... No. 4582 of 1821.
Coffin .... No. 4815 of 1823.
Fanshawe .... No. 2580 of 1862.
Iodocius .... No. 1751 of 1863.

FLOATS.

Fearn .... No. 2003 of 1859.

FISH-TRAPS.

Allen .... No. 150 of 1860.
Stevens .... No. 1120 of 1860.
Damm .... No. 3548 of 1867.
Leheup .... No. 1425 of 1868.
Engholm .... No. 1276 of 1869.
Leach .... No. 358 of 1873.
Leach .... No. 1281 of 1873.
Lake, W. R. .... No. 4043 of 1876.

FISHWAYS.

Phymoui .... No. 974 of 1858.

HARPOONS AND SPEARS.

Bayles .... No. 1367 of 1783.
Congreve and Colquhoun .... No. 4563 of 1821.
Lance .... No. 8541 of 1840.
Ackerman .... No. 10914 of 1845.
Rechten .... No. 125 of 1856.
Tindall .... No. 1110 of 1857.
Roys .... No. 2301 of 1857.
Roys .... No. 2340 of 1857.
Walker .... No. 450 of 1861.
Roys and Lilliendahl .... No. 550 of 1865.
Welch .... No. 3312 of 1867.

FISH-HOOKS.

Andrews, John .... No. 1719 of 1789.
Bell, William .... No. 2063 of 1795.

Poole, Moses .... No. 11520 of 1847.
Bainbridge, A. F. .... No. 533 of 1852.
Huddart, G. A .... No. 889 of 1852.
King, R. J. N .... No. 2902 of 1853.
Box, William H .... No. 923 of 1857.
Newton, William E .... No. 1135 of 1859.
Hackett, William A .... No. 428 of 1865.
Baylis, Charles .... No. 3177 of 1865.
Warner, Jos .... No. 413 of 1866.
Gedge .... No. 150 of 1867.
Welch .... No. 1765 of 1867.
Morrall, A .... No. 2714 of 1867.

NETS.

Stuart and Stuart .... No. 1872 of 1859.
Henry .... No. 3099 of 1860.
James .... No. 492 of 1861.
Hector .... No. 1871 of 1863.
Hurn and Hurn .... No. 3334 of 1865.
Bryson .... No. 193 of 1866.
Mack .... No. 2008 of 1866.
Wilkinson .... No. 1146 of 1867.
Hallett .... No. 1331 of 1867.
Hallett .... No. 1332 of 1867.
Hallett .... No. 1333 of 1867.
Johnson .... No. 2140 of 1867.
Brabazon .... No. 323 of 1869.

OYSTER CULTURE.

Ayckbourn .... No. 2930 of 1863.
Crofts .... No. 1040 of 1864.
Bert .... No. 1316 of 1870.
Michel .... No. 4103 of 1874.
De Lagillardaie .... No. 3506 of 1875.
Jennings and Anderson .... No. 2470 of 1876.
Mewburn, J. C .... No. 1447 of 1877.

PARALYZING BY ELECTRICITY.

Baggs .... No. 2644 of 1863.
Bennett and Ward .... No. 3228 of 1868.

PISCICULTURE.

Johnston, N .... No. 2594 of 1854.

REELS.

Curr, John .... No. 3157 of 1808.
Kenton .... No. 1956 of 1856.

Moore ....No. 932 of 1862.
Mure ....No. 1806 of 1868.
Heaton ....No. 388 of 1873.
Corbet ....No. 3283 of 1876.
Corbet ....No. 655 of 1877.

RODS.

Dawson and Restell ....No. 2017 of 1853.
Bennett, John ....No. 461 of 1857.
Porecky, A ....No. 1553 of 1858.
Britten, B ....No. 3528 of 1868.
Moultray, James D ....No. 1648 of 1874.
Jack, Alexander ....No. 1840 of 1874.
Holroyd, E. A ....No. 1806 of 1876.
Aston, J ....No. 1553 of 1877.

TACKLING.

Cobb ....No. 881 of 1767.
Hector ....No. 361 of 1858.
Frère ....No. 1755 of 1866.

SINKERS.

Corrin, R ....No. 792 of 1877.

PRESERVATION OF FISH.

FOOD.—PRESERVATION OF FISH.

Porter, T., and White, J ....No. 278 of 1691.
Cockburn, A ....No. 793 of 1763.
Batley, B ....No. 2441 of 1800.
Batley, B ....No. 2465 of 1801.
Granholm, L ....No. 4150 of 1817.
Siegnette, L. E ....No. 7036 of 1836.
Benjamin, H., and Grafton, H ....No. 9240 of 1842.
Fitch, M ....No. 10322 of 1844.
Fatio, A. M., and Verdeil, F ....No. 231 of 1854.
Hervé, J. L ....No. 70 of 1855.
Acres, E ....No. 965 of 1855.
Cooke, M. J ....No. 1320 of 1855.
Bethell, J ....No. 1559 of 1855.
Demait, F. M ....No. 2223 of 1855.
Warriner, G ....No. 1982 of 1856.
Vasserot, C. F ....No. 2976 of 1856.
Gorges, E. V. J. L ....No. 14 of 1857.
Davies, G ....No. 223 of 1858.
Osler, S ....No. 1198 of 1858.
Hughes, E. J ....No. 536 of 1859.
Brooman, R. A ....No. 1202 of 1859.

Clark, W	No. 1404 of 1860.
Clark, W	No. 309 of 1861.
Brade, A. G	No. 576 of 1861.
Bouvet, J	No. 2569 of 1862.
Morgan, John	No. 713 of 1864.
McCall and Sloper	No. 2794 of 1864.
Nicoll, D	No. 1349 of 1866.
Medlock, H., and Bailey, W	No. 1707 of 1866.
Clark, W	No. 379 of 1867.
Brooman, C. E	No. 1200 of 1867.
Somervell, J	No. 105 of 1868.
Lake, W. R	No. 3194 of 1868.
Richmond, E. H	No. 1381 of 1869.
Beanes, E	No. 2345 of 1869.
Gamgee, J	No. 60 of 1870.
Gard, W. G	No. 2096 of 1870.
Highton, H	No. 2568 of 1870.
Gard, W. G	No. 3105 of 1870.
Mariotti, L	No. 3321 of 1870.
Durand, F	No. 460 of 1871.
Henley, T. F	No. 1233 of 1871.
Gedge, W. E	No. 2698 of 1871.
Vazquez and Rosenberg	No. 3387 of 1871.
De Malortie and Woods	No. 3803 of 1872.
Clark, C. S	No. 1837 of 1874.
Herzen, A	No. 2032 of 1874.
Hönck, J	No. 3645 of 1874.
Fryer, C. E	No. 3652 of 1874.
Fryer, C. E	No. 3653 of 1874.
Weatherby, C. P. N	No. 351 of 1875.
Melburn and Jackson	No. 933 of 1875.
Hyatt, T	No. 2307 of 1875.
Ermatinger, J. H	No. 3073 of 1875.
Brewer, E. G	No. 3120 of 1875.
Jensen, P	No. 3729 of 1875.
Brewer	No. 160 of 1876.
Jenson	No. 1492 of 1876.
Knott	No. 1920 of 1876.
Knott	No. 2064 of 1876.
Pols	No. 2432 of 1876.
Newton	No. 2479 of 1876.
Grier	No. 3107 of 1876.
Knott	No. 4446 of 1876.
McKay	No. 4605 of 1876.
Tongue	No. 4612 of 1876.
Harvey	No. 4624 of 1876.
Radcliffe	No. 4920 of 1876.

AMERICAN PATENTS.

MANURES, FISH.

12,480. Demolon and Thurneyssen....March 6, 1855.
26,548. Hall, W. D....Dec. 20, 1859.
33,706. Hyde, J. B....Nov. 12, 1861.
43,639. Glover, W. H. H....July 26, 1864.
46,847. Hall, W. D....March 14, 1865.
88,223. Smith, A....March 23, 1869.
97,939. Lugo, O....Dec. 14, 1869.
99,251. Smith, A....Jan. 25, 1870.
99,452. Lugo, O....Feb. 1, 1870.
99,673. Hooper, Hooper and Lugo....Feb. 8, 1870.
99,896. Hooper, W. J. and T....Feb. 15, 1870.
3,840. Lugo, O....Feb. 15, 1870.
99,924. Lugo, O....Feb. 15, 1870.
102,689. Lugo, O....May 3, 1870.
104,327. Lugo, O....June 14, 1870.
112,653. Taylor, T....March 14, 1871.
125,939. Deering, J. M....April 23, 1872.
152,921. Shepard, S. D....July 14, 1874.
208,224. Crowell, A. F....Sept. 24, 1878.

ENGLISH PATENTS.

MANURES, FISH.

Evans, J....No. 10806 of 1845.
Barker, E....No. 11924 of 1847.
Mitchell, B....No. 12023 of 1848.
Bethell, J....No. 12250 of 1848.
Brooman, R. A....No. 14255 of 1851.
Elliot, G....No. 546 of 1853.
Chisholm, J....No. 1375 of 1853.
Oxland, R....No. 2027 of 1853.
White, William....No. 2181 of 1854.
Bachelard and Harvey....No. 648 of 1855.
Gill and Sheridan....No. 744 of 1855.
Theroulde, F. A....No. 1079 of 1855.
Manning, J. A....No. 1579 of 1856.
Osler, S....No. 1198 of 1858.
Jervell, J....No. 684 of 1861.
Dewar, J....No. 1351 of 1863.

Millward, J .... No. 3775 of 1868.
Hamilton, E. C .... No. 2114 of 1873.
Rawson, Sillar, Slater & Wilson .... No. 2662 of 1873.
Crookes, W .... No. 2790 of 1873.
Metcalte and Massingham .... No. 3329 of 1873.
Wise .... No. 4176 of 1874.
Firman .... No. 2380 of 1876.

CANADIAN PATENTS.

FISHING.

Including FISH-TRAPS, FISHWAYS, FISH-HOOKS, HARPOONS, NETS, PISCICULTURE.

FISH-TRAPS.

Cottingham .... No. 2621 of 1873.

FISHWAYS.

Brewer .... No. 2386 of 1873.
Brewer .... No. 4720 of 1875.

FISH-HOOKS.

Harper and Smith .... No. 2110 of 1873.
Skinner .... No. 3067 of 1874.
Smith *et al* .... No. 6666 of 1876.

HARPOONS.

Pelletier .... No. 2383 of 1873.

NETS.

Morin, M .... No. 53 of 1840.
Morin, M .... No. 84 of 1845.
Lemoine, L .... No. 113 of 1847.

PISCICULTURE.

Holton and Green .... No. 2220 of 1873.
Clark .... No. 4927 of 1875.
Wilmot .... No. 6054 of 1876.

II.—ABSTRACT OF PATENTS ISSUED IN GREAT BRITAIN UP TO THE YEAR 1878, HAVING REFERENCE TO THE PURSUIT, CAPTURE, AND UTILIZATION OF THE PRODUCTS OF THE FISHERIES.

By R. G. Dyrenforth, M. D.,
Examiner, United States Patent Office.

DECOYING FISH.

No. 59 of 1632.—Grent.—Provides the net, spear, or hook with a looking-glass to lure the fish. Provisional. No drawing.

No. 295 of 1692.—Williams and Marwood.—Fish lured by means of lights burning upon or under the water. Provisional. No drawing.

No. 4582 of 1821.—De Chabannes.—*Attracting and catching fish.*—Lamp under water having one or more communications with the atmosphere to feed the flame and allow the smoke to escape; mirrors connected with traps or nets to lure the fish; living fish surrounded by glass or other protection in or about the nets, further to lure. Drawing, Plate I.

No. 4815 of 1823.—Coffin.—*Catching fish.*—Bait tossed overboard to bring schools of mackerel about the vessel; hooks are then used weighted by brightened lead. Drawing, Plate VIII.

No. 2580 of 1862.—Fanshawe.—For decoying or for seeing when a sufficient number of fish have been collected in a net, employs a stationary or movable submerged electric or other light; phosphorized oil, or other luminous fluid; or submerged reflectors reflecting light from above.

A globe of plain or colored glass, covered by strong wire net, contains the light and is supplied with air by flexible tubing; or a lantern is employed, constructed with a double roof that, the air therein becoming rarified by the heat, a current may be produced and the lantern rendered self-supplying with air.

The illuminating apparatus is lowered through a well near the center of the boat; or lights are sustained by buoys around the vessel.

When lines are used for cod, salmon, and other fishing, places a small wire-protected glass globe filled with a luminous fluid near the bait, on a horizontal line, and supplies air for combustion through a flexible tube. Drawing, Plate II.

No. 1751 of 1863.—Jodocius.—Ordinary electric light to lure the fish into a cage or net. Drawing, Plate III.

FLOATS.

No. 2003 of 1859.—FEARN.—A double-cone buoy formed of staves is made air and water tight by means of tongues and grooves. The heads are provided with stays to prevent bulging or collapse, and have a bracket for attaching the rope. Drawing, Plate I.

FISH TRAPS.

No. 150 of 1860.—ALLEN.—*Bait-can.*—For a bait-can which shall occupy less space when out of use, provides a flexible bag open at the top, where the edges are connected to a metal ring serving as a flange to a metal covering-plate. In the center of this plate is an opening provided with a perforated lid, and the plate has a handle capable of being turned down. The metal top is connected by jointed uprights to a shallow metal pan, which forms a bottom for the can. The sides being flexible, the uprights being jointed in the center and the pan at the bottom being larger than the top-plate and ring, the can may be folded when not in use. Small sleeves slip over the joints and keep uprights straight when the can is in use. Provisional. No Drawing.

No. 1120 of 1860.—STEVENS.—*Keeping fish alive.*—Supplies oxygen to the water either by injecting air into it or by throwing the water upwards into the air. Uses a box having a false bottom, in which a number of small holes are pierced. Air is supplied by means of a bellows. The water may be raised into the air by endless chains and buckets or by paddle-wheels. No drawing.

No. 3548 of 1867.—DAMM.—*Fish-tank.*—Supplies air to the water in the vessel containing the fish. A is the tank, having a pipe leading from the pump C, fitted with a piston-rod, handle, E, and a toothed sector, G, whereby reciprocating motion may be imparted to the pump and the water forced through pipes H and I. The pipe I and head *i* are fitted with short pipes, *a*, open to the atmosphere to force the water to the external air. As the stream enters the tank it impinges against a disc to prevent undue disturbance of the water therein. Drawing, Plate IV.

No. 1425 of 1868.—LEHEUP.—*Fish-boxes.*—Employs wooden screws to secure the parts together, or a pin having at one end an eye or hole to receive a pin or bolt; or employs a pin with a T-shaped head, the shank being clinched or riveted; or employs an L-shaped pin or bolt. No drawing.

No. 1276 of 1869.—ENGHOLM.—*Retaining caught fish in life.*—The water drawn from the cistern or tank is forced through pipes provided with air-induction nozzles down into the cistern again. An air-pump may be used. Provisional. No drawing.

No. 358 of 1873.—LEACH.—*Apparatus for hauling in nets.*—To facilitate the labor of hauling in great lengths of rope, netting &c., the boats are provided with auxiliary removable screw-propellers, driven by a steam-engine, which may also pump air or water into the fish-well. The water in the well communicates with the water outside of the boat, so that when the latter is in motion a current is established. The lines are coiled around a barrel, which is rotated by the engine. The capstan is mounted on the same shaft with the barrel, and both are connected to the shaft at will by a double-acting clutch. On the lower end of the spindle which carries and actuates the barrel and capstan is mounted a large bevel-wheel, driven by bevel pinion mounted on a horizontal shaft connected with the engine crank-shaft. The bevel pinion is capable of motion endwise on the shaft, so that it may be thrown out of gear with the large bevel-wheel. The other end is provided with a clutch-coupling, so that the propeller shaft may be driven when required. The barrel is provided with a traveling guide to lay the rope evenly when wound thereon. The guide is provided with grooved pulleys mounted on a vertical double-screw shaft, which is driven at a varying speed, regulated by the coiling-barrel, by means of chain and pulley on the lower end of the hauling-barrel. No drawing.

No. 1281 of 1873.—LEACH.—*Fishing.*—Upper part of the boiler projects through the deck and is provided around its upper edge with a circular rail or guide which supports the barrel of the capstan, which latter forms a cap for the boiler, and is provided with anti-friction wheels which run on the circular rail or guide on the top of the boiler. To the lower edge of the capstan barrel is adapted a horizontal guide-wheel, with a V-shaped edge, which runs against V-shaped guide-wheels mounted in bearings secured to the deck. The capstan barrel also carries a large toothed wheel which is driven by a pinion on the upper end of the vertical shaft mounted in bearings attached to the side of the boiler. The large toothed wheel and large horizontal guide-wheel are attached to a band-break-connection to the capstan. No drawing.

No. 4043 of 1876.—LAKE.—*Apparatus for catching fish.*—A sliding shank or bar terminates in a disc, in the center of which is pivoted a latch. Hooks are secured to the lower ring of a crown and pass through openings in the said disc. The crown is composed of a ring, disc, and connecting ribs, this disc having an oblong slot for the passage of the shank. A spiral spring surrounds the shank, and has a tendency to push the first-named disc in the direction of the points of the hooks and cause the latter to close. A bar or dog is fastened to the crown disc, and is provided with a lip which engages with the short end of the latch when the device is "set."

The bait-holder terminates in a hooked end, being pivoted at its opposite extremity to the first-named disc. This holder is bent to form an angle so as to engage with the latch.

To set the trap, compress the spring by bringing the disc and ring towards each other until the lip passes over the lower end of the latch. The other extremity of the latch rests in the angle of the bait-holder. As soon as a fish displaces the holder by taking the bait, the latch is liberated from engagement with the lip of the dog, and the spring causes the disc and ring to move apart, bringing the points of the hooks together firmly securing the fish, which, in most instances, will be instantly killed. A ring is provided to which the line may be fastened. No drawing.

FISHWAYS.

No. 974 of 1858.—PHYMONI.—*Apparatus for Catching Fish.*—A trench cased with brick, wood, or stone, provided with a cover.

Operation: Raises the cover by means of cross-bars a few inches to allow the water and fish to pass in, when the fish will secrete themselves within the trench. When necessary, places bait in the trench to allure or decoy the fish. Also, places along the sides of the trench, against the apertures, a lattice of wire so hung as to rise or fall and close the aperture against egress of the fish. No drawing.

HARPOONS.

No. 1367 of 1783.—BAYLES.—"A triangular instrument which, when struck, pushed, or thrown, and falling on its point on large fish, will penetrate, cripple, kill, and hold such fish; or, if falling on its side, its withers will grapple and take the fish." No drawing.

No. 4563 of 1821.—CONGREAVE and COLQUHOUN.—Force the barbed instrument into the fish by means of a rocket. Rocket may be used alone, or before or behind the harpoon, or to carry a line, or a shell which, bursting within the body of the whale, will kill it, and may besides fill it with gas and prevent its sinking. No drawing.

No. 8541 of 1840.—LANCE.—The rod is provided with a screw-thread running its entire length and working into a ferule in the head of the harpoon; to the reverse end is attached a fly, which forces the harpoon into the fish as he moves forward. Attached to the harpoon is a buoy, to assist in impeding the fish's progress.

Figure 4 is a varied form of the instrument, made with triangular-cutting edges toward the point, and furnished at its base with barbs which turn on pivots. Drawing, Plate V.

No. 125 of 1856.—RECHTEN.—The barb of the harpoon is made in one piece, turns on a center at the end of the shaft, and is retained by a peg, so that when the barb enters the fish the peg will be broken and the barb will place itself at right angles to the shaft. The shaft consists of two bars welded together; the head is made tubular a short distance, and a hole extends transversely through it. The harpoon is shot from a gun. Drawing, Plate II.

No. 1110 of 1857.—TINDALL.—To obtain precision in firing, non-liability to derangement, to hold the fish in whatever position the harpoon is lodged, and to increase the efficiency:

The muzzle of the gun is of smaller diameter than the rest of the barrel, for fitting on a collar contained in the center of the harpoon-head when the harpoon is to be discharged. Collar has on each side a short barbed harpoon-piece, with the barbed ends turned backwards. Line attached to harpoon-pieces by means of a thimble connecting it with a shackle, jointed to the harpoon-pieces. Gun loaded with a long cylindro-conoidal missile, a short piece of the after end of which is made to fit the bore of the gun, whilst the rest is of a reduced diameter, terminating in a cone or point. Mouth of harpoon collar fits exactly to smaller forward portion of the cylindrical ball, and is turned out internally to fit to the angle forming the connection between the larger and smaller diameters of the ball. Ball in barrel close up to powder of charge, and collar with its duplex harpoon being placed upon the muzzle, it follows that when the weapon is discharged the ball enters the collar, but being caught therein at the part behind exactly fitting the bore, it carries away the duplex harpoon and the line with it. Ball and harpoon thus enter the fish together.

With duplex harpoon just described may use an expanding harpoon or harpoon-shot, the two harpoon pieces or arms which it carries being folded down and retained by the barrel of the gun. A spring throws them out when the harpoon is discharged, and pull of the fish expands them fully. Arms of this harpoon are at right angles to those of the main harpoon. The expanding harpoon has studs to guide it and secure the correct relative position of the two sets of harpoon arms.

In another form, shown in figure 6, the head is of the common form, but fitted with a solid cylindrical shank passed through a cross-piece which fits upon the muzzle of the gun at a distance to allow the escape of the compressed air when the gun is fired. To this cross-piece is attached the shackle for connection with the line and rings. The after end of the harpoon shank has a stud or collar piece, and when the gun is fired this stud or collar piece strikes against the arm through which the shank is passed, thus carrying the line along with it. Drawing, Plate VI.

No. 2301 of 1857.—ROYS.—Rocket with an explosive charge in its head is attached to a feathered shaft provided with barbs. The rope or chain is attached near the rocket-head and is also provided with barbs. Drawing, Plate VII.

No. 2340 of 1857.—ROYS.—A small tube has barbs projecting from the sides, and to these a bridle is attached, which is connected to a cord or chain, and upon this a cross-bar is fixed to prevent the shell passing through or entering too far into the fish. No drawing.

No. 450 of 1861.—WALKER.—The harpoon has jointed barbs at its point and a case containing the explosive charge. The opposite end is formed in the shape of a hook and is provided with a spring catch to secure the link and line. Drawing, Plate III.

No. 550 of 1865.—ROY and LILLIENDAHL.—The spear is provided with jointed barbed arms which open outwardly. Drawing, Plate II.

No. 3312 of 1868.—WELCH.—To the head of the shaft are secured a shell, B, and barbs, C, the latter hinged upon a pin and kept closed by an elastic band. The shaft is bored out and the igniting fuse is placed therein. E is a wire ring to which the line is attached. In figure 4 *b* is a metallic tube around the fuse. Drawing, Plate III.

HOOKS.

No. 1719 of 1789.—JOHN ANDREWS.—The fish-hooks are first formed of steel wire in the usual way and then hardened, tempered, polished, and completed.

The hooks, cold, are placed upon thin cap paper and covered with yeast to prevent the fire from penetrating too quickly into the steel, injuring the beard and fine point; they are then placed upon an iron plate and put into an iron case and placed in a slow fire until red-hot, after which they are removed and placed in a tub of milk-warm water for the space of one minute, and finally in fine emery, where they are heated until dry. The hooks are then brightened by agitating them in a barrel containing a mixture of water, castile-soap, and emery, after which they are again dried by being brought in contact with dry ash sawdust, and put into leather bags and agitated by hand.

In order to temper the hooks, hour-glass sand made hot and the fish-hooks placed loosely therein are kept constantly agitated, by which means, also, the hooks will become dark blue. The hooks are then removed, placed in a leather bag, and agitated, and afterwards they are put up in steel paper. No drawing.

No. 2063 of 1795.—WILLIAM BELL.—Casts hooks from steel or common fusible iron. No drawing.

No. 11520 of 1847.—MOSES POOLE.—The hook is so constructed that by the aid of instruments combined therewith the holding of the fish will be more certain.

a, the hook; *b*, the retaining instrument which is attached to the stem of the hook at *c*, there being a stop at *d* to prevent the hook and retaining instrument from coming too close to each other. *e* is a spring which has a tendency to keep the hook and instrument closer together. The spring *e* is affixed to the stem of the hook and connected to the retaining instrument by means of the short link *f*. In the retaining instrument a notch is made, into which the end of the sliding bolt *h* enters, such sliding bolt being slotted and attached by studs to the stem of the hook *a*. Drawing, Plate IV.

No. 533 of 1852.—A. F. BAINBRIDGE.—*Flies.*—Attaches the wings, fins, &c., to the hook by means of elastic, flexible, and controllable thread; such as vulcanized rubber. Drawing, Plate VI.

No. 889 of 1852.—G. A. HUDDART.—Artificial-fly wings resembling he natural wings of flies, made from rubber, gutta-percha, or analogous, compounds, by molding, the molds having lines or markings corresponding to or in imitation of the markings of the wings of the natural insect. No drawing.

No. 2902 of 1853.—R. J. N. KING.—*Artificial bait.*—Artificial bait for fish, in the form of a minnow, made of brass or other metal, formed square inside. Side hooks are fastened to the minnow. A brass block is fitted inside the minnow, to which steel springs are riveted, and hooks which extend from the tail of the minnow are soldered to these springs. No drawing.

No. 923 of 1857.—WILLIAM H. BOX.—First, electroplates fish-hooks; second, attaches the hook directly to the swivel-box by means of a knob or pin-like head, and not to an eye, as usually practiced. Drawing, Plate V.

No. 1135 of 1859.—WM. E. NEWTON.—"Sockdologer" fish-hook, rendered perfect and sure in its operation and less dangerous to be handled while baiting, by arranging between the main hooks, and connected to the same by two arms, D, D^1, a bait-hook, F, in such manner that by forcing the two arms to a horizontal position the main hooks are spread open or set. The top portion of the main hooks is made elastic, so that by the action of this portion, together with the power obtained by two additional springs, the main hooks spring together as soon as the slightest strain on the bait-hook disturbs the horizontal position of the two arms. Drawing, Plate V.

No. 428 of 1865.—WM. A. HACKETT.—The fins, wings, or vanes, made of metal, horn, bone, or other material, are secured to the fishing-hook in such manner that when the said hook is drawn through the water or held in a running stream it will be made to spin or twirl. The hook is swiveled to the line. No drawing.

No. 3177 of 1865.—CHARLES BAYLESS.—Makes at the junction of the stems of the two hooks a spring by coiling or bending the wire of which the double hook is made. Around the stems of the double hook a band or ring is placed capable of sliding upon said stems. No drawing.

No. 413 of 1866.—JOSEPH WARNER.—Makes at or near the end of the shank of the hook an eye, *b*, and side-grooves or depressions *c*, into which the said eye opens and by means of which eye or eye and grooves or depressions the line or gut is readily and securely attached. Drawing, Plate V.

No. 150 of 1867.—GEDGE.—A double fish-hook, composed of two parts or branches.

The first branch, *a*, is bent at right angles at its upper part, where it is pierced with a hole, *e*, the edges of which are rounded for the better passage of the line *l*. At its lower part it has a hook on which the bait is placed.

The second branch, *f*, is attached to the first at *b*, and is held in a fixed position by a piece, *c*, the sides or cheeks of which form springs nipping and retaining this branch, which is further provided with a ring, G, and a hook.

To use this fish-hook, open the branch *f* and fix it to the collar or nipping piece *c*, pass the line through the hole *e*, of the first branch, then attach it securely to the ring G, of the second branch, and then bait the hook on the first branch *a*. Drawing, Plate VIII.

No. 1765 of 1867.—WELCH.—*Swivel for fishing tackle.*—The box or case of the swivel is made of metal tubing open at both ends. To one end attaches, by soldering, a loop. The shank or stem of the swivel, which works in the box, is made of wire, and is provided with a knot at one end, the knot being situated within the bow or loop of the case when the stem is placed within said case. The opposite end is fashioned into a loop. By this construction the shank or stem is irremovable from the case, and at the same time free to rotate within it, while the case is free to rotate upon the shank.

In applying the swivel to spinning bait, solders the tube constituting the case to the back of this spinning bait; but instead of soldering the loop or bow to the box or case itself, it may be soldered to a tube or portion of a tube soldered to the spinning bait, within which last-mentioned tube or portion of a tube the box or case of a swivel is fixed. The hooks are attached to the loop or end of the shank or stem, and the end of the fishing-line or swivel connected with the fishing-line is attached to the loop or bow of the spinning bait. Drawing Plate, VIII.

No. 2714 of 1867.—A. MORRALL.—A piece of wire is bent into or caused to assume such a shape that when in its normal condition it nearly represents the letter **V**. The ends of the wire are formed into barbed hooks. When baited the hooks are pressed together. Drawing, Plate VIII.

NETS.

No. 1872 of 1859.—STUART.—Manufactures nets of single yarn. No drawing.

No. 3099 of 1860.—HENRY.—The net submerged has a rubber tube carried around and attached to it, which communicates with an air-pump for inflating it with air, causing the net to float. Empty casks are also attached as floats. A weight is fastened to the bottom of the net by hooks, which may be opened at pleasure by a cord operated from the boat to allow the net to be disengaged from the weight and float to the surface of the water. Drawing, Plate IX.

No. 492 of 1861.—JAMES.—The nets are buoyed by balloons inflated with air, and having proper ballast (water). The nets are drawn in by a windlass or by other suitable means. Drawing, Plate X.

No. 1871 of 1863.—HECTOR.—The net is formed with compartments and closing apertures to retain the fish in the net. Drawing, Plate X.

No. 3334 of 1865.—HURN.—Forms the nets of continuous pieces of tanned leather. No drawing.

No. 193 of 1866.—BRYSON.—Maintaining artificial light under water for the purpose of attracting or decoying fish.

First. Has a copper vessel with two compartments, into one of which oxygen is forced or condensed, while the other contains hydrogen, also forced or condensed. Each compartment is provided inside with a valve to regulate the escape of the oxygen and hydrogen to the point of combustion, so that these gases are not allowed to mix except at that point, whereby explosion is prevented. Light having been applied to the gases, the apparatus is lowered into the water and the gases continue to burn until entirely consumed, the presence of atmospheric air to keep up combustion not being necessary. The light is covered with a water-tight glass globe, provided with a receptacle to contain the water generated during the combustion. The light, when introduced under the water, attracts the fish thereto. Lime or other substance may be employed in contact with the gases.

Second. Has nets for catching the fish attracted by artifical lights.

Two nets, the upper being larger than the lower; attaches weights to the circumference of the upper net, and floats at and near its center, the lower net being placed under the light. The upper net is let down over it, which, owing to the weights and floats, descends somewhat in the form of an umbrella, and incloses all the fish within its reach, which are thereby caught between the two nets. Drawing, Plate IX.

No. 2008 of 1866.—MACK.—To the buoys and net anchors are attached. The rope from the anchors serves as guide lines to the net. Lights are also used to attract or lure the fish.

a, anchors; *b*, bottom of sea; *c*, main line to the buoys; *d*, stretching-tackle for bottom of net; *e*, stretching-tackle for upper part of net; *f* and *g*, stream stopper-tackle; *h*, stretching-tackle to the surface of the water; *k*, net; *l*, *m*, *x*, bait; *n*, electric light; *o*, *p*, bags filled with herring; *v*, buoy. Drawing, Plate XI.

No. 1331 of 1867.—HALLETT.—The nets have a platform of triangular or semi-circular shape. To the mouth of the platform attaches weighted chains, also a lightly loaded line through the center of the platform; intersects the platform at intervals with cod-lines or light roping by lacing. To the sides of the platform by lacing a weighted line or chain, attaches a "walling" or "leader" of net work, which runs around all sides of the platform except the mouth. No drawing.

No. 1332 of 1867.—HALLETT.—The net is made cone-shaped and the upper part of the mouth is well supported by buoys. The under part of the mouth is made with a deeply curved margin, bordered by a ground-rope or chain, which is heavily weighted, or the chain is tarred or covered with oakum or hemp, or it is galvanized. The net is fitted with two pockets, one on each side, made by lacing the upper and under parts, beginning at the outer edge and gradually working toward the middle and small end of the net; attaches to the sides of the mouth leaders of net-work with sole and back ropes, the sole-ropes being loaded and the back-ropes being corked. To the bottom of the net attaches chains loaded. At the rear end of the net attaches a heavy weight; at the front end a buoy. No drawing.

No. 1333 of 1867.—HALLETT.—The same as No. 1331 of 1867, and having pockets similar to those used in beam trawls, except that they are not covered with net-work at the surface of the water, and have in place thereof two small buoy ropes. No drawing.

No. 2140 of 1867.—JOHNSON.—Twists the twine while in the process of netting and makes the net with a looped knot. Forms the net of twines, alternate ones being larger than the others. Drawing, Plate VIII.

No. 792 of 1877.—CORRIN.—Fittings for sinking fishing nets. Weights of oval form have a hole through the center through which is run a line about two feet in length; to this line at each end of the weight another line is spliced, and both are fastened to the bottom of the net, so that the line through the central hole is on a level with the bottom line or rope of the net, and only half or a portion of the weight then projects below the net. No drawing.

OYSTER CULTURE.

No. 2930 of 1863.—AYCKBOURN.—Tiles, &c., preferably concave, arranged in groups upon each other and at right angles, catch the spawn as it floats. The tiles are coated with common clay of the consistency of thick cream, and on this is laid Portland cement, to which the young oysters attach and grow. When it is desired to remove them, the oysters and clay are broken off together. No drawing.

No. 1040 of 1864.—CROFT.—Tanks are so constructed that water will pass freely among the oysters placed in rows, inclined with their mouths upwards, and supported by long narrow tiles indented, to be easily broken, and provided with projections for holding them up. Over the tiles is placed material to induce the spat or brood of the oyster to adhere. Provides water-tanks, the temperature and water-supply of which may be regulated and varied (fresh or salt), but the water should be acclimatized gradually. Has star collectors, formed of wheels on spindles, connected by rollers armed with bunches of bristly substances or hooks. Has also spat collectors, tile perforated and filled with cork. Drawing, Plate XII.

No. 1316 of 1870.—BERT.—The walls of the inclosure are built of masonry, their foundation below the lowest tides and their heights above that of the highest tides, to prevent loss of eggs. The inner surface is irregular at a height of three feet. The bottom of the basin has a declivity toward the center when the trench is formed, in order to drain the water from the basin and suppress the deposits of mud. A torus two inches high is built, perforated at different places to allow the diluted mud-deposits to pass, but prevent the oysters from following.

A, walls; B, rough inner surfaces; C, inner walls; E, trenches; G, toruses; F, oysters; I, platforms; J, stone supports for platforms; L, pillars which support cross-pieces and there movable ceiling. Pl. XVI.

No. 4103 of 1874.—MICHEL.—*Hives for Breeding Oysters.*—First. Moulds from cement rectangular vessels open at top and having their bottoms provided with numerous holes.

Second. Forms an open rectangular trough without perforations through its bottom. The ends of this trough project downward, forming a stand.

A pair of these vessels, the lower one perforated, and the upper one placed on it as a cover, form a hive; numbers of which are placed side by side on the beach or on timber sleepers, to form a breeding and rearing bed.

The lower perforated vessels receive the spat, and are kept clean by the wash-water running through them. The upper vessels, besides serving as covers, form nurseries, and protect the oysters against enemies and changes of weather. Drawing, Plate XII.

No. 3506 of 1875.—DE LAGILLARAIE.—*Breeding or Cultivating Oysters.*—Secures the oysters to a wire and suspends them in the water in a manner that renders them easily accessible for inspection or removal. The oysters thus wired are secured to trellises. The oysters may be placed in cages provided with hooks, said cages being made buoyant and moored to a submerged chain, held by floats. Drawing, Pl. XVI.

No. 2470 of 1876.—JENNINGS and ANDERSON.—*Propagation of Oysters.*—Place the oysters and spat in water-tight receptacles made of earthenware or other material and covered over at top. The site upon which the receptacles are placed may be enclosed by protecting walls, which, whilst affording protection, will allow of the passage of water as the tide rises or falls; or the receptacles may be placed in water-tight wooden troughs. Drawing, Plate XII.

No. 1447 of 1877.—MEWBURN.—*Apparatus used in Breeding Oysters.*—A box made of any suitable material communicates with the exterior by water and air supply cocks, or has the upper part open without an air-cock, when the apparatus is situated away from the sea and fed artificially, or is closed hermetically when the apparatus is placed so that it emerges at each tide. In this apparatus the collecting hives and the breeding oysters are placed at spawning time, or spawn or spat obtained elsewhere may be put therein. A piece of silk, the meshes of which are too fine to allow the spawn to pass through, is then placed over each water inlet and outlet, so that no spawn or spat can escape when the water is renewed.

Whether the apparatus emerge or not, the collectors should offer the greatest possible surface for the deposit of the spat, and be fed by water, sufficiently clarified by filtering, to insure and maintain the cleanness of the collectors. The water must be renewed without sensible agitation, its introduction be regulated by means of cocks, and a slow current obtained, continuous or intermittent, according to the position of the apparatus. The water is renewed sufficiently often to put the spat in the best conditions of vitality, filtered water being always employed.

The apparatus consists of a box provided at its lower part with a cock which communicates with a filter situated a little below. The cover of the box, which closes it hermetically with the aid of strong bolts or screws, is provided with two India-rubber tubes having valves, one of which opens outward to allow the escape of air at the rise of the tide, while the other opens inward to allow air to enter at the fall of the tide or when the apparatus empties. The box is firmly secured in the water by stakes, to which it is only pinned, to allow of its being removed when required, and the India-rubber tubes are placed on rigid pipes long enough to give the apparatus time to fill before the water reaches their top or the waves pass over.

Inside the apparatus is placed a hive composed of a series of frames containing wire or other gauze, and placed one above the other. The frames are filled with a layer of shells, broken and sifted, and a certain number of parent oysters placed at intervals. There is fixed at the side of and below the box a filter of any suitable kind, through which the water will pass before its introduction into the box. The water arriving at the rising tide will pass into the filter and be deprived of all foreign matter, and finally enter the apparatus above, driving out the air through one of the valves, and carry sustenance to the breeding oysters placed among the frames. The water in retiring allows the second valve to open and admit air to the apparatus.

The spawn or spat, in rising in the apparatus will traverse layers of collectors, and the water in retiring will filter through the layers of broken shells and deposit thereon the spat which it holds in suspension. The apparatus can be inspected daily and the hive withdrawn when sufficiently filled and replaced by a fresh hive. A series of apparatus may be supplied through one filter. No drawing.

OYSTER DREDGES.

No. 2906 of 1860.—ENNIS.—An open-mouth cage so formed and fitted that when lowered into the water and drawn over the beds it will first disturb and then gather up the oysters.

Fitted to and extending across the mouth of the case, are two scrapers, set parallel to each other and on opposite sides of the cage. In front is a rake held in position by binding-screws forming a double rake. The drag-iron having an eye is secured to the sides of the cage, and to the eye the rope is attached for hauling the dredge. Drawing, Plate XIV.

No. 323 of 1869.—BRABASON.—The links of the dredge are runners like those of a snow-sleigh, turned up at the ends and connected by an iron ring or link which rests on them and presents a smooth under surface which carries it along the bottom of the sea, passing over the spat or brood of oysters without injury to them. Provisional specification. No drawing.

OYSTER RAKES AND TONGS.

No. 2171 of 1866.—JOHNSON.—Two converging rakes are provided with handles united by a rivet, one handle being provided with a spring which pulls upon the upper end of the other. The rakes are held open by a lever having its fulcrum on the projecting end of one of the handles, one end of the lever taking into a catch upon the other handle and a lowering rope or chain being attached to the opposite end of the lever. The weight of the apparatus will cause the catch to hold as the apparatus is lowered while open. When the apparatus rests upon the bottom, the weight being relieved, the lever unlatches and the rakes are drawn together by the spring. Pulling upon the rope then also draws the rakes together. The teeth are removable. Guards at both the ends and back of the rakes prevent anything falling out. Drawing, Plate XIII.

No. 1438 of 1867.—JOHNSON.—Two rakes with scoop-shaped or open heads, each attached to a handle and the handles connected by a joint, near the middle, the rake teeth coming toward each other. Drawing, Plate XIV.

ELECTRICITY, PARALYZING FISH, &C.

No. 2644 of 1863.—BAGGS.—Wires or hooks and lines, nets, &c., connected to a galvanic battery. The fish receiving an immediate shock is paralyzed or killed. Drawing, Plate XV.

No. 3228 of 1868.—BENNETT and WARD.—*Electricity, galvanism or magnetic electricity.*—Places in the boat a galvanic battery with its coils and necessary accompaniments, insulated wires, in connection with and passing from the *opposite* poles of the battery and inside or along the line, to which the harpoon or other instrument is attached. No drawing.

PISCICULTURE.

No. 2966 of 1853.—BOCCIUS.—A vessel having apertures in its top and bottom of such a size as to prevent the eggs or spawn passing through, yet sufficiently large to allow water to percolate into the apparatus when immersed. It is made of two or more sections or parts, one part fitting into or on the other. In the interior are placed trays, one over the other, if required, upon which the eggs are placed. Drawing, Plate XVI.

REELS.

No. 3157 of 1808.—JOHN CURR.—*Method of applying ropes for catching and detaining whales.*—The boat is provided with a reel, brake, brake-lever, guide for the line, and rollers to keep the line near the center of the boat. When the whale has nearly run out the line of one boat, it may be joined to the line of another boat. Drawing, Plate XV.

No. 1956 of 1856.—KENTON.—Makes the reels of papier maché and passes a metal axis through the reel and connects the winch by which the said reel is turned to the axis. The case in which the reel works is made of a tubular piece of papier maché closed by discs, through the center of which the axis of the reel passes. Drawing, Plate XV.

No. 1806 of 1868.—MURE.—The carrier, to facilitate the winding of several lines and hooks simultaneously on the winder and to permit one line or hook to be withdrawn at a time without entangling the rest or breaking them, consists of a tablet having at each end a cap or ferrule, the center of which is hollow to receive a piece of cork in which to insert the barbs of the hooks. On each side are a number of openings reached by a sloping duct, through which the lines are successively and alternately passed in a hank till completely wound up, the hooks being fixed in a similar manner, openings being made for their reception on both sides of the apparatus. The hooks are protected by a metal band attached to the cap of the ferrule, one end working on a hinge and the other fixing with a clasp. May have a simple line-winder, which is of the same construction, less the cap for bearing the hooks. This apparatus is furnished with a cover, having an opening at one end through which to draw the lines. Another form consists of an endless band of pasteboard in the shape of a flat cylinder, and the lines are drawn around this, being passed through openings at the sides as before. Hook-carriers consist of a wooden plate with ends terminating in a metal cap having notches for the hooks, which are protected by hinged metal bands, the opposite ends being furnished with hasps. Another form is a bobbin or reel with a cover or case at the bottom, and in the interior is a circular shaft, the upper end of which terminates in a button or cork for the hooks, and the line is passed around small projecting knobs on each side of the shaft at the top and bottom. No drawing.

No. 388 of 1873.—HEATON.—Recesses one of the side disks and places on its axis a pinion facing outward towards the recess and connects the pinion to a hollow axis working through the corresponding plate for winding the line on, and over the recess. On the outside of the disk applies a plate with a rim around its interior, having teeth cut on the inner edge, which gear with an intermediate wheel acting on the pinion. This outer plate works on an independent axis passing through the axis on which the line is wound and is secured in position by the spreading head of a screw on the other side. A handle is applied to give motion to the plate, and from the teeth on the plate acting on the intermediate wheel, motion is given to the pinion and winding axis, the multiple of motion being governed by the size of the pinion and intermediate wheel. No drawing.

No. 3283 of 1876.—CORBET.—*Attaching reels to rods.*—Instead of attaching the reel to the outer surface of the rod in the ordinary manner, provides a frame consisting of two side plates and two tubular or hollow ends. Into one of these tubular ends inserts the top part of the rod, and into the other the lower portion or handle of the rod. No drawing.

RODS.

No. 2017 of 1853.—DAWSON and RESTELL.—The joints are constructed in the shape of a segment of a circle; that is to say, the cross-section of the joint will represent the segment of a circle, or that of a tube bent into the shape of a trough or gutter. The joints are permanently connected, and each joint lies in the joint next preceding it when the rod is inserted in the butt, which is made cylindrical and to resemble a walking-stick. No drawing.

No. 461 of 1857.—JOHN BENNETT.—Upon the ends of the rods to be joined rigidly together, attaches a metallic tube or ferrule, the tube on one end being larger than that on the other, so that one of the tubes may enter and slide in the other tube. The larger tube has in its axis a smaller one, so that the tube which enters the larger tube or ferrule slides between two concentric tubes. The two rods are joined together by means of a cylinder of India rubber (which acts as a joint), and the smaller tube or ferrule is provided with a bayonet joint, in which a pin in the larger tube engages. No drawing.

No. 1553 of 1858.—A. PORECKY.—Strips of whalebone, horn, tortoise-shell, or other corneous matter, or the artificial imitations, are formed by any suitable process into plain, fluted, ornamental, flexible, or rigid tubes. They may be either wholly or partially filled with a suitable composition or material, or may be left unfilled, thereby allowing of their sliding, one over the other, like a telescope. No drawing.

No. 3528 of 1868.—BRITTEN.—Instead of wood, for those parts which are required to be pliant, makes use of steel or other suitable metal that will secure the requisite strength with lightness and elasticity. These parts are formed into tubes, which are tempered in the usual way. The ring-fittings for guiding the line are on little bands of spring metal, which go round the rod and clip it, and are movable along the rod. The whole rod complete can be inclosed in the lower joint by a cap and ferrule. The several lengths of rod are connected by short pieces of thin tubular metal soldered to the ends inside the larger and outside the smaller following lengths, and are made slightly conical in opposite directions to fit tightly when drawn out. No drawing.

No. 1648 of 1874.—MOULTRAY.—In the socket, covered with brass, there is made an incision sufficiently large to receive a spring. By means of two metal pins driven into the wood and covered by brass, the metal spring is fastened. The upper end of the spring is bent upward and placed in the incision at its upper end, so that by coming in contact with the brass covering of the socket it may be prevented from coming out of the incision. In the ferrule which receives the socket and at a place thereon corresponding to the place which the spring will occupy in it when the socket is inserted in the ferrule, an opening is cut to allow the spring to work freely. When the socket is placed in the ferrule the spring holds them fast. No drawing.

No. 1840 of 1874.—JACK.—Fixes the stem part of each joint to its corresponding socket-piece by forming a very small, short projection, or pin, on the large part of the stem-piece, which, when inserted into its socket-piece, enters a small recess cut in from its outer end and is then to be turned round into a groove cut or embossed into the inner circumference of the socket. No drawing.

No. 1806 of 1876.—HOLROYD.—Forms the joint-fitting in the usual way, with the addition to the lower end ferrule of a projecting collar and screwed spigot, both situated at the upper part of such lower end ferrule. Also forms the upper ferrule with an enlarged screwed socket to receive and hold the lower ferrule and its screwed spigot, forming an air and water tight joint when the two are fixed together. Drawing, Plate X.

No. 1553 of 1877.—ASTON.—Takes wood or cane sections, having at each end the usual metallic ferrule, but having a slot, into which is placed a plate of metal pierced at each end, there being a corresponding piercing in the ferrules. Into this secures a pin, thereby forming a joint

4582 of 1821.

P. 17.

DECOYING.

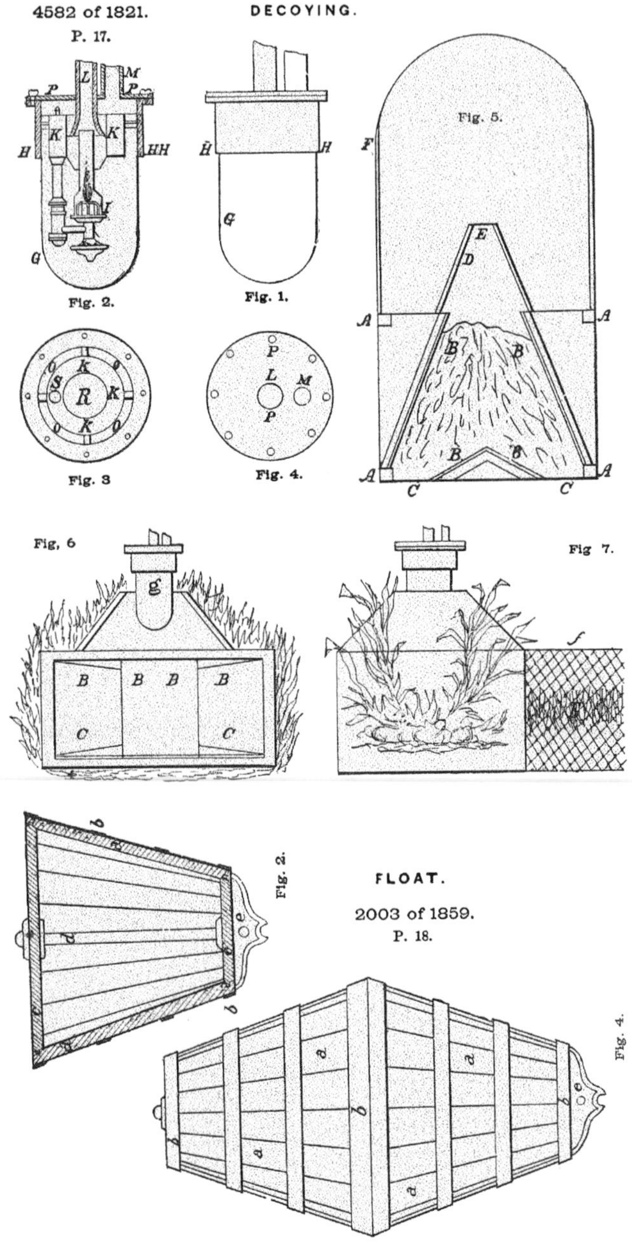

FLOAT.

2003 of 1859.

P. 18.

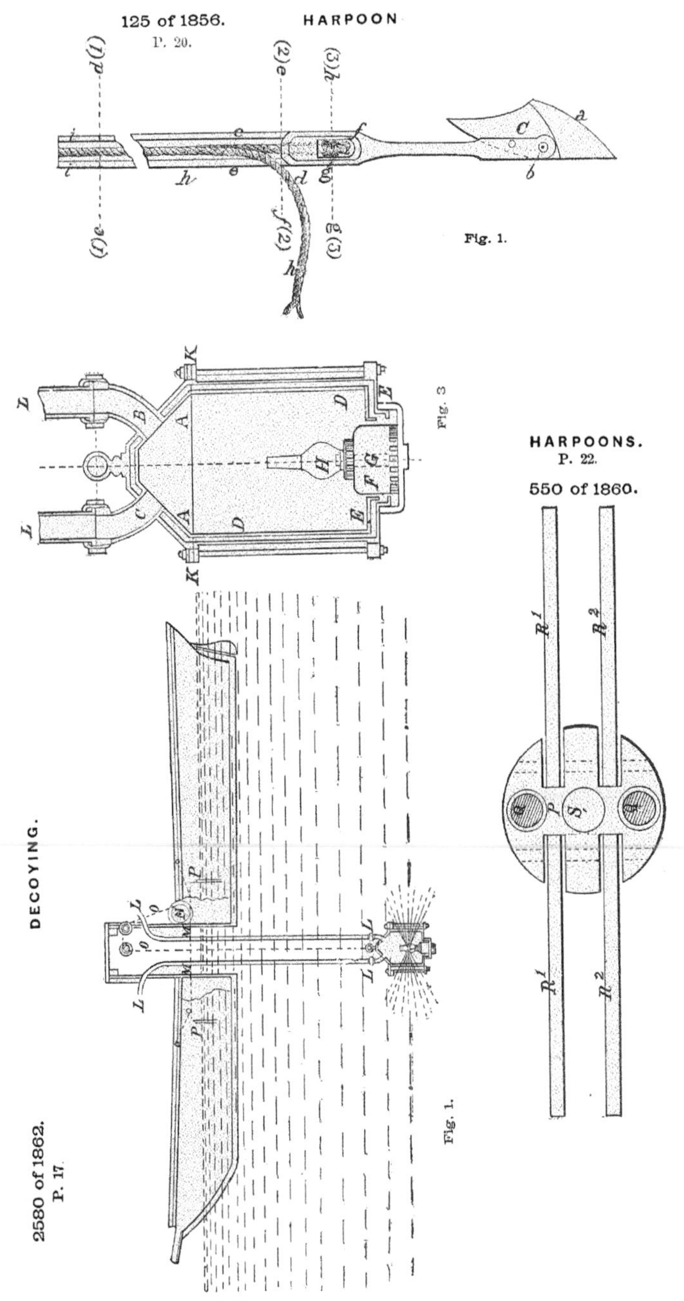

125 of 1856.
HARPOON
P. 20.
Fig. 1.
Fig. 3
HARPOONS.
P. 22.
550 of 1860.
DECOYING.
2580 of 1862.
P. 17.
Fig. 1.

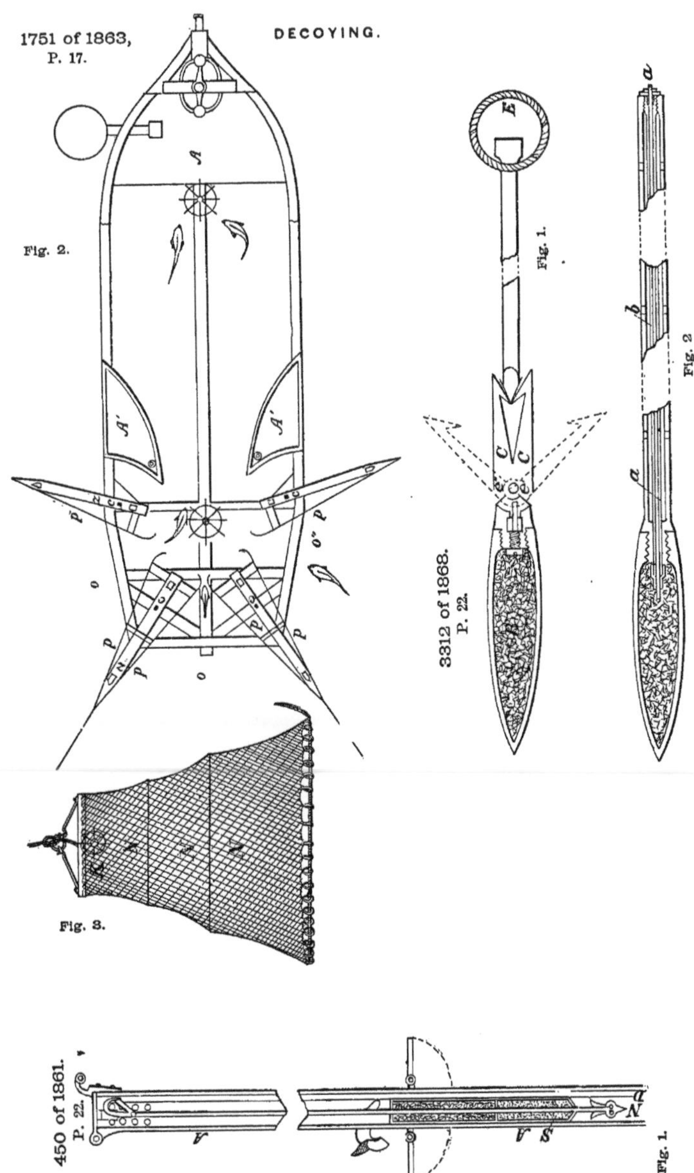

450 of 1861.
P. 22.
Fig. 1

3548 of 1867.
P. 18.

FISH TRAPS.

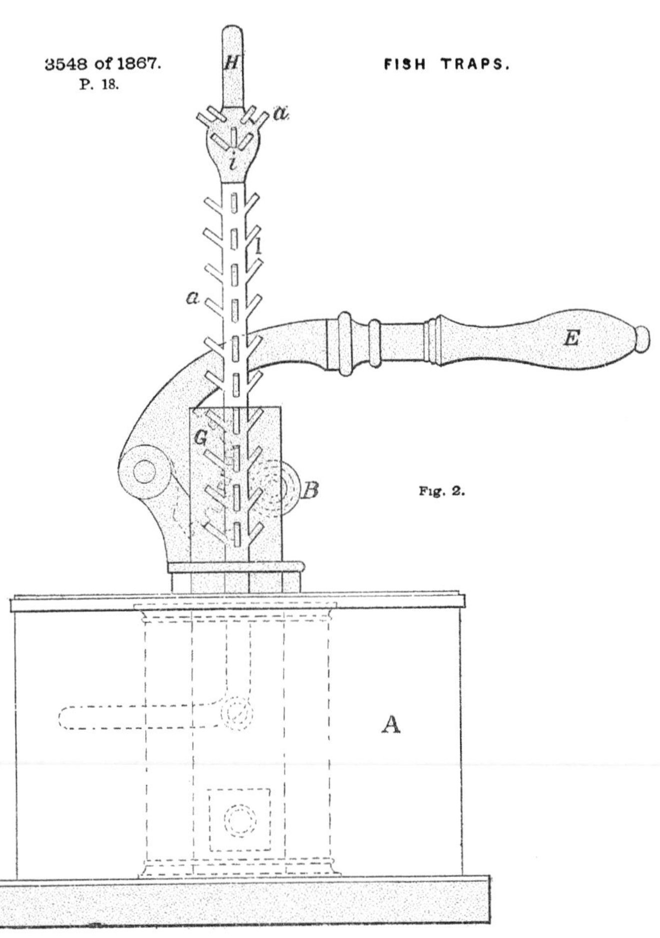

Fig. 2.

11520 of 1847.
P. 22.

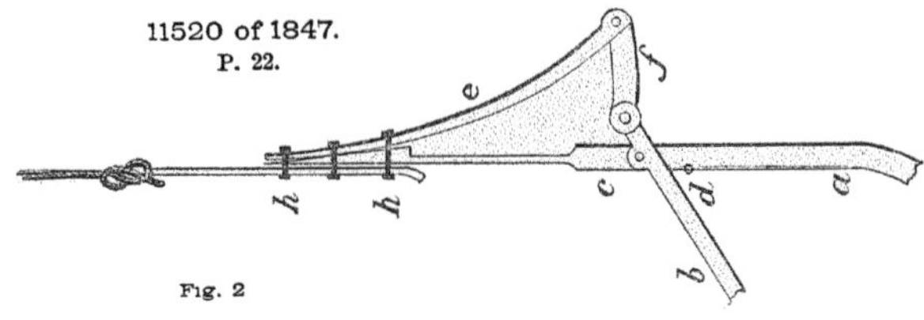

Fig. 2

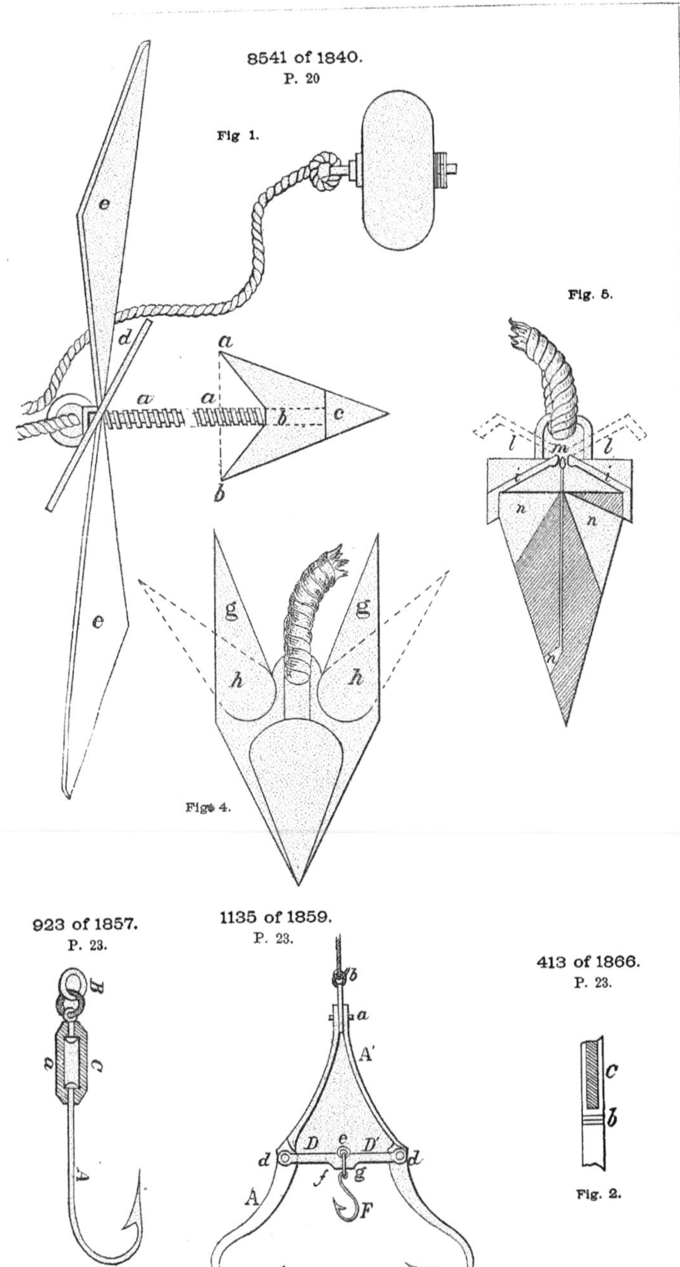
8541 of 1840.
P. 20
Fig 1.
Fig. 5.
Fig. 4.
923 of 1857.
P. 23.
1135 of 1859.
P. 23.
413 of 1866.
P. 23.
Fig. 2.

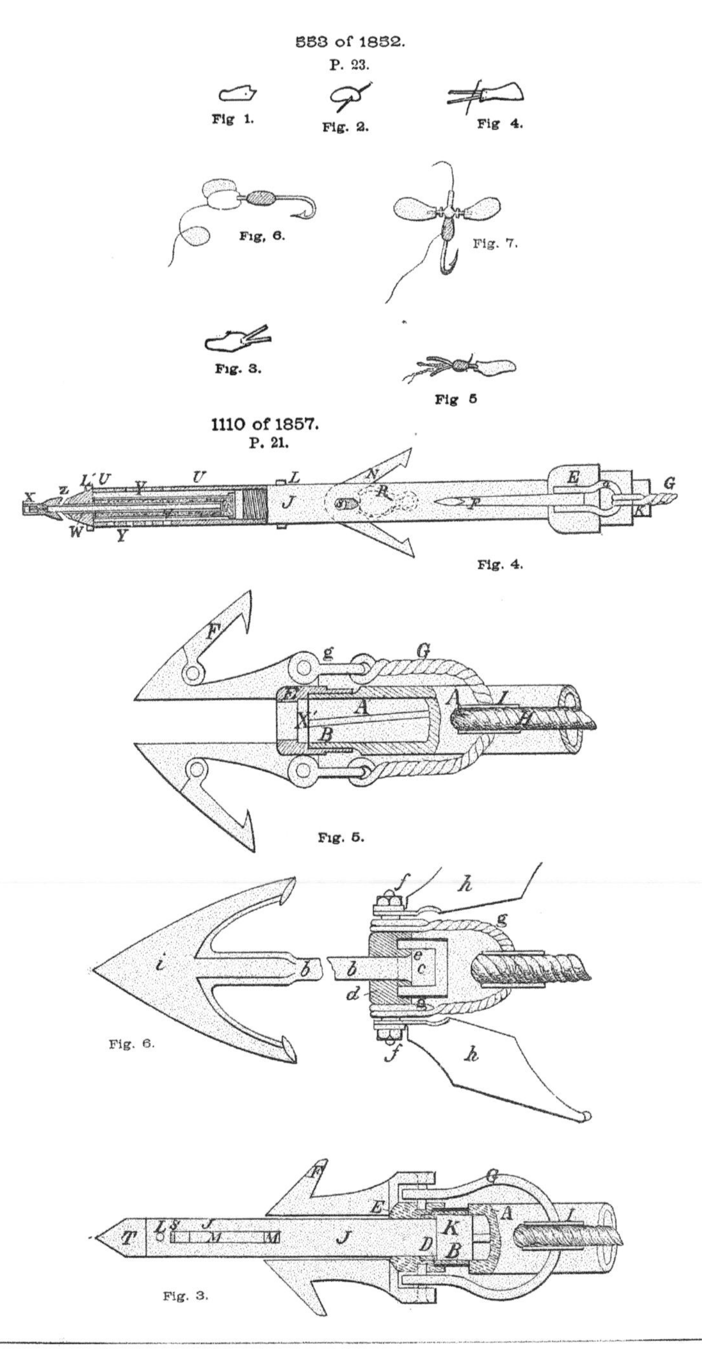

553 of 1852.
P. 23.
Fig 1.
Fig. 2.
Fig 4.
Fig, 6.
Fig. 7.
Fig. 3.
Fig 5
1110 of 1857.
P. 21.
Fig. 4.
Fig. 5.
Fig. 6.
Fig. 3.

2301 of 1857.
P. 21.

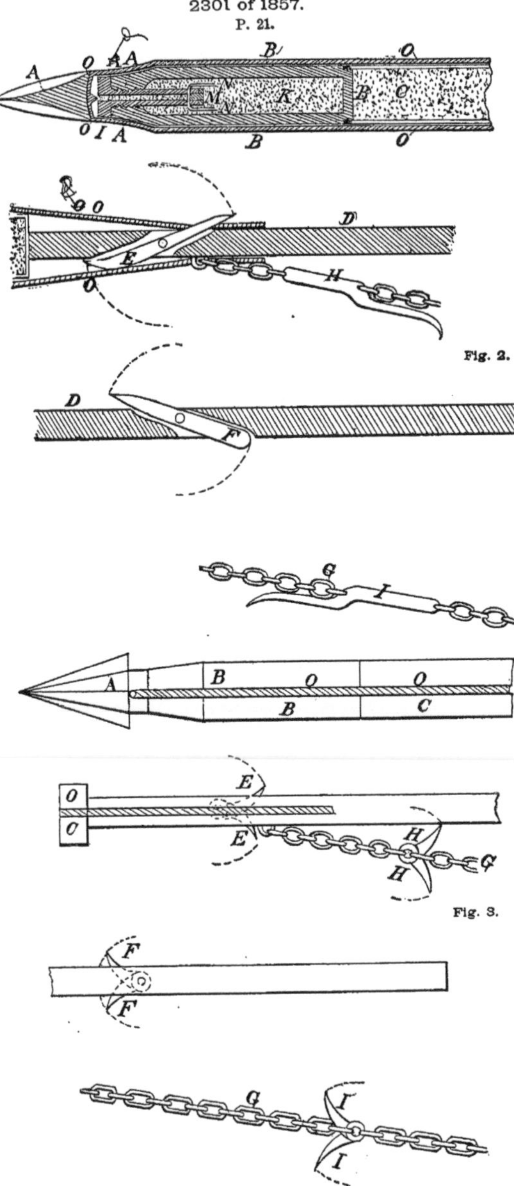

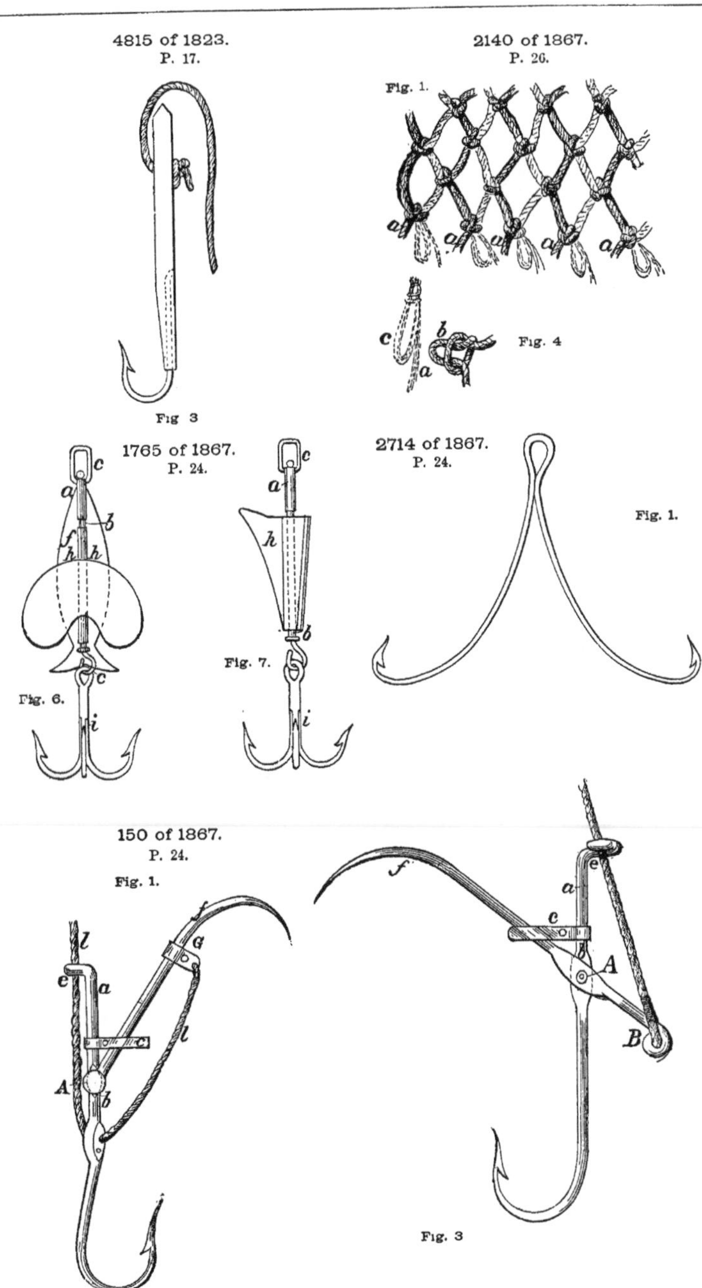
4815 of 1823.
P. 17.
Fig 3
2140 of 1867.
P. 26.
Fig. 1.
a
b
c
Fig. 4
1765 of 1867.
P. 24.
c
a
b
f
h
i
Fig. 6.
Fig. 7.
2714 of 1867.
P. 24.
Fig. 1.
150 of 1867.
P. 24.
Fig. 1.
l
e
a
G
f
c
A
b
B
Fig. 3

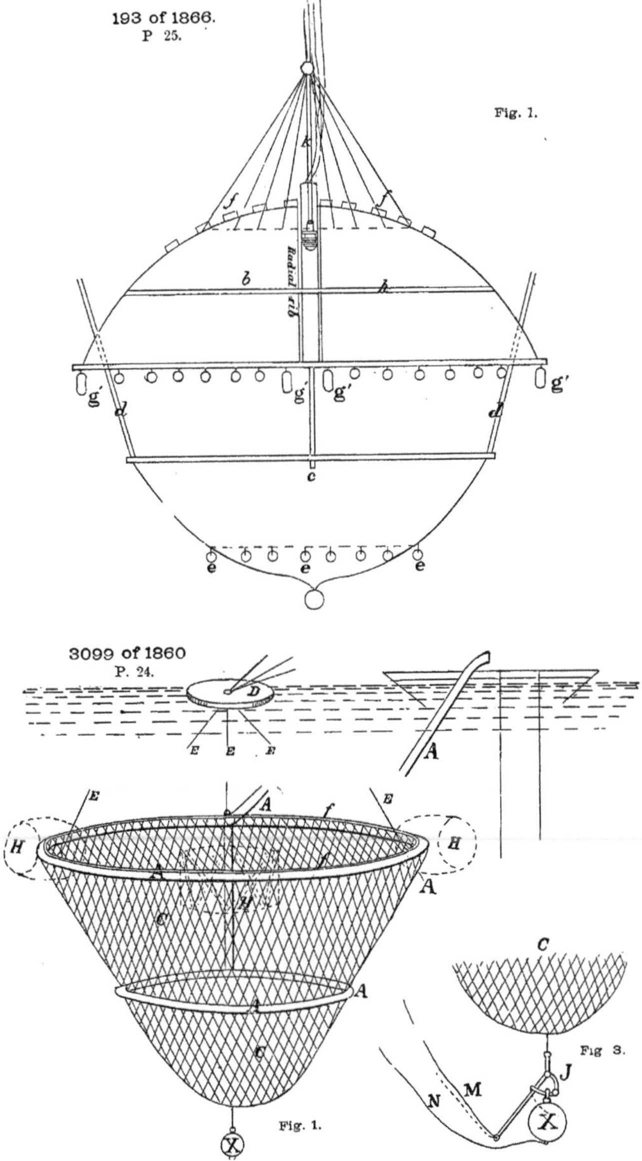
193 of 1866.
P 25.
Fig. 1.
Radial rib
3099 of 1860
P. 24.
Fig. 1.
Fig 3.

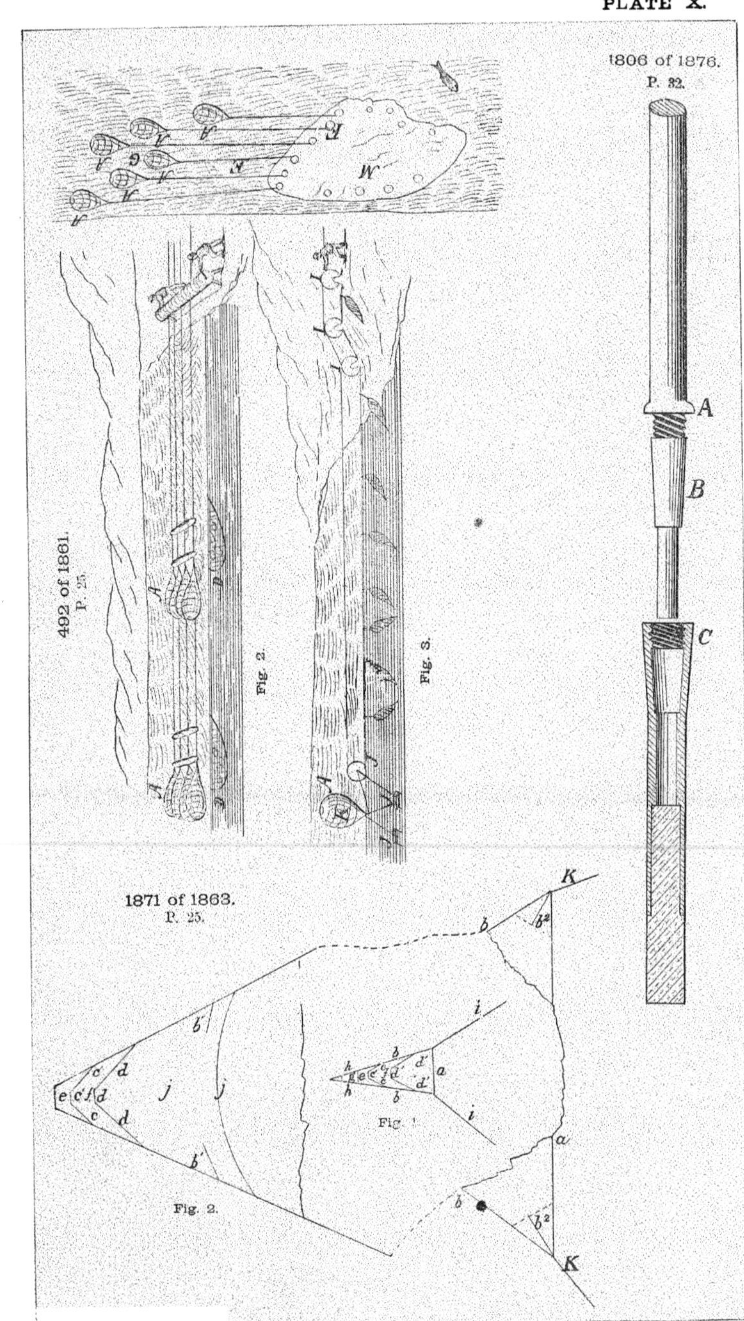
1806 of 1876.
P. 32.
A
B
C
492 of 1861.
P. 25.
Fig. 2.
Fig. 3.
1871 of 1863.
P. 25.
Fig. 2.
Fig. 1

2008 of 1866.
P. 25.

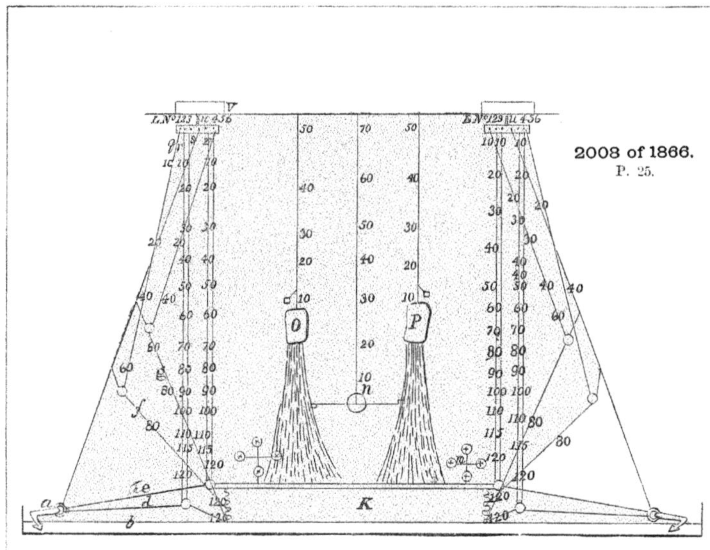

1146 of 1867.

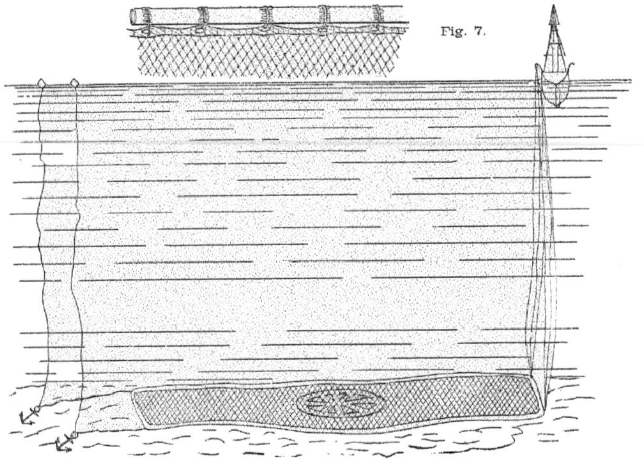

Fig. 8.

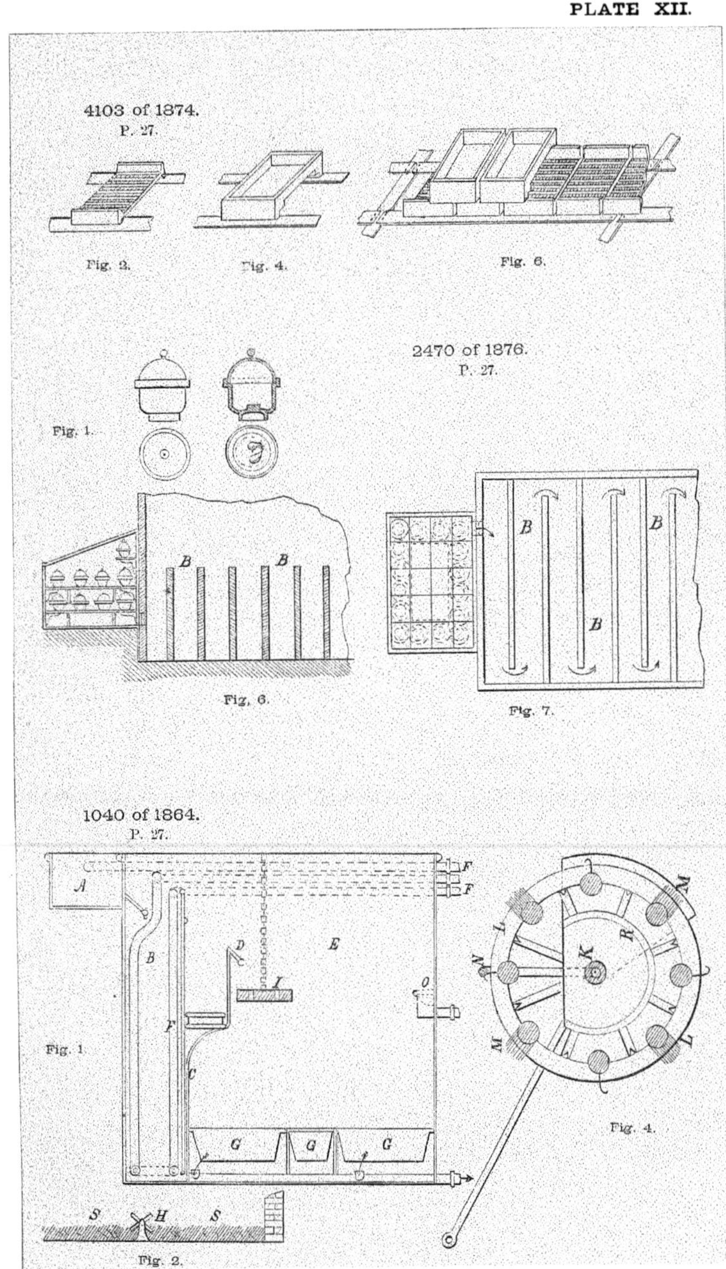
4103 of 1874.
P. 27.
Fig. 2.
Fig. 4.
Fig. 6.
2470 of 1876.
P. 27.
Fig. 1.
B
B
Fig. 6.
B
B
B
Fig. 7.
1040 of 1864.
P. 27.
A
F
F
B
D
E
I
O
F
C
Fig. 1.
G
G
G
S
H
S
Fig. 2.
M
L
N
K
R
M
L
Fig. 4.

2171 of 1866.

P. 29.

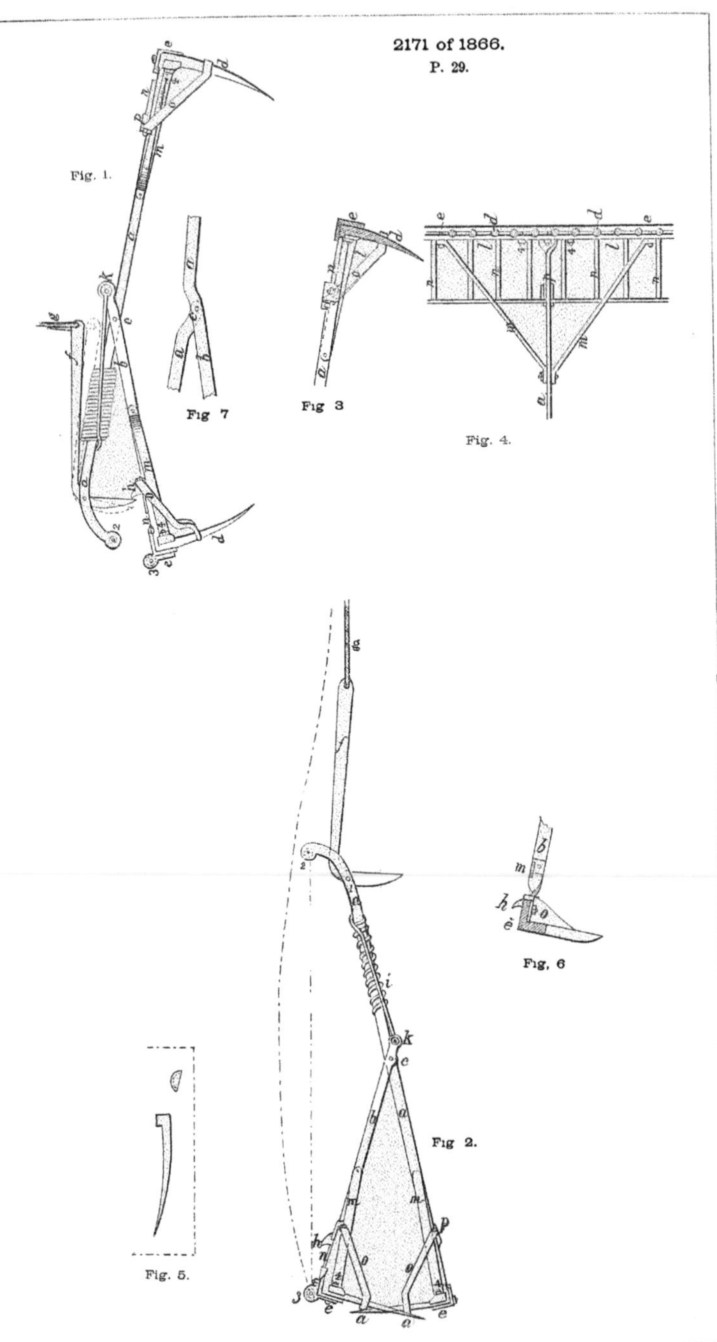

2906 of 1860
P. 29.

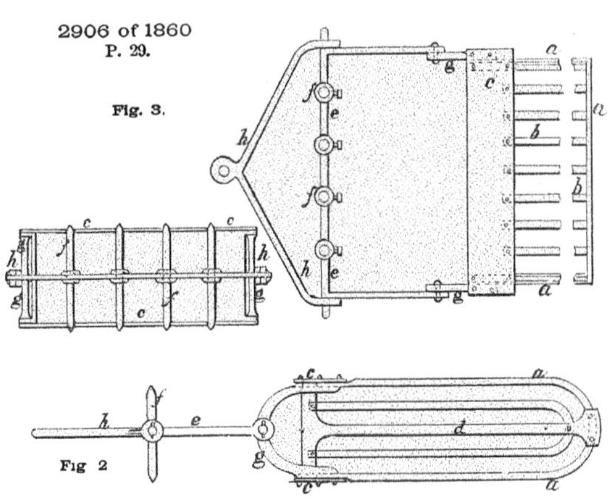

Fig. 3.

Fig 2

1553 of 1877.
P. 32.

1438 of 1867.
P. 29.

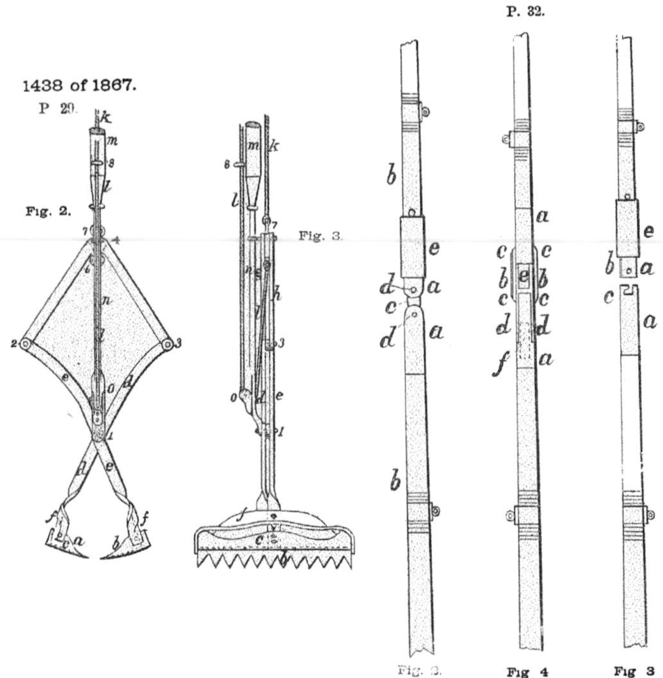

Fig. 2.

Fig. 3.

Fig. 2.

Fig 4

Fig 3

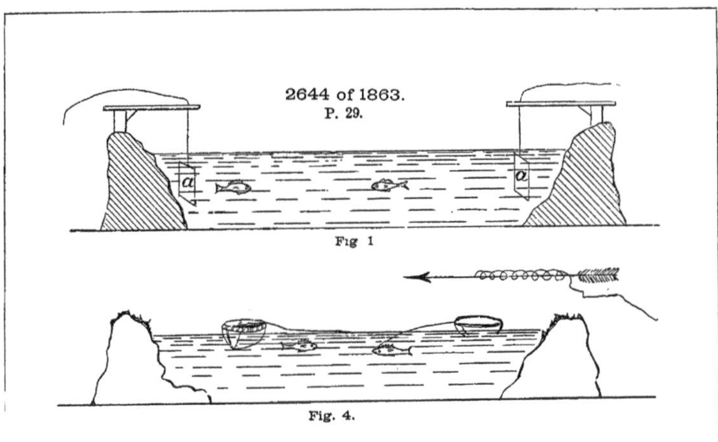
2644 of 1863.
P. 29.
a
a
Fig 1
Fig. 4.

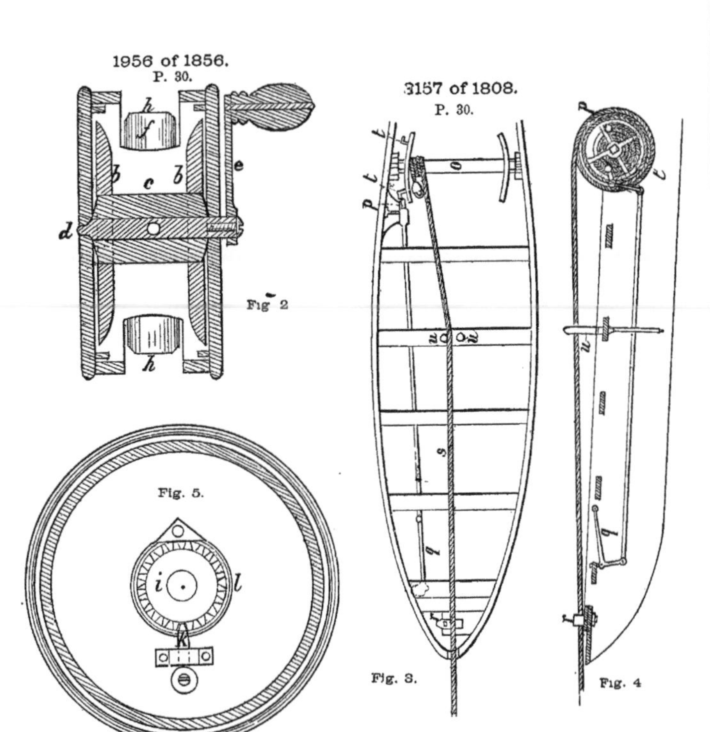
1956 of 1856.
P. 30.
h
f
b
c
b
e
d
h
Fig 2
3157 of 1808.
P. 30.
p
t
o
u
s
q
Fig. 3.
Fig. 4
Fig. 5.
i
l
k

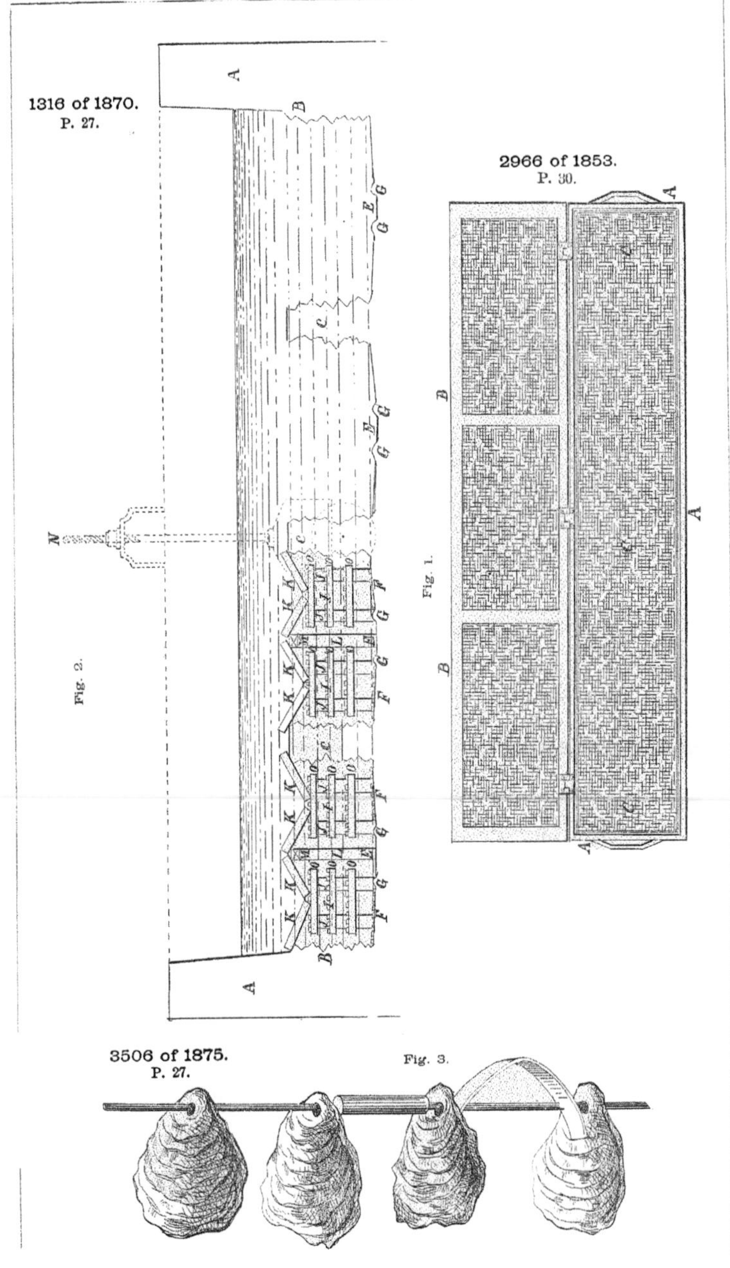
1316 of 1870.
P. 27.
Fig. 2.
2966 of 1853.
P. 30.
Fig. 1.
3506 of 1875.
P. 27.
Fig. 3.

and allowing the rod to be folded up for portability when not required for use; but when required, the rod is unfolded and a tapered metallic tube which is over each joint is moved downwards, making it permanent or secure. The plate of metal may be constructed with a slot at one end in an oblique direction; in this case it would be hooked on to the pin, so that in case of one section of the rod being broken, another can with facility be substituted. Drawing, Plate XIV.

TACKLE.

No. 881 of 1767.—COBB.—Long lines; small lines for snoods with hooks fixed thereon and small weights to sink the same. Small buoys prevent the hooks and bait touching the bottom of the sea, that the fish may readily discover the bait. No drawing.

No. 361 of 1858.—HECTOR.—Galvanized wire rope or lines and galvanized wire instead of the ordinary hemp ropes and materials: No drawing.

No. 1755 of 1866.—FRERE.—Barbed hooks connected to line attached to a spring whereby the line will be capable of yielding when pulled by the fish, and when his power is expended the fish will be drawn up. No drawing.

III.—A PETITION TO THE FISH COMMISSION OF THE UNITED STATES, SIGNED BY THE FISHERMEN OF BLOCK ISLAND, JUNE 12, 1877.

A PETITION.

STATE OF RHODE ISLAND, NEWPORT COUNTY,
New Shoreham, Block Island, June 12, 1877.

To the honorable Fish Commission of the United States:

We, your petitioners, citizens of the United States and of Block Island, natives and fishermen, believe that the catching of cod by means of trawl lines is diminishing the cod on our grounds, from which we believe that fish will eventually be driven if that mode of taking them is continued, because—

First. At any time the cod will bite the bait upon a small hook more readily than upon a large hook, because the small hook is hidden. The hand-fisherman uses a large hook, for he must, in pulling, keep his line taut, and the sharp struggles of the fish would break a smaller. Small hooks are used on trawl-lines, which are never very taut, and hence are quite elastic. During December, January, February, and in early March, when these fish deposit their spawn, they are very shy, and only a few will bite the large hook on the hand-line. In the past they have been secure in depositing their spawn, thus securing their propagation. Now that the trawl-lines have come into use, the fish, biting at the baited hooks, are taken in large quantities at a time when their destruction involves the destruction at the same time of myriads of eggs, thus directly tending to prevent the perpetuation of their species.

Second. It very frequently happens that a trawl-line with from 600 to 1,000 hooks attached becomes so chafed by the rocks on the bottom that, when loaded with fish, it breaks, and only a portion is secured. The remainder, with hundreds of struggling fish, is carried hither and thither until the fish die. Their struggles frighten the other fish very much more than when caught on a hand-line, for then the captured fish is immediately taken from the water, and but little of his distress is seen by other fish. Moreover, when the fish die, their bodies, becoming putrid, effectually rid the ground of cod until the bodies rot from their fastening and drift away, leaving the water pure again, or soon to become so.

Third. Oftentimes, the fish caught on the trawl-hooks are left to struggle for hours before the trawlers get time to take them on board; and, indeed, this is usually the case, for they use so many lines that they can haul them only a few times in a day. Meanwhile, many of the fish get clear. Suffering from the lacerations of the hook, and fearful of becoming again entrapped, they communicate their fears to their sympathizing companions. It is very noticeable that since the trawls have been used on our grounds, the fish have been much more shy than formerly, and that they struggle much harder when caught. Frequently they are caught with the small hooks of trawl-lines still in their mouths; or, without hooks, but with mouths still bleeding where torn in escaping. *The foregoing sentence is true of fully one-sixth of the fish we have caught since trawls were used on our grounds.*

Fourth. The hand-fisherman returns home each night and dresses his fish on the land, the offal being used for manure. The trawler remains at sea for days, and sometimes for weeks, even—remains until his vessel is loaded. In cleaning the fish he throws overboard the offal, which sinks to the bottom and there decays. This putrid matter drives the fish away so long as it remains. The trawler works frequently on the tideward side of the grounds, so as to catch the fish as they come in. Thus the offal is thrown over at places whence, as it is moved at all by the tide, which is usually not very strong, it is swept slowly over the whole fishing grounds, poisoning them for the time throughout their entire extent.

We further believe that trawl-fishing tends directly to injure the hand-fisherman; to injure the markets for fish; to injure the standard of fish as an article of food; and, eventually, to the great injury of all concerned in the business of catching cod, trawlers included, and for the following reasons:

First. These trawl-lines, stretched for miles on the fishing grounds, and running in parallel lines as close together as is profitable, make a barrier which, while in theory it only takes equal chances with hand-fishermen, in fact monopolizes the ground to the irreparable injury of the man with a single hook; for the trawl-hooks are placed so close together that comparatively few fish cross the line, being either caught or deterred by the struggles of others captured; and the trawlers can so place their lines in succession or in parallels as to occupy all the most desirable parts of the ground, one trawler requiring the space of one hundred men with hand-lines.

Second. Trawling, to be successful, requires larger boats and more expensive gear than are within the means of hand-fishermen. The boats must be larger in order to carry the fish caught, and also in order to better withstand the effects of storms; for the trawler generally remains all night at sea, while hand-fishermen return home every night.

Third. Trawlers sometimes catch 8,000 or 10,000 pounds in one or two days. They are unable to dress so many fish, as they sometimes catch, which, not seldom, are greatly damaged before they can be taken to market. So large a quantity will glut any available market, and the fish often spoil before they can be sold. Cod become very cheap, and the dealer is led to hold the fish on ice or in pickle in the hope of better prices; for prices of course fall with the greatly increased abundance of the fish, and the chances are great that he who buys must lose if he sells at once. Meanwhile the fish are injured more or less, even on ice or in brine.

Fourth. As a few days of good trawling may give those fishermen an enormous quantity of cod, especially where from thirty to sixty vessels are trawling; and as no one can tell when such good fishing may come, the dealer buys with great caution, lest he shall have on his hands a large amount while the market is becoming more and more crowded every day. Serious losses of this kind have made dealers very cautious in buying.

But the trawling is good and the trawler is impatient to go to the ground again, and anxious to get rid of fish which he fears will spoil on his hands; so he sells at any price. Thus the difficulty is aggravated, and the amount of damaged fish thrown on the market is greatly increased. When cod were caught only by hand-lines the supply was much more uniform and more susceptible of correct anticipation, while at the same time the fish were sold to the consumer at prices very reasonable, compared with other food staples. At the same time each man could care for what he caught, and thus a much better article of food was produced.

Fifth. When, a few years since, trawling was begun at Gloucester, Mass., cod were caught in large quantities near the port. So close were they taken that they were sent ashore as soon as a large dory was loaded, and in such quantities that the dories were generally loaded and sent ashore three times in a day. Soon the fish could not be caught there, having been either taken or scared into deeper water. The trawlers kept on until now the fish are not caught in quantities, on some of the grounds east of Massachusetts, nor except in one hundred and twenty fathoms of water. As the grounds failed, the trawlers extended their operations along the coast, spoiling the grounds wherever they went. Where they have fished long the hand-fisherman can no longer get his living by catching cod. Now they have come to our grounds. They are nearly all strangers, men of Portuguese birth, and we can already see the baneful influence they are exerting. Nor are the causes of the direct injury to us difficult to find nor hard to understand. The fish are driven into deeper water and so far from shore that we, returning home every night, can compete only at great disadvantage, for we can be on the ground only a few hours at best; and the farther we must go, the fewer hours we can remain. It is remembered well here that four of our best fishermen went seventeen times in one season to Coggeshall's Ledge, nearly twenty miles from Block Island, and caught an amount of cod reckoned at four quintals per trip to each man. Since trawlers have been there we cannot catch enough to pay for the trip.

We would further respectfully call your attention to the following facts:

First. In former years our government, in the fostering care which it exercised for the fishing interests of the country, offered bounties to encourage men to engage in fishing for cod. A certain sum was allowed for each bushel of foreign salt used by vessels of more than fifty-five tons registered tonnage. Block Island was at that time entirely destitute of a harbor in which vessels could ride in a storm. We were compelled in stormy weather to haul our boats above high-water mark to save them from the fury of the waves. From this cause we could not use vessels large enough to claim the bounty offered; hence we were compelled to compete at great disadvantage with the fishermen of more favored localities.

Second. When, in 1870, the breakwater at Block Island was begun, one of the principal reasons given by the government for so large an expenditure of money was that the work would greatly foster the fishing interests of the island. If our fishing grounds are permitted to be spoiled by the use of trawl-lines, then has the breakwater become of no avail to benefit the fisheries of Block Island.

We, your petitioners, almost without exception, are owners of small tracts of land, from which we derive a small income, insufficient, it is true, for our support, but which, when added to the amount realized from cod-fishing, has hitherto given us a comfortable living. On these tracts of land we have built us houses. Here are our homes. Now that our fishing threatens to fail us, we are very apprehensive that we may be compelled to leave our homes; for if this business shall be ruined, our island cannot support more than one-half of its present population. We are, almost to a man, too poor to own boats large enough to trawl successfully; but in our own boats we have been successful. We would point with pride to the fact that while from most of the ports of the coast many vessels and men are lost every year in the fishing business, not one Block Island boat has been lost within the memory of man.

Therefore we do beseech your honorable commission to urge the passage of a law prohibiting the catching of cod by trawl-lines in the waters of the Atlantic Ocean between Montauk Point on the west and No Man's Land on the east, and thus insure the return and continuance of the good fishing we formerly enjoyed, lest by the ruin of this fishery we shall be compelled to seek our subsistence in other pursuits and in other localities, away from the island endeared to us as the place of our birth and the home of our childhood.

That you will thus secure to us, on our native island, the opportunity to obtain our subsistence by that honorable toil to which we have been accustomed from our youth, we will ever pray.

William Dodge.	Darius B. Dodge.	Seabury A. Mitchell.
Welcome Dodge.	Uriah B. Dodge.	C. C. Holmes.
Aaron W. Dodge.	John Thomas.	R. W. Thomas.
Edward P. Littlefield.	James E. Rose.	Joshua T. Dodge.
Charles A. Paine.	Herman A. Mitchell.	William J. Steadman.
George C. Sprague.	Lorenzo B. Mott.	Joseph H. Willis.

Caleb Wescott.
Enoch Steadman.
William T. Dodge.
Benjamin Rose.
L. Steadman.
Whitman W. Littlefield.
Rufus A. Willis.
Giles P. Dunn.
Samuel A. Dunn.
Joshua D. Dunn.
John Ray Littlefield.
Daniel Mott.
John E. Dunn.
Samuel R. Littlefield.
E. C. Smith.
Barzillia B. Dunn.
Amos Mitchell.
Samuel G. Mitchell.
Charles Sprague.
Leonard Mitchell.
Charles Littlefield.
Oliver D. Sprague.
Willard Sprague.
J. R. Sprague.
Stanton S. Allen.
Lemuel B. Rose.
James H. Mitchell.
Asa R. Ball.
Henry Ball.
John Rose.
John E. Ball.
George W. Braymon.
Elisha Dickens.
Noyse Ball.
Charles Ball.
Edward C. Allen.
John A. Mitchell.
H. B. Steadman.
N. L. Willis.
Edgar C. Sprague.
Nathaniel Sprague.

Silas N. Littlefield.
John E. Willis.
Charles Hall.
William P. Dodge.
Halsey Littlefield.
Lewis N. Hall.
James M. Dodge.
Charles W. Willis.
Ransford A. Dodge.
William M. Rose.
Walter R. Littlefield.
Lemuel A. Dodge.
William R. Mitchell.
Edwin A. Dodge.
William Card.
Lorenzo Dodge.
Nathaniel Lathan.
James N. Latham.
George W. Conley.
Ezekiel Mitchell.
Benjamin T. Coe.
Lloyd E. Ball.
O. F. Willis.
Andrew J. Dodge.
Leander A. Ball.
Aaron W. Mitchell.
George E. Thomas.
Elihu W. Rose.
Seneca Sprague.
William Sprague.
Lewis E. Thomas.
Joshua Dodge.
Horace J. Negus.
C. Negus.
M. Negus.
George W. Willis.
George A. Hull.
William J. Greene.
Samuel P. Dodge.
Horatio N. Milikin.
Elias Littlefield.

Nathan C. Dodge.
John P. Steadman.
Ray W. Dodge.
Marcus M. Day.
Erastus Rose.
Willial Rose.
Alfonso Perry.
Solomon Dodge.
Joshua Rose.
Clarence Rose.
Halsy C. Littlefield.
Gideon P. Rose.
William C. Littlefield.
Richard A. Dodge.
John C. Dodge.
Samuel Ball.
A. N. Sprague.
N. B. Wescott.
Edwin Dodge.
Thomas H. Mott.
Hermanza Rose.
Martin V. Ball.
James A. Dodge.
Freeman Mott.
Howard C. Mott.
Lartis Steadman.
William L. Milikin.
Welcome Dodge, 2d.
Simon Dodge.
John W. Milikin.
S. D. Willis,
Andrew V. Willis.
Samuel L. Hayes.
C. W. Dodge.
Robert Rose.
Hezekiah Mitchell.
Charles F. Sprague.
Edgar Dickens.
Eleander Dodge.
Benjamin Sprague.

APPENDIX B.

FISHERY EXPOSITIONS.

IV.—ABSTRACT OF AN ARTICLE FROM THE "NORDISK TIDSSKRIFT FOR FISKERI," 1878, ENTITLED "OBSERVATIONS ON FISHERY EXPOSITIONS," &C.*

By A. Feddersen.

In a former article, Mr. Feddersen had declared himself in favor of fishery expositions, independent of agricultural or other expositions; not including, however, great international exhibitions, where every branch of human activity is represented. Wherever the fishery exhibit is only an appendix to some other exhibition, it will be neglected and not awaken sufficient interest. In spite of this, there was a good fishery exhibit at the agricultural and industrial exposition in Vibarg (Denmark) in 1875, and at Norrköping (Sweden) in 1876. This result was owing entirely to the arduous labors of a few zealous and enthusiastic men. These successes were, unfortunately, not followed up—at least as far as Denmark was concerned—at the Svendborg (Denmark) exhibition, and at the great international exhibition held at Paris in 1878.

There are many who doubt the usefulness of such exhibitions, saying that the practical result is by no means commensurate to the efforts made. It cannot be denied that there is a field for "humbug" in all exhibitions, and that there is a tendency to make a special show for the occasion. But as exhibitions are the order of the day, and probably will be for some time to come, people should not stay away because they are not in every respect as perfect as they might be. The object should be to get the greatest possible good of them and endeavor to keep away from them everything which savors of "humbug," and surely some good will come of it; new inventions will become wider known and goods of sterling quality will find a larger market. Exhibitions, if properly managed, can certainly be highly instructive. As an instance of this, the Bergen exhibition of 1865 may be mentioned, the beneficial effect of which is felt to this day.

It is probable that the direct benefit derived from fishery exhibitions will be greater than that of other exhibitions, as there will be less chance for mere show. Although great progress has been made with regard to fishing apparatus, there will always be a chance to make them still more effective, so that they can be handled with greater ease, or be manufactured of better and more durable material. In this connection

* Nogle Bemærkninger om Fiskeriudstillinger og andre Fiskeritorhold. Abstract made by Herman Jacobson.

we will only point to the increased use of cotton for fishing apparatus since 1865. And as regards the proper treatment, preservation, utilization, and shipping of the products of the sea, there is still a vast field for improvement. Fishery exhibitions will finally prove very beneficial to fishing legislation, as they will afford an opportunity to competent and practical men to meet and fully discuss this important subject. There is still so much ignorance with regard to this last-mentioned matter, and there are so many different and widely divergent views, that it is extremely desirable, both for the government and those specially interested in the fisheries, to arrive at some well-established principles which may form the basis of suitable legislation. There is, therefore, ample reason to encourage and uphold fishery exhibitions.

As far as Denmark is concerned (and perhaps other countries), it must be said that fishery exhibitions will prove failures unless a strong and direct appeal is made to fishermen and to fish-dealers, as well as to manufacturers of fishing apparatus, to take an active part in such exhibitions. Astonishment has sometimes been expressed that our fishermen do not take a greater interest in these exhibitions; but it should be borne in mind that, with few exceptions, fishermen are very conservative, both as regards apparatus and methods, and that it requires a special effort to stir them up. If this matter were seriously taken in hand by some zealous and energetic men, we would see a great improvement in our fisheries brought about by exhibitions. The better preservation of fish, and the increased knowledge of the best markets for their goods, would certainly be a practical benefit to the fishermen; gradually improved apparatus, boats, &c., will be introduced, and make the fishermen more and more independent. Until this result is brought about, the fishermen should be assisted by the government, which unfortunately has hitherto done too little in this respect. A knowledge of improved methods and apparatus should be spread, practical and theoretical instruction should be given, and money should be liberally appropriated for furthering the fishing interests. Since the fishery exhibition at Svendborg in 1877, and our fishery exhibit at Paris in 1878, were failures, owing chiefly to lack of interest and the entirely inadequate sums appropriated for the purpose by our government, a strong effort should be made to be well represented at the international fishery exhibition to be held at Berlin in May, 1880. As Germany will always be the chief market for our fishing products, this should certainly not be neglected.

In conclusion, the following suggestions are submitted, in the hope that at some fishery congress they may assume the form of definite resolutions:

1. The fishery exhibit at agricultural or industrial exhibitions is abolished.

2. In its stead meetings of fishermen and all persons interested in the fisheries are held annually in different parts of the country, where fishing products, apparatus, &c., are exhibited.

3. From time to time—say every three or five years—larger fishery expositions are held, at which all those nations are represented which fish in the same waters as we.

The government should of course give liberal aid to the fishing interests, and a society should be formed for promoting the fisheries in all their different branches.

V.—REPORT ON THE DEPARTMENT OF FISHERIES IN THE WORLD'S EXPOSITION IN PHILADELPHIA, 1876.

BY JOAKIM ANDERSSEN, *Juryman.**

I.

THE FISHERY EXHIBITION IN PHILADELPHIA IN 1876.

COMPRISED ACCORDING TO THE "GROUPING FOR THE JUDGES' WORK."

(GROUP V.)

Fish and productions of fisheries, fishing implements, &c.:

1. Aquatic mammals: seals, whales, &c.; living specimens in aquaria; specimens stuffed, salted, or preserved in other ways.
2. Fish, alive and preserved.
3. Fish in brine, and parts of fish used for food.
4. Crustacea, echinoderms, bêche de mer.
5. Mollusks, oysters, clams, &c., used for food.
6. Shells, corals, and pearls.
7. Whalebone, shagreen, fish-glue, isinglass.
8. Fishing implements: nets, baskets, hooks, and other apparatus used for fishing.
9. Pisciculture: aquaria, hatching-pools, vessels for transporting roe and young fish, and other apparatus used for hatching, raising, or preserving fish.

The judges, whose number for the whole exposition was 250 (half of that number being Americans) in the department of fisheries, were, in the beginning—

1. Professor *Spencer F. Baird*, director of one department of the government exhibit and Assistant Secretary of the Smithsonian Institution in Washington, chairman.

2. Mr. *Seth Green*, Superintendent of Fisheries of the State of New York, from Rochester, N. Y.; and

3. The editor of this report.

Later, when Mr. Seth Green, partly on account of his health and partly on account of his many other engagements, could not attend, and was therefore discharged at his own request, his place was filled by a man of similar experience, Mr. *T. B. Ferguson*, Commissioner of Fisheries for the State of Maryland. Both Professor Baird and Mr. Ferguson were so overwhelmed with work and were so much occupied in various ways that they had scarcely any time to attend to their jury work. In consequence, Mr. *B. Phillips*, secretary of the government exhibit, a

*Beretning om Fiskeriafdelingen ved Verdensudstillingen i Philadelphia i 1876, af Joakim Anderssen, Jurymand. Aalesund, 1877. Translated by H. Jacobson.

highly intelligent gentleman and well-known correspondent of one of the great New York papers, was chosen to act for Professor Baird and Mr. Ferguson, in conjunction with the editor, as permanent member of the committee. It ought to be mentioned that, as in the beginning so few objects were exhibited in the department of fisheries, three judges were deemed sufficient for this group; but as objects continued to come in all the time, these judges had enough to do, so that this group could not finish its work until the very last moment, July 30. It must likewise be mentioned that at the Philadelphia Exposition the verdict of the judges was not final, but that it had to be submitted to the criticism of the Centennial Commission and had to be approved by them; the verdict thereby became more impartial and dignified.

EXHIBITORS AND OBJECTS EXHIBITED.

A.—NORWAY.

Norway, which, both with regard to its products and implements, was best represented (excepting of course the United States, whose exhibit was very full), numbered thirty-two exhibitors, who are mentioned below, together with the objects exhibited by them.

1. *Commercial Union of Aalesund.*—An almost complete collection of all the fishing implements used in Norway, as well as models of fishing-boats.

2. *Carl E. Rönneberg & Sons, of Aalesund.*—Dried codfish, dried and pressed codfish in tin cans.

3. *Laurids Madson, of Aalesund.*—Sides of codfish, prepared in the Scotch manner.

4. *H. A. Helgesen, of Aalesund.*—Preserved fish-cakes, fresh salmon in tin cans (lobsters were promised but were not sent).

5. *Patent Twine Manufactory (Kraasbye Brothers, of Aalesund).*—Fishing-lines of silk, flax, and hemp.

6. *The Bergen Museum.*—Large and instructive collection of fish in spirits of wine, stuffed fish, models of vessels and boats, lodging and ice houses.

7. *The Bergen Commercial Union.*—A complete collection of fish dried and salted, salt herring and roe, and several kinds of fish-oil.

8. *The Bergen Smoking Establishment.*—Smoked herrings, four different sizes (great herrings, merchants' herrings, middle herrings, Christiania herrings).

9. *Fagerheim's net factory (Gottlieb Thomson, Bergen).*—Salmon and herring nets, mackerel and codfish nets, made of hemp and cotton.

10. *Peter Mohn, Bergen.*—Salt fat herrings in tin cans.

11. *Peter Egidens, Bergen.*—Fine herrings and anchovies.

12. *Thomas Erichsen, Bergen.*—Different kinds of fishing-hooks.

13. *H. Dons, Christiania (Christiania Preserving Company.)*—Small smoked herrings in oil, and fish-cakes in hermetically sealed tin cans.

14. *Christiania sail factory.*—Nets and lines made of hemp.

15. *Hadeland glass manufactory.*—Glass floats and glass balls for nets and lines.

16. *Falck-Ytter, Christiania.*—Fishing-sleighs.

17. *W. Nordrock, Christiania.*—Anchovies.

18. *Mrs. Gina Smith, Christiania.*—Anchovies.

19. *Mrs. Rina Tellefsen, Christiania.*—Anchovies.

20. *Zeorg Lund, Christiania.*—Anchovies.

21. *C. C. Just, Christiania.*—Anchovies.

22. *Christian Johnsen, Christianssund.*—Dried codfish.

23. *Jens O. Dahl, Havó.*—Codfish and herring nets, codfish-lines.

24. *Borderich & Co., Lyngvœr.*—Fish-flour (codfish chopped fine), white caviar (made of codfish roe), isinglass.

25. *Norwegian Preserving Company, Mandal.*—Fresh preserved flounders, mackerel, and anchovies.

26. *C. A. Thorne, Moss.*—Anchovies in oil, fresh lobster and salmon in hermetically sealed cans.

27. *Stavanger Preserving Company.*—Preserved fish-cakes in wine and other sauces, fresh lobster in hermetically sealed cans.

28. *C. Stórmer, Svolvœr.*—White caviar.

29. *Svend Fohn, Tónsberg.*—Four different kinds of spermaceti.

30. *Anton Rosing's Widow.*—Cakes of fish-flour.

31. *L. B. Soyland, Flekkefiord.*—Preserved skate.

32. *F. Hjorth, Frederikstad.*—Anchovies.

(The articles which were to be sent from the *Bergen Glass Manufactory* did not arrive.)

Of these exhibitors the first thirty, whose articles were found to be in excellent condition, were recommended for awards, and the Centennial Commission also adopted this recommendation. The two last mentioned exhibitors, however, received no award, because the articles exhibited by them were spoiled, probably because the cans had not been properly sealed.

B.—Sweden

had the following exhibitors:

1. *The Royal Swedish Commission, H. Widegren, superintendent of fisheries, Stockholm.*—A complete collection of fresh-water fishing implements, also nets, lines, &c., used by Swedish fishermen on the banks in the North Sea and the Kattegat, models of bank-fishing vessels and boats, especially of a boat for transporting live fish, salt Gottland and Blekinge herrings, eel, and codfish, dried codfish, and anchovies (pickled spratt), and finally a collection of fish from the Swedish lakes and coasts in glass jars filled with spirits of wine.

2. *Gustav Andersson, Tjellbacka.*—Skinned and boned herrings, anchovies and sardines in hermetically sealed tin cans.

3. *H. C. Bergstróm, Lysekil.*—Anchovies and herrings in hermetically sealed tin cans.

4. *N. O. Erickson, Lysekil.*—Anchovies and herrings in hermetically sealed tin cans.

5. *J. J. Hallgren, Gullholmen.*—Anchovies and fine skinned and boned herrings.

6. *August Lysell, Lysekil.*—Anchovies and fine skinned and boned herrings.

7. *Edward Nilson, Grebbestad.*—Smoked mackerel in oil in hermetically sealed tin cans.

8. *N. Wikström, Stockholm.*—Preserved salmon.

All the first mentioned seven exhibitors were recommended for awards; the salmon exhibited by N. Wikström, although good, was found wanting in freshness of flavor.

(C. M. Amundson, Udevalla, and Leidesdorf's manufactory of fishing implements, Stockholm, had announced several objects for exhibition, which, however, were not sent.)

C.—France

had of course the largest, finest, and best arranged exhibit of all kinds of sardines, sardels, and anchovies hermetically put up in oil. The exhibitors were principally from Nantes, L'Orient, Paris, Bordeaux, and Belle Isle, and most of them were recommended for awards.

D and E.—Spain and Portugal

exhibited common sardels and other food-fish boiled in oil, and besides a large number of preserved cuttlefish and mollusks boiled in oil with tomatoes, truffles, &c., and finally dry-salted and hard-pressed sardines in small wooden boxes, for poor people and therefore comparatively cheap. The Spanish exhibits came from Corunna, Bermeo, Pontovidero (Galicia), Loredo, Velva-Grove, Nevera, and Barcelona; and the Portuguese exhibits from Lisbon, St. Ybes, Oporto, Faro Alga, and Cambra Balbura.

F.—Italy

had likewise exhibited several kinds of sardines in oil, anchovies, and pickled eel from Leghorn, Genoa, and Bologna.

G.—Russia.

Preserved whitefish and carp from St. Petersburg and Astrachan sturgeon-caviar, isinglass, and seal-oil.

H.—Turkey's

only exhibit consisted of a kind of caviar of whole fish-roe, in the form of sausages, called "botargo," having a most delicious flavor.

I.—Austria.

Preserved skinned, strongly salted, and pressed small herrings (a kind of sardels) in oil.

J.—THE NETHERLANDS

had a fine exhibit of excellent herring-nets, floating nets, codfish-lines, trawl-nets, a model of a large herring-seine, models of fishing-vessels, and boats, both ancient and modern, from A. G. Maas in Scheveningen; pickled herring from M. I. Suries & Co. in Rotterdam; smoked salmon from T. E. Novenhays in Amsterdam, and models of cotton nets from Arntzenius, Jamnek & Co. in Goor.

K.—GREAT BRITAIN.

Pilchards in oil, fresh preserved salmon, from C. Freyer and Crosse & Blackwell, London; fresh-water fishing implements from W. & J. Ryder, Birmingham; and fish-hooks from Joseph Buchanan, Glasgow.

L.—GERMANY.

Samples of nets made of cotton, flax, and hemp from "Mechanische Netzfabrik und Weberei-Actien Gesellschaft," Itzehoe, Holstein, and Igler, Bohemia.

M.—CHILI

had a comparatively large exhibit of different kinds of preserved fish and mollusks (12 kinds), prepared algæ and sea-grass, in oil, in the Spanish-Portuguese way.

N.—JAPAN.

Models of fishing implements, fine nets of silk and flax, fish-poles with the line running inside the pole, lobster and eel traps, different kinds of fish baskets and boxes for keeping live fish, strongly salted and smoked salmon with head and tail, sewed in coarse cloth, dried sea-weed (sea-weed-tengusa), of which an insipid but cooling dish is prepared by mixing a large quantity of sugar with it, isinglass, and prepared fish-skin (shagreen).

O.—CHINA.

Models of fishing implements, isinglass, dried crustaceans and fins.

P.—BRAZIL.

Isinglass, preserved fish, turtle-oil, and butter.

Q.—ARGENTINE REPUBLIC.

Fish-baskets of straw and fishing-lines.

R.—LIBERIA.

Nets made of the fiber of trees (ramee).

S.—CAPE OF GOOD HOPE.

Fresh crayfish preserved in tin cans.

T.—VICTORIA.

Shell-fish, crabs, and stuffed fish.

U.—TASMANIA.

Stuffed fish (salmon and trout), and different kinds of shell-fish.

V.—BERMUDA.

Corals, shell-fish, beautiful live salt-water fish in aquaria.

X.—CANADA

had exhibited a large quantity of preserved fish and other marine animals, chiefly lobster, mackerel, and salmon (Canadian Meat and Produce Company, Richibucto), besides salt salmon, trout, and herring (also smoked), dried codfish, pollock (Halifax, Nova Scotia), fishing implements, especially for fresh-water fishing, and winches for hauling in long lines.

Y.—THE UNITED STATES

had, as was to be expected, the largest and most complete collection of all kinds of fish and fishing implements, boats, models of boats and vessels, &c., and everything pertaining to the fisheries and fish trade, all under one and the same roof, in the United States Government's building, in which there were catalogued specimens of all the products of the United States—grain, fruit, minerals, coal, land and marine animals, industrial products, both ancient and modern, war materials, &c. It would lead us too far to enumerate everything pertaining to fish and fisheries, and in fact does not come within the scope of this pamphlet, while we shall make brief mention of those exhibits which the individual States had made in their respective "State buildings" or in the main buildings.

1.—*Massachusetts.*

The Commission of Fisheries of this State had exhibited a complete and beautiful collection of boats and vessels, especially mackerel-schooners, fishing-boats, &c. The city of Gloucester, at present the largest and most important fishing station in America, had exhibited a complete and very instructive collection of everything belonging to bank-fishing, as well as models of ancient and modern boats and vessels floating on a pond specially arranged for this purpose, on which seine-fishing was illustrated by models, and on whose banks there were models of different establishments for receiving and preparing fish. Among the apparatus peculiar to America we must mention the so-called "bait-mills," by which suitable pieces of bait are cut very rapidly from the raw material, salt herring, swordfish, cuttlefish, &c., and likewise the "ice-crushers" and other implements for breaking the ice, used for keeping fresh fish, in small pieces. W. K. Lewes & Brothers and W. Underwood & Co., Boston, had a separate exhibit of salmon, lobster, codfish, mackerel, and shell-fish, preserved fresh; and the Gloucester Isinglass Company and Norwood & Son, Ipswich, had an exhibit of isinglass.

2.—*Maine.*

Extract of fish (juices) resembling Liebig's extract of meat, suitable for nourishing soups, put up hermetically in tin cans, a new invention by J. G. Goodale, Saco; mackerel and lobster put up fresh by J. Winslow Jones, Portland; oysters put up fresh by the Annapolis Packing Company; preserved lobsters and large mackerel by the Portland Packing Company; preserved oysters and clams by Bunham & Morrell, Portland; fishing-poles by Charles E. Wheeler, Farmington.

3.—*Pennsylvania.*

A large selection of fishing-poles of split bamboo, and other fresh-water fishing implements, from Fox, Shipley, and John Krider, Philadelphia.

4.—*New York.*

Preserved salmon, mackerel, oysters, clams, and isinglass, from Kemp, Day & Co., New York; whalebone, salt eel and salmon, anchovies, sardines, and caviar, from Max Ams, New York; live salt-water and fresh-water fish and green turtles, in aquaria, from Eugene Blackford, New York.

5.—*Maryland.*

Models of oyster vessels and boats; oyster scrapers and tongs, with the winches belonging to them, used in the Chesapeake Bay; samples of oysters from different depths; model of the hatching-house and apparatus in Druid Hill Park, Baltimore; model of a floating fishing-battery near Baltimore; fish in spirits of wine; large quantities of preserved oysters (Murray & Co., Baltimore.)

6.—*Ohio.*

Stuffed fresh-water fish (Cuvier Club, Lake Erie).

7.—*Chicago.*

Isinglass made of sturgeon.

8.—*California.*

Fresh salmon, put up in cans, from Columbia River.

9.—*Oregon.*

Fresh salmon put up in tin cans, and salt salmon in one-half and one-quarter barrels, from the Oregon Packing Company.

Being on the jury, I had, of course, an opportunity to see and examine all the above-mentioned exhibits, and all my observations led me to the decided conclusion that America stands very high as regards its salt-water and fresh-water fisheries; as high as in many other branches of industry. This will become clearer from the following description of—

II.

THE FISHERIES OF NORTH AMERICA.

Besides the observations made by me at the exhibition, I had another opportunity of gaining further knowledge of this subject by a journey to New York, Long Island, Boston, and Gloucester, which I made in company with Mr. F. Wallem.

I shall first speak of the *fish markets*. Of these, the Fulton Market, in New York, is the largest and best arranged. It consists of a series of large connected buildings, situated partly along the East River and partly along some of the streets of New York, and contains convenient places for wholesale and retail fish-dealers, for offices, and packing; also a library and reading-room, as well as kitchen and restaurants. At the exhibition, and later in his establishment in Fulton Market, I made the acquaintance of one of the great fish-princes, Mr. Eugene Blackford, whose magnificent and well-arranged establishment contains numberless live fish, and fresh fish on ice of every kind, and, as a specialty, soft crabs, which in New York are considered a delicacy during the period when they change their shell, and are therefore eaten in enormous quantities, shell and all, both boiled and broiled, lobsters and green turtles, which are brought weekly from the West Indies, and are from New York sent to other cities, and frogs. I suppose that Mr. Blackford is the only one of these fish-dealers who himself supplies his market with live and fresh fish. For this purpose he keeps a little steamer, furnished with a purse-seine, which twice a week makes trips between the mainland and Long Island, and generally returns with a considerable quantity of fish. Nearly all the fish-dealers have their own fishing-schooners, or have at least an interest in one. Whenever there is no sale for fresh fish, or the prices are very low, the fish are placed in large and well-arranged ice-cellars, where they freeze, and are kept till they can be sold to greater advantage. Although the sales were quite good at the time I visited New York, I nevertheless found in Mr. Blackford's ice-cellar a large quantity of frozen fish, especially large salmon, which had fallen a few cents in price. Whilst in the street the temperature was 108° Fahrenheit, it was 40° in the cellar, which made it necessary to put on warm woolen clothing before descending into it.

This large fish-market supplies Philadelphia and many other cities with fresh fish or fish on ice nearly all the year round. The kinds of fish which are most common are: Codfish, flounders, mackerel, salmon, brook-trout (*Salmo fontinalis*), bass (*Perca atraria, Labrax lineatus*), blue-fish (*Scomber saltator*), shad (*Alosa* or *Clupea sapidissima*), turbot, pompano

(*Trachynotus*). The prices of fresh fish in a market as large as this one, of course, vary considerably. The following are the average prices: Codfish, 5 to 6 cents; flounders, 8 cents; common mackerel (*Scomber scombrus*), 8 to 10 cents; Spanish mackerel (*Cybium maculatum* and *regale*, *Scomber maculatus* and *regalis*), which is esteemed as highly as salmon, 25 to 30 cents per pound. The value of all the fish annually sold in Fulton Market is about $2,000,000. Of these, Mr. Blackford sells about one-tenth, or $200,000 worth.

In Boston the fish-trade is carried on in a little different manner from New York. Besides a large common market, where all kinds of fresh, salt, and smoked fish are sold, there are special markets for fresh fish, where the fish are received from the fishing-schooners and placed on ice, and where the vessels that are going out are furnished with bait and ice. It must be mentioned here that all the fishing-schooners which are sent out from Boston and supply the Boston markets with fish are always furnished with a considerable quantity of ice for keeping the fish fresh. These vessels usually make one trip every two weeks, and generally return to Boston with a full cargo of fish. Their fishing implements consist of lines, with which they chiefly catch codfish and flounders, but also small swordfish, which are quite common on the coast between Cape Cod and Cape Ann, where the Boston fishermen have their station. The prices of fish in Boston are generally a little lower than in New York.

In Gloucester, the most important fishing-station in America, which possesses more than 500 well-furnished and beautiful fishing-schooners, there is no fish market like in New York and Boston, but a large number of very considerable fish-establishments, which supply a great portion of the United States and Canada with fresh fish, and more especially with salt herring, mackerel, dried codfish, and smoked halibut. As the fish-trade of Gloucester is very extensive, a more detailed description will not be out of place.

Gloucester, situated on Cape Ann, in the State of Massachusetts, a few miles from Boston, was founded by Englishmen about two hundred years ago, and, after many ups and downs, it has, from an inconsiderable fishing-village, by energy and pluck, risen to be an important city, with a population of about 17,000, who live almost exclusively by the fisheries, which, as I mentioned before, employ about 500 schooners with an aggregate tonnage of 30,000, besides a large number of boats engaged in coast-fishing. The result of these fisheries for a single year (1875) are, according to a volume published in 1876 entitled "The Fisheries of Gloucester," as follows:

		Value.
Bank (Newfoundland) cod....	177,473 quintals.........	$998,628
George's cod.................	185,758 quintals.........	1,021,669
George's halibut	2,462,364 pounds	172,365
Bank halibut	7,248,423 pounds	507,389

"Hake" (lyr)*	4,257 quintals	12,774
Cusk (*Brosmius vulgaris*)	2,349 quintals	7,047
Pollock	9,417 quintals	32,964
Herrings	38,292 barrels	153,168
Coast fisheries by "the dory fishermen":		
Fresh fish		89,738
Prepared fish		185,697
Fish-oil		8,945
Mackerel	18,172 barrels No. 1	327,112
Mackerel	7,065 barrels No. 2	184,780
Mackerel	21,763 barrels No. 3	174,104
Mackerel	4,039 barrels No. 4	24,205
Total		$3,900,586

Furthermore:

Pickled fish:		
31,750 herrings, valued at	$13,494	
163 barrels cod, 40¼ barrels swordfish	1,097	
410⅝ barrels trout, and 76¾ barrels fins	4,042	
21⅞ barrels salmon, 250 barrels tongues, &c.	2,282	
Clams, &c	10,000	
All other fish	8,000	
Oil, not mentioned above	100,000	
		138,915
Grand total		$4,039,500

These figures will show the prices of the commonest American fishing products. It will be noticed that roe is not mentioned, as but few roe-fish are caught, and as nearly all the roe is used for bait.

Next in importance to the Gloucester fisheries are those of New York and Boston, then those of Portland, Me., Baltimore, California, and Oregon. The coast fisheries carried on along the whole coast of the United States, especially of menhaden, shad, porgy, bluefish, lobsters, oysters, and clams, yield considerable sums of money every year. The value of the salt-water fisheries of the United States has never been accurately calculated, but I think that it amounts to about $20,000,000. With regard to the value of the fresh-water fisheries of the United States, the uncertainty is still greater; but it is certain that even now it is very considerable, and hopes are entertained that in future it will be still greater, owing to the energetic measures which are being taken to further and to encourage these fisheries. These hopes are well founded, to judge at least from the happy results which so far have been obtained by pisciculture. Whilst lobsters of unusual size are caught in great numbers on the coasts of Maine, Massachusetts, and New York, the largest

* "Lyr" is the Danish for pollack (*Gadus pollachius*); but the "hake" of our East coast are species of *Phycis*.—T. H. B.

number of oysters is caught in Maryland (Chesapeake Bay) and Delaware (Cape May and the Delaware Bay). The salt-water fisheries are most highly developed in Massachusetts, whilst the most important fresh-water fisheries are in New York and Maryland, where the most excellent measures have been adopted for increasing and developing the fisheries. In the following I shall give a brief description of these fisheries, chiefly from the official reports of the State superintendent of fisheries for the State of New York, Mr. Seth Green, and the commissioner of fisheries for the State of Maryland, Mr. Ferguson, but partly also from my own personal observations.

Every State which takes an interest in the fresh-water fisheries has its own fish-commission and a superintendent of fisheries responsible to this commission. The State makes an appropriation so as to enable him to carry on his work, which consists in increasing the number of fish by every possible artificial and natural means, and in encouraging and furthering pisciculture. At the head of the fisheries stands a United States Commissioner of Fisheries, at present Prof. Spencer F. Baird, of the Smithsonian Institution at Washington, a naturalist of great fame.

The magnificent and costly hatching-houses and fish-ways which have been established in the States of Maryland and New York show the ingenuity, the practical manner, and the extent to which pisciculture is carried on in these States. The results obtained by these establishments are truly astonishing, as millions of fish of every kind are called into existence, filling the large rivers and lakes. It is especially those kinds of fish which are considered the best food-fish, *e. g.*, salmon, trout, and shad, that form the principal objects of pisciculture, but other fish, if they are of any value at all, are not neglected. The populous cities of America are therefore as a general rule well supplied with fresh fish, which form an important article of food. American pisciculture, which includes a system of protection with carefully framed regulations for protecting the young fish, has attracted great attention especially in Germany, where of late years hatching-houses on the American plan have been established under the superintendence of a gentleman from New York. The result of these experiments is not known to me. I cannot say with absolute certainty how many millions of fish are annually hatched and placed in the many lakes and rivers of America, but their number must be very considerable.

Of the hatching-houses those invented by Mr. Seth Green, of Rochester, N. Y., and by Mr. T. B. Ferguson, of Maryland, deserve the greatest attention, constituting the New York State hatching-house in Caledonia, N. Y., the Druid Hill hatching-house in Maryland, and the fish-ways near the Great Falls of the Potomac. Recently so-called "ponds" for keeping live fish and for protecting young fish have been established in the Detroit River. These "ponds" are sheets of water hedged in with poles joined by boards in such a manner that the water can circulate freely, and that the young fish can easily slip through the openings. At one end of these ponds there is a movable gate fastened at the bottom to a mud-sill and protruding obliquely about one foot over the surface of the water. Whenever a haul is made near the "pond," the net with the fish is pulled through the gate by pushing the gate down with a pole; and the fish are emptied into the pond without being touched by human hands, and without leaving the water, so that the fish with the young reach the pond in an entirely fresh condition. Another arrangement, partly for making net-fishing easier and partly for keeping the young fish, is found near Havre de Grace (not far from Baltimore), and consists of floats or "batteries" with movable aprons on three sides, on which the net is hauled in such a manner that the fish and their young go direct from the net into a fish-pond in the float, from which the young fish can through small openings pass easily into the open water. These "fish-batteries" are only used in shallow water and in places where net-fishing could not well be carried on without some similar arrangements, and are moved from one place to the other wherever it is thought that there will be good fishing. The idea of the three movable aprons, which touch the bottom when the float lies still and can be raised up by means of chains whenever the float is to be moved, is this, that the influence of the current may be avoided by hauling the net in on that side where the current will not interfere with it. On the float are long poles, which are stuck into the bottom when the float is to lie still, and are raised when it is to move again. The aprons are then raised so high above the water that their corners can rest on frames and thus be held up until the next anchoring place is reached. There are also on these floats winches worked by horses for hauling in the nets, and dwelling-houses and sheds for keeping and preparing the fish. These floats are generally manned by 60 to 70 fishermen. The fourth side, which has no apron, is used for taking the fish on shore. Such an arrangement modified according to local demands would prove very useful in our country wherever nets cannot be hauled on shore in the usual manner.

It is impossible for me to give in this place a detailed description of the hatching-houses and fish-ways; all the more so as it would require drawings to make it perfectly clear. I therefore refer the reader to the above-mentioned Reports for 1875 and 1876, published by Mr. Seth Green and Mr. Ferguson, which contain plates. These Reports may be obtained by addressing Prof. Spencer F. Baird, Smithsonian Institution, Washington.

Great exertions have been made of late years to stock the American rivers and lakes with foreign and domestic fish by transporting live fish and impregnated fish-eggs by railroad from California, and from Europe by steamers, in boxes specially constructed for the purpose. These endeavors have partly succeeded beyond all expectation, so that at present eastern waters contain not only the highly-prized California salmon

(*Salmo quinnat*), but also trout and carp from Germany, and these fish seem to be in as flourishing a condition as in their proper home. A number of practical laws, partly local and partly applying to the whole country, have been made for the better protection and encouragement of the fisheries.

No less than to the development of the river and other fresh-water fisheries, have the Americans given their attention to the improvement of vessels, boats, and implements used in coast and ocean fishing, and a closer examination of this subject shows the high rank to which the American salt-water fisheries have attained.

The well-known American fishing-schooners, especially the Gloucester mackerel-schooners, are as beautifully constructed and as comfortably arranged and fitted up as a pleasure-yacht, and cost from $6,000 to $10,000 and $12,000, fully equipped, including fishing implements and boats. These schooners, which sail very well, have a tonnage of 60 to 130, and a crew of 9 to 14, according to the size of the vessel and the character of the fishing in which it is to be used. The schooners used for bank-fishing either near Newfoundland (Grand Bank) or George's Bank, are only furnished with long lines like the Swedish and Norwegian bank-fishing vessels, but instead of the large and heavy boats used by the latter, they have small flat-bottomed boats, so-called "dories", which are considered unusually good and safe, and are handled a great deal easier. Every bank-schooner has about 6 to 8 of these, arranged according to their size, three to four on each side of the deck. Whilst fishing is going on there are generally only two men in every dory, and single lines with a few hundred hooks are used, not as in Sweden and Norway a long row of lines tied together, with as many as 2,000 hooks, which latter arrangement, of course, involves a much greater risk in stormy weather. Nor do the Americans use the glass floats so common with the Swedish and Norwegian fishermen, probably because the lines can be handled a great deal easier without such floats, and are also more independent of the various currents. The nature of the bottom near the American coast is probably also more favorable for keeping the bait in place. The Norwegian half-moon-shaped weight for sinking the lines is not known in America, where a weight shaped like a plummet or a cylinder with a thick brass wire stuck through it is used. The large hooks or prongs used in Norway, by which the fish is frequently torn to pieces without being caught, are known from olden times, but have long since been abolished as unpractical and barbarous. The mackerel and herring schooners which are engaged in fishing either along the coast or in the Bay of Fundy during summer, or on the coasts of Newfoundland and Labrador chiefly during autumn, and in the Gulf of St. Lawrence during winter, are now all furnished, not with lines, but with the so-called "purse-seine", a new invention which of late years has become the favorite American fishing implement in all waters except on the banks, where wind and current forbid its use.

The idea of catching fish with seines in the open sea has been entertained long since, but as far as I know it has only been carried out in America by the introduction of the purse-seine, with which large numbers of mackerel, herring, shad, menhaden, and other coast fish are caught. Whenever a school of fish makes its appearance it is quickly surrounded by the seine, by one boat rowing in a circle whilst a dory lies still with the one purse-string and the one pulling-string (?), until the whole net is out in the water, whereupon all four strings are brought together on the boat, on whose railing there is a stationary arrangement for drawing the net together, which, when closed, forms a complete sack or purse, from which it derives its name, and which holds the fish, often amounting to several hundred barrels for a single haul. By means of large hooks the fish are then hauled up into the vessel, which lies ready to receive them. The length of these seines is from 150 to 220 fathoms, their depth in the middle from 15 to 30 fathoms, but at the ends only from 6 to 10 fathoms. The size of the meshes varies. The central portion forming the purse contains the smallest meshes (about $2\frac{1}{8}$ inches between the knots), made of the strongest cord; then follows on each side of the centre a portion with larger meshes and thinner cord, and still larger meshes and thinner cord at both ends, all calculated to make the handling of the seine as easy as possible. Such a seine (about 200 fathoms long and 25 to 30 fathoms deep), fully equipped with good cork floats (about 700) and nut-shaped weights weighing about seven-eighths of a pound each fastened at short intervals along the bottom rope, with the exception of the centre piece, which generally is without weights, so it can quickly be pulled together, and with either galvanized lead or brass rings (weighing 2 to 3 pounds) through which the two pulling-ropes pass from end to end, costs from $800 to $900. Only the best cotton thread or the finest hemp cord is used for these seines. The boats, which are well built and constructed in the most practical manner, cost about $300 apiece. Instead of bark or catechu, tar is used for fixing up seines which have been in use for some time. After having been put in boiling tar, they pass between rollers to make them pliable and to squeeze out the superfluous tar.

Many difficulties had to be overcome before the idea of the purse-seine was carried out practically; but all these difficulties were conquered by American energy and perseverance. These difficulties consisted chiefly in making the seine as light as possible and having the purse of a suitable size and shape in proportion to the seine itself, and in the method of drawing the net together. At first the following method was followed: Before the seine was set a large and heavy leaden weight, with two blocks of iron, was, by means of ropes, lowered to the bottom in the place where the seine was to be set; then the seine was gradually rowed out into the water, pulled together at the bottom, and finally hauled up into the boat. All this process consumed considerable time and labor, and during the pulling together at the bottom many fish were lost, whilst now the process is quick and safe. The greatest drawback was this, that the seine must reach the bottom, and wherever the water was too deep for doing this it could not be used. The sack-nets, used with us for catching pollock, or the net invented by Mr. Kildal, in Nordland, for catching codfish, are only incomplete realizations of the above-mentioned idea, and cannot compete in practical usefulness with the purse-seine, whose use is not confined to certain localities, as the sack-net, nor to certain portions of the bottom, like Kildal's net, but which can be used everywhere unless hindered by strong waves or currents. Purse-seines would prove extremely useful in our country, where schools of herring often keep in the middle of the fiords, where they cannot be reached by common nets, or in any other way except occasionally by a floating net.

Convinced of the incalculable profit which would accrue to our country from a more general use of the purse-seine, I have everywhere recommended its introduction; and my efforts in this direction seem to be rewarded, as Fagerheim's mechanical net factory near Bergen has given serious attention to the matter, and has already received several orders for purse-seines. But besides carefully manufactured seines on the American plan, and light boats, some practice will be necessary before the introduction of the purse-seines will yield full results. After having in Boston procured models of purse-seines and boats, which are to serve as guides to our manufacturers, I still desire that some American purse-seine fishermen could be engaged to instruct our fishermen in the use of these seines, so that our experiments might not prove failures, but lead to a speedy adoption of these seines. I may say here that negotiations have been opened with an American fisherman, which, so far, however, have not led to any definite results.

I have been somewhat lengthy in my description of the purse-seine, but the great importance of having it introduced with us will serve as an excuse. It seems strange that this seine, which has been in use in America for almost twenty years, and which in fact has become the principal American fishing implement, has not yet been introduced in the Scandinavian countries, from which so many good sailors have emigrated to America, and have there become experienced fishermen; but, as far as I can ascertain, it is a fact that this seine has not been introduced in a single European country, and is only known by name.

In bays and along the coasts the Americans very frequently use another somewhat expensive fishing implement, which is unknown with us, the so-called "pound-net," a sort of self-acting trap, something like our self-acting salmon-traps, only considerably larger, with which all kinds of fish are caught. As this net seems peculiarly adapted to the American coasts, with their great wealth of fish, and on account of its high price (about $400) does not seem suited to our circumstances, I will not give any further description of this very ingenious contrivance.

The well-known trawl-net and floating net of the Dutch, French, and English is, as far as I could ascertain, not used in America, where the purse-seine fully supplies its place. For the smaller fisheries in bays and mouths of rivers our common nets are likewise used, only with this difference, that the floats are not fastened to the net itself, but swim on the surface of the water, fastened at short intervals to the strings connected with the net. Common casting-nets are also used, and purse-seines are sometimes used in this way simply by taking the pulling-ropes off.

An implement peculiar to the American bank-fisheries are the so-called "nippers," rings made of cotton yarn, used instead of gloves when handling the ropes. A furrow or groove runs all along the outside of these rings, and the ropes, whilst being hauled in, rest in this groove.

A bank-schooner generally makes three to four trips every summer, and, if the market is good, often realizes from $10,000 to $12,000 a season. The codfish and halibut, which are prepared and salted on board, are divided in about the same manner as in the Swedish and Norwegian bank-fisheries: the owner of the schooner, who furnishes the lines and other implements, receiving one-half and the crew the other half of the net yield. The result of the mackerel and herring fisheries varies more than that of the bank-fisheries. The average sum realized by mackerel-schooners is $8,000, and by herring-schooners $5,000 to $6,000, which is divided in the same manner as the result of the bank-fisheries. The small schooners which carry on line-fishing along the coast, and sell their fish fresh on ice, realize, on an average, $4,000 to $5,000 annually, which sum is distributed in different ways, but generally, as with our small cod-fisheries, in such a manner that the owner receives one-fourth and the crew the remaining three-fourths of the net income (the owner's risk being, of course, considerably smaller).

As the continent of North America, comprising the United States (now including California and Oregon), extending from the Atlantic to the Pacific, and the Dominion of Canada, consumes nearly all the fish which are caught by American fishermen, the fish are only prepared with a view to rapid consumption. They are therefore nearly all shipped fresh on ice, or sprinkled with salt and then dried or smoked a little. Codfish does not, therefore, undergo the long drying process as with us, and in Iceland, Nova Scotia, and Newfoundland, which supply distant markets, *e. g.*, Spain, Portugal, Italy, the West Indies, and Brazil, where only well-dried fish can be sent. When the fish have been taken out of the brine, either just as they are taken from the schooners or from large barrels where they have been kept in brine, they are dried on poles stretched a few feet from the ground, for three days, without being turned and pressed, are packed in large boxes and shipped inland by railroad, selling at from 5 to 6 cents per pound. Fish prepared in this manner will of course not shrink much, and weigh heavy. Some kinds of fish, as for instance the cod and pollock, after having been dried in the above-mentioned manner, are skinned, boned, and cut in narrow strips, put up in small boxes weighing from 35 to 50 pounds each, and sent farther inland by railroad, selling at from 6 to 8 cents per pound. Mackerel and also to some extent herrings are not treated in quite so summary a manner. They are prepared very much in the same way as with us, by being split and salted down in barrels which are made by machinery and do not look very solid. In order to keep better, the largest and fattest, mackerel and herring which from August to November are caught on the coast of Labrador are cut open and their entrails are taken out. The common herrings, resembling our spring herrings, which during the spring and summer months are caught on the coasts of America and Newfoundland, undergo a peculiar process by being salted in the holds of the schooners, from which at the end of the trip they are taken to the warehouses, where they are transferred to barrels and shipped inland at a price of $3 per barrel. Such herrings are of course of an inferior quality, and they cannot be used as with us, but must be soaked in fresh water (or milk) and then either boiled or smoked. During the year 1876 several cargoes of salt herring of different size and quality were shipped from America to Sweden. The herring-fisheries on the coasts of Labrador and Newfoundland might be developed much more than they are at present. But so far the herring has not been much esteemed in America, and the herring-fisheries have consequently been somewhat neglected.

That the shipping of herring from Norway to Montreal and Chicago has paid, must be in part ascribed to the desire of the Scandinavian emigrants to have this genuine Norwegian article of food, and in part to the difference of quality between the Norwegian and American herring, the former having a more delicate flavor than the larger Labrador herring.

The Americans also make use of their fisheries in many ways unknown to us. The finer portions of the halibut, of which large numbers are caught on the banks and near the coast of Greenland, are prepared and smoked like salmon and sold at a comparatively cheap price (8 to 12 cents per pound), whilst a number of other fish, *e. g.*, the menhaden (*Brevoortia*), and a sardel-like fish are used for making oil and guano. (The Pacific Guano Company uses enormous quantities of fish for these purposes.) Oil for medicinal purposes is, as far as I could ascertain, not made in America.

As I have mentioned before, great quantities of lobsters, oysters, and clams are caught on the eastern coast of North America, are sold at a cheap price, and therefore form a very common article of food, partly raw (oysters and clams) and partly cooked, oyster and clam soup being a very common, cheap, and delicious dish. Of late years many oysters are put up in hermetically-sealed cans, and find a ready market, partly for ships going out on long voyages and partly in Europe. The American oyster has a somewhat different shape from ours, as well as from the French and English oyster, being somewhat longer and more fleshy than ours. In America oysters are not raised artificially, as in France and England, as their natural rate of increase seems sufficient to supply the demand; but wherever the oyster trade forms a large source of income, as in Maryland, certain regulations are enforced for protecting the oysters. At Cape May and in the Delaware Bay oysters are caught and eaten all the year round. The same also applies to clams and lobsters. Fish-meal and fish-balls put up in hermetically-sealed cans are not known in America, and the well-flavored articles of this kind in the Norwegian exhibit, therefore, attracted much favorable attention.

The common mackerel are, in America, sorted according to four sizes and qualities—Nos. 1, 2, 3, 4—all differing in price. The so-called Spanish mackerel is a fish resembling the trout in its beautiful appearance, with red spots, and almost as large as a small salmon, and sometimes fetches as high a price as salmon. Whilst ling is very scarce near the American coasts, nearly all our common fish are found in large numbers; *e. g.*, codfish, hake, pollock, haddock, cusk, flounders, halibut. The haddock reaches a much larger size than with us. It is highly esteemed, and is sold at a tolerably high price, either fresh or slightly smoked. It is well suited for being shipped on ice on account of its thick skin and its firm flesh. A fish not known with us, but very common in America, is the so-called "sheepshead" (*Sparus* or *Sargus ovis*), which gets its name from the peculiar resemblance of its head to that of a sheep.

To enumerate the many different kinds of fish found near the coasts of North America would be of but little interest to the general public. To those who take a special interest in the matter I would recommend the following works: "Report of the Commissioners of Fisheries of Maryland, 1876," and "Classification of the Collection to illustrate the Animal Resources of the United States," also published in 1876 by Prof. G. Brown Goode, M. A., in which all the North American food-fishes are enumerated and classified.

The American method of freezing whole cargoes of herring and mackerel, so that they keep fresh in the hold for two or three months, has yet to be spoken of. Unfortunately, I had no opportunity during my stay in America to witness the working of this method, but I was told that it was very practical. An American gentleman had promised me some written information on the subject, but so far I have waited for it in vain.

III.

BRIEF ACCOUNT OF THE FISHERIES OF NOVA SCOTIA AND NEWFOUNDLAND.

In Nova Scotia, whose capital (Halifax) I visited on my return from America, and in Newfoundland, whose capital (St. John's) I likewise visited in order to make myself acquainted with the fisheries, I found that they resemble ours rather than the American fisheries. There is more of a regular custom trade, or, in other words, the fishermen are more dependent on the merchants than in the United States, where they are generally their own masters with regard to the way in which they wish to dispose of their fish. The fishermen of Nova Scotia are, to some extent, and those of Newfoundland nearly altogether, equipped by the merchants, to whom they thus become debtors, gradually paying off their debt by fish. It therefore often happens, as with us, that in poor fishing seasons the merchants run a risk of not having the money laid out by them refunded, or that they sometimes are cheated out of the whole or a portion of the fish on which they had calculated, by fishermen (in order to raise cash) selling their fish at a higher price than they would receive from their merchants to foreign merchants or to the French fishermen's colony at St. Pierre, southwest of Newfoundland.

The Nova Scotia cod-fisheries hold about the same relation to those of Newfoundland as the Söndmöre and Nordmöre cod-fisheries to those of Loffoden and Finmarken. These fisheries are carried on very much in the same way as with us. The coast-fishery is chiefly carried on by large, well-manned boats, with nets and lines, and the bank-fishery by swift schooners fitted up like the Gloucester ones and furnished with dories and boats, only not quite so handsome in their appearance. The French fishing-vessels belonging to the above-mentioned French colony of St. Pierre have a different shape and different rigging (generally cutters or large sloops like the English lobster-vessels); they are also manned and equipped differently from the Newfoundland bank-fishing vessels. The French fisheries near Newfoundland differ altogether from those of other nations, especially with regard to the preparing of the fish and roe, and the distribution of the fish. The state stands at the head of the fisheries, and has a governor at St. Pierre who superintends the whole. The present governor, M. Boubert, with whom I had the pleasure to travel from Halifax to Liverpool, told me that the French Government had sent him a number of Norwegian fishing implements, especially codfish nets and lines with floats; but as he had not yet had any opportunity to try them, he could not say anything regarding their practical use in these waters.

A Norwegian who has lived for a number of years in New York as partner in a large business establishment which failed during the civil war, and who now lives in Halifax, as Belgian consul, Mr. C. E. Rónne (son of the late Danish Doctor Rónne of Christianssand), who was educated at the Norwegian Naval Academy, but had to leave the naval service on account of his weak eyes, a man of position in Halifax, with an unusual knowledge of languages and great experience, was likewise the companion of my voyage from Halifax to Liverpool. From this interesting gentleman I obtained a great deal of information regarding New England and Newfoundland in general, and their fisheries in particular, for which I herewith express my heartfelt gratitude. I made copious notes regarding the leading features of the Nova Scotia and Newfoundland fisheries, but to complete these there are still wanting, the manner in which the fish are prepared, statistical data, &c., and I shall give these more from oral than from written accounts, or from my own personal observation, as my stay in both places was only very short.

In Nova Scotia as well as in Newfoundland the fish are prepared in a more conservative manner than in the United States, viz, with the view of obtaining an article that will keep well and may be sent to tropical climates, especially the West Indies, Brazil, Spain, Portugal, and Italy. The codfish, which forms almost exclusively an article of export to foreign ports, is first treated in the usual manner, viz, as soon as possible after having been caught it is opened and all the entrails are taken out; it is then salted in the hold of the vessel, generally with white Cadiz or Liverpool salt. When the vessel returns from her trip the fish are taken to the "drying-places", where they are washed and cleaned of all superfluous salt and all impurities. They are then laid in small heaps, and afterwards, when the weather is favorable, side by side on scaffoldings, which, in Newfoundland, are very high, so as to let the air pass through freely and let them dry thoroughly, in just the same manner as we dry our codfish. The washing process, however, is somewhat different, for the fish are not left so long in the water as to get soft and lose some of their flavor. The thin black skin is also left on the Newfoundland codfish, as people do not think it worth while to take it off. As the climate of Newfoundland often prevents the rapid drying of the fish, there are on every drying-place, close to the scaffolding, small huts where the fish are placed in rainy or damp weather. A number of fish are nevertheless damaged during the drying process, and turn sour and dark; such fish are then sprinkled with thin lime, which makes them look white enough. The fish which are bought fresh from the boats are of course prepared and salted in sheds, and then after some time treated in the above-mentioned manner. The fresh liver is melted by steam to oil for medicinal purposes exactly as we do it in Norway; and the old livers which cannot be used in this way are made into brown train-oil. The roe of the codfish is treated in the same manner as with us, and is either shipped to France or sold to the French colony at St. Pierre. But the Newfoundland cod-liver oil and roe are by no means esteemed as highly as our Norwegian oil and roe, which is probably caused by the different mode of preparing it, and by the different food on which the codfish live in these parts.

Besides cod-fisheries, which are the most important, Nova Scotia and Newfoundland carry on extensive herring and seal fisheries in the sea extending between Nova Scotia, Newfoundland, and Labrador. Whilst the herrings from Nova Scotia are chiefly shipped to Canada and the United States, the Newfoundland herrings nearly all go to England, chiefly to Liverpool. These herrings are prepared in the usual manner. The quality of the Newfoundland herrings, however, is by no means very good.

The average quantity of dried fish shipped from Nova Scotia is 36,000,000 pounds, and from Newfoundland 108,000,000 pounds. I am not able to state the quantity of oil, roe, herrings, and seal-skins which are shipped from these two countries.

The prices do not vary much from those of other countries, and are subject to the same fluctuations which are caused by the varying results of the fisheries and by the state of the markets. The dried codfish are shipped from Nova Scotia and Newfoundland to the West Indies and Brazil, partly in boxes and partly in baskets or a sort of tubs holding about 108 pounds each, and to Europe by merely placing them loose in the hold of the vessel. The shipping of new dried codfish generally commences towards the end of August or in September.

It may be known what an influence the cuttle-fish (calamare) has on the Newfoundland fisheries, as it is a most excellent bait, which can scarcely be replaced by any other. Just like the capelin in Finmarken, the cuttle-fish at certain times visits the coast of Newfoundland in large schools, and large numbers are then caught to be used as bait in the cod-fisheries, which commence in May and last till the end of September or October. The cuttle-fish are either kept fresh or salted, and their price varies considerably. With regard to the bait-herring, everything is exactly as with us.

IV.

POSTSCRIPT.

After having made myself acquainted with the character of the American fisheries, it was my object to apply to our fisheries all the practical American improvements as far as this might be possible. I therefore undertook my journey to New York, Long Island, Boston, and Gloucester, accompanied by Mr. F. M. Wallem, a newspaper correspondent of great and varied knowledge; the special object of this journey being to become still better acquainted, from personal observation, with the American fisheries, and particularly with the working of that most excellent fishing implement, the purse-seine, whose introduction into Norway we both considered of great importance, especially for our mackerel and herring fisheries. We were everywhere received with the greatest politeness and readiness to be shown all that was to be seen. I must here make special mention of Mr. A. A. French, in New York, head of a branch office of the American Net and Twine Company, of Boston, who showed us about on Long Island, where we saw the most important fishing implements and the way in which they are used, and of Mr. Eugene Blackford, of New York, who took us to Fulton Market and showed us everything of interest in that vast establishment. In Boston we were very kindly received by the representatives of the firm of A. A. French & Co. (the American Net and Twine Company), and in Gloucester we were fortunate enough to meet with the same kind reception from the head of the firm of Procter, Trask & Co., which last year shipped herrings from Gloucester to Sweden, and also from one of the partners of the firm of

Cunningham & Thompson, Mr. Thompson by birth a Swede, who for a number of years carried on bank-fishing and halibut-fishing with a vessel of his own, and is now solidly established in his present business; and finally from another Swede, Mr. Joseph Simpson, who, after having been at sea for many years as a bank and mackerel fisherman, has, with an American, Mr. Maker, established a repair shop at Gloucester under the firm of Maker & Simpson, employing 10 to 12 men exclusively in the repairing of purse-seines. This gentleman explained to us the arrangement and use of the purse-net, and undertook to get us models of the same. Other valuable and interesting information we received from Capt. H. Allan, of the fishing-schooner Bonanza, a new and elegant vessel of 137 tons, equipped for catching herring with a purse-seine near the coast of Labrador, which had just returned from there with a full cargo (about 1,800 barrels), and from his son, Mr. John Allan, who was commissioner of the Gloucester fishery exhibit in Philadelphia. To all these gentlemen we herewith express our gratitude.

We also visited the well-known boat-builders, Messrs. Higgins & Gifford, in Gloucester, whose workshops annually produce several hundred dories and a large number of purse-seine boats, and ordered a model of the last mentioned boat.

In making a few concluding observations in connection with the improvements in our fisheries, which in my opinion might advantageously be introduced in our country from America, I do this not only with the wish and hope of benefiting our fisheries, but also with a consciousness of the vast importance of the subject and of my inability to do full justice to it. However, I will boldly write the following, hoping that it will at any rate induce people to give the matter some thought and awaken some discussion which may further the interests of the great and important subject of the fisheries.

In the first place, especially with the view of developing our great herring-fisheries, the purse-seine should be introduced, and for our coast and bank fisheries light and quick-sailing fishing-vessels in connection with the above-mentioned dories. The usefulness of such vessels cannot be doubted, as a great deal in the fisheries depends on being swift in reaching the fishing-place, in catching the fish, and in bringing them home in a fresh condition.

Purse-seines, of different sizes and prices, are now made to order by Fagerheim's mechanical net factory near Bergen (A. G. Thomsen). On account of their being easily handled I would recommend purse-seines of first-class cotton thread, not longer than 150 fathoms, and 25 fathoms deep. The price of such a purse-seine will probably range from 2,400 to 2,800 crowns ($643 to $750).

I would not advise changing common herring-nets of thick hemp thread to purse-seines, as they would be too heavy and the change would involve a great deal of labor. Models of purse-seine boats, dories, and swift-sailing fishing-vessels may be obtained from Messrs. Higgins & Gifford, in Gloucester, Mass., U. S. A.

After what has been said above it might seem practical to use steamers instead of sailing-vessels for the fisheries, but in the present condition of the fisheries there would be much in the way of carrying out this idea. I have some experience in this matter, for a few years ago I, in company with Mr. Frederik Haussen and Mr. Jens Sahl, built and equipped a fishing-steamer of about 20 horse-power and a tonnage of about 500. This steamer Erknö, which otherwise was a perfect success in every respect, and which was intended for the spring-herring fisheries and the bank-fisheries, was, after a few failures, chiefly occasioned by the stoppage of the spring-herring fisheries, sold to the Söndmöre-Romsdal Steamship Company. The expenses of a tolerably strong and not too small steamer like the Erknö are too great in proportion to the advantage accruing from its use. If enterprises of this kind are to pay, they ought to be carried on on a large and well-devised plan, requiring considerable capital.

The attempts which have been made to use fishing-steamers both in the Loffoden and in Finmarken have proved the correctness of this view. Even in America the use of steamers in the fisheries (excepting of course the seal and whale fisheries) is very limited.

The use of floating nets deserves attention next to the use of the purse-seine, as has been fully proved by the experiences of the Stavanger fishermen during the last year. Practical and well-equipped fishing-boats, as well as good and strong fishing implements will always pay in the long run, although the first outlay may be considerable.

The necessity of having suitable laws for protecting the fish and their young, and of having systematic arrangements for restocking our rivers and lakes with fish, partly by artificial hatching and partly by transferring fish and their young from one water to the other, deserves our fullest attention, and we may learn a great deal from America in this respect.

The method of preparing fish for the trade is likewise a question of great importance to our country. Every one acquainted with these matters knows what a loss is involved by drying salt codfish too little, to soak it, as is the custom with us, and to dry it lying in an oblique position. By this wrong way of treating the fish, not to mention the great carelessness in killing and cleaning it, and in letting too much time elapse before salting it, it of course loses much of its nourishing quality and its juiciness, becomes softer, lighter in weight, and will not keep as well. The drying process with us is generally carried on too slowly and in too careless a manner, as the fish remain unnecessarily long piled up in heaps without being turned; they consequently begin to ferment and turn dark and sour. The consequence of drying codfish on rocks which are heated by the sun, is that the fish are often burned and get shrivelled. The drying of codfish on scaffoldings or on small stones has this advantage, that it does not burn so easily and that it dries better, the warmth and the air acting evenly on both sides of the fish. To salt the dry codfish in close boxes or barrels, lets the salt penetrate it much better and makes it heavier and juicier than when it is merely salted on the scaffolding. I think it would therefore be an advantage if those vessels which sail to the Loffoden and Finmarken to buy raw fish, had close boxes or bins in which the fish could be salted, with an arrangement to let the brine flow out from time to time. The so-called Scotch method which recently has been adopted by several large fishing-houses is therefore to be highly recommended. I convinced myself of this at the Philadelphia Exposition by examining the codfish exhibited by the firm Lauretz Madson, in Aalesund (prepared in the Scotch manner), which not only looked very fine, but had kept well and were very heavy.

The old law prohibiting the salting of codfish in boxes or barrels was doubtless well meant, but it has now become clear how senseless this law was.

I can also recommend the new American method of treating gently dried codfish, viz, by taking off the skin, cutting out the backbone and breastbones, then cutting it lengthwise in narrow strips and packing it hermetically in small wooden boxes; this method will prove of special advantage where the saving of freight is an object, when fish have to be sent to distant countries, and in keeping the fish in a good state of preservation. The advantage will be evident if a good price can be obtained, as the freight on that part of the fish which is valueless as food, is saved, as the fish itself will keep better even during the longest journey, and as the skin and bones may be used as fertilizers and will readily sell to guano manufacturers.

I think it would be worth while for our fish-merchants, besides shipping whole codfish in boxes to the West Indies and Brazil, to ship such skinned and boned codfish not only to these countries but also to Buenos Ayres, Montevideo, Peru, Australia, and possibly to Japan, China, and the interior of Germany, especially at times when whole codfish do not find a ready market.

The making of isinglass from fish-maws forms a considerable branch of American industry, as a good deal of isinglass is used for making beer and wine clear. I must strongly recommend the putting up in hermetically-sealed tin cans of stuffed crabs (*Cancer pagurus*), which has so successfully been begun by Mr. C. Wiese, of Osmundsvaag, as well as of fish-balls in brown sauce, convinced that these articles will always find a ready market, especially on board the large transatlantic steamers. I think that halibut smoked in the same manner as salmon would also sell very well with us, but would not be so well suited for the foreign trade, as fish which is not thoroughly smoked does not keep as well as strongly-smoked fish. Gently-smoked anchovies or small herrings, hermetically sealed and put up in oil, like those which Mr. Henry Dons, of the Christiania Preserving Company, had exhibited in Philadelphia, are sure to sell well.

If we consider what large quantities of shellfish, especially clams and muscles (*Mytilus edulis*), are eaten, both raw, with vinegar and pepper, and boiled in milk, in America, France, and also in Spain and Portugal, and what a cheap and healthy food they are, it seems that we, who have so many muscles, ought certainly to follow the example of these countries. I sincerely hope that the freezing of herrings and mackerel in the holds of vessels will also soon be introduced in our country as the most suitable way of preserving large quantities of fish for a long time, and, likewise, that the shipping of fresh fish on ice will become more common with us than it is now. I think that so far the lack of proper means of communication has prevented our adopting these improved methods of shipping fish. It must not be forgotten, however, that in order to ship fresh fish on ice with any reasonable hope of success, the fish must really be fresh when it is put on ice; as fish which is several days or even hours old does not answer the purpose. If, therefore, those fish which are caught in the open sea are to be shipped on ice, it is absolutely necessary that the fishing-vessels either have ice-boxes or regular fish-boxes, where the fish can be kept alive. It is very important that both live and killed fish should be brought to market as soon after they are caught as possible, as it is well known that a fish loses much of its wholesome, nourishing quality when it dies a natural death and the blood cannot flow off. Small fish-ponds for keeping those fish alive which have been caught in nets will, therefore, in connection with fish-boxes on board the vessels, prove extremely useful, and deserve to be introduced wherever it is possible.

I must, in conclusion, mention quite a new rowing-apparatus, invented by Mr. William Lyman, of Middlefield, Conn., which was patented during the Philadelphia Exhibition. This so-called "bow-facing rowing gear" consists in having the oar divided in two parts, which are connected by double galvanized-iron hinges, which move in the form of a parallelogram, and are fastened to the boat by small balls in bronze caps, which fit exactly in two pieces of board screwed firmly to the boat, from which, therefore, the motion proceeds towards both sides. The rower sits with his face towards the prow of the boat and uses the handle of the oar in the usual manner, whilst the oars themselves move in the same direction and drive the boat forward when the rower draws the oars towards himself, and backward when he pushes them away; therefore exactly the reverse of the usual mode of rowing. To a person unaccustomed to it, this way of rowing looks very strange, as the boat seems to move the wrong way. The advantage is this, that the rower can always sit and look in the direction in which he is going; but I think that this invention will never be of much practical use, except in harbors, or for hunting and fishing. As far as I know, two samples of this rowing gear were bought during the exhibition, besides the one which I bought, and brought to Norway by Mr. A. Brun and a ship-builder, Mr. Brönlund, who therefore are able to give further information regarding this curiosity.

APPENDIX C.

THE SEA FISHERIES.

VI.—REPORT ON THE AMERICAN FISHERIES*.

By Fredrik M. Wallem

Note by Translator.—*The prices of fish given in section VI are in large part incorrect.*

INTRODUCTORY REMARKS.

In the United States of North America the traffic in fresh fish is of the greatest importance, whereas in salted and dried fish it is of comparatively secondary importance; and the development in all fish-traffic in the Union augurs that this condition will strengthen and advance till fresh fish to a greater extent than now will become the chief product of the fisheries and will employ the greatest capital. The American fish-dealers enlarge their field of operations with remarkable energy and ability. They extend their fishing-grounds along the east coast both south and north, so that in a twinkling they pass from the Gulf of Mexico to a considerable distance up on the Greenland coast, traversing a coast-stretch of 600 to 700 geographical miles. At the same time they extend their field eastward by going farther to sea and employing steamers for fishing-vessels; while the catch inland in the great lakes and streams takes a decided advance, which result is based upon systematic protection and artificial propagation. Halibut, for instance, they fish for off the Greenland coast (since the year 1870), and sell in the fresh state some hundreds of miles inland after having been sent in ice by rail from the landing-place; nay more, the American fishermen have attempted to bring halibut from the Iceland coast.† It will soon be attempted to send salmon caught in California to markets in Europe in a fresh state, just as American oysters in the shell have been, for some time already, to a few in England. The Americans will soon catch fish a thousand miles from home, if they continue to be eaten fresh as latterly; they will buy fish on one side of the globe and sell them on the

*Om de Amerikanske Fiskerier. Indberetning til Departementet for det Indre fra *cand. jur.* Fredrik M. Wallem om en af ham med offentligt stipendium foretagen Reise til Philadelphia-udstillingen i 1876. Udgivet efter Foranstaltning af Departementet for det Indre. Christiania. 1878. Report on the American Fisheries, by Fredrik M. Wallem, of a journey undertaken by him at the public expense to the Philadelphia Exhibition in 1876. Published by direction of the Department of the Interior. Christiania. Bergh & Ellefsen's printing office. 1878. Translated from the Norwegian by Tarleton H. Bean.

†The first attempt at halibut-fishing off Iceland was made in 1873, but failed, it is said, on the ground of a bad choice of season; it was the experimenter's purpose to repeat the attempt.

other after skillfully cooling and preserving them with ice. The art of preserving fish fresh for a long time and transporting them over great distances is an object of much study, and has called forth many experiments, while salting is little heeded.* And so this national pursuit may advance to this result, that the traffic in fresh fish will become more and more important. The numerous fishery-inspectors in the United States have, besides other duties, to give the coast-extent, lakes and other waters, detailed and continuous advice about food-fishes, to protect them and promote their increase where they already are found, and at the same time to provide barren or depleted waters with a new stock of the kinds of fish which are best adapted to furnish suitable and healthy food for the people. To assist the fishermen and promote extension, the natural as well as the artificial, has become a business which employs many scientifically-cultivated men and many industrious and skillful public functionaries besides. And the fruit of this whole united effort is available with that practical and quick grasp which is peculiar to the people. With regard to good implements, boats and ships, the American fishermen appear to be equally apt to profit by what they have and to invent improvements and new things to the utmost limit. The fishermen consist as a rule of clever people, of whom not a few are from nations of Europe most actively engaged in fishing, so that it is probable that the most of the improvements from Europe are known to them.

The fisheries take a place nearest in the class with the Norwegian so-called "great fisheries"—cod and herring—which is an acknowledgment of the second rank for the United States. In the cod-fisheries on the banks—George's and Newfoundland—the Americans certainly participate with a great number of vessels, but other nations fish here perhaps with a greater number, and compete with them in the world's market. Herring-fishing about Labrador, New Brunswick, and Newfoundland and thereabouts, is not an important business, though partly a comparatively new industry; some of the American vessels fish for herring when they cannot on account of the season prosecute any other fishing, but some buy fresh or frozen herring from the shore fishermen, either to use them, or to sell them fresh in the large coast towns. Of the good (fat) herring the major part are consumed in the United States themselves; the inferior, thin ones they export, in the latter part of the year, to Europe. The oyster and lobster fisheries on the United States coast are, compared with the European, of great importance, and supply not only suitable and very agreeable food for all classes in the Union, but also a tolerably important article of export for the world's market, especially canned.

With this short survey finished, I shall now give an account of the results of my journey.

*In the years 1861-'73 were issued twenty-five patents for preparing and preserving fish and bait; only one for salted fish.

I.

THE FRESH-FISH TRADE.

The exhibits of the different countries represented had greater or smaller divisions for their fishery-industry. In the matter of implements there was something, but in the line of products there was much to see. The American division was richest in the first-named respect, while their products were not present in great variety in the exhibition itself, but in the fish-markets. As the products in many respects give the American fisheries a peculiar character and well merit the serious attention of other nations, I set myself as a special task to examine this matter. I adopt therefore as a suitable introduction for my report the fresh-fish traffic.

By the distinguished favor of the Norwegian juror, Consul Joakim Anderssen, I was introduced to some of the more prominent business men in this branch, and in company with him I went through Fulton Market, New York, where an important trade in fresh fish is carried on, and in whose market-building they have their local association, assembly-room, library, &c. Later we continued our examination in Gloucester and Boston until the consul's return home in August, after which I alone went to sea in a fishing-vessel and followed mackerel-catching for two weeks, to become acquainted with its practical working.

I shall not now undertake to mention the fish-merchants' association, its organization, with prices-current, &c., but immediately proceed to the business itself in Fulton Market, in New York City.

Fresh fish are sold here from stands, not boats or ships, and are said to be used at all seasons of the year with ice, to keep them fresh as long as possible. I visited the place the first time in August, in very warm weather; the last time I was there was in February. Then the streets were covered with some feet of snow and ice. Both times the fish were partly hard frozen, partly packed in ice. No fish were sold living; the only approach to living "fish" were the large turtles, which are brought in by steamers from the West Indian and Florida coasts. The retailers were not "fish-wives," but young men, merchants' clerks.

The number of kinds of fish offered for sale was great, and the prices at different times of the year varied greatly, without, however, fluctuating much from one day to another.

Here merchants of moderately large capital carry on the traffic, part of whom have their own fishing-vessels, giving the business a character something like the Norwegian trade in salted and dried fish and herring; I mean that it is free from the mean and dirty market traffic which one as a rule associates with the sale of fresh fish. The retailers in Fulton Market have, in part, marble counters, neatly-arranged stands, a private office where the owner of the stand may note his sales. The large merchants exercise supervision, and the young attendants, dressed in long aprons, handle the fish and deliver them at fixed prices. The fish-merchant to whom we were introduced, Mr. Eugene G. Blackford, was not only a capable business man but also a highly accomplished gentleman. He was able to give us not only information on all things concerning the fish-traffic, but also scientifically-founded communications on the natural history of fishes. As president of that great society, the American Fish Culturists' Association, he was identified with all the prominent scientific men in that branch, and with the large staff of fishery commissioners in all parts of the United States. We could not have been introduced to any one who was better fitted to be our cicerone and our living lexicon. A great portion of the information which I acquired on the fresh-fish trade I owe to his favor and intelligence.

II.

KINDS AND PRICES* OF FISH.

As I remarked before, the kinds of fishes in the market are numerous, and the prices, naturally, different in different seasons. I shall name such fishes as are generally used for domestic purposes, and the prices of some.

Haddock (similar to our hyse) is one of the commonest and best fish. It is taken on the coast as well as on the Great Banks. They are sold fresh in small quantities from 6 to 8 *öre** per pound; they are employed also as stock-fish.

Pogies or menhaden (*Alosa menhaden* of the herring family) are likewise a good small fish; they are taken in great masses with steamers and purse-seines along the coast and some distance out to sea. A great portion is made into oil and guano, the manufacturing of which is done in large factories on the coast. By the barrel, which average 300 to 500 each, they are sold from 44 cents to 88 cents. In the fish-market they are sold fresh, and retail for 6 to 8 and 10 *öre* per pound.

Thin herring are found in the fish-markets, especially in winter, and then frozen and fresh. These come from the Newfoundland and New Brunswick coasts, where they are either caught or purchased for $1.09 per barrel. They sell them fresh at retail for 6 to 8 and 10 *öre* per pound. They are also salted, and then principally whole. Another fat kind of herring, most like the Norwegian great herring, which is found in the fall on the coasts named, and off Labrador, they generally work up into a very choice salted article for the West; lastly they salt the thin herrings for export to Europe.

Some more esteemed and higher-priced kinds of fish are: butterfish (*Poronotus triacanthus*), catfish (species of *Amiurus*), flounders, sheepshead (*Archosargus probatocephalus*, a sea-carp), sturgeon, swordfish, rockfish (*Roccus lineatus*, a sort of sea-perch). Cod I did not see; they are

*FOOT-NOTE.—The *öre* equals $\frac{27}{100}$ or about $\frac{1}{4}$ of a cent.—TRANSLATOR.

found now and then at fair prices. The common mackerel are sold in season for 40 to 50 *öre* apiece. Here mackerel is the object of an important fishery which lasts from spring till fall on the coast and at sea, the prices varying greatly. A large portion are salted and sold in barrels at prices from $4.91 to $15.29, or even $20.47, for the largest and finest.

Halibut vary greatly in price according to the scarcity or abundance of fish; they send them also by rail many hundred miles inland, whole, partly filled and partly surrounded with ice, in boxes of from 350 to 400 pounds. The prices in the fish-markets fluctuate greatly according to the magnitude of the catch, and when they prosecute the fishery far out at sea on the banks and even along the coast of Greenland no one can constantly have a correct opinion as to where the fishing will be the August, 1876, in New York was 40 *öre* per pound wholesale, 60 to 75 *öre* per pound retail, and were noted later in a fishing-port near Boston 50 per cent. cheaper; moreover, the price may vary from 9 *öre* to 60 *öre* per pound in large lots.

Halibut are to some extent salted (especially the heads); some parts also are smoked (especially the backs and the bellies). The cheeks are considered a delicacy. From the heads, also, oil is expressed. Salted halibut heads are sold for $4.91 to $6.28 per barrel. With regard to the assorting of halibut I shall only remark, that white-naped halibut bring as high as 100 per cent. more than the black-naped.

Salmon fluctuate also greatly in the city markets—from 45 *öre* to 50 *öre*—but decline in the height of the season to 23 *öre* per pound—in July for instance; they may as early as August advance to 90 *öre*, and in November, in the hard-frozen state, they may bring 33 cents per pound. Speculation in frozen salmon is considerable, for the accumulation in the winter months is often great, and as a consequence of over-speculation the holders may be obliged in January and February to sell their stock at a rather low price, to prepare for the arrival of the fresh fish in the market. Smoked salmon is not uncommon; the price varies from 50 to 90 *öre* per pound.

The dearest and most esteemed fish are a fresh-water species and the pompano (*Trachynotus carolinus*, a member of the mackerel-family); this delicacy is taken in the South, the champion of the sub-tropical waters, and commands as high as $1 per pound. Another much hunted fish is the Spanish mackerel (*Cybium maculatum*); we tested it in a restaurant in Fulton Market and found it fat, delicate, and savory.

The commonest fish in general use is the shad (*Alosa sapidissima*, of the herring-family), which often tastes a little mawkish, but in other respects is a fine, though bony fish. Whether the Americans, among other things, have a decided taste for fish will be seen from what follows.

III.

A CULINARY FISH-DINNER WITH INTERNATIONAL DISHES.

Some time after the jury of the exhibition had given its award upon the fish-products from every quarter of the globe, the well-known society, the American Fish Culturists' Association, gave a fish-dinner, at which the choicest international delicacies and rarities of fish-preparations were served at a meeting of connoisseurs. This was in reality a higher jury, which was here to pronounce judgment upon the fish-food of all nations. The whole selection was made by the associations' most capable fish-experts, and as special caterer was engaged "that culinary artist," Mr. M. Sudreau, which was the highest official guaranteee upon the bill of fare.

It is naturally not my purpose to give anything in reference to the feast; I shall concern myself only with the official portion of the affair—the bill of fare. This gives through its contents a clear statement of what this "higher jury" considered specially worthy to be served. And in this statement lies an award which shows more clearly than the jury itself what belongs to the choicer fish-preparations, as the fish-products of every country securing premiums were brought into a single collection. To the whole was given a humorous coloring, as an example of which, to a portion of the current American fish-preparations was given a special name after this or that scientific man or matador or functionary in fishery branches; not, however, preventing the attentive specialist from studying the serious side of the affair. No Norwegian fish-dealer can read that bill of fare without observing what a part the Norwegian fish-products were assigned at this fish-dinner. And if he intend to speculate in the American market with fine products, the bill of fare will doubtless give him many useful hints. I shall therefore give an epitome of it.

The repast was begun with genuine turtle-soup or green turtle *à la* Blackford. This is not a costly article in America (a plate costs, as a rule, in the restaurants, 80 *öre;* while in England one must pay 82 cents to $1.09). After soup, was served lobster salad, "Seth Green's style." Among the extra selected warm entrées were crayfish salad, roast oysters, and roast crawfish (*Cambarus*). After these, in small part savory preliminaries, came the pith of the affair—the international dishes in selection.

Of American fish-products were served: Striped bass (*Roccus lineatus*, a perch); pompano (*Trachynotus carolinus*, belonging to the mackerel-family—a costly delicacy, which brings as high as a dollar a pound in the fish-market); bowls of terrapin (*Malacoclemmys palustris*); deviled crabs (whole small crabs, which are eaten shell and all; in shedding, the fact is that the shell is quite soft); turbot, filet of sole, and frog or toad salad.

Baked American fish-dishes were represented by sheepshead (*Archo-*

sargus probatocephalus, belonging to the sea-carps), and bluefish (*Pomatomus saltatrix*). And these were served cold: Eels in jelly, crayfish, salmon, lobster salad, caviar from California, and oolachans from Alaska.

Norway was represented by: Mackerel in oil, halibut, stewed fish, baked mackerel, and preserved mackerel, together with salmon.

Sweden's representation was: Anchovies and mackerel.

From Portugal were served: Sea-eel, sardines in oil, ling in oil, cuttlefish in oil, soles in oil, mackerel in oil, and swordfish.

From Spain: Sea-eel with tomato sauce, *mixillon*, sardines in oil, sardines in vinegar, and baked bass (a perch).

From Italy: Sardines.

From Holland: Salmon.

From France: Sardines, tunny, and anchovies.

From Russia: Caviar and *poisson au blanc.*

From Turkey: Botargo (roe of *Mugil* sp.) in the form of caviar.

From China: Fins of a kind of shark, white-shark fins, dried *Octopus* egg, and dried fish-stomachs.

From Japan: Shark and dried salmon.

From Africa: Crayfish from the Cape of Good Hope.

For dessert were served, among other things, pudding *à la* Neptune and Neapolitan ice-cream. And for "decoration pieces" were given, besides other things, *Bateau de Pêcheur à la* Roosevelt, and *Kan-Ten*, a Japanese seaweed, *à la* Sekezawa Akekio.

There was also a rich selection, especially of mackerel, eels, and sardines, both from different countries and in different modes of preparation. These food-fishes were served fried, in oil or in vinegar. Salmon also was well represented, and, so far as concerned a single dish, certainly in a rather new form, namely, as dry-fish from Japan. Shark-fins and cuttlefish in oil seemed more curiosities than the actual fish-dishes of foreign countries. The edible seaweed from Japan excited much attention on account of its quality as a refreshing food; not the least because it also represented an important industry in Japan—a kind of tillage of the ocean bottom. Of the modes of preparation, that "in oil" was especially conspicuous, and it has thereby gained a special recommendation.

To the Norwegian manufacturers of fish-products it will at once appear strange that in this selection of "the whole world's" fish-products the common wares from the great Norwegian fisheries were not represented, though both Italian preparations of Norwegian dried fish and Spanish preparations of Norwegian split cod appear to have been obliged to pass in among the dishes prepared with oil. The culinary artist, Mr. Sudreau, had the opportunity to offer the guests Norwegian as well as Canadian split cod, Norwegian and American salted herring, &c., but he has probably found that such things are not according to the American taste. The single exception made in the manner was in serving "stewed fish from Norway." This was prepared from chipped dried cod (exhibited by Bordewich & Co., in Lyngvær) together with "Japanese dried fish," composed of dried salmon. Neither dainty herring nor fat herring from Norway were served, nor Norwegian anchovies, although both Swedish and French anchovies were. Perhaps the Norwegian specimens were spoiled in the strong summer heat; of this, however, I have no certain information. If, notwithstanding this, a small market be found in America for the Norwegian fish-products here named, it will happen in this way, that the strongly mixed population, especially in the Western States, contains many families from countries in Europe where the Norwegian fish-wares are current articles. These families become customers for the Norwegian as well as for the corresponding American wares; also as supplies for different European ships' crews small lots of Norwegian fish-products may find some sale. But, taken in the mass, the population of North America will not become customers for Norwegian dried cod, split cod, and pickled herring.

This committee, conversant with the subject of fish-dishes, confirmed me also in another assumption with regard to the Norwegian manufacturing. As before remarked, there were served both large and small fish in oil—not fewer than seven dishes were in oil—among them Norwegian and Portuguese mackerel, sardines (both Spanish and Portuguese), eels, &c. In restaurants in the great cities in North America one will scarcely find highly spiced herring or anchovies in the way that the North-European taste demands them, but almost exclusively oil-prepared articles. This, I assume, is due to a culinary principle, that it is not desirable to serve up strong articles, with which particularly should be classed brandy and beer or ale, in a dry and warm climate. The Americans have, in this point, appropriated the South-European taste for oil-prepared articles without liquors. Naturally, here, also, exceptions are found, as before mentioned, concerning dried cod and split cod, especially in the Western States; in the communities strongly interspersed with German, Scandinavian, and Irish in the West even highly-spiced herring, sausage, and pickled meats are staple articles; they are served up as "free lunch" in eating-houses, because the strong seasoning makes it necessary for the customer to drink beer to quench the burning thirst which these articles produce and gradually augment. I have tried these things. They are, according to my taste, a very disagreeable food, and the traffic itself with this sort of "free lunch" is ill-esteemed as an ugly, rumseller's speculation.

The bill of fare, moreover, regarded from a culinary standpoint, has interest in this, that prepared fish-roe, other than Russian caviar itself, must be able to find a market, forasmuch as it was adopted to be served up with this dinner. It is true, only the Turks and Chinese supplied these delicacies, but for a manufacturer in Norway this might well be almost a matter of indifference and no serious hinderance from imitation. As for the rest, the Norwegian exhibitors, Bordewich & Co. and Störmer (in Svolvær), had caviar, the first of cod-roe. These articles, however, were not served up with the dinner. For Norwegian manufactured fish-roe to succeed in a European or American market, the manufactured article must still doubtless be given a stronger agreement with the universal taste, just as it manifests itself in the Russian caviar. At the same time it should be admitted that Russian caviar does not suit all tastes and that a change in manufacturing it might insure success. A comparatively new mode of preparing or pickling was sardines in vinegar (from Spain). I am of the opinion that Norwegian herring in vinegar, or pickled like English pickles, or merely in vinegar and onions with seasoning of pepper, just as they often are served latterly in the west-country families (in Norway), might become a salable article. ("Herring in jelly" resembles somewhat an article which was experimentally introduced into the market from Norway.)

Taken as a whole, I think that the Norwegian manufacturers of fish-products, especially of dishes for the table, will be able to extract useful hints from the bill of fare mentioned, which certainly was made up under the direction of persons conversant with the subject, and with every regard to refined culinary skill. Another "complimentary dinner" which was given by the same association on the 14th of February, 1877, in New York, I had the opportunity of studying, but I found nothing which I have not already mentioned above.

IV.

NORTH AMERICA AS A MARKET FOR IMPORTED FISH-PRODUCTS.

(A FEW STATISTICS.)

From the foregoing remarks on the common kinds of fishes in America, the prices and taste, it is evident, so far as I see, that North America cannot become a great market for Norwegian fish-products, and that for many reasons, any one of which is sufficient to decide the matter. That the most important Norwegian fish-products, as a rule, do not suit the taste of Americans, since neither herring nor dried cod nor split cod are used in households or are served up at any meal, is the principal condition which prevents the sale of these products in America, taken as a whole. In the next place, the kinds of fishes which Americans are most fond of are either not found at all on the coasts of Norway, or sparingly, and therefore, as a rule, will be too dear after transportation across the Atlantic, which is one of the main causes which prevent Norway from supplying the articles most common in America. But whether the Norwegian fisheries themselves can procure the proper kinds of fishes for America, or whether the Norwegian fish-products will be manufactured according to American taste, as they now are occasionally, the main condition of the trade will be to furnish fish-products either fresh in ice or fresh hermetically sealed (canned).

I believe that no one in Norway is at present able to fulfill these conditions; and, in the next place, at the same time that one would find himself able to overcome the difficulties which interpose, he would have, in the Canadian fisheries and kinds of fishes, which are closely related to the Norwegian, too powerful competition.

In the mean time one cannot hope at present or even in the near future to find Americans as customers for Norwegian fish-products; so the business is not ended, on the contrary it is scarcely begun, and it is a matter of considerable importance for Norway, because the relation hinges quite naturally on this: cannot the Norwegians compete with the Americans in the American markets as well, perhaps, as the Americans with the Norwegians in the Norwegian markets? They do so already, and will certainly, year by year, become more dangerous in competition. American salted herring has already been introduced into Sweden, Germany, and Russia, Norway's best customers for the articles mentioned. No doubt many believe that the American article is not dangerous to the Norwegian traffic, because it is carelessly prepared and of inferior quality; but the Americans will hardly fail to make themselves familiar with the mode of preparation which the new customers' taste demands, whether it refers to their great herring or their fat herring. They have almost as much material as the Norwegian, they do not lack the occasion, and certainly there is no want either of inclination or ability to enter into competition. Split cod from the American waters compete with the Norwegian both in Europe and South America; why not also herring from the same places?

It seems to me that it would be very appropriate if the consuls concerned had their attention directed to this matter, and should, through their reports, give the mercantile class of Norway the necessary information on the American competition with Norwegian fish-products in the different markets, together with the result from season to season. I have not been able to get any collected official report concerning the fisheries in America or the exports of fish-products. The United States of North America have no fishery statistics, and the data I have obtained and found respecting them are partly the estimate of private individuals conversant with the subject, partly a digest of many different statistical tables which I have been obliged, under various difficulties, to revise. At the same time, I think that our knowledge of the American export and import of fish-products is so small that any contribution thereto will be received with thankfulness, and not the least from those who are as greatly interested therein as the Norwegian fish-merchants. I shall therefore, here communicate an epitome of the results which were obtained in the way indicated above.

If I estimate the yearly profit of the United States fisheries at fully $27,300,000, I think that would come as near as possible to the truth. This estimate is founded partly on the estimate of private individuals on the consumption of fresh fish of all kinds in the great cities, partly on the official reports from fishery inspectors on the catch in some States, and partly on the official statistics of commerce and navigation. In the $27,300,000 is naturally not included what foreign nations capture on the banks in America, nor what the fisheries of Canada yield. If one should take both these factors into the calculation the amount mentioned may perhaps be increased by one-half, because the French fisheries alone on the Newfoundland banks have a yearly profit of $1,365,000 to $1,638,000, and the Canadian fisheries yield $10,920,000 to $12,285,000 yearly.

I shall not undertake to state more definitely how the sum of $27,300,000 arises, because it would simply be to render one series of estimates and another series of data, which would not help to make the matter clearer. So much of the statistics shall I, however, particularize as to mention two chief divisions, namely, the profit of the salt-water fisheries at about $20,475,000 and the profit of the fresh-water fisheries at about $6,825,000.*

With regard to America's exports and imports of fishery-products, that is a matter more easily substantiated. The following summary of the official statistics for 1875 gives an instructive survey:

IMPORTED INTO THE UNITED STATES.

a. Fish-products free of duty:			
All kinds of fresh fish	amounting to	$351,889	
Salted herring	" "	288,590	
Salted mackerel	" "	584,283	
All other kinds of fish-products	" "	928,344	
Total duty free			$2,153,106
b. Fish-products paying duty:			
Pickled herring	amounting to	226,494	
Pickled mackerel	" "	553	
Sardines and anchovies in oil and otherwise	" "	526,179	
Other fish-products	" "	102,283	
Total paying duty			855,509
			3,008,615

The preceding year the importation of the items here named was $3,208,527; the articles free of duty amounted, however, to only $1,800,000, but the duty-paying imports were greater, namely, $1,400,000. The importation of sardines and anchovies especially was of greater importance, amounting to about $1,000,000.

* For comparison it perhaps may be instructive to state that the Norwegian marine fisheries may be estimated at $12,285,000 to $13,650,000 yearly and the French at $15,015,000 to $16,380,000.

According to the same official statistics for 1875 the exports of fish-products from the United States of North America were:

Dried and smoked fish	$710,121	
Fresh fish	69,448	
Pickled fish	359,669	
Fish otherwise prepared	1,855,550	
Whale and fish oil	455,236	
Oysters	170,277	
		$3,620,301
Besides those used in transit:		
Fresh fish	3,895	
Herring	11,722	
Mackerel	10,254	
Sardines and anchovies	23,296	
Oil of all kinds	11,236	
All other fish products	157,053	
		217,456
		3,837,757

While importation in the last year has been diminished, exportation seems to have increased, by which one may well conclude that the fisheries are in constant advancement. The American fish-dealers' exchange with foreign countries amounts also to about $7,000,000; but an account more in detail as to the countries with which this exchange occurs may perhaps be of great interest to the Norwegian fish-merchants, wherefore I shall compile an abstract of tables relating to the subject.

The fresh fish, amounting to $352,000, which were imported duty free, were almost exclusively from Canada; the same was the case with the $584,000 worth of mackerel and about half the quantity of herring, quoted at $289,000. On the other hand, the importation of the remainder of the herring, $226,000, together with sardines, anchovies, and all other fish, was from the following countries:

	Valued at—	From—
Pickled herring, 14,243 barrels	$154,302	Holland.
5,675 "	61,459	Germany.
1,278 "	6,854	Newfoundland and Labrador.
218 "	2,897	England.
126 "	480	Quebec and Ontario.
24 "	298	Scotland.

As the prices quoted are invoice-prices, it is seen that the herring sent from Europe were invoiced at over $10 to $12 per barrel, while herring from the Canadian coasts were only $4 to $5 per barrel. Direct from Norway and Sweden they are imported for $2; but I am informed that some of the herring imported from Hamburg are Norwegian.

Sardines and anchovies in oil reached a value of $1,000,000 in 1874, but in 1875 they were imported only to a little over half the amount, or $526,179, distributed among the following countries:

From France	to the amount of....	$445,022
" England	" " "	56,518
" Germany	" " "	11,072
" Holland	" " "	8,028
" Italy	" " "	2,448
" Quebec and Ontario	" " "	1,894
" Spain	" " "	789
" Sweden and Norway	" " "	152

In the importation of anchovies it also holds good that some Norwegian (and Swedish) wares go by way of Hamburg to America.

The great item of import—"all other fish-products"—represents over $1,000,000, and is due mainly to the following countries:

From Canada, &c.	about...	$900,000
" China	"	48,295
" Hong-Kong	"	243
" Germany	"	22,822
" France	"	12,337
" Holland	"	5,750
" Cuba	"	2,635
" Sweden and Norway	"	955
" Mexico	"	806
" Italy	"	795
" England	"	518

The cities and ports to which the greatest portion of the importation came are the following:

Boston and Charlestown, Mass.,	to the value of	$1,298,921
New York,	" " "	754,884
Portland and Falmouth, Me.,	" " "	369,816
Passamaquoddy, Me.,		158,586
San Francisco, Cal.,		101,152
New Orleans, La.,	" " "	94,104

Next come the cities near the great inland lakes, which are supplied with $100,000 worth of fresh fish. To New York, New Orleans, and San Francisco are imported the greatest quantity of anchovies and sardines, while Boston imports most of the salted herring and mackerel.

With regard to America's exportation of fish-products, the greatest items in American wares in 1875 were as follows:

DRIED AND SMOKED FISH.

71,489 cwt.,	valued at..	$450,655	to Hayti.
18,005 "	" ..	64,514	" Hollandish West Indies
12,089 "	" ..	49,628	" French West Indies.
7,565 "	" ..	38,133	" Cuba.
5,546 "	" ..	20,075	" San Domingo.
3,823 "	" ..	24,419	" Brazil.
1,972 "	" ..	11,453	" English West Indies and Honduras.
1,923 "	" ..	14,264	" United States of Colombia.

A smaller portion of items go to China, Japan, France, the Azores, and Madeira, the Spanish colonies in Africa and elsewhere. Nothing is exported to Spain, Portugal, and Italy, where, on the contrary, Newfoundland split cod has so great a market.

Of fresh fish, as above noted, are exported $69,000 worth, of which about $68,000 worth are shipped from Key West, Fla., to Cuba.

Pickled herring are exported to about the same markets as the dried and smoked fish, and, besides, to the French colonies, the English colonies of Australia, Liberia, the Sandwich Islands, Porto Rico, and Venezuela.

The $2,000,000 worth of "all other fish-products" went for the greatest part to England, the English colonies, and Hong-Kong, because the most important items are the following:

	Value.
To England	$974, 673
" English colonies in Australia	298, 280
" Hong-Kong	216, 522
" Germany	74, 998
" Cuba	46, 924
" France and her colonies	37, 733
" United States of Colombia	33, 461
" Hayti	30, 032
" Peru	28, 183
" Chili	7, 441
" China and Japan	8, 826

The exportation of oil, as previously stated, amounted to $455,000; the quantity was 896,000 American gallons, or about 30,000 barrels, which were shipped in the following principal items:

To England	304,605	gallons, valued at	$125, 583
" France	241,161	" "	123, 937
" Scotland	197,891	" "	80, 670
" Canada	119,007	" "	65, 999

The remainder went to the English possessions in Australasia, Cuba, Hong-Kong, Mexico, and elsewhere.

The great exportation of oysters, which is said to be still in its infancy, amounts to $170,000; the following were the most important markets:

Canada	$70, 114
England	38, 661
English possessions in Australasia	11, 639
Germany	10, 798
Argentine Republic	6, 609
Uruguay	6, 458
Cuba	4, 388

Next come Brazil and Chili, Mexico and the Sandwich Islands, Venezuela and various states in South America, Japan, China, &c.

The ports from which are exported the American fish-products are only partly the same which receive the greatest import of such wares; exportation, for example, goes on especially by way of San Francisco, which exports to the value of $1,500,000; New York to the amount of about $1,000,000; Boston about $700,000, and Key West a little over $100,000.

It appears from this abstract that Americans get their greatest supply of fish in oil from France and England, pickled herring from Canada, Holland, and Germany, while all other fish-products come mainly from Canada, China, and elsewhere. And exportation occurs chiefly to England, the West Indies, Australasia, Eastern Asia, and South America.

As it is probable that Norway cannot compete with America in these foreign markets—the West Indies and the east coast of South America excepted—so also it is likely that she will not readily be able to satisfy America's demand for fresh fish, fish in oil, and the rest of the chief articles. But it is more nearly certain that America can act in opposition to Norway as well in Europe as in other countries. It is of much interest to know this.

V.

PRESERVING FISH WITH ICE, AND THE SIGNIFICANCE TO NORWAY OF THIS MODE OF TREATMENT.

The fact that the Americans are able to furnish salted fish-products so cheaply that it may become a serious apprehension that they will compete with the Norwegians even in the markets of Europe, arises partly from this, that the American fishermen sell their fresh fish so readily and profitably, partly because the fishermen are better equipped and more skillful in their calling than the Norwegians, taken as a whole, and finally for the reason that the kinds of fishes concerned abound in the American waters, and besides they are comparatively little in demand for the consumption of the country's own population. I shall next treat of the factors here indicated.

The first—the traffic in fresh fish—I have previously treated in its own chapter, in which I remarked that the preserving of fish with the help of ice and frost plays a principal part in the business. But I shall now enter more fully into this matter, because I will communicate all the information in regard to preserving fish with the aid of ice which I believe at present may have practical interest.

I have read with much attention the articles which have been published on this subject in the Norwegian newspapers, and I have specially noted the interesting information which the consul-general, Mr. W. Christophersen collected. I hardly mistake when I think that the *summa summarum* of the public discussion of the subject has been this, that for the present it is neither practical nor advisable to encourage Norwegian fish-dealers to attempt the exportation of fresh fish in ice from our great fisheries.

Against this result I shall venture to make some objections, because I will point out the weakness of the conclusions which have been advanced.

They have, in the first place, so far as I can judge, confined themselves to *England* as a market and to *cod* as an article of export. Just as I, on the one hand, regard this limitation of the matter as inadequate, so, on the other hand, I consider the views which have put themselves forward in this region; and I may say, further, that the question being limited thus, the answer must be what it is, namely, that for the present the attempt to send *cod in ice to England* should be discouraged. That this should be the answer arises simply from the fact that the English importers of fish desire *living cod*. With it the affair is decided. Closer investigation as to what kinds of cod are most esteemed in the English market might, therefore, at present seem superfluous. And to institute calculations as to how great expenses will attend the carrying of fresh dead cod in ice from Lofoten or Kristianssund to Hull or London may be unnecessary. However, the subject may acquire interest later on, and therefore I shall not retain certain information communicated to me on the prices of cod in England:

Living North Sea and Iceland cod are sold for	$19.11	to	$27.30	per score.
Ekerö cod, wet-salted, " " "	.98	"	1.17	" "
Lofoten cod, " " " " "	.79	"	.98	" "

The Norway-coast cod, as well as the Lofoten, are thus seen to rate at very low prices; yes, lower than one as a rule can secure in the fishing-places in Norway itself. Cod in ice command only one-fourth to one-sixth as much as cod in the living state, and are said, under the present management, to have great difficulties to overcome. These difficulties consist chiefly in this, that the wholesale fish-dealers in England (the whole of England's and Scotland's trade in fresh fish is in the hands of ten to twelve wholesale dealers) antagonize the importation of all other fresh fish than those which they themselves get. They have their own vessels with wells for keeping living fish, and every attempt to compete with them in this or in the traffic generally they contend against so recklessly that a fresh attempt hardly occurs.* One may even in Norway have a little experience of this. But should the opposition from the wholesale dealers themselves be relinquished, yet will the fact that the cod is a kind of fish which easily loses its fresh taste in freezing, always render competition with the Englishmen's own living cod the more difficult. After all, one will naturally prefer the living cod.

Of halibut, salmon, and mackerel—three kinds of fishes which are well adapted for sending in the frozen or iced condition—there have gone, on the contrary, for many years, a not inconsiderable quantity from Norway

* When the Americans began to export to England fresh meat in ice the English butchers raised a strong opposition, and only after sundry conflicts, which cost much money, the Americans, with the aid of the press and the people, won admission. However, a time was selected when a single English butcher, in order to spite the Americans and their meat, called all his worst meat "American," and sold it cheap.

to England, which demonstrates practically, indeed, that the fish business treated of can be prosecuted with England and certainly succeed besides, because it is carried on by nearly the same Norwegian firms summer after summer, year after year.

But the chief question was, not whether fish from the so-called great fisheries—herring fishery and cod—could be exported in ice instead of in pickle, or salted and dried. In England, the main question was, "Will I buy dead cod instead of the living?" And the answer was "No!" But in South and East Europe the inquiry was different, namely, "Will I buy fresh fish instead of salted and dried?" And it is this question which I believe cannot be regarded as capable of being answered in the negative in reference to the investigation of the English business-relations. My personal opinion of the matter is this, that herring as well as cod can be transported without too great expense to Norway's present great customers in East and South Europe. The difficulty consists not in this but in the fact that one of those markets must be prepared to preserve these frozen or iced wares in this condition, even in the manner of loading. But this difficulty itself can gradually be overcome, provided the traffic only can endure the expenses attendant upon it, which I can have no opinion of, since I do not know the proportions of the ice business in the countries concerned. That the whole may be amended without exorbitant expenses, so that fresh cod and herring may be delivered in South and East Europe, I am confident, from the fact that similar kinds of fishes are delivered in America to markets as far distant from the fisheries. At the same time it is indeed clear that all depends upon whether those customers really desire *fresh* fish from Norway. If they do not, then the matter is thereby decided. But until one has experience of this by some experiments, the question cannot be regarded as satisfactorily answered, and it was only this I wished to take exception to in the results which the public discussion appears to have desired to establish, and that immediately.

For a clearer understanding of the matter, I had some time before examined the relations of this business in America, the fish concerned as well as other articles of food, and I shall now communicate what I learned.

With regard to the treatment of fishes with ice, a distinction must be made between merely *packing* fish in ice and *freezing* them for storing. The first—packing in ice—is employed in ordinary transportation, as from the sea or fishing-banks to ports, and from the ports (generally after replenishing with fresh ice) to the selling-places or markets in cities. The other ice-treatment, by which the fish are frozen, is employed mainly in storing fish, as, for example, to enable the fish-dealer in question to preserve fish from summer until winter or from one season to the next.

Preserving in ice appears to require no skill, yet it demands some judgment to answer the purpose entirely, as to the quantity of ice in proportion to the quantity of fish, the convenient size of the pieces of ice, the material and shape of the box—all are things of importance, and not the least elements of a practical economy.

Greater judgment, however, is demanded in freezing for a subsequent storing of wares. This is truly a new art, on which a patent was taken ten or twelve years since in America, and it will become very important. I shall, therefore, venture to describe a so-called freezing-apparatus or frost-vault.

VI.

THE AMERICAN REFRIGERATOR.

A refrigerator must not only keep the article cold, but it must keep its temperature near zero or below the freezing-point; and to do this is required not only a constant supply of ice, but also such an effect of ice as will produce and maintain intense cold.

Most persons have seen a common ice-chest for household use; its purpose is only to keep articles of food cool or cold. But no matter how much or how often one may fill it with ice, the provisions will not generally freeze; partly because the ice-compartment is too small, partly because the mass of ice cannot of itself send out sufficient cold over the provisions to freeze them hard; besides the distance from the ice is so great that the intervening air makes freezing (in mild weather) nearly impossible.

The largest and best furnished freezing-vault which I saw was one belonging to the above-named Mr. Eugene G. Blackford. It was, practically speaking, a cellerage 80 feet long, 11 feet wide, and 10 feet deep, fitted up as a small store-room. The outer frame of the vault was much like a ship's deck, tight, and composed of planks. A trap-door led down to a room where it was dark and the temperature some degrees below the freezing-point; on the day on which I inspected it there was a strong summer heat of 35° C. (=95 Fahr.). The ceiling itself was double, and the lining was partly sawdust (to prevent the influence of warm air), partly ice mixed with a certain proportion of salt (to send the cold in the ice out over the room, "liberate" it, as it is called in physics). Along the sides and at the ends and across the vault itself were constructed large conduits or rather long reservoirs for ice mixed with salt, to act on the air in all parts. The situation and shape of this reservoir form a very important part of the apparatus. The art is to get the cold in the ice, which is "liberated" with salt, to operate so that it will be of the greatest benefit and also most valuable with least expense. As cold or cold air "falls," these ice-reservoirs are placed highest up in the top; the cold air must also fall right through the whole vault and cool the entire room in its wandering. Were they, on the contrary, placed on the sides or bottom, the cold liberated from the ice would simply fall along or down the reservoir itself. Side-reservoirs would also merely form a cold wall or ice-belt around the vault. Such a cold wall they have also, but the ice of the reservoir is best utilized, as indicated, by allowing the cold to fall out in the room; therefore the wall-reservoirs, which appear as closed conduits, are given an oblique position, so that the cold from the uppermost ice may fall out freely and operate as a corresponding ice-blast from the ceiling. The floor is least liable to an invasion of warmth; the ice-receiver here is, therefore, most poorly supplied with ice.

To protect the provisions against ice-water drip from the ice-box, there are placed conduits and conductors. As for the rest the oblique position of the reservoir will contribute toward carrying the ice-water away without dripping. When once the warm air is expelled and the ice has supremacy, it requires less to maintain freezing for a year or as long as one wishes.

This vault of 10 by 80 by 11 feet, or 8,800 cubic feet, could easily hold 100,000 pounds of fish. It was divided by walls and ice-pipes into four parts, each part furnished with an endless number of shelves of moulding. To keep the temperature in the whole apparatus some degrees below the freezing-point, 1,400 pounds of ice is required daily in the heat of summer. The price of the ice was about $4 per ton of 2,000 pounds. The expenses in this respect were therefore only about $2.80 per day in the heat of summer and proportionally less during the remainder of the year. In winter are required for the apparatus only about 1,000 pounds of ice per week, about 29 cents per day. In the cooler country of Norway an apparatus of this size would require a smaller quantity of ice, and the ice would be had at a much cheaper rate, in Bergen City for instance, for $1.64 to $2.18 per ton, which would amount to only 82 cents to $1.09 per week to preserve 100,000 pounds of fresh fish in winter; truly, a cheap preserving! Even a smaller quantity of fish could profitably be stored here when the prices in midsummer are very low, and it could be calculated upon with safety that they would advance later in the autumn, as happens with the price of salmon, which rises from 10 and 12 cents to 25 and 30 cents; yes, as high as 50 cents a pound toward Christmas. But it will readily be perceived that a quantity of fish, say only 1,000 pounds, would not support the increased expenses of an apparatus so large for a longer time.

Not every kind of fish can endure such cold, but like many kinds of provisions some lose much in flavor. And even the articles of food which endure it must be treated in a peculiar manner. Among other things it must be observed carefully that fish intended for storing in cold must be frozen stiff immediately, or as fresh as possible. Delaying the stiff-freezing some time, or, under certain circumstances, any time, is a source of great loss. Also, in thawing, measures must be taken that nothing shall be spoiled. Thawing must, by preference, take place in cold air so as to "repel (*skyde*) the ice," as it is called in the Norwegian household language, before further dressing or cooking can occur.

The kinds of fishes which best endure storing in cold are salmon and halibut; next very fat herring and mackerel; but cod and flounders and other common fishes will lose their flavor; also oysters. The fatter fish are, the less will their agreeable taste suffer; the poorer, the greater such danger.

Experiments have been made with all kinds of provisions, drinks, and fruit. The result so far reached is this, that every kind endures a certain low temperature, below which one cannot venture without injuring the flavor. Especially fruit and most kinds of flowers are very sensitive in this respect. To treat them directly with ice-cold air will, I think, not succeed; but perhaps the indirect method, or the now common cold-air method for cooling off the preserving-room will do better. Pork is also somewhat difficult to handle, but for a different reason: it requires such intense cold to be entirely protected. Again, it has been found that the more uniform the cold is kept the better.

My informant states that the use of ice as a means of preserving these articles dates back only twelve years. A patent for freezing salmon had yielded the patentee $30,000 to $40,000; but while the patent was respected concerning salmon, competitors soon learned that the patent-right was not infringed upon by employing freezing for other provisions, and after a lawsuit respecting it the patentee lost, as the tribunal declared that freezing other articles of food than salmon was free or not patented. At the same time the use of ice is considered yet to be in its infancy, or taken as a whole only to have acquired importance in the treatment of salmon and some other fish-wares, together with fresh meat. With regard to fruit no one has gone beyond experiment; but as soon as the proper ice treatment is discovered one must admit that America will export to Europe large quantities of fresh fruit,* just as is already the case with fresh meat, and experimentally with fresh salmon.

VII.

TWO KINDS OF REFRIGERATORS ON BOARD PACKET-STEAMERS FOR CARRYING FRESH MEAT.

As the transportation of fresh meat from America to Europe in steamers has attracted marked attention, and as a similar transfer of fresh fish will perhaps in time become an example for the Norwegian fish and game dealers, I undertook a journey across in a steamer which was furnished with cooling-off apparatus (refrigerators), and then remained many days in Liverpool to investigate the condition of the meat after the voyage, and to make myself better acquainted with all the details of the arrangement.

So far as my experience goes, the method employed in the transportation of fresh meats across the Atlantic is copied from a rather common method of refrigeration used by the pork-butchers in the West. This last-

* Fruit-growing is a great industry in the United States. Much is exported in the form of conserved and preserved fruit. Of all kinds of apples alone were exported in 1875, as officially advised, more than $1,000,000 worth.

named method I shall, therefore, treat of first, also for the reason that it may possibly be interesting to the public.

The great pork-slaughtering houses operate chiefly in winter; the summer killing is regarded as of an inferior quality (judging from the recommendatory advertisement, "winter-cured" American pork). The reason is, that heat will prevent, in a greater or less degree, the successful salting of pork; in the cold of winter one may be more certain that the pork will be put in salt sufficiently early, or that the air will not influence the flesh before the salt has begun to operate. During a visit which I paid to a pork-butchery I convinced myself of the significance of refrigeration for the trade. The body of the hog, as soon as the animal is killed, is thrown into a large scalding-trough filled with hot water, where it remains some minutes, to be scalded, or sufficiently long for the whole carcass to become so thoroughly hot that it takes a long time before it is cold enough for salting and packing. It is important, now, to shorten this interval, and also to prevent any hurtful influence from the action of the air. In winter this is not so dangerous. In summer or in warm weather it is hurtful. A sudden cooling-off by putting the pork in ice, it is thought, will injure its flavor; and even a natural cooling-off in wintry air is not to be recommended. It is important, therefore, to devise a method by which refrigeration could proceed to a proper length of time and uniformly, independent of the weather, and in such a way that the right point of time for salting could be determined with safety. They have here, among other things, constructed for a refrigerator a large box with a blowing contrivance or pump. Inside of this box, large enough to hold a small drove of hogs, the carcass is hung up; thereupon the air-pump is set in motion, first to pump out the warm air given off by the carcass, and then to blow more and more cool air in until the air corresponds with cold water, and cold air from an ice-chamber follows. The current of air travels constantly the same way, flowing along the ceiling or the upper side of the box towards one end, from which it goes backward along the bottom to the other end, when it carries the heat from the carcass up and out. When the cooling is somewhat advanced, they do not allow the current of air to escape into the outside air, but into an ice-chamber, and thereby it becomes colder and colder until it is everywhere uniform and the cooling of the carcass is finished. To give the current of air greater cold, ice is mixed in the ice-room with salt in such proportion as will give out the degree of cold desired.

This last part of the refrigeration forms the basis of the method of treating fresh meat in its transportation over the Atlantic in steamers. The meat is brought on board soon after killing, and is quickly cooled off and sewed in muslin; thereupon it is hung up in the inclosure arranged in the room for transportation. It is important now to maintain a low temperature until the vessel reaches the European port. How this is accomplished I shall seek to explain as I describe the refrigerators which I saw in the American steamer which carried me across.

The steamer had three rooms furnished with cooling-apparatus (refrigerators); of these, two were on the second deck, a little forward of midships; the third was astern in the peak, and extended from the keelson up through the two decks. The first were kept cold by water, the last by air. Each of the rooms was 9 to 10 feet high, 12 to 14 feet wide, and 40 to 45 feet long. They were perfectly empty inclosures. The cold-water rooms, which I shall mention first, were provided with doors for loading and unloading the meat. The doors opened out into a stairway by which one reaches the deck. Immediately outside the door was the steam-engine which forced the cold water; close by was the ice-box, through which the water is led; also, finally, two separate zinc-covered and well-lined ice-storerooms—one for each room—5 to 6 feet wide, 10 to 12 feet long, and 9 to 10 feet high. A sufficient quantity of ice was kept for the voyage. The inside of the refrigerating-rooms was provided with close series of hooks, on which the carcasses were hung, sewed in muslin. They must, as far as possible, hang clear of one another, partly to prevent chafing during the rolling of the ship, partly to give the cold air free access to all parts of each carcass. When the rooms are filled there is not room for a boy, hardly enough for a cat, to creep along the floor. The fuller they are the easier it is to maintain the cold. All sides of the rooms were encircled by lead pipes 2½ inches in diameter, 16 coils on each side. The pipes were fastened directly to the wall. This was the furnishing of the room.

The work of refrigeration I shall next briefly mention. The problem is to keep the temperature uniform and low—usually 37° Fahr., or, more accurately, never above 40° Fahr., nor below 35° Fahr. To this end the rooms were first completely inclosed, made tight, then hermetically sealed. In the pipes, which with an aggregate length of 1,700 to 1,800 feet encircled all the walls, was pumped cold fresh water by a little steam-engine, which worked a pump for each room. The water circulates through the whole length of the pipes in a room, returns to the starting-point, holds, then, a temperature near 29.3° F., is conducted over ice and salt, forced again into the pipes, and produces and maintains a temperature of about 35.6° to 37.4° F. This current of water is continuous, since the circulation is uninterrupted; the quantity of water is increased by the melting of the ice; the surplus finds an outlet. Fresh ice is supplied constantly in the ice-boxes, and each ton of ice is mixed with nearly a quarter as much salt. The temperature of the room is controlled through a very small window, inside of which a thermometer is hung, and by that the temperature of the body of water is regulated. To protect the meat, the pumping is continued until the last carcass is unloaded.

The work or oversight of the work of the pumps is taken in charge by two men who watch alternately. This superintendence, as well as the engines, pumps, pipes, &c., is at the expense of the shipper of the meat; the steam, on the contrary, is furnished free by the ship. The freight amounts to so much per ton of the whole room occupied; this trip, about 30*s*. per ton. The apparatus employed returns free in the steamer, and the vacant room is used by the vessel for other freights.

The third refrigerating-room had cold air instead of cold water, wooden air-tubes 14 to 15 inches wide instead of the leaden water-pipes of 2½ inches diameter. The current of air which is to cool off the room and keep the temperature low is set in motion by a blowing-arrangement which is driven by a little steam-engine of ½ horse-power. The blower which stands upon the first deck makes 180 to 210 strokes per minute. This drives the air through a box of ice and salt down into the room, which is everywhere penetrated by it; it is led back in a similar way through a box filled with ice and salt, and comes from this up again under the fan, which forces it downwards in an endless round as long as there is meat to be preserved in the room. The temperature, which is here controlled within the blower, is in the room about as in the cold-water rooms, but often 3° Fahr. lower. The principle is the same, and no one is able to point out any difference of advantage of one over the other system. Both are considered excellent. This cold-air room was 30 feet long, two decks high, and for the rest extended alongside over the whole peak. This accommodated 250 whole beef carcasses, which hung in two layers, one for each deck's height. The box in which the fan worked was only 5 feet high, 2¼ feet long (alongships), and 1¼ feet broad (athwartship). The expenses were met as in the other room, and the freight was paid, as there, only for the room occupied.

This, then, was the information I had to convey, and which I believed would be a matter of interest. One may see herein why the ice treatment of provisions is employed in America in the different industries and for different purposes. That something similar might be practiced in Norway appears to be a conclusion not without foundation; it is important only to overcome the difficulties and regulate the mode of treatment which the Norwegian traffic must involve. But to point out more in detail the practical mode of laying hold of the matter may be found to lie somewhat outside of my present task; it will be enough to indicate incidentally that the wealth of our country in game, fish, and other provisions which are mainly used fresh, might be made serviceable to a great extent if the tradesmen interested should follow the example of the Americans. The many fast steamers along the whole coast and going to foreign countries appear to me, for instance, well enough fitted as a field of experiment for such a refrigerating-room; yes, merely having ice-boxes on board a large steamer for chartering for the use in question might perhaps be found both a good speculation for the owners and a convenient and noteworthy arrangement for the consignors and consignees in our large cities and in foreign countries.

It remains to prove by an adequate trial of such enterprise what I, for my part, consider for the present as certain, namely, that the using of ice is a practical method of preserving and transporting provisions in the fresh state for a long time; also, for ourselves, I think it will not require a long time before great advantage will be found in using ice for our great fisheries, even to a greater degree than is now common in America. I think, also, that our fat-herring fisheries will find a valuable adjunct in ice-using, when, for instance, the fishing occurs far from a salting-station, and it is important to preserve herring in summer until salt, barrels, and sufficient help can be secured. In America they freeze whole cargoes of herring. Bait, too, for the cod and coal-fish fishery might be preserved by the proper amount of ice where now it is wasted or cannot be had.

The matter is first brought into practical operation for the salt-water fisheries, so we should next investigate how far our numerous rivers, lakes, and fiords, well adapted for fishing, ought to derive advantage from the use of ice, and be made to furnish a great quantity of goods for sending abroad. As matters now stand, our inland and fresh-water fisheries rank rather low, and the yield can scarcely be conveyed to market, except in a half-spoiled condition, in summer. The experiments, the experience in the practical business itself, which one acquires with reference to this fishery in America, have brought me to believe that much might be accomplished by us.

The fresh-water fisheries of the Americans, inland and on the great lakes, are prosecuted so largely by using ice and storing frozen fish, that the traffic in these fish-products is a comparatively great business. On this I shall venture to communicate more detailed information.

VIII.

FRESH-WATER FISHERIES IN AMERICA.

That the fresh-water fisheries have an entirely different signification for America from that in other countries was shown in the Exhibition in many ways. While the fishery-division of most countries consisted mainly of fishery-products from the sea-fishery, and only occasionally had some fine implements for lake-fishery, or a portion of the products of the fresh-water fishery, the main strength of the American division lay in this direction. One noticed instantly that this branch was the object of special partiality on the part of the American functionaries concerned.

In Agricultural Hall there was a series of aquaria in which various fishes disported themselves, surrounded by all the comforts which belong to a happy domestic life in fresh-water depths; the handsome, sprightly aquarium-tenants were not only well fed, but were also shown all other attentions which fish delight in; their home was furnished with beautiful sea-weeds, shells, corals, and muscles; in shady grottoes of curious stones reposed friendly-disposed sea-creatures and crabs of exquisite beauty; fresh water bubbled forth continuously and diffused an agreeable coolness which the outside spectators certainly envied them in the oppressive summer heat; and, finally, they were constantly surrounded by a dense circle of spectators, especially children and ladies, who felt themselves very fortunate to see them, admired them, and in their delight gave them rare dainties. Nearly as much attention was bestowed on the fishes cast in plaster and painted with great skill. This collection (the cost of which is stated to be $27,300), was larger than that, and had a place in the Government Building, where also a collection of photographs (with the scale of measurement attached) and color sketches of all sorts of fishes were exhibited. Here was found a great variety of implements, hatching-apparatus, and the requisites for the transportation and planting of young fishes.

The products of the fresh-water fisheries were also exhibited in great abundance, chiefly all forms of preserved salmon. In a couple of refrigerators were shown fresh fish, which, in spite of the oppressive heat, kept their fresh, delicate appearance many months. This was both an exhibition of fish and of a mode of preservation, and the two parts combined were an illustration of the advanced development of the fish-traffic. Beside California salmon lay pompano from the Gulf of Mexico, always equally fresh and sweet. The great prominence which they had thus given the fresh-water fisheries must naturally excite one's interest in them; but since, as before remarked, no detailed statistics of the fisheries are to be found, one is cut off from access to an easy mode of furnishing himself clearer information on their extent and importance as a public industry. At the same time there are found so many local statements respecting them, and also copious individual estimates concerning this and the proportions of the fisheries in other directions, that one with their help can form a tolerably good conception of the part they play.

In a more detailed account of the relations of the great lakes it is seen that the whole fishery-traffic here is entirely similar to the great coast and sea fisheries. The author, one of the esteemed functionaries of the Government of the United States, had visited most of the great lakes and procured statements regarding the scientific as well as the practical interest in this connection. The annual supply of fresh-water fishes which fifteen cities alone take near the borders of the lakes, is found to amount to not less than 32,250,000 pounds, of an aggregate value of $2,184,000, the local expenses not included. Nearly two-thirds were furnished in the fresh state, and consisted of 14,000,000 pounds of salmon, trout, pike, whitefish, and fresh-water catfish; over 4,000,000 pounds (or 20,000 to 25,000 barrels) of fresh-water herring were supplied, but of these 10,000,000 pounds more were delivered salted.

The prices were not stated for the individual kinds, but the total of $2,184,000 for 32,250,000 pounds gives an average price of 27 cents per English pound, or about $2.73 for 36 Danish pounds, and this must be called a good price. But if one take properly into consideration that here are mentioned fish so fine and well flavored as salmon and trout it must be granted that the prices were very low. And without low prices on these inland fish-products it would not be practicable for the Americans to furnish so considerable a quantity for export to Europe and to almost the whole world. In reality the state of things is this, that Americans furnish, in the European markets, fresh salmon, for example, cheaper than many countries in Europe themselves can get them from their own fishermen, and that, notwithstanding the fact that the Americans have considerable expenses on their goods besides a long transportation.

The explanation of this peculiar circumstance must be exclusively sought neither in the abuse or overdoing by the countries concerned of the fisheries in their waters by which the abundance of the fishes diminished and the prices enormously advanced, nor should the explanation be sought in the natural wealth of fishes in American rivers. None of these explanations will strike the right point in the matter. The only satisfactory reason, in my opinion, is this: the Americans latterly prosecute their inland fisheries with a deliberation on the basis of practical scientific measures, which in the course of time must bring and already has brought it to pass, that these fisheries will increase and year by year make their competition with all other countries quite overpowering. Americans can overdo the fishing in a water-course just as well as others, and in reality the complaints about such abuse are not few. But at the same time that they seek to prevent this abuse they do not restrict themselves to protective laws and prohibitory enactments against fishing at certain times or with certain destructive implements. They do not confine themselves to passive measures; they do not depend exclusively on nature's own assistance to counterbalance the fishermen's want of judgment and selfish efforts. The American fishery commissioners have a hold on much more effective measures, active measures: they transfer young fishes to the best water-courses; they see that the barren rivers and waters again get a supply of edible fishes; they erect great hatcheries for the "cultivation" of new material for the fisheries, hatching out both the fish themselves and the subordinate fish on which they feed. And all these active endeavors take place on a grand scale and with a generosity on the part of the people, which at first thought might seem exaggerated, but which upon closer consideration will be found to contain a wise economy combined with practical truth and correct apprehension of what is for the best. As I next have to treat of the fishery itself I shall defer until the next chapter mentioning the hatching operations and give here only some few presumably very instructive particulars on the "cultivation of fish-material" for the fisheries.

In the year 1872 the United States Congress voted $15,000 to defray the expenses of transporting shad (*stamsild*) to the Pacific States and the States bordering on the Gulf of Mexico and the Mississippi River, and to transport salmon, whitefish, and other edible fish to the waters in the Union which were best adapted for them. The same year was voted anew $10,000, and the next year $17,500. But this was not the beginning of these operations, nor was it a solitary series of concessions; but it was, so is it stated in the reports, the natural outgrowth of what so many individual States already had voted, and so many private individuals had already performed. From another statement it is learned that in thirteen States was voted on an average $2,730 yearly for many years for the development of the inland fisheries, and individual States had in the last eight or ten years also applied $27,300 to $40,950 for restocking their water-courses with fish. On the Pacific coast there is a river, the Columbia River in Oregon, in which nearly 7,000,000 salmon-fry were let loose, to make the fishery flourish. In the same river they are also caught quite extensively; because one has estimated that of the catch there in the last year will be produced $3,013,000 to $3,276,000 worth of canned salmon in tin boxes to export to the States and England. Salmon of the year are planted as well as the fry. They spare no pains to aid the producing-power of nature itself; the stream, productive and rich in fishes, is by combined scientific and technical aid made still richer, still more productive. They will take care that the great fishery shall become still greater, and under no circumstances less. As there are naturally many fish here, it must be a suitable place for them; so they plant millions of fish because there is plenty of room.

The inland fishery is carried on with no small employment of capital in implements, boats, ice-houses, steamers, &c. Large stationary nets and traps are used, which have an average value of $546 to $819. Of boats valued at $546 there are hundreds on the great lakes, but $109 is a common value. A peculiar kind of boat called "Norwegian," and so described that I must believe it to be related to the "*listerbaaden*," is, however, considered clumsy for rowing by Americans, wherefore only fishermen from Norway, Sweden, and Denmark use this form. On Lake Michigan the fishery is prosecuted also with small steamers; they cost, as a rule, $1,092. Not fewer than 100 ice-houses for keeping fish fresh are found throughout the extent of the lakes named. In several places a considerable quantity of salmon and trout are stored until a later time, when the fishing is ended, and they sell at a higher price. Also in the fishing-vessels themselves they have apparatus for hard-freezing fish, and they sail from one lake to another with a cargo frozen in this manner; on board the vessels in Lake Superior alone are frozen 270,000 pounds. The greatest inland market for the fishery-products of the great lakes is Chicago, to which city, in 1872, were conveyed about 4¾ million pounds of salmon, trout, pike, &c., and a total of 7½ millions of fresh-water fish, at a value of $4,641,000.

It will be a long time before the fishery of any Norwegian water-course will reach results which can in any manner stand by the side of those here mentioned. But there is certainly no insurmountable obstacle in the way of increasing the profits of all the inland fisheries of Norway to such considerable amounts as will endure the comparison. Norway has so many famous fish-streams and rivers, that one should have in them an invitation to the attempt, especially since one has so practical and satisfactory an example to follow as the Americans present. But as matters now are the good waters yield only indifferently well to a rational fishery; one sees the profits diminish and the cost of carrying on an antiquated fishery increase, the less the occasion for old methods becomes.

IX.

HATCHING YOUNG FISHES FOR THE SAKE OF THE FISHERIES.

As in most other countries where the fishery is a greater industry, some persons also in the United States have sometimes complained that the fishing is falling off; and as elsewhere so here one has for some time heard complaints of uncertainty as to where the cause was to be sought for. Some were of the opinion that it was over-fishing or with too many implements; others were just as sure that the destructive quality of certain appliances was the cause. That the fish were scared away by the noise of steamers, cannonading, or by the bad smell of decayed offal from manufactories, I have, however, not seen advanced.

The question of the decrease of the number of fishes on the coast arising in earnest in the years 1860 to 1870, was, in the spring of 1871, intrusted to Prof. Spencer F. Baird to investigate the matter. He received a commission to learn how far the fishing on the coast, as well as in the fresh waters, was diminished; what cause had occasioned the decrease, and what expedients in the form of law and otherwise ought to be employed to furnish the fishing-grounds with a new supply of fish, and in other ways make fish diet *cheaper for the people.* The professor immediately set his hand to the work. With the assistance of a number of practical men and men of scientific acquirements he begun investigations of the temperature of the water at different depths, its varying transparency, its chemical composition, the influence of currents in the depths and at the surface, food-supply for the edible fishes—in short, examinations of everything on which the success of the fisheries must depend. This was the first year; the next year saw a rapid advance, and Professor Baird associated then with himself a corps of thirty-seven specialists, more than half of whom were professors, teachers, and students in zoölogy and natural history, the rest being fishery inspectors and similar functionaries from eight different States and from British America; so great was the interest with which they participated in the investigations. A fund was established for the propagation of one or another special kind of fish, as, for example, the salmon from the Atlantic coast; and for the people at large, who ought to become interested in the mysteries of development, a large aquarium,* the importance of which for scientific study was especially set forth. It was not long before they found out what should especially be done; they resolved to devote great energy to the propagation of young fishes, or, to speak perhaps more correctly, to the protection of young fishes. Instead of leaving the young to their own fate, they secure the spawn and the milt in time, allow the whole process of development to proceed in the publicly established apparatus, to set free therefrom the young as soon as they can take care of themselves. For many years have a portion of the fishermen followed the business of hatching spawn, raising the young in order to sell the grown fish later; and so skillful had individuals become in this achievement that the scientific men learned from them and took some of them into the service of the state as superintendents of the hatching operations throughout the Union. This thing, that individuals had acquired a considerable fortune by hatching and rearing fish, contributed naturally towards giving the whole matter the powerful impulse in a purely practical direction, which it gained very early.

In a book on trout-culture, written by a practical breeder, it is said, quite significantly for the stand-point which has already been reached, that it "pays better to rear trout than hogs"; and every one knows what the pork business is for America. All that I have learned indicates that this assertion has gained general acceptance both among the common people and the learned; and it is said to be assured that in the art of rearing fish the Americans surpass all others.

Partly to control the fisheries themselves, and partly in great measure to carry on the hatching operations, there were by degrees appointed in eighteen different States, taken together, fifty-two commissioners or fishery inspectors, besides a regular staff of subordinate officials. There was established by the United States Government, besides, a fish commission, whose chief is the well-known Prof. Spencer F. Baird. With this complement of special practical cultivated officials and talented scientific men the effort was untiring to produce and distribute young fishes from the mountain to the sea in small rivers and lakes, in the great streams, and in the great bays, both of fresh-water and salt-water fish.

Of large hatching-establishments there are many, public as well as private. A more detailed description of them would certainly be interesting, but without illustrative drawings such a thing would be hard to understand. Besides, the public is much too little informed about the matter to regard anything other than the practical results as particularly interesting. I shall, therefore, here devote my attention to throwing light upon what I will call the great enterprise of the hatching operations in America.

* This aquarium was set up after the example of the previously named society, the "American Fish Culturists' Association." It was established in New York on a rather large scale, and had at one time even a living whale to exhibit, which, however, unfortunately died, after the expiration of some time, from consumption, according to the statement of the doctors.

One of the most popular kinds of fishes in America is the shad, or "*stamsild*," which forty or fifty years ago was caught by the million in many bays and mouths of rivers. As soon as there was talk of restoring the depleted fisheries to their former magnitude, it was resolved, among other things, that they should also try the shad, both because it is a favorite article and because it multiplies rapidly. An experienced cult-urist, Mr. Seth Green, was taken into the service of the government and began the experiments. At the first hatching the eggs yielded only 2 per cent. of young; later it advanced to 70 per cent., and increased to 99. This was regarded as a triumph, because this high percentage of young hatched out was greater than ever was seen or hoped for, and it was also among the first great attempts with the herring kind. During a period of twenty days 40,000,000 of young shad were liberated at the mouth of the Connecticut River. No one knows in what time this fry will become mature or return to its nursery. Some think it will be about five years; others, a much shorter time. The next summer they began to hatch a new brood, and after an attempt of twenty days the summer heat became too strong for the eggs (they had already gained experience as to what temperature the young should have), and they concluded their work after having liberated 60,000,000 of living shad-fry. The following year they wished to await the result of the 100,000,000 before intrusting much money to the sea. The fourth year the people were highly surprised. It was three years since the first young shad were set free, and as early as the close of May it was reported that for twenty years such shoals of shad had not been seen approaching the land, and vessels which had come through the neighboring sound reported also great schools which stood towards the mouth of the river. The next day there was reported from five different fishing-places an unusually large catch of shad and from the rest that the fishing was remarkable. It continued to be great the whole fishing-season through, and the fish were large and good. By a comparison of this year's catch with that of previous years, it is found that this year's greatest haul yielded about 60 per cent. larger profits than ever before reported. The State's (Connecticut's) fish commissioner did not stop with this; the result had indeed been good, and they continued to set free the young and then capture the adult. In the two following years 156,000,000 eggs were taken and impregnated. Many States followed the example, and there will be planted yearly millions of shad-fry, carefully hatched, for the enriching of the fishery of the coast and in the rivers.

An interesting experiment was undertaken with shad in 1871. This fish was never seen in California or on the Pacific coast. In the year mentioned upward of 20,000 young shad were sent there and set free at Sacramento. They wished to introduce shad into a sea entirely new to them, and submit them to the experiment as to whether they would live there, and whether they would return to the place where they were liberated. The 20,000 shad disappeared in the deep in 1871; two years later a few shad were seen here and there in the neighborhood, and in 1874 a number were caught. They had reached a good size, weighing as high as 3½ pounds. This with many other experiments testified that shad require three years to become adults, that the shad will return to its feeding-place or where it was set free, and that the shad can thrive in entirely different seas from that in which it is thought to be a native.*

With a similar herring-species, the alewife, many experiments have been made also, which have testified how easy the alewife is to hatch out, how quickly it grows to an edible size (three years), and how rapidly it multiplies. It is estimated that every shad with spawn has 50,000 to 80,000 eggs; and that the "fresh-water herring" or alewife has, in proportion to its size, four times as many eggs. This naturally great capacity for reproducing itself is what one takes advantage of when one attempts with hatching-apparatus to protect the young from falling a prey to its natural enemies. Even a small number of herring will, provided their reproduction is protected, be able to increase to a great shoal in the space of a few years, and with this consideration in view it becomes an entirely practical economy to liberate millions of young every year and allow the fishermen to catch the adults for the fish-market. From one fishing-place it is stated that they yearly set free 4,000,000 to 5,000,000 young shad, and that the fishing steadily improves, so that now they catch between 300,000 and 400,000 pounds of mature shad.

It was after a little practice in the art that they succeeded in hatching out this herring-species, which now appeared so greatly increased. Before they had confined themselves mainly to salmon and trout, whose hatching was an ancient and well-known matter in Europe. And after the successful experiments with the shad itself they did not neglect these species of fish, because salmon and trout are far more valuable articles, and the California salmon especially is celebrated for its agreeable flavor. They applied themselves very diligently to their multiplication and distribution to new waters. For the illustration of this work I shall merely call to mind the operations on the Columbia River, in Oregon, and only add here that in 1875 11,000,000 of salmon-eggs were collected at the establishment of the United States Government, which were sent eastward to the waters of different States; this shipment amounted to not less than 20,000 pounds, including the packing. From another river were sent 5,000,000 eggs, packed in boxes of 50,000 each. On this scale they prosecute the work now with unabated vigor and with the mutual co-operation of the different States.

The economical question will indeed be of great interest for Norway.

---

*In a report printed in January, 1878, and which I have just received, it is stated that it is considered certain that the "adult shad will return to the place where it was set free when young." And by the fish-commissioner of California it is reported: "Shad were in 1876 and in 1877 right abundant in the Sacramento River." . . . . "There can be no doubt that the first shad which were brought from the Hudson River in 1871 have been out in the depths of the sea and have now returned and spawned."

In the numerous reports which I have read I have, however, found very little concerning it, so I cannot state in dollars what it will cost to hatch out some millions of salmon or shad. At the same time I have seen that the gathering of salmon-eggs, according to a statement, is becoming less expensive each year, and that the profit steadily increases. The same was still more the case with the shad, which yields so remarkably many more eggs. In the large establishments it is managed so that they succeed in one of them in hatching out a million shad for $1.09, including all expenses. This million of shad one cannot naturally keep under his care until they become salable fish. They set free the whole mass, and it disappears in the deep to return in about three years to the shore where it was liberated. It is believed that one-fourth part return from their wandering about in the sea, where they are persecuted by so many fish. But even with so small a portion in safety, it must indeed turn to good account. Because, if one estimate, for instance, that an establishment hatches one billion in three years at $1.09 per million, then would this outlay of $1,092 in the space of the next three years yield a shoal of two hundred and fifty million shad. If only 2 per cent. of these are caught, one will then get five million shad, which would amount to from ten million to fifteen million pounds of fish, worth at least 11 cents per pound, which is $273,000 to $409,500 gross receipts. This calculation will seem, however, so exaggerated that, perhaps, no one will accept the result. One can, therefore, take off, deduct freely, and reckon on getting merely 10 per cent. profit, and then it reaches $27,300 to $40,950 gross profits from the young shad hatched out for $1,092.

I have been informed that this shad (and the alewife also) might be particularly adapted to our Norwegian fiords. Lately I have come to the belief that an experiment in this direction will be worth the trouble. If one could, along the Norwegian coast, for an outlay of several thousand crowns, cause some shoals of shad to visit the coast annually, there would be a possibility that an attempt would be made with our own herring; perhaps it would multiply just as rapidly. With the American experiments in sight, I cannot, for my part, consider it a bold idea to contemplate the possibility that one, by skillful hatching, might be in condition to make up for the vanished spring-herring. But what will it cost? With the above calculation as a clew, the price would not appear to be great. But the point for the Norwegian herring concerned is this: that no one yet understands hatching it, scarcely how to use an apparatus; yes, indeed, no one knows how long it takes before it will mature or become salable. The prospects so far are indeed uncertain, and it is perhaps far too early to suggest the idea. It is also with a certain risk that I at present entrust it to paper and—publicity. However it is allowed to stand as my calculation.

With regard to the expense of hatching salmon and trout, they have in Europe varied experience in these branches. But the Americans maintain, as previously remarked, that they have carried the business much farther, and hatched them both more safely and cheaper—uniformly cheaper. It will, therefore, be interesting to see the prices of salmon-eggs. I shall first, however, premise the remark that, according to the statement, salmon of both sexes must be bought from the fishermen and then brought to the establishment to spawn. The expenses of this plan naturally become considerably greater than with the little shad, because salmon ready to spawn are costly. The table refers to an establishment in Penobscot, Me.:

Year.	Number of eggs per fish.	Price per 1,000 eggs.
1871	651	$16 25
1872	2,268	4 25
1873	3,560	2 73
1874	5,151	2 00

In the last-named year 3,039,000 salmon-eggs were gathered, assorted, and packed for the price of $2 per thousand eggs; the whole operation cost $6,000. A similar condition obtains with the trout; but, notwithstanding, this is so profitable that this hatching is carried on as an industry by fishermen and countrymen. And so great importance do they attach to the general distribution of salmon and trout in all waters, that the fish-commissioner of the State of New York has decided that the State establishment, which can now produce the young indefinitely, shall deliver to every owner of small streams or lakes as many young as he desires to plant in his waters, whereas hitherto only the great waters were supplied. They desire that the edible fish shall spread to all waters, even to the smallest streams and ponds.

In connection with this benevolence and generosity with which the fishery-inspectors are animated, to oblige all interested, it should be mentioned that they instruct the people in hatching and rearing young fishes. And the work enlarges so as to include more and more kinds of fishes, so as to succeed gradually by study and experiments in learning the peculiarities of the fishes concerned, and what there is to notice with regard to the temperature of the water, the nature of the bottom, articles of food, &c. For this study material is collected from the whole world; even from the interior of China information is seen to be brought on breeding; and it was, therefore, not surprising to see that the Norwegian spring-herring question also, and the dispute between Axel Bœck and Ossian Sars about their new herring-theory has found a place in the official reports. In a fishery-meeting in New York in 1877, to which I was invited, they wished even to have an account of the Norwegian legal provisions for the preserving of fish, on which occasion I, on a special summons from the directors, was obliged to come forward with a discourse. A fishery-inspector from Holland also, who, by chance, was present, was questioned on his country's fishery-relations.

Besides the proper fish hatching and rearing they have also applied themselves to cultivating and raising oysters, lobsters, frogs, eels, &c. The so-called oyster-culture, with which is next intended systematic preserving and capture, is a great business, without a parallel in any country.*

From what I have here communicated in extracts and brief summaries it will presumably be evident that the American example contains a stirring invitation also to Norway to develop her fresh-water fisheries, which are now greatly neglected.

X.

LIFE ON BOARD A FISHING-SCHOONER AT SEA.—MACKEREL-CATCHING WITH THE PURSE-SEINE.

After having waited some time for an opportunity to go to sea, to witness the business out there, I succeeded in getting a promise of a place on board the schooner William Baker, Captain Pearce. It was an old vessel, but a good sailer, and the captain was recommended to me as an experienced, enlightened, and generous man, who would take much interest in communicating to me all the information he could give. He had carried on the herring-fishing at Labrador, halibut-fishing off the west coast of Greenland, and was now determined to prosecute mackerel-fishing in the sea north of Boston.

Late on a rainy evening I was informed that the vessel was now ready to sail, in Gloucester Harbor, and that I could come on board. Neither the weather nor the vessel particularly invited one out in the dark, foggy night. But after being shown a tolerably good bunk astern, where besides myself four of the crew had quarters, I soon found myself adjusted and anxious to get under sail. Early in the morning we cast loose and the vessel hauled out into the channel. But the wind was still; we could make no headway. While we waited for the wind a portion of the crew passed away the time by taking a bath and swimming out in the deep. Their invitation to me to swim a race with them I was in the notion of accepting, when the signal was given to make sail and get under way. All came on board, took off their swimming-clothes, put on dry clothes, and caught hold at the anchor-breaking and later at the hauling-out so that it was a pleasure to see them. The brutal execution of discipline, so often censured on American merchant-ships, did not exist on board here. The whole crew were native Americans, active and experienced fishermen. They associated with one another with good-will, eating at a table common to us all, and carried on their work with mutual satisfaction. Neither beer nor whisky is found on board; but warm

* In one of the last official reports to the United States Government is found printed a full description of the oyster-industry in the United States. Here, according to the older statements, the whole oyster-trade is estimated to amount to the sale of 4,000,000,000 oysters, worth about $69,250,000. To this may be added the profit of other shellfisheries, and of the oyster-shell, which is burned into lime.

coffee and tea can be had from five in the morning to six o'clock in the evening. In other respects the victuals were good and nourishing, consisting mostly of beef, pork, all kinds of fresh fish, different kinds of pie and pudding, sometimes vegetables, with eggs occasionally; in short, about as in a frugal municipal family in Norway. The men were not hired, but had half the weight or a half share of the profits, which were calculated after the expenses of salting, packing, barrels, &c., were deducted. The cook, who besides the usual work in catching had to prepare the meals, got besides for his part $27.30 per month. The trip just ended had been made in three weeks and had yielded a profit of a little over $81.90 net per man, which is considered a very good trip. The crew in herring-fishing, on the contrary, is generally hired. They ship at $10.92 to $13.65 per month for young boys; $32.76 to $40.95 for able-bodied fishermen.

After being under sail a couple of days we saw a whole fleet of mackerel-schooners. There were between 80 and 90 sail, some of which had made a light catch. We passed some steamers which carried on menhaden-fishing, and which went into port every evening to unload. These vessels were quite recognizable by this, that they had two men on the lookout in the cross-trees, on the foremast, and the rest in boats prepared for fishing. Among other sail we passed also an old-fashioned French-rigged vessel, which carried on mackerel-fishing with trolling-lines; the vessel was belayed and went before the wind. A crew of ten men stood to leeward, and each man with at least two hand-lines, which he incessantly cast out and hauled in, while two men ground bait (in a hand-mill) and threw out "chum." The catch was poor, and the mode of catching, itself, most fishermen had given up for capture with purse-seines, which have superseded all other implements.

While crossing back and forth we often saw mackerel-schools playing in the water, but they vanished suddenly. The folks told me that one could smell mackerel as well as menhaden when the large schools of fish were in the water. I entertained doubt of the truth of this information, but in the following week I became perfectly convinced of its authenticity. Early one morning one of the crew said that he smelt menhaden, and went immediately aloft on the foremast to discover the school. A half hour later we saw a school playing on the surface of the water; it was large mackerel. In haste everything was made ready; the purse-seine, which lay on the after hatch coiled up in a bundle and wet with brine (to prevent rotting), was quickly flung down in the seine-boat, which was kept constantly in tow; next two small boats set out, so-called "dories", flat-bottomed, light-rowing boats, half "*sjægte*" and half lighter. These are said to be the best fishing-boats known. When all was ready I leaped down into the boat, and away it went. After a half-hour's rowing the seine-boss found that the time had come to row in on a large school, which played quite delightfully. In three minutes the seine, of about 250 fathoms, was rowed out and cast in a circle around the school. After ten minutes more the seine was pursed, and in it they estimated that they had caught 300 barrels of mackerel. A dory was sent with a message on board the schooner, which was managed by the captain and a small boy; after a little manœuvering the schooner sailed close to the seine, got a portion of the cork-line on board, so that the catch was brought between the vessel's side and the seine-boat. A large dip-net, with tackle and a long handle, was made ready, and in a few minutes the living mackerel were thrown upon the deck by the half-barrel. The captured fish in the seine became, in the mean time, very uneasy, and rushed from one side of the seine to the other; suddenly the seine burst in many places; they sought to haul in the seine, both in the boats and on board, and after much trouble they succeeded in retaining and heaving on board a total of about 50 barrels. The rest of the mackerel escaped through the large holes in the seine.

As soon as the catch was secured on board they took in all sail and began on the preservation of the mackerel. With small dip-nets the fish were thrown in a square trough, and from this, as they were cut and washed, they were assorted according to size. The fish was split or cut in the belly (?) about as cod which are manufactured into split fish. While sprinkling them with salt they give them some slight cuts in the back to make the flesh swell and give the fish a fatter and fuller appearance. This is a trick which both the sellers and buyers understand. Twelve men are engaged at a time with the salting, while the remaining two men examine and repair the seine. In the space of two and a half hours 47 barrels stood ready salted. For every four barrels of fish was required one barrel of Liverpool salt, the packing being done later on shore.

The next morning at 6 o'clock we were again in the boats, made a new cast, but caught nothing. At 8 o'clock we were again out, went half way round with the seine, when the whole school sank to the bottom. We caught nothing. The mackerel were wild and shy, played a little at the surface of the water, but vanished quickly, to emerge again farther away. The mackerel-schools were very large this day; for as far as we could see on all sides they were playing on the surface of the water. We made no catch notwithstanding. The captain's attempt to entice the schools with "chum" also failed. At 11 o'clock we again made a cast, but took only eight mackerel. At 2 o'clock p. m. we made the fourth cast and got a couple of barrels; at 4 o'clock, another cast, and took 10 barrels, but small mackerel; and at 6 o'clock we made another, but got almost nothing save small fish. This was a laborious day; but, notwithstanding, we were the next day, at 5.30 in the morning, again in the boat, made a cast, but instantaneously the school turned around and made their escape. We saw many schools also this day, and at 6.30 we were again on the way after one. This time we were successful. They estimated the catch in the seine at 400 to 500 barrels. They were very large and fat mackerel. After some trouble, the vessel sailed to us, got a portion of the seine on board, and the taking in the fish was about to begin. But the fish were very uneasy in the seine; sunk to the bottom with such force that the boat was on the point of capsizing, although we placed eight men on the other gunwale to counterbalance the mackerel. At one time all went smoothly enough to haul in on the seine and make the purse smaller and smaller to prevent the frantic rushing of the mackerel. But suddenly they sank again to the bottom, careened the boat over so that we took in a quantity of water. We were scarcely ready to place ourselves on the other gunwale when we felt that the boat suddenly righted itself and lay still. The most knew what had happened; it was that the mackerel succeeded in breaking the old seine. Through a large hole, which became larger and larger, about the whole school escaped; and although we in all haste hauled in on the fragments and tried to form a new purse, we succeeded in saving not more than five in the whole 500 barrels.

At 9.15 we set sail for the nearest port; it was considered useless to attempt to mend the more than half worn-out seine. After a day's quick sailing, we reached Boothbay late in the evening. In the harbor lay a schooner just arrived, which was filled to the rail with fresh-caught mackerel. The crew worked the whole night in preserving them. With resignation our crew saw this work. Had we had a better seine, we would also have had remunerative night-work in salting some hundred barrels of mackerel. The next morning I left the vessel, to return to Gloucester by steamer and railroad.

In this way 600 to 700 schooner-rigged vessels carry on mackerel-catching out in the sea, and almost exclusively with purse-seines. They are of from 120 to 150 tons burden, and 10 to 14 men for crew. I have seen the statement that one, as a rule, can calculate that each schooner during the summer catches 1,000 barrels, at $10.92, which would give $6,552,000 or $7,644,000 as the aggregate profit. The catch, however, is quite variable; some successful vessels have caught many thousand barrels in a season. The fishing begins in April or May, far south on the coast. Then the mackerel are fat. The fishing-fleet follows them northward week after week, and in July or August they have advanced as far north as Nova Scotia. Thereupon they bear southward again. The mackerel have now become very fat and large; the catch is then at its best. In September the schools are in full retreat to the warmer waters again; and in October and November the fishing closes.

Salted mackerel are submitted to public inspection. They are assorted into three numbers, the prices of which in August, 1876, are subjoined here:

Length of fish.	Number per barrel.	Price per barrel.
No. 1, 13 English inches	About 140	$16 33
No. 2, 11 English inches	About 240	7 64
No. 3, under 11 English inches	About 350 and over	6 00

For No. 1's with the heads cut off is obtained $5.56 more than for common 1's. For packing, barrels, and inspection, is estimated on the average $1.91 per barrel.

The same schooners which prosecute mackerel-catching go partly also into the herring-fishing; the best fat-herring fishing begins in the middle of August and continues at Newfoundland and Labrador until about mid-winter. In the winter months are caught also large fat herring. Alternating with this fishing the same vessels carry on in part cod-fishing on the Great Banks, and halibut-fishing there and off Greenland. These fisheries I had not the opportunity to make myself more familiar with, as it would have taken a long time to follow the vessels out on the banks and see them. But with regard to the profit in general I can state that it is about as in mackerel-fishing, with the difference, however, that the herring-fishing yields something less than $5,560 to $8,190 to a schooner for the season, while bank-fishing for cod and halibut yields something more, namely, as much as $10,920 to $13,650. However, the halibut-fishing fluctuates greatly, it is said; since it may sometimes yield a far greater profit, of which one has an illustration in this, that a vessel has brought its owners about $27,300 profit in a year, and that skippers have earned from $1,638 to $5,560 for their share.

A description of the universal implement for the capture of mackerel, menhaden, and herring, namely, the purse-seine, I have already sent to the honorable department in a printed letter. Besides I have treated of the purse-seine in a couple of articles in the "Bergen Post" last summer, in which I gave an account of a trial trip with a Norwegian purse-seine of hemp thread. This trial trip, at which I was present by request, after four days' sojourn at home on my return from Philadelphia, was made from Stavanger and required 14 days' time in the beginning of June. The result arrived at was briefly this, that even a large, heavy purse-seine of hemp thread may at a pinch be used for the capture of herring, but hardly of mackerel. However, we saw only herring, but not a single mackerel, which could hardly be expected either, since the weather was cool and partly stormy near the coast as well as twenty miles out in the North Sea. That a large purse-seine of about 200 fathoms is not suitable arises from this, that it is cumbersome to handle; the thread alone in a hemp seine is about 50 per cent. heavier than in a cotton seine, and the heavier the seine is, the more cork and lead must it have; from this it follows again, that heavier twine must be used; one gets also a far heavier implement, and for its management is required a larger crew, which again involves a larger boat—in short, step by step one departs from the chief qualifications for the purse-seine's cardinal virtue, *facility of management* united with strength, by which its whole cost as well as the expenses of working it are not so inconsiderably greater at the same time that the profit in general must be diminished. The fact that the purse-seine in question later in the summer caught mackerel partly shows that it should not be entirely rejected.

Another Norwegian purse-seine of hemp thread, procured in Bergen, was used north of Doore in the summer-herring fishing; it was, though more nearly perfect, yet larger and more massive; but it caught a few herring. Some errors and inaccuracies I had occasion to point out and partly remedy in both of these. In the southern spring-herring district in the winter a partial attempt was made with more or less unpractical imitations of purse-seines, as the common herring-seines were fastened together; concerning these attempts I think it should be said that they surely injure rather than benefit the matter; because they must as a rule bring disappointments and indirectly weaken the desire to make the attempt with a proper and easily managed implement. In Sweden and Germany also they wish now to experiment with American purse-seines of cotton; thither were sent, after the receipt of orders, many small models with descriptions from Norway.

The purse-seine is, however, fully discussed in Consul Joakim Andersen's interesting communication on his operations as a juryman at the Exhibition. I shall therefore not occupy myself further with it, but take leave of it by closing with a little note, which has its special interest: The American purse-seine is arranged on exactly the same principle as that which forms the basis of an implement of capture for herring invented by Berent Chr. Vedeler, of Bergen, now deceased, on which he, by a supreme resolution of the 12th of March, 1859, received a patent for five years. A drawing and model of Vedeler's purse-seine are found in the Polytechnic Journal for 1864 (the time registered by engineer N. H. Brun, *m. fl.*), pages 123 and 124. From this it is seen that it differs very slightly from the American; the difference is essentially this, that Vedeler allowed the pursing-rope to run in rings along three sides of the seine, while the Americans more practically let the pursing-rope run only along the bottom line. Moreover Vedeler decided that his seine ought to be only 40 fathoms long and 10 fathoms deep, also only a sixth part of the usual size among the Americans for large seines. Since Vedeler said nothing about the thread, I assume that he has used hemp thread, which, as before mentioned, is not used in America, where cotton thread is considered far more suitable and the only proper thing for purse-seines. It is not so unlikely that Vedeler's patent is the first entirely original invention, and that his invention, like so many remarkable ones, has found its way to America, and there received the merited appreciation and such a practical adaptation that it has become the most important implement for a very considerable business. In any case it seems to me that Mr. Vedeler's invention deserves mention and his talent to be commemorated at this time, since his original idea returns to its native land in improved form now to find in all probability full appreciation.*

* The purse-seine was in use in America in its present form at least as early as 1855.—Translator.

XI.

CONCLUDING REMARKS.

As one will have seen, I have treated only of what is especially peculiar in the American fishery business. I have only occasionally and quite hastily touched upon what there is in common to the Norwegian and American relations. The task which I set myself was to find out in what direction the development of America's fisheries went, as I believed that therein would be found the cause of the growing superiority of the Americans as fishermen and fish-dealers. If I have succeeded in showing this, and that the track in which development proceeds there must lead to great profits above the present average, then I may flatter myself that I have given those most interested in Norway's fisheries some useful hints. That I have not been able to exhaust everything, is evident; I have not dared to extend my treatise over the entire field, but was obliged to confine myself to several points so as to be able to give something collected and complete. Since one for so many years frequently has heard complaints, for instance, about the management of the herring, the herring-barrel's capacity, the handling and drying of split cod, &c., I consider it useless to repeat or support these current complaints in the survey of the state of things in America; besides, there is doubt whether the Norwegians, just in this respect, had so much to learn from the Americans.

What Norway needs first and foremost is an enlarged market for its products. So far as the herring are concerned, this may in part be attained by using better barrels, which will endure longer transportation without allowing the herring to become dry of pickle. Beyond this there is indeed nothing new for the management of herring. But for the fishery-products in general more is required; they must have more varied preparation and not be exclusively salted, pickled, or dried.

These modes of preparation will no doubt be demanded by the most important markets for Norwegian wares; but it is certain that both fresh and oil-prepared articles may also find profitable markets. And indeed the more variety one employs in the treatment of the raw material the less will one be liable to suffer from overfishing and overproduction. When we employ new modes of preservation we will find new markets, and when we secure more markets we will make more profitable sales.

In discussing the American traffic with fresh fish in ice, their storing of fish in a freezing-apparatus, their import and export, preserving articles and treating them with oil, I have sought to point out how the fish-traffic is developing in America, and thereby also indicate in what direction, in my opinion, we also ought to go to work in Norway. In treating of the American hatching-operations and what is therewith associated, I have wished to point out what means they have at hand for increasing the country's profits from the lesser fisheries, yes, perhaps, partly even for creating new and quite important fisheries. In finally discussing the arrangement and use of the purse-seine, I have sought to direct attention to an implement of capture which we stood greatly in need of, namely, an instrument with which one can fish in the open sea, and among other fishes catch also the herring, which will not resort to the shore. It seems to me that much might be accomplished with all the implements here named. The fisheries in our country, it is true, are associated with so many ancient traditions and continue in many parts in so little developed proportions, that it could not be expected that new ideas should be accepted at once, to say nothing of a complaisant reception; but perhaps on that very account one should labor the more diligently to extend in this field an acquaintance with improvements and new methods of work.

VII.—SHORT INTRODUCTION TO THE PROPER CARE AND MANAGEMENT OF THE BALTIC FISHERY.

By H. Widegren, *Stockholm*, 1874.*

THE FAUNA OF THE BALTIC.

The fish living in the Baltic are either such as live and propagate exclusively in salt water, *e. g.*, the herring, the small herring, the codfish, the flounder, and others, or such as must properly be considered as fresh-water fish, but which can live and propagate in the but slightly briny water of the large and small bays and inlets of the Baltic. The fresh-water fish which are also found on the coast of the Baltic and form the principal objects of the fisheries, are the perch, the pike, the roach, the bream and other carp-like fish, as well as the burbot. Besides these, there are found in the Baltic several kinds of fish, which, like the salmon, the grayling, the gwiniad, and the eel, must chiefly be considered as migratory fish, staying sometimes in salt water, and at other times, principally during the spawning season, in fresh water.

In consequence of the character of the fish-fauna of the Baltic, the fisheries carried on in its waters are not only sea-fisheries in the proper sense of the word, but also such fisheries as are carried on in our Swedish lakes, with which many of the inlets of the Baltic—especially those which only through narrow and shallow sounds are connected with the outer coast—show the greatest similarity in regard to those conditions which have an influence on the fisheries.

As we are going to give a short introduction to the care and management of the Baltic fishery, we will first consider the fishery carried on in the inner bays and inlets of the Baltic, and which, to distinguish it from the herring and other sea fisheries, is by the fishermen called "rock-fishery" or "coast-fishery." In this treatise we shall chiefly make use of the latter name as more expressive of the idea, as this name mentions the locality where this fishery is chiefly carried on, viz, the coasts of bays and sounds.

---

* Kort Vägledning för Östersjö-Fiskets rätta vård och bedrifvande af H. Widegren. Stockholm, 1874. Translated by Herman Jacobson.

I.

THE COAST-FISHERY ON THE COAST AND IN THE BAYS OF THE BALTIC.

In order to carry on the coast-fishery in the proper manner, the first thing required is an exact knowledge of the mode of life of those fish which are to be caught; and, secondly, an intimate acquaintance with all the fishing-implements employed at different seasons of the year. To do full justice to the subject would require more space than can be allowed in a short introduction; we shall, therefore, only give the most important points which should be known and observed by a successful fisherman.

Kinds of fish caught by the coast-fishermen.—The principal fish found on the Swedish coast of the Baltic and in its bays and sounds are, the perch, carp, crucian, tench, roach, chub, *Cyprinus vimba*, bream, pike, salmon, trout, gwiniad, *Salmo albula*, burbot, and eel. As useful either for bait or as food for larger fish we must mention the ruff, the stickle-back, the minnow, the bleak, and the smelt.

Mode of life of the above-mentioned fish.—All the above-mentioned fish have this in common, that at certain seasons of the year they visit certain places on the coast and the coast-waters, and that every year during the spawning season they go to such places as seem specially suited for propagating. Although all of them may in a certain respect be more or less called migratory fish, it has been observed that each one of them in those waters in which it lives confines its migrations to certain limits, that, for instance, the perch and the bream, &c., each only visit their certain bay (and this a bay near to the deep water), and that they scarcely ever extend their migrations beyond this, the limits of their wandering and spawning, unless the spawning-places are disturbed or other natural causes lead to a change. This characteristic trait of the fish, to confine its migration to certain limits, and each kind and school to select that spawning-place where it was born, is especially striking with the migratory fish properly so called, the salmon, carp, gwiniad, and others, which generally go up certain rivers and streams for the purpose of spawning. By marking young salmon which were on the point of leaving the river where they were born, it has been proved that these fish, which of all our fish wander away farthest from their regular place of sojourn, nevertheless return to it regularly. It has moreover been proved that if a whole school of fish, for example, gwiniad, &c., having its spawning-place in a small stream, is caught, no fish of this kind will ever return to this stream, although the nature of the water has remained the same, and although they will continue to go in streams close by, where the fishing has not been of so destructive a character. On the other hand it has been found that fish have left such streams which, by draining, cultivation, or other agencies, had their natural character changed so as no longer to offer a suitable place of sojourn. Every experienced fisherman knows, moreover, to what a degree the placing of fagots and the preparing of artificial spawning-places attracts fish and induces them to spawn in a certain place.

Rules for carrying on the fisheries, made in accordance with the character of the different kinds of fish above mentioned.—The experience which has been gained regarding the migrations of fish, their extent, and the conditions under which they are undertaken, is of the greatest importance to the practical fisherman.

In the first place it must not be expected that the fisheries will be equally productive every year, unless, especially during the spawning season, fishing is carried on in such a manner as always to leave a certain quantity of fish in the water, so that the propagating process may go on undisturbedly. It is wrong, therefore, as is sometimes done with us, to use large seines and catch the entire school of fish coming to a certain bay, in the hope that other schools from other parts of the Baltic will soon replace it. Such a change from their regular route is entirely at variance with the nature and habits of fish. From what has been said above, it will be seen that in order to count on continued good fisheries, the nature of the water should be kept unchanged as much as possible, and in fact it should in every way be made still more suitable for the various kinds of fish. Care should therefore be taken not to disturb vegetation in those places where fish spawn in spring, and as regards the fish of the salmon family, which spawn in streams during autumn, it will be necessary to keep the gravelly bottom, which these fish like, free from mud, shavings, &c. As for keeping the natural conditions undisturbed, it must be mentioned that by excessive fishing—which unfortunately is too often practiced with us—certain smaller kinds of fish, *e. g.*, the bleak and the smelt, are not entirely destroyed, but that larger and finer kinds of fish are thereby deprived of their natural food, and are thus forced to eat their own fry, which of course seriously endangers the future of the fisheries. If a man wishes to improve his fishery, and does not to a certain degree spare the small fish which are of no use for the table, he would make the same mistake as he who stocks water with fine fish without supplying them with the necessary food. It is moreover well known that in spring the fish generally go on grassy bottoms and in small brooks and streams for the purpose of spawning; that after this they go in deeper water, and later in summer stay at a certain depth; that in autumn they again seek sandy or grassy bays, and finally in winter either gather in certain deep basins of the sea or near currents. From this knowledge it follows, that in order to make the fisheries successful, one should attentively follow the migrations of the different kinds of fish all the year round, observe the exact time of their wandering from one place to the other, and finally examine the nature of the bottom and the depth of the sea in different parts of the fishing-waters, because a person not acquainted with all these conditions cannot know with absolute certainty in which places the fish may be found at different seasons of the year.

A farmer or mechanic who only occasionally engages in fishing, therefore, runs the same risk as a bird which, born in a cage, suddenly gains its liberty. It starves to death on account of its lack of the faculty of observing and its ignorance of those places where food may be found. To this must be added the circumstance that a good farmer, accustomed to handle the plow and spade, does but rarely possess the necessary skill in using lines and hooks or nets, and will, consequently, not be a very successful fisherman. Fishing should, therefore, only be carried on by persons who have been able to gain some practice in it, and who can devote their whole attention to it; and those who have such small fishing-waters that it would not pay to keep a special person to take care of them would, therefore, do best to club together with some of their neighbors and let out their fishing-waters to skilled fishermen.

After having thus given the most important rules which should be observed with regard to the mode of life of fish, and after having likewise pointed out the way in which our fisheries could best be furthered and protected, we will briefly mention the way in which a fisherman should go to work, the methods of fishing, and the fishing-implements which can and should be used at different seasons of the year.

Various ways in which the fisheries may be improved.—Rarely, or perhaps never, do we find a sheet of water which is so favorable to the propagation of the different kinds of fish living in it that its condition could not in any way be improved, that is, made more convenient and suitable for the spawning of the fish. Just as the farmer must be very careful to water, to plow, dig, and fertilize his ground, because, being left to itself, it will be overrun with weeds and will not yield the produce which, with some care, might be expected from it, thus the proprietor of fishing-waters must take care of these waters and aid nature by artificial propagation, and make the water a suitable dwelling-place for the young fish, and protect these as much as possible against their enemies. The propagation of fish fortunately goes on under such conditions as to enable man to extend considerable aid to nature; and to give this aid should be the first duty of every proprietor of fishing-waters who has his true interest at heart.

Of our common fish, the perch, the pike, bream, roach, and other carp-like fish spawn in spring or early summer, whilst the salmon, gwiniad, char, and burbot spawn in autumn and winter. Most of the fish which spawn in spring lay their roe on pieces of wood, aquatic plants, algæ, grass, reeds, &c., to which the roe remains sticking until the young have slipped out. This is the case with the perch, for example, which lays its eggs in bag-like heaps on pieces of wood or on reeds, as also with the roach, whose roe is in separate grains, fastened to pieces of wood, stones, or aquatic plants found near the shores of lakes. The pike, the bream, and the tench and other carp-like fish lay their eggs on grassy bottoms or among aquatic plants. These fish love to lay their eggs in places where there is a current, as in small streams, the mouths of brooks, &c. The fish of the salmon kind, on the other hand, lay their eggs free, not fastening them to any object, and select for this purpose gravelly and stony places in brooks and rivers, on whose free bottom they lay their eggs. In order to protect the eggs which have been thus laid the fish of the salmon kind beat the bottom with their tails, in order thus to cover the roe with gravel and sand.

Every one who intends to further the propagation of fish and thus to improve the fisheries must, in the first place, ascertain how those fish of which an increase is desired spawn in nature, and then to arrange his course of action in accordance with the knowledge gained. As regards the fish of the salmon kind, whose roe generally takes a longer time for developing, experience has shown that their number can best be increased by protecting the fish during the period it stays in the streams and is occupied in spawning, and also by introducing artificial propagation, that is, impregnating the roe in an artificial manner and keeping it in special establishments until the young fish are large enough to take care of themselves. As there can be no question of establishing hatching-places for fish of the salmon kind on the Baltic, as these would have to be made in brooks and streams, this is not the place for describing the arrangement of such establishments; but we shall here mention the various means by which fish-waters may be improved.

If it is the intention to increase the number of those fish which spawn in spring and whose roe is fastened to branches and other objects in the water, the owner of the water must, first and foremost, see to it that such objects are found in the water. This is all the more important as through the destruction of the forests, the draining of the marshes, and the gradual rising of the Scandinavian peninsula, the natural spawning-places, at least in certain localities, are diminished or deteriorated. By placing in the spawning-places, a short time before the spawning-season commences, fir branches two to three yards long, or fagots, or by laying pieces of sod on the bottom of smaller sheets of water, or by planting aquatic plants which it is known that the bream and other kinds of fish prefer to lay their roe upon, natural spawning-places may be much improved. If such care is to be further extended to the young fish special ponds may be dug and their sides clothed with fagots or suitable aquatic plants. A number of fish which are about to spawn are then placed in these ponds, where they lay their eggs on the fagots or plants. The branches, full of roe, are taken out every day and placed in a smaller pond 2 feet deep under the water, which by a hedge of fagots is separated from the outer sea, so that large fish and crabs may be prevented from entering the pond and destroying the roe.

The bream, pike, and ide, which last-mentioned kind loves to lay its eggs in flowing water on a grassy bottom, may be inclosed in separate smaller basins at the mouth of brooks and streams where the above-mentioned natural conditions are found or can be artificially procured. After the fish have laid their eggs they are caught and taken out, whilst the roe is, of course, left on the grassy bottom until it is hatched. The young fish are then allowed to go free if such is their desire. By these and other similar means the number of fish in a bay may be considerably increased. It is evident, however, that these means will help but little unless measures are taken to prevent the young fish from being caught in nets with small meshes or with other fishing-implements. Care must also be taken that the schools of fish in one and the same sheet of water are not, by excessive fishing, diminished to such a degree as no longer to be able to propagate their species at the rate necessary for keeping up their numbers. He who cuts down the tender blade will never reap any grain, and he who only sows one-tenth of the seed which his field ought annually to yield, will never have a full harvest. It is evident, therefore, that the owner of fishing-waters must not only employ the above-mentioned means for increasing his number of fish, but must also see to it that the spawning process is not disturbed and that the tender young fish are properly protected. With the view of obtaining this end and in view of the fact that fish will wander from one sheet of water to the other—thus making it possible that one owner of fishing-waters may disturb the fisheries of another—the common interests of the proprietors of fishing-waters imperatively demand that all carry on their fisheries in a manner suited to the nature of the fish and the peculiar condition of the water. He who desires to reap a full harvest from his fishing-waters must, therefore, not only himself carefully observe all the rules necessary for preserving and protecting his fisheries, but he must likewise see to it that his neighbors do the same. Wherever such rules have not yet been adopted it will be in the interest of the owners of fishing-waters to introduce them as soon as possible, as only after this has been done will there be any reasonable hope that the measures for improving the fisheries will be successful.

With regard to the nature of the fauna of our Swedish coast, there are chiefly two rules which ought to be observed in fishing, and these are, not to fish with nets during the spawning-season, and not to use nets whose meshes are shorter than one decimal inch, except in cases where bait is to be caught. Just as important as it is in spring to prepare suitable spawning-places for the fish, it will be to see to it that the above-mentioned rules are not transgressed during the year. Where all the possible means have been employed for aiding the propagation of fish, and where only suitable fishing-implements are used, the owner of fishing-waters, like the farmer, must not miss the harvest-time. Human ingenuity has, fortunately, in course of time invented so many methods of catching fish, that he who is well versed in these methods may derive a benefit from his fishing-waters nearly all the year round without using methods of fishing by which the young fish are destroyed, or by which the future of the fisheries is undermined.

Seasons and implements for the Baltic coast fisheries.—As the pike begins to spawn earliest in spring, and as it is a voracious fish-of-prey which should not be spared too much, it should form the first object of fishing. For this purpose it is necessary, before the ice is completely gone, to close the entrances to the larger inlets by brushwood. As the pike in most parts of the country begins to spawn much earlier than other fish spawning in spring, nets may be used, at any rate in the beginning of this fishery. Towards the end of April or the beginning of May, when the perch, the roach, the flounder, the bream, and other fish commence to spawn, net-fishing must, of course, be stopped, and stationary nets should be used, placed at right angles with the shore, and so as not to close the entrance to the smaller inlets where these fish usually spawn. Whilst the ice lasts, traps should be set for catching bream. Bundles of brushwood are also laid in May with traps for catching roach. Those roach which are caught during the spawning-season should be kept in marshy waters in a convenient place, so that during summer they may be used for bait. In the same manner smelt and bleak are also caught with large nets, and are used for food or for bait. It must be mentioned that fish caught during the spawning-season will live much longer in marshy waters than those caught when the spawning-season is over. A wise fisherman, therefore, will supply himself with as much bait as possible during the spawning-season. Besides nets and traps, wicker-baskets are used during the spring spawning-season for catching perch, bream, pike, &c., and are placed as deep as possible, as also the so-called Hertzman's nets, which are used at some fishing-stations, and with which generally a good many fish are caught.

After the spawning-season has closed, fish may be caught during June and July, either with fishing-lines in deep water or with nets in the fishing-waters and other places suitable for this implement. Different kinds of fish, of course, require different kinds of bait, live fish, fry, or worms, according to the kind of fish you wish to catch, whether pike, perch, bream, or other fish. At midsummer-time, fishing with hooks and lines properly commences. In July, immediately after the bream has done spawning, this fish is caught with smaller nets, which have a purse with large meshes. These nets, which are chiefly used in the province of Skåne (Southern Sweden), are let down from two boats in deep water, and in favorable weather a good many fish are caught in them. These nets only cost from \$4 to \$5.50 (American money) apiece. In July and August fishing is carried on with seines, common nets, and hooked poles. Casting-nets are also during summer thrown out among the reeds, and are used for catching all kinds of fish, with or without poles. During the autumn months seines and nets should chiefly be used, especially in those places where the bleak and smelt spawn. But even during this season a good many fish may be caught in deep water with deep-water nets. During winter, traps are set in streams and the mouths of brooks and in the spawning-places of the burbot, for catching this kind of fish. In deep bays nets are set during winter, towards which the fish are driven by poles. Under the ice roach are caught in traps and are then used as bait for pike, which kind of fish is even caught under the ice with hooked poles or hooks and lines. By using the methods of fishing mentioned above, at the different seasons, a thrifty and energetic fisherman may derive a good income from his fishing-water all the year round.

II.

THE FISHERIES IN THE OPEN BALTIC.

a. THE HERRING-FISHERY.

The different kinds of herring which are found in the trade and on the coast of Sweden.—It is well known that in the sea which surrounds the Scandinavian peninsula there are found different kinds of herring, varying in size and fatness, which on certain portions of the coast are caught, and prepared in different ways reach the great markets under different names. Nearly all over Sweden the following kinds are found in the trade: Norwegian herring, gråben herring, lodd herring, fat herring, Gottenburg or Bohuslän herring, Kulla herring, Bleking herring, small herring, anchovies, skarp herring, spiced herring, &c. All these different kinds are prepared from only two kinds of herrings, viz, the herring proper (*Clupea harengus, L.*)—in the Baltic called "strömming"—and the sprat (*Clupea sprattus, L.*), of which the former both in nature and in trade occurs in far greater numbers than the latter, which is only caught and prepared to a comparatively small extent, mostly as anchovies. As the "strömming" is nothing else but a variety of the herring proper, as I intend to show later, the term "herring" used in this treatise is understood to mean both the herring of our western coast and the "strömming." The sprat can easily be distinguished from the herring proper by its smaller head and by the circumstance that its ventral fins are nearer the head than with the herring proper. The sprat, moreover, on its lower side ends in a sharp edge somewhat resembling a saw, which is not the case with the herring.

The herring, which on certain coasts forms a rich source of income, has its proper home in the North Sea and the Atlantic Ocean, but is also found in the seas connected with the above, the Kattegat and the Baltic. Like other fish the herring has also in course of time undergone certain changes regarding size, fatness, &c., according to the different seas or fiords where nature has placed it. These changes have chiefly been caused by a difference of food not only in the Atlantic Ocean, the Kattegat, and the Baltic, but even in different portions of the Western Sea and the Baltic. We therefore find that every portion of the sea and even certain bays have, so to say, their own race of herrings, which certainly are not a different species from those found on other neighboring coasts, but which, nevertheless, can easily be distinguished as a different variety owing to the surrounding nature. Thus, for example, a larger kind of herrings is at certain seasons of the year found in some bays of the Baltic and can easily be distinguished from those herrings which live and spawn on the outer coast; and the herring found on the coast of Bohuslän and in the Christiania fiord differ in size, &c., from the herring found on the western coast of Norway. These differences have not only given rise to different ways of preparing the herring and to different names under which the herring comes in the market, but from them certain conclusions may be drawn regarding the mode of life of the herring, from which, again, important lessons may be derived regarding the protection and the improvement of the herring-fisheries. Even at this day there are many fishermen who entertain the opinion, which before science had spread more light was quite common, that the herring only accidentally came from remote portions of the ocean to those coasts where it is caught, and therefore these fishermen thought to do right by using these accidents and catching as many herrings as possible; in other words, to fish with the most destructive implements, even those by which a whole race of fish would be destroyed. But since experience has shown that Norwegian herring are never caught on the coast of Bohuslän, nor Kulla herring on the coast of Bleking, nor Gottland herring on the eastern coast, &c., and since the time and place have been discovered where the herring spawns; as well as the mode of life of the tender fry, its place of sojourn, &c., it has been ascertained that the herring—like the salmon and other fish—has certain limits to its migrations, certain places where it spawns, &c. If good herring-fisheries are to continue on certain coasts they must be carried on in such a manner as not to catch all the fish which come to a certain place either to spawn or to live. Care should also be taken to spare the young fry, because if this is not done the race of fish on the coast in question may be destroyed, since no new race can be expected to come here, and thus a large source of income will be lost, whilst if the young fish are spared good fishing may be expected every year.

In several places on the Baltic and the Western Sea carelessness with regard to the preservation of the race of herrings and the protection of the young fish has been severely punished. The investigations which have been made for several years, have shown conclusively that careless and destructive fishing has contributed not a little to the cessation of the great Bohuslän herring-fisheries, which unfortunately have not yet recovered, chiefly because as soon as a school of young herrings shows itself on that coast it is immediately caught with nets that have small meshes. Near Bresund, in Norway, the herrings used to come to the coast for many years, but ceased to come when people began to use nets with small meshes. To give instances from nearer home we will mention that not so long ago herrings came to the coast near Bråviken and to the mouth of the Motala River, as well as near Lósingsskär and Botilshäst, where large quantities were often caught. But people commenced to use nets during the spawning-season, by which this entire race of herrings was caught; and since that time fishing has entirely ceased in those places. Similar instances might be given from many other places on the Baltic. With these experiences fresh in our remembrance it will be evident to every one how important it is to carry on the fisheries in accordance with certain well-defined rules based on a thorough knowledge of the nature and mode of life of the fish, if the future of the fisheries is not to be seriously endangered.

To enable the fisherman to judge for himself what is best for the improvement of the herring-fisheries in every case, besides those rules which may possibly have been laid down by a law of the state, it will be necessary to give some further information regarding the nature and mode of life of the herring.

Natural history of the herring.—The herring is a gregarious fish and is generally found in large schools, especially at the time when it approaches the coast, which it does regularly at certain seasons of the year, partly to spawn and partly to seek food in calmer waters both before and after the spawning-season.

During winter the herring lives in the deep water outside those coasts on which it has its spawning-places; but even during this time it visits the deep fiords, and therefore moves about in the same way as during summer, which is shown by the fact that in the Baltic herrings may be caught during the winter with nets placed under the ice at different depths (from 5 to 24 fathoms), and even with drag-nets in bays and inlets. During its migrations to and from the coast as well as during its stay in the deep waters of the open sea, the herring is sometimes near the surface and at other times near the bottom; and these changes of place are thought to depend partly on the temperature of the water and partly on the currents and other natural causes, concerning which, however, experience has not yet taught us such certain lessons as to draw from them reliable conclusions regarding the depth at which the herring is found at different seasons of the year. Fishermen had, therefore, best make experiments by setting nets at different depths.

The spawning-season varies among the herrings found in one and the same sea, and even the different schools or tribes have different spawning-seasons, and even in one and the same school all fish do not spawn at the same time, probably owing to difference of age or to slower or more rapid growth, &c.

In the Baltic the herring spawns either in spring or in summer, and is accordingly called either spring herring or summer herring. In the Southern Baltic the herrings continue to spawn till the middle of October, whilst in the northern portions of this sea the spawning-season closes in August. The fish spawn either outside the coast on raised bottoms at a depth of 13 to 15 fathoms, or in the bays running inland, mostly in places where the bottom is overgrown with algæ. The spawning is done very quickly, as soon as the school has gathered in its spawning-place, the whole process probably occupying not more than five or six hours. The roe is laid on aquatic plants, stones, pebbles, &c. The development of the roe occupies a shorter or a longer time, according to the different temperature of the water.

In May, when the water is colder, it takes 14 to 18 days to hatch the roe, but in August and July, when the water in the spawning-places generally has a temperature of 14 to 15 degrees (C.) (57.2°–59° F.), it only requires six to eight days. The newly-hatched young fish, which are smaller and more transparent than most other young fish, and are, therefore, hard to distinguish, are a little over ¼ inch long, and have, for eight days after the hatching, a bag attached to their body, which hinders the young herring from being very brisk in its movements during the earliest part of its life. Only after the young fish has lost this so-called umbilical bag it begins to swim about, gather in schools, and seek food. It is difficult to ascertain with absolute certainty the growth and size of the young herring at certain periods of its life, especially as not all the young fish have the same ability to gather food, on which circumstance their growth of course depends.

Attempts have been made to raise young fish by placing them in small basins and feeding them regularly, but so far these attempts have proved unsuccessful, as the young fish did not live longer than five weeks, at which time they have reached a length of about ½ inch. During the whole first year of their life the young fish may be found in the spawning-places, both on the outer coast and in the inner bays. Young fish, measuring 1 inch in length, may be supposed to be about two months old; at the age of three months they measure about 1½ inch in length, all the fins are completely developed, and the color of the body resembles that of the grown herring, so that they may be easily recognized as the young of this fish, which formerly could not have been done. After examining young fish found in the spawning-places one has felt justified in concluding that the young herrings measuring about 3 inches in length, which in spring are found in the spawning-places, are those fish which have been hatched earliest during the preceding year, and are, therefore, about a year old. The young fish measuring 5 to 6 inches in length, which are often caught in nets, are therefore supposed to be only two years old. When a fish has reached this size the roe and milt begin to develop rapidly, and when it has reached a length of 8 inches it is capable of propagating, and may then be supposed to be about three years old.

The food of the young herring, as well as of the full-grown herring, consists chiefly of small crustaceans scarcely discernible with the naked eye, which are found in large quantities in the water both in shallow and deep places. By towing in the sea-water with a net made of fine gauze large numbers of these little animals may be caught. They are more or less plentiful at different times and under different conditions of weather, and at different depths. This may possibly explain to some extent the fact that the herrings are not always found at one and the same depth. In summer these small crustaceans are found nearer the surface, and the herrings at this time likewise go nearer the surface. Like other fish, the herring abstains from food for some time before and after spawning, and its stomach is then generally empty, but after spawning it begins to take food again, and gradually recovers the strength and fatness which it had lost during the spawning process. This explains why the herring is fat at one time of the year and lean at another.

About two months before spawning commences the herring may, as a general rule, be said to be fattest and best. This fatness it retains almost to the end of the spawning-season, when it begins to get lean, and when it is not fit to be caught. The herring, after having done spawning, usually goes into deeper water in order to seek food, and does not return until it has entirely recovered its strength. That the herring, like other fish, returns to the place where it was born as soon as it has become capable of propagating is proved by the fact mentioned above that certain schools or tribes of herring spawn at the same place every year. That the number of fish is one year larger in one place than in another is doubtless caused by changes in the weather, currents, &c. Similar causes may even produce an almost total failure of the herring-fisheries in some locality. Cold and unfavorable weather during the spawning-season doubtless often kills large numbers of the young fish of some school, which of course will affect the herring-fisheries for several years to come. These and other circumstances on which the herring-fisheries depend have so far been so little explained that not much can be said regarding them; but it is fully known and understood that man may destroy the herring-fisheries in some portion of the sea not only by using nets which will catch both old and young fish, but also by disturbing the spawning-places.

It has been mentioned before that certain tribes of herring, especially the larger ones, spawn near the coast on bottoms overgrown with algæ. If this bottom is made unfit for spawning by pulling out or otherwise destroying the algæ by dragging nets along the bottom or in any other way, the herrings are forced to seek other and more suitable places for spawning, and they consequently leave these waters which they used to visit regularly. Experience gathered in Bohuslän and other places has shown that the herring is very sensitive in this respect, and leaves its old spawning-place entirely if its nature is changed or disturbed. Every one, therefore, who wishes to protect his fisheries should be very careful not to change the nature of the spawning-places, either by disturbing the growth of the algæ or other aquatic plants or by throwing refuse or impure matters in the water.

Different methods of catching herrings.—From what has been said regarding the nature and mode of life of the herring, it will be seen that in order not to destroy the whole tribe by catching the young fish or by disturbing the spawning-places, it will be best not to use nets during the spawning-season of the herring, but only use them in autumn and winter, when the herring visits the deep waters of the inner bays. Fishing with nets having large meshes may, however, be carried on at every season of the year.

On the coasts of Skåne, Bleking, and Gottland the herrings are not caught with stationary nets, but with so-called floating nets, which method of fishing is in many respects very advantageous, for which reason I shall briefly describe it.

After the usual number of nets, 27 to 30, have been well arranged and placed in a boat furnished with all the necessary apparatus and provisions, three men enter the boat and go out to sea. The time for leaving the shore depends on the wind and on the distance from land at which the herring is just then supposed to be, because the nets should be cast during dusk. When the casting is to begin the sail is lowered, one man places himself at the prow, another in the middle, and the third at the stern of the boat. The one at the prow takes hold of the oars and rows with the wind, the one in the middle loosens the floats and the weights, and the one at the stern casts the net. In this manner the whole net gets in the water with the exception of one end, which is hanging over the edge of the boat. To the last loop of the net a weight is attached by a rope of a certain measured length, with its float, which is thrown out, and then the whole net is carefully laid while the boat is rowed forward. When the first net has been set, the second one is taken, the loops are joined by a strong knot, to which again a weight and float are fastened, and this is continued until all the nets have been set, in such a manner that the largest floats are in the center of the whole stretch of nets, because otherwise the net would sink in the middle if a very large number of fish should happen to be caught. Finally, when all the nets have been set, there is attached to the last hoop, besides the weight and the float, the so-called floating line, a rope 30 fathoms long, to which at about a fathom's distance from the net a stone of the size of a fist is attached, so the nearest net might not be raised too high, especially if the weight has gone down deep. If the depth is only one or two fathoms no stone is used. The floating line is then cast out and finally fastened to the fore part of the boat. Boat and nets are then allowed to drive with the wind and current, and once every hour the nearest net is examined, to see whether the herring "takes," as the Gottland fishermen say. If you happen to fall in with a large school and the current is not too strong the net must generally be hauled in after two to three hours, so as not to catch more fish than the boat can carry. The Gottland boats carry about 300 "hvlar," besides the net and other apparatus. But in order to derive the full benefit from the herring-fisheries, it is not only necessary to take the proper care of them, so there is always a sufficient quantity of fish, but a thorough knowledge of the different ways of preparing fish for the trade is likewise required.

As fishermen very often are not able to sell the fish they catch for a reasonable price whilst in a fresh condition, it is very important for them to know the best methods of preparing them, especially in these times, when the improved means of communication enable people to get even necessary articles of food from a distance. Every one should therefore endeavor to obtain and retain a good market for his fish by preparing them well and by constantly improving his goods.

The improved means of communication and intercourse between different parts of the world make it possible that the Baltic herrings may now be advantageously sold both at home and abroad, whilst formerly scarcely any were exported. The methods of preparing the herring have to be varied according to the different markets for which it is destined, as different countries have different tastes.

The preparation of the herring for the trade.—The methods of preparing the herring for the trade, at present in vogue, are the following:

1. Salting the herring (common Baltic salt herring) for home consumption or the German ports on the Baltic.

2. Preparing the herring after the Norwegian or Dutch method (so-called "*delikatess-sill*," *i. e.*, delicious or delicacy herring) for home consumption.

3. Spicing the herring (spiced-herring) for home consumption and for the foreign market.

The choice between these three methods will chiefly be determined by the fatness and general condition of the fish; but also by the greater or less ease with which markets for the differently-prepared fish are reached, and other similar circumstances, which may best be considered by the fisherman himself. The fat herrings which are sometimes caught during autumn and mid-summer on certain portions of the coast, are of course best suited for a finer article of goods—"delicacy-herring," or spiced-herring—whilst the common herring is best salted, taking care, however, that by salting a superior article of goods is obtained.

General rules for preparing fish.—The first and foremost rule is to bring the fish as soon as possible after it is caught in contact with the preserving element, viz, salt. Great care should be taken that the fish before being salted is not exposed too much to the sun, because it will in that case easily spoil or rot. During summer every boat should therefore be provided with a sufficient quantity of tarpaulin, so the fish may be kept well covered during the homeward voyage. It will also be found very useful to have on the boat a box with broken ice, in which the fish are laid as soon as caught, and are thus kept fresh until salt can be applied. Fish which have been brought to market fresh, and have for a time been exposed to the warmth, should never be salted, because such fish are frequently a little spoiled. Another very important rule which should invariably be observed is, that everything should be done in as neat and cleanly a manner as possible. Fish-refuse, or any other refuse, should therefore never be tolerated in the salting-houses, or in the vessels used for salting. Nor should old brine ever be used, as it contains slime, blood, &c., and does not salt the fish thoroughly, but is apt to give it an impure and disagreeable flavor. The quality of the salt is also of great importance. It is not only necessary to use loose, strong, and hard salt, which is the best for salting fish, but also to obtain the best quality of the kind of salt needed. Salt which has been damaged by sea-water, or which contains impurities, should of course never be used.

If one has fresh and good herrings just taken from the water, good salt, and clean and ample vessels, all the necessary conditions are fulfilled for preparing a first-class article, following one of the methods given below.

1. *Method of preparing common Baltic herrings for home consumption and for the German ports on the Baltic.*

Two mistakes are often made in salting herring as this process is at the present time carried on by the fishermen on most of our coasts, viz, salting it too much and pressing it too hard. It is highly important to prepare the fish in such a manner that it may for a long time be preserved in good condition. It is of course also important, both for the buyer and seller, that the barrels should be well packed. Both these objects may be obtained without having the fish salted too strongly, and without pressing it almost flat, so it loses all its natural fatness and tastes of nothing but salt. In many places the fish are pressed so hard that they form a lump, from which the brine flows off without penetrating, which makes the fish dry and rancid and by no means agreeable as an article of food. Even if such fish were to find a market in some places, this method of preparing it must be condemned. Although it is of course impossible to lay down rules for preparing fish which would hold good in every case, or satisfy every taste—especially as one buyer cares little for the flavor or fatness of the herring, but only for its weight, whilst another cares nothing at all for the latter—most buyers nowadays endeavor to obtain an article having a good pure flavor, and being at the same time carefully packed. To prepare such an article the following directions are given, which may of course be modified to suit the different tastes, &c. These directions have for several years been followed in the best salting-houses in Gottland and on the southern Baltic coast, and fish prepared in this way will never lack buyers.

In preparing the common herring St. Ybes, Lisbon, or other strong kinds of salt should be used; but Cagliari salt, as well as some looser kinds of English and French salt, may likewise be used, especially if the fish are intended for speedy consumption. The salt should be crushed so that the larger crystals also melt in the brine, and the salt comes in the greatest possible contact with the flesh of the fish.

As salt herring are generally shipped to distant places, and are thus during the voyage exposed to the pressure of other goods, or whilst being transported by railroad or wagons run the risk of being handled carelessly, they should always be packed in carefully made tight barrels, with good strong hoops, so they can stand a long journey without the brine running out. It should be remembered that herrings from a leaky barrel are not worth one-fourth as much as those in a good barrel. As soon as the herring has been taken from the net they should immediately be thrown in small vessels filled with pure and clear brine. There should never be so many herrings laid in a vessel that the lower ones are pressed too hard by the upper ones, but if the number of fish is very large a greater number of vessels should be used. After the herring have thus been immediately brought in contact with salt, they are taken out by degrees to be cleaned and gutted, care being taken that all the entrails are taken out, but not the roe and milt. The practical way of doing this is well known to every fisherman. As soon as the herring have been cleaned they are laid in another vessel also filled with pure brine. When the whole lot has been cleaned, or even while the cleaning is going on, the cleaned herrings are taken out of the brine and rinsed in fresh and clean sea-water, whereupon they are for awhile placed in small baskets or kegs with a perforated bottom, so the water may flow off. When this has been done the fish are placed in tight barrels, which are kept in readiness for the purpose, and sprinkled with dry salt. The sprinkling is done in the following manner: The fish are laid loose in a barrel with crushed salt, 3 kappar to the barrel; whenever a layer has been finished the fish and salt are stirred so they may mingle thoroughly. After 24 hours the fish are taken out and again placed in baskets or kegs with perforated bottoms, so the brine may flow off. After this has been done, which generally takes an hour, the fish are regularly packed and salted in tubs. The fish are placed in layers with their backs downward. Between every layer of fish there is a layer of crushed salt, at the rate of 5 kappar to every barrel. After the tub has been thus filled, a light weight is placed on the top, merely to keep the fish under the brine, and not press it too hard, which makes the fat and the juice of the fish run out into the brine, thus destroying the delicate flavor of the fish. After the tubs have been thus filled they are allowed to stand open for several days, and as the mass of fish gradually settles down, new layers are added to every tub. When after some days the fish do not settle any more, the tubs are closed. They ought then to be rolled gently and turned upside down every two weeks, so the brine may thoroughly penetrate all the fish. Whenever the herrings are to be shipped, the tubs are looked after once more; if they have settled any, they are filled up for the last time, and are then considered ready for the market. The brine which flows over from the tubs and that which is obtained after every salting, may be put in those vessels in which the fish are kept immediately after being caught and whilst they are being cleaned. It is important, however, to see to it that this brine is changed as soon as it has been used more than once or twice and becomes mixed with impurities. To use 9 kappar salt to the barrel, as is done on the coast of Oestergötland, is not advisable, because the fish is pressed too hard and gets too salty. After the fish has been dry salted, no more than 5 kappar salt to the barrel is needed, whereupon the fish should immediately be placed in tubs and not be pressed more than is necessary for filling the tubs properly. In Norrland they let the herring lie uncleaned in brine for 24 hours, and moreover in brine which has been used many a time before for the same purpose. It will easily be seen from what has been mentioned above that this custom should be abolished and that the herring should be cleaned as quick as possible.

On the coast of Karlskrona they dry-salt the fish with only 1 kappar salt to the barrel, and then salt it with 7 kappar to the barrel. This method cannot be recommended whenever the fresh fish should have a chance to soak in a sufficiently strong brine, whilst, if this is done, it does by no means require as large a quantity of salt.

Baltic herring prepared in the above-mentioned way finds a ready market not only at home but also in foreign ports on the Baltic. The price of herring varies very much in different years, and is dependent partly on the result of the fisheries in each year, but also on the price of Norwegian and other foreign herring. In some years when the herring-fisheries have been successful, both in Norway and Sweden, the Swedish fishermen can scarcely obtain a price which fully pays them for their trouble. It may, therefore, be advantageous to seek foreign markets, and prepare the fish for these. Salt herrings may, at certain times, find a ready market in the ports of Northern Germany, Stettin, Stralsund, and other places. The best time for this trade is from mid-summer till the beginning of September. Herrings which are intended for the German market ought to be prepared in the above-mentioned manner, but should be very carefully packed in strong tubs, not holding as much as the Swedish barrel (about 220 pints), but in tubs of the same size as those used in Bornholm and on the German coast, which only hold about 193 pints each. Such tubs, if they are well packed and the fish are in good condition, fetch from $3.50 to $5.50 each in the Stettin market, a price which many a year may prove very acceptable to the fishermen of Southern Sweden, especially if one takes into consideration the fact that these tubs are much smaller than the Swedish and therefore contain fewer fish.

2. *Method of preparing the so-called "delicacy-herring" for home consumption.*

It is well known that every year considerable quantities of Dutch herrings and Norwegian fat herrings are imported into Sweden, partly in large tubs, but mostly in small barrels or kegs, and that these fish are mostly consumed by the better classes. Experiments have shown that the large and fat Baltic herring, which is caught in several places, can easily be prepared in the same manner, and make a better and particularly fine domestic article, which comes very near to the foreign "delicacy-herring," and therefore finds a very ready market at good prices all over Sweden; all the more as the Swedish "delicacy-herring" can be furnished for a much lower price than the foreign.

The very name "delicacy-herring" shows that it is not intended for every-day use. It is therefore generally kept in smaller kegs than the common herring. It is evident that the "delicacy-herring" should not be salted as much as the common herring, as it thereby loses its delicious flavor. As it therefore must be salted with finer and looser salt, it follows that it cannot be kept as long as the common salt herring. In preparing "delicacy-herring," finer and looser kinds of salt should be employed, *e. g.*, Liverpool salt, Lüneburg salt, Cagliari salt, &c. The Lüneberg salt is said to be the best for this purpose.

Norwegian method of preparing delicacy-herring.—As soon as the herrings are caught they are put in pure brine, whilst the cleaning process is going on. Some only take out the stomach, but it will be best, as is done in preparing the common herring, to take out both the stomach and the entrails. As soon as the herrings have been cleaned they are immediately laid in small tubs or kegs, in regular layers with the back downward; salt is placed between every layer at the rate of 6 kappar to the barrel, and salt is also placed on the top. As the herrings during the first days settle in the tub, new layers are added. After about six days an opening is made with a stick between the herrings and the side of the tub, which is filled up with salt, whereupon the tub is closed. Before being shipped every tub is examined and if necessary filled up, as was done with the common herring. If sufficient brine should not form in the tub, a little hole is made in its side with a gimlet, and pure brine is poured in, whereupon the hole is closed. It is very advisable to turn and roll the full kegs as often as possible. Herring prepared in this manner has kept entirely good and fresh for six months.

Dutch method of preparing Baltic herring.—Fresh and fat Baltic herrings are, as soon as they come out of the water, placed in small kegs, and are for at least an hour stirred with fine Lüneburg salt. Then the fish may be cleaned in the usual manner, or also, without being cleaned, be placed in kegs with fine Lüneburg salt between every layer. After the kegs have been filled they are closed and examined and filled again in the manner described above. The herring which has not been cleaned does not keep quite as long as that which has been cleaned. Fish prepared in this manner at Herba, in Gottland, has kept fresh and good for more than a year.

Swedish herrings, prepared in the Norwegian or Dutch manner, have fetched a good price both in Stockholm and other cities of Sweden.

3. *Method of preparing spiced herring.*

So-called spiced herring is an article of trade which, like anchovies, is kept in glass jars or very small kegs. It may be prepared from any kind of herring, and is esteemed as highly with us as in some cities of Northern Germany. But its preparation can so far not be said to form a special trade, but must rather be considered as experiments made by housewives in order to give some little variety to their meals, but especially to the lunch-table. But as these herrings might be in demand in some places, and might possibly fetch a good price in the foreign market, I shall here give the receipt for preparing them.

Freshly caught herrings are immediately laid in vinegar, adding one-fourth part water and some salt. After twenty-four hours the herrings are taken out and the vinegar is allowed to flow off. The fish are then placed in a tub or keg, with the following spices in the following quantities to every hval of herrings: 1 pound dry fine salt, 1 pound powdered sugar, 1 ounce pepper, 1 ounce laurel leaves, 1 ounce saltpetre, ½ ounce sandal, ¼ ounce ginger, ¼ ounce Spanish hops, ¼ ounce cloves. Others use the following spices: 1 pound salt, ½ pound sugar, 2 ounces pepper, 2 ounces allspice, 1 ounce cloves, 1 ounce Spanish hops. The herring should remain in this mixture for two months before being used. Some lay the herrings immediately in vinegar which has not been weakened with water or salt, and after twelve hours they are taken out and treated in the above-mentioned manner. If the spiced herring should after a while be without brine, good brine of Lüneburg salt should be poured in, and then they will keep for years.

b. THE COD-FISHERY.

Of the many fish belonging to the cod family, *e. g.*, the codfish proper (*Gadus morrhua*), the pollock (*Gadus virens*), the haddock (*Gadus aeglefinus*), the ling (*Molva vulgaris*), the hvitling (*Gadus merlangus*), &c., which live in salt water, and which, in the Kattegat and the North Sea, form the object of those extensive fisheries by which many inhabitants of the Norwegian and Bohuslän coasts make their living, there is found in the Baltic only the common codfish (*Gadus morrhua*, L.), at least in such quantities as to repay the trouble of catching it. In the Sound and the portions of the Baltic adjoining it haddock (*Gadus aeglefinus*), glyskoljan (*Gadus minutus*), hvitling (*Gadus merlangus*), pollock (*Gadus virens*), and blanksej (*Gadus pollachius*), are frequently caught, but nowhere in the Baltic proper are they found in such numbers as to form the object of special fisheries. From these its relatives, the codfish proper is distinguished by its upper jaw projecting over the lower jaw, by having a beard on the lower jaw, by having its side bent near the center of the middle dorsal fin, and by having such small eyes that their diameter is much less than the distance from the corner of the eye to the tip end of the nose. The haddock, the glyskoljan, and the hvitling have, it is true, a projecting upper jaw also, but can easily be recognized: the haddock by having almost straight sides and a black spot on each side about under the middle of the first dorsal fin, the glyskoljan (*Gadus minutus*) by the circumstance that the diameter of its eyes is larger than the distance from the corner of the eye to the tip end of the nose, and the hvitling (*Gadus merlangus*) by its not having the beard which is found on the lower jaw of the codfish, the haddock, and the glyskoljan. The pollock and the blanksej (*Gadus pollachius*) both have the lower jaw projecting farther than the upper jaw, and are thereby distinguished from the above-mentioned fish found in the Sound and the Baltic, which likewise belong to the cod family. The pollock is again distinguished from the blanksej (*Gadus pollachius*): the former has a cloven caudal fin, its sides are almost straight, and its color gradually changes from a dark sea-green on the back to silver-gray on the belly and sides; the latter has a caudal fin, which is but little indented, its sides are sharply bent, and the brownish-black color of the back is clearly defined from the silver-gray of the sides.

The codfish proper never reaches the same size in the Baltic as on our western coast or the coast of Norway. Whilst in the North Sea and the Western Ocean it grows very large, and often reaches a weight of 40 pounds, the Baltic cod seldom weighs more than 15 pounds. Like the herring, it gets smaller and smaller the farther north in the Baltic it is found. The average weight of the codfish found in the Southern Baltic and the Sound varies from 3 to 6 pounds, whilst near Gottland it is only 2 to 3 pounds, and on the coast near Stockholm only 1 to 2 pounds.

The color of the codfish varies considerably, owing chiefly to the difference of food and the different bottoms on which it lives. Generally the upper parts of its body have an ashy-gray or olive color, thickly dotted with round spots of a yellow or brownish hue, decreasing in number towards the sides; the lower part of the body is whitish, without any spots. The varieties which are found most frequently are the so-called "Berg"-cod, in Bohulsän, which has a reddish color, thickly covered with spots, and having reddish or grayish-brown fins and back; the "Pall"-cod, near Gottland and the "Berg"-cod of the Southern Baltic, whose whole body is of a dark color with but few spots. The full-grown codfish prefers deep water, either on the outer coast or in large bays and inlets, and only during the spawning-season it temporarily goes into shallow water. It spawns at different times on the different coasts: in the Sound in March, on the coasts of Skåne and Bleking from the middle of March till the end of April, near Gottland and on the coast of Stockholm during April and May. When the spawning-season approaches, the codfish ascends from the deep to shallow waters, either on the outer coast or in bays and inlets. There is this peculiarity about the roe of the codfish, that it does not adhere to aquatic plants and stones like that of other Baltic fish, but, according to observations made by the Norwegian naturalist G. O. Sars, floats about freely near the surface of the water. Even with a low temperature of the water the eggs are hatched after 18 days, and with a higher temperature, even in a shorter time. After being hatched, the young fish continue to float about near the surface of the water at least as long as they still have the umbilical bag which most young fish carry for some time after being hatched. This bag serves as the food of the young fish; and as soon as it is consumed the fish requires other food, and seeks places where suitable food can be obtained, and where it can find protection against the attacks of the numerous fish-of-prey which eagerly devour the young fish. Such places are the algæ-covered bottoms near the shore, where small crustaceans, scarcely discernible to the naked eye, are found in profusion and form the first food of the young fish. As the young codfish grows, becomes stronger and larger, and is able to defend itself against its enemies, viz. fish-of-prey of every kind, not the least dangerous among them being the old codfish themselves, it goes into deeper waters, where it finds larger crustaceans, worms, and snails, which at a more mature age form its favorite food. When fully grown the codfish is a voracious fish-of-prey, devouring almost everything coming in its way, young fish and fish of every kind. It therefore prefers the deep waters, where it feeds on the large schools of herrings, and often visits the banks where the herring spawns, and devours its spawn and young. Since the roe of the codfish does not adhere to plants or stones, but floats about freely near the surface of the water, it depends on current, weather, and wind to what coast it will float, and a large portion of it is consequently very often cast ashore and lost. And as it is well known that the codfish, like other fish, when fully grown, revisits the coast where it was born, it is impossible to calculate on seeing again, as full-grown fish, the young codfish which were born on a certain coast. For the roe laid near some coast may, by current and wind, be carried to distant parts, and the home of the young fish will be the coast to which the roe has been carried accidentally; and this coast will be revisited by them when they are fully grown, when, after having closed their annual regular visits to the deeper waters, the time comes for them to seek shallow waters.

Nature has thus arranged it so that even with the greatest care and protection it is impossible to calculate with absolute certainty on a successful cod fishery on any given coast. Experience has shown that on certain coasts no codfish have come to spawn for several years, although fishing had by no means been carried on in a destructive manner, and although the natural conditions continued as favorable as during the time when the codfish annually visited those coasts in large numbers. It is supposed, and probably correctly, that the cause why enormous cod-fisheries have for many years been carried on uninterruptedly in some localities, *e. g.*, the Loffoden, the Norwegian coast, the Shetland Islands, Iceland, &c., must be found in the fact that these coasts or groups of islands are so favorably situated near or in deep waters, that even when current and wind are comparatively speaking less favorable, a sufficient quantity of roe and young fish is carried into the bays and sounds to insure good fisheries. It must also be remembered that the cod-fisheries carried on in the Loffoden and other large fishing places on the Atlantic Ocean, although carried on near the coast, have altogether the character of ocean fisheries, fishing going on mostly in the open sea and at a considerable depth, 50 to 100 feet and more.

It is evident from all that has been said that the cod is a kind of fish which prefers the deep waters or banks in the open sea, and that one cannot calculate on its coming to a certain coast every year, and on catching it with the apparatus usually employed in coast-fishing. It has been shown by actual observations that the Baltic also contains a considerable number of codfish on those bottoms and banks which extend almost along our entire coast. Fishermen who expect annual productive codfisheries must therefore possess the necessary apparatus for deep-water fishing. We shall now give a few brief directions how to carry on these fisheries.

As the banks on which the codfish stay the greater part of the year are situated at a considerable distance from the shore, it is evident that the fisherman should have a good vessel, strong and large enough to reach the shore when a storm should spring up. For this purpose the boats used for salmon-fishing on the coast of Bleking, known by the name of "Blekings-ekor," are well suited. These boats are large enough to offer ample protection for the fish which have been caught, so the fisherman runs no risk of having his fish spoiled before he comes home.

Cod-fishing on banks may be carried on with so-called "hand-lines" or "codfish-lines," with "angling-lines," and with nets. As the "hand-line" is so well known, it will not be necessary to describe it.

The "angling-line" (long line or trawl-line), with which fishing in the open sea can and should be carried on, resembles a common long fishing-line, only with this difference, that it is furnished with floats which keep the bait from the bottom, as otherwise it would be eaten by different marine animals. The line should be made of such strong material that it will not tear when being hauled in; when laid, it should of course be steadied by weights sufficiently heavy to prevent its being driven away in stormy weather. To mark the place where the line is laid, a buoy is used, with a flagstaff and flag large enough to be easily seen when the fisherman comes to haul in his line. As bait may be used, pieces of herring or other fresh fish, worms, snails, and muscles. The line may be laid either in the morning or in the evening, and in favorable weather the fisherman should so arrange it that he can stay at the fishing-place until it is time to haul in the line. During this time of waiting, the crew may employ themselves by fishing with "hand-lines."

Net-fishing in deep water or in the open sea should be carried on with common codfish-nets, which, however, should be a little deeper than those used in coast-fishing. Each set generally has 24 nets. When the nets are to be set, they are fastened to two ropes of about the same length as the depth of water where the nets are to be set. To the ends of these ropes an anchor is attached; to this is fastened another rope reaching to the surface, and having at its end a buoy to indicate the place where the fisherman has to look for his nets. These, which have been laid in the boats in good order, are then set in the manner shown in Figure 1, Plate I. When several fishermen set nets close to each other, it becomes necessary to attach an anchor also to the end of the row of nets, with a rope reaching up to the surface and having a buoy attached to it, to show in what direction the nets have been set, so that other fishermen may not set their nets across the former, and thus produce confusion and make the hauling-in difficult. If the net has been set so far from the coast that the fisherman can no longer see it, he must either cast anchor and remain in the fishing-place till the net is taken up, or he must when leaving the coast mark some object on it, and then by the aid of his compass row or sail for some time in a certain direction, so that he can easily find the place where his nets are set, even if stormy weather should oblige him to seek the coast before his nets are taken up. In cod-fishing one should have two sets of nets, so the one may dry whilst the other is in the water.

Methods of preparing the codfish.—To prepare a good article of codfish, it should never lie in the boat without being cleaned for any length of time, as it may easily spoil. A careful fisherman carrying on cod-fishing on a large scale should therefore always have in his boat small boxes or kegs in which the fish may be laid in salt. The crew should also be large enough, that two or more persons may immediately commence to kill the fish, so the blood may flow off, and, if possible, clean and salt them. It is likewise important that the fish should not be bruised or trodden on, as thereby their flesh becomes loose, full of holes, and its appearance is not very inviting. The Baltic codfish may be prepared either as so-called "brine-cod" (*Kabeljo*) or so-called "dry-cod" (*Klippfisk*). The Baltic codfish may of course also be prepared as "common dried cod" or so-called *lutfisk*, although by its small size it is not very well suited for this method of preparing it. Whether the fish is to be prepared as "brine-cod" or "dry-cod," it must first be cleaned thoroughly, so that no blood is found near the backbone; the entire skin is carefully removed from the whole lower part of the fish. In large codfish the backbone is taken out, whilst in smaller ones it is allowed to remain; the head is cut off, and the fish is then ripped open, so that it presents the appearance shown in Figures 1 and 2, Plate II. After the fish has been ripped open, cleaned and washed, the water is allowed to flow off, whereupon it is laid in layers in barrels and salted, the outer side downward, and with sufficient salt between each layer to keep the fish from spoiling. After the fish has remained in brine for about eight days, and its flesh has become firm, it should be taken out. If it is to be used for "brine-cod," it is again placed in barrels with enough fine white salt between the layers to keep the fish from spoiling; whilst if it is to be used for "dry-cod" (*Klippfisk*), it is treated in the following manner: The fish are taken from the brine, and laid in rows on slanting boards, so the brine may flow off. Whilst being taken up they are washed in the brine and brushed carefully, so as to remove all impurities. After the fish have laid on these boards for a night, they are spread out to dry either on flat rocks or on a sort of lattice-work placed in a drying booth. It is best to lay the fish on a lattice-work to dry, as the rocks often get very hot and therefore cause the fish to shrivel. When the fish are laid out to dry, the air should not be damp, nor should the fish be exposed too long to a hot sun. In the evening, as soon as the air gets damp, the fish should be piled up in heaps and be again spread out in the morning. This is continued till the fish gets half dry, when the pressing commences, which is done in the following manner: The fish are piled in large heaps, covered with boards, and on these stones of a suitable weight are placed (Figure 4, Plate III). Whilst being pressed the fish should again be spread out for drying, if the weather is favorable, but should likewise, when night comes or when damp weather sets in, be piled up and pressed, and the sides of the pile covered with matting or tarpaulin so as to keep the moisture out. This is continued till the fish gets so dry that when pressed with the thumb no impression is made, showing that the flesh has become quite hard. The fish are then packed in wooden boxes and are ready for the market. Fish prepared in this manner find a ready sale not only at home but also abroad, in England and Germany, where "dry-cod" fetches a higher price than "brine-cod."

c. SALMON-FISHING WITH LINES.

The salmon is a kind of fish which lives half the time in fresh water and half the time in salt water. Its nature compels it during summer to seek swiftly-flowing streams, where during autumn it deposits its roe among pebbles and rocks. Observations have shown that its roe loses its vital power as soon as it comes into contact with salt water. The young salmon hatched in the streams stay there about two or three years, and generally during the rising of the streams in spring return to the sea or to large lakes, where the easier access to food makes them grow rapidly. The young salmon, when they have reached the sea, as well as the full grown salmon, live on small fish, *e. g.*, herring, launce, smelt, &c. When the salmon has become capable of propagating, after a stay of one or two years in the sea or some lake, it returns to the stream where it was born, deposits its roe, and goes back to the sea; and thus its life continues to be a regular change of its place of sojourn until it is either caught or meets with its death in some other way. Fishermen living near the coast can, therefore, not expect good salmon-fisheries unless the salmon are protected in the streams during the spawning-season; nor can the fishermen living along the streams hope to see the salmon again unless the coast-fishermen carry on fishing in such a manner as not to prevent the salmon from going up the streams. Both classes of fishermen have, therefore, an equal interest in having the salmon-fisheries regulated in such a manner as to suit the nature and mode of life of the salmon; for if this is not done, both the coast waters and the streams will soon lose their wealth of salmon. The laws for protecting the salmon-fisheries therefore prescribe that no salmon are to be caught in the streams during their spawning-season in autumn, and that no nets shall be set at or near the mouth of streams in such a manner as to hinder the salmon from reaching their spawning-places in large numbers. Experience has shown that wherever these regulations have been carefully observed, the salmon-fisheries have very soon improved considerably. Every fisherman, therefore, who has the true interest of the salmon-fisheries at heart, should, above everything else, see to it that the fishery-laws are carefully observed in his neighborhood.

Salmon-fishing in the Baltic is chiefly carried on with nets. As this method of catching salmon is well known, it needs no further description. But salmon may also be caught with lines in the open sea near the outer coast. This is done on the southern coast of Sweden, and we shall therefore briefly describe this method of fishing.

It is well known that the salmon stays in the sea during the latter part of autumn, winter, and spring. Whilst the young salmon which spend their first year in the sea prefer to stay near the mouths of rivers, or, at any rate, not far from the coast, the older ones generally spend the cold season of the year in deep water, following the schools of herrings which are found there. The fishermen on the coasts of Skåne and Bleking make use of this circumstance, and during winter and spring, whenever the sea is free from ice, and stormy weather does not interfere with fishing, catch many salmon, which at this time are sojourning in deep water.

As was mentioned above, hooks and lines are used in this fishery. The hooks are laid in the open sea, and the lines are kept in position by anchors or heavy weights in the same way as is done in the cod-fishery in the open sea. The line is not, as is generally done in other fisheries, sunk to the bottom, but is kept floating near the surface by means of large cork-floats (Figure 5, Plate I). The line must of course be strong and the weight heavy, so the fish, which are generally large and powerful, may not carry everything away with them. As a very long line would doubtless get entangled during the winter-storms, only short lines are used, measuring about 20 fathoms in length, with no more than three to four hooks on each line. To make up for this deficiency a large number of lines is set, each with its separate weight. Cheap and very suitable weights may easily be obtained by inclosing large pieces of rock in a triangular lattice-work of wood, with sharp sticks of wood projecting on all sides.

For bait, herrings are generally used, which are attached to the hooks in the manner shown in Figure 5, Plate I. The hook should be of strong galvanized-steel wire, of the size and shape shown in Figure 6, Plate III.

Fishermen who use hooks and lines for salmon-fishing should of course be provided with a sufficient number of lines, so they may set new lines when going out to sea for the purpose of examining those which have been set for some time. As soon as the warm weather sets in, salmon-fishing with hooks and lines ceases, partly because the salmon then go up the rivers, and partly because the warm temperature of the water makes the bait spoil too quickly, so that it becomes entirely useless.

VIII.—THE SALT-WATER FISHERIES OF BOHUSLAN AND THE SCIENTIFIC INVESTIGATIONS OF THE SALT-WATER FISHERIES.

BY AXEL VILHELM LJUNGMAN.*

I.

THE NECESSARY BASIS FOR CARRYING ON THE BOHUSLÄN SALT-WATER FISHERIES AND THE SCIENTIFIC AND PRACTICAL INVESTIGATIONS AND EXPERIMENTS REQUIRED FOR OBTAINING THIS BASIS.

§ 1. Every state ought to consider it as its duty to make scientific investigations, at any rate within its own limits.

In order that a comparatively poor, extensive, and thinly populated country may do its duty in this respect, it is doubtless necessary that the work be done systematically, according to a well-matured plan, if the object in view is to be attained, *i. e.*, a thoroughly scientific knowledge of one's own country. Societies or individuals may, in this respect, do as they deem best--their work and their sacrifices will in any case do some good—but the *state must act according to a distinct plan, so that from want of means one portion of the investigation may not suffer, which, by a wiser and more systematic use of all the means at the command of the state, might have led to good results without thereby injuring any other part of the investigation.*

It is always cheapest to do everything systematically, and is the surest way to reach one's object, and it is almost indispensable at a time when so considerable a portion of the public revenues must be devoted to the defense of the state against foreign enemies.

Wealthy states (especially those which possess colonies) can and ought to extend their scientific investigations also to uninhabited and uncivilized portions of the world. In this way we shall, in course of time, attain to such a complete scientific knowledge (physico-geographical, geological, mineralogical, botanical, zoological, ethnographical, linguistical, and archæological) of our world as our rapidly progressing time demands.

§ 2. The great services which science has rendered to agriculture, mining, and industry, as well as to nearly all our trades, and the losses which a lack of theoretical knowledge has frequently occasioned, show the absolute necessity of following the only certain guidance of science. In all branches of human activity a desire is manifested at the present

* *Bohusläns Havfisken och de vetenskapliga Havfiske undersökningarna. Af Axel Vilhelm Ljungman. Gottenberg,* 1878. Translated by Herman Jacobson.

time to utilize all the results of science, to abandon old prejudices and all endeavors which are only based on accidents.

The great importance of the fisheries has shown the urgent necessity of scientific investigations, so that they may be carried on in the proper manner and protected from injuries caused by ignorance and greed, a necessity which the government has recognized by making appropriations and by instituting a course of investigations.[1]

§ 3. A suitable fishery legislation and administration of the fisheries can likewise only be based on a careful scientific and practical investigation. It must be remembered that both with regard to the fisheries and other industries it is of importance that the state does not meddle more than is necessary, for by making too many rules more harm than good is often done, as, contrary to all calculations, such rules may frequently hinder the free development of any trade. To find the medium in this respect presupposes a thorough knowledge, both theoretical and practical, of the whole trade, and a well-matured plan based on this knowledge, which, without too great difficulties, may be carried out in such a manner that the results can be calculated beforehand with some degree of certainty. Without sufficient knowledge of a trade or industry it is not possible to gain any firm basis for legislation or calculate any of the possible results.

§ 4. Scientific investigations are, as will be shown below, necessary, not only for gaining a theoretical basis for legislation and for successful administration, but also for the furthering of the fisheries themselves, as even with regard to these they may bring to light facts which may prove extremely useful[2]. A common objection to this view is this: that a trade does not need the aid of science, but that it is best to let it develop freely. It is well known that science has proved useful to the fisheries as well as to agriculture and other trades, chiefly by showing the way in the making of experiments and thus facilitating any improvements or new inventions, although the great mass of people who gradually reap the advantages of such inventions hardly ever think about the scientific work which necessarily had to precede them.

Although it is best to keep the fisheries free from too much legislation and meddlesome interference of the government, a complete knowledge of the entire natural history of fish is both useful and necessary, as well as of the proper method of preparing fish for the trade; and in all these respects science may extend considerable aid.

§ 5. This whole field should be investigated scientifically even if no-

[1] This paragraph, like some of the following, is taken from former articles of the author, viz: "Report on an expedition for examining the salt-water fisheries in the Skagerak and Kattegat, made during the summer of 1871, on board the royal gun-boat Gunhild," Upsala, 1873, (partly given in "*Nordisk Tidsskrift for Fiskeri*," II. Copenhagen, 1874, p. 1-14), and "*Preliminär Berättelse*," &c.—Preliminary report on the herring and herring-fisheries on the western and southern coast of Sweden. Upsala, 1875.

[2] *J. MacCulloch "On the herring"* (Quarterly Journal of Science, Literature, and Arts. XVI. London, 1824), pp. 210-211, 216, 222.

material advantage should result from such investigations in the immediate future. The history of natural sciences and trades furnishes numerous examples how a science may be studied and worked up for many years, in some cases even for a century, without yielding any practical result, until all of a sudden some grand invention surprised the public. How long, for instance, was electricity considered by the great mass of the people as a useless matter, good enough perhaps for the learned to know something about, but of no practical value, until its practical application produced a sudden and radical change in public opinion. Science ought to be cultivated conscientiously and perseveringly for its own sake, and sooner or later its results will prove useful in practical life.

§ 6. As the scientific and practical investigations which come into question here are to give us that knowledge which is indispensable for obtaining the necessary basis for carrying on and administering the fisheries in the best possible manner, as well as for useful legislation on the subject, in fact for a final solution of the whole fishery-question, it will be evident that these investigations must extend to everything concerning the fisheries. These investigations must, therefore, not be confined to technical, law, and administrative questions, but must extend to questions of economy and natural science. All the different points from which the fisheries may be viewed must be considered if any good result is to be obtained. For the omission of one of these may essentially change the results. A most thorough and complete treatment of the whole subject is absolutely necessary.

§ 7. It is well known that the so-called inductive method is the only one both in natural history and in a trade which will lead to a reliable general knowledge. From many agreeing facts a deduction is made regarding a general law, which will gain in probability in proportion as the induction is complete in all its parts. This shows the necessity of making as many observations as possible at different times and places, and of comparing these with older observations handed down to us by reliable writers. The necessity of making numerous observations during a long period of time increases, as there are very frequent exceptions from general rules which cannot always be considered as abnormal, and as a lack of agreeing facts with regard to even one or two points may make it very difficult to reach any certain conclusion. It is, of course, not possible to obtain in this way that degree of completeness which would lead to absolute certainty.

With regard to our present subject—the fisheries—one must be careful to avoid the very common mistake of making hasty observations or facts which have not been fully established[3] the basis of more or less preten-

[3] Under this head comes the use of entirely accidental methods of explanation, which is but too frequent. From an accident anything may easily be explained, but then such an explanation may be utterly worthless. Anything that is accidental has in itself something inexplicable, and stands without its proper causal connection, and it is much more difficult to assign its cause than to understand that fact which it is intended to explain.

tious and imposing scientific systems. It must be remembered that in science, as little as anywhere else, is there a royal road to reach one's object, and it is often nothing but vanity which has led persons to build such air-castles and call them scientific achievements.

§ 8. The difficulty of finding the causal connection is also much greater in that portion of natural history which is most important for our purpose, that is, the so-called physiology of relations, than in anatomy or the history of natural development. The anatomist, after having dissected a few specimens and found them to agree, may generally be certain that he has ascertained their normal condition, and that any deviations from this which may possibly be discovered in the future must be considered abnormal. The anatomist can and must often be satisfied with examining only a few specimens, and may from these draw a tolerably certain conclusion; but this would not answer in the physiology of relations with its many changes and irregularities. Here it is necessary to employ every means at our command for taking the greatest possible number of observations, and then, after critically examining their reliability, and instituting the most careful comparisons between them, and by using every method of induction, analogy, and hypothesis for reaching a conclusion, to obtain the most probably or at least approximately correct view. Thus the demonstrative certainty gradually decreases in the physico-mathematical sciences in proportion as we depart from the abstract, outward forms of objects, or from general laws or component parts, and enter the domain of organic nature, which becomes more difficult for the naturalist the more life itself comes into question.

§ 9. Regarding the general view of nature and the different methods of explaining its phenomena, it must be said that a really scientific explanation, going back to final causes, is scarcely possible, as soon as from general views we enter upon details. An explanation from absolutely certain causes, carried through consistently, must always move in a circle, because the world is a whole, developing systematically, and as the various phenomena of nature mutually depend upon each other, so that one phenomenon may depend upon another which follows it, whilst we from our youth up are accustomed to draw a conclusion by advancing from a *post hoc* to an *ergo propter hoc*. The aim of natural science is, therefore, to be as free as possible from teleological and mechanical prejudices and methods of explanation, and to endeavor to show the actual connection between the different phenomena, and not to draw philosophical *a priori* conclusions as to their absolute necessity.

On the material which has been acquired in this manner every one must, according to the best of his ability, base his more or less philosophical theories, being careful, however, to keep these latter strictly separate from the facts.

§ 10. The sources of knowledge to which we are directed as regards fish, fishing-waters, and fisheries, are: literature, the experience of fishermen and superintendents of fisheries, and direct observations and experiments.

As regards literature we must have reference not only to that more scientific portion of it contained in book form, but also to that more scattered information found in newspapers and periodicals[4] or in the various official documents in city or state archives. In order to make full use of this source of information so very important for the herring-fisheries, it is of course necessary to consult the more important foreign works and documents (for example, the Danish, Dutch, and British). The importance of studying the special literature of a subject will be self-evident if we remember the well-known fact that in those fields of human knowledge which have been cultivated most, no one, whatever his natural talents may be, can, through his own exclusive endeavors, obtain that knowledge which is stored up in literature for the benefit of posterity, much less carry his knowledge very far beyond this limit. A thorough acquaintance with all the facts which others have brought to light on a certain subject must form the starting-point for those endeavors from which the greatest possible results may be expected.

But it is not only that knowledge which is stored up in literature which must be taken into consideration, but everything which has been collected, preserved by tradition by the fishermen and the superintendents of the fisheries. The gathering and working up of this very heterogeneous material is connected with considerable difficulty, and presupposes a good deal of experience obtained by direct personal observations, as well as a varied knowledge of all the literature on the subject. The great mass of the people are often more inclined to be influenced and even prejudiced by anything coming from foreign parts; the experience of foreign countries is doubtless also in many respects richer and more varied than our own and more fully corroborated by experiments and scientific investigations. The chance of increasing our stock of knowledge by studying the experience of foreign countries should, therefore, not be neglected. In doing this, however, it should be remembered that there are great differences of nature, law, and economy between our own and other countries, which point is but too frequently lost sight of.

The material obtained from literature and the experience of fishermen must be critically sifted, and for this purpose as well as for extending our knowledge beyond this limit, direct personal observations and experiments are necessary. Experiments are moreover required to prove the correctness of opinions that have been advanced, and of hypothetical explanations.

A full description of the best means for utilizing these various sources of knowledge will be given below when each portion of the fishery investigations will be treated separately.

§ 11. Fishing is a trade which absolutely requires special experience. This experience embraces the proper use of the different fishing-appa-

[4] See *A. Boeck*, in "*Nordisk Tidsskrift for Fiskeri*" VII. Copenhagen, 1872, p. 7.

A. V. Ljungman, "*Preliminar Berattelse*"—Preliminary report for 1873–1874, on the investigations regarding the herrings and herring-fisheries on the western coast of Sweden. Upsala, 1874, p. 70.

ratus, the preparation of fish, and the use of fish in the household. All these points must be worked up carefully and thoroughly so as to give all the necessary hints. But as all this experience cannot be gained by one man engaged in the fishing-trade—*ars longa vita brevis*—as much as possible of it must be gathered both from old documents and from the fishermen themselves, and must then be proved by personal observations, which must also be made with the view of developing and increasing the material. In order to make this material as valuable as possible it is of course highly necessary to become acquainted with the more prominent foreign fisheries, especially those which excel ours.

The technical investigations must therefore relate to the existing condition of the fisheries but also to their further development and to any possible improvements. Among the subjects which in this respect deserve special attention, the large periodical Bohuslän herring-fisheries doubtless occupy a prominent place. With regard to these fisheries we need, above everything else, a brief but complete review of all the experience gained in the course of centuries concerning the herring-fisheries, the preparation of herring, and the herring-trade; for as it is well known that these, our largest fisheries, are periodical, and cease entirely for many years at a time, we cannot expect that all the experience we need is handed down to us by tradition.[5]

The technical investigations ought moreover be specially directed to the scientific treatment of everything relating to the different methods of preparing and preserving fish.

Finally, it would be necessary to carry out according to a well-matured plan all the different experiments required to corroborate our knowledge, to try new fishing-laws, new apparatus, methods, and other improvements. In order to make such technical investigations and experiments really valuable and useful, they ought to be committed to persons who can devote their whole time to it; for of the young naturalists who are generally detailed for such investigations, it cannot be expected that they should have that undivided interest, that local knowledge, and that practical experience which are absolutely required to make such investigations truly successful.[6]

§ 12. As it is the object of the fisheries, as well as of agriculture, to utilize the productive powers of nature for human purposes, and this not only with regard to the quantity and quality of nature's productions, but also to their preservation and possible development, the chief condition of success will be a complete and reliable knowledge of the nature of these productions, of the causes which create and sustain them, and of the mechanical or chemical aids and apparatus by which they may be investigated and utilized. By comparing this knowledge with that experience which practical fishing constantly supplies, we obtain the so-

[5] *P. Dubb, "Anteckningar om sillfisket i Bohuslan" (Kgl. Vetenskaps Akademiens Handlingar* för år 1817), p 32, 33.

[6] *A. V. Ljungman, "Om fiskerilagstiftningen för bohus-länska skärgården," IV* (Göteborgs-Posten, 1875, no. 78).

called knowledge of the trade, or, in other words, the science of the trade.

The science of the fisheries, *i. e.*, the knowledge of how to carry on the fisheries, fully developed and arranged as a whole, forms the science of fishing, properly so called, and in proportion as it is really scientific, fishing becomes a branch of natural science.

In order to make the necessary technical investigations, a good knowledge of natural science is required as well as a special knowledge of those branches of natural science which form the theoretical foundation for the special science of the fisheries.

§ 13. As fishing requires a knowledge of the mode of life and other characteristics of fish as well as of the fishing-waters, so fishing carried on as a trade requires a knowledge of the laws of economy. We therefore need an economy of the fisheries just as much as their practical and scientific knowledge, although the latter is certainly an essential condition of the former.

A wise administration must never favor one trade at the expense of another which is just as important or perhaps even more so, thereby bringing about a conflict of interests which cannot in any way be beneficial to the state. The fisheries must therefore be considered in their relation to other trades and occupations, especially agriculture and navigation, the general welfare of the state, the means of communication, &c. Special regard should also be had to the changed circumstances of our times in case the great herring-fisheries should again be revived; which event, strange to say, has for nearly half a century been rather considered as a curse than as a blessing for Bohuslän.[7] The *social question* of our coast will, therefore, likewise have to be considered—a question whose solution may puzzle our wisest men. All this becomes the subject of a special branch of knowledge, which might be termed "the economy of fisheries," whose aim would be to promote the fisheries by working up the various scientific methods and corroborating their usefulness by practical experiments, always considering the economical value of the fisheries for the public as the foundation on which all improvements in the fishery-laws and the administration of the fisheries should be based.

The fisheries should therefore form the subject of the most thorough and exact treatment even from an economical point of view, and this all the more as the want of such treatment has doubtless been the chief cause of the insufficiency of our fishery-legislation.

§ 14. The investigations must, furthermore, extend to the whole subject of law and administration; for since their purpose is to gain that knowledge which is necessary for carrying on and superintending the fisheries in the most efficient manner, attention must not be diverted from this object or extended beyond its limits. As such a course would only delay

[7] *A. V. Ljungman: "Några Ord om de stora bohus-länska sillfiskena."* Gottenburg, 1877, p. 28.

the results or prevent us from reaching them,[8] the investigations must be carried on with a special regard to the needs of a good administration, which ought, therefore, to be clearly specified.

A complete knowledge of everything pertaining to law and administration is certainly just as necessary for the legislator as a technical, economical, and scientific knowledge. This last-mentioned knowledge is necessary for proving the practicability of the legislative and administrative measures; and in order that the full significance of these measures may be understood, a sufficient knowledge of local and technical conditions is required. A satisfactory solution of the whole fishery-question, to serve as a basis for systematic investigations and for a reform of fishery-legislation, can therefore only be reached by placing each separate part of the question in thoroughly competent hands. The lack of such preliminary and preparatory measures is doubtless one of the chief causes of deplorable defects in our fishery-legislation and of practical mistakes springing from them.

What we need, therefore, is a complete and systematically arranged review of all the laws relating to the Bohuslän salt-water fisheries from the oldest times down to the present. Such a review should, as far as possible, give the causes of every amendment to these laws and tell us how the amended laws worked; the laws should be examined with the view of testing their applicability to the changed circumstances of our times, and they should finally be compared with the experience of other countries. If all this were done in a most thorough manner, we might look for truly beneficial results.

§ 15. In order to fully understand the Bohuslän salt-water fisheries, some purely historical investigations are necessary, which may yield some material of great value which could not have been obtained in any other way. As an illustration of this assertion we may quote the example of *Axel Boeck*, who, in his well-known work, "*Om Silden og Sildefiskerierne*" (On the herring and the herring-fisheries), has based his whole treatment of the important questions, "why the great periodical Scandinavian herring-fisheries have ceased"[9], and "what influence is by outward physical conditions exercised on the migrations of the herrings" on such investigations.[10] Besides, how could we without such historical investigations ever settle the question regarding the nature, spawning-time, &c., of the old Bohuslän herrings?[11] or corroborate or disprove the

[8] Thus it has often been the case that *practical* objects have been used as a bait for carrying through purely *theoretical* measures, which in no way could prove a benefit for the trade.

[9] *A. Boeck*, "*Om Silden*," &c., I, Christiania, 1871, p. 82–119.

[10] " " " " I, p. 72–82.

[11] The supposition regarding the relation of the old Bohuslän herring to the present herring forms, as is well known, the basis for all our fishery-legislation since 1852; and the opponents of this legislation, therefore, chiefly direct their attacks against this supposition. (See "*Handlingar rörande sillfisket i bohusländska Skärgården*," Stockholm, 1843, p. 71–73, 156, 172. "*Götebarg's Handels-och Sjöfarts-Tidning*, 1853, No. 147, supplement. "*Nya Handlingar, &c.*," I, Gottenburg, 1874, p. 22–24, 29–32, 63–66.)

assertion that a severe winter with much ice has a beneficial influence on the herring-fisheries, an assertion which has been so strongly made by Professor *Nilsson* and others?[12] or how could we, in any other way, ascertain the practical working of the older fishery-laws and decide in what respect and in how far the fishing-trade has improved, remained stationary, or gone down?

It will scarcely be necessary to say any more regarding the great value of historical investigations in themselves and their absolute necessity for gaining as complete as possible a knowledge of our fisheries.

As regards our largest and most famous fisheries, the great periodical herring-fisheries, we must in the first place not only examine the possible causes (real or supposed) of the periodical cessation of these fisheries, which has played such a prominent part in our fishery-legislation, but also the causes of the somewhat regular changes in the course and location of the fisheries, and in how far these changes depended on physical conditions, &c. By making these investigations we obtain a great mass of historical facts which will amongst the rest serve as a guide in framing laws in case these great fisheries should revive.[13] Much information would also be gained in this way, throwing light on many a dark point in the natural history of the herring and aiding in solving the difficult question of the migrations of the great herring-schools. The literature referring to these questions amply proves how important a complete historical review of the Bohuslän herring-fisheries is for reaching a satisfactory solution of the whole herring-question.[14]

It is evident that the history of other fisheries will also be very interesting and may furnish valuable material. The historical part of these investigations should, therefore, by no means be omitted or neglected.

§ 16. The investigations must finally embrace all the facts of natural science, which will enable us to master the theoretical conditions "of a successful carrying on and administration of the fisheries" in a much higher degree than it is possible for the uneducated fisherman. A scientific knowledge of nature forms the theoretical basis on which alone a wise and beneficial management of the fisheries can be built up. We

---

[12] "*Handlingar rörande, &c.,*" p. 64, 67, 74, 77, 156, 163.

A. W. Malm, Naturhistoriska studier i det fria och: Rammuren. Gottenburg, 1860, p. 5.

[13] The regular changes in the course of the great herring-fisheries, and the consequent changes in the quality of the herrings require a legislation to suit these changes. Legislation should, therefore, be guided by the experience of former centuries. The first part of an approaching fishing-period cannot be considered as the standard for the whole time it lasts. In all probability it will in its course undergo considerable changes, and we must, consequently, look for this standard in the corresponding part of former fishing-periods. That this way of judging is correct is proved from the last fishing-period (1747–1808). For many of the regulations made during the latter portion of this period, and based on the experience of the first portion, proved to be antiquated and of little use, and in some cases were even more hurtful than helpful. The fishermen themselves, therefore, often demanded changes in the administration and in the laws. It must be said, however, that many of these changes were demanded from purely egotistical reasons.

[14] *Preliminär berättelse* (Preliminary report) for 1873–'74, p. 71–72.

need therefore a complete and thorough knowledge of the true object of the fisheries, of the nature of fish, and of the nature of the water in which the fish live, and finally of all those conditions on which the propagation, development, and life of fish depend.

The scientific investigations which are of importance must in the first place refer to an increased knowledge of the anatomy, physiology, development, characteristics, and varieties of fish, as well as to their distribution, and the probable causes of their appearing and remaining on different bottoms, and consequent upon this to their varying food, their isolation, hydrological relations, &c. They must also relate to the spawning of fish (time, place, whether near the surface or on the bottom), to their growth, difference of age, food, enemies, sickness, &c., dependent on physical conditions, their daily life, their regular annual migrations (caused chiefly by their desire either to seek food or to propagate the species), to their sudden appearance or disappearance, and to its causes, &c., as also to the nature of the fishing-waters, to the plants and animals contained in them as well as to their physico-geographical character. In the following we shall endeavor to point out the methods which should be followed in gathering all the material which is needed for a thorough knowledge of our salt-water fisheries.

§ 17. For a thorough study of the physiology, development, anatomy. &c., of fish it is doubtless necessary, if its results are to answer the increased demands of our times, that a person should have leisure, so that he can devote his whole time to it, and that he should be in possession of all the material and scientific apparatus which are required for such investigations. For some of these investigations, well-arranged aquaria will be of special value. With regard to these studies the author recommended, guided by the experience of foreign countries,[15] in his preliminary report,[16] the establishment of a complete station for scientific investigations of the sea[17] in a suitable place on the Bohuslän

[15] In France chiefly gained by the work of *Coste* and later by that of *Lacaze-Duthier* and in England, France, and Germany by the great public aquaria, as well as in Italy by the zoological station in Naples founded by *A. Dohrn* and subsidized by the government. (See: *Bulletin de la société impérial zoologique d'acclimatation*. 1862, p. 107-114; 1863, p. XLVII-LXIII; 1864, p. 261-269; 1865, p. 533-541; 1872, p. 164-167, 268.—*Archives de zoologie expérimentel et générale*. III, p. 1-38.—*Preussische Jahrbücher*. XXX, p. 137-161—*Zeitschrift für wissenschaftliche zoologie*. XXV, p. 457-480.—*H. Beta, Die Bewirthschaftung des Wassers und die Erndten daraus*. Leipzig, 1868, p. 236-248.—*J. G. Bertram*, The harvest of the sea. 3d edition. London, 1873, p. 293-296.—*F. Buckland*, Familiar history of British fishes. London, 1873, p. XI.—On the organization and progress of the Anderson school of natural history at Penikese Island. Cambridge, Mass., 1874.)

[16] *Preliminär Berättelse* for 1873-'74, p. 71. United States Commission of Fish and Fisheries. III. Washington, 1876, p. 166.

[17] In the above-mentioned report the author has expressed the opinion that the necessary special hydrological investigations should be carried on by persons specially detailed for the purpose, who should have their headquarters at the same station. This idea, although not without its advantages, might, however, meet with considerable difficulties when carried out practically.

coast,[18] which ought to be under the supervision of competent zoologists and botanists and furnished with all the necessary scientific apparatus.

It will be self-evident that such an institution would not only further the study of this part of natural science in our country, but would be almost indispensable for such a study; nor can there be any doubt that its activity can and ought to extend to a much larger field than the mere investigations of the fisheries.

§ 18. In order to gain a sufficient knowledge of the mode of life of fish, of their migrations, &c., in a certain given region, it will be necessary to make uninterrupted observations for a number of years with all the means at one's command (especially by fishing at all seasons of the year); and in order to make such observations truly valuable they should be carried on simultaneously in different parts of the given region; for local differences in the physical conditions will also produce differences in the appearance, mode of life, &c., of the herrings; and by observing the herrings only in one place we would just as little gain a general knowledge of this fish and its mode of life as we would obtain a knowledge of the meteorological conditions of a large country by observations gathered at a single point. In order to gain the true value of phenomena observed in a certain place, a more general knowledge is required, which can only be obtained by comparative studies.

From the foregoing it will be seen how difficult it is to arrive at absolute certainty, and how necessary, therefore, to base our knowledge on the greatest possible number of systematically gathered facts. It will also be evident that both the time during which, and the number of places where, these facts are gathered should be increased in proportion as the kinds of fish which are the objects of investigation are in the former case liable to appear at longer intervals, and in the latter case are subject to more local changes.

§ 19. In order to facilitate and to accelerate the acquiring of the desired knowledge, it will be necessary to have recourse not only to historical researches, but also to the experience of fishermen. Regarding the value of this last-mentioned source of information, it must be borne in mind that the information furnished by our fishermen on the mode of life and the migrations of fish, &c., are very much on the style of the predictions of our old-fashioned weather prophets. These old signs are frequently of just as little value for the fisheries as for agriculture. But although meteorologists have long since shown the worthlessness and fallibility of such predictions, people continue to believe in them, for uneducated persons are apt to remember the few times when such predic-

[18] See also *Preliminär Berättelse* for 1874–1875, p. 18. The most suitable place for such a station would doubtless be the mouth of the Gullmar fiord in the neighborhood of Fiskebäckskil. Farther north the station could not well be, if it should answer its purpose also with regard to the investigations of the expected large periodical herring-fisheries.

tions proved true, and to forget the many times when they were not fulfilled. Such persons will never think of comparing the cases when such signs were without any significance whatever with those cases when they were followed by certain results, nor do they weigh the probability of the one or the other. They are always inclined to follow a *post hoc* by an *ergo propter hoc.* The opinions of fishermen are also often at variance with each other, even with regard to the influence of outward circumstances on the fisheries. It must finally not be forgotten that the nature of the fisheries themselves requires great caution in applying results gained by positive experience. Fishing is generally carried on with very insufficient apparatus and only at *that* time and in *those* places where the greatest gain may be expected. There is a great difference between the occurrence of fish in a certain place and the occurrence of fisheries in the same. The fish may, for instance, come in a certain way which makes it impossible to catch them with the only apparatus on hand, and the fisheries, therefore, come to an abrupt end, although there are plenty of fish. The fishermen are, moreover, frequently governed by prejudices and actually cease to fish before the most profitable period of the fishing-season has arrived, simply because they think they have noticed some adverse signs. All the information gathered from fishermen must, therefore, be sifted in the most critical manner, and the most extensive fishing must be carried on by the observer himself with every imaginable kind of apparatus, in order to corroborate or disprove the statements of the fishermen.

§ 20. In order to gain more reliable and more complete knowledge than can be obtained from fishermen or through historical researches, it is, as I remarked above, absolutely necessary to make direct personal observations in a number of places. There should be a separate observer in every place, who, following a well-devised plan, would make daily observations on the fisheries, &c., which would serve as a basis for a natural history of fishes, and for historical and statistical fishery-reports (annals of fisheries). The superintendents of the fisheries would certainly be able to render much valuable assistance in making these observations.

§ 21. The necessity of comparing the course of the fish, their mode of life, and their migrations, with the meteorological and hydrological conditions, for the purpose of increasing our knowledge of their natural history, has long since been recognized, and caused the Royal Scientific Society at Gottenburg, as early as the beginning of this century, whilst the last great herring-fisheries were still going on, to set a prize for the best treatise on "The influence of the currents on the Bohuslän herring-fisheries." When the fisheries ceased, Dr. P. Dubb, the most unprejudiced and learned of our older authors who have given attention to the Bohuslän herring-fisheries, also expressed the opinion that meteorological and hydrological causes occasioned the periodical coming and going of the herrings on the coast of Bohuslän, and proposed that the state should make an appropriation for a scientific investigation of these causes.[19]

It is clear, however, that any scientific investigation which intends to ascertain in how far there is any periodicity in the coming and going of the herrings, and whether such periodicity applies to our herring-fisheries, and, in case this is so, what laws govern this periodicity, ought to extend over at least a century. This length of time need not frighten any one, for long before the century has come to a close such investigations will have yielded results which will amply repay for all the time and trouble.

§ 22. In order to obtain reliable results from combined observations of the fish and fisheries and of meteorological and hydrological facts, it will be necessary, as I have already said in my above-mentioned preliminary report,[20] to have as complete as possible a series of simultaneous observations. This requires a number of persons placed at suitable stations, whose observations are collected in one report, as is done, for example, with regard to the investigations of the sea made on the coast of North Germany.[21] Without such exact, reliable, and uninterrupted observations of the fisheries and their physical conditions made during a longer period and for the purpose of comparison, it will be utterly impossible to reach any higher degree of probability or certainty.

§ 23. All the necessary meteorological observations had best be made by the stations of the Royal Meteorological Central Institute, which have been established on the western coast of Sweden; but for hydrological observations, as well as for observations of the fish and fisheries, we have as yet no stations for making continuous observations[22]. As the application of hydrological data to the natural history of fish and the course of the fisheries absolutely requires that these observations should be *uninterrupted* and go on all the year round, especially during the cold season, when the principal fisheries are carried on, it will be self-evident that all hydrological observations which have been made hitherto chiefly during the summer months cannot be of any very great value. This is not said to detract from the generally acknowledged value of one or the other portion of purely theoretical hydrology, such as we possess in the investigations of the Swedish waters made by *Forchhammer*, *Edlund*, *Meyer*, *Mohn*, *F. L. Eckman*, and others; but what we

[19] *P. Dubb, Anteckningar om sillfisket i Bohuslän* (*Kgl. Vetenskaps Akademiens Handlingar*, 1817), p. 46. Similar investigations have for nearly the same purpose been made in several foreign countries, and some of our own writers have acknowledged their importance.

[20] *Preliminär berättelse*, 1873–1874, p. 70; 1874–1875, p. 17.

[21] See: *Ergebnisse der Beobachtungs stationen an den Deutschen Küsten über die physicklischen Eigenschaften der Ostssee und Nordsee und die Fischerei.* 1873–1876. Berlin, 1874–1877.

[22] Since this was written the Nautical and Meteorological Bureau has been established, which doubtless will supply this want, and furnish the necessary hydrological observations in the shortest time and with the least outlay.

have said merely refers to hydrology as an aid in the study of ichthyology and the fisheries.

§ 24. But the separate hydrological conditions, such as the different currents of the sea, its temperature, the proportion of salt and gas contained in the water, &c., must be studied, not only in themselves and in their relation to the fisheries, but also with regard to their influence on the vegetable and animal life of the sea. The scientific investigations of the sea must therefore endeavor to find the connection between the different organic forms, both in a general way and more especially with regard to those that are of economical value for man. Thus, with regard to the herring, science ought to find out the influence which the diatoms and other animalculæ exercise on it, and the conditions on which their occurrence and distribution depend.[23] The vegetable and animal life of the sea must therefore be studied, not only from a morphological, physiological, descriptive, and physico-geographical point of view, but also with regard to the position which each individual form occupies in the great household of nature.

§ 25. The bottom of the sea must not be forgotten, but must be made the subject of a thorough scientific investigation, both as regards its orographical and geognostical character, not only in itself but chiefly with regard to the influence which it exercises on the currents of the sea and on its vegetable and animal life.

§ 26. The extent of the different scientific investigations is thus clearly given by the very character of the study of natural history, which, the more scientific it is, the more it should be a *comparative* study, because nature forms a continuous whole where one link of the chain is connected with and depends on another, so that no satisfactory result can ever be obtained if one branch is studied as a specialty to the exclusion of those with which it is connected.

§ 27. As has been mentioned above, all such investigations of the sea and the fisheries, if they are to lead to the desired result, must be carried on simultaneously in as large a number of places as possible; for in no other way can a deep insight be gained into the hydrological conditions of the nature of the fisheries themselves, and of their connection with meteorological causes. These investigations must, therefore, as is already done in meteorology, be made by the united efforts of several nations. The investigation of the nature of those small seas round which so many of our modern civilized nations dwell—Englishmen, Dutch, Germans, and Scandinavians—and which, more than any other seas, are full of fishing and sailing vessels, should certainly be of such interest for science, the fisheries and navigation, that there should be no delay in making them. Germany has also in this respect made a beginning, by sending out expeditions, and by having daily observations taken at

[23] *J. MacCulloch "On the herring"* (The Quarterly Journal of Science, Literature, and Arts. XVI. London, 1824), p. 219.

a number of stations.[24] This example is now, to some extent, followed by Norway, where the government has, since the year 1861, instituted, at its expense, a series of investigations of the herring and cod fisheries, and where, at the present time (1876), the chiefly zoological investigations have not only been made in that portion of the open sea where, according to Prof. *G. O. Sars*, the Norwegian herring has its proper home, but also in more distant portions of the ocean. The meteorological societies of Utrecht and Edinburgh have, the former from 1856 to 1864, the latter since 1873, directed their attention to this subject, although they have, so far, at least, not published any of their results.

§ 28. In Sweden, this special branch of natural science has till quite recently been somewhat neglected by the great mass of our people, although it cannot be denied that several of our naturalists have, by their self-sacrificing labors, produced very important and valuable scientific works in some very closely related subjects. Thus there are very few parts of the world whose aquatic fauna and flora are so well known (with regard to the different species) as Bohuslän. All these labors, which certainly must aid the investigation of the fisheries, have been made for an entirely different purpose, and, therefore, as a general rule, pass by the most important points for *our* subject.

The proposition which has been made several times to institute a more or less exhaustive scientific investigation of the fisheries, has, therefore, not yet led to any positive result, the cause of which must chiefly be found in the little importance of our fisheries to the state, an importance which possibly has been somewhat undervalued. The most extensive, and doubtless the most valuable of these older propositions, is the one which Prof. *C. J. Sundevall* made with regard to the scientific and technical sides of the salt-water fisheries, more than twenty years ago.[25]

§ 29. As I have endeavored to show in the foregoing, all these investigations will be most successful and yield the fullest scientific results by, *first*, establishing a complete station for scientific investigations of the sea on the coast of Bohuslän;[26] and, *second*, by appointing a sufficient number of observers to gather continuous data regarding the fisheries at the more important fishing-stations;[27] and, *third*, by instituting a special meteorological and hydrological investigation of the sea.[28]

[24] *Jahresbericht der Commission zur wissenschaftlichen Untersuchung der deutschen Meere in Kiel.* I–III. Berlin. 1873, 1875.

[25] *Stockholm's läns Kgl. Hushållnings-Sällskaps handlingar* (Transactions of the Stockholm Economical Society), VI, p. 211–212.

[26] *Preliminär Berättelse,* 1873–1874, p. 71; 1874–1875, p. 18. See above, § 17. Since this was written, such a station for zoological investigations has been established by the liberality of Dr. *A. Regnell*, at the instigation of Prof. *S. Lovén*, at Christineberg, in the district of Skaftöland; but we do not know in how far this station will make those investigations, which are of most importance for the fisheries.

[27] *Preliminär Berättelse,* 1873–1874, p. 74; 1874–1875, p. 17–18.

[28] *Preliminär Berättelse,* 1873–1874, p. 70; 1874–1875, p. 17–19. This last-mentioned wish may be said to have been fulfilled by the establishment of the "Nautical and Meteorological Bureau," which has been placed under the direct supervision of the Royal Navy Department.

It ought to be an object of special interest for us to obtain an accurate knowledge of our own waters, especially since such a knowledge would be of as great practical and scientific importance to our fisheries and navigation as the geological investigations have been to our agriculture and mining. The proposed investigations of the sea are no less necessary, and will certainly prove just as useful.

§ 30. With regard to the arrangement of these investigations, it may be well in this place to add a few remarks concerning their scientific portion. It has, at least of late years, become a custom with us to put all such investigations into the hands of a committee of older scientists. Although cases could be mentioned where such an arrangement was not only not hurtful but proved of absolute benefit, such cases must certainly be considered as exceptions. The most extensive scientific investigation ever undertaken in Sweden, viz, the geological survey, has been arranged on a totally different plan, which, most assuredly, is the only one which deserves to be followed. In foreign countries, such investigations have, as far as known, hardly ever been placed in the hands of a committee. In Norway, where the fisheries are of much greater importance, and where, consequently, the investigations must be much more extensive, not only the making of a plan for such investigations regarding the herring and the herring-fisheries, but the whole management of the investigations has been placed in the hands of quite a young man, who had not even finished his course at the university.[29] In Germany, these investigations have certainly been entrusted to a commission, but its members do all the principal work themselves.[30] In the United States of North America, the investigations of the fisheries, both as regards their arrangement and their execution, have been placed in the hands of one and the same man.[31] If no person or persons can be found to whom the investigations may be entrusted, it is not worth while to make any; for a committee, even if its members are fully aware of the object of the investigations, can scarcely reach any valuable results through others, unless these possess the faculty of acting for themselves.

In case such a committee should, however, be considered indispensable, it will be important to place at its head a man who will not be led astray by any interest foreign to the proper object of these investigations.

If any investigations of this kind are to be truly useful, their result must be laid before the public just as it is, without any additions or emendations.

§ 31. It has already been mentioned in the foregoing, and is really self-evident, that a well-devised and detailed plan is absolutely necessary, so the object may be reached with the greatest possible saving of labor, time, and money, and to avoid the danger of entering other fields which

---

[29] *Nordisk Tidsskrift for Fiskeri.* VII. Copenhagen, 1872, p. 8.

[30] *Jahresbericht der Commission zur wissenschaftlichen Untersuchung der deutschen Meere in Kiel.* I-III. Berlin, 1873, 1875.

[31] United States Commission of Fish and Fisheries. 1-III. Reports of the Commissioner for 1871-1872, 1872-1873, 1873-1874, and 1874-1875. Washington, 1873-1876.

have no connection with the point in question. Many a practical and scientific investigation has by its result proved the truth of this assertion.

It is, furthermore, necessary to have as complete as possible an outfit of all the required apparatus of the best construction; for as the expenses are small compared with the result, one should not, through negligence or foolish saving, run the risk of obtaining incomplete or incorrect results. No expense, labor, or care should therefore be spared to make the apparatus as complete and as efficient as possible; for the result of the investigations to a large extent depends on this.

We need not mention again, in conclusion, that such numerous and exhaustive investigations must be continued without interruption for a long period of time, and that no grand results must be looked for after a *few* years, or expected from the work of *one* person; for the different portions of the investigation ought to be distributed among a considerable number of persons.

§ 32. A rich field is thus opened for scientific investigations and for practical experiments, which must be worked up in all its parts before that knowledge and experience can be gained which is the essential condition of an entirely satisfactory arrangement and management of our salt-water fisheries. Even if the time when this point shall have been reached, as regards our knowledge of the nature of fish and of the sea, is far distant, we should not hesitate to make use of the little knowledge we possess for improving the condition of our salt-water fisheries and further their success by legislative and administrative reforms, always bearing in mind, however, the incomplete and insufficient character of that knowledge on which our reforms are based. The consciousness of the small extent and insufficiency of our knowledge should not make us indifferent or indolent, and we must not forget that a clear knowledge of difficulties is a good step forward towards overcoming them.

If a basis has to be found for legislation on which the welfare of thousands of human beings depends, no mistakes should be made; least of all such as could easily be avoided by uprightness and a little self-criticism. Our actual knowledge of the subject is unfortunately still so limited that there can be no question of demanding one or the other axiom, which has been by no means fully proved an infallible basis for legislation; but it is rather a duty to point out the defects in our knowledge and endeavor to remedy them as soon as possible. In saying this we do not mean that our knowledge, insignificant in itself, could not serve as a basis for *some* improvements in our legislation; for it must not be forgotten that as long as perfection cannot be reached we must endeavor to make the best use of what we possess. It is, under all circumstances, the duty of the scientific investigator, in a field like the fisheries, to give all the facts just as they are, and not, from a desire to appear as a great discoverer, or from fear of censure, to hide the naked truth.

II.

BRIEF REVIEW OF OUR PRESENT KNOWLEDGE OF THE MODE OF LIFE AND THE MIGRATIONS OF THE HERRING, AND THEIR PHYSICAL AND BIOLOGICAL CAUSES.

1. The rich herring-fisheries which took place on the coast of Bohuslän at the end of last year (1877) and the beginning of the present year (1878) have caused me to publish a brief statement of the present condition of the scientific investigations regarding the mode of life of the herring and its annual and other migrations, as well as of those natural conditions which may be their causes. I have done this with the special object of making the scattered scientific material accessible for the general public, as this certainly would be of practical use in case the herrings should again regularly visit our western coast in any considerable numbers. The amount of knowledge we possess is small; but if brought within the reach of a larger public it may nevertheless have a beneficial influence on the fisheries. So far, we do not possess any scientific collection of all the material prepared with a view to the special needs of our coast; just as little any general and comprehensive epitome of it. In order, therefore, to make this treatise as timely as possible I have collected all the facts in my possession, and by comparing them with the results of the most recent foreign investigations I have endeavored to make them as complete and reliable as possible, which, however, has not materially changed my previous views. These more or less strictly scientific investigations have so far not produced any result which could lead to a complete and much-needed reform in this branch of human knowledge; but in most points our knowledge is just as incomplete and as little critically sifted as it was a hundred years ago.

2. The object of this entirely preliminary review of our present knowledge of the mode of life and the migrations of the herring is, *first*, to give in as brief and as clear and systematic a form as possible all the material which has gradually accumulated in the course of time, in order to make it more accessible both to scientists and to fishermen; *second*, to give a review of the historical development of the more important points in our knowledge, in order, if possible, to prevent mistakes in the future, and to facilitate a more correct understanding of all the circumstances; and, *third*, to show the necessity and to point out the course and possible results of continued scientific investigations. An entirely satisfactory scientific review of all our knowledge presupposes an ample amount of material critically sifted, consisting in observations made uninterruptedly during a long period of time at a large number of places; for, as in meteorology, all truly scientific knowledge must be based on a collection and comparison of as large a number of observations as possible, made by reliable persons during a long period of time and in many different

places. The most extensive and most complete number of observations will bring our knowledge as near actual truth as possible. It will be self-evident that whenever, as in the case of the fisheries, we have to do with periodical changes embracing long periods of time, there will be special need of many and long-continued observations.

3. But as such entirely sufficient material is for the present not accessible, and could not possibly be procured during the comparatively short time which I have been able to devote to the investigation of the herrings and the herring-fisheries on the western coast of Sweden,[32] even if I had had ever so many able assistants, I cannot do better than to use the material on hand, which, certainly from a scientific point of view, is insufficient, but which, nevertheless, possesses some practical value. Even the most careful preparation of this heterogeneous mass of material must, however, be defective, because no satisfactory result can be reached, unless we possess a thorough knowledge of all the points based on the most comprehensive scientific material. Although, in a work like the present, it may be important to give in each case the exact source from which the information has been drawn, I have thought fit to deviate from this generally observed rule, chiefly because it is my intention to treat the whole subject more fully in a larger work which I am preparing, the title of which will be "On the Herring and the Herring-Fisheries," and also to reduce the time[33] and cost of this brief review as much as possible. I nevertheless hope that this little work may be of some use, and prove a help to those of our writers on the fishery-question who by different circumstances are confined to the observations of others.

4. Natural science, considered as a systematic review of all nature, is constantly growing more many-sided and more complicated, in proportion as it develops and as it is simplified by having many different facts condensed into general axioms ("*Les sciences progressent en se simplifiant*," *Leibnitz*). Nature forms a continuous whole, all parts of which are connected by an indissoluble causal connection both among themselves and with the constantly developing universe; the scientific investigation of a natural object or a natural phenomenon can reach completeness only in proportion as it is many-sided. It is a very common mistake to view the different phenomena isolated from others; and science suffers in consequence, entire systems being built on such incomplete views, which may for the time being satisfy at least the less scientific and critical portion of the public, but which are very hard to root out, all the harder

[32] Since July, 1873, I have (commissioned by the government) been busy in collecting such scientific data regarding the natural history of the herring and the herring-fisheries as I deemed necessary for improved legislation on our herring-fisheries.

[33] The greatest difficulty has been experienced and almost insurmountable hinderances placed in the way of gaining the necessary time for a work like the present, by the necessity of carrying on simultaneously the investigations regarding the other portions of the biology of the herring, and the many other different points in this whole herring question, the legislative, administrative, and economical parts of which have taken up the greater portion of my time spent on the coast of Bohuslän.

when they are propped up by the strong pillars of authority. The different changes in the development of science follow entirely different methods, and the investigations of the period immediately preceding ours, devoted more to special branches, have doubtless, through their results, furnished a very necessary basis for the more many-sided tendencies of our present science.

5. Only through long periods of time the human race can gradually reach a higher and more complete knowledge. The history of sciences shows in every branch of knowledge an exceedingly slow development, quicker at times, but at other times slow, stationary, and even retrograding. Even the gaps in the historical material, or the just as common and easily-explained custom of directing attention from the less important to the most important representatives of scientific development, lead a less experienced and thoughtful man to entirely different points than those in question. It is therefore best, as a general rule, not to look for too much "that is new in science" in an author who has in his special line of scientific investigation been preceded by many and prominent writers who have used all the old material, and who, having solved those problems which were of easy solution, have left the most difficult questions unanswered.[34] But even of the most unassuming author we may demand that he shall treat his subject from the present scientific point of view.

6. A remarkable misconception of the aim of scientific work, which even in our time is not altogether rare, is the idea that the object of the naturalist consists chiefly in increasing the scientific material by making rich collections of objects and observations, by describing everything as minutely as possible, and by combining all the facts to a whole, which is then occasionally termed in a somewhat contemptuous manner "a compilation." A higher view of science must, however, disapprove of such a lowering of that most important kind of scientific work, which is far different from mere compilation, which only aims at arranging the works of several authors systematically in one work. In saying this I do not mean to deny the value or necessity of scientific collections or compilations, but merely to raise a faint protest against those who overestimate these compilations and undervalue combined scientific activity. Science doubtless needs as complete and as critically sifted material as possible, but this material in itself is not yet science. The object of the naturalist, therefore, does not only consist in the constant accumulating of observations in order to reveal hitherto unknown facts or to corroborate known facts, but in combining the material gathered by himself and

[34] If we therefore go over the works of older authors on this subject in chronological order, we soon find that their views only very gradually become clearer and more distinct, and that the more critically sifted and arranged axioms which we intend to give we by no means owe altogether to our predecessors. Only by joining the various facts and arranging them systematically they become important, and not least by thereby showing their insufficient character if viewed from a truly scientific point of view.

others into a whole. The last-mentioned work is no less important than the former, and is actually the really scientific portion of the work.

7. Even in biology there are many questions which cannot be satisfactorily answered by *one* person, but which demand the systematic work of *several* naturalists, just as is the case in meteorology. That this is specially necessary with regard to our subject will be evident when we think of the complete series of observations which are needed—observations referring not only to biology, but also to meteorology, hydrology, &c.; for only by making full observations in *all* these directions can we arrive at any satisfactory result. And in order to have these observations as full and systematic as possible there should be a number of stations working according to one and the same plan.[35]

Although the time is certainly very far distant when sufficient observations will have been collected to furnish the necessary material for a satisfactory scientific solution of our problem, we shall see from the following that, as far as Bohuslän is concerned, a complete meteorological station would be extremely useful during a rich fishing-season. Those who devote themselves to the herring-fisheries would gain an increased knowledge of the herring, its mode of life, and its migrations, and their dependence on the changes of the weather, and such a knowledge would certainly be of great practical use to them.

8. In order to gain a correct view of the causes of the irregularities in the mode of life and the migrations of the herring, it is specially necessary to get as complete a knowledge as possible of the influence which physical and biological causes exercise on the herring.

In treating this very difficult and but little known subject it must not be forgotten that hitherto fishing, carried on as a trade, has been almost the only means of observing the influence of such causes; and as fishing is only carried on at those seasons and places and with those implements which promise the greatest success, it will be clear how incomplete and unreliable our knowledge must be. Great caution is required in gathering and receiving information, for mere fanciful and hypothetical theories lead us only farther away from our true object. This part of my work had chiefly to be confined to the collecting and arranging of all the data found in literature.

9. We must first consider the influence which the *sun* and the *moon* exercise. The sun produces day and night and the four seasons. The changes produced by the turning of the earth round its axis and round the sun act not only through the greater or smaller quantity of light, but in a still higher degree through the differences of temperature and the general changes in the weather which they produce. Attempts have

[35] Those who have to do this preparatory work should be fully impressed with the fact that the result of their labor will chiefly benefit the scientists of the future. They must, therefore, sacrifice their own scientific vanity and the hope of reaping the fruits of their labor, but be satisfied to know that it will form part of the foundation of the science of the future. This self-sacrificing work may well be said to ennoble him who engages in it, and who deserves the esteem and gratitude of humanity.

even been made to prove that the sun produces considerable periodical changes in the weather by the regularly changing number of the solar spots. The moon, on the other hand, exercises an important influence through the difference of light which her changes produce,[36] through the tides regulated by her, and the different currents occasioned by the tide, which are chiefly caused by the attraction of the earth. In this last-mentioned respect the sun exercises a similar but weaker influence, which more or less modifies that of the moon.

10. As regards the influence of *sound* on the herring, it is well known how easily it is frightened by any unusual noise. There is no doubt that the sense of hearing is not very strongly developed in the herring, although the old opinion, which was held even by *Linné*, that fish are entirely deaf, has long since been proved to be erroneous. The herring generally returns immediately to the place from which a sudden noise had scared it away. It has also been sufficiently proved that it is highly improbable that a long-continued noise may drive the herrings away from places which they have been in the habit of visiting from time immemorial. But even quite recently it has again been maintained that such has been the case, and steamboats, men-of-war, fortresses, fishermen, coopers, &c., have been accused of having by noise brought the herring-fisheries in some places to a premature end; but *no* satisfactory proof for any such assertions has ever been brought forward. The herrings have disappeared from places where the thunder of cannon has never been heard, and where no steamboats have ever come, whilst they have continued to make their appearance in localities where steamboats and the shooting of guns were frequent. As far as steamboats are concerned we may specially mention the mouth of the Thames, the Firth of Forth, the Sound, the Great Belt, &c. It is certain that thunder causes the herrings to go to so great a depth that they cannot be reached with the common fishing-implements, which will be further proved below by facts; and a violent cannonade may, of course, have the same effect. It is well known that a strong wind blowing towards the coast often hinders or interrupts the approach of the herrings, but it is scarcely probable that the roaring of the waves when dashing against the rocks produces the same effect.

11. Regarding the influence of *electricity*, we still are without the necessary observations on which any certain opinions might be based. It has been known from olden times that during a thunder-storm the herrings seek the deep water, but so far it is impossible to say whether this action is caused by the thunder, the glare of the lightning, the electricity itself, or all these influences combined. *Valenciennes* says that the herrings become violently excited by thunder, and that newly-hatched herrings

[36] There is a deeply rooted conviction among the fishermen that the changes of the moon exercise a most decided influence on the mode of life and the migrations of the herring. In examining the different opinions entertained by fishermen, it was found, however, that all this supposed influence may be reduced to the few above-mentioned cases, and that it is by no means very important.

may suddenly leave the coast from this cause. The Dutch Meteorological Institute at Utrecht says, in its report of 1859, on observations made during the so-called "great fisheries," "that during a thunder-storm accompanied by heavy rain the herrings do not come up to a height where they can be caught till dawn"; and the Meteorological Society of Edinburgh mentions "that according to their observations there will be good fishing the same day when a considerable thunder-storm extends over the greater portion of Eastern Scotland, but that there will be scarcely any fishing on the day following on that portion of the coast-waters which, so to speak, forms the outer edge of the great deep." *G. C. Cederström* believes that he has found "a surprising connection between the course of the fish and electricity." It seems, however, that this "connection" may be ascribed to other causes than the influence of "cosmic electricity."

12. Regarding the influence of *light* we possess a considerable number of observations, and opinions based on them. Light is therefore justly considered as one of the more important causes which exercise an influence on the more or less regular course of the herrings. We shall below give a full account of these observations and of the various opinions based on them. The herring, like many of our salt-water fish, cannot bear a very strong light, or prefers at any rate a dim light. *Neucrantz* therefore supposes that the glare of the lightning drives the herrings into deeper water.[37]

It is well known that both the herring and the small herring stay in deeper water during the day than during the night on account of the light, although the depth, of course, varies according to the greater or smaller intensity of the light. A bright moonlight night is therefore considered less favorable for herring-fishing, because the herring stay in deeper water. It is, however, just as probable that this opinion has been created by the increased difficulty of catching the herrings in a bright light, as by the supposition that in the darkness the herrings find some protection from their pursuers, and therefore seek dark places even during day-time.[38] This probably also causes the influence which the varying degree of transparency of the water exercises on the fish, eries, fishing always being best when the water is less transparent or turbid, or when the rays of light are broken by small waves.

The changes of light and darkness caused by the setting and rising of the sun seem to have a great influence on the herring, exciting them considerably and causing them to come up from the deep; fishing with floating nets is consequently most profitable at those times. If during very dark nights the herrings keep in such deep water that they cannot be reached with the floating nets, it happens occasionally that the light of the rising moon attracts them towards the surface so the nets

[37] To show how sensitive fish are to the influence of light, we may mention the fact, that codfish kept in shallow vessels open to the sun have become blind from the strong light.

[38] When seeking food the herrings are, even in day-time, not unfrequently found near the surface of the water.

can be quickly filled. The oblique direction in which the rays of the sun and the moon strike the surface of the water doubtless has a great influence. In many places the fishermen use torches during fishing-firmly convinced that the herrings are at night-time attracted towards the surface by light. *Valenciennes* mentions an observation made by French fishermen, "that in winter the herrings do not begin to stir until the sun rises, when they come nearer the surface, and that the same also applies to the newly-hatched herrings."

13. Of all meteorological causes, the *temperature of the air* seems to exercise the most decided influence on the herrings, chiefly through its influence on the weather in general, and more especially on the temperature of the water and the quantity of herring-food. The herring (as we shall show farther on) prefers a certain even temperature, and as this is not found near the surface or near the coast, it goes into deeper water or farther away from the coast. We thus possess numerous data showing that both the herring and the small herring have by cold weather been hindered from approaching the coast, or that the fisheries which had already commenced during mild weather have been interrupted from the same cause. During the winter herring-fisheries mild weather is therefore generally considered favorable, although this is certainly not the case under all circumstances. For it is well known from the last great Bohuslän herring-fisheries as well as from the western coast of Scotland, that the remaining of the snow on the mountains and the consequent considerable cold are thought to promise good winter fishing, whilst thawing weather produced by a southeasterly wind is considered unfavorable. Great heat has just as much influence on the fisheries as great cold, and the fisheries which are carried on during the warm season are just as much inconvenienced by the heat as the winter fisheries by the cold. The herrings seem to keep at a certain depth which is regulated by the temperature, so that in moderately mild weather they are nearer the surface, and at a greater depth either in very cold or very warm weather. *Münter* relates, that in Pomerania the wicker baskets for catching herrings must be placed deeper, as in spring the warmth of the sun increases; and it is well known from the large fisheries which during summer and the beginning of autumn are carried on in the northwestern portion of the North Sea, that cooler weather brings good fishing. After a very hot spring and summer the herrings are also said not to come so near the eastern coast of Scotland as they do otherwise. There it has, however, been noticed that a high temperature at the beginning of the fisheries, if immediately followed by a comparatively lower temperature, has been favorable to fishing. It is well known on the Limfiord (Denmark) that a warm summer is generally followed by rich autumn fisheries, so that the saying has become proverbial there, "Many flies, many herrings." Both the herring and the small herring are northern fish which like cool but not entirely cold water, and some of our most important fisheries are therefore carried on during the coldest

season of the year. If the cold is too severe, however, the herrings are thought to become torpid.

The temperature of the air thus seems to exercise its influence chiefly through its extremes, which the herrings cannot well stand, and through its influence on the spawning of the herrings. During the spawning-process the herrings need a certain even temperature, and, therefore, in case of great heat or cold, go to deeper spawning-places. The temperature of the air, therefore, has a much greater influence on the fisheries of spawning herrings than on the common herring-fisheries, whilst the latter are more influenced by wind and current. The influence of the temperature of the air will be more noticeable when fishing is carried on with floating nets than when stationary nets are used.

As different winds produce a different temperature, and thus exercise an essentially different influence on the herrings, we shall, farther on, when speaking of the influence of the wind, give more data regarding the influence which the temperature of the air has on the herring.

After a mild winter, and during a mild, early spring, the spawning of those herrings which spawn in spring and the spring-herring fisheries begin somewhat earlier than otherwise. When in spring the air gets warmer the young herrings seek shallow waters, but when cooler weather sets in they return to the deep water.

14. The *pressure of the air*, as far as known, only exercises an influence on the herrings through the changes which it produces in the weather and in the direction and force of the wind; for all we know regarding the direct influence of the pressure of the air on the sea and thereby on the herrings is the fact that it is not very considerable. Observations which in this respect have been made with regard to other fish are not sufficiently numerous and general to base any certain opinion on them.

The Meteorological Society of Edinburgh has published the results of their comparison between the observations on the course of the fisheries made by the superintendents of fisheries during the years 1867–1872, and the simultaneous meteorological observations, from which it appears that the richest hauls were generally made when the barometer was "high and steady," whilst the fishing was not so good when the barometer was "low." *Frank Buckland*, however, has recently directed attention to an article, by *J. Salmon*, of Lowestoft, in "Land and Water," January 16, 1862, according to which an "unsteady" barometer had during the autumn herring-fisheries in the Southern North Sea (the so-called "Yarmouth fisheries") been favorable to fishing. It is well known that fishing is good when the herrings keep at a moderate depth, and the strength of the wind and the motion of the water are favorable to the use of fishing-apparatus.

In the Skagerack the lower or higher state of the barometer has a considerable influence on the herring-fisheries, both through its influence on the weather in general and more especially on the wind and the currents of the sea.

15. Regarding the influence which *fog* and *haze* have on the herrings, we possess scarcely any information except the observation made by the Dutch floating-net fishermen that foggy weather is not favorable to fishing. Fog and haze probably have a similar influence as a cloudy sky, but doubtless they often injure the fisheries by impeding the sailing near the coast. Nothing definite is known concerning the influence which the different degrees of moisture in the air or the evaporations from the surface of the water may possibly have on the herrings.

16. The *clouds* also must be considered. A cloudy sky is generally thought to be favorable to fishing; this idea probably originates in the circumstance that clouds produce a subdued light, which brings the herrings nearer to the surface and renders it more difficult for them to see and escape the fishing-apparatus. During the night clouds will diminish the warmth, the lower regions of the air and the surface of the water will, therefore, become cooler, and by day-time clouds will prevent the upper portions of the water from getting too warm. The greater or less degree of cloudiness by day or night is therefore of considerable importance for the fisheries, especially during summer.

17. Regarding the influence of *rain*, *snow*, or *hail* on the herrings, it will easily be understood that this chiefly depends on its connection with other meteorological conditions. When it rains or snows the sky is generally covered with clouds, the pressure of the air is lower and the weather milder (the latter caused, among other things, by latent warmth being set free). Rain or snow is therefore generally considered favorable for fishing. Excessive rains, however, followed by floods, are considered unfavorable in Scotland. When rain is accompanied by a thunder-storm or a hurricane, it is of course these last-mentioned phenomena which exercise an influence on the herrings and the herring-fisheries.

That *snow* should scare the herrings away, as *Neucrantz* says, cannot be considered as the general rule, although this may occasionally be the case in consequence of local peculiarities (for example, when fishing is going on at a season of the year when a fall of snow would produce a considerably lower temperature), but the form in which the "falling weather" shows itself (rain or snow) seems to be immaterial. The influence which the fishermen ascribe to the snow remaining on the mountains has been spoken of above.

18. Of all the outward physical causes, the *winds* doubtless have the most important influence on the mode of life and the migrations of the herrings. This influence of the wind may, however, be occasioned by the different temperature, cloudiness, rain, &c., which it produces, by the influence which it exercises on the force, velocity, and direction of the currents by the considerable changes which it brings about in the height of the water, the motion it produces in the upper layers of the water, and the consequent greater or smaller waves, turbid state of the water, or different breaking of the rays of light. This influence of the wind is

still more increased for us by the fact that it either hinders or favors fishing in a greatly varying degree, thus influencing the very sources of our knowledge.

The influence of the wind on the herrings varies considerably as the herrings approach the coast either for the purpose of seeking food or for the purpose of spawning, a circumstance which hitherto has been too much overlooked, which makes the working-up of the material on hand peculiarly difficult.

From the observations at our command it seems to be certain that when herrings approach a coast for the purpose of spawning, they seek quiet waters. Storms often prevent them from reaching their accustomed spawning-places, or compel them, if they can wait no longer or have come close to the coast, to come quite near the land or to go far away from their usual spawning-places which are in deep water, whilst a gentle wind, from whichever quarter it may come, has very little influence on the herrings. Land-wind may, however, as a general rule be considered most favorable to the fisheries during the spawning-season.

Regarding the influence of storms on the herrings during the spawning-season fisheries, we must here give the results of that valuable comparison of facts from the Norwegian spring-herring fisheries, which we owe to the "historical investigations" of *Axel Boeck*. He found that during a violent land-storm the herrings seek the coast, when the best fishing is near the sheltered fishing-places, whilst during violent sea-wind the herrings but rarely go to those places where they are generally caught and which are quite open towards the sea. During long-continued northwesterly and southwesterly storms herrings often visit places where they never come otherwise, and go farther up the fiords. During long-continued sea-wind the spring-herring fisheries cannot be carried on in the otherwise quite regular course from south to north. *Boeck* also relates, of the same fisheries, that during calm weather the herrings often separated into smaller schools, and the chief fisheries did not commence until a southwesterly or northwesterly wind had stirred up the water and mixed the lower and warmer layers with the upper and cooler ones.

On the east coast of Scotland, near the Moray Firth, land-wind is considered favorable to the spawn-herring fisheries; but for those fisheries which are carried on farther out in the open North Sea, sea-wind is considered better, more especially on the banks far from the coast. It is, furthermore, thought that for these fisheries southern winds are better than western winds, and for the herring-fisheries near Yarmouth western winds are most favorable. For those herring-fisheries which in the northwestern portion of the North Sea, at a considerable distance from the coast, are carried on by the Dutch and Germans, northerly winds are considered more favorable than southerly, and westerly than easterly winds. The fisheries in the open sea will, of course, show a considerable difference from the coast-fisheries in regard to the influence of the winds.

Gisler relates that in the Bothnian Gulf storms do not drive the herrings from their places of sojourn far out at sea, and that the herrings, when storms prevent them from approaching the sea-coast, are supposed to spawn out at sea where there are suitable banks.

As regards our present Bohuslän herring-fisheries, they are so insignificant, the spawning-places are all in such sheltered locations, and the fisheries are carried on with so little energy, that it is very difficult to obtain sufficient data on which to found any certain opinion. We have been informed, however, that during the herring-fisheries near South Hisingen, the herrings go farther up towards the mouth of the river during land-wind, when there is good fishing in the neighborhood of Ny-Elfsborg; and when westerly and southerly winds prevail the best fishing is farther out near the coast of Andal and Hästvik.

During that part of the last great-herring fishing-period, when large numbers of herrings spawned near our coast, land-wind was generally considered most favorable to fishing.

As the most important, most profitable, and safest herring-fisheries are those which are carried on during the spawning-season (as during the other portions of the year the fish are not so fat and their course is more uncertain), it will be clear that, although physical conditions exercise a far greater influence on the last-mentioned fisheries, our knowledge of them is much more limited.

When the herrings come to the coast for the purpose of seeking food, wind and especially storm has an entirely different influence on them, and the occurrence of the herrings is chiefly determined by the quantity of herring-food found in a certain place. Thus the herrings often approach a coast with that wind which drives in large quantities of herring-food, and leave the coast with that wind which drives the herring-food away. This explains why during the last great Bohuslän herring-fishing period the herrings did not always appear in the same numbers during one and the same direction of the wind; for during that period when the herrings spawned near the coast land-wind was more favorable than sea-wind, whilst during that period when the great mass of the herrings only approached the coast for another purpose, a strong sea-wind often caused the herrings to enter the fiords and thus brought about the beginning of the fisheries.

In discussing the causes why the great-herring fisheries came to an end, people made the mistake of supposing that all manner of refuse floating in the sea and various noises kept the herrings on the outer coast and hindered them from coming nearer until the storm drove them in. Both from our last great-herring fishing period and from our later much less important herring-fisheries it is well known, however, that a south-easterly wind drives those herrings which have come for the purpose of seeking food away from the coast, because it blows from the land in such a direction that it both increases and accelerates the Skagerack current on the outer coast and directs it farther away from the coast, carrying with it the great mass of herring-food.

If we examine the observations made in different places, we shall soon find that one and the same wind will exercise a different influence, just according as in one place it blows from and in another place towards the coast. It is, for instance, reported from the Firth of Forth that during the winter of 1863 unusually long-continued westerly (*i. e.*, land-)winds hindered the small herrings from going to their usual places in the inner portion of the fiord, and caused some of them to seek shelter near the Granton breakwater, in the middle portion of the fiord. Fishing was consequently not very successful, whilst during the more severe winter of 1866 violent land-winds drove the herrings and small herrings far up into the fiord. As in that region land-wind is more common during a severe winter, such a winter is generally considered more favorable to the fisheries than a mild one, just the contrary from what it is in Bohuslän, where westerly wind and consequent milder temperature and higher water are decidedly more favorable to the fisheries than land-wind with cold temperature and low water. The same difference will become apparent when we compare Bohuslän with Norrland.

By combining all the observations made in different places, we get a small number of general axioms regarding the influence of the wind on those herrings which come to the coast for the purpose of spawning, and these would be the following: Sea-wind is favorable to the fisheries, as it causes the herrings to seek the coast; violent storms compel the herrings to seek deep water or shelter in the calmer fiords, but a brisk sea-wind is generally favorable. Small herrings often seem to seek shelter near the land, and in Bohuslän it has been observed near Hisiugen that in summer during easterly wind the young herrings like to go towards the mouth of the river. As regards the Skagerack herring fisheries, it must not be forgotten that in judging of the influence of the wind one must take into consideration its direction and force, not only near the coast of Bohuslän but also out in the North Sea. In order to form a correct judgment it will therefore be necessary to have synoptic observations of the wind to refer to, and not to forget that the effect frequently will not be noticed till long after the cause.

We furthermore possess different observations and opinions about the influence of the wind on the herrings, which are of a less certain and scientific character, and of which we do not know whether they have been gathered during those fisheries when spawners were caught, or during those fisheries when only fish coming to the coast for food were caught, or from both. We finally possess some data concerning the direct influence of storms on the herrings, for example, that "immediately after a storm they keep near the surface, and are actually giddy and so weak that some are cast on shore or come near the land," and, "after the lapse of a few days, if the storm continues, get quite soft, as if their flesh was melting" (*Gisler*); that storm compels the herrings to keep closer together, and that a violent storm tends to mix the herrings and the small herrings, whilst otherwise they always keep in separate schools.

Thus we know from Scotland that the herrings, immediately before a storm, come near the surface; and in several places in Bohuslän, especially on the coast of Fjellbacka, it has been observed that the herrings go away from the coast against storm and weather, although some time after the storm there is generally again good fishing. It must be remembered that a storm is often preceded by commotion in the water and other causes which may influence the herrings.

19. Regarding the *influence of the weather in general*, it should be noticed that the herrings evidently prefer even and mild weather, free from all extremes; and such weather is, in most cases at least, an essential condition of successful herring-fisheries. The herrings especially dislike all sudden changes in the weather, and it is even asserted that they are so sensitive in this respect that by their actions they show a distinct foreknowledge of changes in the weather, to which circumstance we shall have occasion to refer again.

It must be borne in mind that it is chiefly the general character of the weather[39] which here comes into question, and that in order to understand its influence it is necessary to have reliable and as far as possible complete synoptic weather observations covering a much larger area than a few fishing-stations. It must likewise not be forgotten that the spawning herring is much less influenced by the weather than the herring which comes to the coast to seek food, so that the fishing for the former is much more certain than that of the latter. Fishing with floating nets near a coast is much more dependent on the weather than fishing with stationary nets.

I have on the coast of Bohuslän often heard the failure of the potato crop brought in connection with good herring-fisheries, and with a certain show of reason, as weather which is favorable for the autumn and winter herring-fisheries, is generally unfavorable for the growth of potatoes. This agrees with *Gisler's* observation from the Norrland herring-fisheries, where the fishermen say that when the corn grows well on the land there is not much fishing in the sea, and when the sea is full of fish there is little corn on the land. Similar observations from the last century we find in the works of *Schöning*, *Ström*, and *Lybecker*, and there is doubtless a good deal of truth in them, especially in former times when agriculture was not so advanced as it is now, and to a great extent depended on the state of the weather.

It should finally be mentioned here that there is supposed to be a centennial periodicity in the weather caused by the relative number of solar spots, according to which larger and smaller solar-spot periods have been spoken of, and by the changing position of the moon towards

[39] The general character of the weather is best seen in the winds as the fullest expression of all its determining features, and as having the greatest influence on the mode of life and the migrations of the herrings. The general character of the weather must, however, be taken into consideration, not only during the fishing-season, or the days and weeks immediately preceding it, but also during the different seasons of the year, and for longer and shorter periods, each comprising several years.

the earth, by a change in the direction of the currents of the sea, and especially of the Gulf stream, or by the influence of the polar ice and its floating towards southern regions, &c. If future scientific investigations should prove the existence of such a periodicity, it is quite probable that light will be thrown on many a dark point in the biology of the herring.

20. Among the hydrological phenomena which for the better understanding of the mode of life and the migrations of the herring require our attention, the more or less agitated condition of the water caused by the winds is the most prominent. The *motion of the waves* seems to exercise an influence on the herrings partly by aërating the water, partly by mixing water of different temperature, and finally by the occasionally very violent agitation which is produced even in the lower regions of the water near the bottom. The last-mentioned effect of the motion of the waves, which shall form the subject of special investigations, has a considerable influence especially on the spawning herrings, which need calm waters for spawning, heavy waves often compelling them to seek spawning-places which are sheltered or in the deep water where the motion of the waves is not so perceptible. *Gisler* says that violent storms tend to weaken the herrings when near the coast, and numerous observations corroborate the fact that during such storms the herrings leave the coast or seek sheltered places; and even the heaving of the sea preceding and indicating the approach of a storm seems often to have the same effect. The direction in which this heaving takes place forms, therefore, one of the more important signs, from which the fishermen predict the future of the weather and of the fisheries. In the open sea, far from the coast, the motion of the waves seems to have no or little influence on the herrings; the Dutch so-called "large" herring-fisheries are therefore not at all influenced by it. Smaller surface waves seem always to have a favorable influence on the fisheries, probably because they break and therefore subdue the light.

21. We must also take into consideration the different height of the water which is caused by wind and tide. On coasts where the tide is very perceptible, it doubtless exercises a very considerable influence, especially on the fisheries,—much more so than on the life of the herring. *Perley* says that in the Bay of Fundy, during the spring tide, in early summer, herrings which have come to the coast to seek food are easily caught even during day-time; and at Yarmouth the richest hauls are made when the tide comes in during the three hours before and the three hours after midnight. According to *Ström* all kind of fishing is more successful at Söndmör when the tide is in than when it is out. I have been informed that on the west coast of Scotland the herrings are nearer the surface during slack tides than when the tide is high. On the coast of Bohuslän the tide is not very perceptible during the season when the principal herring-fisheries are going on; and this whole matter has been studied so little that not much can be said regarding the influence which the tide has on the herrings.

In speaking of the influence of the wind, it has already been said, that when it blows *towards* the coast and consequently produces high water it is favorable to the fisheries, whilst when it blows *from* the coast and produces low water it is unfavorable. Very high water, however, is, according to *Gisler*, not favorable to fishing on the coast of Norrland. On the coast of Bohuslän it is considered a general rule that steady and fine weather and high water are best for the fisheries. Very high water is, with us, only caused by violent winds blowing from the sea, which, of course, often interrupt the fisheries. *G. C. Cederström* says that the herrings are more lively when the water is *moderately* high than when it is *very* high.

22. Of all the hydrological causes, the *currents of the sea* doubtless exercise the most important influence on the mode of life and the migrations of the herring. This influence seems chiefly to depend on the herring-food which these currents carry, through the temperature and the nature of their water, and through the aid which they render to the migrations of the herrings.

That the currents influence the herrings in the choice of their spawning-places is chiefly caused by their influence on the temperature of the water and their carrying the necessary food for the young herrings. According to *Eckström*, it is also quite probable that the herrings in moving to a distant spawning-place take advantage of the ease with which the currents carry them towards their destination. This opinion seems to be corroborated by the place where and the direction in which those herrings which spawn in autumn came to the coast of Bohuslän during the last great herring-fisheries. On the course of the spawning herring during the spawning-season the current seems, as *Boeck* already has said, to have but little influence, as the herrings go to their spawning-places both *with* and *against* the current. This does not coincide, however, with the views of other naturalists, according to whom the herrings always go against the current. As land-wind was during our last herring-fishing period considered favorable as long as those herrings which spawn in autumn came to the coast to spawn, and as most of their spawning-places were on the southern coast, it seems that the herrings generally went against the current. The circumstance that fishing for spawning herrings near South Hisingen, at the mouth of the Göta River, is best when land-wind prevails is explained by the fishermen in this way: that the land-wind accelerates the current of fresh water which is going out and increases the intensity of the under-current of salt water with which the herrings are supposed to come in.

With those herrings which come to the coast for the purpose of seeking food, all this is different; for they are chiefly influenced by the occurrence of this food, which is again dependent on the current.[40] Thus

[40] One must be careful not to draw too rash a conclusion that spawning herrings will soon come to a coast in great quantities because many herrings come to that coast to seek food.

during the latter part of our last great herring-fisheries the herrings came with those currents that were going *towards* the coast, and currents going *out* to sea were consequently considered unfavorable to the fisheries, especially when they took their direction from a southeasterly wind. Regarding the coming to our coast of those herrings which occur in the greatest number, and especially of the so-called "old sea-herrings," there is a remarkable agreement between the place where and the order in which they came, and the direction of those currents of the sea which come from the North Sea and the Kattegat along the coast of Bohuslän and in the Skagerack. The current coming from the North Sea goes north of Skagen towards the Pater-noster Rocks, near which it is met by the current from the Kattegat going north; thereupon they both follow the coast, and after having passed Lindesnäs finally go into the North Sea in a westerly direction. The fishing for the herrings coming from the open sea has generally commenced near Tjörn and the Marstrand Islands, from which point the herrings spread towards the north and south, in the former case following the current; and as the current north of Sotenäs turns away from the coast, the herring-fisheries on the northern coast have generally been less certain and less important than those on the central coast.[41]

The young herrings often go with the current, and therefore often undertake comparatively long voyages, of which we have a proof, unfortunately hitherto overlooked, from the coast of Bohuslän, on whose southern portion especially the young of those herrings which during autumn spawn in the Kattegat are often seen.[42]

The herring delights in going with its head against the stream, especially when in search of food, and near the coast it prefers those places where there is a rapid current. The herring is a fish which likes flowing water; but this does not mean that it is driven about by the waves like a piece of wood. According to ancient and modern observations the herring goes just as easy with as against the stream, and when pursued flies as rapidly against it as with it.

During our last great herring-fisheries, and especially towards their end, it was asserted by persons opposed to the boiling of fish-oil that the refuse from the oil-refineries, which was thrown into the sea, prevented the herrings from coming near the coast, whither they were only driven by violent storms and currents. This view, however, was strongly opposed by many fishermen; but *G. C. Cederström* seems still to lean towards the view of the great power of the current over the herrings,

[41] This was probably during the last great herring-fisheries also caused by the circumstance that those herrings which spawn in autumn, as far as known, chiefly spawned on the central and southern coast, whilst those herrings which spawn in winter were far more frequent on the northern coast. With the small herring this is entirely different; for the most extensive small-herring fisheries have always been carried on on the northern coast.

[42] The coast of Bohuslän offers much better protection to the young fish, and is probably in other respects, too, a much more suitable place of sojourn for them than the open coasts of the Kattegat, which are nearly void of organic life.

and maintains that they must give way to storms and strong currents, adding, however, that in that case they either seek shelter or go into deep water.

From observations made during the English and Scotch fisheries we know that the tide, especially in narrow waters, by the regularly changing currents which it produces, exercises a considerable influence on the herring-fisheries. The richest hauls are made when the current is swiftest, because the floating net is then carried over a greater area. The rising of the tide is generally considered more favorable than its falling, and the herrings have often been observed to swim towards the tide.

The greatest difficulty in utilizing our observations of the influence of the currents on the herrings, both for science and for the fisheries, is this, that these observations almost exclusively refer to the surface-currents, although there is reason to suppose that deeper currents have often had an influence on the herrings.

23. Closely connected with and often directly dependent on the currents, at least in the North Sea, is the *color of the water*. A large number of observations made during the so-called "great" herring-fisheries in the open North Sea show that more fish are caught when the water is green than when it is blue. The green color of the water also indicates this in the northern waters, which are richer in "herring-food" and in fish; and the blue color in the southern waters, where there is less herring-food and consequently also fewer herrings. Prof. *G. O. Sars's* observations, made during the summer of 1876, show, however, that the relation between the "herring-food" and the color of the water may be exactly the reverse.

It has already been said above that thick or turbid water is better for fishing than very clear and transparent water.

24. Regarding the influence of the greater or less *saltness* of the water on the herrings there are a number of opinions, some of them directly opposed to the one held by me. Thus it has been supposed that the herrings when spawning sought less salty waters. *H. A. Meyer* believes that those herrings which are found in the western part of the Baltic and which spawn in spring, prefer sea-water mixed with a good deal of fresh water, and mentions various instances from the Schlei-fiord and other places on the Baltic. But on closer examination this does not seem necessarily to follow from these observations; but they seem rather to lead us to this conclusion, that the herrings when about to spawn only look to the convenient location of the spawning-places, and that neither a small degree of saltness, as in the Schlei-fiord, nor great saltness, as on the east coast of Scotland, prevents them from selecting a place. Because the sea-water in the fiords is in many places less salty than in the open sea, it does not follow that the herrings during the spawning-season come to the coast on account of the smaller degree of saltness of the water.[43]

[43] *Neucrantz's* hypothesis (in which he follows *Pliny*) seems more plausible: that the herrings are by their instinct led to spawn near the mouths of rivers, as these localities possess great advantages for the young fish, principally plenty of food and shelter.

In this as in many other respects the spawning herrings are less sensitive, when impelled by a natural instinct—in this case the strong desire to spawn.

The fishermen in the Scotch fiords, however, say that great floods caused by continuous rain often produce failures of the fisheries.

It is well known that herrings, especially young ones, when seeking food come to the mouths of rivers, and this fact has been explained in different ways. *Parnell* thinks that it is caused by the increased temperature produced by the mingling of fresh and salt water, whilst *Duhamel du Monceau* supposes that it can only be caused by violent storms, or that at any rate the herrings do not show the least preference for fresh water. *Boch* [possibly intended for *Boeck.—Translator's note*], again, believes that only violent persecutions by their enemies compel the herrings to seek the mouths of rivers. *G. C. Cederström* thinks that this is caused by the circumstance "that the herrings' need of acid is easier satisfied in fresh water than in salt water." Other authors think the herrings seek the mouths of rivers because they find more food[44] or better shelter from their enemies in the less salty, more turbid, and calmer water. I cannot deny that even spawning herrings are occasionally found in the mouths of rivers, but as we know so little concerning it it may well be supposed that it is caused by extreme conditions of the weather. *Valenciennes*, however, remarks that the herrings do not enter the mouths of rivers until they have spawned. As far as the Bohuslän coast is concerned it must be said that the coast-herrings peculiar to it are caught most frequently near the mouths of the Göta and Glommen Rivers.[45] Near the mouth of the Göta River, especially, it has been observed that the young herrings during summer when east wind prevails like to come near the mouth of the river. It may be mentioned, as a peculiarity, that during the last great herring-fisheries the herrings in the year 1752 accidentally went so far up the Göta River that they were caught near Tingstad, six and a half (English) miles up the river. In the year 1733 the herrings are said to have gone very far up the river Oder, in Germany.

25. It is quite probable that the quantity of *air contained in the sea-water* has some influence on the herrings; and several authors have referred to it. So far we have no direct observations of the influence which the greater or less quantity of air contained in the sea-water has on the herrings; but it is natural to suppose that coast-waters which have been well aërated by a strong sea-breeze attract a larger number

[44] It is scarcely in accordance with actual truth, as *Buckland* supposes, that the small herrings devour the refuse from the sewers of great cities, which is so frequently found near the mouths of English rivers. This refuse, doubtless, produces a large number of lower microscopical organisms, which either directly, or by serving as food for small crustaceans, &c., benefit the herrings.

[45] It must not be forgotten that in these places the herrings can easily be sold at a comparatively high price and are consequently more sought after.

of herrings, and that, therefore, after the breeze has gone down there will be good fishing in those waters.

26. Among the hydrological causes the *temperature of the water* doubtless, next to the currents of the sea, exercises the greatest influence on the herrings. There is good reason to suppose that the herrings prefer a certain even temperature of the water, and that they consequently avoid too warm or too cold water. This degree of temperature, however, differs greatly according to the different locations, fisheries, and races of herrings. The fishing for spawning herrings is, for example, on the east coast of Scotland going on at a season of the year when the temperature of the water is very high (from the middle of July till the middle of September), or very low (January to March). The observations of the Scotch and Dutch Meteorological Societies made during the Scotch and Dutch summer herring-fisheries have shown that the temperature of the water most favorable to these fisheries is about 13° C. (55.4° F.). During the Scotch winter-fisheries, however, the temperature of the water ranges from 4°.5 to 5°.5 C. (40.1° to 41.9° F.), and during the Norwegian spring herring-fisheries it only ranged from 3° to 4° C. (37.4° to 39.2° F.). But our observations are still so incomplete and relate so exclusively to the spawning-herrings, that it is impossible to say anything with absolute certainty excepting the fact that the herrings, when the temperature of the surface waters is either too high or too low, go to deeper waters. But as the changes in the temperature of the water are chiefly caused by the much better known and more fully-observed temperature of the air and by the currents of the sea, we refer the reader to what has been said above (13 and 22) regarding their influence on the herrings. It will be clear that the former has a more decided influence during summer and the latter during winter.

As has been mentioned above (18), the agitation produced in the water by strong wind is favorable to fishing, by mixing the upper and lower layers of the water and by thus equalizing its temperature.

The preference shown by the herrings for an even temperature of the water, has led to attempts to explain thereby the apparent irregularity in the occurrence of the herrings.

27. It is well known, from olden times, that the *formation of ice* on the sea has an influence on the herrings and the herring-fisheries, although we do not possess sufficient observations on this point to form any certain scientific opinion. From Professor *Edlund's* observations regarding the formation of ice in the sea, we know that the formation of bottom-ice will drive the herrings away. It is more than a mere supposition that the formation of bottom-ice not only drives the herrings away but also destroys their eggs and young ones, and on those coasts—for instance, of the Baltic—where ice is frequent in winter, the herrings do not spawn during that season. It is likewise well known that on the coast of Canada much floating ice keeps the herrings and other fish away from the coast. On the coast of Bohuslän, however, it has been observed

that the herrings occasionally go under the ice which has formed for some time,[46] and that there is good fishing when the ice has broken.

28. Regarding the influence of the *depth of the water* and the *pressure of the water* on the herrings we do not possess sufficient observations to form a definite scientific opinion. The scientific authors who have recently treated the biology of the herring have arrived at essentially different views regarding the question whether the herring must be considered as a fish specially fitted for a life near the bottom of the sea in the deep basins outside the coast, or whether its nature fits it better for a life near the banks in the open sea or comparatively nearer the surface. It is clear that the migrations of the herrings must in the latter case go on near the surface, whilst in the former case the herrings would, both in coming and going, seek deep waters.

The Dutch fishermen who use floating nets have observed that the herrings often change very suddenly from one depth to another, but it is not known whether these changes only extend to a few fathoms or to a greater depth; nor do we know whether these changes are made on account of the greater or less pressure of the water, though the last-mentioned hypothesis is certainly highly improbable.

A French naturalist, *Carbonnier*, has expressed the opinion that the herrings, like other fish, are, immediately after having done spawning, compelled to seek deeper waters to obtain the greater pressure "which has become necessary on account of their empty belly"; but although we certainly possess a number of observations all tending to show that the "empty" herrings go into deeper waters than the "full" herrings, we do not know enough on this point to justify us in embracing *Carbonnier's* opinion; for herrings have been known to come to the surface immediately after spawning, and data are not even wanting which go to show that "empty" herrings lived nearer the surface than "full" ones.

29. We must also take into consideration the influence exercised on the herrings by the *character of the bottom*, both as regards its formation and composition and its vegetation. Concerning its orography, we know, from observations made during the Scandinavian herring-fisheries, that the herrings, when approaching the coast, often follow the valleys of the bottom, probably because there they find calmer and more sheltered waters and a more even temperature, not excessively cold in winter nor very warm in summer. Thus the herrings seem, during the last great herring-fisheries, and even several times since, to have approached the southern coast of Bohuslän by way of the so-called "great furrow" or valley; and the depression of the bottom, which, from its northern end, extends towards the Marstrand fiord, has evidently something to do with the circumstance that the great fisheries generally commenced near the islands at the mouth of this fiord. Similar easily distinguished valleys

[46] This phenomenon might possibly be explained by the supposition that a layer of ice prevents the sea-water underneath from getting any colder, as the soil keeps warmer under a cover of snow.

running crosswise from the great valley direct towards the coast, which doubtless have exercised an influence on the herring-fisheries, are found on that part of the coast of Bohuslän lying south of Soten. On the northern coast it has also been observed that both the herrings and the small herrings follow the deep valley between the outer and inner coast.

Hans Ström and, recently, *G. O. Sars* in a more scientific form have mentioned that the current is always stronger near the edge of a bank rising from the depth of the sea, and that in such places there is always a greater abundance of fish-food and of fish. It is quite probable, therefore, that such banks have an influence on the migrations of the herrings, especially if we remember that all the great herring-fisheries are carried on near such banks, and that the North Sea chiefly owes its wealth of fish to such banks.

Inner coast-waters protected by islands, rocks, or rising ground from the violence of the sea, where the herrings can remain undisturbed for a great length of time (by which, however, they become almost worthless), likewise require our attention. All fishing for those herrings which have come for the purpose of seeking food is carried on in such waters, which, moreover, afford excellent shelter for the young herrings.

Regarding the geognostical character of the bottom, it is well known that the herrings when spawning like a hard and firm bottom, and avoid a muddy or clayey bottom, or a sand-bottom whose upper layer is easily stirred up by the waves. It is also well known that the herrings when spawning prefer a bottom covered with plants.

30. We must not overlook the importance of the *geographical location* of a coast in influencing the herrings, although this influence has certainly been overrated by older authors. In this connection we have to consider the greater or less distance of a coast from the poles, its location on the eastern or western side of a continent, and its greater or less distance from the open sea. The herrings do not extend their migrations, in any considerable number, at least, farther south than those regions where the fauna has a decidedly boreal character, or farther north than that point where the sea is filled with polar ice. The location of a coast on the eastern or western side of the Atlantic Ocean is highly important; and herrings are found in great numbers on the west coast of Europe in degrees of latitude (for example, in Nordland and Finmarken) in which on the east coast of America (as in Greenland) they are comparatively scarce. The nearness of the Polar Sea, with its enormous wealth of "herring-food," does not only increase the number, size, and the quality of the herrings, but also influences their selection of spawning-places. Sometimes, however, as is the case in Northeastern America, the Polar Sea may prove hurtful in its influence by large masses of floating ice preventing the herrings from coming near the coast. The greater or less distance of coast-waters from the open sea is of great importance, for experience has shown that waters, such as the Baltic, which are far

from the ocean, are never visited by any of the great schools of herrings, and, therefore, offer no opportunity for any really "great" herring-fisheries. To this circumstance Bohuslän doubtless owes, to some degree at least, the comparatively short duration of its fishery periods, and the long intervals between these periods. In addition to this, it must not be forgotten that the coast of Bohuslän is not near as much laved by northern waters containing much "herring-food" as the coasts of the North Sea.

31. Regarding the influence on the herring of *biological* causes, it will be clear that, in one respect at least, viz, the satisfying of the herrings' demand for food, they exercise a very decided influence, and also that they entirely depend on climatical, hydrological, orographical, and geognostical conditions. From the foregoing it will be clear that the "herring-food," both by its quantity and by the depth in which it is found, will have an influence not only on the herrings but also on the herring-fisheries. Although the herrings certainly do not take any food whilst spawning, the occurrence of "herring-food" has, nevertheless, also an influence on the spawning herrings' course near the coast; since they still need a little food, and all the more, the longer before the commencement of the spawning-season, they come near the coast. It is self-evident that the quantity of "herring-food" in certain coast-waters will determine the size of the herrings living in these waters; and even the greater schools of herrings living in the open sea seem to a great extent to be dependent—at least as far as their young ones are concerned—on the quantity of food found near the spawning-places. As the quantity and occurrence of "herring-food" are dependent partly on the above-mentioned physical conditions and partly on the quantity of food and of organic matter necessary for its formation,[47] it will easily be understood how important it will be to obtain an accurate scientific knowledge on this subject, based on the horizontal and vertical distribution of the "herring-food," with a special view to its dependence on physical conditions; and as the acquiring of such a knowledge involves much trouble and time, very little has so far been done in this direction, so that our knowledge

[47] See: *K. Moebius, Das Thierleben am Boden der Deutschen Ost- und Nordsee,* Berlin, 1871, p. 9, and: *Die Auster und die Austern wirthschaft,* Berlin, 1877, p. 83, in which last-mentioned place he says: "Every cenobitic region possesses in every period of generation the highest degree of life which it is capable of forming and sustaining. All the organic matter contained in such a region is, therefore, completely absorbed by the beings produced there. It is probable, therefore, that in no part of the earth capable of producing life any organic matter is left for spontaneous generation." But is it not possible that exceptional conditions in one place may destroy a species which, also, without limiting the other species belonging to this cenobitic region, could exist there? And might there not be places where a species might flourish and live in large numbers, no matter whether it was formed there or brought there from other places, but where, owing to the want of the conditions necessary for utilizing the organic matter found there, such matter is stored up for the future? And does not geology furnish many similar instances? And do not the Polar regions owe their extraordinary wealth of organic matter to some such process of storing up during milder climatic conditions?

is consequently somewhat incomplete.[48] It will, therefore, scarcely be necessary in this place to adduce further proof of the necessity of soon acquiring such knowledge, which, in the future, will be of great practical use.

The "herring-food," which is chiefly composed of small crustaceans, depends, as to its occurrence and numbers, on lower and smaller organisms, through which it indirectly absorbs all the organizable matter in its neighborhood. Without an exact knowledge of these organisms and the conditions under which they are found we shall never attain to a correct view of the causes producing the "herring-food" and the herrings. It has long since been known that among these microscopic lower organisms, diatoms occupy a prominent place. On the coast of Bohuslän, especially near Fjellbacka, the fishermen have observed that the tender young herrings generally stay among the greenish "slime," which is chiefly found where fresh-water courses empty into the sea; and as they had not observed the spawning of the herrings they drew from this the remarkable conclusion that the young herrings originated from this "slime," chiefly consisting of diatoms. The more frequent occurrence of diatoms in the mouths of rivers is probably also one of the causes of attraction which draws especially the young herrings thither. In the Christiania fiord Prof. *G. O. Sars* has observed a similar organic "slime," consisting chiefly of diatoms, in great quantities, early in spring or immediately after the breaking of the ice; and in the open Skagerack I, myself, whilst examining the so-called "Koster Grounds" during the summer of 1871, found large numbers of diatoms in the current going along the outer coast of Bohuslän.[49] In the Polar seas *Scoresby* has already observed large numbers of diatoms, and his observations have been corroborated by the Swedish Arctic expeditions. The cenobitical and practical importance of the diatoms, as well as their development from a "formless organic slime" ("free indefinite protoplasm," "formless indefinite organic matter") has recently been discussed by Prof. *Youle Hind* and Prof. *G. O. Sars.* As this "sea-slime" chiefly occurs in the Polar seas, especially where there is Polar ice, and is by the currents driven farther south, Professor *Sars* very beautifully remarks that "in the inhospitable Polar Sea, filled with ice, we find the last causes of the inexhaustible wealth of the more temperate seas." The influence of the Polar regions and the Polar ice on the herring-fisheries

[48] Prof. *K. Moebius*, of Kiel, a member of the commission appointed to investigate the German seas, says very truly: "We are still woefully ignorant regarding the physical conditions of a cenobitic region and their relation to the plants and animals of such a region, as also regarding the mutual influence of the plants and animals existing there at one and the same time. So far we know but very little regarding the cenobitic life of the different and distant regions of land and water."—*Zeitschrift für wissenschaftliche Zoologie, XXX.* Supplement, p. 376.

[49] As even at that time I was convinced of the importance to the fisheries of these microscopic organisms, I have repeatedly pointed to the necessity of making them an object of special scientific investigations.

has thus again been brought into prominence, but in a different form and free from all those fanciful ideas which are characteristic of the last century.

32. Among the biological conditions which exercise a great influence on the herrings we must mention the *persecutions of their enemies.*

From the observations of the fishermen it is well known that fish-of-prey, especially the codfish, can, by sudden and determined attacks, scatter whole schools of herrings; and on the west coast of Norway the small schools which either go by the side of the larger ones or follow them are supposed to be chased by codfish.

Whales and seals seem not to have such an influence on the schools of herrings, and the first mentioned owe it chiefly to their colossal size that they are reported as capable of driving the herrings wherever they please. Even in the old Norwegian laws it was forbidden to kill whales which drove herrings towards the coast during the herring fisheries, and even at much later times various authors have spoken of the whales "as commissioned by Providence to lead or rather drive the herrings to those coasts for which our Lord had destined this blessing"; and have also considered it as a special providence "that the whales and fish-of-prey again cause the herrings to fly to their proper home; the eternal ice of the Polar seas, whither their enemies could not follow them and disturb them in the peaceful enjoyment of their rest." Even in our days the Norwegian fishermen not unfrequently ascribe the failure of the fisheries to the circumstance that the herrings have not been properly chased by the whales and fish-of-prey. That the whales and fish-of-prey follow the herrings instead of leading them, for the sole purpose of devouring them, has long since been proved by *Martin*, *Jessen*, *Bloch*, *Nilsson*, and others. It is quite probable, however, that the whales and fish-of-prey cause the herrings to keep closer together.

As regards the other larger enemies of the herrings, we know of none which can exercise the slightest influence on their migrations; but they may, as is especially the case with birds, indicate the place where herrings are, and thus be of great importance to the fishermen.

Among the enemies of the herring we must undoubtedly also count *man*. But man's influence on the mode of life and the migrations of the herring is very insignificant in comparison with the above-mentioned larger animals and fish-of-prey. The chief means by which man persecutes the herrings is the different fishing-apparatus, among which only the net influences the course of the herrings to any degree.

The herrings keep somewhat shy of the nets, and this is the reason why darkness, turbid water, or a surface agitated by the wind is most profitable for net-fishing. When the water is turbid, fishing with floating nets may occasionally be successful even by day time. If the water is too much agitated, fishing with floating nets will generally be unsuccessful. When the herring keep near the surface of the water, but seem unwilling to enter the floating net, they may occasionally be driven into it by making a noise. Fishing with stationary nets is of course somewhat different, because such nets are always set in such a manner that the herrings must strike them in moving from one place to the other.

When a school of herrings during their migration strike a net they are not thereby hindered in their progress, but they go either above or below it, and after having passed it again pursue their course at their usual depth. This makes it possible by placing several nets in a row to catch the whole school. It is generally thought, however, that by placing the nets too close together the herrings are hindered from entering the fiords.

During the spawning-season the herrings are not afraid of the net, even in broad daylight, but rush blindly towards it, seemingly with the intention of squeezing themselves into its meshes, and this in such a furious style that they frequently push down the net entirely. Fishing by day-time with stationary nets or with drag-nets held by anchors may generally be carried on only during the spawning-season or when the water is very turbid. This proves that the herrings are much less afraid of their enemies when animated by the propagating desire than when merely seeking their food. *Kröyer* says very truly with regard to those annual visits which the herrings pay to the coast for the purpose of spawning: "If we consider how little the herrings are disturbed in their course, and how calmly they allow themselves to be caught or devoured by other fish, we must become convinced that fear does not put them to flight and that noise does not scare them, but that their instinct guides them on the way they must follow."

Farther on I shall have occasion to speak of the influence which the enemies of the herrings exercise on their periodical visits (55, 56, 60.)

33. The different outward *conditions of nature* must, however, as regards their influence on the herrings, be considered not only separately, but combined and connectedly. But as different effects spring from the same cause, owing to difference of the seasons, different local circumstances or different objects of the herrings' visits to the coast, and as fishing with different apparatus produces very different results, it will be necessary, in comparing observations from different places and times and from different kinds of fisheries, first to combine those that are more closely connected, so as to obtain an exact knowledge of every kind of fishery during every season of the year, before one can draw general conclusions. Both in collecting and arranging the observations too little regard has in general been paid to the above-mentioned points, or to the mutual relations of the various meteorological, hydrological, orographical, and geognostical data and their relation to biological facts. This has also made it very difficult for me to give a complete review of the observations and opinions of myself and others.

Thus, for example, the physical investigations of the herring-fisheries made by the Dutch and Scotch Meteorological Societies have been made with too exclusively a practical object, and thus only furnished information regarding the conditions of the weather under which the herrings make their appearance in a manner favorable to the floating-net fisheries, whilst they leave us entirely in the dark with regard to many other interesting points.

The great majority of all the observations we possess relate to fishing during the spawning-season; and it is well known that these fisheries are both richer and more certain than those carried on at other seasons, and have therefore generally been considered more important. The catching of those herrings which have come to the coast to seek food has only recently become more important; and we therefore do not as yet possess a sufficient number of observations on these fisheries, which is to be deplored, as the herrings when seeking food are much more dependent on outward natural conditions than when they are spawning.

The physical conditions prevailing in certain waters (among them principally the differences of saltness and climate as being dependent on the weather and the currents of the sea), the geographical location and the orographical and petrographical character of the bottom, will of course exercise a great influence on all the organic beings found in these waters, or, in other words, on their whole cenobitic life; it will be clear therefore that only a complete knowledge of all these conditions, both in the present and in former times, will satisfactorily explain all the phenomena presented by the races of herrings belonging to these waters, such as the different spawning-seasons, the varying degree of fatness, flavor, &c., as well as the more or less regular periodical changes in the migrations of the herrings. Unfortunately our knowledge of all these matters is so far very limited; we thus neither possess very exact knowledge regarding the influence of the weather on hydrological conditions, nor regarding the influence of these last-mentioned conditions on the different biological conditions. Such knowledge, in order to answer its purpose, should not be confined to one locality, but should extend to a large number of fishing-stations, which would enable us to gain more general and satisfactory opinions regarding the combined influence of physical and biological causes on the herrings and their migrations and consequently on the herring-fisheries.

In order to obtain such knowledge it will be necessary to have access not only to good orographical and petrographical maps, as well as to synoptic weather statistics, but also to synoptic tables of hydrological and biological observations.

34. After having considered as fully as my limited time would allow, the influence of physical and biological causes on the herrings and the herring-fisheries, I will go over to a more direct representation of the migrations of the herrings, referring, of course, to all the foregoing observations and opinions.

In order to reach a sufficiently distinct terminology and a more complete knowledge of the whole subject, it will be necessary by way of introduction to give a brief systematized review of fish as to their place of sojourn.[50] The place where fish are found may be considered by itself, or as a basis for dividing the fish into different groups. With regard to the former we can distinguish between the uninterrupted and a more or less accidental or periodical, that is, occasionally interrupted occurrence of fish. The periods may be daily, yearly, or extend to a longer space of time. With regard to the second point, the fish may be divided in the different groups mentioned below, using as a basis either the character of the water where the fish are found, on the geographical location, or the degree of regularity with which fish make their appearance, the extent of time and space of their periodical appearance, and the greater or less stability as to their place of sojourn. It will be clear, however, that these groups cannot always be distinctly defined, but that the lines of demarcation are often somewhat vague, especially between the sub-divisions coming under one and the same head.

A.—With regard to the character of the water, fish may be divided

1. *as to the saltness of the water*, into
   - *a.* fresh-water fish, and
   - *b.* salt-water fish (or sea-fish).

     Between these two groups there is, however, a sort of neutral territory, some kinds of fish living in either water. There are also some "sea-fish" which ascend the rivers for the purpose of spawning, their young ones returning to the sea (so-called "*anadroms*") in order to grow to maturity, or in order to satisfy their craving for food, in which case the fish which have grown to maturity in fresh water go into the sea for the purpose of spawning (so-called "*katadroms*").
2. *as to its degree of agitation* (flowing or stagnant water), into
   - *a.* river or brook fish, and
   - *b.* sea or lake fish.

     Among the lake-fish there are likewise some which occasionally go up into the rivers.

B.—With regard to the geographical location, fish may be divided

1. *with regard to their horizontal distribution*, into
   - *a.* *littoral* or coast fish,[51] that is, fish which always stay near the coast,
   - *b.* *pelagian* or sea fish, that is, fish which always, or at least the greater portion of the year, live in the open sea.

---

[50] An entirely different subject, foreign to my present investigation, is the question of the general geographical distribution of fish over the globe, and their geological distribution throughout the ages.

[51] *J. R. Lorenz, Physikalische Vehältnisse und Vertheilung der Organismen im Quarnerischen Golfe.* Vienna, 1863, p. 332.

2. *with regard to their vertical distribution*, into
   a. *bottom-fish*, that is, fish which always, or the greater part of the year, live near the bottom. With regard to the character of the bottom, we can again distinguish among these *fish which live on clayey* bottoms, and *fish which live on rocky or stony* bottoms, on *banks* or on the slopes extending from the coast to the great deep,[52]

and

   b. *surface-fish*, that is, fish which generally live near the surface of the water.

C.—WITH REGARD TO THE DEGREE OF REGULARITY WITH WHICH FISH MAKE THEIR APPEARANCE, AND THE EXTENT OF TIME AND SPACE OF THEIR PERIODICAL APPEARANCE, FISH MAY BE DIVIDED, INTO

1. *stationary fish, which live always in the same locality.*
2. *migratory fish, fish which only periodically appear in a place with a certain degree of regularity.*[53]

   These migratory fish may be subdivided

   a. with regard to the direction of the migration or its local object into
      α. fish moving chiefly in a horizontal direction and seeking other waters. These are therefore principally found among the coast-fish,
      β. fish moving chiefly in a vertical direction and seeking deeper or shallower waters,
      γ. fish moving both in a horizontal and vertical direction seeking not only a different depth but entirely different waters.
   b. with regard to the local object of the migration within a certain given region, into
      α. fish which limit their migrations to this region, and
      β. fish which occasionally extend their migrations farther.
   c. with regard to the chief object of the periodical migration, into
      α. fish seeking spawning-places, that is, fish which leave their accustomed dwelling-places principally for the purpose of spawning, and
      β. fish seeking food,[54] that is, fish which migrate chiefly to seek food, and which, therefore, are less regular, both as to the length and course of their migrations, because the occurrence of food depends on changeable physical conditions. For this reason those fish which visit certain localities for the special purpose of seeking food are occasionally classed in one group with the other migratory fish under the general name of "roving fish"

[52] This would be the place to draw attention to the division proposed by *Lorenz* (in the above-mentioned place) of the littoral fish into "stationary bottom-fish" and "roving bottom-fish."

[53] It is evidently nothing but gross ignorance which has caused a few of our writers on the fishery-question to use the term "wandering fish" (from the German) instead of the old Swedish term "*flytt fish.*"

[45] The term is taken "*a parte potiori*," which is assuredly the desire for food, which is doubtless stronger than a desire for rest and quiet well-being.

(*strykfiskar*[55]). They may, under the influence of less common physical conditions occasionally appear in places where they are not found otherwise, and where they must therefore be considered as accidental visitors.

d. with regard to the season when these migrations take place, into "winter-fish" and "summer-fish," &c.

e. with regard to the number of periodical visits paid to a coast during the year, into fish which come *once* a year and fish which come *several* times a year.

f. with regard to the steadiness of the visits to a certain locality, into

α. resident fish, and

β. periodical fish.

3. *Erratic fish, that is, pelagian fish which roam about irregularly and only visit a coast accidentally.*[56]

D.—WITH REGARD TO THE GREATER OR LESS STABILITY IN THEIR PLACE OF SOJOURN, the fish may finally be divided into

1. *fish which, on account of the torpor of winter or summer, by sucking themselves fast to objects resting at the bottom or floating about in the water, or from other causes, are generally in a state of rest.*
2. *fish which are more or less in motion, which, with many, assumes the character of a* REGULAR DAILY MOTION. (*Even those fish which generally are in a state of rest may occasionally be classed in this group.*)

35. After giving the above outline of the way in which fish may be divided into different groups, we must ascertain what position the herring holds with regard to these different divisions and subdivisions.

The herring is most decidedly a *salt-water fish*, although it certainly also occurs occasionally in water whose saltness is very limited, for instance, in the northern portion of the Gulf of Bothnia; and for short periods, whilst spawning or seeking food, it will also enter bays and mouths of rivers whose waters contain very little salt.

36. The herring is both a *littoral* and a *pelagian* fish. When young it generally stays near the coast, but begins comparatively early to follow the currents of the sea and go some distance from the coast. As a general rule, however, the herring is more of a littoral fish when young, and a pelagian fish when older. Very small shoals of herrings may sometimes be altogether littoral, the individuals composing them, as far as known, scarcely ever going any distance from the coast. The larger shoals, however, generally spend the greater part of the year out in the open sea, and the great schools are altogether pelagian in their character, visiting the coast only during comparatively short periods of the year.[57]

[55] *S. Nilsson, Förnyad underdånig berättelse om fiskeriarne i Bohuslän.* Stockholm, 1828, p. 15. *Handlingar rörande sillfisket i bohuslänska Skärgården.* Stockholm, 1843, p. 37.

[56] *S. Berthelot, Oiseaux voyageurs et poissons de passage.* II. Paris, 1878, pp. 99 and 125.

[57] Instances are not wanting, however, when such pelagian herrings have, under peculiar circumstances, remained near the coast for a longer time.

It has even been supposed that some schools of herrings spawn on the banks far out in the open sea, without ever approaching the coast for that purpose.

Those herrings which remain stationary near a coast, or which only go a very short distance from it, will best be called "coast-herrings," to distinguish them from the more pelagian or "sea herrings." This difference, although only a relative one, is certainly one which has a considerable influence on the fisheries.

With regard to the place of sojourn and its influence on the character of the herring, a number of different opinions have been advanced in course of time. It is well known from the herring-fisheries in the western part of the North Sea, and especially from the Dutch fisheries, that the herrings, both before and after spawning, are found in large numbers at a considerable distance from the land; and that the herring-fisheries on the eastern and southern coasts of Great Britain proved successful at certain regular intervals; the supposition therefore seemed highly probable that it was the same school of herrings touching the English coasts on their southward journey, and people seemed naturally inclined to ascribe to the herring a decidedly pelagian character, and from these originally popular opinions *Anderson*, and, later, *Gilpin*, developed their strange theories of the migrations of the herring. On the western coast of Scandinavia people had certainly not been in a position, like those on the eastern coast of Great Britain, to base their views on the course of the floating-net fisheries; but the opinion had gradually gained ground (probably through observations made by seamen and fishermen) that the herrings during the time they were not near the coast lived out in the open sea in a northwesterly direction from the region which they used to visit for the purpose of spawning. Gradually, however, *Anderson's* migration theory gained adherence here and there among the educated classes. Thereupon this theory was gradually opposed by *Bloch*, *Lacépède*, *MacCulloch*, and *Nilsson*, the last-mentioned one specially endeavoring to prove the entirely littoral character of the herring, an opinion which, though strongly opposed by fishermen, gradually gained numerous adherents among the naturalists, but which nevertheless is only correct in part, only being applicable to comparatively small schools of herrings. In opposition to this too one-sided opinion of Professor *Nilsson*, *Axel Boeck* has maintained the old distinction made by the fishermen from time immemorial between "sea-herrings" and "coast-herrings," but has not gone so far as to ascribe to the former a thoroughly pelagian character. This has been done, however, on truly scientific grounds by Prof. *G. O. Sars;* and, finally, *G. Winther* and myself have more in detail developed the views which have here been presented regarding the chief place of sojourn of the herrings.

37. The opinion that the herring is a *surface-fish* has only recently begun to gain ground in scientific circles, although, strange to say, it had for a long time been quite prevalent among the fishermen. Herring-

fishing has principally, and in olden times exclusively, been carried on with apparatus that could only be lowered to a comparatively insignificant depth both in the open sea and near the coast; and as these fisheries were going on at different seasons of the year, and as herrings were occasionally seen by seamen, it will be easily understood that the herring was first considered as a surface-fish, none of the older writers on the herring-question having apparently entertained any other opinion. It was *Anderson*, always inclined to the wonderful, who first pronounced another opinion, viz, that the proper home of the herring was the "bottomless deep," under the polar ice, where sharks and codfish could not breathe and disturb the herring in its "proud repose." Although *Anderson's* theory had many adherents, and for nearly a century enjoyed almost universal popularity among naturalists, but little attention seems to have been paid to the question whether the herring was a bottom-fish or not. *Nilsson*, however, pronounces a more distinct opinion on this subject. He supposed that the herring was, properly speaking, a deep-water fish, which, in his opinion, was proved by the fact that herrings are found in the stomachs of codfish, but he most emphatically opposes *Anderson's* view that the herring could only live in very deep water.[58] This view has since then been embraced and further developed by *Axel Boeck*, who, however, went much further than Professor *Nilsson*, and thus, for example, placed the proper home of the Norwegian so-called "spring herring" at the bottom of the deep valley which extends along the coast of Norway; and in proof of his assertion, has mentioned the fact that in the stomach of herrings caught immediately on their arrival remnants of small crustaceans had been found which only live at a very great depth. A different opinion, however, was soon after advanced by Prof. *G. O. Sars*, who at first considered the herring as a "bank-fish," like the codfish,[59] but later as a surface-fish, like *G. C. Cederström*, who, though inclining to the opinion that the herring, like the eel, sometimes concealed itself on the bottom, nevertheless raised some well-founded objections to *Axel Boeck's* assertion, and his mode of proving it.[60] The proofs which have been brought forward in support of the theory that the herring was specially formed for a life at the bottom of the great deep, have been thoroughly refuted by the two above-mentioned authors, and are in no wise re-established by the direct observations made on the west coast of Norway, through which we know that the spawn of herrings, though seldom, is still found as deep as 60 to 100 fathoms, and that herrings are occasionally caught with stationary nets at a depth of 50 to 60 fathoms.

[58] *Nilsson*, strange to say, mentions the unusual pressure of the water, to prove the unreasonableness of *Anderson's* opinion.

[59] Quite recently this opinion has been modified by saying that the herring, although a "surface-fish," nevertheless showed a decided preference for the banks where the codfish live, on account of the stronger current generally found there.

[60] Even *Cederström* mentions the strong pressure of the water, "exercising a hurtful influence on the gills," as the principal cause why the herrings did not go into deeper waters.

From what has hitherto been known, it will be evident that the herrings can certainly go as deep as 100 fathoms, but that they must nevertheless be considered as "surface-fish," which, according to accidental physical causes, change the comparatively short distance from the surface at which they generally live. The circumstance that the herrings, when near the coast, often go into deeper water, might possibly be ascribed to the milder winter temperature and to the cooler summer temperature, as well as to the greater calm and shelter which the deep waters doubtless offer. Whilst spawning, the herrings must, of course, go towards the bottom.

38. Although there may be very small schools of coast-herrings, composed of stationary fish, the herring must, in a general way, be considered as an entirely *migratory fish.*

From the fact that the herring is a surface-fish, it almost necessarily follows that the migrations of the herring generally go in a *horizontal* direction, an opinion, however, which science has only reached quite recently; for, according to *Anderson*, *Nilsson*, *Axel Boeck*, and several other naturalists, the migrations of the herring go in a vertical direction, having for their object a different depth of water with varying pressure and temperature.

With regard to a certain region, the migrations of the herring may be specially directed towards this region, or they may only pass through it, in which latter case the herring would only be a *fish-of-passage* as far as that region is concerned.

The herrings which visit a coast are, with regard to the object of this visit, either herrings which seek a spawning-place or herrings which seek food, in which latter case their coming and going is less regular and more dependent on physical conditions.[61]

With regard to the season when the herrings visit the coast, they are divided into winter-herring, spring-herring, summer-herring, or autumn-herring.

The sea-herrings generally visit the coast only once a year, but sometimes also twice.

With regard to the steadiness of the herrings' visits to a certain coast the herrings must be considered *regular* migratory fish, as far at least as the extent of coast is not too much limited, and not too much regard is paid to the irregularities of those herrings which come in search of food; but the very large schools of sea-herrings may also, as will be shown below, be considered as *periodical* visitors to the coast—such periods extending over eighty to one hundred years.

39. With the exception of those parts of the day when the herring is

[61] This circumstance has given rise to the often quoted and misunderstood saying of *MacCulloch* that the herring is "an apparently most capricious fish." (Quarterly Journal of Science, Literature, and the Arts, XVI, No. XXXII, London, 1824, p. 214.) Most of the instances of the capriciousness of the herrings seem to have been taken from the herring-fisheries on the western coast of Scotland, which have for their object only herrings which have come in search of food.

supposed to be asleep it is in constant motion, and its *daily course*, especially whilst near the coast, is entirely regular.

It is well known now that the herrings generally keep quiet during the middle of the day and the middle of the night, but are in motion mornings and evenings, and that they go into deeper waters by day-time and near the surface by night. The herrings are, therefore, undoubtedly influenced by the changes of light, especially when the rays of light strike the water in a very oblique direction at the rising and setting of the sun or moon, which seems to waken the herrings to new life and cause them to seek those depths which are best for fishing. The principal changes in the daily course of the herrings are doubtless caused by the varying occurrence of the "herring-food" and by the herrings' desire to find shelter from their enemies.

It is also well known that the herrings go near the coast towards sunset and return to the deep about sunrise. According to *Axel Boeck* the Norwegian spring-herrings during the spawning-season go to the spawning-places at nightfall and leave them in the morning, but towards the close of the spawning-season they also come to the coast during the day, so that the fishermen generally consider rich day-fisheries as an indication that the fisheries are approaching their close, a prediction which, however, is not always fulfilled. *G. C. Cederström* says that in the Baltic the autumn-herrings often go into the deep at night, and come nearer the coast towards morning, but that the reverse may also occasionally be the case.

40. Regarding the *annual migrations* of the herrings to and from the coast, a number of different opinions have been advanced in course of time. Some of these I have briefly hinted at when speaking of the character of the herring as a littoral or pelagian fish (36).

Older writers, and the fishermen themselves, seem not to have entertained any other opinion than that the herrings coming from the ocean approached the coast at certain times of the year, generally in a direction from north to south. The idea that the proper home of the herrings might be the Polar Sea, near the North Pole, never entered the mind of the older writers on the fishery-question, who knew that the herring-whales keep farther south than the great whales of the ice-filled Polar Sea; that these last-mentioned whales lived on entirely different food than herrings, and that no herrings had been seen near Spitzbergen, or, as a general rule, farther north than the North Cape in Finmarken. The herring-fishermen, with their limited geographical knowledge, were scarcely able to form or entertain a Polar-migration theory.

The herring-fisheries on the coasts of Shetland, Scotland, and England gradually go farther south in proportion as the spawning-season comes later during the year. The English at the spawning-time generally fished near the coast, and the Dutch had their principal herring-fisheries only in the North Sea. Their knowledge of the herring was consequently limited and led them to suppose that it was one and the same great school of herrings which coming from the north went all round Great Britain and thus produced the different fisheries. This opinion soon became a generally-received axiom, and is as such given in the older works on the herring-question (for example, the works of *Camden*, *Schoock*, and *Ionston*).

This was the origin of the great migration-theory which was by later authors advanced in the *Atlas maritimus et commercialis* published in London in 1728,[62] according to which the herrings were supposed—from want of room and food—to come from the north in such enormous masses that in passing between Greenland and the North Cape (which large space of sea was for them only a narrow sound), they had to keep very close together in order to pass. It was also supposed that after having passed this sea the herrings divided into two schools, the one (again divided in two smaller schools by Ireland) going west and the other east of Great Britain, not to be seen again after they had reached its southern coast. According to this opinion the herrings were supposed to propagate not only near the North Pole but also near the coasts of Great Britain.

Fascinated by these bold conjectures *Johan Anderson*, always inclined towards fantastic opinions, determined to work them up in detail; and he did not forget the Scandinavian countries with their separate divisions of the great school of herrings, which, as he supposed, did not only have their proper home in the "bottomless deep" under the Polar ice, but also sheltered from the persecutions of men and fish-of-prey, increased to such an extent that an enormous number of herrings was forced every year to leave their home and visit those coasts which Providence intended to bless in a special manner. It will not be necessary to dwell any longer on this strange and fanciful theory, especially as nearly every one who has written on the migrations of the herrings has devoted far more time to it than it deserves, even to the exclusion of really important scientific questions. This theory, worthy of *Münchausen*, may serve, however, as an example of the credulity and the total want of critical discernment prevailing not only among the great mass of the people, but also among naturalists, some of whom give this opinion in their works as if it were a scientific truth.

But as another fish—popularly called "herring," and by many considered identical with the herring proper—comes to the eastern coast of North America from the south, it became necessary to modify the migration-theory with regard to America. And this was done by *John Gilpin*, who let the herrings follow the declination of the sun and annually wander in an elliptic course between the Polar circle and the Tropic of Cancer all round the northern part of the Atlantic Ocean, thus constantly avoiding both excessive heat and excessive cold. It will not be necessary, either, to give much time to this fantastic theory, which,

[62] *Dott* according to *M. E. Bloch*, *Oeconomische Naturgeschichte der Fische Deutschlands*, I, Berlin, 1782, p. 188, a statement which, however, seems to be doubtful.

although it has never become general, has been fully described in the well-known works of *Kröyer* and *Axel Boeck*.

Anderson's migration-theory, which finds adherents to this very day, has, in course of time, undergone various changes. Thus some suppose that the herrings go into southern waters for the sake of propagating, and then return with their young to the Polar Sea, which offers an abundance of food (*Pennant* and others), because the herrings were seen to come near the coast full of roe or milt and leave it empty; and because *Anderson's* explanation of this fact seemed insufficient, one began to think of analogous facts in the life of birds and other migratory animals, or was forced to the opinion that there were inconsiderable changes in the herrings' visits to the coast in the direction of their journey, &c.

41. *Anderson's* migration-theory was subjected to a thorough and annihilating criticism by the distinguished ichthyologist *Bloch*, whose opinion has been shared by *Noël de la Morinière*, *Lacépède*, and *Quensel*. With more originality *MacCulloch* has also followed *Bloch's* opinion, and has directed attention to the impossibility of making *Anderson's* theory agree with the evident irregularities in the course of the herrings. Some years before *MacCulloch*, *Couch* had opposed the migration-theory and had described the character of the herring as a "local fish" on the coast of Cornwall. *S. Nilsson* has with great emphasis pronounced himself in opposition to the theory of a central school of herrings near the Pole, and has specially mentioned the physical impossibility of the young herrings developing in the great deep of the Polar Sea; he has also opposed the opinion that every coast should have its special race of herrings distinguished by outward marks and a separate spawning-season, being, consequently, more local and littoral in its character.[63] Professor *Nilsson*, therefore, not only opposed the theory of a central race of herrings near the North Pole, but of such a central race altogether. The dispute caused by Professor *Nilsson's* writings on the regulation of the Bohuslän herring-fisheries caused the Rev. *O. Lundbeck*, pastor of the church at Klädesholmen, to advance the theory of a central race of herrings probably living in the North Sea, to which we owed the great herring-fisheries, and from which, in course of time, the smaller races of coast-herrings had separated, a theory which might possibly be harmonized with the views advanced in *Bloch's* criticism of *Anderson's* theories,[64] but which is in direct opposition to the facts and opinions given by Professor *Nilsson*. *Lundbeck's* hypothesis found no adherents, and seems to

[63] Professor *Nilsson* went so far in his zeal to give to every coast its special local race of herrings, as to entirely deny the possibility of two or more different races occurring "on one and the same coast and under exactly the same natural conditions." This one-sided and doubtless erroneous opinion has recently found an adherent in Prof. *G. O. Sars*.

[64] *Bloch* believes that the time of spawning depends on age and temperature, and from this opinion it may easily be deduced that the herrings which spawn in the North Sea during autumn, and which are actually somewhat smaller than the common herrings, are only the young of the Norwegian spring-herrings.

have been entirely consigned to oblivion. *MacCulloch's* opinion, however, has met with great and universal favor, and has been shared by *Yarrel*, *Parnell*, and others, whilst Professor *Nilsson's* opinion is held by *C. S. Sundevall*, *Ekström*, *Valenciennes*, *Mitchell*, *Berthelot*, and others.

Recent theories regarding the annual migrations and coast-visits of the herrings chiefly differ from each other in this, that the migration is thought to extend over a greater or less territory, just in proportion as the herring is considered a more littoral or more pelagian fish, and in ascribing various natural instincts as the causes of these migrations.

42. We must finally mention the theory advanced in explanation of the fisheries of new herrings on the western coast of Norway, that the herrings do not, as is generally supposed, spawn every year, but only every other year.[65] This theory was in the beginning only used in explanation of the exceptional occurrence of a small number of so-called "herrings-of-passage," but was more generally applied when people began to take into consideration the unusual and frequent occurrence of so-called "new herrings" or "winter herrings." If this theory is correct, the same herrings would, as a general rule, visit the same coast only every other year for the purpose of spawning, and the annual fisheries of spawning-herrings must, therefore, principally be caused by herrings which alternate in their years of spawning. The frequent visits which herrings pay to the coast between the spawning of every other year, but during the spawning-season, must, therefore, be considered as exceptions occurring at the end (or the beginning) of a period of fisheries (for spawners). After having spawned the spring-herrings would not occur among the summer-herrings during the following summer and grow fat, which, as we know, takes place quick enough, but would remain lean for more than a year. Still less is there any cause why the so-called "new herrings" should appear so seldom during the fishing-season and so frequently after its close. This theory* may be convenient for explaining the above-mentioned phenomena, but it cannot be fully accepted unless it can be harmonized with other phenomena, which are the principal ones to demand an explanation.[66] The investigation of the subject only becomes more complicated through such theories, whose value is, therefore, very doubtful.

43. It has been mentioned before (22) that the young herrings begin to wander about at an early age, chiefly to seek food or shelter from their enemies, or possibly more agreeable places of sojourn. It has frequently

[65] Although it is generally supposed that herrings can spawn several times in succession, we have no positive proof of this, and this question seems actually never to have been examined scientifically.

* The theory mentioned in the beginning of 42.—Translator's note.

[66] I do not mean to imply by this that every herring capable of propagating the species must spawn every year, but merely that when possessed of full health and strength every herring will generally do it. It must, moreover, be remembered that the individual fish composing one and the same school do not all become capable of propagating the species at the same age.

been observed that the young herrings, as they grow up, leave the shallow waters near the coast and go into deeper waters farther out towards the ocean, whence, after a while, they return to the coast in company with the older herrings. The knowledge of the details of these migrations is, like our knowledge of their physical and biological causes, so limited that very little can be said regarding them.

Regarding the coming of the herrings from the sea to the coast we only know that during the spawning-season they generally approach the spawning-places in dense schools, coming from the north, and that when visiting the coast for other purposes the schools are smaller and more scattered, extending over a larger stretch of coast, and come both from the north and the south. Those herrings which come to seek food generally remain for some time in the outer waters before they come near the coast, and their visits are neither as regular nor as long as when they come to spawn. But even the great mass of herrings does, during the spawning-season, not remain near the coast longer than one or two months, exceptions from this rule being very rare indeed. Herrings which have thus remained near the coast over their regular time become almost entirely worthless. During the last great Bohuslän herring-fisheries this seems to have occurred more frequently.

In approaching the coast the herrings generally begin at a certain point, spreading from it either to the left or right or in both directions, influenced in this by the weather, the currents of the sea, and the nature of the bottom. The herrings do not like to visit the place where they have spawned, a second time. It has also been noticed that the large herrings do not go as high up the fiords as the small ones, and that when the spawning-season comes in winter or spring the large herrings spawn before the small ones, whilst when the spawning-season comes in summer or autumn the small or younger herrings spawn before the larger and older ones. After spawning, the herrings have often been observed to go nearer the coast than before spawning; fishing with drag-nets may therefore be carried on long after fishing with stationary nets has ceased, as the "empty" fish (those that have spawned) do not easily enter a stationary net.

The going-out of the herrings is generally a much quicker process than their coming-in, and as it is more difficult to catch herrings whilst they are leaving the coast, we know very little about it. After the herrings have left the coast, they do not stay outside any length of time, but immediately go out to sea to seek food and enjoy the greater protection which the deeper water affords. When the herrings have been to the coast for the purpose of spawning, they generally leave the coast in a northerly direction.

With regard to the extent of the annual migrations of the herrings, I have already mentioned the different opinions, and I will only add here that the larger a school of herrings is, the greater will be the extent of territory where they must seek their food, and the farther from the coast must they extend their migrations. It is not known from direct observations how far the largest schools of herrings extend their migrations, but certainly much farther than *MacCulloch*, *Nilsson*, *Boeck*, and their followers assert.

44. The annual migrations of the herrings may be influenced by physical causes both as regards their time and their direction. It is well known that favorable, mild weather accelerates, whilst bad weather retards the approach of the herrings to the coast,[67] and that wind and current may bring a much greater number of herrings to one part of the coast than to another near it. The general rule, however, is that the herrings, when coming in to spawn, visit the place where they were born. When the herrings come in to seek food, they will generally go to those waters where they have been accustomed to find food in the greatest abundance; those physical causes, therefore, which have an influence on the occurrence of food will also influence the direction of the herrings' migrations, as I have had occasion to remark before.

45. The annual migrations of the herrings are chiefly caused by the desire to propagate the species and to seek food. For spawning, the herrings need a suitable bottom for depositing their eggs, a bottom which also must contain a sufficient quantity of food for the young herrings and afford shelter for them. All these requirements are only met near a coast. Even if herrings, as has sometimes been said, not without a show of reason, spawn on the Dogger-Bank or other still more distant banks in the North Sea, this does not disprove our assertion, for it is doubtless only the greater ease with which the young fish can reach the coast from these banks which has made it possible for the herrings to spawn there.[68]

The grown herrings must again go to the ocean to seek their food, which they chiefly find in the currents and those waters which come from the Polar Sea. In some places, however, they find the required food during some part of the year near the coast; and thus there may be fishing towards the end of summer and the beginning of autumn, as on the western coast of Norway, or during autumn and winter, as on the coast of Bohuslän. The influence which the desire for food exercises on the annual migrations of the herrings has sometimes been overrated, so that it has occasionally been considered as the chief cause, even in cases when the desire to propagate was undoubtedly the principal cause.

As the spawning herrings, on account of their being packed more closely together and on account of the steady course which they pursue, are more exposed to the persecutions of their enemies, and as this danger of course increases the nearer they get to the coast, they generally go

[67] See 13. This is applicable chiefly to those herrings which spawn in spring (that is, after the close of winter when the ice is breaking up). On coasts like those of Norway and Scotland, laved by warm currents of the sea, this is less noticeable.

[68] It is therefore not improbable that the young herrings which in such large number are found near the western coast of Norway are at least in part the offspring of herrings which have spawned on the North Sea banks.

into deep water immediately after having spawned, in order to find the necessary shelter, and leave the coast much quicker than they came. The larger herrings seem likewise to thrive better in the open sea than near the coast, and consequently do not stay there longer than is absolutely necessary. *Neucrantz*, however, goes too far when he supposes that the herrings leave the coast only to escape unpleasant physical conditions, for instance, cold or violently agitated water. It has already been mentioned that want of space or the persecutions of enemies have in former times by some been considered as the chief causes of the annual migrations and regular coast-visits of the herrings. Such opinions are, however, no longer entertained, and therefore cannot claim our attention.

46. The great periods (eighty to one hundred years) of the large races of sea-herrings have long since been known, as far as certain points on the coast of Bohuslän are concerned, but have not formed the subject of scientific investigations till the present century. In olden times this phenomenon, as peculiar as it was important from an economical point of view, was connected with religious ideas or with some superstitious notion of the period, and it was only *Ström*, *Lybecker*, *Dubb*, and *MacCulloch* who spoke of these almost inexplicable facts in a scientific manner, From the last-mentioned author we have the expression, often quoted in season and out of season, that the herring is an entirely "capricious" fish.

Nilsson, who had set himself the special aim to find the causes why the Bohuslän herring-fisheries came to an abrupt end in the year 1808, for the first time examined the question regarding the long periodical visits of the so-called "old" or "genuine sea-herrings" to the coasts of the Skagerack in a truly scientific manner. The result of it was, that their conformity to natural laws was totally denied, and the periodicity of our great herring-fisheries was explained by the herrings having been driven away by man, enough young fish, however, having been left every time to gradually produce new fisheries, to be followed in turn by the final expulsion. This opinion, which was stubbornly opposed by the fishermen who in *Lundbeck* had found a literary spokesman, who maintained that it was the nature of the herring "to change its place, and that its visits to our coasts were periodical," was generally shared by the naturalists of that time, such as *C. J. Sundevall*, *S. Lovén*, *W. von Wright*, *Ekström*, *Malm*, *Widegren*, and others. Even *Kröyer* shared this opinion to some extent, as in these migrations of the herrings continuing for many years and then ceasing all of a sudden he could see nothing else but the changes to which all sea-fisheries are subject; at a later time he chiefly ascribed the undeniable fact of these migrations to the increase in the number of birds and fish-of-prey, changes in the weather, the character of the bottom, the sea-water, and excessive fishing with destructive apparatus.

In direct opposition to this view supported by the most influential scientific authorities, *Löberg* and *Axel Boeck*, sustained by popular opinion and by the history of the herring-fisheries of Western Scandinavia, have endeavored to prove that those races of herrings which visit the coasts of Bohuslän and Western Norway change their spawning-places periodically, although they could give no reason why it should be so. *Axel Boeck*, following in part *H. Ström* and other older authors, also showed that there are several tolerably regular changes in the course of the herring-fishery during the great fishing periods. Later *G. O. Sars* has made an ingenious attempt to explain one phase in these changes, viz, the arrival of the herrings at different times during the fishing period, by the varying occurrence of the "herring-food" supposed to depend mostly on meteorological and hydrological conditions; in the beginning, however, he seemed inclined, like *Nilsson*, to deny the periodicity and to suppose that the visits of the herrings continued without interruption unless checked or hindered by man's interference, but later entertained an opinion which agreed more with that of *Boeck*. *G. Winther* has also shared *Boeck's* view in describing the analogous Sound fisheries, whose changes, however, are less marked. Finally I have attempted to explain the periodicity partly by the tendency of the school of herrings to become great through the influence which its size must exercise on the cenobitic conditions in the disproportionately small spawning-places, partly by the greater distance from the spawning-places at which the herrings just on account of the size of the school must seek their food, which depends on supposed periodical changes in the meteorological and hydrological conditions.

47. The great migration periods of the large schools of herrings are marked by very regular changes both in the time of the herrings' annual arrival at the coast and in the locality where they arrive. It must be remembered, however, that all the knowledge we possess of these changes is derived from the fisheries, and that the too exclusive use of apparatus only suited to coast-fishing has made the coast-fisheries more prominent than they would have been otherwise. But wherever herring-fisheries are carried on not only near the coast but also with floating nets at a considerable distance from the land, such as is the case, for example, near the east coast of Scotland, or with purse-nets as on the northeastern coast of North America, the changes are much less marked and have therefore hitherto almost entirely escaped attention. *MacCulloch* has some observations on this point chiefly with regard to the fisheries on the eastern coast of Scotland, but it is only recently that I among others have directed attention to the fact that the herrings on the eastern coast of Scotland have changed their chief place of visit to a point about 100 English miles farther south, and have left the Moray Firth, and that they have commenced to come near the coast earlier during the season, so that the September fisheries are very insignificant now compared to what they were formerly. On the coasts of Bohuslän and Norway, where fishing is chiefly carried on with seines and stationary nets, such changes have been known from time immemorial.

48. Thus Prof. *Hans Ström* in Norway observed that the herrings

during the period they visited the coast of Söndmör (1736–1756) came later and later every year, and predicted, in accordance with an old tradition and the experience had at Stat, that the herring-fisheries of Söndmör would come to an end. This really took place in Bohuslän, where it had been observed already towards the middle of the last great fishery-period, that the herrings came to the coast later and later every year, which led people to fear that as in times of old the herrings might again gradually leave the Swedish coasts. Somewhat later (1782) *Ström* compared the Bohuslän fisheries with those of Norway, and, basing his opinion on their evident similarity, predicted that the end of the Bohuslän fisheries was near at hand.

About ten years later *Lybecker* expresses himself more distinctly, as follows: "If with prophetic eye we could see the future and predict the fate of the fisheries, we might say with a great degree of probability that a change will take place soon. We know from history that when herrings or other fish-of-passage arrive near the coast later and later, and at the same time keep farther and farther away from the coast, this means a change in the migrations of the herrings, and may even point to their leaving the coast entirely. This has been the course of the Norwegian herring-fisheries, and even of the Swedish herring-fisheries during their older periods, and in fact with all those fisheries where fish-of-passage are the principal object, with the only exception of the Scotch and English fisheries. * * * If we take into consideration the roving nature of the herrings and the examples from olden times, it is highly probable that the herrings will come later every year and finally leave our coast altogether."

It had frequently been maintained that too much fishing, and fishing with destructive apparatus, were the proper causes of the herrings coming later every year, and might even lead to the complete cessation of the fisheries; and people therefore made futile attempts to obviate this danger by legislation. As the ominous predictions regarding the herring-fisheries were, however, not immediately fulfilled, they were almost forgotten; but when the herring-fisheries came to an end in the year 1808 people imagined that the herrings arriving later and later every year fully proved the assertion that they had been driven away by the imprudent action of the fishermen. It was said that refuse thrown into the water, and noise, had prevented the herrings from coming near to the coast, that they had spawned in the open sea, and had, then, in consequence of the languor and weakness following the spawning, been driven towards the coast by storms.

During the more recently closed Norwegian spring-herring fisheries it was (according to *Löberg*) noticed, not without anxious forebodings, that the herrings, which in the beginning of the fishing-period did not come near the coast till early in February, gradually came earlier and earlier, so that finally the fisheries commenced before New Year; and that this change was followed by another, the herrings again coming later and later, till the fisheries did not commence before February. This peculiarity, however, was thought to be a consequence of the irregularity with which the herrings visited the same places on the coast. It was not till *Axel Boeck* began to investigate the matter that this whole question was treated from a more scientific standpoint. He showed that the coming of the herrings to the coast at different times during the period was subject to certain rules, and that this regularity in the movements of the herrings was observed not only during the Norwegian spring-herring fisheries of the seventeenth and eighteenth centuries, but also during those herring-fisheries which were going on on the coast of Bohuslän during the second half of the fifteenth and seventeenth centuries. This peculiar phenomenon has therefore become far more important than it was thought to be in former times; and it may well be said to contain the key to the question of the periodicity of the great Scandinavian herring-fisheries. *Boeck* was not able to assign any cause for these entirely regular changes in the time of the herrings' visits to the coast. This has been attempted, however, by *G. O. Sars* and myself, and an account of these attempts will be given below.

49. At a very early period of the last great Bohuslän herring-fisheries it had been observed that the herrings came to the coast a little farther north every year. This became so noticeable that it was mentioned in the Parliamentary Fishery-Commission's report of January 15, 1770. These changes took place in the following order: the fisheries commenced on the central (or as it was then called "northern" coast),[69] but soon after turned to the southern coast, and during the years 1760–1765 went as far south as the northernmost part of the Holland coast, although the coast near Elfsborg and Marstrand was the principal fishing-place. Up to the year 1780 the herrings gradually left the southern coast and chiefly visited the central coast, going as far north as Strömstad from 1773 on, and making their appearance near the Hval Islands in the southern part of Norway from 1778. These changes also attracted the attention of foreigners, and *Lybecker* speaks of them as sure signs that the Bohuslän herring-fisheries were approaching their end.

When the herring-fisheries actually came to a close, and people began to argue about the causes of this misfortune, those who ascribed it to imprudent and destructive fishing saw in these changes a proof of their assertions. They maintained that as soon as the southern coast began to be covered with salting-houses the herrings left this coast and came to the northern coast, and when this too began to be filled with similar establishments, "the herrings seemed disturbed and scared, and came in smaller schools, approaching both the northern and the southern coast in those places where there was least noise and where least refuse was thrown into the water." (*Svensson.*)

[69] *Axel Boeck's* assertion that during the last great Bohuslän fisheries fishing first commenced on the southern coast (*Om Silden og Sildefiskerierne*, p. 106) is therefore not correct.

Even on the west coast of Norway it had (according to *Löberg*) been observed during those spring-herring fisheries which had been going on there since the close of the Bohuslän herring-fisheries in 1808, that the herrings changed the places of their annual visits, and many attempts were made to explain this phenomenon. None of these attempts, however, found much favor, and *Löberg* therefore maintained that probably these changes were caused by the influence of wind and current.

New interest began to be taken in this question when *Axel Boeck* proved that these changes were to some extent regular, and had been shown to be regular not only during the older fisheries, concerning which our sources of information were very meagre, and during the last Norwegian spring-herring fisheries, but also during the great Bohuslän herring-fisheries of the sixteenth and eighteenth centuries.[70]

Boeck has not assigned any natural cause for this regularity in the changes of the herrings' visits, and I believe that I am the first who has made any attempt to find the causes of this phenomenon. I supposed that during *that* part of the fishing-period when the herrings came to the coast for the purpose of spawning, they preferred its northern portion, because the temperature of the water was higher and more even during the later part of the season when they came there, whilst the southern coast would again offer peculiar advantages of temperature during the earlier part of the season when they came there. During that part of the fishing-period, however, when the herrings came to the coast for other purposes than spawning, their choice of a place would chiefly depend on current and wind; fishing on the central and northern coasts was therefore more certain than on the southern coast. More will be said farther on (60, 63) concerning these attempts to explain the changes in the migration of the herrings.

50. Among the peculiar phenomena of the latter part of the last great Bohuslän fisheries, attention has been drawn to the unusual occurrence of small herrings among the larger ones during the last thirty years. This phenomenon has also become more significant since *Axel Boeck* has shown that something very similar took place prior to the close of the last Norwegian spring-herring fisheries, thus seemingly being an indication that the fisheries are approaching their end. During the above-mentioned herring-fisheries none but large herrings were caught, and on the coast of Bohuslän, for example, it was only immediately before and after the annual fisheries that small herrings were caught among them. The Norwegian spring-herring fisheries generally begin every year with straggling herrings and are mostly followed by smaller herrings.

The case which *Boeck* mentions from the Stavanger coast and from the year 1766 does evidently not belong here, as it only proves a less productive local autumn-herring fishery, when herrings of different size are generally caught.

[70] It is highly probable that the same was the case during the Bohuslän fisheries of the second half of the sixteenth century, as the fisheries came to an end much earlier in the southern than in the other parts of the coast.

51. In order to get a more correct idea concerning this peculiar mingling of great and small herrings towards the end of a fishing-period, it will be necessary to consider another phenomenon which seems to be connected with it, and which has hitherto been overlooked. It is known from the last great Bohuslän herring-fisheries that during the last thirty or forty years (therefore during more than half the period) the herrings came to the coast for entirely different purposes than spawning, and that the herrings, though not exactly being a mixture of great and small fish, differed greatly in size, fatness, and general quality.[71] It then became customary to call the full-grown herrings—whose number was small—by a characteristic name, "select herrings" or "fat herrings." It was thought that impure water and noises had caused the herrings to stay in the open sea, until after spawning they were in so weak a condition that a strong wind would drive them towards the coast.

A similar phenomenon has during the last ten years been observed in the Norwegian spring-herring fisheries, so that instead of spawning herrings ("genuine spring-herrings") an inferior kind of herring has been caught, which is called "mixed herring" or "new herring," the number of full-grown herrings being exceedingly small; their spawning-season seems moreover to come somewhat later than that of the genuine spring-herring, which spawns in winter, and they might therefore possibly belong to a small race of coast-herrings which spawn in spring. *Boeck* considered this phenomenon as a dark and mysterious enigma; *G. O. Sars* was the first who—as far as the Norwegian spring-herring fisheries were concerned—examined the whole question from a scientific point of view. As regards our (the Bohuslän) fisheries, it was scarcely possible to suppose that the so-called "new herrings" were spring-herrings which only visited our coast after having spawned, as the well-known characteristics of the "new herrings" prevented their being considered as spring-herrings which had but recently done spawning. It was therefore supposed that they were old and young fish which would not spawn till the following winter, and which during the preceding autumn would keep nearer the coast than the spring-herrings, which latter would, when going to their new spawning-places in the outer deep coast-waters, drive the "new herrings" towards the coast. But *Sars* has failed to explain why such a "driving-in" of great masses of "new herrings" did not take place during the preceding period when the herrings came to the coast for the purpose of spawning. It is tolerably certain that these so-called "new herrings" are, to a great extent at least, such fish as have not yet reached the age when they are capable of spawning; but as this would not apply to the great mass of the herrings, the supposi-

[71] See *H. Ström* "*Sammenligning imellem de Norske og Svenske Fiskerier*" (Comparison between the Norwegian and Swedish fisheries) in *Dansk Museum*, January 1782, p. 7, 9–11, where he supposes that the above-mentioned Bohuslän herrings are the young of the spring-herrings which have emigrated from the Norwegian coast, and are therefore the same as those which at that time were in Norway called "winter-herrings."

tion gained ground that the herrings only spawned every other year, an opinion to which I cannot give positive assent, at least to such an extent as would be necessary.

In comparing the above-mentioned Norwegian and the Bohuslän fisheries (the latter having for their object herrings which come to the coast for a totally different purpose than spawning), it will soon be found that the phenomena are very similar, and that the spawning-herring fisheries are immediately followed by a longer or shorter period of new-herring fisheries; and I have even been led to suppose, basing my opinion on the development of the last great Bohuslän herring-fisheries, that all great herring-fisheries, at least in Bohuslän, are not only followed but also preceded by a similar period of "new-herring" fisheries. By this term as well as by the term "period of spawning-herring fisheries," I understand, of course, only separate portions of one and the same great period of herring-fisheries; and as the intervals between two such great periods on the coast of Bohuslän generally last from sixty to one hundred years (an average of seventy), these intervals would be shorter on coasts which are nearer that part of the ocean where the herrings chiefly find their food, for instance, the western coast of Norway, and possibly in very favorable localities almost imperceptible. It is evident that the "new-herring fisheries" are much less certain than the "spawning-herring fisheries," which is very noticeable on the west coast of Norway.

It is also my opinion that the "new herrings" on the west coast of Norway belong to an entirely different race of herrings from the Norwegian spring-herring,[72] and that they may possibly be identical with those herrings which periodically visit the coast of Bohuslän. The circumstance that the "new herrings" were not generally seen during that part of the fishing-period when spawning-herrings were caught, such as was the case during the Norwegian spring-herring fisheries, is said to be owing to the weaker "new herrings" having been chased by the stronger spawners to those regions which these had formerly occupied themselves. According to this supposition it would seem impossible that great spawning-herring fisheries could be going on simultaneously on the coast of Bohuslän and on the western coast of Norway; but quite likely that those herrings which during autumn have visited the coast of Bohuslän for the purpose of spawning, visit the west coast of Norway later in winter as "new herrings" or "winter-herrings."

These suppositions of mine do not claim any higher scientific value, but may nevertheless prove useful by possibly directing attention to the necessity of collecting and combining facts relating to the history of the herring-fisheries much more than has been done hitherto.

52. After having thus briefly mentioned the different theories regarding the migrations of the large races of herrings and the phenomena which characterize the large fishery-periods, we must mention the nu-

[72] Compare, however, the above-mentioned entirely different opinion regarding the relation of the "winter-herring," communicated by *H. Ström*.

merous attempts to find the *causes* of these migrations; this is probably the most difficult and most violently disputed part of the whole herring-question. As these fishery-periods have been most distinctly marked by long intervals on the coast of Bohuslän, as there they have caused the greatest economical revolutions, and as consequently they are better known, having at an early period been made the subject of scientific investigations, a brief review of the successive views regarding the causes of the cessation of the last great Bohuslän fisheries will be in place here.

When the great herring-fisheries came to an end in the year 1808, and many men experienced heavy losses, causing great want and suffering on the coast of Bohuslän, it was quite natural that in Bohuslän, at least, people began to think seriously about the causes of this great misfortune and about the possibility of retrieving it. When by direct observations it had been ascertained as an undeniable fact that the so-called "old herrings" had really left the Skagerack, the opinion gained ground among the more educated classes that the herrings had been chased away by destructive fishing, by noise, and by the great quantities of refuse from the oil-refineries which had been thrown into the sea; this opinion was publicly expressed in a pamphlet published in 1822 by *Mr. Svensson*, the proprietor of large salting establishments. This as well as the repeated demands for subsidies from the state to promote the fishing interests and help the impoverished fishermen finally induced the government to order a scientific investigation of the whole matter. This investigation was entrusted to Prof. *S. Nilsson*, who during the summer seasons of 1826, 1827, 1832, and 1833 visited the coast of Bohuslän. In his reports he gave his above-mentioned opinion as to the cause why the herring-fisheries had come to an end. But when he proposed, in order to help the Bohuslän herring-fisheries, that fishing with close nets should be prohibited and in its place fishing with stationary nets having wide meshes should be introduced, his general views began to be opposed, especially the one that the herrings should have been driven away by too much fishing, which last-mentioned idea people thought they could trace in his report for 1828. Professor *Nilsson* replied that his expressions had been entirely misunderstood, that he had never "either entertained or expressed" such an "unreasonable idea." During the conferences with a number of fishermen which were held in the year 1833, the opinion that the herrings should have been driven away by too much fishing, by noise, or by impure water was strongly opposed. When the above-mentioned causes no longer found favor, the opinion gained ground that the fisheries had come to an end through the use of close nets, an opinion which found some support in an "ominous" expression of the distinguished ichthyologist Mr. *Bloch*. This opinion was not directly submitted to the criticism of the coast population, and consequently remained in favor for some time, but was finally also abandoned.

Thus one opinion followed the other, and finally it was maintained in a somewhat dictatorial manner that in all these causes which had been assigned for the cessation of the fisheries there was at any rate some particle of truth.

53. In other places, likewise, where the herring-fisheries had ceased or had decreased, the question as to the causes of this phenomenon had become the subject of discussion, and various explanations were attempted, all of which were also applied to Bohuslän. None of these explanations, however, gained general favor; they were, nevertheless, subjected to a thorough criticism by *MacCulloch*, *Kröyer*, *Löberg*, *Axel Boeck*, and others. In spite of this they continued in one shape or the other to be believed and contradicted, and even in our own times attempts have been made to solve the problem by following some of these old-fashioned ideas. There are, besides, quite a number of modern explanations or suppositions which explain the phenomenon by purely natural causes, whose value cannot yet be determined, because these natural causes are not fully understood. Explanations have, however, been attempted, not only by such more or less probable causes, but even by myths or entirely accidental circumstances. The desire to find the causes of all natural phenomena is deeply implanted in human nature, and when science or ingenuity is unable to find these causes people will resort to accidents and myths. Only those persons, however, who are of a strictly critical and thoughtful nature, and who, consequently, both appreciate the difficulties and are but too well acquainted with the defects and the limits of human knowledge, will feel inclined, at times at least, to give up all hope that the question will ever be satisfactorily answered.

54. In examining, however, all the causes which have been assigned for the cessation of the fisheries, we find that they may be arranged under three heads. The *first* of these embraces mythical or accidental causes; the *second*, causes produced by human agencies; and the *third*, biological and physical causes.

A.—*Mythical and accidental causes:*

1. God's wrath on account of the abuse of his gifts, human godlessness and ingratitude, Sunday fishing, refusing to pay tithes to the clergy, &c., or dissatisfaction with the laws and regulations made by the government, &c.;
2. Magic;
3. Spilling of blood;
4. Cruelty shown towards the herring;
5. Using herrings as manure;
6. Occurrences which accidentally took place at the same time when the herrings disappeared, such as conflagrations on the coast, the erecting of new light-houses, &c.;
7. Neglect on the part of the whales and other so-called "herring-hunters" to drive the herrings towards the coast;

8. The capriciousness of the herring or its innate instinct independent of outward circumstances.

B.—*Causes produced by human agencies:*

1. The gradual destruction of the herrings by:
   a. too much fishing, and more especially by catching young herrings in close nets,
   b. preventing the herrings from reaching suitable spawning-places,
   c. destroying the spawn, or
   d. destroying the places where the young fish are accustomed to seek food and shelter;
2. The gradual expulsion of the herrings from the coast-waters by:
   a. noise,
   b. too many fishermen,
   c. disturbing methods of fishing, especially fishing with floating nets too early in the season,
   d. disturbing the spawning-process,
   e. disturbing the spawning-places by fishing or throwing refuse into the water,
   f. leaving dead herrings on the bottom, or throwing guts and gills of fish into the water,
   g. making the water impure by refuse from oil-refineries, &c.,
   h. hindering the herrings from going undisturbedly to their spawning-places,
   i. insufficient and delayed fishing and the consequent scarcity of food for the herrings.

C.—*Biological and physical causes:*

1. Gradual destruction of the herrings by unfavorable weather, an unusual increase in the number of fish-of-prey and birds-of-prey, lack of food, &c.;
2. The forced expulsion of the herrings from the coast-waters by:
   a. the increasing number of fish-of-prey and birds-of-prey,
   b. the lack of food,
   c. a change in the nature of the coast-bottoms, making them unfit for spawning (these changes may be brought about by bottom-ice or floating ice or by the changed nature of the local fauna and flora of the sea),
   d. changes in the meteorological and hydrological conditions or in their periodicity,
   e. the herrings having to go too far from the coast in search of food.

55. It will scarcely be necessary to say that frequently a greater or less number of the above-mentioned causes have been combined in order to produce a greater effect. Nor will it be necessary to give much attention to the causes mentioned under the first heading (A), although they have for a long time met with great favor among the common people. As regards the causes mentioned under the second heading (B), they might practically be divided into two subdivisions different from

those mentioned above, the first embracing all those causes based on the idea that the herrings were either destroyed or driven away by human agencies, and the second embracing all those causes based on the idea that the herrings had been forced to leave the coast from lack of food. The causes mentioned under the first heading have, generally speaking, not met with universal favor, many of their defenders being led by ill-concealed feelings of envy; and the causes mentioned under the second heading are generally in direct opposition to the first. Strange to say, the method of explanation which has recently been adopted by *G. C. Cederström* has seemingly met with some opposition by the knowledge which we have gained concerning the great Bohuslän herring-fisheries, that those fishing-periods lasted longest during which fishing was carried on with the greatest zeal, whilst those were shortest during which fishing was neglected.[73] It ought scarcely to be necessary to refute this theory, and as far as the above-mentioned theories of explanation are concerned, we may point to the, generally speaking, reliable opinions of the authors mentioned before (53). It must be granted that the influence of human agencies on small fisheries may be noticeable; but their influence on the great herring-fisheries is doubtless exceedingly small, and can in no wise be the cause of such phenomena as the cessation of the great herring-fisheries. At the present time it is very rare to find any scientist who still holds to the old and fully refuted opinions.

56. The biological and physical causes doubtless deserve more attention. With regard to them a distinction may be made between the theory that the herrings are periodically destroyed and that they leave the coast during long intervals. *Kröyer* has mentioned that if a school of herrings is by unfavorable weather compelled to spawn in unsuitable places for several years in succession, it may be entirely destroyed or at least be diminished to such a degree that the fisheries must come to an end. Later, *G. C. Cederström* has thrown out the hint that unfavorable outward conditions had towards the end of the last great Bohuslän fishery-period decimated the herrings and thereby brought about the end of the fisheries. All the suppositions, however, cannot explain the periodicity of the great herring-fisheries; for these fisheries, as, among the rest, has been said by *Nilsson*, *Löberg*, and *Boeck*, have come to an end, not from lack of herrings, but because the herrings left those regions where they had been accustomed to come. If this were not the case a gradual decrease in the number of herrings ought to have been noticed towards the end of a fishery-period, but nothing of the kind has ever been observed. There is far greater probability in the supposition that from some outward causes the herrings have been induced to periodically leave those regions which for a long time they had visited regularly. The most prominent among them is this, that the herrings should have

[73] This supposition is by some people harmonized even with the actual deterioration in the quality of the herrings which undoubtedly takes place towards the end of a fishery-period.

been driven away by the increased number of fish-of-prey and birds-of-prey. This originally popular explanation is quite old, and has been mentioned in a somewhat fault-finding manner by *Dubb*, and has been attacked by *Axel Boeck*, but has, nevertheless, quite recently (in the "Book of Inventions") found a scientific champion in Prof. *F. N. Smitt*. He expresses himself regarding the causes of the periodicity of the herring-fisheries as follows: "In all probability it is chiefly to be sought in the common occurrence that when a race of animals which serves as food for others, under peculiarly favorable circumstances increases in a very marked degree, it also attracts more enemies, which increase in number in proportion as the quantity of their food increases. The weaker gives way to the stronger; the herrings, therefore, seek new spawning-places which afford better protection. When on the other hand the fish-of-prey and birds-of-prey do not find the same quantity of food, they diminish in number. If, therefore, a new race of herrings comes to the old spawning-place and again finds its condition favorable, they may increase at a very rapid rate." According to this explanation all herring-fisheries ought to be periodical, for there is scarcely a region where the herrings are not exposed to enemies; but such a complete periodicity as is here spoken of will only be found with very few herring-fisheries. Nor do we find in any fishery-period an uninterrupted increase in the number of the enemies of the herrings. Thus there were rich shark-fisheries on the coast of Bohuslän immediately before and in the beginning of the great herring-fisheries of the sixteenth century; and it is well known that in Scotland and other places the sharks and other powerful enemies of the herrings are very irregular as to the number in which they occur; this is easily explained, as they cannot for their food rely entirely on the herrings, which only visit the coast for a short time every year, because they need rich food all the year round. Very erroneous ideas seem to be entertained quite generally regarding the occurrence of fish-of-prey during coast-herring fisheries, and their dependence on such fisheries. These fish-of-prey, which otherwise are scattered over a large area, gather in dense schools during the herring-fisheries, and are, therefore, noticed more than at other times. Some of these fish-of-prey chiefly depend for their food on the fisheries, and the herrings are by no means as easy a prey as is generally supposed. It will, therefore, be clear, that according to this theory the enemies of the herring ought to increase in proportion as the mass of herrings increases, whereby the herrings would again decrease. This generally takes place, so that the unusual increase of one or the other kind of fish is soon neutralized again. If, therefore, an increase in the number of fish-of-prey were the cause of the herrings moving away from the coast, some cause ought to be assigned explanative of the very strange disturbance of the natural balance between the number of herrings and that of their enemies. And this cannot be done, at least if Professor *Smitt's* supposition is correct, that when the herrings under favorable circumstances increase very rapidly, the same should also be the case with the fish-of-prey. But on the other hand we seem justified in supposing with *Kröyer* and *N. W. Malm*, that a decided increase or decrease of fish-of-prey may cause a temporary decrease or increase of the herrings at least in some of the smaller herring-fisheries.

57. Lack of food has likewise been considered as a cause why herrings have gradually left a coast. *Leeuwenhock* already has considered the varying quantity of food as the principal cause why herrings changed their place of sojourn; but, as far as I know, this idea did not become general until the question of oil-refuse was discussed during the last great Bohuslän fisheries;[74] and when the herrings had ceased to come to the coast of Bohuslän, a gradual decrease in the quantity of food was assigned as one of the causes of this misfortune. This last-mentioned opinion has, so far as the Bohuslän herring-fisheries are concerned, been embraced by Prof. *G. O. Sars*. If we now suppose, with Professor *Smitt*, that the revival of the great herring-fisheries is owing to the accidental arrival of a new "race of herrings," which increased at a rapid rate, it is reasonable to suppose that this rapid increase produces lack of food, and this explanation will seem more plausible than an increase in the number of fish-of-prey from the same cause. But even then it will be difficult to explain why not all herring-fisheries are periodical, which is certainly the case only with a few. This periodicity ought also to be particularly noticeable with those herrings which come to the coast for the purpose of seeking food, which is by no means the case. The theory that the periodicity of the herring-fisheries is dependent on the varying quantity of "herring-food," has been further developed by Prof. *G. O. Sars*, who supposes that the herrings are obliged to seek their food in a certain regular order at a greater or less distance from the coast. By means of this supposition, he endeavored to prove that the Norwegian spring-herring fisheries are not periodical in the proper sense of the word, but that the occasional decrease of these herrings, or their staying away entirely, is caused by the circumstance that at times these herrings had to seek their food so far out at sea that they could only come to the coast late in the season. They would, consequently, have to spawn immediately on the very outermost bottom. The fisheries would, therefore, be short and insignificant.[75] The circumstance that the Norwegian summer-herrings continue to be very flourishing has also induced Pro-

[74] About the same time, Prof. *H. Ström* had directed attention to the circumstance that the "herring-food" may be found in a place one year and stay away the next, and that the herrings would consequently have to follow it up. *Ström* also mentions that the small crustaceans, which principally compose the "herring-food," prefer the currents of the sea, and that the varying direction of those currents may also cause the crustaceans to change their place, and consequently produce new migrations of the herrings. The wind may also have a good deal to do with all these changes.

[75] According to *Axel Boeck*, it is an old opinion in Norway that the herrings, in the intervals between the great fishery-periods, have not left the coast, but have only transferred their spawning-places to inaccessible bottoms. This opinion has, however, been almost entirely abandoned at the present time.

fessor *Sars* to oppose the general opinion that a period of spring-herring fisheries had recently come to an end. Although it has not been sufficiently proved by actual observations that the spring-herrings do no longer spawn in their usual places, this seems scarcely probable; and this explanation would in no wise be applicable to the great Bohuslän herring-fisheries, which, as far as known, agree with the Norwegian spring-herring fisheries in all essential points. From the circumstance that the summer-herring fisheries continue to be just as productive (and occasionally even more so) as during those years when there were still spring-herring fisheries, no such conclusion as the one mentioned above can be drawn with regard to the latter; for it does not follow that, because the spring-herrings have left their old spawning-places, the summer-herrings should also leave the western and northwestern coasts of Norway; nor does the fact that the summer-herrings remain prove that the spring-herrings must do the same.[76] Professor *Sars* seems also to be somewhat undecided with regard to his theory, for he has at a later time, in accordance with a very general opinion in Norway, expressed the idea that there is a direct connection between the Norwegian spring-herring fisheries and the great herring-fisheries. The above-mentioned opinion of Professor *Sars* may, however (as will be shown below, 63), be developed so as to become more generally applicable; and it is, therefore, not impossible that this very opinion contains the germ of a final solution of the problem regarding the migrations of the great shoals of herrings.

58. Intimately connected with this question is the explanation of these migrations from physical causes. The opinion is very old that changes in the physical conditions are the probable cause of the periodicity of the herring-fisheries. The learned and thoughtful Prof. *H. Ström* began already to see the error in the usual method of explaining the periodical cessation of the herring-fisheries by human agencies, and endeavored to explain the greater or less quantity of herrings, and even the fact of their leaving the coast entirely, by physical causes.[77] He mentioned, for instance, that the rich spring-herring fisheries which took place during his stay at Söndmör occurred at a time when the weather was very unfavorable to agriculture, causing a total failure of the crops, and that such failures are generally indicated beforehand by the frequent occurrence of a fish—the horngädda—which generally lives in more southern regions. Dr. *P. Dubb* likewise supposes that changes of weather and current are the true cause of the periodical coming and going of the genuine "sea-herrings" on the coast of Bohuslän. *Ekström* has explained the circumstance that on the coast of Södermanland the herrings are more

[76] See 45; also, *H. Strom, Söndmör*, I, p. 468; *Dansk Museum*, January, 1782, p. 3–4; *A. Boeck, Om Silden*, p. 130; *A. V. Ljungman, Preliminär berattelse* for 1873–'74, p. 6.

[77] *H. Strom, Dansk Museum*, January, 1782, p. 3–9. In this passage he points out that changes in the condition of the ice near the North Pole probably cause the periodicity in the migrations of the herrings.

frequent than on the coasts of Stockholm and Oestgöta, by the different direction of the wind prevailing during the time when the herrings come near the coast. In the seventeenth volume of the Encyclopædia Britannica (last edition) the irregularities in the visits of the herrings to the west coast of Scotland have, in accordance with the opinion of *Pennant* and *MacCulloch*, been explained by well-known changes in the Gulf Stream, which changes should cause the herrings, which always seek an even temperature, to change their old places and seek new ones. This opinion has recently been taken up by *Frank Buckland*. Prof. *G. O. Sars* has finally endeavored to prove that periodical changes, connected with the movement of the great herrings towards the north, probably take place in the currents of the sea on the northern coast of Norway. He thinks that these changes are indicated by the periodical occurrence of wood, &c., washed ashore from foreign countries, and maintains that his theory of the "herring-food" being found at different times at different distances from the coast presupposes regular periodical changes in the currents of the sea.

59. After having given the above historical review of the different theories regarding the biological or physical causes of the periodicity of the herring-fisheries, it remains for me to indicate the manner in which I have further developed these theories during the last five years. I started with an investigation of the question, "Which fisheries are periodically regular, and which not"; and I have found that only very *large* fisheries carried on near the coast and dependent on the propagating instinct of fish are periodically regular. From this I have drawn the conclusion that it is the *number* of a race of herrings which chiefly causes them to periodically change their spawning-places near the coast. Other fisheries show irregularities with regard to the coming of the herrings, but no periods marked by a complete cessation of the fisheries and by regular changes. Thus the Norwegian summer-herring fisheries owe their existence to one or several large races of herrings; but, as far as known, they are not periodical. This seems also to be the case with those fisheries which are carried on in the open sea at some distance from the coast, where the fishermen follow the herrings to their spawning-places. Our knowledge of these herring-fisheries, however, is very incomplete, for we know very little regarding the quality of the herrings and the possible changes of time and place.

The races (or schools) of herrings may nevertheless practically be divided into large and small ones, the line between the two not being very sharply drawn. And the very fact of certain races of herrings being large has led me to explain the periodicity of the herring-fisheries in two different ways, which I shall give below.

60. The enormous numbers in which the large races of herrings make their appearance must doubtless (especially when they select a limited extent of coast for their spawning-place) produce a great change in the natural condition of the coast-waters, both by their furnishing food to

numberless marine animals, and by their consuming a large quantity of food. This change may finally become so marked as to make these waters unfit for spawning, at least for a large race of herrings. The only, and nearly always sufficient, protection of the herring in its combat for existence is its fecundity; and although we must acknowledge, with *Kröyer*, that "danger does not put the herrings to flight, and that noise does not scare them away, but that their instinct points out the way they have to follow," this very instinct would naturally lead them to leave spawning-places which are no longer fit for spawning and seek new ones. It will therefore be clear that in proportion as the extent and nature of the spawning-places no longer correspond with the size of the race of herrings, the influence of this size will make itself more and more felt, and produce a change of time and place in the fisheries.

In order to judge the probability of this theory it will be necessary to find out how the above-mentioned regular changes of time and place of the herrings' visits to the coast can be explained by it. When a large race of herrings is compelled to seek another and distant spawning-place (in the case of Bohuslän, a more easterly one), they will, in consequence, come later in the year; but if they extend their "hunting-excursions" so as to come a little nearer, or the spawning in a still undisturbed spawning-place occupies less time, they may come earlier, and after having spawned, return earlier to their former district. Thus the herrings would gradually come earlier and extend their visits to other parts of the coast (in Bohuslän farther south) until they have brought their "hunting-district" as near the coast as possible. As this was really the case, and as the farthest spawning-places (in Bohuslän those on the southern coast) were disturbed, the herrings were compelled to seek their spawning-places nearer and nearer to the point where they first approached the coast (in Bohuslän farther north). The search for spawning-places took up some time; the herrings consequently came later and also left the coast later. They therefore also arrived later at their "hunting-grounds," and left the grounds later for the purpose of spawning. In proportion as they reached the "hunting-grounds" later, they would have to advance farther (that is, farther north) into these, because they would arrive in a more starved condition, and therefore require more food, which could only be obtained by scouring through a larger extent of water. The circumstance that during the last great Bohuslän fishery-period the herrings irregularly visited the southern, central, and northern coast, is easily explained by the fact that they did not come to the coast for the purpose of spawning, and that they always waited for some time outside the coast before coming nearer.

61. This theory of the successive disturbance of the spawning-places may possibly also explain the more incomplete periodicity which, as an experienced fisherman informed me, is noticed in the Sound and in the Great Belt, where the herrings seem to alternate between eastern and western spawning-places, so that one year there is good fishing in the

Sound and the next year in the Great Belt.[78] A similar alternation, although, of course, on a much larger scale, might well be supposed to take place between the eastern and western shore of the North Sea.[79]

62. At the first Scandinavian Fishery-Exposition held at Aalesund in 1864, *Axel Boeck* is said to have advanced the opinion that the end of the last Bohuslän fishery-period was contemporaneous with the recommencement of the Norwegian spring-herring fisheries, because the Bohuslän herrings had transferred their spawning-places to those banks in the North Sea which the spring-herrings had been accustomed to visit during those years when the spring-herring fisheries had ceased. The spring-herrings, therefore, on finding their spawning-places taken up had returned to the west coast of Norway. Afterward *Boeck*, it seems with good reason, abandoned this opinion, but the attempt to connect the migrations of two great races of herrings with each other nevertheless deserves attention. The same opinion has been entertained by other writers both before and after *Boeck*. When during the last winter a race of herrings, similar to the "new herrings," visited the coast of Bohuslän, I expressed the opinion that these herrings had been forced to give way to the Norwegian spring-herrings, which about ten years ago had begun to leave their old spawning-places on the west coast of Norway. However this may be, it cannot be doubted that the movement of one race of herrings has an influence on that of other herrings, although this influence may by no means be instantaneous.[80] It is clear that the coming in of larger masses of herrings in one and the same place, though at different seasons of the year, will essentially increase the influence of the disproportionately large races of herrings on a limited extent of coast-waters. It may also be possible for a race of herrings to be driven from its territory by a larger and stronger race, especially if the latter finds its territory too limited in proportion to its size.

This explanation has also opened out new views by applying it to the distinction made between the "new-herring fisheries" and the "spawning-herring fisheries" properly so called, for to some extent, at least, it may explain the fact that "new-herring fisheries" both precede and close a large fishery-period. It also facilitates the explanation of the regular changes of time and place in the visits of the herrings during a fishery-

[78] Although it has not been fully proved that such a mutual periodicity exists between the herring-fisheries in the Sound and the Great Belt, this whole matter deserves attention and ought to be investigated.

[79] A fact which may well be connected with the migrations of the herrings from the western to the eastern part of the North Sea, is the cessation of the otherwise regular whale-fisheries near the Faroe Islands from 1754–1776. But this fact, like the great migrations of the herrings in the North Sea, may be explained by supposing that the fish moved in a northern and southern instead of an eastern and western direction.

[80] Even herrings of different age, though belonging to one and the same race, may thus have to give way to each other, and the proposed method of explanation may be applied to the different theories regarding the relationship and maturity of the "new herrings" advanced by *H. Ström*, *G. O. Sars*, and myself.

period. The chief difficulty consists in finding a "*primus motor*" or the original cause which makes the great races of herrings move; and until a better explanation is found I would assign as this cause the change of biological conditions produced by the great size of a race of herrings, and a supposed periodicity of meteorological and hydrological causes, and possibly a combination of both.

63. The other and perhaps simpler way of explaining the periodicity of the herring-fisheries by the size of the race of herrings, may be reached by considering the very evident effect of this cause, viz, that the herrings are compelled to seek their food on a larger territory, farther from the coast, and more dependent on the changes of weather and current; and here Prof. *G. O. Sars's* theory regarding the visits of the herrings at different times during a fishing-period would come in well. In accordance with this theory it might well be supposed that the herrings would finally have to seek their food at such a distance that they could not reach their old spawning-places at the right time, but would have to select other spawning-places which were within easier reach. But as the herrings chiefly live on small crustaceans floating about in the water, we must, in following this theory, suppose a change in the occurrence of this "herring-food," which could scarcely be explained except by a periodicity of the currents and by the changes in the weather which principally produce this periodicity. No one has so far, however, been able to show the existence of such a periodicity, although it has been supposed to exist, and although there are facts which point in this direction. This hypothetical explanation is, therefore, nothing but a further development of the old opinion that the periodicity of the herring-fisheries is caused by physical changes, and its chief merit consists in indicating by the very point from which it starts the cause why not all herring-fisheries are periodical in consequence of these changes.

It will be clear, however, that this explanation can easily be harmonized with the regular changes of time and place in the so-called "landing" of the herrings, and this consideration should by no means be lost sight of. When the herrings are near the coast they can also land sooner and go farther along the coast (in Bohuslän and Western Norway farther south) than when they are far from the land in the open sea. Regular changes in the one will, therefore, also produce regular changes in the other. It will be more difficult, however, to explain in this way the exceptions from this regular course of changes in the fishery during a fishery-period. And such exceptions have occurred both during the last Norwegian spring-herring fisheries and during the latter part of the last Bohuslän fisheries. This theory may also be further developed by combining it with the other theory that the one race of herrings has to give way to the other so that the great races of herrings would be uninterruptedly moving backward and forward.

64. If, as I have supposed, two great herring-fisheries should be inti-

mately connected with each other, it must also be supposed that the regular changes of time and place are likewise connected. The later arrival of the herrings in a more northerly place than usual would indicate the near end of a fishing-period in one case as an earlier arrival in a more southern place in the other.

65. The favorable conditions on which the development of a great race of herrings depends are only found on a coast which is near the open sea. The great race of herrings which has periodically visited the coast of Bohuslän can scarcely be supposed to have developed there (at least not under conditions like the present), and the greater distance from the sea (and more especially from the Polar currents with their abundance of "herring-food") is doubtless the chief cause why the Bohuslän fishery-periods are more distinct, shorter, and separated by longer intervals than, for example, the fishery-periods of Western Norway. The same cause might also explain the fact that the sea-herrings for a number of years came sooner to the western coast of Norway than to the coast of Bohuslän, and that the space of time between the earliest and the latest arrival of the herrings near the coast was so much greater during the last Bohuslän than during the last West Norway fishery-period.

Another cause of the relative shortness of the Bohuslän fishery-periods may be found in the circumstance that, as the herrings belonging to the coast of Bohuslän spawn in spring, this season is the most suitable for spawning on this coast, whilst in the Kattegat, the Sound, and the Belts autumn is the more favorable season. As the sea-herrings which visited the coast of Bohuslän during the great fishery-periods chiefly spawned in autumn, it must be supposed that during their visit to the Skagerack they were compelled to spawn under comparatively unfavorable conditions, especially as regards the newly hatched young ones. This may, to some extent, have induced them to seek other spawning-places sooner than would have been the case otherwise. It is also quite likely that the coast of Bohuslän, towards the end of the fishery-period, when the herrings did not come in till December, was less inviting (at least for those herrings which spawned during winter). This may also have been caused by unfavorable weather. If, as *Axel Boeck* has shown, a temperature of the water of + 3° C. (37.4° F.) is not injurious to the herrings, it does not follow that this is not the case with a lower temperature accompanied by the formation of bottom ice. As most of the spawning-places on the coast of Bohuslän are located in shallow water, the cold must produce far greater changes in the temperature of the water than in the spawning-places on the western coast of Norway, which are located in deeper waters and are exposed to a much more powerful current of the sea with a far more even temperature. Too little attention seems to have been paid to the great injuries which several closely following severe winters must have inflicted on the spawning-places of the herrings. This unsuitableness of the coast of Bohuslän as a spawning-place for a great race of herrings which are accustomed to spawn in winter, would be another reason for opposing the idea that the Norwegian spring-herrings had alternated their visits between the coast of Bohuslän and the west coast of Norway.[81]

66. For the sake of comparison and completeness, we must also give an account of those circumstances which have been assigned as the causes of the disappearance or diminution of fish in localities where they have been observed for a long time. Among the causes which have been mentioned, the following are the principal ones: *Epidemics among the fish*, *changes in the nature of the water or of the bottom* by volcanic eruptions or by the accumulation of mud (caused, among other things, by the increased denudation of the coast consequent upon the destruction of the forests), and *steamboat traffic*.

In carefully examining the rich literature on this subject, it will be found that of all the causes which have been mentioned as having an injurious influence on the fisheries, only the following have been more generally accepted: *Excessive fishing*, *fishing with destructive apparatus*, destroying the vegetation of the bottom, the eggs, and the young ones, *preventing the fish from reaching their spawning-places*, *impure or turbid water*, *fish-of-prey*, and, finally, *lack of food* (which may be caused by human agencies).

Among those fish which, like the herrings, have either entirely or to some extent left places where formerly they have been very common for longer or shorter periods, we may here mention the following North Sea fish: The codfish, the haddock, the mackerel, and the shark. On the northeastern coast of America there are a number of fish of which the same is known. It thus appears that just the most important sea-fish are quite irregular in their coming and going, and, unfortunately, our knowledge of the causes of this phenomenon is exceedingly limited. Not only our scientific but also our historical knowledge of these fish, and especially of the herrings, is so limited that at present there is very little hope of having this scientific problem satisfactorily solved in the near future. Such a work requires not only a most extensive biological and physical knowledge of our waters, but also a thorough acquaintance with the history of the different fisheries. It is greatly to be lamented that *Axel Boeck's* premature death put an end to the important study of the history of the Norwegian herring-fisheries, to which he had devoted himself for several years, and that the rich material which he had collected for a history of the Scotch and Dutch herring-fisheries has not been worked up. It is quite likely that this material, properly worked up in a scientific manner, would furnish many and important contributions towards the solution of the problem regarding the migrations of the great races of herrings.

---

[81] It is entirely different, however, with those sea-herrings that visit the coast of Bohuslän or the west coast of Norway during winter for other purposes than spawning.

From the little that is known regarding the periodicity of the great herring-fisheries, it will be clear that the periodical coming and going of the herrings, which on the coast of Bohuslän has been observed for six successive centuries, cannot possibly be purely *accidental*, although the causes of this phenomenon can so far only be guessed at. All attempts to explain this phenomenon from accidental causes must therefore be classed with the mythical explanations.

67. In briefly recapitulating the different scientific theories regarding the migrations of the herrings, it will be found that they may all be arranged under the following heads:

A.—*The theory of a central race of herrings*, according to which all herrings which are in the world belong to one great central race, from which all kinds of herrings, both great and small, come. This theory is varied as follows:

1. This central race of herrings is supposed to live in the Northern Polar Sea, from which large schools emigrate every year to those coasts where herring-fisheries are carried on (*Anderson*, *Pennant*, and others).
2. This central race of herrings is constantly moving through the Northern Atlantic Ocean in a circle, whose extent is regulated by the declination of the sun (*Gilpin*).
3. Besides this great central race of herrings living in the Northern Atlantic Ocean, causing the great herring-fisheries, smaller local races have separated in course of time, causing the smaller coast-fisheries (*Lundbeck*).

According to the first two of these three theories there would be no regular migrations, whilst such would take place according to the third.

B.—*The theory of separate races of herrings*, according to which the different fisheries are caused by separate races of herrings, each having its own locality. This theory is varied as follows:

1. The theory of a *coast-race of herrings*, considering the herring exclusively as a bottom-fish. This may again be subdivided:
   - *a.* Some suppose that there is only one local race of herrings in every place, which, if not driven away by human agencies, always stays near the coast. There is consequently no difference between coast-herrings and sea-herrings, and there are no regular migrations (*Nilsson*).
   - *b.* Others think that more than one race of herrings may occur in one and the same place. There is consequently a difference between coast-herrings and sea-herrings, and there are regular migrations; but the proper homes even of the sea-herrings are the deep valleys on the bottom of the sea near the coast (*Axel Boeck*).
2. The theory of a *sea-race of herrings*, considering the herring as a surface-fish. This theory is also subdivided:

a. Some deny the occurrence of more than one race of herrings in one and the same place, the difference between coast-herring and sea-herrings (littoral and pelagian herrings) and regular migrations (*G. O. Sars*).

b. Others maintain that there is a relative difference between coast-herrings and sea-herrings, that more than one race of herrings may be found in one and the same place, and that the great schools of herrings migrate regularly (*A. V. Ljungman*).

68. It is doubtless necessary from time to time to give a review of the results of the scientific investigations, so as to obtain a suitable starting-point for new and systematic researches. As the scientific material has been considerably increased during the seven years since *Axel Boeck* published his well-known work "*Om Silden og Sildefiskerierne*" (The herring and the herring-fisheries), a new review of this material had become necessary to show the extent and character of our present knowledge, and to present the different opinions on a subject of such vast economical importance as the herring-fisheries. In endeavoring to contribute *my* share towards the solution of this important problem, I have invariably directed attention to the necessary theoretical premises, which, unfortunately, has often been neglected. In doing this one is less exposed to the danger of being led astray by the false hope of having hit the right thing in every case; a clearer view of all the points is gained, as well as a knowledge of the difficulties and of the insufficient character of the means by which these difficulties can be overcome. This will make us more humble and more cautious, and help us to avoid the proud certainty which is so often found in older and less critical works. As it cannot be expected that this in itself most difficult and practically most important question regarding a part of the biology of the herrings can be quickly solved by the labor of *one* person, it will be necessary that naturalists who have given attention to this subject should freely communicate to each other their different theories and the results of their investigations. The historic review of these investigations shows how beneficial and indispensable this interchange of ideas has been. The mere gathering of facts does not lead to any great result. These facts must be compared and combined into more or less developed theories, always, of course, remembering that there is a chance of these theories leading to no results or only to partial results. Theories are frequently apt to mislead; but even from the most erroneous theories some useful truth may finally be developed by constant work and by a continued purifying and eliminating process. Mistakes play an important part in the development of human knowledge and lead to greater caution and thoughtfulness in the future. The very knowledge that something is not as we supposed at first, is a step forward; in order that an erroneous opinion may be refuted, such an opinion must first have been advanced. *My* opinion on this point may also serve as an explanation of the circumstance that both in this and in former articles I have given

theories with whose errors I was well acquainted, errors chiefly owing to the fact that all the conditions for a satisfactory development of these theories had not been fulfilled.

If by the foregoing treatise I have succeeded *in making the accumulated scientific material more accessible, and in directing attention to the absolute necessity of more complete and more exhaustive scientific investigations* regarding the important fishery-question, my principal object has been reached.

IX.—THE GREAT BOHUSLÄN HERRING-FISHERIES.

By Axel Vilhelm Ljungman.*

Among the causes which have exercised a powerful influence on the material and moral well-being of the population of Bohuslän, the *large periodical herring-fisheries* doubtless occupy a very prominent place. They are called periodical, because, as far as known, they have only lasted from twenty to eighty years, with intervals of sixty to one hundred, or of an average of seventy years, when the North Sea herrings did not enter the Kattegat and visit our coasts. We will, for the present, not inquire into the causes of this phenomenon, but observe what influence these fisheries have had on the well-being of our country, and endeavor to draw certain lessons for the future. We intend to show that these fisheries have exercised an influence not so much through the enormous income derived from them as by the corruption and immorality which they brought in their train, and by the sudden and radical changes which they occasioned in the quiet and even development of the resources of the province.

In considering each one of the fishing-periods separately we arrive at the remarkable and unexpected result that in course of time these troubles rather increase than decrease. To prove this we shall give a brief account of the older fisheries in chronological order, and give a fuller account of the more recent fisheries.

The oldest date we have regarding the Bohuslän herring-fisheries is from the reign of *Olof "the Saint"* (A. D. 1000–1028). The war between him and *Olof "Lap-king,"* which prevented the Visigoths from drawing their usual supply of herrings and salt from Bohuslän,[1] caused the bold utterances of *Thorgny* at the general assembly at Upsala in 1020. The old northern sagas relate that there were considerable herring-fisheries in Southern Norway, which undoubtedly means the coast of Bohuslän during the reign of King *Sverre*, that is, during the second half of the eleventh century. The data regarding the herring-fisheries which we have from those remote times are, however, so incomplete and so vague

Några ord om de stora bohus-lanska Sillfiskena. Af *Axel Vilhelm Ljungman.* Göteborg, 1877. Translated by Herman Jacobson.

The province of Bohuslän has an area of 1,952 square miles and a population (December, 1876) of 248,024. The chief city is Gottenburg, with 70,000 inhabitants, the second city of Sweden.—*Translator's note.*

[1] It must be remembered that in 871 the Norwegian king, *Harold "Haarfagre"* (Fine hair), had united Bohuslän with Norway, in whose possession it remained till 1658, when it was ceded to Sweden.

that all we can gather from them is the fact that even then the herring-fisheries were carried on only at long intervals, for if they had been going on uninterruptedly they would certainly have assumed such proportions as to attract the attention of the authors of the sagas. The conditions under which such fisheries can be carried on in Bohuslän are so favorable, compared with the northern and western coasts of Norway—and Norway was at that time powerful enough to protect the fisheries during the short fishing-season—that it would be unreasonable to suppose that such had not been the case, or that the herrings should have visited this coast steadily without any considerable fisheries being carried on. But as in all probability the herrings did not come regularly every year, the Bohuslän herring-fisheries must have been less important in those times. This will appear still more probable if we take into consideration the smaller population, the constant disturbances occasioned by civil and foreign wars, and the limited knowledge of proper fishing-implements and of the proper way of preparing fish for the trade.

The first Bohuslän herring-fishery of any importance has, therefore, probably been the one which attracted so much attention during the first part of the thirteenth century, and which is supposed to have begun towards the end of King *Håkon Håkonsson's* reign, that is, in the beginning of the second half of the twelfth century. King *Håkon "the Old"* died in 1262, and in 1257 it seems the fisheries had not yet commenced.[2] It was only under the reign of this king that the outer coast of Bohuslän was occupied and cultivated by man, which certainly was the most essential condition of the development of the sea-fisheries as a trade independent of agriculture.

Bohuslän, like the rest of Norway, had, during the twelfth century, reached a degree of civilization, wealth, and population which scarcely found its equal in the other portions of Scandinavia. The city of Marstrand dates its existence from that period (although its excellent harbor had before that time become a favorite meeting-place of merchant-vessels); and the establishment of wealthy convents[3] near Dragsmark seems likewise to point to the importance and development of this coast. With the convent founded by King *Håkon* there was combined a sort of high school, one of the first of its kind, if not in the whole of Scandinavia, at least in that portion. This school, during the following two centuries, became of considerable importance for the whole country.[4] The constant intercourse kept up by a flourishing commerce between Norway and foreign countries, the great interest taken in litera-

[2] Concerning these fisheries see *A. Boeck's* article: *"Det bohuslänska sillfiskeriet's historia,"* in *"Nordisk Tidsskrift for Fiskeri,"* I, Copenhagen, 1873, pp. 1–3; and *A. E. Holmberg: "Bohuslän's historia och beskrifning."* Uddevalla, 1842–'45. I, p. 80; II, p. 84; III, p. 194, 406. Second edition, Oerebro, 1867, I, p. 96, 279; II, p. 314; III, p. 191. The opinion advanced by the first-mentioned author that these fisheries had commenced long before the end of the twelfth century seems, however, somewhat doubtful.

[3] Even long before this time there was a similar convent at Konungahella.

[4] *Holmberg: Bohuslän's historia och beskrifning,* III, p. 102–108; 2d ed., II, p. 227.

ture and art, evinced by translations of the master-works of the literatures of Western Europe, and by beautiful buildings, of course contributed their share towards the mental development of the people. And as Bohuslän, on account of its favorable location, its wealth, and population, had formed an important portion of the monarchy ever since the beginning of the tenth century, where the kings often resided and where the nobility gathered, it cannot, as under changed circumstances was the case at a later period, have been neglected over other portions of Scandinavia. This period was, in many respects, the golden age of Bohuslän, and it reached a degree of wealth and political power which even the party-warfare of centuries could not destroy entirely. These civil wars resulted in the strengthening of the royal power and the establishment of a well-ordered government, which Sweden, for instance, did not obtain till the reign of *Gustaf Vasa* (1523–1560). All this caused the rich herring-fisheries, which are supposed to have begun about 1260, to be carried on with energy, in order to utilize the vast masses of herrings which came to the coast. These herring-fisheries continued without any considerable disturbance by war till far into the thirteenth century, and it is not impossible that there was good fishing off and on till the year 1341. If our suppositions regarding the beginning and end of these fisheries are correct, which, however, cannot be said with absolute certainty, this fishing-period must have extended over eighty years or more, and would, therefore, have been the longest fishing period on record. From this fishing-period we have the first account of foreigners being allowed to participate in the herring-fisheries, a measure by which first the herring-trade and then all the other trade of the country gradually got into the hands of foreigners, by which Bohuslän lost much of its importance, which, to a great extent, depended on its shipping and commerce. The foreigners who visited these fishing-grounds had certainly to pay a tax for the herrings which they took away, but it is not known that any such tax was demanded from the citizens of the country, either for exporting herrings or for the privilege of participating in the fisheries.

Under the reign of *Magnus Håkonsson*, the son of the above-mentioned king, *Håkon Håkonsson*, in the year 1274, the old provincial laws of Norway were revised and collected in a common law, which is the oldest law that was ever in force in Bohuslän. Its regulations regarding the fisheries are, in the main points, in force in Norway to this very day. As regards the herring-fisheries, the regulations are evidently taken from the older laws of Northern and Western Norway,[5] and only relate to the so-called spring-herring fisheries, which are carried on during the winter, but not to those fisheries which are carried on during the milder season of the year, and consequently not to the Bohuslän fisheries.

At the end of the above-mentioned fishing-period there followed a longer interval during which those kinds of herring which chiefly form

[5] *Håkon Håkonsson's* Law XV, 5. Compare L. M. B. Aubert: "*De norske Retskilder og deres Anvendelse.*" Christiania, 1877, p. 36.

the object of the great fisheries do not seem to have visited our coast. About the middle of the fourteenth century we find very flourishing herring-fisheries, and on the 15th of July, 1453, Pope *Nicholas V* urged the archbishop of Lund, the bishop of Skara, and the abbot of the convent of Hovedö to protect the right of the priests in the diocese of Oslo to receive a tithe of all the herrings that were caught against any interference of the secular powers.[6] But we possess a much more distinct and important proof that rich herring-fisheries occurred during this period, in a deposition[7] made at the district court of Askim, on Tuesday after St. Botulph's day, that is, the 22d of June, 1496, which says, "that Hvinge and other coasts have from olden times belonged to the Swedish empire and to the district of Elffnesborg, no one can question; and those who caught herrings there paid a tax at Elffnesborg, as is well known." We thus must conclude that herring-fisheries had been going on there which were still remembered, and that the authorities taxed the people for the privilege of participating in these fisheries. This change of our periodical herring-fishery to a kind of government fishery[8] is something entirely new and unknown in former legislation. In all probability it may be traced to the foreign tendency and the constant impecuniosity of the union kings, who came from German stock. In Denmark, where the union kings mostly resided, the rule had been established that everything which did not belong to an individual or to a community belonged to the king. A circumstance which caused the introduction of such a herring-tax, or at any rate facilitated it, may have been the popular notion that the herring-fisheries were a special gift of Providence, for which gratitude should be expressed to the king as to the representative of divine power. This last-mentioned idea is doubtless derived from the pagan idea that the highest priestly power belongs to the king. In accordance with this idea the Norwegian law granted to the king a certain portion of the whales which from time to time were driven against the coast of Norway. Although we have no positive proof that the tax on herring-fisheries was also introduced in Bohuslän, it is highly probable that this was the case sooner in Bohuslän than in Vestergotland, unless the crown claimed this tax at one and the same time in both provinces, which is the most probable supposition. The royal power was at that time much more developed in Norway than in Sweden, and it is scarcely probable that the union kings should have introduced such a tax in Vestergotland and not also in Bohuslän. The herring-fisheries of Western and Northern Norway were not treated in the same manner, because they were of a different character, there being no periods when these fisheries stopped entirely, like those of Bohuslän; although something similar takes place in the winter fisheries, the summer and autumn fisheries have generally continued without interruption.

---

[6] *Boeck* in "*Nordisk Tidsskrift for Fiskeri*," I, p. 3-4.

[7] *J. Oedman:* "*Bohusläns beskrifning*," Stockholm, 1746, p. 378-380.

[8] *C. G. Styffe,* "*Framställning af de s. k. Grundregalernas uppkomst och tillämpning i Sverige*," p. 266.

We have no data regarding the time when this Bohuslän herring-fishery of the fourteenth century began or when it ended. We know that King *Erik*, "*the Pommeranian*" (1389–1439), asked the English king, *Henry V*, in the year 1415, to prohibit the inhabitants of several English sea-towns from fishing for herrings on the coast of Norway, which privilege had been granted to them in the year 1294; but it is uncertain whether this prohibition referred to actual fishing or to general trade—more especially to the exportation of lumber—which was going on under the pretext of herring-fishing.[9] Nor is it absolutely certain that this prohibition referred to Bohuslän, although this is quite probable. The herring-fisheries at that time formed a pretext for Englishmen, and still more for Dutchmen from the province of Zealand, as well as for Germans, to get a great portion of the trade of Scandinavia, "and especially Denmark," into their hands; and the lumber trade was at that time, and even till Bohuslän was united with Sweden, one of the principal sources of income of this province. Since King *Erik* at the same time prohibited the Germans from fishing in Skåne, whilst he granted still greater privileges to the Dutch, it almost seems as if he had intended that the Dutch should monopolize the Scandinavian herring-fisheries. They were doubtless less dangerous than the Germans, who occasionally attempted to acquire these rights by force.

During this fishing-period the city of Marstrand obtained similar privileges from King *Christopher*, "*the Bavarian*," in July, 1442, and its church, in the year 1460, was granted a tithe of the herring-fisheries.[10] We mention these facts merely to prove that this city must at that time have been in a flourishing condition, which could not have been brought about by anything but the herring-fisheries. Uddevalla had probably at this time likewise gained importance and become a city. Its oldest privileges, however, date from the end of the fourteenth century.

To this herring-fishery of the fourteenth century the old historian *Peter Claussön* doubtless refers in his well-known work "*Norriges ocomliggende Oers sandfaerdige Bescriffuelse*" (True Description of Norway and the surrounding Islands), when he says: "It is said that in former times there have been very extensive herring-fisheries in Viksidan,[11] which have disappeared by magic, bad men having sunk a copper horse in the sea and thereby driven the herrings away from the coast; but this is only a fable; for it was the wickedness of men, their abuse of God's good gifts and their godless life, which caused the fisheries to cease. This was also the case during the last herring-fishery which the merciful God gave to Viksidan in the year 1556." *Peter Claussön* lived during the second half of the sixteenth century.

Concerning this last-mentioned herring-fishery, which commenced about

[9] See *Holmberg*, "*Bohuslän's historia och beskrifning*," II, p. 14; 2d edition, I, p. 211.

[10] *Holmberg*, "*Bohuslän's historia och beskrifning*," III, p. 406–408, 429; 2d edition, III, p. 192–193, 212.

[11] At that time Bohuslän was called "*Viksidan*," to distinguish it from the southeastern coast of Norway, which was called "*Agdesidan*."

the middle of the sixteenth century,[12] and which continued without interruption till 1590, we possess much more information. The change of these fisheries from public to crown fisheries in Vestergotland, which is actually proved, whilst the same is *supposed* to have taken place in Bohuslän, has now been fully proved also with regard to this last-mentioned province. It was carried out in a manner which exercised a great influence on the fishing-trade and on the physical and moral welfare of our coast. On the 19th of April, 1561, King *Frederick II* sent a letter to the Bohuslän authorities in which he says: "Since we have learned that the fisheries are very much increasing in Marstrand, we ask you to see to it that the royal tax on fishing is regularly collected."[13]

Further information concerning this right of the crown we find in the "Rules and Regulations for the Bohuslän herring-fisheries," published on the 12th of July, 1561, which are said to be the oldest regulations of this kind, and which at that time were generally known by the name of the "Marstrand law-books." We must here give the introduction to these "Rules and Regulations" as showing the claims of the king and the great influence of this fishery on the material welfare of our province. It reads as follows: "We, *Frederick II*, make known to all men, that since we have learned how the Almighty God has extended his favor and blessing to our kingdom of Norway by letting the herrings come to its coast, we decree for the benefit of all those who are engaged in the fisheries, *first*, that all foreign merchants attending the fisheries must stay in Marstrand and nowhere else,[14] and there pay to us a tax of 320 herrings per season and one-half dollar for every 12 tons of herring exported from the kingdom.[15] *Second*, that no foreign merchants visiting these fishing-stations shall bring with them any butter, skins, tallow, or any other goods, except what they absolutely need

[12] As already in the year 1557 the citizens of Oslo, Tönsberg, and Sarpsborg, had obtained the privilege of trading with Marstrand, Kungelf, and Udevalla, in exchange for certain rights granted to the inhabitants of these towns, it seems that even at that time the fisheries had become so extensive as to attract attention; and the herrings must certainly have come to this coast several years previous to 1557. For it often takes a long time till good herring-fisheries become known, which has been fully proved by our fisheries during the seventeenth century.

[13] The extracts from the royal letters and regulations concerning the herring-fisheries of the fifteenth century are taken from *Axel Boeck's* above-mentioned treatise on the history of the Bohuslän herring-fisheries in "*Nordisk Tidsskrift for Fiskeri*," I, p. 5–27, to which we refer those who desire further information regarding this fishing-period.

[14] This regulation was occasionally dispensed with, for instance in 1566 for the Lubeck merchants; but in 1573 and 1580 it was decreed that foreign merchants could only trade in other places by special permit from the Marstrand tax-gatherers.

[15] This tax was finally raised to one dollar for every 12 tons; the consequence was that many foreign merchants made false entries, giving the names of Swedish merchants in other towns as the owners of the herrings; or bought the herrings from the fishermen on the outer coast, and thus paid no tax at all. By a Royal Decree of 1580 it was strictly forbidden to export any fresh herrings.

during the time they stay here.[16] Any one transgressing this rule forfeits his goods. Our own subjects shall be allowed to fish wherever herrings are found, and pay their annual tax in Marstrand.[17] They need not pay any tax on those herrings which they salt for their own use, but on herrings salted for exportation to foreign parts they shall pay the same tax as foreigners.[18] Both citizens and foreigners are commanded to faithfully observe all these regulations." These so-called "Law-books" must not be considered, however, as having introduced any new or permanent law, or as having changed any of the general laws of the country, with the exception of regulations which were occasionally made just for one season.[19] Transgressions of these laws and regulations did not come before the common courts, but before the royal tax-gatherers, who in fact superintended the entire fisheries. It appears from the introduction to the oldest of these so-called "Law-books," that the inhabitants of Bohuslän, like all others who participated in these fisheries, had to pay a certain tax, and that the same right of fishing was given to the king's subjects in Denmark and Germany as well as to those in Norway. It is evident that such a law as that which the kings had made with regard to the Marstrand coast, and later also with regard to the more northern portions of the coast, allowing foreigners to participate in the fisheries on very much the same conditions as their own subjects, drew a large number of people to our coasts during the fishing-season, and *Peter Claussön* also reports that every year several thousand vessels and boats came from Denmark and Holstein as well as from other countries. As the king of course desired the greatest possible revenue from his fisheries, and as this revenue was paid partly in an annual quantity of herrings, and partly in a certain sum of money on those herrings that were exported, it was of course desirable to draw a large number of fishers to the coast and export as many herrings as possible. It appears,

[16] As there were constant complaints that the foreign merchants injured the home-trade, a decree was published in 1569, that they should only be allowed to trade from Michaelmas (September 29) till the first Sunday in Lent. And by further decrees of 1573 and 1580, the privileges of foreign merchants were limited still more.

[17] The tax in herrings must always be paid in the largest and best herrings, and was measured in a separate vessel holding about one-third of a ton. Every fisherman must sell to the government a boat-load of the first herrings at the "usual" price. The government moreover had the first right to buy the best herrings—until all the royal salting-houses were supplied—any one who made a higher bid than the tax-gatherers being heavily fined. The tax-gatherers, however, must pay the highest price which could reasonably be demanded.

[18] In 1580 the tax was raised to one dollar for 12 tons, also for citizens if they exported their fish in foreign vessels.

[19] The assertion made by some people, that several regulations contained in these "Law-books" have been handed down to our own times by popular tradition and are still observed by the coast population of Bohuslän, shows only complete ignorance of the present state of affairs. The most complete of these "Law-books" is mentioned in *Th. Boeck's Oversigt over Literatur, Love, Forordningar, Rescripter m. m. vedrörende de norske Fiskerier* [Review of the literature, laws, regulations, decrees, &c., of the Norwegian Fisheries], Christiania. 1866, p. 3–8.

however, that the inhabitants of Bohuslän did not look favorably upon fishermen coming from abroad, and *Peter Claussön* relates that the Bohuslän people hated all those who came there to fish, and frequently cursed them and beat them, so that it was not safe for any fisherman to go among them, unless they were well armed and a number of them went together. Gradually, however, the number of foreign fishermen grew so large that they became all-powerful, and the principal part of the herring-trade passed into their hands. In order to become a source of national welfare, the herring-fishery must, as was the case with the Dutch, be not so much an object in itself as a means of increasing commerce and navigation. From the time (1612 and 1620) when the Dutch resolved to carry on the herring-fisheries according to strict rules for their own sake and as an object in itself, they began to go down, and gradually lost their importance, so that this so-called "gold mine" of the Netherlands dwindled down to nothing, and had finally to be supported by considerable government subsidies.

The method of fishing likewise led to trouble, for in order to catch the greatest possible number of fish, nets were used exclusively, requiring a large number of people, who were thus taken away from other occupations, and being crowded together, occasioned disturbances and immoral practices. The cause why fishing was carried on with nets must doubtless be sought in traditions and in the regulations of the above-mentioned "Law-books," as well as in the manner in which the royal tax was collected.[20] Even the preparing of the herrings for the trade, which consisted chiefly in drying, required by far too many men.

That these fisheries were very considerable may be inferred from statistics given in a pamphlet published at the time, according to which Marstrand annually salted, dried, and exported 600,000 tons of herrings.[21] We must here give *Peter Claussön's* brief description of these fisheries, as throwing a good deal of light on this whole question. He says: "Several thousand people from the neighboring countries, Norway, Denmark, and Holstein, had come here with their wives and children, and had built themselves houses on the coast. Noblemen, as well as merchants and peasants, had erected large and beautiful houses, some of them two or three stories high; some of these were so large that 168 tons of herrings could conveniently be hung up and dried at one and the same time. Extending for 50 to 60 miles along the outer coast, there were many thousand houses and huts, and numberless people lived on every bay and fiord and island. Thousands of vessels arrived annually from Denmark, Germany, Holland, England, Scotland, and France for the purpose of buying herrings and shipping them to distant countries." Marstrand, which was the centre of the fishery and the trade connected

---

[20] See "Law-book" of October 26, 1575, §§ 8–11, 13; *Holmberg*, "*Bohuslän's historia och beskrifning*," II, p. 88; 2d edition, I, p. 283.

[21] *Holmberg*, "*Bohuslän's historia och beskrifning*," II, p. 84–85; III, p. 408; 2d edition, I, p. 280; III, p. 193.

with it, rose to considerable importance; it had two mayors, a chief of police, a syndic, ten aldermen, &c., which certainly is an evidence of prosperity.[22] Another evidence is found in the amount of war-taxes which Marstrand had to pay during the war with Sweden, and which was as high as that of the large city of Bergen. The inhabitants of Marstrand complained about this, and succeeded in having the taxes lowered; but this fact shows indisputably that at that time Marstrand was the second city of Norway, and ranked higher than Trondhjem, Oslo, and Tönsberg. Uddevalla seems also to have derived considerable benefit from these fisheries.[23]

We will now see what influence these rich fisheries exercised on the moral condition and true well-being of Bohuslän. The king, in order to increase the revenues of the crown, desired to draw as many fishermen as possible to the coast of Bohuslän, and succeeded in seeing his desire fulfilled. The consequence was, that all sorts of people came to Bohuslän from Norway, Denmark, and the German provinces of the Danish crown, many of them by no means persons of high moral character. *Peter Claussön*, in the above-mentioned pamphlet, gives us some idea of the character of the coast population, when there we read of the "godless life which the people led, drinking, gambling, whoring, murder and quarrels being every-day occurrences" in the cities of Marstrand and Udevalla.[24]

As the number of bloody frays increased, it became necessary to appoint special surgeons. It is highly characteristic of those times that, whenever the surgeons had to dress fresh wounds, they had to announce the fact to the tax-gatherer, so the king might not lose the fine which was imposed on frays of this kind. The "Law-books" imposed very heavy fines on the transgression of any of their regulations; still disorder and vice were not much diminished.[25] The tax-gatherers, in whose hands great power was laid, were frequently guilty of violent extortions. The revenues which the crown derived from these fisheries proved of little benefit to the country, for they were chiefly employed for carrying on a useless war against Sweden.

Towards the year 1590 the fisheries began to decrease, as was generally supposed, on account of the godless life led by the fishermen, and the abuse of God's gifts; and the last "Law-book," which was issued in 1589, therefore recommends an earnest reform, saying: "As all good gifts come from Almighty God and His Divine Majesty, thus our kingdom of Norway has richly experienced His favors, its inhabitants as

[22] *Holmberg*, "*Bohuslän's historia och beskrifning*," III, 408–409; 2d ed., III, p. 193–194.

[23] *Holmberg*, "*Bohuslän's historia och beskrifning*," III, p. 113; 2d edition, II, p. 239.

[24] *Holmberg*, "*Bohuslän's historia och beskrifning*," II, p. 93, note; III, p. 113, 409; 2d edition, I, p. 287–288, note; II, p. 239; III, p. 194.

[25] *Holmberg*, "*Bohuslän's historia och beskrifning*," II, p. 93, note; 2d edition, I, p. 287–288, note.

well as others having been blessed with successful herring-fisheries. But since there is danger that God may withdraw His blessings on account of the great sins and vices of the inhabitants of the coast, our tax-gatherers, each one in his district, shall see to it that people in the fishing-stations lead good and Christian lives; that there is preaching every Sunday, and people are exhorted to lead a godly life, so that God may be moved by the prayers of good Christians to extend His blessings to us also in the future." A short time after this "Law-book" had been issued, the herrings entirely disappeared from the coast of Bohuslän; and *Peter Claussön* relates that "many hundred merchants and fishermen went to great expense, but all in vain."

It would naturally be supposed that after the close of the fisheries the coast population were suffering from great poverty and want; but as nothing of the kind is handed down by tradition or by writings from those times,[26] it must be supposed that the evil consequences were in some measure diminished or warded off in such a way as not seriously to influence the whole province.[27] It is possible that many of the inhabitants of the coast moved to other parts, or found some other employment. The land-owners of Bohuslän were at that time well-to-do and independent, having other sources of income; whilst in the cities, among the rest in Marstrand, the considerable commerce had produced a state of well-being. Although the herring-fisheries exercised a great influence on the population of Bohuslän through their demoralizing tendencies, and through the poverty consequent upon their sudden cessation, Bohuslän suffered less than it did two hundred years later when the same occurrence took place. One reason was certainly the shorter duration of the fisheries in the sixteenth century as well as the very limited freedom of trade.

About seventy years later, when Bohuslän, after having for eight centuries formed a province of Norway, was incorporated with Sweden, the herrings again visited our coast; and there would certainly again have been large fisheries if the sanguinary war between Denmark and Sweden, which lasted from 1675 to 1679, had not prevented all fishing. Moreover, the conditions for drawing together on this coast a large number of experienced fishermen were not so favorable as when Bohuslän still belonged to Denmark-Norway.

In order to give to the herring-fisheries some legal sanction, a royal decree was published, October 13, 1666, concerning a regulation which was to be observed during the herring-fisheries. In this regulation certain ports are mentioned, viz, Gottenburg, Kalfsund, Marstrand, Mollösund, Gullholmen, and Lysekil, in which alone herrings might be taken ashore and be prepared for the trade, and where inspectors, endowed with the necessary authority, should supervise the fisheries and see to

[26] *Holmberg*, "*Bohusl. hist. o. besk.*," II, p. 100, 101, note.

[27] It is quite probable, however, that these demoralizing herring-fisheries have left traces on the central coast of Bohuslän, especially near Tjörn, which may be felt even in our days; for otherwise it would be difficult to explain the low moral state of the population on that coast, of which *Holmberg* and other authors speak.

it that only good herrings got into the market. People had learned wisdom by experience, and henceforth only Swedish subjects were allowed to engage in the herring-fisheries, whilst foreigners had to acquire this privilege by special compact. But no such compact or treaty with foreigners is ever mentioned. Foreigners were also forbidden to buy herrings in the ports. There was no tax on herring-fishing, and it is supposed that the Swedish Government by granting this freedom intended to make the population of the newly-acquired province more favorably inclined towards itself. The tax on herring exported to foreign countries was lowered considerably. To maintain good order among the fishermen a so-called "port-law" was published the 10th of May, 1669.

As the inhabitants of Bohuslän henceforth carried on the fisheries almost exclusively under a comparatively mild government, they derived considerable benefit from the fisheries. *Holmberg* mentions as a proof of this that most of the church ornaments in Bohuslän date from this period.[28]

We have no data regarding the exact time when this fishing-period came to a close. But about the year 1670 the herrings seem to have ceased to come to the southern coast of Bohuslän, and according to the most reliable authorities fishing seemed to have closed in 1679 or 1680 also on the central and northern coast. According to an old tradition there is said to have been occasional fishing till the commencement of the great Northern war, under *Charles XII*, in the year 1700.[29]

In the foregoing it has been said that the law of *Magnus Håkonsson* is the oldest law of Bohuslän. This law had been examined, however, and its language changed a little under the reign of King *Christian IV* when it was printed in the year 1604. As regards the fisheries, however, the regulations of the old law remained almost unchanged. This law of 1604 remained in force in Bohuslän till the winter of 1682, when the Swedish law was introduced.[30] The regulations of the Swedish law regarding fishing were fewer in number and shorter, as the fisheries were not so important for Sweden as they had been for Norway; but they changed nothing regarding the privilege of fishing on the sea-coast, for coast-fishing was at that time, in Sweden as well as in Norway, with few exceptions, open to all inhabitants of the country.[31]

About sixty or seventy years after the great herring-fisheries of the sixteenth century came to a close, the last great Bohuslän fishing-period commenced, concerning which all the inhabitants of this province

[28] *Holmberg*, "*Bohusl. hist. o. besk.*," I, p. 135; II, p. 85; III, p. 115, 346, 411; 2d edition, I, p. 148, 280; II, p. 241; III, p. 136, 196.

[29] *Lundbeck*, "*Anteckningar rörande Bohuslänska fiskerierna, i synnerhet sillfisket*" [The Bohuslän fisheries, especially the herring-fisheries]. Gottenburg, 1832, p. 35–36.

[30] *Holmberg*, "*Bohusl. hist, o. besk.*," I, p. 135; 2d ed., I, p. 148.—*Aubert*, "*De Norske Retskilder*" [Sources of Norwegian Law], p. 397–406.—"*Nytt juridiskt Arkiv*" [New Law Archives], 1876, II, No. 12, p. 1–9.

[31] Among these exceptions the more important are the so-called "crown fisheries," near the royal domains, parks, or islands, where fishing can only be carried on by special permit of the government authorities.

have heard, and which, according to unanimous testimony,[32] began in the year 1747. Fishing seems to have commenced in the neighborhood of Tjörn and the Marstrand Islands, but soon after seems to have extended along the whole southern coast to the boundary of Holland. Later the herrings chiefly came to the coast between Marstrand and Lysekil, and after the year 1773 also to the northern coast. In 1778 occasional herrings are said to have been seen near the Hval Islands, in that part of Southeastern Norway which bounds Bohuslän. On the northern coast the herrings advanced a little farther north every year, whilst their quality had already begun to deteriorate. In speaking of the northern coast in those times, the coast north of Sotenaes is not counted in, but this term only applies to the coast between Marstrand and Lysekil. Towards the end of this fishing-period, however, large quantities of herring again came to the southern coast; but this was considered an exceptional case. The Norwegian naturalist *Axel Boeck* has shown that a similar change has taken place, both in the Norwegian spring-herring fisheries and in the Bohuslän fisheries during the sixteenth century, and we are therefore justified in expecting that this will also take place in the future, in case the herrings should again come to our coast. This fishing-period came to an end in 1808, after having lasted sixty-two years, and this event was foreshadowed by the moving of the fish in a northerly direction, by the later and later appearance of the fish, "finally only about Christmas time," and by its being mixed with small herring during the last year of this fishing-period. The value of these indications for the future is increased, since the above-mentioned Norwegian naturalist has shown that the same took place at the close of the Bohuslän fisheries in 1590 and of the Norwegian fisheries in 1787 and 1870.[33]

The fisheries, however, grew in importance only very gradually, for Sweden could not, as Norway had done formerly, send a sufficient number of experienced fishermen to the coast, but these had to be educated by degrees. From the Dutch the Swedes learned the proper way of preparing the herrings, and soon movable nets were adopted instead of stationary ones. As a great many more herrings were caught than could conveniently be salted and smoked, people in the year 1760 began to make oil of those that were left over. All this was easier, for both the new method of fishing and the manufacture of oil required only a comparatively small number of men, which as early as during the fisheries of the sixteenth century had been considered a great advantage. Foreigners were excluded from the fisheries and from the trade in fresh herrings, although the last-mentioned regulation was not strictly enforced

[32] The year 1752, which in some works is mentioned as the time when these fisheries commenced, is probably the year when the herrings commenced to approach the Gottenburg coast in any considerable number.

[33] *A. Boeck*, "*Om Silden og Sildefiskerierne*," &c. [On the herring and the herring-fisheries, &c.]. I. Christiania, 1871, p. 102–118; *Göteborg's och Bohusläns Hushållning-Sällskaps Quartalsskrift* [Quarterly Review of the Gottenburg and Bohuslän Economical Society], October, 1870, p. 36–39 and 44–54.

in the case of the Danes and Norwegians. Foreigners who intended to become Swedish citizens were for three years freed from all personal taxes and enjoyed the same privileges as the natives. The government asked no tax for the privilege of participating in the fisheries, but even paid a subsidy for furnishing large seines, and also in other ways encouraged the fishing-trade. The government also endeavored to draw people to the coast to engage in fishing by giving them free building-lots, lumber from the royal forests, freedom from military service, &c. From 1756 till 1787 the government even permitted Swedish subjects who had fled from Sweden on account of minor offences to return without being punished if they would settle on the coast and engage in fishing. In 1765 a decree was published permitting people who lived in the most distant provinces on the Gulf of Bothnia to go to Bohuslän by sea free of expense if they would engage in the herring-fisheries. The number of those who came to the coast of Bohuslän every year during the fishing-season in order to be employed in fishing or in the preparation of fish for the trade was, during the most flourishing period, estimated at 50,000, not counting in the stationary population of the coast.[34]

Besides holding out inducements for people to engage in fishing, endeavors were also made to further the fishing-interests by improved and more complete laws, for which purpose during the period 1767–1772 a special parliamentary "fishing-commission" was appointed to which all questions concerning fishery-legislation were referred. During the period 1774–1778 special reports on the subject were ordered by the government. The result of the work of the above-mentioned commission was a general fishery-law, which, for the time when it originated, must be considered as possessing considerable merit, and a special law for the North Sea fisheries, which afterwards also included our herring-fisheries. In the former law, which in all essential points is the same as our present fishery-law of the 29th of June, 1852, the privilege of fishing on the inner coast was limited to the proprietors of the coast,[35] which rule in

[34] See *S. Nilsson*, "*Handlingar rörande Sillfisket i Bohusländska Skärgården*" [The herring-fisheries on the coast of Bohuslän]. Stockholm, 1843, p. 11.

[35] In those places where persons having the privilege of fishing had been in the habit of catching fish on "each others' coast," everything should remain in *statu quo*, and such fisheries should be in common to all proprietors of a certain extent of coast, a regulation which rightly understood might prove very useful. (See "*Nya handlingar rörande Sillfisket i Bohusländska Skårgården,*" I. Gottenburg, 1874. Appendix, p. 15–16, § 12.) The granting of the exclusive privilege of fishing to the owners of the coast was likewise done with the view of promoting the fishing-interests, as it was thought that they would be in the hands of those who for their own advantage would carry on fishing in the most approved manner. It was moreover only the logical development of those principles of law which gradually had obtained in Sweden as well as in other Germanic countries. It is an error to suppose that the general fishing-law of 1766, as well as its explanation published in 1771 regarding the western coast of Sweden between the Sound and the Norwegian frontier, had been entirely abrogated by the law of 1774 "for the North Sea fisheries and the salting-houses in the districts of Gottenburg and Bohuslän"; for this was certainly not the intention. Such a change would have required a resolution of Parliament sanctioned by the king. (See §§ 2, 40, 42 of the constitution of 1772.)

former times held good only in exceptional cases, rather because it had been in force from time immemorial than because of any royal decree. An exception, however, was made with regard to our great periodical herring-fisheries, or as the law terms them "the great North Sea herring-fisheries," from which no Swedish citizen could be excluded, on whatever coast it might be, even on those coasts where, prior to 1766, the proprietors had had the exclusive privilege of fishing.[36] By thus distinguishing from a legal point of view the periodical herring-fisheries from the other fisheries, the three hundred year old claim of the crown to the former was formally established. To further the herring-fisheries, the privilege was granted to catch herrings in nets even on those parts of the coast which were held by private owners. From the decrees which were published from time to time (from the year 1748), and from the "Complete Regulations for the North Sea fisheries," published in 1774, it appears that it was the intention of the government that henceforth the fisheries should no longer be hindered by granting royal privileges or monopolies to individuals, as formerly had been done several times (for example, in the year 1745.)[37] To preserve order in the ports and at the different fishing-stations a new and improved set of "Regulations for ports" was published in 1771, by which a number of special officers were appointed, who were to superintend the fisheries and maintain order; thus 1772–1774 a "chief superintendent," 1783–1791 a "superintendent," which office was to be filled by the chief pilot of each district, and finally, 1791–1821, a "superintendent of herring-fisheries."[38]

In accordance with the economical views of those times the government, especially during the first half of this fishing-period, endeavored to encourage the fishing-trade by a high premium on exported herrings, by which the owners of large establishments were certainly benefited, but which otherwise proved no advantage.[39] Such a rich and natural trade as the Bohuslän herring-fisheries of that period ought certainly to have supported itself without any premiums. If the large sums which now only benefited a few capitalists had been used for maintaining good order and morals in the fishing-stations, the whole fishing-trade and

---

[36] In the law of 1852 the expression "the great North Sea herring-fisheries" is changed to "such salt-water fish as approach the coast in large schools," which change, however, was of no practical consequence. Any positive change in the fishery-laws would have to be made in accordance with certain rules laid down in § 87 of the constitution of 1809.

[37] As late as 1778 we find in a "report on the herring-fisheries," a suggestion that no new exclusive personal privileges might henceforth be granted, and no old ones renewed.

[38] Regarding the Bohuslän fishery-legislation, see *Sjöberg, A.*, "*Om den Svenska Fiskerilagstiftning*" [On the Swedish fishery-legislation]. Lund, 1866; and the same author's articles in the "*Göteborgsposten,*" 1875, Nos. 47, 52, 59, 61, 78, and 1877, No. 102.

[39] It is highly characteristic of those times that it was a frequent occurrence that those sums of money which had been appropriated for premiums, to a considerable extent found their way back to the authorities who had granted them, in the shape of bribes.

especially the fishermen themselves would have been benefited and the future of the herring-fisheries would have been very different from what it is now.

The great importance of these herring-fisheries will best be seen from the fact that during the decade 1770–1780, the average annual quantity of herrings amounted to about a million barrels—1,100,000 barrels in 1787—and later this quantity was doubled, and, according to some authorities, even trebled, upwards of three million barrels having been realized in one year.[40] It is probable, however, that this last-mentioned figure is somewhat exaggerated, or at any rate is an exceptional case.[41] The number of herrings which came to the coast was so large that the quantity caught only represented a very small portion of the whole number; and the fisheries had generally come to a close, not because there were no more herrings, but because every one being supplied with herrings they fetched no price at all. It will be self-evident that such extensive fisheries put large sums of money into circulation. And if we take into consideration the number of people employed in preparing and transporting herrings as well as in those trades which are dependent on the fisheries, we will be able to get an idea of the great direct and indirect economical value of these fisheries. The coast of Bohuslän, and especially the district of Elfsyssel, was at that time densely populated, and possessed numerous salting-houses and oil-refineries. The following statistics are gathered from official documents: in 1787 there were in Bohuslän 338 salting-houses and 429 oil-refineries, with a total of 1,812 boilers, using 40,986 barrels of herrings per day. The number of large seines was 358 and of boats 2,100.[42] These figures even increased considerably

[40] *Holmberg*, "*Bohusl. hist. o. besk.*," II, p. 85–86; 2nd ed., I, p. 280–281.

[41] *P. A. Granberg*, "*Staden Göteborgs historia och beskrifning*" [History and description of the city of Gottenburg], Stockholm, 1814-1815, II, p. 158, 159, 216–217. *P. Dubb*, "*Anteckningar om sillfisket i Bohuslän*" [Remarks on the Bohuslän herring-fisheries] in the Transactions of the Royal Academy of Sciences, 1817, p. 33. According to the first-mentioned authority the highest annual export of herrings from Gottenburg was 190,000 barrels prepared herrings and 50,000 aumes herring-oil, whilst according to *Dubb* the export from the whole coast was upwards of 350,000 barrels herrings and 120,000 aumes herring-oil. Twenty barrels of herrings of medium quality were required to produce one aume of herring-oil.

[42] "*Handlingar och Protocoller rörande Kgl. Majts. i nåder förordnade Beredning öfver Sillfiskeri-Handteringens närmare reglerande* [Official Reports on the herring-fisheries], Gottenburg, 1789, p. 29–30, 37, 43, 54, 89–90, 109, 146, 177, 178, 180, 186. In order to give an idea of this flourishing period and explain the fact that even to this day people are sighing for a return of those halcyon days, we will quote the following from *O. Lundbeck*, *Anteckningar rörande bohusländska-fiskerierna* [The Bohuslän fisheries], Gottenburg, 1832, p. 42–43: "He who knew the coast of Bohuslän 25 years ago, and now sees it again, will scarcely be able to refrain from tears. Then it presented an imposing appearance. From the sea itself rose massive walls and pillars supporting immense salting-houses and oil-refineries. Farther inland rich warehouses and busy workshops might be seen, as well as palatial residences of the merchants and neat cottages of the fishermen and workingmen. The coast was crowded with a busy throng and the sea studded with sails. Every night it looked as if there were a grand illumination,

during the following years. The cities, of course, derived the greatest benefit from the fisheries, and it may well be said that the cities and those capitalists who owned the large establishments, were really the only ones who had any positive profit from the fisheries. Gottenburg, especially, must here be mentioned, which, according to "*Granberg's historia*," owed its flourishing condition chiefly to the East India trade and the herring-fisheries. *Granberg* says that the exportation of herrings and more especially of herring-oil gave a new impetus to commerce in general and exercised a decided influence on all trade.[43] The cities of Bohuslän likewise flourished considerably during this period. Marstrand almost quadrupled its population, and Uddevalla rose to importance as well as the city of Kungelf.[44] The great ease with which in those days money was made in Bohuslän and the many chances offered, especially to persons of the working-classes, to lead a joyous and careless life, of course attracted large numbers from all parts of the kingdom, and, as might be supposed, generally persons of low morals. The above-mentioned decree allowing persons who had been convicted of minor offences to return to Sweden if they would engage in the fisheries, actually made Marstrand from 1775–1794 a kind of free port or harbor of refuge for criminals, and did not serve to raise its general standard of morality. If, furthermore, we take into consideration the fact that the owners of salting-houses and oil-refineries were licensed to keep groceries and retail liquor-stores, the consequences may easily be imagined. The greater portion of the male population of the coast were scarcely ever sober, as enough money was earned during the fishing-season to keep them going all the rest of the year. The grog-shops were, therefore, crowded all the year round with the exception of Christmas Day and Good Friday. Fishermen and workingmen, coming from other parts, generally arrived a month before fishing commenced, and this season of idleness was almost exclusively devoted to drinking and carousing. As a natural consequence of such a life the majority of the coast population had no thoughts for the future and no moral strength to bear reverses. Sanguinary quarrels were not so frequent as during the sixteenth century, but all other vices prevailed.[45] The authorities never thought of stemming the tide of corruption, their whole attention being occupied with the prevention of smuggling, for which purpose a number of small men-of-war were in

many thousand lights shining from the windows and from the numerous lamps along the quays, and being reflected in the waves. Everything was life and bustle, and tons of gold changed hands. Now nothing is seen but ruins, only here and there a dilapidated fisherman's cottage, awakening melancholy thoughts in the heart of the visitor. Would that soon these glorious times for which thousands are sighing might return." This was written in the year 1831.

[43] *Granberg, Goteb. hist. och beskr.*, I, p. 65; II, p. 153, 176. The enormous sums which this exportation yielded were used as capital in starting or supporting important home-industries, and their want was painfully felt when in 1808 this source of wealth ceased.

[44] *Holmberg, Bohusl. hist. o. beskr.*, III, p. 120–121, 349–350, 415–421, 425; 2nd edition, II, p. 246; III, p. 138–139, 200–204, 207.

[45] *Holmberg, "Bohusl. hist. och beskr.,"* II, p. 82, 92–93; 2d edition, I, p. 277, 287.

1774 stationed on this coast. For settling difficulties among the fishermen an enlarged and improved code of "port regulations" had been published, according to which certain judicial and police powers were entrusted to some of the fishermen, but all this did not improve the moral character of the population. The coast of Bohuslän gradually became a sort of vast poor-house, all sorts of homeless and shiftless people congregating there in addition to those who through their debaucheries had lost all they had earned during the fishing-season. Since all the better class left Bohuslän every year at the close of the fishing-season, and finally for good, when the fisheries came to a close, and took all their earnings with them, Bohuslän reaped all the evil consequences of the fisheries without enjoying any of their benefits. The poverty and misery on our coast when the fisheries totally ceased in 1808 actually beggars description. But it was not only the coast which suffered; the agricultural interests of the province had been totally neglected from want of men willing to work on farms and from the general degeneracy of the times. Strange to say, the enormous fortunes which had been made and remained in the hands of a few, disappeared quickly or passed into other and worthier hands. It is not to be wondered at that under these circumstances large herring-fisheries, such as those of the eighteenth century, began gradually to be considered as a curse rather than as a blessing, which opinion was, among others, expressed by the historian of Bohuslän, *Axel Emanuel Holmberg*,[46] and by its zealous and highly-honored governor, Count *C. G. Lövenhjelm*.[47] It must be granted, however, that all the evil consequences of great herring-fisheries might be avoided, or at least greatly diminished, by proper precautions, and that such fisheries, if properly managed, might greatly further the material development of Bohuslän.[48] Regarding the last great fisheries it must be said that their evil consequences are chiefly to be ascribed to wrong management on the part of the authorities, who sacrificed the interests of the fishermen, the workingmen, the coast, and the whole province to those of a few large exporters. This mistake was caused, to a great extent, by the wrong economical principles prevailing in those times and by the want of education and enlightenment among our coast population. No petitions were, therefore, ever made to the government authorities or to the Parliament to remedy existing evils, and no improvement could, therefore, ever be looked for.

The great changes for the better which, during the last thirty or forty years, have raised both the rural and the coast population of Bohuslän to a very respectable height of intelligence and well-being, may serve to indicate the way which should be followed if great herring-fisheries

[46] *Holmberg*, "*Bohusl. hist. och beskr.*," II, p. 91–94; 2d edition, I, p. 286–288.

[47] *Göteborg's och Bohuslans Kgl. Hushållnings-Sällskaps Handlingar* [Reports of the Economical Society of Gottenburg and Bohuslän], for 1847, Gottenburg, 1848, p. 27–28.

[48] Concerning the hopes of a return of the great herring-fisheries see the author's article in the "*Göteborgsposten*," 1876, No. 216, and in the *Bohusläns Tidning*, 1876, No. 77.

should again occur.[49] As regards the coast, the great cholera epidemic of 1834 must be mentioned, which carried off the greater portion of the worthless population. A general improvement then took place by the more perfect means of communication, the increased commerce, and the constantly-increasing prosperity consequent upon this, as also by the change in the fish-trade, fewer fish being salted and a great many more being sold fresh;[50] the extension of the bank-fisheries by the introduction of more suitable vessels, which enabled the fishermen to undertake voyages to more distant and richer fishing-banks (during the first half of the fourth decade of this century as far as the Jæder, and in the beginning of the sixth decade as far as Storeggen); by subsidies from the government, the Economical Society, and private individuals, and last but not least by the truly benevolent *liquor-law* of 1855, which marks an epoch in the history of Bohuslän. This last-mentioned law shows conclusively how much good may be accomplished by wise legislation, and how uccessary it is that the government should take a firm stand in suppressing all those evils and disorders which are caused by a low standard of morality or by too many opportunities for satisfying the sensual appetites. With regard to the aid which the state may extend to the different trades and industries, it is now generally acknowledged that nothing is more hurtful to the best interests of the country than the creating of a state within the state by establishing and encouraging monopolies. It is to be hoped, therefore, that our coast will, in the future, be spared that kind of encouragement by the government which was given to it during the eighteenth century. May our people, on the contrary, learn more and more to rely upon their own exertions, and may all material progress serve to further good order, sobriety, and morality, and thus advance the true welfare of our province. The future historian will then be able to give our population a better character than that which *Axel Emanuel Holmberg*[51] was obliged to give them thirty or more years ago, even if he could not record as large and flourishing herring-fisheries as *Holmberg*. And then let all "exceptional" or "monopoly" legislation be done away with, always introduced under the false pretence of benefiting the "poor fishermen." The population of Bohuslän need no longer be the charity-child of the government.

To further the true welfare of our coast we need not only a wise and enlightened government, but above everything else energy and enlightenment among our own people. It is not enough that they obey the laws and are skilled and diligent in their various occupations, but in order to make real progress they must take a warm and active interest in all public affairs. If a community is to develop to the highest point of material and moral prosperity there must be a good deal of public

[49] Great weight must be attached to the circumstance that the coast of Bohuslän does no longer, as was the case during the last great fishing-period, form the rendezvous for all the loafers and good-for-nothing people from every part of the kingdom.

[50] *Holmberg*, "*Bohusl. hist. och beskr.*," II, p. 88, 99–101; III, p. 191, 192.

[51] *Holmberg*, "*Bohusl. hist. och beskr.*," II, p. 39–40; III, p. 192; 2nd edition, I, p. 233.

spirit; and probably our province is lacking a little in this respect. May, therefore, the indifference which is still too prevalent give place to a burning zeal for the public welfare, and mere egotistical interests be more and more placed in the background; and truly our province may boldly meet all the storms of time.

Our great Bohuslän herring-fisheries, and more particularly the last great fishing-period, give us many a useful hint for the future. They ought to furnish convincing proof that it is not always a large income, or, as it is erroneously termed, "an excessive share of God's gifts," which contributes most largely to the well-being of individuals or nations. They show that what is easily gained is also easily lost, especially if nothing is done to put the gain to a proper use; they show that in order to further trade and industry something more is needed than money subsidies, and that man, even in his material endeavors, must have some higher object than the mere making of money, and that *good order and enlightenment* are essential conditions for attaining to true and permanent welfare, and finally that it is a great and grievous mistake to think that such welfare can ever be reached by nothing but money.

X.—SOCIETY FOR PROMOTING THE NORWEGIAN FISHERIES.*

[From "Bergensposten"—a daily paper published at Bergen, Norway, Tuesday, March 4, 1879.]

SPECIAL NOTICE.—In view of the fact that our fisheries need further development in nearly every direction, the undersigned have agreed to form a society for the purpose of promoting the Norwegian fisheries. These fisheries, which form one of the most important sources of income of our country, have at different times attracted the attention of the government, and not without exercising some beneficial influence. But since the fisheries in other countries have progressed, and the utilization of their products has been more and more developed, it has become apparent that our country has remained behindhand with regard to its fisheries and all the various industries connected therewith.

If it is true that Norway is one of the greatest fishing-countries in the world, her citizens should consider this as a strong incentive to develop and utilize this vast source of income to its greatest extent; with regard to this matter our country should not be excelled by any other.

With the example of other countries and with their experience to guide us, this society will endeavor to aid and develop our fisheries by hatching and raising fish, by improving the methods of fishing and the fishing-apparatus, and by utilizing to their fullest extent all the products of the fisheries; it will in fact be the object of this society to aid every endeavor to further the fishing-interests.

We therefore invite our fellow-citizens in town and country to become members of this society. The annual contribution of each member will probably be 5 crowns ($1.34), but we hope that there will be many public-spirited citizens who have both the desire and the means to pay more.

JOHAN AMELN,
and sixteen others.

BERGEN, *February* 27, 1879.

EDITORIAL.—As will be seen from a notice in our issue of to-day, a number of the most prominent and intelligent citizens of Bergen have started a society whose object it is to promote the fishing-interests.

It will be said of this movement, as of so many others started at the right time and by the right men, that it should have been begun long

* *Selskabet for de Norske Fiskeriers Fremme.* Bergensposten, Tirsdag d. 4 de Marts 1879. Translated by Herman Jacobson.

ago, and that it is astonishing that no one has thought of it sooner. We will not discuss this question; suffice it to say that such societies could not be started at a better time and with better prospect of success. There is a general depression of all trades and industries, and every attempt to aid a great industry will awaken sympathy with the great mass of the people, because all feel that they have an interest in the matter. And there is no industry which will appeal so strongly to the sympathy of our people as the fisheries; a society which undertakes to develop the fisheries will meet a very general demand not only in our city but in every fishing-station from Cape Lindesnaes to the North Cape. We can say without exaggeration that never has a society been started in our country with a more timely, practical, and patriotic object.

The notice which has been published starts with a well-known and deplorable fact, namely, that our country is far behind other countries with regard to the fisheries. And the society makes it its object to place our country where it belongs in this respect. Every thinking person must have found out long ago that as regards our fisheries our country has not kept step with other countries; for nearly every journal has year after year informed us how many important improvements have been made in other countries, whilst we have done little or nothing. Our fisheries have certainly not gone down, for the value of their productions has been constantly on the increase; but the fisheries of other countries have increased more rapidly and have made important progress, which has thrown us in the shade. This is a serious matter for Norway and more especially for the city of Bergen. If we are outstripped in the competition with other countries in that industry which on our entire coast from Christianssund to Vardö is the chief source of income of the whole population, the future will look dark in many other respects and the general development of the country will be hindered or retarded.

The signers of this notice, who are fully agreed as to the great importance of this matter, are not saying too much when they maintain "that our fisheries need further development in nearly every direction." And as this deplorable fact is the cause of publishing this notice, every one should consider it a strong incentive to join the society and thus aid a good cause. There cannot possibly be any doubt on this point. But the question which will arise first is, How shall our fisheries be aided? where shall we begin? and what shall be done first?

We anticipate that this society will gain many members in town and country, and that both the state and city authorities will subsidize this important undertaking. We also hope that the society will possess as much common sense and scientific ability, as is represented by the signers of the notice. And with such anticipations we ask the important question, "What shall be done?"

The signers of the notice mention "the hatching and raising of fish", "improving the methods of fishing and the fishing-apparatus" and "utilizing to their fullest extent all the products of the fisheries", in short the aiding of "every endeavor to further the fishing-interests." Such a programme means the establishment of an institution to work in each of these directions, and the organization of special societies or committees for the better carrying-out of all these objects. It will therefore be of great importance that this society should be joined by men of all classes throughout the country, so that no special interest or no special knowledge be excluded.

The whole question therefore becomes one of great importance to the entire country; and it is quite natural that the programme has not entered into any details as regards the carrying-out of the many different objects of the society. The society will probably organize very soon, and when it has once become an accomplished fact, we have no doubt that the plans for work will be laid out with that efficiency which may be expected from the signers of the notice and the classes of society which they represent. The work will then progress with that energy and caution which the great and national importance of the question demands.

In pointing once more to this notice and its important object, we consider any further recommendation superfluous, and would merely say in conclusion, that this question concerns a matter of the most vital interest to our whole country.

XI.—STATISTICS OF THE LOFFODEN FISHERIES FOR 1878.

(From the official report of the Superintendent, First Lieutenant in the Navy, Niels Juel.)

Number of fishing-stations	56
Extent of fishing-district in nautical miles	$12\frac{1}{12}$
Highest number of vessels in the district	722
Highest number of boats in the district	4,912
Highest number of men in the district	27,350

Number of telegrams.

Months.	Received.	Sent.	Total.
January	1,003	1,707	2,710
February	2,175	3,288	5,463
March	5,666	7,578	13,244
April	3,880	4,915	8,795
Total	12,724	17,488	30,212

Number of boats in the district at the end of each week.

Week ending January 19	
January 26	
February 2	2,227
February 9	2,549
February 16	3,087
February 23	4,003
March 2	4,180
March 9	4,357
March 16	4,400
March 23	4,670
March 30	4,673
April 6	3,010
April 13	

Number of men, boats, &c., grouped according to the different kinds of apparatus.

	Fishermen.	Crews.	Boats.
Net-fishing	13,168	2,154	*2,430
Line-fishing	7,258	1,689	1,977
Deep-bait fishing	2,297	844	†844
Hired men	3,311		
Total	23,034	4,687	5,251

* 269 of these also occasionally used lines. † 701 of these used no lines and 143 used lines

There was an *increase* from last year of 2,542 in the number of net-fishermen, and an *increase* from last year of 417 in the number of deep-bait fishermen, and a *decrease* from last year of 1,504 in the number of line-fishermen.

The percentage of fishermen using different apparatus was as follows: 58 per cent. used nets, 32 per cent. used lines, 10 per cent. used deep-bait.

Number of vessels at the Loffoden March 16, 1878.—Steamers 5, schooners 59, sloops 26, yachts 376, other vessels 202, total 668, with 3,111 sailors, and a tonnage of 342,620.

Cargoes of these vessels.—Not specified 479, dry goods 19, groceries 18, flour 27, ham, lard, &c., 10, bait 14, notions 34, wooden ware 9, with a total value of $70,400.

Number of days when no net or line fishing could be carried on.

Districts.	January.		February.		March.	
	Nets.	Lines.	Nets.	Lines.	Nets.	Lines.
Skroven	6	6	10	10	8	8
Svolvær	6	6	9	7	6	6
Vaagene	6	6	11	8	7	7
Hopen	6	6	11	11	10	8
Henningsvær	7	7	14	10	11	8
Stamsund	......	......	14	13	10	9
Ure	......	......	15	13	10	6
Balstad	......	......	19	12	13	8
Sund	......	......	14	14	12	9
Sörvaagen	......	......	14	12	11	8

Temperature of the air at the station of Svolvær (in degrees Fahrenheit).

Week ending—	Average temperature.		Highest temperature.	Lowest temperature.
	At noon.	Average lowest temperature.		
January 19	34.16	31.64	39.20	28.94
January 26	31.64	24.44	39.92	21.02
February 2	....	26.24	35.06	23.
February 9	31.64	27.68	39.02	21.02
February 16	29.84	25.34	35.06	19.94
February 23	34.88	30.74	39.92	27.14
March 2	28.94	23.18	37.04	17.96
March 9	32.	22.46	37.94	17.06
March 16	32.36	25.88	41.	19.04
March 23	37.76	27.14	....	19.94
March 30	33.44	21.02	39.38	15.08
April 6	42.62	33.08	46.04	32.00
April 13	44.06	32.	48.92	28.94
Average for the season	34.16	26.60	....	.....

The average temperature last season was 33.08 F., and the average lowest temperature 24.98 F. It has therefore been a little warmer this year. January and the first part of February on the other hand were considerably colder. The average temperature from January 19 to February 9 was last season 35.60°, and this season 31.82 F. The cause of the early fishing this season can therefore not be the higher temperature, as some have supposed. During the fishing-season I have not been able to discover any connection between the fishing and the temperature of the air. At Hopen and Henningsvær there was good fishing all through February and March, no matter whether the thermometer rose or fell. The few good fishing-days which the East Loffoden fishermen had during February were from the 12th till the 16th, when the average temperature was 28.94 F. From the 11th March to 16th March on the other hand fishing at the East Loffoden was very poor, although the temperature had risen to an average of 32.36 F. Again there was good fishing from the 18th to the 23d of March, when the highest average temperature of March, 37.76 F., was reached. I have therefore come to the conclusion that the fisheries are entirely independent of the temperature of the air, and if Professor Sars and others think they have observed the contrary, they must have taken an exception for the rule.

Temperature of the water at the station of Svolvær in Fahrenheit.—The temperature of the water in the harbor of Svolvær has been observed every day from January 26 to March 2, both at the bottom and at a depth of 6 fathoms (bottom). The temperature at the bottom varied from 32 to 40.10 degrees F., and was generally higher in proportion when the temperature of the air was lower. The instrument used was a Casella-Miller thermometer.

Week ending—	Air.		Water.	
	Noon.	Lowest.	Surface.	Depth of 6 fathoms.
February 9	32.36	27.68	35.69	37.04
February 16	29.84	25.34	35.35	37.78
February 23	34.88	30.74	33.94	36.35
March 2	28.94	23.18	35.06	38.84

The fact of the water at the surface being colder when the temperature of the air was 34.88 F., than when it was colder, is probably caused by the melting of the snow-water. It seems less probable, however, that the effect of the snow-water should be felt at a depth of 6 fathoms, especially as no stream worth the name falls into the sea at this place, and as far as our knowledge goes nothing of the kind has been observed. Between the temperature of the water and the wind there seems to be a certain connection. We could not ascertain which was cause and which was effect as the weather was very changeable. The tide seemed to have no effect on the temperature.

Deep-water temperature.

Depth.	January—		February—							April 8.
	24.	26.	12.	14.	16.	20.	23.	25.	26.	
Surface	34.25	32.45	36.95	37.40	34.25	35.60	35.60	35.15	38.30	35.60
10 fathoms	35.60	35.60	38.30	38.30	35.60	36.95	36.50	41.90	43.25	36.05
20 fathoms	......	......	40.10	......	......	36.50	37.40	45.50	46.40	36.50
30 fathoms	37.40	39.20	42.80*	43.70	37.40	37.85	38.30	50.90	46.85	36.95
40 fathoms	......	......	......	......	......	38.30	39.20	51.80	42.80	......
50 fathoms	......	......	......	45.50	......	......	......	......	......	39.20
60 fathoms	40.10	41.00	......	......	......	38.75	40.10	52.70	44.60	40.10
70 fathoms	......	......	......	46.40	39.20	......	......	......	......	......

*Observations stopped on account of a storm.

All these observations were taken at the same place, about one-third mile south-southeast of the Svolvær light-house. The instrument was unfortunately broken on the 1st of March, and no further observations could be taken. The observations on the 8th of April were taken with another instrument, which, however, had no indicator, and the result was approximately calculated from data gathered at former observations.

As a general rule the temperature at a depth of 10 fathoms has been 35.60–36.50 F., at 30 fathoms 37.40–38.30 F., and at the bottom 39.20–41 F. There were, however, so many exceptions from this that there is every reason to doubt the correctness of the observations. No fault could be discovered in the instrument; when brought to the surface from the different depths it always fell to the same point at which it stood when lowered. The observation of the 14th of February corroborates the one of the 12th, and that of the 26th makes that of the 25th quite probable. If on the 25th February a current of 52.70 F. degrees warmth really entered the west fiord, it has during the twenty-four hours which elapsed whilst the observations were being taken, mingled with the cold water in the fiord, and this comparatively warm mixed water has risen to the surface; on the 26th it had reached to within 10 to 30 fathoms from the surface, whilst the temperature at the bottom had sunk to 44.60 degrees F. The temperature in the harbor of Svolvær at a depth of 6 fathoms also strangely favors the probability of the observations being correct, as on the 12th, 14th, 25th, and 27th February it was 39.65 to 40.10 degrees, a warmth which was only reached once during the course of the winter, viz, on the 3d February. It is therefore quite possible that there are really very strange currents. No conclusions, however, can be drawn as to their influence on the fisheries until the existence of these currents is fully proved. If instruments can be obtained, the observations will be continued next year.

Proportion of spawners and milters.

Apparatus used.	February.			March 1-18.			From March 19.		
	Average percentage.								
	Spawners.	Maximum.	Minimum.	Spawners.	Maximum.	Minimum.	Spawners.	Maximum.	Minimum.
Bottom-nets	54	66	47	46	54	40	52*	70*	35*
Floating-nets	.....	.....	.....	43	46	40	40	43	36
Bottom-lines	64	77	43	48	59	36	43*	50*	40*
Floating-lines	63	69	57	58	70	46	45	67	31

*Observations of 1877.

As far as the bottom-apparatus is concerned this year's observations agree with those of last year, and may therefore be considered closed. They agree, moreover, with those natural conditions which the superintendent considers as the cause of the statistical data as far as the spawners are concerned. The floating-lines show during the first half of March a larger percentage of spawners than the bottom-lines. More accurate observations, however, would have but little practical significance as far as the statistical data of spawners are concerned, and they will therefore not be continued. The discrepancy between the results of the two kinds of apparatus may possibly be caused by the small number of observations (only 80), as there is no cause to suppose that there is any difference between the bottom and the floating apparatus as regards the spawners and milters. As Professor Sars has made entirely different observations and has given his reason for them, I shall endeavor briefly to refute his views.

Professor Sars says in his report (p. 55): "Those boats which had used bottom-nets had almost exclusively caught milters, whilst more spawners had been caught by those who had employed floating nets or lines. That the proportion between spawners and milters must be such, I could, even without palpable proof, have told people beforehand from my observations of the full-grown roe." The observations made by me and my assistants show that the majority of fish caught in bottom-apparatus may be milters, but they likewise show that occasionally the majority may be spawners, and that on an average an equal number of both kinds are caught. It is both improbable and impossible, that milters should be caught exclusively in bottom-apparatus and spawners in floating-apparatus. It is improbable, because the floating-net is 12–40 fathoms from the bottom, and if the professor's observations were correct a similar extent of water must intervene between the milters and spawners. It is impossible, because floating-nets are not used very extensively, viz, only by the fishermen at the stations east of Sörvaagen, and where should the large proportion of spawners come from, if the bottom-nets caught only milters?

Number of livers to the barrel.

Week ending—	Skroven.		Svolvær.		Vaagene.		Hopen.		Henningsvær.		Stamsund.		Ure.		Balstad.		Sund.		Sörvaagen.		Værö.	
	Nets.	Lines.	Nets.	Lines.	Nets.	Lines.	Nets.	Lines.	Nets.	Lines.	Nets.	Lines.	Nets.	Lines.	Nets.	Lines.	Nets.	Lines.	Nets.	Lines.	Nets.	Lines.
Jan. 26	....	....	....	....	250	300	360	420	....	400	....	300	....	....	....	380	....	...	360	480	...	...
Feb. 2	....	....	....	400	250	300	360	420	300	400	300	350	....	....	....	380	300	350	360	480	...	...
9	300	400	300	400	250	300	360	420	300	400	300	350	....	....	....	380	300	350	360	480	...	...
16	300	400	300	400	300	380	350	450	300	400	300	350	....	....	300	360	300	350	360	480	...	...
23	300	460	360	500	300	380	350	450	300	400	300	350	....	....	....	480	300	350	360	480	250	300
March 2	300	460	360	500	300	380	430	520	300	400	350	400	300	350	460	500	350	400	360	440	250	300
9	360	500	360	500	400	500	430	520	350	450	380	430	300	400	460	500	350	400	360	440	350	450
16	360	500	400	500	450	580	480	580	400	500	400	460	300	400	460	500	350	400	300	400	350	450
23	400	500	400	600	400	550	480	580	500	600	400	460	300	400	500	740	350	500	360	480	350	450
30	400	500	450	650	450	580	480	580	500	650	430	600	450	600	500	740	350	500	360	480	450	600
April 6	360	400	450	650	430	580	480	580	500	650	500	650	300	600	500	740	350	450	360	480	400	500
13	....	....	....	....	....	....	450	600	....	....	....	....	....	....	460	700	....	...	550	700	...	...

Highest number of livers from line-fishing, week ending April 13.. 700
Highest number of livers from net-fishing, week ending April 13.. 550
Lowest number of livers from line-fishing, week ending January 26.. 300
Lowest number of livers from net-fishing, week ending January 26.. 250

Both in 1876 and last year there was a period from the middle of February till March when the fish increased in fatness; and as the number of fish generally increased about the same time, I concluded that new schools had come in. This year no increase of fatness was observed except about the middle of March in the district of Sörvaagen, but on the other hand it was observed that this year the fish were fatter than last year from the beginning of the fisheries. Fishing also commenced early and promised well at all the stations. From the middle of February the fatness gradually decreased, so that probably no new schools arrived after that time with the exception of the district extending from Henningsvær to Ure, where the proportion of liver kept unchanged longest, and where there was steady and good fishing all the time. Although the proportion of liver kept unchanged for a long time west of Nufsfiord, there was no fishing from the middle of February till the middle of March, and no net-fishing till the end of March. The unfavorable weather was probably the cause of this, as the fishermen could not reach their usual fishing-places as often as was desirable.

Price of fish (not stated per what quantity).

	Net-fishing.		Line-fishing.		Deep-bait fishing.	
	Absolute.	Average.	Absolute.	Average.	Absolute.	Average.
Highest	$6 24	$6 03	$5 62	$5 36	$7 77	$5 09
Lowest	3 75	5 00	3 48	4 55	4 02	4 55

Price of liver, roe, and bait (per barrel), and of heads (per 100).

Fresh liver	per barrel..	$4 28 to 7 50
Old liver	do.	4 28 to 4 82
Roe	do.	2 14 to 5 36
Fresh herring for bait	do.	2 54 to 4 28
Salt herring for bait	do.	2 14 to 4 82
Cuttlefish for bait	do.	3 21 to 6 43
Muscles for bait	do.	2 14 to 4 28
Heads	per hundred.	8 to 32

This year's fishery has been the second largest ever known; and if the weather had not been so very unfavorable during February and March, the number of fish would—in spite of the total failure of the fisheries in April—have been as large as last year; for it is my opinion that the schools on the banks were much larger. In January and February 4,500,000 were caught; in April, 2,500,000; and in March, 17,750,000; whilst last year (1877) the number was 16,000,000 in March. The first of the two following tables shows the result of the Loffoden-fisheries for every week from 1869 to 1877, and the second shows the result per month, calculated for the last and the first week of the month according to the days when fishing was going on and the number of fish caught during the week. According to these data an average Loffoden-fishery ought to yield about 20,500,000, of which 4,300,000 (21 per cent.) are taken in January and February, 12,300,000 (60 per cent.) in March, and 4,000,000 (19 per cent.) in April. This year (1878) the percentage was as follows: January and February, 18.2 per cent.; March, 71.7; and April, 10.1.

Month.	1869.		1870.		1871.		1872.		1873.		1874.		1875.		1876.		1877.	
	Date.	Thousands.	Date.	Thousands.	Date.	Thousands.	Date.	Thousands.	Date.	Thousands.	Date.	Thousands.	Date.	Thousands.	Date.	Thousands.	Date.	Thousands.
February															1	200		
			5	1,000											5	600	3	130
			13	1,500	13	2,000							13	2,500	11	1,500	10	500
			19	2,400			18	3,500	15	2,000	15	1,000	20	3,750	19	2,500	17	1,500
	27	*4,200	27	3,600	26	2,500	25	5,000	22	2,000	22	2,000	28	6,000	26	4,200	24	3,000
March									1	3,500	1	3,700					3	4,750
	7	8,400	6	6,600	5	4,000	3	6,250	9	5,500	8	5,000	7	9,000	5	7,500	10	8,250
	14	11,400	12	9,600	12	6,000	10	8,000	16	8,500	17	7,500	14	13,500	12	9,000	17	11,500
	21	16,800	20	13,200	18	8,000	17	10,500	23	10,500	22	12,000	21	16,500	19	11,750	24	17,000
	27	18,000	27	16,200	26	11,500	24	12,000	30	14,000	29	14,000	28	18,500	26	15,500	31	20,250
April	3	19,200	3	18,000	2	14,500	1	13,000	6	18,000	5	15,000	4	21,000	2	20,250		
	11	20,400	10	20,400	10	16,800	7	15,000					11	23,000	9	21,250	7	24,250
	16	20,700	14	21,600	14	17,500	14	17,000	14	19,500	14	16,000			15	22,000	14	28,000
Corrected according to medicinal tax		20,500,000		22,000,000		17,500,000		20,000,000		20,000,000		20,000,000		23,000,000		23,000,000		28,000,000
Caught after April 14th						1,000,000		250,000		60,000		55,000				500,000		1,500,000

*These figures indicate the number of fish caught up to the date mentioned.

Month.	1869.	1870.	1871.	1872.	1873.	1874.	1875.	1876.	1877.
Caught in January and February	4,200,000	3,600,000	3,100,000	5,700,000	2,500,000	3,700,000	6,000,000	6,000,000	4,000,000
Caught in March	14,100,000	13,200,000	9,400,000	7,300,000	12,100,000	11,000,000	13,400,000	13,600,000	16,250,000
Caught in April	2,400,000	4,800,000	5,000,000	4,000,000	4,900,000	1,300,000	3,600,000	2,400,000	7,750,000
Total	20,700,000	21,600,000	17,500,000	17,000,000	19,500,000	16,000,000	23,000,000	22,000,000	28,000,000

TOTAL RESULT OF THE FISHERIES, 1878.

Number of fish caught	24, 660, 000
Number of heads sent to the guano factories	16, 500, 000
Liver (barrels)	53, 150
Medicinal oil (barrels)	3, 044
Roe (barrels)	26, 130

Upwards of 14,000,000 fish were caught with nets, 9,250,000 with lines, and 1,250,000 with deep-bait.

Gross receipts of the Loffoden-fisheries, 1878, $1,742,000.

Number of fish and quantity of roe, liver, and oil per week.

Week ending—	Total number caught.	Salted.	Number caught during each week.	Liver.	Medicinal oil.	Roe.	Fishing-days.	
	Thousands.			Barrels.			Whole.	In part.
January 26	160	... ...	158	400	13	350	4	....
February 2	360	....	200	900	20	750	1	1
February 9	700	....	340	1, 900	55	1, 500	1	2½
February 16	2, 500	....	1, 800	6, 800	350	5, 200	2	3
February 23	3, 000	....	500	7, 800	480	6, 000	....	2
March 2	5, 500	3, 500	2, 500	15, 000	925	11, 200	3	....
March 9	8, 500	6, 000	3, 000	22, 000	1, 650	16, 500	3	1
March 16	11, 750	9, 000	3, 250	28, 000	2, 000	19, 000	....	....
March 23	17, 750	14, 500	6, 000	39, 000	2, 700	24, 000	4	1
March 30	22, 250	18, 750	4, 500	48, 000	3, 000	25, 000	....	....
April 6	23, 500	19, 750	1, 250	50, 000	....	....	2	2
April 13	24, 750	21, 000	1, 250	53, 000	....	....	5	1
Total	......	......	......	....	......	....	25	*13½

* Against 32 and 21 in 1877.

The largest number of fish caught, in proportion to the number of fishing-days and the number of men engaged, was in the week March 16 to March 23. The result all through March was unusually even. It is estimated that after the 14th of April 120,000 fish were caught. About half a million of fish were consumed during the fisheries, as well as 1,000 barrels of liver. About 20,000 cod were salted in barrels for sale. The quantity of bait used was as follows: Nine hundred barrels fresh herring; 12,000 barrels salt herring; 3,500 barrels cuttlefish; 900 barrels muscles; with a total value of upwards of $67,000. Most of the salt bait was prepared by the fishermen at their homes.

The treasurer of the "Medical Fund" reports that the taxes on fish in the districts of Nordland and Tromsö in 1877 amounted to $43,682, distributed as follows:

Dried codfish	$16, 913
Salt codfish	12, 946
Oil	7, 431
Roe	615
Herrings	5, 777

Result of the fisheries in the districts of Nordland and Tromsö, 1875, '76, *and* '77.

Fisheries.	1875.				1876.				1877.			
	Fish.		Oil.	Roe.	Fish.		Oil.	Roe.	Fish.		Oil.	Roe.
	Salt.	Dried.			Salt.	Dried.			Salt.	Dried.		
	Millions.		Barrels.		Millions.		Barrels.		Millions.		Barrels.	
Loffoden-fisheries	15¼	7½	35,000	21,000	17	5½	35,000	22,000	25¼	4¼	40,000	29,000
Other winter fisheries	1¼	6	13,000	8,500	2½	6	14,500	12,000	2½	8¾	19,000	16,000
Summer and autumn fisheries	....	9½	8,000	1,000	½	9¼	8,000	1,000	½	12¾	10,000	1,000
Total	16½	23	56,000	30,500	20	20¾	57,500	35,000	28¼	25¾	69,000	46,000

FISHERMEN'S EARNINGS.

Gross average earning for each *fishing-day* in February	$1 87
Gross average earning for each *fishing-day* in March	4 35
Average receipt per day, counting *all* the days of the season	93
Highest total sum earned by a net-fisherman	214 40
Lowest total sum earned by a net-fisherman	48 24
Highest total sum earned by a line-fisherman	120 60
Lowest total sum earned by a line-fisherman	32 16
Highest total sum earned by a deep-bait fisherman	85 76
Lowest total sum earned by a deep-bait fisherman	42 88

Hired men earned from $32.16 to $40.73 besides board and lodging.

APPENDIX D.

DEEP-SEA RESEARCH.

XII.—REPORT ON THE NORWEGIAN DEEP-SEA EXPEDITION OF 1878.

By Prof. G. O. Sars.*

I.

Hammerfest, *July* 10, 1878.

Mr. Editor: As has already been announced in the plan of the expedition heretofore published in your paper, Hammerfest will be our chief station during the present year. At this place the expedition is supplied with coal and other necessities for its various cruises into the Arctic Ocean. Three such cruises are mentioned in the plan: one toward the east, one toward the west, and one toward the north. The first of these, which principally concerned the so-called East Sea (Östhav) has now been completed, and I will improve the time while we are lying here at Hammerfest equipping ourselves for our second cruise toward the west, to make good my promise and send your paper something about the expedition and about what it has accomplished so far.

The scientific investigations were begun, as you have already learned from telegrams, in the West-fjord, where we chose a point a considerable distance up the fjord about directly opposite Tran-isle. The West-fjord here has, according to previous soundings, its greatest depth, namely, 350 fathoms. A series of careful observations of the temperature were made in this place, whereby the remarkable fact heretofore observed farther out in the sea, namely, that at a certain depth (here only 40 fathoms) can be found a temperature considerably lower than that found in both lower and higher water-strata, could be established with perfect certainty by the use of instruments improved in many important respects. A cast of the dredge was also made in this place, whereby various curiosities were brought up from the deep. The weather was here, as on our whole cruise, brilliant, and we most thoroughly appreciated the summer breezes, well knowing that we before very long should have to exchange this beautiful sunshine for the rough climate of the Arctic Ocean. In Tromsö, where we stopped only a few hours, we took on board a pilot who is to accompany us on our cruises this summer. He is an experienced Arctic seaman who has spent no less than thirty-five summers in the Arctic seas about Spitzbergen, Jan Mayen, and Nova Zembla, hunting seal and walrus. He has made the

* Translated by Prof. R. B. Anderson of the University of Wisconsin, Madison, Wis., from a series of letters to "Dagbladet" by Prof. G. O. Sars.

impression of being a man of rare reliability and intelligence, and will doubtless prove a great help to the expedition, especially when we come to Spitzbergen, where he seems to be nearly as well acquainted as at his birth-place, Tromsö.

Before we reached Hammerfest we visited the Alten-fjord, where explorations were made at two points. Here, too, we found an intermediate minimum of temperature, though not before reaching the depth of 100 fathoms. The fauna of the sea-bottom showed, as might be expected, a more marked Arctic character than in the West-fjord, where it still was perfectly Atlantic.

In Hammerfest, where we arrived on Saturday the 22d of June, early in the morning, the members of the expedition were most cordially received by the city authorities, and the two days we spent here afforded us ample evidence of the rare hospitality and kindness for which this most northern town of our mundane sphere is famous.

On Monday night, the 24th of June, we weighed anchor and directed our course to the north and east in the usual steamship route. On our way we examined two of the large Finmark-fjords, namely, Porsanger-fjord and Tana-fjord. In both a series of careful observations of the temperature were taken, which did not, however, show any such intermediate minimum as was found in the West-fjord and Alten-fjord, undoubtedly on account of the greater shallowness of the water. The fauna was likewise examined, both with the dredge and the trawl-net, whereby its genuine Arctic character could be established. Among the hauls made here it is necessary to make special mention of the one made in the Tana-fjord with the beam-trawl, under the supervision of Captain Grieg. A richer haul we zoologists have hitherto scarcely seen. The trawl-net was brought up containing more than two barrels of loose mud, out of which protruded large beautiful sea-anemones and variegated star-fishes, and wherein we saw tumbling about a number of fishes (sea-perch, flounder, and skate). We were here thoroughly convinced of the superiority of the beam-trawl over the common dredge, especially after we had made various important improvements, not only in the net but also in reference to the arrangement of the weights which are to hold the runners in the right position against the bottom. But the instrument being very large, it is also difficult to manage, and hence it can as a rule be used only in calm weather, and in a comparatively smooth sea. We have since had occasion to test it with excellent success in the open sea, and desire all the more to make use of it hereafter, since it has been found that even the most active animals and fishes can be secured in this manner.

In fine, calm, bright weather we doubled the barren and exposed coast of East Finmark and arrived during the night on June 25th at Vardö Isle, where we remained during the following day to complete our equipment and to determine more accurately the geographical position of this point. Early on the morning of the 27th of June, we weighed anchor and turned the prow to the east for our first ocean cruise. The fair weather still continued for some time, so that we dredged with excellent results on the same day at a distance of about forty miles from the coast, where the water was found to be 148 fathoms deep. But the rapidly falling barometer warned us to look out for drizzly weather, which was not long in coming. Toward evening it began to blow from the west, and during the night we had a perfect gale with chopping sea, which made our ship roll in a most disagreeable manner, so that it finally was found expedient to lay the stem of the vessel against the waves and thus await a change in the weather, a method of working the ship with which we had become only too familiar on our first expedition. This state of things continued not only during the remaining portion of the night, but also all of the next day, and while it lasted no kind of investigations could be thought of. Under these circumstances time naturally hung heavily on our hands, and we kept looking anxiously at the barometer to see whether no change might be expected. But the barometer appeared very capricious. Now it would rise a little and make our hopes rise correspondingly, then it fell again without having advanced more than a few millimetres. The following day the weather had cleared up a little, but a pretty stiff breeze was still blowing, and it was so damp and cold that we fairly dreaded the idea of going on deck. Meanwhile we had been able to throw the lead early in the morning and take a series of observations of the temperature, whereby it became evident that we had already advanced within the cold area. Having thus found the eastern boundary line between the Polar and the Atlantic currents, and the manner in which they pass into each other having been more carefully examined, we stopped here, and after having undertaken a dredging we continued northward for the purpose of further tracing the above-mentioned boundary line. The weather continued drizzling and cold, very much like winter, with interchanging showers of snow and sleet, and a temperature that fell even down to 33.8° Fahr. Thus it was not to be wondered at that we who came from summer warmth of more than 68° Fahr., found it disagreeable, and, though being in the midst of summer, we were obliged to put on winter clothes from head to foot whenever we desired to breathe fresh air upon deck. During the four following days the Polar current's boundary was followed accurately. In so doing we first sailed to the north, then to the west in the direction toward Beeren Island. On the way we also undertook a couple of dredgings, whereby we gained a tolerably correct idea of the character of the fauna, though the weather threw many obstacles in the way of these investigations. On the evening of the 3d of July we saw the first ice, which appeared in the form of quite small detached blocks, of the most fantastic shapes, but later in the form of connected floes. Birds increased in number and kind as we advanced. Auks, fulmars ,and gulls flocked everywhere, and in the horizon were seen a number of high columns of smoke as from a fleet of steamships. They were whales that were gormandizing on the abundant fauna of the Polar waters, and which as we drew nearer exhibited their broad, black backs above the surface of the water, after having emptied their lungs with a rumbling noise and whirled the water up in the form of a high column of smoke. A glittering illumination spread itself over the sea filled with floating blocks of ice. Late in the evening a shout came that land was in sight ahead. It was our experienced pilot, whose keen eye first had discovered Beeren Isle through the fog. The rest of us in vain strained our eyes; all we could see was fog and the sea. Tired of looking, and knowing from the chart that we still were a considerable distance from land, the majority of us resolved to retire to our berths in order to enjoy with quickened energies on the following morning the sight of Beeren Isle and if possible undertake to land.

Our waking the following morning gave us the agreeable sensation of a perfectly smooth sea. We had not known the ship to lie so quietly for a long time, and as the screw only now and then throbbed, we soon understood that we were near the shore. We were therefore not slow in donning our clothes and springing upon deck in order to get a more perfect idea of the situation. At about the distance of a quarter of a mile from the ship lay the barren rocks of Beeren Island before us, partly shrouded in fog. We were in the lee of the land, east of its most southern point, doubtless the most picturesque part of the island. A high promontory, with sharp weather-beaten crests, extends precipitously into the sea, and in front of it rises again, in the form of a beautiful obelisk, a high, wonderfully shaped, isolated rock, called the Stappe. When the fog rose a little from the land in the rear of these rocks, extensive connected masses of snow were seen, interchanging with steep precipices and barren, gravelly flats. Further toward the north lifted itself out of the fog Mount Misery, which is 1,700 feet high, and around the summit of which winds a peculiar precipice, looking like an artificial breastwork. Between this highest mountain on Beeren Island and the south point lies the so-called South Harbor, where we intended to land, in order to put ashore the mail entrusted to us for the Dutch expedition, which, on its passage to the east, had determined to touch the same point somewhat later. According to our instructions the place was situated near the so-called Mayor's Gate (Borgermesterport), a wide gateway in the rock, through which in calm weather one can row a large yawl. By the aid of our glass we soon discovered the portal, minutely described and represented by an illustration in the Report of Nordenskjold's Expedition, in 1863, and therefore, when we had approached as near to it as we dared with the steamship, we dropped our anchor, whereupon two yawls were put into the sea and furnished with men.

We rowed without accident through the famous and grand Mayor's Gate, which was guarded by a multitude of noisy gulls, and landed on a gently sloping, sandy beach, where we, without the slightest difficulty, and perfectly dry-shod, planted our feet on the ground of Beeren Island.

Directly to the right of the Mayor's Gate, and a few paces up from the harbor, lies an old deserted Russian hut, the point indicated to us by the Hollanders. The roof was partially dilapidated, and here and there the floor was torn up and drenched with snow-water; but the walls were well timbered and had resisted the destructive influences of the wintry blasts and of the snow tolerably well. The plain and simple interior arrangement, a couple of bedsteads and a rudely-fashioned table, gave us an insight into the dreary existence which its occupants must have experienced during the long wintry nights, while the storm howled without and the snow gathered in towers round about the hut. That time had hung heavily on their hands was also sufficiently evident from the numerous inscriptions and carvings which covered the walls and bedsteads. With an industry and exactness that partially made up for the lack of artistic talent, we here found carved with a jack-knife ships of all sizes and descriptions, the cordage and yards represented as minutely and accurately as possible. In the rear of the house lay parts of the skeleton of a polar bear, which undoubtedly had been altogether too impertinent to escape with his life. The time required to dispose of the mail was occupied by a part of us for the purpose of taking a short stroll into the interior part of the island. Nothing more melancholy and dreary can be imagined. Even Jan Mayen seemed to us a garden in comparison with these barren flats, strewn with nothing but pebbles and gravel.

After having taken this invigorating exercise on shore, we returned to our ship, where we weighed anchor and proceeded westward, in order to determine more accurately the slope toward the great deep outside. At the distance of about forty miles from the island we cast the dredge at a depth of 35 fathoms. The bottom here consisting chiefly of coarse sand, the harvest was comparatively insignificant. On the other hand, the surface of the water was here filled with pelagic animals; our surface-net especially yielded enormous quantities of peteropods (*Limacina arctica*), and many of the specimens were of quite unusual size. The surface temperature was, as might be expected from the proximity of the ice, very low, and the sea-water was filled with a peculiar sea-slime, which on our former expedition had been observed under similar circumstances. But at a somewhat greater distance from the island a very sudden change took place, the temperature of the water rising at once to from 35.6° to 41° Fahr., while the color changed from a greenish to a dark-blue hue. It was the warm water of the Atlantic that here met the Polar water, without being able, however, as it seems, ever to get over to the coasts of Beeren Island, where the Polar current seems to be as dominating as at Jan Mayen. At somewhat long intervals the lead was thrown as we progressed outward, showing first 115, then 457, and then 750 fathoms, without the discovery of any abrupt descent anywhere. At the last-named station a complete series of observations of the temperature were taken, which showed 32° Fahr. to be situated much deeper than we had expected to find it so far north, namely, between 400 and 500 fathoms. A dredging undertaken at the same place gave comparatively little return, the bed of the sea being so soft that the mouth of the dredge undoubtedly became filled up too soon with the tough clay, without being able to catch anything after being so filled.

We now bent our course to the south, then again to the west, in order to find the depth of 1,000 fathoms. This having been accomplished, we turned our prow toward Norway, making soundings at suitable intervals, in order to determine the ascent at this place from the deep. The ascent was here much more abrupt than further north, it being between 500 and 300 fathoms, which seems to indicate the existence of a real precipice between Beeren Island and Norway. We could not, however, devote much time now to establishing the details in regard to this precipice, our coal and water supply diminishing to an alarming extent, and we having still to examine the fauna at this point. Two hauls with the beam-trawl at different depths gave exceedingly interesting zoological results. In the first haul, which was made at a depth of 447 fathoms, with a temperature at the bottom of 33.08° Fahr. we got, among other things, a specimen of a species of halibut (*Hippoglossus pinguis*) more than a foot long. This species of halibut is not known on our coasts, and belongs to the far north. In the second haul, which was made in a depth of 190 fathoms, we also secured some fish of the *Cottoid* family, among which there apparently was a new variety, and besides we got an extraordinary amount of lower animals, which gave to us, as zoologists, abundance of work, even long after we arrived at Hammerfest.

We anchored in the harbor of this town on Monday noon, July 8, upon the whole well satisfied with the results of our first cruise into the Polar Sea, and with the brightest expectations in regard to the two cruises yet to be made before the expedition is completed.

II.

HAMMERFEST, *July* 27, 1878.

MR. EDITOR: We are again lying here well moored in Hammerfest's Harbor, after having once more plowed the waves of the Polar Sea, and I shall avail myself of the opportunity, while we are resting after our work done, of sending you some brief account of our last cruise, continuing my story where I left off in my previous letter.

After stopping about four days in Hammerfest, which was necessary for taking on board coal and other prerequisites, we weighed anchor on the morning of the 13th of July, and proceeded northward through the South Island Sound in beautiful, calm sunshine. On the so-called Bond Island Ridge, one of the most celebrated fishing-grounds in this locality, we stopped for a short time, whereupon our fishing-tackle was brought out. In a short time we hauled up several fine-looking codfish, which differed in no material respect from the so-called winter codfish (*Skreid*), and in size scarcely were inferior to the common Lofoden codfish. The contents of the stomach were examined carefully, both in these and in other species of fish caught at the same time. It appeared as usual that the codfish had not been very delicate in the choice of his food, which was very mixed, and consisted partly of crabs and mollusks, and partly of small fishes. In one of the stomachs we found a wolf-fish a span long, and this was yet so fresh that it could be preserved as a specimen in spirits. The coal-fish seemed to have been far more delicate in the choice of his food, which consisted exclusively of cuttle-fish, and, upon further examination, this proved to belong to the well-known Arctic form, the *Gonatus amœnus*, of which there heretofore has been found only one specimen on our coasts. This last discovery was of no little interest to us, partly because by it the appearance of this cuttle-fish in large numbers on the coasts of Finmark could be established, partly because the remarkable change in coal-fish fishing that recently has been observed in these regions could be naturally explained by the very appearance of this peculiar food.

A westward course was now taken and the jagged mountains of South Island soon disappeared from above the horizon, while we still had a glimpse of the loftier, snow-covered plateaus of Seiland. An indistinct land-line was yet seen for a short time in the southeast; then all vanished, and we had nothing but the boundless sea on all sides around us in the horizon. On the same day soundings and a series of observations of temperature at a depth of 95 fathoms were taken. At twelve o'clock in the night the lead was thrown out again, indicating 630 fathoms, and on the evening of the following day we found 1,110 fathoms, whence it appeared that the descent toward the deep here was tolerably gradual. At the last-named place we stopped, and the trawl went to the bottom, accompanied by our best wishes. In the morning the apparatus was hauled in good condition on deck, and it brought up from the deep a draught larger than any we ever had gotten before. In the net were found, in addition to a great variety of lower animals, no less than five specimens of a rare Arctic fish (*Lycodes*), one of which measured more than a foot in length and seemed to be full-grown. The excellent qualities of the trawl were still further demonstrated by this successful haul, and for the time being the dredge heretofore used fell wholly into discredit with us. As we progressed westward the air steadily grew colder. During a part of the time we had been surrounded by dense fog, and on the following morning the weather was so disagreeable that we had to put on a complete suit of winter clothing when we went on deck. We sounded in the forenoon, finding 1,200 fathoms, and took with great care a series of observations of the temperature, finding 32° Fahr. at only 30 fathoms' depth. Here we sent the trawl down again; but although the greatest precautions were taken both in letting it down and in the further maneuvering of it, it soon appeared, upon the hauling in of the trawl, by the insignificant stretching of the accumulator, that the apparatus for some reason or other had not followed the bottom. In spite of this fact the net had caught in the stratum of water nearest above the sea-bed two specimens of the sea fauna of such extraordinary interest that they abundantly repaid the trouble and care we had given to this haul. One of these was a fish, the other a cuttle-fish, both alike remarkable and interesting. The fish was of a brilliant scarlet color, with extraordinarily far-projecting, thread-like ventral fins, and belonged to a hitherto entirely unknown genus and species of the cod family.* The cuttle-fish likewise proved to be a new species of the remarkable and hitherto but little known genus *Cirroteuthis*. Of both a drawing in colors was immediately made, which will be of valuable service in the preparation of the final report.

On the following morning we were surrounded on all sides by a dense fog so that we could see scarcely more than a few fathoms from the ship. The wind changed successively from north to northwest and west, which, in connection with the position of the barometer, indicated that we were just passing the north side of a tornado and that we in all probability soon would be outside of its range. By the heavy swell setting in from the southwest we were also informed with sufficient certainty that there was at no very great distance from us to the south a storm, and that, too, one of the very worst sort, so that we deemed ourselves fortunate that we on this occasion found ourselves so far into the Arctic Sea. Toward evening the sea became remarkably smooth, and an icy cold filled the atmosphere. In the west was seen toward the horizon a peculiarly clear glimmering in the air, which we already, from our former cruise, recognized as being ice-blink. At 9½ o'clock "Drift-ice ahead!" was shouted, and a piece of ice, much worn and perforated by the sea, came slowly floating past our ship, the first messenger from the Greenland ice. This was followed by still another, then by more and more, and finally the sea was filled on all sides with blocks of all possible sizes and of the most fantastic forms. Colossal mushrooms with hollow, beautiful, bright, green stems; swans, with far-extended necks; boats, with full crews; wonderfully jagged pillars stooping or leaning in various directions; flats sloping irregularly and half hid in the sea, crowded one on the top of the other—in short, the most extravagant forms passed in review before our ship as we progressed. Far out in the horizon was discovered a snow-white irregularly winding line, from which single bluish tops reared their heads, and over which a clear ice-blink, not unlike a sort of aurora borealis, appeared. Here the ice seemed to be more continuous, and our experienced pilot, who was sent aloft in order that he might be able to form a more accurate estimate of the situation, advised us that further progress westward was impossible. We had reached the Greenland ice, and although this had taken place somewhat sooner than calculated in our plan, we had to submit to the inevitable

Rhodichthys regina Collett.—T. H. B.

and change our course, steering northward and following the apparent direction of the edge of the ice. The following night we reached the northernmost station, where we cast the lead, finding a depth of about 2,000 fathoms, the greatest depth we had yet observed. We now turned our course to the east, casting the lead at suitable intervals, and usually taking a series of careful observations of the temperature in order to determine accurately the curve which indicates the modifications of the temperature at various depths. Thereby we established, among other things at various stations, the interesting fact that at a certain depth below the surface there is found a minimum of temperature, then again a more or less distinct rise of the temperature, below which the usual gradual decrease toward the bottom was observed.

The following day the trawl was sent down to a depth of 1,200 fathoms, and although it was evident that it had gone down on the wrong side, that is, with the beam down and the runners up, it still contained several interesting objects from the deep, and among them two specimens of the same Arctic fish (*Lycodes*) which we had caught before at a somewhat less depth. In the night we sounded again, finding a depth of 1,500 fathoms, and the series of observations of temperature taken showed that we were already outside of the actual limits of the Polar current, as we did not find 32° Fahr. before reaching the considerable depth of 400 fathoms.

The tornado we had touched recently had now evidently passed us entirely. On the following day the weather was calm, and the sea was so smooth that even microscopic examinations of the peculiar sea-slime, which on this expedition was observed in various places in the ocean, could be made successfully. We were now rapidly approaching a point where the depth, according to a sounding made during the voyage of Gaimard, was recorded as being 260 fathoms, and which, therefore, was to constitute the extreme points of the bank extending between Beeren Island and Spitzbergen. It being of importance to investigate more accurately the ascent from the deep in this place, the lead was cast at short intervals, but the depth did not decrease in any marked degree. Exactly at the point indicated on the chart where Gaimard's lead had been cast, we still found a depth of 1,060 fathoms, which presupposes a considerable error in the chart of the soundings heretofore made at this place. Not before we had made three or four soundings further to the east did we find the real ascent to the bank, and thus we had an opportunity of making here a not unimportant correction in the sketching of the depth-curves. At a depth of 650 fathoms, that is, on the very slope of the bank, the trawl was let down and brought up again late in the forenoon of the following day to be placed in good order on the deck. It contained about three barrels of mud. An exceedingly plentiful harvest of zoological specimens was secured, and, as usual, there was found, in addition to lower animals, a considerable number of fishes in the net, among which were several of great importance. Upon the bank the trawl was sent to the bottom again at a depth of 180 fathoms, but the net was here torn to pieces by the sharp stones, so that we secured only what had accidentally been caught in the meshes, chiefly hydroids and polyzoa. The temperature both of the atmosphere and of the water had meanwhile sunk to so considerable an extent that it was evident that we had again come within the boundaries of the Polar current. We therefore regarded it as our duty to proceed a little further to the east, in order to investigate the physical and biological conditions in this stretch of the ocean, which for the most part of the year is filled with ice. We cast the lead at short intervals, finding a gradually ascending slope until we reached a depth of only 21 fathoms. At the surface the water was perfectly ice-cold, that is, 31.64° Fahr., the lowest temperature we had observed at the surface. It being presumable that the bottom here was stony, the dredge was sent down instead of the trawl, and it brought up from the bottom a considerable amount of coarse sand mixed with stone. Between and on these stones, and entirely covering the tangles, were found enormous masses of hydroids, many of which were very beautiful. Whole forests of these must cover the bottom in this locality. We did not find it suitable to our present purpose to make investigations further to the east, and so we turned our prow toward Beeren Island, which appeared in sight on the same day about noon. First rose above the horizon the peculiarly formed Mount Misery, and after that a long, low stretch of land on the north side of the mountain. It seemed to be endless, and afterwards proved to be the flat and undiversified northernmost portion of the island. A stiff breeze from the north had meanwhile arisen, which increased as we approached the land, and it soon made the waves so tumultuous that we were obliged to give up all thoughts of landing at this time. Off Mount Misery the wind was so violent that the sea was whipped into mist, and the ship careened fearfully. We hugged the shore so closely that we were in sight of South Harbor and the Russian hut, in order, if possible, to find out whether the mail left here by us for the Dutch expedition had been received. By the aid of our glasses we soon spied the signal left by us, and the flag had been removed, which we of course regarded as a sign that the expedition had been there. A letter just received here, probably brought by some fisherman, assures us that everything had been found in good order. The letter is from the chief of the expedition.

Having lain quiet for some time off the south side of the island awaiting if possible a favorable change in the weather, all sails were hoisted at ten o'clock in the evening, the engine was set to work at its utmost capacity, and we turned the stem of the ship southward toward Norway. So long as we still were in the lee of Beeren Island we had comparatively smooth water and a moderate breeze; but as we got further out to sea the wind increased and the waves waxed higher, and finally there blew a perfect gale from the north, the waves dashed high, and we sped forward at the rate of sixty-five miles per watch. It was the first time that we were obliged to scud under bare poles on account of a real gale, and although the ship frequently careened and pitched badly, making boxes and other things tumble about in wild confusion, we still had abundant occasion to praise the excellent qualities of the Voring and pronounce her an excellent sea-going vessel. In spite of the fact that the sea was exceedingly chopping and the surges very bad, the latter rushing against the ship from all quarters, the vessel acted splendidly and did not ship a single sea.

We hastened southward with impetuous speed, and at the expiration of less than twenty-four hours we had made the four hundred miles from Beeren Island to Norway. The first landfall was Ing Island, one of the outmost of the islands of Finmark. A large point of it was seen through the fog and was immediately recognized by our experienced pilot.

Thenceforth we had the clearly marked steamship route before us, and we anchored in the Hammerfest Harbor all safe and sound early in the morning of the 25th of July.

III.

ON BOARD THE VORING, *September* 1, 1878.

MR. EDITOR: The scientific work of the expedition is at length completed, and with the consciousness of having improved the time to the best of our ability, and completely carried out the plan arranged for the expedition, we are now taking a good rest after our exertions, while the Voring is leisurely carrying us southward along the usual steamship route. It is now something more than a month since my last letter, and during this time we have seen so much and had such varied experiences, that I dare not at this time undertake to give you an exhaustive account of our whole cruise. For the present you will therefore have to content yourself with only a part. The continuation will follow as soon as opportunity offers itself.

After a sojourn of four days at Hammerfest, which was necessary for completing our supply of coal and water and of other provisions, we weighed anchor on Monday, the 29th of July, at 6½ o'clock in the afternoon, and after having fired four guns as a farewell salute to the city, we steamed northward along the usual route, out the Sörö Sound past the Ship's Holm out into the ocean. The weather was calm and warm, and the sea was so smooth that scarcely any motion of the ship was discernible. But the atmosphere toward the north was quite hazy, so that the sun, which, during the whole day, had been shining from a perfectly pure and clear sky, later in the evening hid itself behind a heavy bank of fog, and so did not afford us an opportunity of observing the partial eclipse which was just then taking place. The following morning we had already advanced a considerable distance into the ocean, and the fog, so common here, soon wrapped us in its clammy, cold atmosphere, and compelled us to put on again our traveling clothes, which for some time had been stowed away. In the afternoon the trawl was let down to a depth of 223 fathoms, and in the evening it was brought on deck in good condition, containing a large amount of mud, in which were found several siliceous sponges. As usual we also found several fishes in the net: a rare flounder, a specimen of a Greenland species of *Aspidophorus*, and a small cottoid. Besides, the mud contained numerous lower animals, with the examination of which we zoologists were occupied until late the following day.

We were now rapidly approaching Beeren Island; but the fog was so dense that no land could be seen before we had come within a few miles of it, when the summit of Mount Misery appeared among the masses of fog. Meanwhile the wind had begun to freshen, and a pretty decided swell set in from the west, which made the prospects for landing at the point visited by us before but little promising. Still, we continued our course northward along the east side of the island, keeping as near to the coast as we thought advisable. Now and then broken parts of the somber island, by this time well known to us, became visible, but were again wrapped in the fog which gradually accumulated into threatening driving cloud-banks. The rapid falling of the barometer also warned us that a storm was brewing, and as we would in that case be unable to accomplish anything out at sea, we agreed not to proceed any further for the time being, but to worry the storm out in the lee of the island. Nor was it long before the storm broke and began to creak in our cordage, but we had the land to the windward and therefore lay perfectly at ease, tacking back and forth.

The following day brought but little change in the situation. The fog came hurrying over Beeren Island in dense masses which entirely enveloped the summit of Mount Misery, and left only the gloomy strand with its steep weather-beaten precipices in sight. Toward evening the wind settled somewhat, and finding ourselves just then off the flat northeastern side of the island, where, according to former reports, coal-beds and rich fossil-bearing strata of rock were to be found, we deemed it proper to attempt a landing for the purpose of making careful explorations and gathering specimens of various kinds. Having approached the coast as near as we could with our ship, the boats were let down and hastily filled with a crew of the younger members of the expedition; they were furnished with guns, botanical boxes, and other articles of equipment suitable to the occasion. It was then eleven o'clock in the evening. But the night is here at this time scarcely darker than the day, and there was nothing to hinder our postponing our sleep until the following morning.

We steered into a little bay on the coast which we had observed from the ship, and where the breakers were considerably less formidable than elsewhere, and got the boats safely drawn on shore and made fast on a strand evenly sloping and covered with boulders and driftwood, where a small stream, the so-called English River, came trickling down. The place was well chosen, and it appeared on our later investigations that it would not have been possible at any other point to get over the precipice, which from the flat land within descends into the sea. Here the ascent was not difficult, and we soon found ourselves upon the plateau, whence endless flat and barren wastes strewn with boulders, with here and there a little lake, stretched as far as the eye could see. Some rare birds, among which a broad-tailed *Lestris pomerina*, resembling a bird of prey, soon attracted our attention, and three fine specimens were brought down by our guns. We followed the coast northward as far as seven miles from the English River, now making excursions into the interior of the island, now approaching the coast, whence we got splendid views of a couple of wonderfully shaped rocks standing isolated in the sea, the English Block and the North Loaf, where myriads of sea-birds had chosen their places for nesting. But now came the fog, dark, cold, and wet, driving upon us from the interior of the island, and it soon became so dense that it robbed us of every outlook, wherefore we concluded that it was about time for us to make an end to further progress and begin our march back. On the way we had the good fortune to stumble upon some fossil-bearing rocks, which here cropped out, and were so loose and crushed by the ice that we found no difficulty in making a large collection. Some slate-formed layers of coal were also found, of which specimens were gathered.

After a pretty exhausting march through the dense fog, over the sharp stones, we finally reached our landing-place at about four o'clock in the morning, whence we could, through the fog, barely catch a glimpse of the Voring, which, in the mean time, had anchored as near to the shore as possible. On board, the captain, with his usual thoughtfulness, kept in readiness for us a cup of steaming coffee, which tasted excellent after our wearisome excursion, and, together with a glass of good grog, gave our bodies the requisite amount of heat.

Meanwhile, there was taking place near the stern of the ship a scene which kept us awake for several hours longer. A couple of the crew had gotten out the trolling-lines, and in an incredibly short time had brought on board some beautiful codfish. As soon as this became known all was life and activity among the crew, and all the trolling-lines that could be found on board were brought into service in a hurry. Several of the members of the expedition also took part in the fishing, and that with a zeal scarcely less intense than that of the crew. One splendid codfish after another was hauled in over the rail, and soon the deck was strewn with sprawling fishes, so that there was scarcely room to walk without stepping on them. About two hundred codfish were in the course of a few hours drawn up, which showed conclusively what a wealth of fish there must be around this island, in other respects so desolate and inhospitable. At six o'clock in the morning we at length tumbled into our berths and soon fell asleep, and did not awake again before noon.

We were still in the lee of the island, for a new storm had broken out, which soon compelled our captain to weigh anchor and keep moving, as before, back and forth along the island. In the evening the prospects were very dark and melancholy. The showers came down from Mount Misery howling and creaking through the cordage of the ship, and whipping the sea into foam. The swells of the sea had also increased considerably, and made the ship, as soon as we got ever so little further from the land, pitch and roll terribly, by which we could easily understand what rough weather there must be further out at sea. Meanwhile, it was our intention at the first perceptible improvement in the weather to leave Beeren Island without delay, with which we by this time, to tell the honest truth, were thoroughly disgusted.

The following morning we were already on our way northward. The wind had quieted considerably, the barometer had risen, and the atmosphere had cleared. But the storm during the previous two days had thrown the ocean into so violent a commotion that our ship, having the seas on the beam, rolled with more violence than ever. Later in the day the sea quieted little by little, and a breeze from the north made it settle still more rapidly. When we had advanced to about midway between Beeren Island and Spitzbergen we stopped, the trawl was sent down to a depth of 123 fathoms, and was hauled on board full of specimens of the fauna of the deep. Not less than twenty-eight fishes (the most of them small, it is true) were secured by this haul, besides a multitude of lower animals, among which were some of great interest. We now directed our course to the westward, in order to determine the descent of the Beeren Island Bank toward the deep outside. In three successive soundings we found down along the bank, first, 444 fathoms, then 795 fathoms, and, finally, 1,149 fathoms. At all these stations careful series of observations of the temperature were taken, both with the usual Casella-Miller thermometer and with the improved Negretti-Zambra, the result going to show that in this stretch of the sea there is found a considerably confused distribution of temperature in the deep. The course was again changed and directed northward to Spitzbergen. On the way the trawl was sent down on the declivity of the bank, but came up in disorder, the net, probably on account of the severe ground-swells, having been wound around the beam.

The following day, toward noon, we got the first landfall of Spitzbergen, but the land was for the most part covered with fog, so that we only here and there caught glimpses of immense masses of ice and snow that shimmered through the fog. We sent a dredge down on the bank, where the water was only 70 fathoms deep. But we were unsuccessful again, the sack of the dredge being so torn asunder by the sharp stones on the bottom that only what accidentally stuck fast to it and to the tangles could be secured. In the evening of the same day we doubled South Cape, with the low island off it, and a pretty stiff north wind having meanwhile set in, we steered toward the southeast point in order to get in lee of the land, and at the same time get, to begin with, some idea of Spitzbergen's grand natural aspects. From a broad valley, completely filled with snow, a mighty glacier extends far into the sea, having abrupt edges and floes at the base. Above it rises a beautiful, dome-like mountain, which bears the name Keilhan's Mountain, so called after our celebrated countryman Keilhan, whose explorations in these northern regions form the basis of geological knowledge of this Arctic land. East of Keilhau's Mountain we got a glimpse of a considerable portion of the east coast facing Storfjord. It lay illuminated by the clear light of the midnight sun, while the west side was enveloped in dense masses of fog. Having made some physical and zoological investigations right by the foot of the glacier, we set our course to the southeast out into Storfjord, until we found a depth of 150 fathoms, where a careful series of observations of temperature was taken, and a dredging made, which gave us a tolerably complete idea of the fauna of the sea-bed. The stem was then turned to the west again in order to complete our first passage between Spitzbergen and the Greenland ice.

The same day we reached the ledge, where a couple of Greenland shark-hunters were seen lying at anchor. The lead here showed a pretty abrupt descent toward the deep, and at a comparatively short distance from the edge we had a depth of 750 fathoms. Here the trawl was sent down, but it was brought up with the net completely torn asunder, which was the more to be regretted, since, from the animals still sticking fast in the meshes of the net, it could be seen that the fauna here must be extraordinarily plentiful. But the bottom was evidently here of such a nature that it would involve a great risk to make another haul, wherefore we proceeded westward, throwing the lead and taking careful observations of the temperature at suitable intervals.

On Thursday, the 8th of August, at noon, we were warned that ice was in sight. And it was found that to the north of us, at the horizon, appeared a white line of considerable length, from which blocks of ice came drifting with the current southward. But the sea being free from ice to the west, we continued our course in that direction. In the evening we passed longitude 0° without our having met with any ice, wherefore the ice previously seen clearly must have been an isolated collection of drift-ice. In the night the trawl was sent down to a depth of 1,700 fathoms, and we awaited with great suspense the result of this haul, as we had never before tried the trawl at so great a depth. But, in hauling it up on the following morning, the rope suddenly snapped on account of the great weight, and the whole trawl, together with 2,000 fathoms of rope, was lost. This was a misfortune greatly to be regretted, and for the time being it could not help depressing our spirits, not only because we had given much time and toil to the maneuvering, but also because we had sent the trawl down with great expectations. As matters now stood there was nothing else for us to do than to move on, and meanwhile see to getting a new trawl ready as soon as possible. As ice was now seen in the horizon in various directions, we did not deem it advisable to press further forward toward the west, and so we changed our course to the northeast, trying as far as practicable to follow the edge of the ice. The further north we came the more ice we saw, and at last we were surrounded on all sides, wherever we turned our eyes, with large and small floes of ice of the usual bizarre forms, but still with sufficient space between them to allow the ship to be maneuvered further in the above-mentioned direction, provided proper care was taken. At six o'clock in the evening we at last came out of the belt of drift-ice, and had a clear and open sea before us. The weather was brilliant during the whole time, the sun shone bright, and the sea between the ice-floes was as smooth as in a harbor.

The following morning we had already advanced up to the next passage, and when we had established a depth here of 1,640 fathoms, our course was directed to the east again toward Spitzbergen. In the evening we sounded again, finding 1,333 fathoms, and an accurate series of observations of temperature was also taken by which it became evident that we had already gotten out of the Polar current, 32° F. not being found before we reached a depth of 400 fathoms.

Meanwhile we had made a new trawl, with a new rope and other belongings, and although the depth was considerably less than at the last station, it still was so great that a successful haul with this apparatus would be of great interest in a biological respect. Hopeful, we then let the trawl sink down, trusting that the new rope would stand the test this time. But when we came to haul the trawl in, the same unusual strain appeared on the accumulator as the previous time; its strings were stretched to thrice their length, although the trawl was raised from the bottom. On our former expeditions, further south, we had several times used the dredge at a similar depth, without anything like this happening, and hence we were in the greatest suspense to get at a satisfactory explanation of this yet inexplicable phenomenon. After much work and considerable anxiety in regard to our apparatus the trawl finally came up, and with it came the key to the problem. The net contained not only, as we had been wont to find, theretofore, the usual biloculina-clay, but, together with this, large, round stones, of which one was estimated to weigh about 300 pounds. The beam holding the runners apart from each other was broken in two by the great weight, and it must be regarded as a wonder that the net, too, was not torn to pieces. The whole sea-bed here seems to be literally covered with small and great stones lying loose in the mud, and they must, without any doubt, come from the icebergs that during the summer season constantly break loose from the glaciers on Spitzbergen and then melt here under the influence of the warm Atlantic current and unload the stones which by the action of the glacier are brought upon the ice. The further examination of the materials brought up brought to light several interesting forms of animals; nor were fishes wanting, three specimens of a beautifully banded species of *Lycodes* being secured in good condition and preserved.

Toward evening of the same day we caught sight of land ahead. It was off the northwest coast of Spitzbergen, and proved to be the long and yet but little known Prince Charles' Foreland, the sharp pinnacles of which first lifted their tall heads above the horizon. The following morning the weather was fair and the sun shone clear and bright. We were then only about thirty miles distant from land, and the mountains on Prince Charles' Foreland and around Icefjord lay perfectly clear with their mighty masses of snow and ice. At 125 fathoms' depth we sent down a dredge which brought up a considerable portion of loose mud, containing the usual Arctic animal forms.

It was determined that the next dredging was to take place at a depth of about 400 fathoms, or where the bank declined toward the deep. So we steered to the west, to the point where according to the previously sketched contour-lines we could expect this depth. Upon casting the lead we found, however, to our surprise, that the depth was only 97 fathoms. We were here clearly on a sharp edge, and the soundings made immediately afterwards further out also showed an unusually steep descent toward the deep. At 416 fathoms the dredge went down and came up again with its net full of gravel and stones. On the tangles were hanging beautiful specimens of a sort of Medusa-head (*Astrophyton*) up to two feet in diameter, together with a few specimens of the same beautiful branched sponge which we on our first expedition had caught in the Umbellular region; and from among the gravel was separated a large number of other marine animals, some of which were of great interest.

According to our plan two passages more were to be made between Spitzbergen and the Greenland ice further to the north. Meanwhile from the experience now gained it was thought sufficient to make one cruise to the north, and in this manner considerable time would be gained. The stem was therefore turned to the northwest and then to the north, in order if possible to determine the point where the Atlantic current meets the Polar current. Meanwhile the wind had begun blowing from the northwest, with a chopping sea, so that the ship on account of the constant consumption of coal lay a great deal too high in the water and consequently made extremely slow progress. But fortunately we were sailing with the current, and were thus helped along a little more rapidly than we had expected. The weather remained unchanged all the next day, and the ship lay the whole time fighting the chopping sea, while the propeller was lifted by every heavy swell entirely out of the water, and beat about wildly in the air, without being able to push the vessel forward with its usual force. Still we *did* go forward, and on the following morning we observed the first floes of ice. We had then advanced to the eighteenth degree. The depth was 450 fathoms, and a series of careful observations of temperature was taken at this point, whereby it appeared, however, that we had not yet by far reached the real Polar current. As it could be judged by these observations that the northern limit of the Atlantic current must be looked for at a considerably higher latitude far in among the drifting ice, and as such a cruise was not really a part of our plan, we decided to stop here and turn our course toward the north side of Spitzbergen. Before this was done we made a cast with the trawl, which gave us ample specimens of the fauna of the sea-bed, among which were several specimens of rare fishes. On the surface of the water were found enormous quantities of the peculiar ocean slime, which we on our previous expedition had observed, and renewed accurate microscopic examinations were made of the same at this time. Dense fog now came drifting in upon us from the ice, and at length so diminished our horizon that the ship seemed to float in the midst of a boundless sea of fog. But suddenly, as if by enchantment, we came in the afternoon out of the dense bank of fog into bright sunshine and had before us at the distance of about forty miles the jagged northwest coast of Spitzbergen, with the so-called seven Ice Mountains. Somewhat nearer the shore, at a depth of 250 fathoms, another haul was made with the trawl, which likewise gave a plentiful zoological harvest, which kept us zoologists busy for a long time. Meanwhile nature claimed her dues, and weary from the day's work we sought our berths, while the Voring directed her course toward the Norse Isles in order to anchor there and take in ballast and water.

The following morning, the 15th of August, we lay well anchored at the place determined upon, and here awaited us the surprise of seeing ourselves in company with no less than four Norse fishermen, and among them the sloop so well known from Nordenskjold's expedition, the Ice Bear. These vessels had already been lying here at anchor for some time while the crews were busily engaged in catching cod in the immediate vicinity. Fog still partially covered the surrounding mountains, so that we were enabled to orient ourselves only piece by piece as the fog rose. We were lying in a quite broad sound, in which the current rushed on with considerable rapidity, carrying with it blocks of ice of various forms and sizes. East of us we had the real so-called Norse Isle. West of us was White Island. Both were barren partially snow-clad masses of rock from which weather-beaten grayish heaps of stones extended down toward the strand. Directly north of us arose out of the fog a peculiarly formed mountain, the so-called Clowen Cliff, and further to the west we caught a glimpse now and then between the fog of the most northwestern of these islands, that is, the Amsterdam Island. South of us we had the northwest point of Spitzbergen's mainland, which extended toward White Island. But the main channel toward Red Bay remained enveloped in a compact mass of fog out of which small and large masses of ice now and then came forth, sailing through the sound past our ship.

In the afternoon a couple of the members of the expedition, accompanied by our pilot, who was well acquainted in these regions, undertook an excursion in a boat southward toward the mainland. On our way we passed numerous large and small ice-floes, which came floating in from the constantly ice-filled sea east of the Norse Islands. Between the ice-blocks were swimming large flocks of auks and black guillemots, of which a few became an easy prey to our guns. At one point where the mountain sides seemed less steep, we landed to take a look at the island. After having passed a high mound of gravel and boulders, among which a few alpine plants eked out a miserable existence, we came into a valley of some width surrounded by steep mountains. The major part of the valley was occupied by a lake of fresh water. But the small amount of summer heat had been able to keep only a small strip nearest to the mound open, while all the rest was covered with eternal ice. The water was carefully examined by the aid of the apparatus which we had brought with us. The only living things we could discover were a couple of specimens of the larvæ of a species of gnat. The round stones strewn everywhere over the bottom of the lake were covered with a close, dirty, greenish crust, which seemed mainly to be formed from a species of alga, of which we took specimens. Over the water flew a pair of solitary gulls. Otherwise everything here seemed so barren and desolate that we were glad to get back to our boat again and pass on further. We rowed north to the other side of White Island and landed again on a flat holm (rocky island), which on account of its somewhat more greenish hue seemed to give promise of a thriftier vegetation. On the sandy strand a few eider ducks tumbled about with their recently-hatched young, but quickly absented themselves when we arrived, plunging dexterously into the sea, one after the other, and they did not come to the surface again before they had gotten outside of the range of our guns. On White Island itself we gathered a few plants, and from its highest point we had a brilliant view of the mighty mountains and glaciers in the so-called Fair Harbor. We returned by way of the north side of White Island. But dense fog soon deprived us of every outlook, so that we only now and then caught a glimpse of the gray, weather-beaten strand of White Island and of one and another iceberg sailing by us. At eight o'clock in the evening we were on board again, where we zoologists were engaged for some time longer in investigating the fauna of the sea-bed in the immediate vicinity of the ship.

On the following morning the fog lifted a little so that we could see a little more of our somber surroundings. Through the sound came, as usual, one floe of ice after another drifting with the current. One of these, which was not observed in time, turned against our bow with so great force that it shook the whole ship as if we had struck bottom, and it warned us sufficiently that it would not have been a mere joke if our ship, at full speed, had collided with one of these compact masses almost as hard as stone. About noon the boats belonging to the fishermen came sailing in from the mouth of the sound, all loaded full to the gunwale. The fishing on the previous night had been quite unusually abundant, and so there was here an unexpectedly convenient opportunity for studying the Spitzbergen codfish and the conditions attending the catching of it in these waters. In order to form a more accurate estimate of the vast amount of fish caught here at this time we give the following reliable figures: On three boats, each having a crew of two men, were caught from 10 o'clock in the evening until 4.30 o'clock in the morning, eleven hundred and fifty-three codfish. After having dressed these fish and rested a short time, the same six men went out again at 8 o'clock the same morning and came back at 1.30 o'clock with eleven hundred codfish. Each man had thus in the course of twelve hours hauled up three hundred and seventy-five fish, which makes one fish every other minute.

At four o'clock we had taken in water and ballast, wherefore we weighed anchor and stood to the north again. According to our plan a few physical and biological observations were to be made on the banks directly north of the Norse Islands. But as it kept freshening with a breeze from the southwest as we came further out, and as the fog was very dense, we changed our plan and directed our course, instead, southward into the "Smeerenberg." The fog, which out at sea stood like a dense, dark wall, had as yet but partially gotten in here, and so we got during our passage through this channel, celebrated from former expeditions for its beautiful mountains and glaciers, a most excellent opportunity of getting acquainted with the mighty and grand natural features of Spitzbergen. Views, each more picturesque and surprising than the preceding one, opened before our eyes as we advanced. Every valley and ravine is here filled with a mighty glacier, which with abrupt walls shoots out into the sea; and above the glaciers tower, further into the interior, beautiful mountains abounding in the boldest peaks and precipices. The straggling masses of fog drifting over the land from the sea, between which the evening sun shed its clear rays of light, spread over all a peculiar mystic halo which added, in a marked degree, to the brilliancy and grandeur of this scene. In the innermost part of the bay unfolded itself before our eyes in the clear light of the evening sun a glorious panorama of mountain peaks, which, with the most fantastic, jagged forms, rose from a valley completely filled for many miles with snow, and from this extended a mighty glacier—the largest one we hitherto had seen—out into the sea. From the greenish blue, shimmering, abrupt end of the glacier came one iceberg after the other, floating with the current out of the fjord. One of these, of mighty dimensions, crowned with glittering peaks, passed close by our ship and was immediately sketched. That our sketch-books did not rest during the remainder of our passage through this interesting channel, is a matter of course. Every one of us that knew how to use a pencil, with some sort of practice, certainly has some view or other in his sketch-book from that glorious sail, and they who were not bold enough to record on paper what they saw, will, at least, long preserve in their memories a vivid and lasting impression of the imposing and solemn scenery that here, for the first time, met our eyes.

Through the so-called South Gate we once more directed our course to the sea, where we again met the dense threatening bank of fog, accompanied by a fresh breeze from the southwest. As the weather looked anything but promising to the seaward we agreed to run into Magdalene Bay, a bay entering Spitzbergen's plateau south of Smeerenberg, and likewise renowned for its imposing scenery. It was then quite late in the evening, and the fog partially concealed the surrounding mountains. But between these show forth distinctly the mighty glaciers, the number of which is really extraordinary. In the innermost part of the bay, in the lee of a peninsula, joined to the mainland by a flat isthmus, and surrounded on all sides by majestic mountains and glaciers, we cast anchor, and it being already late in the night the most of us sought our berths in order to be able on the following morning to study with refreshed strength the grand scenery of Spitzbergen, and to make some physical and biological observations at this interesting point.

The morning brought calm weather, but the fog still enveloped to a great extent the mighty mountains which here on all sides lift their jagged peaks to the skies. The sea was everywhere filled with blocks of ice of all sizes and forms, from quite small fragments to respectable icebergs, and presented a peculiar greenish color on account of the constant melting of glacial ice. Enormous numbers of the Arctic pteropod (*Limacina*) were seen moving about on the surface of the water, and among them were also a few specimens of the crystal-clear whale-food (*Clione*), and at some distance from the ship a pair of seals were seen inquisitively lifting their heads up to look at the ship. At this last sight our hunters grew lively. Rifles and ammunition were quickly brought out, and three specimens, two ringed seals and one young large cub, had to give their lives as a penalty for their curiosity. Later in the day some of the members of the expedition made a trip ashore to the peninsula lying before us, which seemed for ages to have been used as a burying-ground. Numerous graves bore ample testimony of the sufferings of men who, impelled by love of knowledge or by greed of gain, had been obliged to leave their bones here. But the graves did not really deserve the name. The soil consists chiefly of stone, so that it had been only possible to cover the rudely timbered coffins as well as circumstances would permit with stones. Now the most of them lay exposed to the air, open and broken asunder by the ice, possibly also interfered with by the polar bear and by other beasts, and in the coffins were found only scattered portions of the skeletons. As if the beasts of prey really had shown respect to the noblest part of the human body, the skulls were still, we found to our astonishment, in the most of these graves lying in their places. From the greatly varied forms of the craniums it could be determined that the deceased had belonged to various nationalities, and a skilled cranologist would undoubtedly be able to point out easily the Dutchman, the Russian, the Norseman, &c. From the highest point of the peninsula we had a splendid view of the south side of the bay. Glacier upon glacier here extended, one beyond the other, as far as the eye could see toward the ocean. The ice here really had the appearance of being the predominating part, and the visible mountain peaks seemed in fact to be nothing more than the boundary lines between the different glaciers. In some places the ice had even been crowded over the mountain peaks and formed peculiar masses suspended, as it were, in the air, and it seemed that they must every moment fall down from the precipitous mountain sides. In the east or at the head of the bay the first one of the glaciers shoots out into the sea. No less than four glaciers here gather themselves into a mighty mass of ice which constantly under the influence of the summer warmth sends out into the sea icebergs of all sizes and forms. One of these, which laid itself right athwart the bow of our ship, gave us considerable trouble when we were to weigh anchor, about eight o'clock in the evening.

In order to investigate the temperature in this bay, constantly filled with ice, we slowly and with all possible care approached the head of the bay where the above-mentioned immense glacier shot out into the sea. Here, surrounded on all sides by floating masses of ice, we sent our lead and our thermometer to the bottom. The depth was 60 fathoms and the temperature at the bottom 28.4° Fahr., the lowest temperature we had observed up to this time. A little further out, where we were less encumbered with ice and could better maneuver the ship, the trawl was sent down and brought up ample specimens of the fauna of the bottom. Not only various lower animal forms, but even fishes were found here and seemed to thrive remarkably well in this ice-cold water. Particularly did we in this haul bring on board numerous specimens of a sort of small codfish, the so-called ice-roach (Ismort, *Gadus polaris*), of which we heretofore had obtained only one specimen.

We now steamed toward the outlet of the bay to the sea, in order finally to make some investigations on the bank and on its declivity west of Prince Charles' Foreland. The wind had entirely subsided and the sea was smooth, but dense fog soon hid the land entirely out of sight. On the following morning we were at our station. The weather was still and calm as on the preceding day, and the fog had so far lifted that the lower parts of Prince Charles' Foreland could be seen. We cast out the lead here, finding 500 fathoms' depth, and hence we were already on the slope of the bank. A little farther out the trawl was sent down at a depth of 110 fathoms and was brought up covered all over with a species of the elegant feather-star (*Antedon*), of which several beautiful and perfect specimens were secured and preserved. We were now nearly through with our investigations in the open sea, and therefore directed our course southward along Prince Charles' Foreland, in order to run into the Ice-fjord, where we had made up our minds to lie still for a few days and make the necessary examination of the ship's engine, which now had been in almost constant use ever since we left Hammerfest.

The following morning we had already reached the inlet of the Ice-fjord. The north side of the fjord still shows the grand scenery characteristic of Spitzbergen; from the northwestern point, where the Dead Man and the Auk Horn lift their heads, there are, as far as eye can see toward the interior, splendid mountain views separated from the mighty masses of ice and snow. On the other hand, the south side has a totally different appearance; the mountains are here less high, and their summits usually form plateaus, while the sides slope gradually and show a regular arrangement both of the horizontal layers and of the glacial rivers that are dug out in the vertical clefts. There is nothing picturesque in the general effect. About in the middle of the fjord we sent the trawl down, but it brought up nothing of interest excepting a young specimen of a peculiar spiny Arctic fish (*Cyclopterus spinosus*). At noon we directed our course toward Advent Bay, which was intended for our station, and outside of which a couple of fishing-vessels lay at anchor. A long flat strip of land which extends into the sea from the western shore forms here a natural mole, within which there is an excellent harbor, in which the largest fleet might be able to find a convenient anchorage and abundant protection against the storms. Here we let our anchor drop. We were now at our goal, where we were to spend at least three or four days. But nature in this locality has nothing of the grand and imposing features that characterize Spitzbergen. The mountains around the bay have the same monotonous character and dismal grayish hue as those we had seen on the entire south side of the fjord. The splendid glaciers which so beautifully diversified the landscape are here wholly wanting, and in lieu of these the valley ascends gradually from the sea, forming a slope, with the same grayish-brown tiresome color that characterizes the mountains. So far we were not a little disappointed in our expectations, but still we had a few nice evenings when the mountains and glaciers on the north side of the fjord blazed in the evening sunlight, casting a beautiful reflection athwart the fjord to our anchoring place. Meanwhile our time was spent in the most profitable manner possible. While our captain was engaged upon the hydrography and chart of the bay, we zoologists industriously examined the sea-bed with our dredges and made excursions on land for collecting plants and whatever else of interest we might find. Nor was the noble art of hunting neglected, and a party was organized of the best and most skillful marksmen to undertake a reindeer chase. But the reindeer is, at this season of the year, very shy, and usually keeps itself farther from the coast than at other times; and not until we had made several and repeated efforts and accomplished miles of fatiguing marching did we succeed in killing a very small young deer, whose exceedingly fine and sweet meat was a welcome addition to our mess.

After having remained three days and three nights in Advent Bay everything was ready for the home passage. The boiler had been carefully examined, a new supply of water had been taken on board, and the bay had been mapped. At six o'clock in the afternoon, on the 22d of August, we weighed anchor, and after having made a haul with the trawl at the outlet of the bay, which, however, gave us but a small return, we directed our course out of the Ice-fjord to the sea. We had only enough coal left to last eight days, so that a longer stay at Spitzbergen, for this reason if for no other, could not be looked upon as advisable. But as Bell Sound, a place famous for the beauty of its scenery, lay directly in our way, we agreed among ourselves that, in case we obtained favorable weather, we would, as a sort of leave-taking ceremony, make a short trip in there, in order to be able to bring home with us a perfectly fresh impression of the imposing scenery of Spitzbergen. The evening was still and the sky cleared, so that we retired filled with the fairest hopes of being able on the following morning to enjoy the sight of Bell Sound's celebrated mountain peaks and glaciers. But we were deceived in our expectations. Dense fog on the following morning enveloped the land and hid all the mountain peaks from sight. Under such circumstances we would scarcely gain anything by running into Bell Sound, and as it was out of the question to spend any time waiting for clear weather the trip was abandoned. So the stem of the ship was turned to the south again, and every trace of Spitzbergen soon vanished in the fog. About half way between Spitzbergen and Beeren Island we finally took a series of careful observations of the temperature, in order to get one more factor in the complicated problem of establishing the conditions of temperature in this belt of the ocean. And herewith our investigations were at length completed. Instruments and apparatus were packed away, and what we now had to do was to get southward to Norway as rapidly as possible.

The weather, which up to this time had been unusually still, showed on the next day all signs of changing for the worse. The barometer fell rapidly, in the horizon appeared threatening cloud-banks, and the wind began to blow from the east. Toward evening the breeze had increased into a gale, but it fortunately blew from the northeast, and hence it was favorable to us. The studding-sails were set, and, as if the Voring herself now was longing to get home, she sped on with unusual velocity, so that we were making much more rapid progress than we from the beginning had calculated. As we got farther south the waves became higher, and the ship, which now was uncommonly light, now and then tossed about so violently in the night that we were several times in a rather disagreeable manner awakened from our sleep. But we had already tested the Voring once before, under similar circumstances, and knew that she would ride the waves securely and bring us all the sooner home across the Arctic seas, and we were all intensely homesick.

At eight o'clock of the next day we got the first landfall of Norway. Far out in the horizon we got a glimpse of something of a deep-bluish hue, which at some times scarcely could be distinguished from the atmosphere, but which gradually became more distinct and defined. In this we finally recognized with certainty the outmost island in the Loppe Sea. It was Bird Island, toward which our course had been directed during the whole time. Still we were a considerable distance from land, and it being late in the night we retired to our berths with the happy consciousness that we should soon be within the skerries in smooth water. When we came on deck the following morning we were just entering Gröt Sound. For the first time for many weeks we again looked upon green fields and trees, and soon the charming Trom Island, with its cultivated fields, its beautiful forests, and its smiling villages, lay before us in its complete summer dress. At twelve o'clock we lay safely moored at anchor in the harbor of Tromsö, and we all soon had the pleasure of receiving by post and telegraph fresh and glad tidings from home.

After stopping a couple of days at Tromsö, which was necessary in order to increase our supply of coal, we weighed anchor on Thursday, the 29th of August, at two o'clock in the morning, and steamed southward along the usual steamship route. The weather was brilliant, and it was a source of great relief to us, after having been tossed about so long on the billows of the Arctic Ocean, to be able to take our ease in smooth water within the skerries. On the evening of the same day we passed West-fjord in perfectly calm, beautiful weather. On the next day our progress was checked somewhat by foggy weather; but the third day was clear and warm as summer, and gave us another opportunity to rejoice at the sight of the glorious mountains and fjord scenes down along the coast of Nordland. On Wednesday, the 4th of September, we swung into the harbor of Bergen, where we were greeted by a general display of flags, and after having given and received a salute we anchored in the usual place near the Sugar-house Wharf. Three of the members of the expedition Chief Physician Dr. Danielssen, Mr. Friele, and Cand. Thorne, here bade us good-bye. The rest of us, after spending two days in Bergen, passed with the Voring to Christiania, where we, after a most delightful voyage, arrived on Monday last, the 9th of September, at four o'clock in the afternoon.

XIII.—ON THE SCIENTIFIC INVESTIGATION OF THE BALTIC SEA AND THE GERMAN OCEAN.*

By G. Karsten.

The following lines are designed to call attention to the investigations of the Baltic Sea and German Ocean, which have been in progress several years, and which are of especial interest in opening up a new field of observation, to which hitherto but little attention has been paid. The extensive experiments of the Americans and Englishmen have increased our knowledge of the physics of the ocean and its organisms, but not being made with the view to continued systematic investigations, they have resulted only in discovering for the time being certain relations for given points; the variations, be they of a periodical or a non-periodical nature, for one and the same locality could not be ascertained during the rapid passages through the ocean. But just these variations are of special significance, since upon them depends our knowledge of the phenomena of the currents and of the relations between the physical conditions and the phenomena of life, as in meteorology, where the final conclusions are not drawn from a few isolated observations but from a knowledge of the limits between which the variations take place. With this view, Dr. H. A. Mayer made extensive investigations of the physical conditions of the western part of the Baltic, hoping thereby to gain information regarding the variable character of the organic world, a fact established by his own observations as well as by those of R. Möbius. The observations of Dr. Mayer have shown that the western portion of the Baltic offers fluctuations in all the physical elements—in the height of the water, its temperature, and the proportion of salt—fluctuations which vary with the seasons and likewise in different years. These observations gave a sufficient explanation of the character of the currents, but in order to properly fix the laws of these currents the co-operation of a number of savans was found necessary, since only by simultaneous observations at many points the enterprise could be made a success.

The impulse to make the present investigation was given by the German Society for Fish-culture, which, fully comprehending its importance, requested the Prussian Government to have the work established. The government acted in accordance with this petition, and entrusted

* Translated by Dr. Oscar Loew.

the work to a Commission stationed in Kiel. This Commission took into consideration the following points, viz:

(*a*) Depth, height of the water, condition of the bottom, quantity of salt and air of the water, and temperature.

(*b*) Flora and fauna of the sea.

(*c*) Distribution, propagation, and migration of the useful animals—problems which required for their proper discussion and settlement observations at numerous stations along the coast, as well as on the high sea. From the preliminary observations of Mayer it was known that in the western part of the Baltic the variations in the water are analogous to those of the climate; further, it was established as a fact that the physical conditions of the eastern and western portions differed—for instance, the variations in the amount of salt show smaller differences in the eastern than in the western part. This is in another relation also found in the German Ocean. The work was commenced by establishing a number of stations along the German as well as the foreign coast, one station being at Heligoland. Two expeditions have thus far been sent out, one in 1871, another in 1872, the former to the Baltic, the latter to the German Ocean. The following is a brief summary of the work of the Commission, the mode of observation, and the results, with the exception of the part relating to organisms, for which the reader is referred to the publications of the Commission.

The most important points for determination were the *amount of salt* and the *temperature*. As variations in these depend upon the currents, and these again upon various causes, as the height of the water, the direction and velocity of the wind, the duration of ice-formation, the amount of rain and snow in the drift regions, &c., it was evident that a thorough study would require a long series of observations. The amount of air contained in the water could not be well determined on account of the want of methods sufficiently simple for the several stations. The proportion of salt is important for several reasons:

First. The difference in the amount of the saline substance is one cause of the currents, the heavier salt water having the tendency to flow to the deepest place. In this manner two currents may be produced—a vertical one when from some cause or other the upper strata become more concentrated, and a horizontal one when two strata of different densities lie side by side. The latter currents predominate in the German Ocean as well as in the Baltic Sea.

Second. The strata in motion will also have temperature of their own. For the waters in question this can be easily shown. The under current of the heavier water of the German Ocean can readily be recognized by its temperature upon its entrance in the Baltic, and the same is the case with the light upper current issuing from the Baltic.

An analogous difference is found in comparing the different strata of the German Ocean with the waters of the Atlantic. Furthermore, there is a certain relation between the amount of salt and carbonic acid contained in the water, and also between this and the organic life. The salt determinations are made with the hydrometer kept at the stations. It may be objectionable that the method used is not absolutely exact on account of the unequal composition of the salt in different parts of the sea. It is not necessary, however, to take this into consideration here, since the currents mainly depend upon the density of the water, which is indicated to a great degree of exactness by the hydrometer. At the stations there is no method simple enough for a trustworthy determination of the air contained in the water; the air collected from the water on expeditions, however, has been subjected to a chemical analysis.

The formula adopted by the Commission was well founded, being the result of the observation that an increase of the specific gravity of 0.0001 corresponded to an increase of 0.0131 per cent. of salt. The results thus far obtained are the following:

The specific gravity increases with the depth. In consequence of the German Ocean containing less salt than the Atlantic, and again the Baltic Sea less than the German Ocean, an under-current of heavier water can be traced flowing from west to east, and a lighter upper current flowing from east to west. This is shown by the current-meter as well as by the densities of the water and the difference in temperature. In very narrow straits, however, as, for instance, in the "Little Belt" and the "Alsensund," the different currents become turbulent and more or less mixed.

The intensity of the currents is variously changed by climatic influences, of which the wind is the most powerful. Prevailing westerly and southwesterly winds drive heavy currents from the Atlantic into the German Ocean and thence to the Baltic, at the same time retarding the light upper current. Easterly and northeasterly winds act reversely, diminishing the heavier under-current and increasing the lighter current on the surface. In accordance with the climatic conditions is the fact that the most salty water enters the Baltic in fall and winter, and the least in spring and summer. The amount of salt in the spring is reduced by the melting of the ice in the north and east; plentiful rains produce a like effect in summer. As the height of the water depends much upon the strength and direction of the wind, the percentage of salt will show a certain relation to the height. This relation, however, is not a simple one. Continuous west winds will not only increase the whole bulk of water in the Baltic, but also produce local differences between the height of the water on the eastern and western coasts, as above mentioned. We have then in the eastern part of the Baltic an increase of the mean level and a decrease of salt, while in the western part, the height of the water decreases and the amount of salt becomes greater.

Taking into consideration certain climatic relations, especially the wind, and the change of these relations in different years, it is clear that not only every stratum of water of a certain locality will show a difference in the amount of salt, but also that there will be deviations in different years; therefore only a prolonged period of observations can determine the average amount of salt of a certain locality. A glance at the following table will give an idea of the great variations taking place. Although the numbers are mere approximations, the differences are, nevertheless, considerable.

TABLE.—

Locality	Surface.				Depth.				
	Maximum.		Minimum.		Fathoms	Maximum.		Minimum.	
	Specific gravity at 63°.5, Fah.	Percentage of salt.	Specific gravity at 63°.5, Fah.	Percentage of salt.	1 fathom=6 feet.	Specific gravity at 63°.5, Fah.	Percentage of salt.	Specific gravity at 63°.5, Fah.	Percentage of salt.
I. BALTIC SEA.									
Helsingör*	1. 0190	2. 51	1. 0062	0. 81	16	1. 0259	3. 39	1. 0086	1. 16
Korsör*	1. 0208	2. 73	1. 0088	1. 19	20	1. 0250	3. 27	1. 0154	2. 02
Friedericia*	1. 0203	2. 66	1. 0092	1. 22	9	1. 0220	2. 88	1. 0104	1. 36
Svendborgsund*	1. 0184	2. 45	1. 0085	1. 15	7	1. 0187	2. 49	1. 0095	1. 26
Sonderburg	1. 0211	2. 76	1. 0092	1. 22	10	1. 0243	3. 20	1. 0095	1. 26
Eckern förde*	1. 0174	2. 30	1. 0079	1. 05	10	1. 0204	2. 67	1. 0121	1. 59
Friedrichs Ort	1. 0201	2. 63	1. 0043	0. 58	8	1. 0219	2. 87	1. 0078	1. 04
Kieler Hafen	1. 0177	2. 34	1. 0000†	......	16	1. 0196	2. 58	1. 0122	1. 60
Fehmarnsund	1. 0135	1. 77	1. 0072	0. 96	6	1. 0147	1. 95	1. 0090	1. 20
Travemünde‡	1. 0161	2. 11	1. 0093	1. 24	5	1. 0163	2. 14	1. 0093	1. 24
Pöl‡	1. 0160	2. 10	1. 0097	1. 29	4	1. 0169	2. 22	1. 0108	1. 42
Warnemünde‡	1. 0098	1. 28	1. 0063	0. 83	5	1. 0128	1. 68	1. 0072	0. 96
Darser Ort	1. 0133	1. 74	1. 0066	0. 86	5	1. 0152	1. 99	1. 0069	0. 91
Lohme, Rügen	1. 0094	1. 25	1. 0032	0. 42	10	1. 0095	1. 26	1. 0050	0. 66
Neufahrwasser	1. 0081	1. 10	1. 0019	0. 25	3	1. 0086	1. 16	1. 0035	0. 46
Hela	1. 0066	0. 86	1. 0014	0. 19	....	....	......	....	......
II. GERMAN OCEAN.									
Ellenbogen, Sylt	1. 0255	3. 34	1. 0208	2. 73	7	1 0258	3 38	1. 0215	2. 82
Wilhelmshafen	1 0266	3 48	1 0220	2 88	8	1 0268	3 51	1 0222	2 91
Borkum	1. 0276	3. 63	1. 0210	2. 75	13	1. 0277	3. 65	1. 0219	2. 87
Heligoland§	1. 0287	3. 80	1. 0244	3. 22	4½	1. 0288	3. 81	1. 0249	3. 28

* Observed by A. H. Mayer.

† Consequence of ice.

‡ Not embracing a whole year.

§ Values probably too large in consequence of instrumental error.

With regard to the details of the currents and their relation to the wind the reader is referred to the publication of the Commission.

The temperatures.—Both bodies of water, the German Ocean and the Baltic Sea, show, in general, different relations as to temperature, but as they intercommunicate by way of the straits of Skagerak and the Kattegat, they exercise some reciprocal influence in this point of view. The condition of the Baltic with regard to climatic influences is almost that of an inland sea, owing to the insignificant extent of its junctions with the German Ocean and its greater outflow. Only in the vicinity of its junctions, and under certain circumstances at some distance from them also, there is a considerable influence brought to bear upon the temperature of the Baltic by the entering under-currents. The temperature of the Baltic varies greatly with the respective temperature of the air, the changes decreasing, of course, with the depth. The unequal temperatures of increasing latitudes will be equalized by the perpetual motions of the waters.

The German Ocean shows much smaller variations of temperatures, and undoubtedly will present different conditions at different points, being connected with the Atlantic in the North by a wide and in the South by a narrow channel, and again by a narrow channel with the Baltic in the East.

While throughout the year currents of but little variations enter the channels from the Atlantic, those from the Baltic are of various temperatures. This, together with the greater depth of the German Ocean, will suffice to show that it requires years of observation and prolonged study to determine the exact relations of temperature. The observations hitherto made for different strata can only be considered as initiating a closer study, the former investigations relating mainly to surface-temperatures. The observations along the coast not having to be made in great depths, the thermometer of the Commission could be of a simple construction. The thermometer for ascertaining the surface-temperature was very simple, reading to .2 of a degree; the temperature could be read either directly in the ocean or in a large quantity of water freshly drawn. For the observations of the temperature in deep water the thermometers were surrounded by a thick layer of India rubber, a poor conductor of heat. The instruments were compared, and the time necessary for each to indicate a change in temperature was noted, as well as that during which they marked the temperature of the water after being exposed to the air. An hour was ascertained to be the average time for each instrument to indicate the temperature of the depth, and fully five minutes that between the removal of the instrument from the water and any perceptible change. The thermometers always remained at the desired depth for one hour before the temperature was read.* During the expeditions upon the open sea this instrument could be used only when the ship or boat lay still or was anchored; in all other cases Casella's maximum and minimum thermometer was used. The results gained are the following:

The temperature of the surface-water of the Baltic, and with diminished extremes also that of the depths, varies with the temperature of the atmosphere. As an example the observations at two stations, Sonderburg and Kiel, are here given:

*These thermometers, surrounded by India rubber, can be had at Steger's in Kiel.

Month.	Sonderburg.			Kiel.			
	Atmosphere.	Surface of water.	Ten fathoms.	Atmosphere.	Surface of water.	Five fathoms	Ten fathoms.
1869.							
January	34.82	34.9	34.9	34.19	36.7	36.9	39.9
February	40.60	38.7	38.3	40.54	38.7	39.9	40.3
March	35.31	37.4	36.7	35.47	37.8	39.0	41.0
April	48.87	44.4	40.8	48.44	48.2	44.4	42.1
May	51.26	51.8	47.5	51.92	53.6	39.3	43.2
June	54.90	53.4	48.0	54.90	57.2	56.1	41.4
July	63.10	59.7	55.2	62.74	65.7	61.7	42.1
August	58.90	51.6	60.3	59.10	63.3	63.0	43.5
September	57.36	59.4	58.3	56.34	60.3	59.4	44.8
October	46.74	52.2	51.8	46.90	53.6	54.9	48.6
November	38.64	42.3	42.8	39.00	46.6	47.5	48.9
December	35.17	41.4	40.3	35.60	39.8	43.2	43.7
Year	47.23	48.2	46.2	46.92	50.2	49.7	43.2

The correspondence of the temperature of the surface-water with that of the atmosphere is evident at a glance. In Sonderburg, however, the periodicity can be traced to a depth of 10 fathoms, while in Kiel a marked decrease is noticeable at a depth of 5 fathoms, and at 16 fathoms a shifting of the seasons and a great diminishing of the extremes. The wider distribution of heat at Sonderburg is produced by the strong current of the Alsensund, mingling different strata, while at a depth of 16 fathoms at Kiel, motion is produced only by the inflow of heavy currents, or by strong winds.

The temperature of the surface-waters is greater than that of the atmosphere; for the temperature of the latter is taken in the shade, while the surface-waters are greatly influenced by solar radiation. As it is hardly possible to recognize the law of the changes in temperature in one year, the following average values of six years of observation in Kiel will give a clearer expression of the retardation of the heating influence of the season:

Kiel, average of six years. (*Degrees Fahrenheit.*)

Month.	Atmosphere.	Surface.	Five fathoms.	Sixteen fathoms.
January	31.40	34.08	36.12	37.60
February	32.20	34.02	34.80	37.23
March	37.17	35.98	35.76	36.27
April	44.62	43.60	40.55	36.80
May	51.70	51.50	47.44	40.18
June	58.00	60.80	55.07	41.96
July	62.60	65.59	61.80	43.60
August	62.58	65.30	62.90	47.45
September	55.42	55.74	59.58	52.70
October	47.87	53.10	54.34	54.34
November	39.44	45.39	47.07	49.44
December	35.40	38.52	40.55	43.02
Six years	46.54	48.67	48.00	43.40

The annual period here enters regularly into the greatest depth. In the air and on the surface July is the warmest month; at 5 fathoms the heat of August predominates, while at 16 fathoms October is the warmest and March the coldest month. The variations of the average values between this table and the previous one, for 1869, demonstrate the inequality of the course of the temperature during the isolated years, and, further, that in the water, as in the air, the climatical differences are represented. Herein the extreme values of the temperatures of the air play a significant part. The following numbers may give an idea how the cold year of 1871, with its low temperatures, influenced the temperatures of the water even at great depths:

Kiel.

Month.	Atmosphere.		Surface.		Five fathoms.		Sixteen fathoms.	
	Max.	Min.	Max.	Min.	Max.	Min.	Max.	Min.
1871.								
January	37.4	5.0	32.9	33.3	32.0	27.9	32.7	30.4
February	50.0	−6.7	34.2	32.0	32.4	31.5	34.2	32.0
March	56.3	25.2	42.1	35.4	38.3	34.9	35.4	34.2
April	55.8	29.7	46.8	38.7	39.8	37.1	37.2	34.9
May	74.3	34.9	54.5	45.5	51.1	41.0	41.0	37.6
June	79.2	41.4	65.7	53.4	59.0	51.1	43.9	41.0
July	76.8	51.8	68.0	59.0	61.2	55.6	43.9	42.1
August	81.5	50.0	73.6	61.2	64.6	59.0	54.5	44.3
September	77.0	40.1	65.7	56.7	61.2	57.4	56.7	55.6
October	61.2	30.4	54.5	47.7	55.6	50.0	56.7	54.5
November	45.3	23.4	47.7	39.8	50.0	41.0	54.5	47.7
December	39.2	5.7	37.6	34.2	38.7	35.4	46.6	38.7

The low temperatures of the air during a severe winter, therefore, reduce the temperatures of all strata considerably below the usual average. These lower temperatures are maintained in the depth for a long period; in the following autumn they again suddenly increase. If then a mild winter follows, the higher temperatures remain for a longer time and keep the lower strata above the average values. The lowering of the temperatures of all the strata below 32° F. finds an explanation in the fact that the maximum density of the water of 2 per cent. salt, as at Kiel, is in the region of the point indicating 29°.8 F., and, therefore, vertically descending currents of the heavier water will effect a rapid equalization of temperature. This will be the more easily possible when, like in the winter months, all water-strata have an increased percentage of salt, with little variation between the different strata. If, however, the lower strata are considerably richer in salt than the upper, then the communication of the lower temperature of the latter will be slow, since the increased densities of the cooler upper strata do not reach the densities of the lower strata, notwithstanding the higher temperature of the latter; violent winds, however, would soon cause a thorough intermingling. The above-mentioned sudden changes of temperature in August and in autumn may be attributed to two different causes: *either* the strata mingle thoroughly, whereby the percentage of salt of the lower strata will be diminished and the temperature of the upper strata communicated to them; as, for instance, in 16 fathoms at Kiel, September 8, 1870, the specific gravity was 1.0167; temperature, 50° F.; September 13, specific gravity was 1.0155; temperature, 61°.2 F.; *or* a sudden entry of heavy underwater from the German Ocean, with its own higher temperature, takes place, replacing the lower cold strata of the Baltic. In the latter case the increase of temperature is combined with an increase of specific gravity. This was observed at Kiel on the 16th of August, 1871, when the specific gravity was 1.0118; temperature, 50° F.; while on the 21st of the same month the specific gravity was 1.0140; temperature, 56°.75 F. These sudden variations are very singular, for the reason that usually the changes in the depths are very slow and gradual, and often for weeks are scarcely perceptible. Such simultaneous changes of temperature and percentage of salt have been noticed at all observing stations along the Baltic. Thus, for instance, at Sonderburg it was observed that the temperature, which from the 13th of December, 1872, to the 22d of January, 1873, at 10 fathoms was lower than 41° F., suddenly changed to 42°.1 F., while at the same time the specific gravity of 1.0195 increased to 1.0243, owing to a powerful current of warmer and heavier water from the German Ocean. If, on the one hand, the Baltic in summer furnishes an excess of heat to the German Ocean by the upper current, the latter in winter, on the other hand, by the under current effects a rise of temperature in the former. This source of heat for the winter is of especially good service in the western portion of the Baltic, and is certainly an important climatic element. It is not yet established with certainty how far to the east this under current extends. The smaller the percentage of salt the greater the maximum of density; hence it is probable that, notwithstanding the lower temperature of the winter in the north and east, the water in the greater depths never cools to the extent of that in the west. Experience, however, on this point is wanting.

Regarding the relations of temperature in the German Ocean, but few observations have thus far been made. Stations were not established previous to 1872. The facts observed, however, are (1) the annual period of the temperature of the water decreases towards the west; (2) the difference of temperature between the strata of different depths is smaller than in the Baltic; and (3) a decrease to the freezing-point never occurs. During the expeditions currents from different sources could easily be traced by the thermometer and hydrometer; for instance, the currents of the Elbe and Baltic. The following table from Mayer's work contains some older observations on the average temperatures of the Baltic, Kattegat, and Irish coast, which show very distinctly the decrease of the differences in the annual periods:

Month.	Baltic.	Kattegat.	Irish coast.
January	35.8	....	46.6
February	36.3	....	45.7
March	36.9	36.3	45.9
April	44.6	42.6	48.4
May	52.7	49.5	51.8
June	59.0	54.7	55.4
July	64.6	60.8	58.7
August	64.8	62.1	60.1
September	59.9	57.8	59.4
October	53.4	52.0	55.2
November	44.6	42.8	49.1
December	39.9	....	48.2

But only by a long series of observations it can be determined to what extent these differences decrease at the different points of the German Ocean in contradistinction to the temperatures of the Baltic, what influence is brought to bear by the currents in different seasons, and what relations in greater depths the period of higher temperatures will show.

It will be seen from the following series of one year's observations at Heligoland that in the German Ocean, the temperatures of the waters of all strata are subjected to greater changes than the temperatures of the waters of the Irish coast although these changes are much smaller than those of the Baltic.

Month.	Air.			Surface.						Half fathoms.					
				High tide.			Low tide.			High tide.			Low tide.		
	Mean.	Max.	Min.	Mean.	Max.	Min.	Mean.	Max.	Min.	Mean.	Max.	Min.	Mean.	Max.	Min.
1872.															
September	59.02	76.10	45.50	62.64	64.6	58.5	62.51	64.6	58.5	63.20	65.1	59.0	63.10	65.0	59.0
October	51.13	61.47	46.85	55.90	58.3	54.0	55.88	58.1	54.0	56.81	59.0	54.9	56.78	59.0	54.9
November	45.25	54.50	32.67	50.74	54.0	49.1	50.74	54.0	49.1	51.30	54.9	49.5	51.30	54.9	49.5
December	37.20	47.75	25.47	44.61	48.6	41.9	44.61	48.6	41.9	45.00	49.1	42.1	45.00	49.1	42.1
1873.															
January	38.35	45.50	25.47	41.90	41.9	41.9	41.90	41.9	41.9	42.18	42.1	41.9	42.20	42.1	41.9
February	33.98	42.12	18.50	39.44	39.8	38.7	39.44	39.8	38.7	39.77	50.4	38.7	39.73	40.5	38.7
March	37.70	50.00	28.62	39.59	40.5	38.7	39.61	40.6	38.7	40.44	41.4	39.4	40.44	41.4	39.4
April	42.82	55.62	32.90	43.10	44.6	41.0	43.05	44.4	40.7	44.10	46.8	41.7	44.10	46.8	41.7
May	48.42	58.55	40.77	46.71	48.6	44.6	46.71	48.6	44.6	47.47	48.6	46.8	47.45	48.6	46.8
June	59.02	72.72	48.20	53.90	57.2	48.9	53.90	57.2	48.9	53.23	55.4	49.3	53.21	55.4	49.3
July	63.14	74.07	52.70	61.37	64.4	57.2	61.37	64.4	57.2	60.82	63.9	55.4	60.84	63.9	55.4
August	62.39	72.95	54.30	64.85	65.1	62.6	64.85	65.1	62.4	63.95	64.8	62.4	63.99	64.85	63.0
Year	48.2	....	....	50.3	....	....	50.3	....	....	50.7	....	....	50.7	....	....

The difference of 29° in the temperature of the atmosphere between the coldest and the warmest month is lowered to 24°.5 in the surface-water, and at a depth of 4½ fathoms to 24°. The absolute extremes, however, are as follows: In the air, 57°.6 F.; in the surface-water, 26°.1, and in deep water, 24°.3. The average volume of the heat of the water in all strata exceeds that of the heat of the air, and it is therefore probable for Heligoland that a source of heat must be looked for in the entering heavier under-current of ocean water. Although the temperature of the air falls considerably below the freezing-point, the temperature of the water at Heligoland in all the strata remains above this point, and, without doubt, this difference causes one of the principal differences in the development of organisms which cannot be explained by the differences in the percentage of salt.

For further information the reader is referred to the report of the Commission.

AUXILIARY APPARATUS.

For the various observations many auxiliary implements are required, such as lead-lines, hoisting-apparatus, sounding-cups, water-bottles, &c. A few words may be said in relation to these, since the usefulness of areometric determinations depends upon the well-working of the machinery.

To secure specimens of bottom water a simple arrangement, first used by Dr. H. A. Mayer, is employed. A strong, well-corked flask is lowered to the desired depth, when it is uncorked by a sudden jerk of the line; the drawing up may be easily done, as experience has shown, without a noticeable change in the quality of the water. This arrangement, however, cannot be used for any but moderate depths, having, like all other means for this purpose, the disadvantage of not permitting a gas-analysis, the air of the bottle becoming partially absorbed by the entering water. An examination of waters thus obtained shows the presence of irregular quantities of the permanent gases, partially derived from the air of the bottle. With the view of exact areometric determinations and gas analyses the Commission has made use of various instruments, those of Professor Dr. Jacobsen, in Rostock, and of Dr. H. A. Mayer proving the best. The apparatus of Jacobsen consists of an India-rubber bag partially filled with mercury and freed from air by pressure. The cork is self-regulating, opening as soon as the bottom is reached and closing again when the bag is drawn up. The apparatus of Mayer consists of a wide, open metallic cylinder, with bottom valves, and permits the bringing up of water from any depth. For further details, especially regarding the amount of carbonic acid, reference is made to the annual reports of the Commission.

Of the other apparatus used during the investigation we only mention the current-meter. The instruments permitting of determination of the direction and intensity of the currents are very defective. From an anchored ship it is comparatively easy to determine the surface-current, but not when the vessel is in motion, especially as regards the determination of the under current. On a firm position or from an anchored ship, the Commission made use of a simple apparatus consisting of two metallic plates combined crosswise and fastened by a fine wire. The current pressing against the plates shifts them from a horizontal position, and thus the strength of the current is approximately determined by angle of deviation. This instrument, however, is not sufficiently sensitive for a weak current, and does not admit of an exact determination of the velocity. Floating bodies combined with the plates also worked unsatisfactorily for the under currents, the upper currents interfering with the indication of the instrument; it was found to be perfectly useless when the ship was in motion or when drifting with the current. As the determinations of velocity and direction of these currents are of great importance, the invention of a good current-meter is very desirable.

Deep-sea investigations proper have been made by the Commission only in a limited sense; the greatest depths investigated were in the Baltic amounting to less than 200, and in the German Ocean to less than 400 fathoms. In such depths it is not difficult to manipulate the instruments. The determination of depth can be made by simple means; the bringing up of bottom may be done by dredging. Should the Commission have occasion to make more extended investigations of the German Ocean currents, or the Navy enter upon such scientific labors, it would be desirable to introduce improvements in the measurement of depths, the apparatus thus far employed not being sufficiently accurate. For trustworthy determinations an apparatus operated by the pressure of the water is required. Some instruments have been constructed upon the principle of Mariotte's law of compressed air, but they are not sufficiently sensitive. It would probably be best to make use of spring-manometers constructed for high pressures of the ocean depths. The knowledge of temperatures and percentage of salt will be highly valuable only in connection with exact measurement of depth.

The Commission hopes that the government will continue the means for carrying on the investigation in question. When the expeditions become more frequent and numerous, and the Navy and commercial ships participate, science cannot fail to be considerably enriched by important results. There will then be occasion to solve many more interesting problems not thus far studied, for instance the changes in the mean level of the ocean, or the secular changes in the level of the coast, the question of the intensity of light in various depths of the water, etc., etc.

APPENDIX E.

THE NATURAL HISTORY OF MARINE ANIMALS.

XIV. REPORT ON THE MARINE ISOPODA OF NEW ENGLAND AND ADJACENT WATERS.

By Oscar Harger.

The following paper includes the species of Isopoda at present known to inhabit the coast of New England and the adjacent regions, as far as Nova Scotia on the north and New Jersey on the south. These limits have been chosen from the fact that nearly all the marine collections of this order made by the Fish Commission have been from the New England coast, except those from the Nova Scotia coast in 1877, while the commission had its headquarters at Halifax. Previous to the work of the Fish Commission extensive collections had also been made, mostly by Professors A. E. Verrill and S. I. Smith, of Yale College, in the Bay of Fundy and at other places along the coast as far south as Great Egg Harbor, in the southern part of New Jersey. The collections thus obtained, and others in the museum of Yale College, have, through the kindness of Professor Verrill, been used in the preparation of this article. As there has not yet been sufficient opportunity for the study of the *Bopyridæ*, only a list of the known species of that family is included, and for this I am indebted to Professor S. I. Smith. The species of the remaining families are described at length, and nearly all figured in more or less detail in the plates accompanying the article. Throughout the article especial reference will be had to the Isopoda of our own coast, and many peculiarities of structure, not found in our genera, will be more or less completely disregarded. As the *Oniscidæ* are a terrestrial family, only a few species, found usually, or only, along the shore are here included.

ISOPODA.

This group is an order of Crustacea, so named from two Greek words, ἴσος, equal, and πούς, a foot, from the general similarity of the legs throughout, all being thoracic. The order belongs to the *Tetradecapoda*, "fourteen-footed," called also *Edriophthalma*, or "sessile-eyed" Crustacea. All of these terms, however, require modification when applied to the animals included in this order, since in the genus *Astacilla* the anterior pairs of legs are quite unlike the posterior, in *Gnathia* there are never more than twelve feet, or legs, in six pairs, and lastly in *Tanais* and its allies the eyes, when present, are not sessile, but articulated with the head, or stalked, as in the higher Crustacea. It may, however, be stated that the relations of the *Tanaidæ* with the rest of the order are remote, and it is perhaps doubtful whether they should be retained among the *Isopoda*, especially as this family differs from the rest of the order in its mode of respiration, as will be explained hereafter.

Although this order is not a large one its representatives are perhaps more widely distributed than in any other order of Crustacea. Every one is familiar with "sow-bugs" or "pill-bugs," which are found even in damp houses and in cellars, as well as under leaves in woods or under almost any pile of rubbish among decaying vegetable matter. These terrestrial species do, indeed, become rare in the colder parts of the world, but are found as far north as Greenland. Other species less familiar, but perhaps hardly less abundant, inhabit ponds and streams of fresh water, and others are found along the shores of all oceans; yet others abound among the marine vegetation of the shallow waters, or fix themselves upon the bodies, or within the mouths of fishes and other marine animals. Species are found swimming free in the open ocean, and others are brought up from the greatest depths to which the dredge has yet penetrated.

It will be convenient to give here a brief general account of the structure of the animals composing this order, and an explanation of the terms used in their description. Most of our marine species have a greater or less number of the segments at the posterior end of the body coalescent, but in the genus *Cirolana* they are distinct; the animals are, moreover, of large size and very abundant in some localities; reference will therefore be constantly made to the figures of *Cirolana concharum*, on plates IX and X, in illustration of the parts of the animal and of the terms used. A few specimens of this animal will help materially in gaining a knowledge of the structure of the group; or, if specimens of *Cirolana* cannot be obtained, a common "sow-bug" (*Oniscus* or *Porcellio*) may be substituted.

The body appears to consist of fourteen segments, of which the first is the head; the next seven form the thorax, or pereion of Spence Bate, and the last six the pleon, sometimes called the abdomen. Returning to the head we find, looking from above, a pair of eyes—each consisting of a group of ocelli—and two pairs of antennary organs. Of these the upper pair, or antennulæ (pl. X, fig. 60), consist on each side of three comparatively large basal segments, which, together, are called the peduncle, or peduncular segments, and support a more slender and tapering flagellum or lash, composed of a considerable number of short segments, decreasing in diameter toward the tip, and each, usually, bearing a fascicle of setæ, which are called by Fritz Müller olfactory setæ, from their supposed function. The antennulæ are very small and rudimentary in "sow-bugs" and their allies. Below the antennulæ are the antennæ properly so called (pl. X, fig. 61 *a*), which are also composed of a peduncle and flagellum. The five basal segments constitute the peduncle, and the following, usually much shorter and smaller segments, are flagellar.

Underneath, the mouth is seen to be protected by a pair of organs called maxillipeds (pl. X, fig. 62 *a*), with which, for convenience of dissection, we shall commence the description of the parts of the mouth. The five terminal segments of the maxillipeds in *Cirolana* (numbered 1 to 5 in the figure) constitute the palpus, but this number varies in the different genera. They are articulated to the external surface of the large basal segment (*m*), usually proportionally much larger than in *Cirolana*, as in *Idotea phosphorea* (pl. V, fig. 28 *b*, *m*), or in the "sow-bug" where the palpus is greatly reduced. The basal segment of the maxilliped is, in general, produced internally beyond the origin of the palpus, and furnished with strongly plumose or pectinated setæ at the tip. Frequently along its inner margin one or more short styliform organs are attached, as in *Jæra albifrons* (pl. I, fig. 5), while along its basal margin is a more or less distinct suture, indicating the epimeral segment of this organ, which will be further explained. The basal segments of the opposite maxillipeds meet along the median line, where their margins are nearly straight, and to the base of the outer margin is attached a more or less triangular external lamella (pl. X, fig. 62 *a*, *l*). The name "maxilliped" is frequently used for the basal segment only, which is often, as in the "sow-bugs," much larger than the rest of the organ and serves to cover and protect the other organs of the mouth.

When the maxillipeds are removed we find two pairs of maxillæ, the outer and inner; of these the outer, or second pair (pl. X, fig. 61 *b*), are in general of a delicate texture, and three-lobed at the tip, the two outer lobes being articulated to the basal piece, and all three lobes ciliated on their inner margins. The inner, or first pair of maxillæ are of a less delicate texture than the outer, and are hardly of the ordinary form in *Cirolana* (pl. X, fig. 61 *c*); reference may, therefore, be made to *Synidotea nodulosa* (pl. VI, fig. 35 *c*), where the two unequal lobes are shown, the inner comparatively small, and supported on a slender peduncle, curved inward, truncated at the tip, and bearing stout, curved, pectinated setæ; the outer much more robust and larger, similar in general outline to the inner, but armed with stout, curved, denticulated spines at the tip.

The mandibles (pl. X, fig. 61 *d*) are usually toothed at the apex, the teeth being supported on a dentigerous lamella, which may be double on one mandible, usually the left, and receive the lamella of the opposite mandible between the two; below this lamella is often a comb of pectinate setæ, and, generally, a molar process, as in *Janira alta* (pl. III, fig. 12 *b*, *m*). In many genera a three-jointed palpus (pl. X, fig. 61 *d*, *p*) is articulated to the external surface of the mandible, and, usually, the terminal segment of the palpus is more or less semicircular, or curved, and bears on its inner margin a very regular comb of setæ (pl. III, fig. 12 *b*), apparently of service in cleansing the organs of the mouth. This comb may be continued or repeated on the second segment, as in *Cirolana* (pl. X, fig. 61 *d*, *p*). In the "sow-bug" and many other genera the mandibles are destitute of palpi. The oral opening between the mandibles is defended by an upper and lower lip, or labrum and labium, which are, however, median, and not paired organs, like the other parts of the mouth.

The seven thoracic segments are of firm texture above, but softer underneath. The dorsal surface is in general more or less rounded, and in *Cirolana* is continued well down at the sides, where, except in the first segment, it is crossed by a suture cutting off a quadrate, or somewhat triangular piece, called an epimeron, or, in the plural, the epimera. The epimera are well shown in the side view of *Cirolana concharum* (pl. IX, fig. 58). They belong to the legs, and form a portion of the large proximal segment called the coxa. Usually, however, the legs are figured as in pl. X, fig. 62 *b*, without this segment, which adheres strongly to the body; often, as in the first segment of *Cirolana*, the suture separating it disappears. The remaining six segments of the legs are more slender, and are called respectively, beginning with the segment following the coxa, the basis, ischium, merus, carpus, propodus, and dactylus, the last being usually slender and curved, often bearing a curved spine or claw at the tip, and, especially in the first pair, capable of flexion on the propodus, so as to form a prehensile hand. In the *Tanaidæ*, as in many of the higher Crustacea, the propodus may be prolonged into a digital process, against which the dactylus closes, forming a chela (pl. XIII, fig. 85), or chelate hand, as in the lobster. In the *Ægidæ* and the *Cymothoidæ* a greater or less number of the dactyli are strongly curved or hooked, for the purpose of retaining firm hold of the host, on which these parasitic species live. Legs thus constructed are called ancoral, as in *Livoneca ovalis* (pl. XI, fig. 67 *d* and *e*).

Of the seven pairs of legs attached to the thorax or pereion, the first three have in general a resemblance to each other, and are often more or less prehensile, while, as in *Chiridotea* (pl. IV, figs. 16 and 20), the last four are more strictly locomotive organs; but to this condition of things there are many exceptions, especially in the development of the first pair of legs, which are quite variable throughout the order, being not even pediform in the males of the *Gnathiidæ*, but two-jointed, in our species, and lamelliform (pl. XII, fig. 76 *d*). Except in this family, however, no confusion arises from speaking of the thoracic appendages as the first to the seventh pair of legs, or thoracic legs, and in general these terms will be used except where it may be necessary to use the technical terms, gnathopods or gnathopoda and pereiopods or pereiopoda, for these organs, as proposed by Spence Bate, according to whose system the first and second pairs are called the first and second pairs of gnathopoda* or gnathopods, and the remaining five pairs the first to the fifth pair of pereiopoda or pereiopods. When necessary these terms will be added as explanatory, having the merit of scientific accuracy as well as applicability to other groups of Crustacea, where a

* See also Edwards, Ann. Sci. nat., III, tome xvi, p. 221-291.

marked distinction of structure and function frequently occurs between the organs homologous with the second and third pairs of legs in the Isopoda.

In the adult females of this order there is commonly formed, on more or less of the under surface of the thorax, an incubatory pouch for the reception and development of the eggs. The outer surface of the pouch is usually formed by four pairs of lamellæ attached just within the origins of the second, third, and fourth, together with the first or fifth pairs of legs, and in the females of many genera, *Sphæroma* and *Asellus* for instance, these lamellæ may be observed in a rudimentary condition on the under surface of the thorax when not actually in use carrying eggs or young. In *Asellus*, and in some other genera, they are found upon the first to the fourth segments, instead of the second to the fifth. In *Anthura* the incubatory pouch extends over only three segments, the third, fourth, and fifth; and in *Astacilla* it is confined to a single segment, being composed of a single pair of elongated plates attached to the fourth segment. In *Tanais* a further remarkable variation occurs, and the eggs and young are carried in sacs attached to the under surface of the fifth thoracic segment, while in the closely allied genus *Leptochelia* the form of the incubatory pouch is normal. In the *Gnathiidæ* and *Anthuridæ*, according to Spence Bate and Dohrn, the incubatory pouch is formed by the splitting of the integument of the inferior surface of the thoracic segments in the females, and for the discharge of the young the outer lamella thus formed further divides into scales, one pair for each segment of the pouch. In *Jæra*, *Epelys*, and probably other genera, a similar mode of development seems to occur.

The six segments of the pleon are smaller than those of the thorax, often much smaller, and frequently more or less united, sometimes consolidated into a single piece with scarcely any trace of division above, but the number of pairs of appendages is generally six, showing the composite nature of the apparently simple organ. Of these six pairs of appendages or pleopods, the first five are more or less concealed beneath the pleon, and consist on each side of a basal segment bearing two lamellæ (pl. IV, fig. 19 *c*), of which the outer is the anterior when they overlap. These lamellæ, at least the anterior pairs, are usually ciliated along more or less of their distal margins with long slender plumose setæ. In the males of most of the genera, the inner lamella of the second pair bears, articulated near the base of its inner margin, a slender stylet (pl. IV, fig. 19 *b*, *s*). This stylet seems to afford, in many cases, specific and even generic characters.

The last segment, sometimes called the telson, has its pair of appendages specially modified, and called the uropods (pl. X, fig. 63). They consist in general like the pleopods of a basal segment bearing two lamellæ, or rami, not being always lamelliform, and in the *Tanaidæ* they are more or less segmented (pl. XIII, fig. 86). One of these rami may disappear, as in *Sphæroma* and in some of the *Idoteidæ* (pl. V, fig. 25 *c*), where a further modification takes place, and the uropods are so articulated to the inferior surface of the pleon as to fold together like a pair of cupboard doors, forming an operculum for the protection of the more delicate pleopods. Except in the *Tanaidæ*, respiration is carried on by means of the pleopods.

In the *Asellidæ*, *Idoteidæ*, and some other families two or more of the segments of the pleon are united, so that, seen from above, the pleon, like the head, may appear to consist of a single segment, as in *Jæra albifrons* (pl. I, fig. 4), but the number of pairs of its appendages, usually six, remains as evidence of this consolidation. In like manner the head is to be regarded as composed of several segments united, and the number of such segments is indicated by the number of pairs of appendages. In the *Tanaidæ* and many of the higher Crustacea, the eyes, more or less distinctly stalked or articulated with the head, are seen to be of the nature of a pair of appendages, which may be regarded as belonging to the first cephalic segment. The antennulæ and antennæ represent, respectively, the second and third cephalic segments, and, in like manner, the mandibles and two pairs of maxillæ represent the fourth, fifth, and sixth segments of the head. A seventh segment is indicated by the maxillipeds. This segment is regarded by Huxley as properly thoracic* instead of cephalic, but, for purposes of description, the segment and its appendages will be regarded as belonging to the head, and the next segment considered the first thoracic.

This segment, like the following thoracic segments, is usually free, and has the dorsal region well developed, but in the adult *Gnathia* it is united with the head, and still more closely so in the *Tanaidæ*. The seventh thoracic segment is the last to develop, and in young Isopoda, taken from the incubatory pouch, only six pairs of legs are commonly found. In *Gnathia* this condition prevails through life, and in the adults the first pair of legs are also modified, especially in the males, so as to quite lose their pediform character, leaving apparently only five pairs of legs. Further modifications of structure will be described in the families and genera in which they occur.

The nomenclature adopted, as explained above, corresponds nearly with that proposed by Mr. C. Spence Bate in his Report on British *Edriophthalma*, and used by the authors of the British Sessile-eyed Crustacea.

The length of an Isopod, in the present article, is given as the length of the body, exclusive of appendages, and is measured from the front of the head to the tip of the pleon. When, as in *Janira*, the head is produced medially into a "rostrum" (see pl. II, figs. 9 and 10), the measurement is taken from the tip of the rostrum, which is a part of the head, and not properly an "appendage."

Among the *Edriophthalma* or sessile-eyed Crustacea, the *Isopoda* may in general be characterized as follows: Body depressed rather than compressed; respiration carried on by means of the pleopods, of which the last pair only are modified into uropods.

---

* Huxley, Anat. Inv., Am. ed., p. 276.

The body is said to be depressed, or flattened from above, in distinction from the form usually seen in the *Amphipoda*, where it is in general flattened from side to side. An important exception to the ordinary mode of respiration occurs in the *Tanaidæ*, as has already been mentioned. In this family respiration takes place in two lateral cavities, situated beneath the integument of a large cephalothoracic shield, covering the head and first thoracic segment. In general, as the name of the order indicates, the legs are similar in structure and function throughout, as in the "sow-bug," but may differ considerably, as in the *Arcturidæ*, the *Munnopsidæ*, and the *Tanaidæ*.

The arrangement of the families in the present paper can only be regarded as tentative, and no higher grouping will be attempted further than to indicate briefly the relationships of a few of the families to each other.

The *Oniscidæ* may, on account of their aërial respiration, be regarded as standing quite distinct from the remaining families, and should, perhaps, be further divided as proposed by Kinahan. As they do not, however, come within the proper scope of this article, I have not attempted to subdivide the family. The *Bopyridæ* have been placed near the *Oniscidæ* in deference to the opinions of Dr. Fritz Müller. Having made no study of this family myself I do not express any opinion as to the propriety of separating it so widely from the *Cymothoidæ*, with which it has usually been associated. The *Asellidæ* and *Munnopsidæ* are closely allied to each other. The *Idoteidæ* and *Arcturidæ* form a group distinguished especially by their operculiform uropods. The above families correspond nearly with the "marcheurs" or walking Isopoda of Edwards, and more nearly with the "gehende Asseln" of Müller. They usually have the antennulæ much less developed than the antennæ, and the uropods terminal or inferior, that is, attached to the end of the last segment, or in the last two families to its inferior surface.

The *Sphæromidæ* and *Limnoriidæ* are closely allied, and perhaps ought hardly to be kept separate as families. The *Cirolanidæ*, *Ægidæ*, and *Cymothoidæ* form another group embracing a wide diversity of forms, from the active predatory *Cirolana* to the sedentary and distorted *Livoneca*, and yet apparently connected by easy gradations. The remaining families are generally regarded as aberrant, and form the "Isopoda aberrantia" of Bate and Westwood. They do not present any very evident relationships with the preceding. Of these the *Anthuridæ* have usually been associated with the *Idoteidæ* or the *Arcturidæ*, or with both. Except an elongated form, however, they do not appear to have much in common with either of these families. According to Dohrn's observations they are related to the *Gnathiidæ* in the structure of the incubatory pouch. The *Gnathiidæ* have the head united with the first thoracic segment, as in the *Tanaidæ*, but this last family is widely separated from the others, and doubtless ought to be regarded as forming a distinct suborder, according to the views of Dr. Fritz Müller.

The arrangement of the families adopted, and to a certain extent their affinities, are indicated in the subjoined table, in which, however, as throughout the article, special reference is had to the representatives of the order in New England waters, extralimital species, genera, and even higher groups, *Apseudes* and the Serolids, for example, being disregarded. The arrangement will be seen to considerably resemble that of Dr. Fritz Müller. I have placed the *Tanaidæ* at the other end of the order, partly, however, from the necessity of a lineal arrangement.

SYNOPTICAL TABLE OF FAMILIES.

I. Respiration pleonal; legs not furnished with a chelate hand.
 1. Legs in seven pairs.
  a Antennulæ small or rudimentary; antennæ longer, often much elongated.
   † Uropods terminal, sometimes rudimentary, rami mostly styliform.
    Legs ambulatory; antennulæ rudimentary; respiration aerial. I. ONISCIDÆ, p. 305
    Legs prehensile; sexes very unlike; adult forms degenerate; parasitic .... II. BOPYRIDÆ, p. 311
    Legs ambulatory or prehensile; segments of pleon united; antennæ with a multiarticulate flagellum .... III. ASELLIDÆ, p. 312
    Last three pairs of legs natatory; segments of pleon united; antennæ with a multiarticulate flagellum .... IV. MUNNOPSIDÆ, p. 328
   †† Uropods inferior, operculiform.
    Legs prehensile or ambulatory, not ciliated .... V. IDOTEIDÆ, p. 335
    First four pairs of legs ciliated; last three pairs ambulatory. VI. ARCTURIDÆ, p. 361
  b Antennulæ and antennæ subequal; body not elongated.
   † Uropods lateral, with one ramus obsolete or subrudimentary.
    Antennulæ and antennæ well developed; pleon of two segments; uropods with one movable ramus .... VII. SPHÆROMIDÆ, p. 367
    Antennulæ and antennæ short; pleon of six segments; outer ramus of uropods small .... VIII. LIMNORIIDÆ, p. 371
   †† Uropods lateral, distinctly biramous; rami mostly lamelliform.
    Mouth carnassial; legs not ancoral; antennulæ exposed in front; pleopods ciliated .... IX. CIROLANIDÆ, p. 376
    Mouth suctorial; first three pairs of legs ancoral; antennulæ exposed in front .... X. ÆGIDÆ, p. 382
    Mouth suctorial; legs all ancoral; antennulæ concealed at base by the projecting front; pleopods naked .... XI. CYMOTHOIDÆ, p. 390
  c Antennulæ and antennæ subequal, or antennulæ much the largest in the males; body cylindrical, elongated.
   † Uropods lateral and superior.
    Legs ambulatory or prehensile .... XII. ANTHURIDÆ, p. 396
 2. Legs in the adult in six, apparently only five, pairs.
    Five pairs of legs ambulatory; antennulæ and antennæ subequal. XIII. GNATHIIDÆ, p. 408
II. Respiration cephalothoracic; first pair of legs terminated by a chelate hand.
    Legs ambulatory and prehensile; head united with the first thoracic segment; antennular flagellum single .... XIV. TANAIDÆ, p. 413

I.—ONISCIDÆ.

Antennulæ rudimentary; legs ambulatory; pleon of six distinct segments, of which the last is small; mandibles without palpi; uropods terminal.*

This large and important group of Isopoda being terrestrial in habit, only a few species are mentioned in this paper. They inhabit moist situations, and are commonly known as "sow-bugs," "pill-bugs," "wood-lice," &c. Several species may often be found under an old board or pile of rubbish. The genus *Ligia* Fabr. inhabits sea-shores, above tide-level, and a few other genera are found under heaps of seaweed, or burrowing in the sand along the shore. Three such species, belonging to as many genera, are here described and figured, but are less fully treated of than the marine species that follow in the other families. Other species, especially of the genus *Porcellio*, may be found in similar situations.

The family may be at once recognized by the apparent possession of only a single pair of antennæ. These are the antennæ properly so called, the antennulæ being minute and rudimentary. This is generally regarded as a character indicating a high degree of development, and causes them to somewhat resemble externally some of the shorter myriopoda, which, like other insects, have but a single pair of antennary organs. The maxillipeds are large and operculiform in this family, with short and few-jointed palpi. The mandibles are destitute of palpi.

The legs are rather weak and fitted only for walking, and usually more or less concealed by the projecting epimeral regions of the thoracic segments. The pleon, in our species, has its segments distinct and decreasing rapidly in size to the last, which bears the more or less exserted uropods. These organs may not, however, project beyond the general outline of the pleon, as they scarcely do in *Actoniscus*, while in *Armadillo* they assist in forming the very regular outline of that part of the body, which closes against the head when those animals, as is their habit, roll themselves into a ball on being alarmed.

This family is placed by Bate and Westwood in a separate "division," the "Æro-spirantia," on account of their aërial respiration. The air, however, requires to be saturated with moisture, and in some of the genera the respiration is, in part at least, aquatic. On this subject the reader is referred to the publications of Duvernoy and Lereboullet and of Nicholas Wagner.

Philoscia Latreille.

Philoscia Latreille, Hist. nat. des Crust. et des Ins., tome vii, p. 43, "1804."

Head rounded in front, not lobed; antennæ with its segments cylindrical, flagellum three-jointed; pleon suddenly narrower than the thorax; uropods exserted, basal segment broad, rami elongate.

* The above diagnosis would not include the genera *Tylus* Latreille nor *Helleria* Ebner, which perhaps ought not to be regarded as belonging to this family, although closely allied to it.

This genus may be recognized among our *Oniscidæ* by the rounded head without lobes, and the conspicuously narrowed pleon. Only a single species is as yet known from New England.

Philoscia vittata Say.

Philoscia vittata Say, Jour. Acad. Nat. Sci., vol. i, p. 429, 1818.
Dekay, Zool. New York, Crust., p. 50, 1844.
White, List Crust. Brit. Mus., p. 99, 1847.
Harger, This Report, part i, p. 569 (275), 1874; Proc. U. S. Nat. Mus., 1879, vol. ii, p. 157, 1879.

PLATE I, FIG. 1.

This species may be recognized, among our terrestrial Isopoda, by the absence of the usual antero-lateral processes on the head, in front of the eyes, and by the sudden contraction of the body at the base of the abdomen or pleon.

Body oval, smooth; about twice as long as broad; head nearly twice as broad as long; eyes large, occupying the antero-lateral regions of the head. The antennulæ are minute and concealed from above. Antennæ minutely hirsute, especially on the last three, or flagellar, segments, inserted below the inner margin of the eyes; first segment short; second about twice as long as the first; third equal in length to the second, clavate; fourth longer cylindrical; fifth longest, slender, cylindrical, straight; flagellum slender, three-jointed, longer than the fifth or last peduncular segment; first flagellar segment about one-half longer than the second; third longer than the second, tapering, tipped with a short transparent filament.

The first thoracic segment is longer than the following ones, which are of about equal length. The anterior angles of the first thoracic segment are somewhat produced at the sides around the head; the posterior angles are broadly rounded. The second and third segments have their posterior angles less broadly rounded, but not at all produced backward. In the fourth segment this angle is scarcely produced, but in the fifth, and still more in the sixth and seventh, it becomes produced and acute. The legs increase in size and length from the first to the seventh pair, and are well armed with spines, especially upon the inferior surfaces of the meral, carpal, and propodal segments. The spines on the latter segment are, however, much smaller than those on the merus and carpus.

The pleon is at the base about two-thirds as wide as the seventh thoracic segment. In the first two segments of the pleon the coxæ, or lateral lamellæ, are short, small, and nearly concealed by the seventh thoracic segment, but in the third, fourth, and fifth segments they are evident and acute but not large. The sixth segment is acute but not prolonged behind, and extends beyond the end of the basal segment of the uropod, which is broad and bears the two rami nearly on the same transverse line. The outer ramus, seen from above, is narrowly and obliquely lanceolate in outline, tapering to the tip, and surpasses by less than half its length the more slender, styliform inner ramus. The uropods, the legs and antennæ, and the segments of the pleon, along their margin, are very minutely hirsute.

The color of these animals is dull and somewhat variable, usually brownish or fuscous, with lighter margins and two broad dorsal vittæ. Length 8^{mm}, breadth 4^{mm}.

This species has been found under rubbish and stones from Great Egg Harbor,! N. J., to Barnstable,! Mass. All the specimens that I have seen have been from the coast, although Say states that it is "very common under stones, wood, &c., in moist situations."

Specimens examined.

Number.	Locality.	Habitat.	When collected.	Received from—	Number of specimens.	Dry. Alc.
1222	Somers and Beesley's Points, N. J.	Shore	—— —, 1871	A. E. Verrill and S. I. Smith	25	Alc.
1911	Stony Creek, Conn	..do		A. E. Verrill		Alc.
2146	Vineyard Sound, Mass	..do	—— —, 1871	U. S. Fish Com	8	Alc.
1910	Barnstable, Mass	..do	Aug. 30, 1875	......do	3	Alc.

Scyphacella Smith.

Scyphacella, Smith, This Report, part i, p. 567 (273), 1874.

Antenna composed of eight distinct segments, with a geniculation at the articulation of the fourth with the fifth segment; terminal portion, or flagellum, composed of three closely articulated segments besides a minute apical one; mandibles slender; exposed portion of the maxillipeds formed of only two segments.

The genus *Scyphacella* was founded by Professor S. I. Smith, in part I of this Report, for the reception of the following species, the only one yet known. In regard to the relations of the present genus with *Scyphax* Dana* Professor Smith says: "This genus differs from *Scyphax* most notably in the form of the maxillipeds, which in *Scyphax* have the terminal segment broad and serrately lobed, while in our genus it is elongated, tapering, and has entire margins. In *Scyphax*, also, the posterior pair of thoracic legs are much smaller than the others, and weak; the last segment of the abdomen is truncated at the apex, and the articulations between the segments of the terminal portion of the antennæ, are much more complete than in our species. The general form and appearance of the genera are the same, and the known species agree remarkably in habits, the *Scyphax*, according to Dana, occurring on the beach of Parua Harbor, New Zealand, and found in the sand by turning it over for the depth of a few inches."

Scyphacella arenicola Smith.

Scyphacella arenicola Smith, This Report, part i, p. 568 (274), 1874.
Verrill, This Report, part i, p. 337 (43), 1874.
Harger, Proc. U. S. Nat. Mus., 1879, vol. ii, p. 157, 1879.

PLATE I, FIG. 2.

The small size, nearly white color, and peculiarly roughened surface of this Isopod will in general serve for its recognition, and the presence

* U. S. Exploring Expedition, Crustacea, p. 733, pl. 48, fig. 5.

of eyes will further distinguish it from *Platyarthrus*, which is often found inhabiting ants' nests, but would hardly be likely to occur in the sand of the beach.

Body elliptical, pleon not abruptly narrower than the thorax, dorsal surface roughened throughout with small depressed tubercles each giving rise to a minute spinule. Head transverse, not lobed; eyes prominent, round; antennæ longer than the breadth of the body; with the first and second segments short; third, fourth, and fifth successively longer and of less diameter; flagellum shorter than the fifth segment, composed of three closely articulated, successively smaller segments, and a very short somewhat spiniform but obtuse terminal one; all the segments, except the minute terminal one, beset with small scattered spinules.

First thoracic segment scarcely embracing the head at the sides; second, third, and fourth segments each about as long as the first, but increasing in breadth; fifth, sixth, and seventh diminishing in length and the last two also in breadth. Posterior lateral angles of the first three segments not at all produced, hardly perceptibly produced in the fourth segment; fifth, sixth, and seventh with the angles increasingly produced but not acute. Legs increasing somewhat in size posteriorly, armed, especially on the inferior surface of the meral, carpal, and propodal segments, with short stout spines.

Segments of the pleon with the coxæ but little developed. Terminal segment slightly rounded at the end, not attaining the end of the basal segment of the uropods, which are robust, with the basal segment spinulose, tapering to the base of the short, stout, outer ramus, and bearing the more slender inner ramus much nearer its base. The inner ramus is actually longer than the outer, but being inserted much lower down does not attain the tip of the outer ramus; both are tipped with setæ.

"Color, in life, nearly white, with chalky white spots, and scattered, blackish dots arranged irregularly. Eyes black." Length 3.4^{mm}.

This species was "found at Somers and Beesley's Points, on Great Egg Harbor!, New Jersey, in April, 1871, burrowing in the sand of the beaches, just above ordinary high-water mark, in company with several species of *Staphylinidæ*," and has also since been found by Professor Smith at Nobska Beach, Vineyard Sound!, Mass., in 1871, and by Mr. V. N. Edwards, on the beach at Nantucket Island!, December 6, 1877. It will doubtless be found at other points along the coast and toward the south.

Specimens examined.

Number.	Locality.	Habitat.	When collected.	Received from—	Number of specimens.	Dry. Alc.
2136	Great Egg Harbor, N J	Sandy beach	Apr. —, 1871	S I. Smith		Alc.
	Nobska Beach, Mass	....do	Aug 18, 1871	......do	2	Alc.
	Nantucket	....do	Dec. 6, 1877	V.N. Edwards	1	Alc.

Actoniscus Harger.

Actoniscus Harger, Am. Jour. Sci., III, vol. xv, p. 373, 1878.

Eyes small; antennæ geniculate at the third and fifth segments; flagellum four-jointed; terminal segments of maxillipeds lamelliform, lobed; legs all alike; basal segment of uropods dilated and simulating the coxæ of the preceding segments of the pleon; rami both styliform.

This genus resembles *Actæcia* Dana* MSS., considered as the young of *Scyphax ornatus*, and found with it on the beach at New Zealand. Professor Kinahan,† on the other hand, regarded the genus as indicating a distinct family. The present genus differs from the description and figures of Professor Dana as follows: The flagellum of the antennæ consists of only four distinct segments instead of about six; the terminal segment of the maxillipeds is less distinctly lobed; the inner ramus of the uropods surpasses the outer, instead of falling far short of it; the outer ramus is styliform instead of being enlarged and subequal to the produced and enlarged outer angle of the basal segment.

Actoniscus ellipticus Harger.

Actoniscus ellipticus Harger, Am. Jour. Sci., III, vol. xv, p. 373, 1878; Proc. U. S. Nat. Mus., 1879, vol. ii, p. 157, 1879.

PLATE I, FIG. 3.

This species may be at once recognized by the pleon, which appears to have four pairs of coxæ produced at the sides instead of three, as in *Oniscus* and other genera of this family. The last pair are, however, the basal segments of the caudal stylets, which are of peculiar form in this genus.

The body is oval in outline. The head appears triangular as seen from above, and is angularly produced in a median lobe, but the lateral lobes are also large and divergent, and broadly rounded. The eyes are small, oval, black, and prominent. They are situated at the sides of the median triangular part of the head, and at the base of the lateral lobes. The antennulæ are minute and rudimentary. The antennæ have the basal segment short; the second enlarged distally, especially on the inner side; the third forming an angle with the second, and clavate; the fourth flattened-cylindrical, longer than the third; fifth longest, slender, bent at base and forming an angle with the fourth; flagellum shorter than the last peduncular segment, tipped with setæ and composed of four segments, of which the second and third are equal and longer than the first, while the last is the shortest, and presents indications of another minute rudimentary terminal segment. The maxillipeds have the basal segment nearly twice as long as broad; the terminal segment elongate triangular, ciliated and somewhat lobed near the tip.

---

* U. S. Expl. Exped. Crust., part ii, p. 736, pl. 48, fig. 6 *a–h*.

† Natural History Review, vol. iv, Proc. Soc., p. 274, 1857.

The first thoracic segment is excavated in front for the head, admitting it about to the eyes. The next five segments are each a little longer than the first, but the last thoracic segment is the shortest. The first segment is dilated at the sides to about twice its length on the median line. The second, and in an increasing degree the succeeding segments are produced backward at the sides. The legs are rather small and weak and of nearly equal size throughout.

The first two segments of the pleon have their lateral processes, or coxæ, obsolete as usual in the family, but the third, fourth, and fifth segments are produced laterally into broad plates, which are close together, and, at their extremities, continue the regular oval outline of the body with scarcely a perceptible break between the thorax and the pleon. This outline is further continued by the expanded basal segments of the uropods, which are even larger than the adjacent coxæ of the fifth segment. At the extremity of the pleon both pairs of rami are visible, the inner springing from near the base of the basal segments below, the outer from a notch near the middle of the inner margin of the basal segment. The rami are tipped with setæ, and the inner just surpass the outer, which, in turn, surpass the produced portion of the basal segments.

Length 4^{mm}, breadth 2^{mm}. Color in life slaty gray.

This species was collected by Professor Verrill, at Savin Rock, near New Haven!, and also at Stony Creek!, Long Island Sound, in company with *Philoscia vittata* Say.

Specimens examined.

Number.	Locality.	Habitat.	When collected.	Received from—	Number of specimens.	Dry. Alc.
2137	Savin Rock, Conn...	Shore....................	—— —, 1874	A. E. Verrill.......	2	Alc.
2138	Stony Creek, Conn..	...do....................	....	.......do............	1	Alc.

The genus *Ligia* Fabricius* is recorded by Gould† from the timbers of a wharf, probably in Boston, and by Dr. Leidy,‡ with some doubt, from Point Judith, R. I., and the characteristics of the genus are therefore here briefly inserted, as follows:

Antennæ with a multiarticulate flagellum; basal segment of uropods exserted bearing two elongated cylindrical rami.

They are found usually in rocky places and under stones just above high-water mark. They are common on our southern coast, and are probably, at least occasionally, transported by accident within our limits. I have seen no specimens from nearer than Fort Macon, N. C.

* Suppl. Ent. Syst., p. 296, 1798.
† Invert. Mass., p. 337, 1841.
‡ Jour. Acad. Nat. Sci., II, vol. iii, p. 150, 1855.

II.—BOPYRIDÆ.

This family has not been studied, and only a list of the species, furnished by Professor S. I. Smith, is included. They are parasitic on Crustacea, and at maturity, the females especially, are generally much distorted and degenerate, often losing a great proportion of their appendages. The males are much smaller than the females, and of a more normal form, and they and the young forms must therefore be relied upon to indicate the affinities of this group to the rest of the order. According to Dr. Fritz Müller these forms indicate a relationship to the *Oniscidæ*, and especially to the genus *Ligia*, and in deference to his authority I have inserted them at this place.

Cepon distortus Leidy.

Cepon distortus Leidy, Jour. Acad. Nat. Sci., II, vol. iii, p. 150, pl. xi, figs. 26–32, 1855.
Harger, This Report, part i, p. 573 (279), 1874; Proc. U. S. Nat. Mus., 1879, vol. ii, p. 157, 1879.
Leidya distorta Cornalia and Panceri, Mem. R. Accad. Sci. Torino, II, tom. xix, p. 114, 1861.

"From the branchial cavity of *Gelasimus pugilator*, Atlantic City, New Jersey." (Leidy.)

Gyge Hippolytes Bate and Westwood (Kröyer).

Bopyrus Hippolytes Kröyer, Grönlands Amfipoder, p. 306 (78), pl. iv, fig. 22, 1838; Monog. Fremst. Slægten Hippolyte's nordiske Arter, p. 262, 1842; Voy. en Scand., Crust., pl. xxviii, fig. 2, 1849.
Edwards, Hist. nat. des Crust., iii, p. 283, 1840.
Stimpson, Proc. Acad. Nat. Sci. Philadelphia, 1863, p. 140.
Gyge Hippolytes Bate and Westwood, Brit. Sess. Crust., vol. ii, p. 230, 1868.
Buchholz, Zweite deutsche Nordpolfahrt, p. 286, 1874.
Metzger, Nordseefahrt der Pomm., p. 286, 1875.
Miers, Ann. Mag. Nat. Hist., IV, vol. xx, p. 64, (14), 1877.
Smith in Harger, Proc. U. S. Nat. Mus., 1879, vol. ii, p. 157, 1879.

Massachusetts Bay !, off Salem, on *Hippolyte spinus*, 30 fathoms, sand and mud, August 4, 1877; on *H. Fabricii*, 22 fathoms, gravel, August 4, 1877; on *H. securifrons*, 90 fathoms, soft mud, August 14, 1877. Casco Bay !, on *H. polaris* and *H. pusiola*, 1873. Bay of Fundy !, on *H. spinus* and *H. pusiola*, 1868, 1872. Off Halifax, Nova Scotia, 43 fathoms, September 27, 1877. Gulf of Maine !, 40 miles east of Cape Ann, Massachusetts, on *H. securifrons*, 160 fathoms, soft mud, August 19, 1877; also near Cashe's Ledge, on *H. spina*, 27 and 40 fathoms, rocks and gravel.

East side of Smith's Strait, north latitude 78° 30′ (Stimpson). "Discovery Bay," north latitude 81° 44′, Greenland (Miers). British Islands (Bate & Westwood). Scandinavian coasts (Kröyer *et al.*). Spitzbergen (Kröyer).

Phryxus abdominalis Liljeborg (Kröyer).

Bopyrus abdominalis Kröyer, Nat. Tidsskr., vol. ii, pp. 102, 289, pls. i, ii, 1840; Monog. Fremst. Slægten Hippolyte's nordiske Arter, p. 263, 1842; Voy. en Scand., Crust., pl. xxix, fig. 1, 1849.
Phryxus Hippolytes Rathke, Fauna Norwegens, p. 40, pl. ii, figs. 1–10, 1843.
Phryxus abdominalis Liljeborg, Œfvers. Kongl. Vet.-Akad. Förh., ix, p. 11, 1852.
Steenstrup and Lütken, Vidensk. Meddelelser, 1861, p. 275 (9).
Bate and Westwood, Brit. Sessile-eyed Crust., vol. ii, p. 234, 1868.
Norman, Rep. Brit. Assoc., 1868, p. 288, 1869; Proc. Royal Soc., London, vol. xxv, p. 209, 1876.
Buchholz, Zweite deutsche Nordpolfahrt, p. 287, 1874.
Metzger, Nordseefahrt der Pomm., p. 286, 1875.
Miers, Ann. Mag. Nat. Hist., IV, vol. xx, p. 65 (15), 1877.
Smith in Harger, Proc. U. S. Nat. Mus., 1879, vol. ii, p. 158, 1879.

Massachusetts Bay!, off Salem, on *Pandalus borealis*, *Hippolyte spinus*, and *H. securifrons*, 48–90 fathoms, soft mud, August 13 and 14, 1877; also, on *Pandalus Montagui*, 35 fathoms, mud and clay nodules, August 10, 1877. Cashe's Ledge!, Gulf of Maine, on *Hippolyte pusiola*, 27 and 39 fathoms, rocky, September 5, 1874. Halifax!, Nova Scotia, on *Hippolyte pusiola*, 18 fathoms, fine sand, September 4, 1877; also, on *H. spinus*. About 30 miles south of Halifax!, on *Hippolyte securifrons*, 100 fathoms, fine sand, September 6, 1877.

Grinnell Land, in north latitude 79° 29′; and "Discovery Bay," north latitude 81° 44′ (Miers). Greenland (Kröyer *et al.*). British Islands (Norman *et al.*). Scandinavian coast! (Liljeborg *et al.*). Spitzbergen (Miers).

Dajus Mysidis Kröyer.

Dajus Mysidis Kröyer, Voy. en Scand., Crust., pl. xxviii, fig. 1, 1849.
Lütken, Crustacea of Greenland, p. 150, 1875.
? G. O. Sars, Arch. Math. Nat., B. ii, p. 354 [254], 1877 ("*D. Mysidis?*").
Smith in Harger, Proc. U. S. Nat. Mus., 1879, vol. ii, p. 158, 1879.
Bopyrus Mysidum Packard, Mem. Bost. Soc. Nat. Hist., vol. i, p. 295, pl. viii, fig. 5, 1867.
? *Leptophryxus Mysidis* Buchholz, Zweite Deutsche Nordpolfahrt, p. 288, pl. ii, fig. 2, 1874.

Labrador (Packard). Greenland (Kröyer, Buchholz). ? Off west coast of Norway (G. O. Sars).

Bopyrus, species.

Bopyrus Leidy, Proc. Acad. Nat. Sci., 1879, pt. ii, p. 198, 1879.
? Smith, Trans. Conn. Acad., vol. v, p. 37, 1879.

A species of *Bopyrus* is mentioned by Dr. Leidy as "a parasite of the shrimp, *Palæmonetes vulgaris*," occuring in the summer of 1879, at Atlantic City, N. J.

III.—ASELLIDÆ.

Antennæ elongated with a multiarticulate flagellum; legs ambulatory or prehensile, not strictly natatory; pleon consolidated into a scutiform segment, bearing terminal uropods, which may be nearly obsolete.

This family is represented on our coast by four species belonging to three genera, and a species of another genus (*Asellus communis* Say) is common in the fresh-water ponds and streams of New England. The genus *Limnoria* Leach has been regarded by modern writers as belonging to this family, but will be found in the present article in the *Limnoriidæ* (p. 79). There remain then to be considered the genera *Asellus* Geoffroy,* *Jæra* Leach, *Janira* Leach, and *Munna* Kröyer, which, as represented in our waters, may be further characterized as follows:

The head is well developed, and in *Munna* is of large size; the body is usually depressed or but slightly arched, except that the pleon is vaulted in *Munna*. The eyes are present in our species though not throughout the family. The antennulæ beyond the basal segment are slender and are always much shorter than the antennæ, which are elongated and composed of a five-jointed peduncle and a slender multiarticulate flagellum. The first three peduncular segments are short; the last two elongated. The parts of the mouth are protected below by a pair of maxillipeds with large external lamellæ and five-jointed palpi. Within the maxillipeds are two pairs of maxillæ of the ordinary form; the outer or second pair delicate and three-lobed at the tip; the inner lobe being formed by the projecting basal segment, while the two outer lobes are articulated; all three lobes are provided with curved spiniform setæ. The inner, or first, pair of maxillæ present two narrow lobes; the outer lobe broader and more robust than the inner, and armed with robust curved spines, while the inner is tipped with much weaker setæ. The mandibles (see fig. 12 *b*, pl. III) are provided with one or two acute dentigerous lamellæ (*d*) at the tip, usually a comb of setæ and a strong molar process below (*m*), and a triarticulate palpus (*p*). This latter organ is, however, wanting in the genus *Mancasellus* Harger† from the Great Lakes and other fresh-water localities of North America.

The seven segments of the thorax are distinct from the head and from each other, and differ but little in general appearance throughout. The legs are mostly slender and elongated, except that the first pair may be more robust and better fitted for prehension. In our marine species the dactylus, at least behind the first pair of legs, is short and armed with two small claws or ungues, while the propodus is capable of considerable flexion on the carpus.

The segments of the pleon are united into a single piece, which is scutiform above, flattened or but little arched, except in *Munna*, and bears, at or near the tip, the biramous uropods, which are, however, nearly obsolete in *Munna*. The pleon often shows more or less trace of its compound character in imperfect transverse sutures on the dorsal surface near the base, and below it is excavated for the pleopods, the posterior pairs of which are delicate and branchial in their nature, while the anterior pairs

* "Hist. des Ins. t. II" (Edw.). For information in regard to the common European form of this genus the reader should consult the admirable work of G. O. Sars, Hist. nat. des Crust. d'eau douce de Norvège.

† Am. Jour. Sci., III. vol. xi, p. 304, 1876. See, also, op. cit., vol. vii, p. 601, 1874, and This Report, part ii, p. 659, pl. i. fig. 3, 1874.

are variously modified in the different genera and in the sexes, so that much confusion has been introduced into the family by mistaking sexual for generic modifications of these organs. The branchial pleopods are usually protected by a thickened anterior pair, which, especially in the females of our marine species, may be consolidated into a single opercular plate, as will be further described. The incubatory pouch in the females does not appear to extend farther back than the fourth thoracic segment, and it may be confined to the second, third, and fourth segments.

In the last-mentioned, as well as in many other characters, this family is closely related to the next, and perhaps the *Munnopsidæ* may yet require to be united with it. Our species of the two families are at once distinguished by the last three pairs of legs, which are ambulatory in the *Asellidæ* and natatory in the *Munnopsidæ*. Our *Munnopsidæ* are, moreover, like the other known species of that family, destitute of eyes, while the marine *Asellidæ* have evident or conspicuous eyes, but the fresh-water genus *Cæcidotea* Packard* is blind, as are also certain foreign species referred to the present family. The relations of the *Asellidæ* with families other than the *Munnopsidæ* are less evident. They were associated by Professor Dana† with his *Armadillidæ* and *Oniscidæ* to form his subtribe *Oniscoidea*, and, *Limnoria* being excluded, the group appears to be a natural one.

Asellus communis Say, confined to fresh waters, and the only known New England representative of the genus, was described and figured by the present author, in Professor S. I. Smith's "Crustacea of the Fresh Waters of the United States," published in part II of this report (page 657, plate I, figure 4). Our marine representatives of the family may be most easily recognized by the consolidated pleon, ambulatory or prehensile legs, none of them natatory, and the slender, elongate antennæ. The genera may be distinguished by means of the following table:

Pleon	flattened above; uropods	short, subrudimentary	JÆRA, p. 314
		well developed	JANIRA, p. 319
	vaulted; head large		MUNNA, p. 325

Jæra Leach.

Jæra Leach, Ed. Encyc., vol. vii, p. "434" (Am. ed., p. 273), "1813–14."

Antennulæ short, few-jointed; antennæ moderately elongated; mandibles with palpi; first pair of legs similar to the following pairs; lateral margins of the thoracic segments projecting over the bases of the legs; uropods short, rami subrudimentary; pleon protected below in the females by a subcircular plate.

The short uropods and projecting lateral margins of the thoracic segments serve to distinguish this genus from its allies, and other characters of generic importance could doubtless be drawn from the pleon and its appendages, as well as from other parts of the structure, but, as it

*American Naturalist, vol. v, p. 751, figs. 132, 133, 1871.

† Am. Jour. Sci., II, vol. xiv, p. 301, 1862.

is represented in our limits by a single species, I have not been able to separate the generic from the specific characters with confidence, and have therefore described the species without attempting it.

Jæra albifrons Leach.

"*Oniscus albifrons* Montague MSS." (Leach).

Jæra albifrons Leach, Ed. Encyc., vol. vii, p. "434" (Am. ed., p. 273), "1813–14"; Trans. Linn. Soc., vol. xi, p. 373, 1815.

Samouelle, Ent. Comp., p. 110, 1819.

Desmarest, Dict. Sci. nat., tome xxviii, p. 381, 1823; Consid. Crust., p. 316, 1825.

Latreille, Règne Anim., tome iv, p. 141, 1829.

Edwards, Annot. de Lamarck, tome v, p. 267, 1838; Hist. nat. des Crust., tome iii, p. 150, 1840; Règne Anim., Crust., p. 204, 1849.

Moore, Charlesworth's Mag. Nat. Hist., n. s., vol. iii, p. 294, 1839.

Thompson, Ann. Mag. Nat. Hist., vol. xx, p. 245, 1847.

White, List Crust. Brit. Mus., p. 97, 1847; Brit. Crust. Brit. Mus., p. 69, 1850; Pop. Hist. Brit. Crust., p. 231, 1857.

Lilljeborg, Öfvers. Vet-Akad. Förh., Årg. viii, p. 23; 1851; ibid., Årg. ix. p. 11, 1852.

Gosse, Man. Mar. Zool., vol. i, p. 136, fig. 243, 1855.

M. Sars, Christ. Vid. Selsk. Forh., 1858, p. 153, 1859.

Bate, Rep. Brit. Assoc., 1860, p. 225, 1861.

G. O. Sars, Reise ved Kyst. af Christ., p. (29), 1866; Christ. Vid. Selsk. Forh., 1871, p. 272, 1872.

Norman, Rep. Brit. Assoc., 1866, p. 197, 1867; ibid, 1868, p. 288, 1869.

Bate and Westwood, Brit. Sess. Crust., vol. ii, p. 317, figure, 1868.

Metzger, J. B. Naturhist. Ges. Hannover, xx, p. 32, 1871; Nordseefahrt der Pomm., 1872–'3, p. 285, 1875.

Parfitt, Trans. Devon. Assoc., 1873, p. (18), "1873."

Stebbing, Jour. Linn. Soc., Zool., vol. xii, p. 149, 1874; Ann. Mag. Nat. Hist., IV, vol. xvii, p. 79, pl. v, figs. 5–6, 1876; Trans. Devon. Assoc., 1879 p. (7), 1879.

Meinert, Crust. Isop. Amph. Dec. Dan., p. 80, "1877." (*Iaira.*)

Harger, Proc. U. S. Nat. Mus., 1879, vol. ii, p. 158, 1879.

Jæra Kröyeri Zaddach, Syn. Crust. Pruss. Prod., p. 11, "1844" (*J. Kröyeri* Edwards ?).

Jæra baltica Fried. Müller, Arch. Naturg., Jahrg. xiv, p. 63, pl. iv, fig. 29, 1848.

Jæra copiosa Stimpson, Mar. Inv. G. Manan, p. 40, pl. iii, fig. 29, 1853.

Packard, Canad. Nat. and Geol., vol. viii, p. 419, 1863.

Verrill, Am. Jour. Sci., III, vol. vii, p. 131, 1874; Proc. Amer. Assoc., 1873, p. 369, 1874; This Report, part i, p. 315 (21), 1874.

Harger, This Report, part i, p. 571 (277), 1874.

Jæra nivalis Packard, Mem. Bost. Soc. Nat. Hist., vol. i, p. 296, 1867 (*J. nivalis* Kröyer ?.)

Asellus Grönlandicus Packard, loc. cit. (*not* of Kröyer).

Jæra marina Möbius, Wirbellos. Thiere der Ostsee, p. 122, 1873; Ann. Mag. Nat. Hist., IV, vol. xii, p. 85, 1873. (*J. marina* Fabricius ?.)

Jæra maculata Parfitt, Trans. Devon. Assoc., 1873, p. "253" (18), "1873."

Stebbing, Trans. Devon. Assoc., 1879, p. (7) 1879, (*albifrons*).

PLATE I, FIGS. 4–8.

This species is at once distinguished from the other marine Isopoda of our coast by the short uropods, arising from a notch in the end of the subcircular pleon. From the terrestrial forms, which it somewhat resembles, and in company with which it may sometimes be found, the above-mentioned character, joined with the multiarticulate flagellum of the antennæ, will serve to distinguish it.

The body is oval and flattened, a little more than twice as long as broad. The head is transverse, broadly excavated on each side over the bases of the antennulæ, sparingly ciliated on the lateral margins, with short scattered spine-like unequal cilia or setæ, which occur in a similar manner along the entire borders of the animal behind the front margin of the head. The eyes are prominent and black, situated near the posterior margin of the lateral regions of the head. The antennulæ are five-jointed, and do not surpass the fourth segment of the antennæ; the basal segment is large and separated from its fellow of the opposite side by about twice its diameter; the second segment is about as long as the first, but of much less than half its diameter; third segment shorter than the second, fourth still shorter, fifth tapering, tipped with setæ. The first three segments of the antennæ are short; the fourth is robust, and about as long as the first three together; the fifth is longest, and is followed by a slender elongated flagellum. The maxillipeds (pl. I, fig. 5) have the external lamella (*l*) short and broad, nearly straight on the inner margin, broadly rounded at the end, and somewhat swollen on the external side; the palpus (*p*) is five-jointed; the first three segments flattened, first short; second dilated internally and ciliated; third ciliate in the inner margin and narrowed to the base of the fourth segment, which is cylindrical; fifth short, conical. The terminal lobe of the maxilliped bears two rows of cilia near the apex, and on the inner side a row of short styliform organs. The outer maxillæ (pl. I, fig. 6 *a*) consist of a semioval portion, broad and ciliated at the tip, bearing above the middle two articulated lobes, armed with strong curved setæ at the tip. The inner maxillæ (pl. I, fig. 6 *b*) are armed with short stout spines, which are strongly spinulose on their inner curved side; inner lobe about half the diameter of the outer. Mandibles with a very much projecting molar process, a comb of pectinated setæ, and a dentigerous lamella, or two of them on the left side.

The first three thoracic segments are of about equal length along the median line, and are together nearly equal in length to the last four, which are also subequal along the median line, but the fifth segment appears shorter than the others on account of its short lateral margin, which has both its anterior and posterior angles strongly rounded. The epimeral region of the segments projects at the sides so as to cover the bases of the legs, and is squarish in the first three segments, rounded in the fourth, and still more so in the fifth, and obtusely angulated behind in the sixth and seventh. The legs are similar in form throughout, but increase in length to the last pair. They have the basis rather robust; the ischium shorter and flexed on the basis; the merus subtriangular, and tipped with spines; the carpus and propodus cylindrical, subequal in length, but the carpus of larger diameter than the propodus; the dactylus short, cylindrical, and provided with two terminal hooklets. There are a few scattered spinules and setæ on the segments, especially the merus, carpus, and propodus. In the males the merus and carpus of the sixth and seventh pairs of legs are provided on their inferior margins with close-set slender curved hairs, which extend nearly the whole length of the carpus and over the distal half of the merus.

The pleon is proportionally broader and shorter in the male (pl. I, fig. 8) than in the female (pl. I, fig. 7). It is broadly rounded behind, continuing the outline of the body without break, and is notched at the tip for the insertion of the uropods, which scarcely project beyond the general outline of the body, and consist on each side of a short, stumpy, cylindrical basal segment, a little oblique at the end where it bears two almost rudimentary rami, the inner about twice as large as the outer, and both tipped with a few short setæ. The lateral margin of the pleon, like that of the body generally, is beset with short, scattered, unequal setæ or spinules. Underneath, the pleon is excavated for the branchial pleopods, which are covered and protected below in the females (pl. I, fig. 7) by a large subcircular plate, sparsely minutely ciliated on the margin. In the male (pl. I, fig. 8) the under surface of the pleon presents on each side a small oval plate, with its inner margin overlapped by a median elongated plate, divided by a central suture, which is open distally. This plate is broad at the base, then narrows toward the middle, after which it expands much more rapidly into an outwardly curved and pointed lobe on each side, ciliated at the tip. Between these two lobes the plate is terminated by two transverse, subquadrate and elongated lobes, which are broadest internally where they are separated along the median line. They are excavated on the anterior margin and less so on the posterior margin, sparsely ciliated behind, and conspicuously so with divergent cilia at the outer short, straight margin. In the females the incubatory pouch appears to be confined to the second, third, and fourth segments.

In size as well as coloration this species varies greatly, females being often found with eggs when less than half the size of the specimen figured. They attain a length of 5^{mm} and a breadth of 2^{mm}, but the males are at least one-third smaller and somewhat narrower than the females, the sides being more nearly parallel. In color there is also much variation. A common color is a dark, slaty gray, with dots or small blotches of yellowish, this color prevailing along the anterior margin of the head. Very frequently darker or lighter shades of green occur, and the incubatory pouch of the females is often bright green. Some specimens are very light colored or nearly white, often with two or more transverse dark bands, with considerable contrast in color; others are reddish brown throughout.

I am unable to separate the American form, *Jæra copiosa* Stimpson, from the common English and European species, although they have hitherto been regarded as distinct. I have had no males from any European locality, but through the kindness of the Rev. A. M. Norman I have had an opportunity of comparing females from Oban, Scotland, with our species, and have found no specific differences. The description and figures given by Rev. T. R. R. Stebbing in the Annals and Magazine of Natural History, IV, vol. xvii, p. 79, pl. v, figs. 5 and 6, show a substantial correspondence in the male; also, so that I have regarded the species as common to both coasts. Whether the Greenland species *J. nivalis* Kröyer, and the Southern species *J. Kröyeri* Edwards, are also identical with *J. albifrons* or not, I am unable to determine, in the absence of specimens for comparison. M. Sars says that he has seen specimens of *J. albifrons* Leach from Trieste, but regards the Greenland species as distinct. Möbius regards the species as identical from Greenland to the Mediterranean, and unites them under the name *J. marina*. Metzger, following Bate and Westwood, is more conservative, using the name *J. albifrons* Leach. Bate and Westwood regard *J. nivalis* Kröyer and *Oniscus marinus* O. Fabricius as doubtfully identical with *J. albifrons*, and *J. Kröyeri* Edwards as distinct. *J. Kröyeri* Zaddach = *J. baltica* Friedrich Müller appears to be, without doubt, identical with this species, as it is separated by that author from *J. albifrons* Leach only by the position of the eyes, which were incorrectly described by Dr. Leach as close together. I have, therefore, referred these two names to *J. albifrons* as synonyms, as has been done previously by Lilljeborg and others. *J. maculata* Parfitt, a species based almost wholly on color markings, I have referred to *J. albifrons*, following Stebbing, who believes that he is "in accord with the author of the species" in so doing.

This species is common, and in suitable localities abundant, on the whole coast of New England!, and extends as far north as Labrador! at least, where it was collected by Dr. Packard, who regarded it as identical with *J. nivalis* Kröyer. It is found among rocks, algæ, and rubbish along the shore, often nearly up to high-water mark, where it may be associated with some of the *Oniscidæ*, to which it has a certain resemblance in form. It occurs "probably" all around the coast of England (Bate and Westwood). I have examined specimens from Oban!, Scotland. It extends to Finmark, on the coast of Norway (M. Sars), and is common on all the coasts of the North Sea (Metzger). It is recorded by Möbius in the Baltic among stones and algæ down to a depth of 18½ fathoms. According to M. Sars this species extends to Trieste on the Adriatic, but without specimens I have not attempted to decide in regard to the synonymy of the Mediterranean species.

Specimens examined.

Number.	Locality.	Fathoms.	Bottom.	When collected.	Received from—	Specimens. No.	Specimens. Sex.	Dry. Alc.
1921	New Haven, Conn.			May 1, 1871		20		Alc.
1917	Stony Creek, Conn.					8		Alc.
1916	Noank Harbor, Conn.			—— —, 1874	U.S. Fish Com	25		Alc.
1915	Vineyard Sound, Mass.			—— —, 1871	....do	1	♀	Alc.
1914	....do	L. w.	Under stones	—— —, 1871	....do	30		Alc.
1920	Provincetown, Mass			—— —, 1872	....do	50		Alc.
	....do		Shore	—— —, 1879	... do	00	♂ ♀	Alc.
	....do		...do	Aug. 13, 1879	....do	00	♂ ♀	Alc.
	....do	L. w.		Aug. 13, 1879	....do	15	♀	Alc.
	... do		Eel grass	Aug. 23, 1879	....do	6	♂ ♀	Alc.
	Gloucester, Mass		Algæ	—— —, 1878	....do	30	♀	Alc.
	....do		Tide pools		....do	7	♀ ♂	Alc.
	Casco Bay			—— —, 1873	....do			Alc.
1919	Eastport, Me.	L. w.	Under stones	1868–1870	A. E. Verrill	7		Alc.
1918	Eastport, Me., Dog Island.		Tide pool	—— —, 1872	U. S. Fish Com.	5		Alc.
1912	Indian Tickle, Labrador.				A. S. Packard	7		Alc.
519*	Hopedale, Labrador.		Stone		....do			Alc.
	Oban, Scotland			—— —, 1877	Rev. A. M. Norman.	4	♀	Alc.

* *Asellus grönlandicus* Packard, MSS.

Janira Leach.

Janira Leach, Edinb. Encyc., vol. vii, p. "434" (Amer. ed., p. 273), "1813–14".
Asellodes Stimpson, Mar. Inv. Grand Manan, p. 41, 1853.

Body loosely articulated as in *Asellus;* antennulæ slender, with a multiarticulate flagellum; antennæ elongated, with a spine, or scale, on the second segment and with a long multiarticulate flagellum; mandibles palpigerous; lateral margins of the thoracic segments not completely covering the bases of the legs; first pair of legs prehensile; the carpus thickened, and the propodus slender and capable of complete flexion on the carpus; dactylus short and armed with two small ungues, as in the succeeding pairs of legs; uropods well developed, biramous.

This genus is represented on our coast by two species, one of which was originally described by Stimpson under the name *Asellodes alta.* It does not, however, seem to present any generic differences from *Janira maculosa* Leach, the type of the present genus. Stimpson's generic description appears to have been drawn from the male, as he says: "External pair of natatory feet having each two laminæ, like the others, but broader and hardened, so as to perform the office of an operculum." The two inner of these laminæ are, however, united along the median line nearly to the tip, as will be seen below.

Our species of this genus may be further characterized as follows: The body is elongate oval in general outline, between two and three times as long as broad. The eyes are distinct. The head is produced medially into a distinct rostrum, and the antero-lateral angles are also produced, but in the typical species (*J. maculosa* Leach) the head is rounded anteriorly. The basal segment of the antennulæ is enlarged; the second is more slender and cylindrical; the third is short, cylindrical, or slightly clavate, and is followed by a short subglobose segment having the appearance of a fourth peduncular segment. Beyond this, is a slender multiarticulate flagellum, composed of about twenty to thirty segments, the segmentation becoming indistinct toward the base. These segments are provided, except toward the base, with slender "olfactory setæ." The first three segments of the antennæ are short and robust, and the second bears, near its distal end, on the external side above, a triangular scale, or spine, articulated with the segment and directed forward, outward, and somewhat upward; the third segment is comparatively short and small; the fourth and fifth segments are slender and elongated, and the flagellum tapers from the base and is composed of many, 80 to 120 or more, segments. The maxillipeds (see pl. III, fig. 12 *a*) are broad, with a rhombic-ovate external lamella (*l*), and a five-jointed palpus (*p*), of which the first three segments are flattened and expanded internally, where the second and third segments are also ciliated. The last two segments of the palpus are cylindrical, and bent inward toward the median line. The outer maxillæ are rhombic in outline, ciliated and spiny along the inner margin and at the tip, as are also the two slender, curved, articulated lobes. The inner maxillæ consist of the usual curved lobes, armed at the tip with denticulated spines, which are larger, stronger, and more numerous on the outer large lobe. The mandibles are strong, and furnished with an acute dentigerous lamella on the right side, received between two such lamellæ on the left mandible; below is a comb of setæ and a strong molar process. The palpus of the mandible is composed of three subequal segments, the last furnished with a comb of setæ.

The thoracic segments do not greatly exceed the head in transverse diameter, and are subequal, the second, third, and fourth with a lateral emargination. The legs are slender and elongated, ambulatory, or the first pair subprehensile and somewhat shorter than the following pairs. In this pair the carpus is slightly swollen and the propodus is capable of complete flexion upon it. The dactyli are short in all the legs, as compared with the propodi, and capable of only incomplete flexion. They are armed at the tip with two robust unguiform spines.

The pleon is broad and flattened above. The uropods are well developed and consist of a cylindrical or slightly clavate basal segment bearing two rami of which the inner is the larger and longer. The under surface of the pleon is excavated, and in the females is protected beneath by a subcircular operculum, but in the males of *J. alta*, and probably in both species, the thickened opercular plates are three in number, viz, a pair of semi-oval plates at the sides and a more slender median plate presenting traces of a suture along the middle.

In the females, the incubatory pouch is formed of four pairs of plates attached to the coxal segments of the first four pairs of legs. These plates may usually be easily seen when the females are destitute of eggs, being then small, elongate, oval, and lying near the under surface of the thoracic segments.

Janira alta Harger (Stimpson).

Asellodes alta Stimpson, Mar. Inv. G. Manan, p. 41, pl. iii, fig. 30, 1853.
Verrill, Am. Jour. Sci., III, vol. vi, p. 439, 1873; vol. vii, pp. 411, 502, 1874; Proc. Amer. Assoc., 1873, p. 350, 1874.
Janira alta Harger, Proc. U. S. Nat. Mus., 1879, vol. ii, p. 158, 1879.

PLATES II AND III, FIGS. 9, 12, AND 13.

This species may be at once distinguished from the following by the absence of spines in the dorsal and lateral thoracic regions, from all the other known Isopoda of the coast, by the flattened, scutiform and consolidated pleon, bearing well-developed, exerted, biramous uropods, which are, however, fragile. It is more slender than the following species.

The body is elongated oval in outline, nearly three times as long as broad. The head is produced in front into a prominent but short, acute, median spine or rostrum, and the antero-lateral angles are also acutely produced, but are shorter and less acute than the rostrum. The eyes are prominent and black, situated on the upper surface of the head, near the lateral margins. They are elliptical in outline, with the long axes converging toward a point near to, or beyond, the tip of the rostrum. The basal segment of the antennulæ is shorter than the rostrum; the flagellum consists of about thirty segments and does not attain the tip of the fourth antennal segment. The scale on the second segment of the antennæ is short and triangular, does not surpass the following segment, and is tipped with a few slender setæ. The maxillipeds (pl. III, fig. 12 *a*) have the external lamella (*l*) obtusely pointed at the apex and angulated on the outer side, otherwise they resemble the same organs in *J. spinosa*, as do the outer maxillæ, the inner maxillæ, and the mandibles (pl. III, fig. 12 *b*).

The thoracic segments are but little broader than the head, the first three and the last two segments are about equal to each other in length; the fourth and the fifth are somewhat shorter. The lateral margins of the segments do not cover the epimera from above, and none of them are produced at the sides into acute and salient angulations, as in the next species. In the first segment the lateral margins are rounded and the epimera project as an angular tooth on each side in front. In the second, third, and fourth segments the emargination is behind a prominent but narrow lobe at the anterior angle of the segment and the epimera are two-lobed. In the fourth segment the posterior angle is nearly included in the emargination, and in the last three segments the posterior angle is elided and the epimera occupy its place. The legs are elongated and armed with spines, especially on the carpal segments.

The pleon is rounded-hexagonal in outline, minutely and sharply serrate at the sides behind the middle, and undulated over the bases of the uropods on the posterior margin. The uropods are slender, easily detached, and liable to escape observation. They are nearly alike in the two sexes, and consist on each side of an elongate, somewhat curved and clavate basal segment, bearing at the end two rami, of which the inner is nearly as long as the basal segment, the outer somewhat smaller and shorter. The rami are slightly flattened, and, like the basal segment, armed with setæ, especially at the tip. The branchial pleopods are protected in the female by a subcircular operculum (pl. III, fig. 13 *a*). In the male, the inferior surface of the pleon (pl. III, fig. 13 *b*) presents on each side a nearly semicircular plate (*b*), with its inner margin overlapped by a median, elongated, and narrow plate (*c*), marked along the median line by a suture. This plate is broadest near the base, then contracts on each side to beyond the middle, after which it expands slightly. The median suture is open near the tip, and, on each side, is a rounded lobe, separated by a sinus from the produced external angle.

Length of body, exclusive of the antennæ and uropods, 8^{mm}, breadth 3^{mm}. Color in alcohol usually pale or brownish, with small black dots on the upper surface. The under surface is lighter, as are the legs and antennæ, especially toward their distal extremities.

This species is at once distinguished from the common European *J. maculosa* Leach by the form of the head, which is rostrate, and has also the antero-lateral angles strongly salient, while in *J. maculosa* the anterior margin of the head is nearly straight and the angles are not produced. From *Henopomus tricornis* Kröyer,* as described and figured by that author, it differs in the elongated uropods.

This species has not been found south of Cape Cod. Dr. Stimpson's specimens were "dredged in soft mud in 40 f. off Long Island, G. M.," in the Bay of Fundy. It was dredged in Massachusetts Bay! in from 54 to 115 fathoms mud, sand, and stones in 1878. In many localities given below in the Gulf of Maine! from 35 to 115 fathoms in 1873, 1874, and 1877, and 120 miles south of Halifax!, N. S., in 120 fathoms gravel and pebbles in 1877. It has also been obtained from several localities in the Bay of Fundy!, in one case at low water on Clark's Ledge, near Eastport, Me. A specimen was collected in 1879, by Mr. Charles Ruckley, of the schooner 'H. A. Duncan,' thirty miles east of the Northeast light on Sable Island, adhering to a specimen of *Paragorgia*, from a depth of 160 to 300 fathoms.

* Naturhist. Tidssk., II, B. ii, p. 380, 1847; Voy. en Scand., Crust., pl. xxx, figs. 2 *a-q*, "1849."

Specimens examined.

Number.	Locality.	Fathoms.	Bottom.	When collected.	Received from—	Specimens.		Dry. Alc.
						No.	Sex.	
	Gulf of Maine, ESE from Cape Ann 29-30 miles.	85	Mud, sand, stones	—— —, 1878	U S Fish Com.	1	♀	Alc.
	Gulf of Maine, ESE. from Cape Ann 30-31 miles.	110–115	Mud, stones	—— —, 1878	... do	1	♀	Alc.
	Gulf of Maine, SE ½ S from Cape Ann 6-7 miles.	54–60	Sand, mud	—— —, 1878	... do	2	♂ ♀	Alc.
1934	Gulf of Maine, SE from Cape Ann 14 miles.	90	Soft mud........	—— —, 1877	... do	1	......	Alc.
1923	Gulf of Maine, E. from Cape Ann 140 miles.	112–115	Sand and gravel	—— —, 1877	... do	1	♂	Alc.
1935	Between Cape Ann and Isles of Shoals	35	Clay, sand, mud.	—— —, 1874	... do	1	......	Alc.
1924	Gulf of Maine, S. of Cashe's Ledge.	90	Rocky	—— —, 1873	....do	...	......	Alc.
1925	Casco Bay, Me......	....	....	—— —, 1873	... do	1	......	Alc.
	Banquereau	....	....	—— —, 1878	Capt Collins ..	1	♀	Alc.
1927	Bay of Fundy, Me ..	....	....	—— —, 1872	U.S Fish Com	3	......	Alc.
1928	Bay of Fundy, Clark's Ledge.	L. w.-30	Rocky	—— —, 1872	... do	...	......	Alc.
1929	Bay of Fundy, Buckman's Head	....	....	—— —, 1872	....do	...	......	Alc.
1930	Bay of Fundy, off Todd's Head.	....	....	Aug. 27, 1872	.. do	...	......	Alc.
1932	Bay of Fundy, Eastport.	....	....	—— —, 1870	A. E Verrill ..	1	......	Alc.
	Thirty miles east of Northeast light on Sable Island.	160-300	On *Paragorgia*..	—— —, 1879	Mr. C. Ruckley	1	♂	Dry.
1933	South of Halifax 120 miles.	190	Gravel and pebbles	—— —, 1877	U S Fish Com.	1	......	Alc.

Janira spinosa Harger.

Janira spinosa Harger, Proc. U. S. Nat. Mus., 1879, vol, ii. p. 158, 1879.

This species is well marked among our known Isopoda, by the double row of spines along the back and the acute laciniations or angulations on the lateral margins of the thoracic segments.

The body is robust, the length but little exceeding twice the breadth. The head is broad, and produced in the median line into a prominent acute spine, or rostrum, about as long as the head. The antero-lateral angles are also produced and very acute, but do not extend as far as the rostrum. The eyes are rounded semi-oval, with the long axes converging toward a point near the base of the rostrum. The basal segment of the antennulæ is less than one-third the length of the rostrum. The second segment is about as long as the first, but of only about half its diameter. The flagellum equals, or slightly surpasses, the third antennal segment, and consists of about twelve segments. The scale, or spine, on the second segment of the antennæ is slender and considerably surpasses the third segment. The external lamella of the maxillipeds has the outer angle prominent, though not acute.

The thoracic segments are produced laterally into one or two acute angulations, giving a sharply serrated or dentated outline to the thoracic region. The first segment is shorter than the second; the second, third, and fourth are about equal in length; the fifth is about the length of the first; the sixth and seventh each a little longer. The first segment is acutely produced at the sides, around the sides of the head, and bears, near the middle of the anterior margin, two short spines, situated about half as far apart as are the eyes, and directed upward and somewhat forward. The second segment has both lateral angles produced into triangular acute processes, of which the anterior is more slender than the posterior and directed more strongly forward. The dorsal spines on this segment are a little farther apart and larger than in the first segment. In the third segment the lateral angulations are more nearly equal than in the second segment and directed less strongly forward. In the specimen figured the third segment bears, on the left side, a single broad angulation, apparently representing the posterior, while the anterior is only indicated by a slight irregularity in the outline. Malformations of this kind appear to be common. The dorsal spines on the third segment are much as in the second. On the fourth segment the anterior angulation is longer than the posterior, and both are directed nearly outward. The dorsal spines on the fourth segment are slightly smaller and nearer together than on the third; but, as in all the preceding segments, they are near the anterior border of the segment. The last three segments are acutely produced at the sides into a single angulation, which is directed more and more backward to the last segment. The dorsal spines on the fifth segment are situated nearer together than on the anterior segments, and rather behind the middle of the segment; they are also smaller than on the preceding segments. On the last two segments they are near the posterior border of the segment, and become somewhat smaller and nearer together on the last segment. The legs are armed with but few, and rather weak, spines.

The pleon is broadest near the base and tapers posteriorly, where the angles are acutely produced; between these angles the margin is rounded and arched over the bases of the uropods, which are about as long as the pleon and less spiny than in *J. alta*. The lateral margin of the pleon is armed with very minute acute spinules, and under a higher power the margins of the thoracic segments and of the head are seen to be similarly armed, especially where most exposed.

Length 8^{mm}, breadth 3.8^{mm}; color in alcohol, white.

This species is near *Janira laciniata* G. O. Sars,* but is distinguished by the double row of dorsal spines, whereas Sars says of that species, "Superficies dorsalis medio leviter convexa spinis singulis tenuibus ornata."

The only specimens yet known are two females, which were taken adhering to the cable of the schooner 'Marion', by Captain J. W. Collins, at Banquereau, August 25, 1878.

---

*Chr. Vid. Selsk., 1872, p. 92, 1873.

Munna Kröyer.

Munna Kröyer, Naturhist. Tidssk., B. ii, p. 615, 1839.

Form of the female dilated oval, of the male elongated sublinear; head very broad (about twice as broad as long), in length equal to one-fourth or one-fifth the length of the animal; eyes occupying the postero-lateral angles of the head, prominent, as if pedunculated but not movable; antennulæ inserted above the antennæ and partly covering their bases, short, a little longer than the head, with a four-jointed peduncle and a few-jointed flagellum; antennæ elongated, equaling or surpassing the length of the body, with a multiarticulate flagellum; mandibles with a three-jointed palpus; maxillipeds with a five-jointed palpus; legs all armed with two terminal ungues; first pair shorter and more robust than the others, with a prehensile hand formed of the propodus and the dactylus; the remaining pairs ambulatory, increasing gradually in length, so that the last pair equal or surpass the body in length. The segments of the pleon are united into a single vaulted segment, and its inferior surface is covered, in the females, by a single opercular plate, while in the males the operculum is composed of three parts, as in the preceding genera.

The generic description as given above is in part taken from Kröyer, the author of the genus. The specimens hitherto obtained do not appear to be separable from his species *M. Fabricii*, to which I have therefore referred them, although differing somewhat from each other. The material has unfortunately been, most of it, in poor condition, many of the specimens having been dried and much broken.

Munna Fabricii Kröyer.

Munna Fabricii Kröyer, Nat. Hist. Tidssk., II, B. ii, p. 380, 1847; Voy. en Scand., Crust., pl. xxxi, figs. 1 *a-q*. "1849".
Reinhardt, Grönlands Krebsdyr., p. 35, 1857.
M. Sars, Christ. Vid. Selsk. Forh., 1858, p. 154, 1859.
Lütken, Greenland Crust., p. 150, 1875.
Harger, Proc. U. S. Nat. Mus., 1879, vol. ii, p. 159, 1879.
Munna, species, Verrill, Am. Jour. Sci., III, vol. vii, p. 133, 1874; Proc. Am. Assoc., 1873, p. 371, 1874.
? *Munna Bœckii* G. O. Sars, Arch. Math. Nat., B. ii, p. 353 [253], 1877. (*M. Bœckii* Kröyer?)

PLATE III, FIG. 14.

This species may be at once distinguished from anything else known on our coast by the prominent, as if pedunculated, but immovable, eyes, on the posterior lateral angles of the large head, together with the elongated and slender ambulatory legs in seven pairs, the first pair only being somewhat shorter.

The first specimens obtained in a recognizable condition were small and differed somewhat from later specimens, especially in size and proportions; the differences, however, do not appear to be necessarily other than what might be due to age and size, and are such as are described by Kröyer in his specimens of *M. Fabricii*. The legs in the small specimen figured are considerably shorter than in larger specimens obtained in 1878, and the flagellum of the antennulæ consists in the small specimens of a single segment, or with traces of subdivision into two, while in the large specimens it is four-jointed, with a rudimentary terminal segment.

The body is in the female elongate oval, tapering posteriorly, and broadest at the third thoracic segment, where the breadth is equal to about half the length. The males are more slender, and are not dilated behind the head. The head forms about one-fifth of the total length, and is nearly twice as broad as long. Its anterior portion between the bases of the antennulæ and antennæ is comparatively narrow on its upper surface, and is rounded or obtusely angled in front. Behind the bases of the antennulæ it is suddenly much dilated at the sides, and a little behind the dilation are the prominent, strongly convex and laterally projecting eyes, immediately behind which the head contracts suddenly in width, and is then slightly rounded behind. The antennulæ arise in a deep sinus on the antero-lateral region of the head. They consist of a four-jointed peduncle followed by a four-jointed flagellum of about the same length as the peduncle. The basal antennular segment is stout, and subtrigonal in form; the second is more slender and cylindrical, while the third and fourth are subequal, quite short and small, together not over half as long as the second segment, and should perhaps rather be regarded as flagellar segments. The four flagellar segments are of a little less diameter than the last two peduncular segments, and are long and cylindrical, the fourth being tipped with a rudimentary segment bearing two strong terminal setæ. The antennæ are much larger and stouter than the antennulæ and are about two or three times as long as the body. They are composed of a five-jointed peduncle and a slender multiarticulate flagellum. They arise nearly in front of the antennulæ and their first three segments are short and stout, not longer taken together than the first two antennular segments. The fourth segment of the antennæ is only about half the diameter of the first three segments, but is greatly elongated, nearly or quite equaling in length the head and thorax taken together, and is cylindrical, and provided with a few short setæ, especially at the tip. The fifth, or last peduncular, segment is slightly more slender and elongated than the fourth, and is followed by a slender tapering flagellum composed of about seventy-five segments, or, perhaps, in perfect specimens, of a greater number. The maxillipeds are large and broad, as required by the large head, and are furnished with a five-jointed palpus, with the basal segment short, the second and third flattened and expanded internally, where they are also ciliated; the fourth narrow; the fifth short, and both provided with scattered setæ, especially toward the tip.

The first thoracic segment is a little shorter than the second, which is about equal in length to the third and the fourth; the last three segments progressively decrease in length and width, and the seventh is somewhat concealed at the sides by the swollen base of the pleon. The basal segments of all the legs are much alike in form, and differ but little in size throughout. They are cylindrical or slightly clavate, the first pair perceptibly shorter and smaller than the second, from which they increase very slightly to the sixth, which is the largest, the seventh not being larger than the second. The legs disarticulate easily at the end of the basal segment, and in the specimens examined nearly all are broken off at this point. Beyond the basal segment the first pair are comparatively short, about half the length of the body. The ischium of the first pair is robust, and a little longer than the merus; the carpus is subtriangular and armed with strong short spines on its palmar margin; the propodus is about as long as the ischium, slightly swollen, and armed with a few spines; the dactylus is short and armed at the end with two stout curved claws, of which the outer is about twice the length of the inner; between the claws is a slender bristle. The second and following pairs of legs are much more elongated than the first pair, the elongation being principally in the carpus and propodus, and, in a less degree, in the ischium and merus, while the dactylus is comparatively but little elongated. In the second pair of legs the propodus is not longer than the carpus, but it becomes proportionally, as well as absolutely, longer in the following pairs until, in the sixth pair, it may be nearly or quite as long as the body and form about two-fifths the whole length of the leg. The dactyli are, in all the legs, comparatively short, often less than one-tenth the length of the propodus, and armed with two unequal claws, of which the longer is about two-thirds as long as the dactylus itself, and the shorter is more than half the length of the longer. In all the legs the ischium is armed with a few short curved spinules, and the elongated propodal segments are furnished with scattered, slender and elongated, straight spines, each with a minute bristle near the apex.

The pleon is remarkably swollen near the base, and is somewhat pear-shaped; posteriorly it is deep, and bears the uniarticulate uropods in shallow grooves near the end. On the upper surface are a few straight slender spines, and below it is covered in the females by an ovate, obtusely-pointed opercular plate, and in the males by a trifid operculum, the median portion being slender, with nearly parallel sides and a central suture, and the two lateral portions slender, semi-ovate and pointed behind. The pleon appears to be carried habitually, during life, flexed upward at a considerable angle.

The length of the specimen figured, by Mr. Emerton (pl. III, fig. 14), is 1.2^{mm}, breadth 0.7^{mm}; but specimens obtained in 1878 measure 3.1^{mm} in length, 1.5^{mm} in width, in the female, and 1.1^{mm} in the male. The pleon measures in length 1.1^{mm} and in width 0.8^{mm} in the larger individuals.

A single much mutilated specimen of this species was dredged in 12 fathoms, South Bay, Eastport!, in 1872, by the United States Fish Commission, and two more specimens, both females, were obtained on eel-grass in Casco Bay! in 1873. Five specimens were obtained adhering to dried specimens of *Acanella* from 150 fathoms, Western Bank!, in 1878, and a sixth, in 53 fathoms, on Brown's Bank!, in lat. 42° 50′ N., lon. 65° 10′ E., by Captain J. Q. Getchell, of the schooner 'Otis P. Lord,' in the same year. In 1879 a specimen was obtained adhering to *Acanthogorgia armata*, by Captain George A. Johnson and crew of the schooner 'Augusta H. Johnson,' on Western Bank!, in lat. 43° 15′ N., lon. 50° 20′ E., 200 fathoms. These specimens were, as has been mentioned, considerably larger than those at first obtained. Kröyer's specimens were from a depth of 50 fathoms, at Godthaab, Southern Greenland, and according to M. Sars the species is abundant on the coast of Finmark among Hydroids in the coralline zone. G. O. Sars records *M. Bœckii* Kröyer, which he regards as scarcely differing from this species, at the harbor of Reikjavik, Iceland.

Specimens examined.

Number.	Locality.	Fathoms.	Bottom.	When collected.	Received from—	Specimens.		Dry. Alc.
						No.	Sex.	
2144	Casco Bay, Me	....	Eel-grass	— —, 1873	U.S Fish Com.	2	♀	Alc.
1936	Bay of Fundy, Me ..	12	....	— —, 1872	.. do	1	♀	Alc.
	Brown's Bank	53	....	— —, 1878	Capt. J. Q. Getchell.	1	♀	Dry.
	Western Bank......	150	On *Acanella*	— —, 1878	....	5	♂ ♀	Dry.
	Western Bank......	200	On *Acanthogorgia armata*.	— — 1879	Capt. G. A. Johnson.	1	♀	Dry.

IV.—MUNNOPSIDÆ.

In this family the body consists of two more or less distinct divisions, the first consisting of the head and anterior four thoracic segments, and the second of the last three thoracic segments, and the pleon, which is consolidated into a single segment, convex above. The eyes are wanting. The antennulæ are much shorter and smaller than the antennæ, and have their basal segment lamelliform. The antennæ are much elongated, with a five-jointed peduncle, of which the first three segments are short and the last two elongated and tipped with a long multiarticulate flagellum. The maxillipeds have their basal segments flattened and operculiform, covering the other mouth parts, and furnished with a large external lamella and a five-jointed palpus. The first pair of legs are shorter than the three following pairs and imperfectly prehensile. The next three pairs are ambulatory and usually greatly elongated. The last three pairs of legs, or at least the fifth and sixth pairs, are different in form from the preceding, and fitted for swimming, with some of the distal segments flattened and provided with marginal cilia or spines. The pleopods are protected by a thickened opercular plate, and the uropods are short and simple or biramous. The incubatory pouch in the females is beneath the first four thoracic segments.

Of this family, two species have been found on the New England coast, and a third, from the Gulf of St. Lawrence, is here included. The specimens obtained have been mostly in poor condition, and one of these, belonging apparently to an undescribed species, is so imperfect that I have decided to await the collection of better specimens before attempting a specific description. In the family characters given above, as well as in the following generic and specific descriptions, I have availed myself largely of the admirable works of M. Sars and his son G. O. Sars, the distinguished Norwegian naturalists, to whom science is indebted for the discovery and characterization of the present group.

The *Munnopsidæ* of our coast may be easily recognized as belonging to the family by the structure of the last three pairs of thoracic legs, which are fitted for swimming by being more or less flattened and ciliated; the last pair, however, may return to the more normal type of leg, so that the fifth and sixth pairs only may be natatory. The three genera which appear to be represented are distinguished as follows: Body suddenly constricted and slender behind the fourth thoracic segment in *Munnopsis* (p. 329); pretty regularly oval in form, with three pairs of flattened natatory legs in *Eurycope* (p 38); suboval but deeply incised behind the fourth segment, in *Ilyarachna* (p. 40), in which genus the last pair of legs are scarcely at all flattened or ciliated.

Munnopsis M. Sars.

Munnopsis M. Sars, Christ. Vid. Selsk. Forh., 1860, p. 84, 1861; Christ-fjord Fauna, p. 70, 1868.

Anterior division of the body dilated, posterior suddenly much narrower and linear. Antennulæ with the basal segment large and flattened, the flagellum elongate and multiarticulate; antennæ very long and slender, many times longer than the body; the last two peduncular segments greatly elongated; the flagellum about equal in length to the peduncle; mandibles subtriangular, entire and acuminate at the apex, without a molar process; the palpus slender with the last segment thick at the base and curved in the form of a hook; penultimate segment of the maxilliped not dilated inwardly; last segment very narrow and linear. Four anterior thoracic segments excavated above, obtusely rounded at the sides; the three following subcylindrical with short acuminate lateral processes; first four pairs of thoracic legs six-jointed (beyond the coxal segment), the first pair short; the second pair not much longer, rather robust and subprehensile in the males; the two following pairs greatly elongated and very slender, many times longer than the body; but with the basis, ischium, and merus very short; last three pairs of legs natatory, all alike, six-jointed, being destitute of dactyli, with the last two segments, the carpus and propodus, foliaceous, margined with long, slender, delicately plumose setæ. Pleon elongate, much longer than broad; abdominal operculum large (nearly covering the whole under surface of the pleon), suboval, simple in the female, but consisting of three distinct segments in the male, one median and very slender, and two lateral, and furnished within with a peculiar curved organ, terminated behind with a much elongated seta; uropods slender uniramous.

Munnopsis typica M. Sars.

Munnopsis typica M. Sars, Chr. Vid. Selsk. Forh., 1860, p. 84, 1861; Christ. Fjord. Fauna, p. (70), pl. vi–vii. figs. 101–138, 1868; Chr. Vid. Selsk. Forh., 1868, p. 261, 1869.

G. O. Sars, Chr. Vid. Selsk. Forh., 1863, p. 206, 1864; Reise ved Kyst. af Christ., p. (5), 1866; Christ. Fjord Dybvands-fauna, p. (44), 1869; Chr. Vid. Selsk. Forh., 1872, p. 79, 1873; Arch. Math. Nat., B. ii, p. 353 [253], 1877.

Whiteaves, Ann. Mag. Nat. Hist., IV, vol. x, p. 347, 1872; Deep-sea Dredging, Gulf of St. Lawrence (1872), pp. 6, 15, 1873; Am. Jour. Sci., III, vol. vii, p. 213, 1874; Further Deep-sea Dredging, Gulf of St. Lawrence (1873), p. 15, 1874.

Buchholz, Zweite Deutsche Nordpolfahrt, Crust., p. 285, 1874.

Heller, Denksch. Acad. Wiss. Wien, B. xxxv, p. (14) 38, 1875.

Norman, Proc. Royal Soc., vol. xxv, p. 208, 1876.

Miers, Ann. Mag. Nat. Hist., IV, vol. xx, p. 65, 1877.

Harger, Proc. U. S. Nat. Mus., 1879, vol. ii, p. 159, 1879.

PLATE II, FIG. 11.

This species is easily recognized among the known Isopoda of our coast by the form of the body, which suddenly diminishes in diameter behind the fourth thoracic segment, so that the last three thoracic segments, bearing the ciliated, swimming legs, are only about half as broad as the anterior part of the body.

Anterior division of the body depressed, posterior subcylindrical; breadth of body less than half the length. Head small, with the length and breadth about equal, equaling the two anterior thoracic segments in length, but of much less breadth, truncate in front and without a rostrum, bearing near the posterior dorsal margin two minute conical tubercles. The eyes are wanting. The antennulæ in the female, when reflexed, extend to the third thoracic segment, in the male to the fourth, with the flagellum longer than the peduncle, pectinate or furnished with a longitudinal series of long setæ, multiarticulate; segments in the female, 23 to 28; in the male, 65 to 66. The antennæ are greatly elongate, about five times as long as the body, very slender; peduncle more than twice the length of the body, the last two peduncular segments beset with numerous short spinules, arranged in longitudinal rows; flagellum nearly as long as the peduncle, composed of about 130 segments. The external lamella (*l*) of the maxillipeds (pl. II, fig. 11 *b*) is narrowed in front with the external margin convex.

The four anterior thoracic segments are subequal, short, about five times broader than long; last three segments broader than long, less than half the width of the preceding segments, bearing near the anterior dorsal margin two small conical tubercles; pleon slightly longer than the three preceding segments together, but not narrower, forming somewhat more than one-fourth the length of the body, elongate-suboval, the breadth scarcely equaling half the length, with a median, rounded, dorsal crest, but little elevated, and bearing in front of this near the anterior margin a small conical tubercle.

Propodus shorter than the carpus in the first pair of legs, equal to it in length in the second pair, which in the males (pl. II, fig. 11 *e*) have the carpus thickened, and armed, on the inferior margin, with stronger spines than in the females; third and fourth pairs of legs about thrice the length of the body, with the three basal segments, basis, ischium, and merus, very short and robust; the last three very much elongated and filiform; the propodus longer than the carpus, both armed with many short spinules arranged longitudinally; dactylus about one-fifth as long as the propodus, slightly curved, naked, very minutely serrulate along the convex margin. Last three legs (pl. II, fig. 11 *f*) with the carpus and propodus elongate-subelliptic, both segments strongly ciliated, the propodus a little shorter than the carpus.

Abdominal operculum in the female (pl. II, fig. 11 *g*) with a longitudinal, elevated, acute median crest, flattened medially in the males. Uropods slightly more than one-third the length of the pleon, composed of two subequal segments. Laminæ of the incubatory pouch in the females attached to the anterior four thoracic segments; the three posterior pairs large; the third and fourth suborbicular; the second elongate; the first much smaller, bifid at the apex.

Length 8–10^{mm}; antennæ 40–50^{mm}; third and fourth pairs of legs 24–30^{mm}. Color, light yellowish, or grayish, in alcohol; lighter below.

The specimens that I have had an opportunity of examining were all more or less imperfect, and I have therefore, in both the generic and specific descriptions given above, made free use of the admirable and exhaustive description of this genus and species by M. Sars,* and the figures of the species on plate II were copied from the same author, having been drawn by his not less distinguished son, G. O. Sars.

This species like its allies is an inhabitant of deep water on muddy bottoms. Three specimens, the only ones that I have personally examined, were taken by the Fish Commission in the Bay of Fundy! between Head Harbor and the Wolves, in 60 fathoms muddy bottom, August 16, 1872. It has been dredged by Mr. Whiteaves in the Gulf of St. Lawrence in 125 to 220 fathoms; by the Valorous Expedition in Baffin Bay in 100 fathoms (Norman); in 25 to 50 fathoms off Cape Napoleon, Grinnell Land, by the Arctic Expedition (Miers); between Norway and Iceland in from 220 to 417 fathoms; Christiania fiord, 200 to 230 fathoms (G. O. Sars); Christiania Sound 50 to 60 fathoms,

* Bidrag til Kundskab om Christiania-fjordens Fauna, 1868, pp. 79–95, pls. vi–vii. (Nyt Magazin.)

whence the species was described by M. Sars; off Storeggen, 400 fathoms (G. O. Sars), and northward among the Loffoden Islands, 250 fathoms; the coast of Finmark, Spitzbergen (Buchholz), and the Arctic Ocean about Nova Zembla (G. O. Sars.)

Eurycope G. O. Sars.

Eurycope G. O. Sars, Chr. Vid. Selsk. Forh., 1863, p. 208, 1864.

Body depressed, subovate as seen from above; about equally attenuated before and behind. Head of medium size, more or less produced between the antennulæ; antennæ very slender, two to four times as long as the body; flagellum longer than the peduncle; mandibles robust, quadridentate at the apex, and bearing below a series of rigid setæ and a strong molar process; mandibular palpus well developed, with the terminal segment enlarged at its base and curved. Four anterior thoracic segments subequal, short; three posterior segments large not suddenly narrower than the anterior segments; the first pair of legs shorter than the next three, with the dactylus short; the next three pairs elongated, and with elongated and slender dactyli; three posterior pairs of legs distinctly natatory, with the carpus and propodus strongly flattened and provided with numerous plumose marginal setæ; dactylus of the ordinary form. Pleon rather large, broader than long, obtusely rounded behind; operculum subpentagonal with rounded angles, much smaller than the pleon. Uropods short, biramous, rami uniarticulate. Dorsal surface of the body smooth and shining.

For the characterization of the genus, as given above, I have depended largely upon the work of G. O. Sars, having had myself, for examination, only the following species:

Eurycope robusta Harger.

Eurycope robusta Harger, Am. Jour. Sci., III, vol. xv, p. 375, 1878; Proc. U. S. Nat. Mus., 1879, vol. ii, p. 159, 1879.

PLATE III, FIG. 15.

This species may be recognized by the flattened and ciliated swimming legs, in three pairs, on the last three thoracic segments, which are not, as in the preceding species, suddenly of much less diameter than the anterior four segments.

Body oval with the length equal to, or slightly exceeding, twice the breadth. Head, behind the bases of the antennulæ, longer than the first thoracic segment, produced medially into a short rostrum about half as long as the basal antennular segment. Antennulæ (pl. III, fig. 15 *a*) attaining the middle of the fourth segment of the antennæ in the females, surpassing the middle of this segment in the males; basal segment subquadrate, spinulose at the distal angles, somewhat narrowed from the base, bearing the second much smaller segment a little beyond the middle of its superior surface; third segment longer and more slender than the second; flagellum of more than twenty articulations, which become indistinct near the base, and are furnished with terminal setæ. Antennæ about thrice the length of the body in the female, somewhat shorter in the male, the sexes differing in the fourth and fifth segments, which, in the females, are subequal in length and, together, as long as the body, while in the male the fifth is shorter than the fourth, and the two segments together are about two-thirds as long as the body. The flagellum is long, slender, and multiarticulate. Maxillipeds (pl. III, fig. 15 *b*) with the external lamella sub-rhombic, emarginate on the exterior distal side; palpus five-jointed, first segment short, produced externally into a very acute angle; second and third segments broad and flattened; fourth narrow with the inner angle produced and rounded; fifth short, oval. Maxillæ of the ordinary form, outer pair with slender lobes. Mandibular palpus elongated, last segment strongly curved.

Thorax widest at the fourth segment; first four segments forming about one-third its length on the median line, last segment longest, all with their antero-lateral angles produced, the anterior four with the epimera projecting as an acute process below, and in front of, the angle. First pair of legs (pl. III, fig. 15 *d* and *d'*) about three-fourths the length of the body; dactylus short; propodus shorter than the carpus; slightly hairy, especially on the propodus with slender hairs. Next three pairs of legs longer than the body, subequal, but increasing a little in length to the fourth; dactyli slender and acicular; propodi and carpi subequal, spinulose along their inner margins in the second pair, but not in the third and fourth. Last three pairs of legs with the carpus strongly dilated and flattened, subcircular as seen in pl. III, fig. 15 *f*, where the sixth pair is represented; propodus also much flattened and dilated; both segments strongly ciliated with plumose bristles, as is also the ischium, or second segment along the outer dilated margin; dactylus about half the length of the propodus instead of less than one-third its length, as in *E. cornuta* G. O. Sars, the species most resembling the present.

Pleon much broader than long, broadly rounded behind. Operculum also broader than long, strongly roof-shaped. Uropods (pl. III, fig. 15 *g*) with the basal segment shorter than the rami, which are uniarticulate, cylindrical, of equal length, obtuse and tipped with a coronet of short spines. The inner ramus is more robust, but not longer than, the outer.

Color in alcohol, honey yellow; length 4.5^{mm}; breadth 2.2^{mm}.

This species appears to approach *E. cornuta* G. O. Sars,* but may be readily distinguished by its greater size, by the shortness of the rostrum, the equal rami of the uropods, and the shape of the external lamella of the maxillipeds, which he describes in that species as "versus apicem dilatata et emarginata utrinque acute producta." In the third and fourth pairs of legs, moreover, the carpus and propodus are not armed with spines as in that species according to Sars' description.

* Chr. Vid. Selsk. Forh., 1863, p. 209, 1864.

This species was dredged by Mr. J. F. Whiteaves in the Gulf of St. Lawrence! at a depth of 220 fathoms muddy bottom, and has not yet been found on the coast of New England. It is introduced here from the probability that it will yet be discovered in the deeper parts of the Bay of Fundy, where the allied *Munnopsis typica* M. Sars has already been found, or even in the Gulf of Maine.

Specimens examined.

Number.	Locality.	Fathoms.	Bottom.	When collected.	Received from—	Specimens.		Dry. Alc.
						No.	Sex.	
1938	Gulf of St. Lawrence	220	Mud		J.F.Whiteaves	10	♂ ♀	Alc.
1939	Gulf of St. Lawrence	220	....do		....do	8	......	Alc.

Ilyarachna G. O. Sars.

Mesostenus G. O. Sars, Chr. Vid. Selsk. Forh., 1863, p. 211, 1864.
Ilyarachna G. O. Sars, Christ. Fjord. Dybvands-fauna, p. (44), 1869.

Body scarcely depressed, subpyriform as seen from above, narrowed behind; its anterior division separated from the posterior by a deep constriction. The head is large and broad and without a rostrum. Antennulæ short, with a flagellum composed of but few segments. Antennæ exceeding the body in length, with a multiarticulate flagellum. Mandibles short and strong, entire at the apex; molar process armed with a few setiform spines; palpus either small and three-jointed or wanting. Four anterior thoracic segments short, excavated above and furnished with lateral processes directed forward; the three following convex above and destitute of lateral processes; the antepenultimate scarcely narrower than the anterior segments and deeply emarginate behind. First pair of legs nearly as in the preceding genus; second pair unlike the others and usually more robust; the following two subequal and commonly much elongated; fifth and sixth pairs of legs much as in *Eurycope;* the last pair unlike the preceding, long and slender, with the segments scarcely flattened, and armed with a long curved claw. Pleon narrowly triangular, pointed at the apex. Abdominal operculum large, covering nearly the whole of the under surface of the pleon, provided with a median crest and numerous marginal setæ. Uropods simple, appressed to the pleon.

For the generic description given above I have depended almost entirely upon the work of Dr. G. O. Sars, who originally described the genus under the name *Mesostenus.* That name being preoccupied he subsequently changed it to *Ilyarachna.*

Ilyarachna species.

A single imperfect specimen of a species apparently belonging to this genus was dredged in 106 fathoms, gray mud, 21 miles east of Cape Cod Light!, September 18, 1879. The species is probably yet undescribed, but, in view of the very imperfect condition of the only specimen yet known, I have decided to await the collection of better specimens before attempting to make out its characters. It may yet be found to represent an undescribed genus, but I am at present inclined to regard it as a species of *Ilyarachna.*

V.—IDOTEIDÆ.

Antennulæ consisting of four segments, of which the basal is more or less enlarged and the terminal clavate; mandibles not palpigerous; thoracic segments subequal in length; pleon with more or fewer of its segments consolidated into a large, scutiform, terminal piece; uropods inferior, transformed into a two-valved operculum protecting the pleopods.

The *Idoteidæ* are represented on the New England coast by ten species; another, found near our northern limits, is included, making eleven in all, belonging to five genera. The family may be further characterized, so far as regards our species, as follows: The body is depressed, and varies in its proportions of length to breadth from about two to one in *Chiridotea cœca* to nearly six to one in *Erichsonia attenuata.* The head is quadrate in outline, except in *Chiridotea.* The eyes are present and usually lateral, but may not be conspicuous. The antennulæ are four-jointed and similar in form throughout the family; they may or may not surpass the head in length, but are usually short and small. The basal segment of the antennulæ is more or less enlarged and usually subquadrate; the second segment is clavate; the third longer and less distinctly clavate; the fourth, or terminal, segment, corresponding with the flagellum of the antennulæ, is nearly straight along its outer, or in the natural position posterior, margin, while the opposite margin is gently curved from near the base, and rounds over more sharply at the tip; along this margin, especially toward the tip, are tufts of short setæ at regular intervals, indicating an approach toward segmentation. The antennæ have a five-jointed peduncle, varying little in form throughout the family; the first of these segments is short; the second is much larger and deeply notched on its under side; the third, fourth, and fifth segments are longer, but more slender and cylindrical or somewhat clavate. The flagellum of the antennæ may be articulated with many or few segments; it may consist of a single segment, or may be rudimentary. The maxillipeds are operculiform and cover the other parts of the mouth below. They consist, on each side, of a large semi-oval plate, with a straight interior margin, meeting its fellow of the opposite side, and bearing on this margin a short, curved, styliform organ. They are provided at the tip with stout pectinate setæ, and along the basal portion of the outer margin lies, on each side, the large external lamella. The palpi of the maxillipeds are flattened and ciliated along their inner margins, and the number of segments may be reduced to three by the coalescence of the last two and of the preceding two. The maxillæ vary but little in the family; the second or outer pair bear as usual three delicate ciliated plates; the first or inner pair are armed with stouter setæ and spines. The mandibles are robust, acutely toothed at the apex, armed with a more or less powerful molar process, and are destitute of palpi.

The thoracic segments are distinct and subequal in length, but may differ considerably in width, and are not united with the head nor with the pleon. The legs, except in the genus *Chiridotea*, are nearly similar in form throughout, and, in the first three pairs at least, are terminated by a prehensile or subprehensile hand, formed by the more or less complete flexion of the dactylus upon the propodus. The first pair of legs is usually shortest and has a triangular carpus. The anterior three pairs of legs are, in general, directed forward, and the posterior four pairs are directed backward and are less perfectly, or not at all, prehensile, a distinction that reaches its highest development in *Chiridotea*. The seventh pair of legs are absent in the young taken from the incubatory pouch, and do not generally attain quite as large size as the sixth pair.

The pleon, seen from above, consists in great part, or entirely, of a large, convex, usually pointed, scutiform piece, representing the consolidated terminal segments. As many as four of the anterior segments may, however, be more or less completely separated by articulations or indicated by lateral incisions or sutural lines. Underneath, the pleon is provided with a structure peculiar to and characteristic of this family, and the next, viz, a two-valved operculum, formed by the specially modified uropods,* or appendages of the terminal segment, closing like a pair of cupboard doors and protecting the delicate pleopods, which are lodged in a vaulted chamber excavated in the under surface of the pleon. This operculum consists, on each side, of an elongated basal plate, often strongly vaulted, angulated externally near the base, where it is articulated with the terminal segment of the pleon, and bearing at the tip one, or sometimes two, small lamellæ. One of these lamellæ usually disappears, but two are present in *Chiridotea*, as also in the foreign genera *Cleantis* and *Chætilia*. When both are present the opercular plates differ only in proportion from the ordinary form of uropods, consisting of a basal segment and two rami. Within the cavity enclosed by the opercular plates lie the usual five pairs of pleopods, each consisting of a basal segment

* In the last edition of the Encyclopædia Britannica (vol. vi, p. 641), these organs are described as the "anterior" abdominal appendages. They are anterior only in position, being in fact the appendages of the posterior segment.

supporting two lamellæ, and two or more of the anterior pairs are ciliated with fine plumose hairs. The inner lamella of the second pair of pleopods bears, in the adult males, a slender style articulated near the base of the inner margin and varying in length and structure in the different genera and species. The pleopods, besides their branchial office, are also of importance in locomotion, being used for swimming, which is a frequent mode of progression in this family, and is often performed with the back downward.

The females are usually broader than the males and carry their eggs and young in a pouch, on the under surface of the thorax, formed of four pairs of plates, attached to the coxal segments of the second, third, fourth, and fifth pairs of legs, and overlapping along the median line.

The known Isopoda of this family on the coast may be most easily recognized by the presence, underneath the pleon, of a two-valved operculum, opening like a pair of cupboard doors, and by the first three pairs of legs being more or less prehensile. Our genera may be distinguished by means of the following table:

Flagellum of the antennæ	articulated, legs	dissimilar, last four pairs not prehensile		CHIRIDOTEA, p 337
		alike, prehensile; epimera	evident above	IDOTEA, p. 341
			not evident above	SYNIDOTEA, p. 350
	not articulated, clavate			ERICHSONIA, p 354
	short and rudimentary			EPELYS, p. 357

Chiridotea Harger.

Chiridotea Harger, Am. Jour. Sci., III, vol. xv, p. 374, 1878.

First three pairs of legs terminated by prehensile hands, in each of which the carpus is short and triangular, the propodus is robust and the dactylus is capable of complete flexion on the propodus; antennæ with an articulated flagellum; head dilated laterally; abdominal operculum vaulted, with two apical plates.

The two species of this genus found on our coast agree further in the following particulars: The body is short, the length being only about twice the breadth, and the outline of the head and thorax together is subcircular. The anterior part of the lateral margin of the head is produced and deeply lobed, the eyes thus appearing dorsal instead of lateral; posteriorly the head is deeply received into the first thoracic segment. The antennulæ are proportionally large, equaling or surpassing the peduncle of the antennæ. The external lamella of the maxillipeds (see pl. IV, figs. 18 and 21) is large and broad and the palpus consists of only three segments, of which, however, the last two are each composed of two coalesced segments, that are separate in the European *Ch. entomon.* Of the two segments thus formed, the terminal is quadrate or rhomboid in outline, with rounded angles and is smaller than the preceding, which expands distally toward the articulation between the two.

The thorax is deeply excavated, in front for the head and behind for the abdomen, so that the thoracic segments are much longer at the sides than along the back, when measured parallel with the axis of the animal. The epimera are separated by sutures, except in the first segment, and have their posterior angles acute. The first three pairs of legs have the dactylus capable of complete flexion upon the propodus, which is more or less swollen and supported by the short triangular carpus. In the last four pairs of legs the three corresponding segments are nearly cylindrical and the dactylus is incapable of complete flexion on the propodus.

The pleòn, or abdomen, is convex throughout and pointed at the tip, and is composed, apparently, of five segments, of which the first three are separated by complete sutures, but the last two are united in the dorsal region, the sutures separating them being visible only at the sides. The opercular plates consist, on each side, of an elongated, vaulted, and attenuated plate, regularly rounded at the anterior end, truncate at the apex, and bearing just within the apex, on the inner side of the organ when closed, two ciliated, ovate or triangular plates. Of these the internal plate, or the one next the median line is much smaller than the outer; the outer also overlaps the inner, a disposition similar to that which prevails in the branchial plates or pleopods. The basal plate of the operculum is ciliated along its anterior and inner margin with bristles, which are plumose except in the region nearly opposite the articulation of the plate, where they become stouter and spine-like. The stylet on the second pair of pleopods in the males is long and slender, more than twice the length of the lamella to which it is attached.

Chiridotea cœca Harger (Say).

Idotea cœca Say, Jour. Acad. Nat. Sci. Phil., vol. i, p. 424, 1818.
Hitchcock, Rep. Geol. Mass., p. 564, 1833. (*I. cœca?*)
Gould, Rep. Geol. Mass., 2d ed., p. 549, 1835; Invert. Mass., p. 337, 1841.
Edwards, Hist. nat. des Crust., tom. iii, p. 131, 1840.
Guérin, Iconog., Crust., p. 35, 1843.
Dekay, Zool. New York, Crust., p. 42, 1844.
White, List Crust. Brit. Mus., p. 94, 1847.
Verrill, This Report, part i, p. 340 (46), 1874.
Harger, This Report, part i, p. 569 (275); pl. v, fig. 22, 1874.
Chiridotea cœca Harger, Am. Jour. Sci., III, vol. xv, p. 374, 1878; Proc. U. S. Nat. Mus., 1879, vol. ii, p. 159, 1879.

PLATE IV, FIGS. 16–19.

This species is at once distinguished from the following by its larger size and short antennæ, which surpass the antennulæ but little, if at all. Among the other known Isopoda of the New England coast, it may be recognized by the broad, subcircular thorax, joined with an articulated flagellum of the antennæ and a two-valved abdominal operculum. The eyes are, moreover, light-colored and inconspicuous, whence the name.

The head is but slightly excavated in front for the bases of the antennæ, and there is a more or less open notch at the sides extending nearly to the eyes. The antennulæ (pl. IV, fig. 17 *a*) are longer than the peduncle of the antennæ and have the second segment strongly clavate; the third cylindrical; the last with about a dozen tufts of short setæ; the peduncular segments are bristly, as are also those of the antennæ. The first segment of the antennæ (pl. IV, fig. 17 *b*) is very short, the second about three times as long, longer than any of the following segments; the third is longer and more slender than the fourth, which is nearly as broad as long; the fifth, or last peduncular, segment is more slender than any of the preceding, slightly clavate, about twice as long as broad, and longer than any except the second. The flagellum slightly exceeds the last two peduncular segments in length and consists usually of about seven segments, each bearing a tuft of short hairs near its extremity, except the first, which is much the longest, bears two such tufts, and is, apparently, composed of two segments united.

The breadth of the thorax is greater than its length along the median line. The first pair of legs (pl. IV, fig. 18 *b*) are a little shorter than the next two pairs, and the propodus or penultimate segment is a little more swollen. The carpus becomes slightly more elongated in the next two pairs. The last four pairs of legs are alike in form and increase in size to the sixth pair, which is the largest. The legs are bristly hairy, especially on the ischial, meral, and carpal segments, where they are provided with stout setæ curved at the tip. The basal segments bear longer and more slender plumose hairs. The epimera are ciliated on their external margins as are the lateral borders of the head and first thoracic segment and the tip of the pleon.

The operculum (pl. IV, fig. 18 *c*) is also ciliated with very fine hairs along its postero-external margin; the larger of the apical plates is broader than in the following species, the width being to the length as 6 to 10. The stylet on the second pair of pleopods in the male (pl. IV, fig. 19 *b*) considerably surpasses the cilia and is curved and acute at the tip. Adult males and females seem to be comparatively rare, and a common form of the second pair of pleopods (pl. IV, fig. 19 *a*) presents an acute stylet, imperfectly separated from the lamella and but slightly surpassing it in length, strongly ciliated like the lamella on its margin.

Length 12–15^{mm}; breadth 6–8^{mm}. The color in life is variable but usually dark grayish, much like the wet sand in or on which it is commonly found. It may be more particularly described as usually of a dark leaden gray on the top of the thorax, sometimes with a central spot, which may be bright pea-green, probably from the contents of the digestive cavity showing through. This dark color is continued in an arrow-shaped, or halberd-shaped, spot occupying most of the upper surface of the head. At the sides of the head and body is a mottling of light yellowish gray, darker again on the edge. The under surface of the body and the legs are pale and generally uniform in color. In alcohol the colors usually fade to a uniform straw color, with fine blackish dots, which are less conspicuous in life.

According to Say this species extends as far south as Florida. It is common on sandy beaches at many localities on the coast of New England, as at New Haven! and other localities on Long Island Sound!,

Vineyard Sound!, Nantucket!, Provincetown!, and Nahant, Mass.! It appears to be very rare, or perhaps does not occur in the northern part of the Gulf of Maine, where it is replaced by the next species; it reappears, however, on the coast of Nova Scotia, having been collected at low water by the U. S. Fish Commission in 1877, at Halifax!. It is usually found on sand below high tide, or burrowing just under the surface, but also swims with facility.

Specimens examined.

Number.	Locality.	Fathoms.	Bottom.	When collected.	Received from—	Specimens.		Dry. Alc.
						No.	Sex.	
	New Haven		Sand			00	♂ ♀	Alc.
1944	Vineyard Sound, Mass		do	—— —, 1871	U.S. Fish Com.			Alc.
1945	Off Nantucket			Sept. 8, 1875	do	1		Alc.
1946	Provincetown, Mass.		Sand	—— —, 1872	do	1		Alc.
1947	Nahant, Mass	L. w.			A. E. Verrill	3		Alc.
1948	Halifax, N S	L. w.		—— —, 1877	U.S. Fish Com.			Alc.

Chiridotea Tuftsii Harger (Stimpson).

Idotea Tuftsii Stimpson, Mar. Inv. G. Manan, p. 39, 1853.

Verrill, Proc. Am. Assoc., 1873, p. 362, 1874; This Report, part i, p. 340 (46), 1874.

Harger, This Report, part i, p. 569 (275), 1874.

Chiridotea Tuftsii Harger, Am. Jour. Sci., III, vol. xv, p. 374, 1878; Proc. U. S. Nat. Mus., 1879, vol. ii, p. 159, 1879.

PLATES IV AND V, FIGS. 20–23.

This species is distinguished from the preceding by its smaller size and longer antennæ, which are about twice as long as the antennulæ and bear a slender flagellum. The eyes are also more conspicuous than in *Ch. cœca.*

The head is excavated in front above the bases of the antennæ; and the incision in the produced lateral margin is nearly closed by the overlapping of the anterior lobe. The antennulæ (pl. V, fig. 23 *a*) are slender and do not surpass the peduncle of the antennæ, the second segment as well as the third is cylindrical, and the last segment bears about nine tufts of short hairs; the peduncular segments bear also a few bristles. The antennæ (pl. V, fig. 23 *b*) have the first segment short; the second, third and fourth about equal in length and more than twice as long as the first; the fifth as long as the third and fourth together, but more slender and cylindrical; the flagellum longer than the peduncle, composed of about twelve segments and tapering from the base. The maxillipeds (pl. IV, fig. 21) have the external lamella (*e*) longer than broad.

The first pair of legs (pl. V, fig. 23 *c*) are somewhat less robust than in *Ch. cœca.* They are a little shorter than the second and third pairs, and have a much more robust hand. The fourth and succeeding pairs of legs (pl. V, fig. 23 *d*) are much as in the preceding species but less spiny and with a greater proportion of plumose hairs.

The external apical plate of the operculum (pl. V, fig. 23 *e*) is slender and twice as long as broad. The stylet on the second pair of pleopods in the males (pl. IV, fig. 22 *s*) does not surpass the cilia, is dilated towards the tip and obtusely pointed.

Length 9^{mm}; breadth 4.5^{mm}. The color is usually light reddish brown, speckled with darker, or marked with dark transverse patches, or bands. A specimen obtained during the summer of 1879, from a clear sandy bottom in 17 fathoms, Stellwagen's Bank, is thus described from life by Professor Verrill: "Color whitish, more or less speckled with salmon on the sides above, the specks more regular and distinct on the head, some lines and specks of flake-white on the middle of the back above the greenish stomach; base of telson salmon brown, its posterior half white; legs marked with salmon."

Dr. Stimpson's specimen "was dredged on a sandy bottom in 10 fathoms off Cheney's Head" in the Bay of Fundy. It occurs in Long Island Sound, where a specimen was taken by Dr. T. M. Prudden off New London! in 1872. The species was, however, considered rare on the coast until 1878, when it was taken in considerable abundance in Gloucester Harbor,! Massachusetts Bay, in seven to eight and a half fathoms, sand and red algæ. It has also been collected at Casco Bay,! Maine, in 1873; at low water in Prince's Cove,! Eastport, in the Bay of Fundy, in 1872, and at Halifax, N. S.,! in 18 to 25 fathoms, sand, September 5, 1877; a single specimen in each case. Three additional specimens were obtained in 1879, as detailed below.

Specimens examined.

Number.	Locality.	Fathoms.	Bottom.	When collected.	Received from—	Specimens. No.	Specimens. Sex.	Dry. Alc.
1953	Off New London....	....	....	—— —, 1872	T. M Prudden	1	♂	Alc.
	Gloucester Harbor, Massachusetts Bay	8½	Sand............	—— —, 1878	U.S Fish Com	10	......	Alc.
	....do	7½	....do	—— —, 1878	...do	00	......	Alc.
	Stellwagen's Bank..	17	Coarse sand.....	Sept. 6, 1879	... do	1	♀	Alc.
	Off Boston Harbor..	16	Speckled sand ..	Sept. 13, 1879	....do	2	♂ ♀	Alc.
803	Casco Bay, Me	....	Sand............	July 12, 1873	....do	1	......	Alc.
1952	Bay Fundy, Prince's Cove.	L. w.	....do	—— —, 1872	....do	1	♀	Alc.
1951	Halifax, outer harbor.	18-25	....do	Sept. 5, 1877	....do	1	......	Alc.

Idotea Fabricius.

Idotea Fabricius, Suppl. Ent. Syst., p. 297, 1798.

Flagellum of the antennæ articulated; legs all terminated by a prehensile hand; epimeral sutures evident above except in the first thoracic segment; pleon composed apparently of four segments, of which the last two are consolidated in the dorsal region; operculum with a single apical plate.

The species to which I propose to limit the name *Idotea** may be briefly characterized as above, and, of these, the three found on our coast agree further as follows: The body is elongated, its length being from three to four times its breadth, and the sides are nearly parallel. The head is quadrate and not produced at the sides. The eyes are lateral. The antennulæ are small and short, hardly surpassing the third segment of the antennæ. The basal segment of the antennæ is very short; the second segment much larger and deeply incised on its under surface; the third, fourth and fifth segments increase in length but decrease in diameter; the flagellum is more or less distinctly articulated, the number of articulations increasing with age. The palpus of the maxillipeds is four-jointed, the last segment being composed of two segments united, as is indicated by a notch near the tip.

The thorax is moderately arched, with the sides but little dilated in the males, somewhat more so in the females. The epimera are conspicuous and separated from their segments by a suture above, except in the first segment, but may not occupy its entire lateral margin. The legs differ but little in form throughout, being all more or less perfectly prehensile, but in the first pair only is the carpus triangular.

The pleon or abdomen appears, when seen from above, to consist of four segments, of which the first two are separated by complete sutures, but the third and fourth by sutures at the sides only. The uropods, forming the abdominal operculum, consist on each side of a flattened, elongated plate, with the anterior end rounded, the sides nearly parallel for most, or all, of its length and bearing at its truncated apex a much shorter more or less tapering or triangular plate. Neither of these plates is strongly ciliated in our species, but a stout, densely plumose bristle springs from the basal plate, on the inside, near the outer end of the articulation between the two plates. The stylet on the second pair of pleopods of the males is not elongated and may not surpass the lamella to which it is attached. The incubatory pouch is conspicuous in the females.

Our representatives of this genus may be recognized among the other known Isopoda of the coast by the following characters: The pleon appears to consist of four segments, the first three short and the third united, in the dorsal region, to the large, more or less vaulted, terminal segment; underneath the pleon is the conspicuous two-valved operculum and, in the antennæ, the flagellum consists of several segments. The three species may be distinguished by the form of the tip of the pleon, which is more or less tridentate in *I. irrorata* (p. 343), pointed in *I. phosphorea* (p. 347), and truncate in *I. robusta* (p. 349).

* The orthography adopted is that of Fabricius, the author of the genus.

Idotea irrorata Edwards (Say).

Idotea entomon Leach, Edinb. Encyc., vol. vii, (Am. ed., p. 243, pl. ccxxi, fig. 7), "1813–14"; Trans. Linn. Soc., vol. xi, p. 364, 1815 (*not Oniscus entomon.* Linné.)
Templeton, Loud. Mag. Nat. Hist., vol. ix, p. 92, 1836.
Moore, Charlesworth's Mag. Nat. Hist., vol. iii, p. 294, 1839.
Stenosoma irrorata Say, Jour. Acad. Nat. Sci., vol. i, pp. 423, 444, 1818.
Hitchcock, Rep. Geol. Mass., p. 564, 1833.
Gould, Rep. Geol. Mass., 2 ed., p. 549, 1835 ; Invert. Mass., p. 338, 1841.
Dekay, Zool. New York, Crust., p. 43, pl. ix, fig. 42, 1844.
Idotea tricuspidata Desmarest, Dict. des Sci. nat., tom. xxviii, p. 373, pl. 46, fig. 11, 1823; Consid. Crust., p. 289, pl. 46, fig. 11, 1825.
"Roux, Crust. Medit., t. 29, f. 11, 12," (B. & W.)
Latreille, Regne Anim., t. iv, p. 139, 1829.
Gould, Rep. Geol. Mass., 2 ed., p. 549, 1835 (*tricuspidata ?*).
Edwards, Hist. nat. des Crust., tom. iii, p. 129, 1840.
Œrsted, Naturhist. Tidssk., B. iii, p. 561, 1841.
Zaddach, Crust. Pruss. Prod., p. 10, "1844."
Lucas, Expl. Algérie, tom. i, p. 60, 1849.
White, List Crust. Brit. Mus., p. 94, 1847; Brit. Crust. Brit. Mus., p. 65, 1850; Pop. Hist. Brit. Crust., p. 223, pl. 12, fig. 2, 1857.
Hope, Cat. Crost. Ital., p. 26, 1851.
Lilljeborg, Öfvers. Vet.-Acad. Förh., Årg. 9, p. 11, 1852 (*Idothea*).
M. Sars, Chr. Vid. Selsk. Forh., 1858, p. 151, 1859 (*Idothea.*)
Bate, Rep. Brit. Assoc., 1860, p. 225, 1861.
Norman, Nat. Hist. Trans. Northumb., vol. i, p. 25, 1865; Rep. Brit. Assoc., 1866, p. 197, 1867; op. cit., 1868, p. 289, 1869.
G. O. Sars, Reise ved Kyst. af Christ., 1865, p. (28), 1866 (*Idothea*).
Heller, Verh. zool.-bot. Ges. Wien, B. xvi, p. 728, 1866 (*Idothea*).
Marcusen, Arch. Naturges., Jahrgang xxxiii, B. 1, p. 360, 1867.
Bate and Westwood, Brit. Sess. Crust., vol. ii, p. 379, figure, 1868.
"Sænger, Fauna of Baltic, Imp. Soc. Nat. Sc. Mosc., viii, 1869."
"Münter und Buchholz, Carcin. Fauna Deutschlands, 1869."
Czerniavski, Zoog. Pont. Comp., pp. 83, 129, "1870."
Metzger, J. B. Naturhist. Ges. Hannover, vol. xx, p. 32, 1871; Nordseefahrt der Pomm., 1872–'73, p. 285, 1875.
Möbius, Die Wirbellosen Thiere der Ostsee, p. 121, 1873. Ann. Mag. Nat. Hist., IV, vol. xii, p. 85, 1873.
Parfitt, Trans. Devon. Assoc., Sess. Crust., p. (19), 1873.
Bos, Bijd. ken. Crust. Hed. Nederl., pp. 34, 67, 1874.
M'Intosh, Ann. Mag. Nat. Hist., IV, vol. xiv, p. 273, 1874.
Stebbing, Jour. Linn. Soc., vol. xii, p. 148, 1874.
Catta, Ann. Sci. nat., Zool., VI, tome iii, p. 30, 1876.
Stalio, Cat. Crost. Adriatic, p. 206, 1877.
Lenz, Wirbellos. Thiere, Trave. Bucht, p. 15, 1878.
Idotea Basteri Audouin, Descr. Savigny's Egypt, Crust., pl. 12, fig. 6, "1830."
Guerin. Iconog., Crust., p. 32, pl. xxxi, fig. 1, 1829–43.
"Roux, Crust. Mediterr., t. 29, f. 1–10," 1830 (B. & W.).
"Rathke, Fauna der Krimm, p. 380," 1830 (Edw.).
"*Idotea variegata* Roux, Crust. Mediterr., pl. 30, fig. 1–9," 1830 (B. & W.).
Idotea (*pelagica ?*) Latreille, Cours d'Ent., Atlas, p. 12, pl. xviii, figs. 20–30, 1831.
"*Armida bimarginata* Risso, Hist. nat. Eur. merid., 5, 109" (B. & W.).
Idotea irrorata Edwards, Hist. nat. des Crust., tome iii, p. 132, 1840.
White, List Crust. Brit. Mus., p. 94, 1847.
Stimpson, Mar. Inv. G. Manan, p. 39, 1853.
Leidy, Jour. Acad. Nat. Sci. Phil., II, vol. iii., p. 150, 1855.

Idotea irrorata—Continued.

Harger, This Report, part i, p. 569 (275), pl. v, fig. 23, 1874; Proc. U. S. Nat. Mus., 1879, vol. ii, p. 160, 1879.

Verrill, Am Jour. Sci., III, vol. vii. pp. 131, 135, 1874; Proc. Amer. Assoc., 1873, pp. 369, 371, 373, 1874; This Report, part i, p. 316 (22), 1874.

Whiteaves, Am. Jour. Sci., III, vol. vii, p. 217, 1874; Further Deep-sea Dredging, Gulf of St. Lawrence, p. 15, "1874."

Idothea tridentata Rathke, Fauna Norw., Nov. Act. Acad., B. xx, p. 21, 1843 (*I. tridentata* Latreille?).

Grube, Ausflug nach Triest, p. 126, 1861.

? *Idotea tricuspis* Dekay, Zool. New York, Crust., p. 42, pl. 9, fig. 35, 1844.

Oniscus Balthicus (*Ideotea marina*) Dalyell, Powers of the Creator, vol. i, p. 228, pl. lxiii, figs. 5-9, 1851 (*O. Balthicus* Pallas?).

Oniscus (*Ideotea*) *entomon* Dalyell, op. cit. vol. i, p. 229, pl. lxiii, fig. 10, 1851 (*not O. entomon* Linné.).

Idothea pelagica, M. Sars, Chr. Vid. Selsk. Forh., 1858, p. 151, 1859 (*not* of Leach).

"*Idotea acuminata* Eichwald, Fauna Caspio-Caucasia, p. 232-233, tab. xxxvii, fig. 6, 1842" (Czerniavski).

Idothea balthica Meinert, Crust. Isop. Amph. Dec. Daniæ, pp. 21, 228, etc., "1877" (*Oniscus Balthicus* Pallas?).

PLATE V, FIGS. 24-26.

Adults of this species are at once distinguished from the other species of the genus on our coast by the tridentate abdomen, or pleon, and young individuals, which often resemble *I. phosphorea*, may be distinguished by the epimeral sutures, which extend quite across the second and succeeding thoracic segments. For character separating them from the other Isopoda of the coast, see at the close of the generic description.

The body is smooth, not tubercular nor roughened. The head is nearly square, narrowing but slightly behind. The eyes are small. The antennulæ (pl. V, fig. 25 *a*) are short, hardly surpassing the third segment of the antennæ. The flagellum of the antennæ (pl. V, fig. 25 *b*) is longer than the peduncle, distinctly articulated, slender, and composed of from twelve to sixteen segments in the adults. When reflexed it reaches the third thoracic segment. The external lamella (*l*) of the maxillipeds (pl. V, fig. 26 *a*) is about twice as long as broad, and is obliquely truncated.

Thorax with the external margins, as seen from above, forming in the adults, a pretty regular curved line, the segments being marked by incisions instead of by serratures as in the other species. In the second and third, as well as in the posterior segments, this margin is formed wholly by the epimera.

The first three segments of the pleon terminate in acute teeth at the sides. The fourth, or last segment, has its lateral margins straight, and is more or less tridentate at the tip, the middle tooth being much the largest. In the operculum (pl. V, fig. 25 *c*) the basal plate is about three times as long as the terminal one, which is broadly truncate at the apex. The stylet (*s*) on the second pair of pleopods in the males (pl. V, fig. 26 *b*) is usually shorter than, or, in smaller specimens, about as long as the lamella to which it is attached, and is abruptly bent toward the lamella at the apex and very obliquely truncated. It is minutely serrulate toward the tip on the side opposite the lamella.

The males of this species sometimes attain a length of 30^{mm} to 38^{mm}, with a breadth of 8^{mm} to 9^{mm} but the females are smaller, rarely, if ever, exceeding 20^{mm} in length, with a breadth of 6.5^{mm}, and are found with eggs when not over 7.5^{mm} in length. The color varies greatly. Frequently it is of a nearly uniform light or dark green, or brownish with minute blackish punctations. It is often longitudinally striped with light color, or nearly white on a dark background, and the stripes may be marginal only, or accompanied, especially in the males, by a median dorsal stripe. More rarely the colors are arranged transversely in bands or blotches, and specimens thus marked are easily mistaken for the next species. The females are usually darker than the males, and often with a light lateral stripe, which may be very narrow or broken into a series of blotches.

A comparison of specimens from both sides of the Atlantic does not seem to furnish any characters by which to separate this species from the common European form, *I. tricuspidata* Desm., and as Say's trivial name has priority I have adopted it. *I. tridentata* Rathke appears to be the same species, but *I. tridentata* Latreille* is described by that author as having antennæ as long as the body; further, Desmarest, just before his original description of *I. tricuspidata* says: "M. Latreille fait observer que cette idotée [*I. entomon*] est bien différente de celle que M. Leach a décrite sous le même nom, * * * * cette dernière qu'il nomme Idotée tricuspide," &c. It would not therefore appear that Latreille was at that time aware that this species had a name, much less that he had himself named it *I. tridentata*. Again, in his Cours d'Entomologie, where he copies figures, doubtless of this species, from Savigny's Egypt, he applies to them the name *Idotea* (*pelagica*?), not recognizing them as his own species. Bate and Westwood quote *I. tridentata* Latreille as a synonym of *I. tricuspidata* Desm., and their quotation† appears intended to refer to a work nearly twenty years older than that of Desmarest. They do not, however, give their reasons for deviating from the ordinary rules of priority, but, perhaps, considered as sufficient the authority of Edwards, who does the same thing. Edwards' description of *I. tricuspidata* Desm. contains, moreover, an evident error, the species being placed in a section of the genus which he thus describes: "§ 2 Espèces dont l'abdomen se compose de trois articles parfaitement distincts (le second étant composé de deux anneux soudés ensemble sur le milieu du dos, mais séparés par une scissure sur les côtés)." *I. irrorata* is included in the same section, but under a subsection, thus correctly characterized: "*aa* Le second article de l'abdomen simple; le troisième offrant près de sa base une fissure de chaque

* Gen. Crust. et Ins., tome i, p. 64, 1806.

† Brit. Sess. Crust., vol. ii, p. 380. The quotation reads, "*Idotea tridentata* Latreille, Con. Crust. et Ins. 1, p. 64," and was doubtless intended for Gen. Crust. et Ins., [tome] i, p. 64, [1806].

côté." No species of *Idotea* that I have seen has the second segment of the pleon composed of two segments, united along the back but separated by an incision at the sides, as described in the parenthesis above, and two certainly of the other species included by Edwards in the section with *I. tricuspidata* agree with it in the structure of the pleon as described in *I. irrorata*. Meinert unites this species with *I. pelagica* Leach under the name *I. Balthica* (Pallas), and in this he may be right, but not being able to consult Pallas' work, I have preferred to use the earliest name that I could certainly connect with the species, rather than to introduce further confusion by adopting a name of the applicability of which I could not satisfy myself. M. Sars also regarded *I. pelagica* Leach as synonymous with *I. tricuspidata*, and says it is found as far north as Tromsoe and southward to the Mediterranean, from which statements I conclude that he intended the present species.

This species is found along the whole coast of New England! and extends southward along the coast of New Jersey at least as far as Great Egg Harbor! and northward to Nova Scotia! and the Gulf of St. Lawrence, where it has been collected by Mr. J. F. Whiteaves. From Cape Cod southward it is abundant, but toward the north it is, mostly replaced by *I. phosphorea*. It is commonly found among sea-weed along the rocky shores of bays and sounds or among the rocks, where its variety of colors affords it protection. It is also found far from land, attached to floating sea-weed, and was thus taken by Professor S. I. Smith and the writer on George's Banks!, September 14 and 15, 1872, at about 41° N. lat., 65° W. lon. One of these specimens was quite large, measuring 38mm in length, but most of them were of moderate size or small. Young individuals are often taken at the surface. According to European authors it is common on the shores of Great Britain and Ireland (B. & W.); on all the shores of the North Sea (Metzger *et al.*); (*I. pelagica*) as far north as Tromsoe (M. Sars); in the Baltic, the Mediterranean, the Adriatic (Heller, Stalio, *et al.*), the Black (Czerniavski *et al.*) and the Caspian ("Eichwald") Seas, and, as with us, is of variable color and varies also somewhat in the shape of the termination of the pleon, which is, however, more or less three-toothed.

Specimens examined.

Number.	Locality.	Fathoms.	Bottom.	When collected.	Received from—	Number of Specimens.	Dry. Alc.
1078	Fire Island Beach, L. I.	....	....	—— —, 1870	S. I Smith	50	Alc.
1079	....do..............	....	....	—— —, 1870	....do	9	Glyc.
1954	New Haven, Conn.....	....	....	Nov. —, 1874	A E. Verrill ..	1	Alc.
1955	Stony Creek, Conn.....	....	....	Oct. 23, 1874	....do	00	Alc.
1958	Lyme, Conn..........	....	....	....	D. C. Eaton....	2	Alc.
1963	Long Island Sound, off Saybrook, Conn.	4	Sand..........	Aug. 3, 1874	U.S. Fish Com.	2	Alc.
1964	Off Stonington, Conn..	5	Sand and gravel	Aug. 14, 1874	....do	12	Alc.
1959	Noank Harbor, Conn...	....	Surface	July 13, 1874	... do	00	Alc.

Specimens examined—Continued.

Number.	Locality.	Fathoms.	Bottom.	When collected.	Received from—	Number of specimens.	Dry. Alc.
1960	Noank Harbor, Conn.	....	Eel-grass	Aug. 28, 1874	U. S. Fish Com.	3	Alc.
1961	Fisher's Island	....	....	—— —, 1874	....do	8	Alc.
1962	Watch Hill, R. I.	....	....	Oct. —, 1872	D. C. Eaton	2	Alc.
1965	Vineyard Sound Mass.	Sf.	....	—— —, 1875	U. S. Fish Com.	1	Alc.
1966	... do	....	....	—— —, 1875	....do	7	Alc.
2153	....do	....	....	Oct. 24, 1875	....do	00	Alc.
1968	Provincetown, Mass	L. w.	....	—— —, 1872	Smith & Harger	2	Alc.
	....do	....	Shore	Aug. —, 1879	U. S. Fish Com.	00	Alc.
	....do	L. w.	....	Aug. —, 1879	....do	00	Alc.
	....do	....	Eel-grass	Aug. —, 1879	... do	00	Alc.
	....do	Sf.	....	Sept. 4, 1879	....do	10	Alc.
426	Beverly, Mass	....	....	....	A. E. Verrill	8	Alc.
	Gloucester, Mass	....	Tide-pool	—— —, 1878	U. S. Fish Com.	2	Alc.
	Gloucester, Mass., Outer Harbor.	7-10	Sand, red algæ	—— —, 1878	....do	00	Alc.
	Between Boon Island and Matinicus Rocks.	....	....	—— —, 1878	Capt. G. H. Martin.	5	Alc.
	Casco Bay, Me	....	....	—— —, 1873	U. S. Fish Com.	11	Alc.
1975	Casco Bay, Ram I	L. w.	....	—— —, 1873	....do	4	Alc.
2150	George's Bank	Sf.	....	Sept. —, 1872	Smith & Harger	6	Alc.
1977	Bay of Fundy	L.w. & sf.	....	—— —, 1872	U. S. Fish Com.	2	Alc.
1978	Off Halifax, N. S	....	....	—— —, 1877	....do	1	Alc.
1979	Nova Scotia	L. w.	....	—— —, 1877	... do	1	Alc.
	Durham coast, England	....	....	....	Rev. A. M. Norman.	4	Alc.
	St. Vaast, la Hogue	....	....	....	Jardin des Plantes.	1	Alc.

Idotea phosphorea Harger.

Idotea phosphorea Harger, This Report, part i, p. 569 (275), 1874; Proc. U. S. Nat. Mus., 1879, vol. ii, p. 160, 1879.

Verrill, Am. Jour. Sci., III, vol. vii, pp. 43, 45, 131, 1874; Proc. Amer. Assoc., 1873, pp. 362, 367, 369, 1874; This Report, part i, p. 316 (22), 1874.

Whiteaves, Am. Jour. Sci., III, vol. vii, p. 218, 1874; Further Deep-sea Dredging, Gulf of St. Lawrence, p. 15, "1874."

PLATE V, FIGS. 27–29.

This species may be distinguished from the others on this coast by the pointed abdomen or pleon. Young individuals sometimes resemble the young of *I. irrorata*, but may still be distinguished by the epimeral sutures of the second and third thoracic segments, which do not entirely cross the segment, but allow more or less of the posterior part of the edge of the segment to form a part of the margin of the animal as seen from above. From *Synidotea nodulosa* it may be distinguished by the evident epimeral sutures and by the three acute teeth at the base of the pleon on each side, instead of a single obtuse tooth, as in that species. For characters separating it from the other Isopoda of the coast see at the close of the description of the genus.

The body, especially of the young, is rough and tubercular along the median line and often also laterally. Older specimens are much smoother, losing their large median tubercles but never becoming as smooth as in the preceding species. The head is narrowed behind. The eyes are of moderate size. The flagellum of the antennæ (pl. V, fig. 28 *a*) is shorter than the peduncle, and consists of about ten to fourteen segments. The maxillipeds (pl. V, fig. 28 *b*) have the external lamella (*l*) broader than in the preceding species, with its inner margin straight and its outer margin curving pretty regularly to a slightly attenuated tip.

The epimera of the second, third, and fourth pairs are rounded behind, and those of the last three pairs are less acute than in *I. irrorata.*

Pleon ovate, a little constricted near the middle and pointed, its three proximal segments rather less acute than in the preceding species. The basal plate of the operculum (pl. V, fig. 28 *e*) tapers toward the end, and the terminal plate is triangular, a little longer than broad. The stylet on the second pair of pleopods in the male (pl. V, fig. 29 *s* and *s'*) is slender, nearly straight, surpasses the lamella to which it is attached, and is obliquely truncate.

Length 25^{mm}; breadth 7^{mm}. The color is very varied, usually dark green or brownish, with patches of yellow or whitish, transversely or obliquely arranged. I have never observed a striped pattern of coloration, so common in *I. irrorata*, and it must occur very rarely if at all. The color is usually darker than in that species.

This species is found associated with the last among rocks and seaweed along the entire coast of New England! and extends northward to Halifax!, Nova Scotia, and the Gulf of St. Lawrence!. It appears to be a more northern species than *I. irrorata*, as it is comparatively rare south of Cape Cod, while it is abundant in Casco Bay, Maine, and in the Bay of Fundy.

Specimens examined.

Number.	Locality.	Fathoms.	Bottom.	When collected.	Received from—	Specimens.		Dry. Alc.
						No.	Sex.	
1980	South End, New Haven, Conn.			Nov. —, 1874	A. E. Verrill			Alc.
1981	Stony Creek, Conn.			Sept. 23, 1874	....do			Alc.
1983	Off Saybrook, Conn.	4	Sand	Aug. 3, 1874	U. S. Fish Com.			Alc.
1984	Long Island Sound			Aug. 10, 1874	....do	3	y.	Alc.
1985	South Fisher's Island.	9		Aug. 21, 1874	....do	1		Alc.
1987	Vineyard Sound, Mass.			—— —, 1871	....do			Alc.
1986	....do			—— —, 1872	....do			Alc.
2147	....do				....do	1		Alc.
1990	Off Nantucket	15		Sept. 8, 1875	....do	1	y.	Alc.
	Cape Cod Bay	14	Green mud	Sept. 15, 1879	U. S. Fish Com.	2		Alc.
	....do	7	Coarse yellow sand.	Sept. 15, 1879	....do	3		Alc.
144	Nahant, Mass				A. E. Verrill	3		Alc.
	Gloucester, Mass	7–10	Sand and algæ	—— —, 1878	U. S. Fish Com.	00	♂ ♀ y.	Alc.
	Ten Pound Island, Gloucester, Mass.			—— —, 1878	....do			Alc.
	Annisquam, Mass	⅓		—— —, 1878	A. Hyatt	20		Alc.
1991	Casco Bay, Me	St.		Aug. 4, 1873	U. S. Fish Com.	1	y.	Alc.
1992	....do			—— —, 1873	....do	4		Alc.
1993	Casco Bay, Ram Island.	L. w.		—— —, 1873	....do	4		Alc.
1994	Bay of Fundy			—— —, 1872	....do	1	♀	Alc.
	Bay of Fundy, Whiting River.			—— —, 1872	....do	00	♂ ♀	Alc.
	Off Halifax, N. S	18		—— —, 1877	....do	2		Alc.
	Egmont Bank, Gulf of St. Lawrence.			—— —, 1873	J. F. Whiteaves	1		Alc.

Idotea robusta Kröyer.

? *Idotea metallica* Bosc, Hist. nat. des Crust., tom. ii, p. 179, pl. 15, fig. 6, 1802.

Idothea robusta Kröyer, Naturhist. Tidssk., II, B. ii, p. 108, 1846; Voy. en Scand., Crust., pl. 26, fig. 3, "1849."

Reinhardt, Grönlands Krebsdyr, p. 35, 1857.

Stimpson, Proc. Acad. Nat. Sci., 1862, p. 133, 1862.

Verrill, Am. Jour. Sci., III, vol. ii, p. 360, 1871; This Report, part i, p. 439 (145), 1874 (*Idotea*).

Harger, This Report, part i, p. 569 (275), pl. v, fig. 24, 1874; Proc. U. S. Nat. Mus., 1879, vol. ii, p. 160, 1879 (*Idotea*).

Lütken, Crustacea of Greenland, p. 150, note, 1875.

PLATE VI, FIGS. 30-32.

This species is easily recognized within the genus by the pleon, which is broadly truncate at the apex and not at all pointed. The pleon is also large and more swollen above than in the other species. For characters separating it from other Isopoda, see near the close of the generic description.

The entire upper surface, except perhaps that of the pleon, is somewhat rugose. The head is nearly square, with the eyes large and prominent. The antennæ (pl. VI, fig. 31 *a*) have the second segment large, the flagellum short, usually of less than ten articulations. Under a sufficient power these organs are seen to be clothed with a very fine close pubescence, which also occurs in a less degree upon the legs. The maxillipeds (pl. VI, fig. 32 *a*) have the external lamella (*l*) short and oval.

The legs are robust and spiny. The epimera, projecting, give a serrated appearance to the sides of the thorax, as seen in figure 30, plate VI, and the dorsum is more convex than in the other species.

The pleon is large and convex, its sides are nearly parallel beyond the middle, and it is broadly truncate, or even somewhat emarginate, at the apex. The basal plate of the operculum (pl. VI, fig. 31 *c*) is elongated, with parallel sides; the terminal plate less than one-fourth as long and nearly square, but tapering slightly and somewhat broader than long. The male stylet on the second pair of pleopods (pl. VI, fig. 32 *c*, *s*) reaches the end of the lamella, to which it is attached, and is slightly curved and rounded at the tip.

Length of male 28^{mm}; female 22^{mm}; breadth 9^{mm}. Color bright blue or green above when alive, becoming darker and dull in alcohol, without the markings of the other species, but often with metallic reflections, when seen in the water, where it is commonly taken swimming free or among masses of floating sea-weed.

It is thus found in mid-ocean, and was described by Kröyer from specimens taken in about 60° north latitude between Iceland and Greenland. It was taken in considerable abundance at Fire Island Beach!, on the south shore of Long Island, by Professor S. I. Smith in 1870; also by the U. S. Fish Commission at Vineyard Sound!, Mass., often in company with *I. irrorata* Edw.; at George's Banks!, September, 1872, small specimens, 5^{mm} in length; between Boon Island and Matinicus Rocks, near the

Isles of Shoals!, by Capt. G. H. Martin, of the schooner 'Northern Eagle,' in 1878, and at Halifax!, Nova Scotia, by the U. S. Fish Commission in 1877, whence it extends to at least 60° north latitude.

The figure and description of *Idotea metallica* given by Bosc correspond well with small specimens of this species such as were taken by Professor S. I. Smith and the writer on George's Banks, and the locality he gives, "the high seas," corresponds also with the habit of this species, so that I am inclined to think that his name ought to be restored. I have, however, retained Kröyer's name, since he so thoroughly described and so well figured the species as to leave no doubt of its identity.

Specimens examined.

Number.	Locality.	Habitat.	When collected.	Received from—	Specimens.		Dry. Alc.
					No.	Sex.	
1080	Fire Island Beach, Long Island	Surface	— —, 1870	S. I. Smith	46	♂ ♀	Alc.
1998	Vineyard Sound, Mass	Surface	— —, 1875	U. S. Fish Com.	1	......	Alc.
1999	....do	Surface	....	....do	00	♂ ♀	Alc.
2002	....do	Surface	July 14, 1871	....do	2	......	Alc.
2003	....do	Surface	Oct. 24, 1875	V. N. Edwards	00	......	Alc.
2004	....do	Surface	Nov. 16, 1875	....do	00	......	Alc.
2000	George's Bank	Surface	Sept. —, 1872	Smith & Harger	4	y.	Alc.
2001	Halifax, N. S	Surface	— —, 1877	U. S. Fish Com	1	......	Alc.

Synidotea Harger.

Synidotea Harger, Am. Jour. Sci., III, vol. xv, p. 374, 1878.

Antennæ with an articulated flagellum; epimeral sutures not evident above; pleon apparently composed of two segments, united above but separated at the sides by short incisions; operculum with a single apical plate; palpus of maxillipeds three-jointed.

Of the two species that I had referred to this genus I had been able to examine only the first when this paper was placed in the hands of the printer. Two specimens of the second species were collected during the summer of 1879, and an examination of their characters leaves no doubt of their generic affinity. Except in the particulars above specified the description already given of the genus *Idotea* will in general apply also to the present, but the species are characterized by a firmer and more solid structure, the segments being more closely articulated and the integument having a somewhat shelly appearance. The pleon is further consolidated than in that genus, the only trace of its composite nature, as seen from above, being a slight incision on each side near the base and running up somewhat obliquely toward the dorsal surface. The well-developed and distinctly articulated flagellum of the antennæ serves easily to distinguish the species from those of the following genera of the family.

Synidotea nodulosa Harger (Kröyer).

Idothea nodulosa Kröyer, Naturhist. Tidssk., II, B. ii, p. 100, 1846; Voy. en Scand., Crust., pl. 26, fig. 2, 1849.
Reinhardt, Grønlands Krebsdyr, p. 34, 1857.
Lütken, Crust. Greenland, p. 150, "1875."
Synidotea nodulosa Harger, Am. Jour. Sci., III, vol. xv, p. 374, 1878; Proc. U. S. Nat. Mus., 1879, vol. ii, p. 160, 1879.

PLATE VI, FIGS. 33-35.

This species may be recognized most easily by the pleon, which is entire, except for a slight incision near the base on each side, and tapers to a blunt but not at all bifid point. The articulated flagellum of the antennæ distinguishes it from *Erichsonia*.

The head and body are roughened and tubercular, having a prominent median row of tubercles and coarse rugæ along the sides of the thorax. The head has a median notch in front, and immediately above this a prominent tubercle directed forward, and succeeded on the median line by two less prominent tubercles. In front of each eye is a still larger tubercle, directed forward and projecting over the anterior margin of the head; behind and within, there are two smaller oval tubercles. The eyes are large, convex, and very prominent. The peduncular segments of the antennæ (pl. VI, fig. 34*b*) increase gradually in length from the first and decrease in diameter from the second, which lacks the lateral incision seen in *Idotea*. The flagellum is distinctly articulated, with about nine segments, of which the last two are very minute. The maxillipeds (pl. VI, fig. 35*a*) have the external lamella (*l*) of an irregular shape, emarginate on the inner side and obtusely pointed. The outer maxillæ (pl. VI, fig. 35*b*) are armed on their external lobe with strong, curved, pectinated setæ, which become much elongated and stout at the tip of the lobe. The inner maxillæ (pl. VI, fig. 35*c*) resemble these organs in other members of the family.

The first four thoracic segments have their external margins rounded. In the last three the margins are more nearly straight, but with rounded angles. The first pair of legs (pl. VI, fig. 34 *c*) are much shorter than the second, and the propodus in the first pair is bristly on what is, in the ordinary position, the upper side.

The pleon is short, and tapers from the base. It is convex, bears two or three small tubercles on the median line near the base, and an impressed transverse line in continuation of the short lateral incisions. The basal plate of the operculum (pl. VI, fig. 34*d*) is oblique at the base with rounded angles, and is somewhat vaulted, with an oblique elevation extending from the articulation to the inner distal angle. The inner margin is straight, and the outer parallel with it to near the end. The terminal plate is slightly oblique at the base, and is elongated triangular, about twice as long as broad. The free margins are finely ciliated, except at and near the base, and the inner margin of the basal plate bears also scattered stouter hairs. The stylet of the males on the second pair of pleopods (pl. VI, fig. 35*d*, *s*) is longer and stouter than in any of our species of *Idotea*. It is nearly twice the length of the lamella, to which it is attached, and of an elongated spatulate form tapering to an obtuse point. The lamellæ are provided with but few cilia, which extend less than half the way from the end of the lamella to the end of the stylet.

Length 10.5^{mm}.; breadth 3.5^{mm}. Females proportionally broader; length 8.^{mm}; breadth 3^{mm}. Color in alcohol gray, often with brownish transverse markings.

This species seems to agree with *Idotea nodulosa* Kröyer, from Southern Greenland, as described and figured, except that the epimeral sutures are not evident above; the lateral margins of the segments are, however, somewhat thickened and prominent with rugæ, as shown in his figure, and I have no doubt that it is the same as his species. It was dredged off Halifax! by the Fish Commission at several localities in the summer of 1877, in from 16 to 190 fathoms on sandy and rocky bottoms, with red algæ at one locality. A specimen was brought from George's Banks! by Mr. Joseph P. Schemelia, of the schooner 'Wm. H. Raymond,' in the summer of 1879, and Mr. J. F. Whiteaves has sent to the Museum for examination two specimens collected by Mr. G. M. Dawson, in 111 fathoms, Dixon Entrance!, north of Queen Charlotte Island, British Columbia. The range of the species would therefore be, as at present known, from George's Banks to Greenland and the Arctic Seas, and southward on the Pacific coast as far as British Columbia..

Specimens examined.

Number.	Locality.	Fathoms.	Bottom.	When collected.	Received from—	Specimens. No.	Specimens. Sex.	Dry. Alc.
	Dixon Entrance, Q. C. I.	111	....	....	J. F. Whiteaves	2	......	Alc.
2006	Off Halifax, N. S....	16	Stones, sand, red algæ.	—— —, 1877	U. S. Fish Com.	2	......	Alc.
2007	South of Halifax, 120 miles.	190	Gravel and pebbles.	Sept. 1, 1877	....do	1	♀	Alc.
	Halifax, outer harbor	18	Sand, stones	Sept. 4, 1877	....do	2	......	Alc.
2008	....do	16	Rocks, nullipore	Sept. 4, 1877	... do	1	......	Alc.
	George's Banks.....	....	....	—— —, 1879	J. P. Schemelia	1	♀	Alc.

Synidotea bicuspida Harger (Owen).

Idotea bicuspida Owen, Crustacea of the Blossom, p. 92, pl. xxvii, fig. 6, 1839.
Streets and Kingsley, Proc. Essex Inst., vol. ix, p. 108, 1877.
Idotea marmorata Packard, Mem. Bost. Soc. Nat. Hist., vol. i, p. 296, pl. viii, fig. 6, 1867.
Whiteaves, Further Deep-sea Dredging in Gulf of St. Lawrence, p. 15, 1874.
Idotea pulchra Lockington, Proc. Cal. Acad. Sci., vol vii, p. 45, 1877.
Synidotea bicuspida Harger, Proc. U. S. Nat. Mus., 1879, vol. ii, p. 160, 1879.

This species may be most easily recognized among the known Isopoda of our coast by the form of the pleon, which is nearly triangular in shape, marked by a slight incision at each side near the base, and distinctly bicuspid at the tip.

The body is rather more robust than in the last species, the length being only about two and a half times the breadth, and is peculiarly marked above by depressed and mostly curved lines, varying in length but mostly short, and confined principally to a region on each side of the median line and extending across the head but not the pleon.

The head is broadly emarginate in front, with a median notch, and its antero-lateral angles are prominent. The eyes are at the widest part of the head, and are strongly convex. The posterior outline of the head is nearly in the form of three sides of a hexagon. The antennulæ attain about the middle of the fourth antennal segment. The antennæ are about one-half as long as the body. The first two antennal segments are short and apparently articulated so as to admit of but little motion; the third segment is a little longer than the first two taken together, and is the largest of the antennal segments in diameter; the fourth segment is somewhat longer than the third, and the fifth or last peduncular segment is the longest, and is followed by a flagellum, a little shorter than the peduncle and composed of about fourteen segments. The last three peduncular segments of the antennæ are somewhat bristly hairy. The maxillipeds are nearly as in the preceding species. The outer maxillæ are destitute of the elongated, pectinate setæ found in that species.

The thoracic segments vary but little in length measured along the median line, but the fifth, sixth, and seventh are slightly shorter than the preceding ones, and this difference is still greater measured along the margins of the segments, where the first is longest, the next three about equal, and the last three shorter. The legs are robust, the first pair shortest, and all more or less bristly hairy. The lateral margins of the segments are much less rounded than in *S. nodulosa*.

The pleon is short, the length being scarcely greater than the breadth at base; above, it is nearly smooth, the impressed lines, so conspicuous in the lateral region of the thorax, being continued for but a slight distance upon its surface. The incision at each side near the base is continued upward and forward by a depressed line on each side; the lateral margins are gently convex to near the tip, which is distinctly bicuspid. The basal plate of the operculum is traversed obliquely by a longitudinal ridge on the external surface, and is rounded in front, slightly narrowed behind, and bears a short, triangular, terminal plate, its length being but little greater than its breadth.

Length 15.5^{mm}; breadth 6^{mm}. Color in alcohol grayish, with white cloudings. Lockington says: "When recent, the coloration of this species is very beautiful, consisting of red cloudings on a lighter ground."

There seems to be no doubt in regard to the synonymy of this species as published by Streets and Kingsley, adopted by the writer in a previous publication, and given above.

The only specimens that I have examined were two, brought from the Grand Banks!, in the summer of 1879, by Mr. Charles Ruckley, of the schooner 'Frederick Gerring, jr.', Capt. Edwin Morris. Dr. Packard's locality is "Sloop Harbor, Kynetarbuck Bay [Labrador], seven fathoms on a sandy bottom." Whiteaves records the species from Orphan Bank, Gulf of Saint Lawrence. Lockington's specimens were collected on the "west coast of Alaska, N. of Behring's Strait, by W. J. Fisher, naturalist of the U. S. S. Tuscarora, Deep-Sea Sounding Expedition." Owen's locality is "the Arctic Seas."

Erichsonia Dana.

Erichsonia Dana, Am. Jour. Sci., II, vol. viii, p. 427, 1849.

Antennæ six-jointed, the terminal or flagellar segment not articulated, clavate; palpus of the maxillipeds four-jointed; legs all nearly alike, prehensile or sub-prehensile; pleon with its segments consolidated into a single piece.

This genus is represented within our limits by two well-marked species, which further agree in the following characters: The head is quadrate, with the eyes lateral. The antennulæ are short, not surpassing the third segment of the antennæ. The antennæ are well developed, more than half as long as the body, with a very short basal segment articulated with little or no motion to the second segment, which is two or three times as long as, and of greater diameter than the first. It is, as usual in the family, incised at its distal end on the under surface. The next three segments are nearly cylindrical. The last or flagellar segment is the longest, and is slightly clavate.

The legs are all terminated by a prehensile or sub-prehensile hand, the dactylus being capable of considerable or complete flexion on the more or less swollen propodus. This flexion is most complete in the first pair. The first two pairs of legs arise near the anterior margin of the segments to which they belong. The place of attachment to the segment moves gradually backward in the following pairs until the last two pairs arise near the posterior margin of the last two segments. The epimera are more or less evident from above, at least in the last two segments.

The pleon constitutes about one-third the length of the body, and is consolidated into a single piece; it bears a more or less evident tooth on each side near the base, and is dilated and obtusely triangular at the apex. The basal plate of the operculum is oblique at the anterior end and abruptly narrowed posteriorly, where it bears a densely plumose bristle, as in *Idotea*; the terminal plate is triangular. The stylet on the second pair of pleopods in the males is well developed, surpassing the cilia; it is minutely denticulated or spinulose near the end and very acute.

The two species found on our coast have but a slight external resemblance to each other, and may be distinguished at a glance, as will be seen from the specific descriptions, and from the figures (pl. VI, fig. 36, and pl. VII, fig. 38). The long, clavate terminal segment of the antennæ distinguishes them at once from young specimens of *Idotea*, especially *I. phosphorea*, which sometimes resemble *E. filiformis*. This character of the antennæ serves, indeed, to distinguish the two unlike representatives of the present genus from all the other Isopoda of our coast.

Erichsonia filiformis Harger (Say).

Stenosoma filiformis Say, Jour. Acad. Nat. Sci., vol. i, p. 424, 1818.
Edwards, Hist. nat. des Crust., tom. iii, p. 134, 1840.
Dekay, Zool. New York, Crust., p. 44, 1844.
Idotea filiformis White, List Crust. Brit. Mus., p. 95, 1847.
Erichsonia filiformis Harger, This Report, part i, p. 570 (276), pl. vi, fig. 26, 1874; Proc. U. S. Nat. Mus., 1879, vol. ii, p. 160, 1879.
Verrill, This Report, part i, p. 316 (22), 1874.

PLATE VII, FIGS. 38–41.

This species may be at once distinguished from the following by the strongly serrated outline of the sides, as seen from above. The clavate terminal segment of the antennæ distinguishes it from the other known Isopoda of our coast.

The body is slender and elongated, but less so than in the next species, the sides are nearly parallel and there is a median row of prominent tubercles, one, large and bifid, on the head, and one upon each thoracic segment. The eyes are prominent. The antennulæ (pl. VII, fig. 39 *a*) surpass the middle of the third antennal segment. The first segment of the antennæ (pl. VII, fig. 39 *b*) is very short; the terminal segment is bristly hairy toward the apex. The external lamella of the maxillipeds (pl. VII, fig. 41 *a*) is emarginate on the outer side toward the apex.

The thoracic segments each bear a prominent median tubercle near their posterior margins, and the first bears also a smaller tubercle near its anterior margin. In the first two segments the posterior external angles are salient and much elevated. The angulated epimera are evident from above in front of these projections. In the third and fourth segments both lateral angles are salient but not elevated. In the last three segments, only the anterior angles are produced, but the epimera fill the places of the posterior angles. This arrangement gives the appearance of fourteen teeth upon each side of the thorax, and the prominent divergent tooth on the pleon makes, in all, fifteen.

The operculum (pl. VII, fig. 39 *d*) is a little more vaulted than in the next species and shorter; the basal plate is less than three times as long as broad; the terminal plate is triangular. The stylet on the second pair of pleopods in the male (pl. VII, fig. 41 *b*, *s*) is slightly curved, finely spinulose near the apex on the side toward the lamella, and minutely and sharply denticulate on the opposite side at the apex, as shown in the enlarged figure (*s*′) of the distal portion of the stylet.

Length 11^{mm}; breadth 3.4^{mm}. The color is a usually dull neutral tint without bright markings, but sometimes more or less variegated with brown or reddish, fading in alcohol.

This species was originally described from Great Egg Harbor, New Jersey, where Say found it in company with *Idotea irrorata*. It is not uncommon along the shores of Long Island Sound! and as far east as Vineyard Sound, Mass.! but has not yet been found north of Cape Cod. It is usually found in tide-pools or among eel-grass and algæ, and has been taken from a depth of 7 fathoms.

Specimens examined.

Number.	Locality.	Fathoms.	Bottom	When collected.	Received from—	Number of specimens.	Dry. Alc.
2010	Long Island Sound..	....	....	....	....	00	Alc.
2011	Thimble Islands ..	....	....	....	A. E. Verrill...	....	Alc.
2012	Long Island Sound, Fisher's Island Sound	7	Sand and shells .	—— —, 1874	U. S. Fish Com.	1	Alc.
2013	Long Island Sound..	4½	Sand and gravel.	—— —, 1874	.. do	2	Alc.
2014	.. do	....	....	—— —, 1874	....do	1	Alc.
2015	... do	....	....	Sept. 10, 1874	...do	2	Alc.
2016	Noank	....	Eel-grass	—— —, 1874	....do	2	Alc.
2017	Vineyard Sound	....	....	—— —, 1875	....do	2	Alc.

Erichsonia attenuata Harger.

Erichsonia attenuata Harger, This Report, part i, p. 570 (276), pl. vi, fig. 27, 1874; Proc. U. S. Nat. Mus., 1879, vol. ii, p. 160, 1879.
Verrill, This Report, part i, p. 370 (76), 1874.

PLATES VI and VII, FIGS. 36 and 37.

This species is at once distinguished from the preceding by its slender form and regular outline; the clavate antennal flagellum distinguishes it from other Isopoda.

The body is smooth throughout and about six times as long as broad, without prominent irregularities and narrowly linear in outline. The eyes are small and black. The antennulæ (pl. VII, fig. 37 *a*) are short, slightly surpassing the second antennal segment. The antennæ (pl. VII, fig. 37 *b*) are stout and smoother than in the preceding species. The external lamella of the maxillipeds (pl. VII, fig. 37 *c*, *l*) is oval and regularly rounded at the tip.

The thoracic segments increase in size to the third, which is equal to the fourth, and the last three are of a gradually decreasing size. The epimera are nowhere conspicuous, but may usually be seen from above, especially in the posterior segments.

The pleon presents only slight traces of a lateral tooth near its base and is but little dilated toward the tip. The operculum (pl. VII, fig. 37 *d*) is longer than in the preceding species, the basal plate is more than three times as long as broad, the terminal plate elongated triangular and obtuse. The male stylet on the second pair of pleopods (pl. VII, fig. 37 *e*, *s*) is nearly straight, hardly surpasses the cilia, and is minutely denticulated near the acute apex.

Length 15^{mm}; breadth 2.5^{mm}. Alcoholic specimens are of a light grayish yellow, with minute black punctations.

It was abundant in eel-grass at Great Egg Harbor, New Jersey! in April, 1871, and has also been found at Noank, Conn.! on eel-grass, but is not common. It has not been found north of Cape Cod.

Specimens examined.

Number.	Locality.	Fathoms.	Bottom.	When collected.	Received from—	Specimens.		Dry. Alc.
						No.	Sex.	
1226	Great Egg Harbor, N. J.	....	Eel-grass	Apr. —, 1871	S. I. Smith ...	00	......	Alc.
2018	Noank, Conn	....	....do	— —, 1874	U. S Fish Com.	1	♀	Alc.

Epelys Dana.

Epelys Dana, Am. Jour. Sci., II, vol. viii, p. 426, 1849.

Antennæ shorter than the anntennulæ and with only a rudimentary flagellum; palpus of the maxillipeds three-jointed; legs all terminated with prehensile hands; pleon consolidated into a single segment with a basal lobe on each side.

Two small and closely allied species from this coast have been referred to this genus. They resemble each other very closely and may be at once recognized by their depressed ovate form, very short antennæ, and generally dirty appearance. The form of the body and absence of powerful mandibles distinguish them from the male *Gnathia*. The length of the body is between two and three times its width. It is marked by a depressed line on each side, running from the posterior part of the head, across the thoracic segments, nearer to their lateral margins than the median line, except perhaps in the last segment, thence continued to inclose a prominent hemispherical protuberance on the anterior part of the pleon, giving the animal somewhat the appearance of a trilobite. The body is slightly roughened under a lens, or sometimes minutely hirsute. The head is slightly dilated at the sides, with the anterior angles produced, and bears a pair of broad, low, triangular tubercles on its anterior part, and a curved posterior depression. The eyes are lateral and prominent, the antennulæ are longer than the head, surpass the antennæ, and have the basal segment but little enlarged. The antennæ (pl. VIII, fig. 45 *b*) are shorter than the head, not surpassing the third antennular segment, the segments increasing in length to the fourth; fifth as long as the fourth, but more slender, bearing a minute, slender rudiment of a flagellum, which is setose at the tip.

The thoracic segments have thick evident margins; first segment smallest, somewhat embracing the head; third and fourth largest;

last segment curving around the base of the pleon. The epimera are not evident from above. The legs (pl. VIII, fig. 46 *a*) are slender and all terminated by a slender prehensile hand, of which the finger, or dactylus, becomes almost acicular in some of the posterior pairs. All the legs are more or less hairy.

The pleon bears on each side, near its base, a rounded lobe, which is separated from the large posterior portion by a more or less evident incision. Dorsally it is convex, and presents two hemispherical elevations, the proximal more convex than, but only half as large as, the distal. They are separated by a broad and deep groove, and the distal convexity is continued upon the obtusely-pointed apex of the pleon. The operculum (pl. VIII, fig. 46 *b*) is vaulted; its basal plate is rounded anteriorly, carinate near its inner margin, contracted externally for the distal third of its length and truncate at the tip, where it bears a stout elongated-triangular finely ciliated terminal piece. The basal plate is coarsely ciliated on its inner margin, and bears a few plumose hairs along its outer free margin. The stylet on the second pair of pleopods in the males is short and stout, surpasses the lamella but not the cilia, and is spinulose just below the blunt apex.

Both species are of a dull neutral color, and commonly covered with particles of mud or other foreign matter. They occur on piles, or under stones, in muddy places, and are dredged on muddy bottoms.

Epelys trilobus Smith (Say).

Idotea triloba Say, Jour. Acad. Nat. Sci. Phil., vol. i, p. 425, 1818.
Edwards, Hist. nat. des Crust., tome iii, p. 134, 1840.
Dekay, Zool. New York, Crust., p. 43, 1844.
Leidy, Jour. Acad. Nat. Sci., II, vol. iii, p. 150, 1855.
Jaera? triloba White, List Crust. Brit. Mus., p. 97, 1847.
Epelys trilobus Smith, This Report, part i, p. 571 (277), pl. vi, fig. 28, 1874.
Verrill, Am. Jour. Sci., III, vol. vii, p. 135, 1874; Proc. Amer. Assoc., 1873, p. 372, 1874; This Report, part i, p. 370 (76), 1874.
Harger, Proc. U. S. Nat. Mus., 1879, vol. ii, p. 160, 1879.

PLATE VII, FIGS. 42 and 43.

This species may be recognized among our Isopoda by its appearance when seen from above, recalling the form of the trilobites, the flattened dorsal surface being marked, as in those animals, by two lateral longitudinal depressions. The pleon is consolidated into a single piece and the antennæ have only a rudimentary flagellum. It closely resembles the next species, but is smaller and most readily distinguished by the lateral margin of the thorax, which is, especially in the anterior part, nearly even instead of zigzag from the projecting angular segments. The anterior angles of the head are also less produced.

The pleon is shorter and broader, its breadth being to its length as six to ten. The deep transverse groove across the pleon is continued to the margin, with only, at the most, traces of a tubercle at each side. The stylet on the second pair of pleopods of the male (pl. VII, fig. 42 *b*, *s*, and *s'*), is a little less elongated than in the next species, not attaining the middle of the cilia.

Length 6^{mm}; breadth 2.3^{mm}. The color is uniform, dull, usually obscured by the adhering particles of dirt.

This species was described by Say from Egg Harbor!, New Jersey, where specimens were also collected by Professors Verrill and Smith, in April, 1871, among eel-grass. It has also been found at Savin Rock!, near New Haven, and Noank Harbor!, on piles and among eel-grass; at Vineyard Sound!; Mass., at Provincetown!, Mass., near Cape Cod in 1879; sparingly near Gloucester! Mass., in 1878, and even as far north as Quahog Bay!, about thirty miles northeast of Portland, Me., where it was taken by the United States Fish Commission, in 1873, along with *Venus mercenaria* and other southern forms.

Specimens examined.

Number.	Locality.	Fathoms.	Bottom	When collected.	Received from—	Number of specimens.	Dry. Alc.
1227	Great Egg Harbor, N. J	....	Eel-grass	Apr —, 1871	S I Smith.....	7	Alc.
2019	Savin Rock, New Haven	L. w.	....	1871–1872	A. E. Verrill ..	2	Alc.
2020	Noank Harbor	....	....	Aug. 12, 1874	U. S. Fish Com.	12	Alc.
2021	....do	Sf.	....	July 13, 1874	... do	7	Alc.
2022	....do	....	On piles	July 27, 1874	.. do	1	Alc.
2024	.. do	....	Eel-grass	—— —, 1874	....do	00	Alc.
2023	Watch Hill, R I	....	....	Apr. —, 1873	A. E. Verrill...	4	Alc.
2025	Vineyard Sound	....	....	—— —, 1871	U. S. Fish Com.	2	Alc.
2026	... do	....	....	—— —, 1871	....do	1	Alc.
2027	.. do	L. w.	Sand..........	—— —, 1871	... do	2	Alc.
	Provincetown, Mass..	L. w.	....	Aug. —, 1879	U. S. Fish Com.	2	Alc.
	....do	½	Eel-grass........	—— —, 1879	....do	1	Alc.
	....do	....	Shore...........	—— —, 1879	....do	9	Alc.
	Gloucester, Mass	....	Tide-pools	—— —, 1878	....do	2	Alc.
2028	Quahog Bay, Me	L. w.	Muddy	—— —, 1873	....do	3	Alc.

Epelys montosus Harger (Stimpson).

Idotea montosa Stimpson, Mar. Inv. G. Manan, p. 40, 1853.

Epelys montosus Harger, This Report, part i, p. 571 (277), 1874; Proc. U. S. Nat. Mus., 1879, vol. ii, p. 161, 1879.

Verrill, Am. Jour. Sci., III, vol. vii, p. 45, 1874; Proc. Amer. Assoc., 1873, p. 367, 1874; This Report, part i, p. 370 (76), 1874.

Smith and Harger, Trans. Conn. Acad., vol. iii, p. 3, 1874.

Whiteaves, Further Deep-sea Dredging, Gulf St. Lawrence, p. 15, "1874."

PLATE VIII, FIGS. 44–47.

This species closely resembles the preceding, and may be recognized among our Isopoda by the characters mentioned under the former species, from which it is distinguished by the following characters: The eyes are prominent; the anterior angles of the head salient. The tubercles on the head are more prominent than in the former species. The lateral margins of the thoracic segments, especially the second, third, and fourth, are angulated and salient. The pleon is more elongated than in the last species, its breadth being to its length as 5.5 to 10, and the depression crossing it is partially interrupted at each side by a tubercle which often projects, as seen from above, just behind the basal lobe, forming a shoulder to the large terminal lobe. The stylet on the second pair of pleopods in the males (pl. VIII, fig. 47, *s* and *s'*) attains about the middle of the cilia.

Length 10^{mm}; breadth 4^{mm}; color, as in the preceding, dull, and usually much obscured by adhering dirt.

A few specimens were collected in Whiting River, near Eastport, Maine, in 1872, which are much more decidedly hirsute than is usual, both on the upper surface and on the legs as well. In other respects they appear to be referable to this species, although the posterior thoracic segments are rather less angulated at the lateral margin. They may be worthy of a variety name *hirsutus*.

Dr. Stimpson's specimens were "taken in deep water on sandy and muddy bottoms" in the Bay of Fundy, and this species usually replaces the last in the northern localities. It has, however, been taken as far south as Block Island Sound !, near the eastern end of Long Island Sound, in 18 fathoms, sandy bottom, and in 29 fathoms Vineyard Sound !. North of Cape Cod it is more common. It was dredged in 25 fathoms on St. George's Bank !, at Stellwagen's Bank ! in 20 to 40 fathoms, rocky and sandy bottom; Casco Bay !, 16 to 17 fathoms mud; Bay of Fundy !, at many localities, usually on muddy bottoms, and in 16–18 fathoms mud and stones, off Halifax !, Nova Scotia, by the Fish Commission, and in 14 fathoms off Richibucto, in the Gulf of Saint Lawrence, by Mr. J. F. Whiteaves. The greatest depth positively recorded is 29 fathoms, but it may very likely have come also from a depth of 40 fathoms near Stellwagen's Bank.

Specimens examined.

Number.	Locality.	Fathoms.	Bottom.	When collected.	Received from—	Number of specimens.	Dry. Alc.
2029	Vineyard Sound	29		Sept 14, 1871	U. S. Fish Com.	8	Alc.
2030	Block Island Sound	18	Sand	—— —, 1874	....do	1	Alc.
	Off Boston Harbor	16	Speckled sand, shells.	Sept. 13, 1879	....do	2	Alc.
	Gloucester Harbor, Mass.	7–8½	... do	—— —, 1878	....do	30	Alc.
2032	George's Bank	25		—— —, 1872		2	Alc.
2031	Stellwagen's Bank	20–40	Rocks and sand	—— —, 1873	A. S. Packard	1	Alc.
2033	Casco Bay	16	Mud	July 12, 1873	U S. Fish Com.	00	Alc.
2035	... do	17	....do	Aug. 30, 1873	... do	00	Alc.
	....do			—— —, 1873	...do	00	Alc.
2038	Bay of Fundy, Eastport.			—— —, 1872	....do	6	Alc.
2039	....do			—— —, 1872	....do	2	Alc.
2040	Bay of Fundy, Whiting R.	2	Mud	—— —, 1872	....do	6	Alc.
2041	Seal Cove, Grand Menan.	8–10		—— —, 1872	... do	10	Alc.
2042	Off Halifax, N. S	16	Stones, sand, red algæ.	—— —, 1877	....do	4	Alc.
2043	....do	18	Mud, fine sand	Sept. 15, 1877	....do	2	Alc.

VI.—ARCTURIDÆ

Form elongated; antennæ large and strong; first four pairs of legs directed forward, ciliated, last three pairs ambulatory; segments of pleon more or less consolidated; uropods operculiform.

This well marked family is as yet represented on our coast by a single species of the genus *Astacilla* Fleming, *Leachia* or *Leacia* of Johnston and other authors. The family can be easily recognized by the four anterior pairs of legs, which are directed forward and strongly ciliated on their inner margins with long slender hairs. The form of the body is elongate and may be very much so, as in our species the length of the body in the male is twenty times as great as its diameter at the middle; in the female eight times. The head is of moderate size and the eyes prominent. The four-jointed antennulæ have the basal segment large and swollen. The antennæ are large and powerful organs, approaching or even surpassing the body in length, with the first two segments short, the second deeply incised below as in *Idotea*, the next three segments elongated, and the flagellum varying in the genera, being multiarticulate in *Arcturus*, and composed of not more than four segments in *Astacilla*. The mouth parts resemble, in general, those of the *Idoteidæ*. The fourth thoracic segment is more or less elongated. The last three pairs of legs are ambulatory, differing much from the first four pairs. The segments of the pleon are more or less united, and the uropods are modified, as in the preceding family, to form an operculum for the more delicate anterior pleopods. They are wholly inferior, and consist on each side of a large basal segment, straight on the median line, where it meets its fellow of the opposite side, and bearing, in our genus, two small terminal plates at the apex.

This structure of the pleon and its appendages, together with the structure of the antennulæ, antennæ, and the parts of the mouth, point to a close relationship between this family and the *Idoteidæ*. With the *Anthuridæ*, however, with which they have often been associated, they seem to have little in common, except, perhaps, the elongate form of body. Even this feature is approached also in the *Idoteidæ*, in *Erichsonia*, for example.

Astacilla Fleming.

Leacia (*Leachia*) Johnston, Ed. Phil. Jour., vol. xiii, p. 219, 1825 (*non* Lesueur).
Astacilla Fleming, Encyc. Brit., 7th ed., vol. vii, p. 502.
Johnston, Loud. Mag. Nat. Hist., vol. viii, p. 494, 1835.

Antennal flagellum short, not more than four-jointed; fourth thoracic segment elongated, and, in the females, bearing the incubatory pouch on its inferior surface.

The characters given above seem sufficient to warrant the separation of this genus from *Arcturus*, notwithstanding the fact that the young of some species, and probably of all, have the fourth thoracic segment no longer than the others as noticed by Johnston*, and later by Stebbing†, who draws from the fact an argument against the validity of the genus. I fail to see, however, why the argument would not be equally valid against the use, among mammals, of characters drawn from the horns and teeth. Nothing is more common, in case of a genus or family possessing a special development of some organ or set of organs, than to find that the young of such a group resemble the adults of less specialized groups. If, however, as may be possible, a gradation can be established between forms which, like *Arcturus Baffini*, have the fourth thoracic segment large but only slightly elongated, and forms like *Astacilla longicornis* or *A. granulata*, in which this segment is much elongated, equaling or surpassing the other six in length, there would then be, perhaps, no sufficient reason for retaining both genera. For the present it seems desirable to keep them separate, and to the characters given above we may add the following:

The head is produced at the sides around the bases of the antennulæ, and is united dorsally with the first thoracic segment, the sutures being evident only at the sides where the segment is produced around the hinder part of the head. The flagellum of the antennæ consists of three, or sometimes only two, distinct segments and a terminal spine, which is perhaps to be regarded as a third or fourth segment. The maxillipeds (pl. IX, fig. 52 *a*) are robust and operculiform, with a thick external lamella and a five-jointed palpus, but little flattened. The mandibles are destitute of palpi.

The first three thoracic segments are subequal and short; the fourth much elongated in both sexes; in the males it is slender and cylindrical; in the females it is more robust, and bears on its inferior surface the incubatory pouch. This pouch is thus confined to a single segment, and is composed of a pair of elongated lamellæ, attached along their outer margins, and overlapping widely along the ventral surface. It occupies nearly the entire inferior surface of the segment. The last three thoracic segments are short and subequal, and the articulation at the posterior end of the fourth segment is capable of considerable motion, and, in our species, is usually flexed backward nearly at a right angle. The first pair of legs (pl. VIII, fig. 49 *b*) have the basis directed backward and the remaining segments ciliated and turned forward, and is more robust than the three succeeding pairs, which are slender, of nearly equal size, and consist of only five segments, which are turned forward from the basis and held beneath the head. They are strongly ciliated, especially on the last three segments. One of the fourth pair of legs is shown on plate VIII, figure 50. The last three pairs of legs are of entirely different structure, being robust and prehensile with strong short dactyli.

The pleon is consolidated into a single segment, which, however, shows traces of its composite nature. It is vaulted above and excavated on

---

* Loud. Mag. Nat. Hist., vol. ix, p. 81, fig. 15, 1836.

† Ann. Mag. Nat. Hist., IV, vol. xv, p. 187, 1875.,

its inferior surface for the delicate pleopods, which are protected by the operculiform uropods. Both rami of the uropods are present in our species, but the outer is much the larger and conceals the delicate inner ramus in an exterior view. The outer ramus only is thickened and of functional importance as an operculum.

The habits of these animals are described by Goodsir in the Edinburgh New Philosophical Journal, vol. xxxi, p. 311. He says, "With the dredge I have procured specimens * * * * alive, and have kept them in glass jars of sea-water with sand and corallines, and have thus been enabled to watch their habits closely.

"Under the circumstances just stated, each individual will select a branch of coralline, will keep that branch exclusively to itself, and will defend it with the greatest vigor against all intruders. It fixes itself to its resting-place by means of its true thoracic feet, and seldom uses these for progression. When it falls to the bottom of the vessel, it fixes its long pointed antennæ firmly into the sand, and, with the assistance of the true feet, drags and pushes itself forward. This, however, may not be a natural mode of progression, but may be adopted in consequence of the artificial circumstances in which the animal is placed.

"Swimming is the natural mode of progression. It is amusing to see one of these animals resting, in an erect posture, on a branch of coralline, by means of its true thoracic feet, waving its body backwards and forwards, throwing about its long inferior antennæ, and ever and anon drawing them through its anterior fringed feet, for the purpose of cleaning them. It frequently darts from its branch, with the rapidity of lightning, to seize with its long antennæ some minute crustaceous animal, and returns to its resting-place to devour its prey at pleasure.

"In this manner the antennæ are the only organs employed in seizing and enclosing the prey, which they drag to the anterior thoracic feet, which hold it while it is being devoured."

I have discarded Johnston's name *Leachia*, or according to his orthography *Leacia*, proposed in 1825, as being preoccupied by Lesueur,* in the Mollusca in 1821. *Astacilla* is used by Fleming in the 7th edition of the Encyclopædia Britannica; 1842 is given as the date in the copy of the seventh volume of the Encyclopædia that I have seen, but Johnston refers to Fleming in 1835 as authority for the name, quoting the Encyclopædia. Fleming says in the Encyclopædia (vol. vii, p. 502): "The genus was instituted by the Rev. Charles Cordiner of Banff in 1784 for the reception of a British species which has been denominated *Astacilla longicornis*." I have not been able to find whether Cordiner published the name at that early date or whether it was a manuscript name only. If actually published in 1784 it would have many years' priority over *Arcturus*, and the author who would unite the genera should use the name *Astacilla*. Even if not published until 1835 it appears to have the best claim to recognition as the generic name of the type here treated of.

*Jour. Acad. Nat. Sci. Phila., vol. ii, p. 89, 1821.

Astacilla granulata Harger (G. O. Sars).

Leachia granulata G. O. Sars, Arch. Math. Nat., B. ii, p. 351 [251], 1877.
Astacilla Americana Harger, Am. Jour. Sci., III, vol. xv, p. 374, 1878.
Astacilla granulata Harger, Proc. U. S. Nat. Mus., 1879, vol. II, p. 161, 1879

PLATES VIII AND IX, FIGS. 48–52.

The elongated fourth thoracic segment distinguishes this species at once from all the other Isopoda of our coast.

The body is in the female eight times and in the male about twenty times as long as broad, the breadth being measured across the fourth thoracic segment. It is roughened and tuberculated throughout. The head is produced at the sides in front beyond the middle of the basal segment of the antennulæ, and is tuberculated above and crossed by two transverse grooves, the first between, and the second behind the eyes, while a third similar groove evidently marks the place of the suture between the head and the first thoracic segment. The eyes are lateral, prominent, round-ovate, broadest in front. The antennulæ in the female slightly surpass the second segment of the antennæ, in the male they nearly attain the middle of the third segment, the flagellar segment being elongated in the male, longer than the three peduncular segments together (pl. VIII, fig. 48 *a*). The second and third segments of the antennulæ are in both sexes short and slender. The antennæ are fully three-fourths as long as the body; the first segment is shorter than that of the antennulæ, being surpassed at the sides by the lateral processes of the head and thus concealed in a lateral view; the second segment is large, scarcely longer than broad, and presents below a deep angular sinus in the distal margin, as in *Idotea;* third segment about as long as the head; fourth segment longest, slightly exceeding the fifth, which is equal to the first three taken together. The flagellum* (pl. VIII, fig. 49 *a*) is less than half the length of the last peduncular segment and usually consists of three distinct segments, of which the first is as long as the other two; the second is equal in length to the third, which is tipped with a terminal spine or claw, probably to be regarded as a fourth segment. Sometimes, however, only two distinct segments exist in the flagellum besides the claw. The flagellar segments are finely and sharply denticulate along the margin which is inferior when the antennæ are straightened. The character of this denticulation is shown in figure 49 *a′* on plate VIII, where a small section of the margin is shown enlarged 100 diameters. The maxillipeds (pl. IX, fig. 52 *a*) are robust and cover the other parts of the mouth; the external lamella (*l*) is ovate and in the figure is somewhat bent outward from its natural position. The palpus of the maxillipeds is five-jointed and but little flattened, strongly ciliated along the inner margin. The terminal lobe

* The figure of the animal (pl. VIII, fig. 48,) was sent to the engraver before I had seen any specimens except the imperfect ones collected in 1877, and the flagellum of the antennæ was dotted from the young specimens. Fig. 49 *a* on plate VIII was made from a specimen obtained in 1878.

(pl. IX, fig. 52 *a*, *m'*) is quadrate, scarcely ciliated at the apex, and distinctly articulated with the maxilliped. The outer maxillæ (pl. IX, fig. 52 *b*) are three-lobed and strongly ciliated. The inner maxillæ (pl. IX, fig. 52 *c*) are two-lobed, the lobes robust and short, the outer armed with short spines at the apex, the inner with three slender curved setæ.

The thoracic segments are coarsely granulated or tuberculated; the first is produced at the sides around the head nearly to the eyes; the others have their anterior and posterior margins transverse. The fourth segment in the female is a little less than three times as long as broad, and is longer than the other six segments taken together, but is only four-fifths as long as the last three segments together with the pleon. It is tuberculated, especially above, but bears no prominent tubercles or spines, and is subcylindrical. In the male this segment (pl. VIII, fig. 48 *b*) is more elongate and much more slender, exceeding in length the three following segments with the pleon. In the ordinary position the thorax is geniculate at the posterior articulation of the fourth segment, forming nearly a right angle with the rest of the body. The last three segments have their epimeral regions angulated and salient. The first pair of legs (pl. VIII, fig. 49 *b*) are of moderate length and, beyond the basal segment, flattened; the basal segment is directed backward but the leg is bent upon itself at the ischium and the remaining segments are directed forward and applied to the under surface of the head. The ischium and merus support but few cilia, and these mostly along their inner margins, but the carpus, propodus, and dactylus are not only ciliated on the inner margin with slender simple cilia, but also bear on the side toward the body stout scattered spinulose setæ, which are specially abundant on the propodus. The opposite side of the leg is nearly smooth. The second, third, and fourth pairs of legs are five-jointed and similar to each other, except that the basal segments of the second and third are somewhat shorter than in the fourth (pl. VIII, fig. 50). The second pair is shorter than the third, and the fourth is a little the longest. All these legs are directed strongly forward and habitually held nearly in the position shown in the figure, under the anterior surface of the body and the head. The last three segments are furnished with elongated setæ along their inner margins. These setæ are inserted in two rows and so placed as to diverge at an open angle. The dactyli appear to be obsolete in these legs. The fifth, sixth, and seventh pairs of legs are of quite a different and more ordinary structure. They contain the full number of segments, and are terminated by robust, slightly curved dactyli. A young specimen obtained has only two pairs of legs of the ordinary form, the last or seventh pair being represented only by rounded tubercles, one on each side of the seventh segment.

The pleon is elongate-ovate, narrower in the male (pl. VIII, fig. 48 *c*). Dorsally it is strongly convex, especially in front. It is two-thirds as long as the fourth thoracic segment in the female, and three-fifths as long as that segment in the male. It is provided with rather coarse tubercles in front, which are arranged transversely in three rows, and behind the third row is a deep transverse groove, behind which the tubercles are less prominent and more of the character of granulations. On each side before the middle is a prominent, sub-acute tooth, directed outward and backward immediately above the articulation of the uropods. The tip of the pleon is not spiniform, but only slightly attenuated and obtuse. The pleopods are delicate in structure, and the anterior pairs are ciliated. The uropods or opercula are more than nine-tenths as long as the under surface of the pleon (pl. VIII, fig. 48 *c*), but cannot be seen from above. They consist on each side (pl. VIII, fig. 51) of an elongated, semi-oval, basal, lamellar segment, thickened and vaulted externally, with the anterior end rounded, and bearing a salient semi-circular process on the outer margin near the anterior end, for articulation with the pleon. Posteriorly this plate is tapering and it is broadly truncated at the tip, where it bears two lamelliform rami. Of these the external is thick, like the basal segment, and is of an elongate triangular form and completes the operculum behind, while the inner ramus is a small and delicate oval plate, articulated to the basal segment near its inner distal angle, and completely covered and concealed by the outer ramus when the operculum is closed. The inner ramus is sparingly ciliated at the tip. The pleopods are very delicate, and the anterior pairs are ciliated.

In the females the lamellæ forming the incubatory pouch are thickened and tuberculated or granulated along the outer edge where they are attached to the segment. The thickened area is bounded by a longitudinal ridge, beyond which the lamella is thin, smooth, and translucent, permitting the eggs to be seen through it, and the thin portion of the right lamella (in the specimen examined) overlaps its fellow of the opposite side so far as to bring its edge along the base of the ridge bounding the thickened portion of the opposite lamella. Near the anterior end and on the outer side is a rounded lobe in the margin of the lamella for articulation with the segment.

Length of female 10^{mm}; male 11^{mm}; diameter of fourth thoracic segment, female 1.2^{mm}; male 0.52^{mm}; color in alcohol, nearly white.

This species was described by the writer without having seen Sars' description of *Leachia granulata*. The volume containing his description has since been obtained by the Yale College Library, and a careful comparison of our specimens with his description leaves little doubt that the species is identical with his. His specimens were somewhat larger than ours, females measuring 14^{mm} and males 17^{mm}. The females in *A. longicornis* Sowerby are much larger than the males, and the reverse relation of size in this species appears to be unusual in the genus.

Specimens were first collected on this coast on George's Bank !, in the summer of 1877, and the three then obtained were found adhering to *Primnoa*, and had been dried and somewhat broken. Better specimens were collected adhering to the cable of the schooner 'Marion,' at Banquereau!, by Capt. J. W. Collins, August 25, 1878, and a fine specimen was obtained in seven fathoms off Miquelon Island!, south of Newfoundland, by Capt. C. D. Murphy and crew of the schooner 'Alice M. Williams,' July 3, 1879. Sars' specimens were collected between Norway and Iceland at stations 18 and 48, of which the respective localities as given by him are latitude 62° 44.5′ north, longitude 1° 48′ east, in 412 fathoms, clayey bottom, and latitude 64° 36′ north, longitude 10° 21.5′ west, in 299 fathoms, clay and sand.

Specimens examined.

Number.	Locality.	Fathoms.	Bottom.	When collected.	Received from—	Specimens.		Dry. Alc.
						No.	Sex.	
2045	George's Bank......	....	....	——, 1877	U. S. Fish Com	2	♀ y	Alc.
2046	...do	....	....	——, 1877	...do	1	♂	Alc.
	Banquereau, N. S...	250	Rocky	——, 1878	Capt. J. W. Collins.	3	♀	Alc.
	Off Miquelon Island.	7	....	July 3, 1879	Capt. C. D. Murphy and crew.	1	♀	Alc.

VII.—SPHÆROMIDÆ.

Body short and convex; head transverse; antennulæ and antennæ multiarticulate, with evident distinction into peduncle and flagellum; mandibles palpigerous; epimera united with the thoracic segments; anterior segments of the pleon short, united and articulated with the large terminal segment; uropods lateral with only one movable ramus.

This family is sparingly represented on the eastern coast of the United States, and within our limits only a single species is found, belonging to the typical genus *Sphæroma*. The animals are usually of small size, and have the body short, broad, and convex. The head is transverse, and both pairs of antennæ are inserted near together below its anterior margin. These organs are much better developed than in the following family. The epimera are faintly indicated in the thoracic segments by impressed lines. The anterior segment of the pleon is similarly marked with transverse sutures indicating the segments of which it is composed. The last segment is large, and one or more of the posterior segments may be notched, tuberculated, spiny, or variously modified, as occurs in many foreign genera. Below, the pleon is much excavated for the pleopods, which, as usual, are in five pairs, the anterior three ciliated. In the males a slender stylet is articulated near the base of the inner lamella of the second pair, and lies along its inner side, so that in the natural position they lie close together on opposite sides of the middle line of the body. These pleopods, though received into a cavity in the under surface of the pleon, are not protected by any operculum nor opercular plates, as in most of the preceding families, nor is the external pair thickened, as in the *Anthuridæ*.

Sphæroma Latreille.

Sphæroma Latreille, Hist. nat. des Crust. et des Ins., tome vii, p. 11, 1804.

Body contractile into a sphere; antennulæ and antennæ short or of moderate length; maxillipeds with a five-jointed palpus; legs all ambulatory; dactyli short and thick; uropods short, ramus and basal segment subequal.

The name of this genus is derived from the peculiar habit of many of the species of rolling themselves into a ball when alarmed. The body is so constructed as to facilitate this operation, the antennulæ and antennæ being received into a groove at the side of the head; the epimeral regions of the thoracic segments behind the first are narrowed nearly to a point and project well downward so as to meet very close together and still leave room for the included legs, while the uropods, shutting together like a pair of scissors, fold also partly under the large terminal segment of the pleon and fill the crevice between the pleon and the head. The maxillipeds in this genus are provided with a long densely ciliated five-jointed palpus. The maxillæ are much as in the *Idoteidæ*, the outer pair three-lobed and strongly ciliated, the inner two-lobed with the inner lobe small and tipped with pectinate setæ, the outer larger and armed with curved denticulated spines. The mandibles have a strong molar process, a dentigerous lamella armed with acute teeth, and a three-jointed palpus.

The legs are rather weak and nearly alike throughout, all ambulatory. The pleon is scarcely narrower than the segments of the thorax and appears to consist of two* segments only, of which the first is much like the last thoracic segment, but more strongly produced at the sides than is that segment and marked with impressed lines. It is articulated with considerable motion to the large scutiform terminal segment, which, in this genus, is rounded and entire at the tip, and not strongly tuberculated nor spiny. Anteriorly, the angles of this segment are produced downward into a rounded lobe in front of the shoulder from which arise the uropods. These organs are not greatly elongated; the basal segment is produced into a plate about equal in size to the single ramus.

Sphæroma quadridentatum Say.

Sphæroma quadridentata Say, Jour. Acad. Nat. Sci. Phil., vol. i, p. 400, 1818.
Dekay, Zool. New York, Crust., p. 44, 1844.
White, List Crust. Brit. Mus., p. 102, 1847.
Harger, Am. Jour. Sci., III, v., p. 314, 1873; This Report, part i, p. 569 (275), pl. v., fig. 21, 1874; Proc. U. S. Nat. Museum, 1879, vol. ii, p. 161, 1879.
Verrill, This Report, part i, p. 315 (21), 1874.

PLATE IX, FIG. 53.

The outline of the body when extended is a pretty regular ellipse, but the animal, when disturbed, rolls itself into a ball with facility, and by

* The pleon is inadvertently described by Bate and Westwood in the British Sessile-Eyed Crustacea, vol. ii, p. 401, as "having all the segments fused together."

this habit may be distinguished from the other marine Isopods of our coast.

The head is rounded in front with an elevated margin, and a slight median projection between the bases of the antennulæ. The eyes are small and sub-triangular, widely separated. The antennulæ and the antennæ are inserted on the inferior surface of the head, and, when the animal contracts, they are received into a groove along the margin of the head and anterior thoracic segment. The antennulæ (pl. IX, fig. 54 *a*) have the basal segment large, the second segment small and conical, the third slender, cylindrical; the flagellum about ten-jointed, ciliated, shorter than the peduncle. In the antennæ (pl. IX, fig. 54 *b*) the peduncular segments decrease but little in diameter, and increase in length from the first to the fifth, and are followed by a flagellum about as long as the peduncle, tapering from the base, with the basal segments strongly ciliated along their inner or anterior distal margins. The antennæ are separated at the base by a triangular, somewhat projecting epistome, which also partly separates the bases of the antennulæ. The maxillipeds have the basal segment short and somewhat triangular, with plumose setæ at the acute apex, and a five-jointed palpus, of which the first segment is short and smooth, and the following segments strongly ciliated along more or less of their inner margins. The outer maxillæ are terminated by three ovate rather acute lobes, which are strongly ciliated. The inner maxillæ have the inner lobe tipped with four pectinated curved setæ, and the outer armed with strong denticulated spines. The mandibles are robust and bear on their external surface at the apex a dentigerous lamella, or usually two such on the right mandible, receiving the lamella of the left between them; below the lamella is a strongly ciliated ridge supporting the dentigerous lamella and connecting it with the molar process, which is large and strong. The mandibular palpi are slender, with the last segment sub-semicircular, bearing at its apex a few serrated spines, and below a comb of straight setæ; the middle segment bears a similar comb with stouter spiny setæ at the ends.

The first thoracic segment is longer than the others, and much elongated at the sides, embracing the head as far as its anterior margin. Above this lateral expansion on each side the segment is excavated for a projecting lobe of the head behind the eye. The second, third, and fourth segments are somewhat shorter than the first and longer than the fifth, sixth, and seventh. The margin of the last segment bends slightly backward at the middle. In the thoracic segments behind the first the epimeral sutures are indicated by a faint depressed line, below which the lateral margin of the second segment tapers to an obtusely rounded point, the third is more acutely pointed, the fourth oblique and acute behind, the fifth and sixth also oblique but less acute, and the seventh rounded. The legs are weak, hairy, and much alike throughout, formed for walking, and none of them chelate. The dactylus in all is short and robust, armed with a stout curved spine or claw at the tip, and a smaller straight spine below it. In the first pair of legs the carpus is short and triangular, the ischium and merus bear on their upper margin a row of long slender plumose hairs. In the second and third pairs of legs these hairs are also found, and the carpus is longer. The fourth pair of legs are robust, the following pairs more slender to the seventh. All are well provided with slender hairs, with a few stouter ones intermixed.

The anterior segments of the pleon are consolidated into a single piece somewhat resembling the last thoracic segment, but marked at the sides by depressed lines, indicating sutures, as shown in pl. IX, fig. 53. At the sides this segment is broadly rounded and projects much below the seventh thoracic when the animal is contracted. The large terminal segment has a similar lobe in front of the bases of the uropods. At the insertion of the uropods the segment is considerably contracted laterally, but is rounded and strongly margined behind. Its anterior lobe, all the thoracic segments, and the head are also margined by an elevation running completely around the animal except where it is interrupted by the uropods. The uropods extend nearly to the tip of the telson, and consist on each side of a basal segment continued backward into a narrow oval plate with entire margins, flattened below, where a similarly-shaped ramus is articulated near its base, the two shutting together like the blades of a pair of scissors. The articulated plate bears four more or less acute serrations on its exterior margin, whence the specific name. The pleopods are ciliated, and the second pair (pl. IX, fig. 54*c*) bears, in the male, on the inner lamella, a slender curved stylet, longer than the lamella, and articulated near its base.

Length about 8^{mm}, breadth 4^{mm}. The color, as usual in shore species, is variable; some are of a uniform slaty gray, many are marked on the dorsal surface with a whitish, cream color, or rosaceous patch, bordered more or less with dark or black. This patch has commonly a longitudinal direction, and is usually symmetrical, and may be broad or much narrowed in the middle. On the dark or barnacle-covered rocks, where these animals are often found, the colors are evidently protective, but they are imperfectly preserved in alcohol.

This species was described by Say, who "found these animals very numerous on the beach of Saint Catherine's Island, Georgia, concealing themselves under the raised bark, and in the deserted holes of the *Teredo*, &c., of such dead trees as are periodically immersed." He also gives East Florida as a locality, and there are specimens in the Yale Museum from Florida! It extends as far north as Provincetown, Mass.! near the extremity of Cape Cod. It is common on the southern shore of New England!, and is usually found among algæ or rocks.

Specimens examined.

Number.	Locality.	Fathoms.	Bottom.	When collected.	Received from—	Number of specimens.	Dry. Alc.
2054	Florida				Smithsonian Inst	00	Alc.
1224	Great Egg Harbor, N. J.			April, 1871	Smith & Verrill	.....	Alc.
2053	New Haven, Conn				S. I. Smith	5	Alc.
	Savin Rock, New Haven	L. w.	Rocky			00	Alc.
2052	Stony Creek, Conn	L. w.	Rocky			00	Alc.
2049	Vineyard Sound, Mass			——, 1871	U. S. Fish Com.	3	Alc.
2050	....do	L. w.		——, 1871	....do	1	Alc.
2051	... do			——, 1875	....do	5	Alc.
	Provincetown, Mass	L. w.		Aug.—, 1879	....do	00	Alc.
	....do	½	Eel-grass	Aug.—, 1879	....do	1	Alc.

VIII.—LIMNORIIDÆ.

Body compressed; antennulæ and antennæ short, subequal; mandibles palpigerous, formed for gnawing; feet not prehensile, all similar, with short, robust dactyli; epimera united with the thoracic segments; pleon of six distinct segments; pleopods similar in form throughout; uropods lateral, biramous.

This family as constituted above contains the single genus *Limnoria* Leach, which appears also to contain but few, or perhaps a single, species* of wide distribution. This genus was placed in the tribe *Asellotes homopodes* with the *Asellidæ* by Edwards, without, however, having examined the animals himself. He has been generally followed in this arrangement by later authors. Previous authors had associated the genus, as it appears to me more justly, with *Sphæroma* and the *Cymothoidæ* in the wide signification of the latter term. White, in his List of British Crustacea, used the name *Limnoriadæ* to include this genus with the *Asellidæ*. I have preferred to constitute a new family for the genus, which has, however, evident relations with the *Sphæromidæ*, and perhaps should yet be united with that family.

Under the circumstances family characters can scarcely be separated with certainty from those of generic or even of specific value only, but for the purpose of comparison with other families certain important characters may be here stated. The body is somewhat depressed dorsally, but is also compressed at the sides, and when extended is subvermiform. It is nearly capable of being rolled into a ball, as in the genus *Sphæroma*. The head is of moderate size and strongly rounded above, as in *Sphæroma*, and the eyes are widely separated and on the sides of the head, a condition not usual in the *Asellidæ*. The antennulæ are short and stout and the basal segment is but little larger than the second; the flagellum

*It is perhaps hardly necessary to remark that *L. xylophaga* Hesse, Ann. Sci. nat., tome x, p. 101, pl. ix, 1868, is not an Isopod. According to Prof. Smith it is *Chelura terebrans* Phillipi, a boring amphipod often found associated with *Limnoria*. See an article by that author in the Proceedings of the U. S. National Museum, 1879, vol. ii, pp. 232–235.

consists of a single, almost rudimentary segment. The antennæ differ widely from any in the *Asellidæ*, since they are less robust than the antennulæ, and but little longer; the peduncular segments are all short, having almost the same proportion to each other as in *Sphæroma* (see pl. IX, figs. 54 *b* and 56 *b*), the last two being together about equal in length to the first three, instead of far surpassing them as in the *Asellidæ;* the flagellum is short and few-jointed, mostly made up of a tapering basal segment, and not at all resembling the slender multiarticulate flagellum of the *Asellidæ*. The mandibles are adaptively modified in accordance with the boring habits of the species, but the other mouth parts do not seem to present characters from which comparisons need be drawn with other families.

The legs are somewhat similar to those seen in many *Asellidæ*, being furnished with short dactyli, each armed with a strong curved claw, and a shorter spine below. A similar form of leg is, however, seen in *Sphæroma*. The epimera are united to the lateral margins of the thoracic segments almost precisely as in *Sphæroma*, an arrangement that does not prevail in the *Asellidæ*.

The pleon has all its six segments well developed and perfectly separated from each other, while in the *Asellidæ* they are united into a single scutiform segment, or at most, the basal segment only is more or less distinct. The pleopods are of the normal number and similar in form and texture throughout; the anterior pairs are ciliated. Each pair of pleopods consists of a basal segment, bearing an inner narrow lamella and an outer oval one, which, except in the fifth pair, are well ciliated. In the male the inner lamella of the second pair bears, on its inner margin, a stylet, as in *Sphæroma* and many other genera of Isopoda. In the *Asellidæ* the branchial pleopods are in fewer than five pairs, and are protected in front by a simple or compound operculum of firmer texture than the other pleopods. Dr. Coldstream * fell into an error in describing the respiratory organs as consisting of "six pairs of scale-like bodies, pendant from the anterior segments of the tail, * * arranged in three rows, in an imbricated manner, one of each kind ('oval' and 'nearly quadrangular') being articulated together on a common peduncle on either side." He further describes, loc. cit., p. 324, "two vesicular bodies of an oval form" behind the branchiæ. These organs were without doubt the external lamellæ of the fifth pair of pleopods, as shown by his figure. There are, however, four instead of three ciliated pairs anterior to the last pair, one of which was overlooked by Dr. Coldstream, and in this error he has been followed by Bate and Westwood. † If the observations of Dr. Coldstream had been correct, an affinity might have been indicated with the *Asellidæ*. The terminal segment is flattened and scutiform, in shape resembling that of *Jæra*, but the uropods are strictly lateral, being attached at the broadest part of the segment and in front of the middle.

* Edinburgh New Phil. Journal, vol. xvi, p. 323.

† Brit. Sessile-Eyed Crustacea, vol. ii, p. 350.

The relations of the present family with the *Sphæromidæ* appear to be more close, but the structure of the mandibles and perhaps also that of the maxillipeds, the fully segmented pleon and the biramous uropods seem to be characters of family value, which, however, a fuller investigation of the boring *Sphæromidæ* might go far to break down.

Limnoria Leach.

Limnoria Leach, Edinburgh Encyc., vol. vii, p. "433" (Am. ed., p. 273), "1813–14."

Mandibles with a nearly even chisel-like cutting-edge at the tip and no molar process; maxillipeds elongate, with a well-developed external lamella and a five-jointed palpus; first thoracic segment large; uropods with the outer ramus very short and almost obsolete.

The above characters differ from those by which Leach separated this genus from *Cymothoa* and the *Sphæromidæ*, with which he associated it.

Limnoria lignorum White (Rathke).

"*Cymothoa lignorum* Rathke, Skrivt. af Naturh. Selsk., v. 101, t. 3, f. 14, 1799" (White).

Limnoria terebrans Leach, Ed. Encyc., vol. vii, p. '433' (Am. ed., p. 273), "1813–14"; Trans. Linn. Soc., vol. xi, p. 371, 1815; Dict. Sci. nat., tome xii, p. 353, 1818.

Samouelle, Ent. Comp., p. 109, 1819.

Desmarest, Consid. Crust., p. 312, 1825.

Latreille, Règne Anim., tome iv, p. 135, 1829.

Coldstream, Edinb. New Phil. Jour., vol. xvi, pp. 316–334, pl. vi, 1834.

"Hope, Trans. Ent. Soc. Lond., vol. i, p. 119" (B. & W.).

Thompson, Edinb. New Phil. Jour., vol. xviii, p. 127, 1835; Ann. Mag. Nat. Hist., vol. xx, p. 157, 1847.

Templeton, Loud. Mag. Nat. Hist., vol. ix, p. 12, 1836.

Moore, Charlesworth's Mag. Nat. Hist., n. s., vol. ii, p. 206, 1838; ibid., vol. iii, pp. 196, 293, 1839.

Edwards, Annot de Lamarck, tom. v, p. 276, 1838; Hist. nat. des Crust., tom. iii, p. 145, 1840; Règne Anim. Crust., p. 197, pl. 67, f. 5, 1849.

Gould, Invert. Mass., pp. 338, 354, figure, 1840.

Fleming, Encyc. Brit., 7 ed., vol. vii, p. 502, 1842.

Dekay, Zool. New York, Crust., p. 48, pl. ix, fig. 33, 1844.

"Kirby and Spence, Int. Entom., 5th ed., p. 238; 6th ed., p. 203" (White.)

White, List Crust. Brit. Mus., p. 96, 1847; Brit. Crust. B. Mus., p. 68, 1850.

Dalyell, Powers of the Creator, vol. i, p. 241, pl. lxv, figs. 7–15, 1851.

Leidy, Jour. Acad. Nat. Sci. Phil., II, vol. iii, p. 150, 1855.

Gosse, Man. Mar. Zool., vol. i, p. 136, fig. 242, 1855.

Steenstrup and Lütken, Vidensk. Meddel., II, vol. ii, p. 275, 1861.

Hesse, Ann. Sci. nat., Zool., V, tome x, p. 113, 1868.

Jones, Trans. Nova Scotian Inst. Nat. Sci., vol. ii, pt. iv, p. 99, 1870.

Verrill, Proc. Am. Assoc., 1873, p. 367, 1874.

Macdonald, Trans. Linn. Soc., II, Zool., vol. i, p. 67, 1875.

Andrews, Q. Jour. Mic. Sci., II, vol. xv, p. 332, 1875.

Limnoria lignorum White, Pop. Hist. Brit. Crust., p. 227, pl. 12, fig. 5, 1857.

Bate, Rep. Brit. Assoc., 1860, p. 225, 1861.

Bate and Westwood, Brit. Sess. Crust., vol. ii, p. 351, figure, 1868.

Limnoria lignorum—Continued.

Norman, Rep. Brit. Assoc., 1868, p. 288, 1869.
Möbius, Wirbellos. Thiere der Ostsee, p. 122, 1873.
Parfitt, Fauna of Devon, Sess. Crust., p. (19), 1873.
Verrill, Am. Jour. Sci., III, vol. vii, pp. 133, 135, 1874; Proc. Am. Assoc., 1873, p. 371, 1874; This Report, part i, p. 379 (85), 1874.
Harger, This Report, part i, p. 571 (277), pl. vi, fig. 25, 1874; Proc. U. S. Nat. Mus., 1879, vol. ii, p. 161, 1879.
M'Intosh, Ann. Mag. Nat. Hist., IV, vol. xiv, p. 273, 1874.
Stebbing, Trans. Devon. Assoc., 1874, p. (8), 1874. Ann. Mag. Nat. Hist., IV, vol. xvii, p. 79, 1876.
Whiteaves, Further Deep-Sea Dredging, Gulf St. Lawrence, p. 15, "1874."
Metzger, Nordseefahrt der Pomm., p. 285, 1875.
Meinert, Crust. Isop. Amph. Dec. Daniæ, p. 77, 1877.
Smith, Proc. U. S. Nat. Mus., 1879, vol. ii, p. 232, fig. 2, 1880.

Limnoria uncinata Heller, Verh. k. k. Zool. bot. Ges. Wien, B. xvi, p. 734, 1866.
Stalio, Cat. Crost. Adriatic, p. 211, 1877.

PLATE IX, FIGS. 55–57.

This species may in general be recognized by its habits, being usually found burrowing in submerged timber, to which, notwithstanding its insignificant appearance, it often proves very destructive.

The body is subcylindrical, tapering slightly at each end and covered above with short hairs to which more or less dirt usually adheres. The head is narrower than the first thoracic segment. The eyes are lateral and consist of about eight ocelli, one central and the others around it. The antennulæ (pl. IX, fig. 56 *a*) are short and seem to arise from near the middle of the front of the head. The basal segment is the largest; the second and third are of slightly decreasing size; the fourth or flagellar segment is much the smallest, and tipped with setæ. The antennæ (pl. IX, fig. 56 *b*) are more slender than the antennulæ, and arise just below their bases and a little farther apart. The first two segments are short; the third slightly longer; the fourth and fifth increasing somewhat in length; the flagellum is not longer than the last two peduncular segments, and consists of a tapering segment, followed by a few short terminal segments provided with a terminal brush of setæ. The maxillipeds (pl. IX, fig. 56 *c*) are slender; the external lamella is semi-ovate, with the inner margin nearly straight, acute, and ciliated at the tip; the palpus is five-jointed but short, with the segments flattened, and all but the first ciliated along their inner margins. The outer maxillæ (pl. IX, fig. 56 *d*) are slender, three-lobed, and ciliated at the tip. The inner maxillæ (pl. IX, fig. 56 *e*) are also slender, the inner lobe tipped with pectinate bristles, the outer with robust spines. The mandibles (pl. IX, fig. 56 *f*) are somewhat elongate, but of a simple form, being curved inward, flattened and chisel-shaped at the tip; below there is a slight tubercle, apparently the rudiment of the molar process; externally, above the origin of the palpus, is a prominent tubercle; the palpus is short, of three subequal segments, the last furnished with a rather imperfect comb of setæ.

The first thoracic segment is about twice as long as any that follow; it is crossed by a broad, shallow depression, and is rounded at the sides.

The second and third segments are each about half the length of the first. The epimeral sutures are evident, and the epimera are rounded behind in the second segment, but a little more prominent in the third, becoming acute and increasing in size and extension backward to the seventh. The fourth segment is slightly shorter than the third, and perhaps a little broader; the last three are short, decreasing in length to the seventh, but maintaining about equal width. The legs are short and rather robust. The first pair have the carpus triangular, but this segment becomes more elongate in the succeeding pairs. The dactyli are robust, and are armed with a strong curved spine or claw at the tip and a smaller one below it. The merus, and usually the ischium and carpus, bear a few spiniform tubercles on the lower surface except in the last pair, which are also more elongated and slender than the others.

The pleon is scarcely narrower than the thorax, and tapers but little; the first four segments are of equal length; the fifth is longer with a median elevation and a transverse depression on each side. The last segment (pl. IX, fig. 57*a*) is transversely oval or subcircular, broader than long, with the anterior margin raised, especially at the middle, where the elevation is continued a short distance on the segment, but posteriorly it is flattened. The posterior margin is ciliate with hairs of various lengths. The uropods (pl. IX, fig. 57*b*) are attached just in front of the middle of the segment at its widest part. They consist on each side of a somewhat wedge-shaped basal segment, ciliated and bluntly denticulated distally on the outer side, and supporting two rami, between which it is produced below into a strong tooth-like process. The outer ramus is very short and curved outward; the inner is not as long as the basal segment, and is ciliated externally and at the tip. Underneath, the pleon is much excavated for the pleopods, which are strongly ciliated. The first pair (pl. IX, fig. 57*c*) consist on each side of a short basal segment bearing two lamellæ; the inner lamella is almost four times as long as broad, with nearly parallel sides, ciliated at and near the tip; the outer, which is also in front of the inner, is sub-oval with the outer margin more convex than the inner, ciliated near the tip and along most of the outer margin, and inserted a little obliquely upon the basal segment. The next three pairs of pleopods are similar to the first pair on each side, except that in the males the second pair (pl. IX, fig. 57 *d*) bears a stylet (*s*) articulated to the inner margin of the inner lamella about the middle. The posterior pair of pleopods are smaller than the others and not ciliated.

Length 4.5^{mm}; breadth 1.5^{mm}; color light grayish.

Much has been written upon the destructive habits of the *Limnoria* or "gribble" and the means of preventing its attacks on woodwork, for which the reader may consult especially the publications of Leach, Coldstream, Hope, Thompson, Moore, Gould, Bate and Westwood, Verrill, and Andrews, who has observed it attacking the gutta-percha of submarine telegraph-cables.

It is found boring in submerged wood along our coast from Florida! to Halifax!, N. S., and the Gulf of St. Lawrence. It occurs above low-water mark, but does not usually live far below that line; it has, however, been found by Professor Verrill at a depth of 10 fathoms in Casco Bay, and was dredged by the U. S. Fish Commission in a depth of 7½ fathoms, Cape Cod Bay!, Mass., in the summer of 1879. It is abundant, according to European authors, in many localities on the coast of Great Britain and in the North Sea. *L. uncinata* Heller, from Verbosca, in the Island of Lesina, Adriatic Sea, appears to be the same species, as the differences pointed out by Heller do not really exist, but were doubtless suggested by the incorrect figures that have been published representing the uropods with rami composed of two or more segments. The form of these appendages, as shown on plate IX, fig. 57 *b*, corresponds well with Heller's description. It was found by Heller associated with *Chelura terebrans*. *Limnoria* is said also to occur in the Pacific Ocean, and from its habits might be expected to have a wide distribution.

Specimens examined.

Number.	Locality.	Habitat.	When collected.	Received from—	Number of Specimens.	Dry. Alc.
2048	Florida	Boring in wood		Smithsonian Inst.	6	Alc.
	Provincetown, Mass	...do	Aug., 1879	U. S. Fish Com	00	Alc.
2047	Casco Bay	...do	——, 1873	...do	30	Alc.
	Bay of Fundy	...do	——, 1872	...do	00	Alc.
	Halifax, N. S	...do	——, 1877	...do	00	Alc.

IX.—CIROLANIDÆ.

Front formed of the approximate basal segments of the antennulæ, which are not covered by an anterior projection of the head; antennulæ and antennæ presenting an evident distinction into peduncular and flagellar segments; maxillipeds with a five-jointed palpus; mandibles formed for biting, palpigerous; legs all terminated by nearly straight dactyli; epimera distinct behind the first thoracic segment; pleopods at least the anterior pairs, ciliated; uropods biramous, the rami flattened and ciliated.

This family is represented on our coast by two closely allied species apparently belonging to the typical genus *Cirolana*, although approaching the allied genus *Conilera*, to which I formerly referred them. They have been hitherto usually referred to the following family, but the differences in the structure of the mouth parts, first pointed out by Schiödte, seem to warrant their separation as a distinct family. The mandibles are formed for biting, being armed with long and powerful teeth, which, closing together like the blades of scissors, are well adapted for lacerating the flesh of fishes on which they feed. The first three pairs of legs are fitted for prehension, but they are destitute of the strongly curved dactyli found in the *Ægidæ*, and still better developed in the *Cymothoidæ*. In the *Cirolanidæ* the propodus, in the first three pairs of legs, is somewhat curved and the dactyli are nearly straight, so that while the first three pairs of legs are powerful organs of prehension, they are also capable of letting go preparatory to the seizure of another victim. The posterior pairs of legs are ambulatory or fitted for swimming by their form and armature of bristly hairs. The ciliated pleopods are also powerful swimming organs, so that these animals are well fitted for the predatory life they lead. The epimera are well separated by sutures in all the thoracic segments behind the first. The pleon is scarcely narrower at base than the last thoracic segment, and is composed of six distinct segments, of which the last is much the longest, but not broader than the preceding segments, and tapers posteriorly. The uropods are lateral, articulated near the base of the last segment and distinctly biramous.

The mouth-organs of this and the two following families have been the object of special research by J. C. Schiödte, whose papers in the Naturhistorisk Tidsskrift have been in part translated in the Annals and Magazine of Natural History. He regards *Cirolana* as representing "the highest development of the crustacean type among the Isopoda," and even hints that *Cirolana* and *Æga* should be removed to opposite ends of the series of Isopoda. The same author would closely unite the *Bopyridæ*, *Æga*, and the *Cymothoidæ* into a single group, the *Cymothoæ*, while acknowledging that the young of *Cymothoa œstrum*, "according to the classification hitherto current, * * * would rather be allied to *Cirolana* than to *Cymothoa*." His classification, however, appears to be based almost entirely upon the structure of the mouth, disregarding the totality of structure upon which alone morphological classification can securely rest. In deference, however, to his views I have here regarded *Cirolana* as the type of a distinct family, which must still be considered as closely related with the two following families, on the principle that it is "more important that similarities should not be neglected than that differences should be overlooked."

Among the more important of the similarities by which these families seem to be united may be mentioned the following, as exemplified by our species. The segments of the thorax and pleon are all distinct from each other, so that the body, in the adults, appears to consist of thirteen segments behind the head, although in the genus *Ourozeuktes* Edwards* the segments of the pleon are consolidated. The epimera are distinct in all the segments behind the first thoracic. The pleon may or may not taper from the base, but it is terminated by a large scutiform segment, sometimes more or less sculptured, and bearing at the sides, near the base, a pair of uropods, in which the basal segment is more or less oblique distally and the rami lamelliform, though one of them may be narrowly so. The pleopods are unprotected by any form

* Hist. nat. des Crust., tome iii, p. 275, 1840.

of operculum and the anterior pairs are ciliated in the young of all three families, but this ciliation, as well as that on the uropods, may be lost in the sedentary adults of the *Cymothoidæ*. In all our species the dorsal surface is smooth throughout, or minutely punctate under a lens, but destitute of distinct roughness, tuberculation or sculpture, except that the telson may be faintly grooved or sculptured, and in some foreign species more distinctly so.

Cirolana Leach.

Cirolana Leach, Dict. des Sci. nat., tome xii, p. 347, 1818.

Thoracic segments subequal; eyes small, well separated; mandibles armed with strong acute teeth; dactyli straight, or but slightly curved; pleon of six distinct segments; basal segment of uropods with the inner angle produced.

Two closely allied species are found on this coast, which I formerly referred to the genus *Conilera* Leach. Further consideration induces me to refer them rather to the present genus, although they have some features which point toward *Conilera*, and are perhaps between that genus and the typical forms of *Cirolana*. From *Conilera*, as described by Bate and Westwood, our species differ principally in the more robust four posterior pairs of legs, in the produced angle of the basal segment of the uropods, and in the structure of the first pair of pleopods, which are not operculiform either in size or texture. Of these two species one is abundant and is described at length. The description will, however, apply almost equally well to the other except in the few points mentioned in the appropriate place. The characters given, though slight, appear to be constant, and I have therefore retained the two specific names.

This genus differs from *Æga* in the structure of the legs, and was placed by Professor Dana in a separate subfamily. In *Cirolana* the first three pairs of legs are strong, and armed with minute spine-like claws at the tip of the nearly straight dactyli; the propodi in these legs are robust, spiny, and somewhat curved, and some of the preceding segments are also armed with spines. These legs thus form powerful organs for seizing living prey, and are not, as in the *Cymothoidæ*, and, in a less degree, in *Æga*, merely fitted by their curved dactyli to retain the hold of the animal upon its host in a parasitic existence. The last four pairs of legs are well ciliated and capable of use either for walking or swimming, and these animals are thus fitted for their active and predaceous life.

Cirolana concharum Harger (Stimpson).

Æga concharum Stimpson, Mar. Inv. G. Manan, p. 42, 1853.
Lütken, Vidensk. Meddel., 1859, p. 77, 1860.
Conilera concharum Harger, This Report, part i, p. 572 (278), 1874.
Verrill, This Report, part i, p. 459 (165), 1874.
Cirolana concharum Harger, Proc. U. S. Nat. Mus., 1879, vol. ii, p. 161, 1879.

PLATES IX AND X, FIGS. 58–63.

This species may be most readily recognized among our Isopoda by the distinct thoracic and abdominal segments, the small lateral eyes, and the evident distinction, in both antennulæ and antennæ, of peduncle and flagellum. From the next species it is distinguished by the tip of the telson, which is truncated, or slighty emarginate, and grooved on the median line above near the end.

The body is, when extended, about three times as long as broad, and is smooth and polished throughout. The head is quadrate, a little broader in front than behind, and embraced at the sides by the first thoracic segment. The eyes are triangular, with the angles rounded, and are often partially covered below by the projecting anterior lobes of the first thoracic segment. They are separated by about three times their longest diameter. The antennulæ (pl. X, fig. 60) are robust, with their basal segments in contact; the first segment is short and sub-spherical; the second also short; the third cylindrical and as long as the first two taken together and followed by a robust, but short, tapering flagellum, consisting of about fifteen segments, of which the second is as long as any other two, but the rest are all short. The flagellar segments beyond the first are provided each with a tuft of "olfactory setæ." The antennæ (pl. X, fig. 61 *a*) are longer and more slender than the antennulæ, and are separated at their bases. The first four peduncular segments are robust; the first two short; the third and fourth each about twice as long as the first or second, and the fifth or last peduncular segment slightly the longest and much the most slender. The fourth and fifth segments bear along the distal portion of their outer margins long bristle-form hairs. The flagellum is slender and composed of from 15 to 18 segments, each bearing a few short bristles. The maxillipeds (pl. X, fig. 62 *a*) are elongated and almost pediform but flattened; the external lamella is small and subtriangular, rounded and hairy at the tip; the palpus is five-jointed, with the last four segments broad, flattened, and well ciliated; the tip of the maxilliped, nearly concealed by the large palpus, is provided with very densely plumose bristles. The outer maxillæ (pl. X, fig. 61 *b*) are short and robust; the two articulated lobes narrow ovate, rounded at the tip, armed, especially the inner one, with spines and plumose or pectinated bristles. The inner maxillæ (pl. X, fig. 61 *c*) are robust, with the outer lobe armed with strong smooth spines; the inner lobe rounded at the end and bearing three straight rather blunt spines, densely covered toward the tip with soft hairs. The mandibles (pl. X, figs. 61 *d*) are robust and horny at the tip, armed with one strong acute tooth, and in the right mandible with one acute and one obtuse tooth along a cutting edge, while the left mandible has three less acute teeth along this edge. Each mandible is, moreover, provided with a molar process or area (*m*), on its inner surface set along its interior and upper margin with spines. A narrowly lanceolate leaf-like appendage is attached just below the molar area. This appendage is furnished with a few bristles near the base, and its upper edge is armed with minute denticles; it is movable and ordinarily concealed behind the mandible. On the external surface, just above the origin of the palpus, each mandible bears two elevated, conical, obtuse tubercles. The palpi are slender, the second segment longest and hairy on the margin beyond the middle, the last segment slender and curved, with the usual hairs or slender bristles along the inner curvature.

The second and third thoracic segments are a little shorter than the others, which are of about equal length. The fourth and fifth segments are widest. The first segment is produced at the sides around the head so as to very nearly attain the anterior lateral angles of the head, and often so as to obscure the lower margin of the eyes. The epimeral sutures are scarcely distinguishable in this segment, but evident in the following segments. The epimera are rounded behind as far as the fourth, but the fifth is slightly angulated, and the sixth and seventh acute and produced backward beyond the margin of the corresponding segment. The first pair of legs are short and stout, and well armed with spines and bristles; the basis is of the ordinary form; the ischium is nearly triangular, having the upper margin much produced in the distal portion and bristly; the merus is expanded in a somewhat similar manner, but the angle is bent forward beyond the short carpus over the base of the propodus; the opposite or lower margin of the merus is armed with short stout spines; the carpus is short and small and possesses but little motion on the propodus, which is robust, somewhat curved, and bears a strong short dactylus. The second and third pairs of legs resemble the first pair, but the carpus increases somewhat in size, and there is more motion in its articulation with the propodus. They are directed forward, while the remaining pairs are usually directed backward and are more flattened. The fourth pair of legs are short like the first three (pl. X, fig. 62 *b*), but, except in size, resemble the following pairs. They are well provided with bristles in tufts, and along the margins of the segments, and especially the merus and two adjacent segments, are armed with long stout spines. The propodus is straight and much more slender than the carpus. The fifth and sixth pairs of legs increase in size, and the propodus especially becomes more elongated, but the seventh pair are a little smaller than the sixth.

The pleon is scarcely narrower at base than the last thoracic segment, and the first segment is often nearly concealed by the last thoracic. The fifth segment is longer on the back but shorter at the sides than the preceding segments. The last segment, or telson, is triangular with the ciliated apex truncated and emarginate or notched at the end of a short median furrow at the tip. The uropods (pl. X, fig. 63) slightly surpass the telson and are strongly ciliated; the inner ramus bears also a few spines near the tip; the basal segment has the inner angle produced along the margin of the inner ramus, which is broad and expanded distally, with a notch at the external angle; the outer ramus is slender and tapering, slightly surpassing the inner.

Length of large specimens 32^{mm}, breadth 10^{mm}, but usually smaller; 22^{mm} long, 7^{mm} broad. The ground color in life is yellowish, with reddish brown on the anterior margin of the head and on the posterior margins of the segments, especially in the dorsal region, where the segments are also marked with black dots. In life the body is somewhat translucent in the thinner parts. In alcohol the translucence disappears and the color fades to a nearly uniform yellowish or buff with black dots.

This species was described by Stimpson from Charleston, S. C. Most of the specimens in the collection are from Vineyard Sound!, where it occurs sometimes in great abundance, and is common especially during the winter. It is found swimming about in shallow water, and may be taken in a scoop-net, and is found also in lobster-pots. It was dredged in 45 fathoms off Block Island!, near the eastern end of Long Island Sound, in 1874, but has not yet been found north of Cape Cod.

Specimens examined.

Number.	Locality.	Fathoms.	Bottom.	When collected.	Received from—	Number of specimens.	Dry. Alc.
2061	Off Fishers Island			May —, 1875	J. H. Latham	100+	Alc.
2060	Vineyard Sound			Mar. —, 1874	V. N. Edwards	10	Alc.
2065	... do	S. f.		Aug. 25, 1875	U. S Fish Com.	1	Alc.
2062	Eel-pond, Wood's Holl		Muddy	July 23, 1875	....do	100+	Alc.
2063	Off New Shoreham			Aug. 19, 1874	... do	1	Alc.
2064	Off Martha's Vineyard	18	Sandy	Sept. 20, 1875	....do	1	Alc.

Cirolana polita Harger (Stimpson.)

Æga polita Stimpson, Mar. Inv. Grand Manan, p. 41, 1853.

Lütken, Vidensk. Meddel., 1859, p. 77, 1860.

Verrill, Am. Jour. Sci., III, vol. v, p. 16, 1873.

Conilera polita Harger in Smith and Harger, Trans. Conn. Acad., vol. iii, pp. 3, 22, 1874.

Verrill, Am. Jour. Sci., III, vol. vii, p. 411, 1874.

Cirolana polita Harger, Proc. U. S. Nat. Mus., 1879, vol. ii, p. 161, 1879.

This species so closely resembles the preceding, that a full description would be little else than a repetition of that given above. It appears, however, to differ constantly from the form already described, by its somewhat more elongated and cylindrical body; in the eyes, which are "elongate trapezoidal in shape, narrowest anteriorly," and in the tip of the telson, which is regularly rounded or slightly pointed at the tip without any truncation, much less any emargination, and is not at all grooved above.

Length 25^{mm}, breadth 6.5^{mm}; color much as in the preceding species.

Dr. Stimpson's specimens were "found on the fine sands at low-water mark on High Duck Island," in the Bay of Fundy, and the specimens that I have examined are from Cape Cod Bay!; from near Salem!, Mass.;

George's Banks!, and east of Banquereau!, or Quereau, latitude 40° 36′ north, longitude 57° 12′ west, where seven fine specimens were taken from a halibut (*Hippoglossus*), June 2, 1879, by Capt. J. W. Collins. It appears to replace the preceding species at the north.

Specimens examined.

Number.	Locality.	Fathoms.	Bottom.	When collected.	Received from—	Number of specimens.	Dry. Alc.
	Cape Cod Bay.............	7	Coarse, yellow sand	Sept 15, 1879	U. S. Fish Com.	2	Alc.
1314	George's Bank, lat 41° 40′ N., lon. 68° 10′ W.	25	Sand..........	——, 1872	Smith and Harger.	1	Alc.
1399	George's Bank, lat 42° 11′ N, lon 67° 71′ W.	150	Soft, sandy mud.	——, 1872	Packard and Cooke.	1	Alc.
	Salem, Mass..............	....	....	——, 1878	J. H. Emerton.	1	Alc.
	East Quereau	190	....	June 2, 1879	Capt J. W. Collins.	7	Alc.

X.—ÆGIDÆ.

Front formed of the approximate basal segments of the antennulæ, which are not covered by an anterior projection of the head; antennulæ and antennæ presenting an evident distinction into peduncular and flagellar segments; maxillipeds operculiform; mandibles formed for piercing, palpigerous, mouth suctorial; first three pairs of legs ancoral, last four ambulatory; epimera distinct behind the first thoracic segment; uropods lateral, biramous, ciliated, and flattened.

This family was represented within our limits by a single species of the typical genus until the summer of 1879, when a single specimen was collected of a second genus belonging to the *Ægidæ*, but having evident relations with the next family, and in many characters intermediate between *Æga* and the *Cymothoidæ*. The two genera by which the family is at present represented on our coast may be further characterized as follows: Both the antennulæ and the antennæ are directed laterally, the former arising near together on the anterior margin of the head and forming part of the outline of the animal as seen from above. They, as well as the antennæ, present an evident distinction into peduncular and flagellar segments. The maxillipeds are operculiform, and have the palpus armed with short hooks for adhesion to the surface of the fish on which they may be feeding. The mandibles are armed with a horny point, but not toothed as in the *Cirolanidæ*, and, while fitted for piercing, are not capable of lacerating and biting off pieces of flesh as in that family.

The first three pairs of legs are ancoral, or armed with strong curved dactyli, which, once implanted in the body of a victim, retain their hold without effort—a structure which attains its fullest development in the following family. The remaining pairs of legs are fitted for walking. The thoracic segments are subequal in length and have the epimera well separated, except in the first segment.

The pleon may or may not be suddenly narrower than the last thoracic segment, and, in our species, is composed of six distinct segments, of which the last is large and scutiform. The uropods are composed of a basal segment, oblique at the apex with the inner angle more or less produced, and bearing two flattened, ciliated rami; they are distinctly lateral, being inserted high up on the sides of the last segment.

This family contains our largest Isopod, *Æga psora*, and to it should probably be referred the huge *Bathynomus giganteus* A. Edwards, from the Gulf of Mexico, measuring more than eleven inches in length. It has usually been regarded as embracing the *Cirolanidæ*. I have already given my reasons for separating them, but have to regret my inability to examine many types of genera apparently more or less intermediate in position between *Æga* and, on the one hand *Cirolana*, and on the other *Cymothoa* and *Livoneca*. I have therefore retained the old classification rather than to unite the following genera with the *Cymothoidæ*.

Our two genera are most easily distinguished as follows: Eyes large and approximate, *Æga*, p. 89; eyes wanting, *Syscenus*, p. 93.

Æga Leach.

Æga Leach, Trans. Linn. Soc., vol. xi, p. 369, 1815.

Eyes large; palpus of maxillipeds five-jointed; three anterior pairs of legs terminated by strong curved claws; posterior pairs slender, with slender nearly straight dactyli; pleon not suddenly narrower than the thorax; pleopods ciliated.

This genus is represented within our limits by a single species, which may be easily distinguished by its large approximate eyes. The basal segments of the antennulæ are flattened and the flagellum is comparatively slender. The maxillipeds have a five-jointed palpus, which is short and flattened and bent around the oral opening, and the inner margins of the three terminal segments are provided with a row of strong hooked spines, which are also found upon the outer maxillæ, thus forming two rows of short hooks on each side of the mouth, by means of which the opening of the mouth can be closely applied to the fish on which these animals prey. The inner maxillæ are slender and styliform and armed with sharp curved spines at the apex, and the mandibles are also acute and fitted for piercing. The body is moderately convex, and the last four pairs of legs are nearly alike ambulatory and of moderate length, the last pair, when extended, scarcely surpassing the telson. The pleon is composed of six distinct segments, and the basal segment of the uropods is strongly produced at its inner angle, as usual in the family. The pleopods are ciliated in the adults as well as in the young.

Æga psora Kröyer (Linné).

Oniscus psora "Linné, Fauna suecica, ed. ii, 1761"; Syst. Nat., ed. xii, tom. i, p. 1060, 1767.
"Pennant, Brit. Zool., vol. iv, pl. 18, fig. 1, 1777 (certe)" (B. & W.).
O. Fabricius, Fauna Grœnlandica, p. 249, 1780.
Mohr, Islandisk Naturhistorie, p. 110, 1786.
Æga emarginata Leach, Trans. Linn. Soc., vol. xi, p. 370, 1815; Dict. Sci. nat., tome xii, p. 349, 1818.
Samouelle, Ent. Comp., p. 109, 1819.
Desmarest, Consid. Crust., p. 305, pl. 47, figs. 4, 5, 1825.
Griffith and Pidgeon, Nat. Hist. Crust., p. 218, pl. viii., fig. 3, 1833.
Edwards, Hist. nat. des Crust., tome iii, p. 240, 1840; Regne Anim., Crust., pl. iv, fig. 4, and pl. lxvii, fig. 1, 1849.
Gould, ?Rep. Geol. Mass., p. 549, 1835; Invert. Mass., p. 338, 1841.
Gosse., Man. Mar. Zool., vol. i, p. 134, 1855.
Æga (*Oniscus psora*) Kröyer, Grönlands Amfipoder, p. 318, 1838.
Æga psora Lilljeborg, Ofvers, Vet.-Acad. Förh., 1850, p. 84, and 1851, p. 24.
Lütken, Vidensk. Meddel., 1858, pp. 65, 179, 1859; ibid., 1860, p. 181 (?) 1861; Crustacea of Greenland, p. 150, 1875.
Schiödte, Ann. Mag. Nat. Hist., IV, vol. i, p. 12, 1868.
Bate & Westwood, Brit. Sess. Crust., vol. ii, p. 283, figure, 1868.
M. Sars, Chr. Vid. Selsk. Forh., 1868, p. 261, 1869.
G. O. Sars, Hard. Fauna, Crust., p. 275 [32], 1872.
Verrill, Am. Jour. Sci., III, vol. v, p. 16, 1873.
Smith and Harger, Trans. Conn. Acad., vol. iii, p. 22, 1874.
Whiteaves, Further Deep-Sea Dredging, Gulf St. Lawrence, p. 15, "1874."
Metzger, Nordseefahrt der Pomm., p. 285, 1875.
Meinert, Crust. Isop. Amph. Dec. Daniæ, p. 89, "1877."
Miers, Ann. Mag. Nat. Hist., IV, vol. xix, p. 134, 1877.
Harger, Proc. U. S. Nat. Mus., 1879, vol. ii, p. 161, 1879.
Æga entaillée Latreille, Règne Anim., tome iv, p. 134, 1829.

PLATE X, FIG. 64.

The present species is the largest Isopod, and indeed the largest Tetradecapod known on the New England coast, reaching a length of nearly or quite two inches and a breadth of one inch, and has even attained to the dignity of a popular name, "salve-bug", by which it is known among fishermen. It may be further distinguished by its large approximate eyes, covering a large proportion of the upper surface of the head, and by the possession of ancoral legs in three pairs only, the last four pairs of legs being fitted for walking.

The body is oval, broadest at the fourth and fifth thoracic segments, where the breadth is about half the length. The dorsal surface is moderately convex and smooth except for minute and rather scattered punctations, which occur also on the legs, especially on the basal segments, on the antennulæ, the uropods, and even the pleopods. The head is transverse and sub-triangular, salient in front between the bases of the antennulæ. Much of the upper surface of the head is covered by the large oval or somewhat reniform eyes, which do not quite meet on the median line. The antennulæ when bent backward nearly or quite attain the anterior margin of the first thoracic segment, and have their first two segments large and flattened, and wedge-shaped in front; of these the basal segment is quadrate in outline, as seen from above, and nearly as broad as long; it closely approaches its fellow of the opposite side in front, but is separated from it behind by a median process of the head; the second segment is triangular in outline, as seen from above, with the apex of the triangle extending beyond the origin of the third slender cylindrical segment, which is followed by a tapering flagellum of about a dozen segments. The antennæ when reflexed extend beyond the first thoracic segment and have the first two segments short and compressed, the third somewhat longer, the fourth and fifth longer and nearly cylindrical, followed by a tapering flagellum about as long as the peduncle and composed of fifteen to twenty segments. The maxillipeds have a short triangular external lamella and a five-jointed palpus, of which the first segment is short and transverse; the second is triangular and bears, on its inner apex, a few slender hooked spines; the third segment is broad and flattened, with the inner margin short, and armed with about three robust hooked spines; the fourth segment is flattened and transverse and armed along its inner margin with about six similar spines; while the fifth segment is small, sub-oval, and armed with much more slender curved spines. The outer maxillæ are provided with curved spines at the apex much like those of the maxillipeds. The inner maxillæ are rod-like and terminate in sharp somewhat curved spines placed close together. The mandibles support a slender palpus of three segments, of which the middle one is much the longest, and the last is robust and sickle-shaped, with a comb of short spines along the inner curve. This segment lies, in the ordinary position, just at the base of the antenna of the same side.

The first thoracic segment is, at its anterior margin, scarcely broader than the head, but expands rapidly backward. It is excavated in front for the eyes, which project somewhat beyond the posterior margin of the head. The second, third, and fourth thoracic segments are each a little shorter than the first; the fifth and sixth are somewhat longer; the seventh is shorter than the sixth. The epimera of the first thoracic segment are not separated by suture, but in the second and following segments they are so separated, and, especially on the anterior segments, marked with two oblique depressed lines. The epimera of the second, third, and fourth segments are rounded or truncate behind, but in the posterior segments they become acute and extend beyond the angles of the segments to which they are attached. The first three pairs of legs are short and armed with strong hooked dactyli. The propodal segments are also curved, and the carpus is short in the first pair but somewhat longer in the second and third pairs. The merus is almost crescent-shaped in the first pair of legs, its horns embracing the carpus above and below, but it becomes more elongated in the succeeding pairs; in all three pairs its inferior margin is armed with a few short, stout spines. The fourth and succeeding pairs of legs are of quite a different type from the first three. The four segments following the first or basal one are straight, cylindrical, or slightly compressed, armed with short spines, especially below and at the distal end, subequal in length but decreasing in diameter to the propodus, which bears in each pair a short, slightly curved and comparatively weak dactylus. The seventh pair is only imperfectly developed in the young specimen figured, but never quite attains the size of the sixth pair, which is the largest.

The pleon is scarcely narrower than the last thoracic segment and tapers but little to the fifth segment. The last segment is triangular, with the sides but little dilated, and is pointed at the tip without grooves or carinations. The uropods scarcely surpass the telson; the basal segment has its inner angle long and spiniform, extending the whole length of the inner margin of the inner ramus and ciliated toward the tip; the rami are flattened, the outer elongate ovate, obtuse; the inner with the inner margin straight, the outer curved and emarginate near the tip. Both rami and the posterior part of the telson are ciliated.

Length 16–50^{mm}, breadth 7–25^{mm}; color in alcohol light brown, darker toward the head; eyes black.

Linné's description of *Oniscus psora* is too indefinite to be certainly recognizable, and in using his trivial name I have followed the authority of Lütken and others. Our specimens agree well with the description of *O. psora* by O. Fabricius, and are undoubtedly identical with that species, which he describes as infesting the cod. They appear to correspond also with Bate and Westwood's figure and descriptions, although those authors make no mention of Fabricius under *Æ. psora*. As Kröyer referred the species to its proper genus, I have adopted his name as authority for the combination.

The specimen figured was dredged in the summer of 1872, a little to the northeast of St. George's Bank!, in latitude 42° 11′ north, longitude 67° 17′ west, in 150 fathoms, soft sandy mud with a few pebbles, and is young, as shown by its size and imperfectly developed seventh pair of legs. Adults may surpass the size of the figure, but the specimen drawn was enlarged three diameters. Adult specimens were obtained from the Provincial Museum, Halifax, Nova Scotia, labeled as found on the cod, and were probably from the fishing banks of that region, or from the Banks of Newfoundland. During the summer of 1879 a considerable number of specimens were received by the Fish Commission through the Gloucester fisheries, of which only a few are included in the table of specimens examined. These specimens were parasitic on the cod (*Gadus morrhua*), and on the halibut (*Hippoglossus*). Specimens have also been obtained from the skate (*Raia*). Whiteaves records this species from a halibut, on the north shore of the Gulf of St. Lawrence. Fine specimens were obtained by Mr. N. P. Scudder from off Holsteinborg, Greenland, in Davis' Straits!, parasitic on the halibut, and collected in July and August, 1879. It extends to Iceland (Edw. *et al.*); the British Isles (B. and W.); the North Sea (Metzger); Finmark (Sars), and Spitzbergen (Miers).

Specimens examined.

Number.	Locality.	Fathoms.	Parasitic on—	When collected.	Received from—	Number of specimens.	Dry. Alc.
1398	George's Bank, lat. 42° 11′ N., lon. 67° 17′ W.	150	....	— —, 1872	Packard and Cooke	1	Alc.
2139	....	....	....	....	Colonial Mus., Halifax.	2	Alc.
	George's Bank........	....	....	— —, 1878	Schooner Alice G. Wonson.	3	Alc.
	......do	....	Codfish	May 8, 1879	J. P. Shemelia	3	Alc.
	......do	....	... do	May 15, 1879	Capt. J. Q. Getchell	9	Alc.
	N. E. George's Bank..	47	....do	Nov. 29, 1878	J. P. Shemelia.....	3	Alc.
2154	Gulf of Maine........	....	Skate (*Raia*)..	— —, 1878	U. S. Fish Com'n ..	20	Alc.
2156	Banquereau........	....	Halibut........	— —, 1878	Schooner Marion ..	1	Alc.
2157	....	40–50	Codfish	— —, 1878	Schooner Rebecca Bartlett.	1	Alc.
2158	Grand Menan Bank...	100	....	— —, 1878	Schooner Peter D. Smith.	3	Alc.
2155	......do	100	....	— —, 1878	U. S. Fish Com'n ..	1	Alc.
	Brown's Bank.	52	Codfish	Dec. 19, 1878	Mr. Isaac Butler ..	2	Alc.
	......do	....	....do	Feb. 13, 1879	Capt. J. Q. Getchell	2	Alc.
	......do	30	... do	May 1, 1879	....do	8	Alc.
	Lat. 43° 25′ N., Lon. 60° W.	180	Halibut	Aug. 21, 1879	Capt. S. W. Smith and crew.	......	Alc.
	Davis's Straits........	....	....do	— —, 1879	Mr. N. P. Scudder.	10	Alc.

Syscenus* gen. nov.

Eyes wanting; palpus of maxillipeds two-jointed; sixth and seventh pairs of legs elongated; pleon suddenly narrower than the thorax; pleopods naked.

This genus is unfortunately represented in the collection by a single specimen. It differs from *Æga* by characters that point toward the *Cymothoidæ*, as in the reduction of the segments of the palpus of the maxillipeds, the sudden constriction at the base of the pleon, and the naked pleopods. The absence of eyes, although a conspicuous character can hardly be regarded as of great taxonomic value. It is separated from the *Cymothoidæ* by the form of the head, which is not produced over the bases of the antennulæ but merely projects slightly between them. The antennulæ moreover are composed of three peduncular segments and a flagellum; the basal segments are much smaller than in *Æga* and less flattened, but still form a part of the anterior outline when seen vertically. The last four pairs of legs differ from the first three, and are more or less elongated and fitted for crawling. The uropods are distinctly ciliated.

Syscenus infelix sp. nov.

This species may be recognized among our Isopoda by the possession of the full number of segments, the ciliated uropods, naked pleopods, and the absence of eyes.

* Σύσκηνος, a messmate.

The body is more than twice as long as broad and only moderately convex. The head is small and as seen from above is transversely somewhat diamond-shaped with rounded angles. It presents in front a slight prolongation between the antennulæ, and on each side of the short median process its outline is excavated above the bases of the antennulæ. The posterior margin is curved, but near each end is a faint indication of a lobe, projecting backward like the ocular lobes in *Æga*, but the eyes are wanting. The antennulæ arise near together on each side of the front and are short, extending when reflexed but little beyond the lateral margins of the head and only slightly surpassing the fourth antennal segment. They are readily distinguishable into peduncular and flagellar segments, the first three segments being of comparatively large size and about equal length; the second segment much flattened below against the antennæ; third more slender than the first two and followed by a short, tapering six-jointed flagellum. The antennulæ are in their natural position reflexed, the second segment being articulated at an angle with the first. The antennæ are considerably longer than the antennulæ and, when reflexed, slightly surpass the posterior border of the third thoracic segment. They are inserted below and a little outside of the antennulæ. The first segment is short and flattened below; the second is also short, the two together being hardly longer than the basal antennular segment; the third segment is about as long as the first two together, and the fourth is a little longer than the third, but of slightly less diameter; the fifth is more than one-half longer than the fourth, but is more slender and is followed by a slender, tapering flagellum of about twenty-four segments. The last two peduncular segments bear a row of elongate bristly hairs along the margin which, when reflexed, is brought next the body, and the row is continued, though with shorter hairs, along the flagellum. The palpus of the maxillipeds is composed of two segments of which the first is nearly square and armed at the inner distal angle with a minute hook; the second is bluntly triangular and armed at the apex, which is directed inward, with three hooklets. The external lamella is small and subcircular. The outer maxillæ are armed with short hooks at the tip; the inner with minute denticles. The mandibles are flattened and denticulate at the tip and bear a three-jointed palpus of which the three segments decrease in size to the last.

The first thoracic segment is twice as long as the second; its anterior margin is adapted to the head; its posterior margin is nearly straight above and rounded at the sides until the epimeral region is reached, when a short, pointed projection juts backward, being the tip of the epimeron on each side, here united with the segment. The next three —second, third and fourth—thoracic segments are of about equal length, and each a little over half the length of the first segment; their posterior margins are nearly straight above and rounded at the sides; the third segment is broadest. The fifth and sixth segments are each a little longer than the second; the seventh about as long as the second. The last segment, and in a less degree the sixth and fifth segments, have their posterior margins excavated along the back; all have their lateral angles rounded, although the angles of the seventh segment are but slightly so. The epimera are short and pointed; those belonging to the second and third segments are larger than the following ones, and are applied directly to the lateral margin of the segments; the posterior four pairs of epimera are shorter and smaller, and are separated from the lateral borders of the segment by a fold of the integument cutting off a portion of the anterior lateral angle and increasing in size to the last segment.

The first three pairs of legs are alike, distinctly ancoral and directed forward. In each the basis is much the longest segment; the ischium is strongly flexed upon it; the merus is expanded distally around the base of the carpus and bears a few bristles at the outer angle; the carpus is short, less than half as long as the propodus, and the dactylus is strong and curved. The fourth pair of legs, like those that follow, is directed backward; the basis is the longest segment and the ischium is strongly flexed upon it and of more than half its length; the merus, carpus and propodus are each about two-thirds as long as the ischium, and all four segments are armed distally with a whorl of spines around the articulation with the succeeding segment; the dactylus is slender, sharp and curved. The fifth pair of legs is longer than the fourth by a little more than the length of the dactylus, the elongation being in the segments from the ischium to the propodus inclusive. The sixth pair is the longest, being, when extended, as long as the thorax and pleon together. This elongation is confined also to the four segments above indicated, and of these the ischium is about as long as the basis; the merus falls a little short of the ischium in length; the carpus and propodus are of equal length, and are as long as the ischium; all these segments are slender and slightly curved, and are armed distally and along their inner side with short spinules. The dactylus is slender and curved. The seventh pair of legs resembles the sixth but is shorter by about half the length of the propodus. The fifth pair does not attain the middle of the carpus of the sixth.

The pleon is of less diameter than the last thoracic segment and about as long as the last five thoracic segments. Its transverse diameter increases slightly to the base of the last segment, where it is broadest; the fifth segment is a little longer than the preceding one, and the last segment is of a broad ovate form, acuminate and ciliated at the tip, truncated at the base and smooth above, except for a faint transverse impression on each side near the base, and a still more faint impressed median line toward the tip. The uropods attain the tip of the telson but do not surpass it; they have the basal segment oblique but not produced at the inner angle, and bearing two elongate-elliptical rami, tapering at the base and ciliated, the inner about one-third longer than the outer. The pleopods are quite naked and destitute of cilia.

Length 23^{mm}; breadth, 9^{mm}; breadth of pleon 4^{mm}; length of head 3^{mm}; breadth 4.2^{mm}.

A single specimen of this species was dredged by the U. S. Fish Commission, about fifteen miles northeast of Cape Cod!, in 130 fathoms brown mud, September 10, 1879.

XI.—CYMOTHOIDÆ.

Head produced anteriorly over the bases of the antennulæ; maxillipeds few-jointed, operculiform; mandibles palpigerous; mouth suctorial; legs armed with strong curved dactyli; epimera distinct behind the first thoracic segment; telson large and flattened; pleopods not ciliated; uropods articulated near the antero-lateral angles of the last segment, and composed of a more or less flattened basal segment bearing two flattened rami; habit parasitic; body often unsymmetrical by distortion in the adults.

This family is represented within our limits by three genera and as many species. They are parasitic in habit, usually on fish, and fix themselves by their strongly-curved claws to their host, often within the mouth, or about the branchial cavity, and frequently become distorted when fully grown. In all our species the head is small, and has the anterior margin produced, concealing the bases of the antennulæ and the antennæ. The head is three-lobed behind, and the first thoracic segment is adapted to it. The antennulæ and antennæ are both short and tapering, without very evident distinction into peduncular and flagellar segments. This distinction is, however, usually more or less evident on examination.

The epimera are well separated, except in the first segment, and may be projecting and conspicuous. The legs are of nearly the same form throughout, but increase in length and become more slender posteriorly.* The basal segments are in some genera enlarged and flattened, but not in ours; the joint between the basis and ischium is strongly flexed, and the segments, at least beyond the ischium, to the dactylus, are short and capable of but little motion on each other. The dactylus is strongly curved and admirably fitted for firm attachment to the host on which the animal may be living. In our species the legs, in the natural position, are concealed in a dorsal view beneath the body of the animal, to the under surface of which they are appressed, the first three pairs being directed forward, and the last three backward, as represented in plate X, fig. 66.

The pleon in our species is not suddenly narrower than the thorax, as it is, however, at least in the adults, in some genera belonging to this family. The segments of the pleon are distinct, the last one scutiform

* In *Artystone* Schiödte the seventh pair of legs "reach to the extremity of the tail and are slender, compressed crawling legs, with a small, almost rudimentary, straight claw."

and of moderate size, not being greatly enlarged. The pleopods are destitute of cilia in the adults.

This family is evidently closely related to the preceding and may yet have to be united with it, or even be extended so as to include also the *Cirolanidæ*. Our representatives of the three families are so few that I have had little opportunity to study the genera, and as before stated, I have separated the *Cirolanidæ* principally in deference to the opinions of Schiödte. *Alitropus* Edwards, *Syscenus* Harger, and *Ægathoa* Dana may be mentioned as genera pointing toward a transition between the *Ægidæ* and *Cymothoidæ*, and it is evident that the latter family is made up of forms degraded by parasitism. They have thus exchanged the ambulatory legs of the *Ægidæ* for strictly ancoral legs, for the most part in seven pairs, and have lost the natatory cilia of the pleopods. Their antennary organs are also much less perfect than in that family. All these modifications are in the line of the sedentary life of a parasite.

The interesting observations of Mr. J. F. Bullar have shown that in certain genera of the *Cymothoidæ* (*Cymothoa*, *Nerocila*, *Anilocra*) a peculiar form of hermaphroditism occurs, the young at a certain stage of development being males with well developed testes and external organs, but possessing at the same time ovaries with the oviduct ending blindly. As development proceeds the male organs are lost by molting, the oviduct obtains an external opening, the incubatory pouch is developed, and the animal becomes a female. Mr. Bullar's statements provoked considerable discussion, but they have recently been verified by Mayer, who has, however, shown that self-fertilization does not occur.

Three genera of *Cymothoidæ* are represented within our limits by as many species, and a fourth species, *Cymothoa prægustator* Say* (Latrobe) may yet be found, being not a rare parasite in the mouth of the menhaden (*Brevoortia menhaden* Gill) in southern waters. The projection of the front of the head over the bases of the antennary organs, and the strongly hooked or ancoral legs are characteristic of the family, and the genera may be distinguished by means of the following table:

Uropods	ciliated, eyes large conspicuous,		*Ægathoa*, p. 393	
	naked; body	symmetrical; posterior epimera elongated,	*Nerocila*, p. 391	
		unsymmetrical; epimera short,	*Livoneca*, p. 394	

Nerocila Leach.

Nerocila Leach, Dict. Sci. nat., tom. xii, p. 351, 1818.

Body oval; head small; eyes of moderate size; posterior thoracic segments and epimera angulated or spiniform, giving a sharply serrated or dentated outline to the thorax; first two "abdominal epimera" also spiniform; pleon of six distinct segments.

Our species of *Nerocila* has the characters of the genus much less pronounced than some foreign ones, as the posterior epimera are nearly

* Jour. Acad. Nat. Sci. Phila., vol. i, p. 395, 1818.

or quite concealed from above by the projecting angles of the segments, and the "abdominal epimera" are mostly concealed beneath the pleon. These organs are the much elongated inferior angles of the segments, which in allied genera, as *Ægathoa*, are short and not produced. In a lateral view they considerably resemble the posterior epimera, giving the appearance of two additional pairs. The specimen first described is smaller than others that have since been obtained.

Nerocila munda Harger.

Nerocila munda Harger, This Report, part 1, p. 571 (277), 1874; Proc. U. S. Nat. Mus., 1879, vol. ii, p. 161, 1879.

Verrill, This Report, part i, p. 459 (165), 1874.

PLATE X, FIG. 65.

This species may be recognized among our Isopoda by the projecting posterior epimera, and the two pairs of spiniform "abdominal epimera" beneath the pleon.

The body is oval, twice as long as broad, smooth, polished, and moderately convex. The head is flattened, broader than long, narrowing anteriorly, broadly rounded or subtruncate in front, three-lobed behind, with the middle lobe largest. The eyes are black and consist of an irregularly rounded patch of small indistinct ocelli, and are visible both above and below. The antennulæ are about as long as the head, and composed of eight segments, of which the first is short, the second is the longest, and the remaining six decrease pretty regularly in size to the last. The antennæ are a little longer and more slender than the antennulæ and have the first segment short, the second subglobose, the third, fourth, and fifth cylindrical, and a little larger than the segments of the flagellum, which are about five in number. The mandibular palpi are longer than any three segments of the antennæ, and the first segment is large, the second elongate conical, the third shorter, cylindrical.

The first thoracic segment is much longer than the succeeding ones and adapted to the head in front. It is slightly produced at its lateral angles behind, or rather appears so from the union of the epimera, which really constitute the projecting angles to the segment. In the second, third, and fourth segments the posterior angles are but little produced, and are equaled or slightly surpassed by the epimera, but in the last three segments the posterior angles are acutely produced much beyond the epimera of the corresponding segments, the angle of the sixth segment nearly attaining the end of the seventh epimeron. In a lateral view, only the last two epimera are decidedly acute, while those of the second and third segments are obtuse and rounded behind. Seen from below, the posterior angles of the epimera are acute throughout. The first pair of legs are slightly more robust than the second and third; the last four pairs are still more slender, the last pair longest, and the last two pairs armed with a few short spinules.

The pleon is shorter than the thorax and much narrower, though not suddenly so and tapers but little posteriorly; the telson is flattened, and regularly rounded behind. The "abdominal epimera" are acute, the second smaller and more slender than the first, but their extension backward varies with the state of contraction of the pleon. The uropods (pl. X, fig. 65 *a*) surpass the telson, and have the inner angle of the basal segment sharply produced. The rami are flattened; the external one twice the length of the basal segment, narrowly ovate or lanceolate, sometimes slightly curved, and surpassing the telson by half its length. The inner ramus is narrowly oval, obliquely truncate behind and about three-fourths as long as the outer.

The length of the specimen figured, which was the one first described, is 15^{mm}, breadth 7^{mm}, but specimens measuring 25^{mm} in length have since been collected; color brown or greenish, with two narrow dorsal bands of lighter color, most evident at the extremities.

The original specimen was obtained on the dorsal fin of *Ceratacanthus aurantiacus* at Wood's Holl!, Vineyard Sound, in 1871, and two more specimens of larger size have since been obtained, also from Vineyard Sound!, Mass.

Ægathoa Dana.

Ægathoa Dana, Am. Jour. Sci., II, vol. xiv, p. 304, 1852.

Body elongate oval; pleon not suddenly narrower than the thorax; head large, subtriangular; eyes large; legs nearly alike throughout, with strong curved dactyli; epimera of moderate size or small; pleon long and large, composed of six distinct segments; pleopods not ciliated; uropods more or less distinctly ciliated, rami subequal.

This genus is represented in our fauna by a species parasitic in the mouth of a squid. The large, granulated eyes remind one of *Æga*, and the ciliated uropods also indicate the approximation of this genus to the preceding family. The ciliation is, however, nearly rudimentary in our species, and is present, at least in the young, of other members of the *Cymothoidæ*.

Ægathoa loliginea Harger.

Ægathoa loliginea Harger, Am. Jour. Sci., III, vol. xv, p. 376, 1878; Proc. U. S. Nat. Mus., 1879, vol. ii, p. 161, 1879.

PLATE X, FIG. 66.

The legs all armed with strong curved claws, the large conspicuous eyes and the slightly ciliated uropods serve to distinguish the present species from the other Isopoda of our coast.

Body elongate oval in outline, nearly four times as long as broad, slightly dilated near the posterior end. Head broadly rounded in front, subequally, but not deeply, trilobed behind. Eyes large, with evident facets, lateral, semi-hexagonal, visible from below, covering nearly half the area of the head above, projecting posteriorly beyond the middle lobe of the head. Exteriorly they form about two-thirds of the lateral margin of the head. Their interior boundary is in the form of three sides of a hexagon, separated at their nearest points by a little more than the transverse diameter of the eye. The antennulæ are about as long as the head, composed of eight segments and separated at the base. The first segment is short and stout; the next two a little longer, but scarcely distinguishable from the following five flagellar segments, which decrease in size to the last. The antennæ are composed of ten segments. They are more slender than the antennulæ, and surpass them by about two segments. The first two segments are broader than the following three, which are also somewhat larger than the five flagellar segments.

The first thoracic segment is shorter than the head, but much longer than any of the succeeding segments, which to the sixth are of equal length, each about one-third shorter than the first. The seventh segment is about one-third shorter than the sixth. The fifth and sixth are broadest, each being about one-third broader than the first. The epimera do not project behind the angles of the segments to which they are attached. The legs differ but little throughout. The first pair are shortest, and the first three pairs are somewhat stronger than the last four, which are armed with a few scattered short spinules. The seventh pair are the longest.

The pleon is a little longer than the seven thoracic segments. The fifth segment is broader behind than in front, and the last segment is as broad at the insertion of the uropods as the third segment, and is rounded behind. Anterior pleopods with the basal segment nearly square. The uropods are unlike on the opposite sides in the specimen figured. The normal form is probably seen in the right uropod, which surpasses the telson by less than half the length of the outer ramus. This ramus is longer than the inner, narrow, with nearly parallel sides and is obliquely truncated at the tip. The inner ramus is somewhat diamond-shaped. The ciliation is nearly rudimentary and might be overlooked. The basal segment is alike on the two sides and has the inner distal angle acute and but slightly produced.

Length 13^{mm}, breadth 3.6^{mm}; color in alcohol yellowish, with minute black specks most abundant on the pleon; eyes black, conspicuous.

The specimen was obtained June 1, 1874, by Mr. S. F. Clark, at Savin Rock!, near New Haven, from the mouth of a squid (*Loligo Pealii*), whence the specific name. Two specimens "parasitic on young mullet" are in the Yale College Museum, collected at Fort Macon!, N. C., by Dr. H. C. Yarrow, which appear to belong to this species, showing that it is not confined to the squid.

Livoneca Leach.

Livoneca Leach, Dict. Sci. nat., tome xii, p. 351, 1818.

Head small, projecting in front over the bases of the antennulæ, which, like the antennæ, are short; legs all alike and armed with strong curved dactyli; body broad, oval, often obliquely distorted.

This genus is represented by a single species, in which the body is of a broadly oval form and depressed. All the legs are short and armed with strongly curved dactyli, and, in the natural position, are closely appressed to the ventral surface, which, however, is more or less exposed below along the middle.

Livoneca ovalis White (Say).

Cymothoa ovalis Say, Jour. Acad. Nat. Sci. Phil., vol. i, p. 394, 1818.

Dekay, Zoöl. New York, Crust., p. 48, 1844.

Livoneca ovalis White, Cat. Crust. Brit. Mus., p. 109, 1847. (*Lironeca*).

Harger, This Report, part i, p. 572 (278), pl. vi, fig. 29, 1874; Proc. U. S. Nat. Mus., 1879, vol. ii, p. 162, 1879.

PLATE XI, FIG. 67.

The broadly oval, more or less distorted and unsymmetrical form of this Isopod serves to distinguish it from any other species yet recognized within our limits.

Body broad, oval, usually oblique, and not, as represented in part I of this report, pl. VI, fig. 29, with the sides of equal length. The legs, moreover, in that figure are in an unnatural position, as they are, during life, concealed beneath the body of the animal and appressed to the ventral surface, the first three pairs directed forwards and the last four pairs backward. The dorsal surface is moderately convex. The head is small, rounded in front, trilobed behind, the middle lobe much the largest, the two lateral lobes extending beyond the eyes, which are not conspicuous, small and broadly separated. Antennulæ (pl. XI, fig. 67*a*) widely separated at the base, with the first segment short and stout; the second longer and somewhat tapering; the third about as long as the first. These peduncular segments are somewhat flattened. The flagellum is longer than the peduncle, tapering and five-jointed, curved backward in the natural position, each segment bearing a row of short blunt setæ, near the distal end, on the inner curve. The antennæ (pl. XI, fig. 67*b*) are about as long as the antennulæ, with the first two segments short and stout, the next three more slender; flagellum three or four jointed, with the last segment imperfectly divided and tipped with a few short setæ. The maxillipeds are narrow, with the outer lamella partially united to the basal segment and the palpus tapering and two-jointed, tipped with a few short curved setæ, at least in young individuals. The mandibles are pointed; their palpi (pl. XI, fig. 67 *c*) tapering from the base and composed of three segments of about equal length, the first subquadrate, the second tapering, the third nearly cylindrical.

The first thoracic segment is longest; the next three a little shorter and about equal; the fifth and sixth still shorter; the seventh shortest measured along the median line, which is usually a curved line except in young specimens. The anterior margin of the first thoracic segment is adapted to the posterior margin of the head and presents three sinuses, the middle one largest, for the median lobe of the head, and two smaller ones for the ocular lobes. The posterior margin of this segment is strongly convex backward throughout. In the succeeding segments this convexity rapidly diminishes so that the fourth has nearly a transverse margin and the last three segments become concave behind in an increasing degree. The epimera are narrow and obtusely pointed behind, and do not surpass the posterior angle of the segment to which they are attached except in the last two segments. The first pair of legs (pl. XI, fig. 67 *d*) are short and stout, the basal segment large but short; the next three segments short and with little motion on each other; the propodus stout and somewhat curved; the dactylus long, curved, and strong. The second and third pair of legs are much like the first, as are the four succeeding pairs, but somewhat larger and longer. The seventh pair (pl. XI, fig. 67 *e*) have the basal segment about twice as long as in the first pair, and the succeeding segments are also proportionally longer than in the first pair, except the dactylus, which is slightly weaker and not longer than in the first pair.

The pleon tapers rapidly at the sides; its first five segments are subequal in length; the last segment forms about half its length, and is flat and broadly rounded behind. Uropods (pl. XI, fig. 67*f*) surpassing the telson with the basal segment, about as long as the rami and but little produced at its inner angle; outer ramus linear oblong, rounded at the end; inner ramus shorter and broader, oblique at the tip.

Length 17–22^{mm}, breadth 10–12^{mm}. These animals when preserved in alcohol are of a leaden color, with the posterior margins lighter.

They are often parasitic on the blue-fish (*Pomatomus saltatrix* Gill). The details figured on plate XI are from small specimens collected on young blue-fish at New Haven!, by Mr. F. S. Smith. Other localities are Thimble Islands!, Long Island Sound; Vineyard Sound!, Fish Commission 1871, one specimen among scup (*Stenotomus argyrops* Gill). A specimen was sent to the Museum in 1878, collected by Dr. T. H. Bean, from the gill of *Micropogon undulatus* caught at Norfolk!, Va., July 9, 1878.

Specimens examined.

Number.	Locality.	Parasitic on—	When collected.	Received from—	No. of specimens.	Dry. Alc.
......	Norfolk, Va	*Micropogon*	July 9, 1878	T. H Bean	1	Alc.
2071	New Haven	Blue-fish		F. S. Smith	15	Alc.
2072	Vineyard Sound	....do	—— —, 1871	U. S Fish Com	1	Alc.
2073	....do	Scup	Aug. 17, 1871	....do	1	Alc.
2074						Alc.
2075	Vineyard Sound	Blue-fish	Sept. 2, 1871	U. S Fish Com	1	Alc.
2076				F. H. Bradley	1	Alc.

XII.—ANTHURIDÆ.

Body elongate, cylindrical; mouth suctorial; legs ambulatory and prehensile, the first pair enlarged; first pair of pleopods thickened and crustaceous, protecting the following pairs; uropods articulated at the sides of the last segment, standing in a more or less vertical position and forming with the telson a sort of cup or flower at the end of the body.

This family is represented within our limits by three species belonging to as many genera, which, in addition to the characters given above, agree further in the following particulars: The body is elongated and vermiform, often more than ten times as long as broad, and of nearly uniform size throughout. The head and thoracic segments are all distinctly separated from each other, and the head and last thoracic segment are shorter than the intervening segments, which are subequal. Both pairs of antennæ are approximate at their bases, and the lower pair or true antennæ are short, not greatly surpassing the head in length. These organs have the basal segment short, the second segment flattened internally and adapted to its fellow of the opposite side, while above and externally it is excavated for the basal segment of the antennulæ. The mandibles are palpigerous, and the mouth parts are fitted for piercing and for suction.

In the first pair of legs the first, second, and penultimate segments are enlarged and thickened; the two intervening segments, merus and carpus, are short; the dactylus forms a curved finger tipped with a stout spine and capable of complete flexion on the robust propodus. In one or two of the succeeding pairs of legs the propodus may be slightly enlarged. The first three pairs of legs have the carpus, or antepenultimate segment, triangular, and their basal segments are directed strongly backward. In the last four pairs the carpus may be short, but is not triangular, and always distinctly separates the merus from the propodus; they are so articulated to the body that their basal segments are directed forward. The first three pairs of legs are articulated to the anterior part of the segment to which they belong, the next three near the middle of the corresponding segments, and the last pair near the posterior margin of the last segment.

The pleon is short, with the segments more or less consolidated, and the pleopods are of the normal number and form. The "operculum" is not formed as in the *Idoteidæ* and *Arcturidæ* of the uropods, but is nothing more than the enlarged and thickened first pair of pleopods, the greater part of it being formed of the external lamella, while the uropods have an entirely different and peculiar structure. They are biramous, and consist on each side of a more or less elongated, flattened, basal segment, so articulated as to lie alongside the telson, and bearing at the apex a terminal plate, the inner ramus, in the same plane with itself, while, on its upper side near the base, stands a more or less perpendicular, oval plate, the outer ramus. The telson is directed obliquely downward, and, with the uropods, forms a ciliated cup-like or flower-like termination of the cylindrical body, whence the name *Anthura*, from the Greek ἄνθος, a flower, and οὐρά, a tail.

The structure of the mouth in this family has been investigated by Prof. J. C. Schiödte, to whose original papers in the Naturhistorisk Tidsskrift I have not had access. The paper on *Anthura* is translated and partly condensed in the Annals and Magazine of Natural History, where that author states that "next the *Cymothoidæ*, though as a type of a separate family, the genus *Anthura* must be placed."

The species of this family may be at once recognized by the peculiar cup-like termination of the body. This cup or "flower" is formed by the telson below, and the uropods at the sides and above; the outer rami of the latter organs being placed nearly vertically, and approaching each other on the median line above, where, however, the "flower" is more or less imperfect. Our three genera may be distinguished as follows: First five segments of pleon consolidated above, *Anthura* (p. 104); segments of pleon distinct, antennæ and antennulæ subequal, *Paranthura* (p. 108); segments of pleon distinct, antennulæ greatly enlarged in the male, *Ptilanthura* (p. 111).

Anthura Leach.

Anthura Leach, Ed. Encyc., vol. vii, p. "404" (Am. ed., p. 243), "1813–'14."

Antennulæ and antennæ short, subequal; thoracic segments not separated by constrictions; pleon with the five anterior segments consolidated above and resembling the last thoracic segment.

Our species of *Anthura* appears to agree in all generic characters with *A. gracilis* Leach upon which the genus was founded. In *A. polita*, however, the consolidated portion of the pleon is seen at the lower part of the sides to be composed of five consolidated segments, and bears the normal number of pairs of pleopods, while Bate and Westwood* say that "the four anterior segments are soldered closely together" in *A. gracilis*, and that "the pleopoda consist of, at least, four pairs of oval plates, strongly ciliated, on each side of the ventral surface of the basal segments of the tail." They had not, however, fresh specimens of the species, which is evidently closely related to ours.

The incubatory pouch of the females in the genus is confined to the third, fourth, and fifth segments, and is composed of three pairs of lamellæ, which overlap from behind forward, while the anterior margins of the first pair are united to the anterior part of the third segment.

Anthura polita Stimpson.

? *Anthura gracilis* Dekay, Zool. New York, Crust., p. 44, pl. ix, fig. 34, 1844 (*not* of Montagu and Leach).
Anthura polita Stimpson, Proc. Acad. Nat. Sci. Phil., vol. vii, p. 393, 1856.
Harger, Proc. U. S. Nat. Mus., 1879, vol. ii, p. 162, 1879.
Anthura brunnea Harger, This Report, part i, p. 572 (278), 1874.
Verrill, This Report, part i, p. 426 (132), 1874.

PLATE XI, FIGS. 68 and 69.

This species is distinguished among its allies on our coast by the nearly complete union of the basal segments of the pleon, which have together the appearance of an eighth thoracic segment. The cup or "flower" at the end of the body serves to distinguish it from other Isopoda.

* British Sessile-Eyed Crustacea, pp. 157 and 160.

The body is smooth, shining and flattened above and broadly keeled in the males below. The head is a little broader than long, deeply excavated on each side of the front for the bases of the antennulæ, and produced at the sides. The eyes are small and lateral but distinct, and are placed on the outer side of the anterior prolongations of the head, about on a line with the bases of the antennulæ. They are too indistinct in the figure, and the eye was even omitted on the right side by the engraver. The antennulæ (pl. XI, fig. 68 *a*) consist of a tapering three-jointed peduncle and a very short flagellum. The first peduncular segment is the largest, and is flattened above and on the inner side; the second segment is smaller, cylindrical, and provided with a comb of hair-like setæ along its outer side; the third is smaller and shorter than the second; the flagellum consists of a single very small segment, with indications of a rudimentary second segment at the end, where it is also tipped with setæ. The antennæ (pl. XI, fig. 68*b*) consist of a five-jointed peduncle, and a short flagellum much like that of the antennulæ. The basal segment of the peduncle is short; the second segment is the largest and is of peculiar shape, being excavated on the outer side to adapt it to the antennula, which lies in the groove thus formed, while the segment is bent upward and inward, and exposes a slender triangular area with the point backward, between, and on a level with, the antennulæ; the next three segments are sub-cylindrical and diminish in size, and are followed by one or two small flagellar segments tipped with setæ.

The maxillipeds (pl. XI, fig. 69*a*) are thick and strong, and are composed of a basal quadrate segment, a little longer than broad, with its proximal external angle elided for the short, sub-triangular external lamella, and bearing two segments representing the palpus. Of these segments the first is but little smaller than the basal segment and is sub-quadrate, tapering a little at the sides beyond the middle. The terminal segment is straight at its articulation with the preceding, and nearly so along the inner side, then rounded in the remainder of the outline. The segments of the palpus are finely ciliated along their margins, except along the external margin of the first segment, where the ciliation nearly disappears; they are also provided with coarse setæ, a few of which occur on the maxilliped, near the outer distal angle. The inner maxilla (pl. XI, figs, 69 *b* and *b'*) is rather robust, and terminated by a strong tooth or spine, below which, on the inner side, is a row of smaller curved teeth. The mandibles are terminated by a horny tooth, below which is a serrulated lobe; the mandibular palpus is robust; the second segment much the longest and provided with stout setæ; the last segment with a comb of rather short setæ. The maxillipeds are of much firmer texture than the other parts of the mouth.

The first thoracic segment is the longest, and is closely adapted to the head behind so as to allow but little motion. The second segment is shorter but somewhat broader than the first, and is rather freely articulated with it, and still more freely with the third; it is carinated below, but its articulations are much less free than in the next genus. The third, fourth and fifth segments are each about the length of the second; the sixth and seventh are progressively shorter. The first pair of legs (pl. XI, fig. 68 *c*) are quite robust and have but little freedom of motion, being directed forward under the head and hardly capable of further lateral extension than is shown in the figure of the animal. The basis and ischium are large and articulated so as to form a curve, bringing the legs forward; the merus is short; the carpus is triangular and extends along the side of the thickened propodus for about half its length, projecting like a tooth at the end; the propodus is ovate, much thickened and armed with a tooth near the middle of the palmar margin, along which it is ciliated, as is also the carpus; the dactylus is short and stout and tipped with a slender, curved, chitinous claw about as long as the dactylus itself. The figure (pl. XI, fig. 68 *c*) represents the inner surface of the leg, the merus being much less conspicuous on the outer side. The second and third pairs of legs are nearly alike and much more slender than the first pair. One of the third pair is represented on plate XI, fig. 68 *d*. In both these pairs of legs the carpus is small and triangular and wedged in between the merus and propodus, which meet above; the merus is a little larger in the second than in the third pair, and in both pairs it is provided with a few setæ at the upper distal angle and along the opposite or palmar side, where the carpus is also armed with setæ; the dactylus bears a few very short setæ. The remaining pairs of legs are rather more slender than the second and third, and the merus is separated from the propodus above by the carpus, which is, however, short. These legs are somewhat hairy, like the preceding pairs.

The anterior part of the pleon (pl. XI, fig. 68 *g*), consisting of the first five segments consolidated, appears much like an eighth thoracic segment a little longer than the seventh; traces of the sutures between the segments can be seen at the sides. The last segment is distinctly articulated, a little elevated dorsally, where it is also somewhat hairy; at the lower part of the sides it is covered by a slightly projecting lobe of the preceding segment, which extends over the proximal part of the basal segment of the uropods. Distally the terminal segment is depressed at a steep angle, and is in the form of a plate, ovate and ciliated at and near the tip, where it is obtuse; the sides are nearly parallel, and it is surpassed by the uropods, which consist, on each side, of a large basal segment, carinated on the outer side and toothed at the articulation with the outer ramus, obliquely truncated at the end, where it bears a short, obtusely-triangular, ciliated, inner ramus, or lamella, in the same plane as the basal segment. The outer ramus, or lamella, forms nearly a right angle with the basal segment, and stands upon its superior outer margin. This ramus is elongate reniform in outline, being notched below for the tooth on the basal segment, and is ciliated along its free superior margin. The first pair of pleopods (pl. XI, fig. 68 *e*) are composed on each side of a short, quadrate basal segment supporting two rami, of which the outer is, like the basal segment, of firm texture, and acts as an operculum; in shape it is semi-oval, with the inner margin nearly straight, and is ciliated distally, and along the outer margin. The inner ramus is much smaller than the outer and of delicate texture, and, in the natural position, is covered and concealed by the outer ramus; it is slender, with nearly parallel sides, rounded at the tip, and not ciliated. In the males the second pair of pleopods (pl. XI, fig. 68 *f*) bears, near the middle of the inner margin of the inner ramus, a slender stylet, slightly surpassing the lamella to which it is attached.

The lamellæ forming the incubatory pouch of the females are of considerable antero-posterior dimensions, and the posterior widely overlap the anterior ones, while the anterior border of the first lamella is united with the third thoracic segment, to which the lamella belongs.

Length 15–18^{mm}; breadth 1.8–2^{mm}. The color is brownish above, mottled with yellowish or honey color, lighter underneath.

This species was described as new by the present author in the first part of this report under the name *A. brunnea*, but there appears to be no sufficient reason for regarding it as distinct from Dr. Stimpson's *A. polita*. It is apparently closely related to *A. gracilis* Leach, although sufficiently distinct according to Bate and Westwood's* description and figures. Those authors, however, seem to have had but very poor and imperfect material on which to base their work. They figure and describe the telson and uropods as truncated and crenulated, and Montagu,† in his original description of the species, says that "the body is terminated by five large caudal appendages truncated at their ends."

Kröyer's‡ descriptions and figures of *A. carinata* approach much more closely to the present species. His figure of the antennula considerably resembles ours, but in his description he gives as the relative lengths of the four segments composing it 11, 4, 3, 5. In our species the last or flagellar segment is much the shortest, as may be seen by the figure, plate XI, fig. 68 *a*. He further speaks of the telson as crenulated, while it is entire in *A. polita*, and his figure (Voy. en Scand., pl. 27, fig. 3 *n'*) shows no tooth-like projection or angle on the basal segment of the uropods, as seen in a lateral view, and the corresponding margin of the outer or superior plate is destitute of the notch shown in the lateral view of these organs on plate XI, fig. 68 *g*. The inner ramus or lamella of the first pair of pleopods is also figured as much larger and more expanded distally than in our species, for which see plate XI, fig. 68 *e*. Unfortunately I have had no European specimens for comparison.

* Brit. Sess. Crust., vol. ii, p. 160, 1868.

† Trans. Linn. Soc., vol. ix, p. 103, pl. v, f. 6, 1808.

‡ Naturhist. Tidssk., II, B. ii, p. 402, and Voy. en Scand., Crust., pl. xxvii, fig. 3 *a–o*, 1849.

This species was described by Dr. Stimpson from specimens taken at Norfolk, Va., and has since been collected by Professors Smith and Verrill at Great Egg Harbor!, N. J., in 1½ fathoms shells and mud; by the U. S. Fish Commission in Long Island Sound!, especially at Noank Harbor!, among eel-grass (*Zostera marina*) and mud; off Block Island! in 17 to 19½ fathoms sand, mud, and stones; at Vineyard Sound!, at low water and in sand, and in 1878 at Gloucester!, Mass., in mud and among algæ.

Specimens examined.

Number.	Locality.	Fathoms.	Bottom.	When collected.	Received from—	Number of specimens.	Dry. Alc.
	Great Egg Harbor, N.J.	1½	Shells and mud	Apr. —, 1871	Smith & Verrill	......	Alc.
2077	Noank Harbor, Conn	....	Eel-grass	Aug. 28, 1874	U. S. Fish Com.	2	Alc.
2078	....do	....	Mud and eel-grass	Aug. 29, 1874	....do	2	Alc.
2079	....do	....	Mud	Aug. 28, 1874	....do	2	Alc.
2080	Vineyard Sound	L. w.	Sand	Sept. 8, 1871	....do	2	Alc.
	Squan Estuary, Gloucester, Mass.	....	Mud	—— —, 1878	....do	2	Alc.
	Gloucester, Mass	....	Mud and algæ	—— —, 1878	....do	1	Alc.

Paranthura Bate and Westwood.

Paranthura Bate and Westwood, Brit. Sess. Crust., vol. ii, p. 163, 1866.

Pleon articulated, composed of six segments; thorax deeply constricted at each end of the second segment; antennulæ and antennæ subequal; palpus of maxillipeds three-jointed; inner maxillæ acicular.

The first character given above is the only one given by Bate and Westwood, who, however, mention that the pleon bears the normal number of pleopods; a character that would not distinguish our species from the other genera. The distinctly articulated flagellum of the antennulæ is provided with a partial whorl of bristles, which, however, forms only the most rudimentary approach toward the structure of those organs in the males of the following genus. The segmentation of the pleon is indistinct in the dorsal region, but is apparent at the sides when seen from above, and the pleon does not at all resemble an additional thoracic segment as in *Anthura*. Both pairs of antennæ are provided in our species with a distinctly articulated flagellum, and are of nearly equal length.

Paranthura brachiata Harger (Stimpson).

Anthura brachiata Stimpson, Mar. Inv. G. Manan, p. 43, 1853.

Verrill, Am. Jour. Sci., III, vol. v, p. 101, 1873; ibid., vol. vii, pp. 42, 411, 502, 1874; Proc. Am. Assoc., 1873, pp. 350, 357, 1874; This Report, part i, p. 511 (217), 1874.

Whiteaves, Am. Jour. Sci., III, vol. vii, p. 213, 1874; Further Deep-sea Dredging, Gulf of St. Lawrence, p. 15, "1874."

Harger, This Report, part i, p. 573 (279), 1874.

Smith and Harger, Trans. Conn. Acad., vol. iii, p. 16, 1874.

Paranthura brachiata Harger, Proc. U. S. Nat. Mus., 1879, vol. ii, p. 162, 1879.

PLATE XI, FIG. 70.

The deep constrictions, by which the second thoracic segment is separated from the first and third, serve to distinguish this species from the allied forms on our coast, and the "flower" at the end of the pleon distinguishes it from other Isopoda.

Body moniliform, with evident segments; head narrower than, and about half as long as, the first thoracic segment, flattened and quadrate above, with a groove behind a raised anterior border, wedge-shaped below, deeply emarginate on each side of the projecting front above for the bases of the antennulæ; eyes lateral, not conspicuous, extending behind the emarginations. Antennulæ (pl. XI, fig. 70 *a*) with the first segment large but longer than broad, flattened above; second and third segments cylindrical; flagellum of twelve or more segments in adult specimens, with the first segment short, second twice as long and the longest segment of the flagellum, which tapers from the second segment and bears on the distal end of each segment an imperfect whorl of hairs. The antennæ (pl. XI, fig. 70 *b*) slightly surpass the antennulæ. They have the first segment short; the second flattened on the inner side, where it is usually in contact with its fellow of the opposite side, and excavated on the outer side above to accommodate the basal segment of the antennulæ; the third segment is short; the fourth and fifth longer and cylindrical. The flagellum consists of about twelve segments, tapers from the base, and is somewhat hairy. Both the antennæ and antennulæ are a little less developed and have one or two less segments in the females. The maxillipeds (pl. XI, fig. 70 *c*) are elongated, with a short, oval external lamella, and a two-jointed palpus. The large basal segment of the maxilliped projects on the inner side nearly to the end of the first segment of the palpus. The palpus has its segments of about equal length and provided with a few scattered bristles. The inner maxillæ (pl. XI, figs. 70 *d* and *d'*) are evident at the tip in an under view of the head; they are elongate and acicular, and minutely and sharply retro-serrate toward the tip. The three-jointed palpus of the mandibles is also conspicuous below; all three of its segments are short, and the last, which lies ordinarily between the bases of the antennæ, is flattened, oval, and provided with the usual comb of setæ.

The thorax is somewhat flattened above, carinate anteriorly below, and has the last segment much the shortest. The first segment is wider than the head and about twice its length, and is more closely united with it than are any of the thoracic segments with each other; it is strongly carinate below, especially on its anterior part, where the carina ends in a prominent tubercle; a much more slender carina bounds the flattened dorsal portion laterally. The second segment is separated from the first by a deep constriction, and is articulated so as to allow considerable motion, especially in a vertical plane; its antero-lateral angles are prominent in the form of low, rounded tubercles, and between them are two less evident tubercles on the front margin of the segment; the dorsal surface tapers behind, and is bounded laterally by carinæ; below, the segment is wedge-shaped, but not carinated; behind, it is separated from the third segment by a constriction not quite as pronounced as that in front. The third segment presents two rather more evident median tubercles in front on the dorsal surface, which is defined laterally by carinæ, fading away at about the middle of the segment; below, it is wedge-shaped and carinate in the males, but membranous along the median line in the females, as are the remaining segments more widely in that sex. In the males they are hard and chitinous throughout, rounded and scarcely wedge-shaped. The fourth segment is slightly longer than any of the others, and bears, near the anterior end of its dorsal surface, an oval depression with slight elongated elevations at each side. A similar structure occurs on the fifth and sixth segments, which are of decreasing length. The seventh is much the shortest thoracic segment, not being longer on the median line than the head; it is somewhat produced laterally.

The first pair of legs (pl. XI, fig. 70 *e*) are not as stout as in *Anthura polita*, and are more flexible; the carpus is the shortest segment, and is triangular, broader than long; the preceding segment, or merus, shows but little in an external view, but is more evident in an inner view, as shown in the figure, and is much broader than long; the propodus is much swollen proximally on its anterior or upper side; immediately in front of the end of the carpus it bears a stout tooth; the dactylus is strong, and tipped with a curved claw. In the second and third pairs of legs the carpus is triangular, but in the posterior pairs it is more elongated so as to distinctly separate the merus from the propodus.

The pleon is short, the telson triangular, acute at the apex. Uropods with the basal segment strongly carinate externally, terminal plate acutely triangular, proximal superior plate oval, curved and attached by its side, nearly meeting its fellow of the opposite side above. First pair of pleopods (pl. XI, fig. 70 *f*) with the external ramus semi-oval; internal ramus less firm in texture, ligulate, ciliated distally. Second pair of pleopods in the males (pl. XI, fig. 70 *g*) furnished with a slender stylet articulated at about the middle of the inner, posterior, lamella, and extending beyond its end. Both the lamellæ are crossed by a transverse suture just beyond their middle, at the point where the stylet is attached to the inner one.

Length 28^{mm}; breadth 2.2^{mm}; females about one-third smaller. The color is usually light yellowish brown, or sometimes somewhat darker, but not as pronounced as in the other members of the family, and nearly the same throughout.

From *P. norvegica* G. O. Sars* our species is distinguished by the eyes, which, though inconspicuous, are present. It lacks the tubercle de-

* Chr. Vid. Selsk. Forh., 1872, p. 88, 1873.

scribed and figured by Heller on the head of *P. arctica*,† and the flagellar segments of both pairs of antennæ distinguish it from *P. costana* Bate and Westwood.‡

This species was dredged by Dr. Stimpson "on a shelly and somewhat muddy bottom in twenty fathoms off the northern point of Duck Island," Bay of Fundy. It is rare south of Cape Cod, but was taken in Vineyard Sound! by the Fish Commission in 1871; also on St. George's Bank!, in 110 fathoms, mud and sand; Gulf of Maine!, down to 115 fathoms; Bay of Fundy!, down to 80 fathoms on muddy, shelly, and sandy bottoms; and off Nova Scotia!, 59 fathoms, pebbles, sand and rocks, and at other localities as detailed below. It was dredged by Mr. Whiteaves in 200 fathoms in the Gulf of Saint Lawrence, between Anticosti and the mainland of Gaspé.

Specimens examined.

Number.	Locality.	Fathoms.	Bottom.	When collected.	Received from—	Number of specimens.	Dry. Alc.
	Vineyard Sound........	... ..	....	—— —, 1871	U.S. Fish Com.	......	Alc.
2081	Gulf of Maine, east from Cape Ann 140 miles.	115	Gravel..........	—— —, 1877	...do	2	Alc.
	Gulf of Maine, southeast ¾ east from Cape Ann 13 miles	53	Mud and stones.	—— —, 1878	... do	1	Alc.
2082	Gulf of Maine, near Brown's Bank.	82	Rocks and barnacles.	—— —, 1877	....do	2	Alc.
1365	George's Bank	110	Brown mud....	—— —, 1872	Packard and Cooke.	2	Alc.
2083	Gulf of Maine, off Portsmouth 22 to 28 miles	80-92	Soft mud	—— —, 1874	U.S. Fish Com.	3	Alc.
2084	Gulf of Maine	65	Mud, sand, and gravel.	—— —, 1874	....do	2	Alc.
2087	Casco Bay, 20 miles southeast of Cape Elizabeth.	68	Mud	Aug. 12, 1873	... do	1	Alc.
2088	Gulf of Maine, 27 miles off Portland.	90	....	Aug. 26, 1873	....do	1	Alc.
	Casco Bay	....	....	—— —, 1873	....do	10	Alc.
2086	Gulf of Maine, 17 miles southeast of Monhegan Island.	72	Brown mud.....	—— —, 1873	.. do	3	Alc.
2095	Eastport, Me............	....	....	—— —, 1870	A. E. Verrill..	1	Alc.
2097	......do	....	....	—— —, 1872	U.S. Fish Com	4	Alc.
2091	Bay of Fundy, between Head Harbor and Wolves	60	Mud	Aug 16, 1872	....do	8	Alc.
2092	Off Head Harbor	75-80	Sand and shells.	—— —, 1872	.. do	8	Alc.
2093	Bay of Fundy...........	....	....	—— —, 1872	....do	1	Alc.
2094	.....do	77	....	Aug 16, 1872	... do	1	Alc.
2096	Bay of Fundy, Grand Menan, New Brunswick	....	....	—— —, 1870	A. E. Verrill..	3	Alc.
2098	Southeast from Cape Sable 18 to 22 miles.	56-59	Sand, gravel, and stones.	—— —, 1877	U.S. Fish Com.	2	Alc.

Ptilanthura Harger.

Ptilanthura Harger, Am. Jour. Sci., III, vol. xv, p. 376, 1878.

Antennulæ with the flagellum remarkably developed in the male, multiarticulate; second and succeeding antennular segments provided

† Denkschrift, Acad. Wiss. Wien., B. xxxv, p. [14] 38, pl. iv, figs. 9-12, 1875.

‡ Brit. Sess. Crust., vol. ii, p. 165, 1866.

with an incomplete, very dense whorl of fine slender hairs; pleon segmented, elongated; palpus of maxillipeds one-jointed.

The most important character of this genus is doubtless found in the structure of the antennulæ in the male sex. In the females the antennulæ are small, and the flagellum consists of a few slender rapidly tapering segments. They thus bear considerable resemblance to young specimens of *Anthura polita*, and being collected with them, were at first mistaken for them. They are distinguished by the larger and more conspicuous eyes, and by the more elongated and distinctly segmented pleon. In the presence of eyes our species differs from a form described by G. O. Sars, *Paranthura tenuis*, from near Stavanger, Norway, in which the males have a well-developed, eight-jointed and densely hairy or setiferous flagellum on the antennulæ.

Ptilanthura tenuis Harger.

Ptilanthura tenuis Harger, Am. Jour. Sci., III, vol. xv, p. 377, 1878; Proc. U. S. Nat. Mus., 1879, vol. ii, p. 162, 1879.

PLATES XI and XII, FIGS. 71-74.

Males of this species are at once recognized by the greatly developed antennulæ, resembling miniature bottle-brushes; females may be distinguished from the young of the other species by the conspicuous eyes; they are much smaller than the adults of the other species.

The body is smooth, flattened above, narrow at the middle, broadest at the base of the pleon. Head broader than the first thoracic segment and nearly as long, on the median line; longer than broad, narrowing to a point in front and much less acutely behind. The eyes are prominent, black, situated within the margin of the head and visible both above and below. The antennulæ in the males (pl. XII, fig. 74 *a*), when reflexed, attain the third thoracic segment; the first segment is large, but not longer than the second; the third is shorter than the second and followed by a short, subtriangular segment, which must be regarded as the first segment of the flagellum, although resembling the last peduncular segment much more than it does the succeeding or second flagellar segment; this segment is small at its base, but expands rapidly above and below and on the side which is next the body in the ordinary reflexed position of the antennula, and on these sides it bears, at its distal end, a fine and dense fringe of long slender hairs, which attain, when appressed, about the fifth following segment. Similar segments, to the number, in some specimens, of eighteen or twenty follow, forming an organ resembling a minute bottle brush or plume, whence the generic name. On one side, however, of the organ, which corresponds nearly with the outer or anterior side, according as the antennula is more or less reflexed, the whorl of hairs is interrupted. In the females (pl. XI, fig. 73) the antennulæ are shorter than the antennæ, with a short flagellum consisting of a small basal segment and a minute terminal one tipped with a few setæ. The antennæ (pl. XII, fig. 74 *b*) are short, differing little in the sexes, hardly surpassing the peduncle of the antennulæ in the males, with a short three or four jointed flagellum bearing a few hairs near the tip. The maxillipeds (pl. XI, fig. 71*b*) have a quadrate basal segment, somewhat emarginate externally for the subtriangular external lamella, and bearing a single suboval terminal segment, or palpus, somewhat truncate and ciliated at the tip. The inner maxillæ (pl. XI, fig. 71*c*) are five-toothed, one tooth being strong and terminal and the other four lateral. The mandibles bear a single-jointed palpus.

The thoracic segments are subequal in length except the last, which is but little over half as long as the others, though broader behind than any of them. They are slightly narrower than the head and margined laterally with a somewhat raised ridge. The third, fourth, and fifth have an elongate oval depression on the median line near the anterior margin. The first pair of legs (pl. XI, fig. 72) have the segments well separated, the carpus nearly equilaterally triangular, the propodus moderately thickened, and the dactylus strong and tipped with a stout claw; the carpus and propodus are bristly on their palmar margins. The remaining pairs of legs are slender and nearly equal in size.

The pleon is about as long to the tip as the last three thoracic segments. The first five segments are consolidated along the dorsum, but distinct at the sides. Each segment rises into a low broad tubercle on each side of the median line. The last segment is about as long as the preceding five, and is elongate-ovate, and obtusely pointed behind. The basal plate of the uropods is about half as long as the telson; the terminal or inner lamella is triangular-ovate, and about equals the telson. The proximal or superior lamella is narrowly semi-ovate, with an emargination on the upper side near the tip. The first pair of pleopods (pl. XI, fig. 71*d*) are shorter than the abdomen, and have the outer plate semi-obovate and the inner shorter, with nearly parallel sides. The second pair of pleopods (pl. XI, fig. 71*e*) bear, in the males, a slender straight stylet, articulated below the middle of the inner lamella and slightly surpassing it. The outer lamella is imperfectly articulated near the middle.

Length 11^{mm}; breadth 0.9^{mm}; females about one-third smaller; color brownish and more or less mottled above, lighter beneath, margined with translucent at the sides, extending on the sides of the head as far as the eyes.

This species is rare on the coast. It has been taken by the United States Fish Commission, on muddy bottom, in Noank Harbor, Long Island Sound!; off Watch Hill!, R. I., in 18 fathoms, sand; and off Block Island!, in 17 to 19½ fathoms, sand, mud, and stones; at Waquoit, Vineyard Sound!, in sand, at low water, September 8, 1871; in Casco Bay!, sand and mud, from 9 fathoms, in 1873, and by Prof. A. E. Verrill, at Grand Menan, in the Bay of Fundy! in 1870.

It is nearly related to and doubtless congeneric with *Paranthura* *tenuis* G. O. Sars,* but is at once distinguished by the presence of eyes, from which character, as distinctive, the name *P. oculata* might be applied to our species if a new trivial name should be thought necessary.

Specimens examined.

Number.	Locality.	Fathoms.	Bottom.	When collected.	Received from—	Specimens.		Dry. Alc.
						No.	Sex.	
2099	Noank Harbor, Conn	....	Mud	—— —, 1874	U.S. Fish Com	1	♂	Alc.
2104	....do			—— —, 1874	....do	1	♀	Alc.
2100	Off Watch Hill, R. I	18	Sand............	July 31, 1874	....do	1	♂	Alc.
2105	Off Block Island	17–19½	Sand, mud, and stones.	—— —, 1874	.. do	1	♀	Alc.
2103	Vineyard Sound, Mass.	L. w.	Sand............	Sept. 8, 1871	.. do	1	♀	Alc.
2101	Casco Bay, Me	....	Mud	July 16, 1873	....do	1	♂	Alc.
2102	....do	9	Sand and mud ..	Aug. 4, 1873	...do	1	♂	Alc.
2106	Bay of Fundy, Grand Menan.	....	....	—— —, 1870	A. E. Verrill ..	1	♀	Alc.

XIII.—GNATHIIDÆ.

Thorax with only five pairs of legs of the normal form in the adults, and apparently consisting of only five segments; antennulæ and antennæ short, with evident distinction into peduncle and flagellum; mouth organs suctorial in the larval state, more or less aborted in the adult; pleon with its segments distinct, bearing the normal number of pleopods; uropods inserted at the sides of the base of the last segment, biramous and resembling the pleopods but of firmer texture.

This family is represented on our coast by a number of forms, all of which, however, appear to be referable to a single species, in which, contrary to what is ordinarily observed in the order, a considerable transformation occurs, especially in the males, after the young leave the incubatory pouch, and before they reach the adult form. The sexes are very unlike at maturity, but in both the thorax may be seen, by a little inspection, to consist in reality of seven segments, of which the first is united with the head, but separated from it by a sutural line near its posterior margin, while the seventh is small and resembles the segments of the pleon, which appears as if consisting of seven segments. The last thoracic segment does not bear a pair of legs. The head is large in the adult male and armed with a powerful pair of curved jaws projecting strongly forward and curved upward. The antennulæ are short and widely separated at base. The antennæ are inserted nearly below them.

The five pairs of pediform legs are ambulatory and nearly alike throughout; the propodal segments are somewhat elongate, and the dactyli weak. All the thoracic segments except the first are distinct in the male, and all are distinct in the larval forms, but the fourth and fifth

* Chr. Vid. Selsk. Forh., 1872, p. 89, foot-note, 1873.

(third and fourth free segments) are indistinctly separated in the adult females.

The pleon is much alike in both sexes and the young, and consists of six distinct segments, each of which bears a pair of appendages. The first five pairs of these appendages, or pleopods, are carried beneath the pleon and subserve the purposes of respiration, while they are also used in swimming. They consist of a short basal segment supporting two rami, ciliated at the tip in the young. The uropods are directed backward and are of firmer texture than the pleopods. They are ciliated near the tip.

Only a single species has yet been recognized within our limits, and the male, female, and young will be described under the specific name.

The striking sexual differences in this family have caused much confusion, the males having been referred to one genus (*Anceus*), and the females to another (*Praniza*), and even these genera have been referred to different tribes or subfamilies. The true relationship of these forms, long ago suspected by Leach, was first made known by M. Hesse,* who, however, seems not to have stated it very clearly and perhaps did not correctly apprehend it at first. His descriptions, however, of the females of *Anceus* apply to what had previously been regarded as the female of *Praniza*, although he says in the same paper that *Praniza* is only the larval state of *Anceus*, which is true only of the young, or larval forms, or the then supposed males of *Praniza*. This family has been further investigated by Bate, Westwood, and Dohrn, to whose writings the reader is referred. It may be here remarked that Bate and Westwood in their account of the structure of *Anceus*, in the second volume of the British Sessile-Eyed Crustacea, appear to have overlooked the last thoracic segment, and suppose that either the first or second segment must be wanting. Dohrn calls attention to the rudimentary (or embryonic) condition of the seventh thoracic segment as the one missing to complete the normal number, but describes and figures† as "untere" and "obere Mundextremität" ("verwandeltes erstes" and "zweites Gnathopoden Paar") what I regard as the maxillipeds and first pair of thoracic legs, or, according to Spence Bate's terminology, which Dohrn seems to have misapprehended, the maxillipeds and the first pair of gnathopods. The second pair of gnathopods are pediform as usual in the Isopoda, and are the first of the five pairs of legs. Of the five pairs of pereiopods normally present, only four are developed in the Gnathiidae. The family is thus remarkable in the order both for the transformations undergone in its development, and for the retention after all of an embryonic feature.

Having discarded the names *Anceus* and *Praniza* for reasons given below, I have also rejected the family name *Anceidæ* and substituted for it a name, suggested by Bate and Westwood and derived from that

*Ann. Sci. nat., IV, tom. ix, p. 106, 1858.

†Zeit. Wiss. Zool., xx, taf. vii, figures 24 and 25.

of the typical genus. The name *Anceidæ* should perhaps be restored in case Risso's species should not prove to be congeneric with *Gnathia termitoides* Leach, *Cancer maxillaris* Montagu.*

Gnathia Leach.

Gnathia Leach, Ed. Encyc., vol. vii, p. "402" (Am. ed., p. 240), "1813-14."
Praniza Leach, MSS.
Anceus Risso, Crust. de Nice, p. 51, 1816.

Head very large and quadrate in the male, smaller and subtriangular in the female; first pair of legs operculiform in the male, subpediform in the female; pleon much narrower than the thoracic segments, with nearly parallel sides, and a sharply triangular telson.

The name *Anceus* Risso, which has been used by modern writers for this genus, ought, according to all rules of priority, to give way to *Gnathia* Leach, as acknowledged by Bate and Westwood,† who, however, hesitated to restore the name on account of Kirby's coleopterous genus *Gnathium*. While the undoubted priority of the name is a sufficient reason for its re-establishment, it may be worth while to add that *Gnathia* was not restricted by Dr. Leach to either sex alone, as that author had the sagacity to "suspect that *Oniscus coeruleatus* Montagu [*Praniza coeruleata* Desm.] was the female" of *Gnathia*, and, as far as I am aware, did not publish a generic name for the Praniza-form, although the name *Praniza* was used by him as a manuscript name, and as such appears to have been published by Latreille in the Encyclopédic Méthodique, which I have not been able to consult.

Gnathia cerina Harger (Stimpson).

Praniza cerina Stimpson. Mar. Inv. G. Manan, p. 42, pl. iii, fig. 31, 1853.
Packard, Mem. Bost. Soc. Nat. Hist., vol. i, p. 296, 1867.
Verrill, Am. Jour Sci., III, vol. vi, p. 439, 1873; vol. vii, pp. 38, 41, 411, 502, 1874; Proc. Am. Assoc., 1873, pp. 350, 354, 358, 362, 1874.
Anceus americanus, Stimpson, Mar. Inv. G. Manan, p. 42, 1853.
Gnathia cerina Harger, Proc. U. S. Nat. Mus., 1879, vol. ii, p. 162, 1879.

PLATE XII, FIGS. 75-79.

It will be convenient first to describe the male of this species and then the female and larval forms. The powerful and prominent jaws in front of the large quadrate head of the males of this small Isopod serve to distinguish it from any other on our coast.

The shape of the body is well described by Dr. Stimpson, as "regularly rectangular, abruptly narrowed at the commencement of the abdomen, which has the appearance of another very small rectangle set into the first, and of only one-third its width." It is somewhat bristly hairy, and much tuberculated and roughened above, especially on the lateral portions of the head and on the anterior thoracic segments. The head is broader than long, depressed medially in front and produced into a rounded lobe between the projecting upturned jaws. The eyes are small

*Trans. Linn. Soc., vol. vii, p. 65, pl. vi, fig. 2, 1804.
†Brit. Sess. Crust., vol. ii, p. 169.

and placed well forward at the sides of the head. The antennulæ (pl XII, fig. 76 *a*) are shorter than the head and slender, sparingly hairy, with a short, few-jointed flagellum. The antennæ (pl. XII, fig. 76 *b*) are also slender, with the first segment apparently composed of two united; the second segment short; the third and fourth longer, nearly cylindrical and followed by a slender few-jointed flagellum. The jaws (pl. XII, fig. 76 *c*) are elongate and turned upward at the apex, irregularly and bluntly toothed near the base within, and somewhat carinate on the outer side near the middle, the carina ending rather suddenly in a tooth-like process of the jaw as seen from above. The under surface of the head is deeply and broadly grooved longitudinally, and this groove is covered by what appear to be the transformed first pair of thoracic legs (pl. XII, fig. 76 *d*). They are in the form of a semi-oval plate on each side, attached near the base of the external side and strongly convex and ciliated on the inner side, where they overlap. This plate is truncated at the apex, where it bears a small oval lamella; on the surface of the large plate are three large, oval, semi-transparent areas. Within these plates is another pair of organs, consisting of a large basal segment and an articulated series of four flattened ciliated segments. These may be regarded as the maxillipeds, with a four-jointed palpus.

The first thoracic segment is indicated above only by a faint sutural line near the posterior margin of the large head. It is followed by five very distinct segments, of which the first two are perhaps most distinct, short, and strongly tuberculated, especially along their posterior margins. The third free segment is broader than the second, square at the sides, with two broad lateral elevations. The fourth free segment is somewhat rounded in front, with its chitinous integument apparently not calcified along the median line. The fifth free segment is narrower than the preceding and produced at the sides around the small last thoracic segment and the base of the pleon. The legs are nearly alike throughout, somewhat hairy and spiny.

The pleon is slightly dilated at the middle, with the angles of the segments salient. The last segment is acutely triangular, ciliate behind, surpassed by the uropods, which are also ciliated with a few bristles; both rami are slender, the inner a little broader than the outer. The pleopods (pl. XII, fig. 78 *e*) consist of two slender elongate lamellæ, the inner longer than the outer, attached to a basal segment and not ciliated in the adults of our species.

Length 4.4^{mm}; breadth 1.3^{mm}; color dirty yellowish brown above, lighter below. This form is *Anceus americanus* Stimpson.

The adult female (pl. XII, fig. 77) differs from the male principally in the following characters: The body is smooth and tapers behind and before, but is much swollen medially, where the segmentation becomes obscure, and the thoracic region seems converted into a sack for the reception of the eggs, plainly to be seen through the transparent integument. The head is comparatively small and subtriangular, emarginate in front. The eyes are placed farther back, and the large conspicuous jaws are wanting. Under the head, the first pair of legs (pl. XII, fig. 78 *a*) are slender, three-jointed with a minute terminal segment, and lie upon a delicate membranous plate on each side; within these are a pair of organs resembling what I have regarded as the maxillipeds of the male.

The first two free thoracic segments are short and curved somewhat around the head; the next two segments are much enlarged and nearly coalescent, and the fifth free segment is nearly similar in form to that of the males. The last thoracic segment is short and small and, as in the male, resembles a segment of the pleon.

The pleon (pl. XII, fig. 78 *c*) differs little from that of the male, but the angles of the segments are less salient.

Length 3–4^{mm}; breadth 1.5^{mm}. Color "pale yellowish or waxen." Dr. Stimpson was "inclined to consider" this form as the female of *Praniza cerina*.

The larval forms bear a much greater resemblance to the female than to the male but are more slender than either, the thorax being, in the smaller specimens, but little broader than the pleon. The head is broad, with large prominent eyes, and is distinct from the first thoracic segment, its posterior margin being truncated. The antennulæ have a short basal segment to the flagellum, which is followed by an elongate cylindrical segment forming about half the length of the flagellum, but bearing at its end a few short segments. The mouth organs project beyond the head, giving it an acute outline, and are evidently formed for piercing and suction. The large jaws of the adult males are, of course, wanting. The maxillipeds are slender and elongated.

The first pair of thoracic legs (pl. XII, fig, 78 *b*) are elongate, with the normal number of segments, a triangular carpus, and a strong curved dactylus, reminding one of the legs of the *Cymothoidæ*. The first thoracic segment is small and short and well separated from the following segments. The next two segments are quite distinct in all the forms, but usually the fourth, fifth, and sixth segments are united much as in the adult female. These forms appear to be the young females, and were described by Dr. Stimpson under the name of *Praniza cerina;* more rarely, however, specimens are found in which all the thoracic segments are distinct and somewhat resemble those of the adult male, but with their peculiarities less marked (pl. XII, fig. 79).

The pleon resembles that of the adults, but is not suddenly much narrower than the thorax. The pleopods as well as the uropods are ciliated at the tip (pl. XII, fig. 78 *d*).

Both these forms of young were taken from the body of a sculpin in the Bay of Fundy in 1872, and, when fresh, their bodies were bright red. In alcohol they fade to a waxy yellow.

Adult males of this species greatly resemble *Anceus elongatus* Kröyer, but his *Praniza Reinhardi* differs in its proportions of the antennary segments from *G. cerina.*

This species was described by Dr. Stimpson from females "dredged on gravelly and coralline bottoms in 20–30 fathoms in the Hake Bay," and males "dredged on a sandy bottom in 10 fathoms off Cheney's Head," Grand Menan, in the Bay of Fundy. It has been collected by the U. S. Fish Commission in Massachusetts Bay!, off Salem, 22–50 fathoms, gravel and soft mud; Gulf of Maine!, at several localities; Casco Bay!, 50 fathoms; Bay of Fundy!, in many localities, 10 to 60 fathoms, rocks, stones, and mud, and young specimens have been taken adhering to codfish and the sculpin. It was dredged by Mr. J. F. Whiteaves in the Gulf of St. Lawrence!, in 220 fathoms, mud. Further details in regard to localities are given in the subjoined table.

Specimens examined.

Number.	Locality	Fathoms.	Bottom.	When collected.	Received from—	Specimens. No.	Specimens. Sex.	Dry. Alc.
	Massachusetts Bay, 3 miles S. E Nahant.		Mud	Aug 31, 1879	J. H. Emerton.	3	♀	Alc.
2108	Massachusetts Bay, off Salem E S. E. 9 to 11 miles.	22	Gravel, stones	—— —, 1877	U. S. Fish Com	3	......	Alc.
2109	Massachusetts Bay, off Salem E. S. E. 8 to 9 miles	33	Mud	—— —, 1877	....do	1	♀	Alc.
2121	Massachusetts Bay, off Salem E. S. E 6 to 7 miles.	25–26	Gravel, stones.	—— —, 1877	...do	1	♀	Alc.
2110	Massachusetts Bay, off Salem E. S. E. 11 to 13 miles.	45–50	Mud	—— —, 1877	.. do	12	♂ ♀	Alc.
	Gulf of Maine, S E ½ S. from Cape Ann, 6 to 7 miles.	54–60	Sand, mud.....	—— —, 1878	....do	12	♂ ♀ y.	Alc.
2107	Gulf of Maine between Cape Ann and Isle of Shoals.	27–36	....do	—— —, 1874	....do	1	♀	Alc.
2111	Casco Bay	50	....	Aug. 6, 1873	....do	2	♀	Alc.
2112	......do	....	....	—— —, 1873	....do	1	♀	Alc.
2113	....	....	....	—— —, 1873	....do	10	♂ ♀	Alc.
2115	Eastport, Me..........	10–20	Rocky.........	—— —, 1872	....do	3	♀	Alc.
2117	......do	....	....	—— —, 1872	....do	3	♀	Alc.
	......do	....	....	—— —, 1868	A. E. Verrill ..	3	......	Alc.
2114	Bay of Fundy	....	On sculpin, &c.	—— —, 1872	U. S. Fish Com	12	y.	Alc.
2116	......do	25–30	....	—— —, 1872	....do	5	♀	Alc.
2118	.... do	....	....	—— —, 1872	....do	5	♂ ♀	Alc.
2119	Bay of Fundy, off Head Harbor.	40	....	—— —, 1872	....do	6	♂ ♀	Alc.
2122	Bay of Fundy	60	....	1870–'72	A. E. Verrill...	00	♂	Alc.
	Off Sable Island	160	On *Lophohelia*.	—— —, 1878	U. S. Fish Com.	4	♂	Alc.
2120	Gulf of Saint Lawrence	220	Mud	....	J. F. Whiteaves	1	y.	Alc.

XIV.—TANAIDÆ.

Respiration cephalothoracic, taking place in a cavity beneath the walls of the united head and first thoracic segment; eyes, when present, articulated; antennular flagellum single; first pair of legs enlarged and more or less perfectly chelate; pleopods natatory, ciliated, not branchial; uropods, terete, terminal, with at least one jointed ramus.

This family differs widely from all the other Isopoda, and indeed from all the sessile-eyed Crustacea, in the structure of the respiratory organs, and in the fact that the eyes, when present, are articulated with the head, or stalked, though without any proper pedicel.

I have seen species of only two genera, *Leptochelia* Dana and *Tanais* Audouin and Edwards, from within our limits. These genera are, by some authors, united under the name *Tanais*, but there seems to be ample reasons for separating them. While they agree in many characters, they differ widely from *Apseudes* Leach, which should probably be regarded as belonging to a different family not represented on our coast, and is accordingly not included in the above diagnosis.

Our representatives of the *Tanaidæ* may be further characterized as follows: The body is subcylindrical and elongated, from four or five to at least eight times as long as broad. The head and first thoracic segment are covered by the large cephalothoracic shield, which tapers somewhat in front, and is dilated behind. Its postero-lateral regions are occupied on each side by the branchial cavity, opening behind by a vertical slit, and in front by a nearly horizontal orifice. During life a lash-like organ can be seen through the body wall, in constant vibration, propelling a stream of water from behind forward through the cavity. The eyes, when present, are distinctly articulated with the head, and in the males are generally larger and more coarsely granulated than in the females. They are absent in one of our species, as in the one mentioned by Willemoes-Suhm from 1,400 fathoms in the Atlantic Ocean, off the North American coast, obtained by the Challenger expedition. They are described as indistinct in other foreign species. The antennulæ are inserted close together immediately below the vertex of the head and between the eyes. They are robust at base, and in the males may be elongated, but in the females are short, with only three or four segments and a minute rudiment of a flagellum. In neither sex have they any trace of the secondary flagellum seen in *Apseudes*. The antennæ are more slender than the antennulæ, and inserted almost directly beneath them. They are five-jointed, with the first and second segments short, the third larger and longer, the fourth and fifth slender and cylindrical, and, like the antennulæ, with indications of a flagellum. The antennæ, like the antennulæ, are tipped with bristles and bear a few scattered similar bristles on their segments.

The mouth organs are aborted in the males, at least in the genus *Leptochelia*, but in the females the mouth is protected below by a well-developed pair of maxillipeds, of which the basal segments meet at an angle forming a keel on the under surface of the head. The palpi of the maxillipeds are four-jointed, and armed with strong cilia; the last segment is strongly flexed on the penultimate. The inner maxillæ are spiny, and have the outer lobe reflexed and bearing elongated cilia at the tip. The mandibles are strong, destitute of palpi, and armed with one or two dentigerous lamellæ at the apex and a strong molar process.

The first pair of legs are robust, and in the males may be large and much elongated; they are in both sexes of our species powerful organs of prehension, being strongly chelate. Like the remaining pairs of legs, they have only five movable segments, unless an articulated spine at the extremity of the fifth segment is to be regarded as the true dactylus. On the other hand, the basal segment in many specimens presents indications of a short segment at its distal end, as if really consisting of the united basis and ischium. If this latter supposition be the true one, the hand of the first pair of legs is formed, as might be expected, of the propodus and the dactylus; the propodus is thickened and provided with a digital process stronger than the curved dactylus, which closes against it; the digital process bears toward the tip a few stout, bristly setæ. These legs are attached to the under side of the united head and first thoracic segment below the branchial cavity, and are directed forward. They are capable of but little lateral motion, and are nearly in contact below, especially toward their bases, which cover and partly conceal the organs of the mouth and the bases of the antennæ. The second pair of legs are very slender in comparison with the first, and are more slender than those that follow. Their basal segments are flattened, somewhat elongated, and usually bent with the convexity outward, in adaptation to the basal segments of the first pair of legs, which they partly embrace. The last three pairs of legs have their basal segments swollen.

The pleon consists, in our species, of five or six segments, and bears three or five pairs of strongly ciliated pleopods of the ordinary form, and fitted for swimming, and also a pair of uropods, consisting of a large basal segment bearing one or two rami. This ramus, or the inner one when there are two, is articulated and composed, in our species, of from two to six segments. The outer ramus may also consist of more than one segment. Like the antennulæ and antennæ, the uropods are provided with setæ, which are often elongate.

In the young the seventh pair of legs are not developed, and, according to Müller, the pleopods are likewise wanting and the uropods have less than the adult number of segments.

This family has been the subject of special research by Fritz Müller, Spence Bate, Dohrn, and others, to whose writings reference may be had for further description of their anatomy and development. Their proper place among the Crustacea cannot be regarded as settled, though the opinion of Fritz Müller that they represent an ancestral type of Isopoda is probably the best offered as yet. According to Dohrn, they present in their development affinities with *Asellus*, *Ligia*, and *Cuma*. Gegenbaur associates his *Tanaida* with the *Podophthalma* rather than the *Edriophthalma*.

Our species of this family are sharply divided into two genera, for which I have, after some hesitation, adoped the names *Tanais* Aud. and Edw. and *Leptochelia* Dana. I have not been able to see Audouin and Edwards' Résumé d'Entomologie, in which the genus *Tanais* is said to have been established, without description, in 1829. In the Précis d'Entomologie, by the same authors, is a figure (pl. xxix, fig. 1), apparently the same as that in the Résumé, which is there called *Tanais* de Costa. Latreille,* in 1831, characterized the genus, basing it upon *Gammarus Dulongii* Aud., figured by Savigny. Westwood,† in 1832, proposed for the same species the name *Anisocheirus*, without, however, mentioning any characters. In 1836, Templeton‡ described and figured, with evident care and accuracy, a species of this family under the name *Zeuxo Westwoodiana*. This species has, according to his figure, six segments in the pleon. Edwards, in his general work, Histoire naturelle des Crustacés, figures and describes *Tanais Cavolinii* (tome iii, p. 141, pl. 31, fig. 6), and refers the figure in the Précis d'Entomologie to that species. In 1843, Rathke§ described and figured *Crossurus vittatus* as a new genus and species allied to *Apseudes* and *Tanais*, but there do not seem to be any characters of importance to separate it from *T. Cavolinii* Edw., and, indeed, Bate and Westwood are inclined to regard them as identical species. If, however, *T. Dulongii* be regarded as the type of the genus, there appears to be nothing but the clothing of the basal segments of the pleon to separate the two genera, and this character seems of no more than specific value, since *T. Dulongii* is described by Bate and Westwood as possessing the peculiar "branchial appendages" at the base of the fifth pair of legs. These appendages are doubtless incubatory sacs, similar to those of *T. vittatus*.

For the second genus I have hitherto used the name *Paratanais* Dana, on the ground that *Leptochelia* of the same author, although having priority, was founded upon the characteristics of the male sex. The type-species, however, of this genus, *L. minuta*, possesses all the characters of *Paratanais* that could occur in the male. *Leptochelia Edwardsii* Dana, *Tanais Edwardsii* Kröyer, moreover, belongs to the same genus, and I have adopted the name for both sexes.

The minute species, by which this family is represented on our coast, may be readily recognized by the proportionately large and strong chelate first pair of legs articulated to the united head and first thoracic segment. The two genera are distinguished by the number of segments in the pleon, which are five, with three pairs of pleopods in *Tanais* (p. 122), and six, with five pairs of pleopods in *Leptochelia* (p. 126).

Tanais Audouin and Edwards.

Tanais Audouin and Edwards, "Résumé (not Précis) d'Ent., p. 182 (without description, 1829), pl. xxix, fig. 1" (B. & W.); Précis d'Entomol., p. 46, pl. xxix, fig. 1.
Edwards, Hist. nat. des Crust., tom. iii, p. 141, 1840.
Crossurus Rathke, Fauna Norwegens, p. 35, 1843.

Antennulæ and antennæ simple; mandibles without palpi; pleon composed of five segments bearing three pairs of ciliated pleopods below,

* Cours d'Ent., p. 403.
† Ann. Sci. nat., tome xxvii, p. 330, 1832.
‡ Trans. Ent. Soc., vol. ii, p. 203, 1836.
§ Fauna Norwegens, p. 35.

and a pair of simple uropods behind; eggs incubated in sacs attached near the bases of the fifth pair of legs of the females.

This genus is distinguished from the next by the structure of the pleon and the uropods as given above, and the females are, when carrying eggs or young, distinguished from all the other Isopoda by the wart-like, or sac-like, appendages of the fifth thoracic segment. Usually a small wart-like appendage is visible on each side of the inferior surface of the thorax just within the bases of the fifth pair of legs, but the size of these organs varies greatly, and in some specimens they become distended with eggs, extended lengthwise with the body and more or less coalescent, so as to form the large, bilobed incubatory pouch, as figured by Rathke. This pouch is, however, attached only to the fifth segment.

The presence of a peculiar appendage to the fifth pair of legs in this genus has been noted by various authors. Bate and Westwood figure, in the second volume of the British Sessile-Eyed Crustacea, page 122, a leg of the fifth pair with the attached pouch, which they "regard as a branchial sac similar to those existing in the Amphipoda, and consequently affording a proof of the nearer relationship of *Tanais* with that order than is possessed by any other isopodous animal." They remark further that "this appendage is wanting in some specimens, and its variable existence is probably a character of specific distinction in the group." Those authors have not, however, separated *T. vittatus* into two species on this character. Stebbing * mentions a specimen with eggs "as described by Rathke." Macdonald † figures a female with an incubatory pouch, which he briefly describes as "a membranous expansion or saccule under the thorax."

Rathke's original description is as follows: "Beide Exemplare, die ich untersuchen konnte, waren Weibchen und trugen Eier unter dem Thorax. Diese aber, die übrigens verhältnissmässig ziemlich gross waren, lagen nicht, wie bei *Idothea*, *Ligia* und vielen andern Isopoden, in einer zum Theil aus Schuppen bestehende Brüthöle eingeschlossen, sondern bildeten zwei länglichovale, dicht neben einander liegende und an der Oberfläche nur wenig unebene Massen von ziemlich beträchtlicher Grösse. Jede von ihnen war zusammengesetzt aus den Eiern und einer durchsichtigen eiweissartigen Substanz, die um jene herumgegossen war, sie wie ein Kitt zusammen hielt, und sie zugleich auch an die Bauchseite des Leibes befestigte. Es zeigten demnach jene Massen ganz dieselbe Zusammensetzung, wie die sogenannten Eiertrauben der Cyclopiden, Lernæaden und Branchiopoden." Rathke, having had only two specimens, does not appear to have perceived the attachment of these masses at the bases of the fifth pair of legs, and of course had no opportunity to see them in various stages of development. A specimen belonging to this genus and measuring 17 millimeters in length was obtained at Ker-

* Ann. Mag. Nat. Hist., IV, xvii, p. 78, 1876.

† Trans. Linn. Soc., II, Zool., vol. i, p. 69, pl. xv, fig. 1, 1875.

guelen Island by Willemoes-Suhm,* who describes the sacs attached to the fifth thoracic segment and attaining, as the young develop, a diameter of three to four millimeters.

Tanais vittatus Lilljeborg (Rathke).

Crossurus vittatus Rathke, Fauna Norwegens, p. 39, pl. 1, figs. 1-7, 1843.
Tanais tomentosus Kröyer, Naturhist. Tidssk., B. iv, p. 183, 1842; ibid., II, B. ii, p. 412, 1847; Voy. en Scand., Crust., pl. xxvii, figs. 2 a-q, "1849."
Lilljeborg, Öfvers. Vet.-Akad. Förh., Årg., viii, p. 23, 1851.
Meinert, Crust. Isop. Amph. Dec. Daniæ. p. 86, "1877."
Tanais hirticaudatus Bate, Rep. Brit. Assoc., 1860, p. 224, 1861.
Tanais vittatus Lilljeborg, Bidrag Känn. Crust. Tanaid., p. 29, 1865.
Bate and Westwood, Brit. Sess. Crust., vol. ii, p. 125, 1866.
Stebbing, Trans. Devon. Assoc., 1874, p. (7), and 1879, p. (6); Ann. Mag. Nat. Hist., IV, vol. xvii, p. 78, 1876.
Verrill, Am. Jour. Sci., III, vol. x, p. 38, 1875.
Macdonald, Trans. Linn. Soc., II, Zool., vol. i, p. 67-70, pl. xv, 1875.
Harger, Proc. U. S. Nat. Mus., 1879, vol. ii, p. 162, 1879.

PLATE XIII, FIGS. 81, 82.

This species is at once recognized among our Isopods by the pleon, which is beset with bristly hairs at the sides, and crossed by two rows of similar hairs near the posterior margins of its first two segments.

The body, though small, is rather robust, the length being about five times the breadth, which is greatest at the first free, in reality the second, thoracic segment. The head and united first thoracic segment is short, not longer than broad. The eyes are distinctly articulated and much less in diameter than the bases of the antennulæ. The antennulæ are shorter than the head and first thoracic segment, and are composed of three segments, of which the first is longer than the other two together, while the second and third are of about equal length; the third segment is terminated by one or two rudimentary segments, surmounted by a tuft of straight bristly setæ. Similar setæ arise from the terminal portions of the two preceding segments. The antennæ are as long as the antennulæ, but more slender, and consist of a five-jointed peduncle, somewhat setose like the antennulæ, and terminated by a rudimentary flagellum beset with setæ. The basal plates of the maxillipeds are ciliated externally, and meet each other on the median line so as to form a keel narrowing backwards; distally they become thicker and bear a four-jointed palpus, of which the second and third segments are dilated internally and ciliated, and the fourth is spatulate and ciliated at its extremity. The inner maxillæ have one of the lobes of the usual form and position, and armed with short, curved spines at the tip, while the other is bent backward and bears several elongated cilia at the tip, and by its constant motion urges a stream of water through the branchial cavity.

The first pair of legs are much enlarged and extend, in their natural position, beyond the head, and the "hand" is ordinarily directed nearly downward. The digital process of the propodus bears a broad lobe on its inner side, and an acute tooth at its extremity; at the side of the lobe

* Zeit. Naturges., B. xxiv, p. xvii, 1874.

is a row of setæ; the dactylus is strong, with an obtuse tooth on its inner margin. In the second pair of legs the dactylus is rather robust and tapers strongly. In the succeeding pairs of legs the dactyli become curved, and, in the posterior pairs, hooked and armed with a comb of slender teeth, while the three preceding segments are also armed with slender teeth or spines at their distal ends. The constrictions between the thoracic segments are well marked, giving the body a somewhat moniliform appearance. In breeding females, a pair of warts, or sacs of greater or less size are found attached to the under surface of the fifth thoracic segment, and containing eggs or young, according to their stage of development. These sacs often, if not usually, coalesce more or less perfectly before maturity.

The first three segments of the pleon are not narrower than the last thoracic segment, and are strongly margined, or tufted, at the sides with plumose hairs. These hairs are continued in two transverse rows, one upon the first and another on the second segment near their posterior margins, across the back of the pleon. This character is only imperfectly shown in the figure, where the transverse rows of hairs should have been more strongly indicated. The last two segments of the pleon are suddenly narrower than the first three. The last is much longer than the fourth and bears a short tooth at each side near the base. This segment may be composed of two united. The three pairs of pleopods are nearly alike (pl. XIII, fig. 82), and consist of a basal segment bearing two semi-oval lamellæ, which, as well as the basal segment, are strongly ciliated. The uropods are scarcely longer than the last two segments of the pleon, and the basal segment is comparatively small; the second segment is nearly as long as the first, the third about half as long as the second and tipped with setæ, with which the first two segments are also provided.

Length 5.5^{mm}; breadth 1.1^{mm}; color brown, mottled with lighter above; beneath, nearly white.

This species occurred on piles and among algæ and eel-grass at Noank!, Conn., in the summer of 1874, along with *Leptochelia algicola*, but in much less abundance. It was described by Rathke from Molde, on the west coast of Norway, and inhabits also the British Isles, and while the present article was going through the press I received, through the kindness of Rev. T. R. R. Stebbing, specimens from Torquay!, England, which confirm my previous determination of our species as identical with the European form. It has been found by J. D. Macdonald "in the excavated wood of piers, in company with *Limnoria* and *Chelura terebrans*." It is doubtfully identified by Bate and Westwood with a Mediterranean species, *T. Cavolinii* Edw. On the authority of Lilljeborg I have regarded it as identical with *Tanais tomentosus* Kröyer, although differing in the number and proportion of the segments of the pleon, as described and figured by that author. Kröyer's specimens were from Øresund, Denmark.

Leptochelia Dana.

Leptochelia Dana, Am. Jour. Sci., II, vol. viii, p. 425, 1849; U. S. Expl. Exped., Crust., p. 800, 1853.

Paratanais Dana, Am. Jour. Sci., II, vol. xiv, p. 306, 1852; U. S. Expl. Exped., Crust., p. 798, 1853.

Antennulæ and antennæ simple; mandibles without palpi; pleon composed of six segments, bearing five pairs of ciliated pleopods below, and a pair of biramous uropods behind; incubatory pouch of the females of the normal form.

The genus *Leptochelia* was constituted by Professor Dana for a form which Fritz Müller has since shown to be the male of *Paratanais* Dana, and although so far as I know the name has not hitherto been used for any but the male forms, I see no reason why it should not be adopted instead of the later name *Paratanais*. I have therefore adopted it for the four species lately described, from our coast. Dr. Stimpson's *Tanais filum* undoubtedly belongs to the same genus, making five species within our limits, only four of which I have seen. The species that I have examined may be further characterized as follows: The body is of nearly uniform size throughout. The antennulæ are directed forward and have a large basal segment, in contact with its fellow of the opposite side at its origin, and composing about half the length of the organ in the females; but in the males this segment, though absolutely much larger than in the females, may not form more than about a third of the total length of the antennula, which is nine to twelve jointed and terminated by a well developed flagellum. The antennæ differ but little in the sexes, and are five-jointed. The organs of the mouth are abortive in the males, and the oral region is covered below by a pair of subtriangular plates, perhaps the rudiments of the maxillipeds. The second thoracic segment is shorter than those that follow it; the fifth and sixth are the longest, and the seventh is shorter than the sixth.

The pleon consists of six distinct segments, subequal in length or with the last somewhat longer than the others. These segments are smooth above, and the first five bear on their under surface each a pair of pleopods, much like those of *Tanais* (pl. XIII, fig. 82), but not ciliated on the basal segment. The last segment bears a pair of uropods, which consist of a large basal segment bearing two terete rami. Of these the outer ramus is shorter and smaller than the inner, and may consist of a single segment so small and short as to be easily overlooked; the inner ramus is larger and longer, and composed, in our species, of from two to six segments. The number of these segments appears to be of value as a specific character, but not perfectly constant.

In the females the incubatory pouch is formed, as in the order generally, by four pairs of lamellæ attached to the bases of the second, third, fourth, and fifth pairs of legs.

Leptochelia algicola Harger.

Leptochelia Edwardsii Bate and Westwood, Brit. Sess. Crust., vol. ii, p. 134, 1868 (*Tanais Edwardsii* Kröyer?).

Tanais filum Harger, This Report, part i, p. 573 [279], 1874 (*non* Stimpson). Verrill, This Report, part i, p. 381 (87), 1874.

Paratanais algicola Harger, Am. Jour. Sci., III, vol. xv, p. 377, 1878.

Leptochelia algicola Harger, Proc. U. S. Nat. Mus., 1879, vol. ii, p. 162, 1879.

PLATES XII and XIII, FIGS. 80, 83–86.

The large and strong chelate claws, six-jointed pleon, and uropods with a short, one-jointed, outer ramus and a six-jointed inner ramus, will, in general, distinguish the present species from any other Isopod on our coast.

The body is of nearly uniform size throughout, and not constricted at the articulations. The head is narrowed in front. The eyes are conspicuous and plainly articulated, and are large in the males. The antennulæ in the females (pl. XIII, fig. 84 *a*) are shorter than the head and first thoracic segment, and are composed of three segments, of which the first is longer than the second and third together, and the third is slightly longer than the second, and, in some specimens, present traces of a division into two segments. The basal segment bears a short, stout seta just beyond the middle and one or two more near the tip; the second has also setæ near the tip, and the third bears a tuft of half a dozen or more setæ at the tip. In the males (pl. XII, fig. 80) the antennulæ are about two-thirds as long as the body and usually eleven-jointed, but sometimes with one or two segments more or less than that number. The basal segment forms, in this sex, about one-third the length of the organ, and is curved from near the base so as to be convex upward; the next two segments decrease rapidly in length, and are followed usually by eight flagellar segments provided with "olfactory setæ" from two to four or more to a segment. The antennæ (pl. XIII, fig. 84 *b*) in both sexes are short, slender, and decurved, terminated by a tuft of setæ. They appear to vary but little in the family.

The first pair of legs have the merus triangular, bringing the ischium and carpus together. In the female (pl. XIII, figs. 83 and 84 *c*) these legs, in their natural position, extend but little beyond the head; the propodus has a stout, digital process nearly in the line of its axis; this process is broadly notched near the base, then elevated into a slightly serrulate lobe, and bears at the apex a short, stout terminal tooth. Near the base of the lobe are usually two stout setæ. The first pair of legs in the males are much larger and more elongated, especially in the last three segments; the carpus is elongate and cylindrical, extending about half its length beyond the head, and attaining the end of the basal antennular segment; the propodus (pl. XIII, fig. 85) is robust and has a strong, curved, and two-toothed digital process, bearing also two stout setæ near the second tooth; the dactylus is also curved and provided on its inner margin with about seven short setæ springing from the bases of as many serratures; the propodus bears on its inner surface, above the origin of the dactylus, a comb, formed by a row of short setæ, and terminated at each end by a longer one. In the second pair of legs (pl. XIII, fig. 84 *d*) the dactylus, with its terminal spine, is not as long as the propodus, which bears two or three setæ near its tip. The third and fourth pairs of legs are shorter than the second. The last three pairs have their basal segments moderately swollen; the merus, carpus, and propodus of these legs are armed with a few spines near their distal ends; the dactyli are short.

The pleon is slightly broader near its base than the thoracic segments. The first five segments are subequal in length, the last longer and pointed behind. The uropods (pl. XIII, fig. 86) consist of a robust basal segment (*b*) bearing two rami, of which the outer (*o*) is very short and uniarticulate; the inner (*i*) is six-jointed, tapering from the base, with the segments of about equal length and provided with setæ near their distal ends.

Length 2.2^{mm}; breadth 0.33^{mm}; color nearly white.

It is possible that this species may prove to be identical with *L. Edwardsii* (Kröyer) Dana, although differing from Kröyer's description* and figures, especially in the following particulars: The peduncle of the antennula, which, according to his description and figure, consists of a short basal segment, an elongated segment, and a third short segment, has by his description the ratio to the following flagellum of five to four. The basal segment that he describes and figures was probably only the enlarged basal portion of the elongated segment, which, together with the following segment, constitutes only about three-sevenths of the length of the organ instead of five-ninths according to his description. He further describes and figures the uropod as biramous, with the inner elongated ramus composed of seven segments instead of six. Other differences could be pointed out in the proportions of the thoracic segments and the segments of the first pair of legs. Bate and Westwood† figure and describe a species, which they regard as *L. Edwardsii*, although their description and figures differ somewhat from Kröyer's, principally in the fact that they figure and describe the uropods as simple, saying in the generic description: "Pleopoda, five anterior pairs biramose; posterior pair unibranched and multiarticulate;" and again under the species (p. 136), "The posterior or caudal pair of pleopoda consist of a single multiarticulate branch, of which the basal joint is larger than the terminal ones: it consists of nine or ten small articuli." They figure it on page 134 as simple, tapering from the base and *seven*-jointed. These authors express their indebtedness "for this interesting addition to our British fauna to the zeal and research of the Rev. A. M. Norman, who took it during the summer of

* Naturhist. Tidssk., vol. iv, p. 174, pl. ii, figs. 13–19.

† Brit. Sess. Crust., vol. ii, p. 134.

1865 among *Zosteræ* between tide marks in Belgrave Bay, Guernsey," and in the description of *Paratanais forcipatus*, on p. 139, mention in a foot-note a specimen from the same locality, "which has a pair of six-jointed anal filaments with a short one-jointed secondary filament arising from the extremity of the basal joint. Can this be the female of *Leptochelia Edwardsii* fully grown?"

Through the kindness of the Rev. Mr. Norman I have been able to examine a specimen labeled "*Leptochelia Edwardsii*, Guernsey, 1866," and do not find that it differs from our species in any characters that can be regarded as of specific value. The antennulæ have indeed only seven flagellar segments, or ten segments in all, which is also the case in some of our specimens, though eight such segments—eleven in all—is the usual number. The thoracic segments have the same proportion to each other as in our species, and the uropods agree exactly with ours in being biramous, with the outer ramus short and uniarticulate and the inner ramus six-jointed.

This is the form of uropod described and figured by Kröyer in *Tanais Savignyi*, which, as Fritz Müller has suggested, is probably the female of *T. Edwardsii* Kr. That species has, however, according to Kröyer, a five-jointed antennula, the last segment being rudimentary. I have observed among a large number of our specimens two which had the last segment divided, though scarcely longer than in the others. These specimens could hardly be distinguished from *T. Savignyi* Kröyer by any characters that I have observed. In view, however, of the great similarity of the females throughout the genus, as exemplified in the females of this species and of *L. rapax*, with both sexes of which I am familiar, I have concluded for the present to retain the specific name which I recently proposed for this species, and wait until an examination of both sexes can be had to decide the questions of specific identity.

I formerly regarded this species as identical with *Tanais filum* Stimpson, and supposed its range to extend to the Bay of Fundy. In view of the number of species now known to exist on this coast, and in the absence of any specimens from the Bay of Fundy, I now regard that as an error, and have corrected it in the American Journal of Science.

This species is rather abundant among eel-grass (*Zostera marina*) and algæ at Noank! and Wood's Holl!, and has been taken during the past summer (1879) at Provincetown!, Mass., among eel-grass, on a vessel's bottom and in old piles, in company with *Chelura terebrans* Philippi and *Limnoria lignorum* White. The specimen sent by the Rev. A. M. Norman enables me to extend its range to the Island of Guernsey!, in the British Channel.

Specimens examined.

Number.	Locality.	Fathoms.	Bottom	When collected.	Received from—	Specimens.		Dry. Alc.
						No.	Sex.	
2126	Noank Harbor, Conn	....	Eel-grass	—— —, 1874	U. S. Fish Com	00	♂ ♀	Alc.
2127	....do	....	Eel-grass and algæ.	—— —, 1874	... do	00	♂ ♀	Alc.
2128	... do	....	....do	—— —, 1874	....do	10	♂	Alc.
2129	Vineyard Sound, Mass	....	....	—— —, 1871	....do	1	♀	Alc.
2130	Vineyard Sound, Parker's Point ...	....	L. w. on algæ....	—— —, 1875	.. do	....	......	Alc.
2131	....do	....	On piles.........	—— —, 1875	... do	....	......	Alc.
2132	.. do	....	Surface	—— —, 1875	....do	3	♀	Alc.
2133	....do	....	....	—— —, 1875	... do	8	......	Alc.
	Provincetown, Mass.	L.w.	....	Aug 22, 1879	....do	18	♂ ♀	Alc.
	......do	....	Eel-grass	Aug. 23, 1879	....do	24	♂ ♀	Alc.
	......do	½	.. do	Aug. 23, 1879	....do	2	♀	Alc.
	......do	....	In old piles	Aug. 23, 1879	....do	1	♂	Alc.
	......do	....	Vessel bottom ..	Sept 3, 1879.	....do	2	♂ ♀	Alc.

Leptochelia limicola Harger.

Paratanais limicola Harger, Am. Jour. Sci., III, vol. xv, p. 378, 1878.

Leptochelia limicola Harger, Proc. U. S. Nat. Mus., 1879, vol. ii, p. 163, 1879.

PLATE XIII, FIGS. 87, 88.

I have seen only females of this species, and these in general much resemble the same sex in *L. algicola* described above, but differ as follows: The eyes are small and inconspicuous, being less than half the transverse diameter of the basal antennular segment. The second segment of the antennulæ (pl. XIII, fig. 88 *a*) is short, only about half as long as the third. In the second pair of legs the dactylus with its terminal claw or spine is longer than the propodus, and the claw is slender and attenuated. The pleon is not wider than the segments of the thorax, and the uropods have the outer ramus two-jointed and surpassing the basal segment of the inner ramus, which is five-jointed, with the first segment long and imperfectly divided.

Length 2.5^{mm}. Color white in alcohol.

The specimens of this species were dredged in 48 fathoms, soft mud, in Massachusetts Bay!, off Salem, by the United States Fish Commission, in the summer of 1877.

Leptochelia rapax Harger.

Leptochelia rapax Harger, Proc. U. S. Nat. Mus., 1879, vol. ii, p. 163, 1879.

PLATE XIII, FIGS. 89, 90.

Females of this species closely resemble those of the two preceding species, but are distinguished by the following characters: The eyes are larger and more conspicuous than in *L. limicola*. The last segment of the antennulæ is scarcely longer than the preceding, instead of nearly twice as long. In the second pair of legs the dactylus is somewhat shorter, and the terminal spine less attenuated. The external ramus of the uropods consists of a single very short and small segment, shorter than the basal segment of the inner ramus, which is not elongated. The inner ramus is five-jointed instead of six-jointed, as in *L. algicola*, from which species the males are easily distinguished by the elongate and slender antennulæ and chelate legs, and by other characters, as may be seen from the following description and the figures.

The males (pl. XIII, fig. 89) are remarkable for the long, slender hand terminating the first pair of legs (pl. XIII, fig. 90). The body of the male is short and robust, and the segments are well separated by constrictions at the sides. The head with the united first thoracic segment is short and rounded, bulging strongly at the sides just behind the eyes, which are conspicuous, considerably less in diameter than the bases of the antennulæ, distinctly articulated and coarsely faceted. The antennulæ are much elongated, especially in the basal segment, which constitutes nearly half the length of the organ, and is more than one-third as long as the body; this segment is straight, swollen on the inner side near the base, then tapers gradually to the tip; the second segment is a little over one-third the length of the first and cylindrical; the third is again about one-third the length of the second, and scarcely thicker than the following flagellar segments, which vary in number from six to eight, and are usually of about equal length. In case there are eight flagellar segments the first is, sometimes at least, considerably shorter than the others. The last segment is tipped with a rudiment, and bears a few setæ. The whole number of segments, therefore, varies from nine to eleven, and if one of the flagellar segments be taken as a unit of measurement, the length of the first three segments will be approximately expressed by the numbers 9, 3.8 and 1.4. The antennæ when extended do not far surpass the middle of the basal segment of the antennulæ, and are comparatively slender; the first segment is short and somewhat expanded distally; the second is slightly longer and expanded so as to be sub-cordate; the third is short and cylindrical, equal in length to the first; the fourth is the longest segment, being longer than the first three taken together, and is slender and cylindrical, with a few setæ near the tip; the fifth is more slender and but slightly shorter than the fourth, and is tipped with a minute rudimentary terminal segment and a few setæ.

The legs of the first pair are large and much elongated. They vary somewhat in size and proportions, but are commonly, when extended, longer than the body of the animal. In these legs the segments preceding the carpus are robust but comparatively short, while the carpus is about half as long as the body, and the propodus (pl. XIII, fig. 90) is even more elongated than the carpus, and is usually strongly flexed upon it. More than half the length of the propodus is made up of the slender digital process, which bears, near the base on the inner side, a low, obtuse tooth, and a larger and more prominent one near the slender incurved tip. The dactylus (pl. XIII, fig. 90) is more than half as long as the propodus, slender, curved, and pointed, and armed with scattered, weak spinules along the inner margin. The digital process of the propodus bears also a few setæ, especially near the base of the outer tooth. The forceps thus formed are in most cases large enough to close around the body of another individual, but vary in size, being in some specimens at least one-third smaller than in others. The basal antennular segment may also be somewhat shorter than above described.

Of the thoracic segments the second (first free) segment is the shortest, and is also slightly broader than the others, and broader than the head. The third, fourth, and fifth segments increase in length progressively; the sixth is as long as the fifth; the seventh shorter. In the second pair of legs, the dactylus with its terminal claw is about as long as the propodus and nearly straight, as it is also in the third and fourth pairs, but the dactyli of the last three pairs of legs are more curved, and the basal segments somewhat swollen.

The first five segments of the pleon are of about equal length. The sixth is slightly shorter, obtusely pointed in the middle, and emarginate above the bases of the uropods, which are composed of a robust basal segment, bearing a minute outer ramus composed of a single segment tipped with setæ, and a five-jointed inner ramus, also sparingly provided with setæ. Between the uropods and below, a thin spatulate plate projects beyond the extremity of the pleon.

In length the males vary from 2.6^{mm} to 3.8^{mm}, and in breadth from 0.6^{mm} to 0.85^{mm}. The females measure in length about 2.3^{mm}; in breadth, 0.5^{mm}.

About one hundred specimens of this species, three-fourths of them females, were collected by Prof. A. Hyatt and Messrs. Van Vleck and Gardiner, in three feet of water, on muddy bottom, in the summer of 1878, at Annisquam!, Mass., and are the only specimens I have seen.

Leptochelia filum Harger (Stimpson).

> *Tanais filum*, Stimpson, Mar. Inv. G. Manan, p. 43, 1853.
> Packard, Mem. Bost. Soc. Nat. Hist., vol. i, p. 296, 1867.
> Harger, Am. Jour. Sci., III, vol. xv, p. 378, 1878.
> *Leptochelia filum* Harger, Proc. U. S. Nat. Mus., 1879, vol. ii, p. 164, 1879.

"Very minute, slender, rounded on the back, white, looking very much like a short piece of thread. Head small, and rather narrowed in front; first thoracic segment of great length; the second half as long as the third, which is about equal in length with the fourth, fifth, and sixth; the seventh being a little shorter than the sixth. The segments of the abdomen are well defined, the first five equaling each other in length, and the terminal one longer than the fifth, but narrower, and rounded behind. Antennæ short and thick, without flagellæ, with blunt tips crowned with few hairs, as are also their articulations. The inner ones are directed forward, and much the stoutest, especially toward their bases; while the outer ones are more slender and curve outward and backward. First pair of legs exceedingly thickened, with very large ovate hands and strong curved fingers. They are generally closely applied against the breast. The remaining thoracic feet are very slender, terminating in sharp, slender fingers, which in the second pair are very long and nearly straight, and in the other pairs short. The legs of the posterior pair are a little the longest and thickest. The ambulatory feet, in five pairs, are of great length and resemble those of Amphipods. The caudal stylets are in length about four-fifths that of the abdomen, and consist of four or five articles, with few hairs, each article becoming narrower, the last one with a tuft of few hairs at its extremity. Length .15 inch; breadth .02. Dredged among *Ascidiæ callosæ*, in 20 fathoms, in the Hake Bay."

I have seen no specimens corresponding fully with the above description, which is copied from Dr. Stimpson; neither have I seen any specimens of this family from the Bay of Fundy. I formerly regarded the species from Vineyard Sound as *Tanais filum* Stimpson, and that name is used in this Report, part i, p. 573 (279), where also "Bay of Fundy to Vineyard Sound" is given as its range. This error was corrected by the writer in the American Journal of Science in 1878. In the absence of specimens from the Bay of Fundy I am unable to say positively that this species is not the same as my *P. limicola*, although the number of segments in the uropods does not correspond with those of that species, and the outer ramus of the uropods, which is rather conspicuous in that species, is not mentioned at all by Dr. Stimpson. Further investigation is needed to settle this question, but the number of species known to me from the coast seems sufficient warrant for regarding this, for the present at least, as a distinct species.

Dr. Packard states that he has dredged *Tanais filum* Stimpson in the Gulf of St. Lawrence, "at Caribou Island, in eight fathoms, on a sandy bottom."

Leptochelia cœca Harger.

Paratanais cœca Harger, Am. Jour. Sci., III, vol. xv, p. 378, 1878.
Leptochelia cœca Harger, Proc. U. S. Nat. Mus., 1879, vol. ii, p. 164, 1879.

PLATE XIII, FIG. 91.

This species is at once recognized among our Tanaids by the absence of eyes. The enlarged chelate claws joined to the united head and first thoracic segment, and the six-jointed pleon serve to distinguish it as belonging to the present genus.

Body slender, elongated, and rather loosely articulated; head narrow in front, not broader than the bases of the antennulæ; eyes wanting; antennulæ distinctly four-jointed (pl. XIII, fig. 91*a*) in the type specimen, first segment forming less than half the length of the organ, second segment longer than the third, last segment about as long as the second, slender, tapering and tipped with setæ; antennæ attaining the tip of the third antennular segment. The first pair of legs (pl. XIII, fig. 91*b*) are robust, but less so than in the preceding species; they extend forward in the natural position about to the tips of the antennæ; they have the basal segment subquadrate, the hand or propodus less robust than the carpus, with a serrated digital process; dactylus short.

The second, or first free, thoracic segment is about two-thirds as long as the third; this in turn is about equal to the fourth and to the fifth segments; while the sixth and seventh segments are progressively somewhat shorter. The second pair of legs are scarcely more slender than the following pairs, and the basal segments are not curved around the base of the first pair.

The uropods (pl. XIII, fig. 91*c*) are short, and biramous; each ramus two-jointed. The outer ramus is more slender than the inner, half its length, and bears a long bristle at the tip.

Length 2.5^{mm}; color white.

The first specimen of this species was dredged along with *L. limicola* in 48 fathoms, soft mud, Massachusetts Bay!, off Salem, in the summer of 1877, and a second specimen apparently of the same species, though differing somewhat in the antennulæ, was collected on the shore at Provincetown! during the summer of 1879. Unfortunately only a single specimen was obtained in each case, but it is very distinct from the other species of our coast. It does, however, closely approach *Tanais islandicus* G. O. Sars,* but appears to differ in the first pair of legs, which Sars describes as follows: "Pedes primi paris validi, manu sat dilatata, carpo vix angustiore, digitis palmæ longitudinem æquantibus vix forcipatis." These legs are in our species distinctly chelate, and the dactylus is much shorter than the propodus (see pl. XIII, fig. 91*b*). He further says: "Uropoda sat elongata, biramosa, ramis, ambobus biarticulatis, valde inæqualibus, exteriore ne 3^{tiam} quidem interioris longitudinus partem assequente." In our species the outer ramus of the uropod is about one-half as long as the inner.

GEOGRAPHICAL DISTRIBUTION.

The whole number of species enumerated is forty-six, three more than were included in my recent paper on New England Isopoda in the Proceedings of the United States National Museum. Their geographical distribution, especially on our coast, is summarized in the lists below.

The following eleven species have as yet been found only south of Cape Cod:

Scyphacella arenicola.
Actoniscus ellipticus.
Cepon distortus.
Bopyrus *species*.
Erichsonia filiformis.
Erichsonia attenuata.
Cirolana concharum.
Nerocila munda.
Ægathoa loliginea.
Livoneca ovalis.
Tanais vittatus.

*Archiv for Mathematik og Naturvidenskab, Bind ii, p. 346 [246], 1877.

The following nineteen have been found only north of Cape Cod:

Gyge Hippolytes.
Phryxus abdominalis.
Dajus mysidis.
Janira alta.
Janira spinosa.
Munna Fabricii.
Munnopsis typica.
Eurycope robusta.
Synidotea nodulosa.
Synidotea bicuspida.
Astacilla granulata.
Cirolana polita.
Æga psora.
Syscenus infelix.
Gnathia cerina.
Leptochelia limicola.
Leptochelia rapax.
Leptochelia filum.
Leptochelia cœca.

The remaining sixteen are included in the following list as found on both sides of Cape Cod, but the letter N. is used to designate such species as are common north and rare south of the Cape, and S. signifies that the species is common at the south but rare northwards.

Philoscia vittata, S.
Jæra albifrons.
Ilyarachna *species*.*
Chiridotea cœca.
Chiridotea Tuftsii, N.
Idotea irrorata.
Idotea phosphorea, N.
Idotea robusta.
Epelys trilobus, S.
Epelys montosus, N.
Sphæroma quadridentatum, S.
Limnoria lignorum.
Anthura polita, S.
Paranthura brachiata, N.
Ptilanthura tenuis.
Leptochelia algicola, S.

The eleven species included in the following list occur also on the coast of Europe. The British species are marked B.

Gyge Hippolytes, B.
Phryxus abdominalis, B.
Jæra albifrons, B.
Munna Fabricii.
Munnopsis typica.
Idotea irrorata, B.
Astacilla granulata.
Limnoria lignorum, B.
Æga psora, B.
Tanais vittatus, B.
Leptochelia algicola, B.

The number of Isopoda included in the present paper is considerably less than are known to inhabit Great Britain, being only about two-thirds as many as are included in Bate and Westwood's work, together with such additions to that fauna as have come to my knowledge since. As has been seen, eight, or nearly one-fifth of our marine species, are identical with those of Great Britain. The number of genera is much more nearly equal. Thirty-one marine genera are enumerated in the present paper, and of these sixteen are also British. The remaining fifteen do not appear to be represented on the British coast, but their place is filled by perhaps a rather greater number of genera. Of the families, neglecting the *Oniscidæ* as not properly included in the present paper, we come to the *Bopyridæ*, which have as yet been but little studied

* The only specimen yet known is from twenty-one miles east of Cape Cod.

on this coast. Five species only are enumerated here, two of which are also British, while Bate and Westwood enumerate twelve. A closer examination of the group may very likely add considerably to the present list.

The *Asellidæ* and *Munnopsidæ*, which Bate and Westwood would unite, have seven marine species belonging to six genera in our list, and, rejecting *Limnoria*, this number corresponds well with the British list of four genera and six species; one species, *Jæra albifrons* Leach, is identical, as are three of the genera—*Jæra*, *Janira*, and *Munna*. The more typical forms of the *Munnopsidæ* have not yet, as far as I am aware, been recognized in British waters.

The *Idoteidæ* are more numerous on our coast and appear to be more diversified than in Great Britain. I have regarded our eleven species as belonging to five different genera, while Bate and Westwood include the seven British species in a single genus. The most conservative could hardly class our species in less than three genera to one English genus, and, judging mostly from the figures and descriptions, I should be inclined to reckon three, or at least two, English genera to five on our coast in this family. One genus and species, *Idotea irrorata* Edw. (Say), is identical. Of the *Arcturidæ* a single representative has only recently been discovered within our limits, while three species, of the same genus as ours, are mentioned by Bate and Westwood, and Stebbing has since added two more species.

A single species of *Sphæroma* is the only representative on our coast of a family numbering no less than five genera and thirteen species in Bate and Westwood's volume. If the last two of these species be united as sexes of the same, and *Dynamene rubra* and *viridis* be also united, as suggested by Stebbing,* there are still left eleven representatives of this family in England to one on our coast. Our species is closely related to the British *Sphæroma serratum* Leach. *Limnoria lignorum* White is the only known representative of its family on both coasts.

The *Cirolanidæ* and *Ægidæ*, which are classed together under the latter name by most authors, have only four representatives in our limits, belonging to three genera. Two of these genera are also found in Great Britain, where they contain no less than seven species, one of which, *Æga psora* Kröyer, is identical on the two coasts. *Cirolana truncata* Norman is not included in Bate and Westwood, but these authors mention three other species belonging to as many genera in this group, making five genera and ten species from Great Britain to only three genera and four species in our waters. The *Cymothoidæ* are represented in our list by three species belonging to three genera, while Bate and Westwood say of this family, "No specimen has hitherto been satisfactorily determined as having been found in our own seas." The Rev. A. M. Norman, however, in the Annals and Magazine of Natural History for December, 1868, p. 422, mentions and briefly describes *Anilocra medi-*

* Jour. Linn. Soc., Zool., vol. xii, p. 148, 1874.

terranea Leach, taken from a "small fish in rock-pools at Herm in 1865." This genus has not been found on our coast.

Of the three genera and three species of *Anthuridæ* in our list two genera are also found in Great Britain, and it is possible that one species may yet prove identical. The *Gnathiidæ* are more difficult of comparison on account of the confusion that has existed in the sexes, and the larval forms. Our specimens seem to be all referable to a single species, doubtless congeneric with the British species, the number of which may, perhaps, by a liberal estimate, be placed at three.

In the *Tanaidæ*, the genera are the same as in Great Britain, and two of our species, *Tanais vittatus* Lillj. and *Leptochelia algicola* Harger, are found on both coasts. There remain a second species of *Tanais* on the British coast, and two species of *Leptochelia* (*Paratanais* of Bate and Westwood) against four species of *Leptochelia* on our coast, as the remaining representatives of this family. The genus *Apseudes* should probably be taken to represent a family not yet found on our coast.

We have, therefore, the following list of marine familes, with the genera in each, that are identical on our coast and that of Great Britain. The species have been already indicated in a preceding list:

Bopyridæ: Gyge, Phryxus, Bopyrus. Two species.
Asellidæ: Jæra, Janira, Munna. One species.
Idoteidæ: Idotea. One species.
Arcturidæ: Astacilla.
Sphæromidæ: Sphæroma.
Limnoriidæ: Limnoria. One species.
Cirolanidæ: Cirolana.
Ægidæ: Æga. One species.
Cymothoidæ.
Anthuridæ: Anthura, Paranthura.
Gnathiidæ: Gnathia.
Tanaidæ: Tanais, Leptochelia. Two species.

Further details of geographical and also of bathymetrical distribution are presented in the table on pages 139 to 141, in which the first column shows the least depth in fathoms at which each species has been collected on our coast; the second the greatest depth; and the following eighteen columns are for different localities, which may be further explained as follows: The Carolinas include Charleston, S. C., Fort Macon, N. C., and Norfolk, Va.; New Jersey includes Great Egg Harbor and Atlantic City, N. J., and Fire Island Beach, on the south shore of Long Island; Long Island Sound includes Savin Rock, New Haven, Stony Creek, or Thimble Islands, Saybrook, New London, and Norwalk, Conn.; Block Island includes Watch Hill, Block Island Sound, and the deeper water off the island; Vineyard Sound includes also Buzzard's Bay, Nantucket Sound, and off Nantucket Island; Cape Cod Bay includes Provincetown and Barnstable; Massachusetts Bay includes Salem, Nahant, Gloucester, and Annisquam, Mass.; the Gulf of Maine includes all outside of the line of 50 fathoms between Cape Cod and Nova Scotia, and extending seaward to include George's Banks; Casco Bay includes Cape Elizabeth and Quahog Bay; Bay of Fundy includes Eastport Harbor and Grand Menan, while species collected at greater depths than 50 fathoms are reckoned also in the Gulf of Maine, and the same is true of those from that depth off Nova Scotia; Nova Scotia includes also Banquereau or Quereau, Eastern and Western Banks, Miquelon Island, and the Grand Banks. Species occurring on the north shore of the Gulf of St. Lawrence are credited also to Labrador. In the last column of the table the general habitat of each species is briefly indicated.

	Least depth.	Greatest depth.	Florida.	The Carolinas.	New Jersey.	Long Island Sound.	Block Island.	**Vineyard Sound.**	Cape Cod Bay.	Massachusetts Bay.	Gulf of Maine.	Casco Bay.	Bay of Fundy.	Nova Scotia.	Gulf of St. Lawrence.	Labrador.	Greenland.	Arctic Ocean.	British Isles.	North Sea.	ORDINARY HABITAT.
ONISCIDÆ:																					
Philoscia vittata	s				!	!		!	!												Shore, under rubbish.
Scyphacella arenicola	s				!			!													Shore, in sand.
Actoniscus ellipticus	s					!															Shore, under rubbish.
BOPYRIDÆ:																					
Cepon distortus					+																Parasitic on *Gelasimus*.
Gyge Hippolytes	22	160								!	!	!	!	!			+	+	+	+	Parasitic on *Hippolyte*, &c.
Phryxus abdominalis	18	100								!	!			!			+	+	+	+	Parasitic on *Pandalus*, *Hippolyte*, &c.
Dajus mysidis																+	+				Parasitic on *Mysis*.
Bopyrus, *species*			?		+																Parasitic on *Palæmonetes*.
ASELLIDÆ:																					
Jæra albifrons	0	*lw*				!		!	!	!		!	!	!	!	!	?		!	+	Under seaweed, in tide-pools, &c.
Janira alta	0	300								!	!	!	!	!							Muddy, stony, or shelly bottoms.
spinosa														!							
Munna Fabricii	0	200										!	!	!			+	+			Eel-grass, coralline, shells, &c.
MUNNOPSIDÆ:																					
Munnopsis typica	60	220[1]									!		!		+		+	+		+	Muddy bottoms.
Eurycope robusta	220	220													!						Do.
Ilyarachna, *species*	106	106									!										Do.
IDOTEIDÆ:																					
Chiridotea cœca	0	*lw*	+			!		!	!	!				!							In sand at low water.
Tuftsii	0	25				!				!		!	!	!							Sandy bottoms.
Idotea irrorata[2]	0	10			!	!		!	!	!	!	!	!	!	+				!	+	Algæ and eel-grass.
phosphorea	0	18				!	!	!		!		!	!	!	!						Algæ and rocks.
robusta	*sf*				!			!			!						+[3]				Open ocean at the surface.
Synidotea nodulosa[4]	16	190									!			!			+				Rocky, stony, and sandy bottoms.
bicuspida	—	—												!	+	+		+			

	Least depth.	Greatest depth.	Florida.	The Carolinas.	New Jersey.	Long Island Sound.	Block Island.	Vineyard Sound.	Cape Cod Bay.	Massachusetts Bay.	Gulf of Maine.	Casco Bay.	Bay of Fundy.	Nova Scotia.	Gulf of St. Lawrence.	Labrador.	Greenland.	Arctic Ocean.	British Isles.	North Sea.	ORDINARY HABITAT.
IDOTEIDÆ—Continued.																					
Erichsonia filiformis	0	7			+	!		!													Eel-grass, shells, &c.
attenuata	0	1			!	!															Eel-grass.
Epelys trilobus	0				!	!	!	!	!			!									Eel-grass and mud.
montosus	2	40					!	!		!	!	!	!	!	+						Rocky and muddy bottoms.
ARCTURIDÆ:																					
Astacilla granulata	7	250									!			!				+			On *Primnoa*, &c.
SPHÆROMIDÆ:																					
Sphæroma quadridentatum		*l10*	!	!	!	!		!	!												Rocks, tide-pools, &c.
LIMNORIIDÆ:																					
Limnoria lignorum[5]	0	10	!			!		!	!	+		!	!	!	+				+	+	Boring in submerged timber.
CIROLANIDÆ:																					
Cirolana concharum				+			!	!													Swimming free, predatory.
polita										!	!		+	!							Do.
ÆGIDÆ:																					
Æga psora		150									!			!	+	+	!	+	+	+	Parasitic on the cod and halibut.
Syscenus infelix	130	130									!										Parasitic.
CYMOTHOIDÆ:																					
[illegible]la munda								!													Parasitic on file-fish.
Ægathoa loliginea				!		!															Mouth of *Loligo*.
Livoneca ovalis				!	+	!		!													Parasitic on bluefish, &c.
ANTHURIDÆ:																					
Anthura polita	0	1½		+	!	!		!		!											Eel-grass, mud, &c.

Paranthura brachiata	20	115	----	----	----	----	----	!	----	----	!	!	!	!	+	----	----	----	----	----	Rocky, shelly, and muddy bottoms.
Ptilanthura tenuis	0	19	----	----	----	!	!	!	----	----	----	!	!	----	----	----	----	----	----	----	Sand and mud.
GNATHIIDÆ:																					
Gnathia cerina	10	220	----	----	----	----	----	----	----	!	!	!	!	----	!	----	----	----	----	----	Rocks and mud, young parasitic on fish.
TANAIDÆ:																					
Tanais vittatus	0	1	----	----	----	!	----	----	----	----	----	----	----	----	----	----	----	----	!	+	Algæ and on piles.
Leptochelia algicola	0	1	----	----	----	!	----	!	!	----	----	----	----	----	----	----	----	----	!	----	Algæ and eel-grass.
limicola	48	48	----	----	----	----	----	----	----	!	----	----	----	----	----	----	----	----	----	----	Muddy bottoms.
rapax[6]	----	----	----	----	----	----	----	----	----	!	----	----	----	----	----	----	----	----	----	----	Algæ and eel-grass.
filum	20	20	----	----	----	----	----	----	----	----	----	----	+	----	+	----	----	----	----	----	"Among *Ascidiæ callosæ*."
cæca	48	48	----	----	----	----	----	----	!	!	----	----	----	----	----	----	----	----	----	----	Muddy bottoms.

[1] *Munnopsis typica* has been dredged only in 60 fathoms in the Bay of Fundy within our limits, but was obtained by Mr. J. F. Whiteaves from 220 fathoms in the Gulf of St. Lawrence.

[2] *Idotea irrorata* extends also to the Baltic, the Mediterranean, the Adriatic, the Black Seas, and is even said to occur in the Caspian Sea.

[3] *Idotea robusta* was described by Sars from 60° north latitude, between Greenland and Iceland.

[4] *Synidotea nodulosa* extends also to British Columbia.

[5] *Limnoria lignorum* is found also in the Adriatic Sea and is said to occur in the Pacific Ocean, near San Francisco, Cal.

[6] *Leptochelia rapax* was collected at Annisquam, north of Cape Ann, and therefore not strictly within the limits of Massachusetts Bay.

NOTE.—In the above table an exclamation point is used to signify the responsibility of the author for the statement that a species has been found at the locality named. The sign + denotes that a species is said by other authors to occur at the locality under which it is used. The depths are given in fathoms; *lw.*, signifies low water; *s.*, shore; *sf.*, surface. See also page 137.

LIST OF AUTHORITIES.

The present list includes only such works and articles, relating wholly or in part to Crustacea, as have been quoted, or otherwise used, in the preparation of the preceding paper, and is chiefly intended to aid in consultation of the authorities quoted. A few of the titles are necessarily given at second hand, as indicated by quotation marks in the list. The references to these works occurring throughout the article are also inclosed in quotation marks, usually with an accompanying mention of the author from whom they are taken. In all other cases the references have been made directly from the works quoted. A considerable number of authorities have not been referred to, and are omitted from the list, because at present inaccessible, and, for many of the most important works that I have been able to consult, I am indebted to the liberality of Professor S. I. Smith, who has given me the free use of his library and afforded other material aid in the preparation of the article. I have also had free access to the libraries of Professors Verrill, Marsh and Dana.

In this list, as throughout the article, the number of the series of various scientific publications is indicated by Roman numerals in capitals. As far as possible references have been made to the original paging, sometimes with that of the separata added in a parenthesis, and, in the following list, a parenthesis is used to denote that the paging is, or is supposed to be, that of the separata.

Agassiz, Alexander. Letter to C. P. Patterson, Superintendent Coast Survey, on the dredging operations of the U. S. Coast Survey steamer "Blake" during parts of January and February 1878. <Bulletin of the Museum of Comparative Zoology, vol. v, pp. 1-9. Cambridge, 1878.

Andrews, A. [Limnoria terebrans attacking telegraph cable.] <Quarterly Journal of Microscopical Science, II, vol. xv, p. 332. London, 1875.

Audouin, Jean Victor, *and* **Edwards, Henri Milne.** "Résumé d'Entomologie, ou d'Histoire naturelle des animaux articulés, complété par une iconographie de 48 planches. [2 vols.] Paris, 1828–29."

Audouin, Jean Victor, *and* **Edwards, Henri Milne.** Précis d'Entomologie ou d'Histoire naturelle des animaux articulés. Première division, Histoire naturelle des annélides, crustacés, arachnides et myriapodes, complété par une iconographie. [8vo, 70 pages, 48 plates.] Paris, 1829.

Audouin, Jean Victor. Description de l'Égypte ou recueil des observations, et des recherches qui ont été faites en Égypte pendant l'expedition de l'armée Française. Explication sommaire des planches de crustacés de l'Égypte et de la Syrie. Publiées par J. C. Savigny. Histoire naturelle, tome i, pt. 4, pp. 77–98. Paris, "1830."

Bate, C. Spence. On the British Edriophthalma. <Report of the British Association for the Advancement of Science, 1855, Reports on the state of science, pp. 18–62, pl. xii-xxii. London, 1856.

Bate, C. Spence. On Praniza and Anceus and their affinity to each other. <Annals and Magazine of Natural History, III, vol. ii, pp. 165–172, pl. vi-vii. London, Sept., 1858.

Bate, C. Spence. Crustacea. [In] List of the British marine invertebrate fauna. By Robert McAndrew. <Report of the British Association for the Advancement of Science, 1860, Reports on state of science, pp. 217–236. London, 1861.

Bate, C. Spence. Carcinological gleanings, No. ii. < Annals and Magazine of Natural History, III, vol. xvii, pp. 24-31, pl. ii. London, 1866.

Bate, C. Spence, *and* **Westwood, John Obadiah.** A History of the British sessile-eyed Crustacea. [2 vols. 8vo.] London, 1861-1868.

Beneden. *See* **Van Beneden.**

Bos, Jan Ritzema. Bijdrage tot de kennis van de Crustacea hedriophthalmata van Nederland en zijne Kusten. [8vo., 100 pages, 2 plates.] Groningen, 1874.

Bosc, Louis Augustin Guillaume. Histoire naturelle des Crustacés, contenant leur description et leurs mœrs; avec figures dessinées d'après nature. [12mo., vol. ii, 296 pages, 18 plates.] Paris, An x (1802).

Buchholz, Reinhold. Zweite Deutsche Nordpolfahrt "in den Jahren 1869 und 1870, unter Führung des Kapitän Koldewey." B. ii, Part viii, Crustaceen, pp. 262-399. pl. i-xv. Leipzig, 1874.

Buchholz, Reinhold. Mittheilungen naturwiss. Vereins v. Neu-Vorpom. u Rügen, i, pp. 1-40. See **Münter, Julius.**

Bullar, John Follett. The generative organs of the parasitic Isopoda. < Journal of Anatomy and Physiology, vol. xi, pp. 118-123, pl. iv. London and Cambridge, 1876.

Bullar, John Follett. Hermaphroditism among the parasitic Isopoda; reply to Mr. Moseley's remarks on the generative organs of the parasitic Isopoda. < Annals and Magazine of Natural History, IV, vol. xix, pp. 254-256. London, 1877.

Catta, J. D. Note sur quelques Crustacés erratiques. < Annales des Sciences naturelles, Zoologie, VI, tome iii, pp. 1-33, pl. i-ii. Paris, 1876.

Coldstream, John. On the structure and habits of the Limnoria terebrans, a minute crustaceous animal destructive to marine wooden erections, as piers, etc. < Edinburgh New Philosophical Journal, vol. xvi, pp. 316-334, pl. vi, 1834.

Cornalia, Emilio, *and* **Panceri, Paolo.** Osservazioni zoologico ed anatomische sopra un nuovo genre di Isopodi sedentari (Gyge branchialis). < Memorie della Reale Accademia delle Scienze di Torino, II, tom. xix, pp. 85-118, pl i-ii. Turin, 1861.

Cuvier, Georges. Le Règne Animal. *See* **Edwards, Henri Milne,** *and* **Latreille, Pierre Andre.**

Czerniavski, Voldemar. Materialia ad Zoographiam Ponticam comparatam. "Transactions of the first meeting of Russian Naturalists at St. Petersburg, 1868." pp. 19-136, pl. i-viii. "1870."

Dalyell, John Graham. The Powers of the Creator displayed in the Creation. [3 vols., 4to, 145 plates.] London, 1851-1858.

Dana, James Dwight. Conspectus Crustaceorum, &c. Conspectus of the Crustacea of the Exploring Expedition * * continued. Crustacea Isopoda. < American Journal of Science and Arts, II, vol. viii, pp. 424-428. New Haven, 1849.

Dana, James Dwight. On the classification of the Crustacea choristopoda or tetradecapoda. < American Journal of Science and Arts, II, vol. xiv, pp. 297-316. New Haven, 1852.

Dana, James Dwight. Report on the Crustacea of the United States Exploring Expedition, under the command of Charles Wilkes, U. S. N., 1838-42. Washington, Text [4to, two parts, 1618 pages], 1853. Atlas [folio, 96 plates], 1855.

Dekay, James E. Zoology of New York or the New York Fauna. Part iv, Crustacea. [4to, 70 pages, 13 plates.] Albany, 1844.

Desmarest, Ansleme Gaetan. Malacostracés. <Dictionnaire des Sciences naturelles, tome xxviii, pp. 138–425 [56 plates]. Paris, 1823.

Desmarest, Ansleme Gaetan. Considerations générales sur la classe des Crustacés. [8vo, 446 pages, 56 plates.] Paris, 1825.

Dohrn, Anton. Untersuchungen über Bau und Entwicklung der Arthropoden. 4. Entwicklung und Organisation von Praniza (Anceus) maxillaris. <Zeitschrift für wissenschaftliche Zoologie, Band xx, pp. 55–80, taf. vi–viii.—5. Zur Kentniss des Baues von Paranthura Costana. <Tom. cit. pp. 81–93, taf. ix. Leipzig, 1870.—7. Zur Kentniss vom Bau und der Entwicklung von Tanais. <Jenaische Zeitschrift für Medicin und Naturwissenschaft, Band v, pp. 293–306, taf. xi–xii. Leipzig, 1870.

Duvernoy, George Louis. Sur un nouveau genre de l'ordre des Crustacés Isopodes et sur l'espèce type de ce genre, le Képone type. <Annales des Sciences naturelles, Zoologie, II, tome xv, pp. 110–122, pl. iv B. Paris, 1841.

Duvernoy, George Louis, *and* **Lereboullet, Auguste.** Essai d'une monographie des organes de la respiration de l'ordre des Crustacés Isopodes. <Annales des Sciences naturelles, Zoologie, II, tome xv, pp. 177–240, pl. vi. Paris, 1841.

Ebner, Victor von. Helleria, eine neue Isopoden-Gattung aus der Familie der Oniscoiden. <Verhandlungen k. k. zoologisch-botanischen Gesellschaft, Wien, Band xviii, pp. 95–114, pl. i. Vienna, 1868.

Edwards, Alphonse Milne. Sur un Isopode gigantesque des grandes profondeurs de la mer. <Comptes Rendus, tome lxxxviii, pp. 21–23. Paris, 1879.

Translated in the Annals and Magazine of Natural History, V, vol. iii, pp. 241–243. London, 1879.

Edwards, Henri Milne. "Résumé d'Entomologie" *and* Précis d'Entomologie. *See* **Audouin, Jean Victor.**

Edwards, Henri Milne. Annotations in Histoire naturelle des animaux sans vertèbres, par J. B. P. A. de Lamarck, 2me Edit., tome v, 8vo. Paris, 1838.

Edwards, Henri Milne. Histoire naturelle des Crustacés, comprenant l'anatomie, la physiologie et la classification de ces animaux. [8vo, 3 vols. text, 1 vol. plates.] Paris, tome i, 1834, tome ii, 1837, tome iii, 1840.

Published as a part of the Suites à Buffon.

Edwards, Henri Milne. Le Règne Animal distribué d'après son organisation, par Georges Cuvier. Les Crustacés avec une atlas. [Crochard edition, text 4to, 278 pages, atlas with 87 plates.] Paris, "1849."

Edwards, Henri Milne. Observations sur le squelette tégumentaire des Crustacés Décapodes et sur la Morphologie de ces Animaux. <Annales des Sciences naturelles, Zoologie, III, tome xvi, pp. 221–291, pl. viii–xi. Paris, 1851.

Edwards, Henri Milne. Rapport sur un travail de M. Hesse relatif aux metamorphoses des Ancées et des Caliges. <Annales des Sciences naturelles, Zoologie, IV, tome ix, pp. 89–92. Paris, 1858.

Eichwald, Eduard von. "Faunæ Caspio-Caucasiæ illustrationes universæ. <Noveaux Mémoires de la Société Impériale des Naturalistes de Moscou, vol. vii. Moscow, 1842."

Fabricius, Johann Christian. Entomologia Systematica emendata et aucta secundum classes, ordines, genera, species, adjectis synonimis, locis, observationibus, descriptionibus. [8vo, 4 vols., vols. i and iii in two parts]. Hafniae (Copenhagen) 1792–1794. Index alphabeticus. [175 pages]. 1796.

Fabricius, Johann Christian. Supplementum Entomologiæ Systematicæ. [8vo, 572 pages.] Hafniae (Copenhagen) 1798. Index alphabeticus. [53 pages.] 1799.

Fabricius, Otho. Fauna Groenlandica. [8vo, 450 pages, 1 plate.] Copenhagen, 1780.

Fleming, John. Crustacea. <Encyclopædia Britannica, 7th edition, vol. vii, pp. 497–504, pl. clxxx–clxxxi, 4to. Edinburgh (1842).

Fraisse, Paul. Die Gattung Cryptoniscus Fr. Müller (Liriope Rathke). <Arbeiten aus dem Zoologisch-zootomischen Institut in Würzburg, Band iv, pp. 239–296, taf. xii–xv. 1878.

Gaimard, Paul. Voyages en Scandinavie, etc. *See* **Kröyer, Henrik.**

Gegenbaur, Carl. Elements of Comparative Anatomy. Translated by F. Jeffrey Bell, the translation revised by E. Ray Lankester. [8vo, 645 pages.] London, 1878.

Geoffroy, Étienne Louis. "Histoire abrégée des Insectes qui se trouvent aux environs de Paris, dans laquelle ces animaux sont rangés suivant un ordre méthodique. Paris, 1762, 1800, 2 vols., 4to."

Goodsir, Henry D. S. On two new species of Leachia, with a plate. <Edinburgh new Philosophical Journal, vol. xxxi, pp. 309–313, pl. vi. 1841.

Gosse, Philip Henry. A Manual of Marine Zoology for the British Isles. [Two parts, small 8vo.] London, 1855–1856.

Gould, Augustus Addison. Crustacea [List of, in Massachusetts]. <Report on the Geology, Mineralogy, Botany, and Zoology of Massachusetts. 2d edition. By Edward Hitchcock. pp. 548–550. Amherst, 1835.

Gould, Augustus Addison. Report on the Invertebrata of Massachusetts, comprising the Mollusca, Crustacea, Annelida and Radiata. [8vo, 373 pages, 15 plates.] Boston, 1841.

Griffith, Edward, *and* **Pidgeon, Edward.** The Classes Annelida, Crustacea and Arachnida arranged by the Baron Cuvier, with supplementary additions to each order. [8vo, 540 pages, 59 plates.] London, 1833.

Grube, Adolph Eduard. Ein Ausflug nach Triest und dem Quarnero. Beiträge zur Kentniss der Thierwelt dieses Gebietes. [8vo, 175 pages, 5 plates.] Berlin, 1861.

Guérin Méneville, Felix Edouard. Iconographie du Règne Animal de Cuvier. Avec un texte descriptif mis au courant de la science. Crustacés. [8vo, 48 pages, 35 plates.] Paris, 1829–1843.

Guérin Méneville, Felix Edouard. Expedition Scientifique de Morée, Section des Sciences physiques, tome iii, pt. i, Zoologie, section ii. Des Animaux articulés. Crustacés, pp. 30–50. pl. xxvii. Paris, 1832.

Harger, Oscar. The sexes of Sphæroma. <American Journal of Science and Arts, III, vol. v, p. 314. New Haven, 1873.

Harger, Oscar. On a new genus of Asellidæ. <American Journal of Science and Arts, III, vol. vii, pp. 601–602. New Haven, 1874.

Harger, Oscar. This Report, part i, pp. 569–573. *See* **Verrill, Addison Emory.**

Harger, Oscar. Trans. Conn. Acad., vol. iii, pp. 1–57, *and* This Report, part ii, pp. 657-661. *See* **Smith, Sidney Irving.**

Harger, Oscar. Description of Mancasellus brachyurus, a new fresh water Isopod <American Journal of Science and Arts, III, vol. xi, pp. 304–305. New Haven, 1876.

Harger, Oscar. Descriptions of new genera and species of Isopoda, from New England and adjacent regions. <American Journal of Science and Arts, III, vol. xv, pp. 373-379. New Haven, 1878.

Harger, Oscar. Notes on New England Isopoda. <Proceedings of the United States National Museum, 1879, vol. ii, pp. 157-165. Separata, Washington, 1879. [List of Bopyridæ by Prof. S. I. Smith.]

Heller, Camill. Carcinologische Beiträge zur Fauna des adriatischen Meeres. <Verhandlungen der k. k. zoologisch-botanischen Gesellschaft in Wien, Band xvi, pp. 723-760. Vienna, 1866.

Heller, Camill. Die Crustaceen, Pycnogoniden und Tunicaten der k. k. Osterr-Ungar Nordpol-Expedition. <Denkschriften der mathematisch-naturwissenschaftlichen Classe der kaiserlichen Academie der Wissenschaften, Band xxxv, pp. 25-46, taf. i-v. Vienna, 1875.

Hesse, Eugène. Mémoire sur les Pranizes et les Ancées (extrait). <Annales des Sciences naturelles, Zoologie, IV, tome ix, pp. 93-119. Paris, 1858.

Hesse, Eugène. Memoir on the Pranizæ and Ancei (abstract). <Annals and Magazine of Natural History, III, vol. xiv, pp. 405-417. London, 1864.

Hesse, Eugène. Mémoire sur les Pranizes et les Ancées. <Mémoires présentés par divers savants à l'Académie des Sciences de l'Institut Impérial de France, tome xviii, pp. 231-302, pl. i-iv. Paris, 1868.

Hesse, Eugène. Observations sur des Crustacés rares ou nouveaux des côtes de France. 15me article. Description d'un nouveau crustacé appartenant au genre Limnorie. <Annales des Sciences naturelles, Zoologie, V, tome x, pp. 101-121, pl. ix. Paris, 1868.

Hitchcock, Edward. A Catalogue of the animals and plants of Massachusetts, VI, Crustacea. <Report on the Geology, Mineralogy, Botany, and Zoology of Massachusetts, pp. 563-564. Amherst, 1833.

Hope, Frederick William. "Observations on the ravages of Limnoria terebrans, with suggestions for a preventative against the same. <Transactions of the Entomological Society, vol. i, pp. 119, 120. London, 1836."

Hope, Frederick William. Catalogo dei Crostacei Italiani e di Molti Altri del Mediterraneo. [8vo, 48 pages, 1 plate.] Naples, 1851.

Huxley, Thomas Henry. A Manual of the anatomy of invertebrated animals. [8vo. 698 pages.] London, 1877. American edition [12mo. 596 pages], New York, 1878.

Johnston, George. Contributions to the British Fauna. <The Edinburgh Philosophical Journal, vol. xiii, pp. 218-222, 1825.

Johnston, George. Illustrations in British Zoology. <Loudon's Magazine of Natural History, vol. viii, pp. 494-498. London, 1835. Ibid., vol. ix, pp. 79-83. 1836.

Jones, John Matthew. Notes on the marine zoology of Nova Scotia. <Proceedings and Transactions of the Nova Scotian Institute of Natural Science of Halifax, N. S., vol. ii, 1869-70, part iv, pp. 93-99. Halifax, 1870.

Jones, Thomas Rupert. Manual of the natural history of Greenland. *See* **Lütken, Christian Friedrich.**

Kinahan, John Robert. Analysis of certain allied genera of terrestrial Isopoda; with description of a new genus and a detailed list of the British species of Ligia, Philougria, Philoscia, Porcellio, Oniscus and Armadillidium. <Natural History Review, 1857, Proceedings of Societies, pp. 258-282, pl. xix-xxii. Dublin, 1857.

Kingsley, John Sterling. Bulletin of the Essex Institute, vol. IX, pp. 103–108. *See* Streets, Thomas Hale.

Kingsley, John Sterling. Notes on New England Isopoda, [by O. Harger, Notice of.] <American Naturalist, vol. XIV, pp. 120–121. Philadelphia, 1880.

Kirby, William, *and* **Spence, William.** "An Introduction to Entomology, or Elements of the natural history of Insects. 5th and 6th editions. London."

Kröyer, Henrik. Grönlands Amfipoder. <Kongelige Danske Videnskabenes Selskabs naturvidenskabelige og mathematiske Afhandlinger, vol. VII, pp. 229–326, (1–98) pl. I–IV. Copenhagen, 1838.

Kröyer, Henrik. Munna, en ny Kræbsdyrslægt. <Naturhistorisk Tidsskrift, Bind II, pp. 612–616, pl. VI, figs. 1–9. Copenhagen, 1839.

Kröyer, Henrik. Nye Arter af Slægten Tanais. <Naturhistorisk Tidsskrift, Bind IV, pp. 167–168, pl. II. Copenhagen, 1842.

Kröyer, Henrik. Monografisk Fremstilling af Slægten Hippolyte's nordiske Arter. <Kongelige Danske Videnskabenes Selskabs naturvidenskabelige og mathematiske Afhandlinger, vol. IX, pp. 211–360, pl. I–VI. Copenhagen, 1842.

Kröyer, Henrik. Karcinologiske Bidrag. <Naturhistorisk Tidsskrift, II, Bind II, pp. 1–123, 1846, and pp. 366–446, 1847. Copenhagen, 1846–7.

Kröyer, Henrik. Voyages en Scandinavie en Laponie, au Spitzberg et aux Féröe, Zoologie, Crustacés. (Publiées sous la direction de M. Paul Gaimard). [40 folio plates.] Paris, 1849.

Lamarck, Jean P. B. A. de M. de. Histoire naturelle des animaux sans vertèbres. 2me Edit. Revue et augmentée de notes par MM. G. P. Deshayes et H. Milne Edwards.

Latreille, Pierre Andre. Histoire naturelle génerale et particulière des Crustacés et des Insectes. [8 vo, 14 vols. text, 1 vol. plates.] Paris, An X–XIII (1802–1805).

Latreille, Pierre Andre. Genera Crustaceorum et Insectorum secundum ordinem naturalem in familias disposita, iconibus exemplisque plurimis explicata. [2 vols., 16 plates.] Paris, 1806–1807.

Latreille, Pierre Andre. "Entomologie ou histoire naturelle des Crustacés, des Arachnides et des Insectes. <Encyclopédie méthodique. Paris, 1789–1825."

Latreille, Pierre Andre. Le Règne Animal distribué d' après son Organisation par M. Le Baron Cuvier, tome IV, Crustacés, Arachnides et partie des Insectes. [8vo, 584 pages.] Paris, 1829.

Latreille, Pierre Andre. Cours d'Entomologie, ou de l'Histoire naturelle des Crustacés des Arachnides, des Myriapodes et des Insectes. [8vo, 568 pages, with atlas of 24 plates.] Paris, 1831.

Latrobe, Benjamin Henry. A drawing and description of the Clupea tyrannus and Oniscus prægustator. <Transactions of the American Philosophical Society, vol. V, pp. 77–81, pl. I. Philadelphia, 1802.

Leach, William Elford. Crustaceology. <Edinburgh Encyclopædia, vol. VII. Edinburgh, "1813–14."

I have seen only an American edition, in which the article is on pp. 221–277.

Leach, William Elford. A tabular view of the external characters of four classes of animals which Linné arranged under Insecta; with the distribution of the genera composing three of these classes into orders, etc., and descriptions of several new genera and species. <Transactions of the Linnean Society of London, vol. XI, pp. 306–400. London, 1815.

Leach, William Elford. Cymothoadées. <Dictionnaire des Sciences naturelles, tome xii, pp. 338–354. Paris, 1818.

Leidy, Joseph. Contributions toward a knowledge of the marine invertebrate fauna of the coasts of Rhode Island and New Jersey. <Journal of the Academy of Natural Science, II, vol. iii, pp. 135–152, pl. x–xi. Philadelphia, 1855.

Leidy, Joseph. Notices of some animals on the coast of New Jersey. <Proceedings of the Academy of Natural Sciences of Philadelphia, 1879, pp. 198–199. 1879.

Lenz, Heinrich. Die wirbellosen Thiere der Travemünder Bucht. Theil I = Anhang I zu dem Jahresberichte 1874–1875 der Kommission zur wissenschaftlichen Untersuchung der deutschen Meere in Kiel. [24 pages, 2 plates.] Berlin, 1878.

Lereboullet, Auguste. Annales des Sciences naturelles, Zoologie, II, tome xv, pp. 177–240. *See* **Duvernoy, George Louis.**

Lilljeborg, Wilhelm. Bidrag till den högnordiska hafsfaunan. <Œfversigt af Kongl. Vetenskaps-Akademiens Förhandlingar, Årg vii, pp. 82–88. Stockholm, 1850.

Lilljeborg, Wilhelm. Norges Crustacéer. <Œfversigt af Kongl. Vetenskaps-Akademiens Forhandlingar, Årg viii, pp. 19–25. Stockholm, 1851.

Lilljeborg, Wilhelm. Hafs-Crustaceer vid Kullaberg. <Œfversigt af Kongl. Vetenskaps-Akademiens Förhandlingar, Årg ix, pp. 1–13. Stockholm, 1852.

Lilljeborg, Wilhelm. Bidrag till kännedomen om de inom Sverige och Norrige förekommande Crustaceer af Isopodernas underordning och Tanaidernas familj. [8vo, 32 pages.] Upsala, 1865.

Linné, Carl von. "Fauna Suecica, sistens animalia Sueciæ Regni Quadrupedia, Aves, Amphibia, Pisces, Insecta, Vermes, distributa per classes et ordines, genera et species. Editio altera, 8vo. Stockholm, 1761."

Linné, Carl von. Systema Naturæ per Regna tria Naturæ, secundum classes, ordines, genera, species cum characteribus, differentiis, synonymis, locis. Tomus I, Ed. 12 reformata. Holmiae (Stockholm), 1766–1767.

Lockington, William Neale. Description of seventeen new species of Crustacea. <Proceedings of the California Academy of Sciences, vol. vii, pp. 41–48 (1–8). San Francisco, 1877.

Lucas, Hippolyte. Histoire naturelle des animaux articulés. Crustacés. [pp. 1–88, 8 plates.] <Exploration Scientifique de l'Algérie pendant les années, 1840–1842. Sciences physiques, Zoologie, I. Paris, 1849.

Lütken, Christian Friedrich. Nogle Bemærkninger om de nordiske Æga-arter samt om Æga-slægtens rette Begrændsning. <Videnskabelige Meddelelser fra den naturhistoriske Forening i Kjöbenhavn, Aaret 1858, pp. 65–78, pl. 1 A. 1859.

Lütken, Christian Friedrich. Om visse Cymothoagtige Krebsdyrs Ophold i Mundhulen hos forskjellige Fiske. <Videnskabelige Meddelelser fra den naturhistoriske Forening i Kjöbenhavn, Aaret 1858, pp. 172–179. Copenhagen, 1859.

Lütken, Christian Friedrich. Tillæg til „Nogle Bemærkninger om de nordiske Æga-arter samt om Æga-slægtens rette Begrændsning"—Om Æga tridens Leach og Æga rotundicauda Lilljeborg samt om slægterne Acherusia og Ægacylla. <Videnskabelige Meddelelser fra den naturhistoriske Forening i Kjöbenhavn, Aaret 1860, pp. 175–183 (1–9). Copenhagen, 1861.

Lütken, Christian Friedrich. Ibid., 1861, pp. 274–276. *See* **Steenstrup, Japetus.**

Lütken, Christian Friedrich. The Crustacea of Greenland. <Manual of the natural history, geology and physics of Greenland and the neighbouring regions; prepared for the use of the Arctic expedition of 1875, by T. Rupert Jones, pp. 146-165. London, 1875.

Macdonald, John Denis. On the external anatomy of Tanais vittatus occurring with Limnoria and Chelura terebrans in excavated pier-wood. <Transactions of the Linnean Society, II, Zoology, vol. i, pp. 67-71, pl. xv. London, 1875.

M'Intosh, William Carmichael. On the invertebrate marine fauna and fishes of St. Andrews. <Annals and Magazine of Natural History, IV, vol. xiii, pp. 140-145, 204-221, 302-315, 342-357, 420-432, vol. xiv, pp. 68-75, 144-155, 192-207, 258-274, 337-349, 412-425. London, 1874.

Marcusen, Johann. Zur Fauna des schwarzen Meeres. <Archiv für Naturgeschichte, Jahrgang xxxiii, Band i, pp. 357-363. Berlin, 1867.

Mayer, Paul. Carcinologische Mittheilungen. VI. Ueber den Hermaphroditismus bei einigen Isopoden. <Mittheilungen aus der Zoologischen Station zu Neapel, B. i, pp. 165-179, pl. v. Leipzig, 1879.

Meinert, Fr. Crustacea Isopoda, Amphipoda et Decapoda Daniæ: Fortegnelse over Danmarks Isopode, Amphipode, og Decapode Krebsdyr. "Naturhistorisk Tidsskrift, III," pp. 57-248. Copenhagen, "1877."

Metzger, Adolf. Die wirbellosen Meeresthiere der ostfriesischen Küste. Ein Beitrag zur Fauna der deutschen Nordsee. <Zwanzigster Jahresbericht der naturhistorischen Gesellschaft zu Hannover, pp. 22-36. Hannover, 1871.

Metzger, Adolf. Nordseefahrt der Pommerania—" Zoologische Ergebnisse der Nordseefahrt, X. Crustaceen aus den Ordnungen Edriophthalmata u. Podophthalmata, taf. vi. Aus Jahrsbericht der Commission zu wiss. Untersuchung des deutsches Meer, im Kiel, Jahre 1872-1873." Berlin, 1875.

Miers, Edward John. List of the species of Crustacea collected by the Rev. A. E. Eaton at Spitzbergen in the summer of 1873, with their localities and notes. <Annals and Magazine of Natural History, IV, vol. xix, pp. 131-140. London, 1877.

Miers, Edward John. Report on the Crustacea collected by the naturalists of the Arctic Expedition in 1875-1876. <Annals and Magazine of Natural History, IV, vol. xx, pp. 52-66 and 96-110, pl. iii-iv. London, 1877.

Milne-Edwards. *See* **Edwards, Alphonse Milne** *and* **Henri Milne.**

Möbius, Karl. Die wirbellosen Thiere der Ostsee. <"Bericht über die Expedition zur physikalisch-chemischen und biologischen Untersuchung der Ostsee im Sommer 1871 auf S. M. Avisodampfer Pommerania." pp. 97-144. Kiel, 1873.

Möbius, Karl. On the invertebrate animals of the Baltic. <Annals and Magazine of Natural History, IV, vol. xii, pp. 81-89. London, 1873.
Translated by W. S. Dallas from the preceding.

Mohr, Nicholas. Forsøg til en Islandisk Naturhistorie, med adskillige œkonomiske samt andre Anmærkninger. [8vo, 413 pages.] Copenhagen, 1786.

Montagu, George. Description of several marine animals found on the south coast of Devonshire. <Transactions of the Linnean Society of London, vol. vii, pp. 61-85, tab. vi-vii, 1804. Ibid., vol. ix, pp. 81-114, tab. ii-viii. 1808.

Montagu, George. Descriptions of several new or rare animals, principally marine, discovered on the south coast of Devonshire. <Transactions of the Linnean Society of London, vol. xi, pp. 1-26, tab. i-v. London, 1815.

Moore, Edward. On the occurrence of Teredo navalis and Limnoria terebrans in Plymouth Harbour. <Magazine of Natural History, new series, vol. ii, pp. 206-210. London, 1838.

Moore, Edward. Limnoria terebrans in Plymouth Harbour. <Magazine of Natural History, new series, vol. iii, pp. 196–197. London, 1839.

Moore, Edward. Catalogue of the malacostracous Crustacea of South Devon. <Magazine of Natural History, new series, vol. iii, pp. 284–294. London, 1839.

Moseley, Henry Nottidge. Remarks on observations by Capt. Hutton, Director of the Otago Museum, on Peripatus novæ-zealandiæ, with notes on the structure of the species. <Annals and Magazine of Natural History, IV, vol. xix, pp. 85–91. London, 1877.

Moseley, Henry Nottidge. Hermaphroditism in the parasitic Isopoda. Further remarks on Mr. Bullar's papers on the above subject. <Annals and Magazine of Natural History IV, vol. xix, pp. 310, 311. London, 1877.

Müller, Friedrich. Bemerkungen zu Zaddach's Synopseos Crustaceorum Borussicorum prodromus. <Archiv für Naturgeschichte, Jahrgang, xiv, Band i, pp. 62–64, pl. iv. Berlin, 1848.

Müller, Friedrich [Fritz]. Ueber den Bau der Scheerenasseln (Asellotes hétéropodes M. Edw.). <Archiv für Naturgeschichte, Jahrg. xxx, B. i, pp. 1–6. Berlin, 1864.

Müller, Friedrich [Fritz]. Facts and Arguments for Darwin. Translated by W. S. Dallas. [8vo., 144 pages.] London, 1869.

Müller, Friedrich [Fritz]. Bruchstücke für Naturgeschichte der Bopyriden. <Jenaische Zeitschrift für Medicin und Naturwissenschaft, Band vi, pp. 53–73, taf. iii–iv. Leipzig, 1871.

Münter, Julius, *and* Buchholz, Reinhold. "Ueber Balanus improvisus (Darw.) var. gryphicus (Münter). Beitrag zur carcinologischen Fauna Deutschlands. <Mittheilungen d. naturwissensch. Vereins von Neu-Vorpommern u. Rügen, i, pp. 1–40, 2 plates. Berlin, 1869."

Norman, Alfred Merle. Reports of deep-sea dredging on the coast of Northumberland and Durham, 1862–64. Report on the Crustacea. <Natural History Transactions, Northumberland and Durham, vol. i, pp. 12–29, "1865."

Norman, Alfred Merle. Report of the committee appointed for the purpose of exploring the coasts of the Hebrides by means of the dredge. Part ii, on the Crustacea, Echinodermata, Polyzoa, Actinozoa and Hydrozoa. <Report of the British Association for the Advancement of Science for 1866, Reports on the State of Science, pp. 193–206. London, 1867.

Norman, Alfred Merle. Preliminary report on the Crustacea, Molluscoida, Echinodermata, and Cœlenterata, procured by the Shetland dredging committee in 1867. <Report of the British Association for the Advancement of Science for 1867, Reports on the State of Science, pp. 437–441. London, 1868.

Norman, Alfred Merle. On two Isopods belonging to the genera Cirolana and Anilocra, new to the British Islands. <Annals and Magazine of Natural History, IV, vol. ii, pp. 421–422, pl. xxiii. London, 1868.

Norman, Alfred Merle. Last Report on dredging among the Shetland Isles, Part ii, Crustacea, etc. <Report of the British Association for the Advancement of Science for 1868, Reports on the State of Science, pp. 247–336 and 344–345. London, 1869.

Norman, Alfred Merle. Crustacea, Tunicata, Polyzoa, Echinodermata, Actinozoa, Foraminifera, Polycistina, and Spongida in "Preliminary Report of the Biological Results of a Cruise in H. M. S. 'Valorous' to Davis Strait in 1875." By J. Gwyn Jeffreys. <Proceedings of the Royal Society, vol. xxv, pp. 202–215. London, 1876.

Œrsted, Anders Sandöe. Beretning om en Excursion til Trindelen, en Alluvialdannelse i Odensef jord, i Esteraaret 1841, d. 19de Octbr. <Naturhistorisk Tidsskrift, Bind iii, pp. 552–569, tab. viii. Copenhagen, 1841.

Owen, Richard. "The Zoology of Captain Beechey's Voyage * * * to the Pacific Ocean and Behring's Straits, performed in H. M. Ship Blossom * * * in the years 1825–28." Crustacea, pp. 77–92, pl. xxiv–xxviii. London, "1839."

Packard, Alpheus Spring. A list of animals dredged near Caribou Island, Southern Labrador, during July and August, 1860. <Canadian Naturalist and Geologist, vol. viii, pp. 401–429, pl. i–ii. Montreal, 1863.

Packard, Alpheus Spring. Observations on the glacial phenomena of Labrador and Maine, with a view of the recent invertebrate fauna of Labrador. <Memoirs of the Boston Society of Natural History, vol. i, pp. 210–303, pl. vii–viii. Boston, 1867.

Packard, Alpheus Spring. On the Crustaceans and Insects [in] The Mammoth Cave and its inhabitants, by the editors. <American Naturalist, vol. v, pp. 744–761. Salem, 1871.

Packard, Alpheus Spring. On the cave fauna of Indiana. <Fifth Annual Report of the Trustees of the Peabody Academy of Science, pp. 93–97. Salem, 1873.

Panceri, Paolo. Mem. Accad. Sci. Torino, II, vol. xix, pp. 85–118. *See* **Cornalia, Emilio.**

Parfitt, Edward. The fauna of Devon. Part IX. Sessile-eyed Crustacea. [8vo, 25 pages.] "Reprinted from the Transactions of the Devonshire Association for the Advancement of Science, Literature, and Art. 1873."

Pennant, Thomas. "The British Zoology, 4th Edit., 4 vols., with 279 plates, 4to. London, 1777."

Pidgeon, Edward. The Classes Crustacea, etc. *See* **Griffith, Edward.**

Rathke, Heinrich. "Beitrag zur Fauna der Krimm. <Memoiren der kaiserlichen Akademie der Wissenschaften zu St. Petersburg, Theil iii, pp. 291–454, 773–774. 1837."

Rathke, Heinrich. Beiträge zur Fauna Norwegens. <Nova Acta Academiæ Cæsareæ Leopoldino-Carolinæ Naturæ Curiosorum, tom. xx, pp. 1–264c., taf. i–xii. Breslau and Bonn, 1843.

Rathke, Jens. "Jagttagelser henhorende til Indvoldsormenes og Blöddyrenes naturhistorie; med anmärkningar af O. Fabricius. <Skrivter af naturhistorie-Selskabet, vol. v, pp. 61–153, tab. ii–iii. Copenhagen, 1799."

Reinhardt, Johann T. Fortegnelse over Grønlands Krebsdyr, Annelider og Indvoldsorme. <Naturhistorisk Bidrag til en Beskrivelse af Grønland, pp. 28–49. "Særskilt Aftryk af Tillægene til 'Grønland, geographisk og statistisk beskrevnet' af H. Rink." Copenhagen, 1857.

Risso, Antoine. Histoire naturelle des Crustacés des environs de Nice. [8vo, 176 pages, 3 plates.] Paris, 1816.

Risso, Antoine. "Histoire naturelle de l'Europe méridionale, tome v. Paris, 1826."

Ritzema Bos, Jan. *See* **Bos, Jan Ritzema.**

Roux, Jean Louis Florent Polydore. "Crustacés de la Méditerranée et de son Littoral décrits et lithographies. Marseilles, 1829–1830."

Saenger, Nicholas. "Preliminary account of an exploration of the fauna of the Baltic. <Communications of the Imp. Society of Nat. Sc. Anthropol. and Ethnol. of the Univers. of Moscow, vol. viii, pp. 22–34. 1869."

Samouelle, George. The Entomologist's useful Compendium; or an introduction to the knowledge of British Insects. [8vo, 496 pp., 12 plates.] London, 1819.

Sars, George Ossian. [Om en anomal Gruppe af Isopoder.] <Forhandlinger i Videnskabs-Selskabet i Christiania, Aar 1863, pp. 205–221. Christiana, 1864.

Sars, George Ossian. Beretning om en i Sommeren, 1865, foretagen Zoologisk Reise ved Kysterne af Christianias og Christiansands Stifter. [8vo, 47 pages.] <"Nyt Magazin for Naturvidenskaberne." Christiania, 1866.

Sars, George Ossian. Histoire naturelle des Crustacés d'eau douce de Norvège. 1me livraison. Les Malacostracés. [4to, 145 pages, 10 plates.] Christiania, 1867.

Sars, George Ossian. Undersøgelser over Christianiafjordens Dybvandsfauna anstillede paa en i Sommeren 1868 foretagen zoologisk Reise. [8vo, 58 pages.] <"Nyt Magazin for Naturvidenskaberne." Christiania, 1869.

Sars, George Ossian. Undersøgelser over Hardangerfjordens Fauna. <Forhandlinger i Videnskabs-Selskabet i Christiania, Aar 1871, pp. 246–286. Christiania, 1872.

Sars, George Ossian. Bidrag til Kundskaben om Dyrelivet paa vore Havbanker. <Forhandlinger i Videnskabs-Selskabet i Christiania, Aar 1872, pp. 73–119. Christiania, 1873.

Sars, George Ossian. Prodromus descriptionis Crustaceorum et Pycnogonidarum, quae in expeditione Norvegica Anno 1876, observavit G. O. Sars. <Archiv for Mathematik og Naturvidenskab, Bind ii, pp. 337* [237]–271. Christiania, 1877.

Sars, Michael. Oversigt over de i den norsk-arctiske Region forekommende Krebsdyr. <Forhandlinger i Videnskabs-Selskabet i Christiania, Aar 1858, pp. 122–163. Christiania, 1859.

Sars, Michael. [Beskrivelse af en ny Slægt og Art af Isopoder: Munnopsis typica Sars.] <Forhandlinger i Videnskabs-Selskabet i Christiania, Aar 1860, pp. 84–85. Christiania, 1861.

Sars, Michael. Bidrag til Kundskab om Christianiafjordens Fauna. [104 pages, 7 plates.] <"Nyt Magazin for Naturvidenskaberne." Christiania, 1868.

Sars, Michael. Fortsatte Bemærkninger over det dyriske Livs Udbredning i Havets Dybder. <Forhandlinger i Videnskabs-Selskabet i Christiania, Aar 1868, pp. 246–275. Christiania, 1869.

Savigny, Jules Cæsar. Description de l'Égypte ou Recueil des Observations et des Recherches pendant l'Expedition de l'Armée Francaise. Histoire naturelle, Planches, Zoologie, Crustacés. [13 folio plates.] Paris, 1817.

Say, Thomas. An account of the Crustacea of the United States. <Journal of the Academy of Natural Science, vol. i, part i, pp. 57–63, 65–80, pl. iv, pp. 97–101, 155–169, 1817; part ii, pp. 235–253, 313–319, 374–401, 423–441. Philadelphia, 1817–1818.

Say, Thomas. Observations on some of the animals described in the account of the Crustacea of the United States. <Journal of the Academy of Natural Science, vol. i, part ii, pp. 442–444. Philadelphia, 1818.

* In this volume the paging from 200 to 263 is incorrectly printed 300–368. The separata are paged 337–371.

Schiödte, Jörgen C. On the structure of the mouth in sucking Crustacea, part i, Cymothoæ. < Annals and Magazine of Natural History, IV, vol. i, pp. 1-25, pl. i, 1868.—Parts ii, Anthura, iii, Laphystius. < Ibid., vol. xviii, pp. 253-266 and 295-305. London, 1877.

Translated from "Naturhistorisk Tidsskrift III, vol. iv with 2 plates, and vol. x with 5 plates. Copenhagen, 1866 and 1875."

Schiödte, Jörgen C. Sur la propagation et les metamorphoses des Crustacés suceurs de la famille des Cymothoadiens. < Comptes Rendus, tome lxxxvii, pp. 52-55. Paris, 1878.

Translated in Annals and Magazine of Natural History, V, vol. ii, pp. 195-197. London, 1878.

Smith, Sidney Irving. This Report, part i, pp. 537-747. *See* **Verrill, Addison E.**

Smith, Sidney Irving *and* **Harger, Oscar.** Report on the dredgings in the region of St. George's Banks in 1872. < Transactions of the Connecticut Academy of Arts and Sciences, vol. iii, part i, pp. 1-57, pl. i-viii. New Haven, 1874.

Smith, Sidney Irving. The Crustacea of the fresh waters of the United States. <This Report, part ii, pp. 637-665, pl. i-iii. Washington, 1874. [Descriptions of Asellus and of Asellopsis by O. Harger].

Smith, Sidney Irving. The stalk-eyed Crustaceans of the Atlantic coast of North America north of Cape Cod. < Transactions of the Connecticut Academy, vol. v, pp. 27-136, pl. viii-xii. New Haven, 1879.

Smith, Sidney Irving. Proc. U. S. Nat. Mus., 1879, vol. ii, pp. 157, 158. *See* **Harger, Oscar.**

Smith, Sidney Irving. Occurrence of Chelura terebrans, a Crustacean destructive to the timber of submarine structures, on the coast of the United States. < Proceedings of the United States National Museum, 1879, vol. ii, pp. 232-235. Washington, 1880.

Stalio, Luigi. Catalogo Metodico e Descrittivo dei Crostacei Podottalmi ed Edriottalmi dell' Adriatico. [8vo, 274 pages.] "(Estr. dal vol. iii, serie v, degli Atti dell' Instituto Stesso.)" Venice, 1877.

Sowerby, James. The British Miscellany: or coloured figures of new, rare, or little known animal subjects; many not before ascertained to be inhabitants of the British Isles. [2 vols. in one, 8vo, 76 plates.] London, 1804-1806.

Stebbing, Thomas Roscoe Rede. A Sphæromid from Australia, and Arcturidæ from South Africa. <Annals and Magazine of Natural History, IV, vol. xii, pp. 95-98, pl. iii A. London, 1873.

Stebbing, Thomas Roscoe Rede. On a new species of Arcturus (A. damnoniensis). <Annals and Magazine of Natural History, IV, vol. xiii, pp. 291-292, pl. xv. London, 1874.

Stebbing, Thomas Roscoe Rede. A new Australian Sphæromid, Cyclura venosa; and notes on Dynamene rubra and viridis. < Journal of the Linnean Society, Zoology, vol. xii, pp. 146-151, pl. vi-vii. London, 1874.

Stebbing, Thomas Roscoe Rede. The sessile-eyed Crustacea of Devon. [8vo, 10 pages, 1 plate.] "Reprinted from the Transactions of the Devonshire Association for the Advancement of Science, Literature, and Art. 1874."

Stebbing, Thomas Roscoe Rede. On some new exotic sessile-eyed Crustaceans. <Annals and Magazine of Natural History, IV, vol. xv, pp. 184-188, pl. xv A. London, 1875.

Stebbing, Thomas Roscoe Rede. Description of a new species of sessile-eyed Crustacean and other notices. <Annals and Magazine of Natural History, IV, vol. xvii, pp. 73–80, pl. iv–v. London, 1876.

Stebbing, Thomas Roscoe Rede. Notes on sessile-eyed Crustaceans with description of a new species. <Annals and Magazine of Natural History, V, vol i, pp. 31–37, pl. v. London, 1878.

Stebbing, Thomas Roscoe Rede. Sessile-eyed Crustacea of Devonshire. Supplementary list. [8vo, 9 pages.] "Reprinted from the Transactions of the Devonshire Association for the Advancement of Science, Literature, and Art." 1879.

Steenstrup, Japetus, *and* **Lütken, Christian Friedrich.** Mindre Meddelelser fra Kjöbenhavns Universitets zoologiske Museum.—2. Forelöbig Notits om danske Hav-Krebsdyr. <Videnskabelige Meddelelser fra den Naturhistoriske Forening i Kjöbenhavn, 1861, II, vol. iii, pp. 274–276. Copenhagen, 1862.

Stimpson, William. Synopsis of the Marine Invertebrata of Grand Manan; or the region about the mouth of the Bay of Fundy, New Brunswick. [4to, 66 pp., 3 plates.] <Smithsonian Contributions to Knowledge, vol. vi. Washington, 1853.

Stimpson, William. Descriptions of some new marine Invertebrata. By William Stimpson, Zoologist to the U. S. Surveying Expedition to the North Pacific, Japan Seas, etc., under direction of Commander C. Ringgold, U. S. N. <Proceedings of the Academy of Natural Science, vol. vii, 1855, pp. 385–394. Philadelphia, 1855.

Stimpson, William. On an oceanic Isopod found near the southeastern shores of Massachusetts. <Proceedings of the Academy of Natural Science, vol. xiv, 1862, pp. 133–134. Philadelphia, 1862.

Stimpson, William. Synopsis of the marine Invertebrata collected by the late Arctic Expedition under Dr. I. I. Hayes. <Proceedings of the Academy of Natural Science, vol. xv, 1863, pp. 138–142. Philadelphia, 1863.

Streets, Thomas Hale, *and* **Kingsley, John Sterling.** An examination of types of some recently described Crustacea. <Bulletin of the Essex Institute, vol. ix, pp. 103–108. Salem, 1877.

Templeton, Robert. Description of a minute crustaceous animal from the Island of Mauritius. <Transactions of the Entomological Society, vol. ii, pp. 203–207, pl. xviii. London, "1836."

Templeton, Robert. Catalogue of Irish Crustacea Myriapoda and Arachnöidea, selected from the papers of the late John Templeton, Esq. <Loudon's Magazine of Natural History, vol. ix, pp. 9–14. London, 1836.

Thompson, William. On the Teredo navalis and Limnoria terebrans as at present existing in certain localities on the coasts of the British Islands. <Edinburgh New Philosophical Journal, vol. xviii, pp. 121–130. 1835.

Thompson, William. Note on the Teredo norvegica, Xylophaga dorsalis, Limnoria terebrans and Chelura terebrans combined in destroying the submerged wood-work at the Harbor of Ardrossan, on the coast of Ayrshire. <Annals and Magazine of Natural History, vol. xx, pp. 157–164. London, 1847.

Thompson, William. Additions to the fauna of Ireland, Crustacea. <Annals and Magazine of Natural History, vol. xx, pp. 237–250. London, 1847.

Van Beneden, Pierre Joseph. Recherches sur la Faune littorale de Belgique. Crustacés. [4to, 180 pages, 32 plates.] "Extrait du tome xxxiii des Mémoires de l'Académie royale de Belgique." Bruxelles, 1861.

Verrill, Addison Emory. On the distribution of marine animals on the southern coast of New England. <American Journal of Science and Arts, III, vol. ii, pp. 357-362. New Haven, 1871.

Verrill, Addison Emory. Results of recent dredging expeditions on the coast of New England. (No. 1). <American Journal of Science and Arts, III, vol. v, pp. 1-16, Jan. 1873.—(No. 2). <Ibid., pp. 98-106, Feb. 1873.—No. 3. <Ibid., vol. vi, pp. 435-441, Dec. 1873.—No. 4. <Ibid., vol. vii, pp. 38-46, Jan. 1874.—No. 5. <Ibid., pp. 131-138, Feb. 1874.—No. 6. <Ibid.,pp. 405-414, pl. iv-v, Apr., 1874.—No. 7. <Ibid., pp. 498-505, pl. vi-viii, May, 1874. New Haven, 1873-4.

Verrill, Addison Emory. Explorations of Casco Bay by the United States Fish Commission in 1873. <Proceedings of the American Association for the Advancement of Science, Portland Meeting, 1873, pp. 340-395, pl. i-vi. Salem, 1874.

Verrill, Addison Emory. Report upon the invertebrate animals of Vineyard Sound and the adjacent waters, with an account of the physical characters of the region. <This Report, part i, pp. 295-778 (1-478), pl. i-xxxviii. Washington, 1874.

Published also separately with the above title or, Invertebrata of Southern New England, by A. E. Verrill and S. I. Smith. [8vo, 478 pages, 38 plates.] Washington, 1874.

Verrill, Addison Emory, Smith, Sidney Irving, *and* **Harger, Oscar.** Catalogue of the marine invertebrate animals of the Southern Coast of New England and adjacent waters. <This Report, part i, pp. 537-747 (243-453), pl. i-xxxviii. Washington, 1874.

Published also as a part of the above Report upon the invertebrate animals of Vineyard Sound and adjacent waters or Invertebrata of Southern New England, by A. E. Verrill and S. I. Smith, pp. 243-453. Washington, 1874.

Verrill, Addison Emory. Results of dredging expeditions off the New England Coast in 1874. <American Journal of Science and Arts, III, vol. ix, pp. 411-415, vol. x, pp. 36-43, pl. iii-iv, and pp. 196-202. New Haven, 1875.

Wagner, Nicholas. Recherches sur le système circulatoire et les organes de respiration chez le Porcellion élargi (Porcellio dilatatus Brandt). <Annales des Sciences naturelles, Zoologie, V, tome iv, pp. 317-328, pl. xiv B. Paris, 1865.

Wagner, Nicholas. Observations sur l'organisation et le développement des Ancees. <Bulletin de l'Académie Impériale des Sciences de St. Pétersbourg, tome x, pp. 497-502. 1866.

Westwood, John Obadiah. Extrait des recherches sur les Crustacés du genre Pranize de Leach. <Annales des Sciences naturelles, tome xxvii, pp. 316-322, pl. vi. Paris, 1832.

Westwood, John Obadiah. British Sessile-eyed Crustacea. *See* **Bate, C. Spence.**

White, Adam. List of the specimens of Crustacea in the collection of the British Museum. [143 pages.] London, 1847.

White, Adam. List of the specimens of British animals in the collection of the British Museum, part iv, Crustacea. [141 pages.] London, 1850.

White, Adam. A popular history of British Crustacea, comprising a familiar account of their classification and habits. [358 pages, 20 plates.] London, 1857.

Whiteaves, Joseph Frederick. Notes on a deep-sea dredging-expedition round the Island of Anticosti, in the Gulf of St. Lawrence. <Annals and Magazine of Natural History, IV, vol. x, pp. 341-354. London, 1872.

Whiteaves, Joseph Frederick. Report of a second deep-sea dredging expedition [in 1872] to the Gulf of St. Lawrence, with some remarks on marine fisheries of the Province of Quebec. [8vo, 22 pages.] Montreal, 1873.

Whiteaves, Joseph Frederick. On recent deep-sea dredging operations in the Gulf of St. Lawrence. <American Journal of Science and Arts, III, vol. vii, pp. 210-219. New Haven, 1874.

Whiteaves, Joseph Frederick. Report on further deep-sea dredging operations in the Gulf of St. Lawrence [in 1873], with notes on the present condition of the marine fisheries and oyster-beds of part of that region. [8vo, 29 pages.] "Ottawa, 1874."

Willemoes-Suhm, Rudolf von. Von der Challenger Expedition. Briefe an C. Th. E. v. Siebold. II, Sidney, im April, 1874. <Zeitschrift für wissenschaftliche Zoologie, Band xxiv, pp. ix-xxiii. Leipzig, 1874.

Willemoes-Suhm, Rudolf von. On some Atlantic Crustacea from the 'Challenger' Expedition. <Transactions of the Linnean Society of London, II, Zoology, vol. i, pp. 23-59, pl. vi-xiii. London, 1875.

Willemoes-Suhm, Rudolf von. Preliminary report to Professor Wyville Thomson, F. R. S. Director of the civilian scientific staff, on observations made during the earlier part of the voyage of H. M. S. 'Challenger.' <Proceedings of the Royal Society, vol. xxiv, pp. 569-585.—On Crustacea observed during the cruise of H. M. S. 'Challenger' in the Southern Sea. <Tom. cit., pp. 585-592. London, 1876.

Woodward, Henry. Crustacea. <Encyclopædia Britannica, 9th edition, vol. vi, pp. 632-666. Edinburgh and Boston, 1877.

Zaddach, Ernest Gustav. Synopseos Crustaceorum Prussicorum Prodromus. [4to, 39 pages.] Regiomonti (Königsberg), "1844."

TABLE OF CONTENTS.

EXPLANATION OF THE PLATES.

PLATE I.

FIGURE 1.—Philoscia vittata Say (p. 306); dorsal view, enlarged six diameters; natural size indicated by cross at the right.

2.—Scyphacella arenicola Smith (p. 307); dorsal view, enlarged about twelve diameters; natural size indicated by cross at the right.

3.—Actoniscus ellipticus Harger (p. 309); dorsal view, enlarged ten diameters; natural size indicated by line at the right.

4.—Jæra albifrons Leach (p. 315); female; dorsal view, enlarged about ten diameters.

5.—The same; maxilliped from the left side, exterior view, enlarged twenty-five diameters; P, palpus; *l*, external lamella.

6.—The same; maxillæ, enlarged twenty-five diameters; *a*, outer, or second, pair of maxillæ; *b*, inner, or first, pair of maxillæ; *i*, inner, *e*, outer lobe.

7.—The same; inferior surface of the pleon of a female.

8.—The same; inferior surface of the pleon of a male.

(All the figures were drawn from nature by O. Harger.)

PLATE II.

FIGURE 9.—Janira alta Harger (p. 321); dorsal view, enlarged five diameters; natural size indicated by line at the right.

10.—Janira spinosa Harger (p. 323); dorsal view of female, enlarged six diameters.

11.—Munnopsis typica M. Sars (p. 330); dorsal view of male, enlarged about two diameters; *b*, maxillipeds; *m*, basal segment; *l*, external lamella; 2 and 3, second and third segments of palpus of maxillipeds; *c*, outer maxillæ; *d*, inner maxillæ; *e*, one of the second pair of legs of the male; *f*, one of the natatory legs; *g*, abdominal operculum of the female, external view.

(Figures 9 and 10 were drawn from nature by O. Harger; figure 11 is copied from M. Sars, drawn by G. O. Sars.)

PLATE III.

FIGURE 12.—Janira alta (p. 321); *a*, maxilliped; P, palpus of maxilliped; *l*, external lamella; *b*, mandible; *P*, palpus of mandible; *d*, dentigerous lamella; *m*, molar process, enlarged twenty-five diameters.

13.—The same; inferior surface of the pleon, *a* in the female, *b* in the male, enlarged ten diameters; *a*, single opercular plate in the female; *b*, external; *c*, median plate of operculum of male.

14.—Munna Fabricii Kröyer (p. 325); female; dorsal view, enlarged about twenty diameters; natural size indicated by line at the right.

15.—Eurycope robusta Harger (p. 332); female; dorsal view, enlarged six diameters; natural size indicated by line at the right; *a*, antennula, enlarged twenty diameters; *b*, maxilliped; *c*, mandible; *d*, one of the first pair of legs, each enlarged twenty diameters; *d'*, propodus and dactylus of the first pair of legs, enlarged about thirty-eight diameters; *e*, propodus and dactylus of the second pair of legs, enlarged twenty diameters; *f*, one of the sixth pair of legs; *g*, uropod, each enlarged twenty diameters.

(Figure 14 was drawn from nature by Mr. J. H. Emerton, the others by O. Harger.)

PLATE IV.

FIGURE 16.—Chiridotea cœca Harger (p. 338); dorsal view, enlarged nearly four diameters; natural size indicated by the line at the right.

17.—The same; *a*, antennula; *b*, antenna; each enlarged twelve diameters.

18.—The same; *a*, maxilliped from the right side, external view; *l*, external lamella; *m*, maxilliped proper; 1, 2, 3, first, second, and third segments of the palpus of the maxilliped, enlarged twenty diameters; *b*, one of the first pair of legs, magnified twelve diameters; *c*, uropod from the left side, inner view, showing the two rami articulated near the tip.

19.—The same; pleopods of second pair from the right side, anterior views, enlarged ten diameters; *a*, common form in males; *b*, rarer form in male; *s*, elongated stylet, articulated near the base of the inner lamella; *c*, form in the female.

20.—Chiridotea Tuftsii Harger (p. 340); female; dorsal view, enlarged five diameters; natural size indicated by the line at the right.

21.—The same; left maxilliped, enlarged twenty-five diameters; *e*, external lamella; *m*, basal segment; 1, 2, 3, segments of palpus.

22.—The same; pleopod of the second pair, from a male, enlarged twenty diameters; *s*, elongated stylet, articulated near the base of the inner lamella.

(All the figures were drawn from nature by O. Harger.)

PLATE V.

FIGURE 23.—Chiridotea Tuftsii Harger (p. 340); *a*, antennula; *b*, antenna; *c*, leg of the first pair; *d*, leg of the fourth pair; all enlarged twelve diameters; *e*, left uropod, or opercular valve, inner view, enlarged ten diameters.

24.—Idotea irrorata Edwards (p. 343); dorsal view, enlarged two diameters; natural size shown by the line on the left.

25.—The same; *a*, antennula; *b*, antenna; *c*, left uropod or opercular valve, external view; all enlarged six diameters.

26.—The same; *a*, right maxilliped, enlarged twelve diameters, *l*, external lamella; *m*, basal segment; 1, 2, 3, 4, segments of palpus of maxilliped; *b*, pleopod of the second pair from a male, enlarged eight diameters, showing stylet, *s*, articulated near the base of the inner lamella.

27.—Idotea phosphorea Harger (p. 347); dorsal view, enlarged about two diameters; natural size shown by the line on the right.

28.—The same; *a*, antenna, enlarged six diameters; *b*, maxilliped, enlarged twelve diameters, showing, *l*, external lamella; *m*, basal segments; 1, 2, 3, 4, segments of the palpus of maxilliped; *c*, leg of the first pair; *d*, leg of the second pair, both enlarged six diameters; *e*, right uropod, or opercular valve, inner view, enlarged six diameters.

29.—The same; pleopod of the second pair from a male, enlarged eight diameters; *s*, stylet articulated near the base of the inner lamella; *s'*, distal end of stylet reversed and enlarged thirty diameters.

(Figure 24 was drawn by Mr. J. H. Emerton, the others by O. Harger.)

PLATE VI.

FIGURE 30.—Idotea robusta Kröyer (p. 349); dorsal view, enlarged two diameters; natural size shown by the line at the right.

31.—The same; *a*, antenna; *b*, leg of the first pair, each enlarged six diameters; *c*, left uropod, or opercular valve, inner view, enlarged four diameters.

FIGURE 32.—The same; *a*, maxilliped, enlarged twelve diameters; *l*, external lamella; 1, 2, 3, 4, segments of palpus; *b*, maxilla of the outer or second pair; *c*, pleopod of the second pair from a male, enlarged six diameters; *s*, stylet articulated near the base of the inner lamella.

33.—Synidotea nodulosa Harger (p. 351); dorsal view, enlarged four diameters; natural size indicated by the line at the right.

34.—The same; *a*, antennula; *f*, flagellar segment; *b*, antenna; *c*, leg of the first pair from the right side; *d*, right uropod, or opercular valve, all enlarged ten diameters.

35.—The same; *a*, maxilliped from the right side, showing, *l*, external lamella; *m*, basal segment; 1, 2, 3, segments of palpus, enlarged twenty diameters; *b*, maxilla of the outer or second pair; *c*, maxilla of the inner or first pair, both enlarged twenty diameters; *d*, pleopod of the second pair from a male, enlarged twelve diameters; *s*, stylet articulated near the base of the inner lamella.

36.—Erichsonia attenuata Harger (p. 356); dorsal view, enlarged three diameters, natural size indicated by the line at the right.

(Figures 30 and 36 were drawn by Mr. J. H. Emerton, the others by O. Harger.)

PLATE VII.

FIGURE 37.—Erichsonia attenuata Harger (p. 356); *a*, antennula; *b*, antenna, each enlarged twelve diameters; *c*, maxilliped, showing, *l*, external lamella, enlarged thirty diameters; *d*, uropod, or opercular valve, enlarged twelve diameters; *e*, pleopod of the second pair from a male, enlarged fifteen diameters; *s*, stylet, articulated near the base of the inner lamella; *s'*, distal end of stylet, enlarged fifty diameters.

38.—Erichsonia filiformis Harger (p. 355); dorsal view, enlarged five diameters, natural size indicated by the line at the right.

39.—The same; *a*, antennula; *b*, antenna; *c*, leg of the first pair; *d*, uropod, or opercular valve, each enlarged twelve diameters.

40.—The same; *a*, maxilla of outer or second pair; *b*, maxilla of inner or first pair; *c*, mandible, showing molar process, *m*, and dentigerous lamella, *d*, all enlarged thirty diameters.

41.—The same; *a*, maxilliped, showing, *l*, external lamella; *m*, basal segment, and 1, 2, 3, 4, segments of palpus, enlarged thirty diameters; *b*, pleopod of the second pair from a male, enlarged fifteen diameters; *s*, stylet, articulated near the base of the inner lamella; *s'*, distal end of stylet, enlarged fifty diameters.

42.—Epelys trilobus Smith (p. 358); dorsal view, enlarged ten diameters; natural size indicated by the line at the right.

43.—The same; *a*, maxilliped from the left side, enlarged twenty diameters; *l*, external lamella; *m*, basal segment; 1, 2, 3, segments of palpus of maxilliped; *b*, pleopod of second pair from a male, enlarged twenty diameters; *s*, stylet, articulated near the base of the inner lamella; *s'*, end of stylet, enlarged fifty diameters.

(All the figures were drawn from nature by O. Harger.)

PLATE VIII.

FIGURE 44.—Epelys montosus Harger (p. 359); dorsal view, enlarged six diameters, natural size indicated by the line at the right.

45.—The same; *a*, antennula; *f*, flagellar segment; *b*, antenna; *c*, maxilliped from the left side; *l*, external lamella; *m*, basal segment; 1, 2, 3, segments of palpus; all the figures enlarged twenty diameters.

46.—The same; *a*, leg of the first pair, enlarged twenty diameters; *b*, right uropod or opercular valve, enlarged fifteen diameters.

FIGURE 47.—The same; pleopod of the second pair, from a male, enlarged twenty diameters; *s*, stylet, articulated near the base of the inner lamella; *s'*, distal end of stylet, enlarged sixty-six diameters.

48.—Astacilla granulata Harger (p. 364); female; dorsal view, enlarged four diameters, natural size indicated by the line at the right; *a*, antennula of male; *b*, fourth thoracic segment of male; *c*, inferior surface of pleon of a male, showing opercular valves; all the figures enlarged four diameters.

49.—The same; *a*, flagellum of antenna, enlarged twenty diameters; *a'*, portion of inner margin of the same, enlarged one hundred diameters; *b*, one of the first pair of legs, upper surface, enlarged twenty diameters.

50.—The same; one of the fourth pair of legs, enlarged twenty diameters.

51.—The same; inner surface of left opercular plate, or uropod, from a female, enlarged twenty diameters.

(All the figures were drawn from nature by O. Harger.)

PLATE IX.

FIGURE 52.—Astacilla granulata Harger (p. 364); *a*, maxilliped; *m*, basal segment; *l*, external lamella; *b*, outer maxilla; *c*, inner maxilla; all enlarged twenty diameters.

53.—Sphæroma quadridentatum Say (p. 368); dorsal view, enlarged five diameters; natural size indicated by the line at the right.

54.—The same; *a*, antennula; *b*, antenna; *c*, pleopod of the second pair, from a male, showing stylet, *s*, articulated near the base of the inner lamella; all the figures enlarged ten diameters.

55.—Limnoria lignorum White (p. 373); dorsal view, enlarged ten diameters; natural size indicated by the line at the right.

56.—The same; *a*, antennula; *b*, antenna; *c*, maxilliped; *d*, maxilla of the outer or second pair; *e*, maxilla of the inner or first pair; *f*, mandible, all enlarged twenty-five diameters; *e'*, distal end of outer lobe of first pair of maxillæ, enlarged sixty-six diameters.

57.—The same; *a*, last segment of pleon, with attached uropods; dorsal view, enlarged ten diameters; *b*, uropod with dotted adjacent outline of last segment of pleon, enlarged thirty diameters; *c*, first pair of pleopods; *d*, pleopod of the second pair, from a male, showing stylet, *s*, articulated to the inner lamella; both figures enlarged twenty diameters.

58.—Cirolana concharum Harger, (p. 378); lateral view, enlarged about three diameters.

(Figure 53 was drawn by Mr. J. H. Emerton, 55 by Prof. S. I. Smith, 58 by Mr. J. H. Blake, and the others by O. Harger.)

PLATE X.

FIGURE 59.—Cirolana concharum Harger (p. 378); dorsal view, enlarged about three diameters. The natural size is shown by the line at the right.

60.—The same; antennula, enlarged ten diameters.

61.—The same; *a*, antenna enlarged ten diameters; *b*, maxilla of the outer or second pair; *c*, maxilla of the inner or first pair; *d*, mandible from the right side, inner view; *p*, palpus; *m*, molar area; the last three figures enlarged five diameters.

62.—The same; *a*, maxilliped from the right side, exterior view, showing, *l*, external lamella; *m*, basal segment; 1, 2, 3, 4, 5, segments of the palpus; *b*, leg of the fourth pair; both the figures enlarged five diameters.

63.—The same; uropod from the right side; inferior view, enlarged five diameters.

64.—Æga psora Kröyer (p. 384); *a*, dorsal and *b* ventral views of a young individual. The central line indicates the length of the specimen, natural size, which is here enlarged three diameters. Adults attain about the size of the figure.

FIGURE 65.—Nerocila munda Harger (p. 392); dorsal view of the type specimen, enlarged about four diameters. The natural size is shown by the cross on the right; *a*, uropod, enlarged six diameters.

66.—Ægathoa loliginea Harger (p. 393); type specimen; *a*, dorsal, and *b*, ventral view, enlarged four diameters. Its natural size is shown by the line between the figures.

(Figure 59 was drawn by Mr. J. H. Blake, the others by O. Harger.)

PLATE XI.

FIGURE 67.—Livoneca ovalis White (p. 395); *a*, antennula; *b*, antenna; *c*, mandibular palpus; each enlarged twenty diameters; *d*, one of the first pair of legs; *e*, one of the seventh pair of legs; *f*, uropod; each enlarged ten diameters.

68.—Anthura polita Stimpson (p. 398); dorsal view, enlarged four diameters. The natural size is shown by the line at the right; *a*, antennula; *b*, antenna, each enlarged ten diameters; *c*, leg of the first pair; *d*, leg of the third pair; *e*, right pleopod of the first pair, interior view, showing inner ramus without cilia; *f*, pleopod of the second pair from a male, showing stylet articulated to inner lamella; each of the figures *c* to *f* enlarged eight diameters; *g*, lateral view of pleon, enlarged six diameters.

69.—The same; *a*, maxilliped, enlarged twenty diameters; *b*, maxilla, enlarged twenty-five diameters; *b'*, distal end of the same, enlarged sixty diameters.

70.—Paranthura brachiata Harger (p. 402); dorsal view, enlarged about three diameters; natural size shown by the line at the right; *a*, antennula; *b*, antenna, enlarged eight diameters; *c*, right maxilliped, enlarged sixteen diameters; *d*, maxilla, enlarged sixteen diameters; *d'*, distal end of the same, enlarged fifty diameters; *e*, leg of the first pair; *f*, first pleopod from the right side, inner view, showing ciliated inner lamella; *g*, pleopod of the second pair from a male, showing stylet articulated to the inner lamella; figures *e* to *g* enlarged eight diameters.

71.—Ptilanthura tenuis Harger (p. 406); male; dorsal view, enlarged about four diameters; *a*, inferior view of the head and first thoracic segment, enlarged eight diameters; the flagellum of the antennulæ omitted; *b*, maxilliped; *c*, maxilla, each enlarged fifty diameters; *d*, first right pleopod, seen from within, showing ciliated inner lamella; *e*, second left pleopod, showing stylet *s* articulated to the inner lamella in the males.

72.—The same; one of the first pair of legs of a male, enlarged sixteen diameters.

73.—The same; female; dorsal view of the head, enlarged twenty-five diameters.

(Figure 71, excepting *b*–*d*, was drawn by Mr. J. H. Emerton, the others by O. Harger.)

PLATE XII.

FIGURE 74.—Ptilanthura tenuis Harger (p. 406); *a*, antennula; *b*, antenna; each enlarged twenty diameters, from a male.

75.—Gnathia cerina Harger (p. 410); male; dorsal view, enlarged ten diameters.

76.—The same; *a*, antennula; *b*, antenna, each enlarged thirty-eight diameters; *c*, mandibles (*l*, left, *r*, right), enlarged thirty-eight diameters; *d*, first leg or first gnathopod from the right side, enlarged twenty-five diameters; all the figures from the male sex.

77.—The same (p. 411); female; dorsal view, enlarged ten diameters.

FIGURE 78.—The same; *a*, one of the first pair of legs or first gnathopod of a female, enlarged thirty-eight diameters; *b*, one of the first pair of legs in a young, parasitic individual, enlarged sixty diameters; *c*, pleon, with the last and part of the penultimate thoracic segments of a female, dorsal view, enlarged twenty diameters; *d*, pleopod of a young, parisitic individual, enlarged sixty diameters; *e*, pleopod of an adult male, enlarged sixty diameters.

79.—The same; young male; dorsal view, enlarged twenty diameters.

80.—Leptochelia algicola Harger (p. 421); male; lateral view, enlarged twenty diameters; natural size indicated by the line above.

(All the figures were drawn from nature by O. Harger.)

PLATE XIII.

FIGURE 81.—Tanais vittatus Lilljeborg (p. 418); dorsal view, enlarged eight diameters. The transverse bands of hairs on the pleon are not sufficiently distinct.

82.—The same; one of the first pair of pleopods, enlarged thirty diameters.

83.—Leptochelia algicola Harger (p. 421); female; dorsal view, enlarged twenty diameters; natural size indicated by the line at the right.

84.—The same; *a*, antennula; *b*, one of the first pair of legs; both from a female specimen and enlarged twenty-five diameters.

85.—The same; hand, or propodus and dactylus of the first pair of legs, enlarged forty-eight diameters, showing the comb of setæ on the propodus.

86.—The same; uropods of a male, enlarged seventy diameters; *b*, basal segment; *i*, inner six-jointed ramus; *o*, outer ramus.

87.—Leptochelia limicola Harger (p. 424); female; dorsal view, enlarged twenty diameters; natural size shown by the line at the right.

88.—The same; *a*, antennula; *b*, antenna; *c*, leg of the first pair; *d*, leg of the second pair; all from the female sex and enlarged twenty-five diameters.

89.—Leptochelia rapax Harger (p. 424); male; dorsal view, enlarged about twelve diameters.

90.—The same; hand, or propodus and dactylus of male, enlarged sixteen diameters.

91.—Leptochelia coeca Harger (p. 427); type specimen, female; *a*, antennula; *b*, leg of the first pair; *c*, uropod; each enlarged fifty diameters.

(All the figures were drawn from nature by O. Harger.)

ALPHABETICAL INDEX TO THE REPORT ON THE MARINE ISOPODA OF NEW ENGLAND AND ADJACENT WATERS.

[In the following index the first reference for the names of the families, genera, and species here described is to the page on which such description is made. The list of authorities, being alphabetically arranged, is not indexed.]

PLATE I.

FIGURE 1.—Philoscia vittata Say (p. 306); dorsal view, enlarged six diameters; natural size indicated by cross at the right.

2.—Scyphacella arenicola Smith (p. 307); dorsal view, enlarged about twelve diameters; natural size indicated by cross at the right.

3.—Actoniscus ellipticus Harger (p. 309); dorsal view, enlarged ten diameters; natural size indicated by line at the right.

4.—Jæra albifrons Leach (p. 315); female; dorsal view, enlarged about ten diameters.

5.—The same; maxilliped from the left side, exterior view, enlarged twenty-five diameters; P, palpus; *l*, external lamella.

6.—The same; maxillæ, enlarged twenty-five diameters; *a*, outer, or second, pair of maxillæ; *b*, inner, or first, pair of maxillæ; *i*, inner, *e*, outer lobe.

7.—The same; inferior surface of the pleon of a female.

8.—The same; inferior surface of the pleon of a male.

(All the figures were drawn from nature by O. Harger.)

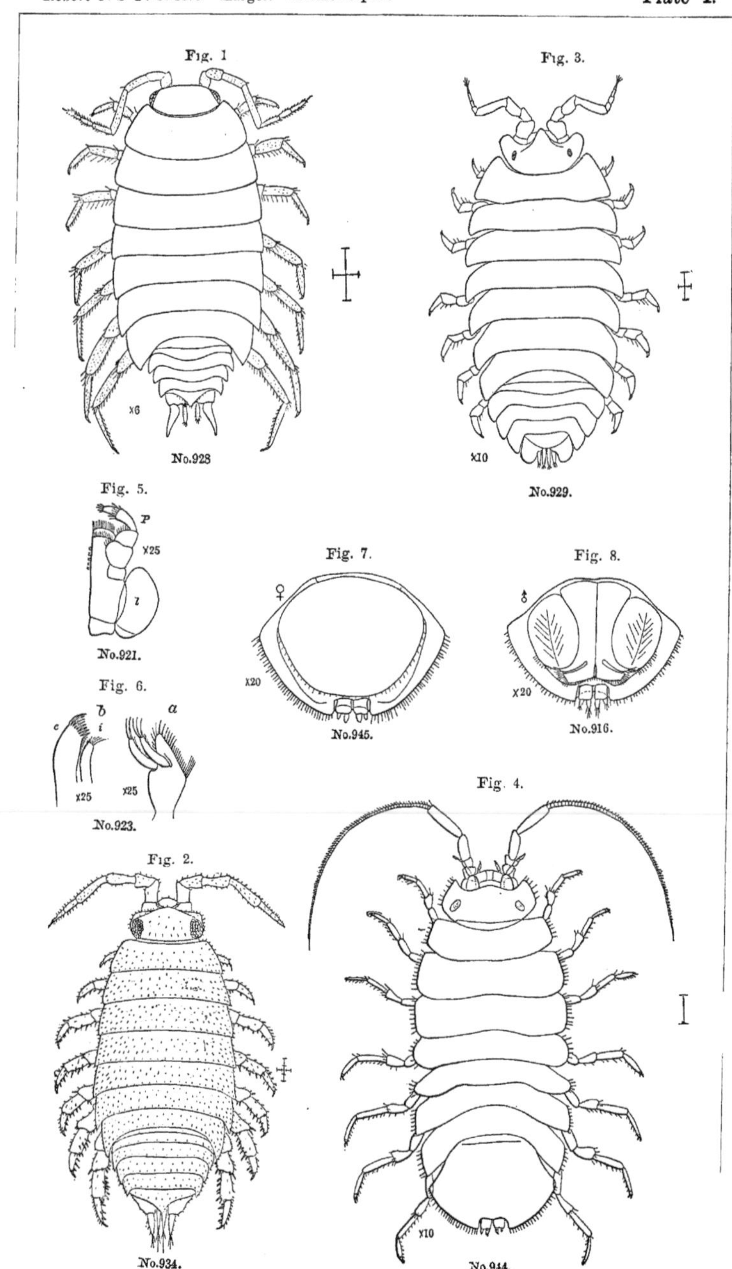
Fig. 1
x6
No. 928
Fig. 3.
x10
No. 929.
Fig. 5.
p
x25
l
No. 921.
Fig. 6.
c
b
i
a
x25
x25
No. 923.
Fig. 7.
♀
x20
No. 945.
Fig. 8.
♂
x20
No. 916.
Fig. 2.
No. 934.
Fig. 4.
x10
No. 944.

PLATE II.

FIGURE 9.—Janira alta Harger (p. 321); dorsal view, enlarged five diameters; natural size indicated by line at the right.

10.—Janira spinosa Harger (p. 323); dorsal view of female, enlarged six diameters.

11.—Munnopsis typica M. Sars (p. 330); dorsal view of male, enlarged about two diameters; *b*, maxillipeds; *m*, basal segment; *l*, external lamella; 2 and 3, second and third segments of palpus of maxillipeds; *c*, outer maxillæ; *d*, inner maxillæ; *e*, one of the second pair of legs of the male; *f*, one of the natatory legs; *g*, abdominal operculum of the female, external view.

(Figures 9 and 10 were drawn from nature by O. Harger; figure 11 is copied from M. Sars, drawn by G. O. Sars.)

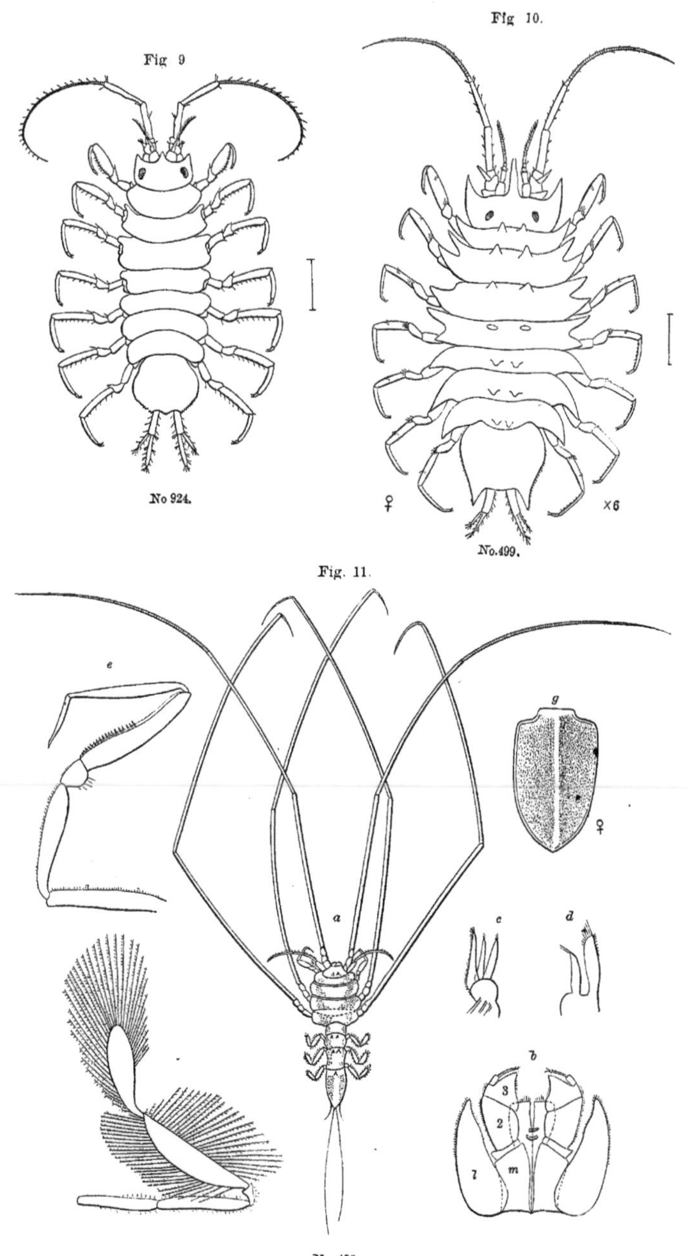
Fig 9
No 924.
Fig 10.
♀
×6
No.499.
Fig. 11.
e
g
♀
a
c
d
b
3
2
1
m
No. 490

PLATE III.

FIGURE 12.—Janira alta (p. 321); *a*, maxilliped; P, palpus of maxilliped; *l*, external lamella; *b*, mandible; *P*, palpus of mandible; *d*, dentigerous lamella; *m*, molar process, enlarged twenty-five diameters.

13.—The same; inferior surface of the pleon, *a* in the female, *b* in the male, enlarged ten diameters; *a*, single opercular plate in the female; *b*, external; *c*, median plate of operculum of male.

14.—Munna Fabricii Kröyer (p. 325); female; dorsal view, enlarged about twenty diameters; natural size indicated by line at the right.

15.—Eurycope robusta Harger (p. 332); female; dorsal view, enlarged six diameters; natural size indicated by line at the right; *a*, antennula, enlarged twenty diameters; *b*, maxilliped; *c*, mandible; *d*, one of the first pair of legs, each enlarged twenty diameters; *d′*, propodus and dactylus of the first pair of legs, enlarged about thirty-eight diameters; *e*, propodus and dactylus of the second pair of legs, enlarged twenty diameters; *f*, one of the sixth pair of legs; *g*, uropod, each enlarged twenty diameters.

(Figure 14 was drawn from nature by Mr. J. H. Emerton, the others by O. Harger.)

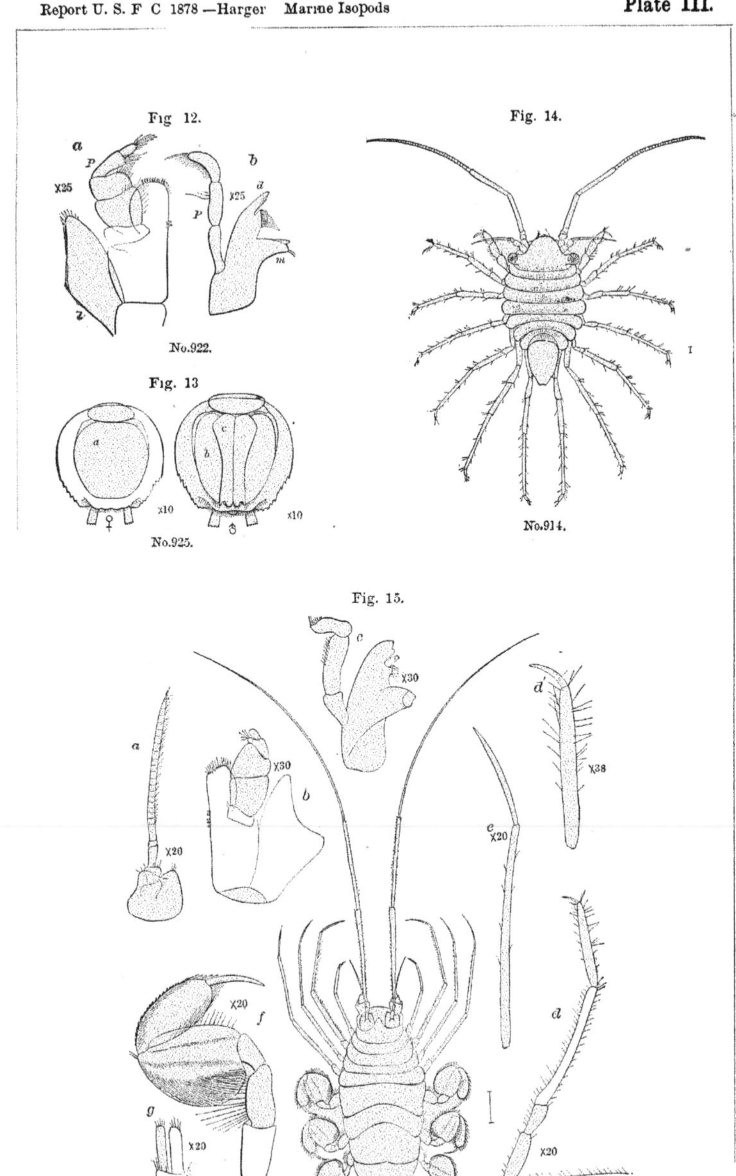
Fig 12.
a
P
X25
b
X25
d
P
m
l
No.922.
Fig. 13
a
c
b
x10
x10
♀
♂
No.925.
Fig. 14.
No.914.
Fig. 15.
c
X30
d'
X38
a
X30
b
X20
e
X20
X20
f
d
g
X20
X20
No.918.

PLATE IV.

FIGURE 16.—Chiridotea cœca Harger (p. 338); dorsal view, enlarged nearly four diameters; natural size indicated by the line at the right.

17.—The same; *a*, antennula; *b*, antenna; each enlarged twelve diameters.

18.—The same; *a*, maxilliped from the right side, external view; *l*, external lamella; *m*, maxilliped proper; 1, 2, 3, first, second, and third segments of the palpus of the maxilliped, enlarged twenty diameters; *b*, one of the first pair of legs, magnified twelve diameters; *c*, uropod from the left side, inner view, showing the two rami articulated near the tip.

19.—The same; pleopods of second pair from the right side, anterior views, enlarged ten diameters; *a*, common form in males; *b*, rarer form in male; *s*, elongated stylet, articulated near the base of the inner lamella; *c*, form in the female.

20.—Chiridotea Tuftsii Harger (p. 340); female; dorsal view, enlarged five diameters; natural size indicated by the line at the right.

21.—The same; left maxilliped, enlarged twenty-five diameters; *e*, external lamella; *m*, basal segment; 1, 2, 3, segments of palpus.

22.—The same; pleopod of the second pair, from a male, enlarged twenty diameters; *s*, elongated stylet, articulated near the base of the inner lamella.

(All the figures were drawn from nature by O. Harger.)

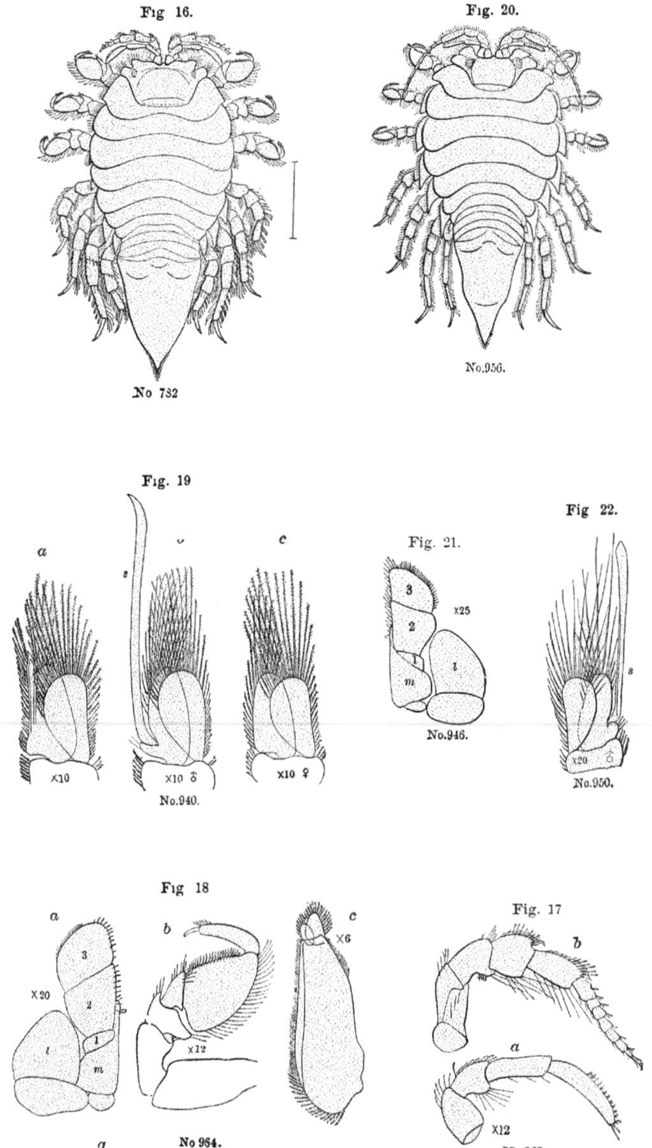
Fig 16.
No 782
Fig. 20.
No.956.
Fig. 19
a
c
s
X10
X10 ♂
X10 ♀
No.940.
Fig. 21.
3
2
1
m
l
X25
No.946.
Fig 22.
s
X20 ♂
No.950.
Fig 18
a
3
2
1
l
m
X20
b
X12
c
X6
g
No 964.
Fig. 17
b
a
X12
No.969.

PLATE V.

FIGURE 23.—Chiridotea Tuftsii Harger (p. 340); *a*, antennula; *b*, antenna; *c*, leg of the first pair; *d*, leg of the fourth pair; all enlarged twelve diameters; *e*, left uropod, or opercular valve, inner view, enlarged ten diameters.

24.—Idotea irrorata Edwards (p. 343); dorsal view, enlarged two diameters; natural size shown by the line on the left.

25.—The same; *a*, antennula; *b*, antenna; *c*, left uropod or opercular valve, external view; all enlarged six diameters.

26.—The same; *a*, right maxilliped, enlarged twelve diameters, *l*, external lamella; *m*, basal segment; 1, 2, 3, 4, segments of palpus of maxilliped; *b*, pleopod of the second pair from a male, enlarged eight diameters, showing stylet, *s*, articulated near the base of the inner lamella.

27.—Idotea phosphorea Harger (p. 347); dorsal view, enlarged about two diameters; natural size shown by the line on the right.

28.—The same; *a*, antenna, enlarged six diameters; *b*, maxilliped, enlarged twelve diameters, showing, *l*, external lamella; *m*, basal segments; 1, 2, 3, 4, segments of the palpus of maxilliped; *c*, leg of the first pair; *d*, leg of the second pair, both enlarged six diameters; *e*, right uropod, or opercular valve, inner view, enlarged six diameters.

29.—The same; pleopod of the second pair from a male, enlarged eight diameters; *s*, stylet articulated near the base of the inner lamella; *s'*, distal end of stylet reversed and enlarged thirty diameters.

(Figure 24 was drawn by Mr. J. H. Emerton, the others by O. Harger.)

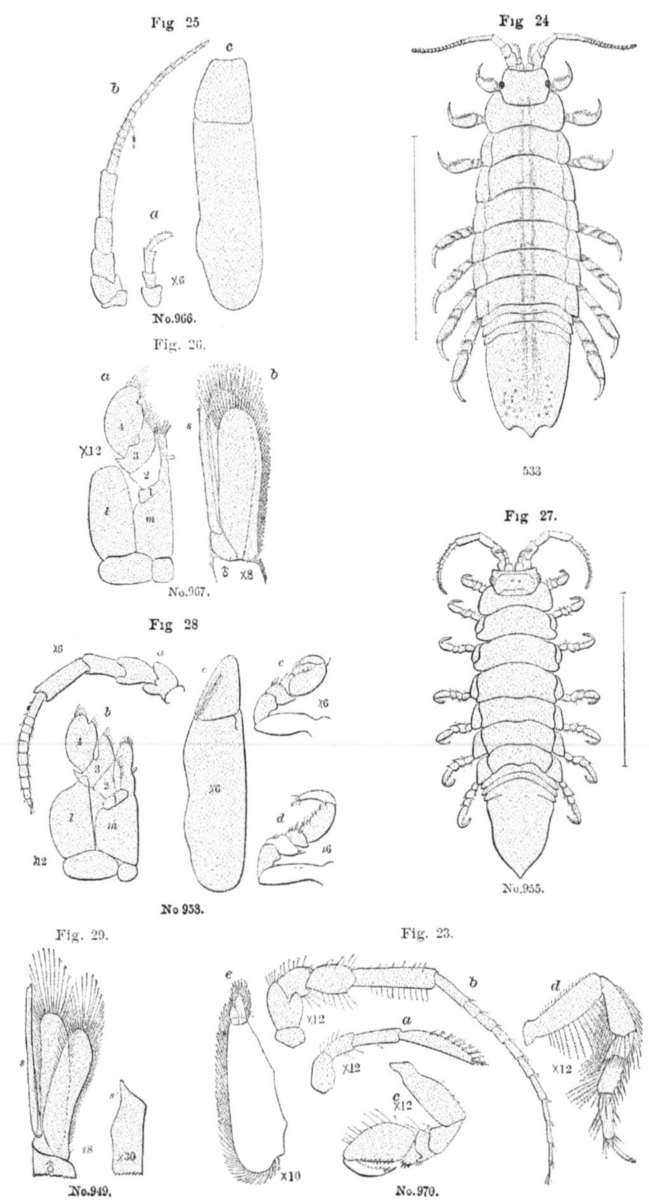
Fig 25
No.966.
Fig. 26.
No.967.
Fig 28
No 953.
Fig. 29.
No.949.
Fig 24
533
Fig 27.
No.955.
Fig. 23.
No.970.

PLATE VI.

FIGURE 30.—Idotea robusta Kröyer (p. 349); dorsal view, enlarged two diameters; natural size shown by the line at the right.

31.—The same; *a*, antenna; *b*, leg of the first pair, each enlarged six diameters; *c*, left uropod, or opercular valve, inner view, enlarged four diameters.

FIGURE 32.—The same; *a*, maxilliped, enlarged twelve diameters; *l*, external lamella; 1, 2, 3, 4, segments of palpus; *b*, maxilla of the outer or second pair; *c*, pleopod of the second pair from a male, enlarged six diameters; *s*, stylet articulated near the base of the inner lamella.

33.—Synidotea nodulosa Harger (p. 351); dorsal view, enlarged four diameters; natural size indicated by the line at the right.

34.—The same; *a*, antennula; *f*, flagellar segment; *b*, antenna; *c*, leg of the first pair from the right side; *d*, right uropod, or opercular valve, all enlarged ten diameters.

35.—The same; *a*, maxilliped from the right side, showing, *l*, external lamella; *m*, basal segment; 1, 2, 3, segments of palpus, enlarged twenty diameters; *b*, maxilla of the outer or second pair; *c*, maxilla of the inner or first pair, both enlarged twenty diameters; *d*, pleopod of the second pair from a male, enlarged twelve diameters; *s*, stylet articulated near the base of the inner lamella.

36.—Erichsonia attenuata Harger (p. 356); dorsal view, enlarged three diameters, natural size indicated by the line at the right.

(Figures 30 and 36 were drawn by Mr. J. H. Emerton, the others by O. Harger.)

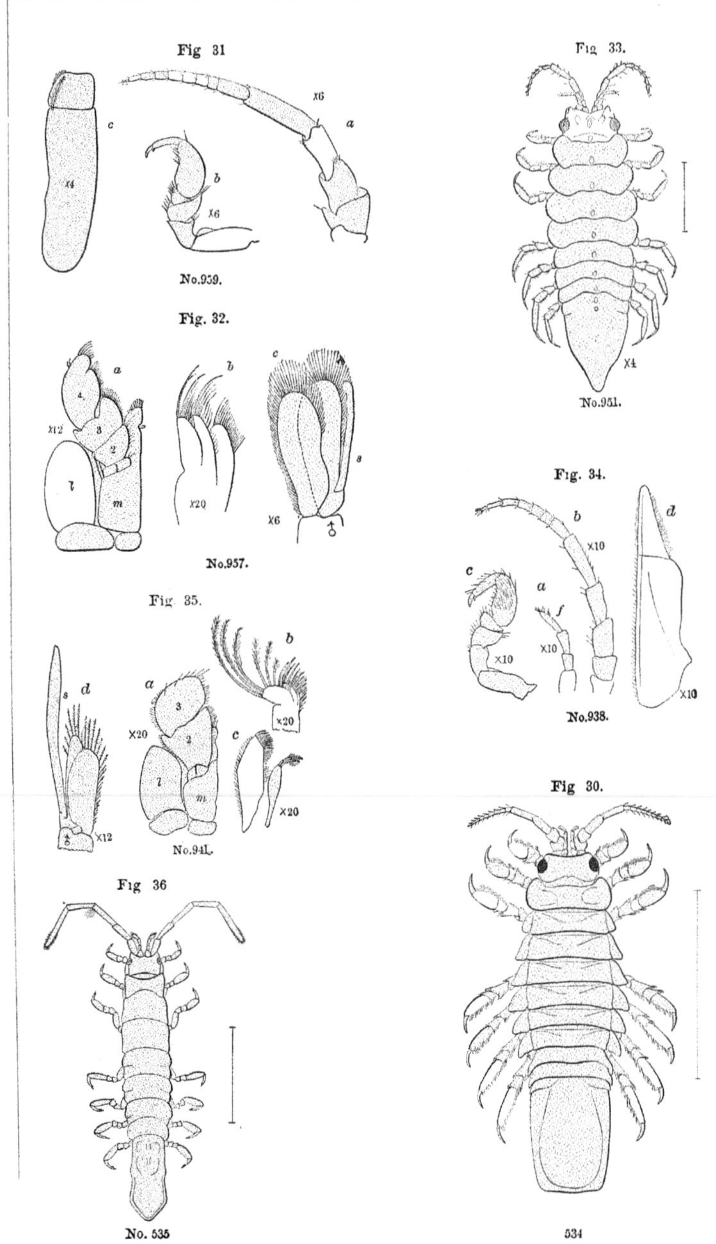
Fig 31
c
x4
x6
a
b
x6
No.959.
Fig. 32.
a
b
c
x12
4
3
2
l
m
x20
x6
s
No.957.
Fig. 35.
b
d
s
a
3
x20
2
x20
l
m
c
x20
x12
No.941.
Fig 36
No. 535
Fig 33.
x4
No.951.
Fig. 34.
d
b
x10
c
a
f
x10
x10
x10
No.938.
Fig 30.
534

PLATE VII.

FIGURE 37.—Erichsonia attenuata Harger (p. 356); *a*, antennula; *b*, antenna, each enlarged twelve diameters; *c*, maxilliped, showing, *l*, external lamella, enlarged thirty diameters; *d*, uropod, or opercular valve, enlarged twelve diameters; *e*, pleopod of the second pair from a male, enlarged fifteen diameters; *s*, stylet, articulated near the base of the inner lamella; *s'*, distal end of stylet, enlarged fifty diameters.

38.—Erichsonia filiformis Harger (p. 355); dorsal view, enlarged five diameters, natural size indicated by the line at the right.

39.—The same; *a*, antennula; *b*, antenna; *c*, leg of the first pair; *d*, uropod, or opercular valve, each enlarged twelve diameters.

40.—The same; *a*, maxilla of outer or second pair; *b*, maxilla of inner or first pair; *c*, mandible, showing molar process, *m*, and dentigerous lamella, *d*, all enlarged thirty diameters.

41.—The same; *a*, maxilliped, showing, *l*, external lamella; *m*, basal segment, and 1, 2, 3, 4, segments of palpus, enlarged thirty diameters; *b*, pleopod of the second pair from a male, enlarged fifteen diameters; *s*, stylet, articulated near the base of the inner lamella; *s'*, distal end of stylet, enlarged fifty diameters.

42.—Epelys trilobus Smith (p. 358); dorsal view, enlarged ten diameters; natural size indicated by the line at the right.

43.—The same; *a*, maxilliped from the left side, enlarged twenty diameters; *l*, external lamella; *m*, basal segment; 1, 2, 3, segments of palpus of maxilliped; *b*, pleopod of second pair from a male, enlarged twenty diameters; *s*, stylet, articulated near the base of the inner lamella; *s'*, end of stylet, enlarged fifty diameters.

(All the figures were drawn from nature by O. Harger.)

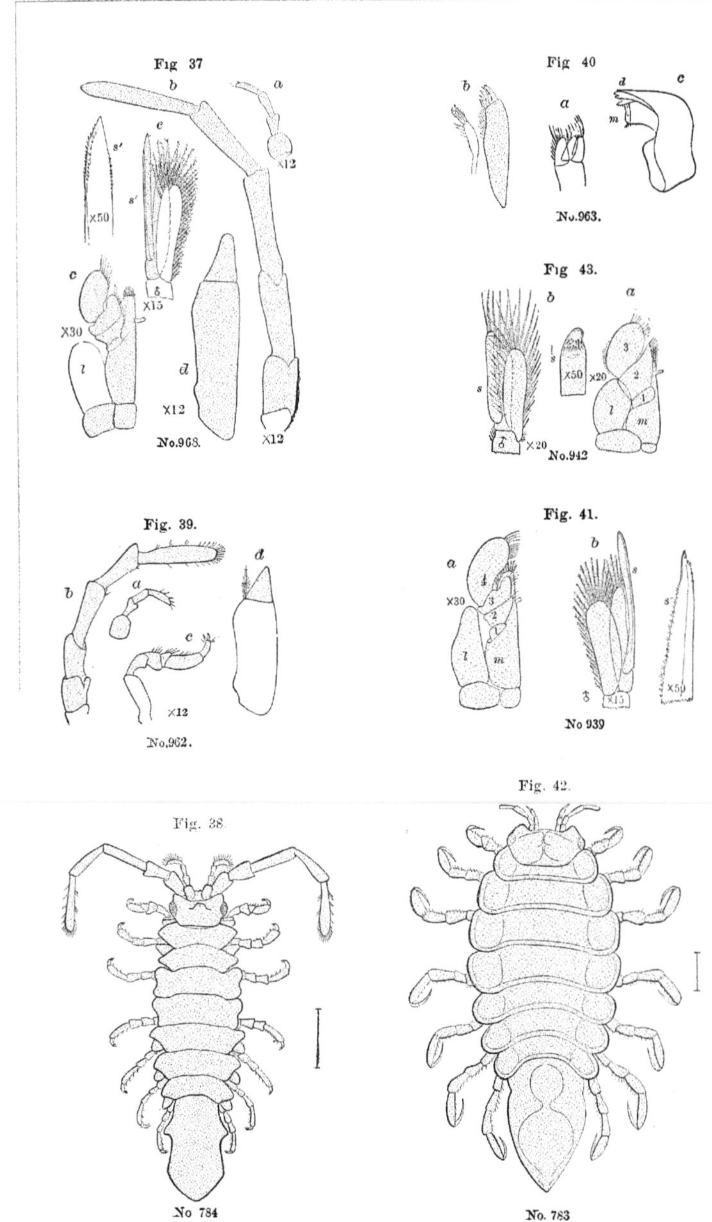

Fig 37
No.968.
Fig 40
No.963.
Fig 43.
No.942
Fig. 39.
No.962.
Fig. 41.
No 939
Fig. 42.
Fig. 38.
No 784
No. 783

PLATE VIII.

FIGURE 44.—Epelys montosus Harger (p. 359); dorsal view, enlarged six diameters, natural size indicated by the line at the right.

45.—The same; *a*, antennula; *f*, flagellar segment; *b*, antenna; *c*, maxilliped from the left side; *l*, external lamella; *m*, basal segment; 1, 2, 3, segments of palpus; all the figures enlarged twenty diameters.

46.—The same; *a*, leg of the first pair, enlarged twenty diameters; *b*, right uropod or opercular valve, enlarged fifteen diameters.

FIGURE 47.—The same; pleopod of the second pair, from a male, enlarged twenty diameters; *s*, stylet, articulated near the base of the inner lamella; *s'*, distal end of stylet, enlarged sixty-six diameters.

48.—Astacilla granulata Harger (p. 364); female; dorsal view, enlarged four diameters, natural size indicated by the line at the right; *a*, antennula of male; *b*, fourth thoracic segment of male; *c*, inferior surface of pleon of a male, showing opercular valves; all the figures enlarged four diameters.

49.—The same; *a*, flagellum of antenna, enlarged twenty diameters; *a'*, portion of inner margin of the same, enlarged one hundred diameters; *b*, one of the first pair of legs, upper surface, enlarged twenty diameters.

50.—The same; one of the fourth pair of legs, enlarged twenty diameters.

51.—The same; inner surface of left opercular plate, or uropod, from a female, enlarged twenty diameters.

(All the figures were drawn from nature by O. Harger.)

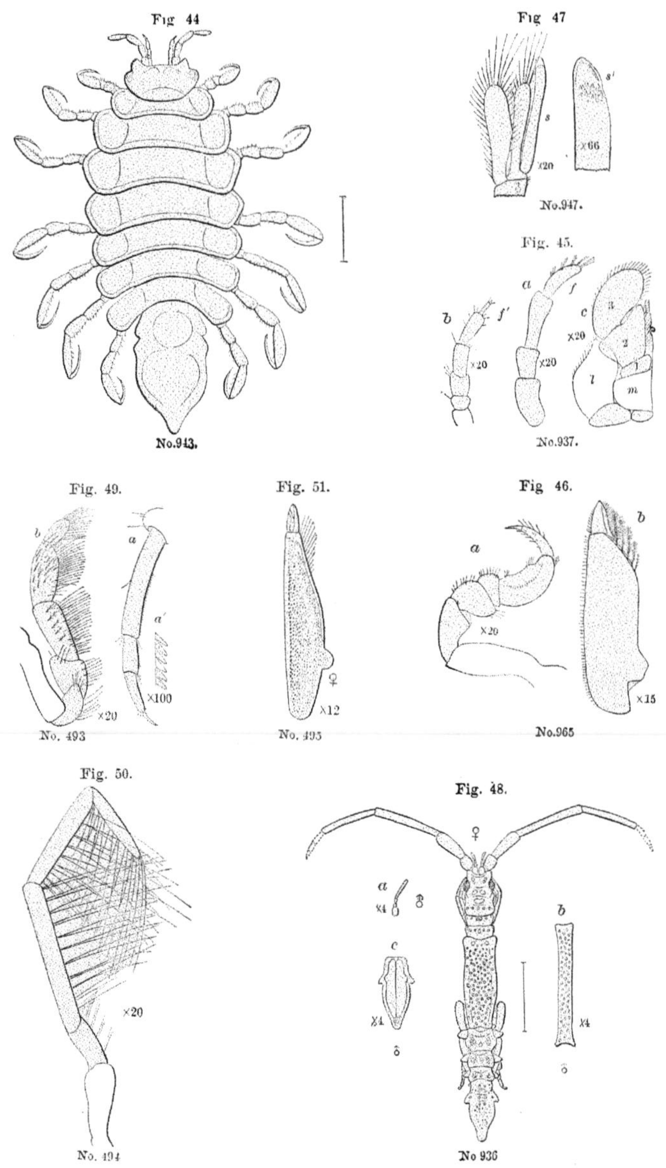
Fig 44
No.943.
Fig 47
s
s'
x20
x66
No.947.
Fig. 45.
a
b
f
f'
c
x20
3
2
1
l
m
No.937.
Fig. 49.
b
a
a'
x20
x100
No. 493
Fig. 51.
♀
x12
No. 495
Fig 46.
a
b
x20
x15
No.965
Fig. 50.
x20
No. 494
Fig. 48.
♀
a
x4
♂
c
x4
♂
b
x4
♂
No 936

PLATE IX.

FIGURE 52.—Astacilla granulata Harger (p. 364); *a*, maxilliped; *m*, basal segment; *l*, external lamella; *b*, outer maxilla; *c*, inner maxilla; all enlarged twenty diameters.

53.—Sphæroma quadridentatum Say (p. 368); dorsal view, enlarged five diameters; natural size indicated by the line at the right.

54.—The same; *a*, antennula; *b*, antenna; *c*, pleopod of the second pair, from a male, showing stylet, *s*, articulated near the base of the inner lamella; all the figures enlarged ten diameters.

55.—Limnoria lignorum White (p. 373); dorsal view, enlarged ten diameters; natural size indicated by the line at the right.

56.—The same; *a*, antennula; *b*, antenna; *c*, maxilliped; *d*, maxilla of the outer or second pair; *e*, maxilla of the inner or first pair; *f*, mandible, all enlarged twenty-five diameters; *e'*, distal end of outer lobe of first pair of maxillæ, enlarged sixty-six diameters.

57.—The same; *a*, last segment of pleon, with attached uropods; dorsal view, enlarged ten diameters; *b*, uropod with dotted adjacent outline of last segment of pleon, enlarged thirty diameters; *c*, first pair of pleopods; *d*, pleopod of the second pair, from a male, showing stylet, *s*, articulated to the inner lamella; both figures enlarged twenty diameters.

58.—Cirolana concharum Harger, (p. 378); lateral view, enlarged about three diameters.

(Figure 53 was drawn by Mr. J. H. Emerton, 55 by Prof. S. I. Smith, 58 by Mr. J. H. Blake, and the others by O. Harger.)

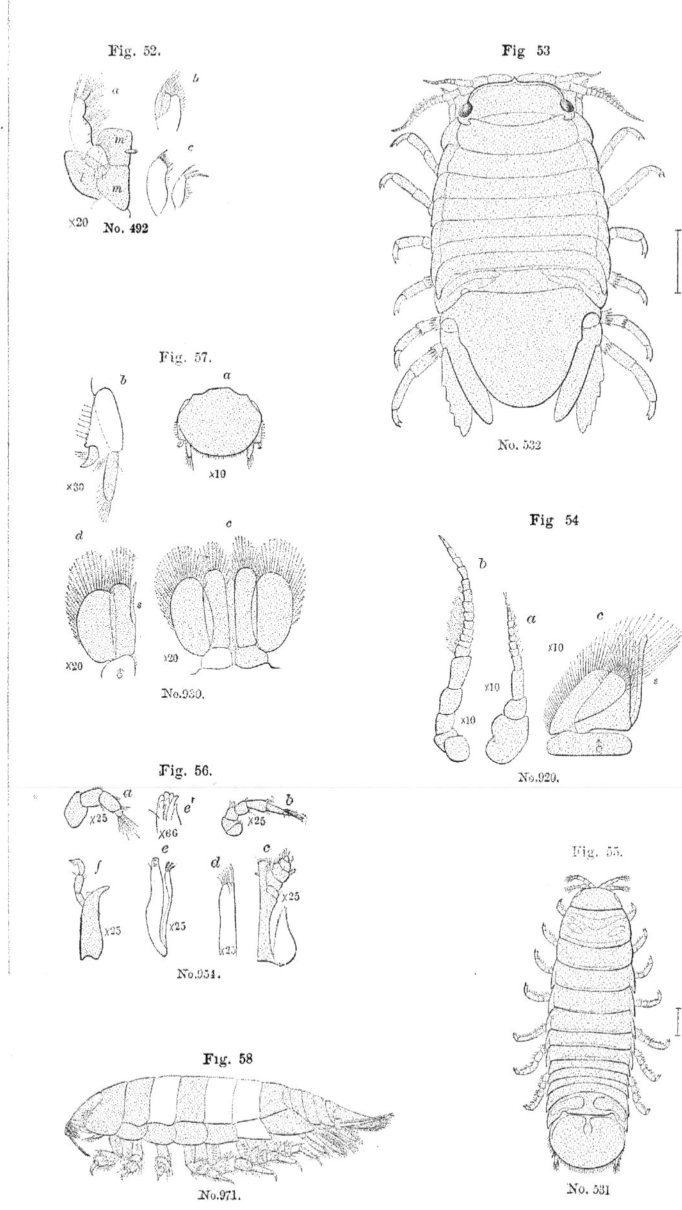
Fig. 52.
a
b
c
m
m
l
×20
No. 492
Fig 53
No. 532
Fig. 57.
a
b
×30
×10
d
c
s
×20
×20
No.930.
Fig 54
b
a
c
×10
×10
×10
s
No.920.
Fig. 56.
a
b
e'
×25
×66
×25
f
e
d
c
×25
×25
×25
×25
No.954.
Fig. 55.
No. 531
Fig. 58
No.971.

PLATE X.

FIGURE 59.—Cirolana concharum Harger (p. 378); dorsal view, enlarged about three diameters. The natural size is shown by the line at the right.

60.—The same; antennula, enlarged ten diameters.

61.—The same; *a*, antenna enlarged ten diameters; *b*, maxilla of the outer or second pair; *c*, maxilla of the inner or first pair: *d*, mandible from the right side, inner view; *p*, palpus; *m*, molar area; the last three figures enlarged five diameters.

62.—The same; *a*, maxilliped from the right side, exterior view, showing, *l*, external lamella; *m*, basal segment; 1, 2, 3, 4, 5, segments of the palpus; *b*, leg of the fourth pair; both the figures enlarged five diameters.

63.—The same; uropod from the right side; inferior view, enlarged five diameters.

64.—Æga psora Kröyer (p. 384); *a*, dorsal and *b* ventral views of a young individual. The central line indicates the length of the specimen, natural size, which is here enlarged three diameters. Adults attain about the size of the figure.

FIGURE 65.—Nerocila munda Harger (p. 392); dorsal view of the type specimen, enlarged about four diameters. The natural size is shown by the cross on the right; *a*, uropod, enlarged six diameters.

66.—Ægathoa loliginea Harger (p. 393); type specimen; *a*, dorsal, and *b*, ventral view, enlarged four diameters. Its natural size is shown by the line between the figures.

(Figure 59 was drawn by Mr. J. H. Blake, the others by O. Harger.)

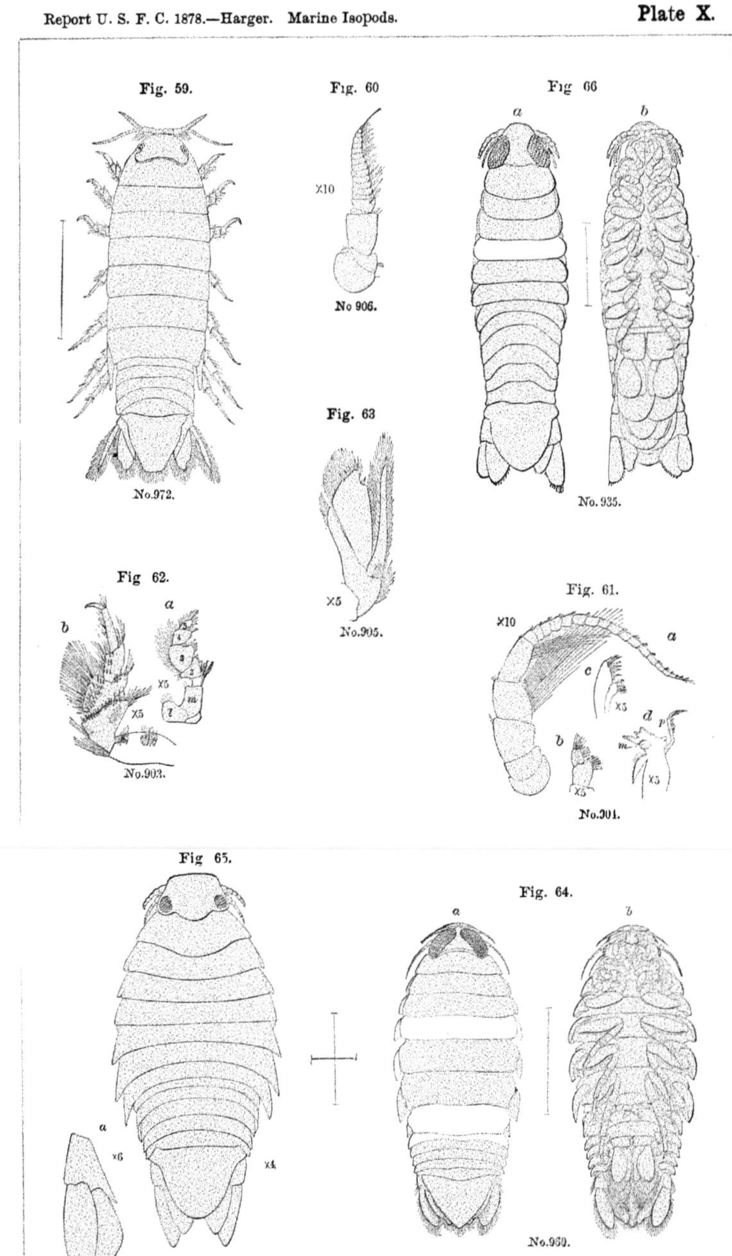
Fig. 59.
No.972.
Fig. 60
X10
No 906.
Fig 66
a
b
No. 935.
Fig. 63
X5
No.905.
Fig 62.
a
b
X5
X5
No.903.
Fig. 61.
X10
a
c
X5
b
d
X5
X5
No.901.
Fig 65.
a
X6
X4
No.927.
Fig. 64.
a
b
No.960.

PLATE XI.

FIGURE 67.—Livoneca ovalis White (p. 395); *a*, antennula; *b*, antenna; *c*, mandibular palpus; each enlarged twenty diameters; *d*, one of the first pair of legs; *e*, one of the seventh pair of legs; *f*, uropod; each enlarged ten diameters.

68.—Anthura polita Stimpson (p. 398); dorsal view, enlarged four diameters. The natural size is shown by the line at the right; *a*, antennula; *b*, antenna, each enlarged ten diameters; *c*, leg of the first pair; *d*, leg of the third pair; *e*, right pleopod of the first pair, interior view, showing inner ramus without cilia; *f*, pleopod of the second pair from a male, showing stylet articulated to inner lamella; each of the figures *c* to *f* enlarged eight diameters; *g*, lateral view of pleon, enlarged six diameters.

69.—The same, *a*, maxilliped, enlarged twenty diameters; *b*, maxilla, enlarged twenty-five diameters; *b'*, distal end of the same, enlarged sixty diameters.

70.—Paranthura brachiata Harger (p. 402); dorsal view, enlarged about three diameters; natural size shown by the line at the right; *a*, antennula; *b*, antenna, enlarged eight diameters; *c*, right maxilliped, enlarged sixteen diameters; *d*, maxilla, enlarged sixteen diameters; *d'*, distal end of the same, enlarged fifty diameters; *e*, leg of the first pair; *f*, first pleopod from the right side, inner view, showing ciliated inner lamella; *g*, pleopod of the second pair from a male, showing stylet articulated to the inner lamella; figures *e* to *g* enlarged eight diameters.

71.—Ptilanthura tenuis Harger (p. 406); male; dorsal view, enlarged about four diameters; *a*, inferior view of the head and first thoracic segment, enlarged eight diameters; the flagellum of the antennulæ omitted; *b*, maxilliped; *c*, maxilla, each enlarged fifty diameters; *d*, first right pleopod, seen from within, showing ciliated inner lamella; *e*, second left pleopod, showing stylet *s* articulated to the inner lamella in the males.

72.—The same; one of the first pair of legs of a male, enlarged sixteen diameters.

73.—The same; female; dorsal view of the head, enlarged twenty-five diameters.

(Figure 71, excepting *b–d*, was drawn by Mr. J. H. Emerton, the others by O. Harger.)

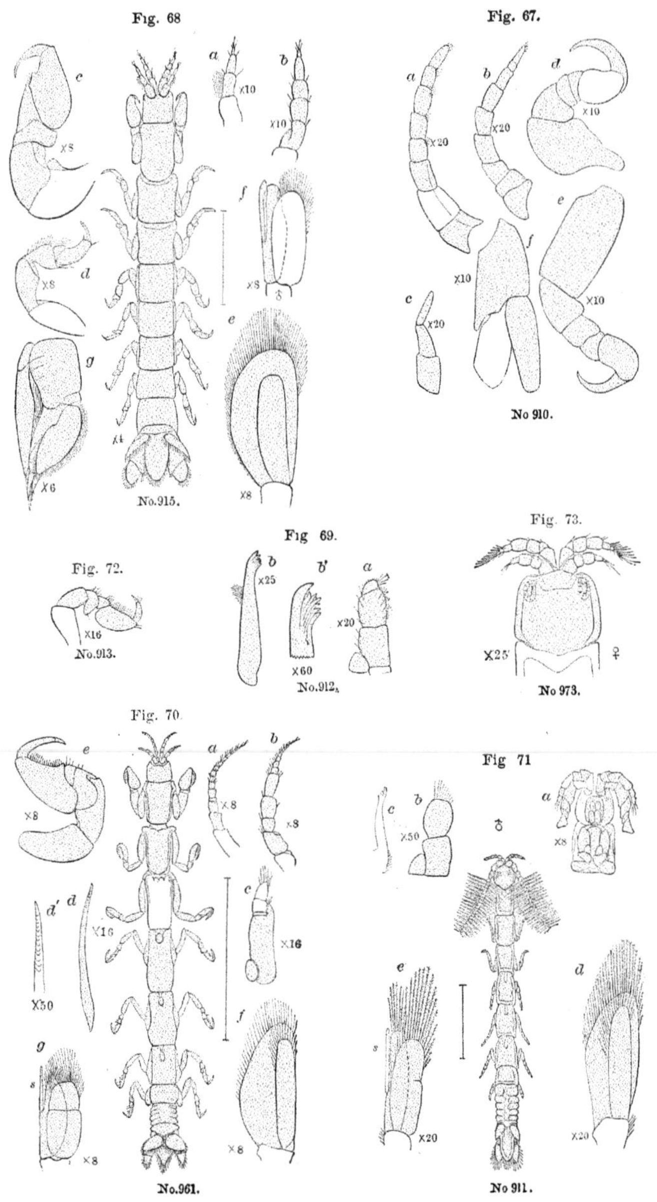
Fig. 68
Fig. 67.
No.915.
No 910.
Fig 69.
Fig. 72.
No.913.
No.912.
Fig. 73.
No 973.
Fig. 70.
Fig 71
No.961.
No 911.

PLATE XII.

FIGURE 74.—Ptilanthura tenuis Harger (p. 406); *a*, antennula; *b*, antenna; each enlarged twenty diameters, from a male.

75.—Gnathia cerina Harger (p. 410); male; dorsal view, enlarged ten diameters.

76.—The same; *a*, antennula; *b*, antenna, each enlarged thirty-eight diameters; *c*, mandibles (*l*, left, *r*, right), enlarged thirty-eight diameters; *d*, first leg or first gnathopod from the right side, enlarged twenty-five diameters; all the figures from the male sex.

77.—The same (p. 411); female; dorsal view, enlarged ten diameters.

78.—The same; *a*, one of the first pair of legs or first gnathopod of a female, enlarged thirty-eight diameters; *b*, one of the first pair of legs in a young, parasitic individual, enlarged sixty diameters; *c*, pleon, with the last and part of the penultimate thoracic segments of a female, dorsal view, enlarged twenty diameters; *d*, pleopod of a young, parisitic individual, enlarged sixty diameters; *e*, pleopod of an adult male, enlarged sixty diameters.

79.—The same; young male; dorsal view, enlarged twenty diameters.

80.—Leptochelia algicola Harger (p. 421); male; lateral view, enlarged twenty diameters; natural size indicated by the line above.

(All the figures were drawn from nature by O. Harger.)

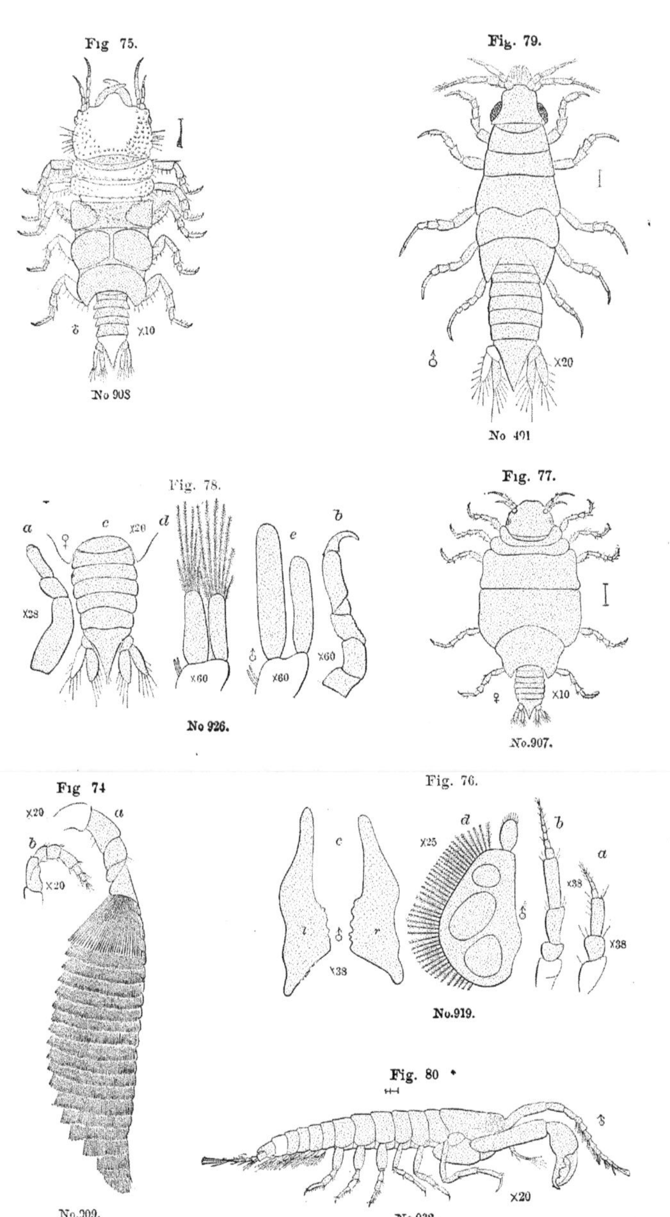

Fig. 75.
No. 908.
Fig. 79.
No. 401
Fig. 78.
No. 926.
Fig. 77.
No. 907.
Fig. 74
No. 909.
Fig. 76.
No. 919.
Fig. 80
No. 932.

PLATE XIII.

FIGURE 81.—Tanais vittatus Lilljeborg (p. 418); dorsal view, enlarged eight diameters. The transverse bands of hairs on the pleon are not sufficiently distinct.

82.—The same; one of the first pair of pleopods, enlarged thirty diameters.

83.—Leptochelia algicola Harger (p. 421); female; dorsal view, enlarged twenty diameters; natural size indicated by the line at the right.

84.—The same; *a*, antennula; *b*, one of the first pair of legs; both from a female specimen and enlarged twenty-five diameters.

85.—The same; hand, or propodus and dactylus of the first pair of legs, enlarged forty-eight diameters, showing the comb of setæ on the propodus.

86.—The same; uropods of a male, enlarged seventy diameters; *b*, basal segment; *i*, inner six-jointed ramus; *o*, outer ramus.

87.—Leptochelia limicola Harger (p. 424); female; dorsal view, enlarged twenty diameters; natural size shown by the line at the right.

88.—The same; *a*, antennula; *b*, antenna; *c*, leg of the first pair; *d*, leg of the second pair; all from the female sex and enlarged twenty-five diameters.

89.—Leptochelia rapax Harger (p. 424); male; dorsal view, enlarged about twelve diameters.

90.—The same; hand, or propodus and dactylus of male, enlarged sixteen diameters.

91.—Leptochelia coeca Harger (p. 427); type specimen, female; *a*, antennula; *b*, leg of the first pair; *c*, uropod; each enlarged fifty diameters.

(All the figures were drawn from nature by O. Harger.)

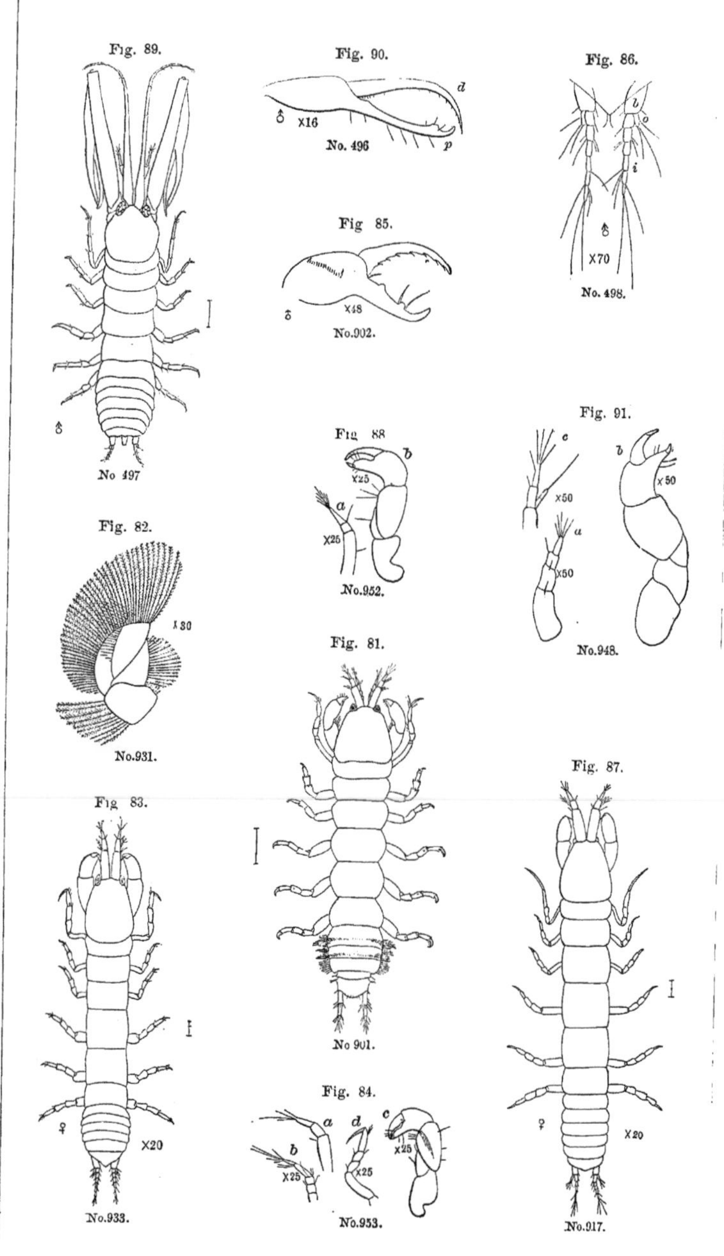
Fig. 89.
No 497
Fig. 90.
No. 496
Fig. 86.
X70
No. 498.
Fig 85.
X48
No.902.
Fig 88
X25
No.952.
Fig. 91.
X50
No.948.
Fig. 82.
X30
No.931.
Fig. 81.
No 901.
Fig 83.
X20
No.933.
Fig. 84.
X25
No.953.
Fig. 87.
X20
No.917.

XV.—REPORT ON THE PYCNOGONIDA OF NEW ENGLAND AND ADJACENT WATERS.

By Edmund B. Wilson.

It is intended to give in this report an account of our present knowledge of the species of Pycnogonida known to occur upon the coasts of New England and Nova Scotia, comprising descriptions and figures of all the forms, and an account of their geographical and bathymetrical distribution. Although the work is mainly systematic, and has been done with special reference to supplying a basis for satisfactory determinations of the genera and species, and their distribution, it has been thought best to give also a brief general account of the structural peculiarities and general natural history of the group. In so doing I have drawn largely from the works of several other writers; and especially from those of Dr. Anton Dohrn, who has made a careful study of the anatomy and embryology of these animals. It should be borne in mind that the structure of the Pycnogonida is, as yet, not well understood, and that further research is needed to fully explain the anatomy and systematic relations of this peculiar and perplexing group. To make the report as complete as possible, it has been made to include not only the collections of the Fish Commission, but also those made by various other parties since the year 1864. The parties referred to were as follows: Expedition of 1864, Professors A. E. Verrill and S. I. Smith; Expedition of 1868, the same with the addition of Professor H. E. Webster and Mr. Geo. A. Jackson; Expedition of 1870, Professor Verrill with Mr. Oscar Harger and Mr. C. H. Dwinelle. The Pycnogonida from these sources, with those of the Fish Commission collections, are at present preserved in the Peabody Museum of Yale College, where they have been studied. I take pleasure in here expressing my great obligations to Professors Verrill and Smith; I am also indebted to Professor Carl Semper for specimens of several European species.

The Pycnogonida form a small and very natural group of articulated animals, which are all marine, have a very wide geographical distribution, and are found at all depths from low-water mark down to many hundred fathoms. Although forming a small and inconspicuous group, they possess a special interest from peculiarities in their structure and development; and though some of the species have been carefully studied by competent observers, opinion is yet divided as to the exact position they should occupy in the zoological system. By some writers they are referred to the Crustacea, by some to the Arachnida, while others place them in a group distinct from both. In some respects they must be regarded as intermediate between these groups; and hence to trace their homologies, especially those of the appendages, is a matter of considerable difficulty. Some of the hairy species bear a close general resemblance to spiders, which has given rise to their common name of *sea-spiders*. Their anatomical structure is, however, very different from that of the spiders, and in their sluggish movements and parasitic habits they are still more unlike those active and predacious animals. Most of the species cling to other animals, such as sponges, sea-anemones, and particularly tubularian and other hydroids; upon these animals they probably in part feed, sucking their juices by means of the large proboscis or rostrum, though their food apparently consists also of more solid matters. They are remarkable, as a whole, for the reduction of the abdomen, and the great development of the legs, which sometimes have an extent equal to nine or ten times the length of the body; the abdomen is always aborted, so as to often appear like a mere tubercle, and, with the exception of one or two forms where it is bi-articulate, it is not divided into segments. The body shows exteriorly four segments, exclusive of the rostrum and abdomen; these segments expand laterally into prominent processes, which may readily be mistaken for the basal joints of the legs, to which they give attachment. The abdomen arises from the posterior segment, from which it is not, as a rule, separated by segmentation. It is usually directed more or less upwards; at its extremity is the anus, usually in a deep cleft.

The most anterior pair of appendages, which are wholly wanting in a few forms (*Pycnogonum*, etc.), are here regarded as antennæ, a view which seems to me to be justified by their position and the origin of their nerves; by many writers they are, however, considered to be post-oral, and as probably representing mandibles. In the higher forms they are three-jointed and usually forceps-like or 'chelate,' in other genera two-jointed, and a recently described genus (*Tanystylum*), with the antennæ composed of a single joint, completes the transition to those forms in which antennæ have quite disappeared. It may be here mentioned that antennæ are invariably present in the larva, so far as known; and that they are then always three-jointed and chelate, their subsequent disappearance in certain forms being apparently a case of "retrograde development." Below the antennæ is the large proboscis or rostrum, at the extremity of which is the mouth; this is triangular in shape, and is sometimes furnished with three denticulated organs not very unlike the jaws of a leech. Within this rostrum is a large cavity, continuous posteriorly with the œsophagus, and containing a complicated apparatus for masticating food; this consists of a great number of chitinous bars lying transversely in the walls of the cavity and giving attachment to numerous setæ, usually bifid at their tips, which extend forward toward the mouth. Posterior to these are found in some species, according to Zenker, horny denticles by which food may be still further comminuted. The homologies of the rostrum are not well understood. It is regarded by Huxley as representing the coalesced cheliceræ and pedipalpi, like that of the Acarina; and Latreille states that in a large species of *Phoxichilus* from the Cape of Good Hope he found longitudinal sutures in the rostrum, so that it appeared as if consisting of the "Labrum, lingua and two jaws united together." Other writers have regarded it as the head, etc. It seems to me that a study of the embryology does not confirm these views, for the rostrum in its early stages presents no trace of sutures or other evidence of its composite nature, but arises as a simple protuberance between the bases of the antennæ. Posterior to the antennæ, and at the sides of the rostrum, are, in many genera, a pair of so-called 'palpi,' which are composed of five to nine joints, and are sometimes furnished with plumose hairs that undoubtedly have a tactile function. The third pair of appendages, which are wanting in the females of certain species, have been termed 'ovigerous legs,' from their office, in the male, of bearing the egg-masses, it having been formerly supposed that the females never possessed them. This term is, however, inappropriate when applied to the female appendages, and it seems preferable to term them *accessory legs*, as certain writers have done, at least until their homologies are better understood. The legs proper are eight in number, and are, as already mentioned, remarkable in many species for their great length. They are composed of nine joints, of which the last, or dactylus, is claw-like and forms, in some species, a sub-cheliform hand with the preceding joint or propodus. In certain genera the dactylus is armed with two movable auxiliary claws, articulated to its upper side near the base; their presence or absence forms a valuable generic character.

The stomach always sends out long prolongations into the legs and antennæ, and sometimes, also, rudimentary ones to the palpi and accessory legs. These diverticula exhibit active peristaltic movements, which drive the food rapidly back and forth and thoroughly distribute it. The movement is plainly visible in some species of *Nymphon*, and is an interesting sight. The stomach-walls contain numerous muscular fibres and are somewhat glandular, but no liver or other special secretory organ is known to exist. The circulatory system is very simple and has been detected in only a few species. It consists of a tubular dorsal vessel, with lateral paired openings for the ingress of the blood. Claparède figures in "*Phoxichilidium*" *cheliferum* a distinct aorta, which divides anteriorly into two trunks, emptying into the body-cavity. I have observed in *Nymphon grossipes* a dorsal pulsating organ, which I believe to be the heart. In the same specimen the perivisceral circulation was also seen. No special respiratory organs have been observed with certainty, though *Phanodemus* and *Oomerus* were described as possessing tracheal (?) openings; it seems improbable that this is their true nature, more especially since the tracheæ which should communicate with them have not been detected. The nervous system is well developed, consisting of a supra-œsophageal ganglion connected by commissures with a series of four large ventral ganglions. The former lies immediately beneath the oculiferous tubercle, to which it sends large nerves; and from it are also derived the nerves of the antennæ, palpi, rostrum, and accessory legs (Zenker). It seems probable, in view of the different origin of their nerves, that the accessory legs are not, as often supposed, branches of the first pair of ambulatory legs, but that they represent a pair of distinct appendages. Moreover, they are sometimes distinctly separated from the first pair, which is notably the case in a peculiar genus from Japan, apparently belonging to the genus *Ascorhynchus* Sars.

The sexes are separate, and the reproductive organs extend far out into the legs; their orifices are upon the lower side of the second joints in all the legs. Reference has already been made to the habit of carrying the egg-masses, followed by the male. These egg-bearing forms were long supposed to be females, but it has been conclusively shown by Cavanna, and subsequently by Dohrn, that they are males. The same fact was also noted in one or two species by Semper and Hoek. I have been able to confirm this in nearly all of our species by examination of the contents of the reproductive organs. In the fourth joint of each leg, in the male, is a large glandular organ, discharging by a number of openings arranged in an irregular row along the inferior side of the joint. Dohrn surmises that the secretion of this organ serves as a cement by which the eggs, when discharged by the female, are glued into a ball and attached to the accessory legs of the male.

Kröyer, Dohrn, and others have carefully studied the embryology. The eggs are collected into round masses upon the accessory legs and thus carried about by the male until after the escape of the embryos so that his body is often covered with the curious young. Segmentation of the yolk is complete. Prominences then appear upon the lower side of the embryo, one of which ultimately becomes the rostrum, and the others form three pairs of appendages, representing the future antennæ, palpi, and accessory legs. The condition of the larval antennæ has been already referred to. In most forms the embryo escapes from the egg with only these three pairs of appendages; but a species of *Pallene*, studied by Dohrn, passed through no metamorphosis, leaving the egg provided with the full number of appendages.

The species of the genus *Phoxichilidium* are remarkable for passing their early larval stages within the digestive cavities of certain tubularian hydroids (*Hydractinia*, etc.), six or eight of them sometimes living together within a single polypite. How they take up residence in the body of their involuntary host has not been observed, but they have been seen to escape by crawling out through the mouth.

The Pycnogonida, as a whole, have never been very carefully studied by systematic zoologists, though the observations of Dohrn, Quatrefages, Zenker, and others have given us a tolerably full knowledge of their anatomy and, in some cases, of their embryology. The systematic work has, with few exceptions, been unsatisfactory and confusing, owing to the paucity of generic and specific characters, the great variation of some species, the difficulty of obtaining large series of specimens, and the want of detailed descriptions. Though the specific characters are well marked, the general resemblance is so close in certain genera (*e. g.*, *Nymphon*, *Phoxichilidium*) as to render close examination necessary for the proper determination of the species. For this reason it is quite impossible to determine, from the descriptions, to what species some of the older names should be applied; and hence, as a rule, only such references are given in the synonymy as refer to figures or full descriptions.

The North American species have hitherto received little attention. Leach described an *Ammothea* from Carolina, and Stimpson another species of this genus from Puget Sound. Thomas Say described, in 1821, from Charleston, S. C., the genus *Anaphia*, of which he had one species (*A. pallida*) represented by three specimens. In 1853, Stimpson enumerated five species in his "Invertebrata of Grand Manan," of which four were described as new. In addition to these, three or four species are mentioned, accompanied in some cases by brief notes, in papers by Professors Verrill, Smith, Packard, and others. The "*Pasithoe*" described by Dr. Gould (Proc. Bost. Soc. Nat. Hist., vol. i, p. 92) is indeterminable. With two exceptions, the species here described were fully figured and characterized in a preliminary paper by the author, entitled "A Synopsis of the Pycnogonida of New England" (Trans. Conn. Acad. Sci., vol. v, pp. 1–26).

The genera known to me are included in the following table, those occurring on the New England coast being indicated by an asterisk. It should be noted that the table is in part compiled from descriptions, some of which are very imperfect. In cases where I have been unable to find the exact characters, an interrogation mark is placed after the name. There is need of a revision of the present genera, which can only be effected by the study of a large collection from all parts of the world.

I have been unable to ascertain the characters of the genus *Gnamptorhynchus* recently described by Böhm, and have therefore not included it in the table.

A. Antennæ present and chelate.
- *a.* Palpi present. (*Nymphonidæ*).
  - *b.* Auxiliary claws present.
    - (1). Accessory legs 11-jointed. Palpi 5-jointed .... **Nymphon* Fabr.
    - (2). Accessory legs 9-jointed. Palpi 8-jointed .... **Ammothea* Leach.
    - (3). Stigmata present (?) Accessory legs 3–4 (?) jointed. Palpi 3-jointed. *Phanodemus* Costa.
  - *bb.* Auxiliary claws wanting.
    - (1). Accessory legs 11-jointed. Palpi 10-jointed .... *Decolopoda* Eights.

aa. Palpi wanting. (*Pallenidæ*).
b. Auxiliary claws present.
(1). Accessory legs 9-jointed ....*Pallene* Johnston.
(2). Accessory legs 5-jointed......................*Phoxichilidium* M. Edwards.
bb. Auxiliary claws wanting.
(1). Accessory legs 11-jointed ....*Pseudopallene* Wilson.
(2). Accessory legs 6-jointed ....*Anoplodactylus* Wilson.
(3). Accessory legs ? — jointed. Stigmata present (?).......... *Oomerus* Hesse ?

B. Antennæ present, simple.
a. Palpi present. (*Achelidæ*).
b. Auxiliary claws present.
(1). Antennæ 3-jointed. Accessory legs 9-jointed. Palpi 8-jointed. *Oicebathes* Hesse.
(2). Antennæ 2-jointed. Accessory legs 10-jointed. Palpi 9-jointed. **Achelia* Hodge.
(3). Antennæ 1-jointed. Accessory legs 10-jointed. Palpi 6-jointed. *Tanystylum* Miers.
(4). Antennæ 1-jointed. Accessory legs 10-jointed. Palpi 9-jointed. *Carniger* Böhn.
bb. Auxiliary claws wanting.
(1). Antennæ 3-jointed. Accessory legs 10-jointed. Palpi 10-jointed. *Eurycide* Schiödte.
(2). Antennæ 2-jointed. Accessory legs 9-jointed. Palpi 5-jointed. *Pariboea* Costa.
(3). Antennæ 2-jointed. Accessory legs 8-jointed. Palpi 9-jointed. *Ascorhynchus* Sars.
(4). Antennæ 2-jointed. Accessory legs 10-jointed. Palpi 9-jointed. *Parazetes* Slater.
(5). Antennæ 2 (?)-jointed. Accessory legs 6-jointed. "Palpi 3-jointed." (1) *Pephredo* Goodsir ?

C. Antennæ wanting
a. Palpi present. (*Pasithoidæ*).
b. Auxiliary claws present.
(1). Accessory legs 9-jointed. Palpi 8-jointed ....*Pasithoe* Goodsir.
(2). Accessory legs 9-jointed. Palpi 7-jointed ....*Endeis* Costa.
bb. Auxiliary claws wanting.
(1). Accessory legs 10-jointed. Palpi 9-jointed. (2) *Rhopalorhynchus* Wood-Mason.
aa. Palpi wanting. (*Pycnogonidæ*).
(1). Auxiliary claws present. Accessory legs 7-jointed ....*Phoxichilus* Latreille.
(2). Auxiliary claws wanting. Accessory legs 10-jointed. **Pycnogonum* Brunnich.

(1). It is impossible to ascertain from Goodsir's original description exactly what are the characters of this genus.

(2). I cannot distinguish *Colossendeis* Jarzynsky from this genus.

The family characters must be regarded as still doubtful. Originally, all the forms were included by Latreille in a single family, the *Pycnogonidæ*. Subsequently those genera with antennæ were separated as *Nymphonidæ*. Dr. Semper has divided the latter into the *Nymphonidæ* with chelate antennæ, and the *Achelidæ* with simple antennæ; and in my "Synopsis" (*l. c.*) those genera with chelate antennæ, but without palpi (*Pallene*, *Phoxichilidium*, etc.), were characterized as *Pallenidæ*. A further division seems to me necessary, in the removal from the *Pyc-*

nogonidæ of those forms which possess palpi; and for this group the name *Pasithoidæ* may be used. The families will then stand as follows:

A. Antennæ present and chelate.
  Palpi present .... *Nymphonidæ.*
  Palpi wanting .... *Pallenidæ.*
B. Antennæ present, simple .... *Achelidæ.*
C. Antennæ wanting.
  Palpi present .... *Pasithoidæ.*
  Palpi wanting .... *Pycnogonidæ.*

This arrangement is, it is true, somewhat artificial, but it affords a convenient division of the genera, and may, for the present, be retained.

Following, is a systematic account of the genera and species.

Family I, PYCNOGONIDÆ.

PYCNOGONUM Brünnich.

Body very broad and stout. Antennæ and palpi wanting. Accessory legs ten-jointed, wanting in the female. Legs stout, dactylus without auxiliary claws.

Pycnogonum littorale (Gröm) O. Fabr.

Phalangium littorale Ström, Söndmör, p. 209, Pl. I, fig. 17, 1762.

Acarus marinus Pallas, Misc. Zool., p. 188, Pl. XIV, figs. 21-23, 1766.

Pycnogonum balænarum L., Syst. Nat., ed. XII, I., p. 1028, 1767.—Chr. Fabr., Ent. Syst., vol. iv, p. 416, 1794.—Latreille, Hist. Nat. des Crust. et des Insectes, Tom. vii, p. 332, 1804.—Gen. Crust. et Insect., Tom. i, p. 144, 1806.

Pycnogonum littorale O. Fabr., Fauna Grönlandica, p. 233, 1780.—Abilgaard in O. F. Müller, Zool. Dan., Volumen 3, p. 68, Pl. CXIX, figs. 10-12, 1789.—Cuvier, Règne Animal, Arachnides, Pl. 21, figs. 1 to 1*d*.—Milne-Edwards, Hist. Crust., vol. iii, p. 537, Pl. 41, fig. 6.—Johnston, Mag. Zool. and Bot., vol. i, p. 376, Pl. XIII, figs. 1-3.—Kröyer, Nat. Tidss., 1ste Bind, 2det Hæfte, p. 126.—Isis, Jahrg. 1846, Heft vi, p. 442.—Voy. en Scand., Laponie, etc., Crust., Pl. 38, figs. 4*a*–*e*.—Norman, Rept. of the Brit. Assoc. for the Advancement of Sci. for 1868, p. 301.—Whiteaves, Ann. and Mag. Nat. Hist., Nov., 1872, p. 347; Rept. of a second Deep-sea Dredging Exp. to the Gulf of St. Lawrence [in 1872], p. 15 [Montreal, 1873?].—Möbius, die wirbellosen Thiere der Ostsee, p. 153, 1873.—Hoek, Niederländisches Archiv für Zool., Band iii, 3tes Heft, p. 236, Pl. XV, figs. 1-3, 1877.—Verrill, Amer. Journ. Sci., vol. x, p. 38, 1875.—Smith and Harger, Trans. Conn. Acad., vol. iii, p. 10, 1874.—Wilson, Trans. Conn. Acad., vol. v, p. 4, Pl. I, figs. 1 *a*–*b*, Pl. II, figs. 3 *a*–*b*, July, 1878.

Pycnogonum pelagicum Stimpson, Invertebrata of Grand Manan, p. 37, 1853.—Verrill, Amer. Journ. Sci., vol. vii, p. 502, 1874.

? *Pychnogonum littorale* Nicolet, in Gay, Historia fisica y politica de Chile, Zoologia, p. 308, Pl. 4, fig. 8, 1854.

PLATE I, FIGURES 1 TO 3.

Body very broad and flat. Lateral processes with scarcely any interval between them. Neck somewhat constricted, but broad and stout. Each segment has a prominent conical tubercle in the median line above, and one or two less prominent ones on each lateral process.

Oculiferous tubercle prominent, broad and rounded. Eyes black widely separated, remarkably small. Abdomen slender, decidedly clavate, truncated at the extremity. Rostrum large, slender, basal half slightly swollen, outer portion attenuated, truncated at the tip. There is a slight constriction near the middle and another near the extremity, which give it a distinctly sinuous outline.

Accessory legs very small and slender, composed of nearly equal short articulations, the first five of which are somewhat shorter than the others; the terminal joint is pointed and slightly curved; the outer joints bear a few small stout spines.

Legs very stout; the three basal joints are short and thick, the first with two or three obtuse prominences above; fourth about twice the second, with one or two prominent tubercles at the distal extremity above; fifth similar, but not so much produced distally; seventh joint or tarsus very short and nearly triangular; eighth (propodus) narrow, somewhat curved; dactylus nearly half as long as the propodus, very stout.

Many of the joints bear very short hairs, which are densely set on the inferior side of the tarsus and propodus. The entire surface of the animal is covered with very small rounded tubercles, which give it a scabrous appearance. Color light yellowish brown to dark brown, the legs often blackish near their extremity.

Length 16 millimeters; extent 38 millimeters.*

This species has a wide range. Phillippi records it from Naples, and it appears to be common along the whole northern coast of Europe. Nicolet described and figured a form from Chili which is certainly very closely allied to, if not identical with, ours, and Mr. Henry H. Sclater informs me that he has received specimens of a variety of this species from Japan. Dr. Böhm reports a single specimen from Kerguelen Island. On our coast it ranges, so far as now known, from Long Island Sound to the Gulf of St. Lawrence (*Whiteaves*), though its occurrence south of Cape Cod is exceptional. In the Bay of Fundy it is not uncommon under stones at low-water mark, and it extends down to 430 fathoms. It is sometimes found clinging to actinias; at Eastport, Me., 17 specimens were taken from *Bunodes stella*, growing on the rocks near low-water mark; and off Cape Sable, N. S., they were found in considerable numbers attached to the base of *Bolocera Tuediæ*.

A comparison of specimens from the Gulf of Maine and from Eastport, Me., with specimens from Valentia, Ireland, received by the museum of Yale College from the Rev. A. M. Norman, leaves no doubt of their identity. Stimpson's *P. pelagicum* is evidently only the immature form.

In my "Synopsis" (*l. c.*) reference was made to Dr. Hoek's observation of the presence of accessory legs in the male and their absence in the female of this species. As

* The length includes the rostrum and abdomen. The extent is the distance from tip to tip of the outstretched legs.

it has been recently shown that in all cases where accessory legs are possessed by only one sex, this is the *male* and not the *female*, this observation was, of course, correct; though Dr. Hoek did not extend it to the other species examined by him.

Specimens examined.

Number.	Locality.	Fathoms.	Bottom	When collected.	Received from—	Specimens. No. and sex.	Dry. Alc.
4762	Long Island Sound; off Race Point Rock.	50	Shells, gravel, rock.	——, 1874	U. S. Fish Com	1 ♂ ○	Alc.
4979	Nantucket Shoals ..	18–21	....	Sept. —, 1874	....do	1 ♂	Alc.
5006	Massachusetts Bay, 1 mile south from Gloucester.	19	Rocks, sand..	——, 1878	....do	1 ♀	Alc.
4913	Gulf of Maine; 13½ miles S. E. from Cape Ann	90	Soft mud.....	——, 1878	....do	1 ♀	Alc.
4761	14 miles N. E from Cape Ann.	35	Stones, gravel	——, 1873	....do	1 ♂	Alc.
4933	29 miles E. from Cape Ann.	110	Soft mud....	——, 1878	....do	4 ♀	Alc.
4934	27 miles E. from Cape Ann	85	Sandy mud, gravel.	——, 1878	...do	5 ♂, 5 ♀	Alc.
4920	Gulf of Maine; Cashe's Ledge.	52–90	Rocks	——, 1873	.. do	1 ♀	Alc.
4764	E. from George's Bank; 41° 25′ N., 65° 42 3′ W.	430	....	Sept. 15, 1872	....do	3 ♂	Alc.
4763	Eastport, Me.; off Cherry Island.	20–25	....	——, 1872	....do	2 ♂, 2 ♀ ...	Alc.
4760	Eastport, Me; Johnson's Bay.	12	Rocks	——, 1872	....do	2 ♂	Alc.
4758	Eastport, Me	20	....do	——, 1872	.. do	5 ♂, ♀	Alc.
4756	... do	L. w.	....	——, 1872	....do	4 ♂, 7 ♀	Alc.
4754	....do	L. w.	....	——, 1870	Expedition '70	6 ♂, 2 ♀	Alc.
4918	....do	L. w.	....	Aug 18, 1868	Expedition '68	4 ♂, 2 ♀	Alc.
4750	Eastport, Me.; on *Bunodes*.	L. w.	Rocks	——, 1864	Expedition '64	17 ○	Alc.
4753	Eastport, Me.......	....	....	——, 1864	Expedition '64	1 ♂	Alc.
4766	W. from Brown's Bank; 42° 49′ N., 66° 19′ W.	82	Hard	——, 1877	U S. Fish Com	2 ♂, 15 ♀ ...	Alc.
4767	About 26 miles S. E. from Cape Sable, N. S.	59	Rocks	——, 1877	....do	2 ♂, 1 ♀ ...	Alc.
4768	About 39 miles S. E. from Cape Sable, N. S.	91	Very fine sand	——, 1877	....do	1 ♀	Alc.
4765	About 44 miles S. E. from Cape Sable, N. S.	88	... do........	——, 1877	.. do	2 ♂, 13 ♀ ..	Alc.

Family II, ACHELIDÆ.

Tanystylum Miers.

Body broad and stout. Antennæ rudimentary, one-jointed. Palpi six-jointed. Accessory legs ten-jointed, present in both sexes. Legs stout, dactylus with auxiliary claws.

Tanystylum orbiculare Wilson.

Trans. Conn. Acad., vol. v, p. 5, Pl. II, figs. 2*a* to 2*f*, Aug., 1878.

?? *Pasithoe umbonata* Gould, Proc. Bost. Soc. Nat. Hist., vol. i, p. 92.

Pallene, sp., Smith in Report on the Invertebrata of Vineyard Sound, p. 550 (544).

Plate III, Figure 11.

Body orbicular, deeply incised between the lateral processes, which are in close contact. Oculiferous segment extremely broad, neck not evident. Oculiferous tubercle large and rounded. Eyes black. Abdomen rather large, tapering, truncated, and slightly bifid at the extremity; it usually projects vertically upward.

Rostrum very large, rounded-conical, little constricted at the base, somewhat shorter than the body.

Antennæ rudimentary, consisting of a single knob-like joint, which is thinly covered with hairs.

Palpi slightly longer than the rostrum; the first, second, fourth, and fifth joints are nearly equal, and about as long as broad; the third and sixth are nearly equal, and about twice the others. The outer joints are somewhat hairy, the terminal one most so.

Accessory legs about half as large in the female as in the male. In the latter the basal joint is somewhat swollen and about as long as broad. The second, fourth, and fifth are nearly equal, and longer than the third; the remaining joints are short, decreasing in size to the last, which is very small. In the male the proportions are nearly the same, but the third joint is proportionally longer, and all of the others are more robust. The three outer joints are nearly globose, the terminal one minute. This joint bears, in both sexes, two spines, one of which is sometimes bifid at the tip. Other spines occur on the four preceding joints and are sometimes bifid.

Legs rather stout, sparsely hairy, the fifth and sixth joints having, above, alternate depressions and elevations, producing a deeply sinuous outline; each of these elevations bears a number of hairs. The three basal joints are very stout and short; the three following are each about equal to the three basal joints united; tarsus nearly triangular, with two or three stout spines below; propodus strongly curved, with a series of stout curved spines on the lower margin, on the upper side hairy; dactylus more than half the propodus, stout and curved; auxiliary claws about half as long as the dactylus.

Color of alcoholic specimens light yellowish brown. Length 1.5 millimeters; extent 6.4 millimeters.

The egg-masses are three or four in number and of a light yellow color. In some specimens the embryos had escaped from the eggs; they closely resemble those of *Achelia*, described on p. —, and the antennæ are large and chelate.

This genus, recently described by Miers from specimens collected at Kerguelen Island, is interesting from the extreme reduction of the antennæ, thus offering a transition from the *Achelidæ* to the *Pasithoidæ*.

T. orbiculare extends from off Martha's Vineyard to Virginia; it is almost invariably found upon Hydroids or Ascidians growing on piles of wharves, etc., and down to 14 fathoms.

Specimens examined.

Number.	Locality.	Fathoms.	Bottom.	When collected.	Received from—	Specimens. No. and sex.	Dry. Alc.
	Off Point Comfort, Va.		Mud	— —, 1877	S. F. Clarke	1 ♂, 1 ♀	Alc.
4916	Brooklyn, N. Y				D. C. Eaton	2	Alc.
	New Haven, Conn		On piles		S. I. Smith	1 ♀	Alc.
4772	Fisher's Island Sound.	10-12		July 13, 1874	U. S. Fish Com	1 ♀	Alc.
5018	Off Stonington, Conn			— —, 1874	do	1 ○	Alc.
4770	Vineyard Sound	14		— —, 1871	do	1 ♂, 2 ♀	Alc.
4774	Off Martha's Vineyard.				do	6 ♂, 7 ♀, 2 ○	Alc

ACHELIA Hodge.

Body broad. Antennæ small, two-jointed, not chelate. Palpi eight-jointed. Accessory legs ten-jointed, present in both sexes. Legs stout; dactylus with auxiliary claws.

Achelia spinosa (Stimp.) Wilson.

Trans. Conn. Acad., vol. v, p. 7, pl. ii, figs. 1*a* to 1*h*, Aug., 1878.
Zetes spinosa Stimpson, Invertebrata of Grand Manan, p. 37, 1853.

PLATE I, FIGURE 4; PLATE II, FIGURE 8.

Body nearly orbicular, deeply incised, segments not apparent. Lateral processes separated by a distinct interval. Neck distinct, but very broad. Oculiferous tubercle large and prominent, acute. Eyes ovate, black. Abdomen very long and slender, bifid at the tip.

Rostrum large, thickest in the middle and tapering to both ends, truncated at the extremity.

Antennæ not quite half the rostrum. The basal joint is about four times as long as broad, somewhat swollen near the extremity, where there are two or three tubercles, each terminated by a stout hair. Second joint rounded and knob-like, with one or two hairs.

Palpi slightly longer than the rostrum; the first, third, and four outer joints are very short, the first somewhat swollen; the second and fourth are much longer and nearly equal; all but the basal joint are hairy, the four outer ones only on the exterior margin.

Accessory legs in the male rather large. The two basal joints are short and stout, followed by three longer and more slender ones. The remaining five are much shorter, the terminal one very small and nearly globular; it bears two very large flattened denticulated spines; each of the two preceding joints has a similar spine; the outer joints are sparsely hairy, most of the hairs pointing backward. In the female this appendage is much smaller and proportionally stouter; the two basal and five distal joints are nearly as in the male, but the third, fourth, and fifth are much shorter and stouter.

Legs rather long; the three basal joints are short and stout, the second longest; the three following joints are nearly equal, each about as long as the three basal joints united; tarsus small, about one-fourth the propodus; the latter is strongly curved and armed below with a series of short stout spines; dactylus about half the propodus, stout and curved; auxiliary claws more than half the dactylus. The entire surface of the legs and body is scabrous with numerous pointed hairy tubercles often tipped with spines; the lateral processes of the body have three or four of these tubercles near the exterior margin; the largest are on the basal joints of the legs; on the other joints they are much smaller. The legs are throughout hairy and most of the hairs are borne on prominent tubercles. Color of alcoholic specimens light brown. Length 2.6 millimeters; extent 8.4 millimeters.

The egg-masses of a male specimen from Eastport, Maine, contain embryos recently escaped from the egg. The antennæ of these are enormously large and strongly chelate. The basal joint bears, at its extremity, on the outer side, a long spine. The two remaining appendages consist of two basal joints and a long, slender, acute terminal one, bearing a spine near its middle. The body is nearly hemispherical and without trace of segmentation. Rudiments of the eyes have appeared. The rostrum is rounded-conical and much smaller than in the adult.

There can be, I think, no doubt of the identity of this form with Stimpson's *Zetes spinosa*. Its most closely allied European representative is *A. echinata* Hodge [Annals and Magazine of Natural History, 3d series, vol. xiii, No. lxxiv, p. 115, pl. xii, figures 7–10, 1864], and I at first thought it was identical with that species. A comparison of *A. spinosa* with three specimens of *A. echinata* from Heligoland, received from Professor Semper, shows the two species to be perfectly distinct. The latter species has a slender, tapering rostrum of a very different shape; the peculiar conical spinous tubercles upon the legs are much more numerous, larger, and more slender; the abdomen is much shorter and stouter. Moreover, in *A. echinata* the second joint, in at least the two posterior pairs of legs, has a very prominent, rounded, hairy tubercle, projecting from the lower and posterior side, which is wanting in our species.

This species ranges from Grand Manan, N. B., to Block Island Sound, though it appears to be peculiarly a northern form, being represented south of Cape Cod, thus far, by a single specimen. At Eastport, Me., it is very common upon Hydroids, Ascidians, and other animals, and under stones near low-water mark; in Casco Bay it is also frequently found under similar circumstances. A single specimen was dredged by the U. S. Fish Commission, off Block Island, August 30, 1874, 34 fathoms, mud, which is the most southern locality recorded, and also the greatest depth.

Specimens examined.

Number.	Locality.	Fathoms.	Bottom.	When collected.	Received from—	Specimens. No. and sex.	Dry. Alc.
4785	Off Block Island....	34	Mud	Aug 30, 1874	U. S Fish Com.	1♀..........	Alc.
	Ram Island Ledge, Casco Bay.	L. w.	Rocks.......	—— —, 1873	....do	17♂, 12♀ ..	Alc.
4775	Eastport, Me	L. w.	....do........	—— —, 1868	Expedition '68	1♂, 1♀	Alc.
	....do	L. w.	....do........	—— —, 1870	Expedition '70	8♂♀	Alc.
4780	....do	20	....do........	—— —, 1872	U. S Fish Com.	1♀..........	Alc.
4781	....do	....	....do........	—— —, 1872	....do	3♂, 4♀, 2○	Alc.

Achelia scabra, *sp. nov.*

Body nearly orbicular, without distinct segmentation. Lateral processes comparatively short and stout, scarcely separated from each other; all except the posterior pair have two prominent conical spinous tubercles on the upper side near the outer margin; there is a similar but larger tubercle on each side of the neck, anterior to the first lateral process. Oculiferous tubercle obtuse, very large and stout; eyes large and conspicuous, black. Abdomen long and slender, constricted in the middle, bifid at the tip; along the sides it is somewhat spinous.

Rostrum large and stout, obtusely rounded-conical.

Antennæ extending to about the middle of the rostrum, very stout; basal joint about two and a half times as long as broad, second joint very short and stout, ovoid.

Palpi nearly as in *A. spinosa.* The hairs upon the exterior margin of the distal joints are very stout and close-set.

Accessory legs also much like those of *A. spinosa*, and presenting similar sexual differences; in the female they are much smaller, and with the third, fourth, and fifth joints much shorter than in the male; the terminal joint is, in both sexes, very minute.

Legs rather long, very rough and tuberculose, so that the outlines, particularly of the outer joints, are very irregular; tarsus very short and small, propodus stout and curved, dactylus two-thirds the propodus; auxiliary claws very slender and small, scarcely one-fifth the dactylus. This latter joint has upon the lower (concave) margin three stout, curved, divergent spines, at the basal angle, followed by an irregular series of smaller ones. The tarsus is also armed, upon its lower side, with a number of spines.

The whole surface of the legs and body is rough and scabrous; many of the larger tubercles upon the legs are tipped with stout hairs or slender spines; but these are nearly wanting on the three basal joints, and are everywhere less numerous and conspicuous than in *A. spinosa.* Color in alcohol, dirty white. Length 2.3 millimeters.

This species, which I at first mistook for *A. spinosa*, is represented by only two specimens, as follows:

Specimens examined.

Number.	Locality.	Fathoms.	Bottom.	When collected.	Received from—	Specimens. No. and sex.	Dry. Alc.
4936	Cape Ann, N. N. W., 15 miles.	23	Gravel stones	——, 1878	U. S. Fish Com	1♂	......
4935	St. George's Banks..	45	....	——, 1873	....do	1♀	......

The following are the most important points in which it differs from *A. spinosa*: The lateral processes are much shorter and in close contact, the three basal joints of the legs almost entirely lack the spinous tubercles characteristic of that species; the antennæ are much stouter, the rostrum stouter and less constricted at the base, and the auxiliary claws are less than half as large; (this last character affords the readiest means of distinguishing the two species.)

Family III, PALLENIDÆ.

PALLENE Johnston.

Body comparatively stout. Rostrum short, rounded. Antennæ robust, three-jointed, chelate. Palpi wanting. Accessory legs nine-jointed, present in both sexes. Legs very long; dactylus with auxiliary claws.

Pallene empusa Wilson.

Proc. Conn. Acad., vol. v, p. 9, Pl. III, figs. 2 *a* to 2 *g*, August, 1878.

PLATE II, FIGURES 5 to 7.

Body robust, smooth, distinctly segmented. Lateral processes well separated. Neck long, very slender at base. Oculiferous tubercle subacute, small, but prominent. Abdomen very small and short.

Rostrum nearly hemispherical, evenly rounded, smooth.

Antennæ sparingly hairy, short and stout. The opposable edges of the second and third joints are coarsely toothed, the teeth evenly rounded, so that the outline is deeply sinuous. There are about seven of these on the second joint, and many more, smaller ones, on the dactylus.

Accessory legs in the male about one-third the legs; the third joint is curved and about equal to the two basal joints united. Fourth joint considerably longer than the third, suddenly expanding at its distal extremity below; the five remaining joints are much shorter and nearly equal; the terminal one smoothly rounded at the extremity; each of the outer four joints is armed with a series of seven or eight spines; these are very broad and thin, with minute slender teeth, which do not extend to the base and are usually terminal; some of the spines are truncated, others smoothly rounded at the extremity. In the female the appendage is considerably smaller, the third and fourth joints are much shorter and stouter, and the latter is not expanded at the extremity.

Legs enormously long, over four times the length of the body, very slender near the base, much stouter distally; the first and third joints are short, the second much longer, about five times the first; the three following are much longer and very stout; the fourth is usually distended by the generative organs; tarsus very short and nearly triangular; propodus nearly straight and very slender; it is very narrow at the base, expanding to two and a half times this width, near the extremity; on the inferior side, near the base, are four or five stout spines, followed by a series of much smaller and more slender ones; dactylus slender, more than half the length of the propodus; auxiliary claws slender, two-thirds as long as the dactylus. The legs bear a few scattered stout hairs, most numerous distally. Length 1.5 millimeters; extent 13 millimeters. Color in alcohol, white.

This interesting species is closely similar to the European *P. brevirostris* Johnston, the type of the genus, and it is possible that a larger series of specimens may prove the identity of the two forms. The Peabody Museum has received, from Professor Semper, three specimens of the European species, collected at Heligoland, in the North Sea; these specimens agree in having a shorter and broader neck than that of our species, and the rostrum is much longer. I think that the species must be kept separate unless a series of specimens show intermediate forms.

New Haven to Vineyard Sound. Several specimens were taken, in 1874, by Prof. S. I. Smith, from tubularian hydroids growing on the bottom of an old ship at Noank, Conn. Professor Verrill notes that the eyes are, in life, of a bright red color.

Specimens examined.

Number.	Locality.	Fathoms.	Bottom.	When collected.	Received from—	Specimens.	Dry. Alc.
						No. and sex.	
5022	New Haven, Conn.	....	....	....	S. I. Smith	1♂ 1○	Alc.
	Off New Haven Light.	....	....	— —, 1865	A. E. Verrill	1	Alc.
4811	Noank Harbor, Conn.	3	Mud	— —, 1874	U. S. Fish Com.	4, 2♀, 1♂	Alc.
4808	Noank, Conn	....	....	— —, 1874	....do	7	Alc.
4810	Vineyard Sound	....	....	— —, 1871	....do	2	Alc.

PSEUDOPALLENE Wilson.

Body robust; neck broad and thick; rostrum more or less acute. Antennæ three-jointed, chelate; palpi wanting; accessory legs eleven-jointed, present in both sexes; legs stout and comparatively short, dactylus without auxiliary claws.

This genus has hitherto been confounded with *Pallene*, and some confusion has thus been caused in the diagnosis of that genus.

In *Pallene*, as described by Johnston (Mag. Zool. and Bot., vol. i, p.

380) the accessory legs are nine-jointed; the neck is constricted and more or less elongated as in *Nymphon;* the rostrum is short and nearly hemispherical; and the dactylus bears two very large auxiliary claws. The presence or absence of these claws is a good generic character; they are always two in number, are movably articulated to the dactylus, and are provided with a special set of muscles by means of which they are moved. It is to be observed, also, that the peculiar spines upon the outer joints of the accessory legs in *Pallene* are very unlike those of *Pseudopallene.*

Kröyer figures three species of "Pallene" in Gaimard's Voy. en Scand., Laponie, etc. (*P. discoidea*, *P. intermedia*, and *P. spinipes*). The first of these is undoubtedly a *Pseudopallene*, and probably also the other two, but, not having examined specimens of them, I have been unable to verify this.

Pseudopallene hispida (Stimp.) Wilson.

American Journal of Science and Arts, vol. xv, No. 87, p. 200, 1878.—Trans. Conn. Acad., vol. v, p. 10, Pl. III, figs. 1 *a* to 1 *e*, July, 1878.

Pallene hispida Stimpson, Invertebrata of Grand Manan, p. 37, 1853.

PLATE II, FIGURE 9.

Body oval, very broad, neck not constricted. Oculiferous tubercle small, rounded. Eyes ovate, light brown. Oculiferous segment half as long as the body. The second and third segments have, above, two prominent conical tubercles, each of which is tipped by a hair. The lateral thoracic processes are very broad and are not separated by any interval; they bear, on the outer margin, two to four acute, hairy tubercles. Abdomen twice as long as broad, truncate, hairy.

Rostrum slightly hairy, acute-conical, as long as the oculiferous segment, with a constriction on each side, below, giving it the appearance of being articulated at this point. The mouth is terminal and surrounded by a rosette of filamentary processes.

Antennæ very stout and swollen, hairy, tipped with amber-color, about twice as long as the rostrum; claws of chelæ blunt and rounded; basal joints enlarged near their attachment; the second joint has, on its lower margin, a prominent rounded tubercle behind which the dactylus closes.

Accessory legs slender; in the female the two basal joints are short, the third longer, the fourth and fifth still longer, sixth about as long as the third; the remaining joints are shorter and decrease in size to the last, which is spine-like and trifid at its extremity; the four outer joints are armed with four or five stout, smooth, curved spines. In the male these appendages are considerably longer and more slender, and the fifth joint has a prominent rim or shoulder at its distal extremity, as in *Pallene empusa*, which is armed below with a few stout spines. The terminal joint is not trifid but simply claw-like; it is attenuated toward the tip and abruptly incurved.

Legs very stout, the three basal joints short and overlapping each other; fourth joint as long as the three basal ones, much distended with the ovaries in the specimen described; fifth as long as the fourth, but much more slender; sixth still longer and more slender; tarsus very short, nearly triangular; propodus tapering from the base, slightly curved, armed on the inferior margin with five or six stout curved spines; dactylus curved, acute, about two-thirds as long as the propodus.

All of the legs bear a number of prominent, conical, spiny tubercles. These are arranged in longitudinal rows on some of the joints, particularly on the fifth and sixth, which thus appear deeply serrate on the margin. The entire surface of the body is rough and more or less hairy.

Color, in alcohol, light brown. Length 3 millimeters; legs 7.5 millimeters; accessory legs 3.7 millimeters.

I have seen only two specimens, namely:

Specimens examined.

Number.	Locality.	Fathoms.	Bottom	When collected.	Received from—	Specimens. No. and sex.	Dry. Alc.
4812	Johnson's Bay, Eastport, Me.	12	Rocks	—— —, 1872	U. S. Fish Com.	1 ♀	Alc.
5005	Grand Manan, N B	50–55	....	—— —, 1872	....do	1 ♂ + ○....	Alc.

Stimpson first obtained this species from deep water off Grand Manan, "on *Ascidiæ callosæ*."

Pseudopallene discoidea (Kröyer) Wilson.

Trans. Conn. Acad., vol. v, p. 12, Pl. III, figs. 3*a* to 3*c*, July, 1878.

Pallene discoidea Kröyer, Nat. Tidss., 1ste Bind, 2det Hæfte, p. 120, 1844; Voy. en Scand., Laponie, etc., Pl. 37, fig. 3*a*—*g*; Isis, Jahrg. 1846, Heft vi, p. 443.

PLATE II, FIGURE 10.

Body oval, somewhat narrower than that of *P. hispida*; lateral processes in close contact. Abdomen pointed, slightly bifid at the tip.

Rostrum obtuse, slightly hairy, outline of sides convex.

Antennæ stout, but not so much so as in *P. hispida;* basal joint not enlarged near the base. Chelæ with the claws acute and finely serrated along the opposable margins, second joint with no tubercles on the inferior margin.

Accessory legs of the female short and stout, all of the joints being broad and short; fourth and fifth joints longest, terminal joint acute; the 7th, 8th, 9th and 10th joints have each a simple spine on the upper side.

Legs nearly as in *P. hispida*, but longer and more slender, particularly in their basal portion, where the joints do not overlap.

The legs and body are armed with conical hairy tubercles arranged nearly as in the preceding species. Color light yellowish brown. Length, 3 millimeters.

This species is represented by two female specimens, of which one was taken with *Caprella* on the tangles in 20 fathoms, rocky bottom, Eastport Harbor, by the United States Fish Commission, August 9, 1872; the other is simply labeled "Eastport Harbor, 1870."

This species is very similar to the last, and a larger number of specimens may show them to be identical. The specimens described present, however, well-marked differences, particularly in the shape and armature of the antennæ, the shape of the rostrum, abdomen, etc. Though not agreeing perfectly with Kröyer's figures of *P. discoidea*, there can be little doubt of the identity of our species with it.

Specimens examined.

Number.	Locality.	Fathoms.	Bottom.	When collected.	Received from—	Specimens. No. and sex.	Dry. Alc.
4922	Eastport Harbor, Me	....	....	—— —, 1870	Expedition, '70	1 ♀	Alc.
4813	....do	20	Rocks	—— —, 1872	U. S. Fish Com	1 ♀	Alc.

PHOXICHILIDIUM Milne Edwards.

Body slender; neck short. Rostrum cylindrical, rounded. Antennæ three-jointed, chelate. Palpi wanting. Accessory legs five-jointed, absent in the female. Legs slender; dactylus with auxiliary claws.

Phoxichilidium maxillare Stimpson.

Phoxichilidium maxillare Stimpson, Invertebrata of Grand Manan, p. 37, 1853.—Wilson, Trans. Conn. Acad., vol. v, p. 12, Pl. IV, figs. 1*a* to 1*e*, July, 1878.
Phoxichilidium minor Wilson, *op. cit.*, p. 13, Pl. IV, figs. 2*a* to 2*b*, July, 1878.

PLATE III, FIGURES 12 to 15.

Body rather stout. Oculiferous segment twice as broad as long. Oculiferous tubercle prominent, acute. Eyes ovate, nearly white in alcohol. Posterior segment much smaller and narrower than the next anterior. Abdomen small and rounded.

Rostrum stout, usually about as long as the oculiferous segment though the length is somewhat variable, nearly cylindrical, rounded at the extremity. It is sometimes slightly constricted a short distance from the tip; in other cases no such constriction is apparent, and the outline of the lateral margins may be slightly convex (*P.* "*minor*").

Antennæ stout, almost destitute of hairs. Claws of the chelæ very strongly curved, quite smooth on the opposable margins; the dactylus projects somewhat beyond the extremity of the preceding joint, and is very thick and strong.

Accessory legs nearly one-third as long as the legs; basal joint stouter than the others; third joint longest; terminal joint strongly curved, smoothly rounded at the tip, armed on each side with six or eight simple spines directed backward, and below, with three or four stouter ones; the other joints have a few scattered hairs.

Legs comparatively stout, remarkably smooth in appearance, though with a very few scattered hairs; basal joint nearly quadrate, about half the length of the second, which is somewhat longer than the third; the three following are nearly equal and longer than the three basal joints united; propodus stout and curved, about four times the tarsus; on its inferior margin are five stout spines followed by a series of very small ones; dactylus stout, more than half the propodus; auxiliary claws small, varying from one-fifth to one-fourth the length of the dactylus.

Color blackish or sepia to nearly pure white. Length of adult specimens 2 to 4.75 millimeters; extent of legs 15 to 30 millimeters.

Most of the specimens from the Bay of Fundy are dark colored and of large size, and differ in several other particulars from those taken in Casco Bay, at Gloucester, Mass., and other southern localities. These differences are so striking that I was led to describe the southern form as a new species under the name *Phoxichilidium minor*. Since the publication of that description, however, a much larger series of specimens has been obtained, which shows conclusively that the two forms cannot be separated, though extreme forms appear very unlike. The southern form is almost always white in color, and very small, even when adult; it further differs in the shape of the rostrum and antennæ, and in being more slender in nearly all respects.

Phoxichilidium femoratum of Northern Europe is closely similar to this species, but is figured as being more slender, of a different color, and with the propodus and dactylus differently armed and shaped. I think it quite possible that they may be shown to be identical, but it seems preferable to keep them separate at present. The so-called "species" of this genus need revision (though in this respect the genus is not wholly without a parallel among the *Pycnogonida*), and undoubtedly a large series of specimens would reduce their number.

The observed range of *P. maxillare* is from Gloucester, Mass., to Halifax, N. S.; and in depth, from low water to 55 fathoms. At Eastport, Me., it is very common under stones at or near low-water mark, and frequently numbers of them cling to each other in a tangled mass.

Specimens examined.

Number.	Locality.	Fathoms.	Bottom.	When collected.	Received from—	Specimens. No. and sex	Dry. Alc.
4937	Gloucester, Mass....	L w.	On piles	— —, 1878	U.S Fish Com.	12 ♂, 13 ♀ ...	Alc.
4939	About 5 miles north from Cape Ann.	....	....	— —, 1878	....do	24 ♂, 25 ♀ ...	Alc.
	Ram Island Ledge, Casco Bay.	L. w.	Rocks	— —, 1873	....do	22 ♂, 39 ♀ ...	Alc.
5026	Casco Bay	D. w.	....	— —, 1873	... do	1 ♂	Alc.
5016	....do	....	....	— —, 1873	....do	8 ♂, 12 ♀	Alc.
5025	Portland, Me	L. w.	On piles	— —, 1873	....do	1 ♂, 3 ♀	Alc.
5004	Grand Manan, N. B .	50, 55	....	— —, 1872	....do	1 ♂	Alc.
	Eastport, Me........	L. w.	....	— —, 1868	Expedition, '68	14 ♂, 6 ♀	Alc.
	....do	L. w.	....	— —, 1870	Expedition, '70	7 ♂, 7 ♀	
	....do	L. w.	....	— —, 1872	U.S. Fish Com.	2 ♂, 6 ♀	Alc.
4795	Halifax, N. S........	L. w.	On piles	— —, 1877	....do	1 ○.........	Alc.

ANOPLODACTYLUS Wilson.

Body slender. Rostrum cylindrical, rounded. Antennæ three-jointed, chelate. Palpi wanting. Accessory legs six-jointed, wanting in the female. Neck elongated, extending forward over the rostrum. Legs slender; dactylus without auxiliary claws.

This genus differs from *Phoxichilidium*, which it otherwise closely resembles, in the number of joints composing the accessory legs, and in the absence of auxiliary claws upon the dactylus. *Phoxichilidium* has been made to include several distinct types, among them a form having eleven-jointed accessory legs (*P. fluminense* Kr.), and "*Phoxichilidium cheliferum*" Claparède, a very remarkable form with the accessory legs ten-jointed and distinctly chelate.

Kröyer's *Phoxichilidium petiolatum* (Voy. en Scand., Laponie, etc., Pl. 38, fig. 3) belongs to *Anoplodactylus*, and probably also *Phoxichilidium virescens* Hodge.

Since the publication of my original description of this genus it has been pointed out to me that Say's genus *Anaphia* (described in 1821) may be identical with it. Say's description was based upon two specimens which did not possess accessory legs and were probably females; hence it is impossible to determine their exact generic characters. Nevertheless, their general agreement with the type of *Anoplodactylus* is so close that I think it probable that they are generically the same; and, if so, of course the name *Anaphia* should be used. To prevent possible confusion, however, the later name is retained until an opportunity is afforded for examination of specimens from the locality where Say's specimens were collected.

Anoplodactylus lentus Wilson.

American Journal of Science and Arts, vol. xv, No. 87, p. 200, 1878.—Trans. Conn. Acad., vol. v, p. 14, Pl. IV, figs. 3 *a* to 3 *e*, July, 1878.

Phoxichilidium maxillare Smith, Report on the Invertebrata of Vineyard Sound, &c., p. 250 [544], Pl. VII, fig. 35, 1874 [*non* Stimpson].

?*Anaphia pallida* Say, Journ. Acad. Nat. Sci. Phil., vol. 2, p. 59, Pl. V, figs. 7 and 7 *a*, 1821.

PLATE III, FIGURES 16 to 18.

Body slender, lateral processes widely separated. Oculiferous segment broad, as long as the two following segments united, not emarginate between the bases of the antennæ. Posterior segment somewhat elongated and very slender, the lateral processes directed obliquely backward. Neck swollen. Abdomen rather more than twice as long as broad, slightly bifid at the extremity. Oculiferous tubercle prominent, acute, placed far forward. Eyes ovate, light brown to black.

Rostrum large, longer than the oculiferous segment, somewhat constricted basally, so as to appear clavate; extremity subglobose.

Antennæ long and slender, hairy, their bases closely approximated; basal joint extending beyond the extremity of the rostrum; chelæ stout, hairy, claws acute, opposable edges smooth.

Accessory legs stout, roughened by minute tubercles, the outer joints with many short stout hairs, most of which are directed backward; the two basal joints are very stout, the first shorter than its width, the second about twice as long; third nearly two and a half times the second, somewhat clavate, suddenly constricted a short distance from the base; fourth half the length of the third, considerably longer than the fifth; sixth much smaller than the preceding.

Legs very long and slender; first and third joints very short; second longer and clavate; the three following joints are much longer, sixth longest; tarsus very short, deeply emarginate; propodus curved, with a rounded lobe near the base bearing five or six strong spines; these are followed by a series of much smaller ones; dactylus stout, about two-thirds the length of the propodus. Entire surface of the body scabrous. Legs with a few scattered hairs, which are most numerous on the outer joints.

The sexes resemble each other closely, but the females do not possess accessory legs; the female is, as a rule, slightly larger than the male. Length 7 millimeters; legs 30 millimeters.

This species is nearest to "*Phoxichilidium petiolatum*" Kr., of Europe. In the latter species, however, according to the figures, the anterior segment is much more slender, and it is emarginate between the bases of the antennæ, which are thus separated by a distinct interval; the posterior segment is represented as stouter and shorter; the rostrum more abbreviated; and the propodus of a different shape. Kröyer figures the accessory legs with seven joints, probably mistaking the constriction near the base of the third joint for an articulation.

Common between tide-marks and down to six fathoms in Vineyard Sound, where it is found on shelly bottoms "clinging to and creeping over the hydroids and ascidians." "It is most frequently deep purple in color, but gray and brown specimens are often met with" (*Verrill*). It is also taken rarely in the Bay of Fundy, there being a single specimen in a vial with *Phoxichilidium maxillare* and *Pycnogonum littorale* from Eastport.

Specimens examined.

Number.	Locality.	Fathoms.	Bottom.	When collected.	Received from—	Specimens. No. and sex.	Dry. Alc.
4807	Long Island Sound			— —, 1874	U. S Fish Com.	1♀	Alc.
4804	Cataumut Harbor	4	Eel-grass	— —, 1875	....do	1♂	Alc.
4800	Vineyard Sound, Mass	3–5	Gravel	— —, 1871	....do	8♂, 16♀	Alc.
4806	....do			— —, 1875	....do	2♂, 3♀	Alc.
4925	....do	8	Hard	— —, 1875	....do	3♂, 7♀	Alc.
5021	Wood's Holl, Mass[1]			— —, 1875	....do	1♂	Alc.
	Eastport, Me					1	Alc.

Family IV, NYMPHONIDÆ.

AMMOTHEA Leach.

Body broad, neck scarcely apparent. Rostrum large, tapering. Antennæ small, three-jointed, chelate. Palpi eight-jointed. Accessory legs nine-jointed; in the female five-jointed (?). Legs slender. Auxiliary claws present.

Ammothea achelioides Wilson.

Trans. Conn. Acad., vol. v, p. 16, Pl. V, figs. 1 *a* to 1 *e*, July, 1878.

PLATE IV, FIGURES 19 and 20.

Body very broad, oval, segments not evident, lateral processes scarcely separated. Oculiferous tubercle prominent, acute; eyes dark; abdomen long and very slender, bifid at the extremity.

Rostrum large, tapering, extremity rounded.

Antennæ about three-fourths as long as the rostrum; basal joint narrowest near the middle, somewhat hairy, with one or two prominent tubercles, each tipped by a slender spine; chela with the claws very slender and strongly curved, armed with a few small spines on the opposable edges.

Palpi slender, longer than the rostrum, sparsely hairy, most so on the distal joints; the first, third, and four distal joints are very short; terminal one shortest; sixth longest; the second and fourth are nearly equal and more than twice the basal joint.

Accessory legs, in all the specimens examined, very short, swollen and pellucid, so that the joints could with difficulty be distinguished. They are composed of five joints; a very short basal one and four other longer ones; the terminal one is tapering, smoothly rounded at the tip. It seems probable that these appendages are either those of the female, or of the immature male.

Legs short, rather slender; the three basal joints are short, followed by three which are nearly equal and about as long as the three basal joints united; tarsus very short; propodus gently curved, with two stout spines on the inferior margin near the base, followed by a few smaller ones; dactylus nearly two-thirds the length of the propodus, rather stout; auxiliary claws two-thirds the dactylus.

The legs are rough and hairy, the hairs usually arising from tubercles or swellings. These tubercles are very large and acute-conical near the outer margin of the body-processes and upon the first joint of the legs; on the outer joints they are smoothly rounded and less elevated, often producing a sinuous outline most apparent on the fourth, fifth, and sixth joints.

Color of alcoholic specimens, light yellowish brown. Length 1.4 millimeters; extent 5.2 millimeters.

Three specimens only, taken in the Bay of Fundy by the United States Fish Commission, in 1872. In general appearance it is closely similar to *Achelia spinosa*. It is apparently nearest to the "*Ammothoa brevipes*," of Hodge, described from the Durham coast, though quite distinct from that species, so far as can be judged from the original figures. The antennæ, in our species, are more slender and with much smaller spines on the chela, and the proportions of the palpal joints are very different; the abdomen is far longer and more slender, and the legs are not so spinous.

Specimens examined.

Number.	Locality.	Fathoms.	Bottom.	When collected.	Received from—	Specimens. No. and sex.	Dry. Alc.
4814	Bay of Fundy........	....	....	—— —, 1872	U. S. Fish Com.	2........	Alc.
4779	Grand Manan, N. B..	....	....	—— —, 1872	....do	1........	Alc.

NYMPHON Chr. Fabricius.

Body slender. Neck distinct. Rostrum cylindrical, rounded. Antennæ three-jointed, chelate. Palpi five-jointed. Accessory legs present in both sexes, eleven-jointed. Legs slender; dactylus with auxiliary claws.

All the species of *Nymphon* are slender, some of them exceedingly so. The antennæ are slender and the claws of the chelæ are armed along their opposable edges with a series of close-set, slender spines. The sexes generally resemble each other closely, the chief differences being found in the accessory legs. These appendages are armed, in both sexes, with a series of flattened denticulated spines, found upon the seventh, eighth, ninth, and tenth joints. The auxiliary claws are usually of small size, and sometimes minute. They are peculiarly deep-water forms, rarely occurring at a depth of less than twenty fathoms, and sometimes extending down to great depths. For this reason almost nothing is known of their habits, though their external development has been well studied. In certain species the specific characters are extremely variable, as described below.

Nymphon Strömii Kröyer.

Nat. Tidss., 1ste Bind, 2det Hæfte, p. 111, 1844; Voy. en Scand., Laponie, etc., pl. 35, figs. 3 *a–f*.—Norman, Rept. of the Brit. Assoc. for the Advancement of Sci. for 1868, p. 301.—Miers, Ann. and Mag. Nat. Hist., 4th series, vol. 20, No. 116, p. 109.—Wilson, Trans. Conn. Acad., vol. v, p. 17, Pl. VI, figs. 1*a* to 1*h*, July, 1878.

Probably *Pycnogonum grossipes* Abilgaard, in O. F. Müller's Zoologica Danica, vol. iii, p. 67, Pl. CXIX, figs. 5–9, 1788.

Nymphon giganteum Goodsir, Ann. and Mag. Nat. Hist., vol. xv, No. xcviii, p. 293, 1845.—Norman, Rept. of the Brit. Assoc. for the Advancement of Sci. for 1868, p. 301.—Whiteaves, Ann. and Mag. Nat. Hist., Nov., 1872, p. 347; Rept. of a Sec. Deep-sea Dredging Exp. to the Gulf of St. Lawrence [in 1872]. Montreal, 1873?—Verrill, Am. Journ. Sci. and Arts, vol. vii, p. 411; vol. vi, p. 439, 1874.

?*Nymphon gracilipes* Camil Heller, Die Crustaceen Pycnogoniden und Tunicaten der K. K. Österr-Ungar. Nordpol-Exp., p. 16, Taf. iv, fig. 15, Taf. v, figs. 1, 2.

PLATE V. PLATE VI, FIGURE 29.

Body very stout, nearly smooth. Neck very short, but deeply constricted. Oculiferous segment large, longer than the two following segments united, stout and swollen anterior to the constriction of the neck. Oculiferous tubercle prominent, smoothly rounded. Eyes very distinct, black, ovate. Abdomen small, tapering toward the extremity.

Rostrum rather large, nearly cylindrical though slightly expanded in the middle.

Antennæ smooth, rather slender; basal joint as long as the rostrum; claws of chelæ remarkably slender and elongated, gently curved, when closed meeting along nearly their whole length; they are armed along their opposable margins with a series of small spines, which are more erect and much more numerous upon the dactylus.

Palpi much longer than the rostrum; basal joint stout, very short; second and third much longer, nearly equal; fourth and fifth a little less and more slender, sparsely hairy.

Accessory legs stout, slightly hairy; the three basal joints are nearly as broad as long; the following three are much longer, the sixth shortest and about as long as the three basal joints united; the remaining joints are much shorter and more slender, the terminal one acute and claw-like, with a row of spines on the inferior edge; the denticulated spines vary considerably and are sometimes nearly smooth.

Legs very long and slender; first and third joints short, about half the second; the three following are very long, sixth longest, fifth shortest; propodus and tarsus slender, nearly equal, hairy; the former is not armed with spines; dactylus long and slender, very acute, about three-fifths the length of the propodus; auxiliary claws very small, about one-fifth the dactylus. Color, when living, light salmon-yellow, the legs often annulated with broad reddish rings. Egg-masses large, two to four in number, bright yellow. Length of largest specimens 15 millimeters; extent 140 millimeters; accessory legs 19 millimeters.

This fine species is not uncommon; it attains its greatest size on muddy bottoms in deep water. Taken at many localities in Massachusetts Bay, off Gloucester and Salem; in the Gulf of Maine off Cape Ann; Casco Bay 50–70 fathoms; Eastport, Me. (Professor Verrill); off Halifax, N. S.; Bedford Basin, Halifax; Orphan Bank, Gulf of Saint Lawrence (*J. F. Whiteaves*). It is found on all bottoms, though, as a rule, it may be regarded as a "muddy-bottom species." The observed range on our coast, in depth, is from 7½ fathoms (Gloucester Harbor, mud and sand) to 115 fathoms (27–31 miles E. S. E. from Cape Ann, gravelly bottom).

I cannot distinguish this species from Goodsir's *N. giganteum*. *Nymphon gracilipes*, of Heller, is also very closely similar to, if not identical with, this species. In his figure, however, the dactylus is represented of nearly the same length with the propodus, and it may be distinct. The "*Pycnogonum grossipes*," figured in the "Zoologica Danica," is certainly not *N. grossipes* Fabr., and it seems to me most probable that it is to be referred to *N. Strömii*.

It is worthy of note that the arch of the upper side of the oculiferous segment, when laterally viewed, is very variable, as is also the length of the constricted portion or "neck."

Specimens examined.

Number.	Locality.	Fathoms.	Bottom.	When collected.	Received from—	Specimens. No. and sex.	Dry. Alc.
4972	Gloucester, N. ¼ W. 6⅓ miles.	45	Mud	— —, 1878	U.S. Fish Com.	2♂, 1♀	Alc.
4837	Salem, W. N. W. 9 to 11 miles.	35	...do	— —, 1877	...do	2	Alc.
4838	Salem, W. N. W. 13 miles.	11–50	Soft mud	— —, 1877	...do	1♂, 1♀	Alc.
4839	Cape Ann, N. W. 14 miles.	90	...do	— —, 1877	...do	4	Alc.
4973	Cape Ann, W. N. W. 30 to 31 miles	110	Soft brown mud	— —, 1878	...do	1	Alc.
4998	Gloucester, N. 5½ to 7 miles.	40–45	...do	— —, 1878	...do	3♂, 1♀	Alc.
4995	Cape Ann, W. by N. 4½ to 5½ miles	57–68	Soft mud, concretions.	— —, 1878	...do	9	Alc.
4845	Cape Ann, N. W. ½ N. 11 miles.	51	Mud, gravel, stones.	— —, 1877	...do	4	Alc.
4966	Cape Ann, N. W. ¾ W. 13 miles.	53	Mud and stones	— —, 1878	...do	1	Alc.
4967	Gloucester, Harbor	7½	Mud and sand	— —, 1878	...do	1♀	Alc.
4968	Cape Ann, N. W. 4 to 5 miles.	42	Mud, clay nodules.	— —, 1878	...do	2♀	Alc.
4960	Cape Ann, N. W. by W. 6 to 7 miles.	54	Sand, clay nodules.	— —, 1878	...do	1♂, 2♀	Alc.
4970	Cape Ann, N. W. by W. 7 miles.	75	Sand	— —, 1878	...do	1♂	Alc.
4997	Cape Ann, N. W. 7½ to 8 miles.	32–36	Sand and pebble.	— —, 1878	...do	3♂, 1♀	Alc.
4971	Cape Ann, W. N. W. 27 to 31 miles	115	Gravel	— —, 1878	...do	8	Alc.
4835	Off Isles of Shoals	35	Clay, mud, and sand.	— —, 1874	...do	1♀	Alc.
4833	Cashe's Ledge, N. 6 to 15 miles	52–90	Rocks	— —, 1873	...do	11	Alc.
4834	Casco Bay, Me	50		— —, 1873	...do	1♂	Alc.
4832	...do	72		— —, 1873	...do	1♀	Alc.
	Eastport, Me			— —, 1870	Expedition '70		
4842	Cape Sable, N. S., N. W. 18 to 22 miles.	59	Pebbles and sand	— —, 1877	...do	1♂	Alc.
4844	Off Halifax, N. S., 8½ miles.	52	Sandy mud	— —, 1877	...do	3♀	Alc.
4843	Halifax, Bedford Basin.	35	Soft mud	— —, 1877	...do	1 ○	Alc.

Nymphon macrum, *sp. nov.*

PLATE IV, FIGURES 21 to 23.

Distinctive characters.—Antennæ extremely slender, with the claws of the chelæ much curved. Accessory legs separated from the first lateral processes by a distinct interval. Terminal joint of palpus very slender. Tarsus longer than the propodus. Auxiliary claws nearly two-thirds the dactylus.

Body very smooth and rather robust, for the genus, with the lateral processes separated by an inter-space about equal to their width. Oculiferous segment constricted, so as to form a short and narrow neck, nearly as in the preceding species; there is a slight, though distinct, interval between the process bearing the accessory legs and that of the first pair of ambulatory legs. Above, there is a distinct sulcus running backward from midway between the bases of the antennæ. Oculiferous tubercle acutish, prominent, situated just anterior to the first pair of lateral processes; eyes large, ovate, light colored, surrounded by dark pigment. Posterior segment narrow with the lateral processes directed backward; abdomen slender and tapering, distinctly bifid.

Rostrum about as long as the oculiferous segment, nearly cylindrical, slightly swollen in the basal half. Antennæ extremely slender—more so than in any other species of the genus known to me; basal joint somewhat longer than the rostrum; chela much elongated; claws, when closed, crossing each other at a considerable distance from their tips. Both claws are armed with a dense row of spines, which gradually decrease in length toward the tips, and finally disappear, leaving the terminal portion bare for some distance; these spines are larger and more crowded on the movable claw; on the other, larger spines alternate with from one to three smaller ones. In one specimen there were 109 such spines upon the movable claw and 184 upon the other.

Palpi with a very short and stout basal joint, about one-sixth or one-seventh as long as the second; the third is considerably less than the second, the fourth still shorter, the terminal one less than the fourth and extremely slender and straight. The entire appendage is slightly hairy.

Accessory legs resembling those of *N. grossipes;* in the male they are considerably longer than in the female, the fifth joint is more slender and elongated, and with a strong s-shaped curvature. The outer joints bear a few scattered hairs.

The legs are very slender, especially the outer four joints, and sparsely hairy; the proportions of the joints are about as in *N. longitarse*, but the tarsus is only about $1\frac{1}{3}$ times the propodus; both these joints have a close and pretty regular series of small, slender spines along the entire inferior margin. Dactylus slender and acute, rather more than half the propodus; *auxiliary claws very large*, nearly two-thirds the dactylus.

Color in alcohol, light yellowish-white. Length of a large specimen 9.3 millimeters. Extent 72 millimeters.

This species is very distinct, and the specific characters appear to vary but slightly. It is, in general appearance, much like *N. Strömii*, but may be at once distinguished by the large auxiliary claws. The interval between the accessory legs and the first pair of ambulatory legs is a feature which I have not seen in any other species of the genus. It has been taken at a few localities in the Gulf of Maine in from 85 (or perhaps less) to 115 fathoms. A single specimen is known from Banquereau, off Nova Scotia. All the specimens, as shown below, are from deep water, and most of them from muddy bottoms.

Specimens examined.

Number.	Locality.	Fathoms.	Bottom.	When collected.	Received from—	Specimens. No. and sex.	Dry. Alc.
4960	Cape Ann, W. N. W. 30 miles.	90–115	Mud, gravel, stone.	—— —, 1878	U. S. Fish Com	4♂, 3♀	Alc.
4962	Cape Ann, W. N. W. 30 to 31 miles.	110–115	Mud and stone	—— —, 1878	....do	8♂, 12♀	Alc.
4963	Cape Ann, W. N. W. 29 to 30 miles.	85	Mud, stone, sand	—— —, 1878	....do	2♀	Alc.
4978	Off Manhegan Island, G. Maine.	5–90	Mud or sand...	—— —, 1874	....do	1	Alc.
4841	Cape Ann, W. 140 miles.	112	Gravel	—— —, 1877	....do	2♀	Alc.
4965	From cable of schooner Marion, Banquereau, N. S.	....	....	Aug. 25, 1878	....do	1♂	Alc.

Nymphon longitarse Kröyer.

Nat. Tidss., 1ste Bind, 2det Hæfte, p. 112, 1844; Voy. en Scand., Laponie, etc., Pl. 36, fig. *2a–b*.—Wilson, Trans. Conn. Acad., vol. v, p. 19, Pl. VII, figs. *2a* to *2h*, July, 1878.—G. O. Sars, Archiv for Mathematik og Naturvidenskab, Andet Bind, Tredie Heft, p. 366, 1877.

PLATE VI, FIGURES 30 and 31.

Entire animal extremely slender. Body smooth. Oculiferous segment produced into a very long slender neck, expanding anteriorly for the attachment of the antennæ. Posterior segment very narrow, lateral process directed nearly backward. Abdomen small, tapering. Oculiferous tubercle rounded, eyes black, ovate.

Rostrum slender, rounded, shorter than the basal joint of the antennæ.

Antennæ very slender, slightly hairy; claws of chelæ very long and slender, their tips crossing when closed; the spines with which they are armed are larger and less numerous than those of *N. Strömii.*

Palpi resembling those of *N. Strömii*, but more slender and with the fourth joint shorter than the third or fifth.

Accessory legs remarkably slender; the three basal joints are very short and nearly equal; fourth nearly twice the length of the first three united; fifth somewhat less; sixth equal to the three basal joints, about twice the seventh; the remaining joints decrease to the last, which is claw-like with a few spines on its inferior margin; spines of the distal joints decidedly curved.

Legs resembling those of *N. Strömii* but much more slender and with the tarsus very long, nearly twice the propodus; both these joints are very slender, nearly straight, and along their entire inferior margin is a regular series of small hairs; dactylus nearly straight, very acute, more than half the propodus; auxiliary claws very small, about one-fourth the propodus. The legs are sparsely hairy, the hairs longest near the outer extremities of the joints, where they often form a semicircle on the upper side. Color, when living, light salmon or nearly white. Sometimes "irregularly and conspicuously striped across the body and legs with bright purple; body clear white" (*Prudden*). Length 7 millimeters; extent 65 millimeters.

This species may be usually distinguished by its extremely attenuated appearance, which is more marked than in any other species of the genus. The neck varies considerably, and in some specimens is much stouter than in others.

Common off Salem and Gloucester, Mass., and at numerous localities in the Gulf of Maine, off Cape Ann; Jeffrey's Ledge; off Isles of Shoals; off Casco Bay; Bay of Fundy; St. George's Banks; off Cape Sable, N. S.; off Halifax. The observed bathymetric range is from 16 to 115 fathoms. It occurs on all bottoms, but is more frequently observed on muddy bottoms. The females seem to be more common than the males

Specimens examined.

Number.	Locality.	Fathoms.	Bottom.	When collected.	Received from.	Specimens.	Dry. Alc.
						No. and sex.	
4849	Salem, W. N. W. 9 to 11 miles.	33	Sand, mud..	—— —, 1877	U. S. Fish Com.	3♀........	Alc.
4851	....do..................	35	Mud, clay nodules.	—— —, 1877	....do	3♂, 1♀....	Alc.
4852	Salem, W. N. W. 8 to 9 miles	33	Soft mud...	—— —, 1877	....do	1♀........	Alc.
4853	Salem, W. N. W. ½ N. 13 miles.	48	....do	—— —, 1877	....do	1♂, 2♀....	Alc.
4850	Gloucester, N. about 5½ miles.	45	Mud........	—— —, 1877	....do	1♂........	Alc.
4947	Gloucester, N. 10 to 13 miles.	45	... do	—— —, 1878	....do	4♀........	Alc.
4948	Gloucester, N. ¼ W. 6⅓ miles.	45	Soft mud....	—— —, 1878	....do	2♀........	Alc.
5023	Gloucester, N. 5½ to 7 miles.	43	....do	—— —, 1878	....do	1♂........	Alc.
4854	Cape Ann, N. W. 14 miles	90	....do	—— —, 1878	....do	1♀........	Alc.
4860	Cape Ann, N. W. ½ N. 12 to 14 miles.	75	Mud........	—— —, 1877	....do	1..........	Alc.
4861	Cape Ann, N. W. ½ N. 11 miles.	51	Mud, gravel, rock.	—— —, 1877	....do	3♂, 1♀....	Alc.
4941	Cape Ann, N. by W. ¼ W. 4½ miles.	38	Mud........	—— —, 1878	....do	1♀........	Alc.
4943	Cape Ann, N. W. 4 to 5 miles.	38-42	Mud, sand, stone.	—— —, 1878	....do	2♀........	Alc.
4945	Cape Ann, N. W. by N. 6 to 7 miles.	73-75	Soft mud ...	—— —, 1878	....do	2♂, 6♀....	Alc.
4989	Cape Ann, N. W. ½ N. 6 to 7 miles.	54-60	Sand, mud..	—— —, 1878	....do	1♀........	Alc.
4990	Cape Ann, W. by N. ½ N. 4½ to 5½ miles.	57-68	Soft mud, concretions.	—— —, 1878	....do	2♀........	Alc.
5015	Cape Ann, W. N. W. 27 to 31 miles.	90-115	Mud, gravel, stone.	—— —, 1878	....do	1♀........	Alc.
4942	Cape Ann, N. W. ¾ N. 4½ miles.	38	Pebbles, sand	—— —, 1878	....do	1♂, 1♀....	Alc.
4992	Cape Ann, N. 8½ miles ..	32	Rock, stones	—— —, 1878	....do	1♀........	Alc.
4993	Cape Ann, N. ¼ W. 10 miles.	28	Sand, stone .	—— —, 1878	....do	1♂, 2♀....	Alc.
5024	Cape Ann, N. ½ W. 7½ to 8 miles.	32-35	Sand, pebbles	—— —, 1878	....do	1♀........	Alc.
4940	Massachusetts Bay	56	Mud........	—— —, 1873	....do	1♂........	Alc.
4976	Jeffrey's Ledge.........	26	Gravel, stone	—— —, 1873	....do	1♀........	Alc.

Specimens examined.—Continued.

Number.	Locality.	Fathoms.	Bottom.	When collected.	Received from.	Specimens. No. and sex.	Dry. Alc.
4848	Off Isles of Shoals			— —, 1874	U. S. Fish Com.	1♂	Alc.
4846	Cashe's Ledge, N. 6 to 15 miles.	52–90	Rocks	— —, 1873	...do	1♂	Alc.
4863	Off Cape Elizabeth	72		— —, 1873	...do	1♀	Alc.
4847	Casco Bay	64		Aug. 6, 1873	...do	1♂	Alc.
4864	...do			— —, 1873	...do	1♀	Alc.
4977	Manhegan Island, E. 2 miles.	48	Soft mud	— —, 1874	...do	1♂	Alc.
4865	Bay of Fundy			— —, 1872	...do	1♂	Alc.
4855	Cape Sable, N. S., N. W. 27 to 32 miles.	88–90	Sandy mud	— —, 1877	...do	3♂, 2♀	Alc.
4856	Cape Sable, N. S., N. W. 18 to 22 miles.	59	Sand, gravel, stone.	— —, 1877	...do	1♂	Alc.
1352	Latitude 42° 44′, longitude 64° 36′.	60	Gravel, stone, shale.	— —, 1872	...do		Alc.
4857	Narrows at mouth of Bedford Basin, Halifax.	16	Shale, stone.	— —, 1877	...do	2♂	Alc.
4859	Chebucto Light, N. W. by W. 8½ miles.	52	Sand, mud	— —, 1877	...do	3♀	Alc.
4858	Outer harbor, Halifax		Rocks	— —, 1899	...do	1♀	Alc.

Nymphon grossipes (L.) Chr. Fabr.

? *Phalangium marinum* Ström, Söndmör, p. 208, 1762.

? *Phalangium grossipes* Linné, Syst. Nat., ed. xii, i, p. 1027, 1767.

Pycnogonum grossipes O. Fabr., Fauna Grönlandica, p. 229, 1780.

? *Nymphum grossipes* Sabine, Suppl. to the Appendix Capt. Parry's First Voyage, p. 225, 1824.

Nymphon grossipes Chr. Fabr., Ent. Syst., Tom. 4, p. 217, 1794.—Latreille, Hist. Nat des Crust. et des Insect., Tom. vii, p. 333, 1804; Genera Crust. et Insect., Tom. i, p. 143, 1806.—Kröyer, Grönlands Amfipoder, S. 92, 1838 [*teste* Kröyer]; Nat. Tidss., 1ste. Bind, 2det Hæfte, p. 208, 1844; Oken's Isis, Jahrg. 1846, Heft vi, p. 442; in Gaimard's Voy. en Scand., Laponie, etc. Pl. 36, figs. 1*a–h*.—Stimpson, Invertebrata of Grand Manan, p. 38, 1853.—Packard, Mem. Bost. Soc. Nat. Hist., vol. i, p. 295, 1867.—Buchholz, Zweite Deutsche Nordpolfahrt, Crust., p. 396, 1874.—Verrill, Am. Jour. Sci., vol. vii, p. 502, 1874.—Möbius, Die wirbellosen Thiere der Ostsee, p. 153, 1873.—Wilson, Trans. Conn. Acad., vol. v, p. 20, Pl. VII, figs. 1 *a–q*, July, 1878.

Nymphon mixtum Kröyer, Nat. Tidss., 1ste Bind, 2det Hæfte p. 110, 1844; in Gaimard's Voy. en Scand., Laponie, etc., Pl. 35, figs. 2 *a–f*.—Norman, Rept. of the Brit. Assoc. for the Advancement of Sci. for 1868, p. 301.—Buchholz, op. cit., p. 397, 1874.—Lütken, Lists * * * compiled for the Brit. North Pole Exp., p. 164, 1875.—Sars, Archiv für Math. og Naturvidenskab, andet Bind, Tredie Hefte, p. 366, 1877.

Nymphon brevitarse Kröyer, Nat. Tidss., 1ste Bind, 2det Hæfte, p. 115, 1844; in Gaimard's Voy. en Scand., Laponie, etc., Pl. 36, figs. 4 *a–f*—Reinhardt, Nat. Bidrag til en Beskr. af Grönland, p. 38, 1857.—Lütken, Lists compiled for the Brit. North Pole Exp., p. 164, 1875.—[=*Nymphum hirsutum* Kröyer, Grönlands Amfipoder, S. 92, 1838, *teste* Kröyer].

? *Nymphon rubrum* Hodge, Nat. Hist. Trans. Northumb. and Durham, p. 41, Pl. X, fig. 1, (1865, *t.* Zool. Rec.).

? *Nymphon gracile* Leach, et auct.

Plate VI, Figures 32 to 37. Plate VII, Figure 42.

Body slender, smooth. Oculiferous segment variable; in some specimens nearly as short and stout as in *N. Strömii*, in others much longer and very slender. Oculiferous tubercle very prominent, conical, very acute. Eyes black, oval or nearly round. Abdomen small, tapering, often bent upward.

Rostrum large, somewhat variable, but usually shorter than the oculiferous segment, slightly swollen at the extremity.

Antennæ slender, basal joint about as long as the rostrum; chela similar to that of *N. longitarse*, but stouter, the claws shorter, slightly hairy.

Palpi slender, with a few small hairs most numerous on the outer joints; basal joint nearly quadrate, about one-fourth the second; third slightly longer than the first two united; fourth less than half the third; fifth longer, slender, tapering, somewhat variable, being stouter in some specimens than in others.

Accessory legs very slender. In the female they are, on an average, about one-eighth the extent of the legs; in the male about one-sixth. The joints have nearly the same proportions as in *N. longitarse*, but the fourth and fifth joints are longer and still more slender.

Legs long and slender, proportions of the first six joints nearly as in *N. Strömii*. Tarsus extremely variable in length (Pl. VII, figs. 1*b* to 1*g*); in young specimens it is less than half the propodus, while in some large adult specimens it is nearly twice that joint; the propodus is armed, on the inferior margin, with a series of slender, slightly curved spines, which are longest proximally; dactylus about two-thirds the propodus; auxiliary claws less than half the dactylus. The legs are sparsely hairy, the hairs often forming, as in *N. longitarse*, a semicircle on the outer extremities of the joints. Color, when living, light salmon-yellow, the legs often banded with reddish or light purple. Length 10.5 millimeters; extent 90 millimeters.

This species is, in most of its characters, extremely variable. Kröyer's *N. brevitarse* and *N. mixtum* are undoubtedly, I think, forms of *N. grossipes*. The former are young specimens, with a short, thick neck, very short tarsus, and abbreviated rostrum; the latter are those having a long slender neck, and with the tarsus from one and a half to two times the propodus. From the large collection in the Peabody Museum I have formed an almost complete series from extreme forms of *N. brevitarse* to undoubted *N. mixtum*, though in none of the specimens of the latter species is the tarsus quite so long as that figured in the Voy. en Scand., Laponie, etc. The palpi, also, vary considerably with age.

The variation is due in part to age, but is not sexual, since male specimens with egg-masses present the same differences. In some specimens the antennæ are tipped with brown, or jet black; in others they are white. The terminal joint of the legs is sometimes similarly tipped with brown.

The following table gives the relative length of the tarsus and propodus in a series of specimens selected to show the variation. The joints measured are, in all but one or two cases, from the second leg of the right side.

	Propodus. mm.	Tarsus. mm.	Ratio of t. to p.
a (*N. brevitarse*)	0.465	0.249	0.54
b	0.498	0.332	0.65
c	0.930	0.670	0.72
d (*N. grossipes*)	1.094	1.094	1.00
e	.999	1.195	1.20
f	1.062	1.328	1.25
g	1.295	1.693	1.315
h (*N. mixtum*)	1.228	1.892	1.541

In Pl. VI, figs. 33 to 35, the variation of the neck is shown. All the latter specimens are adult males.

This and the preceding species are the commonest of the group. The most southerly locality from which I have seen specimens is Long Island Sound (two young specimens, 50 fathoms, off Race Point Rock, 1874); and the most northerly is Orphan Bank in the Gulf of St. Lawrence, dredged by Mr. Whiteaves in 1873; Dr. Packard has recorded it from Labrador. Taken by the United States Fish Commission off Salem and Gloucester, 19 to 48 fathoms; Gulf of Maine, off Cape Ann, 18 to 90 fathoms; off Isles of Shoals; off Cashe's Ledge; off Cape Elizabeth; Casco Bay, common; St. George's Banks, 50 fathoms; common off Halifax, 16 to 101 fathoms; Bedford Basin, Halifax Harbor, 35 fathoms, soft, oozy, offensive black mud. In depth the observed range is from 12 to 110 fathoms. Like the preceding species, it is found upon nearly all bottoms, but it seems to be less of a muddy bottom species, and is more often taken on rocky or gravelly bottoms.

It seems to me not improbable that Leach's *Nymphon gracile* is identical with *N. grossipes*, though none of the descriptions and figures of that species, which I have seen, suffice to identify it with certainty. The species of *Nymphon* from Northern Europe are in considerable confusion, and stand in need of revision.

Specimens examined.

Number.	Locality.	Fathoms.	Bottom.	When collected.	Received from—	Specimens. No. and sex.	Dry. Alc.
4891	Long Island Sound, W. of Race Point Rock.	50	Rock, shells, gravel.	—— —, 1874	U S Fish Com.	2○	Alc.
4893	Salem, W. N. W. 9 to 11 miles.	33	Sand, mud	—— —, 1877	... do	1♂, 1♀	Alc.
4904	......do	35	Mud, clay	—— —, 1877	....do	1♂, 4♀	Alc.
4905	Salem, W. N. W. 13 miles.	48	Soft mud	—— —, 1877	....do	2♂, 4♀	Alc.
4892	Salem, W. N. W. 5 to 7 miles.	22	Gravel	—— —, 1877	....do	2♂, 3♀	Alc.
4894	......do	20	Rocks	—— —, 1877	....do	1♂, 1♀	Alc.
4895	......do	19–20	Gravel	—— —, 1877	....do	2♂, 1○	Alc.
4897	Salem, W. N. W. 6 to 7 miles.	26	Gravel, stone	—— —, 1877	....do	2♂, 3♀	Alc.
4950	Gloucester, N. 3½ to 4½ miles.	33	Rocks	—— —, 1878	....do	1♂	Alc.

Specimens examined—Continued.

Number.	Locality.	Fathoms.	Bottom.	When collected.	Received from.	Specimens. No. and sex.	Dry. Alc.
5011	Gloucester, N. 3 to 4 miles.	25–26	Sand, gravel, stone.	—— —, 1878	U. S. Fish Com.	1♂	Alc.
4952	Gloucester, N. N. W. 4 to 6 miles	19½	Sand, gravel.	—— —, 1878	....do	2♂, 1♀	Alc.
4987	Gloucester, N. by E 2¼ to 4 miles.	19–23	Sand, gravel, stone.	—— —, 1878	....do	2♂	Alc.
5010	Gloucester, N. to N. by W. 5½ to 7 miles	40–45	Soft brown mud.	—— —, 1878	....do	1	Alc.
4955	Gloucester, N ¼ W. 6½ miles.	45	Mud	—— —, 1878	....do	1♀	Alc.
4912	Cape Ann, N. W. ½ N. 11 miles	50	Mud, gravel, &c.	—— —, 1877	....do	1♂	Alc.
4951	Cape Ann, N. W. ¾ W. 13 miles.	53	Rock to mud.	—— —, 1878	....do	1♀	Alc.
4953	Cape Ann, N. W. 4 to 5 miles.	42	Sand, mud, clay nodules.	—— —, 1878	....do	1♀	Alc.
4954	......do	38	Mud to rock.	—— —, 1878	....do	1♂	Alc.
4956	Cape Ann, W. N. W. 30 to 31 miles	110	Soft brown mud.	—— —, 1878	....do	1♂	Alc.
5009	Cape Ann, N. W. by N. 7 miles	73–75	Soft mud....	—— —, 1878	....do	1	Alc.
4906	Cape Ann, N. W. 14 miles.	90	Mud	—— —, 1878	....do	1♀	Alc.
5002	Cape Ann, N. 8½ miles.	32	Rock, stones.	—— —, 1878	....do	1♂	Alc.
5000	Cape Ann, N. E. 2½ miles.	18	Rough rock..	—— —, 1878	....do	1♂	Alc.
4958	Cape Ann, N. N W. 15 miles	23	Stone, gravel, shells.	—— —, 1878	....do	1♂, 2♀	Alc.
4957	Cape Ann, W. N. W. 29 to 30 miles	85	Gravel, pebbles.	—— —, 1878	....do	2♂	Alc.
4880	Cape Ann, S W. 14 miles.	33	Gravel, stone	—— —, 1873	....do	1♂, 2♀	Alc.
4890	Off Isles of Shoals..	35	Clay, mud, sand.	—— —, 1874	....do	2○	Alc.
4930	Cashe's Ledge, N. 6 to 15 miles.	52–90	Rocks	—— —, 1873	....do	1♂, 1♀	Alc.
4885	Off Cape Elizabeth..	68	....	Aug. 13, 1873	....do	1♂	Alc.
4881	Casco Bay	....	....	—— —, 1873	....do	2♂, 7♀	Alc.
4883	......do	....	....	July 17, 1873	....do	2♂	Alc.
4884	......do	34	....	Aug. 27, 1873	....do	1♀	Alc.
4886	......do	18	....	Aug. 27, 1873	....do	1♀	Alc.
4888	Manhegan Island, N. 8 miles.	64	Mud or sand	—— —, 1873	....do	1♂	Alc.
1308	Latitude 41° 25′, longitude 66° 25′.	50	Sand, shells.	—— —, 1872	....do	1♂	Alc.
4877	Grand Manan, N. B.	....	....	—— —, 1872	....do	7♂, 14♀	Alc.
5003	......do	50–55	....	—— —, 1872	... do	3	Alc.
4866	Eastport, Me	....	....	—— —, 1868	Expedition '68.	3♂, 21♀	Alc.
4870	......do	....	....	—— —, 1870	Expedition '70.	6♂, 23♀	Alc.
4868	Eastport, Me., off Head Harbor.	100	....	—— —, 1870	....do	4♀	Alc.
4872	Eastport, Me., Johnson's Bay.	12	Rocks	Aug. 7, 1872	U. S. Fish Com.	3	Alc.
4875	Eastport Harbor, Me.	20	....do	Aug. 7, 1872	....do	1♂, 1♀	Alc.
4873	......do	60	Mud	Aug. 16, 1872	....do	5♀	Alc.
4879	Eastport, Me., off Cherry Island.	20–25	....	—— —, 1872	....do	1♀	Alc.
4876	Eastport, Me	....	....	—— —, 1872	....do	1♂, 8♀	Alc.
4898	Bedford Basin, Halifax.	35	Soft mud	—— —, 1877	....do	1♀	Alc.
4909	......do	26	....do	—— —, 1877	....do	1♀	Alc.
4903	Chebucto Light, N W. by W. 9 miles	53	Mud, fine sand.	—— —, 1877	....do	1♂	Alc.
4899	Halifax, outer harbor	25	Gravel	—— —, 1877	....do	2♂, 3♀	Alc.
4900	Narrows at mouth of Bedford Basin	16	Stone, shells.	—— —, 1877	....do	1♀	Alc.
4901	Chebucto Light, N. by E. 26 miles	101	Fine sand	—— —, 1877	....do	2♂	Alc.
4902	Halifax, outer harbor	16	... do	—— —, 1877	....do	1♀	Alc.
4908	......do	20	Shingly	—— —, 1877	....do	1♀	Alc.
4910	......do	25	Rocks	—— —, 1877	....do	2♂	Alc.
4911	......do	43	....	—— —, 1877	....do	2♂, 2♀	Alc.
4880	Orphan Bank, Gulf of St. Lawrence.	....	....	—— —, 1873	....do	1♂	Alc.

Nymphon hirtum Fabricius.

Ent. Syst., vol. iv, p. 417, 1794.—Kröyer, Nat. Tidss., 1ste Bind, 2det Haefte, p. 113; Voy. en Scand., Laponie, etc., Pl. 36, figs. 3 *a–g*.—Norman, Rept. of the Brit. Assoc. for the Advancement of Sci. for 1868, p. 301.—Buchholz, Zweite Deutsche Nordpolfahrt, p. 397, 1874.—Miers, Ann. and Mag. Nat. Hist., 4th series, vol. 20, No. 116, pp. 108-9, Pl. IV, fig. 3, 1877.—G. O. Sars, Archiv for Mathematik og Naturvidenskab andet Bind, Tredie Hefte, p. 365, 1877.

? *Nymphon hirsutum* Sabine, Supplement to the Appendix, Capt. Parry's First Voyage, p. 226, 1824.

Nymphon hirtipes Bell, Belcher's Last of the Arctic Voyages, Crust., p. 401, Pl. XXXV, fig. 3, 1855.—Wilson, Trans. Conn. Acad., vol. v, p. 22, Pl. V, figs. 2 and 3; Pl. VI, figs. 2 *a* to 2 *k*, July, 1878.

Nymphon femoratum Leach, Zool. Misc., vol. i, p. 45, Pl. 19, fig. 2, 1814.—Johnston, Mag. Zool. and Bot., vol. 1, p. 380, 1837 (*teste* Hodge).

PLATE VII, FIGURES 38 TO 41.

Body very robust, lateral processes scarcely separated. Oculiferous segment broad and stout, neck very thick. Oculiferous tubercle much elevated, slender, rounded. Eyes ovate, black. Abdomen slender, tapering from the middle toward the base and tip.

Antennæ very hairy, rather stout, basal joint slightly longer than the rostrum; claws of chelæ slender, acute, very strongly curved, when closed crossing each other at a considerable distance from the tips. The spines, with which they are armed, are rather long, slender, and not very closely set; toward the base they become strongly curved or even hook-shaped.

Palpi very stout; basal joint nearly quadrate, half the length of the second; the remaining joints decrease regularly to the last. The appendage is densely hairy; on the outer three joints the hairs are densely plumose.

The accessory legs differ considerably in the sexes. In the female there are three short basal joints, followed by two which are considerably longer, nearly equal, and somewhat clavate; the sixth is about two-thirds the fifth, and the remaining joints become successively smaller to the last, which is acute and claw-like, and armed below with a series of spines. In the male the appendage is larger and stouter, the fifth joint is about twice as long as the corresponding joint in the female, and near its outer extremity it is swollen and furnished on each side with a dense tuft of long hairs; the spines of the outer joints are scarcely denticulated and alike in both sexes.

Legs comparatively stout, often distended with the generative organs; first and third joints about as long as broad; second longer, somewhat clavate, longer in the male than in the female; the three following joints are much longer, the sixth longest; tarsus short, half the propodus, which has, below, a series of slender spines; dactylus about two-thirds the propodus; auxiliary claws very small and slender, about one-fifth the dactylus. All the appendages are thickly covered with coarse hairs, which are most numerous on the outer joints. The body is slightly hairy or nearly naked. Color light dull yellow. Adult specimens are very frequently covered with rubbish, and living Bryozoa, Sponges, Rhizopods, etc., are often attached to them. Length 12 millimeters; extent 73 millimeters.

This species has not before been recorded from our coast, though taken in great numbers off Halifax by the United States Fish Commission in 1877. It occurs on rocky, gravelly, or muddy bottoms, down to 50 fathoms. Sept. 24th, 1877, several hauls made off Halifax in 50 fathoms, muddy bottom, brought them up by hundreds, clinging to the meshes of the trawl-net. A single specimen was dredged off Salem, Mass., in 48 fathoms, soft mud. Many of the specimens had egg-masses. In some of these, young were found in various stages of growth. In the earliest stage observed (Plate VII, figure 41) the body is very large and swollen, without a trace of segmentation. The rostrum is short and directed downward. The five anterior pairs of appendages are developed, the posterior one rudimentary. The basal joint of the antennæ bears a long flagellum.

Specimens examined.

Number.	Locality.	Fathoms.	Bottom.	When collected.	Received from—	Specimens.	Dry. Alc.
						No. and sex.	
	Off Salem, Mass	48	Soft mud ...	—— —, 1877	U. S Fish Com.	1...........	Alc.
4816	Chebucto Light, N. W. by W. 9 miles.	53	Mud, rocks.	—— —, 1877	....do	75 + ♂. ♀....	Alc.
4818	Chebucto Light, N. W. by W. 8½ miles.	52	Sandy mud..	—— —, 1877	....do	150 + ♂ ♀...	Alc.
4823	Chebucto Light, N. 9 miles.	57	Mud, sand, gravel, st.	—— —, 1877	... do	3 ♂, 5 ♀	Alc.
4931	Sambro Light, W. by N. 10 miles.	42	Gravel, rocks	—— —, 1877	... do	1 ♂, 1 ♀.....	Alc.
4827	Sambro Light, W. by N. 9 miles.	42	Fine sand...	—— —, 1877	....do	2 ♀	Alc.
4828	West from last......	42	Sand, rocks.	—— —, 1877	....do	4 ♂, 1 ♀.....	Alc.
4822	Outer Harbor, Halifax.	25	Rocks	—— —, 1877	....do	2 ♂, 1 ♀....	Alc.

It is, unfortunately, a difficult matter to know whether the name *hirtum* should really be applied to this form. It is impossible to determine from Fabricius's very brief description of *N. hirtum* from the "Norwegian Ocean"; and hence most writers, including Kröyer, who first fully described and figured the species, have referred to it as "*Nymphon hirtum* Fabr. ?" Our specimens differ from Kröyer's figures in Gaimard's "Voyages en Scandinavie, Laponie," etc., in several particulars, most notably in the form of the antennæ and proportions of the palpal joints; Kröyer's specimens were from Iceland. G. O. Sars, in a recent paper (Archiv for Mathematik og Naturvidenskab, andet Bind, Tredie Hefte, p. 365, 1877), records, from the same region, a form which he identifies with *N. hirtum* Fabr. and with *N. hirtipes* Bell, but which "*vix* = *N. hirtum* Kröyer." It seems to me probable, under these circumstances, that Kröyer's figures are inaccurate, and that, therefore, the name *N. hirtum* must be restored.

The following table is intended to show the general geographical and bathymetrical distribution of the species described in this paper. To indicate those localities from which I have examined specimens the mark of affirmation (!) is used; in cases where the locality is given on other authority, the + sign is used. A few of the species occur in the deeper waters far to the southward of their ordinary limits; this is indicated by the ± sign.

	Geographical distribution.																					Bathymetrical distribution.		
	Virginia.	Long Island Sound.	Vineyard Sound.	Off Martha's Vineyard.	Off Nantucket.	Massachusetts Bay.	Off Isles of Shoals.	Casco Bay.	Saint George's Banks.	Gulf of Maine.	Bay of Fundy.	Off Cape Sable, Nova Scotia.	Off Halifax.	Off Nova Scotia.	Gulf of St. Lawrence.	Labrador.	Greenland.	Northern Europe.	Mediterranean.	South Pacific.	Kerguelen.	Littoral.	Fathoms.	European.
Pycnogonum littorale (*Ström*) *O. Fabr*		±!			!	!			!	!	!	!			+		+	!	+[1]	?+[2]	+[3]	+	0 — 430	
Tanystylum orbiculare *Wils*	!	!	!	!																			0 — 14	
Achelia spinosa (*St*). *Wils*		±!						!			!											+	0 — 34	
Achelia scabra *Wils*									!	!													23 — 45	
Pseudopallene hispida (*St.*) *Wils*											!												0 — 12	
Pseudopallene discoidea (*Kr.*) *Wils*											!												0 — 20	
Pallene empusa *Wils*		!	!																			+	0 — 3	
Phoxichilidium maxillare *St*						!		!		!	!		!									+	0 — 55	
Anoplodactylus lentus *Wils*		!	!								!												3 — 8	
Anemothea achelioides *Wils*											!													
Nymphon Strömii *Kr*						!	!	!		!	!	!	!		!		+	+					7½ — 115	
Nymphon macrum *Wils*										!				!									85 — 115	
Nymphon longitarse *Kr*						!	!	!	!	!	!	!	!				+	+					16 — 116	[4]220
Nymphon grossipes *Fabr*		±!				!	!	!	!	!	!		!		!	+	+	+					12 — 110	[4]416
Nymphon hirtum *Fabr*						!							!				+	+					25 — 57	[4]299

[1] *Teste* Philippi. [2] *Teste* Nicolet. [3] *Teste* Böhm. [4] *Teste* G. O. Sars.

LIST OF WORKS REFERRED TO IN THIS ARTICLE.

The following list of works upon the Pycnogonida referred to in the synonymy or elsewhere in this paper has been appended to render the references more intelligible and easier of access. The list has no pretensions to being considered a complete or even tolerably full bibliography of the subject, but contains only the titles of such works as have been used in the preparation of this report.

Abilgaard, Petrus Christianus. *In* Müller, Zoologica Danica seu Animalium Daniæ et Norvegiæ rariorum ac minus notorum Descriptiones et Historia. Volumen Tertium, p. 68, Pl. CXIX. Havniæ (Copenhagen), 1788–9.

Bell, Thomas. *In* The Last of the Arctic Voyages, being a Narrative of the Expedition in H. M. S. Assistance, under the Command of Captain Sir Edward Belcher, C. B., in Search of Sir John Franklin, during the years 1852–3–4. Account of the Crustacea by Thomas Bell; Pycnogonidæ, pp. 408–9, Pl. XXV, figs. 3–4. London, 1853.

Böhm, Dr. R. Ueber zwei neue von Herrn Dr. Hilgendorf gesammelte Pycnogoniden. <Sitzungsbericht der Gesellschaft der Naturforschende Freunde zu Berlin, No. 4, pp. 53–60, 1879.

Buchholz, Reinhold. Zweite Deutsche Nordpolfahrt in den Jahren 1868 und 1870 [unter Führung des Kapitän Koldeway], Part VIII, Crustaceen, pp. 396–7. Leipsig, 1874.

Cavanna, G. Riassunto di una Memoria sui Pignogonidi. in Bulletino della Societa Entomologica Italiana, Firenze, VIII, pp. 282–297. [Original memoir, Studi e ricerche sui Pignogonidi, Pte-la, Anatomia e Biologia in Publicazioni del R. Instituto di studi superiori practici e di perfezionamente in Firenze, I, pp. 249–264. 1876.

Claparède, A. René Edouard. Untersuchungen über Anatomie und Entwickelungsgeschichte wirbelloser Thiere an der Küste Normandie angestellt, p. 102, Pl. XVIII, figs. 11–14. Leipsic, 1863.

Dohrn, Anton. Ueber Entwickelung und Bau der Pycnogoniden. <Jenaische Zeitschrift für Medicin und Naturwissenchaft, Band V, p. 138, Tab. V und VI. Jena, 1869.

——— Neue Untersuchungen über Pycnogoniden. Abdruck aus den Mittheilungen der Zoologischen Station zu Neapel, I. Band, I Heft, pp. 28–39. 1878.

Fabricius, Otho. Fauna Grönlandica, p. 233. Copenhagen, 1780.

Fabricius, Joh. Christ. Entomologia Systematica emendata et aucta secundum classes, ordines, genera, species, adjectis synonimis, locis, observationibus, descriptionibus. Tomus IV, pp. 416–7. Hafuiæ (Copenhagen), 1794.

Gaimard, M. Paul. *vide* **Kröyer.**

Goodsir, Sir Harry. Description of a new species of Pycnogon. <Annals and Magazine of Natural History, Vol. XV, No. XCVIII, p. 293. 1845.

Gould, Augustus A. Description of a new species of Crustacean. <Proceedings of the Boston Society of Natural History, Vol. I, pp. 92–3. Boston, 1844.

Harger, Oscar. *vide* **Smith** *and* **Harger.**

Heller, Prof. Camil. Die Crustaceen, Pycnogoniden und Tunicaten der K. K. Österr-Ungar Nordpol-Expedition. Pycnogonida, pp. 16–19, Tafel IV & V. Wien, 1875.

Hesse, M. Observations sur des Crustacés rares ou nouveaux des Cotes de France. Mémoire sur les nouveaux Genres Oicebathes, Uperogcos et Sunaristes. <Annales des Sciences Naturelles, Cinquième Série, Tome VII, pp. 199–216, Pl. 4. Paris, 1867.

——— Mémoire sur des Crustacés rares ou nouveaux des Cotes de France. Description d'un nouveau Crustacé appartenant à l'ordre des *Pycnogonidiens* et formant le genre Oomère, Nob. <Annales des Sciences Naturelles, Cinquième Série, Tome XX, Vingt-quatrième article, pp. 1–18, Pl. 8. Paris, 1874.

Hodge, George. Report on the Pycnogonoidea, in Brady, Reports of Deep Sea Dredging on the Coasts of Northumberland and Durham, 1862–4. <Natural History Transactions of Northumberland and Durham, 1865 [*t.* Zool. Record], pp. 41–2, Pl. X, fig. 1.

——— List of the British Pycnogonida with Descriptions of several new Species. <Annals and Magazine of Natural History, III Series, Vol. XIII, pp. 113–117, Pl. XII and XIII. London, 1864.

Hoek, Dr. P. P. C. Ueber Pycnogoniden. <Niederländisches Archiv für Zoologie, herausgegeben von C. K. Hoffmann, pp. 235–252, Tafel XV und XVI. Leiden, Leipsig, Mai, 1877.

Johnston, George. Miscellanea Zoologica. An attempt to ascertain the British Pycnogonidæ. <Magazine of Zoology and Botany, Vol. I, pp. 368–382, Pl. 13, figs. 1–12. 1837.

Krohn, August. Ueber das Herz und den Blutenlauf in den Pycnogoniden. <Wiegmann's Archiv fur Naturgeschichte, Berlin, 1855, pp. 6–8. Berlin, 1855. Also in Annals of Natural History, Vol. XVI, 1855, pp. 176–7. London, 1855.

Kröyer, Henrik. *In* Grönland's Amfipoder beskrivne af Henrik Kröyer. Oversigt af de grönlandiske Kræbsdyr, ledsæget af nogle zoologisk-geografiske Bemærkninger. <Det Kongelige Danske Videnskabernes Selskabs naturvidenskabelige og mathematiske Afhandlinger, Afh. VII, pp. 312–326. Kiöbenhavn, 1838.

——— Bidrag til Kundskab om Pyknogoniderne eller Söspindlerne. <Naturhistorisk Tidsskrift, Anden Rækkes, förste Bind, pp. 90–139, Pl. I, figs. 1 *a* to 1 *b*. Kjöbenhaven, 1844–5.

——— Beytrag zur Kentniss der Pycnogoniden. <Oken's Isis, 1846, Heft VI, pp. 430–447, Taf. II, figs. 1 *a* to 1 *f*. Leipzig, 1846.

——— *In* **Gaimard, M. Paul,** Voyages en Scandinavie, en Laponie au Spitzberg et aux Féröe, Atlas, Pl. 35 to 39. Paris, 1849.

Lamarck, J. B. P. A. de. Histoire Naturelle des Animaux sans Vertèbres. Les Pycnogonides, Tome V, pp. 100–105, 2nd. Ed. Paris, 1838.

Latreille, P. A. Histoire Naturelle, generale et Particulière des Crustacés et des Insectes. Pycnogonides, Tome VII, pp. 330–3. Paris, 1804.

——— Genera Crustaceorum et Insectorum secundum Ordinem naturalem in Familias disposita, Iconibus Exemplisque plurimis explicata. Pycnogonides, pp. 143–4. Paris, 1806.

——— *In* **Cuvier,** Règne Animal distribué d'après son organisation: Les Arachnides; Des Pycnogonides, pp. 85–88. Paris.

Leach, William Elford. Zoological Miscellany, Vol. I, pp. 33 and 43, Pl. 13 and 19. 1814.

Linné, Carl von. Systema Naturæ per Regna tria Naturæ, secundum classes, ordines, genera, species, cum characteribus, differentiis, synonymis, locis. Ed. 12 reformata, Tomus I. 1766–7.

Lütken, Christian F. Lists of the Fishes, Tunicata, Polyzoa, Crustacea, Annulata, Entozoa, Echinodermata, Anthozoa, Hydrozoa, and Sponges, known from Greenland. Compiled for the use of the British North Polar Expedition. Article XV, The Crustacea of Greenland, pp. 143–165, Appendix. 1875.

Miers, Edward J. Report on the Crustacea collected by the Naturalists of the Arctic Expedition in 1875–76. <Annals and Magazine of Natural History, 4th Series, Vol. XX, July No., 1877, Pycnogonida, pp. 27–29. London, 1877.

——— Zoology of Kerguelen Island: Crustacea; Pycnogonida, pp. 12–15, Pl. XI. October, 1877.

Milne Edwards. Histoire naturelle des Crustacés, comprenant l'Anatomie, la Physiologie et la Classification de ces Animaux. Tome III, pp. 530–37, Pl. 41, figs. 6 and 7. Paris, 1840.

Möbius, Karl. Die wirbellosen Thiere der Ostsee: aus dem Bericht über die Expedition zur physikalisch-chemischen und biologischen Untersuchungen der Ostsee im Sommer 1871 auf S. M. Avisodampfer Pommerania. Kiel, 1873.

Nicolet, H. *In* **Claudio Gay,** Historia fisica y politica de Chile. Paris, 1854.

Norman, Alfred Merle. *In* Last Report on Dredging among the Shetland Isles by J. Gwyn Jeffries, Rev. A. Merle Norman, W. C. M'Intosh and Edward Waller; Part II, on the Crustacea, Tunicata, Polyzoa, Echinodermata, Actinozoa, Hydrozoa, and Porifera. By the Rev. Alfred Merle Norman, M. A. <Report of the British Association for the Advancement of Science for 1868, pp. 247–346. London, 1868.

Packard, Dr. A. S. Observations on the Glacial Phenomena of Labrador and Maine, with a View of the recent invertebrate Fauna of Labrador. <Memoirs of the Boston Society of Natural History, Vol. I. Boston, 1867.

Pallas, P. S. Miscellanea Zoologica Quibus novæ imprimis atque obscuræ Animalium Species describuntur et observationibus iconibusque illustrantur, p. 188, Tab. XIV, figs. 21–23. 1766.

Philippi, R. A. Ueber die Neapolitanischen Pycnogoniden. <Wiegmann's Archiv für Naturgeschichte, Vol. IX, pp. 175–182. Berlin, 1843.

Quatrefages, M. A. de. Études sur les Types Inférieurs de l'Embranchement des Annelés: Memoire sur l'organisation des Pycnogonides. <Annales des Sciences Naturelles, 3ième serie, Tome IV, pp. 69–83. Paris, 1845.

———. Observations générales sur le phlebenterisme: anatomie des Pycnogonides. <Comptes Rendues, Tome 19, pp. 1152–57. Paris, 1844.

——— Recherches sur les Pycnogonides. <Extraits des Procès-Verbaux des séances de la Société Philomatique de Paris. 1844. pp. 86–88.

Reinhardt, J. *In* Naturhistoriske Bidrag til en Beskrivelse af Grönland, af J. Reinhardt, J. C. Schiödte, O. A. L. Mörch, C. F. Lütken, J. Lange, H. Rink, Tillæg No. 2, Fortegnelse over Grönlands Krebsdyr, Annelider og Involdsorme. Kjöbenhavn, 1857.

Sabine, Edward. *In* Supplement to the appendix of Captain Parry's [first] voyage for the discovery of the north-west passage, in the years 1819–20. Crustacea, pp. CCXXVII–CCXXXVIII, Pl. 1-2. London, 1824.

Sars, George Ossian. Prodromus Descriptionis Crustaceorum et Pycnogonidarum, quæ in Expeditione Norvegica anno 1876, observavit G. O. Sars. <Archiv for Mathematik og Naturvidenskab, andet Bind, Tredie Hefte, p. 337. Kristiania, 1877.

Say, Thomas. An account of the Arachnides of the United States. <Journal of the Academy of Natural Sciences of Philadelphia, Vol. II, pp. 102–114. Philadelphia, 1821.

Semper, Carl. Ueber Pycnogoniden und ihre in Hydroïden schmarotzenden Larvenformen. <Arbeit aus dem Zoologischen Zootomischen Institut in Würzburg, 1874, pp. 264–68, Taf. 16 u. 17. Also in Verhandlungen der physikalisch-medicinisch Gesellschaft in Würzburg, VII, 274. 1874.

Slater, Henry H. On a New Genus of Pycnogon and a variety of Pycnogonum littorale from Japan. <Annals and Magazine of Natural History, 5th series, Vol. III, Article XXXII, pp. 281–283, 1879.

Smith, Sidney I. *In* Verrill *and* Smith, Report on the Invertebrate Animals of Vineyard Sound and Adjacent Waters with an Account of the Physical Features of the Region. [Extracted from the report of Professor S. B. Baird, Commissioner of Fish and Fisheries, on the south coast of New England in 1871 and 1872.] Pycnogonidea, p. 544 (repaged, 250). Washington, 1874.

Smith *and* **Harger.** Report on the Dredgings in the Region of St. George's Banks, in 1872. <Transactions of the Connecticut Academy of Arts and Sciences, Vol. III. July, 1874.

Stimpson, William. Synopsis of the Marine Invertebrata of Grand Manan or the Region about the Mouth of the Bay of Fundy, New Brunswick, pp. 37–38. Washington, 1853.

——— Descriptions of new species of Marine Invertebrata from Puget Sound, collected by the Naturalists of the North-west Boundary Commission, A. H. Campbell, Esq., Commissioner. <Proceedings of the Academy of Natural Sciences of Philadelphia, 1864, No. 3, p. 159.

Ström, Hans. Söndmör's Beskrivelse. 1762.

Verrill, A. E. Results of recent Dredging Expeditions on the Coast of New England. No. 3, American Journal of Science and Arts, 3rd Series, Vol. VI, pp. 435–441; No. 4, *l. c.*, vol. VII, pp. 38–46; No. 6, *l. c.*, Vol. VII, pp. 405–414; No. 7, *l. c.*, Vol. VII, pp. 498–505.

———. Results of Dredging Expeditions off the New England Coast in 1874, *l. c.*, Vol. X, pp. 36–43. 1875.

Whiteaves, J. F. Notes on a Deep-sea Dredging Expedition round the Island of Anticosti in the Gulf of St. Lawrence. <Annals and Magazine of Natural History, Nov. 1872, pp. 342–354.

———. Report of a second Deep-sea Dredging Expedition [in 1872] to the Gulf of St. Lawrence, with some Remarks on the Marine Fisheries of the Province of Quebec. [Montreal, 1873?]

EXPLANATION OF PLATES.

PLATE I.

FIG. 1. Pycnogonum littorale, 469, male; dorsal view.

2. The same; ventral view.

3. The same; *a*, accessory leg; *b*, ambulatory leg.

4. Achelia spinosa, 473; *a*, general dorsal view; *b*, antenna; *c*, spine from accessory leg.

PLATE II.

FIG. 5. Pallene empusa, 476; dorsal view.

6. The same; *a*, ventral view of rostrum and part of the oculiferous segment; *c c'*, spines from accessory leg.

7. The same; antenna.

8. Achelia spinosa, 473; terminal joints of leg.

9. Pseudopallene hispida, 478; *a*, general dorsal view; *b*, antenna; *c*, spine from accessory leg.

10. Pseudopallene discoidea, 479; *a*, dorsal view of body; *b*, antenna; *c*, accessory leg of female.

PLATE III.

FIG. 11. Tanystylum orbiculare, 471; *a*, general dorsal view; *b*, terminal joints of leg; *c*, palpus; *d*, accessory leg of female; *f f'*, spines from accessory legs.

12. Phoxichilidium maxillare, 480; dorsal view of body.

13. The same (smaller southern form); terminal joints of leg.

14. The same (larger form, from Eastport).

15. The same; *a*, accessory leg; *b*, ova; *c*, antenna.

16. Anoplodactylus leatus, 482; dorsal view of body.

17. The same; antenna.

18. The same; *a*, terminal joints of leg; *b*, accessory leg.

PLATE IV.

FIG. 19. Ammothea acheliodes, 484; dorsal view of body; *r*, rostrum; *a*, antenna; *b*, palpus; *d*, abdomen; *e*, oculiferous tubercle; *s'*, etc., lateral processes; *l*, legs; *b* (smaller figure), accessory leg.

20. The same; *a*, terminal joints of leg; *b*, antenna; *c*, palpus.

21. Nymphon macrum, 487; dorsal view of oculiferous segment; terminal joints of leg; palpus.

22. The same; *a*, antenna; *b*, spines from fixed claw; *c*, spines from movable claw.

23. The same; accessory legs of male and female (both are enlarged to the same amount).

PLATE V.

FIG. 24. Nymphon Strömii, 485; lateral view, natural size.
25. The same; lateral view of body.
26. The same; dorsal view, natural size.
27. The same; *a*, accessory leg; *b*, spine from accessory leg.
28. The same; antenna.

PLATE VI.

FIG. 29. Nymphon Strömii, 485; dorsal view of body.
30. Nymphon longitarse; *a*, dorsal view of body; *b*, terminal joints of leg; *c*, *c'*, *c''*, spines from accessory leg.
31. The same; antenna.
32. Nymphon grossipes, 491; *a*, dorsal view of body; *b*, lateral view of body.
33 to 36. The same; series to show variation in oculiferous segment.
37. The same; series to show variation in length of propodus.

PLATE VII.

FIG. 38. Nymphon hirtum, 495; dorsal view; *r*, rostrum; *a*, antenna; *b*, palpus; *c*, accessory leg; *d*, abdomen; *l*, leg.
39. The same; *a*, fifth joint of accessory leg of male; *b*, corresponding joint of female.
40. The same; antenna.
41. The same; recently hatched larva.
42. Nymphon grossipes; antenna.

INDEX.

PLATE I.

FIG. 1. Pycnogonum littorale, 469, male; dorsal view.
2. The same; ventral view.
3. The same; *a*, accessory leg; *b*, ambulatory leg.
4. Achelia spinosa, 473; *a*, general dorsal view; *b*, antenna; *c*, spine from accessory leg.

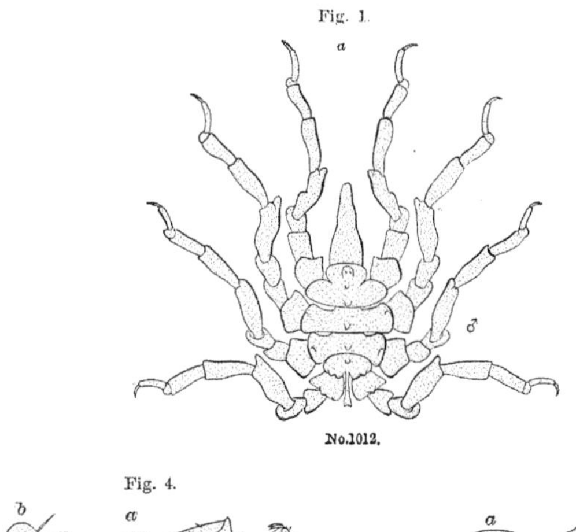
Fig. 1.
a
♂
No.1012.

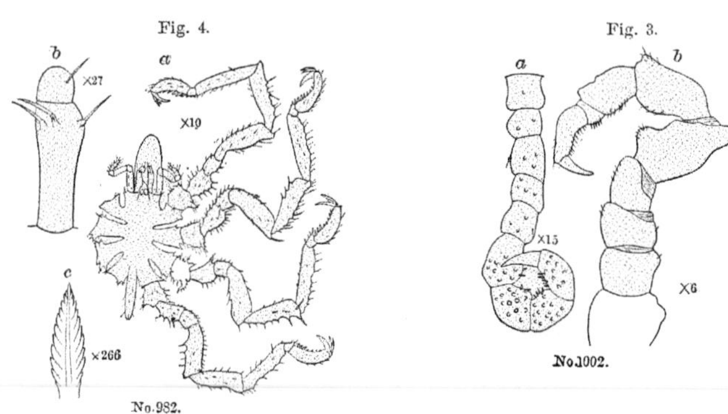
Fig. 4.
b
×27
a
×19
c
×266
No.982.
Fig. 3.
a
b
×15
×6
No.1002.

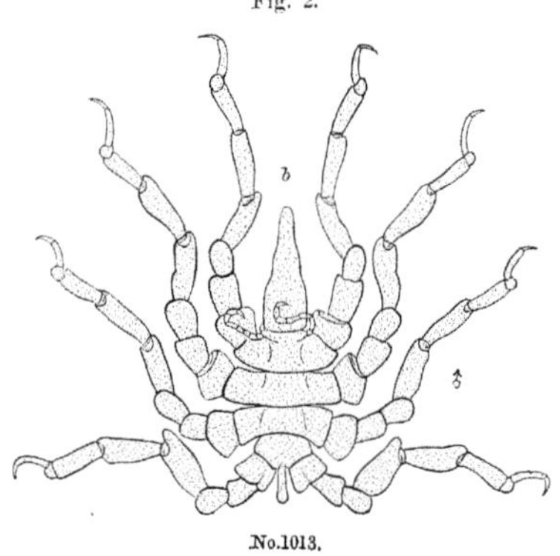
Fig. 2.
b
♂
No.1013.

PLATE II.

FIG. 5. Pallene empusa, 476; dorsal view.

6. The same; *a*, ventral view of rostrum and part of the oculiferous segment; *c c'*, spines from accessory leg.

7. The same; antenna.

8. Achelia spinosa, 473; terminal joints of leg.

9. Pseudopallene hispida, 478; *a*, general dorsal view; *b*, antenna; *c*, spine from accessory leg.

10. Pseudopallene discoidea, 479; *a*, dorsal view of body; *b*, antenna; *c*, accessory leg of female.

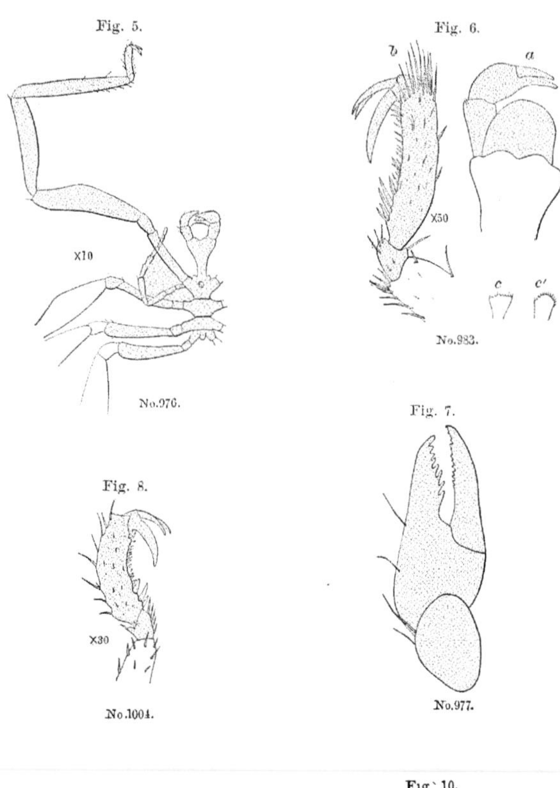
Fig. 5.
X10
No.976.
Fig. 6.
b
a
X50
c
c'
No.983.
Fig. 8.
X30
No.1004.
Fig. 7.
No.977.

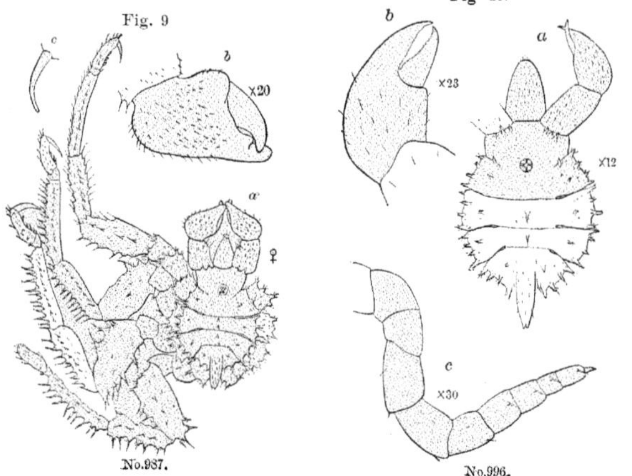
Fig. 9
c
b
X20
a
♀
No.987.
Fig. 10.
b
a
X23
X12
c
X30
No.996.

PLATE III.

FIG. 11. Tanystylum orbiculare, 471; *a*, general dorsal view; *b*, terminal joints of leg; *c*, palpus; *d*, accessory leg of female; *f f'*, spines from accessory legs.

12. Phoxichilidium maxillare, 480; dorsal view of body.

13. The same (smaller southern form); terminal joints of leg.

14. The same (larger form, from Eastport).

15. The same; *a*, accessory leg; *b*, ova; *c*, antenna.

16. Anoplodactylus leatus, 482; dorsal view of body.

17. The same; antenna.

18. The same; *a*, terminal joints of leg; *b*, accessory leg.

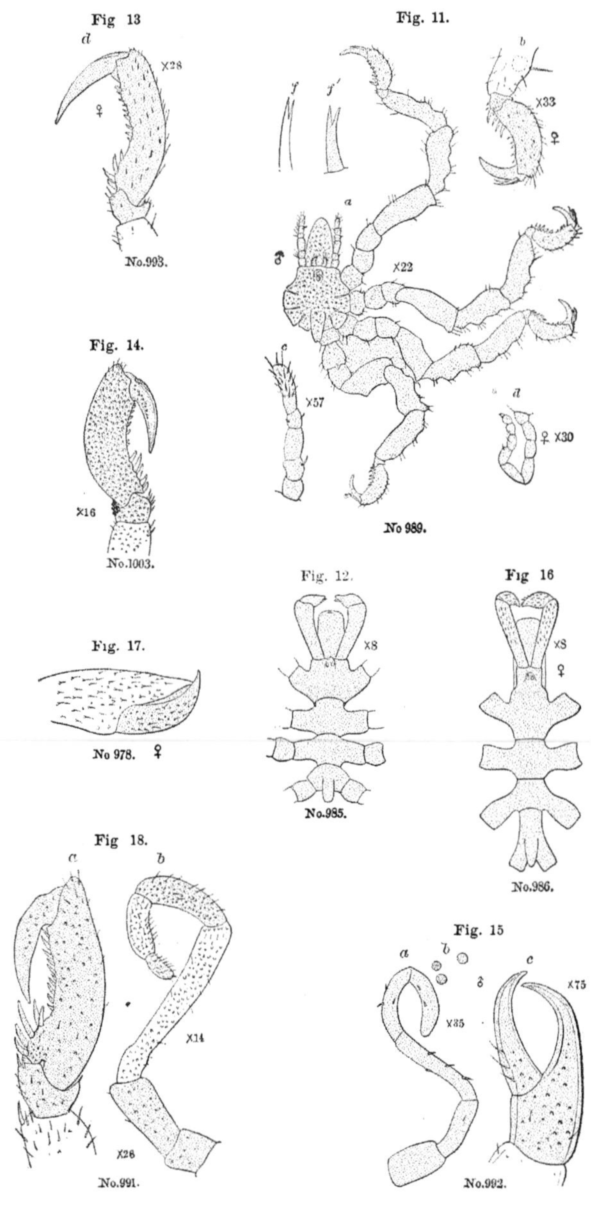
Fig 13
d
X28
No.993.
Fig. 11.
b
X33
a
X22
c
X57
d
X30
No 989.
Fig. 14.
X16
No.1003.
Fig. 12.
X8
No.985.
Fig 16
X8
No.986.
Fig. 17.
No 978.
Fig 18.
a
b
X14
X26
No.991.
Fig. 15
a
b
c
X75
X35
No.992.

PLATE IV.

FIG. 19. Ammothea acheliodes, 484; dorsal view of body; *r*, rostrum; *a*, antenna; *b*, palpus; *d*, abdomen; *e*, oculiferous tubercle; *s'*, etc., lateral processes; *l*, legs; *b* (smaller figure), accessory leg.

20. The same; *a*, terminal joints of leg; *b*, antenna; *c*, palpus.

21. Nymphon macrum, 487; dorsal view of oculiferous segment; terminal joints of leg; palpus.

22. The same; *a*, antenna; *b*, spines from fixed claw; *c*, spines from movable claw.

23. The same; accessory legs of male and female (both are enlarged to the same amount).

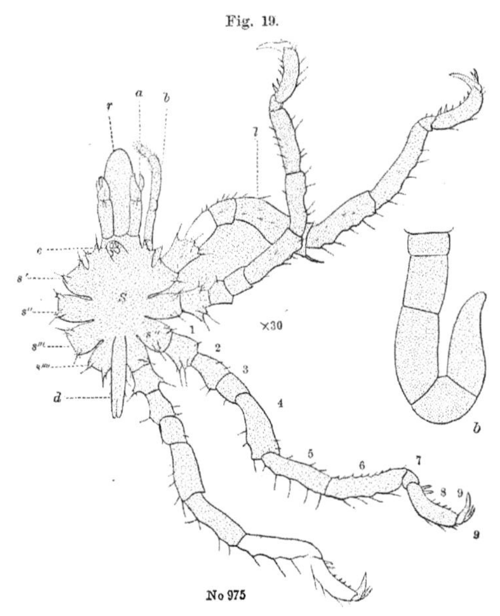
Fig. 19.
r
a
b
7
c
s'
s''
s'''
s''''
S
d
1
2
3
4
5
6
7
8
9
9
×30
b
No 975

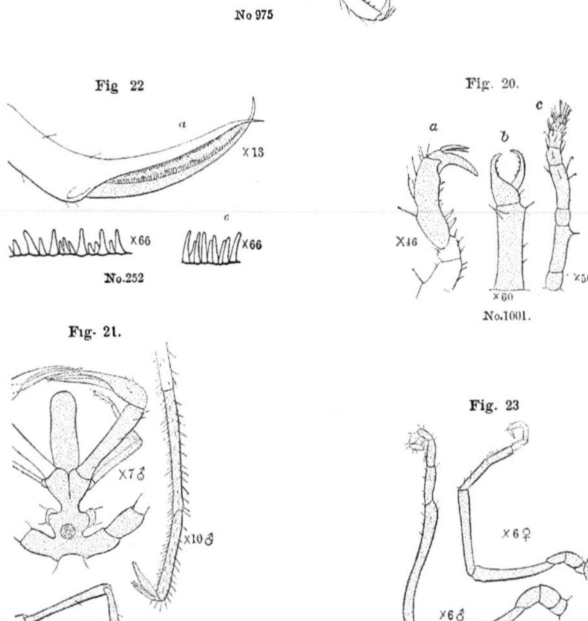
Fig 22
a
X13
c
X66
X66
No.252
Fig. 20.
a
b
c
X46
×60
×50
No.1001.
Fig. 21.
X7♂
X10♂
X10♂
No. 253
Fig. 23
X6♀
X6♂
No. 254

PLATE V.

FIG. 24. Nymphon Strömii, 485; lateral view, natural size.
25. The same; lateral view of body.
26. The same; dorsal view, natural size.
27. The same; *a*, accessory leg; *b*, spine from accessory leg.
28. The same; antenna.

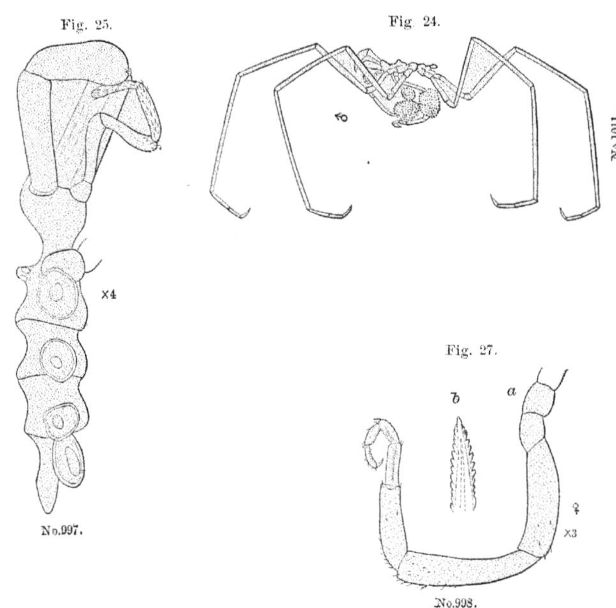
Fig. 25.
×4
No.997.
Fig. 24.
♂
No.1011.
Fig. 27.
b
a
♀
×3
No.998.

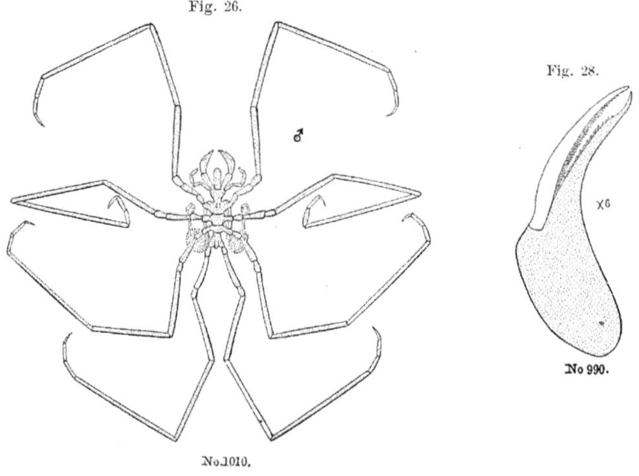
Fig. 26.
♂
No.1010.
Fig. 28.
×6
No 990.

PLATE VI.

Fig. 29. Nymphon Strömii, ·485; dorsal view of body.

30. Nymphon longitarse; *a*, dorsal view of body; *b*, terminal joints of leg; *c*, *c'*, *c''*, spines from accessory leg.

31. The same; antenna.

32. Nymphon grossipes, 491; *a*, dorsal view of body; *b*, lateral view of body.

33 to 36. The same; series to show variation in oculiferous segment.

37. The same; series to show variation in length of propodus.

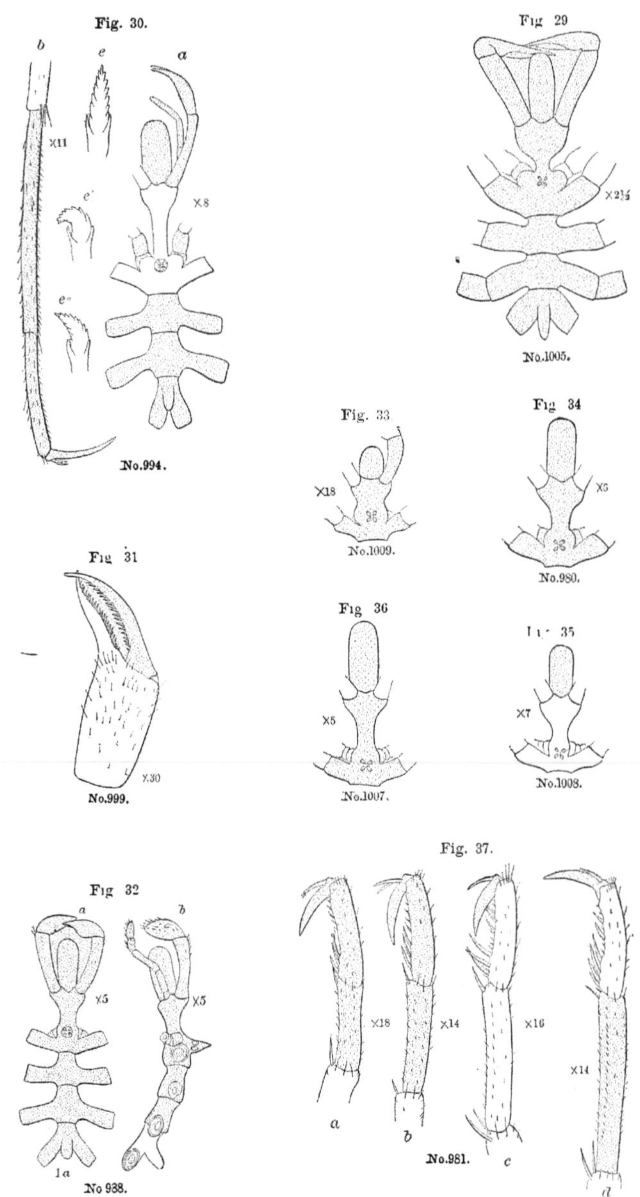
Fig. 30.
b
e
a
X11
e'
X8
e''
No.994.
Fig 29
X2½
No.1005.
Fig. 33
X18
No.1009.
Fig 34
X6
No.980.
Fig 31
X30
No.999.
Fig 36
X5
No.1007.
X7
No.1008.
Fig. 37.
Fig 32
a
b
X5
X5
1a
No 938.
X18
X14
X16
X14
a
b
c
d
No.981.

PLATE VII.

FIG. 38. Nymphon hirtum, 495; dorsal view; *r*, rostrum; *a*, antenna; *b*, palpus; *c*, accessory leg; *d*, abdomen; *l*, leg.

39. The same; *a*, fifth joint of accessory leg of male; *b*, corresponding joint of female.

40. The same; antenna.

41. The same; recently hatched larva.

42. Nymphon grossipes; antenna.

Fig. 39.

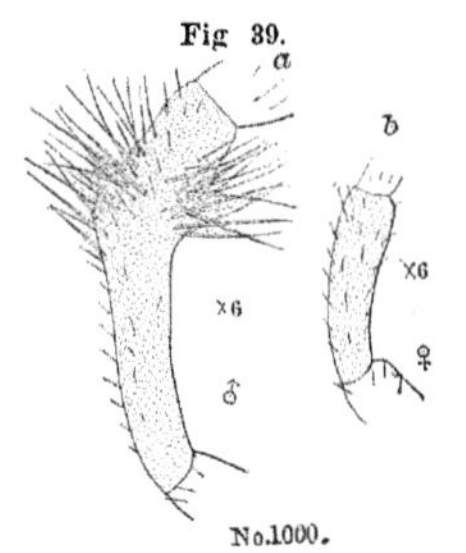

No. 1000.

Fig. 40.

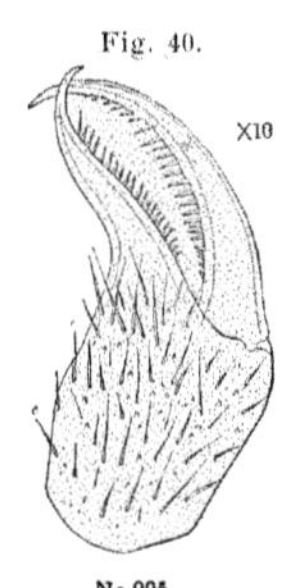

No 995.

Fig. 38.

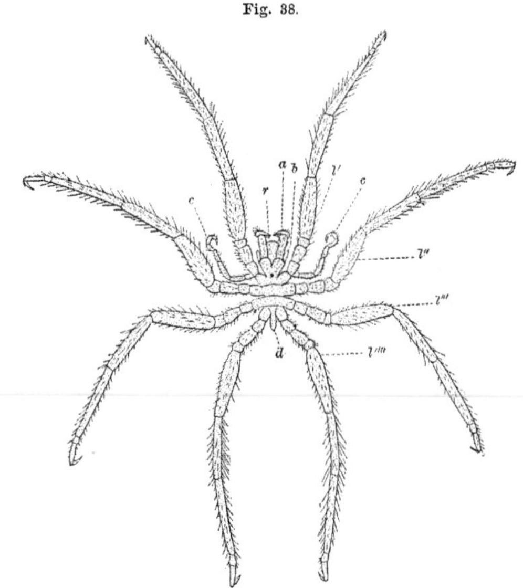

No 974.

Fig. 41.

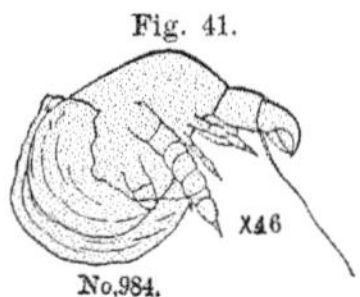

No. 984.

Fig. 42.

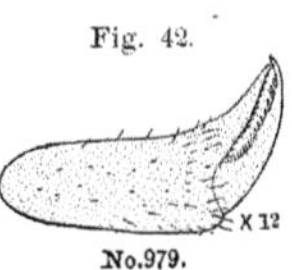

No. 979.

APPENDIX F.

THE PROPAGATION OF FOOD-FISHES.

GENERAL CONSIDERATIONS.

XVI.—THE ENEMIES OF FISH.*

By Baron de la Valette St. George,

Professor of Anatomy and Director of the Anatomical Institute at Bonn.

Read at a meeting of the German Fishery Association, Berlin, March 31, 1879.

[From Circular No. 3 of the German Fishery Association, Berlin, May 14, 1879.]

[Translated by Herman Jacobsen]

War is the watch-word of the whole of organic nature; there is a constant war of all organisms against outward unfavorable circumstances, and there is constant war among the different individuals. The seed-grain which falls into the ground, the worm crawling on the earth, the butterfly hovering over the flower, the eagle soaring high among the clouds, they all have their enemies—outward enemies threatening their existence, and inward enemies eating their life and strength.

Even fish, which claim our special attention, are by no means permitted to spend their life in peace. Plants and animals endanger their very life, and when they have been fortunate enough to escape these, man comes and seeks to catch and destroy them with numberless arts and tricks.

Confined to a special sphere of life, the water, they frequently do not find in it the necessary conditions of existence. In their very cradle, so to speak, that is, in the egg, the tender germs, scarcely awakened to life, are threatened by a dangerous enemy belonging to the lowest grades of the vegetable kingdom. This is the much-dreaded *Saprolegnia ferax;* in an incredibly short time its long threads envelop the egg, choke it, and destroy it.

The best preventive is an ample and continuous supply of cold water of a temperature of about zero, a dim light, and the immediate removal of spoiled eggs. Using a brush only destroys the spurs or threads of the *Saprolegnia* and consumes too much time.

These parasitical plants may prove dangerous even to older fish, for I have observed them on full-grown trout. But, as Dr. *Wittmark* says in his excellent treatise on the enemies of fish, the *propter hoc* and *post hoc* should be well distinguished. I believe that such fungous formations are only found in fish which are worn-out or have been weakened by sickness, and that in such cases it accelerates their death. It is well known to all pisciculturists how important it is to keep all ponds

* *Ueber die Feinde der Fische*, Vortrag des Herrn Freiherrn von la Valette St. George.

or vessels scrupulously clean, and especially keep away all decaying animal matter.

The higher algæ and aquatic plants of every kind may prove hurtful to the fish by limiting the extent of water and hindering the free movement of the fish. This is also the case with the so-called "water plague," *Wasserpest*, *Elodea canadensis*, which, however, does not deserve this name, as it contains much food and develops a great deal of oxygen.

Among the protozoans which form the connecting link between vegetable and animal life we find a small but very dangerous enemy of the fish, namely, the *Psorospermia*. These are round or oval bodies, often possessing a tail, with an internal cellular formation, measuring about 0.005‴, which were first discovered in 1841 by *Johannes Müller* in the socket of the eye of the pike and in small pimples on the skin of the perch, the stickle-back, and several fish of the cyprinoid family. They form the contents of small capsules, measuring $\frac{1}{5}$ to $\frac{1}{2}$‴, which are imbedded in the membranes. They have also been found in the bladder of pike and codfish. Recent investigations have thrown some light on the origin of these beings. They are products by separation of the so-called "Gregarines," which develop an amœba without any kernel, which again changes to a gregarine. *Lieberkühn* has observed the development of the psorosperm into an amœba, and *E. van Beneden* the change from the amœba to the "gregarine," the one in the pike and the other in the lobster. So far it has not been ascertained with absolute certainty in what way the fish are affected by them.

It is certain that these parasites occasion the destruction of the tissue which surrounds them by producing festering sores, and that fish infected by them must gradually die.

Certain formations of a higher group—the *infusoria*—have also recently been accused of being enemies of fish; some of them, such as the *Opalina ranarum*, in the intestinal tube of the frog, and the *Trichodina pediculus*, have long been known as internal and external parasites. They can get in under the outer skin and destroy it. *Livingston Stone* recommends the transfer for a short time of fish affected in this way to salt water. It is said that among the mollusks the *Tichagonia polymorpha* does not disdain the spawn of fish.

We must now turn to a group of animals which has a very bad reputation, and which, belonging to the worms, are comprised in the family of intestinal worms or helminths. This group sends a whole army of animals into the field, some only visible through the microscope, others measuring inches and even yards, and often possessing terrible weapons; they live and find their food in the abdominal cavity, the intestines, muscles, gills, and skin of fish.

The eel contains no less than 25 different kinds of these parasites, the perch 23, the pike 21, the salmon 16, the trout 15, and the carp 12.

The intestinal worms are divided into four classes, viz, *Cestodes*, *Trematodes*, *Acanthocephala*, and *Nematodes*.

The fecundity of some of these worms is considerably increased by the so-called "change of generation," that is, the interpolation of generation without sexual connection between the regular generation.

From one of the innumerable eggs of the tape-worm, for instance, there develops an embryo armed with six little hooks, which finds its way through the animal tissues, settles somewhere, and develops into a tape-worm. In the beginning it is inclosed in a bladder, and in this state is known as the "bladder-worm." Generally not until it has reached the stomach or intestinal tube of some other animal, does the head get free and develop the different joints of the tape-worm, in which eggs develop in the regular manner.

In another kind of the helminths, the *Trematodes*, we also find this peculiar manner of propagation. From the embryo worm-like animals are developed *sporocysts* or *redia*, which again develop little worms with tails, the *cercaria*. These very lively little animals, which frequently have bristles on the head, envelop themselves in a capsule and throw off their organs of motion.

If in this condition they enter the stomach of that animal which is to be their definite place of abode, the cyst, as I have shown by experiments years ago, is digested, and its contents becomes a fully-matured animal of the *Trematodes* kind. Before they get so far they may, however, pass through several different animals.

Tapeworms are very frequently found in fish, often in an entirely undeveloped condition, which shows that they were first eaten by some other animal along with the animal in which they lived.

Among these must be counted the *Ligula*, which is frequently found in the abdominal cavity of our fresh-water fish, such as the bleak, blay, crucian, salmon, trout, pike, and perch. In some parts of Italy this worm is called "Macaroni piatti," and is considered a great delicacy. In storks, herons, gulls, and wild ducks it is found in its more developed form. According to *Van Beneden* it only gets into these birds accidentally.

Very similar to the *Ligula* is the *Schistocephalus*, which often fills the abdominal cavity of the stickle-backs to such an extent as almost to make them burst. When 25 years ago I pursued ichthyological studies here in Berlin, under the instruction of my venerable teacher, Professor *Peters*, I often fed crows and ducks with these worms. These experiments were made in my student's quarters in the third story, and were therefore attended with considerable difficulties. *Von Willemoes-Suhm* has pursued his experiments in a reversed manner, and has raised the embryos with six little hooks from the eggs of the *Ligula* taken from a diver, and from those of the *Schistocephalus* taken from a gull. Among these undeveloped forms we must also mention the *Scolex polymorphus*, which is found in many salt-water fish, and comprises different stages of development.

A very simply-formed tapeworm is found in the intestinal tube of nearly every fish of the genus *Cyprinus;* it is called the *Caryophyllaeus mutabilis.* It has only one joint, closely connected with the head, which develops in the same fish, and forms the connecting link with a long and varied series of maritime forms, which in their youth live in osseous fish, and with these migrate into rays and sharks, where they reach maturity. These are the *Tetraphyllidæ* of *Van Beneden*, which have four sucking disks which are either unarmed or have hooks or stings. To the former belong the *Echinobothrium*, *Phyllobothrium* and *Anthobothrium;* to the latter, the *Acanthobothrium*, *Calliobothrium*, and *Onchobothrium.* The *Echinobothrium* found in the ray has only two sucking disks, but two sharp stings on the forehead, and a neck full of long bristles. Next to this comes the *Tetrarhyncus*, which has four hooked trunks. When quite young it is found in plaice, and when fully grown in rays and sharks. Nearly related to this one is the *Triænophorus nodulocus* with two pairs of three-pronged hooks, which in its undeveloped condition is found in the liver of fish of the genus *Cyprinus*, and when fully matured in the intestinal tube of fish-of-prey.

The *Cestodes* found in fresh-water fish are not so numerous and varied as those of the salt-water fish. Of the latter there may still be mentioned the *Bothriocephalus punctatus* of the plaice, and the *Bothriocephalus rugosus* of the codfish; and of the former, the *Bothriocephalus proboscideus* of the salmon, the *Tænia ocellata* of the perch, the *Tænia osculata* of the "Wels," *Silurus glanis*, and the *Tænia torulosa* of the "Orf," the *Tænia longicollis* of the salmonoids, mostly confined in capsules, and the *Tænia macrocephala* of the eel.

The *Trematodes* may be subdivided into a number of families according to the number and position of their sucking disks. In this way we may distinguish the *Monostoma*, *Distoma*, *Tristoma*, *Polystoma*, *Holostoma*, and *Amphistoma.* The sucking disks of the lower extremity are sometimes furnished with hooks as, for example, in the *Gyrodactylus.*

The last of this kind is the *Diporpa*, which in the middle grows together with another individual, and then forms a peculiar twin-animal, the *Diplozoon paradoxum.* The *monostoma* are rarely found in fish. They are found in the intestinal tube of the stickle-back (*M. carryophyllinum*), and the barbel (*M. cochleariforme*), in the gills of the "brachse" (*M. praemorsum*), and in capsules in the small "Maràna" (*M. Maraenulæ*). *Von Nordmann* found an incredible quantity of a youthful formation (*Diplostomum*) of the *Holostoma* living in aquatic birds, in the eye of the perch, the burbot, fish of the genus *Cyprinus*, and in the last mentioned also in the skin. In the lens of the eye of the burbot, 290 such little animals were counted, whilst the vitreous humor contained about half that number. This must of course make the fish more or less blind.

The *Distoma* are very common in fish. Thus the perch has five different kinds, and the eel ten. In our fish the most frequent are the *Distomum globiporum*, the *D. tereticolle* of the burbot, pike, salmon, and trout, the *D. nodulosum* of the perch, the *D. torulosum* of the "Wels," the *D. ferruginosum* of the barbel, the *D. macrobothrium* and *tectum* of the smelt, the *D. laureatum* and *varicum* of the "Aesche", the *D. ocreatum* of the herring, salmon, and May-fish, which, when young, lives quite free, and is a parasite on larvæ of worms and small crustaceans.

Of the *Polystoma* I will only mention the *Octobothrium* of the May-fish, which lives in the gills like the *Diplozoon* of the cyprinoids, the *Gyrodactylus* and *Dactylogyrus*.

The *Acanthocephala* are round, tube-formed worms, without mouth and intestinal tube, whilst at the head-end they have a trunk with hooks. Their embryos have smaller hooks, with which they pierce the intestinal tube of the animals in which they live, principally crustaceans, then wrap themselves up in a capsule, and in that state are transferred to other animals, fish, birds, or whales.

We thus find the *Echinorhyncus proteus* when young in small crustaceans, and when more developed in the perch, the "Wels," the carp, the salmonoids, and other fish. It is likewise found in many salt-water fish, as the codfish and the plaice; whilst it does not occur in the rays and sharks.*

Different from the *Acanthocephala* is the family of the *Nematodes*, numbering upwards of 1,200 kinds, distinguished chiefly by a more or less developed organ of digestion. These *Nematodes* are very prolific, and exhibit many peculiar phases of development; a change of generation does not, however, seem to occur with them, at any rate it has so far not been observed. It has been noticed, however, that parasitical hermaphrodites have been produced by free individuals of opposite sexes. There may also possibly be a development of the egg without impregnation.

The *Nematodes* are very frequent in fish, sometimes half developed in capsules, and sometimes fully grown.

We thus find the *Gordius aquaticus*, a very long (1 meter) nematode, living in the water, inclosed in land and water insects and their larvæ, as well as in minnows and loaches; and the *Cucullanus* when young in small crustaceans, and when fully grown in the perch and eel. In the swimming-bladder of the salmonoids we find the *Ancyranthus;* in the stomach of the eel, the *Filaria denticulata;* in fish of the *Cyprinus* kind, the *Trichosoma tomentosum;* and in the plaice, the *Heterakis foveola*. The large genus *Ascaris* has also many representatives in fish, for example, *Ascaris truncatula*, in the perch; *A. gasterostei*, in the stickle-back; *A. clavata*, in the codfish, trout, and salmon; *A. mucronata*, in the burbot and pike; *A. collaris*, in the plaice; *A. siluri*, in the "Wels"; *A. acus*, in the carp, pike, and trout; *A. dentata*, in the barbel; *A. hirsuta*, in the smelt; *A. obtusocauda*, in fish of the *Coregonus* kind; *A. cristata*, in

* As far as can be judged from the very incomplete description, the unusual mortality of the crawfish noticed in several places ("*Deutsche Fischerei Zeitung*," 1879, p. 62) might be traced to the *Echinorhyncus*, perhaps *E. polymorphus Br.*, *E. Milarius, Zenker*, *E. Astaci fluvialis v. Siebold.* When young they live in small crustaceans, and when more developed in aquatic birds. These latter would, therefore, transmit the infection.

the pike; *A. adunca*, in the "May-fish"; *A. labiata*, in the eel; and *A. constricta*, in the sturgeon.

Diesing mentions a *Trichina cyprinorum*, but all attempts to develop *trichina* in fish have so far been unsuccessful. Quite recently I have made experiments with goldfish, feeding them with meat which contained *trichina*, but the trichina passed through the intestinal tube. The "fish trichina" which are from time to time spoken of in the newspapers are, therefore, probably myths.

A very dangerous and troublesome parasite is the "fish-leech," which lives on the skin and the gills of fish, often in such numbers as to torment the poor fish. We find the *Piscicola geometra* on fish of the genus *Cyprinus*, the salmonoids and the pike, the *P. respirans* on the barbel, and the *P. fasciata* on the "Wels." The *Branchiobdella* lives on crawfish, the *Histriobdella* on lobsters, while others live on salt-water fish and mollusks.

Also among the crustaceans we find a number of parasites which live on the blood of fish, the so-called "fish-louse," remarkable for a peculiar regressive metamorphosis during their development.

The young are very lively and resemble small crustaceans, but when they have reached their permanent place of sojourn, only those organs remain which are necessary for feeding and propagating.

Of the very large number of these parasites, I only mention the *Ergasilus Sieboldii* on the carp and pike, the *Lamprogena pulchella* on the "orpe," the *Lérnæocera* on the carp, the pike, and the codfish, the *Achtheres percarum* on the perch, the *Tracheliastes polycopus* on the barbel, and the *Argulus foliaceus* on the carp; the last mentioned of which I had frequent occasion to observe in my goldfish ponds. These parasites seem not to do special harm; still I found it advisable to counteract their spreading by draining the ponds from time to time and by removing the parasites.

I do not believe that more developed crustaceans, which form an excellent food for trout, can hurt the fish-eggs, but there are quite a number, such as the *Anceus*, *Cymothoa*, &c., which live as parasites on fish.

Among the *insects* the water-beetles and their larvæ, principally the *Dytiscus*, *Acilius*, and *Calymbetes*, are justly considered enemies of fish. I have seen how a *Dytiscus marginalis* killed a Mexican salamander six inches in length, for whom it was to serve as food, by biting it in the neck. The larvæ of the dragon-fly are also said to hurt the fish.

In passing to the *vertebrates*, we find that the fish themselves are dangerous enemies of their own kind. Not only will it happen that a male trout with an utter lack of gallantry will eat the eggs instead of impregnating them, but many fish, which are considered harmless vegetarians, actually turn cannibals, and, especially at a more advanced age, devour the eggs and young fish. This is the reason why there are so frequently no young fish in goldfish-ponds containing strong and healthy male and female fish, whilst those eggs—few in number—which by the water flowing through the pond are accidentally carried beyond the limits of the pond develop very successfully.

Arrangements may be made accordingly. To take out the old fish after spawning is easier said than done. Although the salmonoids as a general rule, that is, as long as they find a sufficient quantity of insects, crustaceans, worms and snails, are not very dangerous depredators, they love to eat the spawn of fish. Nearly all fish will be guilty of the same offense when tempted by such delicate morsels. Trout which have acquired a taste for such food may prove very dangerous to their younger comrades. The fish-of-prey, properly so called, the pike, the barbel, "Wels," burbot, and eel are well known as such, and they are caught wherever this is possible.

Among the *amphibia*, the water-salamander, the water-toad, "Unke," and the frog should be kept away from the fish-ponds as much as possible, because they may inflict great damage to eggs and young fish.

In order not to pass the *reptiles*, we will also mention among the enemies of fish the crocodile and the water-snake. I do not know whether our common *Coluber natrix* likes fish as well as it does frogs and tritons.

All *aquatic birds* are born enemies of fish. The water-fowl seems to be the most harmless of all, although it cannot be trusted entirely. The stork is decidedly worse than his reputation. But the most dangerous enemies are the herons, which, especially during moonlight nights, do great damage. Among the birds-of-prey the *Circus rufus*, the *Pandion haliætus*, and the *Haliætus albicilla* are the most dangerous and powerful enemies of fish. A most dangerous enemy is the kingfisher, *Alcedo ispida*. The crow, *Corvus corona*, also likes fish, and is remarkably skillful in catching them. The wagtail, *Motacilla flava*, and *alba*, and the water-ousel, *Cinclus aquaticus*, are likewise fond of fish-eggs and young fish.

Among the *mammals*, the enemies of fish are not so numerous, but the few are all the more dangerous. Of the *Cetacea*, we must mention the fin-fish, the narwhal, and the dolphin, and of the *Phocidæ*, the seal. The water-rat, *Hypodæus amphibius*, and the brown rat, *Mus decumanus*, should be kept away from fish-ponds as much as possible. Although the first-mentioned lives on reeds, it does damage by undermining the dikes, whilst the last-mentioned most assuredly hurts the fish. *Brehm* has given detailed reports of the great damage done to fish by the *Crossopus fodiens*, which eats the eyes and brains of living fish, even those of considerable size. The domestic cat does not disdain fish-food, and I have often watched cats lying in wait for fish on the banks of ponds. The best known and most dangerous enemies of fish are the otters, the *Vison lutreola* and *americanus*, the *Enhydris lutris*, and the *Lutra vulgaris*. The reputation of the last-mentioned kind is so bad, that I need not say any more about it, but only recommend the excellent methods of catching them introduced by *Von der*

Borne. Their near relatives, the weasel, *Mustela vulgaris*, *M. erminea*, and the polecat, *M. putorius*, cannot be trusted either.

It must also be mentioned that some members of the *bear* family are fond not only of honey, but also of fish.

We have thus quickly passed through the whole animal kingdom, and have arrived at the last and most dangerous, because most intelligent, enemy of fish, namely, *man.*

Ignorance and covetousness have in many parts of the country reduced the number of fish to a minimum, and nothing but efficient fishery-laws and rational pisciculture can remedy the evil. The *German Fishery-Law* and the *German Fishery Association* have opened out a new era for the German fisheries. On this foundation, which has been laid by the best men of our country, we must all build according to our ability. As our revered president has remarked, "there should be a hatching-box near every water-mill." Only united activity will bring us nearer to our object—to raise the general welfare of our nation.

XVII.—IS SAWDUST AS SERIOUS AN OBSTACLE TO THE ASCENT OF SALMON IN OUR RIVERS AS IS GENERALLY MAINTAINED?

By Prof. H. RASCH.*
[Translated by Tarleton H. Bean.]

That the rivers on which there is considerable cutting of timber gradually become more and more destitute of salmon is an undeniable fact; but while it is asserted that the sawdust introduced into the river from the saw-mills causes the salmon coming from the sea either to forsake its foster stream because of meeting the sawdust, to seek another river not polluted, or else, when the fish attempts to pass through the areas quite filled with sawdust, then this, by fixing itself in the gill-openings or between the gills, causes its death, yet later experience seems to entitle us to the assumption that sawdust neither causes the salmon to forsake its native stream nor produces any great mortality among the ascending fishes. The hurtfulness of the sawdust to the reproduction of the salmon is not so direct, but is exceedingly great in this, that it partly limits and partly destroys the spawning-grounds of the river.

The river Drammen, below Hellefos, has for many years been greatly polluted by sawdust, and the abundance of salmon decreased constantly until the fishermen at Hellefos adopted the so-called artificial method of hatching, whereby they supplied the river each year with a considerable number of fry, which, after wandering to sea, returned to the cataract, although the quantity of sawdust is the same as heretofore; and one cannot see that the ascending fish is in any marked degree affected thereby. The case is different when it reaches a cataract where many saw-mills are situated, and there meets an insurmountable obstacle to its further advancement. Its desperate leap is in vain, and as it is driven down exhausted in the water filled up with sawdust, it will undeniably be liable to get some of it so tightly wedged in the gills that it cannot get rid of it, and death will then sooner or later be the result. To this danger the male salmon will be especially exposed near and at the spawning time, since the increased length of the so-called notches of the lower jaw prevent it from completely closing its mouth. The salmon which are not seldom found dead after the spawning time are nearly always males. That, at the same time, most of the deaths result from violent struggles

*Meddelelser fra norsk Jæger-og Fisker-Forening, 2den Aargang, Kristiania, 1873, p. 56.

between rivals is probable. If one could secure for the ascending fishes an easy passage over the intercepting cataracts and dams, then certainly very few fish would die from getting sawdust in their gills.

That young salmon bred from a race of salmon which has its own river, when they are set free in a strange river and one which is in an unusual degree polluted by sawdust, will not be prevented by this circumstance from returning to this last-named stream after their wandering in the sea, one had a convincing illustration in the great experiment instituted last year by Director A. Hansen. In olden times the salmon-shoal which had its spawning-place in Soli River could ascend to it through the then passable Soli cataract, but when they, for the sake of the increased mill-business, erected above the cataract a dam so high that the salmon could not ascend to their spawning-grounds, this salmon shoal gradually died out entirely. With the consent of the mill-owners Mr. Hansen in 1868 constructed a hatching-apparatus, which in November of the same year was supplied with impregnated salmon-eggs transported from the fishery at Hellefos. On St. John's Night, 1869, the young arising therefrom were liberated from the apparatus into the river, partly above and partly below the dam. Last summer a portion of the planting returned as young salmon, and according to experience gained elsewhere we should wait for the great body of them until the coming summer; because the greatest portion appear to pass the first two years of their lives in the rivers and two years in the sea.

In case one could aid the advance of the salmon around the Sarp cataract or Soli cataract—and perhaps in this way a few less important water-falls—and in connection therewith furnish the Glommen with artificially hatched young, one may now be fully assured that the abundance of sawdust which incumbers both branches of the Glommen, which again unite between Sarpsborg and Fredrikstad, will not prevent the salmon from going up to the falls, where they will then probably soon find access to a rightly constructed salmon-ladder, which would help them up to a portion of the great river freer from sawdust. The result of Mr. Hansen's experiment should therefore be a good support for the watchful action of the management of our association, which will in due time be communicated to the members.

XVIII.—THE PURIFICATION OF REFUSE WATER.

By E. Reichardt, of Jena.*

[Translated by Herman Jacobson.]

It is of great importance, not only for hygiene, but also for pisciculture and agriculture, that this whole matter should be more fully investigated, both theoretically and practically, in order to gain and diffuse knowledge.

I have on a former occasion published a treatise on this subject in this journal (vol. 209, p. 1), but, urged from many sides, I will not hesitate to reproduce, part of this treatise, embodying all the recent discoveries.

It is an undoubted fact that refuse of various kinds is in a reckless manner thrown into public waters, thus doing injury to public health and depriving agriculture of valuable fertilizing matter, and finally depopulating brooks and rivers of fish, which form so valuable an article of food.

The objection that this had been so from time immemorial does not hold good; no bad habit like this could ever be justified even by the usage of ages. Not even old-established manufactures can claim such a privilege, because the progress of industry, based on the progress of chemistry, has taught us to make use of a number of hurtful and unhealthy substances whose refuse flows into public waters. Any such privilege, very questionable in itself, cannot possibly include innovations of every kind which were formerly quite unknown.

Every man ought to be held responsible for any injury to public interests caused by his business.

Since the above principle is often carried out very rigorously with regard to street-cleaning, &c., why should this not be done with regard to public water in brooks, rivers, and ponds? If changes in any branch of industry, even if these changes only mean an enlargement of the business, involve hurtful influences, it can justly be demanded that such influences should be neutralized.

Chemistry is not only one of the strongest levers of modern industry, but it certainly falls within its province to remedy all injuries to public property caused by industry. Chemical knowledge should not only be utilized in a one-sided manner for the benefit of various industries, but

* *Reinigung des Abfallwassers.* Von E. Reichardt in Jena. Archiv der Pharmacie, vol. xii, Halle, 1879.

it should also strive to utilize all refuse matter in as complete and satisfactory a manner as possible.

It is a fact which can be observed everywhere that nature utilizes refuse matter of different kinds in such a manner as not to injure deeper-lying springs, that is, the upper layers of soil or rock absorb the refuse, either changing its character or forcing it to enter other combinations, so that in both cases the lower strata receive but very little of it. This purifying process, which is going on on an extensive scale, is both of a mechanical and chemical nature, and presents the best and simplest starting-point for discussing this whole question.

PURIFICATION OF WATER BY REPOSE.

The success of this purifying process is very clearly demonstrated by the glacier water flowing from lake to lake. In a turbid and milky condition the icy water leaves the mountains on its way to the plain, till it enters a lake often of very considerable depth and extent. Although the same quantity of water leaves the lake to continue its journey towards the plain, it comes out as clear as crystal, whilst long banks of clay or sand gradually mark the entrance of the turbid glacier water. The same observation may be made in rivers. A strong current carries all the floating particles of clay and sand far down the river, while a sluggish current allows them to gather in the bed of the river or on its banks, thus purifying the turbid waters. An attentive observer may watch the same process in every brook; wherever an indentation of the bank delays the rapid flow of the water, numerous particles of mud will gather, and many substances injurious to the life of fish are in this way removed from the water. This natural process of purifying flowing water by allowing it to enter a condition of repose can easily be imitated in an artificial way, and should be adopted wherever turbid water enters brooks and rivers.

This is particularly the case in the neighborhood of mines, quarries, factories, for instance, beet-sugar factories, &c. It will suffice in every case to construct so-called "mud-catchers"—ponds through which the turbid water is led. If the water, as is often the case in mines, flows out with the strength of small brooks, 2 to 3 ponds should be constructed, one by the side of the other, regulating their depth according to the quantity of impure matter in the water. It will also be well to cultivate in these ponds aquatic or floating plants or reeds, and to plant willows on their banks, as such vegetation aids the purifying process in more ways than one. I have often observed that by simply following these rules a single pond proved sufficient to purify completely the turbid water flowing from a mine. In such a pond hardy fish, and even carps, were raised.

From time to time the pond would yield a great quantity of mud, which, when taken out during the cleaning of the pond, proved a valuable fertilizer.

If the flow of turbid water is not very strong it will be sufficient to dig deep water-tight pits, 2 or 3, close together, as shown in the following diagram:

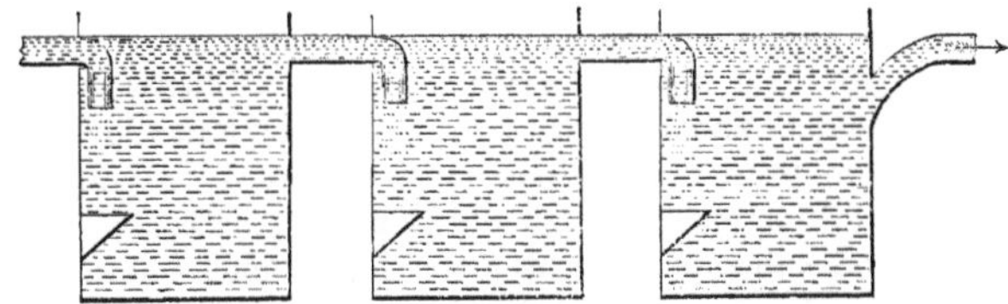

These pits, which should be lined with brick and be cemented or made water-tight by clay, receive the turbid water through drain pipes (of terra cotta) which bend towards the bottom of the pit. The flow of water, however, is broken by a stone projecting from the side of the pit, so that below this stone the water is in repose. The pipe through which the water leaves the last pit is bent upward, so that any particles floating on the water, such as oil, &c., may remain in the pit, from which they are removed from time to time, and are in this way prevented from entering the public water-courses. Such floating particles are specially injurious to fish, because they are in the habit of collecting anything floating in the water; tar, petroleum, &c., may thus prove poisonous.

By repeated personal observations I have become convinced that even lighter organic matter floating in the refuse water has settled at the bottom, and that from the third, or even from the second pit, the water flowed out perfectly pure and clear.

PURIFICATION OF WATER BY CHEMICAL PROCESS.

This method of purifying water will, of course, be influenced by general or local conditions.

In the first place it should be unlawful to introduce any hurtful matter into public waters; and it would be very desirable if the recently appointed inspectors of manufactures were to give some attention to this question of refuse. As soon as there is a doubtful case, it should be submitted to competent chemists or health-officers, making, if necessary, a last appeal to the imperial health officer. German manufactures are but too frequently carried on in a one-sided manner, excluding the chemist who is often the only person capable of giving information or rendering aid, especially as regards the greatest possible utilization of all manner of refuse.

Such utilization is often prevented by the ignorance of manufacturers, who, though well versed in everything pertaining to their special branch of industry, avoid anything which does not seem to come within their immediate province. It is therefore the duty of the government to take this matter in hand by diffusing information and making regulations.

Even large factories simply lead the refuse of soap from the washing of wool, &c., into the nearest river. Such a thing could not happen in England, where experience has taught people to utilize such refuse for the manufacture of soap or gas, &c. It is therefore as great an advantage to industry as to the purification of the water not to let soapy water flow into public water-courses.

It may justly be demanded of every manufacturer who uses large quantities of water that he should purify the refuse water, and allow only such water to flow into the river or brook which cannot do any harm. The water used in working machinery should be kept apart from water which in cleaning dyed substances absorbs the superfluous coloring matter; this last-mentioned water should be purified, while the former may safely be allowed to flow into the river.

It will also be advisable to see to it that not unnecessarily large quantities of water are made impure; this will be comparatively easy, as a great deal of water may be saved by the modern improvements of our machinery.

In the above I have, of course, only given hints, which will have to be specially adapted to every individual case, and which will only be appreciated by competent persons. Here is another field of usefulness for our inspectors of manufactures.

No manufacturer should allow refuse water containing free acid or free alkali to flow into public water-courses; such strong chemical substances should at any rate first be changed to salts, which are far less injurious. Lye may mostly be used again, especially if the first strong lye is at once employed.

The simplest, cheapest, and very generally used purifier is lime, employed either as quicklime or chalk. Nature employs lime and magnesia as purifiers of the soil. Organic substances combine with them; water containing iron loses it; and thus water penetrating the surface soon becomes pure, containing only particles of the stratum through which it has passed. The purifying effect of lime is still further heightened by the circumstance that a large number of coloring substances enter into insoluble combinations, not only with lime, but also with carbonated lime. These combinations have long since been employed in the manufacture of coloring substances. The effect of the lime does by no means reach its end by its chemical combination with acids, but is continued when the lime has assumed the form of carbonated lime. Lime is, therefore, the most effective purifier of all those waters which contain coloring matter. Lime likewise enters into insoluble combinations with albuminous substances, and therefore removes those substances which chiefly cause putrefaction. Alum is frequently employed with the lime, but its addition should be entirely regulated by the local demand.

Lime is used in the same kind of pits as mentioned above. The burnt lime is placed in the first pit. As soon as the pit is filled with refuse water it is stirred a few times, and the water will become clear in a very short time, so that it frequently enters the second pit with but very little coloring or impure matter. The water flowing from the third pit is generally so pure that it contains nothing but the superfluous lime and the soluble salts formed by it.

The quantity of superflous lime is generally very small, as lime only dissolves in 500 parts of water; but it might nevertheless, prove injurious to the fish. It is therefore necessary *to lead off the water in open ditches,* as lime very quickly becomes impregnated with carbonic acid from its immediate surroundings or from the air; the carbonated lime is then separated, as well as the last remnants of coloring matter. The water which has passed through these ditches will thus enter the river in a perfectly pure condition, and it is not even necessary to lét it flow through such open ditches for any considerable distance. It has also been proposed to let the pipe through which the water finally flows into the river open from below, about the middle of the bottom, so that the refuse water might immediately mingle with a great quantity of river water, and thus be deprived of anything of an injurious character which might have remained in it.

In most cases, however, the application of lime is sufficient to purify the water. In my former treatise on this subject I have entered into details, and will here only mention a few instances of purifying water. Soapy or fatty matter will generally be separated from the water by lime. These, however, are special cases, which were mentioned in my former treatise in order to show the value of the method. These lime deposits form very valuable fertilizers, so that, according to observations made in England, the expenses of this purifying process are fully covered. The whole arrangement is not at all expensive, and if once introduced it only requires a little attention to make it pay. Two pits are frequently sufficient to purify the water.

In Saxony the government ordered an investigation* to ascertain the number of complaints of water having been made impure by refuse. In 1877 complaints were made in 140 places and traced to 273 sources. Half of all these cases were traced to the weaving industries, especially dyeing, bleaching, and wool-weaving; 9 per cent. to the manufacture of paper; 8 per cent. to the manufacture of leather; 8 per cent. to mining industries; 6 per cent. to the manufacture of articles of food; 2 per cent. to the manufacture of chemicals. Of the 626 breweries in Saxony only 6 were accused of rendering public water-courses impure.

I found that many brewers have introduced purifying pits in connection with their establishments, and have worked them very successfully, as the mud obtained from these pits forms a most valuable fertilizer. In one case I was enabled to get a better insight into this question and to render aid.

A large brewery was accused of making the water in a large neighboring pond so impure by refuse water that it began to putrefy, thus causing considerable damage to the pond and annoyance to the people living near it. A chemical analysis of some of the most turbid refuse

* GUNTHER, "*Berliner Klinische Wochenschrift,*" 1879, No. 8.

water from this brewery showed large quantities of albuminous matter, in fact substances containing nitrogen, and the search for organic substances showed 1,000 parts in 100,000 parts of water, that is, 1 per cent. The purifying arrangement, containing two pits, which has now been introduced into this brewery lets the water from the second pit run out as clear as spring water, containing only faint traces of nitrogen and only 5 parts organic substances in 100,000 parts of water; in fact, the composition of the refuse water differed but little from that of the original water employed in the brewery; all that could be noticed was a slight increase in the quantity of lime. The water after having been thus purified entered the pond at no great distance from the brewery.

If we compare the experience of this and other countries with the actual condition in manufacturing districts located on small streams and rivers, it cannot be denied that so far little or nothing has been done to prevent public waters from being rendered impure. The above-mentioned cases where water has been artificially purified are entirely isolated, although the above-described purifying apparatus is neither expensive nor difficult to keep in order.

It must be acknowledged that even in Germany more and more attention is paid to the depopulation of the fishing waters, so that in many places suitable regulations regarding the fishing-season have been introduced either permanently or temporarily. But one of the greatest evils has so far been almost entirely ignored, and this is the impure water of many streams and rivers, rendered so by many different kinds of refuse, whose utilization as fertilizers or for industrial purposes is urgently demanded in the interest of economy. It will, perhaps, be difficult to make rules which would apply in every case; but so far nothing has been done to remedy this evil, and it is to be hoped that soon we shall have suitable legislation on this question.

The basis for such legislation will be found in the following points:

No impure or hurtful matter of any kind shall be allowed to enter any public water.

In the different industries the impure water, properly so called, shall be separated from simple refuse water, and the former shall undergo a purifying process of either a mechanical or chemical character.

Both methods should be under the superintendence of the health-officers; investigations are to be caused and directed by the inspectors of factories.

The washing of colored substances in public waters shall be prohibited, as it can be done by suitable machinery in a much more efficient manner; the refuse water should, however, be subjected to a further purifying process.

In most cases it will suffice to use lime for this process, and the sediments should, if possible, be utilized in some manner.

The refuse materials from privies should never be thrown into public water, as they possess a considerable value as fertilizers, and can be much more suitably employed for such purposes.

XIX.—NOTES ON THE FUNGUS DISEASE AFFECTING SALMON.*

By A. B. Stirling.

It is widely known that a destructive epidemic has this spring appeared among the salmon of the rivers Eden, Esk, and Nith. The mortality among the fish has been so great as to cause considerable alarm among proprietors, salmon-commissioners, tax-men, anglers, and the general public.

The newspapers inform us that within three days the watchmen have taken out of the Esk as many as 350 dead salmon. All who have examined the fish carefully agree in referring the disease to the presence of a fungoid growth.

The other fish in those rivers, as the smolts, trout, eels, lampreys, minnows, pike, and flounders, are also said to be attacked in a similar way to the salmon, and fears are entertained that the disease may become thoroughly established in the district.

In these circumstances I have thought it might be interesting to describe the condition of some of the fish which have come under my observation. In March last, my friend, Dr. Philip Hair, of Carlisle, sent me the fin of a salmon which had been affected by the disease, and requested me to state, if possible, its nature. Unfortunately, the fin was in a putrid condition when it reached me, and, as a result of the examination, I could only state to Dr. Hair that the disease was probably a fungoid one. A few days later I received from Dr. Hair a fine specimen of a trout, but it was not stated whether the fish was taken alive or picked up dead. It was, however, quite fresh, and the effects of the disease were painfully exhibited on the carcass. A hurried examination of this specimen enabled me to inform Dr. Hair that the disease was due to what I had previously suspected, namely, a fungoid growth.

While examining this specimen I observed, entangled among the fronds of the fungus, foreign matter of various kinds, namely, torulæ or yeast-fungus, triple phosphates, fecula, human hairs, hairs of the cat and mouse; also desmids, diatoms, shreds of dyed wool and cotton, with other fragments of matter unknown to me. Respecting the torulæ, I, in my letter to Dr. Hair, asked if their presence could be accounted for

* Notes on the fungus disease affecting salmon, by A. B. Stirling, assistant conservator of the anatomical museum in the University of Dublin, communicated by Professor Turner. Proceedings of the Royal Society of Edinburgh, session 1877–1878, Monday, July 1, 1878, vol. ix, page 726.

by bakeries or breweries in Carlisle, whose refuse might have got into the river.

My letter was published by Dr. Hair in the Carlisle Journal of March 29 and in the Field newspaper of March 30, and as worded it might have been inferred that I regarded the presence of bakeries and breweries as the cause of the disease. This was of course not intended. On April 12th I received two salmon and a trout from J. Dunne, esq., chief constable of Cumberland and Westmoreland, all of them in a diseased condition. Mr. Dunne requested me to make an examination of those fish, and hoped, on public grounds, that I might be able to discover the true nature and cause of the disease.

As a result of my examination of those fish, I sent a preliminary report to Mr. Dunne. This report was forwarded to the fishery-inspectors, and was considered of so much importance that it was published in the Times and many of the provincial and local newspapers. Sir Robert Christison had also very kindly supplied me with a number of specimens from the river Nith, all of them affected with this disease. An examination of these has confirmed me in the opinion expressed in the report above referred to. All these fish had the disease in an advanced stage, being more or less affected about the head, chin, branchiostegal rays, and fins in every instance. One salmon had rubbed the chin till the lower jaw had nearly separated at the symphysis, the skin was rubbed off the branchiostegal rays, and the rays broken. A trout had the upper left jaw bare of skin, the bone worn and hanging loosely attached to the cheek, the pectoral fin of the left side in rags, and the rays worn to stumps.

Another salmon had the skin rubbed off the nose and crown, and the matted fungus covered the bare parts; the dorsal fin was quite destroyed, the strong anterior rays being reduced to stumps of half an inch in length, and the remains of the fin bare, bleached, and without membrane. Beneath the dorsal fin on each side were spaces extending 3 inches forward toward the head, and 2½ inches backwards toward the tail, thickly covered with the fungus. Besides these there were other spaces on the sides of the fish from 1 inch to 2 inches in diameter, all covered by the fungus, which gave the fish a spotted appearance.

This fish appears to have been alive when taken, as the skull and brain had been punctured by the fisherman. The greater part of this fish was cooked; it was very firm and fat, and the three persons who made a meal of it pronounced it capital. I tasted a portion of the flesh from a part where the fungus covered the skin, and could not detect anything different in the flavor from an ordinary fishmonger's salmon.

The fungus appears, in the first instance, to attack those parts of the fish that are not covered with scales, as the crown, nose, sides of the head, chin, throat, and the membranous parts of the fins. From those parts the fungus extends by vegetative growth (which seems very vigorous) to those portions of the surface of the body which are covered with scales. On the sides of the fish, where small patches of the fungus were situated on the scales (and no rubbing had taken place), no sore could be detected, and the fungus was easily wiped off with the finger.

I may also mention that all the fish which I received from the Eden River, both trout and salmon, were infested with tape-worms of a large size, the worms being about two yards in length and three-sixteenths of an inch in breadth. One of the salmon had from 60 to 80 yards of those worms in the pyloric portion of the gut. Another salmon had three varieties of worms in various parts of its alimentary canal—first, in the stomach were many round worms, about 4 inches in length, tapering to each end, and as thick as ordinary whip-cord in the thickest part of the body; many of those worms were entangled among the gill-rays, it being their habit to crawl there when the fish dies, and from their presence in this situation they are called gill-worms by the fishermen; second, a small spiral worm, which attaches itself by burrowing in the outer walls of the intestine, in the fat and pyloric appendages; third, tape-worms seated within the pylorus and intestine.

On May 30th I received from Sir Robert Christison a large salmon from the Nith. This fish was believed to have been to the sea after being attacked with fungus, and was captured on its return. The specimen was a female, and had the roe about one-fourth grown; the viscera were very healthy, and no entozoa were found in it. The head of this female is peculiar in having a kip on the under jaw, and a cavity in the upper jaw to receive it, as in the male fish of the species. The right side of the head, including the eyes and nose, was very deeply rubbed and the bones injured, but no fungus adhered to the injured part. The pectoral fin on the same side had no membrane, the rays being bare, broken, and separate from the muscles at their roots. There were several patches on both sides of the fish, from which the scales were rubbed off, but no fungus adhered to the rubbed parts. In several of those rubbed parts, although the skin was unbroken, a portion of the muscle, corresponding in breadth to the external injury, and half an inch in depth, was in a pulpy condition; beneath other rubbed spots the muscle was quite sound. The dorsal, ventral, caudal, and anal fins were all more or less injured by rubbing. No fungus adhered to any of the fins except the anal, the rays here being reduced to stumps of an inch or half an inch in length, on which a thickly matted covering of fungus is seated. The branchiostegal rays are very slightly rubbed, and are the only other part of the fish on which the fungus remains. In my report to the fishery commissioners in April last I stated that the fish did not die of the fungus, but of the injuries they inflict by rubbing, in trying to rid themselves of the pest. As some objection was taken in regard to this statement, I quote, in corroboration of my views, from a letter published in the Field of May 25th last. The letter was written by Commander Duncan Stewart, R. N. He says:

"In regard to the disease from which salmon are suffering in some of our rivers, it may be of advantage that I should mention what I observed in a small river at the head of Castrie's Bay, in Siberia. I found the river rather low, but with plenty of clear running water. But what astonished me was to see thousands of salmon in all stages of disease and death, some darting away, but soon stopping to rub the side on the bottom or on a rock; others were constantly rubbing, others unable to rub. In those last cases large sores, from the size of a shilling to that of a half-crown, of a most filthy appearance, were always present. Fish in which the scales had been rubbed off would try to get out of my way, but I could kill them with a stick; those with the skin gone would rub themselves against my trousers."

Supposing this salmon from the Nith had been to sea, and had while there got rid of the greater part of the fungus with which it was affected, it had returned to the river in such a mutilated condition, and with unhealed sores of such a nature as in all likelihood would have ultimately proved fatal. Besides, the fact that the fungus was not killed by the salt water, but was found in a highly vigorous condition on the parts to which it still adhered, gives but small hope of any permanent benefit to diseased fish from a visit to the sea.

The fungus belongs to *Saprolegniæ*, a natural order of doubtful affinity, said to have the habits of molds and fructification algæ. This order consists of the genera *Saprolegnia* and *Achlya*, which are great enemies of fish and other animals preserved in aquaria.

The filaments of the fungus arise free from the outer surface of the epidermic layers of the fish, having neither branches nor articulations. They are tubes, the walls of which are perfectly translucent, and in their interior, at irregular intervals, are small groups of fine granular matter.

The majority of the filaments are spear-shaped at their upper terminations, and appear to be barren.

The prolific filaments, on the contrary, enlarge at their upper extremities, and form elongated club-shaped chambers, in which granular matter gathers. In the midst of this granular matter small round bodies appear, and, those enlarging, gradually develop into spores. The prolific filaments apparently contain more granular matter, and are of greater caliber than the other filaments. They are evidently destined from the first to be the propagating media.

The spores escape by an opening in the summit of the chamber. This aperture is not an original opening; it is produced in a somewhat remarkable manner. So long as the spores are unripe and unfit for expulsion, a slender continuation of the filament projects from the apex of the chamber in a manner similar to the neck of a bottle. At the point at which this joins the spore sac there is a slight contraction, which goes on gradually increasing in depth. Ultimately, when the spores are fully matured, it drops off, and the aperture is formed. The filaments forming the mycelium of the plant are tortuous and branched; they ramify in the mucous and epidermic layers of the fish; they do not penetrate the corium where there are no scales. In other situations they never reach a greater depth than the outer surface of the scales; they are tubular. The whole plant, being without septa, forms a single individual of apparently indefinite extent. The spores are variously shaped at different stages, ovate and kidney being the commonest forms. They are very minute, and require a power of 450 diameters to observe them well. The cilia are two in number, a longer and a shorter one, and are situated at the long axis of the spore. They are difficult to observe, and always disappear in permanently-mounted preparations, although the spores themselves remain unaltered in all other respects. When the fungus is stained with logwood or picric acid, excellent permanent preparations can be got. It has been stated that the fungus dies with the fish. I have not found this to be the case; on the contrary, all my observations have been made from dead fish. Some of the specimens sent me from Carlisle by Mr. Dunne were missent to Aberdeen, and returned to me on the seventh day after the death of the fish, and yet I have scores of permanent preparations from these specimens which show distinctly the characteristic form of *Saprolegnia ferax*.

I have also found the fungus perfectly identical in all the specimens I have examined, which consist of salmon, sea-trout, and river-trout from the Eden, and salmon and grayling from the Nith.

It has also been said that a salt solution destroys the fungus, "*which melts in the solution like sugar in water.*" On the contrary, salt and water is an excellent preservative of Saprolegnia; masses of it before me as I write have been in a salt solution for two months, and it remains unaltered. Further, the salmon captured in the Nith, which is believed to have gone to the sea in order to get rid of the fungus, had the fungus growing vigorously on several parts of its body. The fungus must either have instantly attacked the fish on its return to the river, or not have been destroyed during its stay in the salt water.

Regarding the cause of the disease, I can offer no opinion further than that some functional condition of the fish seems necessary for the propagation of the fungus. The germs of *Saprolegnia ferax* must exist at all times and in many places; and, if so, there must be a reason why fish are not constantly affected with the fungus and in every river. I am persuaded that the condition of the fish is in some way either suitable or unsuitable for the propagation and growth of the fungus. Whether this arises from too high or too low condition, I am quite unable to say; but I may remark that while some of the fish examined were in the kelt stage, others were in a condition perfectly fit for food.

XX.—ADDITIONAL OBSERVATIONS ON THE FUNGUS DISEASE AFFECTING SALMON AND OTHER FISH.

By A. B. Stirling.*

* * * By the kindness of James Tait, esq.,† I received a common river-trout and a minnow, both of which were captured near Kelso bridge in Tweed River; both specimens were affected with fungus—the *Saprolegnia ferax*. I may here mention that I have noticed several able letters, which have appeared in the Scotsman newspaper from time to time, in which the writer states that the fungus is only a secondary attack, and that a primary disease of an inflammatory kind first affects the head and other parts of the salmon before the fungus can settle upon it. I do not for an instant doubt the fact that the writer saw fish with sores of the kind described by him upon them, when there was no fungus present to cause them. I can only say that, among all the fish which I have received for examination, consisting of salmon, sea-trout, smolts, common trout, grayling, and minnows, I have not seen one with a sore on which this fungus was not present; while on every fish examined there were some patches of fungus which could easily be wiped off, leaving only a slight stain, and in some instances no mark could be discerned, and no loosening and shedding of the scales or ulceration of the subjacent surface. Again, in every instance where the fungus was rank, long-seated, and felted, sores in every degree, from slight abrasion to sloughing, were found under them. With reference to the trout and the minnow before mentioned, the trout had fungus seated upon the gums of both the upper and lower jaws, which involved both the teeth and lips, and had spread upward and backward upon the head, and its destructive progress could be easily traced. First, the skin of the lips was broken in several places, and shreds of it were hanging loose, to which the fungus was adhering; while, as it spread backward over the nostrils and crown of the head, the skin and its pigment spots could still be seen intact where the fungus was seated, a portion of which had been carefully shed aside to expose the skin. On each of the pectoral fins a patch of young fungus was seated, and the mucous coat was seen through the fungus to be quite entire; the same appearance was seen upon the anal fin and scaled parts of the body. The minnow had only

*Proceedings of the Royal Society of Edinburgh, session 1878–'79, x, No. 103, p. 232. Communicate June 2, 1879.

†Of Kelso.

one patch of fungus upon it, which was seated within its mouth on the inner margin of the right lower jaw; it filled the mouth, which was distended by its growth; and every other part of its body was free from fungus or blemish of any kind.

The reason why most of the fish affected with fungus are first attacked by it upon their heads may arise from various causes. All river fish present their heads to the downward current of the water, whether they are swimming or at rest, and as the spores of the fungus are floating down with the stream the heads of the fish are the first parts to come in contact with and be affected by them. Further, the mucous glands are most numerous and active upon the head of the fish, which is also more thickly covered with mucous than other parts of the body, and the spores which fall upon it adhere more readily; and the fins and tail, from their continuous waving motion, are more liable to arrest the passing spores than the parts of the body from which they spring, and, from this cause, are generally affected sooner than the bodies of the fish.

The number of the dead and dying fish of all kinds removed from the river Eden in 1878 by the police, and published by Mr. Buckland in his report for that year, show that there were 1,271 salmon, 140 fresh-water trout, and 40 brandlings or parr, being over 50 of the large fish to every one of the smaller. About 1,000 of the salmon were clean fish, and it may be inferred that the trout and parr were also clean, which goes far to show that the so-called disease is as much a mechanical as a functional one. Further, from documents descriptive of the effects of the disease in the river Tweed, in the lower district, during this season, 1879, which were collected by the police from taxmen and practical fishermen on the river, I find that the proportion of large fish affected, dead, or dying—namely, salmon and sea-trout—is very great compared with the smaller fish which were found to be affected in a similar way. The smaller fish alluded to consist of river-trout, grayling, smolts, perch, and gray mullet.

From observations of the fungus, and of the fish affected by it, I am led to believe that the so called salmon disease does not depend upon a prediseased condition of the fish. It is a true parasitic attack, to which every fish in any affected river seems to be liable, as every kind of fish, irrespective of condition, appears to be a proper nidus for the propagation of the *Saprolegnia ferax* when a living spore from that fungus attaches itself to it. While engaged during the spring and summer in the microscopic examination of the *Saprolegnia ferax*, I observed that as the season advanced many of the patches of fungus seated upon the fish were barren, consisting of spear-shaped filaments only, having no zoosporangia at their apex, and consequently they produced no zoospores. The filaments were long and very thin, and almost void of protoplasmic contents, indicating that the plant was losing its force and in a state of decay.

The *Saprolegnia ferax*, in all probability, is always present in our rivers in more or less active condition. It is believed that this fungus has two modes of reproduction, namely, by oospores and by zoospores. The oospores are few in number, and may be looked upon as ova, and they required sexual impregnation. They are called resting spores, from a belief that they remain dormant in the water for an indefinite period, which may continue for many years; and during this phase of their life they may germinate in limited numbers, providing only for the continued existence of the species. While in this state of abeyance there is no plague of fungus, from the ova only producing neutral or barren plants, which bear no fruit or seed. After a period of longer or shorter duration, a season, or a series of seasons, may follow, during which an unknown influence arises which acts upon the resting spores, by which they are stimulated to great reproductive energy, and the plants they produce being fruitful, the asexual mode of reproduction commences.

The zoospores are produced in pod-like cases called zoosporangia, which are situated at the apex of the filaments, and may be looked upon as fruit or seed. They are the ciliated spores and are the media by which the fungus is communicated to the fish. The zoospores are produced in great numbers, each zoosporangium containing from 100 to 150 of them. The oospores or ova are produced in a globular sack, which forms at the root-ends of the filaments or upon the roots themselves. Those sacks are called oogonia, and each sack contains a few oospores or ova, three or four to nine being the numbers I have observed in the four instances in which I have seen them in the whole courses of my investigations.

Suppose an oospore (resting spore) to be capable of producing, under favorable circumstances, a plant carrying 100 filaments, and each of the filaments to produce 100 zoospores, 10,000 germs would be derived from a single ovum or resting spore, every one of those germs being capable of producing a plant as productive as that from which it derived its existence; a multiplication of innumerable millions would be produced in a few days, the ciliated spores being as plentiful in the water as snow-flakes are in the air during a snow-shower; and in this way the plague of fungus, the so-called salmon disease, is originated.

I obtained in April the living fungus from a grayling caught by Mr. J. Williams, student of medicine, when angling in Keerfield Pool in the Tweed, near Peebles. It had been cut in two halves and the tail portion selected; it was packed in a tin vessel with wet moss, which had preserved the fungus in active vegetative growth, when I received it on the morning after its capture. A pale pink bloom was plainly visible over the whole surface of the matted fungus, and, when it was held up between the eye and the light, a new growth appeared to cover the older fungus on its outer surface to about one-eighth of an inch in height.

When examined under the microscope in water, free ciliated zoospores, which had escaped from the zoosporangia situated at the extremities of the filaments, were observed in motion; they moved in a fitful way, by shorts jerks, not by a continuous movement.

Those zoospores were pyriform in shape during the short time they were observed in motion; on becoming stationary the cilia disappeared, being probably withdrawn into the body of the spore, which then assumed a globular form. This change took place in a very short time—not exceeding ten minutes—and while under observation minute projections became visible on the edge of the spore, which grew into delicate filaments of considerable length. I have succeeded in fixing the development of the fungus in this state, and it can be seen in various stages of growth, all of which were ciliated spores within the space of one hour.

This, the asexual mode of propagation, is remarkable for the rapidity with which it is accomplished. A few of those ciliated spores become attached to any part of either a healthy or a diseased fish; in one hour the cilia will have disappeared and a filament of some length will have sprouted from the spore. Thus, in a single day, a fish on which no fungus could be discerned is to-morrow seen to be affected, and in three days is spotted or patched over with fungus from head to tail.

In the second or sexual mode of production of spores a short pedicle is pushed out from one of the sides of a filament, on which a globular sack—oogonium—is formed, and within this sack a number of oospores are produced, which are spherical in shape and have a cell-wall or envelope, and some are provided with a nucleus in the center. These, after impregnation, escape from the oogonia, and are probably capable of living in the water for an indefinite period, in a dormant or resting state, until the conditions arise which are favorable for their germination.

It may be asked, how does the fungus affect the fish, and do any recover from its effects? The fungus produces a local irritation and inflammation of the integument, as is evidenced by the congestion and even ecchymosis of the true skin, by abrading of the scales, and in the more advanced stages by ulceration and sloughing, affecting the whole thickness of the integument and mucous surface.

Wherever the fungus adheres and spreads, the function of the skin is necessarily interfered with. Light, which is so essential to the fish in promoting its pigmentary secretions, is cut off from a large portion of its skin. Endosmosis, exosmosis, and the secretion of the mucus for lubrication are destroyed, and in this way constitutional symptoms would be occasioned which, if the disease continued, lead to the death of the fish.

The second question, "Do any fish recover from fungus attack?" may now be answered more hopefully. The fishermen and watchmen on the Tweed report having seen several fish with new skin growing over the sores upon their bodies, from which this fungus had disappeared, and I am inclined to believe that this is so. A male kelt has been sent to me by Mr. List, which was taken in tidal water below Berwick bridge. This fish is 2 feet in length, and weighs about 3 or 4 pounds; it is supposed to have been affected with fungus, and to have completely recovered from its effects. No particle of fungus could be found upon any part of its body, and there was only one raw sore. This sore was only five-eighths of an inch in length and three-eighths in breadth. It had evidently been larger, and had a smooth healing border. All the upper surface of the head and snout were covered with skin, but very uneven over its whole surface, from depressions and projections which may have been caused by sores which have been healed over, and the hinder part of the operculum had an irregular cicatrix of considerable size upon it. The breast and belly, from the gill-covers to the vent, were blood-streaked and spotted, and there were brownish marks upon both its back and sides as if fungus had recently adhered to it. All the fins were entire—not one ray was broken—and the fish as a whole looked remarkably well for a kelt, and if it had been affected with fungus, which I fully believe, its recovery has been almost perfect.

A salmon taken at some distance up the river, and which is affected with fungus, has been taken down to Berwick, and placed in a box or corve, and is now anchored in the river, in the tideway, where the water is at all times less or more salt, and at intervals is towed out to sea, where the full influence of the salt water acts upon it; and when I last heard of it considerable improvement had taken place. Mr. G. H. List has paid particular attention to the protection of any fish being affected with fungus disease in any of the coast fishing stations; and, after the most careful inquiry, no trace of any fish in the least degree diseased at any of those stations could be got, nor, as far as any fishermen either knew or heard of, was any salmon with fungus upon it ever seen in salt water.

I have tried to propagate this fungus upon dead flies, spiders, and other small animals, following the directions of Pringsheim, "N. A. A. L. C.," 1851, p. 417,* who says: "All that is required to obtain a living specimen of this singular plant is to allow the body of any small animal, such as a fly or spider, to float for a few days in rain-water exposed to the light. By this method a crop of *Saprolegnia* may be obtained at any season." In this way I got a fungus upon the flies and spiders after an exposure of from 12 to 20 days, which, on examination, was found to be a common mold, exactly similar to that produced upon a solution of gum-arabic, gelatine, and meat infusions. I have tried to propogate the *Saprolegnia* fungus upon minnows, but without success hitherto, doubtless because the method adopted did not provide the proper means, there being wanting the necessary stimulus which exists in the river, or, what is more likely, the life of the fungus itself. The minnows were placed in a large glass vessel filled with town water from the tap. A piece of skin with this fungus adhering to it was taken from a salmon smolt and placed in the water along with them. In three days they had eaten up both skin and fungus, and remained unaffected. Several large patches of this fungus were then taken from the skin of a

* Cited by Dr. Burdon Sanderson in his paper on the "vegetable ovum," Cyclopædia of Anatomy and Physiology, edited by Dr. Todd.

salmon and placed in the vessel along with them. In a few days it had all disappeared, and produced no effect. Another method was suggested by Mr. G. H. List, who also kindly furnished me with material for the trial. Pieces of skin with this fungus growing upon them were cut from the bodies of dead salmon at the river side, and were put into wide-mouthed bottles, which were at once filled with river water, the skin not being allowed to dry. On receipt of the bottles the pieces of skin, along with the water in which they were brought, were emptied into the vessel among the minnows. The water in the vessel was not changed for three days, at the end of which time the minnows were still unaffected. Fresh water was then put in the vessel, and the pieces of skin retained in the water, which was changed every second day for eight days. The minnows were not disturbed by the pieces of skin. They nestled under them and nibbled every morsel of fungus from them, hiding and playing about them until they had to be removed from putridity. All the minnows are still alive and are in beautiful condition, taking food greedily, worms cut small and crystals of sugar being their favorites. They have been kept since 14th May till now (12th July) and are as healthy and lively as when put in the vessel.

XXI.—SICKNESS OF THE GOLDFISH IN THE ROYAL PARK, BERLIN.*

BERLIN, *July* 18, 1879.—A destructive epidemic has broken out among the inhabitants of the large goldfish-pond in the "Thiergarten," which rendered it necessary to remove all the fish from the pond last Friday forenoon. About two weeks ago the park-guards noticed an unusually large mortality among the goldfish. On closer examination it was found that the death of these little fish took place accompanied by very peculiar and regularly recurring phenomena. A gathering formed on the head where it is joined to the body, which soon commenced to fester; the head became soft; the body began to swell considerably; the scales, which in their natural state lie close and smooth on the body, seemed raised up by a festering substance; the lower gills, which are generally white, had a dark-red color, and a blood-like secretion oozed out of the pores. The poor fish evidently suffered from difficulty in breathing, and kept near the surface of the water. This abnormal condition became more intensified, the fish finally lost their scales and died after three days, often after a few hours.

The number of dead fish increased from day to day till at last the number of deaths amounted to sixty per day. An immediate examination of this entirely unheard-of phenomenon was therefore made. Some of the dead fish were taken to the Physiological Institute, where a microscopical examination revealed the fact that these fish contained a large number of living infusoria, which had almost entirely eaten up the liver. In other fish, especially those which were particularly bloated, the whole inside was a spongy substance, and various phenomena strongly resembled those generally observed in dropsy of human beings. The species of infusoria found inside these fish could not be determined, for it is the first time that infusoria of this kind have been observed at the Physiological Institute. After the character of the sickness had been ascertained, its cause had to be found. It was ascertained that most cases of sickness occurred in the northern portion of the pond, where fresh water is introduced. This portion of the pond was carefully examined and found to be full of a slimy substance which, when dried, broke into innumerable fine dust-atoms.

The supply of water was of course stopped at once, in consequence of which measure the water in the pond has sunk one-half meter. Sci-

* "*Krankheit der Goldfische im Berliner Thiergarten.*" [From "Deutsche Fischerei-Zeitung," second year, No. 29, Stettin, July 22, 1879.] Translated by Herman Jacobson.

entific men will now subject the water to a searching examination. This water comes from six springs located near the hippodrome at Charlottenburg, which are under the inspection of Counsellor Hoboecht. From the Hippodrome, the water is conducted through pipes to the "König's Platz," where it supplies the fountains, and from here it flows into the goldfish-pond. If impurities have been introduced into the water (and this is beyond a doubt), it seems natural to suppose that these impurities have originated in the springs themselves, as was the case in the Tegel water-works. A scientific investigation will doubtless explain why such impurities have got into the water just now, after the Hippodrome water-works have been in operation for so long a time. Possibly the increased quantity of water taken from the springs has stirred up waters which had hitherto been left entirely undisturbed. The shutting off of the water has of course also stopped the fountains in the "König's Platz" for the time being. The taking out of the fish from the pond this morning attracted a large crowd. The work itself was troublesome and consumed considerable time. Three fishermen caught the fish in a large net. The fish were examined at once and the healthy separated from the sick. This examination showed the sad result that almost 20 per cent. of all the goldfish in the pond had become infected. Until the pond is thoroughly cleaned and the whole matter has been satisfactorily investigated, the fish will be kept in the ponds back of the nurseries, where hitherto some of them have wintered. If it is any way possible, attempts will be made to cure the sick fish.

XXII.—THE ECONOMIC VALUE OF THE NORWEGIAN LAKES AND RIVERS AS A FIELD FOR FISH CULTURE.

By N. Wergeland.*

[Translated by Tarleton H. Bean.]

INTRODUCTION.

At the public meeting of the Imperial Acclimatization Society, held in Paris February 20, 1862, M. de Quatrefages, vice secretary of the society, delivered the following address:

From Hesiod to Virgil and from Virgil to our day the poets have vied with one another in praising the boundless munificence and maternal goodness of the goddess which watches over the harvest. But, without offense to the beautiful spirit, their commendations have been wrongly bestowed. Ceres is but a nurse, and that a severe one. She resembles Hercules, in that she helps only those who first help themselves. Before she makes the furrow fruitful she insists that the laborer shall water it with his sweat as an offering, and does not always protect it from the scorching or freezing breath of Æolus's children, nor from Jupiter's thunder-showers.

There is on ancient Olympus a much less exacting and a very differently liberal goddess. I refer to Tethys, the old ocean's bride and mother of springs and streams; in other words, the goddess of the water. She proves always a tender mother, gives always without numbering, and without ever requiring a return. Of him who cultivates her domain she demands neither plowing nor harrowing; she excuses him from all labor save that which is necessary to the harvest. It is perhaps on this ground alone that she has been neglected; because mankind has sometimes a strange heart which is inclined to ingratitude. It easily disregards what is acquired without trouble; it forgets a benefactor whose always open hand and heart have anticipated its desires, but holds better in remembrance and higher in esteem one whose benefactions must be extorted. This is doubtless the reason why the ancient Grecian priests lavished upon Ceres the expressions of filial gratitude which rightfully belonged to Tethys.

But one fine morning, as if overtaken with regret, they suffered Venus to be born from the foam of the ocean; Venus the goddess of love, fruitfulness made corporeal. This was at once to repair an injustice and to

* Meddelelser fra Norsk Jæger—og Fisker—Forening, 7de Aargang, 1 ste Hefte, Kristiania, 1878, pp 1–47; 2 de Hefte, pp. 101–172.

acknowledge the eternal truths which the somewhat obscure, somewhat graceful, myths of antiquity so often conceal.

Fruitfulness is, according to universal experience as well as according to the highest apprehensions of science, the chief attribute of water. Without water the richest soil would remain absolutely barren, while the water appears to be sufficient in itself alone to bring forth all kinds of living beings. Wherever it collects and remains, even in small quantity, life manifests itself in a thousand forms; before the spring sun has dried up the water in the ruts of our roads each of these has witnessed generations of microscopic algæ, rotatores, and lower crustacea to be born, grow, and die; the smallest pond is a whole world wherein representatives of the two organic realms and of the four principal divisions of the animal kingdom contend together; but what is this in comparison with the picture which presents itself to our sight when we direct it towards our brooks and rivers?

To see this organizing, life-producing energy which appears to be assigned to water in all its might, one must, however, turn his gaze toward the sea; one will then not merely feel surprise but overpowering amazement. To produce the marvel, one need not go to the tropical zone, concerning whose inconceivable fertility the sea-faring ones can narrate; our own coasts are sufficient for the inquirer.

He will immediately be surprised by a striking fact. In the sea it is not the loose bottom which corresponds to our arable land which proves itself most fruitful, it is the *rock*. The harder and firmer it is, the more impenetrable it is to all that can be called roots, the more living beings of both kingdoms it nourishes. From Belgium to Spain, Brittany's rocky coast is incontestably the richest. It is on its unalterable, impenetrable granite that the uninterrupted belt of sea-weed extends densest and broadest, which gives the soda industry and agriculture an importance sufficient to make up for all others; it is here that all depresions, all little creeks with their bottoms covered with loose stones transform themselves into shady valleys, where algæ of all kinds and all sizes represent the mainland's moss, greensward, thicket, and forest; it is here also where the grass-eating animals, which find the most abundant nourishment in the most luxuriant vegetation, are most numerous and most fruitful, and thereby themselves give the richest nourishment to the greatest number of flesh-eating kinds. But all takes place in the water, all is produced thereby and returns thereto. The soil amounts to nothing, because the starting point in the circle in which life and death follow each other is always *a simple plant fastened on the naked rock.*

This evidence of Creative Power which the water displays in itself, even to its smallest molecule, and which increases with the fluid masses, must kindle the human soul. With this evidence stand in closest relation the cosmogenic speculations of different nations, likewise all the theories of spontaneous generation which different men, of considerable merit in other respects, have attempted to put forward contrary to what experience has established.

When the existing continents rose up from the seas in which they were born, the greatest portion of the soluble substances which could serve for the support of living beings collected with the water in the sea. After this time thousands, perhaps myriads of cycles elapsed, and the land without cessation was washed away by rain. That is to say, distilled water has not ceased to furnish this immense reservoir with materials of the same nature, with organic detritus.

Hereby it becomes explicable how the river water flowing over a great expanse becomes enriched, how the sea water becomes a nourishing bath for the beings which it contains. In this manner is explained the abundance of the products of all kinds which the waters possess, and whose existence seems a paradox; plants without roots nourish themselves solely through their branches or leaves; stationary animals wait for the occurrence of their food, which is never wanting: free-moving animals, which float almost passively, a ball for the wind and waves, which scatter them everywhere, find everywhere that which is required for their nourishment.

But, on the other hand, it holds good in the ocean, also, that where no washing off of the solid land can reach, there also life ceases and death reigns. The fluid plain has its deserts just as the dry land.

Such an one is an enormous area in the southern part of the Pacific Ocean, separated by Humboldt's Stream from the coast of South America, which has been rightly called the Desert Sea. Here the waves rise and fall without moving anything but water; the billow is never traversed by any fish, nor the air by the pinions of any bird. That the sea, at a certain distance from the coast where the organized matter washed down from the dry land sinks to the bottom or is consumed by the multitude of living beings, does not everywhere show this unusual barrenness, which has so greatly astonished the seafaring ones who crossed this region, is so because, by the universal laws which govern our planet, there goes on an incessant mixing of all its parts. Even the revolution of the land produces streams which flow from the equator to the poles and from the poles to the equator, and which carry the waters which have washed the Old World over to the new continent, and the waves which have washed America's coasts back again to Europe. These streams carry, just as our great rivers, with which we have long compared them, elements of all kinds, which are plundered from the dry land; furrowing, in a manner, the ocean in all directions, they distribute, wherever they extend, fertility and life.

As the soil is not fruitful unless it is regularly watered, so also is the water fruitful only by virtue of the elements which it receives and transforms from the mainland. The sea sends the mainland rain and dew which are indispensable, to it; the mainland sends the sea the nourishing materials which it needs. Each of them expects a return for what it gives, and neither the one nor the other has ever refused its return. In this manner, receiving and giving without ceasing, both contribute to nature's wonderful harmony.

When men appeared at last upon the earth, and entered into the great circle of mutual influences, originated conditions which were produced by the nature of the surroundings. In the beginning of communities were found everywhere hunters and fishermen exclusively. They desired of the earth as of the water only what it produced of itself, and as a consequence they required enormous room in which the not numerous tribes might find the uncultivated fruits, the fish, and wild animals which were necessary for their support, and which often failed. In our day still some tribes are in the same condition, and we call them savage.

Mankind became at length herdsmen; that is to say, they collected some useful animals about them; they were raised thereby a round in the ladder of civilization. Less exposed to the cravings of hunger, these tribes increased and became hordes. But to support the animals which they had procured for themselves, men were obliged to move from pasture to pasture. They remained, therefore, nomadic and barbarous.

Finally, they learned how to cultivate vegetables, and trees, and plants, and soon thereafter how to improve them; they abode also in one place, and became agriculturists. But with the new work which they had assumed they were obliged from the first partly to strive against nature, partly to call her to their help. To procure a place for rice, wheat, corn, or potatoes, the weeds had to be removed; to increase the crops and renew the exhausted ground's fertility, manuring became necessary. Agriculture was called to life; it secured a steadily increasing population its daily bread. They performed their labors through centuries, and the experience gained secured steady production; civilized men live plentifully, by the million, in a space where a few thousand nomads, a few hundred hunters, would starve to death.

We all see what has been effected in this direction; but what always escapes the attention of many is that human industry is directed only to the soil, and has forsaken the water. With regard to culture, the hunter has altered his condition; the fisherman has become a savage. In this respect the most refined European races find themselves, with few exceptions, exactly in the same condition as the tribes of the Orinoco or of Australia; the white does not in any respect excel the negro. As his colored brethren have fished he fishes everywhere, always carelessly and without judgment; more numerous and equipped with better implements, the civilized white has fished more than the worse equipped black, and has wound up by exhausting the brooks, lakes, and rivers, as well as the sea, of both small and great fishes. But good is often produced from evil in its climax, and necessity has seldom failed to teach mankind wisdom. The diminution of wild animals led undoubtedly to the taming of our domestic animals; agriculture was instituted, perhaps, in the midst of the pangs of hunger. The decrease of the abundance of fish has brought fish-culture to mind; and at present *aquaculture*, that is, the cultivation of the waters with reference to fish-propagation, is about to win its way to recognition and practice as *agriculture* did thousands of years ago.

I.

GENERAL CONSIDERATIONS.

Few countries possess such a wealth of lakes and rivers of all sizes as Norway. In his work, "The Kingdom of Norway," Dr. O. J. Broch gives the combined areas of these waters as 7,600 square kilometers, or 2.4 per cent. of the whole area of the country. Their situation with regard to elevation above the sea, in connection with the climate and the topographical relations, causes by far the greater portion to be especially adapted as a place of residence for the kinds of fishes which are universally considered the choicest and most valuable, because they contain clear and cool water, in which these fishes thrive best and acquire the finest flavor. Of the waters, only a small portion, lying in the lowest regions, are unsuited to these better kinds, because of their sluggishness and higher temperature; these are, however, well adapted to other less esteemed, but at the same time valuable, species of fishes. Most of these waters, in earlier times, when the population was smaller, were very rich in fish, and the greater ones were therefore regarded as manorial rights, which, as such, were separately liable to taxation. Forty or fifty years ago the greatest portion of the waters situated in the mountain regions proper, and the rivers generally, were what one might call rich in fish, although the abundance, according to the statements of the inhabitants, was even at this time considerably diminished; but latterly the quantity of fish is steadily and rapidly being diminished by the constantly increasing fishery of the growing population, which in this country, as everywhere in Europe, urges on the pursuit, and especially at a time when it is the most injurious to the continuance of the fish supply—the spawning time—because the fish is most readily caught on the spawning-grounds. The steadily diminishing abundance only increases the demand instead of putting a check thereon. Fishing implements were gradually constructed in such manner that the smallest edible fish could never escape, and brooks which were the natural haunt of young fish were swept systematically from one end to the other by fine-meshed nets with careful search, so that only an insignificantly small number could reach the age of reproduction. This was, of course, not so everywhere in like degree; but over a large part of the country, by this mode of proceeding, prosecuted more or less eagerly, has been established a scarcity which in places approaches complete absence of fish of the better kinds, which were the chief object of pursuit, just as in many other European countries.

The sad result brought about in this manner, which in those places had reached its culmination more than thirty years ago, is naturally universally regretted; but the primitive consolation, that the Lord will constantly take care of the continuance of the abundance of fish, and that this gift was inexhaustible, was so rooted in their apprehension that men ascribed the diminution of the fish to the most marvelous causes instead of the real one.* Precisely the same thing occurred with the most valuable of all our fishes, the salmon, which, however, is indigenous in only a small part of the course of our great rivers. In the beginning of the century down to the end of its first twenty years, this abundance was so great that in many places the servants stipulated that they should eat salmon only three days in a week. But this abundance, by the same mode of procedure as was employed for the fresh-water fishes, and owing to other causes arising with the gradually developed industry, diminished to such a degree that the capture of a single salmon had become in many places in Southeastern Norway a rare occurrence, and it fell off to such an extent in many localities that the merchants did not think it worth while to keep the implements of capture, whereas formerly, when the price of the fish was only $\frac{1}{3}$ to $\frac{1}{5}$ of what it had in the mean time advanced to, good and even rich fisheries were a yearly experience.

Such was the state of things in this country, as well as in many other parts of Europe, when in this portion of the world it finally dawned afresh upon the consciousness that man's care, by bringing nature's powers into activity in an intelligent manner, might win from the waters a considerable production of fish, a production which, when the business is prosecuted with the requisite energy and care, might become very considerably greater than one could have any conception of from previous experience. Influenced by his own observation, it occurred to a farmer in the year 1842, in the Vosges, in France, to attempt to hatch out young trout in order to restore them to a depleted river.† The

*I have twice in Aal, in Hallingdal, received the explanation that the sea-worm was the cause of the scarcity. The first time, in 1840, Vatsfjord was the scene of its ravages, and it is said that they had procured castor, with which the water was sprinkled around to poison or drive off the worm, but without avail; the fish were absent and remained away. I have since had a good opportunity to see who does the work attributed to the sea-worm. Near sunset every evening the people assembled from every house in the neighborhood, and swept the water with fine-meshed nets, and they caught therefore only very few fish in the water, whereas the same little flowing river was rich in trout weighing three-eighths of a pound to one-half pound. The last time, in 1872, I heard that Buvandet, below Rensfjeld, had been the scene. A clergyman, one of his assistants, and a couple of farmers, owners of the water, had in partnership sprinkled castor to drive off the sea-worm, naturally with the same success as before.

†It is singular how seldom it happens that men avail themselves of accidental experience. The indication of artificial fish culture is not of rare occurrence here, since in many places the same experience is had as in the following case: In 1841, a perfectly trustworthy man told me that he, some years before, had been fishing and hunting late in the autumn at Gjendinsoset. Impending storms drove the fishermen in the greatest haste away to their country district, 5 to 6 miles distant; the nets were pulled up in the greatest haste, and the boat placed in the boat-house, while they

attempt succeeded, and aroused general attention in France, and thereby an impulse was given to a new industry, which, wherever the natural fundamental conditions are present or can be procured, will bear fruit of particularly great value.

Although, in the last century, this industry has received so little attention in Europe that it might be regarded as entirety forgotten, it has, notwithstanding, been known and practiced for a very long time. The most ancient civilized people of the East, the Chinese, practiced it steadily to a great extent, and have practiced it from time immemorial. One of their proverbs reads, "The more fish a country produces, the more men it produces." Artificial culture is so ancient here that it is considered to have been always prosecuted, and they have many species of fish which are cultivated in every house, in every pond, and which are regarded as belonging as much to the household as other domestic animals. They belong to the great cyprinoid family, are vegetable feeders, and are fed just as regularly as cattle and other quadrupeds. The stock is procured by collecting annually the naturally deposited spawn or naturally hatched young, and this collecting is a distinct industry. In the central provinces, which are drained by the Yangtse-kiang, near Kieow-Kiang, in the province Kiangri, in the month of April, more than 150 junks of a very considerable tonnage are occupied in bringing in cargoes of young, which they transport to and distribute in the interior of the country.

Just as fish culture is carried on everywhere, to a great extent, so the provisions of law have constantly received attention, in order that the abundance of fish in the natural waters may not be diminished by making them the subject of stringent legislation. It is said that 1,222 years before the Christian era, an emperor of the Tscheou dynasty, together with his consort, wished to go fishing; it was in the fourth month, during the spawning season. One of the prime ministers, Tschangsype, cast himself on his knees before him, and submissively called his attention to the fact that he was about to violate one of the most stringent laws of his empire, and that by acting thus he might bring destruction upon one of the most important of the common means of subsistence, whilst he would thereby incur a great responsibility before the tribunal of history. The emperor admitted that the minister was right, and desisted from his intention.

Fishing in all lakes, channels, and brooks which do not immediately

forgot to take out the plug, so that the boat might empty itself. On the following spring, when the boat again was put in the water, little young trout swarmed in the water remaining in the boat, hatched out from the spawn and milt which the imprisoned ripe fish had liberated when they were taken from the nets. Hatching may also occur under peculiarly unfavorable circumstances, for it cannot be doubted that the water must have been entirely frozen for a long time. To be sure, the boat was in the boat-house, which, during the winter, was covered by snow; but the place lies over three thousand feet above the sea, and at this height the cold is considerable and protracted.

flow out of or into a great river rich in fish is absolutely prohibited during six months of the year, from March to September, in order to secure the fishes against the rapacity of their pursuers and to insure propagation. The maintenance of these provisions and political inspection of the waters is presided over by specially selected mandarins and private citizens, to which last the state leases fishing privileges by cantonments. These general lessees, called konau-ho, pay a yearly tax to the state, and are pledged, 1, to appoint times for planting a quantity of young fish in the waters leased by them, corresponding with the extent of these waters; 2, to see that communication between their waters and the rivers rich in fish is always open at the spawning season, so that the fish may come into them to spawn; 3, with a stringent watchfulness to see that no one fishes during the time from March to September, and that nothing is done which can work injury to the thriving of the fish. In compensation, no one is allowed to fish in their cantonments without written permission, which they furnish to companies who carry on fishing according to the regulations created by the lessees. In rivers of medium size the close season is reduced to three months, and in the largest rivers which empty into the sea every one is allowed to fish during the whole year.

Owing to these provisions and the universal household fish-culture, fresh-water fish constitute a very large part of the accustomed food of the people, and so it has been from time immemorial without this source of nourishment ever having threatened to be exhausted.* The ancient Romans likewise carried on systematic fish-culture to a considerable extent, and their methods have not in the flight of time gone entirely into oblivion; but this cultivation was in the main only the enterprise of private individuals in inclosed fish-ponds. Universal legal provisions having in view the preservation of the abundance of fish in the open, generally accessible and public waters, scarcely existed; because if this had been the case it would, like the rest of the Roman laws, without doubt have been observed, at least to some extent, through the lapse of time, and would also doubtless have prevented the universal diminution of the abundance of fish, which in all the most civilized countries of Europe exists even down to the present time. The practice of the Romans is, however, as remarked, not entirely forgotten; they have in most countries continued to a greater or less extent to maintain fish-ponds, and to supply them with young in a manner which may be called artificial, in so far as this supplying goes on under direct human supervision, and is not left entirely to nature's care. It has similarly also been prosecuted here in Norway, at one time or another, by certain rich men; they say, also, that the monks at Storhammer have attempted it. It is presumably this circumstance alone which explains the occurrence in a couple of places of a species of fish, the carp, which does not belong to our northern fauna, and of another which is indigenous in the eastern

* Dabry de Thiersant, French consul-general in China, 1871.

part of the peninsula, the pike, so far as is known in a single place in western Norway, where it ordinarily does not exist. In a single place systematic and artificial fish-culture has been carried on extensively for centuries as the sole means of subsistence of a community consisting of many thousand individuals, namely, in Laguna di Comachio, near the Adriatic Sea south of Venice. But no one thought about imitating this business before last year, though the same or a similar opportunity for such industry is found in many places in France as well as in Italy. The objects of culture here are fish which do not, like the salmon, spawn in fresh water, and afterwards reach their greatest development in the sea, but which, on the contrary, spawn in the sea, while their young at stated times frequent the streams in the shallow lagoons, there to reach their full development, chief among them being the eel, which has the same habit here in the North as in the South.

But the commonly-practiced fish-culture of the Chinese, Romans, and modern Europeans is restricted chiefly to species of fish of particularly great fecundity, which live in sluggish waters of an average higher temperature, which spawn in spring or summer, and whose eggs are hatched without difficulty of any kind in the space of a few days, namely, carp and its kindred genera, together with the pike and perch, to which may be added the eel, whose young may be easily collected in their migrations up the river courses. The choicer species of fish belonging to the many different species of the salmon family, the most of which spawn late in the autumn or in the winter, have been the subjects of artificial culture in very few places.

The artificial fertilization of the spawn of the nobler species of fishes, the salmon, in the manner in which it has been practiced for the last twenty or thirty years, was not, however, entirely unknown; for there are found printed works which describe it dating from the middle of the preceding century and later; but these aroused general attention as little as the practical performance of the operation, which took place here and there. About the year 1842, when, as before remarked, a peasant, Remy, in the Vosges, concluded to attempt the artificial fertilization and hatching of trout-eggs, the affair first was fortunately brought to the knowledge of French scientific men who appreciated its great economical importance to the nation, and many of these have since that time with the greatest zeal labored to bring, and have also succeeded in bringing, the mode of operation to the desired perfection. Among these many scientific men Mr. Coste, professor of embryology in the College of France in Paris, a member of the French academy, is generally regarded as the one who has labored the most and the most successfully for the advancement of this thing, in which duty he was strongly supported by the Emperor Napoleon. Since the year 1852, one may regard the plan of operation to have been brought to perfection, and since that time the business, so far as the salmon-like fishes are concerned, had been carried on to a steadily increasing extent everywhere in Europe and America, where the opportunity is presented.

About the same time the matter awakened attention among us where already in many places the abundance of fish was reduced to a minimum, a reduction which, moreover, has continued for many years with unabated zeal in many places in this country, and even now is continued here and there on no small scale.

Since the year 1848 the legislature has taken into consideration the destruction of salmon-fishing, and has sought by more stringent provisions to control the instinct of prey. Since 1863 the way has likewise been open by it for restraining this mode of procedure in the lakes and rivers, since the necessary increase ought to be gained in all the places interested. For about the same length of time, by the contribution of public funds, artificial hatching has been carried on over the whole country, and public attention has been directed to the matter, while instruction in the art has been given wherever it has been sought.

These measures have borne evident fruit, and the country therefore owes great gratitude to Prof. H. Rasch, who chiefly gave it the impulse, as well as to his indefatigable assistants in its practical execution. But the result has not yet by far reached the extent which it can and ought to reach, and which it probably will reach when the matter is taken hold of with the energy and care which it deserves. That it is not at present greater cannot depreciate the man's services, which hitherto have borne the matter forward; one must much rather wonder that he has succeeded in winning so great victories over deeply rooted prejudices, and the universal reluctance among people to submit to previously unknown restrictions against habitual unrestrained free fishing, the use of which they must first see before they can, perhaps rather will, comprehend them.

But the experience gained through more than twenty years' practice in many countries in America, as well as in Europe, has shown that we now stand very far from the goal which we can and, therefore, ought to seek to reach. We have hitherto in this country confined our operations to placing little barriers against improper rapacity; these barriers ought to be given the necessary dimensions which are required for the attainment of the object in the well-understood interests of all. Having, besides, to some extent provided for the sowing of the field, they will win therefrom increased production; but this care has not been sufficiently great by far, partly because it is limited to the salmon by ill-advised provisions of law, which render difficult, often impossible, that which has the claim of the first requisite, the desired abundance of mature spawn; partly because they have placed their trust in, and, therefore, to an unreasonable degree given their attention to the advancement of, natural culture. Moreover, they considered only the production of the delicate young, and have liberated these, which cannot be regarded as in much better condition to escape the multitudes of enemies than the spawned eggs, in the rivers, to be eaten up in masses before they reach any size. Finally, they have, for the lake fisheries concerned, not at all considered that, just as the farmer by manuring can multiply his crops, so by a suitable mode of procedure one may also increase the maintaining capacity of the water by looking after the increase of the nourishment from which the fish, which are made the object of especial care, get their subsistence. The opinion has been much more generally held that this food is injurious to the rearing of the cultivated fishes, as it consists chiefly of living fish of smaller and commoner kinds, which are regarded as enemies of natural culture, as spawn-eating, in competition with all other fishes, small and great, and with a multitude of other living animals, four-footed as well as winged, and insects of a multitude of species, which culture, as remarked, has attributed to it an importance which by no means can or ought to be attached to it if one wishes speedily to reach the goal, a considerable increase of the abundance of fish.

Since I have chiefly in view to show the value of the lakes and rivers as a field for fish-culture, I shall next take these into consideration by showing what ought to be done, so that the kinds of fishes which should be the objects of cultivation in them may be produced in the greatest possible abundance; and next, what should be done for the fish which belong to both the sea and the fresh water, although these last have already obtained, and probably by more thorough modes of proceeding will further retain, superiority over those in economical respects, and, therefore, ought to stand in the first place. Thus I pass on to—

II.

WHAT FURTHER SHOULD BE DONE AND WHEREFORE.

It is known that in natural fish-culture only a very small fraction of the quantity of eggs deposited are developed; therefore, the Lord of nature has made their fruitfulness great in proportion to the danger of destruction to which the eggs and the young are exposed. I shall here confine myself solely to the chief representative of the fishes, which will certainly be the especial object of culture—the trout. What percentage of naturally spawned trout eggs reach their full development as young it has been found impossible to learn with certainty, just as little as to what age the multitude which come to life as tiny young ones under natural conditions live. Men who have closely studied this matter believe that one out of ten or one out of a hundred eggs develop into young fish. The rest go to ruin, are buried up, destroyed, or eaten up; perhaps the greatest portion disappear in the last mode, for everything that lives in the water, large and small, even the fish that lay the eggs, eat them as the greatest delicacy.* Trout or salmon roe is the most irre-

* An experienced American fish-culturist thus describes what happens at the spawning-place:

"When the spawning time approaches the trout seek a suitable place on a gravelly bottom in shallow flowing water, especially that originating from springs. When they have paired themselves, which takes place only after violent, often deadly battles between the males, both go to the chosen place and lie still there if they are not

sistible lure (bait) for trout and salmon, as well as other fishes. Such is the case in countries with much milder winters than we have. How great a portion of the young are eaten in the space of the first year, when they stay unprotected in their native place, we have naturally no certain knowledge of; it is probable that at least half are eaten or destroyed by the above-named causes under the water, as may frequently happen. With us, at all events, over the greatest portion of the country, we have cold winters, which produce bottom ice, and at least spring drifting of the ice to a considerable degree, which as a rule always comes in contact with and traverses the best spawning places of the trout or other salmonoid fishes, which are besides readily laid bare in the course of the winter.

But suppose that the eggs and the young fish endure being locked up in the ice without dying, they will by no means endure the drifting of the ice or lying bare in the frost. In this way the profit or product of natural culture becomes so uncertain that it cannot be depended upon to give any result which in any way can or ought to be taken into consideration. It seems to me that the confidence in the rapid increase of the abundance of fish through natural culture fully corresponds with the confidence of the farmer who thinks that the garnering of premature corn will be sufficient provision of seed for the coming year's harvest. On the other hand, a long experience has now established that one, if he manages things with proper care in all necessary directions, may safely count upon about ninety-five young from a hundred eggs, and that one can rear these young ones under proper conditions through a year with a loss of only five per cent. The artificial culture also is as safe, the natural as unsafe, as possible.

Since this is the case, one has it also in his power by labor and outlay, which are inconsiderable, to procure all the young that he considers necessary for stocking a fishing-stream with perfect certainty, provided one can procure the necessary quantity of eggs; all regard for

disturbed; but the males are for the most part occupied in driving away rivals, who pry around. It is curious to see a little male by the side of a large female. Under ordinary circumstances the smaller male respectfully gives way to the larger, but at the pairing time the smaller will in an instant attack one three times as large as himself, should he approach to within a few feet of the female; as a rule the male is fully occupied in driving away rivals. When these are numerous the female will often come to the help of her chosen mate. But after the female has selected a consort there is no longer any contest; the disappointed males fly as soon as the consort makes a show of attack; they appear to respect the intimate union. The female meanwhile forms a nest, which consists simply of a shallow depression 6 to 8 inches in diameter and 2 to 3 inches deep. It is constructed in this way: the female thrusts her nose down in the gravel and pushes it aside with her tail as she raises her head again. This work goes on many days until the cavity is large enough for her. After they have laid over the nest for some time the female is ready to deposit a portion of her eggs. The male seems to know this instinctively, because whereas he had been busy expelling rivals, he is then always at the female's side, and the instant she lays her eggs he allows his milt to flow over them.

"When the eggs are deposited the male forsakes the female, who thereupon covers the natural culture becomes superfluous for the fish species concerned, which ought to be the object of cultivation, and absolutely injurious in so far as one, from his anxiety to advance this culture, seeks to eradicate or omit to stock the water with other kinds of fishes which might serve the cultivated fishes as food, even if they in any degree concur with this about other means of food, for the fishes which one cultivates will chiefly be fish of prey, which will develope with a desired rapidity only when they have an abundance of other fishes as food.

The essential condition for abundant production of fish, next to the possession of water, is ability to be able to get fish-spawn in the desired quantity. This ability will always be present in all well-stocked fishing waters if no injurious law regulation places an artificial barrier by prohibiting the capture of spawning fish at the right season, that is, at the spawning time itself. Every trout or salmon yields 2,000 eggs per kilogram (two pounds) of its weight. To procure 1,000,000 eggs there will thus be required fish of the united weight of about 500 kilograms, but few males being required in proportion to the females. After the lapse of a year, one will have at least 800,000 young fishes of one-twentieth to one-tenth of a kilogram each, or 60,000 kilograms in place of 500 kilograms. If, therefore, the mothers and fathers which are taken as spawning fish must be consumed and a portion more are taken under the same pretext, this signifies nothing in the face of the certainty of having brought back the necessary young which, even as yearlings, will weigh fully one hundred times as much.

It was just to prevent the loss of profit in fishing at the close-season that the existing legal enactment for salmon was made. This misunderstanding of real interest, this injurious prohibition will probably disappear when the pending new proposition for a change in the fish law obtains legal validity. Besides, the spawning-fish are always poor food in comparison with what they are at other times of the year. In England such spawning fish are considered inedible, and such will also be the case in this country when a greater abundance of fish no longer

the eggs with gravel by sweeping over them with her tail all that is found near the nest. If the female is not satisfied with the covering she will go into the stream and push suitable stones backward with her ventral fins over the nest until it is completely covered. After a few minutes the male returns to see how the work progresses, eats some eggs if he can find any, and departs again. The female, on the contrary, does not go away, but remains at the place and does not forsake it until all the eggs are spawned, which occurs in many installments and occupies a long time, often as much as three days. The female, as well as the male and all the hangers-on swimming around, have meanwhile eaten as many as they could of the eggs. When the first pair has left the place another comes on the same errand. The female finds a suitable place and begins to prepare a nest. As soon as the first-spawned eggs appear this business is given up and female and male vie with the lookers-on in eating all the roe before they again resume work. If one next takes into account that all kinds of water fowl seek after the spawn with great eagerness, that the tender young fish coming to life in the spring serve in great portion as food for the larger fishes, it is no wonder that there are so few trout in our streams, but a great wonder that any are left."—(Trout-culture, by Seth Green. 1870.)

tempts one to take them as food just at the time when they are the poorest.

When no such prohibition exists there will generally be no difficulty in finding as many spawning fish as may be necessary, provided the water contains the requisite quantity. By proper forming of spawning places one can regularly take on them every fish which makes its appearance for spawning; if it is not quite ready for that when it is taken, it can be set free in the water again; it will again make its appearance at the place when the proper time comes. Or one can in the American manner construct the spawning places so that one without touching the fish, leisurely and at ease in the daytime can collect all the spawn deposited and fertilized the evening before.

That there will be fish in abundance to supply all the spawn which is considered necessary for planting in a stream when it has become properly stocked is certain; that natural culture will take place by the side of the artificial is thus self-evident. But whether this gives any yield of any living fish or not is a matter of entire indifference, excepting in so far that the quantity of spawn deposited in this manner, and the possibly small number of young arising therefrom, plainly increases the nourishing capacity, since, as remarked, the spawn as well as the young will serve as food for a whole multitude of all kinds of fishes.

After—

1. *Artificial hatching* of the multitude of eggs, which is considered necessary for the proper yearly recruiting of the water, or filling up the decrease caused by steady fishing for the fish which have reached a suitable size, is required also—

2. Rearing of the young in an inclosure until they are at least six months old, when they will have reached such a size that they themselves may appear as enemies among a host of enemies whose prey they would have become at an earlier age; there is required in aquaculture, as in agriculture, *proper inclosing of the ground.* This must, in the scheme here proposed, provide for the hindering or destruction of enemies which will divide the harvest with the breeder, and take the lion's share or the whole if they are able. Among these, man stands first. It is, therefore, a matter of course that the laws must secure for the fish-culturist, as well as the farmer, the indisputable right to the fruit of his labor, and thus make the proprietorship and right of fishing in the water of every condition just as clear and fixed as the corresponding right to the ground in question.

As long as the right to free and unrestricted fishing in brooks and rivers is recognized fish-culture cannot pay, because our brooks and rivers are just as important for the rearing of young fishes as they are also chosen waters in which they may nourish themselves until they become full-grown or mature products, which will be indiscriminately fished for, and thereby the result of every effort for the increase of the fish will be brought to naught. What is needed in this direction this is not the place to develop more fully; I suppose it will be done when the realization of the necessity is sufficiently clear to the public. But even if legal enactments contain all the necessary provisions, this will not be enough. Law alone is inoperative; it must have living supporters, particularly when the temptation to break it is especially great, as will be the case when the fishing-streams, rivers and brooks, receive the complete supply of fish which they can accommodate and sustain. Constant and active watchfulness must supplement the protection of the law, and the business will with the greatest ease be able to meet the expense which must herewith be associated. Where a single individual commands an isolated field for fish-culture, the matter is very simple; where the culture must be carried on in partnership by more than one, or by many interested parties, watchfulness becomes necessary over a greater area than for the isolated proprietor's district. He who cultivates for himself alone, independent of others, will see to carrying on the business in the manner most profitable for himself, and act in accordance with that object, and, if he does not do this he must blame himself for the possible damage or loss. Where many persons are interested, the temptation to undertake to feather one's own nest at the expense of the rest is very great, and not every one is able to resist it. In such case the custody must be open for inspection, that even the interested parties observe the rules of the business, namely, that the catching of fish be judicious, so that no improper division may be made, nor the business be injured. But in this case it is not sufficient as hitherto to depend chiefly upon the men seperately appointed for such inspection. If it is important to have fixed legal enactments, or agreements respected, then should every means auxiliary thereto, which costs nothing, be brought also into the greatest activity, and this activity should not as generally hitherto, be paralyzed to the greatest possible degree.

Where numerous keepers of the law or established custom are appointed it is clear that many and very improper advantages can and must escape their observation, which, however unavoidable, must become generally known to other persons among the people more or less interested in the matter. Why not seek to invoke such assistance for the support of the law when it can be procured without cost? In nearly all hitherto established laws of such a nature they have, as it appears, made it possible to prevent the public from interesting themselves in any way for their support. The specially-appointed inspector was to have a share in the fines which were imposed for transgressions; the discoverer and informant, who had no such position, received, on the contrary, nothing. It appears now as if they, with full knowledge, wished to relax the operation of the law to the utmost extent in which, with a little appearance of decency, it could exist. That such maintenance of the law will be secured to a considerable degree by allotting to every informant the same compensation can be subjected to no doubt, and by such a plan alone can such laws fully acquire their intended strength.

The mode of proceeding hitherto followed made the law almost inoperative, and encouraged a license in treating it, which it is, or should be, the design of the law to destroy. "No receiver of stolen goods, no thieves," says the old proverb. No fracture of such a law of any importance can occur without the participation of assistants, since the law itself now does its best to help the receivers by depriving them of that encouragement to the announcement of irregularities which it grants to the specially-appointed inspectors, for whom such encouragement would appear less necessary, since they are paid especially to see that the law is respected. Were such encouragement granted to those who are not inspectors, the transgressions of the law would on that account alone become exceedingly rare, as no one could be certain that a transgression occurring would not be reported, which in ninety-nine out of a hundred cases must come to the knowledge of many persons whose silence could not be depended upon. The objection against encouraging a system of informants, which we have heard mentioned as an argument against the institution of such a reward, will signify nothing, because an occasion for information will then never, or very seldom, arise.

But, besides the encroachment of men, there is no little multitude of animals which will tax the abundance of fishes in somewhat the same degree as beasts of prey on the land and in the air reduce the abundance of game, and this taxation is in reality very much more considerable than people generally have any conception of. *Otters, loons, ducks*—especially fish ducks—destroy a considerable quantity of fish, and should therefore be persecuted with all means one is in possession of. In the same class must be included fish of prey, not merely of the kinds which are not objects of culture, but also those of the cultivated species which have become so old that they increase annually in size but slightly in proportion to the nourishment which they require. The most profitable yield from fish culture will clearly be obtained when the fish is regularly caught as soon as it has reached the age and size at which the quick increase begins to fall off. This size will differ in the different species of fish, likewise in different waters, according to the greater or less wealth of nourishment, and its quality. No general rule in this direction can be given; it can only be acquired through experience in each separate locality.

According to experience, a trout consumes daily animal food equal to $\frac{1}{100}$ of its weight; this has at all events proved fully sufficient feeding for fish maintained in the same way as stalled cattle. They have thrived upon it, grown quickly, and become fat. They could of course have consumed more; but this quantity may be regarded as a proper medium, especially if one does not include very large ones. A common lake-trout which, for example, has reached a size of five kilograms, will thus in the space of a year consume of food of all kinds 365 times 0.05=18.25 kilograms, while at the same size and age it will increase scarcely more than one-half in the space of a year, or 2.5 kilograms, In the first three or four years its own weight doubles yearly with the same amount of food; thus, for instance, a trout during the fourth year which at the beginning weighs about 0.75 kilogram, consumes about 365 times 0.0075= 1.85 kilograms, while it will have gained in weight 0.75 kilogram; the ratio between the food consumed and the increase of weight is also at this age quite particularly more profitable than at a later age, because the increase of weight in the last case bears the proportion to the nourishment consumed of 75 to 185, or 1 to 2.4, and in the first case of 2.5 to 18.25, or 1 to 7. Even if the increase were the same at this age as earlier, the proportion would become as 5 to 18.25, or as 1 to 3.6, or in a considerable degree less profitable than in the younger stages.

One must also strive to catch the older fish as completely as possible, and for this end the spawning time will furnish the best opportunity. Whether one will then eat them or put them into a separate smaller pond, where they will be easy to catch at any time, for preserving and feeding them until they are in better condition, is a matter of taste. That fish of prey of other species which possibly may occur ought to be exterminated by all means is self-evident.

For the attainment of a reasonably large profit it is, moreover, as before mentioned, necessary, in the greatest possible extent, to improve (gjöde) the water which is the object of cultivation. This may occur in different ways, depending on how the circumstances may be varied. It applies to the whole circle of creation that the lower organisms live upon plants, and in their turn serve as food for the more highly organized flesh-eating animals; and it is a settled thing that men, by assisting the operation of nature, can, to a very considerable extent, and in many, if even not in all, directions, promote this activity toward a very considerably increased production. It will everywhere be in the power of the fish culturist, in the same way as is employed for the fishes in question, which are the peculiar object of the breeder's care, to promote the hatching of species of fish which feed chiefly on vegetables, in order that subsequently, when they have reached the proper development, they may serve as food for the choicer fishes.

In the same way one may promote the increase of crustaceans and mollusks, which likewise, to a great extent, serve as food for the nobler fishes. It will, moreover, in many places be an easy matter, by the employment of vegetables, which, in comparison with meat, cost little, such as carrots, peas, meal, and potatoes, to feed a greater multitude of the vegetable-eating fishes and other aquatic animals than the waters of themselves could support. Moreover, one can, if the opportunity offers, to a greater or less extent, provide directly for the nourishment and food of the cultivated fish by the use of all kinds of animal offal, the flesh and entrails of all sorts of fish, birds, and four-footed beasts, which have little or no value.*

By the use of such means, among which cod-roe might, perhaps, be

*Mr. Seth Green says: "It is more profitable to raise trout than hogs."

used with profit, it is in the power of the fish-culturist to eke out the stock of his water to a very considerable degree, which is limited only by the existing access to fresh, running water, or, more properly, on the renewal of the quantity of oxygen absorbed by the water; by an adequate renewal thereof, the number of fish may be increased until the space becomes as closely packed in proportion as a saufjös (sheepfold) usually is. How far one ought to go in this direction experience will in every place quickly teach. If more fishes are placed in the water than it can support, the leanness of the fish will soon attract attention; this, however, will not occur, because in case of need they will mutually eat one another. If the supply of oxygen becomes too small in proportion to the need of the abundance of fish, this will quickly and plainly show itself also in this, that the fish will seek at the surface of the water the wanting vital air and that many will die. Probably only a few waters can be found in this country where it will be possible to carry fish-production to this last extremity, since one observes that the superfluity with regard to the means of nourishment does not exist except in winter and after continued intense cold, which dries up the tributary brooks. The difficulty, and in most cases the impossibility, of furnishing the fishes an extra food-supply during the winter will restrict stocking far below these extremes, which can be reached only by wholly artificial breeding in smaller especially constructed ponds, where the fish may be supported in the manner which may be most closely compared with the permanent stall-feeding of cattle. But even if one, as will generally be the case, finds it most profitable to restrict the stocking of the waters far below this measure, they may still receive and support a quantity of fish which will considerably exceed what would be considered a great abundance.

Further, it is important *to carry on the collecting in an intelligent manner.* This will not be done until one constructs his fishing-implements in such a way that he reaps only the mature fruit; that is to say, that he catches all the fish which have reached the size shown by experience to be the most profitable, which, as before remarked, may, however, be very different, according to the species of the fish which one attempts to produce, and the food one may be able to procure for them. To permit a portion to escape again, in order to reach a greater size, causes, as shown, a greater or less loss in proportion to the shorter or longer time one permits them to live after reaching the most profitable size; to catch them earlier also causes a loss, though of less importance.

But it is important also to harvest the crop at the time of year when it is most fully mature, most savory, and of the greatest value, not merely on account of its flavor but on account of its greater weight; that is to say, in summer from April until the middle of August at the latest for all the autumn and winter spawning fishes. The difference of quality and weight between the summer and winter is remarkably great; it may, in the salmon-like fishes, amount to over the half the weight of the fish in its best condition, and so far as the flavor is concerned the proportion is precisely similar. Professor Rasch states in his book on "The means of improving the salmon and fresh-water fisheries of Norway, 1857," that the Duke of Athole has related the following:

"By consulting my journal I find that I caught this fish marked as a spawner (kelt) on the 31st March, with a rod, two miles above Dunkeld-Broen, and it then weighed exactly ten pounds. Five weeks and two days later I caught it again, and it had, in the short time specified, gained the almost incredible increase of 11¼ pounds, for on its return it weighed 21¼ pounds. The salmon here mentioned was caught and marked nearly 40 English (6 Norwegian) miles from the sea. It had thus in this time wandered this way back and forth, and still had time to obtain the quantity of food which it consumed to produce such an increase of weight. There can be no doubt of the trustworthiness of this fact, because his grace was extremely precise with regard to his marking experiment, and carried for this purpose with him small zinc tags numbered and furnished with the means of fastening them. Thus we find this fish marked number 129, and the date entered in his grace's journal."

This observation refers of course to the salmon and not to the trout, but there can be no doubt that the last species of fish is subject to the same laws as the first, even if not to the same degree.

Hitherto there has at the same time obtruded itself an important hinderance in the way of extensive catching of late fishes in summer, namely the difficulty of preserving them in the best condition for any length of time. To transport them fresh has been possible only for short distances; to salt them so that they will be preserved has also its difficulties; in every case the lake fish loses thereby a good portion of its value as a salable article. These circumstances have certainly had varying influence in restricting the fishery to the spawning time, since the frost has already to some extent made its appearance, although the greater ease then of capturing the fish at the spawning places has of course been the essential motive of the common people for deferring the fishing chiefly to this time.

If the means cannot be found for preserving fish unspoiled in the fresh state for a long time, the profit of systematically prosecuted fish culture will be diminished in no small degree in all places some distance away from the chief means of transit and the trade centers, and the most and best of our fishing streams are thus situated. This is, however, fortunately the state of things; it is in our power to preserve fish perfectly fresh for a long time by a very simple means which is everywhere in this country at hand in more than the necessary quantity, namely, ice. The plan which has hitherto been employed in shipping fresh fish over to England, packing in boxes with loose pieces of ice and sawdust, permits, according to the statement of Americans, the transportation of fish fresh and unspoiled on the railroads to a distance of 800 kilometers, about 490 miles—that is to say, preserves them during such transportation for twenty-four hours; but this will not be sufficient. If one, on the contrary, employs a plan recently introduced into Canada, and freezes the fish in the proper quantity safe and sound in blocks of ice, so that the fish and ice form a compact mass with two to three inches of pure ice between the fish and open air, an operation which can easily be performed in tin boxes of suitable form by packing in a mixture of ice and salt, which gives a cold of 31.45° Fahr., then, if placed in an ice-house, or during shipment packed with care in pieces of ice and bad conductors of heat—moss, which is found everywhere, or sawdust, if this can be had—they will probably be preserved for a long time unspoiled, just as safely as the mammoth which for many thousand years has lain buried in Siberia's ice-fields, and which now and then comes to light in a perfectly unspoiled condition.

There is thus nothing that prevents or discourages fishing—harvesting—from occurring chiefly in summer, when this is associated with the least trouble and the least discomfort from cold and bad weather, and nothing in the way of preserving fish in the most valuable condition—perfectly fresh—as long as it may be found profitable. One has it then fully in his power to bring them to market at the time and the places when and where it will pay best, even if these are far distant and many days are required to reach them. The preservation of the harvest is also just as simple as the preservation of corn and hay.

I shall next briefly mention the habits of the salmon, which in certain particulars differ from those of the lake fishes.

Ancient, fully trustworthy experience has shown that the salmon, like the birds of passage, seek and, with unerring instinct, return to the place where they were born, and equally well whether their birthplace is a mighty stream or a little brook which the salmon in many places cannot penetrate without lying flat on their sides or employing accidental floods in order to traverse the shallower places between the pools where they can find water deep enough. This last phenomenon one has nowadays rarely an opportunity to observe, since nearly all these brooks have long since been fished out, so that for many years no salmon have been born in them. I have, however, personally had the opportunity of observing this fact in the little river or, more properly, brook, which forms the boundary between us and Russia on the south side of Varanger Fjord. It may be forded dry-shod, when it is lowest, at many places, yet salmon are found in most of the pools quite up to the lake in which it has its origin. Every brook, which at least now and then has a supply of water so far uniform that the salmon can swim over the shallow places, can be made a collecting-place for the salmon in any quantity, because one has it in his power to make each of them the point of departure for young salmon in every case with little trouble. The hatching of eggs may take place wherever a spring, or even a brook, supplied with water only in the winter half of the year, is found—in the last event, of course, with a little greater cost of construction, which at the same time will signify nothing compared with the profit. The roe which one gets to begin with he must procure from other places, and from places where good salmon are found; for there are many, sometimes considerably different varieties of the greatest difference in value. After the lapse of two or three years one will be able to get more at a place that he can manage to hatch—under the condition previously mentioned as now existing, the absurd opposition against removal—which it cannot be doubted will occur.

The hatching of salmon-roe is in all respects just as certain as the hatching of trout-eggs; of which it is not necessary further to speak. Only in one direction is the case of the salmon different from that which applies to lake fish. The salmon goes, one or two years after it has been hatched, out to sea, and the nourishing capacity of this is unbounded.* One does not need, so far as the salmon are concerned, in any way to limit the abundance of the eggs which are taken for hatching for fear that the fish will not be able to find food enough for their full and complete development.

But it is not sufficient, as hitherto, to take care of the young till they can scarcely be regarded as fully hatched; one must further protect them until they assume the wandering habit, and instinctively seek the sea. The older method in reality results in destroying at least half, perhaps three-fourths, of the young which one has with care hatched out. Of course the care of the young long continued will involve an outlay for suitable ponds and for the food as well as the tending of the fish; but this outlay will amount to nothing compared with the increased abundance of fish which will spring from it. The mode of procedure heretofore adopted is perhaps the principal reason why the profit of the work hitherto done is so inconsiderable.

It applies to salmon, without doubt, in a still higher degree than to the lake fishes that every attempt to assist natural culture, in order to increase the abundance of fish thereby, and escape the labor of hatching, is perfectly idle work. For the salmon it is evidently the largest rivers that have any importance in this direction; for the lake fish it is essentially small brooks. But it is just in the largest rivers that the unfavorable conditions peculiar to our climate appear most plentifully and with the most destructive power, along with all unfavorable circumstances which in more propitious climates bring it to pass that natural

*Capt. John Ross, who . . . . . . undertook a voyage of discovery to the arctic regions, to find the so-called northwest passage, states, in his report on this expedition, the following: "When spring finally arrived after the first winter, came in sight from the ship a great river, on whose shore Esquimaux gathered, to fish. They thrust unweariedly the whole day their spears at random down in the turbid river, and at every third cast they usually got a salmon. On the vessel there was a fishing-net made of coarse materials, brought along for such a case; this was thrown out, but was broken by the weight of the fish. A new net of coarser material was then quickly knit, and with this 5,500 salmon were taken at one haul. All the empty vessels on the ship were salted full, and a large portion, which could not be accommodated, were presented to the Esquimaux. This describes plainly enough the unbounded nourishing capacity of the sea.

culture in all proportions yields an inconsiderable and even in many cases no profit at all.

It is attempted now, chiefly with the object mentioned, to clear the path through the long extent of our large rivers, to which, partly by nature, partly by art, their approach is obstructed, by the construction of salmon-ladders in different places; however, curiously enough, not in the place most important of all in this respect in the whole country, a place which by itself is more important than most of the other rivers combined, namely Sarpen. It might appear that since no increase of the abundance of salmon has been gained by the introduction of such ladders, and the larger field thereby found for the salmon, owing to the increased natural culture, it will be useless to continue such a scheme. This is, however, by no means the case. Such a plan must on many accounts be considered in a high degree profitable, even if it should cost what one might regard a large sum of money.

Access to the hatching of eggs is nowhere unbounded, while common interest demands that it should be prosecuted on as large a scale as possible. The waters of springs, which may be used for such hatching—and it is only those which have pure water, in the greatest possible degree free from minerals of all kinds in solution—limit the quantity of spawn which can be hatched out, and their occurrence is not particularly frequent. One may, of course, by appropriate arrangement of apparatus, provide for the replacing in the spring-water the oxygen which was gradually consumed during the hatching of the embryos, but one has no means of removing from the water the carbonic acid generated in the place of the oxygen consumed. This will steadily increase by the continued use of the water, and quickly reach such a point that the water will become deadly for the embryos and the young. Long before such a point has been reached the water must be regarded unsuitable for hatching. One cannot generally calculate that people who never see, as adult salmon, any young which they may have hatched out will interest themselves in such hatching. It is vain to expect that the springs which must occur along the upper courses of our rivers will ever be employed for the hatching of salmon to the desired extent as long as the way is not cleared for them to return to their birthplace and they are retained near the mouth of the river for the advantage alone of those who live there.

If one wishes to advance the hatching of salmon in the greatest possible degree, he must, by means of ladders, clear the way to the upper sources of the river, and then hatching on a large scale will not fail to take place along the extent of the rivers wherever the necessary conditions are present. The profit hereof will of course substantially fall to the residents along the lower portion of the river and along the coast outside of its mouth; but some profit will accrue to those resident on the upper waters, and this probably sufficiently great to incline them to regular hatching. This would especially apply to the men on whose rivers sportsmen will be required to buy permission from the owners to catch salmon, a privilege which usually commands a high price. Inland residents have probably a legitimate claim* that they should not be cut off from permission which probably might be granted them to enjoy a portion of the blessing which the sea can give in the form of the best fish which find their way from it up through the streams as far as they are able to advance, and this so much the more since the lower residents under all circumstances will skim the stream and have absolutely the greatest profit from the abundance of salmon which the increased hatching must produce. There is also ample reason for building salmon-ladders wherever it will be practicable to place them, even if regard for the promotion of natural culture can or ought to have no weight. Should it fortunately happen that natural culture in some peculiarly favorable place produce a yield worthy of mention, so much the better, provided only that the dependence upon such a yield do not cripple the work of artificial hatching—the only mode that is perfectly certain—and the work of caring for the young in their tender youth.

In parenthesis I shall here say a few words about the attempt to hatch salmon in lakes, with a view to keeping them there. This is, in my estimation, a complete misconception of the problem. In the first place, one cannot, according to my belief, destroy the wandering instinct of the true salmon (*Salmo salar*) by placing it in a lake. It will certainly find its way out of this into the sea just as surely as out of a river; the one is just as easy as the other, and one cannot destroy instinct. Only by keeping them confined in a basin from which they cannot possibly escape can this instinct probably be controlled, and this is attended with a danger of their leaping out on land and perishing. In the next place, the nourishing capacity of every lake, even the largest, is limited in comparison with that of the sea. To confine sea salmon—wandering salmon—there, even if it were practicable, would also be to subject them to comparative starvation instead of plenty. As regards the Venern salmon, which persons have attempted to introduce into our country, it is, in practical and economical respects, precisely the same as our female trout. The fact that some have been pleased to call it a salmon does not by any means make it a sea salmon, or a true salmon in any respect. So far as flavor and weight are concerned it differs in nothing from the female trout in Mjösen, and the corresponding large trout in our other larger lakes, and it is therefore in my opinion both unnecessary and unprofitable work to introduce it into our waters.

Salmon are born in the river, live there a short time in their tender youth, and then seek the sea, where they grow with astonishing rapidity.

* Already the Norwegian parliament has recognized the legality of such a claim as fully as it could be done at the time, as it prohibits the barring the way of the fish from the beach to the river source. This law, however, in a space of time was forgotten, and industry has been allowed without complaint here and there to block the way of the salmon where it was open before. That this happened without complaint was probably only because the abundance of fish was already diminished to such an extent that this barring of their way made no difference.

When, driven by instinct, they again retrace their way to their birthplace they become on their way along the coast, which they always follow, the object of a fishery of the greatest economical importance, because they are then in the best condition, which they quickly lose after their entrance into the river, where, according to experience, they take little or no nourishment. Residents on the sea-coast are thus in all respects at least as much interested in the hatching of salmon in the greatest possible quantity as the dwellers along the river courses. This community of interest is not yet comprehended in its full extent. Since the coast-dwellers as a rule are cut off from the opportunity of working personally for the production, or even for the preservation, of the salmon supply, they harvest what others have sown, and ought therefore rightfully to be obliged, in proportion to their catch, to share in the expenses which the hatching of the young salmon in the river-courses involves. A law of this kind would be eminently just, and would doubtless in a high degree advance the profit for all, but first and to the greatest extent for the coast-dwellers themselves.

Law has now established a certain size of mesh for the implements for catching salmon. This should gradually be increased so that only full-grown salmon of twelve Danish pounds weight, 6 kilograms, and upwards, can be caught. The salmon is then four years old. It will doubtless be urged as an objection to this that they will thereby lose permission to catch the sea-trout. This will certainly become true so far as the salmon-nets are concerned. But since the sea-trout are of inferior value compared with salmon, and one will gain more by allowing the salmon to become mature than by catching the small ones, the loss of the sea-trout in the salmon-nets cannot be taken into consideration. One must resign the capture to the common set-nets. At the same time the now legal sale of young salmon 8 inches long should be absolutely forbidden; under a size of 16 marks (8 Danish pounds), four kilograms, should no salmon be allowed to be sold. Moreover, the necessary active attention must be exercised in order that such a provision of law, as all others, may be properly respected; because only in this way will all persons interested be able to reap the full share, which, according to the circumstances existing on their soil, rightfully belongs to them. It is evident that the coast population have it in their power to capture nearly all the fish which come into the rivers, just as the residents on the river banks are able to omit the hatching of the young, the plain result of which will be that the coast people can get no salmon, since after a shorter or longer time no salmon will be found in the sea along the adjacent coast-extent. Both categories of proprietors' interests are thus closely dependent each upon the other. It would therefore seem that it would be to the greatest advantage of all the fishery-owners in the naturally-united districts whose interests are thus consolidated, if the whole matter within the district were carried on in partnership, as well with regard to the planting as the harvesting. Without the aid of suitable legislation such an agreement could not be rendered feasible; a majority should have the right to control the minority; because one or more perverse ones are found everywhere, and such associations serving the common use should not be checked by a single person or a few persons, whose interests probably may be of little importance compared with those of most of the others.

In comparison with the cultivation of the kinds of fishes related to the trout, whether they be in the sea or fresh water, the cultivation of the salmon will always be beyond comparison the most profitable, taken as a whole, for all the interested persons in partnership, or for the individual who can secure for himself the greater portion of the profit of the planting of the young which he must hatch and rear and then liberate into the sea. There is of course an important difference in the rapidity of growth between different species of lake as well as salt-water trout; but if one institute a comparison between the species of both kinds of trout, which may be considered to represent the middle class, and salmon, the relation will show itself to be as follows: After being hatched out in the spring the young salmon remain in the river where they are born until the next spring; then about half of them change their markings and go to sea. Whether these are stronger individuals or one of the sexes is not known. The other half remain in the river until the spring following this, then change their dress and go to sea. By continued culture also after the first year a portion of the whole quantity of the young hatched out in the course of the winter will always go to sea. It is also unnecessary to take account as to which year's fish they represent. Those which forsake the fresh water have a length of 6 to 7 inches and a weight of 125 to 350 grams. These young have been marked to find out their subsequent growth, and thereby it has been found that even in the autumn of the same year in the spring of which they went to sea at least a portion of them returned to the river, and had then a weight of 1 to 3 kilograms, an average of 2 kilograms. When they next return from the sea they weigh from 3 to 6, or on the average of 4 kilograms, and are then in their third year. The next or fourth year they reach an average weight of 8 kilograms, and so on, but in a diminishing scale. According to the experience in many different fish-culturists' establishments in Europe and America the increase of the trout with a good supply of food may be estimated for corresponding ages, respectively, at 0.275, 0.650, and 1.500 kilograms. Comparing these numbers, the proportion shows itself to be, for the salmon and trout in question, at the close of—

	Salmon.	Trout.
The second year, average weight.....	2 kilograms.	0.375 kilograms.
The third year, average weight......	4.5	0.750
The fourth year, average weight.....	8	1.500

This shows what a considerable value salmon culture has compared with trout culture. The yield from the first is, after the fourth year, more than five times as much as the last; and even this will yield a good profit, as will be shown later. As is well known, the export of ice has in the last ten years steadily increased in extent and economical importance in this country. In many places along the coast where, naturally, fitting opportunity has offered, ponds are constructed with considerable expense for the production of ice alone for export. Nothing is more natural than to use these ponds for fish culture along with ice production, to which they are all, to a greater or less degree, adapted, according as the supply of running water and its quality may be. The expenses which fish culture will occasion are nothing in comparison with the cost of procuring ponds by means of damming brook-courses; but the yield from fish culture, carried on with care, may be of very great importance, especially if circumstances allow one to select salmon as the object of culture. It will, at all events, give a very good return for the outlay which the apparatus for fish culture and the labor upon it demand.

In the same way as one has found it to answer a good purpose to procure ice-ponds at considerable expense, will one in many places where the opportunity offers be able, with profit, by damming, to construct larger and smaller fish-ponds at such a distance from the sea that the production of ice will have no other importance than as a means of preservation for the harvested fish. Such an opportunity is offered in many places in our mountain districts, where the ground which must be sacrificed to transform it to a lake bottom instead of solid land cannot be considered to have any value in comparison with that which it will acquire by being transformed into a field for fish culture. One has it thus in his power, to an extent which perhaps must be called very considerable, to enlarge the field for this culture, already very great in proportion to other countries, which, as shall be shown later, has a more varied economical importance than agriculture on good ground of corresponding extent.

I shall next briefly mention—

III.

THE FISHES WHICH SHOULD BE THE OBJECTS OF CULTURE; ALSO THE KINDS OF FISHES AND OTHER AQUATIC ANIMALS WHICH SHOULD BE REARED AS FOOD FOR THESE.

Since I, as before remarked, have our lakes and rivers specially in mind, I shall first treat of the species which live exclusively in fresh water, and then of those which live both in fresh and salt water.

Following the prevailing taste among us, as also to a great extent in other countries, the genera and species of fresh-water fishes should probably be taken into consideration in the following order: Trout, red char, gwiniad, grayling, perch, perch-pike, pike, and crawfish, besides those improperly included with the fishes in familiar language. To these some persons would add the bream and lake, which, however, for a reason which I shall mention hereafter, I place out of consideration.

TROUT (*Salmo ogla, ferox, fario, punctatus*) are the most widely distributed of all our fresh-water fishes. They are found from near the level of the sea up to the snow limit, and at elevations over 1,000 metres above the sea they are the only indigenous fish. Just as the localities differ in which they occur with regard to temperature and quantity of water, its mobility, and the nature of the bottom, so the trout, which are considered by naturalists as belonging to the same species, differ in form, size, color of the skin and of the flesh, and in flavor. Although especially well-flavored in places which are suitable for its thriving, in less favorable and turbid rivers it may be of very inferior value compared with other fish, and it suffers at all events from a circumstance produced by its characteristic taste, that one can scarcely find another fish from the sea or the fresh water of which one becomes weary more quickly. In order that the trout may acquire their flavor, clear cold water and a stony bottom are necessary. They thrive of course in streams with a muddy bottom, especially when these have a steady and tolerably strong influx of fresh cold water; but then they seldom have the same flavor if they should become very fat and large. If the influx of water is smaller, and is liable to cease entirely in the summer, so that the water in the lakes is considerably heated, they are not suitable for trout; and this condition is fulfilled as a rule in all regions of country which are situated lower than 250 to 350 metres above the sea-level; also in all smaller lakes. It applies even to the largest rivers rising in the mountain-tops—more especially, though, at an elevation less than 100 metres above the sea-level. Fresh-water trout should, therefore, scarcely be made an object of culture in small streams lower than 150 metres above the sea-level, or at all events only where the water-course is pretty uniformly supplied with cold water the whole year round by inflowing brooks or springs. It may of course live, and through bountiful feeding reach a considerable size, in comparatively small streams or artificial basins in a low country near the sea-level, but it acquires there a flavor far inferior to that which it has in elevated regions.

What particularly influences the flavor of the trout naturalists have not with certainty fathomed. It is assumed generally that cold water with more stone than mud bottom, and rich in insect larvae, and especially in the smaller crustaceans, offers the most favorable conditions for this species of fish, especially when the supply of this kind of food, as is the case in many places, is so great that the rapacity of the fish is either not at all developed or but little developed, as is plainly shown to be the case in various waters by the whole form of the fish, and as the vain fishing, with baiting arrangements of the most tempting kind in other waters, appears fully to confirm. But even where the supply of food from the insect world is smaller, and where thus the preying instinct manifests itself, the trout becomes very well flavored if there is only a sufficient supply of food and the water is clear and especially cold. The best-flavored trout are found therefore in the waters situated the highest, even up to the snow-fields; but where a certain water contains trout of particular appearance and flavor, scarcely two are found alike. The trout spawn, as it appears, exclusively in running water, especially in brooks which empty into a lake or larger river in which they stay, and then press towards the head-waters as far as they are able to advance. There is, however, a theory that they also spawn in still water; but about this there is no positive knowledge.

The RED CHAR (*Salmo salvelinus* and *alpinus*) lives in fresh water from the level of the sea up to a height of about 600 metres above it. In southeastern Norway it is a rare fish, since it seldom appears at the surface of the water, and is not easily caught with the implement usually employed to catch the trout—the artificial fly. In western Norway, on the contrary, it appears to be much more common, shows itself more plentifully, and is caught by the fly in greater numbers, than the trout. How far north this applies is unknown to me, but it is at any rate the case as far as Trondhjen. With regard to flavor the chars vary as much as the trout. In southeastern Norway they have as a rule a finer flavor than the trout; especially in certain waters in western and northern Norway it falls far below this in flavor.

How far the char can thrive is at present, so far as is known, not determined. They appear to thrive best in deep, cool waters, with the uniform afflux from bottom springs or brooks. It is strictly a lake fish; it spawns in the same water and does not ascend the brooks which empty into or flow out from it; at any rate not here in southeastern Norway.

It is a common assumption that the char and the trout do not agree well in the same water. The statement has been made that when the char is introduced into water where only trout were found before, the abundance of these was diminished. Although the char certainly may be regarded as a voracious fish like the trout, it appears, however, to judge from its whole structure, to be so in a less degree than the trout. The theory of a direct war between the adult fishes of both races mutually is thus untenable, neither can this reputed opinion be explained. It has been supposed that the char destroys the trout eggs and young with greater voracity than the trout consumes the char's. This statement appears to me entirely untenable. The trout spawns in brooks wherever it can reach them, and its young remain in them in their tender youth; while the char never frequents the brooks. It can thus not do any harm to these eggs or young of the trout. The char, on the contrary, spawns in the very water where its eggs and young are entirely exposed to the attack of the trout, and it refuses them certainly just as little as other fish eggs and young, its own included. The relation is thus rather the opposite of what is supposed. But whatever may be the relation in this respect, and even if the supposition were well founded, the admitted relation cannot be taken into consideration where energetic cultivation is in question, which does not rely upon natural culture for the preservation or the increase of the abundance of fish. In reality it may possibly be that the perfectly innocent char has been blamed for that for which the sea-worm is held responsible elsewhere, while the real culprit has been man and him alone, who has fished with stupidity and rapacity, and will not comprehend and confess that he has done his utmost to kill both the goose and the goslings, in which he has finally succeeded; because with this explanation it is clear that the more easily accessible trout will be captured and diminished in numbers more rapidly than the less accessible char.

If in possession of water in which it is admitted that the char will thrive and acquire a desirable flavor, this and the trout may unquestionably be raised together. That they will fight for the food at hand is certain and unavoidable; but that the char should diminish the abundance of the trout is plainly inconceivable if the supply of both kinds is regulated by artificial hatching and protection of the tender young in the quantity which may be found suitable for the proper stocking of the water in proportion to its nourishing capacity. It may, of course, happen that one of these kinds of fish will find better conditions for natural culture than the other, and thus increase proportionally more in number, but in such a case one will have it in his power, by limiting the artificial culture of the favorably situated species, to restore the desired proportion.

The char is considered not only here in South-eastern Norway as the finest, best flavored fresh-water fish, but the same value is attached to it also in England, France, and Southern Germany, where it occurs in the natural or cultivated state.

GWINIAD (*Coregonus oxyrhynchus, lavaretus, fera, vimba*) make their home in the less-elevated larger lakes and streams, where they, presumably because of strong persecution alone, for the present reach but a small size. Experience in the attempt of introduction shows that it thrives particularly well in our more elevated mountain lakes, where it reaches considerable size, fatness, and flavor. While it nowadays in this country, where it occurs most generally, seldom reaches a size of two to three kilograms, in Finland it grows to six kilograms and over. It is generally known as one of our best fishes, which one can eat more freely without becoming tired of it than trout or salmon. It is not, like the trout, what is called a gluttonous fish, or a fish of prey, since its mouth is small and not so armed with teeth that it can seize a somewhat developed fish, even though of small size. It lives, therefore, on water-insects of different kinds, but also, like all the others, on spawn and young fishes. Its spawning-time is late in the autumn, and for a spawning-place it chooses a stony bottom in shallow water. It is comparatively more fruitful than the trout, since its eggs are of somewhat smaller size.

This fish is admirably adapted for rearing along with the trout and the char in the same water, since it not merely on its own account is well worth cultivation, but also because its young are excellent food for the larger trout, which the char on its part cannot injure. How high it will thrive we have no experience of; the highest place known to me, where it is planted and thrives remarkably, lies about 900 metres above the sea. It appears thus not absurd to suppose that it will also thrive under conditions otherwise favorable. But if one does not wish to make this a special object of cultivation, and will regularly hatch out and raise a proper number of young in proportion to the area of water which he may have control of, he should, however, not neglect to stock the water with it as a contribution to the food of other fishes. It will in this respect possibly prove to be a kind of fish that will thrive highest above its hitherto common place of resort.

It might, perhaps, be supposed that the planting of the gwiniad in such places where the already indigenous fish, the trout, finds so rich a supply of food that it does not devour all of it, and thus does not need any kinds of smaller fish or young fish as additional food, would effect an unnecessary competition about food between the gwiniad and the trout, and thereby a diminished profit from the fishery as a whole. That this possibly might become so by thickly stocking with both species together cannot be contested. In such a case one obtains restricted planting in the aggregate. But I should, however, think that rearing the gwiniad by the side of the trout, whereby a variety in fish-food becomes possible in places where they have hitherto been confined exclusively to trout, will prove a source of real profit.

The GRAYLING (*Thymallus vulgaris*) has hitherto been confined to the same localities as the red char in the eastern part of the country. It is not found westward of the rivers of Laugen, Vormen, and Glommen, and the tributaries falling into these. It resembles the red char more closely than the trout in form, color, and flavor. It thrives particularly in cold clear waters with afflux of larger brooks or rivulets, with a strong or even rapid current, against which it at certain times of the year advances, on which account also it is called, in certain places, current-grayling. There appears to be good ground for supposing that it will thrive in suitable places as well as the red char; but as far as is known no attempt has been made to introduce it into other lakes or rivers than those in which it naturally exists, neither has it, so far as is known, hitherto been the object of hatching. It has, upon the whole, been less noticed, just as it is comparatively rarer, than the red char, wherefore I have taken it into consideration after this. The flesh of the fish is savory and very wholesome; it takes the artificial fly readily; and, finally, it spawns in the spring, while the fishes previously named spawn in autumn and winter. The first-named quality recommends it as an article of food for cultivation in the same degree as the red char; the other recommends the grayling as a game fish for every one who, along with the profit of fish cultivation, also prizes fly-fishing as a pastime; and the last-named quality makes it just as worthy of recommendation to be bred along with the red char where circumstances permit, since it naturally lays its eggs in the open water, and the fry proceeding therefrom will serve as food for the whole planting of fishes at a time of the year when neither trout or char eggs or young are present; while the artificial hatching which is necessary for the regular stocking, without requiring special apparatus, may take place in May to June in the empty apparatus used for the hatching of trout and char eggs. For the nourishing of the young in their tenderest youth only is required a particular apparatus of little extent and expense. It is just as little a fish of prey as the red char, and possesses nearly the same fecundity.

With these four kinds of fishes the list is complete of the salmonoids, or salmon-like fishes, which ought to be special objects of cultivation as articles of food. There are still a couple small fishes belonging to the same principal division, which also deserve to be taken into consideration as objects for cultivation; but since they are supposed to deserve to be considered as a natural food for the larger and more valuable fishes rather than as food for men, for which, however, to some extent they are used in the places where they now occur in any quantity, and since they, as it appears, will thrive and reproduce only in larger lakes with bottoms of a certain quality, I will refer to them under the kinds of fishes that ought to be hatched as food for the larger and more valuable fishes. Should it happen that by such cultivation they will become so numerous that they might also be considered as food for people, so much the better.

Of another chief division of fishes, the spiny-finned, which inhabit our fresh waters, I have next cited two kinds to be considered above all others, although it is generally supposed that they cannot be compared with the preceding in flavor. About this, opinion may be, and is much divided.

Of these I have placed the PERCH (*Perca fluviatilis*) first, since this species is most widely distributed. It is found generally more or less numerous in nearly all rivers, brooks, and lakes, even up to the spruce limit. Whether it will live and thrive at greater elevations is, so far as is known, not determined, probably because it is readily eaten by the common people in the mountain districts, but is not regarded particularly by the side of the trout, which is more generally distributed there. It bears very different names in the different parts of the country where it occurs: Abor, tryte, skjebbe, &c. It lives and thrives in all kinds of lakes, small and great, with or without constant afflux, and is just adapted, therefore, for cultivation in such lakes in which the previously-mentioned kinds of fishes will not thrive properly. It is decidedly a rapacious fish, and it is supposed, therefore, by fish breeders in Southern Europe that it ought to be excluded from the lakes in which the *salmonidæ* are cultivated. Experience in our country seems, at the same time, to show that it cannot be very dangerous to the trout or the red char, since it occurs up to their present limits, as well as in all waters along with both of these, and without, as far as is known, doing them any serious injury. It is a slothful, sluggish fish, while the trout is lively and particularly quick in its movements. I assume, therefore, that there should be no hesitation in raising the perch along with trout and other salmonoids, if one cares to cultivate this kind of fish and the opportunity to treat it by itself is wanting. The flesh is found by many, when properly prepared, to be peculiarly well flavored, especially as a change from other fish, and it is highly nourishing. With the necessary supply of food the perch reaches a considerable size, up to one and a half and two kilograms; but if one wishes to gain this size, care must be taken that the fish obtain abundant nourishment from other smaller fishes of different kinds. It spawns in spring, May to June, and is peculiarly fruitful.

The genus of fish most closely related to this is the PERCH PIKE (*Lucioperca sandra*), one of our largest lake-fishes, and also one of the most savory; it grows to a size of 12 kilograms and upwards. It is found at present in only a few of our larger, least elevated lakes; it is not found in this country north of Öieren. As far as known, no one has attempted to introduce it to other lakes, although there are a great number of such in which it doubtless would thrive well. Since the fish is very little known, I shall state that its flesh is as white as the cod's, likewise just as free from bones, and it is, although of a somewhat more decided flavor most nearly like that of the turbot—rated by most persons who know it just as highly as the best cod. Its natural place of resort shows that this fish is not adapted to the colder, more elevated lakes, and that on this account it cannot be expected to thrive where the conditions are particularly favorable for the thriving of the trout. Since it is, besides, a predaceous fish, which is considered more voracious than the perch, it should not be bred together with trout or red char. While, on the contrary, the gwiniad remains more at the surface of the lake, while the perch-pike, by preference, remains at the bottom, there can be no objection against planting them in the same lake. That care must be taken to supply sufficient food, if one desires a rapid development of this fish to a suitable size, is self-evident. It spawns in spring, and is very fruitful.

Of the soft-finned fishes, I will state that only one species will be taken into consideration for cultivation, namely, the PIKE (*Esox lucius*), the most voracious of all our fresh-water predaceous fishes. It is comparatively little distributed in our country; generally it can be said to be only in the eastern parts, where, it is true, it ascends in the great rivers to a height of 250 to 300 metres above the sea in the Glommen; but in smaller lakes and in the larger western rivers it occurs scarcely higher than 200 meters above the sea. In the whole southern and western part of Norway it occurs in only a few places, where it appears to have been introduced, while in the eastern and northern portions of the Scandinavian peninsula it is generally distributed far into the interior; it is found thus in many lakes on the boundary between East Finmark and Finland. In the region of the country where the pike is indigenous it is a well-esteemed fish, though, of course, not so highly esteemed as the trout or the gwiniad. In the portions of the country, on the contrary, where it has been introduced, and this applies to the region to the west and north of the valley of the river Drammen, and to this valley itself, in which last it was introduced not many years ago, it is less prized as a means of food. It is, taken on the whole, an indolent fish, which, however, when it sets out to seize a victim, can move itself very quickly. It is thus a dangerous enemy of the trout, and annihilates this wherever they are found in the same lake, in the regions which are favorable for its thriving, namely, the stiller, shallower, and thus warmer portions of the same. Syrifjord, at Ringeriket, offers a good illustration in this connection. In its tributary, Stensfjord, and the shallower portions of Tyrifjord itself, the pike rules, and it is a rare thing to get a trout without much walking. In Holtsfjord branch, on the contrary, it is much deeper and colder, and in its immediate vicinity the trout is the presiding fish; and here it is just as great a rarity to capture a pike as it is to catch a trout in the other places.

With abundant supply of food the pike grows rapidly, and it may reach a very considerable size, 12 to 18 kilograms and more. In all lakes of lower elevation, for instance such as are liable to want supplies in the great drought of summer, the pike may profitably be cultivated along with the perch, which is its usual associate wherever it occurs, and the pike-perch, which is very nearly as voracious a fish of prey. It spawns from February to midsummer, and is, like the preceding species, particularly fruitful; its own offspring, therefore, constitute a large part of its food.

The CRAWFISH, like the pike-perch and the pike, frequents the less elevated watercourses in the eastern portion of the country. Up to a few years ago it was not found west of the valley of the Glommen, and in this region not higher than a hundred metres above the sea, in the smaller and warmer rivulets, brooks, and lakes. In the Glommen proper, it is very rare and does not extend up to Mjösen, probably because of the greater coldness of the water than in the tributaries. About twenty years ago it was transported to Stensfjord. It appears that it has not distributed itself there very much. Although comparatively little distributed, and thus little known in this country, it is liked by many where it is found. It might with ease be transported to many localities on the western low-lying rivers and lakes, and its cultivation and fattening are very easy.

With these I think that the list of fresh-water inhabitants worthy of cultivation for food ought to end. The fishes belonging to the last-named principal division (soft-finned), the extensive family of carp-like fishes, *Cyprinidæ*, of which many kinds are found in this country, and among which the bream in particular is the principal one and the most widely distributed of the largest of this family, I think do not deserve especial cultivation, except as food for those previously named, in places where any of the cyprinoids will thrive. They all stand—at all events those which are now at all abundant in this country—far below all of the previously mentioned in flavor, and are distinguished by an abundance of bones of the finest and most pointed kind, which make it a troublesome, sometimes painful, matter to eat them. The carp proper, which in more southerly countries has been and now is an object of breeding, is found in only two places with us, where it is introduced, although it is said that the attempt of its importation in ancient times was made in many places, but without success. What value the carp may have as a source of food in comparison with the species of fish named, I cannot decide. I am inclined to believe that it does not much surpass the bream in this respect. Should any one wish to undertake to cultivate it, places may be found, of course, where this may be done successfully, but it must take place in waters of entirely different quality than those in which it is said that such culture was attempted before. Low-lying warm waters, deep in places, with muddy bottom are their proper haunts. All carp-fishes are very prolific, and live principally on vegetables partly rotten. They are thus just suited as food for others—carnivorous fishes.

The same reason which appears to disparage undertaking the culture of any genus or species of the *Cyprinidæ** does not, on the contrary, apply to the LAKE (*Lota*), which is a rather generally favorite food in the early months of the year wherever it occurs, namely, in all the great watercourses of Southeastern Norway up to quite a considerable height; in the Glommen as far as Röros. This single fresh-water representative of the cod family is a greedy fish of prey, and is possessed of very great fertility, so that, on the one hand, it is little adapted for the stocking of waters where other more esteemed species of fish are bred, while, on the other hand, its natural fecundity is sufficient to maintain the race in proper abundance; from which it happens, furthermore, that its haunts have such an extent that individuals rarely have any particular interest in their increase. Should we ever proceed so far that systematic cultivation, for the interest of all, of the abundance of fish in the large lakes or larger divisions of our greater water-courses should be practiced, and should one then ever consider it of interest to increase the supply of this species of fish, it will be very easy, by hatching out the spawn, which can be procured with ease in great quantities, to provide for the development of great multitude of young.

It cannot be open to question that one, by artificial hatching, and by protected rearing in their tender youth of the young thus produced, as long as the nature and mode of life of each particular species of the

*If one wishes to describe this great family of fishes by the quality most apparent to people in general, one should style them the fishes crammed with fine bones and with insipid flesh.

cultivated fishes require, will be enabled to increase the abundance of fish in our lakes and rivers to a very considerable degree. All of these waters at one time were much richer in fish than they now are, and it is thus a certain thing that, without any care in this direction, they will supply nourishment for a considerable increase of a number of fishes in the future just as well as they have done in the past. But if one enters into the cultivation of fish, it will be just as unreasonable to stop at the limits which the natural fruitfulness of the water establishes as it would little reward the farmer to neglect to increase the natural fruitfulness of the soil, especially as it is just as easy, perhaps much easier, to produce an increase of the nourishing capacity of the water as to furnish to the soil the materials which are necessary to increase the crop.

Such an increase of the fertility of the water may be brought about in many ways. The simplest, as before remarked, will be, along with the fishes which are the objects of culture, to hatch out, and in the proper degree rear, other fishes, crustaceans and mollusks, as food for them, to the extent which prevailing circumstances in every place will allow; and these conditions will be found to vary to a considerable extent. Along with these means one may naturally also employ artificial—if one will unnatural—food, and thereby augment the yield in an extraordinarily high degree; but the opportunity of procuring this kind of fish-food is in a still higher degree different from increasing living organisms. I will therefore, place it out of consideration.

As fish-food are adapted, chiefly—

ÖREKJYTEN, Gorkjyte, Gorkim (*Phoxinus aphya*), which occurs as far up as 900 metres above the sea. It may be that it, like the gwiniad, will admit of being planted and thrive in still more elevated lakes and brooks than those in which it occurs now. One thousand two hundred to 1,300 metres above the sea-level should be regarded as the limit of the occurrence of the trout, and it is perhaps not impossible to introduce the Örekjyten to this height; if not everywhere, at least in certain places, where the proximity of the "*Snebræernes*" is not too great.

It multiplies rapidly, and since it spawns in or after midsummer, its spawn or young will supply the young cultivated fishes of all kinds at a time of the year when other spawn and young do not occur. It is at all times a cherished food for trout. Where this fish will no longer thrive, on account of the elevation above the sea, one has probably no other species of fish to plant in its place as food for the larger choicer fishes, except young gwiniad (Sik), as far as these will thrive upwards. The minnow is regarded in many places in the country, as before intimated, as injurious, since it is with reason supposed that it destroys the spawn and young of the choicer fishes in great quantities. Thereby it counteracts, just as all other fishes without exception, the natural culture of all fishes; but this circumstance will, as repeatedly stated, lose all importance when we, as a fundamental condition for fish-culture, begin by constantly providing for such a planting of fully-developed young fishes of the kinds which one prizes as objects of especial care, as corresponds with the existing nourishing capacity of the water-area. The more minnows which under such conditions occur, and the better they thrive, the more profitable for the development of the species of fish which one desires to produce as the particular fruit of culture.

A species related to this—

The MÖRT (*Leuciscus rutilus*), occurs in the eastern and southern portion of the country in nearly all the waters and rivulets with a current not too strong and with a muddy bottom, up to a somewhat lower level than the preceding. It will probably also admit of being planted in waters more elevated than those in which it is now found. It is, as the most of the genera belonging to the great cyprinoid family, particularly fruitful. It spawns in spring, in April and May. There is no difficulty in the way of its introduction in great numbers, if it be found necessary in this way to insure that the supply of this food may not diminish too much, which may easily occur with merely natural reproduction, where the spawn as well as the tender young are the defenseless prey of every fish, small and large, of other and even the same kind. At a still lower height occurs another cyprinoid—

LAU or LÖJE (*Alburnus lucidus*), which occurs in large lakes and rivers, but resembles the mört in habits and appearance; the most evident difference is that it has a yellow iris, while the mört has a red one. It spawns on grassy bottoms, at the same time as the mört, and might perhaps be introduced to more elevated waters. Both of these species feed on aquatic plants as well as aquatic insects. It thrives thus in a less degree than other fishes of choicer kinds, if the insect world yields it nourishment.

The KARUDSEN (*Cyprinus carassius*) lives in pools and ponds with muddy bottoms, up to a height of 150 to 200 metres above the sea-level. It will probably easily admit of being transferred to more elevated places in still waters with muddy bottoms. It lives, as the two species previously named, to a great extent on vegetables, occasionally rotten ones. It spawns in May and June, and is very fruitful. The *Karudsen* may reach a very considerable size, even up to one kilogram, and are then considered well flavored.

Of the salmon family occur, as before intimated, in the great Scandinavian lakes, two species, which may be mentioned.

SLOM or NORS (*Osmerus eperlanus*) and LAKE HERRING or SIKLÖJE (*Coregonus albula*) appear abundant in very numerous schools where they are found. Both occur in Mjösen, where the last, at the spawning time in autumn, is now and then caught in great multitudes, many hundred barrels. In Sweden, also, the nors is caught in great abundance at the spawning time, which occurs in March to April. The attempt has been made to introduce the lake-herring into many of our larger lakes, but the attempts hitherto made have, so far as is known, not succeeded. If sufficient care should be exercised, it might be possible to introduce both species to many lakes, and to greater heights than those at which they now occur; and they will then, even if one may not wish to use them as food for mankind, be particularly useful as food for other larger fishes which persecute them with eagerness.

As being closely related to fishes, must another class of animals which may serve as nourishment both for larger and smaller fishes not be left out of consideration, namely, frogs and toads. Just as the grown animals are relished by the larger fishes, so are their eggs and tender young a favored food for smaller fishes of choicer kinds. To transfer these animals from one water to another will, of course, hardly be successful, since they are just as much land animals as they are water animals; but nothing is easier than to transfer their eggs or spawn laid here and there to basins or waters where it may be of use.

As is seen from the foregoing brief representation, the ability to procure natural food for the larger fishes is all the greater and easier the lower the water-basin lies which shall be prepared or preserved for fish-culture. This is a very favorable circumstance, because the lower the level becomes, the higher the water's average temperature, and the less uniform the water supply, the more voracious, as it appears, become the species of fish which may be made the object of successful culture. Even the trout becomes in the less elevated warmer waters more of a predacious fish than at greater heights. Whether this is because of the greater wealth of insect food in the elevated regions than in the waters of the lowland, or owing to other circumstances, we have as yet scarcely any certain knowledge or conjecture about; the fact is in the meantime as stated. A commonly received opinion, especially with regard to the trout, which, with the exception of the red char, is regarded as the best of all our fresh-water fishes, perhaps because it is the most common—is at the same time that fish of the salmon family become all the fatter and more savory the greater the supply they find of insects, crustaceans, and mollusks. This supply is meanwhile, as experience teaches, very different in different waters. It shows, therefore, of itself that also in this direction one ought to provide, as far as possible, for procuring himself a supply of food by stocking the waters in which fish-culture is prosecuted with the species of these kinds of animals which admit of being transferred from one place to another.

The same opinion is stated by Prof. G. O. Sars, in a prize essay on Norwegian crustacea, issued as a publication of the university in 1865: "One will be able to form a still clearer idea of the enormous numbers of these animals if one reflects that they, notwithstanding their small size, constitute the chief food of most fresh-water fishes, a fact which is sufficiently established by Leydig, who, upon the dissection of very many different kinds of fresh-water fishes, always found the stomach's contents almost exclusively to consist of entomostraca and copepoda. It is thus most highly probable that the good quality of fish in different localities largely depends upon the greater or lesser quantity of these small animals, wherefore, even from practical considerations, in the artificial fish-culture so zealously carried on in our time, they certainly ought not to be left entirely out of consideration. They will especially recommend themselves as wholesome and efficient nourishment for the still delicate young fishes, just as, also, certain species deserve to be planted in larger fish-ponds. These small animals play a still more important role in the great economy of nature by filling all the stagnant pools and swamps with their countless multitudes."

The recommendation suggested by the professor concerning the cultivation of these crustaceans as food for fish-fry was about the same, practically, as that attempted by Mr. Sauvedon in the vicinity of Paris in 1864, '65, '66, and '67, and has completely answered the purpose.*

The aquatic insects which, after passing through the larval condition in the water, take on a winged form, it will perhaps be difficult to transport from one place to another; it will, however, be worth the trouble to attempt it. The animals of the extensive crustacean or crawfish family, which live in mud or among the stones at the bottom of the water—which by common people are called Grundaat, because they destroy fishing implements placed on the bottom—easily admit, on the contrary, of being transported from one body of water to another; and since they, with little care, multiply with great rapidity, there will be no difficulty in making a supply of this nourishment ample in the basins which may serve as abiding places for the younger fishes. Since these basins must stand empty at certain intervals, it is easy to provide that a new after-growth of such crustaceans shall take place when the basin again is to be used, if the quantity of these should be improperly reduced by a preceding stock of fishes. How far the free-swimming crustaceans, which sometimes occur in countless quantities even in the most elevated mountain streams and throughout the larger rivers, will admit of being introduced to waters where they do not occur, is a question well worth examination. I have seen the Vormen so full of these animals that it appeared as if there were one to two animals in each cubic inch of water, at least nearest the surface.

They think they have discovered in Scotland that the fatness and the flavor of the fish in certain waters was really to be attributed to a mollusk, a *water-snail*, which lives in these waters in great abundance, and this species of snail has, therefore, already been introduced into this country, but is hardly distributed as far as it is desirable it should be. The introduction of these animals into new waters in which they can live is not difficult, and since their fecundity is very great, a great wealth of food will very quickly be produced for fish whose cultivation may be attempted by their introduction.

Of the species of fish which belong as well to salt as to fresh water there are only three, the SALMON (*Salmo salar*), the SEA-TROUT (*Salmo eriox* and *trutta*), and the EEL (*Muræna anguilla*).

---

* Bulletin de la Société d'Acclimatation, December, 1867.

The two first-named kinds have been previously so far mentioned that I shall in this place merely add a few remarks. Both are born in fresh running water, both remain therein during their earlier youth, and go therefrom to the sea, where they quickly develop, after which they again, driven by the instinct of reproduction, return to the places where they were born. Some have thought that this returning, at all events so far as the salmon are concerned, is also caused by the instinct of freeing themselves in fresh cold water from certain parasites which infest them in the sea in summer; but this appears scarcely probable, since the lake and river trout are also attacked in summer by similar animals. The return to fresh water takes place earlier in the case of the salmon than with the sea-trout, which last, as a rule, first enters the mouths of the rivers towards autumn. Owing to this circumstance, one has a longer fishing season for the first than for the last.

It is thought that the salmon and the trout do not agree well, but conflict with each other about the spawning-places, so that where the trout appears in any great abundance, particularly the large form, *eriox*, there the salmon decreases, and *vice versa.* It may, however, be somewhat doubtful how far this observation is founded upon any mutual antipathy between the races. It may possibly be based upon other circumstances; for instance, conditions accidentally more favorable for the propagation of one species or the other in different years, which plainly will cause one or the other species to occur in greater comparative abundance some years after than was the case earlier.

With regard to flavor, the salmon must generally be preferred to the trout, particularly salmon of the better kinds. It has already been stated that salmon, like the fresh-water trout, vary considerably in quality from one place to another; this is, however, true in a less degree with the ocean-trout. In the salmon this may be the case even in a very high degree with fish from adjacent rivers, a phenomenon which can be explained only by race differences, since fish from the different rivers get their development in the sea, where both find equal conditions for thriving and fatness. At Christiansand we have a striking example of this. While the salmon in Torisdal River is plump and beautiful in form, bright in color, and must be called excellent in fatness and flavor, the salmon of Topdal River is thin, dark in color, and can be styled nothing but indifferent or even bad in plumpness and taste.

In regular culture one has it in his power, among fish as well cattle, to select the best races which are known, and it seems to be beyond question that these, transplanted to a new locality, will retain their peculiarities; because, as remarked, the wealth of the sea is everywhere free for all, and supplies all a like abundance of food. What can produce the existing great difference in adjacent places is not easy to perceive. The only mode of explanation seems to be this, that the rivers possess very different nourishing capacity for the tender young, so that they in one in a manner are checked, while in the other they thrive greatly.

Should this supposition be well founded, which the attempt of a few years with spawn brought from another place will show, then one should, in the places which prove to be less favorable, seek to provide the remedy for the absence of nourishing capacity for the young by direct feeding, and also constantly renew the race by bringing spawn from more profitable places.

But even an inferior race of salmon grows much more rapidly than sea-trout, even the larger form, so that the relation between these with regard to yearly increase of growth will be not far from the same as previously pointed out with regard to the relation between salmon and lake-trout. If, then, the mutual antipathy mentioned between the salmon and the sea-trout may be only a baseless conjecture, this difference in growth from year to year, in connection with the salmon's returning to its birthplace earlier in the year, will cause its culture to become much more profitable than that of the sea-trout, which requires just the same labor and outlay for apparatus. It has been previously stated that salmon-fishing ought to carried on in such a way that one, at all events after hatching has been prosecuted for three or four years, will not catch any but grown fish, that is to say, of about 6 kilograms in weight. The apparatus which will be required for such fishing will be unserviceable for the capture of sea-trout, which only at a comparatively advanced age reach such a size. I should, therefore, consider it best to undertake the culture of only one of the fishes mentioned, and then preferably the more profitable one—the salmon.

For both these species of fish in question, one is clearly free from all care with regard to their nourishment, except during their earliest youth. The wealth of the sea in food which they require is certainly boundless; at all events, men are able, so far as their insight at present extends, to do nothing to increase it, even if it should be considered desirable.

As regards the EEL, it acts just the reverse of the species previously mentioned. It is born, so far as we now know, in salt water alone, and migrates from this in early youth, when of the size of a coarse darning-needle, up along the brooks and rivulets to the fresh waters, where it passes many years of its life. In migrating up into the water-courses it is not easily stopped by any obstacle; if the current is stronger than it can swim against, the young eel takes to the land, and continues its way in compact columns of many thousands on the moist bank; it winds even up the trunks of trees in dams, and thus advances where one would consider it impossible. The migration takes place, as remarked, in dense multitudes of many thousands. Whether or not eels are born in the sea we have no certain knowledge, neither do we know whether they are also born in fresh water.

The journey of the eel-fry up into the water-courses takes place in April, May, and June—in different places at different times, probably according to the condition of the weather. From midsummer and during autumn, especially during the dark autumn nights, the adult eel goes from the fresh waters, where it has passed its preceding life, down into the sea, as commonly supposed, to propagate its species, and thereafter does not return to the fresh waters, but continues its life in the sea, where it is thought to make regular journeys from place to place. No one has ever come across adult eels migrating upwards from the sea. What causes this wandering out to the sea no one knows. The exodus includes both larger and smaller eels at the same time, which do not appear to be of the same age or degree of maturity, and so there are found simultaneously in the sea both small and large eels, as in the fresh waters from which they descend. It is a warm-water fish, and therefore is seldom found more than 500 feet above the sea-level, but thrives in all kinds of waters, large and small, with or without affluents, if they only have a muddy bottom in which they can bury themselves in winter.

It lives exclusively on animal food—insects, snails, fish, and flesh of all kinds until and after it has become putrefied. It is easily caught in many ways; the easiest, however, appears to be unknown in this country, at all events it is employed very rarely—namely, to capture it in so-called eel-traps in its migration towards the sea, which is universal over all Sweden.

It is considered by many a very well-flavored fish, and commands a high price. It grows very quickly. While 1,800 young eels are required to weigh a pound, one-half kilogram, only 30 yearling go to make up the same weight. At the age of three or four years the eel reaches a weight of one kilogram. One can cultivate it by collecting the young during the migration from the sea, and therewith stocking ponds and pools, and if one provides for a supply of food, living or dead, they will grow quickly. The mode of cultivation which, however, will presumably answer the purpose best will be to place the eels which are caught in the river or the sea, fresh or salt, and which may not have reached a suitable size, in a separate pond, and then to feed them well with all kinds of offal of flesh and fish, to which there must be easy access; if it is a little putrefied it will, as remarked, do no harm. They will then grow very quickly, and give a good return for their care and food. One must in the mean time look out that they do not escape from the pond, by surrounding them with an inclosure which the eel cannot climb over or creep through; otherwise they will in the night, during a heavy dew or rain, leave the pond and disappear. A belt of loose sand about the pond will prevent them also from escaping, if it has sufficient breadth. A pond for such purpose ought to have a muddy bottom and hollows in the sides in which fish may conceal themselves and find shelter both in warm and cold weather; if one can conduct running water through the pond, it will become so much the more profitable. In an extent of twenty square meters 150 to 200 eels can be accommodated and thrive well, if they simply obtain sufficient food.

This fish is well adapted for stocking ice-ponds, which, because of a want of a regular water supply or other circumstances, cannot be used for salmon or sea-trout. There is no difficulty in planting them. The young eels will be found in all rivulets which proceed from waters wherein the eel lives, regularly every spring, if one takes the trouble to look for them in May and June. They migrate, as remarked, not singly, but in dense masses of many thousands, and can be collected with ease. If it is necessary to bring them from a very great distance, the transportation is not difficult. The eel's tenacity to life is known, and they are able to live many days in water-plants if they are now and then moistened and kept in a cold place.

IV.

ACQUIRED EXPERIENCE.

In order that one may form an idea by any means clear concerning what profit he may expect from systematic fish-culture prosecuted to a different extent in different localities, it is necessary to know the results which have been reached from such culture in other places. I shall, therefore, communicate below a few instructive illustrations sought among a multitude of experiences in different countries and authentic sources.

In the State of New York, in the vicinity of Caledonia, Messrs. Seth Green, A. S. Collins, and S. M. Spencer, in 1865 to 1866, constructed a fish-farm nearly three-quarters of a mile (1,200 meters) from the source of the Caledonia Springs, a brook which originates from springs in the bottom of its bed, and which at the farm conveys a bulk of water of nearly 80 barrels, or about 10,500 liters,* per second, or about 9,000,000 hectoliters in twenty-four hours; a respectable body of water, which is completely at the disposition of fish-culture. The farm contains an area of about 20 hectares, which, in a length of about 800 meters, is traversed by a brook. Since the ground is quite level, there is no overflow into the brook, whose water, therefore, is perfectly clear; it contains a small portion of sulphur and chalk, but these must be well adapted to the fish, since the brook has been renowned for its trout, which are now numerous in it. The object of the construction of the farm was to rear trout for sale as food, but circumstances have caused the operations to involve chiefly the bringing in of impregnated eggs, newly-hatched young, and one to two years' old fish for sale for stocking other lakes, brooks, and ponds. Many millions of eggs are hatched out annually.†

The selling prices are: For a single thousand impregnated eggs, $10; many thousands and upwards, $5 to $8 per thousand; for newly-hatched young, $30 per thousand; yearlings, $12 per hundred; fishes two years old, $25 per hundred, or a little over the price of dead fish for food. No more old fish are kept and reared than are necessary to procure the de-

* One liter = 1.0362 quarts; one hectoliter = 103.52 quarts.

† Seth Green: Trout Culture, 1870. Leon Soubeiran: Pisciculture dans l'Amerique du Nord, 1871. Raveret Wattell Progrès de Pisciculture aux États-Unis, 1873.

sired quantity of impregnated eggs. If we estimate that only 2,000,000 eggs are sold annually at the lowest price, these yield a gross profit of $16,000, or $800 per hectare ($80 per rood) of the whole property, which can only be in part occupied by them for the necessary fish-culture ponds.

Mr. Green states that a pear-shape basin of 6.3 meters in length, 1.9 meters in breadth about the middle, and from 0.15 to 0.64 of a meter in depth, with a capacity about 3.6 cubic meters, and with a water supply of 1.6 liters of filtered spring water per second is quite sufficient for 6,000 to 8,000 young after they are completely hatched, when they begin to take food of their own accord.

That a similar pear-shape basin of 9½ meters in length, 3.15 meters in breadth at about the middle, and a depth 0.15 to 1.26 meters, with a capacity about 15½ cubic meters, and a water supply of about 37½ liters of brook water in a second, will be sufficient for 5,000 two-year-old fish; while a similar basin 15.75 meters long, 9.5 broad about the middle, and from 0.15 to 1.5 meters deep, with a capacity about 48 cubic meters, and a supply of running brook-water equal to that of the last smaller basin, is fully sufficient for 2,000 fish in the third year; observe, under the condition that the temperature of the water is not under 1° (34.25° F.) and not over 12° R. (59° F). The fish will die if the temperature of the water rises to 16° R. (68° F.), unless there are cold springs in the bottom of the basin which will cool off the water. The rule is, the colder the better down to the limit stated.

The hatching is done in a trough 20 feet long, 18 inches broad, and 6 inches deep, which is divided into compartments 18 inches long by cross-pieces 2 inches high, which are secured to the sides of the trough; other such cross-pieces are used at pleasure, or when it is necessary to make the water deeper. The bottom of the trough is covered with shingle. The spring water is filtered in a trough 6 feet long, in which are placed, at an angle of 45°, three to four screens covered with flannel, to give the greatest possible filtering surface. The loss of eggs in hatching does not exceed 6 per cent.

Later, Mr. Green was appointed superintendent of the fisheries of New York State, and he constructed a State hatching-house on the same brook—Caledonia Springs—in which are hatched out and distributed annually 4,000.000 to 5,000,000 of impregnated eggs and young.*

Count M. de Causans, on the 10th of December, 1858,† the lake be-

* One of the herring family, *Alosa sapidissima* (the shad), which in America is very highly esteemed as an article of food, and which, like the salmon, spawns in fresh water, had been for some years almost fished out. Mr. Green has in the later years annually hatched millions of young of this fish in open water—one year nearly 8,000,000—so that now the abundance in many places has become just as great as it was a hundred years ago. The young have been carried from New York State to streams in California, where they thrive well, although this fish did not before exist on the Pacific coast. A couple of genera related to this fish are found sporadic on our coast. Would it not be worth the trouble to attempt the introduction of the dainty shad to our coasts and rivers?

† Bulletin de la Société d'Acclimatation à Paris.

longing to him, Saint-Front, in the canton Fay-le-Froidi, department Haute-Loire, in France, has a surface of fully 30 hectares, a depth of up to 10 meters, and an elevation of 1,200 meters above the sea. It is fed partly by numerous springs in the bottom of the lake, partly by brooks which flow into it after first traversing the meadows which surround the lake; a larger brook, La Gagne, carries the water from the lake to the river Loire. The lake contains trout which are much esteemed, and which are sold on the spot for 2 francs 50 centimes to 3 francs per half kilogram = 1 Norwegian pound.

Until 1854 they restricted themselves to liberating in the lake a few hundred small trout caught in the brooks; but this autumn they took pains to procure 30,000 to 40,000 impregnated eggs, which were placed in an apparatus constructed for the purpose in the lake itself, to be hatched out. In subsequent years this was annually continued in an increasing scale, just as persons by cleaning out the tributary brooks have attempted the improvement of natural culture. In the space of the year 1857 by net fishing they took regularly from 25 to 30 kilograms of fish of an average size of 300 grams every time. He can now, without interfering with the abundance of the fish, sell at least 15,000 kilograms annually, and his manager, Mr. Millet, thinks that one might, without injury, take 200 kilograms per hectare annually. The lake contains also carp and other cyprinoids, together with an abundance of minnows and frogs, which serve as food for the trout.

The Marquis de Selve has constructed a fish-cultivating establishment* at Godset Villiers, in the communeCerny, in the department of the Seine and Oise, near Paris. It consists of a canal with manifold windings, which in a total extent of 12 kilometers (one Norwegian mile), traverses a nearly horizontal field of 12 hectares, and ends in a larger basin from 10 to 30 meters in diameter, and is 5 meters deep; the canal is 2 meters in breadth, and is fed by water taken from Cerny canal, which from the basin flows out again into the river Essonne. The whole inclination from the beginning of the canal to the outlet from the basin is only 80 centimeters, but the water supply is sufficient to maintain a suitably strong current; various rather strong springs increase at many points the water supply, as they furnish the hatching apparatus with the necessary water for the development of salmon and trout. At the highest portion of the field are constructed smaller canals for the rearing of the delicate young until they have reached such a size that they can be liberated in the larger canals. There is abundance of water which nourishes an endless multitude of small crustaceans, and the bottom contains lime, which is of great advantage for the development of the common large crawfish.

In the spring of 1864 was begun a project of digging a couple of kilometers of the large canal to experiment with trout and the common

*Rapport par Ch. Wallut, 15 Mars, 1867, Bulletin de la Société d'Acclimatation à Paris.

crawfish. Young trout were purchased from a hatching establishment in Paris; but even the results reached in a few months determined the marquis to give his apparatus the full extent intended. In the month of October, 1864, therefore, a multitude of workmen engaged in digging the rest of the 12 kilometers of canal; in April, 1865, the water began to traverse the whole length. The expenses of the scheme (the digging, locks and bridges, buildings, purchase of eggs, fish, crawfish) amounted to fully 150,000 francs, and the expenses of maintaining, for food, watching, repairs, instruments, &c., must be estimated at about 12,000 francs per year.

According to the report the establishment was partly three and one-half years old, and the rearing of crawfish was carried on for the same length of time. Unfortunately, crawfish grow slowly; it requires eight to ten years before they can be offered in the market of Paris. In order to shorten this time, 325,000 crawfish of three to four years old were gradually purchased and liberated in the canal. These have thrived wonderfully, and will admit of comparison, so far as flavor and size are concerned, with the best crawfish from Maas River or from Germany.* They have multiplied, and are found of all sizes from one to three years. The Marquis de Selve estimates the number at 8,000,000 to 10,000,000, which by no means appears overdrawn. Last year already he sold many thousand francs worth; this year he expects to sell over 20,000 francs worth. Since Paris is supplied for the greatest part from Holland and Germany, an interruption may easily occur in the shipping, and a cessation of nearly twenty-four hours brings up the price to double or triple what it was. The marquis, who is close by and can obtain instant information, can take advantage of these accidental pauses, and has therefore constantly a reserve on hand for such favorable opportunities. The marquis, however, cherishes no illusions; as soon as autumn approaches various sacrifices must still be made. To insure the ability to deliver a uniform supply he must still for three or four years stock his canals with crawfish to the amount of 25,000 francs yearly, in order to be properly furnished with spawners. The profit on these purchased crabs is only 12 to 15 centimes (2.2 to 2.7 cents) apiece; when those born in the establishment are grown, the case becomes entirely different. There are sold now about 1,000 crawfish daily.

Besides crawfish are cultivated also trout and other salmonoids. The first planting is nearly three years old. On the 18th of August, 1866, was caught a splendid trout; it was not weighed to avoid injuring it, but from its appearance it must have weighed between 2 and 2½ kilograms at the age of 27 to 28 months; it was sent away living as a present because of its size and beauty. Trout 20 to 24 months old weigh from 500 to 750 grams.

The hatching apparatus is placed in a separate building in connection

* Crabs are almost as necessary a constituent part of a French dinner as meat or fish; if not as a dish proper, then as a means of decoration.

with the smaller canals. Three watchmen with large dogs keep off robbers; besides, many workmen of different kinds are employed. For feeding are used two horses weekly at 25 francs apiece, together with carrots and other vegetables to the amount of 5,000 francs yearly; each adult crawfish consumes one gram of flesh or carrots daily. The marquis has full confidence that the scheme will pay when it comes into full condition and operation.

Mr. Livingston Stone, in Charlestown, North America,* arranges from his own experience the following propositions:

1. Under favorable conditions the increase in weight of a trout in the third year equals a fifth part of the food it consumes.

2. For a trout of this age the daily ration of flesh or fish offal ought to be about a hundredth part of its weight.

3. At this age the weight generally doubles in the space of a year. With care and abundant food one may reach the same result in 6 months. April to September.

4. All kinds of viscera (lungs, liver, &c.) are adapted for food for adult trout. During the first 6 months the cost of the food for the young is next to nothing.

5. Even in summer, fresh trout, packed in ice and sawdust, may be shipped a long distance. Thus treated they endure railroad transportation for a distance of 500 miles (800 kilometers).

The price of trout in the markets in New York varies from 50 cents to $2.25 (1.80 to 4.50 crowns) per pound.

At many establishments in Auvergne in North France,† the following results have been reached. Of salmon, trout, and red char have been hatched out 92 to 98 per cent. of eggs impregnated according to Professor Coste's plan. At the age of 24 to 26 months, trout and red char reach a length of .25 of a meter and a weight of 200 to 300 grams; at the age of three years they reach a length of .30 of a meter and a weight of 500 to 700 grams. Four year olds are .40 of a meter long and weigh from 800 to 1,500 gams. The average loss from different causes is only 5 per cent. yearly although the temperature of the water in many places in summer now and then rises to 24° C. = 19.2 R. (75.2° F.).

In Bayern various fish-cultural establisments have existed for many years. One of these founded in München by a fish dealer, Mr. Küffer, deserves a somewhat detailed description, since it is distinguished as much by the simplicity of its construction and the smallness of the space on which it is built, as by the profitable results which have been reached. It is situated in one of the suburbs of München, and is fed partly by water from the river Iser, partly from a strong spring which comes up on the upper side of the river in the establishment itself. Like all springs

* Rapport sur de le progrès de la pisciculture Americaine par Raveret Wattel, 1873.

† The following information is drawn from Traité de pisciculture par M. G. Bouchon Brandely, secretaire du Collége de France, 1876. (?) The book is the result of a journey made for investigations of the relations with regard to fish-culture in France, Italy, Switzerland, Austria, Bayern, Belgium, Holland, and England, 1874.

its temperature is nearly constant, 46.4° Fahr. The water of the Iser, which is brought in by a little canal, has on the contrary a very variable temperature, and just as variable clearness, as a natural consequence of the extremely changeable climate of Bayern and the river's rising in a high, partly snow-clad mountain region, and its course through easily soluble strata of earth.

The establishment is constructed on a field with little inclination, and occupies a surface of not over 150 square meters (1½ Ar. = 0.15 Maal). Highest, but lower than the spring, are placed a series of stone troughs for the salmonoids of different ages, and for the small fishes which are required for their support, in a series along the wall and covered with a movable screen, to prevent the light from shining down upon them, and at the same time to protect the fish against thieves or beasts of prey. Somewhat lower, different basins are dug out in the bottom for other salmonoids from two to four years old; however, with the exception of one which is intended for the small fry. These water-reservoirs are constructed in the simplest manner possible. The sides are formed of planks driven down into the ground; a gate for entrance and exit keeps the fish confined; and some holes made in the bottom for shelter, constitute the whole of the internal structure, while the arrangement is completed by some boards, which partially cover the basins to produce a little shade and darkness in them. These basins or bowls have a size of 18 to 20 square meters and a depth of 45 to 50 centimeters. The height of the water varies between 30 and 40 centimeters. Immediately above the basin intended for the young is a little larger basin, likewise dug out in the ground, which has been constructed for a variety of salmon peculiar to the waters of Bayern, the Donau salmon.* This basin is so constructed that about half of it is under a little house, and is there covered by a floor, which can be lifted at pleasure, to observe the fishes, catch them, or distribute food to them. In this basin are found only individuals from one to three years old. The above-mentioned basin for the young differs from the rest only in having a supply of water-plants as shelter or cover for the young fish. The bottom in all the rest of the basins is covered with larger and smaller stones, and care is taken to remove all vegetation as soon as it shows itself.

All these troughs and reservoirs are filled from the spring. The troughs receive the spring-water first, and from these it flows down in the lower basins. The basin for the young is fed, however, directly from the spring with perfectly pure water, which has not been used before. The narrowness of the space has made Mr. Küffer attempt fish culture in a manner which comes the closest to stall-feeding. To show this it

* This; the hucho, *Salmo hucho*, is a variety, which like the Venern salmon, *Salmc ogla*, never leaves fresh water; it has white flesh and reaches the enormous size of 50 kilograms and upwards. It has been attempted to transplant it in this country, but without success.

is sufficient to state that in one of the stone troughs 1.50 meters long, 75 centimeters broad, 60 centimeters deep, with a water-supply of .675 cubic meter (= 675 liters) there are six Donau salmon and common salmon together, each of which weighs from 10 to 12 kilograms. One of the Donau salmon measured 1.20 meters. Born in the establishment, it was at the age of a year placed in this little box, and in eight years it reached this enormous size. The length of its comrades varied between 85 and 110 centimeters; they were likewise born in the same establishment. The space was so narrow for them that they for many years had found it impossible to turn or to move from the place. Their long-continued captivity had made them just as tame as other household animals. They were not alarmed when one approached them but evidently expected food from his hand. Every female of this age and size lays annually 16,000 to 18,000 eggs.

This is not the only example of the stall-feeding and packing which Mr. Küffer's establishment furnishes. To the fish-hatching structure he has added an annex in one of the storehouses, where he carries on the fish trade. The fishes which he sells each day for consumption he keeps living in stone troughs, or wooden tanks lined with zinc. The reservoirs are furnished with water from the common water aqueduct of the city, which does not contain the same water as that with which the fish-culture establishment is furnished, but which, however, is perfectly serviceable. In a stone trough of the same measure as that which contains the large salmon there are over 200 trout, weighing from 350 to 450 grams. All these fish were lively, provided with a good appetite, and were apparently in the best condition. One cannot, however, conclude from this that such a packing is to be recommended as a rule.

All the salmonoids in Mr. Küffer's establishment differ in sex and age, so as to prevent them from mutually destroying one another. They are fed with the viscera of the fishes which are sold for consumption in Mr. Küffer's storeroom, and in order that they shall never hunger, the food is supplemented by commoner species of fishes, which he places in the same basin. Mr. Küffer is of the opinion that one must feed to satiety if one wishes to see the fish develop and thrive. He feeds them, therefore, twice in twenty-four hours at least, morning and evening. The fishes which are intended as food for the rest are reared in a separate basin, where they, as it were, get no other food than what they find in the water. Those of the fish which live in the same basin with the fishes whose food they are sooner or later destined to become, live on what these waste.

Mr. Küffer employs for impregnation the French, or Mr. Coste's method. The apparatus for hatching is the same as is used in many other places, but it is entirely covered, so as to completely exclude the light. Sometimes he is obliged to distribute the eggs in the basin intended for the rearing of the young. Under all circumstances he succeeds completely in the hatching. The loss during the whole hatching season is just as little with the eggs which are placed in this basin as with those which are placed in the apparatus especially constructed for hatching. But he does not employ the basin for this purpose unless his hatching apparatus becomes too small to accommodate the whole quantity of eggs.

The reservoirs which are to contain crawfish are entirely covered like the stone troughs, and get the water directly from a spring and from the river. It is highly amazing to see the enormous quantity which Mr. Küffer has packed together in so small a space. In a division which is not longer than 2½ meters and 1½ meters broad, there were heaped upon one another more than 6,000 crawfish. They were so crowded that they lay in many layers, one above another. They were remarkable for their size and weight; some weighed 250 grams and upwards. The males are always separated from the females, and live in a reservoir by themselves. Notwithstanding the narrow space and the little care bestowed upon them, only a very small number die.

In an establishment at Salzburg, which is almost as small as Mr. Küffer's in München, fishes are fed with minced horse-flesh and the commoner kinds of fishes. They feed 30,000 large and small fishes, which are in the establishment, at a daily expense of 44 cents.

In Amsterdam, where a grand establishment is constructed for fish-hatching, chiefly for stocking the rivers of Holland with salmonoid fishes, in the winter of 1873 and 1874, from 250,000 impregnated eggs, they got 238,000 living young; the loss was under 5 per cent. All of these were liberated in the rivers Yssel and Vecht. With regard to the increase of the abundance of fish in the larger rivers by means of the hatching in this establishment during a few years, it is noticeable that the state owns 38 fisheries, which are leased, the amount of whose rental is quite considerable; thus, for example, share No. 2, which before 1873 had been leased for 10,000 florins yearly, at the auction sale of leases the same year was rented for 35,000 florins. Share No. 12, for which previously was given 8,000 florins, was leased for 47,500 florins, and so on. These figures speak for themselves; upon the whole, the rent is more than triple what it was earlier.

As universally applicable results from the observations made, M. Bouchon Brandely lays down as essential the following rules:

The depth and extent of the reservoirs and basins ought to vary according to the number and age of the fishes which are to live in them. For trout of two to three years, is required a depth of 60 centimeters; for older trout, 1 meter. A greater depth will, of course, do no harm, but that mentioned is sufficient. For the young, broad trenches of little depth are better than reservoirs. The bottom should consist of small stones with water-plants here and there, with the borders planted with bushes, which hang out over the water, partly to give shade, partly to supply nourishment, since quantities of insects and larvæ will frequent them and fall into the water, where they with greediness will be captured by the young. The young of different species of fish, even belonging to the same genus, ought to be kept separate; likewise should the young of the same species and the same age be separated according to size. The development of different individuals is considerably different in rapidity, and the larger will quickly attack the smaller if they live together. The most important thing to observe in a structure for the culture of salmonoids is to secure a constant supply of clear water, and the colder this is the better.

For the rearing of the young newly-hatched is naturally required a larger place than for their hatching. The larger the place one gives them, and the purer, the colder the water one has for use, the more certain is one that they will thrive well, and that the loss at this most sensitive age will become the least possible.

These are a few of a great multitude of examples of more or less successful fish culture prosecuted on different scales. That there are plenty of examples of less successful or even quite unsuccessful culture is self-evident. Everything depends in this as in all other practical undertakings, on how far one properly appreciates the conditions, and does *all* that is necessary in pursuance of them, in order that the affair shall succeed; if these things are done, the examples mentioned show that the business can and will yield a good profit. It is particularly the hatching of the salmonoid fishes that is unsuccessful, and the reason is easily given. They have partly lacked a sufficient water supply, but especially the comparatively low temperature of the water in summer, which is an irremediable condition for the thriving of the salmon.

With regard to the quantity of food that fish require in comparison with what warm-blooded land animals need, the fish commissioners of New York state the following:

"The food consumed is used by all animals chiefly for the production of motion and heat, because all animals are physical machines, which must be furnished with fuel if motion is to be developed, since they will wear out with friction if this wear and tear is not compensated for by new nourishment. A man or a horse can only perform his full quantum of work when his body gets the full nourishment required, and on the other hand he can not consume the full quantity of nourishment unless he works. The dry-land animals are warm-blooded and movable; many of them, in natural state, find their food only by hunting it; while fishes are cold-blooded, and although they sometimes undertake long journeys, as a rule they keep themselves still in the same place. A trout will remain by the week in the same place in a pool or a hole. A pike will lie still like a sunken stick between the rushes, except when it between times plunges off after its prey, after which it usually returns to the same place; in the possession of all the senses except hearing, in the fullest development, it resumes its condition—that of motionless, sleepless attention.

"The consequence of this is, that while the warm-blooded animals use a considerable quantity of nourishment to produce the high temperature of their blood, and compensate for the wear and tear occasioned by constant motion, the cold-blooded transform nearly all their food to an increase of their own size, and grow, therefore, in proportion thereto, without using more than a small quantity for maintaining the low temperature of their bodies, and to compensate for the small wear and tear occasioned by their little motion. It is therefore a mistake to feed fishes with the flesh of warm-blooded animals, unless it is entirely worthless offal, because it is contrary to natural economy. Cattle can be fed up and fattened only by a liberal use of the natural means of nourishment, whose procuring demands considerable employment of human labor, whereby a comparatively large portion is used without corresponding increase in growth, while fish, left to themselves or fed in a proper manner, will consume what in itself costs nothing, and what could not in any manner be utilized. They should, indirectly or directly, be fed from the water's domain, whose crop practically is left to perish, without at present being useful for men. Neither is the flesh of the inhabitants of the water, when it can be obtained, less valuable for men than that of land animals. Pound for pound, it contains the same quantity of nourishment and will sustain human life just as well, while under certain conditions it is more profitable. It gives the brain and the nerves phosphorus, which is not contained in all kinds of food, but which is just as essential to complete health as gluten or starch, while it at the same time is so much more easily digested than the flesh of land animals and birds that it is used as a modified form of fasting, and at certain times of the year is more wholesome than the last-named kinds of food. A wise economy of the means of nourishment which nature gives will, therefore, as the population increases, compel us to turn our attention to the harvest which the water can yield."

V.

ESTIMATED PROFIT AND THE ECONOMICAL VALUE OF THE WATER-AREA.

I shall next, with the guidance which the foregoing results of experience furnishes, attempt to show what profit one may expect from fish culture in our rivers and lakes, if this business in the future is attended to with the care which is now bestowed upon the prosecution of an industry of corresponding importance, and this everywhere in proportion to the opportunities offered by more or less favorable conditions for the necessary attention to the business.

From the examples cited is seen what also lies in the nature of the business, that this culture can be carried on in very different ways, alike by artificial rearing and complete feeding in apparatus and basins specially constructed therefor through the whole time which intervenes before the fish can reach the development one desires; also, by what one might call perfect stall-feeding, by inconsiderable cleanings out in the brooks in which natural spawning occurs, and the formation of spawning places, at the time when the spawn is mature; by placing eggs in common boxes in open water, which labors, though somewhat mutually different, must be regarded as of nearly equal importance, and the least to which can be given the name of fish-culture. Between these two extremes there will be a great multitude of variations of more or less extensive cultivation, just as is the case in agriculture, produced partly by local circumstances, partly by the cultivator's greater or less ability, information, and aptness. The different waters will, besides, present just as great variations in natural fertility as the soil. Some localities will be able to be compared with rich wheat-land, while others can be considered as only equal to poor oat-land.

It is worth while also to find out a proper middle course which can be depended upon to yield a probably medium result. As a middle course for the cultivator's greater or less completeness, I will adopt a natural lake of such situation that the hatching in a regular apparatus, and the rearing of the young in separate water-basins through the first summer, will take place, which will not require greater outlay or labor than can be accomplished by any, even the most indigent, owner of fishing waters who interests himself in the business and will attend to it with the necessary care, at the same time that the natural means of nourishment are sought to be increased by the culture of commoner fishes and other aquatic animals which, likewise, any one, even the poorest fish culturist, can perform himself or with the help of his half-grown children.

As a representative of the fishes of medium value, I should select the fish most widely distributed in our country, the common trout, just as one in agriculture employs barley as a common denominator. The red char, in places where it can thrive well and reach its best development, will be the most valuable fish, and represent wheat, just as the pike and the perch will stand in the series with oats; both extremes will be cultivated in nearly corresponding quantities, and mutually balance each other.

The selected representative occurs at all elevations above the sea, even up to the uppermost limits where fish can live, and is there the only kind which occurs. It is a rule, which may be regarded as universal, notwithstanding that, as before remarked, each particular lake may be considered to have its peculiar variety of trout, in the matter of appearance as well as taste, that the higher the lake lies the better flavored are the trout which live therein. Thus it is this very species which should become the object of cultivation above all others; but against this the climate opposes, at these heights, great, in many places insurmountable, difficulties, as it will be impossible for people to establish winter residences at such elevations. It is evident, however, that people are able to live at considerable heights if they find it profitable there, especially after they have learned to use peat as fuel. There are besides only a few lakes of any extent which are worth considering, that lie higher than the stations where, for a long time, people have had dwellings, and have thrived well. When, therefore, at some time the importance of fish-culture has become recognized, it will certainly come to pass that the necessary men for regular cultivation will remove to all lakes of any importance in elevated districts, so that fish-culture, to the extent previously supposed will be carried on, even up to the most elevated lakes which may be worth cultivating; because, even if hatching cannot take place in the lakes themselves, there is nothing to prevent obtaining satisfactory results by bringing to these lakes the required stock of young from less elevated localities in which the hatching can be done with greater ease. Such transportation will, as a rule, be an easy and simple thing, since it will not be long, and, as a rule, can go on in such a way that one, as often as he wishes, can look out for the most thorough renewal of the quantity of water in the vessel in which the transportation occurs.

The next and most difficult question to answer is, how large a planting of fish may be considered proper under such a mode of cultivation as previously provided? With regard to this I shall next consider some of the previously cited data.

Mr. Seth Green states that two basins with an area, the uppermost of 22.5 and lowermost of 55.8 square meters, and with a capacity of 15.5 and 48 cubic meters, respectively, are sufficient for 5,000 individuals of one to two years old, and 2,000 individuals of two to three years old fish, respectively, with a supply of water through both reservoirs, one after the other, of 37.6 liters per second. Since the trout, as a rule, stay on the bottom, and only rarely distribute themselves at different heights in the water, the spaciousness of the area is more to be considered than its cubic contents. To each fish in these basins is thus allotted, respectively, .0045 and .0111 square meter, and cubic space of, respectively, .005 and .024 cubic meter, with 5 and 24 liters of water. For each fish in the first basin, there is also a surface of, for example, 15 centimeters long, and 5 centimeters broad, while the fish at this age can have a length of about 10 and a breadth of about 2 centimeters, more or less; in the other basin, each fish has a space of, for example, 20 centimeters in length and 5.5 centimeters broad, and it will have a length of 15 centimeters, and a breadth of 3 centimeters or upwards. One perceives, also, that fish can live and thrive in a very small space, if the water supply and the food are sufficient. According to M. Bouchon Brandely's statement, each four-year-old fish has .3 square meter; it is, for instance, 75 centimeters long and 40 wide, and its size will be about 25 centimeters long and 5 broad, while to each fish will be allotted a quantity of .3 cubic meter, which equals 300 liters of water. The water supply he does not mention. The relation concerning the surface and cubic room each fish ought to have is thus, according to the two statements, considerably different; there can be no doubt that the larger it is the better.

But in open water, where the fish must live on what they find in it in the form of insects, mollusks, crustacea, and larger or smaller fishes of other commoner, chiefly vegetable-eating kinds, probably a considerably greater surface room and likewise cubic space should be allotted to each one. Since a natural lake, wherein fish can live in our country in winter, must have a considerably greater depth than the basins indicated, if one estimate according to the bottom alone, there will arise a much larger space for each fish; but this increase cannot be taken into consideration except in so far that the commoner fishes, which live in schools near the surface of the water, will thereby get the necessary space. Since I take it for granted that the young are preserved in the greatest possible degree from enemies in the first year only, as long as this can be done without considerable expense for building special breeding ponds, also that one has not the opportunity of procuring other kinds of reservoirs for young fishes than such as will freeze to the bottom in winter, young fishes must be liberated in the water late in the autumn when the cold commences—they are then only about one-half year old as developed fish. Because of the small size at this age, I think that the deaths in the first year become considerably more numerous than experience has shown them to be under more solicitous care, and that likewise the loss from year to year will become much greater than experience has shown it to be with proper care in ponds. I assume, therefore, that of 1,000 young about one-half year old, which are liberated in the water one autumn, the next autumn only 700 one and a half year old fish will live, and of these the following year 550 two and a half year old fish; moreover, of these the following year there will be 450 three and a half years old, while the next year will be found only 400 grown fish over four years old, which thus in the fifth year become the profit of culture. These grown fish I will on the average estimate at a weight only one kilogram, though, according to experience, the average will reach a considerably greater weight.

Experience has sufficiently shown that among the fishes, as in other classes of animals, individuals differ greatly with regard to the rapidity of growth. But this cannot be taken into consideration; one will naturally catch and sell the fish which have reached the proper size, 1 kilogram and upwards, or whatever other size and weight he may find most profitable, without the slightest regard to what age the individuals may have; it is essential that one can calculate with certainty the chosen size reached by the corresponding ages of the number of fish calculated. If this size is reached earlier or by a greater number of fish, the profit will become so much the greater. The thing is, not to calculate the probable profit too high, also the business more profitable than it is likely to become, and this is perhaps done here as well with regard to the sale from year to year, which probably is placed at too high a figure, as with regard to the increase in size and weight of the fishes surviving, which is placed likewise at much too little; both parts in all likelihood exceed the probability.

For each fish that one will calculate to capture in the fifth year of 1 kilogram in weight I assume that one ought to find in the water (450 + 550 + 700 + 1,000) divided by 400 = 7 fishes of younger ages, as likely the fishes of the oldest class, which may be too small, will be compensated for by younger individuals of greater weight from year to year.

If one estimates now for each fish, without regard to age, a surface of 1.5 square meters or 1.5 meters long and 1 meter wide, or for each fish of every age a surface five times as great as M. Bouchon Brandely adopts for four years old, it should probably be assumed that no crowding can be expected to take place; because a lake of the size mentioned may generally be considered to have an average depth of at least 5 meters; each fish, small and great, thus obtains 7.5 cubic meters or 7,500 liters (Potter), or twenty-five times as great cubic space as that considered sufficient for adult fish by M. Bouchon Brandely. In a water area of 3 hectares, which equals 30,000 square meters, can also be estimated 20,000 fish of all ages, and of these one-eighth, or 2,500 adult fishes over four years old, of a weight of 1 kilogram and upwards. There is, as remarked, naturally nothing to prevent taking fish of younger age or smaller size. Many will perhaps even prefer fish of half this weight. I hold, however, to the size and age mentioned essentially so as not to estimate the profit too high and attainable in too short a time, at the same time, because at the age indicated the fish may be considered fully mature, as after this age the common trout, which alone I have here in mind, increases more slowly in size and weight, and also becomes less profitable.

According to the calculated decrease, from year to year, to a number of 2,500 fishes of over four years old, will correspond a number of the youngest class of about 6,500 individuals in the first year which must be liberated in the water to compensate for the decrease of the captured adult fishes. But it is not enough only to calculate on the hatching out of this number. The most delicate age from the time the hatching out can be considered completed with the absorption of the egg-sack until autumn, is, according to all experience, the most dangerous time for the young, since the death rate is greatest, and here it is assumed that the ponds or basins procured for the rearing of the young during the first summer are not likely to be of the best kind, and that the attention, at all events, at the beginning, is incomplete; as a result of this, the death rate must be estimated comparatively high. If one place this decrease at about 50 per cent., which will likely be estimating it high enough, and add at the outside about 10 per cent. for loss during the hatching itself, there will be required at the highest 10,500 eggs each year for hatching.

It is probable that this quantity of eggs will yield a larger number of fish than is necessary for obtaining the stated number of adults; but this can likely involve no harm, because if it should be evident that the abundance of fish will thereby become too great, either with regard to the supply of food or of oxygen, it is an easy matter, by the assistance of natural means or by the capture of a larger number of fish, to remedy the difficulty. It will, probably, under all circumstances, be the best to hatch out a greater quantity of eggs than is actually necessary, if there is opportunity to retain them, and thereby gain experience as to the water's nourishing capacity in both directions, because the larger the planting which can be kept in good condition the greater the profit. To procure 10,500 impregnated eggs, are required female fish of a combined weight of 6 to 7 kilograms, and a corresponding number of male fish, weighing from 2 to 3 kilograms. This quantity of fish, with the appropriate mode of proceeding, will be easy to procure in a not altogether fished-out lake of the size mentioned, or even from far-distant lakes, in which case, however, it must be bought. After the expiration of the second year, at all events, the lake will yield the desired quantity of eggs from the best developed of the young first hatched. For the hatching of this quantity of eggs, is required 3 to 4 square meters of space in a suitable apparatus. Even if one, besides the fish which may be objects of culture, wishes to hatch out some gwiniad, chiefly as food for the trout, and therefore increases the size of the hatching apparatus to 5 square meters of surface, the structure can only become inexpensive if no unnecessary luxury is brought into use in a construction. In case of necessity the hatching may occur in an open field.

Besides this hatching apparatus, as remarked, separate places must be procured for the newly-hatched young. This will, as likewise previously remarked, everywhere be an easy thing, by the construction of canals or small ponds or damming up some little creek, which are arranged so that they may be furnished with water from some brook or spring, or both together. These ought to be prepared the year before they come into use, so that they may be sufficiently stocked with water-insects, and crustacea. This requires only a few days' work and material of small cost.

The hatching of the kinds of fishes which may serve as food for the cultivated fishes proper, and which will thrive in a lake, is not difficult, since it may take place in the lake itself with very simple and inexpensive apparatus, and in a very short time, since their spawning season is short, and the eggs are hatched in a few days. The collecting of the necessary quantity of eggs by the use of likewise very simple and inexpensive arrangements is not at all difficult if the lake only contains such fish. Neither is the transportation from other lakes associated with other difficulties than the transportation of trout eggs; it will be necessary only a couple of times. Even if this may give rise to any outlay, it will be just as well invested as the outlay for seed-corn for fertile soil or the procuring of artificial manure for the land which needs it.

In three hectares of water, should one, from the fifth year, harvest, on the average, 2,500 adult fishes, of 1 kilogram each, or 2,500 kilograms.

The price of the kind of fish which I have chiefly in view here, the trout, was, even thirty or forty years ago, 2.7 cents per mark—which equals one-fourth of a kilogram—in the mountain regions. In the towns it is now considerably higher, especially at certain times of the year, and for fresh fish; I will, however, place it at 2.7 cents for one-fourth kilogram, which equals 11 cents per kilogram, notwithstanding that this article has something near the same nourishment as flesh, and the price of flesh must be placed at at least double. *The value of the mean profit of three hectares of water, about 2,500 kilograms, is thus calculated to be $275 annually, and for each hectare also over $82.50 gross.*

If one estimates that the careful cultivation of the water area mentioned will demand the same outlay for labor, buildings, and implements of all kinds as the cultivation of a corresponding area of land, which ought to be placing it more than high enough; and if one considers, moreover, that the profit of the first year's harvesting of the water's natural culture covers the labor necessary for the hatching, rearing, and care, then will, from the fifth year, the proportion between the yield from fish cultivation and agriculture stand as follows:

In Dr. O. J. Broch's previously-quoted work, "The Kingdom of Norway," which at present may be truly regarded as the most trustworthy statistical work, the area of the cultivated land is given as 2,700 square kilometers, or 270,000 hectares, and the value of the products of the land for the year 1865 at $13,750,000, or not quite $55 per hectare on the average. According to what experienced farmers in the most fertile tracts of our country have stated, this average profit must be considered a little higher than in reality is the average profit from land of all qualities; while, on the other hand, the estimate upon the yield from fish culture places it lower than it, at all events in many places, may be made. The proportion between agriculture and fish culture, also, should be as 2 to 3 in equal areas. But even if it must be admitted that this estimate is too high, and that some reduction ought to be made, so much, however, should be considered established, *that fish culture will yield a profit which fully equals that from agriculture over equal areas.*

As previously stated, nature has allotted to us an area covered with fresh water, more or less adapted to fish culture, of 7,600 square kilometers, which equals 760,000 hectares, which, besides, we have it in our power, with comparatively little expense, to increase considerably, by overflowing with water, districts which are and always will be worthless except in so far as they are put to such use. With the estimated yield per hectare of cultivated water should also—if some time in the perhaps not too distant future fish culture shall be prosecuted with the same perseverance and care which are now vainly bestowed on utterly poor and ungrateful soil, and in the entire extent whereto nature gives the opportunity—the total profit reach as much as $55,000,000 yearly and upwards, if the example chosen can be regarded as a proper representative of the average yield of lakes at different elevations and in different parts of the country. Of this, however, there may be doubt, though one in the lower districts of the country, and in many places in the southern and western portion, will have it in his power, in a very considerable degree, to increase the yield by the use of other means of food for the fish besides living fish and other water-animals, which at the same time are reared in the waters. It is possible that the yield from the most elevated lakes, where cultivation to a considerable extent will be checked by local and climatic conditions which cannot be overcome, will be so far below the stated average that the greater yield of the lakes more favorably situated will not counterbalance the reduction in the total yield hereby caused. So as not to estimate too high, I will, therefore, only place the *medium yield at half of that previously calculated, or at* $27,500,000 *annually.*

But this is, however, only a portion of the profit which carefully-prosecuted fish culture will give; because the *salmon fishery*, which I here only casually and in passing have considered, can—if the opportunity for the increase of the abundance of this species of fish is unbounded, above what it is now, at all events to an extent whose measure it is impossible to indicate, and when the fishery can be prosecuted everywhere along our wide-stretched coast where fishing apparatus can be placed, when the abundance of fish has reached its greatest possible increase—in a comparatively short time by suitable modes of proceeding gain an importance and yield a profit which, in all probability, will not fall below what the lake fishery alone can yield; and possibly give considerably greater profit than this. If one estimate, therefore, the profit of the salmon fishery, like the previously-reduced value of the profit which the fresh waters probably will be able to produce, when the business at some time in the future receives the attention and the labor which it fully deserves, *the probable profit of the cultivation of fishes of the best kinds, which can be born only in fresh water, will be estimated low if placed at* $55,000,000 *gross yearly.*

This industry may also reach such a development that its gross yield may exceed the value of the whole annual import of the country, which in a good year amounts to the sum mentioned, and considerably surpass the combined profit of agriculture and grazing, which, according to Dr. O. J. Broch's tables in the work quoted, amounted in 1865 to $13,750,000 + $22,000,000 = $35,750,000.

Thus the matter presents itself when regarded in its entirety and under the supposition that the work is carried on with the greatest possible economy in the same manner nearly as the mass of farmers carry on their work.

I shall next attempt to set forth how fish culture appears when it is considered as a means of industrial speculation, as a field for the employment of capital, which seeks security and a good interest. In order that a speculation of this species may give good returns, the business must be carried on to such an extent that the necessary assistants can be given, as far as possible, continuous employment thereat, since in the opposite event, by only employing the labor at different times of the year, it will become unnecessarily expensive. As such a size I would consider a water area of 30 to 40 hectares, combined in one or separated into many connected smaller lakes, to be tolerably suitable. I shall select the smallest size, in order to make the expenses proportionally larger than they will be when a larger area can be had for cultivation; because the expenses increase just as little in this, as in other industries, in proportion to the size of the business. According to the previously-mentioned data, should, for the stocking of such a lake, according to the same proportion as estimated earlier, be required yearly 10 × 10,500, which equals 105,000 impregnated eggs. In order to make the case as unprofitable as it is ever likely to be, I will assume that the lake is so depleted that one in the first year must buy a part of the eggs in another place, that no profit be had before the fifth year after the beginning of the business; at the same time that the price of the product in distant markets can be placed at only $3\frac{1}{5}$ cents per one-fourth kilogram, which equals 13 cents per kilogram, although the price here now is two or three times that much; wherewith I shall place the necessary expenses so high that no doubt can be entertained of their adequacy.

I will assume that the business is begun in the month of September, since one must be through handling the quantity of eggs which are to be hatched out, while at this time, also, as a rule, the harvesting of the water's crop will be past. Hereby the 1st of September becomes the proper beginning of the year with reference to fish culture. At this time I calculate as expended:

For hatching apparatus, building of spawning-places, and ponds for the young in the coming year	$412 50	
For a dwelling for two laborers' families	825 00	
Boats, implements, materials, &c	192 50	
Three-fourths of the necessary quantity of eggs, three-fourths of 105,000 = 80,000, which easily and once for all can be procured by the purchase of 60 kilograms of living fish in the course of the summer; price and expenses $2.75 per kilogram	165 00	
The outlay at the beginning of the work in round numbers		$1,650 00
In the first year I assume to be expended:		
For two permanent workmen	550 00	
Implements and materials	55 00	
Loss of interest, 5 per cent. of the previous outlay	82 50	
		687 50

For the second year:		
For two permanent workmen	$550.00	
Implements and materials, about	48.12 5	
Loss of interest, 5 per cent. of previous outlay ($2,337.50)	116.87 5	
		$715 00
For the third year:		
For two permanent workmen	550.00	
Implements and materials, about	35.75	
Loss of interest, 5 per cent. of previous outlay, ($3,052.50)	152.62 5	
		738.37 5
For the fourth year:		
For two steady workmen	550.00	
Implements and materials, about	42.62 5	
Loss of interest, 5 per cent. of previous outlay, ($3,790.875)	189.54 3	
		782.16 8
The combined outlay in advance and loss of interest		4,573.04 3

or, in round numbers, $4,675, which is to be considered as the fixed capital invested in the enterprise.

From the fifth year, when the period of preparation may be considered ended, and when the regular harvesting may first be supposed to begin—though, as repeatedly remarked previously, there is nothing to prevent it from beginning earlier, and that possibly with profit—the annual expenses at a very high estimate may be stated at:

Two permanent workmen	$660.00	
Transient labor in summer	330.00	
Implements and materials	165.00	
Stock of ice	137.50	
$2\frac{1}{2}$ per cent. for repairs of buildings and apparatus of a value highly estimated at $2,200	55.00	
For packing, transfer, and conveniences for selling, 20 per cent. of the gross profit, $3,437.50	687.50	
Combined annual expenses		$2,035.00
Gross profit on 30 hectares equals 2,500 × 10, which equals 25,000 kilograms × .1375		3,437.50
Estimated net profit		1,402.50,

which amount ($1,402.50), *divided by the sum of* $4,675 *invested as fixed capital, gives a yearly interest of* 30 *per cent.*

But the selling price of the article which is here treated is in the larger markets considerably higher than is here calculated, and in a business of the kind which is here in view one must assume that they will look out to sell in places where a high price can be expected without regard to a somewhat prolonged transportation, which only in a comparatively small degree will increase the expenses of transfer. The article stands, considered as a delicacy, far above meat in price, and it commands in Christiania markets 13.75 to 19.25 cents, and often more, per one-fourth kilogram, which equals 55 to 66 cents per kilogram for Venern salmon, which, as a rule, stand far below our best kinds of mountain trout in flavor and value. In France the price varies, according to circumstances and the season, between 3 and 6 francs, or 55 cents and $1.10 per one-half kilogram, which equals $1.10 to $2.20 per kilogram. Nearly the same price as in France applies in north Germany, and probably also in England. It will thus be a long time before the price, owing to increased production, will approach the amount at which it has been previously estimated, because, in order that this may happen, the prices of the necessaries of life, such as meat, must also fall to half what they now are.

In the very near future also a higher selling price than that estimated certainly will be reached, and it can hardly be doubted that the profit previously calculated upon will be obtained from capital which may be invested in such industry as is here treated of, even if the yield become considerably less than estimated; and this so much the more certainly as the business in the extent contemplated will be able to be carried on in due time with far smaller outlay than here estimated.

If one consider the estimated profit as a product equally of the area of water used and of the capital employed in its utilization, for which there is probably entire reason, since the water area is the basis of the possibility of the employment of capital in the industry here in question, the value of the capital which the water demands will advance in proportion to its extent. In such a division the net profit according to the estimate established should be $701.25 on 30 hectares, or $23.375 per hectare annually. The ratio of interest to principal, according to the lawful interest on landed property and fixed possessions, gives this hectare a capital value of $467.50. But notwithstanding that the previously-given calculation is carried out in such a way that the profit arising therefrom must be regarded as a minimum, I will, however, state that this must be considered above the average amount, because not all the waters will possess a like degree of fertility; neither will they, because of existing circumstances, be able to be cultivated as completely as stated, partly because a portion, even if not a large portion, of our fresh waters are adapted only for the culture of the kinds of fishes which are considered inferior, whose price must be estimated lower than that of trout or red char, although at present here in the chief city and the adjacent provinces it is often more than double as high as the previous stated price for the kinds of fishes mentioned. In order to be sure not to estimate it too high, I will therefore reduce the average value per hectare to less than half of that calculated, and place it at $2.20.

But even if one, moreover, must consider this as a maximum measure for the waters in their totality, it is, in consequence of what has previously been shown, only a part of the value which must be taken as the average price when the combined fresh waters shall be taken into consideration. The salmon fishery, which is, as can be seen, smaller or larger simply as the rivers are smaller or larger which flow directly out into the sea, must also be brought into consideration. Its yield under thorough and careful management must, as previously set forth, in the future be counted upon to reach a value at least equal to that which the inland fisheries proper will give. The proper average value of our rivers and lakes in their whole extent will thus amount to at least $440 per hectare, and, for 760,000 hectares, to a sum of over $334,400,000 at such time when the people have learned, in some measure, the necessary extent and the necessary mode of using the source of prospérity which Providence has allotted to them in the rivers and seas. In this way we arrive at the conclusion, that—

The economical value of our rivers and lakes may be increased in the future to a total of over $300,000,000 *if the same labor and attention are bestowed on their cultivation as are now devoted to agriculture and grazing.*

The capital value just estimated for a hectare of water shows what outlay can be made with profit in order to put in condition our larger bodies of water for the use of fish-culture alone, by damming up in places where there may be an opportunity to secure a proper supply of running water. The share of the profit calculated per hectare is $33.375 yearly. If one starts from the standpoint, that in the employment of a sum for procuring a water supply, the half portion of the yield at the same rate—$33.375 per hectare—ought to give 5 per cent. interest on the capital invested, and besides 3 per cent. thereof for the repairs of the necessary dams, about $275 can be invested with profit per hectare of the water which thereby may be dammed up. If, for example, by the employment of a capital of $11,000 there can be dammed up a water area of suitable depth of 40 hectares or upwards, there will in this event accrue a yearly profit of $935, which sum gives $550 for the interest of the invested capital and $330 for repairs, and a surplus of $55 for the increase of the interest or the reduction of the outlay. The remainder of the invested capital for the business of fish-culture will besides obtain at least 20 per cent., since the outlay in this case will not need to be greater than for the previously-calculated work upon 30 hectares of water, or $4,675, to which in this event accrues a share of $935. If one combined both investments the whole amount becomes $15,675, on which accrues a profit of $935 plus $550, which equals $1,485, which is an average of about 10 per cent. If one assume that the same area of water can be dammed for a smaller sum, for example $5,500, which quite frequently ought to be able to be done, the yield will be about 15 per cent. at least of the whole investment. If there is an opportunity to sell the ice which will form on a body of water thus dammed up, the profit will clearly become very large; because if one only reckons about the fourth part of the usually low price, $137.50 per hectare, this gives $5,500 in addition to the income. From this it will be seen that the time may come when comparatively worthless land to a great extent will be transformed into water-reservoirs, to prosecute fish-culture there alone.

It is truly beyond all doubt that the preceding calculations will be considered exaggerated or untrustworthy, though it must be difficult to show that the profit is estimated too high in weight or in price, or that the expenses associated with the business are placed too low in any direction. The bulk of the whole production alone—on the average of a value of $82.50 or 600 kilograms of unsorted salmon and different kinds of fresh-water fish per hectare, for 760,000 hectares equals 456,000,000 kilograms—is so great that its sale for consumption in the interior of the country and for export to foreign countries may appear to be in no small degree improbable. Upon closer reflection, however, this improbability ought to disappear.

There is here in question a means of nourishment which, everywhere in the old and the new world, is placed in the first rank with regard to flavor just as it approaches the first rank in nourishing value in proportion to its weight. With a full supply thereof at a price which is lower than that of most other far less esteemed and valuable means of nourishment, it cannot be doubted that it will be used instead of these, to a very considerable extent, by the whole population of the country of all classes. How great the consumption will become by its increased use alone throughout the country it is difficult to say, but it ought, perhaps, to advance to one-half of a kilogram daily per individual among three-fourths of the population. Without regard to its increase through time, which will be promoted no little by the facilitated access to nourishment which this business will produce, the consumption within the country should amount to one hundred and eighty times one and a quarter millions, which equals 225,000,000 kilograms yearly, or about the half of the entire production.

That there will be found sale in foreign countries at the price calculated of what may not be consumed at home, even if very considerably more than the half may be left for export, there can be no doubt. In the space of two generations Europe has become always more and more out of condition to feed its population with its own products. There have constantly been imported fertilizers and articles of food, especially grain, in large, steadily increasing quantities from other parts of the world without this satisfying the constantly growing demand, whose increase, besides, has just as steadily been counteracted by emigration on a large scale. Besides grain, meat has of late years been imported in different forms from the most remote regions of the earth, and most recently in the fresh state and in immense quantities, without thereby causing any trace of a tendency towards a decline of the prices of articles of food, which, moreover, are advancing just as uniformly as they have been for the past decade. Under such circumstances there is probably not the remotest reason for supposing that the articles here treated of, if they, as remarked, are brought to market in a perfectly fresh state and at a lower price than is paid for other articles of food of greatly inferior value, will be unsalable even if the whole estimated quantity is marketed abroad. In north Germany alone, which can produce but few fishes of this kind and quality, there are many millions of people who would prefer them to other food if they could be had at a reasonable price, which for these millions means four to five times what is here estimated.

But this transportation in the fresh state to a distant place, which, of course, is not entirely unknown, but is practiced in a manner which can give but little security for the perfect preservation of the article during many days or weeks, may perhaps be considered both costly and troublesome. In order to remove all uncertainty in this direction, I shall state that carrying of fresh meat from America to England in steamers specially arranged for the purpose costs 30 shillings sterling per ton; that is, $7,425 per 1,000 kilograms, or seven-tenths of a cent per kilogram; an expense which is nothing in comparison with the increase in the calculated price which is to be expected in a foreign market. This is the cost of both transportation and refrigeration, which requires ice bought where it is high and the use of machinery during the whole time of transportation, together with expensive fittings of the rooms wherein the transportation goes on. The price of refrigeration alone can also not be placed higher than half—seven-twentieths of a cent per kilogram—which equals one-fifth of a cent per pound for an average period of about fourteen days. In all places where fish will be reared, the collection of the necessary quantity of ice costs nothing, as it will be done by the persons occupied in the business as a whole, and paid for it by the whole year. Salt is required for freezing, but it is only changed from a solid, dry body to pure liquid brine, and as in this form it can be used for all ordinary purposes just as well as in the solid food this article also thus costs nothing. The preserving-house and apparatus for freezing are very simple things, which can be provided once for all at an expense which must be a very small fraction of what the arrangement of a ship's room for transportation by refrigeration of fresh meat represents.

The expenses of freezing fish and their preservation for a long time in a refrigerator during a transportation of eight to fourteen days in the frozen state cannot thus possibly be more than the previously-mentioned cost of transportation across the Atlantic Ocean; and this expense, stated to be seven-twentieths of a cent per kilogram, is only a small fraction of the previously-estimated outlay for preserving, transportation, and packing, $2\frac{3}{4}$ cents per kilogram.

The necessary freezing and storing for a long time, and the refrigerating during transportation to places much more distant with the means of conveyance of the future, cannot in any way cause a remarkable increase of price of the product.

There is just as little probability that the increase of the population prevailing in the last decade will in the future become stationary, or in a remarkable degree decline, as there probably is that the prices of the necessaries of life in coming time will decline in any especially considerable degree in the markets of Europe. But even if such a decline should take place, even if it, on the average, and for this article especially, should go down to the half of what may be considered at the present a low estimate, or from 82½ cents to 41¼ cents per kilogram, the product in question must find a sure sale, since, it with an enormous profit against it, as was previously brought into consideration, can be sold far below this price, and thus, as a delicacy of the first class, must become sought in preference to articles of food of inferior account. In order that the product which is here in view may lose its rank and value as a delicacy it must become common every-day fare—that is, be able to be sold at such a price and in such quantities that the great public may have the means and the opportunity to eat it four or five times a week. If one takes into consideration only a small portion of the population of North Europe, say 15,000,000 of people, and assumes that these, on the average, when the article is to be had as cheaply as here estimated, will consume one-half kilogram twice a week, which is by no means improbable if the price becomes as stated—less than half the cost of the commonest kind of meat—for this small fraction of the population of North Europe will be required annually 780,000,000 kilograms at least; that is, ONE-HALF MORE THAN THE WHOLE ESTIMATED PRODUCTION. There is thus certainly no reason to fear that the enterprise will fail from want of sale of the portion of the product which we may be able to offer for sale. The probability is, on the contrary, that many times the quantity will find sale at considerably higher prices than those previously brought to notice, and this so much more surely since there is every reason to suppose that at the same time when an increased fish production may be secured there will be an increase of the population which may be counted on as consumers on nearly the same scale as in recent past time.

With full conviction that the labor bestowed upon the cultivation of the water will pay well, I advise every one who may have the opportunity, to take hold of this cultivation, and that with full energy; because, let it be remembered, half or quarter of the work, according to plans hitherto employed, and universal Norwegian customs, will give just as little profit as importance in this as in other directions. Moreover, I must recommend that this enterprise, with the intervention of legal provisions, be hedged around just as carefully as any other pursuit of similar importance—agriculture, grazing, or cultivation of the woods. So long as the hitherto mentioned license can go on uncensured and unchecked, the enterprise here treated of will come into practice only to a very small extent, and to inappreciably little use for people in general; BECAUSE ONLY BY PERSONAL OWNERSHIP, INCONTESTABLE AND FULLY PROTECTED BY LAW, OF THE PROFIT WHICH CAN BE EXPECTED WILL AFFORD SUFFICIENT ENCOURAGEMENT TO PUT THE ENTERPRISE IN THE NECESSARY VIGOROUS PRACTICE IN THE EXTENT TO WHICH OPPORTUNITY IS OFFERED. THE BEST AND LARGEST FIELDS FOR FISH CULTURE, WHICH TO A GREATER OR LESS EXTENT ARE OWNED IN PARTNERSHIP, WILL, WITHOUT SUCH PROTECTION OF THE LAW, BECOME JUST AS POOR AS THEY AT PRESENT ARE UTTERLY BARREN, BECAUSE NO ENCOURAGEMENT TO ARTIFICIAL AND ENERGETIC CULTURE WILL THEN EXIST. PRODUCTION WILL THEN, AS HITHERTO, BECOME DEPENDENT ALONE ON NATURAL CULTURE; AND THE EXPERIENCE OF A THOUSAND YEARS HAS CLEARLY SHOWN *that there is no natural production, however strong and rich this may be, which can stand against the greed of men, when this is not checked by the personal consciousness of ownership.*

XXIII.—WHAT DOES A FISH COST?*

By Christian Wagner.

After the tiny, jelly-like fish has left its egg it receives nourishment for several days from the adhering yolk-bag, mostly resting on its side upon a blade of grass as near as possible to the surface. Thence forward, however, it preys on living aquatic animalculæ, and though enjoying a life scarcely perceptible by means of a microscope it soon begins to hunt for the mite-like water-fleas, the larvæ of gnats, &c. For some time I have been in the habit of keeping, and not only during the breeding season, small cups and larger dishes for raising different insects, which facilitate the observations.

The goldfish, carp, tench, &c., are quite voracious when only eight days old. They consume in three days about as many of these animalculæ as their own weight. Higher or lower temperatures are at this time of the greatest influence. The warmer the weather the greater is the vitality of the fish and the more rapid its growth, if properly fed.

By continued natural feeding the little fish doubles its size in about eight days, and then, fourteen days old, is looking for larger food, which now mostly consists of larvæ of gnats. The consumption of food equal to one-third of its own weight per day is still continued; but the growth does not progress at the same rate, it only increasing about 100 per cent. in the next fourteen days. A fish of four weeks, with sufficient food, will, perhaps, in four weeks double its size; then in eight weeks, and so forth, at the same ratio, if the winter or other circumstances do not interfere.

Although our predaceous fish, the so-called winter-fish, as pikes (pike-perch), trout, &c., down to the little stickleback, sometimes take food in rather cold weather, the so-called summer-fishes (carp, tench, bream) eat almost nothing in winter time. The colder it is the slower they breathe, and though on warm sunny days they occasionally appear near the surface, they rarely take adequate food. They are always satisfied with the little nourishment contained in the water, which, by breathing, is conducted into the stomach—it is true at the expense of their own bodies, for in spring-time all these fishes appear more or less emaciated. Suppose such a fish, one year old, be it summer or winter fish, to have attained a length of about six centimeters, and to represent, according to its weight, a food value of one pfenning [equal to nearly one-

* Translated by H. Diebitsch from Deutsche Fischerei-Zeitung, No. 46, Stettin, November 12, 1878.

fourth of a cent.—TRANSL.] (fifty of such fishes equal one kilogram), it has in the coming summer one hundred more warm days for growth. It may be taken for granted that the fish in these one hundred days, as in the first year of its existence, will consume every three days as much food as its own weight, which in the second summer amounts to 33⅓ times its own weight of feeding matter, worth 33⅓ pfenninge.

The predaceous fish, as is known, lives on other fishes, even of his own species, if not fed with blood, scraps, &c., and consumes food matter as valuable or nearly as valuable as itself. In fact, up to the second summer the fish represents only 10 per cent. of the value of its feeding expenses, having only attained the weight of 120 grams, and in the third year, consuming at the same ratio, it costs already 10 mark (1 mark nearly 24 cents), and so forth.

Proceeding on this undoubtedly correct basis, calculation would finally arrive to an enormous amount. I shall afterwards give a striking example. If the so-called summer-fish in its third year does perhaps not use up quite 33⅓ times as much food as its own weight, the winter-fish, on the contrary, as it keeps on eating throughout the whole year, consumes a great deal more.

According to my observations, a three-pound pike prefers a one-pound pike to a one-fourth pound one, and a pike one-fourth pound in weight rather takes another one one year old than a smaller fish or an angle-worm, &c. As a special dainty, the frog may be mentioned. In the following I shall state facts, and shall prove by figures how dear a fish can become, though apparently an inexpensive inhabitant of the water.

I had rented a (Grand Ducal) fish-breeding establishment at Varel, and a friend of mine, Mr. Krommelbein, placed in one of the ponds, which had proved especially adapted, 2,000 so-called *Streck*-carp, about as long as a hand; these are fit for propagation in the fourth year.

In order to destroy the too great quantity of fry, it is customary to introduce small pikes; in this case about 20 were added. As formerly done, this pond after three years was to be fished in its turn. When informed by Mr. K., I prepared for the 1,800 carp to be received (10 per cent. of loss), which, after former experience, ought to have attained the weight of 1½ pounds each. You will be surprised to hear the result of the entire proceeding, viz: Two eatable fishes—pike—of 30 pounds* each, a number of two-year-old carp-fry, † one-year-old pikes without value, equally small perches, ‡ and many sticklebacks! The above two pikes had made such astonishing growth in consequence of the abundant food.

The summer, like the winter, fish of the same age often differ in size

* The pikes introduced with the Streck-carp must have been too large, since they actually eat up the carp.—[EDITOR.]

† The Streck-carp must have deposited eggs. Have none of the originally introduced carp been caught, from which those two-year old must have been bred?—[EDITOR.]

‡ Had these also been introduced?—[EDITOR.]

in the ratio of 1 to 4, according to their feeding. A pike of six years weighs on an average about 5 pounds; but if its nutriment is abundant, as was the case in the carp pond mentioned, a triple weight has been obtained. What was the expense for these two pikes? From former results we know that the same pond had yielded 3,000 pounds of eatable fish when stocked with 2,000 carps (without larger pikes), and taking 1 mark as the value of 1 pound, then either of the two pikes costs 100 mark per pound!

How, during three years, so many enemies to the fry could originate is the question still to be answered, and the answer is quite simple. Just as weeds in the field without direct seeding grow in greatest luxuriance, so the hosts of unwelcome destroyers originate in the water-basins. If we, for instance, observe ducks swimming among the deposits of eggs (either on plants as with summer-fish or on the bare shore with others), we will see how they carry on their wings the contents of the water when suddenly frightened, the eggs easily adhering to their feet and feathers. Reaching other, perhaps more quiet, water they dive and get rid of their adhesions, &c. Many kinds of water-fowl, also rats, frogs, and other animals, thus distribute useful seed, as well as that of weeds, if this expression be allowed.

Whoever, therefore, wants to breed a certain species of fish, must know how to calculate and must thoroughly cultivate the water. The water is like the field; where there is no cultivation there is no result in either. My experiences serve me as proofs for the statements made in 1863 (in *Zoolog. Garten zu Frankfurt-am-Main*). The area of my property, if used for agriculture, would scarcely support a laborer and family, while by pisciculture it gives employment to fifteen men, three horses, and a steam-engine. The profit to myself is much greater than any farmer or gardener could make of it, for the water is much richer than the field, if pools are cultivated like land. The field is my school, the water my field!

CHRISTIAN WAGNER.

OLDENBURG.

APPENDIX G.

THE PROPAGATION OF FOOD-FISHES.

APPLICATIONS.

XXIV.—THE PROPAGATION AND DISTRIBUTION OF SHAD IN 1878.

By James W. Milner.

A.—STATION ON ALBEMARLE SOUND.

Encouraged by the success of the apparatus devised by Mr. T. B. Ferguson, Commissioner of Fisheries of Maryland, for hatching shad in tidal waters, it was decided that a renewed experiment should be made in the waters of the Southern States with the improved method. The floating boxes had proved inadequate and unsatisfactory in waters without current.

The region of the numerous fisheries in Albemarle Sound was selected for the point of operations. These, numbering in all about forty, are located on the shores of the sound, principally in the northern and western region and the mouths of the Chowan and Roanoke Rivers, and employ seines from five hundred to twenty-five hundred yards in length. Most of these employ horse-power for hauling in the seines, although several are fitted with fine steam-engines. They are considered as an important part of the plantations and estates upon which they are situated, and a very considerable amount of capital is invested in the enterprise. The preparation of the shore alone is an expensive operation, employing many men, diving experts, and explosives, to clear off the snags, cypress knees, and stumps which are found in the shallower portions of the water.

The fishing for shad and alewives or herring ordinarily begins in March and lasts to the 1st of May, the lower fisheries cutting out several days before the upper ones. Fishing is continued from midnight Sunday evening until midnight Saturday. The fishing in this locality has hitherto been confined almost exclusively to the seines, although of late years pound-nets, or "dutch nets," as they are called in this locality, have been introduced. Pamlico Sound, which is adjacent, is also an extensive fishing region.

The nets visited by our steamers extended from Colerain, on the Chowan, to Drummond Point, on the northern side of the sound, and to Jamesville on the Roanoke, something like thirty fisheries being thus available.

The barges fitted up and used the previous year by Maryland, containing the improved machinery for shad-hatching, were, with the machinery, purchased by the United States commission from the Maryland commission and put in working order. On the 19th day of March the revenue-steamer Thomas Ewing, Capt. Alvan A. Fengar in command, by instructions from the Secretary of the Treasury, took the fleet in tow and carried it as far as Norfolk, Va. One of the barges was left at this port to be fitted up as a second machinery-scow. A large tug was employed to tow the three barges through the canals, and they arrived on March 26 at the headwaters of Albemarle Sound, near the mouth of the Chowan River, and were anchored in a protected position adjacent to Avoca, the plantation of the Capehart family, which occupies an extent of land lying between the rivers Chowan and Roanoke.

The services of the steamer Lookout, belonging to the Maryland Fish Commission, had been obtained for the work of the season. She was sent to Norfolk in December of 1877, to secure a harbor not liable to be obstructed by ice at the time she might be needed, and by permission of the Secretary of the Navy was kept at the navy-yard. She arrived in Albemarle Sound on the 3d of April. She there rendered efficient service under the direction of Maj. T. B. Ferguson, whose hearty coöperation and assistance were of great moment to the United States Fish Commission. Several trips of the Lookout from Avoca to Washington demonstrated the practicability of carrying eggs and young fish with perfect safety, so far as the steadiness of the cones before referred to is concerned. Notwithstanding the roughness of the sea, the gimbals in which the cones are suspended insure the horizontal position of the surface at whatever angle the deck itself may stand.

Major Ferguson's appointment as one of the United States commissioners to the Paris Exhibition made it necessary for him to break up his connection with the United States Fish Commission about the time of closing the work on Albemarle Sound.

Very important assistance was also rendered in Albemarle Sound by a small open steam-launch furnished to the commission by the Secretary of the Navy; indeed, without it, very much less wonld have been accomplished. A second launch, derived from the same source, was in use in addition at Havre de Grace.

On the morning of the 28th about eight thousand eggs were taken and put into some Brackett boxes which we had in our outfit. The large seines in the vicinity were in full operation, the two nearest us being those of Avoca Beach and Scotch Hall. These seines were each about twenty-three or twenty-four hundred yards in length. Both of these beaches are on the great Capehart plantation. To Dr. W. R. Capehart and to his father we are indebted for continued aid and active coöperation, which were essential to the success of our work. The previous year, at an outlay of nearly $650, Dr. Capehart made an experiment with the floating boxes, but because of the lack of current no adequate results were obtained. This year he had a large tank erected, and using one of our pumps with his steam-engine hatched a large number of shad.

Up the Chowan, within a distance of twelve miles, were three or four large seines, and along the northern shore of the sound, in the vicinity of Edenton, N. C., and Drum Point, were some eight or ten more. At several of these fisheries steam-engines are used in drawing the seine to shore. The catch of shad was quite limited, thirty to one hundred being near an average haul, while the alewives were very numerous.

On April 1, a general move was made for obtaining shad ova. On this day two hundred thousand eggs were taken, and from this time on a considerable number of eggs were obtained up to the 1st of May, the number reaching 10,387,000. The largest number of eggs taken in any one night was on the 15th of April, when eighty-two shad were stripped, affording what were estimated at 1,605,000 eggs. On the 17th fifty-nine ripe shad were stripped, and on the 18th, seventy-six.

A singular fact attending the work of gathering spawn was the concentration of the spawning fish upon the Avoca Beach, the one nearest to our station; so constant was this that fully four-fifths of the eggs taken were obtained at this one point, although the fisheries for twelve miles up the Chowan, and for fifteen miles along the northern shore, and five or six miles up the Roanoke River, were visited. As usual the bulk of the spawn was taken at night, the largest amount being brought from the seines coming to shore about an hour after dark; a few are taken in the morning, but it seems to be the uniform law that scarcely any are taken after broad daylight. Albemarle Sound proves to be one of the best localities for shad production that has been tried, as the numerous large seines are continuously hauled from Monday morning (midnight) to Saturday midnight of each week. There is no tide in the vicinity, and the hauling is not delayed at any time, as all hours of the day and night are favorable.

The shipment of fish to remoter points began April 11. Correspondence and a conference with the commissioner of Virginia, Col. Marshall McDonal, and of North Carolina, Col. L. L. Polk, had resulted in their assuming the distribution of young fishes to the waters of their States at a distance from the vicinity. The first shipment was made to Nottaway Mills, Va., into the Nottaway, tributary to the Chowan River, at that date. From that time to May 2, when this work closed, 4,926,500 young fish were distributed; of these, 2,145,500 were put into the immediate waters; 1,039,000 were distributed in other waters within the State, making 3,184,500 young put into the waters of North Carolina; 1,142,000 were sent into Virginia, while in other States 600,000 were planted. The accompanying tables will show the details of the gathering of eggs and of the distribution of the fish from the Avoca station.

Although in every respect the region of operations was most admirably adapted to furnishing a large number of young shad, it proved to be rather inaccessible for railroad travel in the distribution of fish. Two steamboats connect it with Franklin, a point on the Seaboard and Roanoke Railroad, within the Virginia line, and about seventy miles distant from our station; but in shipping fish it was necessary to send them by the steam-launch to an outside pier where they remained in charge of the attendants until the steamer came. No telegraph communication could be had nearer than Franklin.

As already remarked, the seine fisheries of this region are on the most extensive scale of any locality in the country; although a single seine on the Potomac exceeds any one on the Albemarle Sound in dimensions, still the average length of seines in the Albemarle Sound is greater and there are more of them. The system of fishing is a very good one, most of the seine proprietors finding it possible to employ the same gang of hands each succeeding year. This, of course, affords them the advantage of a trained and experienced force, which is a matter of much consequence.

At the Scotch Hall fishery on the Capehart plantation, where steam-engines are used, a system of signals is employed with the steam-whistle, through which the men are called to any point necessary, and the position of the seine can be known by the proprietor, while in his office or at his home, by the special signals given at different intervals.

Where the outlay is large and the labor bill great, as it is at these steam-fisheries, it is found essential to have the material in the outfit of superior quality, so that there may be little liability of delay through breakage or accident. A peculiar line is used at these fisheries, imported from Russia, made expressly for the Russian navy, and said to be used only in two or three industries in the United States, one the Albemarle Sound fisheries and another in oil-well drilling.

Some of these fisheries have proved very profitable to their owners. One is referred to, the sum of the profits from which for nine years was $55,000. A table showing the catch at Scotch Hall fishery for a series of years is appended.

Large shoals of rock-fish or striped bass visit the shores later in the year. A remarkable haul made on one of Dr. Capehart's shores in 1876 yielded 35,000 fish. Many of these weighed 80 and 90 pounds, and 365 of them had a total weight of 23,785 pounds, an average of 65 pounds. This year the run of shad into the sound was very light; only one haul of consequence was heard of, this being when 998 were taken at Avoca Beach on the 17th of April.

It may not be generally known that the waters of Albemarle Sound are entirely fresh from Roanoke Island to the head of the sound, the inlets from the sea being so small that very little salt water is introduced; the large rivers emptying into it also prevent the tides from having much effect upon the water; Pamlico Sound has a larger area of salt water.

Early in April it was announced that large hauls of herring were being made in Pamlico Sound; about the middle of April large catches were made at the lower fisheries in Albemarle Sound, and from that on till the 1st of May the nets were crowded with herring. The run was unprecedented, the older fishermen asserting that nothing equal to it had ever occurred; one of the nets of Mr. Peter Warren took at one haul 400,000. The northern shore of the sound and the Chowan River seemed to be in the direct course of the fish; later they struck the net at Avoca Beach; the best haul made at this point was 165,000. Scotch Hall, about three miles nearer the mouth of the Roanoke River, did not seem to be in the range of the fish; they ascended the Roanoke in much more moderate numbers, though all that the fishermen desired and more could be taken. The herring crowded the waters of the sound to such an extent that they seemed to drive the shad and other fishes away, and the catch of shad became even smaller than it had been. The steamers from Franklin came daily freighted with salt and went back loaded heavily with salted herring; the prices dropped very rapidly until they were offered in some instances at 50 cents a thousand at the beach. The women employed to dress the salted herring worked night and day, and a large increase of the force was made at most fisheries. It was impossible in these immense hauls to take care of all of the fish, and frequently a large surplus was carted away from the beach to spread on the fields as manure.

No satisfactory theory suggests itself for this immense increase in the herring. The seines stopped fishing eight or ten days earlier than usual because of this immense influx of herring, as the prices became lower for every additional hundred thousand salted.

A change in the run of the shoals of fish at different points, from year to year, is observed here as it is in many other places and with other species of fish. It is impossible to say whether this is owing to an alteration in the contours of the bottom from the heavy storms, or to a change in the distribution of the food of the fishes, or to a question of temperature, but it is a fact that certain shores, which in a series of years have gained notoriety for great yields of fish, subsequently diminish in value, and other stations supplant them in this respect.

The facilities for hatching fish which were at our command were, first, the apparatus first used in the season of 1877, namely, a barge, on the outside of which levers protruded from air-ports; from the bows were suspended buckets, an up-and-down movement being afforded these by means of eccentrics, which from their irregular form, with one long side and one short side, produced a sudden drop and a slow rise; this apparatus is the well-known invention of Major Ferguson. Cones were arranged along the sides of the housing which covered the scow; two large casks were raised on a platform to an elevation higher than the top of the cones, and were filled by a pump run by the same engine which propelled the shafting. An improvement was made on the cones at the suggestion of Mr. F. N. Clark, which obviated the continual attention required in skimming off foul matter, shells of eggs, and the like, which continually clogged the perforations in the inner rim, and produced an overflow of eggs and fishes from the cones. By means of Mr. Clark's contrivance the specific gravity of sound eggs, at a properly regulated pressure, caused them to remain some distance below the surface; the refuse matter, including the *Saprolegnia*, constantly flowed off. The engine was worked night and day. When the stock on hand was small the engineers also attended at night to the cylinders and cones, but during the height of the season, when everything was filled, night and day attendants were required. Some little difficulty was at first experienced with the cylinders in time of storm, a very little increase in wind or wave occasioning too much agitation of the water and eggs within the cylinders, and our first warning of this danger was the loss of over 300,000 eggs, by a strong wind and sea setting against the cylinders containing them on one side of the barge. This we were afterward able to prevent in part by stopping the machinery and allowing the wave alone to give the movement to the eggs within the cylinders. A screen or breakwater might, however, easily be devised by placing a frame work outside the cylinders, reaching a little below the water, and nailing or fastening to it either canvas or thin boards; not being prepared with this device during the present season, it was thought best to use the cones to the largest extent before utilizing the cylinders, and sometimes, when bad weather threatened, we took the precaution to remove the eggs and fish from the cylinders to the cones inside the building, if any of them happened at the time to be empty, as was explained in my report of 1877. Very much less loss was experienced by this apparatus than could be expected from any form of floating box where only side currents are to be depended on.

Where a continual river current is found, the cheaper floating boxes may be used quite as efficiently, except for the greater area required; but the larger portion of the shad-spawning grounds being within tidewater and where currents are very slight, the great advantage of this certain and constant agitation of the water is readily appreciated.

As in all artificial propagation of fishes the presence of a skillful expert is necessary; trusting the work to beginners and those who have little experience and ability in fish-hatching will afford as small results as it does with any other apparatus.

A small experiment was made with the Chase jar. This Mr. Oren M. Chase has used at Detroit for the past four years in hatching white-fish. It was found to work with quite as much efficiency in hatching shad, and it is quite probable that with some modifications to suit the different conditions of shad-hatching it would be found to excel everything else in the concentration of space and hatching a very large quantity of eggs in masses contained in quite small vessels.

A device invented by one of the working members fo the corps, Mr. W. T. Wroten, also deserves notice. It embodies the principle of the Chase jar, except that instead of applying the current through a rubber tube and diffusing it from the center of the vessel it is applied through vertical funnels or channels on the sides of the vessel, forcing the water in through a narrow space or slot extending round the bottom. This is an advantage in the fact that the vessel being made sufficiently small—

to contain about fifteen or eighteen quarts—can be carried out in the boats, and the spawn as soon as it is impregnated and "rises" can be immediately turned into this bucket, where it can receive much better care before reaching the hatching station, and the vessel can then be placed where a stream of water can be introduced, and the entire operation, from the time the eggs are impregnated to the time the fish are taken out, can be carried on in the vessel without transferring or moving the eggs.

The experimental device of Mr. Wroten is a little crude in its construction, but with another year's use, and the improvements which will be suggested, it is quite likely to be considered as a valuable acquisition to the apparatus for fish-hatching.

The Maryland yacht Lookout, which was at the station, had also, on its forward decks, six cones which were employed in hatching eggs and in two efforts to transport large quantities of fish to the waters of the Potomac and streams in Virginia and Maryland.

At Avoca station a few experiments were made in taking herring-spawn, with very good success; the variety was the so-called glut or small-eyed herring, which here runs high up the streams; the larger-eyed herring remaining down the bay—an instance of opposed habits in the same species in different regions, as in the Potomac the so-called branch herring runs up the streams, and the glut remains in the open waters. The eggs were handled in precisely the same manner as those of the shad; they had rather a tendency to adhere to the sides of the pan or whatever they touched, but still with a little pains were washed free, and were put into the buckets to hatch the same way as the shad. The young were very minute, and it was impossible to keep them in the vessels, because they were small enough to slip through; the wire-cloth that we had in use being only twenty-four meshes to the inch.

About the end of April the large seines began to cut out, the great flood of herring making it unprofitable to continue fishing, as the shad all abandoned the shores.

On the 29th of April, the revenue-cutter E. A. Stevens reported for duty in affording facilities for our work. As the last seine, the one at Avoca Beach, which had proved so profitable to our work, was to cut out on the 1st of May, and the steamer Lookout was at hand expecting to carry back a large stock of fish and eggs to the Potomac River, it was determined to get rid of all the young fish on hand and have the Stevens tow the barges as far as Norfolk, if not farther. On the morning of the 2d of May the barges were tied behind the steamer, and we started down the Sound, lying over at night in the narrow cut south of the entrance to the Albemarle and Chesapeake Canal. On the morning of the 4th we reached the navy-yard at Norfolk, having had some trouble in getting through the shoal passages and cuts with the long string of barges. We were tied to the anchor-buoys at the navy-yard, and the steamer returned to Albemarle Sound. About four o'clock the same afternoon the revenue-cutter Thomas Ewing arrived with instructions to take one of the barges and a steam-launch to Havre de Grace. Captain Fengar, however, obligingly consented to take two barges instead of one. The wind rising to a gale he put off starting until the next morning; but at an early hour on the 5th the cutter got under way with the two barges and one launch in tow; the remaining barge was taken into the dock and tied in one of the slips at the navy-yard.

On the 7th the steam-launch of the Franklin was sent alongside, Mr. Kullman, a machinist from the steamer, coming with it to act as engineer; on the evening of this day I came on to Washington, leaving the two barges and the steam-launch to be towed up to Havre de Grace on the return of the revenue-cutter Ewing. On her second trip this steamer left Norfolk on the 9th of May and took the remaining barges and the launch in tow for Havre de Grace. Going up the bay they encountered a strong wind and sea, and were for a time in considerable danger; at about midnight the wind rose to quite a gale and the steamer ran into the Great Wicomico River for harbor; the launch, however, parted her hawsers and went adrift; as it was impracticable to turn around with the two scows, she had to drift along until they were anchored, when the steamer went out and found her after some search and took her into the harbor.

B.—STATION NEAR HAVRE DE GRACE, MARYLAND.

The barges arrived on the 11th of May at Havre de Grace. I had been for twenty-four hours awaiting them. They were anchored in the Narrows in about the same place they were last year.

As we were well supplied with steam-launches and facilities for obtaining spawn from the different seines and nets, I concluded that the steamer Lookout could be best utilized by being sent to the Potomac to work the fisheries of that river, as she possessed considerable capacity in the cones on her forward deck for the care of eggs. Captain Chester accordingly left with the Lookout on the morning of the 15th of May.

The first eggs were taken on the 17th of May, the number being 25,000; the total number from that day until the 11th of June obtained at this station was 12,730,000. The greatest number gathered in one day was 1,940,000, on the 29th of May, from 97 spawners. On the 27th 65 good spawners, and on the 28th 71 were taken, this period seeming to be the climax of the spawning season.

The first shipment to a distant stream was made on the 15th of May—150,000 shad—to the Tombigbee River in Mississippi. The total number shipped to other States from this point was 2,535,000; the number put in at this station and in immediate waters, 5,105,000; and the number put in other streams of Maryland and in the Potomac River, 1,705,000, making a total of 9,345,000 fishes.

Mr. Thomas Hughlett, of Easton, Md., State Commissioner of Fisheries, took the responsibility of a considerable portion of the State distribution and afforded efficient aid to our work.

C.—POTOMAC RIVER STATION.

Captain Chester succeeded in obtaining 1,430,000 eggs on the Potomac River, a portion of which were put in the river and two shipments made to other waters of Virginia. The results of his work are shown in the tables.

D.—GENERAL RESULTS.

The total number of eggs taken for the year at Avoca and Havre de Grace stations, and the Potomac River was 24,547,000; from these about 14,521,000 fishes were distributed, or about 60 per cent., which is of course small, the losses before the machinery was complete accounting for most of it.

The success of the work was, however, great enough to meet all special requisitions, and it was deemed unnecessary to establish a station at any point further north. A shipment of 150,000 shad was sent on the 11th of June to Sacramento River, California, going through with great success. This is the fourth shipment made to that river by the United States Commission in co-operation with the State, a previous one in 1871 having been made under the auspices of the State alone.

The results from placing shad into the Sacramento River, where they had no previous existence, are of the most encouraging character, as the number of shad taken has increased yearly, so that in the present year it makes a considerable item in the Sacramento fish market.

The news of continued captures has also been heard from Louisville, Ky., and a few points on the Mississippi River. Shad were taken in the month of March at Wetumpka, on the Coosa River. The run of shad at Louisville began about the 1st of May, and closed about the 20th; the greatest number being taken between the 10th and 18th, as near as could be learned by the Fish Commissioner.

TABLES OF SHAD PROPAGATION AND DISTRIBUTION IN 1878.

Record of shad-hatching operations conducted at Avoca, N. C., on Albemarle Sound, from March 28, 1878, to May 1, 1878, on account of the United States and Maryland Fish Commissions.

Date	Hour.	Temperature of—			Wind.		Condition of—		Ripe fish.		Eggs obtained.
		Air.	Surface water.	Bottom.	Direction.	Intensity.	Sky.	Water.	Males.	Females.	
		°	°	°							
March 28	12 m.	69	60	......	SW.	....	Cloudy..	Muddy..	2	2	16,000
29	12 m.	69	60	......	NE.	....	Fair	..do	....	....	....
30	12 m.	60	60	......	NE.	....	Cloudy..	..do	....	....	....
31	12 m.	75	62	......	SW.	....	Fair	..do	....	....	....
April 1	12 m.	62	63	......	NW.	....	..do	..do	20	12	225,000
2	12 m.	74	61	......	NEE.	....	..do	..do	12	14	188,000
3	12 m.	65	61	......	ENE.	....	Cloudy..	..do	30	18	288,000
4	12 m.	53	60	58	NE.	....	..do	..do	2	1	20,000

Record of shad-hatching operations conducted at Avoca, N. C., on Albemarle Sound, from March 28, 1878, *to May* 1, 1878—Continued.

Date.	Hour.	Temperature of—			Wind.		Condition of—		Ripe fish.		Eggs obtained.
		Air.	Surface water.	Bottom.	Direction.	Intensity.	Sky.	Water.	Males.	Females.	
		°	°	°							
April 5	12 m.	63	58	57	NNW.		Clear	Muddy	2	1	20, 000
6	12 m.	56	57½	57	NW.		Cloudy.	do	9	6	120, 000
7	12 m.	64	58½	57	W.		Clear	do			
8	12 m.	62	61	58	ENE.		do	Clear	16	11	195, 000
9	12 m.	67	61	58½	SEE.		Cloudy.	do		13	240, 000
10	12 m.	64	60	58			Rainy	do		16	320, 000
11	7 a. m.	63	61	58	W.		Misty	do			
	12 m.	68	64	58	NE.		Clear	do			
	7 p. m.	65	63	58	SW.		Cloudy.	do		5	100, 000
12	7 a. m.	54	59	58	WSW.		Fair	do			
	12 m.	77	63	60	W.		Clear	do			
	7 p. m.	64	61	59	Calm.		do	do		8	140, 000
13	7 a. m.	69	62½	59	WSW.		do	do			
	12 m.	79	64	60			do	do			
	7 p. m.	66	64	63	W.		do	do		19	380, 000
14	7 a. m.	64	62	61	NEE.		Hazy	do			
	12 m.	68	66	63½	E.		do	do			
	7 p. m.	60½	65½	63	E.		do	do			
15	7 a. m.	60½	65	63	NE.		Cloudy.	do			
	12 m.	66½	66	62½			do	do			
	7 p. m.	60½	65½	62			do	do		83	1, 605, 000
16	7 a. m.	56	64	61	NNW.		do	do			
	12 m.	56	60½	60½	NNW.	Strong	do	do			
	7 p. m.	53	59	59½	NNW.	do	do	do		59	1, 160, 000
17	7 a. m.	56	59	56	NW.		do	do			
	12 m.	64	60½	57	NW.		do	do			
	7 p. m.	53	63	58							
18	7 a. m.	51	61	58							
	12 m.	55	63	60							
	7 p. m.	55	64	60½						76	1, 465, 000
19	7 a. m.	58	62	59							
	12 m.	72	64	60			Fair	Clear			
	7 p. m.	61	66	60			do	do		27	540, 000
20	7 a. m.	60	61	60½			do	do			
	12 m.	78	64	63			Cloudy.	do			
	7 p. m.	65	64	61½						11	190, 000
21	7 a. m.	69	64	62½			Clear	Clear			
	12 m.	78	71	60½			do	do			
	7 p. m.	68	70	61			do	do			
22	7 a. m.	65	65	62½		Calm	do				
	12 m.	77	71	62½		do	do				
	7 p. m.	75	72	63		Fresh	do			27	540, 000
23	7 a. m.	70	66	62		Light	Cloudy				
	12 m.	79	70	66		Brisk	do				
	7 p. m.	75	67	65		Light	Fair			31	610, 000
24	7 a. m.	72	66	62½		do	Cloudy				
	12 m.	76	67	66		Brisk	do				
	7 p. m.									1	20, 000
25	7 a. m.	64	66	65		Light	Cloudy				
	12 m.	78	67	65		Brisk	do				
	7 p. m.	64	67	64½		Calm	Clear			22	440, 000
26	7 a. m.	64	65	64½		do	do				
	12 m.	80	71	65½		Light	Fair				
	7 p. m.									4	80, 000
27	7 a. m.	67	67	65							
	12 m.	85	76	65½							
	7 p. m.	77	70	68						20	400, 000
28	7 a. m.	65	70	67							
	12 m.	80	76	68							
	7 p. m.	68	70½	67½		Slight	Fair				
April 29	7 a. m.	59	67	67½		Calm	Fair				
	12 m.	71	70	68½		Slight	do				
	7 p. m.	60	69	69		Calm	Clear			42	820, 000
30	7 a. m.	70	70	67		do	Fair	Clear			
	12 m.	81	72	67½		do	do	do			
	7 p. m.	66	72	70		do	do	do			165, 000
May 1	7 a. m.	64	70	69		do	do	do			
	12 m.	83	76	70		do	do	do			100, 00

***Record** of shad-hatching operations conducted at Havre de Grace, Md., on the Susquehanna River, from May 7, 1878, to June 12, 1878, on account of the United States and Maryland Fish Commissions.*

Date.	Hour.	Temperature of—			Wind		Condition of—		Ripe fish.		Eggs obtained.
		Air.	Surface water.	Bottom.	Direction.	Intensity.	Sky.	Water.	Males.	Females.	
		°	°	°							
May 7, 8, 9, & 10										20	385,000
12	7 a. m.										
	12 m.										
	8 p. m.	56	61½			Calm	Cloudy	Clear			
13	7 a. m.	50	58	57		Strong	Hazy	Roily			
	12 m.	59	62	59		Fresh	Fair	do			
	8 p. m.	53	60	58		Slight	Cloudy	do		3	60,000
14	7 a. m.	50	59	57½		do	do	do			
	12 m.	51½	59	57½		Calm	do	do			
	8 p. m.	50	57	57½		Fresh	do	do		1	20,000
15	7 a. m.	46	56	56½		do	do	do			
	12 m.	52	56½	55		do	do	do			
	8 p. m.	53½	57	55		Calm	Fair	do		3	60,000
16	7 a. m.	53	56½	55½		Fresh	do	do			
	12 m.	61	62	57		Slight	do	do			
	8 p. m.	60	59½	57		do	Cloudy	do			
17	7 a. m.	59	60	57		do	do	do			
	12 m.	60	61	57		do	do	do			
	8 p. m.	57	60	57		do	do	do		7	150,000
18	7 a. m.	57	59½	57½		Calm	Fair	do			
	12 m.										
	8 p. m.	62	65½	62		Slight	Clear	Roily		2	40,000
19	7 a. m.	61	60	59		do	Fair	do			
	12 m.	70	64	63		do	Clear	do			
	8 p. m.	60	63½	63		do	Fair	do		3	60,000
20	7 a. m.	60	63½	63		do	Cloudy	Clear			
	12 m.	65½	64	63		do	do	do			
	8 p. m.	64	64½	63		Strong	do	do		26	520,000
21	7 a. m.	66	64½	63		Fresh	do	do			
	12 m.	73	68	63½		Calm	Fair	do			
	8 p. m.									29	580,000
22	7 a. m.	57	63	62½		Fresh	Fair	Clear			
	12 m.	64	65	65		do	Clear	Roily			
	8 p. m.	66½	67½	65		Slight	do	do		31	620,000
23	7 a. m.										
	12 m.										
	8 p. m.									23	470,000
24	7 a. m.	70	67½	65		Calm	Cloudy	Clear			
	12 m.	73	70	66		do	do	do			
	8 p. m.	66½	68	66		do	do	do		44	880,000
25	7 a. m.	65	68	65½		do	do	do			
	12 m.	74	70½	67		do	do	do			
	8 p. m.									27	540,000
26	7 a. m.										
	12 m.										
	8 p. m.									38	760,000
27	7 a. m.										
	12 m.										
	8 p. m.									65	1,300,000
28										71	1,430,000
29										97	1,940,000
June 1	7 a. m.					Gale	Cloudy	Roily			
	12 m.					do	do	do			
	8 p. m.					do	do	do			
2	7 a. m.	70	66	63		Strong	do	do			
	12 m.	72	66	63		Fresh	do	do			
	8 p. m.	70	65½	63		Slight	do	do		4	80,000
3	7 a. m.	69	65	63		do	do	do			
	12 m.	70½	67½	65		do	Fair	Clearing			
	8 p. m.	68	71	67		Calm	Clear	Clear		53	1,060,000
4	7 a. m.	71	69	66		Slight	Cloudy	do			
	12 m.	76	69	66½		do	Fair	do			
	8 p. m.	74	72	68		Fresh	Cloudy	do		40	750,000
5	7 a. m.	72	71	68		do	Fair	do			
	12 m.	77	72	69½		do	Cloudy	do			
	8 p. m.	66	72	69		Strong	Clear	do			
6	7 a. m.	62	68½	66		do	Hazy	do			
	12 m.	66	74	71		Fresh	do	Roily			
	8 p. m.	65	74	71		Calm	Clear	do		20	300,000
7	7 a. m.	66	72	69		Slight	do	do			
	12 m.	73	74	72		do	Fair	do			

Record of shad-hatching operations conducted at Havre de Grace, Md., &c.—Continued.

Date.	Hour.	Temperature of—			Wind		Condition of—		Ripe fish.		Eggs obtained.
		Air.	Surface water.	Bottom.	Direction.	Intensity.	Sky.	Water.	Males.	Females.	
		°	°	°							
June 7	8 p. m.	69	72	70	....	Fresh	Hazy	Clear	....	12	200,000
8	7 a. m.	70	71	68	....	Slight	Cloudy	..do	....	....	....
	12 m.	......	......	......	....	....	....	....	....	....	....
	8 p. m.	72	70½	68	....	Slight	Cloudy	Roily	....	8	120,000
9	7 a. m.	68	70	67	....	Fresh	Cloudy	Clear	....	....	....
	12 m.	72	72	70	....	Slight	Fair	..do	....	....	....
	8 p. m.	65	73	70	....	Calm	Cloudy	..do	....	12	200,000
10	7 a. m.	65	71	68	....	..do	..do	..do	....	....	....
	12 m.	69	72	69	....	Slight	..do	..do	....	....	....
	8 p. m.	......	......	......	....	....	....	....	....	7	125,000
11	7 a. m.	......	......	......	....	....	....	....	....	....	....
	12 m.	......	......	......	....	....	....	....	....	....	....
	8 p. m.	66	70	68½	....	Fresh	Fair	Clear	....	5	80,000
12	7 a. m.	63½	68½	66	....	Slight	Hazy	..do	....	....	....

Record of shad-hatching operations on the Potomac River, from May 16, 1878, to June 3, 1878, on account of the United States and Maryland Fish Commissions.

Date.		Hour.	Temperature of— Air.	Surface water.	Wind. Direction.	Intensity.	Condition of— Sky.	Water.	Seine hauled.	Fish taken. Males.	Females.	Ripe fish. Males.	Females.	Eggs obtained.	Remarks.
			°	°											
May	16	6 p.m.	49	61	NW.	Fresh	Clear	Roily	4 p. m	120					
	17	7 a.m.	59	58½	SSW.	do	Cloudy								Four hauls a day, averaging 100 shad, but none ripe.
	18	7 a.m.	60	59	SSW.	Light	Clear	Clear							
	19	7 a.m.	59½	60	SW.	do	do	do	10 a. m	160	40				
		12 m.	61	61½	S.	Fresh	Cloudy	do	5 p. m	60	30		2	45,000	
		6 p.m.	60½	62	SE.	Light	do	Roily	11 p. m	70	30		1	5,000	
	20	7 a.m.	66½	62	S.	do	do	Rain, fog	11 a. m	150	75				
		12 m.	62	62½	S.	do	do	Roily							
		6 p.m.	65	63½	S.	do	Clear	Clear	6.30 p. m	100	65		1	10,000	
	21	7 a.m.	66½	64	NW.	do	do	do	May 20, 11 p. m.	160			3	60,000	
		12 m.	68	69	S.	do	do	do	6 a. m	110	40				
		7 p.m.			W.	do	do	do	11.40 a. m	175	45		1	20,000	
	22	7 a.m.	64	66	NW.	Fresh	do		7 p. m	80	6				Seine broke and fish escaped.
		12 m.	69	67	NW.	do	do		7 a. m	7			1	20,000	Put 50,000 young shad in Neapsico Creek.
		6 p.m.	65	72	NNW.	Light	do		1 p. m	110	40		1	20,000	
	23	7 a.m.	64	66	NW.	do	do		7 a. m	40	10				Left Keystone Point in steamer at 10 a. m.; went up to Chapman's fishery; found they were not fishing; returned and anchored near Indian Head.
		6 p.m.	67	69		Calm	do	Clear	5.36 p. m	787			19	400,000	
	24	7 a.m.	66	66½		do	Cloudy	do							
		12 m.	73	66½	SE.	Fresh	do	do							
		7 p.m.	68	66		Calm	do	do	6.30 p. m	600			29	500,000	
	25	7 a.m.	67	66	NW.	Light	do	do	8 p. m	200			11	220,000	Moved to Washington, D. C., and anchored off Eighth street wharf.
	29	6 p.m.	63	72	W.	do	Clear	do	10.30 p. m				3	50,000	Taken at Glymont.
														60,000	Taken from gillers.
	30	7 a.m.	66	70	SSW.	Fresh	Clear	Clear							Heavy squalls from north and northwest; over 100,000 eggs lost from wooden boxes.
		12 m.	85	72	SSW.	do	do								
		7 p.m.	69	73	N.	Strong	Rain								
	31	7 a.m.	63	70	NW.	Fresh	Cloudy		1 a. m	240					
		12 m.	67	71	NW.	do	do		11.40 a. m	350					Heavy squall on the river from northwest at 4 p. m.
		6 p.m.	60	71	NW.	do	do	Roily							
June	1	7 a.m.	62	70	NW.	do	do		1 a. m.	100					Last haul of the season.
		12 m.	65	69	N.	Gale	do								
		6 p.m.	64	69	N.	Fresh	do								
	2	7 a.m.	64	68	NNE.	do	do								
		12 m.	65	68	NNE.	do	do								
		6 p.m.	66	68	NNE.	do	do							20,000	Taken from gillers.
	3	7 a.m.	67	67	NE.	Light	do								
		6 p.m.	72	70	E.	do	Clear								

Statistics of fisheries at Avoca, N. C.

Date.	Number days fished.	Shad.			Rock.		
		Number taken.	Best day's work.	Best haul.	Number taken.	Best day's work.	Best haul.
1869.							
March	26	9, 034	March 30, 3,290	March 30, 1,213			
April	26	35, 556					
May	12	6, 238					
Total	64	50, 828					
1870.							
March	25	6, 907					
April	26	22, 082	2,163	737			
May	9	7, 590					
Total	60	36, 579					
1871.							
March	20	9, 509					
April	26	19, 693	April 11, 2,239	April 12, 1,241			
May	6	1, 677					
Total	52	30, 879					
1872.							
March	20	2, 940			1, 073		
April	26	40, 488	April 11, 7,857	April 10, 3,289	8, 681		
May	7	4, 916			1, 653	May 1, 700	
Total	53	48, 344					
1873.							
March	21	8, 793			1, 955		
April	26	40, 670	April 5, 7,640	April 5, 2,550	23, 225	April 8, 2,800	800
May	11	7, 814			7, 430		
Total	58	57, 277					
1874.							
March	23	23, 604					
April	26	46, 842	April 9, 6,243	April 4, 3,210			
May	10	8, 650					
Total	59	79, 096					
1875.							
March	22	16, 171					
April	25	34, 485	2,561	1,224			
May	......	10, 827					
Total	......	61, 483					
1876.							
March	26	6, 968			3, 875		
April	26	22, 350	April 28, 1,610	April 28, 505	19, 029		
May	10	7, 831			62, 878	May 6, 48,100	*35, 000
Total	62	37, 155			85, 782		
1877.							
March	18	7, 499			1, 318		
April	26	12, 950	April 21, 1,365	April 21, 524	6, 781	March 28, 1,175	700
May	15	3, 791			4, 705		
Total	59	24, 240					
1878.							
March	......	4, 744	March 19, 830	March 19, 363	3, 018	March 30, 300	
April	......	5, 543			7, 973		
Total	......	10, 287					

* Many of this catch weighed 80 to 90 pounds; 365 of them weighed 23,725 pounds.

Number of hauls made during one week, 1868, *and number of shad caught.*

Date.	First haul	Second haul	Third haul	Fourth haul	Fifth haul.	Total.
1868.						
April 6	920	1,330	550	693	520	4,013
7	352	384	860	940	1,377	3,913
8	1,341	1,329	1,200	1,941	....	5,811
9	1,678	2,215	1,704	....	....	5,597
10	1,600	2,722	1,850	760	376	7,317
11	1,052	2,900	4,777	1.850	....	10,579
Total shad caught one week						37,230

Record of distribution of young shad made from April 11, 1878, to June 14, 1878, by United States and Maryland Commissions on Fish and Fisheries.

Date of transfer.	Place whence taken.	Number of fish.		Introduction of fish.				Transfer in charge of—
		Originally taken.	Actually planted.	State.	Town or place.	Stream.	Tributary of—	
1878.								
April 11	Salmon Creek, near Avoca, Albemarle Sound, North Carolina (at mouth of Chowan River).	2,500	2,500	North Carolina	Avoca	Salmon Creek	Chowan River	U. S. F. C.
April 11		111,000	111,000	Virginia	Nottoway Mills	Nottoway River	do	W. G. Williamson.
April 11		139,000	139,000	North Carolina	Weldon	Roanoke River	Albemarle Sound	Do.
April 12		100,000	100,000	do	Neuse Station	Neuse River	Pamlico Sound	S. G. Worth.
April 12		150,000	150,000	do	Railroad crossing	Meherrin River	Chowan River	W. G. Williamson.
April 13		144,000	144,000	Mississippi	Vaughn Station	Big Black River	Mississippi River	C. W. Schuermann.
April 13		116,000	116,000	Alabama	Demopolis	Tombigby River	Mobile River	J. F. Ellis.
April 13		115,000	115,000	Virginia	Richmond	James River	Chesapeake Bay	H. B. Nicholas.
April 15		120,000	120,000	North Carolina	The Mill	Salmon Creek	Chowan River	U. S. F. C.
April 20		100,000	50,000	do	Raleigh	Neuse River	Pamlico Sound	H. E. Quinn.
April 21		700,000	100,000	Maryland	Several points	Potomac River	Chesapeake Bay	William Hamlen.
April 22		50,000	40,000	Virginia	Railroad crossing Seaboard and Roanoke Railroad.	South Branch Nansemond River	James River	H. B. Nicholas.
April 22	do	100,000	100,000	North Carolina		Tar River	Pamlico Sound	Thomas Taylor
April 23	do	100,000	100,000	Mississippi	Friar's Point	Sunflower River	Yazoo River	J F Ellis
April 23	do	40,000	40,000	do	Holly Springs	Cold Water River	do	C. W. Schuermann.
April 23	do	40,000	40,000	do	Railroad crossing	Tallahatchie River	do	Do
April 23	do	40,000	40,000	do	Grenada	Yalabusha River	do	Do
April 24	do	100,000	100,000	Virginia	Salem	Roanoke River	Albemarle Sound	W. F. Page.
April 24	do	800,000	800,000	North Carolina	Avoca	Salmon Creek	Chowan River	U S. F. C.
April 24	do	100,000	100,000	do	Lockville	Cape Fear River	Atlantic Ocean	Col L. L. Polk.
April 25	do	60,000	60,000	Georgia	Macon	Ocmulgee River	Altamaha River	H. E. Quinn.
April 25	do	60,000	60,000	do	Albany	Flint River	Appalachicola River	Do.
April 25	do	200,000	200,000	North Carolina	Avoca	Salmon Creek	Chowan River	U S F C.
April 25	do	200,000	200,000	do	Coleraine	Chowan River	Albemarle Sound	Do.
April 25	do	115,000	115,000	do	Scotch Hall	Albemarle Sound	Atlantic Ocean	Do.
April 25	do	85,000	60,000	Virginia	Petersburg	Appomattox River	James River	H B Nicholas.
April 26	do	250,000	250,000	North Carolina	Plymouth	Roanoke River	Albemarle Sound	U S F C
April 26	do	25,000	25,000	do	Avoca	Chowan River	do	Do
April 26	do	70,000	70,000	do	Scotch Hall	Albemarle Sound	Atlantic Ocean	Do
April 26	do	200,000	200,000	do	Franklin	Blackwater River	Chowan River	Thomas Taylor.
April 28	do	300,000	300,000	do	Avoca	Salmon Creek	do	U. S F. C.
April 29	do	18,000	18,000	do	do	do	do	Do
April 30	do	100,000	90,000	Virginia	Milford Station	Mattapony River	York River	H B Nicholas.
May 1	do	100,000	90,000	do	do	do	do	H D. Johnson.
May 1	do	100,000	100,000	do	Taylorsville	Little River	South Anna River	W. F Page.
May 1	do	45,000	45,000	North Carolina	Avoca	Salmon Creek	Chowan River	U S F C
May 1	do	335,000	100,000	Maryland	Potomac Point	Potomac River	Chesapeake Bay	William Hamlen.

May 2	....do	150,000	150,000	North Carolina	Franklin	Blackwater River	Chowan River	Thomas Taylor.
May 2	....do	100,000	100,000	......do	......do	......do	......do	Do.
May 2	....do	200,000	200,000	......do	Avoca	Salmon Creek	......do	U. S. F. C.
May 3	....do	50,000	50,000	Virginia	Seaboard and Roanoke Railroad crossing	South Branch Nansemond River.	James River	Page and Johnson.
May 15	Havre de Grace	90,000	90,000	Mississippi	Fulton	Tombigby River	Mobile River	J. F. Ellis.
May 15	....do	60,000	60,000	......do	Aberdeen	......do	......do	Do.
May 17	....do	100,000	100,000	Maryland	Greensborough	Choptank River	Chesapeake Bay	Thomas Hughlett.
May 18	....do	50,000	50,000	Tennessee	Humboldt	Middle Fork of Forked Deer	Forked Deer River	H. E. Quinn.
May 18	....do	50,000	50,000	......do	Huntingdon	South Fork of Obion River	Mississippi River	Do.
May 20	....do	120,000	120,000	Missouri	Neosho	Shoal Creek	Arkansas River	C. W. Schuermann.
May 23	....do	100,000	100,000	Mississippi	Meridian	Okatibee Creek	Chickasawha River	R. E. Earll.
May 23	Steamer Lookout (Black Warrior).	50,000	50,000	Maryland	Neapsico Creek	Potomac River	Chesapeake Bay	U. S. F. C.
May 23	....do	50,000	50,000	Virginia	Freestone	......do	......do	Do.
May 24	Havre de Grace	100,000	100,000	Maryland	Federalsburg	Nanticoke River	......do	Thomas Hughlett.
May 24	....do	100,000	100,000	Alabama	Pollard	Escambia River	Pensacola Bay	F. A. Ingalls.
May 24	Steamer Lookout	75,000	75,000	Maryland	Glymont	Potomac River	Chesapeake Bay	U. S. F. C.
May 25	Havre de Grace	150,000	150,000	......do	Havre de Grace	Susquehanna River	......do	Do.
May 25	....do	150,000	150,000	Georgia	Albany	Flint River	Appalachicola River	J. F. Ellis.
May 26	....do	350,000	350,000	Maryland	Bull's Mountain	Northeast River	Chesapeake Bay	U. S. F. C.
May 26	....do	60,000	60,000	Kentucky	Somerset	Cumberland River	Ohio River	H. E. Quinn.
May 26	....do	60,000	60,000	......do	[illegible]'s Station	Green River	......do	Do.
May 26	....do	100,000	100,000	Maryland	Havre de Grace	Susquehanna River	Chesapeake Bay	U. S. F. C.
May 27	....do	100,000	100,000	Louisiana	Tickfaw	Amite River	Lake Pontchartrain	W. M. Russ.
May 27	....do	100,000	100,000	Illinois	Farlow	Kaskaskia River	Mississippi River	C. W. Schuermann.
May 27	....do	200,000	200,000	Maryland	Havre de Grace	Spesutie Narrows	Chesapeake Bay	U. S. F. C.
May 27	....do	400,000	400,000	District of Columbia	Washington	Potomac River	......do	Do.
May 28	....do	400,000	400,000	Maryland	Havre de Grace	Spesutie Narrows	......do	Do.
May 28	....do	200,000	200,000	Virginia	Riverton	Shenandoah River	Potomac River	W. F. Page.
May 28	Steamer Lookout	100,000	100,000	Maryland	Fort Washington	Potomac River	Chesapeake Bay	U. S. F. C.
May 28	....do	300,000	300,000	......do	Glymont	......do	......do	Do.
May 29	Havre de Grace	175,000	175,000	......do	Salisbury	Wicomico River	......do	Thomas Hughlett.
May 29	....do	75,000	75,000	......do	Princess Anne	Manokin River	Tangier Sound	Do.
May 29	....do	1,500,000	1,500,000	......do	Havre de Grace	Spesutie Narrows	Chesapeake Bay	U. S. F. C.
May 30	....do	500,000	500,000	......do	......do	......do	......do	Do.
May 31	....do	250,000	250,000	......do	Millington	Chester River	......do	Thomas Hughlett.
June 1	....do	175,000	175,000	Kentucky	High Bridge			F. A. Ingalls.
June 1	....do	50,000	50,000	South Carolina	Railroad crossing	Great Pee Dee River	Winyaw Bay	S. G. Worth.
June 1	....do	50,000	50,000	......do	Columbia	Broad River	Santee River	Do.
June 1	....do	100,000	100,000	Indiana	Elkhart	Elkhart River	Lake Michigan	J. F. Ellis.
June 1	....do	50,000	50,000	Arkansas	Benton	Sabine River	Washita River	H. E. Quinn.
June 1	....do	50,000	50,000	......do	Arkadelphia	Caddo Creek	......do	Do.
June 1	....do	175,000	175,000	Virginia	Shadwell	Rivana River	James River	W. F. Page.
June 1	....do	50,000	50,000	Maryland	Easton	Tread Haven	Choptank River	S. M. Rixey.
June 1	....do	75,000	75,000	......do	Near Easton	Miles River	Eastern Bay	Do.
June 1	....do	500,000	500,000	......do	Havre de Grace	Spesutie Narrows	Chesapeake Bay	U. S. F. C.
June 2	....do	100,000	100,000	Illinois	Rickford	Rock River	Mississippi River	C. W. Schuermann.
June 3	....do	500,000	500,000	Maryland	Havre de Grace	Spesutie Narrows	Chesapeake Bay	U. S. F. C.
June 3	....do	100,000	100,000	......do	Greensborough	Choptank River	......do	S. M. Rixey.

Record of distribution of young shad, &c.—Continued.

Date of transfer.	Place whence taken.	Number of fish.		Introduction of fish.				Transfer in charge of—
		Originally taken.	Actually planted.	State.	Town or place.	Stream.	Tributary of—	
1878.								
June 3	Havre de Grace	50,000	50,000	Maryland	Hillsborough	Tuckahoe River	Choptank River	S. M. Rixey
June 3	...do	100,000	100,000	Kentucky	Bowling Green	Green River	Ohio River	William Russ.
June 4	Steamer Lookout	150,000	150,000	Virginia	Waynesboro	South River	Shenandoah River	W. F. Page.
June 5	Havre de Grace	150,000	150,000	Maryland	Laurel	Patuxent River	Chesapeake Bay	J. M. Donaldson.
June 6	...do	150,000	150,000	...do	...do	...do	...do	David Scott
June 6	...do	500,000	500,000	...do	Havre de Grace	Spesutie Narrows	...do	U. S. F. C.
June 7	...do	50,000	50,000	North Carolina	Salisbury	Yadkin River	Great Pee Dee River	S. G. Worth.
June 7	...do	50,000	50,000	...do	Catawba Station	Catawba River	Santee River	Do.
June 8	...do	100,000	24,000	Iowa	Cedar Rapids	Des Moines River	Mississippi River	B. F. Shaw.
June 8	...do		36,000	Iowa	Logan	Boyer River	...do	Do.
June 8	...do	100,000	100,000	Missouri	Saint Louis	Mississippi River	Gulf of Mexico	F. A. Ingalls.
June 9	...do	125,000	60,000	Ohio	Fremont	Sandusky River	Lake Erie	H. E. Quinn.
June 9	...do		65,000	Indiana	Terre Haute	Wabash River	Ohio River	Do.
June 9	...do	100,000	50,000	Illinois	Charleston	Embarras River	Wabash River	W. H. Hines.
June 9	...do		50,000	...do	Marion	[illegible]a River	...do	Do.
June 9	...do	100,000	50,000	Georgia	Cartersville	Etowah River	Coosa River	J. M. Donaldson.
June 9	...do		50,000	Alabama	Salisbury	Tallapoosa River	Alabama River	Do.
June 10	...do	150,000	150,000	Maryland	Snow Hill	Pocomoke River	Chesapeake Bay	S. M. Rixey.
June 10	...do	150,000	150,000	...do	Railroad crossing	Bush River	...do	W. F. Page.
June 10	...do	100,000	100,000	...do	Havre de Grace	Spesutie Narrows	...do	U. S. F. C.
June 11	...do	185,000	185,000	...do	...do	...do	...do	Do.
June 11	...do	150,000	150,000	California	Tehama	Sacramento River	Pacific Ocean	T. N. Clark.
June 14	...do	80,000	80,000	Maryland	Cockeysville	Gunpowder River	Chesapeake Bay	N. Simmons.
June 14	...do	120,000	120,000	...do	Havre de Grace	Spesutie Narrows	...do	U. S. F. C.
		16,680,500	15,700,500					

XXV.—BIOLOGICAL OBSERVATIONS MADE DURING THE ARTIFICIAL RAISING OF HERRINGS IN THE WESTERN BALTIC.

By Dr. H. A. Meyer.*

PREFACE.

I have published my observations on the spawning-season of the herring, on its growth, the influence of the temperature of the water on the spawn, as well as some other facts relating to the mode of life of the herring, in the Annual Report of the Commission for a scientific investigation of the German seas for 1874–'76, Berlin, 1878.

These investigations I have continued during the years 1877 and 1878, in the direction pointed out in that report. As it will be some time before the next report of the commission is published, and as from a practical point of view it will be of interest to make the newly acquired experience quickly known, I publish the following preliminary report, containing the more important results of the last two years' observations of the development of the herring.

I.—INFLUENCE OF THE TEMPERATURE ON THE DEVELOPMENT OF THE EGGS OF THE HERRING IN SPRING.

In the above-mentioned report of the commission, the various experiments are given, which go to prove that very cold water will considerably delay the hatching of the eggs of the autumn herring. It was still an open question whether the same applied to the eggs of those herring which spawn in spring, which could only be settled by making the necessary experiments. This seemed all the more desirable, as it was a question not only of scientific but also of practical interest. If impregnated eggs of those fish of the herring kind which spawn in spring and summer can be preserved in a healthy state for a longer time by keeping them cold, the artificial raising of fish will thereby be benefited, because then it will become possible to send such eggs to distant countries. As the time which it takes the summer spawn to develop under ordinary circumstances is but short, the attempts to import the spawn of valuable American summer fish have so far not been successful.

* *Biologische Beobachtungen bei Künstlicher Aufzucht des Herings der westlichen Ostsee.* Von Dr. H. A. Meyer, Berlin, 1878. Translated by Herman Jacobson.

I therefore examined the condition of spring spawn when kept in very cold water with the following results:

For this year's experiments I used full-grown fish which had been caught with hook and line near Bappeln, on the Schlei (Duchy of Schleswig), on the 26th of April. Milt and roe were ejected by these fish into porcelain dishes the very moment they were taken out of the water, without exercising any pressure. The temperature of the water at the time these fish were caught was 8°.4 C. (47.12° F.), the saltness midway between the surface and the bottom 1 per cent. (specific gravity=1.0076 at 17°.5 C.=63.5° F.). The artificial impregnation was, as in former cases, accomplished in porcelain dishes, which during the journey to Kiel were often supplied with fresh sea-water, which was kept at the above-mentioned temperature.

The diameter of the eggs, altough they came from different fish, only varied between 1.22 and 1.37 millimeters. On closer examination it was found that nearly every egg had been impregnated and that their normal development had begun.

The following experiments were made with these impregnated eggs:

1. A number of eggs were placed in the open water of the Bay of Kiel, whose temperature at this time was 11–12° C. (51.8°–53.6° F.), and whose saltness near the surface was 1.40 per cent. These eggs were left in the water till the young fish were hatched.

2. Some eggs were likewise hatched in the open water of the Bay of Kiel, but after having been impregnated, they were from the 2d to the 5th day after this had taken place put in water whose temperature was only 2° C. (35.6° F.), in order to learn the influence of cold water on scarcely developed eggs, and likewise to see what would be the result of suddenly placing them in colder water.

3. Eggs which for eight days had been in the water of the Bay of Kiel with a temperature of 11–12° (51.8°–53.6° F.), and whose development had almost been completed, were suddenly placed in water whose temperature was only 2° (35.6° F.), in order to ascertain the power of resistance to cold of eggs which were near being hatched.

4. One-fourth of all the eggs were, immediately on their arrival at Keil, placed in water whose temperature was only 2° (35.6° F.), where they were left, to retard their development as much as possible.

5. Finally, some eggs were placed in still colder water, in order to ascertain the degree of cold which becomes destructive to herring-eggs.

A uniform temperature suitable for these purposes can easily be obtained by using wooden boxes similar to refrigerators, especially if these boxes are carefully surrounded by non-conductors of heat. The water used in these experiments was salt water from the Bay of Kiel. In all the experiments the water was changed once a day and fresh water put in the vessels after being reduced to the desired degree of coolness.

At as low a temperature as 2° C. (35.6° F.) it seems quite easy, even without taking any special precautionary measures, to keep herring-eggs fresh and healthy for a month. Only where a number of eggs had become pasted together did they begin to mold or rot. Wherever the eggs lay at the bottom of the vessel in a single layer it would not even have been necessary to change the water once a day during the normal period of development.

In artificial hatching special care should therefore be taken to distribute the eggs evenly over the vessel. If this is done their sticking to the bottom is no hinderance, but rather an advantage, as the fine sediment which forms at the bottom can easily be removed without injuring the eggs.

In the following review of the results of my five experiments I have omitted many details, because I would only have to repeat what I have communicated in former publications.

Experiment 1.

From the eggs kept in the open water of the Bay of Kiel young fish were hatched in 10 to 11 days.

This agrees with my former statement (Report of the Commission for 1877, p. 240) as regards the time required for the development of autumn eggs in water of the same temperature. These required 11 days at a temperature of 10° to 11° C. (50°–51.8° F., and spring eggs, which formed the subject of my present investigation, required about the same time, having during the first day been in water having a temperature of 8.4° (47.12° F.), which afterwards was changed to 11° to 12° (51.8°–53.6° F.).

The young fish after having left the eggs only differed from those hatched in autumn in being somewhat smaller. Two of the larger ones were measured with the following results (given in millimeters):

Total length	From the end of the head to the umbilical bag	Length of the umbilical bag	From the lower end of the umbilical bag to the "sphincter ani."	From the "sphincter ani" to the point of the tail.
6.6mm	0.84mm	1.10mm	3.17mm	1.48mm
6.4mm	0.46mm	1.08mm	3.17mm	1.37mm

Experiment 2.

Placing eggs at the beginning of their development for three days in water having a temperature of 1° to 2° C. (33.8°–35.6° F.), did not injure them in the least, but retarded the hatching about four to five days. This process of development was therefore not stopped, but merely retarded. This slower development must have continued after the eggs had been replaced in water having a temperature of 11° to 12° (51.8°–53.6 F.).

In this experiment the time from the impregnation of the eggs till the young fish escaped was fourteen to fifteen days.

Experiment 3.

Placing nearly developed eggs from water having a temperature of 12° (53.6° F.), into water having a temperature of only 1° to 2° (33.8°–35–6° F.), does not injure them.

For this experiment eggs were selected whose embryo if left in water having a temperature of 11° to 12° (51.8°–53.6° F.) would have been hatched after two days, whilst now, in water having a temperature of only 2° (35.6° F.) they required twelve days. In comparing this result with that of the preceding experiment, it appears that further developed eggs are more retarded by the influence of cold than those whose development has not advanced quite so far. In the second experiment the hatching was only delayed four to five days in spite of their being exposed to cold water for fully three days; whilst in this third experiment they were delayed twelve days, although the eggs were only exposed to the cold for two days. The whole time consumed from impregnation to hatching was, in this experiment, about twenty days.

Experiment 4.

Those eggs which immediately upon their arrival at Kiel were placed in water having a temperature of 1° to 2° C. (33.8°–35.6 F.) did not develop as evenly as during the first three experiments.

The first young fish left the eggs on the twenty-eighth day after impregnation, the majority between the twenty-ninth and thirty-third day, and a few even later. If we take into consideration, that during the first day these eggs had been in warmer water (8°.4 to 12°=47.12°–53.6° F.) and that if immediately on being impregnated they had been placed in water having a temperature of 1° to 2° (33.8°–35.6° F.) their development would have been a few days slower, it may well be supposed that in that case they would have required thirty-three to forty days.

In my former experiments with autumn-eggs a similar delay occurred at a temperature of 3°.5 C. (38.3° F.). It seems, therefore, that the eggs of the spring fish differ somewhat in this respect from those of the autumn-fish. This difference, however, is not marked enough to draw special conclusions from it. It must be granted that there is a similar delay caused by cold in both cases.

Experiment 5.

The period of development was by this experiment shown to increase in length if still colder water was applied. At first, water was used having a temperature of 0° (32° F.); with this temperature, the first young fish were not hatched till the forty-seventh day, and if we count in the first day spent in warmer water, still later.

The young fish, however, did not seem to be quite healthy, although some of them swam about for days in a lively manner. Many retained a very noticeable deformity of the back. It could not be ascertained whether this was solely caused by the lower temperature; for it is probable that other causes aided in bringing about this result, for example, keeping the eggs for more than one and one-half months in small vessels which could not be thoroughly cleaned; the impossibility of keeping the temperature exactly at 0° (32° F.); and, finally, the change of water, which could only be effected once a day. I do not maintain that it is utterly impossible to produce perfectly healthy fish at a temperature of 0° (32° F.), for, in repeating the experiment with better apparatus, some of the mistakes of the first experiment might be rectified; but a renewed experiment would scarcely seem profitable, because it has been ascertained as a fixed fact that at a temperature of + 1° C. (33.8° F.) the eggs of the herring develop in a perfectly normal manner, whilst repeated experiments have shown that this is impossible at a temperature of only — 0°.8 C. (30.56° F.). At this temperature the yolk becomes opaque, expands, and finally bursts the shell of the egg.

In the water of the Baltic, which is not very salty, the dividing line lies between + 1° (33.8° F.) and — 0°.8 C. (30.56° F.); at any rate, very near to zero.

I have so far not been able to ascertain whether this condition would remain the same in the water of the North Sea, which has a greater degree of saltness, and whose freezing-point is lower.

The fact that the spawn of the herring can stand such a low temperature sufficiently explains why young herrings sometimes make their appearance in the Schlei immediately after the breaking up of the ice. They can, without any risk, lay their eggs even in very shallow water, as no thick cover of ice, which alone might prove dangerous, forms in spring. Even those eggs which have been laid earliest do not fully develop until the water has become somewhat warmer. The autumn herring never spawns in shallow water, but only where there is a current. In the Western Baltic, therefore, the young herrings will scarcely be destroyed by cold. On the other hand, it might be of interest to investigate whether currents coming from the Polar Sea during the spawning-season of the herring could strike the spawn in northern waters, for instance, on the coast of Norway. This would furnish an answer to the mysterious problem why the herrings leave certain coasts which they had been in the habit of visiting for many years. The surface temperature cannot decide the question, but the temperature of the water at the depth at which the eggs are found.

In the above-mentioned report of the commission (p. 241), I have mentioned the fact that in the eggs of the autumn fish the yolk diminishes in size if the season of development is extended on account of the cold. But as the autumn eggs used in former experiments differed in size, and the young fish hatched from them in length. it was impossible to decide whether those embryos whose time of development had been prolonged by the cold had already increased in length whilst in the egg. As the spring eggs used in this year's experiments were all of the same size, it could be proved that the decrease of the yolk is invariably accompanied by an increase of the embryo. The young fish were measured, with the following result:

	Millimeters.
After 7 to 8 days from the time they left the egg	4.7 to 6.0
After 11 to 12 days from the time they left the egg	5.2 to 6.6
After 20 days from the time they left the egg	6.0 to 6.9
After 28 to 35 days from the time they left the egg	6.1 to 7.2

If these results are compared with those which are given in the report of the commission, for young fish hatched from autumn eggs from Korsoer (Denmark), it will be found that these last-mentioned fish were considerably longer. Their length varied between 5.4 millimeters, when the time of development was shortest, and 8.8 millimeters, when it was longest.

II.—INFLUENCE OF NORTH SEA WATER ON HERRINGS' EGGS FROM THE BALTIC.

In order to ascertain the influence of the water of the North Sea—which contains more salt than that of the Baltic—on the eggs of the Baltic herring, I took some eggs which had been impregnated, on the 26th of April, at Cappeln, on the Schlei (a fiord of the Baltic), on the following day to the aquarium of the zoological garden in Hamburg. The North Sea water used in this aquarium at this time only contained 3.25 per cent. salt, and its temperature was 12° C. (53.6° F.). Here the Baltic water was gradually mixed with the North Sea water, so that the eggs were not exposed to the full degree of saltness till after forty-eight hours. On the 7th of May the first young fish were observed swimming about freely, and during the succeeding days they were followed by others. The time of development was therefore very nearly the same as in Baltic water of the same temperature. The day when the eggs were taken to Hamburg was unfortunately very hot, and as the eggs were not evenly distributed, but were placed several layers deep at the bottom of the vessel, the larger number of them spoiled. But the fact that the remaining ones reached their full development about as fast in the North Sea water as in that of the Baltic, shows that the saltness of the water does not exercise any very marked influence. There were no arrangements in the Hamburg aquarium for raising the fish, and this first experiment therefore only proves that the eggs of the Baltic spring-herring can develop in the North Sea, leaving it an open question whether the young fish hatched from these eggs can live and grow to maturity in the North Sea.

III.—RAISING YOUNG HERRINGS FROM ARTIFICIALLY-IMPREGNATED EGGS.

As far as I know, no one has succeeded in artificially raising young herrings. My own numerous experiments in this direction invariably failed, because the eggs began to mold, and because no suitable food for the young fish could be found. They died, if not sooner, at any rate after the yolk had been consumed. The growth of the fish could, therefore, not be observed in one and the same individual, but had to be estimated approximately from a series of measurements made on different fish, which only kept fresh for a short time, and then had to be replaced by freshly-caught fish of the same size.

In the spring of 1878 I at length succeeded in raising young fish, reaching a length of 72 millimeters, from eggs which had been used in the above-mentioned second experiment. My observations confirm, as a general rule, the data regarding the growth of the herring given in the report of the commission. There were, however, some differences in the details which will justify me in giving in this place a full report of this experiment.

As has been mentioned before, the eggs were impregnated at Cappeln, on the Schlei, on the 26th of April. The second, third, and fourth day after impregnation they were kept in water having a temperature of only 2° C. (35.6° F.); the remaining time they were kept in the open water of the Bay of Kiel, having a temperature of 11–12° (51.8°–53.6° F.). The entire season of development lasted 14–15 days. A short time before the young fish escaped from the eggs the dish containing the eggs was placed in an oval wooden vessel, measuring 135 centimeters in length, 95 in breadth, and 77 in height, and holding about 0.7 cubic meter of water. Half of this water was every day replaced by fresh water from the bay, which could flow off slowly, but continuously, through a sponge firmly pressed in a round opening at the bottom of the vessel. This sponge served as a filter, hindering the animalcules which serve as food for the larvæ of the herring from escaping. During the course of the summer the temperature of the water on the surface of the bay increased to 25° C. (77° F.), and in the wooden vessel, which was generally protected from the light and heat by a wooden lid, to about 20° (68° F.). The saltness varied between 1.15 and 2.20 per cent.

When within two days the greater portion of the eggs had been hatched, I did not wait any longer for the remaining ones to be hatched. The number of young fish was anyway very considerable. They always kept together like a swarm of bees, and when the sun was allowed to shine on the water they often came to the surface.

After one to two days many of these young fish already showed a considerable increase in length, the largest measuring 9.2 to 9.3 millimeters. After three days many had lost the umbilical bag entirely and showed a widely opened mouth. After five days food could be recognized in the intestinal tract. In some it consisted of a fine-grained greenish matter, whilst in most it was composed of embryos of gasteropods and bivalves of the smallest kinds of *Rissoa*, *Ulvæ*, *Lacuna*, *Tellina*, *Cardium*, *Mya*, which at this season of the year fill the water of the Bay of Kiel near the shore. These embryos can easily be distinguished by their small shell, and swim about in the water in a very lively manner. Those which were found in the intestinal tract of the larvæ of the herrings had a length of 0.11 to 0.16 millimeter. Sometimes 20 were found in one and the same larva resembling a string of fine beads and filling the whole space from the mouth to the "sphincter ani." The copepods, at first of the Nauplius kind, were not quite so frequent among the contents of the intestinal tube.

When Professor Hensen and I examined some of these larvæ on the 10th day after being hatched, we found a small number of colorless and scarcely visible particles of blood.

After the 10th day the number of our young fish, which had so far enjoyed excellent health, began to diminish in a very noticeable degree. Finally their number dwindled down so rapidly that I was afraid my experiment would be brought to a premature close. The fish did not seem to grow much more in length, although some progress could be noticed in its transformation from a larva to a definite fish shape. But the length of the largest one on the 47th day after impregnation was only 12 millimeters, whilst, according to my observations of young herrings raised in the open water of the Schlei, it ought to have been about 17 millimeters. An increase of only 3 millimeters during a whole month could certainly not be called a normal development. The intestinal tube was nevertheless filled nearly all the time.

Hitherto fresh water had been poured into the oval vessel through a thick cloth, so as to keep out any enemies of the herring. I now made a change in this respect, by pouring the fresh water direct into the vessel, hoping thereby to give to the young fish more and more varied food. I can of course not decide whether the favorable turn which matters took was owing to this change, but I know of only one cause of the sudden growth of the fish, namely, the largely increased number of copepods.

By this increased growth during the third and following months the artificially hatched fish at the end of the fifth month reached exactly the same size as the herrings of the same age living in the open water of the Schlei—which I have mentioned in a former report. This was further corroborated by a number of young herrings raised in the Schlei simultaneously with those kept in confinement.

Age, counted from the impregnation of the egg	Measurements of herrings living in the open water of the Schlei.	Measurements of herrings raised artificially from Schlei-eggs during the spring of 1875.	Growth of the artificially raised herrings per month.
	mm.	*mm.*	*mm.*
One month	17-18	10-11	....
Two months	34-36	17-19	7- 8
Three months	45-50	30-35	13-16
Four months	55-61	48-54	18-19
Five months	65-72	65-70	17-16

Probably it was only the want of suitable food, brought about by filtering the water through a cloth, which detained a large number of fish from their normal growth. The few, however, which passed this ordeal successfully showed in the most unmistakable manner that they knew how to make up for their involuntary fasting. This is an interesting observation, because it shows how much the growth of the herring depends on the quantity of food. A still more striking example of rapid growth was exhibited by a few small sprats which were sent to me from Cappeln on the 20th of August, 1878.*

These fish, measuring 30 to 35 millimeters in length, reached a length of 66^{mm} in thirty-five days, whilst they were confined. We, therefore, observe in a small fish closely resembling the herring a growth of 25^{mm} in thirty-five days, or 22^{mm} during one month. It is well known that the growth of all fresh-water food-fish varies very much. But the fact, proved by actual experience, that salt-water fish, like herrings and sprats, are not only retarded in their growth by want of sufficient food, but will make up for lost time as soon as food is more plentiful, must be considered as another proof that fish of different size may have the same age. This will be a welcome statement to those who could not make the occurrence of herrings of all sizes at every season of the year agree with the supposition of two, or at most three, spawning-seasons. There is at the present time scarcely any doubt that the best-fed fish of the autumn spawning are fully equal in size to fish of the spring spawning—therefore 6 months older—which have been retarded in their growth, but that these last-mentioned fish can at a later time reach the size which belongs to their age.

It is likewise certain that the food of the herring is, as a rule, more plentiful during some months of the year than during the rest; that, for example with us, on the shores of the Bay of Kiel, there is a very noticeable lack of pelagian animals towards the end of winter. Our herrings, therefore, probably grow slower at certain seasons of the year than at others.

All that can be aimed at is, therefore, to ascertain the average growth, more especially as one year may not resemble the other in this respect.

Although this treatise was to be confined to the growth of the embryo, I also made a few observations on the growth of the fins, which may find a place here.

On the 25th day after the fish had left the egg, the dorsal fins began to appear, and on the 33d day 11 distinct rays could be seen. Six weeks after the fish had left the egg, the dorsal and caudal fins in fish measuring 15-19 ^{mm} had assumed their complete shape; but in the former some rays were still wanting, whilst in the latter the smaller half of the rays was wanting. The anal fin likewise showed a form resembling its definite form very closely. The number of rays in different fish of the same age varied very considerably. The shape of the pectoral fins had not changed much, and did not by any means resemble

*Large numbers of these fish appeared about the same time in the Bay of Kiel.

the adult shape. Of the ventral fins there were only very indistinct indications. After ten days more had elapsed—therefore 7 to 8 weeks after the fish had left the egg—many of the larvæ entered the transition period. They now measured 25 to 28^{mm} in length. This is the same length as that of the smallest larvæ, of the same age, living in the open water of the Schlei, whilst with the autumn herring the transition period does not set in until the fish have reached a much larger size. The fish which had been artificially raised in the Bay of Kiel had, therefore, preserved the character of the Schlei spring fish.

The young herrings of the same size which can now be caught in large numbers in the Bay of Kiel differ somewhat from the above-mentioned herrings in the position of their fins, and to some extent resemble the autumn herring. They are, therefore, certainly not hatched from Schlei eggs, but are probably natives of the Bay of Kiel. In my former reports I had to leave it undecided whether the Bay of Kiel contained spawning-places; but now I can answer this question in the affirmative.

On the 5th of May, 1878, and during the following days, there was excellent herring-fishing at the mouth of the river Schwentine.

On the 5th and 6th of May only fully-matured fish containing spawn were caught; on the 7th of May some were caught which had spawned, and from day to day the number of spent fish increased, until no other were caught.

Also at the mouth of the canal, near Holtenau, the same large and full fish were caught.

This confirms a supposition, which I have expressed in another place, that the spring or coast fish have numerous spawning-places in the Western Baltic, although none of these spawning-places are of great importance.

KIEL, *September*, 1878.

XXVI.—THE PROPAGATION AND GROWTH OF THE HERRING AND SMALL-HERRING, WITH SPECIAL REGARD TO THE COAST OF BOHUSLÄN.

BY A. V. LJUNGMAN.*

[Translated from the Swedish by Herman Jacobson.]

The great importance of the herring-fisheries to the Scandinavian countries has led to scientific investigations for the purpose of gaining increased knowledge of all those conditions on which the proper management and administration of these fisheries depend, and with a view of making them as productive as possible. The principal results of these investigations which have been carried on in Norway and Denmark have been given from time to time in the "*Tidsskrift for Fiskeri*" and the "*Nordisk Tidsskrift for Fiskeri*"; and the author of this little treatise hopes to give an account of the Swedish investigations as they are gradually progressing. As such a work may be considerably furthered by observations furnished by those specially interested in the fisheries, who have special opportunities for observing, I have determined to follow the example of two former contributors to this journal, *Axel Boeck* and *G. Winther*, and to work up historically and systematically those portions of the various sciences having a bearing on the subject towards which important contributions may be expected from a large class of people outside of scientific circles, and whose more general knowledge may be of importance for our fisheries. In order to enlighten and benefit this class of people, as well as to derive the fullest possible benefit of their co-operation, it is important that by a short and clear résumé of what has been done hitherto, they should be enabled to gain an insight into the condition of our fisheries.

Among the phases in the life of our salt-water fish whose knowledge may be considerably increased by observations and contributions by the common people, the *visible phenomena accompanying the propagation and growth of the herring* occupy an important place. And with the view of throwing more light on this subject, I wrote this treatise four years ago, and have now reproduced it, embodying all the results of recent investigations, especially those of the German Fish Commission.[1]

* *Om sillens och skarpsillens fortplantning och tillväxt, med serskild hänsyn till Bohusläns skärgård. Af* A. V. LJUNGMAN.—[From "Nordisk Tidsskrift for Fiskeri," 5th year, part 4, Copenhagen, 1879.]

[1] *Jahresbericht der Commission zur wissenschaftlichen Untersuchung der deutschen Meere in Kiel*, IV–VI. Berlin, 1878.

The coast-herring living in the Skagerack spawns during spring, in suitable places on the coast, few of which seem to have been known to those authors who have written on this subject. The cause of this is probably found in the circumstance that net-fishing, which is best suited for catching herrings during the spawning season, is not very common in Bohuslän, as nets can only in exceptional cases be used in those places where the herrings spawn. An even sand-bottom, free from rocks and stones, is very rare on the coast of Bohuslän, and the drag-nets which are the best for use during the spawning-season cannot easily be drawn across a rocky bottom, which forms the most suitable spawning-place. The herrings, however, seem also to spawn on clayey bottoms overgrown with aquatic plants. Good spawning-places are found quite frequently farther out along the whole coast of Bohuslän, and it is certain that the herrings, though perhaps not in very large numbers, spawn in these places every year; at any rate much more frequently than was thought formerly, when such an occurrence was considered a rare exception.[2] There has been, even quite recently, a tendency to underrate the number of our coast-herring which spawn in spring.[3]

With regard to the nature of the bottom on which the herrings spawn the observations made by different authors all agree. As suitable bottoms we generally find mentioned rocky or stony bottom, sand-bottom, bottoms overgrown with algæ or other aquatic plants, whilst it is generally denied that herrings can spawn on soft, muddy bottoms without any vegetation. *Mitchell's* assertion, that herrings cannot spawn on sand bottoms,[4] may find its cause in the circumstance that the waves stir up the sand on the more shallow banks near the coast of Scotland, which would, of course, disturb the eggs; but his assertion, that the herrings lay their eggs also on hard, clayey bottom,[5] cannot be properly substantiated. The assertion that the herrings prefer a bottom overgrown by a "peculiar kind of algæ," which "limits the number of their spawning-places to very few,"[6] is likewise without proper foundation. Although the herring is, therefore, not limited in the choice of its spawning-place, we cannot agree with *Valenciennes*, who says that the herring spawns almost anywhere, in calm weather even out in the open sea.[6] It is on the contrary our opinion that the herring chooses spawning-places which are not only suitable for hatching the eggs, but also for feeding and protecting the young fish. The herring prefers calm water during spawn-

[2] WRIGHT, W. v., "*Handlingar rörande sillfisket i bohuslänska skärgården*. Stockholm, 1843, p. 166.—ECKSTROM, C. U., *Öfversigt of Kgl. Vetenskaps-Akademiens Forhandlingar*, I, 1844, p. 26, 82.—HOLMBERG ,A. E., *Bohuslans historia och beskrifning*. III. Uddevalla, 1845, p. 215.—NILSSON, S., *Skandinavish Fauna*. IV. Lund, 1855, p. 509.

[3] YHLEN, G. v., *Nya handlingar rörande sillfisket i bohuslänska skärgården*. I. Göteborg, 1874, p. 12.

[4] The herring, its natural history and national importance. Edinburgh, 1864, p. 294.

[5] The herring, &c., p. 29 and p. 32.

[6] LOVÉN, S., *Handlingar rörande sillfisket*, p. 160.

[6] *Histoire naturelle des poissons*, xx, Paris, 1847, pp. 79–80.

ing, and when spawning in fiords and sounds generally keeps near the land. The choice between neighboring spawning-places often depends upon the weather, and it has several times been observed that violent and continued storms have compelled the herrings to spawn at some distance from the coast in places which otherwise were not suitable; as likewise, that in too cold or too warm weather they seek deeper spawning-places, for even prior to spawning they prefer an even temperature. According to *Ekström*, they also prefer places where there is a current.[7] It has also been observed that the older and larger herrings prefer those spawning-places which are near to the open sea, whilst the smaller and younger ones go nearer the coast or higher up the fiords.

Besides those herrings which regularly spawn at or near the coast, there seem to be some which generally spawn in the open sea on banks suitable for the purpose, located at a sufficient depth to afford protection against any violent commotion of the waves; and it is an old conjecture, that those herrings which during long periods have in large numbers visited the western coasts of Scandinavia for the purpose of spawning, during the intervals visit such banks in the North Sea.[8] As those portions of the eastern Skagerack near the cost of Bohuslän which are suitable for spawning are limited in extent, are not very well protected, and are generally found to have a comparatively small depth, it will easily be understood why they have not become permanent spawning-places. The supposition that a large race of herrings spawns there regularly every year is therefore not in accordance with the actual facts.

Regarding the depth in which the herrings spawn it seems that they generally prefer a depth of a few fathoms; more recent observations have proved, however, that occasionally they may spawn at a very considerable depth (60 to 100 fathoms), and that the eggs may be hatched there. It is not certain, however, that such a depth is favorable to the raising of the young fish.[11] Along the coast of Bohuslän there are probably few spawning-places deeper than 10 to 15 fathoms[12], most of them, especially those higher up the fiords, being only 2 to 5 fathoms deep.

It has long been known and has been mentioned among others by *Pennant* and *Naël de la Morimère*, that considerable quantities of fish-eggs are found floating near the surface of the open sea; the Dutch fishermen even believe that most of the herrings are raised from such floating masses of fish-eggs ("herring-beds"); but, as we shall endeavor to prove, this cannot be the case, as the eggs of the herring are heavier than water and can therefore not float on the surface. In order that the eggs may be fastened to suitable objects, the spawning process should go on near the bottom, and these so-called "herring-beds" owe

[7] *Die Fische in den Scheeren von Mörkö.* Berlin, 1835, pp. 216, 223.

[8] A. Boeck, "*Om Silden og Sildefiskerierne.*" I. Christiania, 1871, pp. 128-129.—*Tidsskrift för Fiskeri.* VII, p. 39.—*Nordisk Tidsskrift för Fiskeri.* II, p. 263–264.

[11] H. Kröyer, "*Danmarks Fiske.*" III. Copenhagen, 1846, p. 163.

[12] They are nearly all located near the northern portion of the coast, especially from the Väder Islands to Koster.

their existence probably to some species of codfish, to judge at least from the time of the year when they are found.[13]

Regarding the spawning-time of our Bohuslän coast herring it has already been mentioned that it is in spring or from the middle of March till the middle of May, chiefly in April.[14] It must be supposed, however, that the larger herrings, especially on the northern coast, occasionally spawn somewhat earlier, sometimes in February, whilst on the southern coast the spawning season sometimes lasts till June.[15] Hydrological conditions, especially the temperature of the water, exercise a considerable influence on the time when the spawning-season begins; it generally begins earlier on the northern than on the southern coast, on the outer than on the inner coast; earlier also after a mild winter and particularly favorable weather. Cold and especially the formation of ice seem to have a great influence on the time of the propagation of the herring.[16] No race of herrings, as far as known, spawns about the time of the winter solstice, but either so long before this period that the young fish may grow to some size before the hardest winter weather sets in, or so long after it that the newly-hatched fish soon meet with mild weather, or are at least not exposed to the dangers consequent upon the formation of ice. The herrings do therefore not spawn on the coast of Scandinavia during December and the first half of January, nor during the latter half of November.[17] It is probable that the spawning-season of a race of herrings has in course of time been fixed according to the varying occurrences of the food required for the young fish, which chiefly seems to consist of young mollusks and small crustaceans.[18] The spawning-season of the herring in a given locality does doubtless to some extent depend on the propagation of these small marine animals.

As those herrings which spawn in March and April are generally larger than those which spawn in May, these latter are probably herrings which spawn for the first time.[19] The three-year-old herring, which

---

[13] *Verslag van den Staat der Nederlandsche Zeevischerijen over* 1860, p. 29—*Uitkomsten verkregen uit de journalen der haring-schepen. Berigt bij het koninklijk Meteorologisch Instituut over* 1860, p. 6–7.

[14] When speaking of the "spawning-time" of a race of herrings, we always mean the time when *large* numbers of fish are spawning; and no one should be led to consider another season as the "spawning-time" because both before and after that time some herrings will spawn.

[15] P. DUBB, "*Anteckningar om sillfisket Bohuslän.*" Kgl. Vetensk. Acad. Handl. f. år 1817, pp. 35, 44.

[16] Compare Professor KRÖYER'S excellent remarks (*Danmarks Fiske*, III, p. 170) regarding the influence of the ice on the spawning-time of the herrings in the Baltic.

[17] See note 14, as also NILSSON, "*Handlingar rörande sillfisket,*" p. 56. EKSTRÖM, "*Die Fische in den Scheeren von Mörkö,*" p. 220, 223, and: "*Jahresbericht der Commission zur wissenschaftlichen Untersuchung der deutschen Meere.*" IV–VI., pp. 100, 237, 248, 249.

[18] *Stockholmsläns Kgl. Hushållnings-Sällskaps handlingar.* VI, Stockholm, 1855, p. 197.—BOECK, "*Om Silden og Sildefiskerierne,*" p. 15.

[19] The same opinion has already been advanced by A. W. MALM in "*Göteborgs och Bohusläns Kgl. Hushållnings Sällskaps handlingar,*" 1857, p. 21.

do not finish their spawning in spring, are found early in autumn with strongly developed sexual organs, and therefore spawn somewhat earlier than the other herrings of the same age. It is chiefly these herrings, besides the older and larger ones, which also spawn somewhat earlier on the outer coast, and which, therefore, are not so frequently caught in nets, which have given rise to the assertion that autumn or winter spawning herrings occur on the coast of Bohuslän.[20] The sexual organs of the herring develop much slower during winter when food is not so plentiful; for this reason the spring-spawning herrings have their sexual organs developed much longer before the spawning-time than is the case with the autumn-spawning herrings. Whenever, therefore, herrings are observed during autumn with well-developed, firm and hard roe or milt, this is a sure indication that the herrrings will soon commence to spawn. The erroneous opinion, which in a similar case has been advanced by *Ström*, *Malberg*, *Lybecker*, *Nilsson*, *Kröyer*, *Löberg*, *Axel Boeck*, and others, concerning the spawning of the Norwegian summer and autumn herring[21], should be a warning example against hasty opinions based on insufficient observations or data regarding the spawning-season of the herring.[22] It is also well known from olden times that the different age of the herrings has an influence on the varying spawning-season.[23] *Nilsson* thus reports that the young herrings at Skelderviken and the coast of North Holland spawn sooner than the old herrings,[24] whilst in the Sound the old herrings, according to *Winther*, spawn sooner.[25] It seems that those herrings which spawn during winter and spring are, in this respect, the very reverse of those which spawn towards the end of summer and during autumn.

The spawning-time of our Bohuslän coast herrings seems to have remained the same, at least as far as can be judged with any degree of certainty from the more or less distinct notices regarding these fisheries in the "*Trangrums-acten*" (law regarding the refuse from train-oil refineries), in the reports of Mr. *Clancey*, superintendent of herring-fisheries in

[20] See my "*Preliminär berättelse för* 1874-'75," pp. 10-12.

[21] H. Ström, "*Physisk og œconomist Beskrivelse over Fogderiet Söndmör.*" I. Sörö, 1762, p. 308.—C. R. Molberg, "*Afhandling om Saltvandsfiskerierne i Norge*" (Kgl. danske Landhunsholdnings-Selskabs Skrifter, iii, Copenhagen, 1790), p. 370.—I. L. Lybecker, "*Om Fiskene og Fiskerierne; Almindeligħed samt om Silde-Fiskerierne i Särdeleshed,*" Copenhagen, 1792, p. 82.—Nilsson, "*Skandinavisk Fauna,*" iv, pp. 496, 511.—Kröyer, "*Danmarks Fiske,*" iii, p. 170.—O. N. Löberg, "*Norges Fiskerier,*" Christiania, 1864, p. 93.—Boeck, "*Om Silden og Sildefiskerierne,*" p. 122.—See, also, G. O. Sars, "*Indberetning om de i Aarene* 1870-'73, *anstillede praktisk-videnskabelige Undersögelser,*" Christiania, 1874, p, 37.

[22] Nearly all similar opinions regarding the spawning-time of the herring and the small-herring are also based on the erroneous idea that all successful herring-fisheries must necessarily be spawning-herring fisheries.

[23] See *M. E. Bloch*, "*Oeconomische Naturgeschichte der Fishce Deutschlands.*" I. Berlin, 1782, p. 191.

[24] *Handlingar rörande sillfisket*, pp. 56, 58, 60.

[25] *Nordisk Tidsskrift för Fiskeri*, iii, p. 12.

Dubb's "Anteckningar om sillfisket i Bohuslän," "Handlingar rörande sillfisket," and in *Ekström's* and *Malm's treatises.*[26]

As has been mentioned before, the coast of Bohuslän is at long intervals also visited by large numbers of "genuine sea-herrings," whose spawning-season seems to be towards the end of summer or the first part of autumn, as far as can be judged from the reports on the herring-fisheries during the eighteenth century.[27] Among these "sea-herrings" there were found especially on the northern coast a small number of herrings whose spawning seems to have occurred *towards the end of winter or the beginning of spring,*[28] but whose relationship could be ascertained with a greater degree of certainty than even that of the great mass of herrings. It is highly probable that the herrings in question which spawn towards the end of winter and generally in the beginning of spring belong permanently to our coast and to its race of coast-herrings, and are in fact its largest, strongest, and, with regard to the sexual organs, earliest developed representatives; it is likewise probable that it is owing to the sea-herrings coming from the North Sea that so many more herrings were caught in nets during the last fishery-period than later. The masses of herrings coming from more distant parts of the ocean drive those herrings which are nearer the land towards the coast, where both are caught together. The same takes place, though to a less extent, with the rich herring-fisheries which occasionally occur in the beginning of the year, which also explains the prevalent opinion that in these fisheries herrings resembling the so-called "old" herrings are caught.[29] During the latter part of the last great fishing-period no other fully matured herrings were caught but these last-mentioned ones.

Regarding the spawning-time of the herring, it should here be mentioned that the opinion has been advanced that one and the same herring could spawn more than once a year, and that therefore one and the same race of herrings had two distinct spawning-seasons.[30] No con-

[26] *"Trangrums-Acten,"* Stockholm, 1784, pp. 76, 77, 78.—G. C. CEDERSTRÖM, *"Fisködling och Sveriges fiskerier,"* Stockholm, 1857, p. 215.—*Kgl. Vet. Akads. handl.*, 1817, pp. 35, 44.—*Handl. rör sillf.*, pp. 64, 66, 90, 117, 120, 126.—*Nya handl. rör sillf.*, I, p. ix, x.—*Oefvers. af Kgl. Vet. Akads. Förhandl.*, I, 1844, p. 120. C. N. EKSTRÖM, *"Praktisk afhandling om lämpligaste sättet att fiska sill, torsk, långa, makrile, hummer och ostron,"* Stockholm, 1845, p. 8.—A. W. MALM, *"Göteborgs och Bohusläns Kgl. Hushållnings-Sällskaps handlingar,"* 1856, pp. 9–10; 1857, p. 21.

[27] See my *"Preliminär berättelse, 1873–'74,"* pp. 19–21, where *all* the conflicting opinions regarding the spawning-time of the so-called "old herring" are for the first time given in a collected form.

[28] NILSSON, *"Handl. rör. sillf.,"* p. 55.—*Skandinavisk Fauna*, iv, p. 508.

[29] See my *"Preliminär berättelse"* for 1873–'74, pp. 29–32.

[30] M. E. BLOCH, *"Oeconomische Naturgeschichte,"* i, p. 192.—B. G. E. DE LA CEPÈDE, *"Histoire naturelle des poissons,"* v, Paris, 1803, p. 434.—*Handlingar rörande sillfisket*, p. 56.—R. PARNELL, *"Memoirs of the Wernerian Natural History Society,"* vii, Edinburg, 1838, p. 319.—*Evidence on the operation of the acts relating to trawling for herring on the coasts of Scotland*, Edinburg, 1863, p. 30.—A. RUSSEL, *"The salmon,"* Edinburg, 1864, p. 86.—See also, KRÖYER, *"Danmarks Fiske,"* iii, p. 170.—This whole question has already been discussed more than two centuries ago, as may be seen from NEUCRANTZ, *"De harengo exercitatio medica,"* Lubecæ, 1654, pp. 18, 19.

vincing proof, however, has been brought forward for these suppositions, which must rather be considered as unsuccessful attempts to explain the fact that herrings which spawn at different seasons of the year occur on the same coast, without having recourse to the supposition that two different races of herrings live in the same water, exposed to entirely similar influences.

Another opinion has also been advanced, viz, that the herrings only spawn every other year. Although it will be difficult to deny the possibility of such an occurrence, even merely as an exception from the common rule, or owing to special circumstances, and although it must be acknowledged that such a supposition affords a very convenient explanation of the relationship and occurrence of the so-called migratory herring (*stråksillen*),[31] it must, on the other hand, not be forgotten that there is not sufficient proof for the absolute correctness of the supposition, and that it brings in its train numerous other difficulties. Nothing of the kind has ever been noticed in other closely-related species of fish, and it seems perfectly clear that we should not look for any such characteristic in a species of animals whose only shield in the battle for existence is its fecundity.

In the same way it has also been asserted, in order to prove the occasionally quite frequent occurrence of so-called "migratory herrings," that the herrings grow so old that through old age they lose the faculty of propagating the species.[32] But no convincing proof of this assertion has so far been brought forward, although it ought to have been comparatively easy to obtain such proof. It is not known how often the herring can spawn—in other words, how long it retains and uses its propagating faculty. A Scotch fishery commission has, however, expressed the opinion that the herring does not live longer than the time occupied by two to three propagation periods.[33]

Some time before spawning commences the herrings which have hitherto lived rather scattered, begin to gather in large masses, which, with the principal races of herrings, assume gigantic proportions, and form so-called "herring-mountains," which gradually approach the places

[31] BOECK, "*Om Silden og Sildefiskerierne,*" p. 24.—G. O. SARS, "*Morgenbladet,*" Christiania, January 4, 1872, No. 3.—*Indberetning om dei Aarene* 1870–'73, *anstillede praktisk-videnskabelige Undersögelser,*" Christiania, 1874, p. 59.—Compare, also, LÖBERG'S entirely different explanation in "*Norges Fiskerier,*" pp. 23–24, and my explanation in my "*Preliminär Berättelse*" *for* 1874–'75, p. 11.—Although it is natural to suppose that those herrings which finish their first spawning very early are so much weakened by it that they need an extra year to gain sufficient strength for another spawning—an explanation which agrees well with the circumstance that the "migratory herring" is smaller than the large spawning herring—we must, as long as this supposition lacks sufficient proof, and as long as the phenomena which shall be explained by it can be explained in a much less doubtful manner, nevertheless, reject it.

[32] C. U. EKSTRÖM, *Oefvers. af Kgl. Vet. Akads. Förhandl.*" *i*, 1844, p. 26. *Praktisk afhandling*, p. 8.—A. W. MALM, "*Göteborgs och Bohusläns Kgl. Hushållnings-Sällskaps handlingar,*" 1857, p. 20.—*Läsning för Fiskare i Bohuslän, Göteberg*, 1861, p. 17.

[33] Report of the Royal Commission on the operation of the acts relating to trawling for herring on the coasts of Scotland, Edinburg, 1863, p. 28.

where they are going to spawn. Occasionally, however, the herrings arrive at the spawning-places some time before spawning commences; during a portion of the great fishing-periods, this seems to have been the rule, but generally this is not the case, although it happens that at the beginning of the fisheries herrings are caught which are far from being ready to spawn. The various individuals composing a school of herrings do not all get ready for spawning at one and the same time, so that the spawning-season of one school generally extends over nearly a quarter of a year; the number of spawning fish is small at the beginning and at the end, and greatest about the middle of the spawning-season. It is therefore an old experience, gained during the great herring-fisheries in the western portion of the North Sea, that in the beginning the fishermen catch more fat herrings, fewer spawning herrings, and scarcely any herrings which have done spawing; that the number of fat herrings decreases in proportion as the spawning herrings become more frequent, and that towards the end of the fisheries nearly exclusively such herrings are caught which have done spawning, together with a few spawning herrings, but no fat herrings at all. This last-mentioned kind seems to give way before the spawning herring—does therefore not go along to the spawning-place, and is not found there whilst spawning is going on.

The herring generally takes no food during spawning and immediately previous to it, and as the sexual organs develop at the expense of fat, the fish are very lean after spawning. During the spawning-season we therefore find, at least with the sea-herring, only very inconsiderable and entirely indeterminable traces of food in its stomach or entrails. This is not so much the case with the coast-herring, which finds sufficient food even near the spawning places, and which seems to continue to take food farther into the spawning-season.

The approach of the herrings to the spawning-places may certainly be delayed or interrupted by unfavorable weather, but when spawning has once commenced the herring blindly rushes forward towards its object without being deterred or hindered by anything; for instance, the attacks of fish of prey, &c.

It has also been observed that when the herrings begin to approach the spawning-places the overwhelming majority are female fish, while the very reverse is the case towards the end of their visit to the coast; and a predominance of male fish is said to be a sure sign that the fisheries are approaching their end.[34] A short time before the beginning of the spawning-season small quantities of fish composed exclusively of male fish are caught.

The herrings generally approach the spawning-place at the beginning of night and leave it early in the morning immediately after having spawned; but during the great fisheries it also happens that the her-

[34] BOECK, "*Om Silden og Sildefiskerierne,*" p. 26; *Tidsskrift for Fiskeri*, VII, p. 24.

rings come to the spawning-places during daytime, and this is said to take place particularly towards the end of their visit to the coast.[35]

Concerning the spawning-process itself, opinions are divided as to whether it continues uninterruptedly till finished, or (as with the carp) goes on at intervals, the contents of the sexual organs being emptied gradually. The latter opinion is advocated by *Axel Boeck*, who mentions a number of very plausible reasons in its favor, which, however, are not altogether convincing. He even goes so far as to speak of the importance which this gradual spawning process ought to have for the fishermen.[36] According to information received from experienced fishermen, two to three weeks would elapse before a school of herrings had by repeated emissions ejected all the spawn and roe contained in their sexual organs; but this does by no means prove that every individual fish spawns at intervals. The fact that the nets sometimes contain fish whose sexual organs are only half emptied is not a sufficient proof that such fish, if left alone, would have retained the products of their sexual organs till they could find another chance to emit them. It is quite probable that miscarriages take place under violent shocks or when death is threatened. It must also be remembered that by no means all those herrings which at one and the same time approach a spawning-place also spawn together, but that a greater or smaller number come near the coast which are not quite ready for spawning. This circumstance may have led to erroneous or exaggerated versions of actual facts. As far as known the spawning process of the great schools of herrings is one continuous act. It is certain that the herring, when free, does not begin to spawn until the entire contents of the sexual organs are so loose that the least pressure will make them flow out;[37] and even if there are intervals in the spawning process these intervals must be very short.

During the spawning process the herrings are packed in a dense mass, are in constant and violent motion, move their tails rapidly, press and rub against each other or against the bottom, press against the nets, &c., all with the obvious intention to facilitate the emptying of the sexual organs.[38] It has been observed that during the emission of the milt the sea-water assumes a whitish color, that a peculiar odor becomes perceptible, and that many scales which have become loose during the process rise to the surface. In net-fishing it has also been observed that the

---

[35] BOECK "*Om Silden og Sildefiskerierne*," p. 59.

[36] *Om Silden og Sildefiskerierne*, pp. 26, 27.

[37] The products of the sexual organs begin to get loose, especially in the male fish, long before spawning commences. With a practical view *Dr. Heincke* has given an excellent table of the gradual development of the sexual organs in the "*Jahresbericht der deutschen Commission zur wissenschaftlichen Untersuchung der deutschen Meere*," IV–VI, pp. 68, 69.

[38] Professor HENSEN, who has observed the spawning of the herring in the Slifiord (Duchy of Schleswig), says that the roe is freely emitted by the female fish while hurrying to and fro over the spawning-place. (*Jahresbericht*, IV–VI, p. 26.)

female fish generally go nearer the bottom than the male fish.[39] After the spawning process is finished the herring hasten back to the open sea, but according to observations made in Scotland, they first gather for a while near the surface in the spawning-place.[40] Together with the roe a sticky slime is emitted, which soon becomes hard in the water, and by means of which the roe, when it sinks to the bottom, is fastened to rocks, stones, and aquatic plants; sometimes the roe even forms large compact cakes.

As the Skagerack herring spawns during the night, and during the dark and cold season of the year, the Bohuslän coast offers but few opportunities for observing the spawning process. This is probably also the cause of the characteristic ignorance of the spawning process of the herring displayed by the Bohuslän fishermen. The remarks which we propose to make on the phenomena accompanying the spawning of the herrings are, therefore, principally based on observations made by fishermen in more favorable localities. For comparison's sake, we will, however, reproduce here the excellent description of the spawning process of the herring given by *Gisler*, which in some respects must still be considered the best of the kind. In his "*Beskrifning om Strömmings-Fiskets beskaffenhet, Norrbotten*" (Description of the Herring-Fisheries in Norrbotten), he thus describes the spawning process of the herring:[41] "When the herrings approach the coast in large numbers and emit both roe and milt, giving a whitish color to the water, the fishermen say that the herrings 'are shining.' When this takes place the following may be observed: The herrings which have halted, say about one-eighth of a mile from the coast, approach the land in large masses, both male and female, and emit milt and roe. Packed closely together they press forward towards the land, beat their tails against each other, and cause such violent commotion that many scales are torn off and float near the surface of the water; a strong and rank odor (*odor aphrodisiacus*) fills the air, and may be perceived at a great distance. During this time the fish do not heed seines or nets but press against them. In a few moments, about sunrise, the roe and milt will give the water a whitish-gray color, extending far out towards the deep; as soon as the fish have commenced spawning they will go out to sea, seeking those places where several currents meet, ejecting roe and milt all the time, till, when they have reached the deep, they have grown quite light and empty; they scarcely return to the coast that same summer. The roe when emitted is surrounded with a gluey juice, by which it is fastened to rocks, stones, plants, and fishing apparatus; lines which have been left in the water near the bottom are often covered with roe to the thickness of an inch, and it is quite difficult to scrape it off. With regard to its spawning process, the herring

[39] *Nordish Tidsskrift for Fiskeri*, I, p. 38.

[40] HUGH MILLER, according to W. BRABAZON: "The deep sea and coast fisheries of Ireland. Dublin, 1848, p. 31.—J. G. BERTRAM: "The harvest of the sea." London, 1873, p. 170.

[41] *Kgl. Svenska Vetenskaps Academiens Handlingar för* 1748, IX, p. 113–115.

bears a strong resemblance to other small fish; bream, perch, crucian, &c., which also press close against each other and cause a great commotion in the water when they spawn. For some days in the beginning of spring only milters are caught, but as soon as spawning commences milters and spawners are caught containing loose milt and roe. When spawning is over, both milters and spawners seek the deep."

After the eggs of the herring have been laid and have become impregnated, some time elapses before they are *hatched;* this time varies according to the temperature of the spawning-place. The somewhat conflicting observations on this subject seem to point to a varying incubation-season for the different races. *C. J. Sundevall's* opinion is probably correct, that "fish do not have a regulår hatching-time for their eggs like birds."[42] This conscientious observer says that on the coast of Stockholmlän, the eggs of those herrings which spawn there, are generally hatched in about 14 days or a little more, but may, when the temperature of the water is higher (say upwards of 68° F.), be even hatched in 3 days.[43] *Ekström* reports that on the coast of Mörkö hatching takes about 14 days.[44] From the Sli-fiord, which has been examined by the German Fishery commission, it is reported that the hatching of the herring-eggs during springtime, when the temperature was, comparatively speaking, high (18 to 20° C.=64.4°–68° F.) took only about 8 to 10 days.[45] The same observation has been made by *Kröyer* regarding the herrings on the coast of Denmark.[46]

[42] *Stockholms läns Kgl. Hushållnings-Sällskaps handlingar,* VI, Stockholm, 1855, p. 158.

[43] *Kgl. Svenska Vetenskaps Academiens handlingar,* I, 1, 1855, p. 17.—See Stockholms läns *Kgl. Hushållnings-Sällskaps handlingar,* VI, p. 195, where it says: The development of the eggs progresses rapidly. During August they are often hatched in 3, or at most 5–6 days. With a water temperature of 14–15° = 57°–59° F., the eggs have been hatched in 6–8 days. During May it took 6–8 days to hatch them." These observations have been repeated by WIDEGREN in his treatise "*Några ord om sillfiske samt om sillens eller Strömmingens rätta beredning till handelsvara*" (*Kgl. Landbruks Academiens Tidsskrift,* X, Stockholm, 1871.—*Tidsskrift for Fiskeri,* VI, p. 68.—*Circulare des deutschen Fischerie Vereins,* 1872, IV, p. 106.—United States Commission of Fish and Fisheries. Report for 1873–74 and 1874–75, Washington, 1876, p. 186.—Report on the herring fisheries of Scotland. London, 1878, p. 182.)

[44] *Die Fische in den Scheeren von Mörkö.* Berlin, 1835, p. 221.

[45] *Circulare des deutschen Fischerei-Vereins,* 1874, p. 268. Later Professor KUPFFER has given the following results of the above-mentioned comission: "The roe of the autumn-herring, with a lower temperature (48.2°–51.8°F.) and a saltness of about 2 per cent., develops in exactly the same time and shows the same phenomena as the Sli-spring herring with a higher temperature (57.2°–68°F.) and a saltness of only 0.5 per cent.; the development of the eggs in the Western Baltic goes on independent of the temperature of the water and its saltness in about 7 days, counting from the time of impregnation; the majority of the eggs are hatched in 7 days, some of them even in 6 days, although the hatching of an indeterminable percentage of eggs may be delayed a few days." (*Jahresbericht,* IV–VI, p. 31-32).—DR. H. A. MEYER adds the following information, based on more recent and complete observations: "that with a temperature of 38.3°F. the, development of the egg takes about 40 days, with a temperature of 44.6°–46.4°F. about 15, and with a temperature of 50°–51.8°F. about 11 days; but that the influence of the temperature on the roe of the spring-herring does not differ from its influence on the roe of the autumn-herring."—*Jahresbericht,* IV–VI, p. 240.)

[46] *Danmarks Fiske,* III, p. 170.

On the southern coast of Bohuslän *A. W. Malm* made observations during April, 1856, which showed that it took the spring-herrings spawning in that region about 24 days for the development of the roe;[47] and from Norway we have *Axel Boeck's* observations, which fully agree with this.[48] In Scotland the views seem to differ somewhat from ours; thus *Allman* says, that the development of the roe of the winter-spawning herring occupies 25 to 30 days,[49] whilst the fishermen say that it generally takes the roe of the summer-herring 2 to 3 weeks and that of the winter-herring 4 to 8 weeks to develop;[50] *Bertram*, however, computes the period necessary for developing the egg at 10 weeks.[51]

The newly-hatched young herrings vary somewhat in size according to the size of their progenitors, those on the coast of Stockholm measuring only 7 millimeters,[52] whilst those on the west coast of Norway reach a length of 10 millimeters.[53]

The newly-born herring with its long and narrow body bears very little similarity to its progenitors, and therefore has to undergo considerable changes in size and shape until it becomes a genuine herring.[54] The tender young herring grows very rapidly; higher temperature and the larger quantity of food consequent upon it will accelerate its developement.

Prof. *C. J. Sundevall* therefore feels justified, on the ground of his own and Baron *C. J. Cederström's* observations, in stating that the young herring on the coast of Stockholm reach a length of 25 millimeters in about two months, 36 millimeters in three months, 50 millimeters in four months, 75 millimeters in one and 125 to 150 millimeters in two years.[55] *Ekström* says that the young herrings on the coast of Mörkö

[47] *Göteborgs och Bohusläns Kgl. Hushållnings-Sällskaps handlingar*, 1856, p. 10, 11.—*Göteborgs och Bohusläns fauna* (vertebrates). Göteborg, 1877, p. 579.

[48] *Om Silden og Sildefiskerierne*, p. 12-13.

[49] Report of the Royal Commission on the operation of the acts relating to trawling for herring on the coast of Scotland. Edinburg, 1863, p. 24.

[50] Evidence of the Royal Commission on the operation of the acts relating to trawling for herring, p. 21, 33, 34.—MITCHELL, "The herring," p. 340.—BERTRAM, "The Harvest of the Sea." London, 1873, p. 168.

[51] The Harvest of the Sea, p. 171.—BUCKLAND finally mentions observations from Scotland by Captain MCDONALD, according to which the roe of the herring is hatched in about 18 days. (Report on the herring-fisheries of Scotland. London, 1878, p. 182.

[52] C. J. SUNDEVALL, "*Stockholms läns Kgl. Hushålln. Sällsk. handl.*, VI, p. 195.—*Kgl. Svenska Vetensk. Acads. handl.*, I, 1, 1855, p. 18.—The same observation is found in A. W. MALM, "*Göteborg's och Bohusläns fauna*," p. 579.

[53] BOECK, "*Om Silden og Sildefiskerierne*," p. 13.—According to the observations of the German Fishery Commission, the length of the newly-hatched herring varies from 5.2-8.8 millimeters. (*Jahresbericht*, IV-VI, p, 32, 33, 240-248.)

[54] C. J. SUNDEVALL, "*Stockholms läns Kgl. Hush. Sällsk. handl.*," VI, p. 196-197. *Kgl. Sv. Vet. Ak. handl.*, I, 1, 1855, p. 18-19, 21-22., Plate IV. (*Jahresbericht der Commission zur wissenschaftlichen Untersuchung der deutschen Meere*, IV-VI, p. 74, 79, 98, 121, 127, 130, 243.) The overlooking of these facts has doubtless been the cause of several mistakes in distinguishing young herring from young small herring.

[55] *Stockholms läns Kgl. Hushålln. Sällsk. handl.*, VI, p. 105, 196-197. *Kgl. Sv. Vet. Akad. handl.*, I, 1, 1855, p. 18-19.

reach a length of 25 millimeters in one month, 50 millimeters in about three months, and 100 millimeters in one year.[56]

Professor *Münter*, who has observed the life of the herring on the coast of Pommerania, thinks that young herrings caught on that coast measuring 56.7 millimeters were two to three months old.[57] The data furnished by Professor *Kröyer* regarding the growth of the young herrings on the Danish coasts agree entirely with *Ekström's* observations from the coast of Södermanland.[58] Professor *Nilsson* says that the young of the autumn-spawning herring reach a length of 75 millimeters in May, but that near the mouth of the Laga River young fish are found at the same season which were larger and were, therefore, presumably a year older,[59] and that on the coast of Bohuslän, according to assurances given by the fishermen, the herrings have reached a length of 25 millimeters towards the end of May, 50 millimeters in August (about the middle of the month), and next autumn, when the herrings are one and a half years old, 75 to 100 millimeters; [60] which observation has later been somewhat modified, the young herring reaching a length of 75 millimeters during the first summer, and those small herrings which measured about 100 millimeters being called "last year's young ones.[61]

Young herring begin to appear on the coast of Bohuslän already in the beginning of May, and with the increasing warmth grow quite rapidly, measuring 65 to 90 millimeters[62] from the point of the lower jaw to the root of the tail (a total length, therefore, of 80 to 110 millimeters).

[56] *Die Fische in den Scheeren von Mörkö*, p. 221–222.

[57] *Archiv für Naturgeschichte*, XXIX, p. 303.

[58] *Danmarks Fiske*, III, p. 170–171.

[59] *Handlingar rörande sillfisket*, p. 59.—Professor NILSSON seems to have forgotten the spring-spawning race of herrings (*Clupea majalis, Nilss.*), whose occurrence in this region cannot have been unknown to him.

The observations of the German Fishery Commission on the young of the autumn-spawning herring in the western part of the Baltic seem to prove that those fish which are hatched in autumn reach the same or perhaps even a greater length in one year's time than those fish which are hatched in spring. (*Jahresbericht*, IV–VI, p. 248.)

[60] *Handlingar rörande sillfisket*, p. 45.

[61] *Handlingar rörande sillfisket*, p. 130.

See, also, EKSTRÖM, "Praktisk afhandling," p. 10, where he says that "in November or December the young fish, then nearly a year old, have reached a length of 75–100 millimeters."

[62] During the latter part of November, 1873, I measured a great many young herring, which had about that time been caught on our northern coast, and found that the total length of one-year-old fish varied from 78 to 109.5 millimeters. Very numerous and accurate measurements of young herrings, made in the bay of Kiel by the German Fishery Commission, from November 14, 1876, to May, 1877, gave the following minimum of total length, viz: November 14, 84 millimeters; end of November, 90 millimeters; end of December, 100 millimeters; end of January, 110 millimeters; of February, 114 millimeters; of March, 135 millimeters; and of April, 138 millimeters. (*Jahresbericht*, IV–VI, p. 245.)

Young herring having a total length of 85 to 95 millimeters are, by A. W. MALM, considered to be almost two years old. (*Göteborgs och Bohusläns fauna* [vertebrates], Göteborg, 1877, p. 581.)

By measuring a large number of herrings caught on the coast of Bohuslän during the latter part of spring, I have found that the majority of herrings in that region may, according to their size, be divided into three groups, namely, 1, those measuring 100 millimeters (total length 120 millimeters), which must be considered as one-year-old-fish;[63] 2, those measuring 145 to 150 millimeters (total length 170 to 175 millimeters), which must be considered as two-year-old-fish; and 3, those measuring 175 millimeters (total length 200 to 210 millimeters), which are presumably three years old, and have fully-developed sexual organs. Occasionally I found a fish measuring only 160 millimeters (total length 185 millimeters) which had loose roe,[64] as well as fish measuring 160 to 170 millimeters (total length 185 to 200 millimeters), which could not possibly have become ready for spawning that same year. Larger fish, measuring about 200 millimeters (total length 23½ centimeters), are probably four years old.[65] As a considerable portion of the food which the herring eats and assimilates is directly or indirectly used for the formation of the milt and roe, the growth of the herring in size is, of course, a slow process.[66] The circumstance that the spawning season of our coast-herrings extends over a period of at least several months, causing a considerable difference of age among the young fish, and a difference in their ability to seek and obtain food, with other accidental circumstances, must be considered as the cause why in one and the same net fish of every possible size may be caught.[67]

The Bohuslän coast herring seems to spawn for the first time when it is three years old, although this must by no means be understood as if all fish born in one and the same year must spawn at that particular

[63] According to the extensive investigations of the German Fishery Commission, the growth of the young herrings in the southwestern portion of the Baltic is more rapid during spring than my observations on the coast of Bohuslän have shown it to be. In the above-mentioned part of the Baltic the herring are said to reach a total length of 130 to 140 millimeters during the first year, *i. e.* 10 to 20 millimeters more than my observations showed. (*Jahresbericht*, IV–VI, p. 246.)

[64] In reproducing my observations (*Preliminär berättelse*, 1873–74, p. 35) in the "*Jahresbericht der commission zur wissenschaftlichen Untersuchung der deutschen Meere in Kiel*" (IV–VI, p. 247), the circumstance seems to have been overlooked that my measurements do *not* include the caudal fin, which of course increases the length considerably. I have in doing so followed the custom of other writers, amongst the rest AXEL BOECK, who does not count in the caudal fin. Modern writers, as well as some of the older ones, have followed a different course in this respect. Desirable as uniformity in this matter would be, the choice must be left to individual opinions; only for the sake of avoiding mistakes it should always be mentioned what length is meant, and this has, unfortunately, not always been done.

[65] A. W. MALM's observations agree with this, as he says that fish having a total length of 190 to 220 millimeters "are presumably in their fourth year," and that fish measuring 300 millimeters (total length) "are probably upwards of 6 years old" (Fauna, pp. 573, 577).

[66] Judging from the observations of the German Fishery Commission, the later growth of the herring is not quite so slow as I have mentioned above.

[67] See "*Preliminär berättelse*," p. 156.

age, for some do not seem to get ready for spawning till they are four years old.

Opinions have been very much divided, both among naturalists and persons engaged in the herring fisheries, as to *the age at which the herring spawns for the first time.* Professor *Nilsson*, on the authority of "intelligent fishermen," supposed that "no kind of fish spawns in its second year," and that "the herring does not spawn till it is five or six years old."[68]. *Ekström*, on the other hand, thinks that those herring which measure about six inches ("counted from the point of the nose to the caudal fin") are two years old, and that those measuring 10 to 12 inches are about 4 to 5 years old; and also says that the herrings on the coast of Bohuslän do not spawn until they have reached a length 7 to 8 inches (total length).[69] Prof. *C. J. Sundevall*, who has observed the growth of the herring on the coast of Stockholm län, thinks that they are ready to spawn when three or three to four years old, when they have reached a total length of about 8 inches or 200 millimeters.[70] *Axel Boeck* is inclined to believe that "the youngest herring which spawns can scarcely be less than three, and certainly not more than four years old," although he is not able to give sufficiently strong reasons for his opinion;[71] he also says that persons who have long been occupied in fishing have informed him that the herring, when spawning must be six or six to eight years old.[72] Prof. *G. O. Sars* seems to have followed Professor *Nilsson* in trusting the authorities mentioned by him, and at first fixed the age when the herring spawns for the first time at four to five, and more recently at five to six years;[73] although he grants that some favored individuals which have just reached the age of four years (that is, "Christiania herring" of the preceding summer) may, in exceptional cases, be ready for spawning whilst of the five-year-old herrings (the "middle-herring" of the preceding summer) a much larger number have reached maturity.[74]

Of foreign naturalists who have given attention to this question we

[68] *Handlingar rörande sillfisket*, pp. 45, 47, 51, 71.—See, also, the same work, pp. 59 and 60, where it says "when the herring begins to spawn for the first time, it is at least 5 to 6 years old."

[69] *Praktisk afhandling*, pp. 10, 11.—See, also, the same, p. 5.

[70] *Stockholms läns Kgl. Hush. Sällsk. handl.*, VI, pp. 105, 151, 161, 162.

[71] *Om Silden og Sildefiskerierne*, pp. 36, 37.—*Tidsskrift for Fiskeri*, VII, p. 21.

[72] *Om Silden og Sildefiskerierne*, pp. 36, 37.—*Tidsskrift for Fiskeri*, VII, pp. 20, 21.—In the "*Christiania Morgenbladet*" of November 5 and November 20, 1872, *Boeck* gives a full account of the six years' development of the herring furnished him by a man by the name of *Dahl*. According to this authority, herrings are on the west coast of Norway called "*Musse*" when 1 year old; "*Bladsild*" (leaf herring), when 2 years old; "*Christiania sild*" (Christiania herring), when 3 years old; "*Middelsild*" (medium herring), when 4 years old; "*Kjöbmandsild*" (merchant's herring), when 5 years old; and "*Vaarsild*" (spring herring), when 6 years old; distinctions which seem to be of very ancient date in Norway.

[73] *Indberetning om de i Aarene*, 1870–73, *anstillede praktisk-videnskabelige Undersögelser*, pp. 38, 39, 40.—Recently COLLETT has expressed the same opinion (*Norges Fiske*, Christiania, 1875, pp. 191, 192).

[74] *Indberetning*, 1870–73, p. 39.

must mention *A. van Leuwenholk*, who thinks that herring spawn when only one year old,[75] and *Yarrell*, who, in accordance with observations made by himself, maintains that during the first year herrings do not develop sufficiently to have mature milt or roe.[76] We must also mention the British Commission for examining the Scotch legislation regarding the herring seine-fisheries (*Playfair*, *Huxley* and *Maxwell*), which has shown that opinions vary much among Scotch fishermen, some supposing that one, others that three, and others that even seven years must elapse before the herring is ready to spawn. The report of the commission says: "No sufficient proof can be brought forward against the assertion that the herring reaches its maturity when one year old." "There is good reason to suppose that the eggs are hatched in at most two to three weeks after spawning, and that six to seven weeks later (that is, at most ten weeks after spawning) the young fish have reached a length of 3 inches." "Since it is well known that young salmon can leave a river and return to it after twelve months eight to ten times larger than when they left, and since the herring lives on nearly the same food as the young salmon, it seems quite possible that it can also grow in the same rapid proportion." "Under these circumstances nine months ought to be a sufficiently long time to increase the length of the herring from 3 to 10 or 11 inches." "It may well be objected, however, that one cannot draw absolutely certain conclusions regarding the growth of fish by means of analogy, and it will perhaps be best to leave it an open question whether the herring has reached its maturity at the age of 12, 15, or 18 months, and consider the last mentioned figure as the maximum."[77] In North America naturalists seem inclined to the opinion that the herring, like most other migratory fish, does not reach its maturity till it is three years old.[78]

---

[75] *Sesde vervolg der brieven, Delft*, 1697, *sid.* 336–337.

[76] British Fishes, 3d edition, I, London, 1859, p. 107.

[77] Report of the Royal Commission on the operation of the acts relating to trawling for herring on the coasts of Scotland. Edinburgh, 1863, p. 27.—Evidence of the Royal Commission on the operation of the acts relating to trawling for herring on the coasts of Scotland, pp. 8, 9, 17, 23, 33, 34.—*Mitchell*, The Herring, pp. 30, 340.—*Bertram*, The Harvest of the Sea, pp. 169, 170.

[78] *M. H. Perley*, Reports on the Sea and River Fisheries of New Brunswick. Fredericton, 1852, p. 290.—Fourth Annual Report of the Department of Marine and Fisheries. Ottawa, 1872, Appendices of the fisheries branch, p. 131.

The work of the German Fishery Commission has led to still another view of this question, according to which the herring reaches its maturity when two years old. Since it has been ascertained that at the age of one year the herring has a total length of 130–140 millimeters, and that fully-matured herrings have been caught in the western portion of the Baltic, having a total length of 160–200 millimeters, Dr. *H. A. Meyer* felt justified in supposing that the growth of the 60–70 millimeters which were lacking to bring the length to the last mentioned figure and the full development of the sexual organs would not require more than one year. (*Jahresbericht*, IV–VI, p. 247.)

If, however, further investigations should confirm this supposition, which is by no means impossible, I believe, for my part, that such investigations ought to prove that such is the case only with *some* of the fish born during one and the same spawning-season, but that by far the larger number of these fish only reach their maturity at the age of three, and perhaps even four, years.

According to unanimous testimony, the herring continues to grow (though slower) long after it has reached maturity, or the faculty of propagating the species; but the assertion that the herring, as well as other fish, know no other limit of growth but death is probably not well founded.[79] It is not known how old the herring is when it ceases to grow; but it is reasonable to suppose that it has reached its full growth at eight years of age.

It is not known to what age the herring can live, but it is not probable that it reaches a very high age. Nor has the assertion been proved that the herring lives so long that it loses its propagating faculty from old age. Interesting attempts have been made to ascertain the age of herrings and other fish from the layers composing the scales[80] or from the number of vertebræ in the backbone,[81] and future histological investigations will doubtless throw more light on this subject. It must finally be mentioned that the above-mentioned Scotch Commission is of opinion that, owing to the violent persecutions to which the herring is exposed, it will scarcely be possible for it to reach a higher age than three to four years and live through two to three propagating epochs.[82]

The Bohuslän coast herring occasionally reaches a length of more than 300 millimeters; but even specimens measuring upwards of 250 millimeters are comparatively rare, especially on the southern part of the coast, probably owing to the fact that fishing is there carried on more vigorously. The largest specimen which I obtained measured 322 millimeters (370 including the caudal fin), a length which corresponds with that given by *Ekström* as the maximum length of the Bohuslän herring,[83] and which exceeds that mentioned by later authors. *Collett* mentions that the largest herring which he could obtain from the boundary waters between Norway and Bohuslän had a total length of 364 millimeters;[84] and *Lundberg*, in his treatise on the herring, containing numerous measurements of the herrings in the Royal Swedish Museum, says he could not find any specimen longer than 344 millimeters; this was one sent from Strömstad by Baron *C. Cederström*.[85] No large specimens, such as are occasionally found in the northern portion of the North Sea, near the northern coast of Norway, Iceland, or Northeastern North America, are, as far as known, ever caught in the Skagerack. *Buckland* gives 17 inches

[79] A. v. LEUWENHOCK, "*Epistolæ physiologicæ.*" Delphis, 1719, p. 218.

[80] A. v. LEUWENHOCK, "*Epistolæ physiologicæ,*" pp. 401, 402.—According to these accounts the herring can reach an age of at least twelve years.

[81] H. HEDERSTRÖM, "*Rön om fiskars ålder*" (Rgl. Vet. Acads. handl., 1759, XX, pp. 222, 329).

[82] Report of the Royal Commission on the operation of the acts relating to trawling for herring, p. 28.

[83] *Praktisk afhandling*, pp. 5, 10.—During the rich herring fisheries in the beginning of 1878, I observed several herrings having a total length of 375 millimeters, and a maximum height of 75-85 millimeters.

[84] *Norges Fiske.* Christiania, 1875, p. 192.

[85] *Bidrag till Kännedommen om strömmingen i Stockholms skärgård.* Stockholm, 1875, pp. 20, 21.

as the length of the largest herring which, to his knowledge, has ever been caught on the coasts of Great Britain;[86] and from Holland it is reported that the crew of the vessel *De Dankbaarheid*, Captain *Klaas Dorlandt*, in 57° 23′ N. L., on the 23d October, 1863, caught a herring measuring 485 millimeters in length.[87]

For the sake of comparison we will here, after *H. Baars*, give the average size of herring caught on the coast of Norway. The average length of the "great herring" is 350 millimeters; of the "spring-herring," 300–320; of the "merchants' herring," 250; of the "medium-herring," 235; of the great "Christiania herring," 200; and of the little "Christiania herring," 180.[88] According to *Collett*, however, the average length of the "great herring" is 330–340 millimeters, and that of the "spring-herring" about 330, whilst three to four year old "summer-herring," which have not yet spawned, often reach a total length of 270 to 280, and "half-grown two-year-old fish" 170 to 190. The largest specimen of the "great-herring" kind in the Christiania Museum has a total length of 378 millimeters.[89]

Regarding the propagation and development of the *small herring*, I have not been able to find any information in old writers, and my own observations are still so far from complete that I deem it best to defer their publication. This I offer in excuse of the brevity of the following account:

The fishermen, at least in that part of the coast of Bohuslän where herring-fisheries are carried on during spring and summer, are well acquainted with the fact that the "small-herring" has fully developed roe and milt in spring and during the early part of summer, and some fishermen have even observed their young some time after spawning. The spawning of the "small-herring" may be somewhat delayed or accelerated by the weather, but seems as a general rule not to have undergone any change with regard to the time when it takes place. In the reports of *P. Clancey*, superintendent of herring-fisheries to the Royal Board of Trade, we find the following notice, that on the 11th March 1811, small-herring containing both milt and roe were caught,[90] indicating that spawning would begin at most 3–4 months later. From this circumstance we may safely draw the conclusion that it is not necessary to suppose an advance in time of the spawning-season of the "small-herring," in order to explain the statements of *Nilsson*, *Wilhelm von Wright*, *Ekström*, *A. W. Malm*, *E. Uggla*, as well as the opposing statement of *G. von Yhlen*, probably derived from *M. E. Bloch's* ichthyology, or from the supposition that the spawning-season was always contemporaneous with the fishing-season. The above-mentioned writers take *autumn* to

[86] Familiar history of British Fishes. London, 1873, p. 122.

[87] *Verslag van den Staat der Nederlandsche Zeevisscherijen over* 1860, p. 15.

[88] *Die Fischerei-Industrie Norwegens.* Bergen, 1873, pp. 50, 51, 54.

[89] *Norges Fiske*, p. 192.

[90] G. C. CEDERSTRÖM, "*Fiskodling och Sveriges Fiskerier,*" Stockholm, 1857, p. 215.

be the spawning-season of the small herring.[91] Mr. *G. von Yhlen*, superintendent of fisheries, who at first gave the *latter part of autumn* as the spawning-season of the small herring has done so in accordance with the experience of old fishermen, although it is my opinion that there is no reason why he should admit the spawning during autumn merely as an exception.[92]

Prof. *C. J. Sundevall* has ascertained that the small-herring on the coast of Stockholmlän spawn towards the end of June and in July,[93] therefore somewhat later than on the coast of Bohuslän. *Kröyer* reports of his *Clupea sprattus*, that "they generally spawn in August, but begin during the latter half, of June, and sometimes continue till Sepember."[94] and of *Clupea Schoneveldi*, that "in spawners which were caught early in spring he found the milt strongly developed,"[95] which indicates an earlier spawning-season for this last-mentioned variety.

The small herring found on the northern coast of Germany (Pommerania, Holstein, East Friesland) is said to spawn in autumn.[96] On the eastern and southern coasts of Great Britain the small herring are reported to spawn twice a year, viz: during summer[97] and during winter immediately after new year.[98] The small herring found on the coast of Iceland spawns during spring.[99]

My own observations regarding the spawning-season of the small-herring show that on the middle coast it begins towards the end of May or early in June. It is probable that the spawning-season begins a little earlier on the northern and a little later on the southern coast.[100] The small-herring which are caught in autumn and the beginning of winter have never very strongly developed milt and roe (a circumstance which can be fully and extensively proved by the preparation of so-called

[91] NILSSON, "*Prodromus ichthyologiæ Scandinavicæ*, Lunda, 1832, p. 22. *Skandinavisk Fauna* IV, p. 521.—W. v. WRIGHT, "*Handlingar rörande sillfisket*," pp. 167, 175.—EKSTRÖM, "*Praktisk afhandling*," pp. 9, 103.—*Oefvers. af Kgl. Vet. Akad's. Förhandl*, I, 1844, p. 26.—A. W. MALM, "*Göteborgs och Bohusläns Kgl. Hushålln. Sällsk. handl.*" for 1856, p. 10. *Läsning för Fiskare i Bohuslän*, p. 15.—E. UGGLA, "*Göteborgs och Bohusläns Kgl. Hushllån Sällsk. handl.* for 1859, Appendix 2, p. 15.—G. VON YHLEN, "*Göteborgs och Bohusläns Kgl. Hushålln. Sällsk. Quartalskrift*, July 1867, pp. 51, 52, April 1868, p. 45.

[92] G. VON YHLEN, "*Göteborgs och Bohusläns Kgl. Hushålln. Sällsk. Quartalskrift*," July 1871, p. 52, July 1872, p. 50. *Nya Handlingar rörande sillfisket*, I, p. 19. Recently Dr. A. W. MALM has communicated observations from the years 1864 and 1865 which correct his above-mentioned older observations (Fauna, pp. 582, 583).

[93] *Stockholms läns Kgl. Hushålln. Sällsk. Handl.*, vi, pp. 109, 185–187.

[94] *Danmarks Fiske*, iii, p. 191.

[95] *Danmarks Fiske*, iii, p. 201. I have observed similar cases on the coast of Bohuslän.

[96] BLOCH, "*Oeconomische Naturgeschichte*," i, p. 207.—L. WITTMACK, *Circulare des deutschen Fischeri-Vereins*, 1875, p. 119.

[97] YARRELL, "*British Fishes*," 3d ed., i, p. 116.—J. COUCH, "*A History of the Fishes of the British Islands*," IV, London, 1865, p. 110.—E. HOLDSWORTH, "Deep-sea fishing and fishing-boats," London, 1874, pp. 133–135.

[98] HOLDSWORTH, "Deep-sea fishing and fishing-boats," pp. 133–135.

[99] F. FABER, "*Naturgeschichte der Fische Islands*," Frankfurt-am-Main, 1829, p. 180.

[100] NILSSON, "*Skandinavisk Fauna*," iv, p. 521.

"skinless and boneless anchovies"), but are nevertheless not very thin, which shows that they cannot have spawned previous to the beginning of the autumn fisheries. The larger small herring which are caught during spring and summer are generally thinner and in a poorer condition.[101]

Regarding the spawning of the small herring it ought also to be mentioned that Mr. *Holdsworth*, who is thoroughly versed in all questions relating to the British fisheries, in his well known work "Deep-sea fishery and Fishing-boats" has expressed the supposition that the small herring, like the codfish, the mackerel, and (according to *Couch*) the pilchard[102] and other salt-water fish emit their roe on the surface of the water during summer generally in the open sea and during winter nearer the coast.[103]

The young of the small herring are said to appear on the northernmost part of the coast about midsummer or the beginning of July. No information can be found in any writers on the subject as to how fast the small herring grows, and how old it is when it spawns for the first time,[104] and my own observations are not sufficiently advanced to draw any certain conclusion from them. But as on the 18th of March, 1874, I received from Kalfsund several small herrings measuring 96–97 millimeters (from the point of the lower jaw to the root of the tail) which had strongly developed sexual organs, and as the majority of those which I received from Tjörn during the latter part of spring, measured only 100–110 millimeters, it does not seem impossible that the small herring spawns for the first time when two years old, although this will probably only occur with some of the descendants of one and the same spawning-season. It is on the whole more probable that the small herring, like the herring, does not become capable of spawning till it is three years old.

The largest small herring which I ever obtained on the coast of Bohuslän measured 149 (counting in the caudal fin, 172.3) millimeters in length; but even specimens measuring 140 millimeters are rare.

For comparison's sake I will, in conclusion, give a few facts concerning the spawning and growth of some fish which are closely related to the herring.

The most important of these, the American herring (*Alosa præstabilis* or *sapidissima*, the shad), spawns like the salmon, high up the rivers, and

[101] LÖBERG, "*Norges Fiskerier*," p. 97.

[102] Fishes of the British Islands, iv, p. 81.—*Holdsworth*, "Deep-sea fishing and fishing-boats," pp. 31, 132.

[103] "Deep-sea fishing and fishing-boats," p. 135.

[104] A little more than a year ago Dr. *A. W. Malm* gave some information on this point, to the effect that young fish measuring 20–34 millimeters (total length), and obtained between July 5th and August 15th, are said to be young ones of that same year, whilst young fish measuring 42–57 millimeters (total length) and obtained towards the end of July or the middle of August, are said to be a little over a year old. (Fauna, pp. 583–585.) According to Dr. *Malm's* opinion, the small herring, which is only half the size, may reach about the same length during its first year as the herring which is twice as large, and whose young measuring 46–49 millimeters are said to be a year old. (Fauna, pp. 580–581.)

its spawning is thus described: "Gathered in dense schools, the spawners and milters move slowly in a circle, the dorsal fins often protruding above the surface of the water. Suddenly, as if struck by an electric shock, they dart off, and immediately roe and milt are expelled in the water. Wherever there is only one couple they slowly swim in a circle, the milter holding his head close to the pectoral fin of the spawner."[105] Although the roe is not loose, it is only a little heavier than the water of the river, so that by artifical impregnation it can be kept floating in the current; it also differs from the roe of the herring by being entirely free from any sticky substance, by which it could be fastened to any object either at the bottom or near the surface. With a water temperature of 24° C. (75.2° F.) it is hatched in about 60–70 hours, but when the temperature is lower it takes longer, requiring about seven days with a temperature of 62.6°–64°.4 F.[106] For a spawning-place the shad prefers either coarse sand or a bottom with a rich vegetation.[107] The *Alosa præstabilis* does not reach its maturity until it is 3–4 years old, although milters which are only 2 years old are said to be able to propagate the species.[108] When five years old they are considered fully grown.[109]

The American river-herring (*Pomolobus pseudoharengus*—"the alewife") seems to reach maturity at the same age as the *Alosa præstabilis*[110], but its roe when ejected is like that of the herring, accompainied by a sticky slime, by means of which it adheres to any objects found in the spawning-place; it is hatched at the usual temperatare in about 70–74 hours.[111]

Our common *Alosa finta* also spawns up rivers and streams, where it empties its sexual organs with violent muscular exertions, beating the water with its tail, so that during quiet evenings or nights the noise of the spawning may be heard at some distance from the spawning-place.[112]

---

[105] Report of the Commissioners of Fisheries of Massachusetts, for 1869, p. 17.—*Slack*, United States Commission Fish and Fisheries, II, report for 1872 and 1873, p. 460.

[106] Report of the Commissioners of Fisheries of Massachusetts, for 1867, p. 36.—United States Commission, Fish and Fisheries, II, pp. 425, 430.

[107] *Slack*, United States Commission, Fish and Fisheries, II, p. 460.

[108] Report of the Commissioners of Fisheries of Massachusetts, for 1867, pp. 23, 40; for 1869, p. 18; for 1870, p. 5; for 1871, p. 12; for 1873, p. 18; for 1875, pp. 5, 52.

[109] Report of the Commsssioners of Fisheries, of Massachusetts, for 1869, p. 21.

[110] Report of the Commissioners of Fisheries of Massachusetts, for 1868, pp. 7, 8, 9, 23; for 1869, pp. 5, 6, 21; for 1874, p. 8; for 1875, p. 52.—United States Commission of Fish and Fisheries, II, pp. LIX–LXI.

[111] Report of the Commissioners on Inland Fisheries of Massachusetts, for 1873, pp. 8, 9.

[112] YARRELL, "British Fishes," 3d ed., I, p. 130.—KRÖYER, "*Danmark's Fiske*," III, pp. 317, 318.

XXVII.—THE INTRODUCTION AND CULTURE OF THE CARP IN CALIFORNIA.

By Robert A. Poppe.

Carp culture in California owes its beginning to the efforts made by the late Mr. J. A. Poppe, of Sonoma, in the year 1872, and previously. Although yet in its infancy, it promises at an early day to become one of the great sources of profit for the agriculturist and small farmer of that State.

The particulars of Mr. Poppe's visit to Germany in 1872, and his return in the same year with several small carp, which he placed in his ponds on the "Palpuli Rancho," in Sonoma Valley, are facts well known to the majority of people interested in the subject in California. To the readers of the reports of the United States Fish Commission, however, the circumstances are doubtless vague and uncertain, and it is the purpose of this paper to give whatever information there is at hand concerning Mr. Poppe's trip to Europe, and also to throw some light on the question as to the time this well-known European wanderer, the carp, first made its appearance in American waters. The subjoined account can be relied on as being authentic in every particular, although less full and exhaustive than it would otherwise have been had Mr. Poppe been spared and given it the care and revision of his long experience. Concerning Mr. Poppe, it may be said in this connection that he was thoroughly imbued with the future success of carp culture in California. He was busy collecting data concerning the carp for publication for a long time, but his expectations and hopes were cut short by an untimely death.

The "Palpuli Rancho," where Mr. Poppe resided, and where the carp were placed on their arrival from Europe, will first demand our attention. This farm is situated about six miles from Sonoma, in a southerly direction, and contains 684 acres. Of this, 440 acres is high arable land, and the remaining 240 acres is marsh, or what is known in California as "tule" land. Sonoma Creek, quite a large stream, navigable almost to Sonoma, bounds the "Palpuli Rancho" on the east, while on the west there are the foot-hills and small eminences of the Coast Range. The name "Palpuli," if universally understood, would afford an index of the character of the farm which bears its name. It signifies "land of many springs." The number of springs on this place is truly wonderful. Almost every acre has one or more of them, and in many localities of a hundred acres there are at least a dozen. They make their appearance on the surface of the ground; sometimes even they seem to come out of the solid rock; then, after winding through innumerable channels, find their way into Sonoma Creek. No accurate and exhaustive analysis of the water has yet been made to the knowledge of the writer, and consequently it cannot be stated definitely what its properties are. Moreover, it is of different degrees of temperature in the different springs, in some being quite warm (86°) and in others much less so. That in the carp ponds has a uniform temperature of 74°. The water, too, is very soft, and contains the salts of sodium, sulphur, and magnesium in solution.

Mr. Poppe has seven ponds on his farm, all artificially constructed, at a cost of much labor and expense. They follow one another closely, being separated only by a levee or embankment of 8 or 10 feet. The first or upper one is perhaps 200 feet above the level of Sonoma Creek, and is 150 feet square. It covers an area before clothed with springs, and has an average depth of nearly 5 feet. The embankment at the lower end is 10 feet wide and about 6 feet high. The general method employed by Mr. Poppe in making his ponds is substantially as follows: The area was first definitely determined, and then plowed up with a strong iron-beam plow as deep as possible. The soil was then removed with the ordinary road-scrapers, and deposited on the lower end of the projected pond to serve as a levee. An alternate plowing and scraping was continued until the necessary depth was obtained. On account of the presence of the springs, and the consequent accumulation of water, the labor of construction was oftentimes anything but pleasant. After the pond was completed, a main channel through the center was dug out, and an exit under the levee prepared; a board wall was built on the inside to prevent the eating out of the embankment by the water. This was necessary on account of the prevailing west winds driving the water to the east side, and thus destroying it. The water was then allowed to rise, and finally passed off through a broad canal to the other side. Where this last pond terminated the succeeding one began. The land has a sufficient slope to insure a uniform fall of water from one pond to the other.

In a similar manner all the other ponds were constructed. The last three are by far the largest and most elaborate. In addition to these just spoken of, which are breeding ponds, &c., there are two others, used for the temporary reception of the carp prior to their being transported to other localities. These ponds are a great deal smaller than the others, but serve quite as distinct and valuable a purpose. It would not be convenient to let the water off the large ponds every time a quantity of the fish were to be sold; so, on an occasion of this kind, they are taken from the receiving ponds with little or no trouble and inconvenience.

I have now described the carp ponds proper, and yet there remains one concerning which something should be said. Nearer the foot-hills, and considerably higher above the level of Sonoma Creek, there is to be found another pond having a temperature of 86°. The spring which feeds it is not more than 20 yards from the others, yet there is a difference of 12° in the temperature. This pond has been devoted to the raising of gold-fish. They grow remarkably large here, and become very fat.

A bath-house is near this pond, receiving its water from it. For a number of years it has been resorted to by invalids and dyspeptics, who, perhaps, not receiving that rejuvenation promised pilgrims by old Ponce de Lèon from his wonderful fountain of youth in Florida, have yet been much benefited, and received decided aid from the warm spring of "Palpuli Rancho."

For a number of years this Rancho has been used as a dairy farm; from 1853 to 1875 it was controlled by Mr. Poppe himself, and since that time it has been leased to a colony of Swiss dairymen. The soil, however, is admirably adapted for viticulture, as the large vineyard set out some eight or ten years ago will show.

With this statement of the location and general characteristics of the farm where the carp were placed upon their arrival, I turn now to a brief *résumé* of the incidents of Mr. Poppe's voyage to Europe.

Mr. Poppe left Sonoma on the 3d day of May, 1872, and San Francisco on the 5th for New York, going by way of the isthmus of Panama. At New York he embarked on one of the German Lloyd steamers for Bremen, arriving there in the usual time without any noticeable occurrence. A few days were spent in visiting, after a lapse of thirty-three years, the scenes of his boyhood, and in finding some trace of his former friends and relatives. After spending a week or so in this manner, Mr. Poppe set out on the important business of his journey, namely, the procuring of.specimens of the carp.

Until recently it had not been definitely known where Mr. Poppe procured his fish in Germany, he never having given the exact locality during his lifetime. Several days after his death, however, a journal published in Stettin was received by his family which contains the desired information. The writer of the article was a companion of Mr. Poppe during a part of the journey to the locality. I extract the following from the journal *Deutsche Fischerei-Zeitung*, December 16, 1879, page 412: "The young carp which Mr. Poppe, of Sonoma, Cal., took with him to America in 1872 were taken from the ponds of a certain miller of Reinfeld, Holstein, who followed the business of carp culture. This city is on the line of the Hamburg-Lubec Railroad. The writer of these lines accompanied Mr. Poppe at that time to Reinfeld."

Here Mr. Poppe procured 83 carp of various ages and sizes. Three were very large, two feet or more in length; the others all the way from that size to the length of an ordinary steel pen. The large ones, of course, owing to the imperfect accommodations, were the first to die, while only the very smallest endured the long voyage. The were placed on a steamer for New York, in tin vessels of the capacity of twenty-two gallons each. These were of different heights, and so arranged in an ascending scale that the water from the upper vessel flowed into the second, thence into the third. From the third it was dipped back into the first, and this process continued throughout the whole voyage. Mr. Poppe had no assistant, but performed all the labor himself, watching through the long hours of the night that the comfort of the fish should be enhanced. On account of the lack of ice on board the steamer, and Mr. Poppe's consequent inability to keep the water at a sufficiently cool temperature, many of the fish died on the voyage. One after another was found dead on the surface and thrown overboard. Upon the arrival of the steamer near New York, only twenty were alive. A delay of two days was had in New York Harbor, occasioned by some irregularity in the quarantine regulations, during which time a dry, stifling wind arose, and continued with unabated fury the whole of one night. This destroyed twelve more of the carp, and when a landing was effected in New York there only eight left. These were placed as soon as possible in hastily extemporized ponds of the Croton Aqueduct Company and left there for three days to become accustomed to their native element, which, except in a highly impure state, and under very peculiar circumstances, they had not had since leaving their native land. After making all the necessary arrangements with the railroad companies respecting the rapid and careful transit of the fish across the continent, and not forgetting, in this instance, a large supply of ice, Mr. Poppe left New York for San Francisco. In seven days he arrived there with the whole number, not having lost a single one in traveling a distance of over three thousand miles. In San Francisco there was another delay, but on the day following his arrival, being the 5th day of August, 1872, three months since leaving San Francisco, he landed his five puny, and almost dead, carp in his ponds in Sonoma Valley. Two fish died in San Francisco, and one on the sloop to Sonoma, leaving five as the number safely landed. These were placed in one of the ponds already described, and formed the numbers from which all the others sprang.

At the time the carp were placed in the ponds, in August, 1872, they were in a very precarious condition; the journey, if continued but a little while longer, would certainly have killed them. They were about as large as an ordinary steel pen, being the very smallest of the 83 with which Mr. Poppe started from Europe. In the May following, the original five had increased to 16 inches in length, and the young fish had increased to over three thousand. Since that time the increase has been very rapid, but the sales have kept pace with it, so that no overstocking has as yet taken place. The spawning season takes place late in the spring—in the months of April and May. The exact time, of course, is uncertain with them, because observations of the fish in the ponds cannot be very accurate. The spawn of the original five carp, in May following their arrival (as was said before), numbered 3,000. Their growth and development, also, was proportionately rapid. The original carp (two only of which now remain) measure over two feet in length and weigh in the neighborhood of fifteen pounds. The young have been known to increase rapidly also, in one year reaching a length of 12 inches and weighing from six to eight pounds.

The young of the carp have been sold to farmers throughout California and adjacent States, and some have been shipped even to the Sandwich Islands and Central America. Sonoma County, California, where Mr. Poppe resided, has been pretty well stocked with them. Among others who are engaged in the business of carp culture in this county may be mentioned Mr. Levi Davis, of Forestville; Mr. William Stephens, Sebastopol; Sylvester Scott, Cloverdale; J. A. Kleiser, Cloverdale; Mr. Field, Petaluma; H. T. Holmes, Santa Rosa; A. V. Lamotte, Sonoma, and others. Mr. Lamotte, by the way, is the superintendent of the "Lenni Fish Association," a society composed principally of San Francisco gentlemen, who have a great desire to encourage fish culture in California. At the first opportunity I shall ask Mr. Lamotte to prepare an outline of the labor of his society for publication in the forthcoming report of the commission.

The southern portion of the State is likewise well supplied with this fish. Shipments have been made to San Bernardino, Los Angeles, and adjacent counties. From all these localities come reports of the success of the undertaking. The shipment to the Sandwich Islands was to a Mr. Charles R. Bishop, a resident of Honolulu, who has extensive grounds and numerous lakelets on his premises for their reception. Mr. Bishop has not yet written concerning them, and consequently I am unable to say what success has attended their introduction there. Mr. Levi Davis, of Forestville, I believe has sold some of the young of his carp, but in what quantity and for what price I am unable to say. He has occasionally, also, written for the California press an account of his labors.

The carp on Mr. Poppe's farm are usually, and indeed almost wholly, fed with the curd from the dairy. They have, however, repeatedly shown a fondness for barley, wheat, beans, corn, pease, and coagulated blood. Mr. Poppe was accustomed to say "they would eat anything a hog would." In most ponds they find much of their food on the bottom, such as vegetable matter, fungus, and other substances. The item of expense for food is at most very small where the carp are on a farm, for almost anything will do that perhaps but for their presence would go to waste.

As to the probable extent to which carp-culture may be carried in California, and what its probable success will be, no one of course can be able to say. That it can become profitable, I have not the least doubt. The carp, more than any other fish common to us here in the extreme west, is hardy, prolific, and does excellently in our waters. It can be raised with much less trouble and expense than any other, and is as good if not better than any. In the southern portion of the State, which seems destined to become the principal part thereof, it is already introduced, and, relying on the evidence furnished us by those who are engaged in its culture, the business has already become of more than transitory importance. There are hundreds of acres of marsh and waste lands in California which by a little labor could be prepared for the culture of the carp. An acre devoted to this purpose is the most profitable investment a farmer could make. Any one making the experiment will acknowledge beyond a doubt that the money necessary was never invested to better purpose. Besides supplying the market, there is a delicious dish for home consumption.

A prominent writer on viticulture in California is devoting his best efforts towards cultivating a taste for the pure native wines of our State, and tearing down, if possible, the barriers which have impeded its introduction in the East and elsewhere. I myself am acquainted with a noted divine who has labored unceasingly to destroy the demand for whisky, by cultivating a more refined taste for pure native wine. If he can introduce and establish the use of the latter, in moderation, he will be doing a service to his fellow-man. Similarly, anything that tends to man's comfort should be encouraged. If the consumption of carp is calculated to increase man's happiness, let us encourage it by all means. In addition to all this, I have said nothing concerning the beautifying of our otherwise beautiful landscapes by the establishment of ponds all over the State. They certainly add a charm and freshness, and make the desert and barren meadow "blossom like the rose."

XXVIII.—ON CARP CULTURE, CHIEFLY IN ITS RELATION TO AGRICULTURE.*

By EBEN-BAUDITTEN.

[Read at a meeting of the Prussian Fishery-Association at Elbing, July, 1877.]

Whilst our farmers are making the greatest exertions to increase the productiveness of their lands by labor, intelligence, and capital, they generally neglect the sheets of water found in their possession, and it is high time that attention is given to the waters. Carp-culture may be considered as one of the principal means of making the water productive.

In early spring, when all nature awakes from her winter sleep, the carp becomes a source of income to the farmer by the sale of fish two, three, and more years old. In autumn, when the farmer is depressed by cares and anxieties because the summer has brought too much rain or too great heat, so that he has not even been able to work his fields in a rational manner, the carp, which is not influenced by the above-mentioned extremes of the weather, will be the principal source of income.

Whilst it requires a vast amount of care and labor to procure the quantity of feed which the cattle need during seven months of winter, the carp, so to speak, sleeps all through winter, and may therefore well be termed the best domestic animal.

Carp-culture, i. e., the raising of carp, must be strictly distinguished from the *keeping of carp* when young carp are procured from piscicultural establishments. Nearly every farm will offer the necessary conditions for keeping carp, while carp-culture requires a number of ponds, *e. g.*, the pond for the young fry, the pond for the larger (the growing) carp, and the pond for wintering the carp. On this last-mentioned pond the success of the carp-culture mainly depends; whilst the ponds for the young and the growing carp may be shallow, the wintering-pond should be 8 to 12 feet deep, and should have some flowing water even during the severest cold. If the current is so strong that the water is always more or less in motion, the wintering-pond need not be quite so deep.

If these conditions cannot be fulfilled, the wintering-pond will never afford absolute security, and the result will therefore be doubtful. In

* *Ueber Karpfenzucht, hauptsächlich mit Bezug auf unsere Landwirthschaft.* [From the "Deutsche Fischerei-Zeitung," Vol. II, No. 14, Stettin, April 8, 1879.] Translated by Herman Jacobson.

such a case it will be advisable to abandon carp-culture and confine one's self to the keeping of carp. We meet with similar cases in agriculture. Many farmers, *e. g.*, will be prevented from raising lambs by local causes, while the keeping of a flock of sheep may be profitable. If a farmer, therefore, has no wintering-pond, but several small sheets of water which have water all during the summer, he will do best in stocking his ponds (no matter what their size may be) with three-year-old carp and a few pikes, at the rate of about 100 to 300 pounds of carp, (no individual fish to weigh less than one-half pound,) to the acre. The growth and health of the fish depends on the character of the bottom and of the water; and in this respect carp greatly resemble the products of the soil.

Any pond, no matter whether large or small, which is a few feet deep and whose water does not contain too much iron, will in autumn yield a rich crop of carp from those which have been placed in it in spring. But where the circumstances are more favorable, and where wintering-ponds permit the raising of carp one, two, and three years old, carp-culture may be carried on successfully. The first pond required is the hatching-pond, "*Streichteich.*" It should not be too large, from one-quarter to two acres, and in it should be placed two or three spawners and one or two milters, to which should be added a one-year-old milter weighing at least one-half pound. The water should be two to four feet deep in the middle of the pond, and the northern and eastern banks should be wide, shallow, and well protected, exposed to the warm rays of the sun; it will be well to have some reeds and grasses grow in the pond. In our province (Prussia) the carp generally spawn for the first time at the end of May. In warm weather they spawn a second, third, and fourth time at short intervals; and it may be observed, every time, that the older fry make room for the younger in the warm and shallow water, and go into deeper water, until they are again driven into deeper waters by the succeeding fry.

Small hatching-ponds are preferable to large ones for, 1st, a small pond will generally be more sheltered; 2d, in large ponds the wind creates larger waves, which frequently cast the spawn on the shore, where it dries up when the water recedes; 3d, small sheets of water can be better protected from overflowing in violent rain-storms; 4th, large ponds will offer greater attractions to the numerous enemies of the carp, *e. g.*, the eagles, the herons, ducks, crows, &c., not to mention the otter, which is a well-known robber of fish-ponds.

The larger the sheet of water the more difficult will it be to protect the fish against the ravages of birds. The herons are particularly fond of the young fry of the carp, but a careful huntsman will soon be able to keep them, as well as the ducks, away by shooting a number. The eagles and the crows should be caught in traps; but the most dangerous enemy of the carp is the small diver, *Podiceps minor*, all the more so as it is quite difficult to get a shot at him, especially in ponds where there are many reeds. In the stomach of such a diver which had been caught about the end of June last year there were found a large number of young carp about an inch in length. At such a tender age they fall a prey to this dangerous bird, which, in spite of its small size, may cause a total failure of the carp-fisheries. This spring, when the reeds were not as yet covering the banks, 5 divers were shot on a pond of about 8 acres near Bauditten, but 8 escaped and could neither be caught nor shot. Their nests, which are very hard to find, were destroyed several times, and 121 eggs were taken; still they did not cease to build nests and lay eggs.

If every one of these 4 pair hatches 8 young ones, there will be—the old birds included—40 divers. And if every one of these destroy 100 young carp a day, this will make 4,000 a day, or nearly 30,000 a week. Wherever, therefore, no other cause can be assigned why a hatching-pond has proved a failure, the supposition lies very near that divers or other aquatic birds were the cause.

The frogs also must be kept away from the ponds as much as possible. This is best done by drawing their spawn ashore with rakes, and then either burying it or letting it dry. The greatest possible calm should prevail in and about such a pond; no cattle should graze near it, it should contain no pike, and the water should be of equal depth all the year round. After having hatched carp for several years, a farmer will have fish one, two, and three years old, which must be kept separate, carefully arranged according to years. For if small fish are in the same pond with larger fish, the smaller ones will suffer. If a small fish, *e. g.*, catches a worm, the larger fish will immediately take it away; one begrudges it to the other just as it is among men, and the weaker has always to give way to the stronger.

If the hatching ponds contain much food, *i. e.*, if they have good water, a clayey and rich bottom, and are not too much crowded with fish, the carp will under favorable conditions weigh upwards of two pounds in the autumn of the third year. The number and nature of the ponds will plainly indicate whether carp should be kept a fourth year. In the ponds for the growing carp the pike forms an important element. The common idea that it makes the lazy carp move about, and thus gives them the necessary exercise, is certainly erroneous; but the pike certainly prevents the carps in these ponds from spawning, which would only do harm, and destroys useless fish, *e. g.*, bleaks and crucians, which only take the food from the carp. In the hatching-ponds, however, the pike is a dangerous enemy.

Time will not allow me to speak of the cheapest and best way of arranging the ponds, which will go hand in hand with the cultivation of fields and meadows, and I will therefore close with a few remarks on the different varieties of the carp. The following must be distinguished:

1. The "carp proper," *Edelkarpfen.* (So called to distinguish it from the crucian.)

2. The Bohemian "mirror-carp," *Spiegelkarpfen.*
3. The Bohemian "leather-carp," *Lederkarpfen.*

The Bohemian "mirror-carp" has been described as a somewhat inferior fish in an otherwise excellent book; but most critics, even in Silesia, agree that it is a very fine fish, growing quicker, having a more delicious flavor, and a much hardier nature than the "carp proper." This has been provèd by actual experiments here in Bauditten. After several vain attempts to transplant the "mirror-carp" to Bauditten from Pardubitz, in Bohemia, a number of these fish have at last safely arrived here, and last year spawned for the first time. Their rapid growth was quite remarkable, for in autumn these young carp measured already 8½ to 9 inches in length, and when compared this spring with the other carp had grown much faster.

I must finally refer in a few words to one of the most beautiful and perhaps most profitable fish, which may likewise be raised in carp-ponds, viz, the "gold-orfe," *Goldorfe.* Its back is scarlet, whilst the belly and the sides have a silver color. After many experiments we have this year at last succeeded in raising some of these valuable fish in Bauditten.

It grows as fast as the carp, is as peaceful a fish, of a hardier nature than the carp, knows how to avoid its enemies, and always keeps near the surface, thus forming a brilliant ornament of any pond. Its flavor must be very delicious, as it is served on the imperial table as a special delicacy on the birthday of the emperor.

XXIX.—ON THE CARP PONDS OF NETHER LUSATIA.*

By Dr. Edm. Veckenstedt.

Nether Lusatia, though not adorned with great landscape beauty nor blessed by nature with a rich and fertile soil, still presents many remarkable and attractive features. There, on the upper and middle course of the river Spree, only a few miles from the capital city of the German Empire, a strange people has preserved its nationality. Even today the nimble Wendin passes us with foreign salutation in fantastic attire. Our ponds and canals show many an idyllic picture, and the proud high trees are ornaments to our parks such as are rarely found in artificial gardens.

The industry of Nether Lusatia, too, has been more and more developed every year. Our ponds not only enliven and beautify the landscape but their object is essentially practical. Carp-breeding here has obtained great results. In Hamburg, to mention one instance, the carp of the "Spreewald" has outrivaled the Bohemian carp.

Cottbus is the place of meeting for the so-called "Carp Exchange." Every year, on the first Monday of the Cottbus fall market, a busy life develops in the Hotel Ansorge there. The fish-dealers from Halle, Leipsic, Dresden, Magdeburg, Posen—who name all the places and call all the names?—among them representatives of such firms as Kaumann, Berlin, F. J. Meyers, Hamburg, the carp-king Fritsche, &c., have arrived from all parts of the compass to wait for the *carp-barons*. With this name the first-class breeders are designated, as Mende-Dobrilugk, who undisputedly raises the largest carps; von Löwenstein and Faber, with a product of 6–800 hundred-weight each; Berger-Peitz, with at least 2,000 hundred-weight, &c. These gentlemen meet in a separate room as *Fischereiverein*, with the expert Mr. von Treskow as their president, to discuss the questions of the day and to determine approximately the price to be asked for the carps. After this business is finished the sale-contracts proper are made.

The weight of the carps from Upper and Nether Lusatia, represented in Cottbus by their breeders, amounts to 8–10,000 hundred-weight; the number of fishes to 2–300,000.

This simple fact alone might occasion a comparison with the results of the artificial fish-breeding, for which so much interest is shown. It is known that since the publications of Professor Coste, of the College

* Die Gartenlaube, 1877, No. 45. An den Karpfenteichen der Niederlausitz. Dr. Edm. Veckenstedt.

de France, fish-culture has received much attention, and though the splendid success—promising to populate rivers, lakes, and ponds with fishes—has only to a small degree been attained for France, yet the Hüningen *Muster Anstalt*, established by the French Government, diligently continues its labors under the direction of German officers and the *Deutsche Fischerei-Verein*, endeavors to develop in all circles of our population a greater interest in the progress of fish-culture.

Whether the successes will correspond to the expectations, time only can show. So far the artificial breeding of noble fishes (*Edel fische*) only can be considered successful; but it seems that the *Salmonidæ* alone, though quite a desirable enrichment of our rivers, will scarcely ever constitute a cheap food for the people. The *Fischerei-Verein* will have to direct its attention especially to the reform of the laws; for in the fish trade a good many things still depend on the option of the individual. So, for instance, many eels are at present confiscated in Berlin for the lack of lawful size, while in Pomerania, whence 90 per cent. of all the eels in the Berlin market come, the same size of fish is not only considered marketable, but iron eel-traps are rented to anybody for a few cents for catching the fish under the ice, where of course no discrimination can be made as to size.

We will now return to Nether Lusatia and rejoice at the results of the natural fish-breeding, which, quietly and without noise, like everything truly good, in its own way has reached its present height. How many prejudices had to be removed, how many notions to be contended with, how many experiences to be collected before Lusatia could succeed in securing for its *cultur-Fisch*, the carp, the market over all Germany.

Let us now go to these ponds and take a glance at the breeding and capturing. The ponds, about seventy in number, have a surface of about 5,000 Morgen (3,500 acres), and yield at present 2,000 hundred weight per year.

Now we will look at one of the small ponds, a so-called *Streichteich*. In this a definite number of *milts* and *roes* (males and females) deposit their spawn. Here especial care is required in the management of the pond-bottom, which, by antecedent cultivation, is fertilized and has its acids neutralized. It is desirable to have the pond sheltered against the noxious influence of the wind; then it must be kept absolutely free from pikes, and it is of advantage that its water in great part is derived directly from the heavens, for the *Himmelsteiche* are usually the best *Streichteiche*. If now the bottom was not too poor, and if wind and weather were not unfavorable, next spring the strong fry will be transferred into the *Streckteich*, for in fall time this movement is rather dangerous, as young carps will never endure the dangers of winter when in their transfer their scales are injured. This *Streckteich* must be rich in nourishment, so that the fish grow rapidly to be fit for the *Abwachsteich*.

In many cases, however, the fishes have to be placed into a *Streckteich* of the second order because their growth was insufficient. The

Abwachsteich contains all the ponds the greatest water-surface, and only 25–40 carps per Morgen (two-thirds of an acre) are admitted, to which one-twentieth of other fishes are added. In this pond the pike plays its principal part. It is, as is generally known, the *factotum* of the carp-pond. Even if it is legendary that the pike forces the lazy carps to locomotion, so as to give them better appetite, yet it is indispensable for the destruction of wild fishes, &c.

As nearly all creatures have to endure the severest diseases in their youth, so also the carp has to overcome its greatest dangers up to the day it is admitted to the *Abwachsteich*. Here external enemies are rarely dangerous, though otter and sea-eagle claim their victims; yet fishes of prey do not injure it, and swan, ice-bird, ducks and divers, frogs and toads are only dangerous to the spawn and fry.

Diseases, too, occur mostly with young carps only; polypes render the fish unfit for its full development; tape-worms constrict its intestines, make it lean, and finally kill it; lice torment it, and produce dropsy. But the water itself may become noxious; its inlet and outlet must be accurately regulated; a ditch carrying bad water to the pond, its sudden rising after a thunder-storm, a lightning-stroke, &c., have often done considerable damage to the breed.

Yet now the autumn day has come on which the capture of the marketable carp begins, and we go to the *Teufelsteich* near Peitz, the largest of the estate. Three weeks before this day the outflow from this pond commenced. All the time the greatest quiet has to reign at the places deepened for the catching, because otherwise the carps, sensitive to sound and timid, would not descend the deep ditches leading to the places of capture, which would render the operation slow and more difficult.

On the day of the fishing itself the drivers begin to wade along the ditches with loud noise, until the fishes are collected at the place of capture, which has an extent of about one *Morgen*. Then the ditches are closed with (stell) nets and the catching begins. Two *Watnetze*, handled by three fishermen, yield about 100 hundred-weight at every *draw*. The fishes are carried to the scale and spread upon platforms. Pikes, *Karauschen*, *Schleien*, are picked out, and the small *Barsche* (perches?) used for manuring fields and meadows.

Four practiced hands throw the carps from the platforms upon the scale, and when it indicates one hundred-weight the fishes are rapidly transferred to a hogshead standing upon a wagon; three filled hogsheads make a load. In sharpest trot the horses hasten to the *Hammergraben* (hammer-ditch), where the fishes are loaded into *Dröbel*. *Dröbel* are perforated covered boats, the surface of which is even with that of the water; they contain, on an average, 25 hundred-weight, and are shipped by hardy sailors to the *Schwielochsee*. There the fishes are transferred into larger *Dröbels*, containing about 100 hundred-weight; in tow of freight-boats they reach Berlin in about one week; Hamburg,

Magdeburg, &c., however, not before four or five weeks. All this time the greatest attention is necessary. The journey to and through Berlin is dangerous on account of the water being either too low or too high, and, besides, every evening the whole transport has to be carefully examined and every single sick or dead fish to be removed; and it occasionally happens that some *Dröbel* go to pieces.

These astonishing results of the natural fish-breeding are the more to be appreciated the more laboriously they are obtained (the manager of the Peitz pond, for instance, spends about 100,000 marks), yet Lusatia shows still better results from the ponds devoted to the raising of goldfish. Most interesting are those obtained by Mr. Eckart in Lübinchen, the well-known breeder of the great *Madüe-maräne*. His splendidly watered ponds, at present celebrated for the lacustrine pile-dwellings found there, contain in all their brilliancy the goldfish and *Orfe*, the trout and *Elritze*, the leather-carp and *Maräne;* and when now the Americans are enabled to breed the most valuable of all *Maränes*, the *Maräne* of the *Madüe* Lake, it is the merit of Mr. Eckart, who was the first to send embryonized *Maräne* eggs across the Atlantic Ocean. Constant study and continued experiments were necessary for his eminent success, and if we want to stock our rivers, lakes, ponds, or aquaria with numerous and different species of fish we will have to work incessantly, for the conditions most favorable to the several species have in great part yet to be found out.

XXX.—THE CARP-FISHERIES IN THE PEITZ LAKES.*

That "carp in beer" is a favorite dish in Berlin is sufficiently proved by the fact that about 500,000 pounds of this fish are annually consumed in this city. It will therefore not be out of place to give a brief account of the famous Peitz Lakes in Lower Lusatia, which mainly supply Berlin with carp, and which were well known even in the time of Frederick the Great.

The Ural-Baltic plateau, which includes a portion of Lusatia, contains a very large number of lakes and ponds. Of these the Peitz Lakes are the most important. These lakes, 76 in number, and forming a water area of almost 5,000 acres, are a royal domain, and are at present rented to Mr. Th. Berger. They produce a very large number of carps, and the annual fishing days in October, especially that of the Devil's Lake, having an area of about 900 acres, form important and interesting events, genuine popular holidays, not only for the inhabitants of Peitz and the surrounding country and the people of the neighboring city of Cottbus, but, because easy of access, likewise to many inhabitants of the capital. It must, however, be borne in mind that these great fisheries, and each one of the 60,000 or 70,000 carps caught during this season, have a previous history extending over a period of about four years; for those well-fed, golden-scaled government fish, resembling each other in size and shape as much as eggs, have not sprung into existence suddenly like the armed men who rose from the dragon-seed sowed by Cadmus, but it required great work and care and trouble to develop them so far; and in order to understand all this we shall have to become acquainted with the details of this industry, and gain some entirely new ideas with regard to the carp and its life. We here see not a fish rapidly parting the waves with its fins, and in undisturbed liberty now diving into the deep, now rising to the surface, always timid and flying from the terrible fish of prey, but a well-cared-for domestic animal, constantly guarded by and accustomed to human beings; a very peaceful, phlegmatic animal, with a predilection for muddy bottoms and slow-flowing water, growing more comfortable and gentle in its ways by its "education," which has been going on for generations, all this tending to make the fish fat and comfortable looking, and giving to its flesh a most delicious flavor.

The life of the carp, which really may be termed a "jolly sort of im-

*Die Karpfenfischerei in den Peitzer Teichen. From a Berlin daily paper. Translated by H. Jacobson.

prisonment," commences in the hatching-ponds, varying in size from one to ten acres, in which as many as 20 pair of well-developed milters and spawners are placed in spring, there to spawn under the genial rays of the sun. It is characteristic of the slow nature of the carp not to do this spawning business at once like other fish, and so far all attempts at artificial impregnation have failed. But if the water during the spawning season has been kept at an even height, and the frogs do not devour too many eggs, young carps are produced in great numbers, as they are very prolific, one pair alone producing several hundred thousand eggs, from which, even under the most unfavorable circumstances, about 25,000 young fish may be counted on. During their earliest infancy these fish live on infusoria, as their little mouths will not allow any other food to pass. The summer goes by, the new year comes in, and in spring the little one-year-old carps—which at this age are very suitable for the parlor aquarium—are placed in larger ponds (generally covering an area of 30 acres each) at the rate of 360 to 600 fish per acre. After they have stayed in these ponds a year, the fish (now two years old) are placed in still larger ponds (generally covering an area of 400 acres each) at the rate of 180 to 360 per acre. After another year has passed, the fish (now three years old) are placed in the large ponds (generally about 900 acres each), in which they stay another year, and reach an average weight of 2½ to 3½ pounds, and thus attain their maturity.

Loneliness produces melancholy, and in order that the carp may not lead a too idyllic sort of dream-life after leaving those ponds where they spent their first two summers, and which are absolutely free from fish of prey, quite a large number of other considerably smaller fish, such as tench, crucians, pike, and even perch—which have been specially raised for this purpose in separate ponds—are during the third year placed in the same ponds with them. These fish give the carp some idea of life in the great world, and by their constant attacks, which, however, are generally harmless, bring a little life into the quiet society of philosophers, and, to some extent, act the part of shepherd dogs. But there are other enemies of the carp which tend to make the carp livelier, reminding us of those persons in "Gulliver's Travels" who had constantly to use rattles to rouse the Lilliputians from their day-dreams; and these are otters, herons, wild ducks and geese, fish-hawks, and human beings—poachers, who rob the ponds during the night.

Thus the day of harvest comes at last. Three weeks beforehand they begin to let the water flow off, and the carps gradually gather in the deep ruts or holes of the bottom. On the morning of the great fishing-day they are driven into a basin about the size of an acre and about one meter deep. This is done by the fishermen, who, armed with purse-nets, wade, often with half their bodies in the muddy water, and, shouting and yelling, drive the fish before them. Slowly the great mass of fish comes rolling on, making the water of a dark, muddy color, and throwing great quantities of mud-like clouds in the air. No one could tell that these are carp, for the dark, round backs, which in innumerable places become visible among the seething mud and water, rather resemble eels or similar fish. The whole spectacle, which is quiet in the beginning, reminding one of the driving of a flock of geese or a drove of sheep, gradually becomes quite exciting, especially toward the end, where from 60,000 to 70,000 pounds of carps are crowded together in a narrow space scarcely 20 paces square. Two simple nets are nevertheless sufficient to close up the two channels leading into the basin, which now resembles a caldron full of boiling mud and water. In this turmoil the pikes fare worst, for some of the carps, which, like tame steers, seem in the last moment to remember that after all they possess considerable strength of muscle, are continually dealing powerful blows with their tails, which the sensitive and cowardly pikes cannot stand very well, so they endeavor as much as possible to crowd into a distant corner. Now the fishing itself commences, and a number of men with two drag-nets, each holding about 5,000 pounds of fish, slowly haul that quantity on shore. Here everything is activity and bustle. Under an open shed we see a large pair of scales with a 100-pound weight. The carps are uninterruptedly brought up from the pond in immense buckets, each carried by two stout men, and thrown on boards by the side of the scales. With lightning-like rapidity, one fish after another is seized by men standing there for the purpose, counting "one, two, three, four * * *" until the scales are evenly balanced. Thirty-one to thirty-three fish generally make the hundred pounds. The full scale is then immediately seized by two men, while an empty one is being filled, and the fish are placed in large casks on one of the many wagons which hold at a short distance. As soon as the three casks, which every wagon holds, are filled, the wagon is rapidly driven along the turnpike, near which the whole transaction takes place, to the Hammer Canal, distant about one kilometer (3,280.709 feet), where the fish are immediately placed in the holds of boats, which contain water. Each of these boats carries 2,500 pounds of fish.

Thus the carps are within a few minutes transferred five times, without having suffered in the least. Near the scales stands, in his rubber overcoat, a note-book in his hand, Mr. Fritsche, from Frankfort-on-the-Oder, a well known fish-dealer, called the "carp-king," and, with Mr. Berger's agent, calmly notes down the number of fish to every hundred pounds, while Mr. Berger himself is busy arranging things, giving orders, and satisfying the many private buyers, male and female, young and old, farmers and town-people, who have come with bags, sacks, and baskets to buy single fish or small quantities up to 200 pounds. Mr. Berger also attends to the picking out of other fish, such as tench, pike, perch, &c., which have been caught in the net. A large quantity of still smaller fish, so-called "spoon-fish," because they have to be eaten with a spoon, are likewise brought up in these nets, many of them almost mashed by the heavy weight of the carps, and dead a few minutes after they have left the water. These are thrown in large baskets and are viewed with eager and longing glances by the many poor people standing round, who here, for a few cents, might procure more than one good meal, and net Mr. Berger perhaps $24 extra. But woe be unto him if he should dare to sell these fish; the inexorable police-officers would at once refer him to a paragraph of the fishery law, according to which these fish dare not be sold, as not having the required size.

Meanwhile the hour of noon comes, and the ardently longed-for lunch time, doubly welcome on account of the pouring rain and the cold, is fast approaching, and Mr. Berger invites his guests to his house near by. Among them we see, besides some landed proprietors from the neighborhood, men of inexhaustible good humor and unlimited capacity of stomach, the well known Lusatian anthropologist and reporter of the *Gartenlaube*, Dr. Veckenstadt. In the hospitable mansion we are regaled with the products of the chase, snipes, reed-birds, ducks, partridges, &c., and one of the epicurians present makes the remark, which may be taken to heart by all good housewives, that the flesh of the pike becomes infinitely more delicious if it has lain in brine for twenty-four hours. The fishermen and drivers are meanwhile taking their lunch in the sheds near the ponds, and after a short pause the work begins anew until late at night, when about 60,000 pounds of carps have passed through the hands of the weighers. As regards the further transportation of the carps, which are the property of Mr. Fritsche the moment they leave the scales, they first go to the Schwieloch Lake, reaching it in five to fourteen days, going through the Hammer Canals, the Spree, and the Spreewald. The difficulties of their route are considerable, for the water is often so low that the boats have to be placed on rollers and conveyed for short distances in this manner. Arrived at the Schwieloch Lake, the fish are transferred to larger boats, each holding about 10,000 pounds, and, placed in the care of reliable persons, they go down the Spree to Berlin, which place they generally reach after eight days, or they go still further to Hamburg, where they get after a journey of four to five weeks, and other places. The total annual rent of the domain is $12,870; the expenses for salaries, wages, wagons, &c., amount to about $7,150; so that Mr. Berger must make at least $20,000 just to meet his expenses. But it is said that he makes a little more!

XXXI.—MR. CHRISTIAN WAGNER'S ESTABLISHMENT FOR RAISING GOLDFISH, AT OLDENBURG, GERMANY.*

The two most important establishments for raising goldfish in Germany and Austria are the one belonging to Baron Max de Washington, of Poels, near Wildon, Styria, and the one belonging to Mr. Christian Wagner, of Oldenburg, of which we intend to give a brief description.

As the method followed in Oldenburg cannot be understood without some knowledge of the location of the establishment, we must mention that its 120 ponds are all close together, and with their dikes, &c., cover about 12 acres of bog-land near the river Hunte. The water, however, does not come direct from this river, but partly from an artificial stream or canal which on two sides forms the boundary of the establishment, partly from a neighboring factory, and from the ponds themselve. As glance at the accompanying diagram will best show what may be called the veins, arteries, and other vital organs of the establishment.

We intend first to show the manner in which the three channels which supply the water are used. The water which comes from the artificial stream G is, by means of the injector F, pumped into the open channel *a*, and after it has flowed through some or all of the ponds to the right and left it goes through wooden pipes into the ejecting-canals *c*, on each side, and eventually returns to the stream by way of the main ejecting-canal *b*.

In order to furnish the necessary insect life to the water which has thus circulated through a portion of the establishment, it is led by a very circuitous route before it again reaches the injector, and mingles freely with the main stream of the river, which is fed by the drains from the neighboring meadow-lands.

The water from the factory (which recently has proved very injurious) is collected in the reservoir E, and from there flows through subterranean pipes (indicated by dotted lines) into the channel *a*, which feeds the hatching-ponds B, and then goes into the flat "coloring-pond" D. After having mingled with the spring-water of these ponds it leaves the establishment, either through pipes laid in the dike of the pond, marked D, or through the main ejecting-canal *b*. Its temperature (sometimes as high as 100° Fahrenheit) cannot be regulated by Mr. Wagner as well as that of the water which is pumped in, and which, during the warm season, is sometimes raised to a temperature of 123° by means of the steam from a 10-horse-power engine.

---

* "*Die Goldfischzüchterei von Christian Wagner zu Oldenburg.*" [From "Deutsche Fischerei-Zeitung," second year, No. 29, Stettin, July 22, 1879.] Translated by Herman Jacobson.

The water which comes from springs at the bottom of the ponds is chiefly used for supplying the different spawning-ponds, marked A, and the ponds for hardening the skin of the fishes ("*Hauthärtungs-Teiche*"),

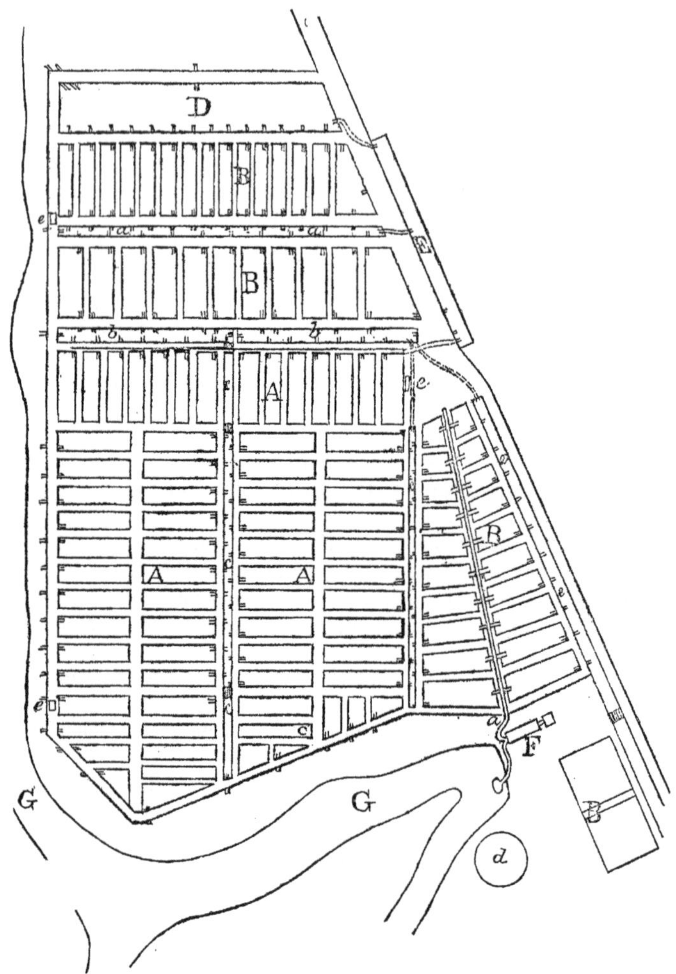

PLAN OF THE OLDENBURG ESTABLISHMENT FOR HATCHING GOLDFISH.

A, A, A. Spawning ponds. B. Rearing ponds C. Ponds for skin-hardening (?). D. Shallow coloring pond E Reservoir. F. Injector and machine-house (contains an engine of ten-horse power, which is connected with a machine of three-horse power). G Artificial stream or canal. *a*. Open supply pipes *b*. Principal discharging canal. *c*. Waste-pipe leading into A. *d*. Eel pond *e*, *e*. Tool-houses.

N. B The dotted lines, as also the short parallel lines, indicate covered pipes.

marked C, for although the spawning-ponds are, by subterranean pipes, connected both with the reservoir and with the hatching-ponds near the machine-house, these pipes are but rarely used. In case of necessity, *i. e.* when the stagnating spawning-ponds require it, a movable wooden pipe is used, through which the water of the stream is pumped into every pond whose water needs stirring up. On its way the water becomes completely saturated with oxygen, and its effect on mature fish is so quick that they often commence to spawn within an hour from the introduction of the fresh water.

As the bottom of the dikes is composed of very porous soil, the water goes from one pond to the other, and the depth of water is about the same in all the ponds, any superfluous water being led out through the channels *c* into the main outlet-canal *b*.

Although at times the depth of water in the ponds is only ½ foot, the average depth is about 2 feet, increasing to 4 feet near the outlets. The extent of surface is of greater importance than the depth of water, the average surface of each pond being about 228 square yards.

The bottom of the ponds is purposely left uneven, and is here and there overgrown with aquatic plants, on which the goldfish love to deposit their eggs.

The dikes between the ponds are generally 6 feet high, while the outer dikes are 8 feet high, 10 feet broad at the base, and 3–4 feet at the top. As the incline is therefore very gradual, and as the grass tends to keep the soil together, the bottom of the dike, though porous, is nevertheless firm.

The chief results of Mr. Wagner's cultivation of goldfish during several successive seasons are as follows: Many fish commence to color at the end of the first year; they are large enough to be sold for aquaria in the autumn of the second year, and they may be made to spawn two or three times a year, as a large number reach their maturity when only twelve months old.

By good feeding and frequent redistribution of the female fish (not allowing the same males and females to be together any very considerable length of time), and by an occasional airing of the water as described above, it has become possible to fix the time of spawning to the very day, and to raise a large number of young fish from comparatively few spawners. Under favorable circumstances the first young fish are raised in March or April, and by adopting the above-mentioned measures a second set of young fish may be raised in July or even earlier, and a third in August or the beginning of September.

It is but natural that fish cannot spawn so often during one season and at so early an age without many of them becoming prematurely barren. These barren fish, which can easily be recognized by a sunken appearance of the parts back of the ventral fin, must of course be separated from the others. They can only be sold as ornamental fish. Even if the spawning process is not hurried too much, it is an exception if a fish is used for spawning more than three years.

Mr. Wagner's average spawning stock amounts to about 3,000 fish, which are continually improved by the introduction of Italian and Portuguese fish, and by adding the finest specimens which he raises every year.

Of fancy fish his ponds contain, besides a few peculiarly colored specimens, the "dolphin," the "head," the "double tail" or "narwhal," and the "telescope-fish." Whenever Mr. Wagner wishes to produce some new fish, he makes some of these monstrosities interbreed, and thus obtains novel specimens.

Although there is no fixed rule, the proportion of females to males in the spawning-ponds is generally as 2 to 1; in sorting them great attention must be paid to their quality, age, &c.

It is likewise important, not only with regard to the old but also to the young fish, that (excepting the winter months) they are properly sorted and distributed, so that fish of the same size are put together, and that sufficient and suitable insect food is supplied for those ponds in which the fish are placed when coming from the spawning-ponds. In order to secure this food the fish are generally placed in ponds which have laid dry for seven or eight weeks; and if it should happen that one or the other of these ponds has less food than usual, it can easily be supplied from one of the neighboring ponds, or in case of necessity from the artificial stream G by applying a double hand-pump.

On this stream depends the supply of water for the twenty hatching-ponds near the machine-house; the suction-pipe of the injector rises or descends according to the depth in which infusoria and other insects are found in the stream. In calm weather they are generally found at or near the surface, and farther down during windy weather. Their exact place of sojourn can always be ascertained by dipping a glass cylinder vertically into the stream, and by observing their position in the column of water.

The insects, however, are not alive when they become food for the fish. Before they reach the twenty ponds they have been killed by the heat of the water—which in summer is often raised to a temperature of 100° Fahrenheit by steam from the boiler. Not satisfied with the effect of the heat—in high temperature fish breathe oftener and consequently take in more food—and this system of what may be called "condensed insect-feeding," Mr. Wagner finds it beneficial to supplement this natural food from time to time with artificial food, using for this purpose blood, small pieces of meat, and occasionally barley which has commenced to germinate (refuse from breweries). This food is not cooked, but simply thrown into the ponds (the blood in small lumps) wherever the water is shallow.

The results of this Oldenburg feeding system, as regards the growth of the fish, are as follows: Some of them double their weight in a week's time, and under ordinary circumstances the young fish have reached a length of 1¼ to 2¾ inches in autumn. When properly colored the largest are then sold as "glass-fish." Most of them, however, do not reach a salable size till the end of the second summer.

The artificial coloring of the fish is just as important as their artificial feeding, and much time and money has been consumed in experimenting until satisfactory results have been obtained. The Oldenburg ponds are very favorably located with regard to the coloring process, for of the three principal ingredients, viz, iron, lime, and tan, the first mentioned is found in considerable quantities, both in the soil and in the water; nevertheless it is not sufficient, and has to be artificially increased from time to time.

The German national colors are in great demand, and a fish which was originally red and white can by proper treatment be transformed into an "imperial fish," exhibiting the national colors, viz, black, white, and red. In spite of the greatest care it will happen that fish are not sufficiently colored when they have reached the size of "glass-fish"; they are then transferred to the large shallow pond D, where they are more exposed to the rays of the sun, which possess a strong coloring power, but are not without danger to the fish, as they often kill them suddenly if the bottom is too bright and shadeless.

In order to make the fish less tender for handling and transferring to the aquaria they are generally for a time placed in the so-called "skin-hardening" ponds, marked C. The peat bottom of these ponds contains little or no sand or clay, but a great deal of iron. The water, likewise, contains much iron; and in ponds of this kind the adding of lime tends to harden the skin of the fish. This method of hardening the skin has made the former slow and wearisome acclimatizing process almost superfluous. By applying lime the same result is obtained with young eels, which are kept in the pond marked *d*, and are also sold for aquaria. These eels are obtained from neighboring waters, into which they come from the Hunte when they ascend that river in May.

By this simple method the goldfish become so hardened that they can be easily handled without suffering injury. Their future welfare (when kept in glasses or aquaria) of course depends on the character of the water and the food. Mr. Wagner recommends spring or pump water, and wherever this cannot be obtained river-water. Rain-water he considers utterly useless. When the goldfish are kept in glasses or small aquaria animal food is almost exclusively recommended by Mr. Wagner; *e. g.*, meat, raw or cooked, scraped very fine, worms, insects, larvæ, ant-eggs, &c. The aquaria should also contain a few aquatic plants at which the fish may nibble. Too much food is injurious, especially in winter, when scarcely any food is required. Mr. Wagner considers it less injurious to give no food for a whole month than too much food. As a rule, no more food should be given than can be at once consumed by the fish.

Before Mr. Wagner ships fish to any considerable distance he lets them fast for a week, and in this way prevents, as far as possible, the water from becoming impure during the journey. The vessel which is generally used for transporting goldfish is an oval tub with a perforated bung at the top. As this tub is not completely filled with water, a certain degree of motion keeps the water pure and fresh, and in favorable weather—cold weather is the best—fish have successfully been sent to Denmark, Russia, England, Southern Italy, America, &c., without a change of water. The journeys are generally not very long, as most of the fish are sold in Germany and Austria. The risk and the difficulties of transportation are therefore considerably diminished.

The price of fish of course varies according to age, size, color, and kind. The most expensive fish are the so-called "telescope-fish," which are sold at $7.14 to $21.42 a pair; next come the "dolphins" and "heads," which sell at $11.90 a pair; then the "double-tail" or "narwhal," which are sold at $4.76 to $2.85 a pair; and finally those fish which are valued on account of their peculiar coloring; these are sold at $2.30 to $23.80 a hundred.

In order to keep up with the constantly growing demand, Mr. Wagner has been obliged to increase and enlarge his ponds from time to time, and a number of ponds which were originally destined for carps have been appropriated for goldfish. In 1874 Mr. Wagner had 56 ponds and raised 99,500 fish; in 1876 he raised 170,000 (50,000 of which he exchanged for imported fish); and in 1877-'78 he had 120 ponds and annually raised 300,000 fish.

That the *Cyprinus auratus* does not bear its name in vain, but produces a golden harvest for its cultivator, is sufficiently proved by the fact that Mr. Wagner has been obliged to constantly enlarge his establishment. At the present time he employs a bookkeeper, a night watchman, an attendant, and fifteen laborers (not counting the men employed in the Berlin salesrooms), all of whom earn a good living, while his own annual profits are very considerable. The same area used for agricultural purposes would scarcely feed a single family.

XXXII.—A REPORT ON THE HISTORY AND PRESENT CONDITION OF THE SHORE COD-FISHERIES OF CAPE ANN, MASS., TOGETHER WITH NOTES ON THE NATURAL HISTORY AND ARTIFICIAL PROPAGATION OF THE SPECIES.

By R. E. Earll.

A.—INTRODUCTION.

The recent inquiry into the decrease of the food-fishes of the east coast of the United States by the United States Commission of Fish and Fisheries, under the direction of the commissioner, Prof. Spencer F. Baird, has led to the establishment of temporary stations at different points along the coast, where special attention has been given to the study of the more important species for the purpose of gathering definite information of their relative numbers past and present, their geographical distribution, and their habits. Of late the commercial importance of what might be styled the great ocean fisheries, together with the complicated questions that are continually arising between our own government and our more northern neighbor regarding them, has led Professor Baird to give particular attention to this subject, with a view to becoming more thoroughly acquainted, not only with the habits and movements of these species, but also the methods employed in their capture and the extent and money-value of the fisheries.

With this end in view, he selected Gloucester, Mass., as the most suitable location for the Commission in 1878, where he arrived with his assistants early in July, and at once began the investigation of the subject. During the summer much valuable information was gathered relating to the extent of the fisheries, and many observations were made on the natural history of the different species. However, as this was not the spawning season for the different members of the cod family, the only obtainable information on the habits of the fish during this period was from the fishermen, who are usually not considered very accurate scientific observers.

After a careful consideration of the subject, it was decided to continue the station through the winter, in order to study the natural history of the spawning fish, that visit the shore in immense numbers at this time, and also to make experiments with the eggs of the cod and other species, with a view to their artificial propagation. Accordingly, the late James W. Milner, deputy commissioner, proceeded to Gloucester, to take charge of the work and to prepare a report on the whole subject. Mr.

Milner arrived late in August, and remained until the preliminaries were arranged and the first eggs had been taken, when the sickness that has so recently resulted in his death compelled him to leave for the South, in order to avoid the cold and stormy weather of the New England sea-coast. The loss of so enthusiastic and experienced a worker, whose efficient labors have aided greatly in bringing the United States to the front in all subjects relating to fish-culture, was a severe blow to the Gloucester work; had he been permitted to remain, the results would doubtless have been more thoroughly satisfactory.

Owing to the absence of Mr. Milner, the writer has been requested to prepare a report from hurried notes made during the winter. Much of the data has been obtained from personal observations and experiments, either in the hatchery, or at the various fish-wharves, or during visits to the different fishing-grounds in the fishing-schooners of the harbor. Much valuable information has also been obtained from the older and more experienced fishermen and from the files of the local papers. In all cases, however, care has been taken to avoid the acceptance of any statements and opinions without being fully convinced of their correctness, and due allowance has been made for the lack of careful and accurate observations on the part of those interviewed. Many questions requiring much more careful inquiry than we were able to make still remain unsolved, and many points have been wholly omitted in the report for want of sufficient evidence either to disprove or confirm them.

The report, then, especially in the portions relating to the natural history and artificial propagation, must be considered as merely paving the way for a more careful and extended study of the subject.

B.—THE SHORE FISHERIES.

1.—ORIGIN OF THE COD FISHERIES OF CAPE ANN.

Of the many different fisheries in the United States yielding remunerative employment to large numbers of men, the cod-fisheries of New England are the most important and extensive. Dating back as they do even beyond the earliest permanent settlement of the country, and being to the struggling colonists often the only unfailing source of supply, they were at this time of vital importance to the people. In fact, the presence of these fish in the waters of New England had much to do with hastening the settlement of the country, and it was doubtless the knowledge of their abundance that led the merchants of the Old World to send their first vessels to our shores.

The following facts, gathered largely from Babson's History of Gloucester and the files of the Cape Ann Advertiser, give briefly the origin of the Cape Ann fisheries and a glance at their condition at intervals to the present time. Apparently the first that was known of the presence of the codfish in this locality was in 1602, when Bartholomew Gosnold, in the ship Concord, while on a voyage of discovery to America, reached the coast of Maine, and sailing southward passed "the mighty headland, which, on account of the great numbers of cod-fish with which the voyagers 'pestered their ships there', then received the name of Cape Cod." From this date foreign merchants, principally those of England, fitted out fishing-vessels for America, these visiting points on the coast of Maine, and meeting with varying success. In 1622 these parties, having found the expenses of the enterprise greater than the catch of fish would warrant, began to devise methods of lessening them. They soon decided upon a plan whereby the vessels should take out a number of men in addition to their regular crews, these to assist in taking the fish, and to be landed on the shore after the trips were secured, where they were to remain during the rest of the year to clear the soil and engage in agricultural pursuits, living chiefly on the natural products of the land; and to devote their time during the fishing season to loading the vessels that were to be sent yearly to the little colony. Accordingly, in 1623, a ship left Dorchester, England, and proceeded to the usual fishing-grounds, coming later into Massachusetts Bay, where she secured the balance of her trip, and, after leaving fourteen men at Cape Ann with suitable provisions, sailed for Europe. The same year a patent of the land was granted to the New Plymouth colony, who in 1624 built a fishing-stage at Cape Ann, the Dorchester fishermen arranging to share the patent with them. The following year a man was sent from Plymouth to build salt-pans at this place, but, the fisheries proving unremunerative, were abandoned by both parties, and the colony was broken up, a part of the Dorchester men returning to England while the remainder removed to Salem.

The next fishing interests at Cape Ann were in 1639, when the general court passed an act for the encouragement of Mr. Maurice Thomson and others, providing for the establishment of a fishery plantation, and granting certain exemptions to fishery establishments, in order to encourage the colonists to engage more extensively in the capture of the different species. This seemed to have a beneficial influence on the fishing interests of the section, and they gradually grew into a more flourishing condition. But it was not until the beginning of the last century that these fisheries assumed important proportions, and then, for the first time, ship-building was extensively carried on, and Cape Ann sent a large fleet to Cape Sable and Sable Island for cod-fish. In 1741 Gloucester owned about 70 sail, and at the beginning of the Revolutionary War she had 80 sail engaged largely in the bank-fisheries, with nearly twice as many chebacco boats fishing along the shore. The effect of the war, together with the small catch of the vessels, resulted disastrously to the fishing interests, and at the beginning of the present century the fleet had dwindled down to 8 sail of more than 30 tons. But while the bank or offshore fleet had been so reduced the smaller crafts had continued to increase, and there were at this time fully 200 chebacco boats, aggregating about 3,000 tons, fishing on the inshore grounds.

In 1819 the fisheries were in such a state of depression that Congress passed "the bounty act" for their encouragement. This seemed to put new life into the business, and in 1825 over 150 sail fitted out for the different banks, and by 1847 the fleet had been increased to 287 sail, with an aggregate of 12,354 tons, or an average of 43 tons, carpenters' measurement, to the vessel. The Cape Ann fishermen first visited the famous George's Bank fishing-grounds about 1830, and by 1850 this locality had become a favorite resort for both the cod and halibut fleets.

In the spring of 1879 there were 39 fishing-firms at Gloucester, and 378 fishing-vessels of over 5 tons burden sailing from the harbor. Of this fleet 174 sail visited the distant banks for cod, 44 engaged exclusively in the halibut fisheries, 66 were provided with purse-seines for catching mackerel, 8 fished for both cod and halibut, 78 fished along the shore for cod, pollock, haddock, hake, and cusk, and the remaining 8 sailed about in search of squid to supply the bank cod-fishermen with bait. Of the 174 offshore cod-fishermen, 130 went to George's and Brown's Banks, and the remainder to La Have, Quereau, Western and Grand Banks. In addition to the above the other towns of the vicinity had each small fleets engaged in some branch of the fisheries; so that the total number of fishing-vessels belonging to Cape Ann at this time reached upward of 415 sail.

Thus the fisheries of Cape Ann have been continuously prosecuted for two hundred and forty years. Small at first, they have met with varying success, reaching their lowest ebb about the year 1800, since which time they have gradually grown in importance, until to-day Cape Ann is the center of the marine fisheries of America; and Gloucester, which from its excellent natural advantages early became prominent, has continually strengthened itself, until it has come to be the great fishery metropolis of the country; and is now, by the aid of laws and business customs, which tend to transfer the business from the fishermen to the capitalists and from the smaller to the larger dealers, gradually absorbing the fishing interests of the State.

With this large fleet engaged in the various branches of the fisheries, and visiting so many different localities, the quantity of fish landed in Gloucester is enormous; the cod-fish alone for the year ending June 30, 1879, reaching 36,665,620 pounds of cured fish, which, at the low average of three cents per pound, would have a total value of about $1,100,000, This quantity of cured fish represents not far from 91,650,000 pounds of round fish, or, on the supposition that the fish average 15 pounds each, over 6,100,000 cod in number. These figures, though not absolutely correct, probably vary but little either way from the actual number landed in Gloucester during the year mentioned. The data from which the calculations have been made were taken partly from the weekly reports of the Cape Ann Advertiser and partly from notes made during my stay in Gloucester.

In looking over the history of the fisheries, we find that when the bank-fisheries have prospered the shore-fisheries have been neglected, but when for any reason the bank-fisheries have been unprofitable the fishermen have resorted to their boats and small vessels for a livelihood. Thus, in 1804, when the bank fleet had been reduced to 8 sail, the shore crafts numbered nearly 200; but when in 1847 the off shore vessels had increased to 287, the chebacco boats numbered scarcely 35, these fishing during a portion of the season only.

We are told that the chebacco boat originated with the fishermen of Cape Ann, and that it derived its name from a river on the north side of the cape, where it was first extensively built. These boats were usually of about 15 tons burden, rigged with two masts but no bowsprit, and had a small forecastle or "cuddy" forward, affording sleeping and cooking accommodations for the four or five men that constituted the crew. The fishermen often ventured fifteen to thirty miles from harbor in them, remaining four or five days before returning to land their catch.

The first small boats extensively used were known as the Hampton boats, from the village where they were first built. These are still used by many of the shore-fishermen of Maine and Massachusetts. They are open lap-streak boats varying from 12 to 20 feet in length, propelled either by oars or by means of two sprit-sails; the masts being movable so that they can be placed in the bottom of the boat when not in use.

The common fishing-dory, now so extensively employed, was little used for fishing purposes prior to 1825. It seems to have had its origin with the boat-builders of Salisbury, Mass., about 1775, being long used as a river-boat, and for lightering purposes, before its seaworthiness became known. It is a flat-bottom, lap-streak boat, with sharp, projecting bow, V shaped projecting stern, and flaring sides, having an average length of 13 to 15 feet on the water-line. Occasionally it is propelled by means of a small sail, but oars are more frequently used by the shore-fishermen.

As early as 1828, a few "pinkies," and "square-stern" vessels of 30 to 60 tons burden engaged occasionally in the shore-fisheries, but it was not until 1843, when the halibut-fisheries began to require this class of vessels, that any extensive winter fishing was carried on. These vessels, after finishing their season's work in the halibut fisheries, began to fish along the shore during the pleasant weather, and it was in this way that the winter shore-fisheries originated. This class of vessels rapidly increased in number, and by 1855 had nearly supplanted the smaller chebacco-boats, though it was not till 1870, or later, that the shore-fisheries began to assume their present important proportions. In the spring of 1879 fully 100 vessels ranging from 10 to 60 tons, with 90 additional dories, engaged in these fisheries, and the fleet landed during the year ending June 30, about 14,475,000 pounds of round cod-fish, besides a great quantity of haddock, pollock, and hake.

fisheries are soon at their height. The vessels are usually provided with dories, taking from three to twelve each according to the size of their crews. Such fishermen as are unable to ship on the vessels now row or sail out in boats. These often endure great hardships, as the wind may rise suddenly and drive them out to sea giving them a hard pull of hours before they can regain the shore, while an occasional unfortunate fails to return.

The pasture-school is composed of fish averaging probably between 12 and 14 pounds, some being much larger while others are quite small. In the falls of 1877 and 1878 the fishing was unusually good until the first of January, the average daily catch per man often reaching 800 to 900 pounds, while an active fisherman at times caught nearly twice that quantity.

At the present time there are but few towns on the north side of the cape extensively engaged in the shore-fisheries, and for this reason little is definitely known about the first appearance of the Ipswich Bay school of cod-fish in that locality. We cannot even feel certain of the month when they reach the grounds, as the fishermen have many and conflicting opinions on the subject. From the best obtainable information it seems probable that cod have visited these waters regularly for many years, and that they were formerly taken in considerable numbers by the boat-fishermen of the section who rowed out from the shore in pleasant weather during the winter months. But for a number of years these grounds were nearly deserted, and it was not until 1877–'78 that the shore-fishermen of Gloucester and Swampscott learned their value.

In January, 1879, after the fish had left "the pasture" several vessels sailed for Ipswich Bay, where they found the cod remarkably plenty, returning in a short time with unusually large fares. The news spread rapidly and soon all the shore fleet were in the bay, while vessels of 60 to 70 tons abandoned the other fisheries and fitted out for this locality. Vessels from other towns along the shore soon joined the fleet, and by the middle of February 104 sail, with upwards of 600 men, were fishing within a radius of five or six miles, and 20,000 to 25,000 pounds of round fish were sometimes taken in a day by the crew of a single schooner.

The above number of vessels was reached only during the height of the season, and several causes operated to reduce the fleet so that at times it was quite small. But allowing an average of 45 sail during the entire four months, each vessel carrying six dories, the trawls averaging 800 hooks each, and we have the enormous number of 216,000 baited hooks spread out upon the sandy bottom to tempt the spawning-fish. It is not surprising, therefore, that the catch reached fully 11,250,000 pounds on this little patch of ground between the first of February and the last of May.

Fishermen are agreed that the individuals composing this school averaged larger than those of any school that had previously visited the shore. There were almost no small ones among them, the great bulk this seemingly accidental variation, that gives every gradation to either extreme, there is a more constant difference in both form and color, due perhaps to the peculiar habits and surroundings of the individual. This difference is so noticeable that the fishermen can easily distinguish the one from the other, and they have come to call the one a school-fish in distinction from the other, which they call a shore-fish or "ground-tender."

The school-fish are supposed to be constantly on the move, remaining usually in the deep water, where they are very active in the pursuit of their prey, consuming such quantities as to keep them in excellent flesh. Such fish are usually very shapely, with small and very distinct dark spots on a light background, and seem to have the head quite small in proportion to the body. On the whole, they are just such fish as would be expected from continued activity and good living. On the other hand, the shore-fish, or "ground-tenders," live constantly among the rocks and sea-weeds along the shore, where the water is less pure and the food less abundant. They seem to lead solitary lives during a greater part of the year, being scattered along different portions of the coast, living upon the little rocky spots, where they feed upon such animals as they chance to find; or at times entering the shoaler water among the sea-weeds, where they feed upon the mollusks and articulates that are often so abundant in such localities. They are generally in poorer flesh than the school-fish, having a relatively larger head in proportion to their bulk, with larger and less distinct spots on a darker background. In addition to these large fish, that for some reason seem to prefer the shore as a feeding-ground, there are many young and immature that have not yet joined the school-fish in their migrations. These fish are the sole dependence of the boat-fishermen in summer, or from June to November, and one must know the grounds pretty thoroughly, and row about from one feeding spot to another, in order to secure any considerable number of them. During the months of June, July, and August, the fishing is quite limited, being confined to a few boat-fishermen who row or sail out daily with hand-lines, returning in the afternoon with from 150 to 300 pounds, which they usually sell at fair prices to supply the fresh-fish trade.

Early in the fall the spawning instincts of the fish cause them to gradually gather from the different parts of the shore to special rocky grounds, where they remain until they have deposited their eggs. At such times, being more numerous in these localities, the fishing becomes more profitable, so that many small vessels and a larger number of boats frequent these grounds, and by the middle of October the daily catch reaches about 400 pounds per man.

Thus far the catch has been composed almost wholly of the young and shore fish; but about the 1st of November the fall school of spawning-fish, known as the "pasture-school," makes its appearance. All the smaller vessels and boats are now pressed into service, and the winter

2.—CHARACTER OF THE FISHING-GROUNDS.

Cape Ann is a prominent headland, dividing the waters of Ipswich Bay on the north from those of Massachusetts Bay on the south. Next to Maine, it has the most bold and rocky shores on the coast of New England, and its rugged granite walls rising to a considerable height above the water, present an inhospitable appearance to the approaching mariner. This granite ridge, of which the Cape is a part, extends some distance from the shore, forming an irregular ocean-bed; and continuing southward, is broken up into a large number of small rocky islands and sunken ledges, separated by deeper channels.

Among these islands and ledges the shore cod, and other species, find a favorite feeding-ground, and the school-fish, though seldom venturing among the innermost islands, come yearly in great numbers to the larger outer ridges where they remain during several months for the purpose of spawning. It is here that the shore-fishermen of Cape Ann find their best fishing during the fall and early winter; the fish being known as the "pasture-school," from the grounds where they are most frequently taken.

Farther east, at a distance of 15 to 20 miles from the shore, and separated from the foregoing by a wide channel of clay and mud, is a ridge of ground about 20 miles long, known as Stellwagen or Middle Bank. This bank lies at the entrance to Massachusetts Bay, between Cape Ann and Cape Cod, with an average depth of 15 to 18 fathoms. The fishermen often resort to this locality when the fish are approaching or leaving the coast, and frequently find good fishing for several weeks.

To the north of the cape is Ipswich Bay, with its low sandy beach and level bottom sinking very gradually until a depth of 25 to 30 fathoms is reached at a distance of several miles from land. The floor of this bay is a vast sandy waste, with only here and there a patch of clay or rocks, the whole supporting but a small amount of animal life, and this limited to a few species. It is essentially a spawning rather than a feeding ground of the cod, and large schools visit the bay for this purpose during the winter, remaining as late as June. The fishermen are just beginning to learn the value of this ground, and in the spring of 1879 over 11,000,000 pounds of round fish were taken, mostly by the Cape Ann fleet.

Farther to the eastward, and extending some distance in a northerly direction, is Jeffry's Bank. This ground is frequently visited by the shore-vessels during certain seasons of the year, and good fares are often secured. It seems more of a feeding-ground for the fish than Stellwagen Bank, and the fishing often lasts during a longer period.

3.—DIFFERENT SCHOOLS.

In examining the cod-fish landed from time to time, one cannot but notice the great individual variation in the species. But in addition to being of uniformly large size with a few very large. Of over 5,000, selected without regard to size at different times during the season, the average weight was 20¾ pounds.

Fishing continued good in Ipswich Bay until the first of June when the school left the shore, being perhaps hurried in their movements by a large school of dog-fish (*Squalus americanus*) that made their appearance in the bay about this time.

After the school-fish leave the shore in summer the fishermen frequently resort to the outer grounds, such as Jeffry's and Stellwagen Banks, when they often secure good fares from what they suppose to be a new school that visit these grounds for the purpose of feeding. We have had little opportunity for examining these fish, but there seems a strong probability that they belong to the school that have just left the shore, and that they remain on these grounds for a few days or weeks on their way to deeper water.

4.—METHODS OF CAPTURE.

Two methods only are extensively used by the cod-fishermen of Cape Ann. Hand-lines have been used from the earliest times, and are still exclusively employed on the rocky ledges during the stay of the "pasture-school" in the fall and early winter. A visit to the harbor at midnight in November, when the fall fishing is at its height, cannot but impress one with the loneliness of the scene, for all is quiet and the region seems thoroughly deserted. But two hours later the rumbling of wheels and the shrill cry of the baitman cause a great and sudden change, for if the fisherman is behind time he frequently finds that all of the "sperling" have been sold. With the first cry of the baitman, lights may be seen in the hands of the fishermen as they emerge from the cabins of the different schooners, and soon the dull thud of oars is heard and boats approach from various quarters, while men and boys come straggling down the different lanes and by-paths from their homes on shore. The night of the fisherman is over. He secures his bait and returns to his vessel, where the other members of the crew are just beginning the work of the day. Soon the measured stroke of the windlass and the hoisting of sails are heard and the fleet is "under way."

If it is calm in the harbor, as it often is at this early hour, one or two boats are "hoisted out," and with lines fastened to the vessel's bow they tow her toward the outer harbor, where her sails catch the breeze and she is off for the fishing-grounds. The boats are now "paid astern," and the rowers join the other members of the crew, who have assembled in the forecastle to eat their morning meal from their private lunch-baskets, all going to the common reservoir for their mug of hot coffee that has been prepared by the schooner's cook. Breakfast over, each gathers his gear in convenient shape, and after filling his bucket with bait, lounges about waiting for the day. The vessel aims to reach the ground just before light, and at the first sight of his land-marks or cross-bearings the captain brings his vessel upon some little spot of ground known to be a favorite resort of the cod. Each crew strives to be first on the ground, which is a small rocky ridge five or six miles east-southeast of Eastern Point, with an average depth of 25 to 30 fathoms. After the anchor has been dropped the jib and foresail are "taken in" and the dories are lowered from the vessel's deck. Soon the men, following each other in rapid succession, are off for their favorite spots, the captain and cook, or at times only the latter, remaining to care for the vessel and fish over the rail. It is indeed a lively scene, with 150 dories and upwards of 40 larger crafts, each striving for the best berth on a little ridge of ground not over 50 by 90 rods in extent.

When the desired knob has been reached, the killick is dropped, and the fisherman seats himself upon the middle thwart with his face toward the stern, his lines and gaff by his side, and his bucket of bait before him. A fisherman uses two lines, each having two hooks, the leads varying in weight from three to five pounds, according to the depth of water and the strength of the tide. The hooks are now baited, from three to six sperling being strung on each, and a line is thrown over on either side, being allowed to run out until the lead reaches the bottom when it is "seized up" five or six feet, so that the lower hook just clears the rocks, with the upper one a foot or two above. The lines are now fastened to the inner braces of the boat, and with one in either hand the fisherman sits expectant, slowly moving his arms back and forth in his endeavor to induce the fish to bite. On hooking a fish, he generally stands in the boat, facing the line, which he proceeds to haul quite rapidly until the fish is at the surface, when with one hand he holds the line, and with the other reaches for his gaff to lift the fish into the boat.

The best fishing usually occurs in the early morning, though the hour may vary and even come in the afternoon; but by one or two o'clock the flag is set in the rigging of the vessel as a signal for the boats to return, and as they come alongside, the fish are pitched into the common pile on deck, after which they are "hoisted in," and the vessel starts for home in order to market her catch before dark.

The quantity of fish taken daily varies greatly, being dependent upon the dexterity of the fisherman, the abundance of fish, and the quality and kind of bait used. The largest hand-line catch in a day, as far as we can learn, was secured off Pigeon Cove in the winter of 1877–'78, when a fisherman landed 2,200 pounds of round fish. About the same time two men, fishing from one boat in the same locality, landed 3,900 pounds; while two other boats with similar crews fell only 100 pounds behind them.*

The method of trawling originated with the fishermen of this region, probably with those of Marblehead, about twenty-five years ago, and has since come into general favor. This method is used almost exclusively when fishing where the bottom is smooth; though it cannot be

*Cape Ann Advertiser.

employed on very rough ground, as the trawl becomes fastened among the rocks, and is often lost together with all the fish that are on it. The trawl consists of a long rope to which are fastened, at intervals of four to seven feet, smaller lines called "gangings," each bearing a baited hook at its free extremity. These gangings are from two to four feet in length, and number from four or five hundred to even fifteen or sixteen hundred, according to the length of the ground-line. The trawl has an anchor, weighing from 8 to 16 pounds, at either end to hold it in position, while buoys, connected with these by means of small ropes, float at the surface to mark their exact location. When using trawls the vessel usually carries a dory for each member of the crew save the captain and cook. On reaching the grounds these boats are "paid astern," and as the vessel sails at right angles to the wind, they are dropped in regular order, each being separated from the other by 30 to 60 rods. Each man now takes his position in the stern of his boat, and, after throwing out the buoy and line and lowering the anchor to the bottom, slowly pays out the trawl as the wind and tide carry him along. When all the trawl is out the second anchor, with another buoy and line, is dropped, and the man is picked up by the vessel. In case neither wind nor tide carry the boat along with sufficient rapidity, the fisherman sculls with one hand while with the other he pays out the trawl; or, where two go in the same boat, one usually rows so that the trawl may be set in any direction regardless of the winds or tides. Thus we have lines often a mile in length stretched out upon the ocean's bottom, with hooks at regular intervals of five or six feet, and the cod cannot pass without being tempted to take the bait. The trawls are sometimes left in the water only a few hours, but more frequently they remain over night, and are often taken up well filled with fish. In hauling, the fisherman first rows to his buoy, and pulls up the anchor with one end of the trawl attached. He then takes his position in the bow of the dory with a trawl-tub before him, into which he coils the trawl as it comes from the water, using his gaff to take the market fish into the boat, and "cutting away" all large but worthless fish, such as sharks and skates.

Another method of fishing with the trawl, known as "underrunning," requires a second buoy-line attached to a small weight on the end of the trawl, the other line being fastened to the larger anchor only. By means of this second line the trawl is brought to the surface, while the anchor remains on the bottom to mark its original position. In underrunning, the man stands, as before, in the bow of the dory with a bucket of bait in place of the trawl-tub, merely passing the trawl over the bow of the boat and again into the water on the opposite side, saving such fish as are found and re-bating any hooks that may require it. This method is often employed for two reasons: first, because it retains for a man his old "berth" where fish may be plenty; and, second, from the fact that in this way the hooks are kept almost constantly in the water, so that no opportunities for fishing are lost.

When trawlers are numerous and there is a disposition to dishonesty or "mooning," as the fishermen style it, which consists in taking the fish from the trawls belonging to another vessel under the cover of darkness, each vessel usually anchors near her own trawls in the evening; but ordinarily, when near home, or during stormy weather, the vessels seek shelter in the harbor for the night, starting out in time to reach their grounds at early dawn. Great quantities of fish are often taken on a single trawl, and when a dory has been filled she hoists her signal, and a boat is sent out from the vessel to lighten her. The largest catch by any shore-vessel during the winter of 1878–'79 was made by the schooner George A. Upton, of Gloucester, from Ipswich Bay. This vessel landed an equivalent of 55,906 pounds of round fish as the result of two and one-half days' fishing with eight dories, the trawls averaging 900 hooks each, selling her trip for $569.56.

The method of catching cod with gill-nets, though so successfully used by the fishermen of Norway, has never been adopted by the fishermen of our coast. Knowing of the profits derived from the use of their nets by these foreign fishermen, Professor Baird, who is ever anxious to introduce among the Americans any methods that will result to their advantage in the prosecution of the fisheries, decided to make experiments with them at Cape Ann, with a view to their introduction among our shore cod-fishermen. Accordingly he secured from parties in Norway a set of these nets and forwarded them to Gloucester, to be thoroughly tested by the employés of the Commission at that place. They reached the hatchery when the pasture-school was on the shore, and were set on the favorite fishing-grounds a number of times. But the strength of the twine had probably been affected in transit, and the nets proved far too frail. The strong tide and rough water caused them to catch among the rocks, where they were badly damaged; while numerous holes indicated clearly that large fish had torn their way through the nets, only such being retained as had become completely rolled up in the twine. The nets were always taken from the water in bad order; but the capture of 800 pounds on one occasion, even under these circumstances, seemed to indicate that nets of sufficient strength might be used to good advantage, at least on the smooth fishing-grounds along the coast.

5.—THE BAIT QUESTION.

With so large a fleet engaged wholly in hand-lining and trawling, the question of obtaining and preserving bait is of the utmost importance to the fisherman, and on its abundance or scarcity depends largely the success or failure of his season's work. Cod-fish, though having the habit of snapping at, and at times swallowing, anything that may come in their way, are on the whole quite dainty fish, and when one expects to be successful in catching them for profit, he must have not only a good quality of bait, but also a kind that the fish are known to prefer. So peculiar are the fish in this particular, that the fishermen have different names for the various schools, derived from the kind of bait on which they live during the fishing season. We often hear them speak of the clam-school, the herring-school, and the squid-school; and when securing bait they will at times pay exorbitant prices for that kind on which the fish are known to be feeding, rather than take an equally good quality of another kind at much lower rates. Thus, when the fish are feeding on squid (*Ommastrephes illecebrosa*) the fishermen secure squid if possible; the same is also true of the herring (*Clupea harengus*), the capelin (*Mallotus villosus*), and other species. But while it is undoubtedly true that during the feeding season the fish take the hook more readily when baited with that particular species that they chance to be pursuing, and while they always prefer fresh to salt bait, yet we think the fishermen in error when they apply the rule with the same fixedness to the schools of spawning fish, and that the shore-fishermen often lose both time and money by so doing. It is quite interesting to watch the effect of this idea upon the shore-fishermen; for they seem fully convinced that, when one kind of bait has been successfully used, it is utter folly to attempt the use of any other kind. Thus in the winter of 1878–'79, when sperling (young herring) became scarce, the fleet waited fully two weeks, hoping that more might be obtained before they would supply themselves with either frozen herring or clams.

In the winter of 1877–'78 the first vessels resorting to Ipswich Bay for cod chanced to be fishing with clams, and, as a result, clams were used by nearly the entire fleet, though frozen herring could be more easily obtained, and were cheaper. Again, in the winter of 1878–'79, the first vessels resorting to the above locality used frozen herring, and the results obtained were entirely satisfactory. Frozen herring were at once announced as *the* bait, and fishermen provided themselves with these only. A few, however, acting on the experience of the previous season, had contracted for clams in advance, and were obliged to use them. These unfortunates, for such they felt themselves to be, frequently received expressions of sympathy from the other fishermen, and it was the general belief that their catch was much smaller than it would otherwise have been. A comparison of the quantity of fish landed by one of these vessels with that of a vessel of equal size using frozen herring, showed that the bait had little effect on the catch, the trips averaging about the same in size, sometimes favoring the one and again the other vessel. Later in the season, when frozen herring could not be obtained, the vessels went south for fresh herring and alewives (*Pomolobus vernalis* and *P. æstivalis*), and it was not uncommon for them at times to refuse the herring, and to spend several weeks in search of alewives, or again to refuse the alewives and search for herring.

The principal kinds of bait used in the cod-fisheries are clams (*Mya arenaria*), sperling or young herring, fresh and frozen herring (*Clupea harengus*), fresh and salt squid (*Ommastrephes illecebrosa*), fresh and salt menhaden (*Brevoortia tyrannus*), capelin (*Mallotus villosus*), and alewives

(*Pomolobus vernalis* and *P. æstivalis*). The shore-fishermen of Cape Ann use principally clams, frozen and fresh herring (including sperling), and alewives.

Clams are used principally during the summer months and at other times when bait is scarce. They occur in considerable numbers in most of the muddy flats along the shore between tide-marks, being small and scattering near the line of high-water, but gradually increasing in both size and number as the low-water line is neared. To these flats the fishermen resort with their clam-forks and baskets during the hours of low-water. When they are plenty, an energetic worker can dig from seven to nine bushels at a single tide, these making nearly two-thirds of a barrel of bait; but in the vicinity of Gloucester the flats have been dug over so frequently that clams are becoming scarce, and the fishermen are often obliged to buy their supply from other places, at an average price of four or five dollars a barrel.

The sperling, now so extensively used by the shore-line fishermen, average from five to six inches in length. They make their appearance in these waters about the middle of September, remaining until driven off by the coldness of the water late in December. We are told that they were first used for bait by the Swampscott fishermen about 1840, and that the demand did not become general until 1866. The supply now comes wholly from Ipswich Bay, where for the past two years the fish have been unusually abundant. They are taken wholly at night, within a short distance of the shore, by means of dip-nets. The men visit the grounds in 20-foot dories, made expressly for the purpose, and as soon as it becomes dark a torch is placed in the bow, and two men row the boat rapidly through the water, while the third stands ready to secure the fish as they are attracted by the light and gather in little bunches, keeping just in front of the boat. A good dipper will often catch half a bucket of them at a single dip. It usually takes but a short time to secure all that can be sold, when the boat returns to the shore, where a wagon is in readiness to carry the fish to market. Ninety men were engaged in this work during the winter of 1878–'79, landing and marketing about 7,000 barrels of sperling, at an average price of $3 per barrel. During the season six men landed nearly a thousand barrels, while a single crew of three men caught 20 barrels in one night.

The fishermen buy only enough bait to last them through the day, getting a fresh supply each morning, as the fish soon become soft, and when in this condition will not stay on the hook. For this reason they are not suitable bait for the trawl, and cannot be used in the offshore fisheries.

Frozen herring usually make their appearance in the Cape Ann markets about the middle or last of December, from which time they are extensively used as bait by all of the fishermen until April, when the weather becomes so warm that they cannot be obtained. The supply comes largely from the coasts of Nova Scotia and Newfoundland, where the fish are abundant during the greater part of the winter. Many of the larger Cape Ann vessels engage in the frozen-herring trade during these months, visiting those points where the herring chance to be most abundant, and bringing large trips to the principal New England markets. Formerly they supplied themselves with nets for catching their own fish, and took full crews of fishermen to assist in the work, but of late they often find it cheaper to buy the fish of the natives, in which case they carry only enough men to work the vessel on the passage. The herring are first frozen on the shore, after which they are thrown, with a little straw, into the hold, and at times even the cabin of the vessel is filled, the crew living in the forecastle. A vessel thus loaded carries from three to four hundred thousand fish. If the trip is to be sold to the fishermen, the vessel is anchored in the middle of the harbor, and a flag set in the rigging as a signal that bait may be obtained. The fishing-vessels are brought alongside of the "baiter" and the herring are counted out, and quickly transferred by the crews to beds of straw or canvas, where they remain in good condition until such time as they are needed. The price varies from 25 cents to $1 per hundred, the average being a trifle under 50 cents. The fish have an average length of nearly 12 inches. In preparing them for bait, they are first slivered, and the head and tail thrown away, after which the balance is cut so as to make about six baits. A vessel carrying eight dories and fishing with trawls requires from eight to twelve hundred herring for a day's fishing.

After the season for frozen herring is over, the fishermen often find great difficulty in securing bait of any kind. In the spring of 1879 shore-fishing was almost wholly suspended for several weeks on this account. About the 1st of May a small school of herring made their appearance in the locality, and the water was soon filled with nets for their capture, but the supply was so small as to afford relief to only a few of the smaller boats. Later mackerel were purchased from the market-fleet when they were cheap, but the price was generally so high that the fishermen could not afford to use them. Again, from the 10th of May until the 1st of June almost no bait could be found in the locality, and the shore-fishing by the small boats was practically suspended. The larger vessels started out to seek it elsewhere, and were often obliged to go as far as Greenport, Long Island, before a supply could be obtained. In this way two weeks were often spent in getting enough for three or four days' fishing. The offshore fleets were also seriously hindered in their work by the scarcity of bait, and usually spent much more time in search of it than they did on the fishing-grounds.

While the season of 1879 has been an exceptional one, owing to the absence of the menhaden (*Brevoortia tyrannus*) from the Gulf of Maine, yet the question of the bait supply has for years been growing more serious, and the difficulty of obtaining it has been constantly increasing. The expense has also been proportionately increased, until it now seriously reduces the profits of the business.

Professor Baird on learning of this difficulty began a series of experiments to throw light on the subject. For this purpose he caused a refrigerator, with a capacity of fully a ton, to be placed in the laboratory of the United States Fish Commission at Provincetown, Mass., where by the use of salt and ice he easily obtained a temperature of 18° F., and found no difficulty in keeping fish for any desirable period. He now suggests a way out of the bait difficulty by the building of large refrigerators in the principal fishing-towns along the coast; these to be filled when bait is plenty and cheap, and the supply to be kept until such time as it may become scarce. He also suggests the use of small refrigerators in the holds of the fishing-vessels, and thinks that by this means bait can be kept as long as desired and as cheaply as by the present method.

The time has undoubtedly come when this question should receive the serious attention of the fishery capitalists of New England, and it only remains for some one actually engaged in the fisheries, or for some enterprising capitalist, to act upon these suggestions in order to bring the plan into general favor.

6.—DISPOSITION MADE OF THE FISH.

The two principal markets for the shore fishermen of Cape Ann are Boston and Gloucester. The former uses the bulk of the fresh haddock, while the latter buys most of the cod, hake, pollock, and cusk.

In former years it was the custom in all of the fishing-towns along the coast for the fishermen to cure their own catch or to land the fish at the wharf of some shoresman who would "make" them at his leisure, charging from 6 to 8⅓ per cent. of their value for his labor. In either case the fish would not be sold till late in the fall, and it was often nearly spring before the fisherman received any money for his season's work. Being usually a man of small means he had no money to carry himself and family through the season, and he was obliged to arrange with the merchant to supply him with goods until the fish could be caught, cured, and sold. In this way the merchant's bills came to have a value largely dependent upon the abundance and price of fish, and, if the season was a poor one, the accounts were often worthless. To protect himself against such losses the merchant came to charge exorbitant prices for his goods, and mutual dissatisfaction was the result.

Many of the towns still do business in this way, but a few have adopted the cash system. Gloucester was among the first to adopt this method, and in this way drew a large number of fishermen into the town, and greatly increased the size of her fleet. The Cape Ann cod-fishermen now receive their money as soon as the fish are landed and weighed, and thus many of the evils of the credit system are overcome.

In addition to this, there is usually so much competition that a fisherman can secure good prices for his catch, and can sell in any way that he thinks most profitable. Early in the fall, when fish are scarce, he usually sells his fish round, but later in the season he often finds it to his advantage to "gut" or even split them before selling. The average price paid for cod during the winter of 1878–'79 was $1 per hundred pounds for round, $1.25 for gutted, and $2 for split fish.

The method of dressing is often quite interesting to the stranger, as the work is carried on with great rapidity. A single dressing-gang consists of three men, each performing a particular part of the work. After the fish have been weighed and pitched into a tub—usually half of a hogshead—the "header," armed with a sharp and pointed knife, seizes the fish by the mouth with his left hand, and rests its back upon the edge of the tub. He then, with one stroke of the knife, severs the attachment between the gill-covering and the belly, and inserting it in the opening thus made, slits the abdomen to the vent. He then makes a cut on either side of the head at the base of the skull, and while its back still rests on the edge of the tub, and his left hand holds its head, he places his right hand upon the body of the fish, and throws his weight upon it, separating the backbone from the skull and tearing the head from the body, cutting away any flesh that tends to hold them together. The fish is now allowed to fall back into the tub, when the "gutter" seizes it and removes the viscera, transferring the livers to one barrel at his side, and the ovaries to another, allowing the remainder to drop down at his feet or to fall back into the tub. He then throws it upon a table, where the "splitter" places its back against a little strip of wood to keep it from slipping, and holding the fish open with his left hand, takes a splitting-knife in his right and cuts along the left side of the backbone to the base of the tail. The fish now lies open on the table, when with a hard stroke of the knife he severs the backbone near its middle, and catching the end thus freed, lifts it slowly, and following along its side with his knife quickly cuts it from the body, sliding the fish from the table into a tub of water, where it is washed before going to the salt-house. Three men will usually dress from two to four thousand pounds per hour, the quantity varying with the size of the fish. When a large quantity is to be dressed, or when dressing on board a vessel, a double gang of seven men is usually employed, the extra man, called the idler, pitching the fish into the tubs and drawing the water to wash them after they have been split.

Two methods are employed in curing the fish. By the first they are placed in butts, with a quantity of salt, and covered with the strongest pickle. Here they must remain for about two weeks in order to become thoroughly "struck," after which they may be placed on the flakes, when, after one or two days' drying, they are ready for the market, though still quite damp and full of salt. This method is employed only on shore, and such fish are known as pickle-cured fish, being inferior in quality to the kench-cured fish, though they find a ready market in all the inland towns.

In "kenching," the fish are salted in piles, either in the hold of a vessel or on the floor of a fish-house. Each fish is placed back downward, so that, as the salt is dissolved by the moisture from the body, the pickle will pass into the flesh and thoroughly preserve it. When properly salted, fish may be kept in kench for fully a year, though they are seldom allowed to remain more than four or five months, as they are liable to grow strong and musty. To prepare these fish for market, they are first thoroughly washed and scrubbed, and then placed on the flakes, where they are allowed to remain until dry.

A new brand of fish, known as the boneless cod, has been introduced within the last few years, and is meeting with a ready sale. By this method pickle-cured fish of different species are taken to the boning-room, where men and boys are employed in stripping off the skin, cutting out the fins and bones, and cutting the flesh into convenient shape for packing in small boxes for the retail trade.

A large number of observations have been made to ascertain the exact loss in weight of different members of the cod family from the time they leave the water until ready for market. From these it is found that the pickle-cured cod loses from 60 to 66 per cent.; the haddock, 62.3; the pollock, 59.8; the hake, 55.5; and the cusk, 50.5. The additional loss of the cod in boning is 21.9 per cent. The details of the above are given in Tables V to XII, inclusive.

In addition to the market-cured fish, that represents the principal value of the cod, other parts of the fish are often saved. Indeed, but little of either the weight or bulk of the fish is thrown away. When considered separately, any one of these parts has a value seemingly insignificant for the individual; but when taken collectively, they have an importance that cannot be neglected in estimating the money value of the cod-fisheries.

The livers, from which both medicinal and tanners' oils are made, are, next to the cured fish, the most valuable. These are always saved by the fishermen, and bring from 8 to 15 cents per gallon, according to the season when the fish are taken. They are in the best condition from July to September, when a thousand pounds of round fish will furnish four or five gallons, yielding from eight to ten quarts of oil, and are poorest from January to May, when only two and three-fourths gallons, yielding but four or five quarts of oil, can be obtained from a like quantity of fish. The livers are usually boiled in large kettles, and the oil thus freed rises to the surface, when it is dipped off and put into barrels for the market. In the bank-fisheries each vessel is provided with butts, where the livers are kept until the oil has been separated by partial decomposition and the natural heat of the sun. This method is known as sun-trying. Much of the oil from the livers of the shore-fish is used for medicinal purposes, and in the crude state brings about 50 cents per gallon, while the sun-tried oil is sold as tanners' oil, at from 26 to 55 cents per gallon.

The ovaries or eggs of the fish come next in importance. During "war times" these brought from $8 to $12 per barrel, and found a ready sale, even at these prices, for use as bait in the sardine fisheries of France. At that time all the fishermen made a practice of saving them when they could be obtained. Of late, owing to a number of different causes, the price has declined, and during the winter of 1878–'79 the fishermen have received only $1.25 for them in a fresh state, and the price when cured for exportation averaged only $3 per barrel. For this reason most of the offshore fishermen refused to save them, and the quantity landed in Gloucester was a trifle under 1,800 barrels, these being mostly brought in by the shore fishermen. A fair average yield during the winter, or from September to April, is about one barrel to every 4,000 pounds of fish; at other seasons they cannot be obtained.

The sounds or air-bladders of the fish have recently become quite valuable, and are now frequently saved. They are prepared for the market in different ways, depending upon the use for which they are intended. By far the greater part are dried, and sold to factories in the locality, together with sounds of the hake, where they are made into marketable gelatine and used principally in clarifying beer. Some are also put upon the market as cooking gelatine, while others are made into glue. Sounds are also extensively eaten, and are by some considered a great luxury. When put up for this purpose they are pickled and mixed with the tongues of the fish, and in this condition bring from $8 to $12 per barrel.

In the offshore fisheries the sounds usually become the property of the vessel's cook, who cuts them from the bone and prepares them for market during his leisure hours. On shore this work is usually done by men and boys, who cut and scrape them for a certain part—usually one-third to one-half of their value.

From Capt. J. W. Collins, of Gloucester, we have the following facts relating to sounds and the sound trade: One thousand pounds of round cod yield from nine to ten pounds of sounds; these, when scraped, weigh about 6½, and after salting, 5¾ pounds. It requires four pounds of green sounds to make one pound of dried. The price paid during the year 1879 for dried sounds was from 22 to 35 cents, the prices received by the fishermen being as follows:

	Cents.
Grand Bank sounds, pickled, per pound	3½
George's Bank sounds, pickled, per pound	4½
Shore sounds, green, per pound	5

Cod tongues are also saved, and find a ready sale in the retail markets, at from 8 to 15 cents per pound. On the offshore vessels these are often cut from the fish as soon as they are taken from the water, the fishermen keeping account of the number of fish caught in this way, and at night each brings his dish of tongues to the captain, who, after counting and crediting them to the fisherman, empties them into the common pile, when they become a part of the general stock and are sold at from 2 to 5 cents per pound. On shore they are cut out by boys and men, each person having half of all he gets. An average yield is about five pounds to a thousand pounds of round fish.

Again, the skins, bones and fins, that come from the places where boneless fish are prepared, usually bring when delivered at the factories an average price of $1 to $2 per wagon-load. I am told that these make the finest quality of isinglass.

The above are the only products that have a market value as merchandise, but in addition many poor people resort to the dressing-wharves during the winter season and cut out the cheeks of the fish, the process being known to the fishermen as "scalping." These are considered very fine eating, and many people are in this way supplied with excellent food, who would be compelled to go without in case they had to purchase. That which now remains after all these different parts have been saved is usually thrown into the sea, though at times the farmers cart it away to manure their land.

Thus we see that though small for the individual, when considered collectively these minor products have an enormous value for the cod-fisheries of the country. But barring those which, though useful, have no market value, and considering the marketable ones only, we find that in the shore-fisheries, when all of the above-mentioned parts are saved, they represent a value equal to 14½ per cent. of the total value of the fish as it comes from the water.

C.—NATURAL HISTORY OF THE COD.

1.—GEOGRAPHICAL DISTRIBUTION.

The cod-fish has perhaps a wider range than any other of our important food-fishes. On the east coast of America it is found from the polar regions on the north to Cape Hatteras on the south, occurring in vast numbers in the vicinity of Labrador and Newfoundland and on the noted fishing-banks lying to the south and west. On the west coast a closely-related species occurs, and the cod-fisheries of the Pacific are being rapidly developed. In the Old World the cod has a wide geographical range, and is very abundant in some localities, the cod-fisheries of Norway being among the most extensive in existence.

It is not our purpose, however, to go into any general discussion of the natural history of the cod, but merely to treat of the subject with special reference to the species as found in the waters of Northern Massachusetts, where it occurs in greater or less numbers, from the shore line to a depth of 90 or 100 fathoms, on all of the rocky spots and ledges, during the entire year. It is also frequently found on sand and clay, but seldom, if ever, remains on muddy bottoms. Cod are most plenty in this locality from November to June, when they visit the shore for the purpose of spawning, during which time they usually remain in from 15 to 40 fathoms of water.

2.—CHARACTERISTICS OF THE COD.

Cod-fish are gregarious in their habits, going in schools of greater or less size, and are governed in their movements by the presence or absence of food, the spawning instinct, and the temperature of the water. When migrating, the schools are quite dense, though by no means like schools of menhaden or mackerel. But when they reach the feeding ground they seem to distribute themselves over a large area, though more or less grouped together in little bunches. This is particularly noticeable on the shore, when the fish are moving about in search of food, and the fisherman soon catches up all that chance to be on one patch of rocks, and must then row to another in order to find a new supply. The same thing is seen on western banks, where a vessel usually carries dories to distribute her crew over different parts of the ground, and often, by setting her trawls in one locality for a day or two, seems to catch up all of the fish, and must then "shift her berth." Fishermen also cite many instances where the fishing is excellent on a few, particular, well-defined spots on different parts of the ground, while almost no fish can be taken in other places.

During the spawning season this tendency to become scattered is less noticeable, for the instincts of the fish seem to bring them nearer together, and great numbers are often taken in one particular locality. Even here, however, the tendency to separate into groups occurs, for some boats find good fishing while others, but a few rods away, catch almost nothing; and in trawling, some parts of the line have a fish on nearly every hook, while other parts take only a scattering one.

In schooling, both sexes are always found together, whether it be on the spawning or feeding ground or on the journey; but the relative numbers of each seem to vary greatly, and we have been able to discover no invariable rule whereby one can predict with certainty the sex that will first appear, or that which will be most abundant at any given time during the season. The fishermen have a commonly accepted tradition that in the spawning schools the females always come first and the males later, but this theory is not supported by facts. Observations were frequently made on the relative numbers of the two sexes landed by the shore-fishermen between September, 1878, and July, 1879. The results showed that during the early fall, or before the school-fish had made their appearance, the fish were nearly equally divided between males and females—first the one and then the other being more abundant. When the school-fish first reached the shore early in November the males were a trifle more plenty than the females for about a week, but from that date until they left the grounds the females were taken in greater numbers, sometimes in the proportion of two to one, and at others in nearly equal quantities. In the Ipswich Bay school during the first two or three days in February there were ten males to one female; by the middle of the month the females composed about 40 per cent. of the catch, and from this date until the 1st of June the males numbered two to one. From reliable fishermen we learned that the same was true of the fish on the offshore banks, and that, though varying greatly in their relative numbers, both males and females were always present.

There is usually a great difference in the size of the individuals taken by the fishermen on the shore feeding-grounds in a single day, for the young and "ground-tenders" remain on these rocky ledges during the entire year, and late in the season the school-fish come in upon the same grounds and are naturally taken with them. But when the school-fish visit a locality not frequented by the young, as they do in Ipswich Bay, there is a noticeable absence of immature fish, and the catch is composed almost wholly of individuals of large size. Thus, in the winter of 1878–'79 many trips of from twenty-five to forty thousand pounds were landed with scarcely a small fish among them, while vessels fishing only a few miles distant found young fish plenty, and there were occasional instances where such vessels caught only small ones. Again, though the school-fish may differ considerably in size, we have not found one, thought to belong to their number, that had not reached maturity. Indications strongly favor the idea that the young remain separate from the school-fish during the first few years of their lives, and we are led to believe that, though they are often taken together, the occurrence is accidental and the young will not follow the old in their migrations until they reach maturity, though after this point is reached they seem to mingle freely without regard to age.

The cod-fish sometimes make long journeys from one bank to another, and, indeed, from one region to a very distant one. It is, of course, nearly impossible to trace their movements at such times, and one can usually only guess at the place from whence they come or the distance traveled.

During the winter of 1877–'78 an unusually large school visited the coast of the United States. At this time cod were more plenty along the shores of New England than for many years. Among the fish captured at Cape Ann and other points were quite a number with peculiar hooks fastened in their mouths. These hooks gave a clew to the movements of the fish, for they differed from any in use by the American fishermen, and proved identical with those used by French trawl-fishermen on the Grand Banks, and indicated that the fish must at some time have been in that locality, as the hooks probably came from no other place. If the above be granted as proven, the fish must have traveled a distance of five to eight hundred miles at least, and, as a portion of the school continued well to the southward, some individuals must have journeyed much farther. Most of the schools that visit the shore have no such tag or mark whereby their former locality may be learned. They are thought to come directly in from the deep water and to depart by the same route, but where they spend the summer months is not known.

Cod-fish are probably governed in their movements by the abundance and migrations of food, the spawning instinct, and the temperature of the water, though the last named seems to exert but little influence. It is generally acknowledged by the fishermen that during the feeding season fish are plenty only where food exists in considerable quantity, and that after "cleaning up" one part of the bank they go to another. They also follow schools of bait for long distances, living upon them until they are broken up or entirely destroyed. Thus they often follow the capelin (*Mallotus villosus*) into the shoal water, and even drive immense numbers of them upon the shore.

The spawning instinct seems to exert a decided influence upon the movements of the fish, for we find them visiting the same locality year after year during the spawning season, often remaining for several months at a time. The fish that visit the waters of Cape Ann during the winter, doubtless come in for the purpose of spawning rather than for food. This seems clear from the fact that they do not arrive when bait is most plenty, nor do they follow any species to the shore. On the contrary, the pasture-school usually arrives about three weeks after the large herring have left the coast, and remains on the south side of Cape Ann, while sperling are abundant in Ipswich Bay. The Ipswich school is also the largest after the sperling have been driven away by the cold weather, and remains on the sand-flats, which supply almost no food. From these facts we are led to believe that food has little influence upon the movements of the fish during the spawning season.

The instinct that leads the spawning fish to seek the shoal water in such great numbers is certainly a wise one, for they generally select spawning-grounds where the tide runs strong and the water is rough, and the large number of individuals is absolutely necessary, that the water may be filled with germs for their successful impregnation. If, instead of schooling in such numbers during this period, they remained scattered over a large area, almost no eggs would be fertilized.

Again, while food is not essential to the spawning fish, it is of vital importance to the young, and it seems a wise provision that these should be brought into being where food is abundant, rather than that they should be hatched in mid-ocean, where almost no suitable food exists.

Cod-fish live at a depth varying from a few feet to over 100 fathoms. They have occasionally been seen schooling or feeding at the surface on the fishing-banks and on the coast of Labrador. In February, 1879, there was good fishing in three fathoms of water within a few rods of the shore in Ipswich Bay; while in May of the same year large numbers were taken in 110 fathoms from "the channel" near Clark's Bank. They seem to prefer a depth of less than 70, and by far the greater numbers are caught in from 18 to 40 fathoms.

In moving from one bank to another, where the intervening depth is much greater, it seems probable that, instead of following the bottom, they swim in a horizontal plane, following a stratum of nearly uniform density and temperature. The fishermen of Cape Ann have often caught them with 70 to 80 fathoms of line, between Brown's and George's Banks, where the sounding-line indicated a much greater depth. The finding of pebbles and small stones in their stomachs is not an uncommon occurrence. The fishermen regard these as an unfailing sign that the fish have either just arrived or are about to leave the bank. These stones may play no small part in adjusting the specific gravity of the fish to that of the stratum of water in which they are to move.

There seems to be a tendency for the large fish to remain in deeper water or nearer the bottom than the small, and usually beyond a certain depth; the deeper one fishes the larger the fish. Formerly, in hand-lining from deck on the banks, the vessels often anchored in 80 or even 90 fathoms, and the catch averaged over two-thirds large; but in hand-lining from dories they seldom fish in over 50 and usually less than 35 fathoms, as they find it difficult to handle so much line, and the catch runs about two-thirds small. The same is true in fishing at different depths at the same time and in the same place. Thus, of two men fishing side by side from the deck of a vessel, the one with his hook on the bottom will catch much larger fish than the other who lets his line but part way down. Larger fish are also taken on the trawl than on the hand-line, for the former lies constantly on the bottom, while the latter may be raised to any distance above it.

The size of the species varies greatly with the different individuals. The boat fisherman visiting the rocks and ledges along the shore in summer catches fish weighing from 2 to 60 pounds, the average being a trifle under 9 pounds. The school-fish run larger, those on the south side of the cape, in the fall of 1878, averaging about 12, and those in Ipswich Bay, later in the season, fully 20¾ pounds. Probably the latter were the largest as a school that have ever visited the shore. On George's Bank, where the largest cod-fish are taken, trips are sometimes landed where the average weight would be fully 40 pounds round, but such cases are exceptional. The largest specimen that we have seen was taken by the schooner Northern Eagle, Capt. George H. Martin, in Ipswich Bay, March 10, 1879, and is now in the collections of the National Museum. It measured 5 feet and 2 inches, and weighed 99½ pounds when landed, probably weighing fully 105 pounds when taken from the water.

We have also authentic record of a specimen captured off Cape Cod in February, 1878, that weighed 107 pounds after being eviscerated, which is equivalent to over 125 pounds round. Other instances have been recorded in the local papers in the vicinity of Cape Ann, from time to time, where cod of unusually large size have been taken; and the Cape Ann Advertiser has very recently noted the capture of one weighing 180 pounds by a vessel belonging to Newburyport, Mass. Whether this was the actual weight or an estimate we have not learned.

Of the many specimens weighed and examined during our stay in Gloucester the average weight of the females exceeded that of the males by nearly $2\frac{1}{2}$ pounds—a difference only partially accounted for by the presence of the eggs. Whether this would be true for other localities or for other years is not known.

The general form of the different individuals varies but slightly, though there is considerable difference in their relative proportions. No external character has been noticed whereby the sexes can be distinguished, and even the most trained observer cannot separate them when green until they have been opened. The difference in the relative proportions is considerable, and the length is not a reliable indication of weight. Some fish are short and thick while others are long and slender. Table No. IV gives the measurements and weights of a number of fish of different sizes, and shows fully the extent of this variation.

The difference in the shape of the shore and school fish seems largely the result of food and habits. The school-fish, moving about in pursuit of food, becomes thick and plump, so that the head appears small in proportion to the body, while the shore-fish, subsisting on such food as can be found on the rocks, grows thin and gaunt, giving its head a larger relative size.

There is a remarkable variation in the color of different individuals of the species. This is doubtless due to surrounding circumstances, or to the character of the bottom on which they live, and to the age of the fish. The young when first hatched are nearly colorless, with the exception of a few dark star-shaped pigment cells, most noticeable in the eyes, and sparingly scattered over the whole surface of the body. These increase rapidly in number during the first few days on certain portions of the body, giving the little fish a very peculiar banded appearance. At the age of six months they are still quite transparent, and the upper parts are well covered with minute black dots, more prominent along either side of the dorsal fins, and gradually shading off into lighter underneath, with the belly nearly white. At this time the fish have a peculiar golden tinge, deepest along the back and sides. Traces of the dark bands still remain, but these are more noticeable on account of the intervening lighter spaces, that seem to extend irregularly downward and backward, giving the fish a blotched or mottled appearance. Gradually the young fish living along the shore come to resemble more nearly the adult in the relative size and distribution of the spots, until at the age of twelve to eighteen months there is a marked similarity between them. But the young continue to live among the rocks and ledges covered with algæ, and soon begin to show a reddish tinge; this increasing and varying with the individual, often giving the fish a deep red color. These red fish are known to the fishermen as rock-cod, from the bottom on which they are taken. They are usually small, having a weight of only a few pounds, but some retain the color until they are of large size, and one specimen was seen during the summer of 1878 weighing 46 pounds. The ordinary shore-fish or ground-tender is quite dark, with the belly of a dirty ash, and the spots usually large and indistinct on a dark background. The regular school-fish, on the contrary, is very light, with smaller and more distinct spots on a lighter background, and has the belly nearly white. Specimens of cod have also been seen in which the whole upper surface of the body was of a uniform straw or lemon color, gradually shading into lighter underneath. A fine specimen of the above was secured in the summer of 1878, and is now in the National Museum.

3.—FOOD OF THE COD.

A list of the stomach contents of the cod would be of little value, except in throwing light on the food that the fish seem to prefer, by showing the relative quantities of the different kinds. A full list, including everything that has been found in the species, would be very long, and embrace nearly everything, whether organic or inorganic, that chanced to come in its way. Any bright or curious object often attracts its attention, and is very likely to be swallowed by it. Thus knives, nippers, and even vegetables lost or thrown from the vessel are frequently found in the stomachs of the fish when they are being dressed. Stones, too, are not uncommon at times, and over a pound has been taken from a single fish. The list of fishes, articulates, and mollusks seems only limited by the size of the individuals or their ability to escape. But while such a variety of food is found in the cod, its principal food is limited to a few species of fish and a small number of mollusks. Among the former the more important are the herring (*Clupea harengus*), capelin (*Mallotus villosus*), lant (*Ammodytes americanus*), and a few others. It often follows these fish in their migrations, feeding upon and destroying great numbers of them, and at times shows great dexterity in their capture. I am told that in the spring of 1879 an immense school of herring made their appearance on and moved slowly across George's Bank, and that with them came the largest school of cod that has been seen in that locality for a long time. The cod remained constantly among the herring, so that, when the latter had passed the fishing fleet, the vessels were obliged to weigh anchor and follow them in order to secure the cod. The cod also drive the capelin into the shoal water, and even upon the shores of Newfoundland and Labrador, in immense numbers, and, when they have reached the shallow bays, fishermen report the water as fairly white from their splashings in their active and eager pursuit of their prey. Among mollusks, squid (*Ommastrephes illecebrosa*) and the common bank-clams are their principal food, the former being preferred to any other species, and the latter often occurring in such quantities in the stomachs of the fish that the French fishermen on Grand Banks frequently catch a large part of their trip on bait secured in this way.

During the spawning season the cod-fish cease to search for food, and give less attention to feeding than at other times, though they will usually take the bait when placed before them. That they do not search for food is shown by the fact that the pasture-school remained within a few miles of a large school of sperling without being drawn after them; and that the Ipswich Bay school was largest after the sperling had left the coast, and remained for a number of months on sandy wastes which supported only three species of invertebrates, *Buccinum undatum*, *Fusus* sp., and *Asterias vulgaris*, in any considerable abundance. The examination of the stomachs of several hundred individuals showed four-fifths of all to be entirely empty, while a greater part of the remainder contained only bait picked from the trawls of the fishermen. A small number contained fish of one or more species that had probably been captured in the locality, while a few scattering invertebrates were found. Of the species mentioned as abundant on the grounds, not a star-fish and but two shells of one species and one of the other were found. But it was clearly shown that the fish would not refuse food, for often the stomachs were well filled with bait picked from the trawl before the fish were hooked. From 10 to 15 pieces were frequently found, and in one case 18 were counted.

The females when fully ripe seemed less willing to feed than at other times, and few were caught with the moving hand-lines; but when the trawl was used, thus leaving the bait motionless on the bottom for hours at a time, they were induced to bite, and many were taken with the eggs running from them. Ripe males seemed to bite readily at any time.

The young fish, as has been remarked, seems to spend the first three or four years of its life in shoal water, among the rocks and algæ. Here its food consists at first of the minutest forms, and later principally of small crustacea, though it often picks up mollusks and worms, and even enters the harbors in summer, where it remains about the wharves, picking up bits of refuse thrown from the fish-houses. The young fish were so plenty in Gloucester Harbor during the summer of 1879 that boys often caught 25 or 30 of them in an hour with hook and line.

4.—ENEMIES OF THE COD.

The cod-fish seems to have few enemies. Among fishes its principal enemy is the dog-fish (*Squalus acanthias*). These fish make their appearance in large schools on the shores of Northern Massachusetts early in May, where they remain until September, moving about from one locality to another, and driving everything before them. They are probably the most pugnacious of any species in the waters of New England, and cod, as well as other fish, are often brought to market bearing marks of their sharp teeth and horny spines. The arrival of a school of dog fish in any locality is the signal for all other species to leave; and in this way the work of the fisherman is often suddenly terminated.

Halibut (*Hippoglossus vulgaris*) are also regarded by the fishermen as enemies of the cod, and many cases are cited where, in former years, they drove them from the fishing banks. In fact, thirty or forty years ago, when the halibut were very abundant in Massachusetts Bay and in the waters about Cape Ann, but had no market value, they interfered greatly with the work of the fishermen, and often good cod-fishing was spoiled by their sudden appearance. At the present time halibut occur in much smaller numbers on these grounds, and no such difficulty is noticeable. Indeed, it is found that cod occur in greater or less numbers with the halibut on the outer banks, where they seem to live peaceably together; and we are led to believe that it was the abundance of the halibut in former times, when they literally covered the ground, rather than any hostilities beween the species, that drove the cod from the banks.

Just how the large cod is affected by the presence of the pollock (*Pollachius carbonarius*) we are unable to say, but the young living near the shore finds in them its most deadly enemy. Young pollock are exceedingly abundant all along the shore during a greater part of the year, often moving in large schools as well as singly, and frequently many barrels are taken in a single day in each of the many traps along the coast. They are especially abundant in the waters off Cape Ann, and being exceedingly voracious, attack and devour almost any small fish that comes in their way. We have often watched their movements in the clear water of Gloucester Harbor, and noted the sudden dispersion of a school of several hundred young cod of six months' growth at the approach of a single pollock seven or eight inches in length.

These little fish show great fear of them, and usually remain near the long kelps and sea-weeds that are growing on the piling of the wharves, and at once dart in among these for protection at the first approach of the pollock, reappearing very cautiously only after the lapse of several minutes. At times a pollock succeeds in approaching unnoticed, when it suddenly darts into the midst of the school and seizes one of the little fish as its prey. Even when of equal size the cod exhibit the same fear, and on putting several of each about ten inches in length into a large tank of water, the cod sought refuge beneath some strips of board that were stretched across one corner, while the pollock swam about freely in the water. On being driven from their hiding place they soon returned, and it was not until the pollock were taken out that they would freely venture from their hiding place.

5.—REPRODUCTION.

Evidence is not wanting to show that cod spawn every year, and that they deposit the entire number of eggs in the ovaries each season. We have examined hundreds of specimens and have failed to find a single instance where the condition of the ovaries did not clearly indicate, to our minds at least, that such was the case. During the first of the season no mature fish were found in which eggs were not present, though they often varied greatly in development from very small to nearly ripe. Again, later in the season, no spent fish were seen with any eggs remaining in the ovaries; and no fish were found during the spawning period in which the condition of the ovaries did not indicate that the eggs were gradually maturing, and would be deposited before the close of the season.

The eggs contained in the ovaries are separated into little irregular conical clusters, each connected with the general mass by a slender thread that expands into a delicate membrane containing minute and diffusely branched blood-vessels. This membrane incloses each of the eggs, and the blood-vessels supply the nutrition so necessary to their future growth and development. As the eggs mature they gradually increase in size, until, when ripe, they become detached from the membrane, and pass down through secondary channels into one main channel leading to the genital opening of the female.

The first ripe female seen during the season of 1878–'79 was found in a lot of fish landed from the shore-fish or ground-tenders September 2. The eggs were noticed to be running from this fish as it lay upon the floor of the fish-house. On opening it was found that it had just begun spawning, for a few eggs only, perhaps five per cent. of the entire number, were transparent, and a small number of these had separated from the membrane and fallen into the channels leading to the genital opening, while the great bulk were far less mature and represented almost every stage of development from green to ripe.

From this date ripe fish, both males and females, were occasionally taken, though they did not become abundant until the middle of October. Early in November, when the school-fish made their appearance on the south side of Cape Ann, the individuals varied greatly in their spawning condition; some were quite ripe and had already thrown a portion of their eggs, while others were so green as to indicate that they would not spawn for several months at least, though in nearly all the eggs had begun to enlarge. By the 1st of December fully 50 per cent. of the catch had commenced spawning, but when driven away, probably by the unusually heavy storms, in January, a few were not quite ripe, and the majority had not thrown all their eggs.

About the 1st of February the fish in Ipswich Bay were found to average fully 90 per cent. males, with the spermaries mostly well developed. At this time there was a great variation in the ovaries of the females; of these not more than one in ten had spawned, while fully 60 per cent. were still green. By the middle of the month the females numbered about 40 per cent., though over half had not commenced to spawn. On March 13, 300 fish from this school were opened, with the following results: 14 per cent. were spent males; 53 per cent. were ripe males; 6 per cent. were spent females; 14 per cent. were females in various stages of spawning, and 11 per cent. were green females. May 10, fully half of the females had not finished spawning, and an occasional green one was noticed. Even in June, when the fish left the coast, a very few, though ripe, had not finished throwing their eggs.

The results of the above observation prove not only interesting, but surprising, for we find the cod-fish spawning during nine consecutive months in the same locality, a period far exceeding that required by any other species of which we have any knowledge.

This fact can be more easily understood when we remember that the individuals do not deposit all their eggs in a single day or week, but probably continue the operation of spawning over fully two months.

That this is true there can be little doubt, for when the females first begin to throw their eggs only a very small percentage of the whole number are ripe, while the balance show every gradation to the perfectly green and immature. By frequent examination of individuals in more advanced stages, it is found that the eggs gradually continue to increase in size as they mature, and that as fast as they become detached from the membrane they pass down through the channels to the opening, and are excluded from the body, either by the will of the parent or by internal pressure caused by the increasing size of the eggs, to make room for others. It would be impossible for a fish to retain all or even a small part of its eggs in the roe-bags until the last had matured, for the increase during the development is very great, and the eggs would come to have a bulk greater than the entire stomach cavity of the fish. The products of the ovaries of a 75-pound fish, after impregnation, would weigh about 45 pounds and measure nearly 7 gallons, equal to over half of either the weight or bulk of the fish.

Another proof that the cod-fish deposits its eggs gradually during a long period is seen in the fact that few can be taken from the fish at any one time. In "stripping the fish," at the hatchery in Gloucester, it was found that only one quart, or less than 400,000 eggs, could be taken from a 21-pound fish in a single day. Allowing the ovaries of this fish to contain 2,700,000 eggs, and the time of spawning to be two months, the fish must deposit in the natural way 337,500, or nearly a quart, each week.

But by the artificial method, where strong external pressure is applied, many more eggs are probably secured at once than would be naturally thrown by the fish. Thus the fish must either gradually deposit more or less eggs each day, during the entire spawning season, or it must deposit at intervals separated by only a day or two at most.

The schools move about but little during the spawning season, except when driven away by enemies or by violent storms. After they reach the waters of Cape Ann, fishing continues best in the same localities and even upon the same spots until they leave. The individuals, too, seem to move about but little among themselves. When the female becomes ripe she remains quietly near the bottom, while the male, a little more active, often swims higher up. This is indicated by the fact that much greater numbers of spawning females are taken with the trawl lying directly on the bottom than with the hand-line a little way above it, while the males are taken on one as readily as on the other.

It may not be impossible that the eggs are fertilized while floating about in the water some minutes after exclusion, and that the strong tides usually found on the spawning-grounds play an important part in distributing the germs, thus making the chances of impregnation more favorable. Indeed it may be possible, and, if the spawning goes on gradually for several months, seems not improbable, that the immediate presence of the opposite sexes during the act of spawning is not necessary, but rather that the eggs are fertilized mainly by accidental contact. Observations would seem to strengthen the probabilities of this theory; for, if the fish went in pairs, they would often be taken on adjoining hooks of the trawl, or one on either hook of the hand-line. Such is not usually the case, however, but on the contrary several of the same sex are more frequently taken together.

The eggs have a specific gravity of 1.020 to 1.025, as indicated by the fact that they float in salt water and sink rapidly in fresh. The oldest fishermen had not the slightest knowledge of this fact, but held to the idea that the females deposited their eggs on the rocks, where they were visited and impregnated by the males, and left to become the food of the various animals so abundant in such localities. They had at times noticed the little transparent globular bodies in the water, but it had never occurred to them that they were the eggs of any fish. They may be found at the surface in common with eggs of the pollock, haddock, and probably other species of the cod family, when the sea is smooth; but when the water becomes rough they are carried to a depth of several fathoms by the current, though the tendency is to remain near the surface.

There are many ways in which these eggs may be destroyed. The principal loss is probably the result of non-impregnation, for unless they come in contact with the milt of the male very soon after being thrown from the parent they lose their vitality. Again, being subject to the winds and tides, they are often carried long distances from the spawning-grounds into the little bays and coves, and are driven upon the shores in immense numbers, or left dry by the tides, where they soon die from exposure to the atmosphere, or during the cold winter weather are instantly destroyed by freezing. Ipswich Bay, the most extensive spawning-ground in the locality, is especially unfortunate in this particular, for the heavy storms from the north and east tend to drive them upon the shore, and each breaker as it rolls in upon the beach must carry with it many millions of eggs.

But such impregnated eggs as escape destruction upon the shores are subjected to the ravages of the myriads of hungry animals living about the rocks and coves, and many are consumed. One day in January we introduced a jelly-fish or medusid, having a diameter of but 1½ inches, into a tray of eggs in the hatching-room, and in less than five minutes it had fastened 70 eggs to its tentacles, often loading them so heavily that they were severed from the body by the weight or resistance of the eggs as they were dragged through the water.

By the aid of a microscope, numbers of *vorticelli* were found upon them, in one case 46 being counted on a single egg; and in addition a peculiar formation, thought to be minute algæ, was often noticed. Just what influence these latter would exert, or whether they would occur in the clear water outside the harbor, is not known. Thus, owing to the many different circumstances that tend to destroy the eggs, probably but a very small number out of a million are successfully hatched, and of the young fish but few reach maturity.

To overcome these difficulties nature has made the cod one of the most prolific of the ocean fishes, and we find not only thousands but millions of eggs in a single female. All members of this family contain large numbers of eggs, but the cod-fish is the most prolific of all.

The exact number varies greatly with the individual, being dependent largely upon its size and age. To ascertain the number for the different sizes, a series of six fish was taken representing various stages of growth from 21 to 75 pounds, and the eggs were estimated. Care was exercised that all should be green, so that no eggs should have been thrown, and that they might be of nearly equal size. The ovaries were taken from the fish and accurately weighed; after which small quantities were taken from different parts of each and weighed on delicately-adjusted scales, and these carefully counted. With this data it was easy to ascertain approximately the number for each fish.

The results obtained are given in Table No. I, appended to this article, showing a 21-pound fish to have 2,700,000, and a 75-pound one, 9,100,000. The largest number of eggs found in the pollock was 4,029,200, and in the haddock 1,840,000. These facts are given in detail in Tables II and III.

When the eggs are first seen in the fish they are so small as to be hardly distinguishable, but they continue to increase in size until maturity, and, after impregnation, have a diameter, depending upon the size of the parent, varying from one-nineteenth to one-seventeenth of an inch. A 5 to 8 pound fish has eggs of the smaller size, while a 25-pound one has them between an eighteenth and a seventeenth.

From weighing and measuring known quantities it is found that one pound avoirdupois will contain about 190,000 of the smaller size, or that 1,000,000 eggs well drained will weigh about 5 pounds. Again, by assuming one-nineteenth of an inch as the standard, or by precipitating a known quantity in chromic acid and measuring, we find one quart, or 57¾ cubic inches, to contain a little less than 400,000, or that 1,000,000 will measure between 2½ and 3 quarts.

With these facts in mind, it will be an easy matter to estimate the quantity of eggs taken for hatching purposes during any given season.

When the little fish first break through the shell of the egg that confines them the fetal curve or crook is still quite noticeable, but they soon straighten, and are then about five-sixteenths of an inch in length. At this time the yolk-sack, situated well forward, is quite large, but so transparent as to escape the notice of the ordinary observer. This is gradually absorbed, disappearing wholly in about ten to fifteen days, and the little fish begins to move about with a peculiar serpentine motion, at times darting quite rapidly, and then remaining motionless, as if resting from its exertions. It now begins its independent existence, and moves about more frequently, apparently in search of food. From this date it is impossible to follow them, for none have been confined, and it is only by catching large numbers at different seasons and carefully recording their weights and measurements that one is enabled to judge of their growth. The habits of the species, that cause them to live near the shore for the first few years, furnish excellent opportunities for such observations, and many were examined during our stay at Cape Ann.

At the outset the problem becomes difficult, in that the spawning period, instead of being limited to a few weeks, as is the case with most species, extends over fully three-fourths of the year, and the difficulty is greatly increased by special causes that affect the rate of growth of individuals hatched at the same time.

The results were what might be expected; for a table of measurements, made late in June, gave an almost continuous series, with only one or two breaks, that could with certainty be taken to represent the non-spawning period of the fish. But though the gaps were so completely closed by the extremes in variation, which seemed to cause even an overlapping, making the last hatched of one season smaller than the first hatched of the next succeeding, yet there was a tendency for the greater number of individuals to be thrown into groups at intervals in the series, these seeming to represent the height of the spawning season for the different years. The break was distinct between the smallest and those of a year earlier, so that, taking the height of the spawning season on the south side of the cape to be December, the large number of young fry ranging from 1½ to 3 inches must have been hatched the previous winter, and were consequently about six months old. The large number of individuals having a length of 9 to 13 inches indicated the normal growth of those hatched a year earlier, or fish of eighteen months to be 10 to 11 inches, and their weight 7 to 8 ounces. The next group, or the fish thought to be thirty months old, measured from 17 to 18 inches, with an average weight of 2 to 2¼ pounds. The fish now begin to increase more in weight than in length, soon appearing in the markets as "scrod," and by the following summer measure about 22 inches and weigh from 4 to 5 pounds.

Beyond this period nothing can be determined, for the variation, constantly growing greater, now gives every size and weight, with no indication of breaks in the series.

But enough has been learned, if the above be correct, to show that the male reaches maturity at three and the female at four years; for the smallest ripe male noticed during the season of 1878–'79 weighed 3½ and the smallest ripe female 5 pounds.

D.—HATCHING OPERATIONS.

1.—OBJECTS OF THE WORK.

Fish-culture, in its crudest forms, was first employed by the Romans and Chinese many hundred years ago; but the fish-culture of the present day, by which such excellent results are being obtained, is a science of recent growth, and it is only within the past few years that it has assumed a thoroughly practical aspect. Its present condition is the result of a continued series of experiments that have given a degree of success far beyond what its most enthusiastic workers had dared to expect.

The present shad-hatching apparatus, that seems so near perfection, is an excellent example; for in this case, though the progress has been rapid, the crude apparatus of a few years ago has been replaced by the new only after the most careful experiments with the eggs of the species. So with other fresh-water and anadromous species; the improved apparatus for successfully hatching them is the result of many experiments and observations.

But, while the above species have been the subjects of careful study, the important marine food-fishes, such as the cod, halibut, and sea-herring, have remained unnoticed. The great importance of these fisheries has led Professor Baird to consider carefully the question of the artificial propagation of several of the principal species, and, after studying the habits and food of the fish for some time, he decided to inaugurate a series of experiments to ascertain what could be accomplished in this direction.

Accordingly, a hatching station was established at Gloucester in the fall of 1878 for the purpose of experimenting with the eggs of the cod, in order to learn how and in what numbers they might be obtained, the kind of apparatus necessary for successfully hatching them, and to what extent artificial propagation might be made practicable. The chief aim was then to study experimentally the whole subject of hatching in its relations to the cod-fish and its eggs, to pave the way for future work, rather than to go into any extensive work for the immediate propagation of the species.

2.—PREPARATIONS FOR HATCHING.

The late James W. Milner, deputy commissioner, arrived early in the fall to take charge of the experiments, and Mr. Frank N. Clark, a professional fish-culturist in the employ of the Commission, came soon after to personally superintend the work in the hatching-room. Mr. Milner remained long enough to see the preliminary apparatus and machinery placed in position and the first eggs taken, when he was obliged to return to Washington on account of his serious sickness. A little later Mr. Clark was called away to look after the interests of the Commission in another State, and Capt. H. C. Chester superintended the work in the hatchery during the remainder of the season.

The building occupied by the Commission during the summer as a scientific station was considered suitable, and was retained for the work. It is situated at a prominent point, on the southwest side of the harbor, on a substantial wharf, with 4 to 6 feet of water at mean low tide. The outer end of the building was set off as a hatching-room, the remainder of the lower part being used as a store-house, while the upper part answered the purpose of an office and laboratory.

A 4-inch pipe was laid from the hatching-room to a point in the harbor at the end of the wharf, and sunk below low-water mark. The outer end of this pipe was fastened to the piling of the wharf, and incased in a box with wire-cloth openings to keep out the animal life of the harbor. The inner end communicated with two 300-gallon tanks, placed in an elevated position in the center of the room, to be used as reservoirs for the salt water. These reservoirs were tapped from beneath by smaller pipes that extended along the walls of the building, at a height of 4 or 5 feet, with faucets at short intervals, from which the water was supplied to the eggs by means of rubber tubing. In one end of the room was an 8-horse-power steam-engine, for working the pump that brought the water from the harbor to the reservoirs in a constant stream, the quantity being regulated by the outflow.

It was of course unknown what hatching-apparatus could be successfully used, as no eggs of the cod had ever been artificially hatched; and indeed it was not then quite clear to the minds of those in charge to which of the three classes, sinking, floating, or adhesive, the eggs of the cod belonged. Cones similar to those used in shad-hatching (figured in the Report of the Commissioner of Fish and Fisheries for 1873–'74 and 1874–'75, p. 376) were selected as likely to give the best results, put up along the sides of the hatching-room, and connected with the faucets by the rubber tubing.

The above constituted the original apparatus of the hatchery, and when it had been properly arranged Mr. Milner turned his attention to the methods for securing the supply of fish and eggs. For this purpose he selected a 5-ton schooner and a 14-foot open market-boat for visiting the fishing-grounds, and, in addition, a small well-boat in which the fish could be brought alive to the hatchery. A little later it was found desirable to build large live-boxes for confining the green fish until they should ripen.

These live-boxes were 12 to 14 feet long, 6 feet deep, and 5 feet wide, made of pine lumber, with 2-inch spaces between the narrow boards, to admit a fresh supply of water to the fish. When finished they were anchored in the harbor beside the wharf, where they remained throughout the season.

3.—MANNER OF PROCURING EGGS.

The supply of eggs was obtained in several different ways during the winter.

The first method employed was to send men to the fish-wharves daily to examine the fish landed and to take the eggs from any ripe females that might be found. This practice was soon given up, as the fish had usually been dead some hours when they were landed, owing to the distance of the fishing-grounds from the harbor, and the eggs had so nearly lost their vitality that they could not be impregnated. Only an occasional lot of fish were found whose eggs could be saved, and few good ones were obtained in this way.

A second method, by which the men went daily in the schooner to the fishing-grounds to take eggs from such ripe fish as they might catch, was pushed vigorously at first. In this case hand-lines were used, as the bottom was too rocky for trawling, and the catch was composed largely of green fish, so that few eggs were obtained.

A third method, which was merely a repetition of the second on a larger scale, was more successful in that more fish were taken, and consequently more ripe ones found. The plan was to utilize the catch of the fishermen by putting spawn-takers on several of the regular fishing-schooners to examine each fish as it came from the water, or as soon as it was brought to the vessel, and to bring the ripe eggs to the hatchery in pans taken out for the purpose. This method was followed during a greater part of the time, and some good eggs were obtained in this way. But here as in the former case hand-lines were used, and spawning fish were not taken in very large numbers. A visit to the fishing-grounds, where trawls were used, later in the season, fully convinced us that as many eggs could be obtained in this way as might be needed, for on a four days' trip to Ipswich Bay, in February, many millions might easily have been secured.

The finding of so many green fish led to the building of the live-boxes at the hatchery, and when these were ready the schooner visited the fishing-grounds daily and brought her catch alive to harbor in the little well-boat, transferring the fish at once to the live-boxes, where they were to remain until they should ripen. These live-cars proved a great success, for the fish kept well and ripened rapidly. In this way many live fish were kept convenient to the hatchery, where they could be carefully watched, and the eggs secured as soon as they had ripened. This method entirely overcame the difficulties of bringing the eggs long distances and of properly caring for them until they could be transferred to the hatching apparatus, and the live-cars soon came to furnish nearly all the eggs.

4.—HATCHING OPERATIONS.

Two spawn-takers visited the live-cars at intervals of one to three days, one taking out the fish with a dip-net, while the other examined them carefully, by pressing gently on the abdomen, to see if they were ripe. If green, they were transferred to an empty live-box floating beside the other; but when a ripe fish was found it was confined in a dip-net and returned to the water until one of the opposite sex could be secured. Thus all the fish were examined regularly every second or third day, and when ripe ones were found they were carried to the hatchery, where the eggs were taken and impregnated.

In "stripping" the fish the spawn-taker usually held its head firmly in his left hand, with its back against his body and its left side uppermost, and, owing to its size and strength, a second party generally held the tail and helped to keep the fish in position, while the spawn-taker, with his right hand, gently pressed the eggs or milt from the abdomen into a large pan placed just beneath to receive them.

The methods employed in impregnating the eggs were similar to those in use with eggs of the shad. They were usually taken in a pan having a little water in the bottom and the milt at once added, after which they were "brought up" in the usual way, by slowly adding water at intervals till the pan was nearly full.

It was found desirable to leave the eggs with the milt for fully half an hour before dipping them out, and at times it took even longer for them to become well hardened. Several other ways for impregnating the eggs were tried, such as taking them in a damp pan and introducing the milt directly upon them before adding the water; and of putting the milt in the water first, and the eggs later; and again, of introducing the two at the same time; but these seemed no improvement upon the ordinary method.

The first good eggs were taken November 13, and when placed in the cones were found to remain constantly at the surface of the water, where they soon clogged the screen through which the waste water made its escape, causing the cones to overflow, and the eggs to be carried over the top with the water. The plan of introducing the water at the top and allowing it to escape at the bottom was equally unsuccessful, for the downward current carried many of the eggs with it, thus clogging the screen as effectually as in the former case. The cones in their original condition were thus rendered useless, and the question of so modifying them as to make them answer the purpose, or of the invention of new apparatus, at once became a very important one; and one difficulty after another had to be overcome before any degree of success could be expected.

Mr. Milner remained long enough, before leaving for the South, to witness this stage of the difficulty, and was the first to suggest an alteration in the apparatus. This consisted in a modification of the inverted cone, so that the water should be introduced through a twisted tube at the apex, thus giving it a spiral motion as it ascended, while the outflow was in the form of a circle surrounding and just above the inflow, in a line with the sides of the cone. On testing, this apparatus was found to clog equally with the other, and was soon abandoned.

In one end of the room was a Clark hatching-trough that had been used in hatching eggs of the herring. This consisted of a long trough about 12 inches square, with numerous partitions dividing it into a number of compartments. The whole was placed at an incline, so that the overflow of water from one compartment would run into the next lower through a little groove at the top of the partition. Mr. Clark introduced into each compartment a small wooden box with a wire-cloth bottom, each to be placed at an angle with the bottom of the trough, with its lower end under the little spout that conducted the waste-water from the compartment above. With the box thus placed in a compartment filled with water, the stream that was kept constantly running would fall into its deepest part, and in this way create considerable current in the water, the surplus gradually passing out through the bottom and up around the sides on its way to the next compartment.

When the cod-fish eggs were introduced into these boxes they were found to have an excellent motion; but of the great amount of harbor sediment and mud in the water much was retained in the boxes by the wire screens and gradually collected on the eggs, causing them to sink to the bottom, where they soon died. This apparatus, though seemingly all that could be desired with clean water, was rendered useless by the fine dirt that could not be kept out.

The writer suggested a modification of the copper cone, so that the water should escape near the top through an intermittent siphon, the end of which should be incased in a large wire-cloth bag, to weaken the strength of the current where it met the screen, and cause any eggs that might be held against the bag while the water was running to fall away when it stopped. This apparatus, like that of Mr. Clark, was rendered useless by the sediment in the water; and in addition, there seemed to be a corrosion of the copper, due to the action of the salt-water, that proved injurious to the eggs.

The Ferguson bucket, which consists of a cylinder of sheet-iron, with a wire-cloth bottom, getting a circulation of water by means of a slow rise and quick drop when partially immersed, was tried, with only indifferent results.

Captain Chester was at this time devising an apparatus which should not only give a certain change to the water, but also partially keep the sediment from the eggs. This apparatus is known as the Chester bucket. It consists of a tin cylinder 18 inches in diameter and 24 inches deep, with four rectangular openings, each 2½ inches wide, extending from near the bottom to within 5 inches of the top. These and the bottom of the cylinder are covered with wire-cloth, to prevent the eggs from escaping and the dirt from entering.

On the outside of the cylinder, along one side of either opening, are placed strips or pockets of tin, at an angle with the side, and extending partially over the openings, so that the adjacent pockets face in opposite directions. As the cylinder rotates on its axis, the water is forced in at the two opposite openings and out at the others.

Beneath the wire-cloth bottom are four more strips of tin, radiating from the center, and placed at such an angle that the rotation of the cylinder forces the water against them, and up through the bottom. The whole is placed in a trough nearly filled with constantly-changing water, and sunk to such a depth that the water nearly fills it. The cylinder turns on a pivot, the power being applied from the engine by means of shafting, to a horizontal arm firmly fixed to its axis, and is kept constantly turning back and forth through an arc of 90°, thus keeping the water changing, and giving the eggs a tendency toward the top center.

When this apparatus had been thoroughly tested, and found to give good results, the cones were taken down, and water-tight troughs placed along the sides of the hatching-room to receive the Chester buckets, and from this date the hatching operations were conducted with a fair degree of success; and while, with pure water, the modified Clark trough or some equally simple apparatus might give excellent results, yet to Captain Chester belongs the credit of having partially overcome the existing difficulties, and of inventing the first apparatus successfully used in hatching floating eggs.

The time required for the development of the eggs of the cod-fish, after they are thrown from the parent, varies greatly, being dependent largely upon the temperature of the surrounding water. Of those taken November 13, the first hatched in 13 days; while of those taken December 17, the last did not hatch until February 5, thus requiring 51 days; giving a difference of 38 days for eggs taken within little over a month of each other. The period of 51 days represents an extreme case, and the circumstances may be worthy of consideration. These eggs, as above stated, were taken December 17 from four fish apparently in good condition, and placed in a bucket in the hatchery. January 28, the bucket was thought to contain too many eggs, and a few were taken out and placed in a floating box, with wire-cloth bottom, anchored in the harbor. The difference in temperature of the water in the two places averaged from one to two degrees.

Of those remaining in the building, the first hatched January 17, and the last on the 23d, making a variation of 6 days.

On January 25, the first fish were noticed from the harbor lot, and from that time they continued to hatch slowly until February 3, when not more than 10 per cent. were out, and 2 days later, when the first fish were 11 days old, and the eggs 51 days from the parent, a few still remained unhatched.

The variation of 11 days for eggs treated in exactly the same manner, suggests the idea that other elements than temperature may enter in to hasten or retard development. One cause, namely, that of the condition of the eggs when thrown from the parent, may considerably affect this period.

We find with the cod, as with other species, that the first fish hatched from a given lot of eggs always seem weak and immature; and again, that the last are usually in the same condition. The first are perhaps from eggs that for some reason have remained in the parent after they should have been thrown; the great majority of healthy fish coming later probably represent eggs in their normal condition; and the weak ones hatching last may be from eggs, that, though not thoroughly matured when taken, had just reached that stage where impregnation became possible. The time elapsing after the eggs leave the fish before they come in contact with the milt may also affect the time of hatching.

Experiments in these lines would be of practical importance in determining how many good eggs could be taken from the fish at one time; how often eggs might be taken from the same individual, and, also, the most desirable time for applying the milt.

A table of temperature observations, showing the condition of both air and water at the first high and low water after 7 a. m., will be found further on. The temperature of the water in the hatchery was always from one to two degrees higher, being raised a little in passing through the pipes. From this table we find that the average time required for hatching eggs, in water of different temperatures, was as follows:

	Days.
In water having an average temperature of 45° F.	13
In water having an average temperature of 41° F.	16
In water having an average temperature of 38° F.	20
In water having an average temperature of 36° F.	24
In water having an average temperature of 34° F.	31
In water having an average temperature of 33° F.	34
In water having an average temperature of 31° F.	50

The water of the harbor reached, and remained for a number of days, at a temperature of 30°, but the eggs in the floating box remained uninjured, even though the little fish in them were well advanced, while the large cod in the live-boxes within a few feet of them were all frozen to death.

Several attempts were made to hasten the development of the egg, by raising the temperature of the water by means of steam-pipes. The time of hatching was frequently shortened in this way, but in all cases the fish seemed premature and soon died. The failure in these experiments may be due to the crude apparatus that could not be regulated so as to keep the temperature constant and avoid fluctuations. These difficulties overcome, it seems not at all improbable that the process of hatching could be materially shortened, and the fish gradually accustomed to cooler water until the natural temperature of the harbor should be reached, when they could be put out.

The problem of hastening or retarding development in the egg is a very important one. Fish-culturists have given some attention to the subject, but none have yet succeeded in the invention of apparatus by which the water can be kept constantly at any given temperature.

Ripe fish were found nearly every time the live-boxes were overhauled, from November 13 to early in January, when the fish were frozen. Forty-three females were "stripped" during the season, and the milt from 60 males was used in fertilizing their eggs. The total number of eggs secured in this way was about nine and one-quarter millions.

It may be a matter of some surprise that so few eggs should be obtained from so large a number of fish; but it must be remembered that the eggs ripen slowly through a period of six to ten weeks at least, and that but few can be secured at any one time. Probably not over 200,000 can be taken from a 10-pound fish in a day, while 400,000 would be a large average for a fish of 20 pounds weight. After the fish were once "stripped" they were allowed to die, as the primary object of the experiments was methods, rather than quantities of eggs. In this way the great bulk and number of eggs were not secured; but when the work shall be resumed for the purpose of increasing the food supply, we see no reason why these spawning fish may not, by exercising care, be "stripped" over and over again until all or at least a greater part of the eggs have been secured, the fish being returned to the live-boxes after each operation. Still the supply of spawning fish seems limited only by the size of the live-cars, and the above method may not become necessary.

About the 1st of January the weather became quite cold, and the temperature of the water on the night of the 3d, for the first time during the winter, fell to 30°. On the morning of the 4th, when the spawn-takers visited the live-cars, they found that all the fish had been frozen to death, and, on examination, considerable ice was noticed in their stomachs. At this time the more important points about the treatment of the eggs having been learned, and the practicability of artificial propagation fully established, it was thought unnecessary to secure a new stock of fish for the live-cars, and it was decided to discontinue operations until such time as they could be resumed on a steamer constructed especially for the purpose. By this means the harbor sediment can be avoided, and the fish followed to any locality where they chance to be most plenty.

The number of fish hatched during the experiments was not far from 1,550,000. At first, while the apparatus remained so imperfect, the loss was great, and nearly or in some cases quite all of the first few lots of eggs were killed. But with the introduction of new methods one difficulty after another was overcome, and the percentage of loss was gradually reduced. The manner of caring for the eggs while hatching soon came to be better understood, and this too had a decidedly beneficial effect; so that, barring the loss resulting from impure water, there was a constant gain in the percentage hatched, and the loss during the last of the season did not exceed 40, and was frequently not over 30 per cent.

When first hatched, the little fish remain nearly motionless, or, at times, indulge in the same spasmodic efforts so noticeable when freeing themselves from the eggs. In a day or two they become more active, darting about for short distances in the water, with a peculiar motion and considerable rapidity. In a few days they begin to absorb the yolk-sacks, and seem quite vigorous, while the pigment-cells increase rapidly, giving them considerable color. When they had reached this stage, they were usually taken to the outer harbor and liberated, to become accustomed to their future surroundings before the yolk-sacks were absorbed, thus giving them the opportunity of seeking their natural food when the first instincts of hunger should lead them to desire it.

The young cod seem more hardy than those of most other species, and may be kept for a considerable length of time with small loss. In one case fully 50 were put in an 8-ounce bottle and kept in a room at a temperature of 50° F., without change of water, for four days, before the first ones died. Early in January a number were sent by express to Professor Baird in Washington, where they arrived in good condition, with no care on the way except that given by the baggage-master on the train.

5.—DIFFICULTIES ENCOUNTERED.

The difficulty of suitable apparatus for hatching the eggs has been fully described. This consisted not only in the invention of something suitable for floating eggs, but an apparatus that could be used in impure water. These requirements, after several unsuccessful attempts, were at last met, and the difficulty partially overcome by the introduction of the Chester bucket.

The greatest source of annoyance, and one that could not be wholly overcome, was the abundance of the harbor sediment or dirt in the water. The trouble from this source was due largely to location. The hatchery was situated at a point of the harbor, with the main channel on one side, and on the other a large cove, into which the refuse of much of the business portion of the city found its way through the street gutters and sewers. The location of the hatchery was then unfortunate, in that it occupied a position where the main current caused by the 11-foot tide, on its passage in and out from the cove, brought a greater part of the dirt and filth of the city directly beneath and beside the wharf, where much of it was pumped up through the pipes into the hatching-room, and found its way to the eggs.

In addition to this, the violent storms caused a heavy undertow to roll in from seaward, and to stir up the mud from the bottom and sides of the harbor, so that at low tide the water was often quite thick. This sediment of course passed up through the pipes, and often resulted in great injury to the eggs. It was not unfrequently the case that a lot of eggs would continue in good condition until the fish were nearly ready to hatch, when a heavy storm would roil the water, and cause the dirt to collect on them to such an extent as to give them a dull brownish color, and from its weight sink them to the bottom, where they soon died.

Every precaution was taken to thoroughly cleanse the water from these impurities before it came in contact with the eggs. Large flannel filters were introduced, and all the water made to run through several of them on its way to the reservoirs. During stormy weather, when the bottom mud was stirred up, the water was often passed through six or seven of these filters, but even then the finer sediment could not be kept back. Frequently the dirt was pumped up in such quantities as to so completely clog the filters that the water would not go through them, and at such times they had to be replaced by clean ones every few minutes during the hours of low water. Other methods of filtering were also tried with no better success.

The new Fish Commission steamer, built expressly for this work from a special appropriation of Congress, will entirely do away with this difficulty resulting from impure water, as she can be safely anchored in the deep water of the outer harbor where no sediment is found.

The corroding action of the salt-water upon the copper and tin of which the apparatus was made, was also the source of considerable trouble. The copper cones were rendered useless on this account, and tin was often eaten entirely through in a few days. This difficulty was partially overcome by thoroughly painting the cones with asphalt, but even then the tin would rust so badly as to seriously injure the eggs. All trouble from this source can be easily avoided in future by making the apparatus of wood or some metal that is not acted upon by the salt-water. Indeed, nickel wire-cloth was used during the latter part of the season for the bottoms of the buckets, and found to answer the purpose admirably.

The fact that the cod cannot live in water colder than 30° F. presents another difficulty, for it is of the utmost importance that a large supply of fish be kept constantly in the live-boxes; and, as the water at the surface of the harbor may reach this temperature at any time for several months during mid-winter, the fish are liable to be frozen. But with a steamer anchored in several fathoms of water in the outer harbor, the live-cars by her side could, at the approach of cold weather, be weighted and sunk to the bottom until the weather should become warmer.

6.—EXPERIMENTS WITH EGGS OF OTHER SPECIES.

While the primary object of the station at Gloucester was for the study of the cod, the question of the reproduction of several other important species received considerable attention, and much valuable information was gathered. Among these species were the haddock (*Melanogrammus æglefinus*), the pollock (*Pollachius carbonarius*), and the herring (*Clupea harengus*).

a. Herring.

Herring visit different parts of the coast from Cape Cod to Labrador at various seasons of the year for the purpose of spawning or feeding, and are abundant in some localities during a greater part of the summer.

In the winter the herring-fisheries of Newfoundland and Nova Scotia are very extensive; and formerly the spring herring-fisheries of the

Magdalen Islands drew a large fleet to that region. The fall fishing is most extensive in the vicinity of Wood Island (near Portland, Me.), and on the south side of Cape Ann, where herring "strike in" along the shore in immense schools, about the middle or last of September, for the purpose of spawning. At such times small vessels, from almost every fishing town between Cape Cod and Eastport, visit these localities with gill-nets, and the fish are sometimes taken in such numbers as to sink the net. At Wood Island alone, in the fall of 1879, the herring fleet numbered over 150 sail.

While preparing for the cod work at the hatchery a small school of spawning-fish arrived in the vicinity of Gloucester Harbor, and it was decided to make experiments with their eggs. Accordingly, the Fish Commission boats were provided with nets, and, for about two weeks beginning with October 12, visited the spawning-grounds daily, setting their nets in the evening and fishing them over at intervals through the night. Ripe males were always plenty, and 50 spawning females were sometimes taken in a single night. Many thousands of eggs were secured in this way, and after impregnation were taken to the building, where large numbers were successfully hatched.

The eggs of this species are adhesive, and when thrown into the water by the fish fasten themselves to the first hard substance with which they come in contact, this being usually the algæ or the rocky bottom. On account of their adhesiveness, when taken from the fish for hatching purposes, they must at once be brought in contact with that particular object on which they are to remain till hatched, as when they have become fastened to any substance it is impossible to remove them without injury. For the purpose of bringing them from the fishing-grounds, a water-tight egg-box was made, with slits or grooves in the sides, to receive movable panes of glass, and keep them in position until they could be transferred to the apparatus in the hatchery.

As soon as the fish were taken from the water the eggs were pressed from them upon these panes of glass, and, after the milt had been applied, were quickly spread over the surface by means of a feather. The glasses were then placed in position in the egg-box and the water was changed at short intervals until they arrived at the hatchery.

A Clark hatching-trough (described on page 37) was arranged with grooves on the sides of the compartments to receive the glasses of eggs, these being three-fourths of an inch apart and placed at an angle with the perpendicular. The glasses were so arranged that every alternate one should rest on the bottom, with the others half an inch above, so that the water must pass over the top of the first pane, under the second, over the third, &c., on its way through the trough, thus giving a constant stream over each pane. A few eggs were taken on wire cloth and others on mosquito netting, but the former rusted so badly as to injure the eggs and the latter collected such quantities of sediment from the water that the results were far from satisfactory. Those taken on the glass did much better, as the eggs could be washed with a camel's hair brush or a feather, and thus kept passably clean.

The development of the eggs was quite marked, and the line of the fish could be distinguished at the end of the third day; the eye could be seen on the fifth, and on the sixth a very slight motion was noticeable. The average time in hatching was about twelve and the shortest ten days.

The greatest difficulty encountered in this as in other cases was from the impure water; but, even under these circumstances, a good many were hatched, and the experiments proved conclusively that the artificial propagation of the species would be an easy matter if at any time it should be thought desirable.

b. Pollock.

Large pollock are absent from the waters of Cape Ann from the middle of January till early in May, the small ones leaving earlier in the fall and returning in April. The young may be taken almost anywhere along the shore, but the large fish seem to confine themselves to definite localities; and though not particularly abundant during the summer at Cape Ann, it is a favorite spawning-ground for the species, and during this period large schools visit this shore.

They begin to grow plenty about the first of October, and by the last of the month are so numerous as to greatly annoy the cod-fishermen by taking the hook before it can get to the bottom.

During this season some of the smaller vessels fish exclusively for pollock, "seizing" up their lines a number of fathoms from the bottom, and at times the fish bite as fast as the fishermen can haul them. Early in November, a crew of four men landed 10,420 pounds, or about 1,100 fish, the result of less than two days' fishing. Owing to a foolish prejudice, the price is always low, at times being less than 30 cents per 100 pounds. The average weight of the fish is about 9 or 10 pounds, and during the spawning season the sexes are taken in about equal numbers.

They seem to spawn while swimming about in the water, and their eggs, being buoyant, are found at the surface with those of the cod; but they may easily be distinguished from the latter by their smaller size. The first ripe female was seen at the fish-wharves October 23. November 11, a few good eggs were taken, and, after impregnation, found to have a diameter of one twenty-fifth of an inch. They were placed in an aquarium at the hatchery, and within forty-eight hours the fish could be distinctly seen, though no pigment cells were visible. This proved that the development of the eggs after leaving the parent was quite rapid, and indicated that they would hatch in five or six days at most, with water of the ordinary temperature.

At the time of taking these eggs no suitable apparatus had been arranged, and we did not succeed in hatching them; and as no others were obtained during the season positive statements cannot be made;

but the eggs were well advanced before they died, and careful observations up to this point fully convinced us that these eggs are as hardy as those of the cod, and that they may be successfully hatched by a similar method.

Table III gives the result of our computation of the number of eggs in individuals of different size, from which it will be seen that a 23½-pound fish has over 4,000,000 of eggs, while a 13-pound one has 2,500,000.

c. *Haddock.*

It is not many years since haddock were very little sought in the markets, and the price averaged only one cent each; but the method of smoking them, introduced into this country by the Scotch, has greatly increased the demand, and now a ready sale can be found for any quantity at good figures. At the present time a large fleet of Gloucester and Portland vessels are engaged in this fishery during the winter months, visiting George's and other offshore banks, and localities further north where the fish are abundant at this season. The vessels are each provided with trawls, and a single crew have been known to take nearly 20,000 pounds in a day.

The fish usually remain on these offshore banks till the winter is over, and they do not reach Cape Ann until just before the spawning season, which for this species begins about the middle of April and continues during nearly three months, the height of the season being in May.

In the spring of 1879 it is thought that two schools visited this coast, the first, composed of fish of large size, arriving early in April and leaving by the middle of May; and the other, composed of smaller fish, reaching the grounds about the 20th of May and leaving gradually after the 1st of July, a few remaining during the greater part of the summer. When the fishing first began, the fish were several miles from the shore, but they continued to "work in," until there was good fishing at the mouth of the harbor for several days, after which they seemed to move back again, and toward the close of the season remained on muddy bottom, when trawls were extensively used in their capture.

Early in May haddock were so plenty that one man caught 1,881 pounds in one day with hand-lines, and about the same time many different fishermen secured over 1,000 pounds. The males were usually a trifle more abundant, though at times the females composed fully half of the catch. The latter average larger than the former, and some days there would be a difference of two pounds in favor of the female.

The first ripe females were noticed on the 23d of April, and in the middle of July an occasional fish had not finished spawning. The first eggs were secured May 5, and others were taken at intervals to June 2, the total quantity being about 250,000. The method of impregnation was similar to that used for eggs of the cod, and the size of the eggs was one-nineteenth of an inch. Though the number contained in the larger individuals of the species reaches over 1,800,000 (see Table II), the quantity obtained for hatching purposes at any one time was quite small as compared with the number taken from the cod or the pollock, and the quantity of milt in the male fish was very much less than in either of the other species.

Different methods were employed in hatching the eggs; among others the Clark trough, and a floating box with wire-cloth bottom placed in the harbor beside the wharf. Those placed in the former were injured by dirt, but the floating box was more successful, and of the eggs placed in this a number were hatched. The line of the fish could be seen when the eggs were three days old, and in five days the fish was fully formed, though no motion could be detected. The shortest time required for hatching was eight, and the average nine days.

7.—CONCLUSIONS.

Up to the time of the establishment of the hatchery at Gloucester, so far as we know, no attempt had been made to impregnate and hatch floating eggs, and the whole subject involving the artificial propagation of so many important species had received little attention from the fish-culturists of the world.

The results of the experiments, during the three or four months of the winter of 1878–'79, were all that had been expected, and gave methods that will be of the greatest value for future extensive work. The principal points involved in hatching this class of eggs are now fairly understood, and most of the difficulties in the way of success have been met and overcome.

That the artificial propagation of the species is not only possible but practicable is proven by the fact that, under the most unfavorable circumstances, a small party succeeded in hatching over a million and a half of young cod during a short season; and that the loss of eggs in hatching was reduced from 100 to only 30 per cent. in about two months.

With apparatus made of suitable material, and placed on the new steamer now being built for the purpose, we see no reason why the work may not be carried on with the utmost success. At Gloucester the steamer can be safely anchored in the deep water of the outer harbor, away from all dirt and sediment, and can, if necessary, be moved to any other place where the fish chance to be more plenty.

With other species hatched by the Commission the great difficulty has been to secure the spawning-fish, from which the supply of eggs could be obtained. This has required a large force of men kept constantly on the fishing-grounds, and even then the quantity of eggs taken has usually been below the desired number, so that the hatching operations have often been limited by the number of eggs that could be secured. Again, with most species the spawning season for any particular locality lasts but a few weeks at most, and the loss of time occasioned by storms and other causes frequently interferes greatly with the success of the work.

With the cod the case is wholly different, for fish are plenty on the New England coast during most of the year, and the spawning season at Cape Ann lasts during eight or nine months.

The supply of spawning-fish can be obtained with little difficulty by a single crew, and brought to the harbor alive from any locality desired by means of an ordinary market well-smack.

These fish can be transferred to the live-cars convenient to the hatchery, to remain until such time as they may ripen. Thus the live-cars can be made a source of almost constant supply, and the hatching operations can be vigorously pushed during fully half the year; while the number of fish that can be hatched seems limited only by the capacity of the hatchery, and hundreds of millions of eggs can easily be secured in a single season.

The young fry seem quite hardy, and can be kept confined a considerable time and transported long distances with small loss; so that it will be an easy matter to carry them to the more southern waters before turning them loose in the sea. In this way it is thought that the range of the commercial fisheries may be somewhat extended, and a large class of people, both fishermen and consumers, greatly benefited. When the subject is regarded from the above standpoint, it is clear that the artificial propagation of the cod, as well as that of several other species, will remove the possibility of the extermination of these species from over-fishing; for the ovaries of 25 good-sized cod-fish, if all the eggs were hatched, would furnish more fish in number than are taken by the combined fleets of cod-fishermen from all the different fishing-ports of the United States during the most prosperous season.

SMITHSONIAN INSTITUTION, *February* 1, 1880.

E.—APPENDIX.

TABLE I.—*Showing the number of eggs in cod-fish of different sizes.*

Number.	Length of fish.	Weight of fish.	Weight of ovaries.	Estimated weight of ovary-walls.	Net weight of eggs.	Number of troy grains weighed out.	Number of eggs in the portion weighed out.	Number of eggs to the grain.	Total number of eggs in fish.
	Ft. In.	*Lbs.*	*Lbs. Oz.*	*Oz.*	*Lbs. Oz.*				
1	....	70–75	8 8	6	8 2	7	1,108	160	9,100,000
1 (*a*)*	....	70–75	8 8	6	8 2	7	1,132		
2†	4 2½	51	7 2	5	6 13	6	1,131	188.5	8,989,094
3	3 8	30	2 8¾	2¾	2 6	6	1,341	223.5	3,715,687
4	3 5	27	2 9⅜	2⅜	2 7	7	1,680	240	4,095,000
5	3 4½	22¾	2 2⅛	2	2 0⅛	6	1,368	228	3,229,388
6	3 3	21	1 15¾	1¾	1 14	6	1,249	208.17	2,732,237

*No. 1 (*a*) represents a second quantity taken from the same ovary the following day, and the greater number may be partially accounted for by the evaporation of moisture during the night.

†No. 2 contained a few ripe eggs.

TABLE II.—*Showing the number of eggs in haddock of different sizes.*

Number.	Length of fish.	Weight of fish.	Weight of ovaries.	Estimated weight of ovary-walls.	Net weight of eggs.	Number of grains (troy) weighed out.	Number of eggs in part weighed out.	Number of eggs to the grain.	Total number of eggs in fish.
	In.	*Lbs.*	*Oz.*	*Oz.*	*Oz.*				
1	28½	9 9/16	9½	5/8	8 5/8	4	1,950	487.5	1,839,581
2	26½	6 13/16	5¾	½	5¼	4	1,479	369.75	849,315
3	26	6 6/16	6¼	¼	6	4	1,457	364.25	856,156
4	24	4 13/16	6¾	½	6¼	5	1,160	232	634,380
5	22	4	5¼	½	4¾	5	970	194	403,132
6	20¾	3 3/16	5¼	¼	4¾	5	960	192	398,976
7	19¼	2 6/16	2 3/8	3/8	2	5	966	193.2	169,050

TABLE III.—*Showing the number of eggs in pollock of different sizes.*

Number.	Length of fish.	Weight of fish.	Weight of ovaries.	Estimated weight of ovary-walls.	Net weight of eggs.	Number of grains (troy) in part weighed out.	Number of eggs in part weighed out.	Number of eggs to the grain.	Total number of eggs in fish.
	Ft. In.	*Lbs.*	*Lbs. Oz.*	*Oz.*	*Lbs. Oz.*				
1	3 3½	23½	2 2	2	2 0	6	1,727	287.8	4,029,200
2	2 8½	13	1 2¼	1¼	1 1½	6	2,043	340.5	2,569,753

TABLE IV.—*Showing the variation in weight of cod-fish of various lengths.**

MALES.

Length of fish.	Condition of spermaries.	Weight of fish.	Length of fish.	Condition of spermaries.	Weight of fish.
In.		*Lbs.*	*In.*		*Lbs.*
16½	Very small	1¼	33		13¼
20		3	33	Very small	13¾
21	Nearly ripe	2¾	33½		9¾
23	Very small	4½	34	Medium	14
24	Small	4	34	Nearly ripe	14
24½	Large	5¼	35		12
25	Nearly ripe	5¼	35		13
26	Very small	6¼	35	Well developed	15½
27	Well developed	8¼	35½	Small	13½
27	Very small	7¼	35½	Well developed	15¼
28	Well developed	7¾	36		12¾
29		8	36	Medium	14¼
30		7	36	......do	15
30	Small	9¼	36½		16½
30		8¾	38		17
30	Nearly ripe	8½	40	Well developed	19
30	......do	10½	40	......do	21
31	Ripe	9¼	40	......do	21½
31	Small	8¾	40½	Medium	20¾
31		7	40½		25
31½		8½	42	Well developed	23½
31½	Well developed	10	42	......do	24½
31½	......do	10¾	43	Medium	25½
33		11	46	Well developed	43
33	Medium	11¼			

*The measurement was to the end of middle caudal rays.

TABLE IV.—*Showing the variation in weight of cod-fish of various lengths*—Continued.

FEMALES.

Length of fish.	Condition of ovaries.	Weight of fish.	Length of fish.	Condition of ovaries.	Weight of fish.
In.		*Lbs.*	*In.*		*Lbs.*
18½	Small	2¼	34	Very small	14
19	Very small	2⅛	34	Small	17½
19½	....	2⅛	34½	....	11
21	Very small	3⅛	35½	Small	14½
22½	......do	4¼	36	Medium	16
23	....	4	36½	....	16½
23	....	4¼	37	Small	14
23	Very small	4¼	39	......do	18
26	......do	5¼	39	Medium	21¾
27½	Medium	7½	39½	Small	18¼
28	......do	8½	39½	......do	20
29	....	7	40	......do	16
30	Very small	7½	40	....	17½
30	....	7¾	40	Medium	20½
30	....	9	40	....	23
30½	....	7¼	41	Medium	23¾
31	....	8½	41	Well developed	32
31	....	9½	41½	......do	27
31½	....	7½	43	....	29½
31½	Small	8½	44	Ripe	31½
31½	Well developed	10	44½	....	35
32	Small	11	45	Well developed	39
33	....	11	48½	Small	31
33	Small	12	50½	Ripe	45½
33½	....	12	57½	Small	54
34	Medium	12¼			

TABLE V.—*Showing the loss in weight of cod from the round to the market-dried fish.**

Number.	Sex.	Weight, round.	Weight, split.	Weight, dried.	Weight of stomach contents.	Weight of ovaries or spermaries.	Length, round.	Length, split.	Date of capture.	Time in curing.	Percentage of loss.
		Lbs.	*Lbs.*	*Lbs.*	*Lbs.*	*Lbs.*	*In.*	*In.*		*Days.*	
1	♀	34 12/16	18 5/16	11 15/16	4/16	4 8/16	48	36½	Jan. 29	22	.655
2	♂	23 14/16	15 3/16	10 2/16	4/16	1 2/16	41	31½	Jan. 29	22	.576
3	♂	16 8/16	10 12/16	7 8/16	Empty.	12/16	37	28¾	Jan. 29	22	.545
4	♂	12 1/16	7 10/16	4 15/16	Empty.	4/16	34¼	26½	Jan. 29	22	.501
5	♂	7 6/16	4 7/16	3 1/16	Empty.	2/16	29½	23	Jan. 29	22	.585

Average loss .6023.

The loss was distributed as follows:

Loss in splitting .... .4044
Loss in pickle .... .1496
Loss on flakes .... .0483

* These fish represent an average dryness for the year.

TABLE V (*a*.)—*Showing the loss in weight of cod from the round to the market-dried fish.**

Number.	Weight, round.	Weight, split.	Weight, dried.	Weight of ovaries or spermaries.	Weight of stomach contents.	Length, round.	Length, dried.	Time in curing.	Sex.	Percentage of loss.
	Pounds.	*Pounds.*	*Pounds.*	*Pounds.*	*Pounds.*	*Inches.*	*Inches.*	*Days.*		
1	40¾	23½	14 12/16	4	Empty ..	47	33¾	37	♀	.641
2	27⅝	14⅞	9⅛	4	..do	41¼	27¾	37	♀	.67
3	18	10½	6⅜	2	..do	37	24½	37	♂	.646
4	14½	7½	4¼	1⅛	½	35	25	37	♂	.707
5	13⅜	7½	4⅜	1⅛	¼	34½	24	37	♂	.679
6	8¾	5	3	⅛	Empty ..	29½	21¾	37	♂	.657
7	5¼	3½	1⅞	⅛	..do	26½	19	37	♂	.643
8	3½	2¼	1⅜	....	..do	22	17	37	♂	.607

Average loss, .659.

* The fish dried as much as in the warmest weather.

TABLE VI.—*Showing the loss in weight of "George's cod" in curing, after being split and salted on the vessel.**

Number.	Length split.	Weight as they come from vessel.	Weight dried.	Percentage of loss.
	In.	*Lbs.*	*Lbs.*	
1	40	$30\frac{3}{16}$	$25\frac{13}{16}$	.145
2	36	$21\frac{7}{16}$	$19\frac{5}{16}$	.099
3	30	11	$9\frac{11}{16}$	.119
4	$30\frac{1}{2}$	$9\frac{5}{16}$	$8\frac{11}{16}$	.067
5	26	$6\frac{1}{16}$	$5\frac{5}{16}$	.124
6	24	5	$4\frac{11}{16}$	.063

Average loss, .115.

* These fish represent a fair average in dryness for the winter season.

TABLE VIII.—*Showing the loss in weight of market-dried cod-fish in boning.*

Number.	Weight dried.	Weight boned.	Percentage of loss.
	Lbs.	*Lbs.*	
1	$25\frac{13}{16}$	$20\frac{13}{16}$	.194
2	$10\frac{13}{16}$	$8\frac{3}{16}$	.243
3	$5\frac{10}{16}$	$4\frac{5}{16}$	.234
4	$3\frac{8}{16}$	$2\frac{8}{16}$	.286

Average percentage of loss, .219.
One quintal of dried fish will therefore make 89 pounds of boned fish.

TABLE IX.—*Showing the loss in weight of pollock from the round to the market-dried fish.**

Number.	Weight, round.	Weight, split.	Weight, dried.	Weight of ovaries or spermaries.	Weight of stomach contents.	Length, round.	Length, dried.	Time in curing.	Sex.	Percentage of loss.
	Pounds.	*Pounds.*	*Pounds.*	*Pounds.*	*Pounds.*	*Inches.*	*Inches.*	*Days.*		
1	$16\frac{15}{16}$	$10\frac{6}{16}$	$6\frac{10}{16}$	$\frac{3}{16}$	$\frac{13}{16}$	37	25	30	♀	.609
2	$13\frac{1}{16}$	$8\frac{12}{16}$	$5\frac{10}{16}$	3	Empty..	$34\frac{1}{2}$	23	30	♂	.569
3	$12\frac{11}{16}$	$7\frac{13}{16}$	$5\frac{4}{16}$	$1\frac{5}{16}$	...do	$33\frac{1}{4}$	$22\frac{3}{4}$	30	♂	.587
4	$10\frac{11}{16}$	$6\frac{3}{16}$	$3\frac{15}{16}$	$\frac{2}{16}$	$1\frac{8}{16}$	$30\frac{3}{4}$	22	30	♀	.632
5	$7\frac{7}{16}$	$4\frac{9}{16}$	3	†0	$\frac{14}{16}$	26	18	30	♀	.597

Average loss, .598.

* The fish represent a fair average in dryness for the year.
† Young.

TABLE X.—*Showing the loss in weight of haddock from the round to the market-dried fish.**

Number.	Weight, round.	Weight, split.	Weight, dried.	Weight of ovaries or spermaries.	Weight of stomach contents.	Length, round.	Length, dried.	Time in curing.	Sex.	Percentage of loss.
	Pounds.	*Pounds.*	*Pounds.*	*Pounds.*		*Inches.*	*Inches.*	*Days.*		
1	$7\frac{1}{4}$	$4\frac{3}{4}$	$2\frac{7}{8}$	$\frac{1}{4}$	Empty	27	$21\frac{1}{2}$	34	♀	.603
2	$4\frac{3}{4}$	$3\frac{1}{8}$	$1\frac{11}{16}$	$\frac{1}{8}$	...do...	25	$19\frac{1}{2}$	34	♂	.645
3	$3\frac{1}{2}$	$2\frac{1}{2}$	$1\frac{3}{16}$	$\frac{1}{8}$	...do...	$21\frac{1}{2}$	17	34	♀	.661
4	$2\frac{1}{16}$	$1\frac{5}{16}$	$\frac{14}{16}$	$\frac{1}{16}$	...do...	18	14	34	♀	.580

Average loss, .623.

* These fish were cured as much as in the warmest weather.

TABLE XI.—*Showing the loss in weight of hake from the round to the market-dried fish.**

Number.	Weight, round.	Weight, split.	Weight, dried.	Weight of ovaries or spermaries.	Weight of stomach contents.	Length, round.	Length, dried.	Time in curing.	Sex.	Percentage of loss.
	Pounds.	*Pounds.*	*Pounds.*	*Pounds*	*Pounds.*	*Inches.*	*Inches.*	*Days.*		
1	$23\frac{12}{16}$	$14\frac{4}{16}$	$9\frac{15}{16}$	$\frac{6}{16}$	Empty	42	32	30	♀	.582
2	$12\frac{6}{16}$	$7\frac{9}{16}$	$5\frac{10}{16}$	$\frac{4}{16}$	...do...	35	$27\frac{1}{2}$	30	♀	.545
3	$5\frac{8}{16}$	$3\frac{12}{16}$	$2\frac{12}{16}$	$\frac{2}{16}$	...do...	$27\frac{1}{2}$	$20\frac{3}{4}$	30	♀	.500
4	$4\frac{11}{16}$	$3\frac{2}{16}$	$2\frac{4}{16}$	$\frac{1}{16}$	...do...	26	$19\frac{1}{2}$	30	♀	.520
5	$2\frac{2}{16}$	$1\frac{6}{16}$	1	0	$\frac{2}{16}$	$20\frac{1}{2}$	16	30	♀	.529

Average loss, .555.

* The fish represent a fair average for the year.

TABLE XII.—*Showing the loss in weight of cusk from the round to the market-dried fish.**

Number.	Weight, round.	Weight, split.	Weight, dried.	Weight of ovaries or spermaries.	Weight of stomach contents.	Length, round.	Length, dried.	Time in curing.	Sex.	Percentage of loss.
	Pounds.	*Pounds.*	*Pounds.*	*Pounds.*		*Inches.*	*Inches.*	*Days.*		
1	$17\frac{14}{16}$	$11\frac{10}{16}$	$8\frac{3}{16}$	1	Empty	$35\frac{1}{4}$	28	30	♀	.542
2	$15\frac{5}{16}$	$10\frac{15}{16}$	8	$\frac{1}{16}$	...do...	$33\frac{3}{4}$	27	30	♂	.478
3	$5\frac{15}{16}$	$3\frac{13}{16}$	$2\frac{13}{16}$	$\frac{8}{16}$	...do...	26	$20\frac{1}{2}$	30	♀	.526
4	$4\frac{1}{16}$	$3\frac{1}{16}$	$2\frac{4}{16}$	$\frac{1}{16}$	...do...	23	$18\frac{1}{2}$	30	♂	.446
5	$3\frac{4}{16}$	$2\frac{7}{16}$	$1\frac{12}{16}$	0	...do...	21	$16\frac{1}{2}$	30	♂	.462

Average loss, .505.

* The fish represent a fair average in dryness for the year.

TABLE XIII.—*Observations on temperature at Gloucester, Mass.*

FOR THE MONTH OF OCTOBER, 1878.

Day of month.	Time of observation.				Temperature of water at surface.		Temperature of water at bottom.		Thermometer in the open air.		Winds.				State of sky.		Time of beginning and ending of rain or snow.
	First high water after 7 a. m.		First low water after 7 a. m.		High water.	Low water.	High water.	Low water.	High water.	Low water.	High water.		Low water.		High water	Low water.	
	Hour.	A. M. or P. M.	Hour.	A. M. or P. M.							Direction.	Force.	Direction.	Force.			
	h. m.		*h. m.*		°	°	°	°	°	°							
18	3 40	P. M.	....	....	56	......	56	......	64	......	SE	Light	....	....	Hazy	....	
19	4 40	P. M.	11 00	A. M.	55	55	56	55	56	57	W	Strong	SW	Strong	Cloudy	Cloudy	
20	5 45	P. M.	12 00	M	53	55	54	54	53	51	W	Light	W	do	do	do	
21	6 45	P. M.	1 00	P. M.	52	53	52	53	54	57	W	Fresh	W	Light	Clear	Clear	
22	7 45	A. M.	1 30	P. M.	52	53	52	53	54	56	N	Light	E	do	do	do	
23	8 40	A. M.	2 30	P. M.	51	51	51	51	54	57	NE	Strong	E	Strong	Cloudy	Rain	Rain, 12 m. to 11 p. m.
24	9 30	A. M.	3 30	P. M.	51	50	51	51	55	57	NW	do	N	do	do	Cloudy	
25	10 25	A. M.	4 25	P. M.	49	50	50	50	53	51	NE	Fresh	NE	Light	do	do	
26	11 20	A. M.	5 20	P. M.	50	51	50	50	56	58	NE	Light	SE	do	Clear	Clear	
27	12 00	M	5 45	P. M.	51	51	51	51	55	56	SE	Strong	SE	Strong	Cloudy	Cloudy	
28	12 45	P. M.	7 00	P. M.	51	51	51	51	60	55	NW	do	NW	do	Clear	Clear	
29	1 45	P. M.	8 00	A. M.	51	50	51	50	49	46	SE	Light	NW	Light	do	do	
30	3 00	P. M.	8 50	A. M.	50	50	50	50	50	49	E	Strong	E	Strong	Cloudy	Cloudy	Rain.
31	3 45	P. M.	9 20	A. M.	50	50	50	50	52	50	SW	Light	SW	Light	do	do	Rain, 9 a. m.; light rain, 4 p. m.

FOR THE MONTH OF NOVEMBER, 1878.

Day of month.	First high water Hour.	A. M. or P. M.	First low water Hour.	A. M. or P. M.	Surface High water.	Surface Low water.	Bottom High water.	Bottom Low water.	Air High water.	Air Low water.	Winds High water Direction.	Force.	Winds Low water Direction.	Force.	Sky High water	Sky Low water.	Time of beginning and ending of rain or snow.
1	4 45	P. M.	10 37	A. M.	50	50	50	50	43	46	NW	Strong	NW	Strong	Clear	Clear	
2	5 45	P. M.	11 45	A. M.	49	49	49	49	50	48	SW	Light	S	do	do	do	
3	6 45	P. M.	12 45	P. M.	48	48	48	48	45	46	NW	do	NW	do	do	do	
4	7 30	A M.	1 20	P. M.	48	47	48	47	46	43	SW	do	NW	do	do	Cloudy	
5	8 15	A. M.	2 20	P. M.	46	46	46	46	32	36	NW	Strong	NW	do	do	Clear	
6	9 00	A. M.	3 00	P. M.	47	45	47	45	34	35	SW	do	W	Light	do	Cloudy	
7	9 45	A. M.	3 45	P. M.	47	45	47	45	35	36	SW	Light	SW	do	do	Clear	
8	10 15	A. M.	4 30	P. M.	45	44	45	44	37	32	NW	Strong	NW	Strong	Cloudy	Cloudy	Snow, 2 a. m. to 3.30 a. m.
9	11 00	A. M.	5 15	P. M.	43	42	43	42	36	35	NW	do	NW	Light	do	do	
10	11 30	A. M.	5 45	P. M.	44	42	44	42	41	39	NW	do	NW	do	Clear	Clear	

TABLE XIII.—*Observations on temperature at Gloucester, Mass.*—Continued.

FOR THE MONTH OF NOVEMBER, 1878.

Day of month.	Time of observation.				Temperature of water at surface.		Temperature of water at bottom.		Thermometer in the open air.		Winds.				State of sky.		Time of beginning and ending of rain or snow.
	First high water after 7 a. m.		First low water after 7 a. m.								High water.		Low water.				
	Hour.	A. M. or P. M.	Hour.	A. M. or P. M.	High water.	Low water.	High water.	Low water.	High water.	Low water.	Direction.	Force.	Direction.	Force.	High water.	Low water.	
	h. m.		*h. m.*		°	°	°	°	°	°							
11	12 00	M	6 15	P. M.	46	47	46	47	46	46	SW	Light	SW	Light	Cloudy	Clear	
12	12 30	P. M.	7 00	A. M.	45	45	45	45	53	49	W	Strong	SE	Strong	do	Cloudy	Rain, 5.30 a. m. to 9 a. m.
13	1 30	P. M.	7 20	A. M.	44	44	44	44	48	45	W	do	SW	do	do	do	
14	2 00	P. M.	7 50	A. M.	44	44	44	44	40	50	NW	do	NW	Light	Clear	Clear	
15	2 45	P. M.	8 40	A. M.	43	43	44	43	40	42	SE	Light	NW	do	do	do	
16	3 45	P. M.	9 45	A. M.	43	43	43	43	49	50	SE	do		Calm	Cloudy	do	
17	4 45	P. M.	10 45	A. M.	43	43	43	43	43	53	NE	do		do	do	Cloudy	Rain, 5.30 p. m.
18	5 45	P. M.	11 45	A. M.	44	43	44	43	43	46	NE	Strong	NE	Strong	do	do	Rain.
19	6 45	P. M.	12 45	P. M.	44	44	44	44	42	48	NE	Light	N. NE	do	do	do	Rain at 7 a. m.
20	7 15	A. M.	1 30	P. M.	45	44	44	44	40	43	NE	do	NE	Stormy	do	do	Rain, 11 a. m. to 5 p. m.
21	8 15	A. M.	2 30	P. M.	45	45	45	45	44	45	SW	do	SW	Calm	do	Clear	Heavy rain.
22	9 00	A. M.	3 15	P. M.	45	45	45	45	44	45	NE	Stormy	NE	Stormy	do	Cloudy	Rain, 5 a. m. to 4 p. m.
23	10 30	A. M.	4 15	P. M.	45	45	45	45	45	44	SSW	Light	W. SW	Light	do	do	
24	11 00	A. M.	4 45	P. M.	45	45	45	45	52	54	WSW	Strong	W. SW	do	Clear	do	
25	12 00	M	6 00	P. M.	45	44	45	44	45	41	WSW	Light	E. NE	do	Cloudy	do	Rain, 1.30 p. m. to 12.30 a. m.
26	12 30	P. M.	6 30	P. M.	45	44	44	44	44	45	NW	do	NW	do	Clear	Clear	Rain at 7 p. m.
27	1 45	P. M.	7 45	P. M.	44	44	44	44	50	42	ESE	do	E. NE	Stormy	do	Cloudy	Rain, 4.30 a. m.
28	2 30	P. M.	8 30	A. M.	44	44	44	44	51	46	S	Strong	S	Light	Cloudy	do	
29	3 15	P. M.	9 00	A. M.	44	44	44	44	39	63	W	Light	W. NW	Fresh	Clear	Clear	
30	4 15	P. M.	10 00	A. M.	43	43	43	43	57	65	NW	do	W. NW	do	Cloudy	do	

FOR THE MONTH OF DECEMBER, 1878.

Day of month.	Hour.	A. M. or P. M.	Hour.	A. M. or P. M.	High water.	Low water.	High water.	Low water.	High water.	Low water.	Direction.	Force.	Direction.	Force.	High water.	Low water.	Time of beginning and ending of rain or snow.
1	5 00	P. M.	10 45	A. M.	43	43	43	43	41	54	NW	Light	NW	Light	Clear	Clear	
2	6 00	P. M.	11 45	A. M.	44	42	44	42	53	44	E	Strong	E	Strong	Cloudy	Cloudy	
3	6 45	P. M.	12 45	P. M.	44	44	44	44	43	61	W	Light	W	do	Clear	Clear	Rain, 1.30 p. m. to 7 p. m.
4	7 15	A. M.	1 15	P. M.	43	43	43	43	40	43	NE	do	NE	Light	Cloudy	Cloudy	Rain, 7 a. m. to 1.30 p. m.
5	8 15	A. M.	2 15	P. M.	44	43	43	43	48	33	W. NW	do	NW	do	do	Clear	Snow, 7.15 a. m. to 8 a. m.

6	9 00	A.M.	3 00	P.M.	42	42	42	42	30	49	W	Strong	W	do	do	do	
7	9 45	A.M.	3 45	P.M.	42	40	42	40	46	38	NW	do	NW	Strong	Clear	do	
8	10 15	A.M.	4 15	P.M.	40	39	40	39	38	41	W	Light	W	Light	do	do	Rain.
9	11 00	A.M.	5 00	P.M.	39	39	38	39	32	32	NE	do	E.SE	do	Cloudy	Cloudy	Snow, 8.30 a.m. to 7 p.m.
10	11 45	A.M.	5 45	P.M.	39	40	39	40	47	50	E.SE	do	E	Strong	do	do	Rain.
11	12 15	P M.	6 30	P.M.	40	40	40	40	53	36	NW	Strong	W	do	do	Clear	
12	1 00	P.M.	7 00	A.M.	41	40	41	40	44	42	W	do	W	Light	Clear	Cloudy	
13	1 45	P.M.	7 45	A.M.	41	39	43	39	47	49	NW	do	W	do	do	Clear	
14	2 15	P.M.	8 15	A.M.	42	41	42	41	49	48	NW	Light	NW	do	do	do	
15	3 15	P.M.	9 00	A.M.	39	39	39	39	49	44	W	do	SE	Strong	Cloudy	Cloudy	Snow, 4.15 a.m. to 6.30 a.m.; rain 10 a.m. to 10 p.m.
16	415	P.M.	10 00	A.M.	40	39	40	39	28	45	NW	do	NW	do	Clear	Clear	
17	5 15	P.M.	11 00	A.M.	38	38	38	38	32	48	NW	do	NW	do	do	do	
18	6 15	P M.	12 00	M.	38	38	38	38	37	39	NW	do	NW	do	do	do	
19	715	P.M.	1 00	P.M.	38	38	39	39	27	44	NW	do	NW	do	do	do	
20	7 45	A.M.	2 00	P.M.	37	37	38	37	24	30	W	do	W	do	Cloudy	Cloudy	Rain.
21	8 45	A.M.	3 00	P.M.	36	36	36	36	45	35	SW	do	E	Light	do	do	Snow, 1.15 to 10 p.m.
22	9 45	A.M.	3 45	P.M.	37	37	37	37	51	35	SW	Strong	SW	Strong	do	do	
23	10 45	A.M.	4 45	P.M.	37	36	37	36	45	25	W	do	W.SW	Light	Clear	Clear	
24	11 45	A.M.	5 45	P.M.	36	34	36	34	34	27	SW	do	SW	Strong	do	do	
25	12 30	P.M.	6 30	P.M.	34	33	33	33	33	36	NW	do	W	Light	do	do	
26	1 30	P.M.	7 30	P.M.	33	32	33	32	30	32	SW	do	SW	do	do	do	
27	2 00	P.M.	7 45	A.M.	32	32	34	32	32	26	NE	do	N	do	Cloudy	Cloudy	
28	2 45	P.M.	8 30	A.M.	33	33	33	33	33	45	NW	Light	NW	do	Clear	Clear	
29	3 30	P.M.	9 15	A M.	33	32	34	32	29	24	NW	do	NW	do	do	do	
30	4 15	P M.	10 15	A.M.	33	32	34	32	30	34	NW	do	NW	do	do	Cloudy	Snow, 7 a.m. to 11.15 a.m.
31	5 00	P.M.	11 00	A.M.	32	32	32	32	33	28	W	do	W	do	do	Clear	

FOR THE MONTH OF JANUARY, 1879.

1	6 00	P.M.	12 00	M.	32	31	33	31	38	42	....	Calm	NW	Light	Cloudy	Clear	
2	7 00	A.M.	1 00	P.M.	32	32	32	32	35	35	NE	Light	NE	Strong	do	Cloudy	Stormy, 7.30 a.m. to 8.30 p.m.
3	7 45	A.M.	1 45	P.M.	31	33	32	33	12	20	W	Strong	W	do	do	Clear	
4	8 15	A.M.	2 30	P.M.	30	30	31	30	20	26	NW	Light	NW	do	do	Cloudy	
5	9 00	A.M.	3 00	P M.	30	30	30	30	25	29	NW	do	W.NW	do	Clear	Clear	
6	9 45	A.M.	3 45	P.M.	30	30	30	30	24	28	NW	do	NW	Light	do	Cloudy	Snow, 12.45 a.m. to 5 p.m.
7	10 30	A.M.	4 15	P.M.	31	30	31	30	27	24	NW	Strong	NW	do	do	Clear	
8	11 15	A.M.	5 15	P.M.	31	31	31	31	35	36	SW	do	SW	Strong	do	Strong	Snow, 8.40 a.m. to 9.30 a.m.; rain, 2.30 p.m. to 3.10 p.m.
9	12 00	M.	6 00	P.M.	32	31	31	31	40	37	NE	Light	NE	Light	Cloudy	Cloudy	Snow, 3.15 a.m.—
10	12 15	P.M.	6 45	P.M.	31	31	32	31	28	27	W	Strong	NW	do	Clear	Clear	—to 2.30 a.m.
11	1 45	P.M.	7 30	A.M.	31	30	31	31	32	21	SW	Light	SW	do	Cloudy	Cloudy	Snow, 6.15 p.m. to 8.30 p.m.
12	2 15	P.M.	8 15	A.M.	31	30	32	30	30	28	NW	do	NW	do	Clear	Clear	
13	3 00	P.M.	9 00	A.M.	33	31	33	31	35	29	SW	do	NW	do	do	Cloudy	
14	4 00	P.M.	10 00	A.M.	32	32	32	32	40	41	W	do	W	do	do	Clear	

TABLE XIII.—*Observations on temperature at Gloucester, Mass.*—Continued.

FOR THE MONTH OF JANUARY, 1879.

Day of the month.	Time of observation.				Temperature of water at surface.		Temperature of water at bottom.		Thermometer in the open air.		Winds.				State of sky.		Time of beginning and ending of rain or snow.
	First high water after 7 a. m.		First low water after 7 a. m.								High water.		Low water.				
	Hour.	A. M. or P. M.	Hour.	A. M. or P. M.	High water.	Low water.	High water.	Low water.	High water.	Low water.	Direction.	Force.	Direction.	Force.	High water.	Low water.	
	h. m.		*h. m.*		°	°	°	°	°	°							
15	5 00	P. M.	11 00	A. M.	30	31	32	31	13	17	NW	Strong	NW	Strong	Cloudy	Clear	
16	6 00	P. M.	12 00	M	29	30	29	30	29	21	NE	do	NE	do	do	Cloudy	Snow, [illegible]m. to 10.30 p. m.
17	7 00	P. M.	1 00	P. M.	31	31	32	31	26	25	W	Light	NW	Light	do	Clear	
18	7 30	A. M.	2 00	P. M.	32	32	32	32	34	40	SE	Strong	NW	Strong	do	do	Rain, 6 a. m. to 7 a. m.
19	8 45	A. M.	2 45	P. M.	31	31	32	32	23	30	NW	Light	NW	Light	Clear	do	
20	9 45	A. M.	3 45	P. M.	31	31	32	31	24	28	NE	do	NE	do	Cloudy	Cloudy	Snow, 2 a. m. to 6.15 a. m.; snow, 10.45 a. m. to 11 15 a. m.
21	10 45	A. M.	4 45	P. M.	32	30	32	30	11	19	NW	Strong	W	Strong	Clear	do	Snow, 6.45 p. m.—
22	11 30	A. M.	5 30	P. M.	32	31	32	31	40	33	SW	Light	N	Light	do	do	—to 1 30 a. m.
23	12 00	M	6 00	P. M.	31	31	31	31	43	40	NW	do	N	Strong	Cloudy	do	Rain, 11.30 a. m. to 12.30 p. m.
24	1 00	P. M.	7 00	P. M.	31	32	32	32	31	32	SW	Strong	SW	do	Clear	Clear	Snow squall, 11.15 p. m.
25	1 30	P. M.	7 30	A. M.	32	32	32	32	39	53	SW	Light	SW	Light	do	do	Snow squall, 9.30 p. m.
26	2 00	P. M.	8 00	A. M.	31	31	33	31	20	10	NW	Strong	NW	Strong	do	do	
27	3 00	P. M.	8 45	A. M.	33	30	33	30	33	20	S	do	SE	do	Cloudy	Cloudy	Rain, 2 30 p. m. to 3 30 p. m.; 6 p. m. to 9.45 p. m.
28	3 30	P. M.	9. 30	A. M.	33	32	33	32	44	33	SE	Light	SE	Light	do	do	Rain, 5 a. m. to 7 a. m.

XXXIII.—REPORT OF OPERATIONS AT THE UNITED STATES SALMON-HATCHING STATION ON THE M'CLOUD RIVER, CALIFORNIA, IN 1878.*

BY LIVINGSTON STONE.

CHARLESTOWN, N. H.,
December 31, 1878.

Prof. SPENCER F. BAIRD,
United States Commissioner:

SIR: I beg leave to report as follows: The winter of 1877–'78 was an extremely rainy one, and in this section of California it rained almost incessantly from the 6th of January till the end of February. In consequence of these rains the McCloud River rose to an unprecedented height, and swept down through the cañon which incloses it with terrible volume and velocity. When it was 14 feet 9 inches above the summer level, it was just even with the floor of the fishery mess-house. From that time till the waters began to subside the fishery buildings were in great danger. The excessive rise in the river brought down drift-wood that had been undisturbed for years, and in immense quantities. This drift-wood coming down with great force in the swift current and composed sometimes of the trunks of huge trees, endangered the buildings to a most serious degree. The water was not high enough to carry away the buildings by the mere force of the current, although it was in itself very powerful, but the momentum of the drift-wood was sufficient to carry everything before it.

During all the time of the high water, the men in charge, viz, Myron Green, Patrick Riley, and J. A. Richardson, together with four or five Indians who helped them, worked with great resolution and courage. During the whole of two days and one night they were in the water, sometimes up to their necks, and often in danger of their lives, guiding the drift-wood so that it would pass through the fishery premises with the least danger. They worked so persistently and skillfully that the houses were saved, but everything else was swept away. All the fences, flumes, chicken-coops, door-steps, hatching-troughs, filtering-tanks, eveything that was on the ground that would float, were carried off. The whole of the interior of the hatching-house was cleared out and left as clean as the dry bed of a river, which indeed it literally became. The damage done to the fishery was so considerable that I applied to the

* The species referred to in the accompanying report is the Quinnat or California salmon—*Salmo quinnat*.

United States Commissioner of Fish and Fisheries for a sum of money for the purpose of making repairs. This being furnished out of the deficiency appropriation voted by Congress in the spring of 1878 for the propagation of food fishes, I went to the McCloud River in May and immediately entered upon the work of putting the fishery in repair. There was an immense deal of work to be accomplished to set things to rights, and to get the place ready for the season's operations in hatching salmon-eggs. The main things to be done were to place the old buildings as they were before the freshet, to build a new building to serve both for a dwelling-house and a post-office, to replace the fences and flumes, to build the spawning-house and the corrals for the parent salmon, to repair the current wheel and the two flat-boats that it rested on and to put them in place in the river, to build a solid wall of rock from the high land to the river to protect the buildings against the force of the current in future floods, to build the rack, &c., and to reconstruct almost the whole of the interior of the lower part of the hatching-house, every portion of which was swept away so clean that not a single thing was left in it, not even the heavy grindstone. In order to make as rapid progress as possible, I put on a large force of men at once, and began work simultaneously on several of the undertakings just mentioned. The getting out of the timbers for the buildings, for the hatching-house floor, for the fences, and for general purposes, occupied the time of most of the men for two or three weeks. As we have no horses at the fishery, it becomes necessary to cut our timbers somewhere on the river above us. The first year that we settled here we found enough suitable trees close by, but each subsequent year we have had to go higher and higher up the river, till this year we found it necessary to go nearly four miles up to find such timbers as we required. This involved the consumption of a good deal of time, not only in getting the timber but especially in floating it down to the fishery, the river being tortuous in its course and very rapid. It was over a month before all the timbers were delivered at the places where they were wanted, and if it had not been for the very efficient help of the Indians, who seemed as much at home in the water as on the land, we probably should not have succeeded in getting the logs down the river at all. As soon as the timbers were ready, we built the bridge and rack across the river to obstruct the ascent of the salmon.

The demand for California salmon-eggs being now very large, I wished to take ten million eggs or more this season, and was, consequently, anxious to get the rack in as soon as possible. The water was still much higher than usual, and the difficult undertaking of bridging the stream was made still more difficult this year by the high water. By the 10th of July, however, it was accomplished, and the river was closed to the upward migration of the salmon. I was the more willing to close the stream as early as this because vast numbers of full-grown salmon, taking advantage of the high water in the Sacramento River, had escaped the nets of the Sacramento fishermen and had already fully stocked the upper waters of the McCloud with spawning fish.

The bridge and rack were hardly completed before the salmon in immense quantities made one of those fierce raids on the rack which I have described in previous reports. For two or three hours thousands of them threw themselves against the rack with all their strength in their fierce but useless attempts to effect a breach in the dam. Finally, finding their efforts ineffectual, they desisted and fell back into the deep pools below.

In the mean time, while the dam was being built, work had progressed very satisfactorily in other directions. On the 20th of June, by the aid of a Spanish windlass, we returned the current wheel and boats to the river. By the 10th of July the post-office building was finished, and the fences, flumes, doorsteps, and most of the smaller things that had been injured or destroyed had been repaired or restored. By the 1st of August the west piazza of the large dwelling-house was finished, together with an additional room. All the buildings had been whitewashed or painted. The large corral for confining the spawning fish was put in place at the fishing-ground, the solid water-wall of rock to protect the fishery-buildings against future floods was nearly finished, the first line of troughs in the hatching-house was laid, the current-wheel and flatboats put in complete repair, the packing-boxes were made, and a new fishing-boat had been built.

During the first twenty days of August we gave our attention chiefly to finishing up the hatching-house and hatching apparatus, building the spawning-house at the fishing-ground, making the smaller nets to catch and confine the parent salmon in while taking the eggs, and in general to perfecting every part of the preparations for taking eggs; and I may add here that never since the United States Fish Commission began work on the McCloud River have the appointments of the fishery and all the arrangements for carrying on operations here been so complete and entirely satisfactory. From the bridge and rack, which are the first steps taken towards securing the season's supply of salmon-eggs, to the minutest points connected with the taking and hatching of the eggs, there was hardly a thing left to be wished for, thanks to the liberal allowance made by the United States Fish Commissioner of Fish and Fisheries for the operations of this station.

On the 20th of August we took the first eggs of the season, numbering 30,000, and from that time till the 5th of October, when the last ice-car was loaded with salmon-eggs for their eastern destinations, our time was taken up with spawning the salmon, taking care of the eggs, preparing the moss for packing, and making the crates for shipping the eggs in.

Having now given a general *résumé* of the work which was done at the McCloud Fishery in the season of 1878, I will mention a few incidents which came under my observation, some of which may be worth recording, and, as they are mostly disconnected, I will take them up in the order in which they occurred.

On the 19th of May, when I arrived at the fishery, the country looked magnificently. All the foliage was fresh and green, owing to the recent heavy rains. Azalias, roses, the beautiful golden poppies of this region, with a thousand other gorgeous California flowers, were in bloom in vast profusion; and so thoroughly saturated with water was the earth, from the excessive rainfall of the winter, that it was long after the usual time when the desiccating influence of the dry season began to show its withering effect upon the vegetation.

On Sunday, May 26, an incident occurred which, though resulting in nothing of importance, seems to illustrate the uncertainty with which life in remote and unsettled regions like this is accompanied. About midnight we were awakened by the dogs barking violently in the direction of the hill behind the house. Upon sending them out to see what was the matter, they went about ten rods to some thick brush, and returned yelping. At the same time we could distinctly hear stones being thrown at them. It was dark. There was only one man in the house besides myself, and we only had one gun between us. With the exception of the hostler at the stage station, a mile distant, there was not a white man within three miles. We were in a country which we knew was often frequented by desperadoes, and where the stage has been robbed six times in a month, and where murders are not of unfrequent occurrence. It might be only one or two burglars in the bushes, but how did we know that they were not a gang of cut-throats who were taking advantage of our weakness to overpower us, and secure the money which is supposed to be at a government station like this. It was impossible to help thinking that if that were the case, how easy it would be for a few determined men to set fire to the buildings, and then to pick us off, one by one, as we endeavored to escape. That has been the fate of a great many persons in unsettled portions of California, and why should it not be ours? I follow out this line of thought merely to illustrate the uncertainty which attends this sort of life. In point of fact the only result was that we remained awake the rest of the night, and in the morning we saw where the men, whoever they were, had thrown the rocks at the dogs. That was all.

A very natural sequel to this incident took place just a week later, and also illustrates the uncertainty which I have just mentioned. About nine o'clock one evening we heard a great deal of noise, accompanied with some quarrelling among the Indians about a quarter of a mile below the house. The noise continuing, two of our men started down the road to see what the matter was, and on arriving at the fishery stable found one or two men engaged in robbing a teamster who was stopping there over night. One or two shots were fired by our party, but the robbers escaped. We found, however, that the rascals had not only robbed the teamster of his money, but had taken from his wagon twenty demijohns of whisky, which they had distributed indiscriminately among the Indians. The result was such as no one can realize who has not been in an Indian country. The Indians were all more or less intoxicated, were very noisy and quarrelsome, and were inciting each other to make a descent on the fishery, and, as they expressed it, "to sweep it clean with the ground." Our men, in the highest degree indignant at this outrageous villany of the robbers, armed themselves for the occasion and determined to give chase to them that very night. They found them about daylight at an Indian lodge, and placing the muzzles of their revolvers close to the robbers' heads, they captured them without resistance. One is now in the State's prison, the evidence against him being conclusive. The other was discharged for want of sufficient proof of his guilt. This furnishes another instance of our insecurity. It is true it resulted in nothing, but had the Indians been sufficiently intoxicated or sufficiently bold to make an attack on the fishery that night, they could have carried everything before them.

On the 21st of June a post-office was established at the fishery, which I named Baird, after Professor Baird, United States Commissioner of Fish and Fisheries.

During the first week in July an Indian named Chicken Charlie called on me and said his father was going to die soon, and he wanted a coffin made. We made the coffin, and after a while, when they supposed the Indian was dead, they put him in the coffin and proceeded to bury him; but before they had finished burying him he came to life again, and they took him out and waited a while longer. The next time he really died, and the following day he was buried over again.

As soon as the dam was completed across the river, the salmon showed signs of being very thick in the river below. On the 11th of July we made a haul with the seine, which confirmed our impressions of the abundance of the salmon, the number taken at this haul being nearly a thousand. About this time the Indians employed at the fishery did some very fine work under water in repairing the rack. We discovered one day that the salmon, by their violent and repeated attacks on the dam, had at last forced a passage-way underneath the rack and were escaping. I immediately put three Indians on the break to repair it. The water was very cold and very swift, and it would have been extremely difficult for white men, unless experienced divers, to do the work; but the Indians, diving down to the bottom of the river and bracing their feet against the dam to resist the force of the current, worked with great skill and perfect self-possession, although remaining sometimes a very unpleasantly long time under water. I will add here that the assistance of the Indians during the work which we have to do in the water is perfectly invaluable. I do not know how we should get along without them, particularly as the snow-water of the McCloud is so cold that white men cannot stay in it any great length of time. The Indians will remain in it till they get so cold that they build a fire when they come out of the water to warm themselves by, as I have often seen them, when the surrounding air is already at 130° Fahrenheit from the natural heat of the sun.

Salmon-jumping.—Soon after the salmon were shut off from ascending the river, I frequently took a boat and went out into the river below the dam to watch the salmon jumping. On the 21st of July I counted 75 a minute (4,500 an hour) jumping in a space perhaps a hundred yards long by thirty yards wide. On the 28th of July I counted 100 a minute (6,000 an hour). On the 31st of July I counted 145 a minute (8,700 an hour). This is the largest number of salmon that I have ever seen jumping in the McCloud River in a minute.

Heat of the sun.—For some unknown reason there are usually one or two days, but no more, during the summer when it is exceptionally hot in the sun. In 1875 this peculiar day came on the 22d of July, when the temperature was 153° in the sun. This year it came on the 26th of July. The thermometer on that day in the sun at 4 o'clock p. m. rose to 149°.

The eclipse of the sun.—On the 29th of July an eclipse of the sun took place. I had told the Indians two months before that it was going to happen, and from that time till the day of the eclipse they came to me every little while to inquire how many days before the "grizzly bear would eat the sun," that being their explanation of the darkening of the sun at an eclipse. When the day arrived, twenty or thirty of them came to the fishery and looked at the sun with the greatest interest through pieces of smoked glass which we prepared for them, and which enabled them to watch the progress of the eclipse much better than they could do in their own way, which is by observing the reflection of the sun in the water. It is a great mystery to them how the white man is able to predict so long beforehand the coming of the "grizzly bear that eats the sun."

On the 25th of March, 1876, an eclipse of the sun occurred, and, at the height of the obscuration, an otter came out of the water in front of the house, looked around, and disappeared. The Indians remembered it, and kept on the watch for the otter during the eclipse this year (1878). No otter came; but it was a singular fact that the next day an otter—the only one we saw during the season—swam down past the house and back again, and disappeared. I think that the Indians who saw these otters will always think that an otter, as well as a grizzly bear, is required to accomplish an eclipse of the sun.

The Indian scare.—On the 21st of July an Indian messenger came in great haste from Copper City, on Pitt River, about eight miles from the fishery, with a letter from the superintendent of the silver mines there, stating that alarming rumors had reached that place about large numbers of northern Indians having been seen on the McCloud, and that the people there had heard that the Indians were meditating an attack on their settlement, and asking if we knew anything about it. About the same time we read in the papers that the Pit River Indians had been making hostile demonstrations on their river. Our McCloud River Indians, who by this time had heard of the alarm at Copper City, were very much excited. We wrote back to the superintendent that we thought there was nothing in it, and that there was no danger. The next morning, however, an Indian squaw told us that the Yreka and Upper Sacramento Indians were coming down to the McCloud to kill the McCloud Indians and what white men there were on the river, meaning ourselves at the fishery. We heard farther that Outlaw Dick, who murdered George Crooks here in 1873, and Captain Alexander, an Indian of very warlike disposition, had urged the northern Indians at a recent council to make a descent upon the McCloud and "clean out," as they expressed it, all the white men and McCloud Indians on the river. To add to the excitement, a Piute chief had visited our Indians the past week to stir them up to make war on the whites.

Three days after, a McCloud Indian came down in hot haste from Alexander's camp and told our Indians that Alexander had gone north to "call" his Indians, and that they would be down next month to make war on the McClouds. Some of our Indians were very much alarmed, and for several days a good deal dejected over this news, and they told us stories of ancient fights that they had had with the northern Indians, and how the Modocs and Yreka Indians had made war on them and burned their children and carried off their squaws. All this occurred just at the time when the SanFrancisco papers were full of the murders and depredations of the Oregon Indians, and we began to think that there might be something serious in the excitement in our neighborhood. At all events, as we had only one rifle at the fishery I thought it prudent to be at least better armed, and accordingly telegraphed for arms and ammunition. The excitement, however, gradually died away. The Piute chief returned to his own tribe; the Oregon Indians began to surrender and come in to deliver themselves up to the soldiers; the McCloud Indians recovered from their alarm, and about three weeks after the first excitement they informed me that Captain Alexander and his Indians had changed their minds and were not coming. This was the end of our Indian scare, and after this we thought nothing more about it. We might not have been in any danger whatever. It is very likely that we were not, and yet when a few white men are in an Indian country where the Indians outnumber them ten to one, as in our case, their very helplessness creates a feeling of uneasiness if there is only the slightest suspicion of danger. We did not know that we were in great danger, but we knew that if we were, with but one rifle among us, we were perfectly powerless to avert it; and that reflection was an unpleasant one in itself.

Hot weather.—Between the 8th and 14th of August, inclusive, we had a hot week, during which the heat was so continuous and excessive that I think it is worth mentioning. The temperature on those days at 3 o'clock in the shade was as follows: August 8, 102°; August 9, 108°;

August 10, 110°; August 11, 110°; August 12, 112°; August 13, 106°; August 14, 102°.

I will also call attention here to the striking contrast between the temperature of the air and that of the water. On the 11th of August the air in the sun was 134°, and the water was 60°, consequently when our men went into the water to work on that and similar days, they experienced a change of temperature of 74°. This is very trying to the health, and some who have worked here in the water have suffered very severely from the effects of it.

Roily water.—About the 10th of August we noticed that the river water was beginning to be turbid, and to look in color like the Missouri at Omaha. This created no alarm, because we had often noticed, after very hot days, that the McCloud water was turbid, the cause being that the unusual heat melts an unusual amount of snow on Mount Shasta, which swells the smaller streams at the head of the river and roils the water. The turbidness of the water, however, continued for several days and increased every day till, on the 15th of August, the water was so muddy that one could not see more than 18 inches below the surface. Then we began to think that there might be some other cause for it than melting snows, and horrible visions of Chinamen mining at the head waters of the McCloud arose in our minds. Every other good salmon-spawning river in California has been spoiled or nearly spoiled for the salmon by mining operations, and to think of the McCloud, the last hope of the Sacramento salmon being ruined in the same way was intolerable. The universal sentiment at the fishery was that if our suspicions were true, "the Chinese must go," and it would not have been difficult to find men enough to carry the decree into execution.

On Saturday, August 17, I decided if the water did not become clearer to send an expedition up the river to ascertain the cause of its turbidness. On Monday, however, it began to get a little clearer, and continued to grow clearer till the 24th of August, when it was about as clear as usual. In the mean time I discovered the cause of the turbidness, which proved to be a very peculiar one at the same time that it entirely relieved the Chinamen from our very unjust suspicions. We discovered that when there is an unusual amount of melting snow on Mount Shasta, the water seeks a new channel through what is generally in summer a dry gulch. This gulch, called Mud Creek, is composed of fine, white, ashy earth, and when the melting snows on Shasta overflow into it, they carry vast quantities of whitish mud into the McCloud. This is what made the river so roily; and the reason that it continued roily so much longer than usual was because there was more snow than usual on Shasta, and the heat for a week was very excessive.

The salmon.—The salmon, as before remarked, were found to be extremely abundant below the dam, and as soon as it was finished they gathered there in vast numbers. Indeed they were more numerous than I have ever known them to be before at that time, viz, the first half of July. This abundance of salmon continued through the season. At first they were very small, smaller than we have ever known them to be before, but about the 13th of August a new run came up of very large fish. This run with the earlier run of small ones made the river swarm with salmon. I have never seen anything like it anywhere, not even on the tributaries of the Columbia. On the afternoon of the 15th of August there was a space in the river below the rack about 50 feet wide and 80 feet long where, if a person could have balanced himself, he could actually have walked anywhere on the backs of the salmon, they were so thick. I have often heard travelers make this remark about salmon in small streams, so I know that it is not an uncommon thing in streams below a certain size, but to see salmon as thick as this in a river of so great volume as the McCloud must, I think, be a rare sight. About this time I kept a patrol on the bridge every moment, night and day, and this precaution, though an expensive one, was well rewarded, for this vast number of salmon continually striking the bridge with sledge-hammer blows were sure, in the course of time, to displace something and effect a passage through to the upper side, and when one did succeed in getting through, the others would follow with surprising rapidity one after another, like a flock of sheep going through a break in a fence. If they were not watched a hundred or even a thousand could easily slip through unobserved, but by the aid of the patrol, who was always provided with material for repairing the dam, a breach was discovered as soon as it was made, and was repaired as soon as it was discovered. This swarm of salmon just alluded to remained at the bridge and kept up the attack at one point or another for three days, and then fell back to the pools below, where, with occasional renewals of their attacks, they remained until they were caught in the seine.

The spawning season.—The spawning season began the 20th of August, with the taking of 30,000 eggs from seven fish. Every haul of the net brought an enormous quantity of salmon. Without our trying to capture many, the net would frequently bring in a thousand at a haul. We found very few ripe fish, however, until the 28th of August, when the spawning season set in in good earnest, and from this date to the last day of taking eggs the yield was very large and remarkably regular.

This leads me to say that the most extraordinary feature about the fishing season this year was that the salmon in the river did not seem to be diminished any by our constant seining. We made enormous hauls with the net every day, spawned a large number of salmon, and gave a large number to the Indians for their winter supply, but always the next day the spawning salmon seemed to be as thick as ever. This abundance of the salmon was a daily surprise to us. Every day we were regularly, though agreeably, disappointed. It was three weeks before we made any impression on the spawners in the river. At last, about the 15th of September, the females with spawn began to fall off a little, but only a little. We had enough eggs by this time, however, and stopped fishing on the 18th of September, not because of any scarcity of salmon, but because we did not want any more eggs. We had in the hatching-house on the evening of that day 12,246,000 salmon eggs, according to our recorded count, though without doubt over 14,000,000 in reality, as our method of counting purposely leaves a large outside margin for emergencies. Had we continued to fish and take eggs till the close of the fishing season, we could probably have taken 18,000,000 eggs, and perhaps more.

It is a fact worth noticing here, that the salmon were smaller this year than usual, the eggs were smaller, and the number of eggs to the fish was smaller. I doubt if the female salmon which we spawned averaged for the season over nine or nine and a half pounds, while in previous years they have averaged twelve or fourteen pounds. Sometimes we spawned twenty salmon in succession, of which not more than three out of the twenty would vary a half a pound from seven pounds. The weights of the salmon which we tagged and set free, given in the table below, are a fair sample of the weights of the females for the whole season.

Table showing the weight of several McCloud River salmon which were tagged with a silver tag and turned loose in the river in September, 1878.

	Weight.		Weight.
	Pounds.		*Pounds.*
No. 1	7	No. 12	7
No. 2	8	No. 13	8
No. 3	7	No. 14	11
No. 4	9	No. 15	9
No. 5	10	No. 16	6
No. 6	6	No. 17	7
No. 7	7	No. 18	6
No. 8	9	No. 19	6
No. 9	7	No. 20	9
No. 10	8		
No. 11	6		153

Average weight, 7.65 pounds.

It will be seen by the above record that twenty salmon, taken indiscriminately, weighed 153 pounds, giving an average weight of 7.65 pounds each. The small size of the salmon in the McCloud River this year was undoubtedly caused, in whole or in part, by the fishing at the canneries on the Sacramento, where the 8-inch meshes of the innumerable drift-nets stopped all the large salmon and let all the small ones through. The eggs when taken proved to be at least a third smaller than those of most previous years, and the average number of eggs to the fish was about 3,500, against 4,200 last year.

I adopted a new and rather unique method this year of driving the fish to the fishing-grounds. As may be readily supposed, the constant drawing of the net over the seining-hole had the effect of frightening the salmon off the ground. Of course it was necessary to get them back again before they spawned, as otherwise we should have lost the eggs. I have hitherto been in the habit of sending a gang of white men and

Indians down the river for this purpose. By going over the fish with boats, by throwing in rocks, by stirring up the holes with long poles, by floating down trees and brush over them, we have usually succeeded in driving back the fish that have *gone down* the river from the fishing-ground. This, however, did not enable us to get at the fish that *went up* the river and that lay in the rapids, and particularly in the deep holes between the seining-grounds and the bridge above. Here vast quantities of salmon collected, which we had never hitherto been able satisfactorily to reach. This year I accomplished it in this way: I had several Indians go up to the bridge armed with long poles. At a given signal three Indians jumped into the foaming rapids below the bridge, and by splashing the water with their arms and limbs and making as much of a disturbance in the water as possible did everything they could to frighten the salmon out of the rapids. On reaching the deep holes, where the fish lay collected by hundreds and perhaps thousands, the Indians dove down in the very midst of the swarms of salmon, and, stirring them up with their long poles, succeeded in driving them out.

In order to co-operate most effectively with the Indian divers, I had the seining-boat, with the boatmen all ready in it, stationed just at the point where the boat starts across the river with the net. On the beach also, where the net is drawn in, the fishermen were stationed at the ropes, seven men at the lower rope, and four men at the upper one, ready to pull in the seine at the proper moment. On the other side of the river, nearly opposite the fishing-boat, was stationed a boatman with a second boat, whose duty it was when the net was payed out to pull down close to the opposite shore where the net itself could not reach, in order to prevent the salmon from skulking there away from the seine. Still lower down on each side of the river were men stationed on the banks to throw rocks into the rapids below, with the intention of driving the fish out of the rapids into the net.

On these occasions the hauling of the seine was quite an exciting event. The Indian swimmers, their dark heads just showing above the white foam, screaming and shouting in the icy waters and brandishing their long poles, came down the rapids at great speed, disappearing entirely now and then as they dove down into a deep hole. As soon as they approached within about four rods of the fishing-skiff, the boat shot out from the shore, the second boat man braced himself and his oars for a quick pull down along the bank. The man at the stern of the first boat began paying out the seine, the fishermen on the beach gathered at their respective ropes, the men on shore began throwing rocks in the rapids, and in a few moments the net was drawn to the beach with an enormous mass of struggling, writhing salmon, often weighing in the aggregate not less than four or five tons. Then the fishermen sprang into the water and examined the fish, taking the ripe ones to the corral and throwing the unripe ones back into the river until the net was emptied. Then all was quiet again and the men proceeded to take the eggs from the ripe fish which they had captured.

I ought to add here that the water is too cold for white men to endure swimming and diving and remaining in it as long as is necessary to drive the salmon from the rapids. Indeed, the mere work of examining and spawning the salmon is altogether too severe an exposure for white men, and almost every one of my men gets more or less prostrated with sickness the first week of the spawning season. And it is not to be wondered at, for we run the seine every night until twelve o'clock, and the water and night air are sometimes 80° colder than where the men have been accustomed to work during the day. For instance, after hewing timbers or building a corral in a sun temperature of 130° in the daytime, they will frequently work in the water and night air in the evening in a temperature of 50°, their clothes wet through all the time. Here the difference in temperature is just 80°. This is obviously exceedingly trying to the most robust constitution, and the result always is that most of the men get sick the first week, though it is also true they usually rally—that is, those who can stand it at all—and are all on duty the next week, attacking their work with renewed zeal and vigor.

The actual spawning of the salmon this year was conducted on the same general plan as last year, except that I made arrangements for doing the work somewhat more systematically, and on a scale corresponding to the great number of eggs which we hoped to take, and which we actually did take. I think other salmon breeders will be inclined to smile an incredulous smile when I say that we frequently took from 700,000 to 900,000 eggs and upwards in one day before four o'clock in the afternoon. Yet this my men actually accomplished several times. The physical exertion required to do it is enormous.

On the evening of the 18th day of September all the eggs were taken and placed in the hatching-houses in good order, the whole work of the spawning season having been done this year, notwithstanding the large number of eggs taken, more smoothly and easily than ever before.

Maturing and hatching the eggs.—The maturing and hatching of the eggs also passed off more smoothly this year than usual. No disasters or drawbacks occurred during the whole season that I remember. Everything worked well, and when the time came for shipping the eggs, there were as fine a lot in the hatching-houses as was ever collected together. There was not an egg shipped, that I am aware of, that had been in the least degree injured by fungus, sediment, insufficient air, or any other cause whatever. All were in a perfect condition of health and vitality.

In confirmation of the above statement I quote below from some of the letters which I received from consignees of the eggs concerning the condition of the eggs on arrival at their destinations.

MADISON, WIS., *December* 20, 1878.

DEAR SIR: Your receipt for freight on California salmon eggs received this day.

The eggs *were very fine; hatched out beautifully.*

Very truly yours,

WILLIAM WELCH,
President Wisconsin Fish Commission.

L. STONE.

MOUNT CARROLL, CARROLL COUNTY, ILLINOIS,
October 16, 1878.

DEAR SIR: The two crates of California salmon eggs, of which you notified me from California, reached me on the 14th instant. They are *in fine condition, only about* 3 *per cent.* being found faulty.

Very truly yours,

SAMUEL PRESTON.

LIVINGSTON STONE, Esq.

GLOUCESTER, MASS., *October* 18, 1878.

MY DEAR SIR: My man writes me of the safe arrival of the salmon eggs *in good condition.* Out of the lot of 250,000 he picked out 6,000 bad eggs, $2\frac{4}{10}$ per cent.

Yours, very respectfully,

FRANK N. CLARK.

LIVINGSTON STONE.

SAINT PAUL, MINN., *October* 28, 1878.

DEAR SIR: The California salmon eggs from McCloud River came to us in the evening of the 14th, and I am glad to say they open up in better order than any we have ever received before. The packing and carriage were a complete success, and up to this time *the loss has not been over* 5 *per cent.*

Very respectfully,

R. O. SWEENY.

Hon. S. F. BAIRD,
United States Fish Commissioner, Smithsonian Institute,
Washington, D. C.

PEMBROKE, ME., *October* 8, 1878.

DEAR SIR: I received the case of salmon eggs that you shipped to me. The eggs were *in good condition,* there being only 823 dead eggs, which is a small percentage.

Yours, respectfully,

LORENZO S. BAILEY.

LIVINGSTON STONE.

TRENTON, N. J., *October* 14, 1878.

DEAR SIR: In accordance with your request of September 23, you are informed that the shipment of salmon eggs for the State of New Jersey, and others (total, 475,000), was received in due time, and that the condition of the eggs on arrival *was most excellent.*

Very respectfully,

E. J. ANDERSON,
Commissioner of Fisheries of New Jersey.

LIVINGSTON STONE, Esq.

ROCHESTER, N. Y., *October* 8, 1878.

DEAR SIR: The eggs arrived at destination October 4. They *were in very good condition.* The first time going over them 4,945 were picked out. They are looking well.

Yours,

SETH GREEN.

LIVINGSTON STONE.

ELGIN, ILL., *October* 12, 1878.

DEAR SIR: The California salmon eggs came *in excellent shape.*

Very truly,

W. A. PRATT.

LIVINGSTON STONE.

PLYMOUTH, N. H., *October* 8, 1878.

DEAR SIR: The eggs arrived here at noon the 7th, *in good condition.*

Yours, &c.,

A. H. POWERS.

LIVINGSTON STONE.

COUNCIL BLUFFS, IOWA, *October* 17, 1878.

DEAR SIR: The 50,000 California salmon eggs shipped me per express were duly received on the 14th instant, and in unpacking the same I find them *in excellent condition.*

Yours, respectfully,

WM. A. MYNSTER.

LIVINGSTON STONE.

The only large loss experienced in the shipment of the eggs this year was in the case of a lot of 500,000 consigned to Hon. Samuel Wilmot, Newcastle, Ontario, Canada. These, as Mr. Wilmot's letters which follow will show, were almost a total loss. There cannot be much doubt that the injury to the eggs occurred on the express car between Chicago and Newcastle. As Mr. Wilmot's eggs were handled at the McCloud River fishery in precisely the same way, were packed in the same way, were shipped in the same way, and, in short, received precisely the same treatment that the other eggs received, from the time of their leaving the parent fish on the McCloud till they were unloaded from the ice-car at Chicago, and as all the other eggs went safely, it does not seem possible that the injury to the eggs could have occurred west of Chicago, because if it had, the same disastrous agency which destroyed his eggs must inevitably have affected some of the other eggs, which was not the case. It will also be seen from Mr. Wilmot's letters that the injury could not well have occurred after the eggs reached Newcastle. The obvious inference then is that the mischief must have taken place between Chicago and Newcastle.

NEWCASTLE, *October* 9, 1878.

Prof. SPENCER F. BAIRD,

United States Commissioner of Fisheries, &c., Gloucester, Mass.:

DEAR SIR: I hasten to inform you, as mentioned in my telegram of yesterday, of the loss of the California eggs that you were kind enough to have sent to me from the McCloud River. The real cause of their death I cannot fully comprehend, but I am inclined to believe that they must have got overheated on the road.

I got a letter from Mr. Stone in September, stating that half a million of eggs would be sent me, and that they would be shipped on or about 28th September, from Redding to Chicago in a refrigerator-car, thence by express to their destination, and that the express company would notify me by telegram when the eggs left Chicago. I also got a postal card from Mr. Stone, dated 23d September, notifying me that 5 per cent. more than the number of eggs ordered would be added to the shipment.

The express agent here, on the morning of the 5th instant, informed me that five crates of eggs had arrived by the morning train and that they were at his office. (This was the first and only notice I received of their coming since receipt of Mr. Stone's letters.) I immediately sent my assistants for them, giving them instructions to handle them carefully and walk the horses slowly from the office to the fishery (about a mile); in the mean time I had my men clean out the hatching-troughs, through which a full flow of water had been running for some time, and also rinse off the trays, to be in readiness for the coming eggs. I was present at this time, and when the eggs arrived I saw them carefully taken off the wagon and carried to the end of the fishery. I then opened the first crate myself. Before doing so I examined the manner in which it was arranged, which was most satisfactory. The outer covering of the inner boxes was well packed with fern leaves, and there was a center chamber dividing the two inner boxes, in which a quantity of fresh ice still remained. This ice must have been put in only a short time before, as some of the pieces were quite large and almost filled up the entire width of the chamber.

Before opening the inner boxes containing the eggs, I pushed a thermometer into the fine moss round and about the eggs, to ascertain its temperature, so that no sudden change would be made in unpacking the ova. The moss gave a record of 54°; the air inside and outside the building (it was a dark, cloudy day) gave a record of 54° also. I then tried the water in the stream and in the hatching-troughs, and found it, after several trials, to show 53° to 53½°. This being so very favorable, I set to work with the most satisfactory anticipations for success.

To make matters as equal as possible, I also sprinkled water on the inner boxes containing the eggs before unscrewing the slats, and allowed the water to percolate through the moss and amongst the eggs. The ova was then taken out by gently lifting each layer with the muslin cloth under them, and immersing the eggs slowly in the hatching-troughs; these were 12 feet long, 12 inches wide, 5 inches deep, with a full flow of water running constantly through them; in opening out the eggs, my assistant, who has been with me several years, drew my attention to the eggs, or the embryo inside, beginning to turn a whitish color; this I noticed clearly. In opening and removing the moss, they presented a healthy appearance, with the usual dark, red color, but almost immediately began to show a faint opaque white streak along the back of the embryo; some showed it more than others. We got through with the operation of unpacking on the evening of Saturday. Some of the eggs were placed loosely on the bottom of the troughs, with a couple of inches of water running over them; others were placed on hatching-trays. A very few, indeed, in removing gave evidence of life or motion inside. I picked out a few dozen that showed life and put them by themselves; these turned out in the course of a few hours just the same as the others, with the opaque white line. As I had a similar loss the previous season, I concluded there was no hope for their safety. I examined the troughs and trays after night, and found the lines of mortality more plainly visible, and on Sunday morning I concluded the result to be almost a total loss. To-day, whilst writing, I can notice a few eggs here and there yet looking as if they were all right, but I fear the white fever, not the *yellow*, has struck them, with no hope of recovery. Every day since Saturday I have picked out a dozen or so that I hoped were healthy, as they gave some signs of life, and put them carefully by themselves, but with the same result—a few hours afterwards, death—and to-day I fear we shall have none left.

Now, the question arises, what has caused this mortality? Has it occurred with other lots sent elsewhere? Or am I alone the unfortunate one? I hope the latter may be the case, as it would be sad, indeed, if a similar fate has befallen all the rest. I am very anxious to learn the fate of the other shipments, and will be pleased to hear from you concerning them. Whether the cause of death took place before reaching Chicago, in the refrigerator-car, or since their redistribution there, I cannot say. The facts are, however, just as I have related, and I feel very sad at the loss, as I had contemplated sending a number of the eggs to our establishment on the Saguenay River, 300 miles below Quebec, from which place I distributed some thousands of the California fry, two years ago.

I am very fearful now that my expectations in reference to the California egg enterprise will be wholly frustrated.

The first lot of eggs I got from you previous to last fall came to hand in the best possible shape, not more than 2 per cent. or 3 per cent. being lost till time of hatching out. How matters have turned out wrong since I cannot tell. I may, however, state my belief that last year's loss was undoubtedly from overheating on the road, as the moss and eggs were steaming hot when they were opened. This year's shipment did not show that state of things on arrival here. Yet the overheating may have taken place before reaching Chicago, and the replenishment of ice may have cooled them off, but the stroke of death did not culminate till the opening out and exposure to the air and water here.

* * * * * * *

THE FISHERIES, NEWCASTLE, ONT.,
November 9, 1878.

LIVINGSTON STONE, Esq.,
Assistant United States Commissioner of Fisheries, &c.,
Charleston, N. H.:

DEAR MR. STONE: I received your favor of 4th instant, in reference to the California eggs which you were kind enough to forward (from your establishment on the McCloud River), to me in October last, and I can assure you that no one can feel more disappointed than I do at the loss of them, for I had set my mind upon going largely into the rearing of these Pacific salmon. However, the misfortune occurred in losing them all save about 1,000, and the question now to be solved is, how did the calamity happen? What was the cause of it?

You ask me certain questions concerning the death of the eggs. These I will answer *seriatim*, and if, from the replies I give you, you can form any correct idea why the loss should have occurred, no one will be more pleased than myself, as it will not only solve the mystery, but will also probably give a clew whereby similar disasters may be prevented in the future in connection with getting California ova from you.

I wrote Professor Baird on the 9th October, giving him particulars of the loss, &c. In all probability he has sent you the letter or a copy of it. I will, however, recapitulate a portion of it by saying, "that I got a letter from you in September saying that half a million of eggs would be shipped to me on or about 28th September," and that, when they arrived at Chicago, the express agent there would notify me by telegram when they would be expressed from that place. I also got a postal card from you, dated 25th September, that 5 per cent. more than the original number would be shipped. To make matters short in this letter, I have concluded to send you a copy of that portion of the letter referring to the loss, written on 9th October to Professor Baird, in which the particulars are minutely given. (See copy attached hereto.) You will observe in it that I did not receive any notice from the express company at Chicago when the eggs were sent on from there. My first knowledge of the ova after your letter and postal card was from the express agent here sending word to me that five crates of salmon had arrived. This notice was on Saturday morning, the 5th October. I will now take up your questions.

Question. At what hour did the eggs arrive at Newcastle?

Answer. The express train from the West arrives at 9.25 a. m.; and very shortly after this time I was notified of the arrival of the eggs.

Question 2. What express company delivered them?

Answer. The Canadian and American Express Company.

Question 3. Was there ice on top of the crates and in the ice-chambers?

Answer. I am not aware of any ice being *on top* of the crates and think there was not, but there was ice in the ice-chambers of the crates.

Question 4. Was the express-car warm in which they were brought to Newcastle?

Answer. This I cannot answer, nor can the agent here tell me, as the cars only stop a moment or so at the station, and no observation was taken at the time.

Question 5. How long after arrival at Newcastle were they unpacked?

Answer. The unpacking commenced between 10 and 11 o'clock a. m., and the work was completed about 4 p. m.

Question 6. Were they likely to grow cooler or warmer in the place where they were kept at Newcastle before unpacking?

Answer. There could be only a very little change, as the day was a very dark, lowery one and pretty cool, the thermometer inside and outside the building ranging at 54°; there were no fires on the premises, neither was there sunshine.

Question 7. Did the eggs appear to be dead on being opened, or was it after they were placed in the water that they showed that they were spoiled?

Answer. At the first glance, when moss and muslin were removed, the eggs looked bright and red, but upon close examination life and motion were only noticed in a few, and my assistant (who has been engaged in the general work in connection with fish hatching, &c., in the establishment for several years) drew my attention to this, stating at the same time that he was fearful that they were going to turn out as those did last year, as he could see a faint whitish line along the embryos in the eggs. I noticed this also. This gave us cause to take extra care in unpacking. A thermometer was put amongst the moss between the layers, which gave a temperature of 54°. The water in the troughs stood at 53° to 53½°, and the air outside and inside the building was 54°. These were very favorable circumstances, and each of us began to remove the eggs, first sprinkling water over the moss in the boxes, then gently removing the upper layer of moss with the muslin, and lifting up the eggs with the muslin underneath them, and carefully immersing the eggs in the trough immediately alongside the packing boxes, so that in each case the eggs in the muslin cloth were not carried beyond 3 or 4 feet before immersing in the water. It was observable that little or no life was noticed by movement of the embryos as is usually the case when handling them, but the faint opaque white line became more apparent when placed in the water. I took part personally with my men in opening two of the crates, noting the above particulars. The opening of the other three crates was performed by my assistants in the same manner and with precisely the same results.

Question 8. Did all the crates open just alike, or were some in worse condition than others?

Answer. There was no perceptible difference in the crates. My assistant thought one slightly better looking, but in the end all proved alike.

Question 9. Did any of the eggs appear to have hatched on the way?

Answer. I may say, no. There were, however, just half a dozen or so that gave signs of premature hatching, but the number was so trifling as hardly to deserve notice.

Having answered your queries as clearly as I possibly can, I hope you may glean something from them that may give a clew to the loss. I must say that I cannot imagine the real cause. What strikes me with great surprise, is how it was that all the other consignments turned out so well and mine so badly. The inference would be that the difficulty must have taken place at Chicago in reshipping, or on the road from that place to this. From what I can learn, the time taken between Chicago and here by express is about 48 hours. At what time the eggs reached Chicago from Sacramento I have not precisely learned, but I think I saw some notice of the arrival of a car load of California eggs at that place about the 2d or 3d of October. If this were the case, and it was the same shipment by which mine came, no time would have been lost between Chicago and here for their carriage.

The next question arises, how many transhipments were there between the places, and could injury have been caused whilst transhipping? Not getting any bill of lading of their shipment at Chicago or upon their arrival here, I cannot particularly answer this; but there would no doubt be a transhipment at Detroit from the American road to the Canadian or Great Western Railway to reach Hamilton and Toronto. At Toronto there would be another transhipment from the Great Western line to the Grand Trunk Railway in order to reach Newcastle. This would make *two* changes of cars (or *three* if a change was made at Hamilton for Toronto), with new express carriers at each change, and from the great monopoly of the express company, and consequent carelessness of many of its employés, roughness of handling the crates, on account of their size and weight, might be the cause of injury, or heated cars (though this could not be the case, as there was plenty of ice in the chambers). It may be that these crates have been tumbled out of the cars like cord-wood, or barrels of pork, or crates of hardware, and the eggs became injured by concussion in falling, and thus killing them. Yet I am doubtful whether this theory will hold good, as it is perfectly astonishing the knocking about that eggs sometimes get and yet receive no injury. If the injury did take place from the last-mentioned cause, it would be impossible to find out where the blame was to be placed, from the many changes in transhipment and no one in particular looking after them. In opening some of the crates the layers were very much displaced, some being quite to one side, as if forced there by some pressure or shock. There were no labels or directions on the crates giving special instructions for "careful handling," or "keeping this side up with care," so that they may have been carried in the cars or in express wagons on their "sides" or "ends." There was a painted address on each, *Sam Wilmot, Newcastle, Ont.*, 105,000 *fish-eggs.*

As you will find in my note to Professor Baird (copy herewith), the crates were brought from the village of Newcastle, which is about three-quarters of a mile from the fishery, in my own wagon, walking the team all the way; they were unloaded in my presence and under my directions, with every possible care. I opened two of them myself and helped remove the eggs, as described, taking, as far as my experience and judgment were concerned, every precaution to prevent any possible injury to the ova; yet the consequences have been as related. I was not present at the opening and laying down of the three last crates, being called away to make the customs entries, &c. My assistants, however, followed the same course I did with the first crates. About 6 p. m. my head man informed me that he was afraid the eggs would all be bad; when I saw them a couple of hours later I came to the same conclusion. On the following morning (Sunday) I saw the white mark on almost every egg. Now and then an egg was noticed with the embryo in it alive, giving rapid, jerky-like motions; these few were picked out and put by themselves, but they died too. During the following few days the men kept close watch and were constantly looking out to find any eggs that might prove sound; and out of the whole half million we managed to get between one and two thousand that had not succumbed to the malady, or whatever else you may call it; these few hatched out in about five or six days after, and we have them yet (looking well) as the last remnant of the *Livingstone* consignment.

In connection with the history of these five *large crates*, and the one *large crate* of last year, it is strange that they should *all* have gone in a somewhat similar way, whilst the former smaller packages of 10,000 and 50,000 in previous years all came to hand in the very best of condition; in fact, the loss in them was extremely trifling. None of these latter-mentioned good consignments hatched out for five or six weeks after being laid down. In the crate of the fall of 1877 there was *not one good egg.*

These, without doubt, were killed from overheating, as the moss and eggs when opened were *steaming hot*. The five crates this fall did not present this steaming or overheated appearance upon opening, yet this opaque white line became visible almost immediately after opening and being put in the troughs, and the one or two thousand that we saved or picked out from the lot, hatched out in a few days after. This, to a certain extent, would show that they must have had more than ordinary warmth for their safety; otherwise they would not have hatched out so prematurely.

In order to get every good or apparently living egg from the large mass on the trays and in the troughs, we kept them on hand as long as we could, in fact till they became unpleasant to the smell; but during this time there was no growth of fungus or byssus upon them. The embryo or young fry inside (which was quite visible in all of the eggs) turned that pallid or opaque white color which always denotes death. I sent a lot of the eggs to Professor Baird that he might examine them; I did not hear of the result.

I have packed and unpacked a very great many fish eggs, sometimes with losses, but as a rule pretty successfully. The loss with these five crates I must confess upsets me; the more so, when you report all the other consignments as unusually good. This being the case, my lot must have come to grief in some one of the following ways, presuming they arrived all safe at Chicago:

1st. By detention or injury received at Chicago before transhipment.

2d. By overheating or exposure, or both, in transitu here.

3d. By rough, improper handling of the crates in transhipment from place to place and on the cars.

There was one thing which struck my attention in opening the first crate, namely, the perfect state the ice was in in the ice chambers, the appearance almost denoting that it had only just been put there; the pieces of ice were large, almost filling up the chamber; in others it was not so apparent. I was under the impression at first that forty-eight hours on an express car would have almost melted any ice put in at Chicago, yet the weather was cool in the beginning of October, and the ferns in the boxes may have kept the ice in the good condition in which it came here.

I must congratulate you upon your success in procuring the immense number of eggs you did this season—some 12,000,000, I believe—and I have much pleasure in acquainting you of my success at the several establishments under my control, the returns from my assistants showing up to the present time upwards of 8,000,000 of salmon eggs laid down. The salmon, trout, and white-fish season being now in its prime, and being busily engaged in collecting the eggs, I cannot yet tell you the result; but I am fearful, the weather having been so very unfavorable, we shall not secure the supply we should like to get.

Let me hear from you, not only on this unpleasant subject of the loss of eggs, but on any other kindred matter in fish culture, in all of which you are so thoroughly conversant.

Excuse my very long and somewhat prosy letter, but when details are to be given, both time and paper must be sacrificed.

Believe me to be yours, very truly,

SAMUEL WILMOT,
Superintendent Fish Culture for Canada.

I may mention here that the supplementary hatching-house did excellent service in helping us to eke out the quota of eggs for the two ice-cars. For illustration, all the eggs going into the first car had to be taken within a period of about a week, because those that were taken before that were in danger of being too far advanced to go in the car, and those taken after that were likely to be not far enough advanced. The supplementary hatching-house, which matured the eggs eight days quicker than the regular hatching-house, by virtue of its warmer water-supply, here came very conveniently to our aid by furnishing the additional half million eggs just when they were wanted.

On the 3d of October the balance of the eggs were sufficiently matured to load the second car. About two millions and a half (2,500,000) still remained in the hatching-house after both the cars were loaded and sent off. These were afterward hatched by Mr. Myron Green and Mr. James Richardson and placed by them in excellent order in the McCloud, Pit, and Little Sacramento Rivers, all tributaries of the Sacramento.

Packing and shipping the eggs.—The packing and shipping of the eggs, as well as the taking, maturing, and hatching of the eggs, passed off more smoothly this year than usual. The packing was done with marvelous rapidity and reflects great credit on all concerned in it, particularly Mr. James Richardson and Mr. Patrick Riley, who placed the layers of eggs in the boxes. Had not the character of the packing, as shown by the way in which the boxes finally opened, been made the subject of unusual commendation from the parties who were engaged in unpacking the eggs at their destination, I should hardly venture to say how rapidly they were packed, lest it might be thought to imply undue haste or want of care. I will, however, under the circumstances, state that the eggs were actually packed at the rate of half a million an hour, and I will add my own testimony also, that I never saw eggs packed with more care, fidelity, and pains, the rapidity with which the work was dispatched being wholly the result of experience and skill and the enthusiasm with which every one employed did the part of the work which fell to his share.

The manner of packing the eggs was in general the same as last year, the only difference being that this year the packing-boxes were made an inch larger both in length and width in order to give more room for the eggs. I, however, took especial pains this year to send large measure, in most instances giving from 5 per cent. to 50 per cent. more than were ordered.

One circumstance must be mentioned here which, though at first it seems unimportant enough, would be attended with the most serious consequences if not provided against. I refer to the diminution of the moss supply. Little by little, each year for seven years, we have encroached upon the supply of moss within our reach. This year we had to go away beyond the Sierra Nevada range to the sage-brush region of Shasta Valley to get our moss, and I am informed by the moss-gatherers that even that source of supply is now exhausted. To a New Englander, at least, the question of the moss-supply would seem trivial enough, and if, as is very unlikely, he could not get moss within a mile he would be willing to go two miles for it if necessary. But the question is not so easily settled in a dry country like California, and it is undoubtedly a fact that there is not within a hundred miles of the United States fishery on the McCloud River an accessible spot where moss can be obtained next year in any considerable quantity. It may, therefore, become necessary next year to meet the subject in some new manner, probably by shipping the moss from the Eastern States or Oregon, or sending an expedition to the neighborhood of Lake Tahoe for it, a distance by the traveled route of about five hundred miles.

I will close this report by making a crude statement of the work which was done at the fishery the last forty days preceding the loading of the second car on the 5th day of October. During this time we caught and examined, one by one, nearly 200,000 salmon. We took and impregnated at least 14,000,000 eggs. We went over almost daily the 14,000,000 eggs and picked out the dead ones. We washed and picked over, almost sprig by sprig, 220 bushels of moss. Our Indians collected and brought in on their backs four tons of ferns for outside packing, sometimes going two miles to get them, and we packed and crated, and loaded into the car at Redding eight or nine million salmon eggs, in addition to making new wire trays, packing-boxes, &c., &c., and doing the thousand little things which are constantly coming up to be done at a place like the fishery. All this work required an average of ten white men and twenty Indians for the forty days referred to.

Supplementary to this report will be found the following tables:

(1.) Table showing the observations taken of wind, weather, and temperature for the season of 1878.

(2.) Table showing the daily number of salmon eggs taken and salmon spawned.

(3.) Table showing the weights of salmon spawned.

(4.) Table showing the distribution of the eggs.

(5.) Catalogue of collection made for the Smithsonian Institution.

LIVINGSTON STONE.

TABLE I.—*Table of temperatures taken at the United States salmon-breeding station, McCloud River, California, during the season of* 1878.

Month.	Air.				Lowest night temperature.	Water.			Wind.	Weather.
	Shade.			Sun.						
	7 a. m.	3 p. m	7 p. m.	3 p. m.		7 a. m.	3 p. m.	7 p. m.	3 p. m.	
	°	°	°	°	°	°	°	°		
May 20	......	......	......	......	......	......	......	......	....	Cold rain.
21	55	56½	52	......	......	50	50	50	....	Do.
22	54	69	59	......	......	51	51½	52	....	Rainy, a. m.; clear, p. m.
23	60	86	66	106	......	52	55	54	....	Clear.
24	62	95	78	120	......	53	56	57	....	Do.
25	58	100	76	124	......	53	57	56	....	Do.
26	56	93	70	116	......	54	57	57	....	Do.
27	63	80	66	96	......	54	55	55	....	Do.
28	53	66	62	......	......	50	51	51	....	Showers, p. m.; rainy, night.
29	52	64	56	......	......	50	51	51	....	Showers, p. m.
30	50	67	57	80	......	49	52	51	....	Do.
31	53	66	64	93	......	50	53	53	....	Clear, a. m.; cloudy and showery, p. m.
June 1	58	76	68	102	......	52	54	54	....	Cloudy, a. m.; clear, p. m.
2	58	89	72	112	......	52	54	54	....	Clear.
3	56	92	69	114	......	52	56	56	....	Slightly cloudy, p. m.
4	56	98	77	120	......	54	57	57	....	
5	58	103	81	122	......	54	56	56	....	
6	72	103	82	130	......	55	58	59	....	Slightly cloudy, p. m.
7	64	92	79	114	......	56	59	58	....	
8	63	99	80	112	......	56	59	58	....	
9	72	103	79	126	......	56	59	58	....	Slightly cloudy, p. m.
10	72	97	80	114	......	56	59	58	....	
11	64	92	......	114	......	56	59	58	....	Cloudy; after 4 p. m., clear.
12	72	100	76	124	......	56	59	58	....	Clear.
13	66	100	76	124	......	56	58	58	....	Do.
14	61	101	77	128	......	56	58	58	....	Do.
15	67	86	76	100	......	56	58	58	....	Cloudy, a. m.
16	62	84	72	106	......	56	58	58	....	
17	67	87	76	87	......	56	59	58	....	Cloudy, p. m.
18	66	96	79	128	......	56	59	58	....	Clear.
19	65	95	78	122	......	56	59	58	....	Do.
20	63	88	78	88	......	57	59	58	....	Cloudy.
21	72	103	81	124	......	56	59	58	....	Clear.
22	64	103	87	126	......	56	59	59	....	Cloudy, p. m.
23	66	92	75	118	......	58	60	59	....	Clear.
24	70	94	80	124	......	57	59	58	....	Do.
25	72	102	80	123	......	56	59	58	....	Do.
26	76	108	81	130	......	56	50	59	....	Do.
27	62	96	83	109	......	57	60	59	....	Cloudy; rain, p. m.
28	56	88	74	108	......	57	59	58	....	Clear.
29	66	97	78	118	......	55	58	57	....	Do.
30	70	105	79	128	......	57	58	57	....	Do.
July 1	64	98	83	122	51	56	59	58	N.	Light clouds; clear.
2	58	85	72	114	47	56	59	58	N.	Light clouds.
3	58	78	68	102	50	56	57	56	N.	Do.
4	60	86	73	112	42	53	57	56	N.	Do.
5	65	95	73	116	53	54	58	57	N.	Clear.
6	58	97	73	120	42	54	58	57	N.	Light clouds.
7	60	97	74	122	50	55	59	58	SW.	Clear.
8	57	100	78	128	45	55	59	58	SW.	Do.
9	62	99	80	127	50	56	59½	58½	SW.	Cloudy.
10	67	85	76	116	57	56	59½	58½	SW.	Clear.
11	62	82	75	112	51	56	59	58	SW.	Do.
12	67	102	78	120	50	56	60	58	N.	Do.
13	59	104	80	123	57	56	60	59	SW.	Do.
14	61	103	84	120	54	57	60	59	SW.	Do.
15	61	98	84	118	67	57	60	59	SW.	Light clouds.
16	62	81	74	90	60	57	58	57	S.	Cloudy.
17	75	78	68	84	58	55	56	56	N.	Do.
18	65	93	73	113	47	55	58	57	N.	Clear.
19	64	95	76	112	46	55	59	58	W.	Do.
20	57	97	77	112	51	56	60	59	SW.	Do.
21	52	95	76	115	54	56	60	59	SW.	Do.
22	57	100	75	118	47	56	59	58	SW.	Do.
23	64	97	76	114	46	56	59	58	SW.	Do.
24	55	96	78	114	46	56	60	59	SW.	Do.
25	52	102	80	126	47	56	60	59	N.	Do.
26	54	104	80	126	54	56	60	59	NE.	Do.
27	56	101	83	123	54	56	60½	59½	SW.	Do.
28	62	102	82	124	58	56	61	60	SW.	Do.
29	64	99	80	118	53	57	60	58½	SW.	Do.

TABLE I.—*Table of temperatures, &c.*—Continued.

Month.	Air.				Lowest night temperature.	Water.			Wind.	Weather.
	Shade.			Sun.						
	7 a. m.	3 p. m.	7 p. m.	3 p. m.		7 a. m.	3 p. m.	7 p. m.	3 p. m.	
	°	°	°	°	°	°	°	°		
July 30	57	104	80	124	52	56	59	58½	NE.	Clear.
31	56	98	80	116	50	56	......	......	N.	Do.
Aug. 1	55	99	78	115	51	56	60	59	E.	Do.
2	55	100	79	118	50	56	59½	59	NE.	Do.
3	55	97	80	114	50	56	60	59	SW.	Do.
4	60	95	79	112	53	56	60	59	SW.	Do.
5	55	91	77	108	51	56	60	59	SW.	Do.
6	55	98	76	121	49	56	60	59	SW.	Do.
7	56	97	76	120	50	56	59	58	SW.	Do.
8	56	102	77	126	51	55	59	58	SW.	Do.
9	58	108	79	128	49	56	59	58	N.	Do.
10	56	110	80	128	50	56	59	58	NE.	Do.
11	59	110	83	134	51	56	60	59	NE.	Do.
12	60	112	83	133	50	56	60	59	SW.	Do.
13	62	106	84	132	52	56	60	60	S.	Light clouds.
14	64	102	88	122	59	57	60	61	S.	Cloudy.
15	75	98	80	121	61	57	59	61	N. & S.	Do.
16	60	94	80	110	50	56	59½	59	NE.	Fair.
17	56	93	76	96	50	56	58	58	NE.	Light clouds.
18	57	99	73	120	49	56	58	58	N.	Hazy and smoky.
19	55	96	78	116	47	55	58	58	......	
20	54	92	77	108	49	57	59	59	......	Clear.
21	58	76	68	80	49	55	57	56	S.	Cloudy.
22	60	83	72	102	56	54	57	57	SW.	Light clouds.
23	52	84	66	104	......	54	57½	56½	......	Clear.
24	47	88	70	114	43	53	57	56	......	Do.
25	51	100	75	121	48	54	58	57	NE.	Do.
26	......	98	78	119	......	......	58	57	SW.	Do.
27	56	102	84	118	52	56	59	58	SW.	Do.
28	58	99	77	105	53	56	60	59	SW.	Do.
29	62	109	78	122	59	57	60	59	NE.	Do.
30	58	109	83	118	54	56	60	59	NE.	Do.
31	60	107	82	126	59	57	59	59	......	Do.
Sept. 1	56	96	76	117	50	56	59	58	SW.	Fair.
2	53	98	76	114	48	56	58	58	N.	Do.
3	64	97	83	108	59	56	58	58	N.	Do.
4	60	99	76	109	55	56	58	57½	N.	Do.
5	52	100	76	120	47	56	58	58	N.	Do.
6	52	97	76	114	47	56	58	58	N.	Do.
7	56	99	78	108	47	55	58	58	N.	Do.
8	52	97	......	104	45	......	......	......	N.	Do.
9	52	......	72	......	44	54	......	56	N.	Do.
10	52	......	71	......	47	54	......	56	N.	Do.
11	52	96	73	96	47	53	56	56	N.	Do.
12	50	93	67	97	44	53	56	56	SE.	Do.
13	45	90	65	95	40	53	56	56	......	Do.
14	45	......	64	......	38	52	......	55	......	Do.
15	41	76	......	82	37	51	55	......	......	Do.
16	49	72	60	72	44	52	54	54	S.	Cloudy.
17	......	76	63	80	......	......	54	54	S.	Do.
18	45	76	64	81	40	52	55	54	......	Fair.
19	44	85	65	90	40	52	55	54	......	Do.
20	46	86	65	93	42	51	55	55	......	Do.
21	46	92	66	101	41	51	55	55	SE.	Do.
22	44	96	67	108	39	51	55	55	......	Fine.
23	46	94	69	111	41	51	55	54	......	Do.
24	44	92	65	103	39	50	54	54	......	Do.
25	44	98	65	112	40	51	54	53	......	Do.
26	......	79	......	79	......	......	53	......	S.	Cloudy.
27	54	73	59	81	48	52	54	53	S.	Do.
28	54	56	54	56	49	51	52	51	S.	Rain.
29	50	52	52	52	49	50	51	51	......	Do.
30	52	60	56	60	47	50	51	50	......	Do.
Oct. 1	48	84	54	98	46	50	53	53	N.	Fair.
2	47	90	64	102	44	50	54	53	N.	Do.
3	50	96	64	110	45	51	54	53	N.	Do.
4	48	96	63	106	44	50	54	53	N.	Do.
5	46	92	64	111	41	50	53½	53	N.	Do.
6	52	100	63	115	48	51	54	54	N.	Do.
7	50	......	......	......	46	51	......	......	N.	Do.
8	......	92	......	108	......	......	54	......	N.	Do.
9	43	79	......	86	37	51	54	......	N.	Cloudy.
10	43	88	......	101	38	50	53	......	N.	Fine.
11	46	79	52	86	39	50	52	52	SW.	Cloudy.

TABLE I.—*Table of temperatures, &c.*—Continued.

Month.	Air.				Lowest night temperatures.	Water.			Wind.	Weather.
	Shade.			Sun.						
	7 a. m.	3 p. m.	7 p. m.	3 p. m.		7 a. m.	3 p. m.	7 p. m.	3 p. m.	
	°	°	°	°	°	°	°	°		
Oct. 12	52	55	50	55	......	49	50	50	S.	Rain.
13	47	52	51	52	44	49	51	51	SE.	Cloudy.
14	48	49	49	49	......	49	49	49	....	Rain.
15	43	72	46	82	37	48	50	49	S.	Fine.
16	53	83	53	96	39	47	50	49	S.	Do.
17	40	84	54	94	37	47	51	50	SE.	Do.
18	41	84	54	103	36	47	51	50	SE.	Do.
19	40	75	64	75	38	48	51	50	S.	Cloudy, a. m.; fine, p. m.
20	39	80	50	91	36	47	51	50	SW.	Fine.
21	46	88	......	105	42	47	51	......	SE.	Do.
22	42	91	57	105	37	47	51	50	SE.	Do.

TABLE II.—*Table of salmon-eggs taken at the United States salmon-breeding station, McCloud River, California, during the season of* 1878.

Date.	Number of eggs taken.	Total number of eggs taken.	Number of salmon spawned.	Total number of salmon spawned.
August 20	30,000	30,000	7	7
22	30,000	60,000	8	15
23	62,000	122,000	19	34
24	54,000	176,000	17	51
26	110,000	286,000	29	80
27	152,000	438,000	46	126
28	302,000	740,000	83	209
29	306,000	1,046,000	82	291
30	444,000	1,490,000	106	397
31	496,000	1,986,000	138	535
September 2	682,000	2,668,000	202	737
3	348,000	3,016,000	112	849
4	374,000	3,390,000	118	967
5	422,000	3,812,000	130	1,097
6	582,000	4,394,000	179	1,276
7	578,000	4,972,000	163	1,439
8	740,000	5,712,000	233	1,672
9	578,000	6,290,000	190	1,862
10	714,000	7,004,000	211	2,073
11	894,000	7,898,000	258	2,331
12	722,000	8,620,000	213	2,544
13	858,000	9,478,000	218	2,762
14	920,000	10,398,000	268	3,030
15	500,000	10,898,000	154	3,184
16	648,000	11,546,000	195	3,379
18	700,000	12,246,000	221	3,600

Total number of eggs taken 12,246,000
Total number of salmon spawned 3,600

TABLE III.—*Table showing the weights of salmon spawned on various days at the United States salmon-breeding station, McCloud River, California, during the season of* 1878.

[The salmon were weighed after the eggs had been taken from them.]

AUGUST 28, 1879.

Number.	Weight in pounds.	Number.	Weight in pounds.	Number.	Weight in pounds.	Number.	Weight in pounds.	Number.	Weight in pounds.	Number.	Weight in pounds.	Number.	Weight in pounds.	Number.	Weight in pounds.
1	6	12	8	22	6	32	8	42	7	52	9	62	7	72	7
2	7	13	7	23	16	33	8	43	6	53	6	63	7	73	8
3	8	14	8	24	14	34	5	44	8	54	6	64	7	74	6
4	8	15	9	25	14	35	8	45	6	55	7	65	8	75	13
5	10½	16	16	26	11	36	7	46	7	56	11	66	7	76	12
6	9	17	6	27	16	37	14	47	7	57	7	67	7	77	7
7	5	18	6	28	14	38	7	48	7	58	7	68	7	78	7
8	6	19	7	29	7	39	9	49	10	59	15	69	9	79	7
9	16	20	14	30	8	40	7	50	16	60	7	70	8	80	6
10	9	21	14	31	8	41	17	51	6	61	7	71	7	81	15
11	15														

81 fish weighed; average weight, 8¾ pounds.

AUGUST 29, 1878.

Number.	Weight in pounds.	Number.	Weight in pounds.	Number.	Weight in pounds.	Number.	Weight in pounds.	Number.	Weight in pounds.	Number.	Weight in pounds.	Number.	Weight in pounds.	Number.	Weight in pounds.
1	16	12	15	23	7	33	7	43	5	53	9	63	8	73	7
2	8	13	7	24	7	34	17	44	5	54	6	64	5	74	7
3	10	14	8	25	8	35	13	45	7	55	6	65	6	75	7
4	9	15	6	26	15	36	8	46	7	56	5	66	8	76	7
5	14	16	8	27	9	37	7	47	8	57	8	67	7	77	6
6	6	17	7	28	8	38	9	48	7	58	7	68	7	78	11
7	12	18	7	29	7	39	14	49	7	59	7	69	5	79	5
8	7	19	7	30	11	40	5	50	7	60	7	70	5	80	7
9	8	20	8	31	14	41	7	51	6	61	10	71	7	81	8
10	7	21	7	32	14	42	17	52	8	62	6	72	6	82	5
11	8	22	17												

82 fish weighed; average weight, 8¼ pounds.

AUGUST 30, 1879.

Number.	Weight in pounds.	Number.	Weight in pounds.	Number.	Weight in pounds.	Number.	Weight in pounds.	Number.	Weight in pounds.	Number.	Weight in pounds.	Number.	Weight in pounds.	Number.	Weight in pounds.
1	14	15	8	29	7	42	6	55	6	68	7	81	5	94	9
2	9	16	7	30	8	43	6	56	12	69	8	82	7	95	8
3	17	17	8	31	7	44	6	57	7	70	13	83	6	96	8
4	12	18	7	32	6	45	8	58	9	71	9	84	8	97	7
5	16	19	7	33	7	46	8	59	8	72	14	85	7	98	7
6	8	20	14	34	8	47	8	60	6	73	8	86	7	99	8
7	8	21	10	35	6	48	5	61	6	74	8	87	8	100	7
8	8	22	11	36	7	49	6	62	11	75	6	88	6	101	6
9	7	23	8	37	7	50	6	63	7	76	15	89	7	102	6
10	8	24	8	38	6	51	10	64	6	77	14	90	6	103	6
11	7	25	6	39	6	52	8	65	11	78	8	91	11	104	8
12	5	26	7	40	6	53	7	66	13	79	9	92	15	105	8
13	8	27	7	41	9	54	6	67	11	80	7	93	7	106	6
14	8	28	8												

106 fish weighed, average weight, 8 pounds.

AUGUST 31, 1878.

Number.	Weight in pounds.	Number.	Weight in pounds.	Number.	Weight in pounds.	Number.	Weight in pounds.	Number.	Weight in pounds.	Number.	Weight in pounds.	Number.	Weight in pounds.	Number.	Weight in pounds.
1	17	19	6	37	9	54	7	71	14	88	8	105	4	122	7
2	5	20	9	38	7	55	9	72	13	89	8	106	5	123	6
3	7	21	6	39	8	56	7	73	7	90	7	107	4	124	8
4	8	22	13	40	17	57	15	74	8	91	9	108	7	125	8
5	14	23	4	41	6	58	7	75	15	92	7	109	5	126	6
6	6	24	7	42	5	59	8	76	7	93	7	110	7	127	7
7	7	25	8	43	9	60	7	77	11	94	8	111	7	128	7
8	7	26	3	44	6	61	6	78	7	95	6	112	7	129	6
9	13	27	6	45	6	62	6	79	13	96	6	113	6	130	5
10	6	28	7	46	7	63	7	80	7	97	6	114	5	131	6
11	7	29	7	47	8	64	7	81	5	98	7	115	6	132	8
12	7	30	6	48	9	65	5	82	16	99	7	116	12	133	17
13	8	31	12	49	7	66	7	83	13	100	6	117	7	134	8
14	8	32	12	50	7	67	9	84	8	101	5	118	6	135	8
15	5	33	16	51	8	68	7	85	6	102	8	119	6	136	8
16	7	34	6	52	6	69	8	86	6	103	6	120	8	137	7
17	9	35	7	53	6	70	7	87	8	104	7	121	6	138	6
18	8	36	5												

138 fish weighed; average weight 7¾ pounds.

SEPTEMBER 9, 1878.

Number.	Weight in pounds.	Number.	Weight in pounds.	Number.	Weight in pounds.	Number.	Weight in pounds.	Number.	Weight in pounds.	Number.	Weight in pounds.	Number.	Weight in pounds.	Number.	Weight in pounds.
1	7	7	7	13	7	19	6	25	9	31	7	36	6	41	8
2	8	8	9	14	13	20	7	26	8	32	7	37	14	42	6
3	6	9	8	15	10	21	6	27	8	33	7	38	7	43	7
4	6	10	6	16	8	22	7	28	6	35	7	39	6	44	7
5	5	11	8	17	5	23	8	29	7	35	8	40	8	45	4
6	6	12	7	18	7	24	6	30	7						

45 fish weighed; average weight 7½.

TABLE IV.—*Table of distribution of salmon eggs from the United States salmon-breeding station, McCloud River, California, during the season of 1878.*

State.	Commissioner or applicant.	Number asked.	Number forwarded.	Destination.
California	B B Redding	2,500,000	2,500,000	Sacramento River and tributaries.
Illinois	Dr. W. A. Pratt	100,000	100,000	Elgin.
Do	N. K. Fairbank	100,000	100,000	Chicago.
Do	Samuel Preston	200,000	200,000	Mount Carroll.
Iowa	B F. Shaw	250,000	250,000	Anamosa.
Do	W. A Mynster	50,000	50,000	Council Bluffs.
Kansas	B F. Shaw	100,000	100,000	Cedar Rapids.
Maine	Lorenzo Bailey	15,000	15,000	Pembroke.
Maryland	T B Ferguson	1,000,000	1,000,000	Baltimore.
Massachusetts	A H. Powers	100,000	100,000	Plymouth, N. H.
Do	E. A. Brackett	100,000	100,000	Winchester.
Michigan	Frank N. Clark	250,000	250,000	Northville.
Do	George H. Jerome	200,000	200,000	Niles.
Minnesota	Dr R O. Sweeny	1,000,000	1,000,000	Saint Paul.
Missouri	B. F Shaw	200,000	200,000	Anamosa, Iowa.
Nebraska	J. G. Romaine	100,000	100,000	South Bend, Ill.
Nevada	H. G. Parker	250,000	250,000	Carson City.
New Hampshire	A H. Powers	250,000	250,000	State hatching-house, Plymouth.
New Jersey	Mrs J. H. Slack	300,000	300,000	Bloomsbury.
Do	West Jersey Game Protective Society.	150,000	150,000	Mrs. J. M. Slack.
Do	Abram S. Hewitt	25,000	25,000	Do.
New York	Seth Green	100,000	100,000	Caledonia.
North Carolina	S. G Worth	350,000	350,000	Henry's Station.
Ohio	Castalia Springs Association	50,000	50,000	Cleveland.
Pennsylvania	James Duffy	150,000	150,000	Marietta.
Do	Seth Weeks	100,000	100,000	Corry.
Rhode Island	C. F. Reed	20,000	20,000	Reedsburg.
Utah	A. P. Rockwood	50,000	50,000	Salt Lake City.
Virginia	Prof. M. McDonald	300,000	300,000	Lynchburg.
West Virginia	C. S. White	500,000	500,000	Romney, W. Va.
Wisconsin	Wisconsin State hatching-house.	100,000	100,000	Madison.
Do	A E. Lytle	100,000	100,000	Geneva Lake
Canada	Samuel Wilmot	500,000	500,000	New Castle.
England	Prof. S F. Baird	100,000	100,000	England.
France	do	100,000	100,000	France.
Holland	do	100,000	100,000	Holland.
Germany	do	250,000	250,000	Germany.
New Zealand	Auckland Acclimatation Society.	200,000	200,000	Auckland.
	Total	10,310,000	10,310,000	

TABLE V.—*Catalogue of Natural History Collection made for the Smithsonian Institution in 1878, by Livingston Stone.*

475° to 500° are all from McCloud River, California.
475°. Trout. September, 1878.
478°. Trout. September, 1878.
479°. Salmon skin. September, 1878.
480°. Salmon skin. September, 1878.
481°. Trout. September, 1878.
482°. Trout. September, 1878.
483°. Salmon skin. September, 1878.
484°. Trout. September, 1878.
485°. Trout. September, 1878.
486°. Salmon skin. September, 1878.
487°. Salmon skin. September, 1878.
489°. Trout. September, 1878.
490°. Trout. September, 1878.
492°. Salmon skin. September, 1878.
493°. Trout. September, 1878.
494°. Salmon skin. September, 1878.
495°. Trout. September, 1878.
496°. Trout. September, 1878.
497°. Trout. September, 1878.
498°. Salmon skin. September, 1878.
499°. Trout. September, 1878.
500°. Rat. September, 1878.
570. Trout. McCloud River, California. July 1, 1878.
571. Trout. McCloud River, California. July 1, 1878.
572. Trout. McCloud River, California. July 3, 1878.
573. Trout. McCloud River, California. July 6, 1878.
574. Trout. McCloud River, California. June 30, 1878.
575. Trout. McCloud River, California. June 27, 1878.
576. Trout. McCloud River, California. June 30, 1878.
577. Trout. McCloud River, California. June 27, 1878.
578. Trout. Dolly Varden (Indian, Wye-dai-deek-it). McCloud River, California. July 6, 1878.
579. Salmon skin (female). McCloud River, California. July 14, 1878.
580. Trout. McCloud River, California. July 3, 1878.
581. Sacramento Pike. McCloud River, Cal. July 6, 1878.
582. Trout. McCloud River, California. July 14, 1878.
583. Trout. McCloud River, California. July 14, 1878.
584. Trout. Dolly Varden (Indian, Wye-dai-deek-it). McCloud River, California. July 14, 1878.
585 to 591. Salmon skins. (Males, 585, 586, 587, 588, 589, 590). McCloud River, Cal. July 15, 1878.
591 to 596. Salmon skins (females). McCloud River, California. July 15, 1878.

596. Trout. McCloud River, California. July 15, 1878.

597. Trout. McCloud River. July 15, 1878.

598. Trout. McCloud River, California. July 15, 1878.

599 to 604. Salmon skins (males). McCloud River, California. July 16, 1878.

604 to 611. Salmon skins (females). McCloud River, California. July 16, 1878.

Jar No. 1. Two Dolly Vardens. Clackamas River, Oregon. Winter 1877 and 1878.

Jar No. 2. Two Trout, one Dolly Varden. McCloud River, California. July, 1878.

Jar No. 3. Five Trout, one Dolly Varden, one Snake. McCloud River, California. July, 1878.

Jar No. 4. Birds. McCloud and Pitt Rivers, California.

611. Trout. McCloud River, California. August 17, 1878.

613 and 614. Trout. McCloud River, California. August 19, 1878.

615 and 616. Trout. Dolly Varden (Indian, Wye-dai-deek-it). McCloud River, California. August 15, 1878.

617. Trout. Dolly Varden. (Wye-dai-deek-it). McCloud River, California. September 1, 1878.

618, 619, 620, 621. Trout. McCloud River, Cal. August 23 and 26, 1878.

622, 623. Salmon heads (male). McCloud River, California. September 3, 1878.

624, 625. Salmon skins (females). McCloud River, California. September 3, 1878.

627, 628, 629, 630, 631, 632, 633, 634. Trout. McCloud River, California. August, 1878.

XXXIV.—REPORT OF SALMON-HATCHING OPERATIONS IN 1878, AT THE CLACKAMAS HATCHERY.

BY W. F. HUBBARD.

CLACKAMAS HATCHERY, OREGON, *February* 4, 1879.

To Professor SPENCER F. BAIRD,
United States Fish Commissioner:

I beg to report to you as follows: The first spawn of last season was taken September 5, 1878, when we took the spawn from one female salmon, the first one we had caught that was ripe. The next was taken September 7, when we took two females. In spawning the fish, sometimes one male would answer for one female; but we almost always used two, and sometimes three. When fishing, we always caught more males than females.

September 9 took the spawn from 4 females.
September 10 took the spawn from 5 females.
September 11 took the spawn from 7 females.
September 13 took the spawn from 12 females.
September 14 took the spawn from 23 females.
September 15 took the spawn from 22 females.
September 16 took the spawn from 19 females.
September 17 took the spawn from 32 females.
September 18 took the spawn from 27 females.
September 19 took the spawn from 38 females.
September 20 took the spawn from 36 females.
September 21 took the spawn from 43 females.
September 22 took the spawn from 35 females.
September 23 took the spawn from 32 females.
September 24 took the spawn from 20 females.
September 25 took the spawn from 27 females.
September 26 took the spawn from 31 females.
September 27 took the spawn from 24 females.
September 28 took the spawn from 21 females.

September 29 the river began to rise, caused by heavy rains, and we were not able to do any fishing, although we took the spawn from four fish which we had in pens built for the purpose of keeping the fish.

September 30 we took the spawn from three fish from the pens.

The river was still rising, and on the night of the 30th it washed away the rack, allowing all the fish that were below to go up the river. After that the river stayed high for two or three days, and when it got low enough for us to fish again all the fish were gone.

The number of eggs taken was 2,081,000.

The number of females spawned was 478.

There were more than twice as many males caught as there were females.

November 7 the dam which supplies the hatching-house with water broke, and we were obliged to take the eggs and young fish out of the house. The company has two flat-boats here, and we fastened them together and made a place between them for the eggs and fish. On the same day we turned into the river 300,000 young fish.

December 9 the river began rising again, and the current was so strong that it killed a good many of the fish, and we saw something must be done or we would lose them all. By this time we had had a good deal of rain, and all the small streams were full of water and we were able to turn the water from one of them, which has plenty of water in the winter, but is nearly dry in the summer, into the hatching-house, and once more the fish were put back into the house; at this time the eggs were nearly all hatched.

December 24 Captain Ainsworth took 3,000 young salmon, which were placed in a land-locked lake in Washington Territory.

December 26 600,000 young fish were turned into the river and Cleer Creek, a stream which runs into the Clackamas below the hatchery.

December 27 150,000 young fish were turned into the river at different points.

January 2, 1879, the last of the young fish were out, 150,000, which were also put in the Clackamas at different points up and down the river.

The total number of fish turned out is estimated at 1,203,000.

The large number of eggs and fish lost is attributed to having to move them from the hatching-house to the river and back, and also to the high-water while they were in the river, which killed a great many,

At the time the rack went out there were a great many fish below it and had it remained two or three weeks longer, we should probably have taken another million of eggs.

W. F. HUBBARD,
Assistant Superintendent.

Respectfully forwarded.

J. G. MEGLER,
Secretary O. & W. F. P. Co.

XXXV.—REPORT OF SALMON-HATCHING OPERATIONS ON ROGUE RIVER, OREGON, 1877-'78.

BY K. B. PRATT.

MR. LIVINGSTON STONE:

DEAR SIR: In accordance with your request, I will endeavor to give you a report of the proceedings at the salmon hatchery at Ellensburgh, Oregon, mouth of Rogue River.

During the summer of 1877, Mr. R. D. Hume, who had just completed a salmon cannery at Ellensburgh, visited the United States fishery on the McCloud River, California, and examined the hatching-house and the work being done there, and decided he would have a hatchery of his own upon Rogue River, in order to keep up the supply of salmon in that stream.

For nearly twenty years salmon had been taken in large numbers and salted, and there was a visible decrease in the number of fish returning to the river each year.

On returning to Ellensburgh, in September, Mr. Hume set about putting up a hatching-house, building it only a short distance from the mouth of the river. About 250 salmon were placed in a fresh-water pond, which had been dug close by the hatching-house, there to be kept until they were ready to spawn, but as there was an insufficient supply of water in the pond, many of the fish died, so that by the time they commenced spawning there were only about 100 left. Of these, 57 were females, from which about 215,000 eggs were taken. Just before spawning time a large cage, with three compartments, was built and sunk in the pond; then the pond was dragged with a net, and the fish placed in the largest compartment of the cage. From there the fish were caught in dip-nets and examined each day, the ripe females put in one division of the cage, and the ripe males in another. Mr. Hume's idea was to handle the fish as carefully as possible when spawning them, and to at once return them to the river, so that they could return to salt water as soon as they chose. With this idea in view, a contrivance for holding the fish while spawning was made, consisting of two pieces of light board fastened together on one side with hinges, and straps extending around the other side at the ends. In this the females were placed on their backs, the straps extending around the shoulders and tail, and with a little care they could not escape. The males were held by one of the men in his arms; so the necessity for taking the fish by the gills was avoided. As soon as the fish were spawned they were marked by cutting a piece from the dorsal fin, and immediately returned to the river, most of them swimming off quite vigorously. The first eggs were taken on November 23, and the last on December 12. The hatching-house was supplied with water from a small stream that was constantly roiled up by the cattle and horses running loose over the country, and then the heavy rains swelled the stream to an unnatural size, and leaves, twigs, and mud would be swept down into the tank and choke up the flannel screens, so that it was necessary to clean them every few minutes, and a watchman was kept on duty all night to see that there was a good supply of water running through the trough all the time. The temperature of the water was sometimes as low as 38°, ranging from that up to 54°, averaging about 47°. The eggs were from 23 to 27 days in showing the eye spots, and from 56 to 60 days in hatching, a few not hatching till 64 days old. After keeping the young fry from three to four weeks they were taken up the river and placed in some small creeks 1½, 6, and 12 miles from the mouth of the river. Owing to the difficulty in keeping a pure supply of water running through the trough, and many other adverse conditions, there was a large loss in hatching the eggs, probably 30 per cent.

Again, in transporting the young fry up the river to the small creeks where they were planted there was a considerable loss, owing to the overcrowding in the tubs and pails in which they were carried. However, at least 100,000 healthy young fry were planted in the streams, and probably many of those that were thought to be suffocated revived after being turned into the stream, for some were seen to swim off after a few minutes.

XXXVI.—REPORT ON AN ATTEMPT TO COLLECT EGGS OF SEBAGO SALMON IN 1878.

By Charles G. Atkins.

1.—HABITAT OF SEBAGO SALMON.

Within the limits of the State of Maine there are known to be four distinct localities inhabited from olden times by fresh-water salmon, commonly called "landlocked salmon." The first of these districts is in the valley of the Saint Croix River, mainly in Grand Lake and connecting waters, on the west branch or Schoodic River, whence the name "Schoodic salmon." The second is Reed's Pond, Union River, Hancock County; the third is Sebec Lake and vicinity, tributary to the Penobscot; and the fourth is Sebago Lake and vicinity, tributary to the Presumpscot River.

Lake Sebago, the principal haunt of the salmon in this district, is the second largest body of fresh water in Maine. It has an area of about sixty square miles. Its depth is known to exceed 100 feet, and is reported to be in places not less than 400 feet deep. Its shores are for the most part sandy, but in some places gravelly and stony, and in a few places the solid ledge comes down steeply to the water's edge. A large portion of the country draining into the lake is also sandy and gravelly, and the streams are generally clear, though considerably discolored by peat swamps.

Though in the midst of a country long since settled, the immediate shores of the lake are almost wholly clothed with forests of recent growth, their sterile character forbidding any extensive attempt at farming.

Sebago Lake discharges its waters into the Presumpscot River, which empties into Casco Bay near Portland. The entire length of this river is about twenty-two miles. It descends rapidly, having a total fall of 247 feet between the lake and the sea, yet in its natural condition there was no impediment to the free passage of fish up and down. There were many rapids which were doubtless resorted to by spawning salmon. For many years, however, the river has been obstructed by many high mill-dams, which have entirely prevented the ascent of fish. The descent is of course still open, and the fresh-water salmon are occasionally taken on all parts of the river.

The principal affluent of Sebago Lake is Songo River, which drains the country lying to the north. Songo River itself is very short, forming merely the connecting link between Sebago Lake and an extensive chain of ponds (so called) above. In a straight line the distance from the lake to the first pond (Brandy Pond) is not over three miles. By the course of the river, which is sinuous to an extraordinary degree, the distance may be twice or thrice as great. A short distance below Brandy Pond the river is crossed by a dam and lock to improve the navigation, which is pursued not only by freight boats, which formerly ran by canal to Portland, but now only across the lake to a station of the Portland and Ogdensburgh Railroad, but also by steamers conveying passengers as far as Bridgton, on Long Pond. Immediately below this lock the Songo receives its main affluent, Crooked River, a stream that rises nearly forty miles to the northward and follows a very sinuous course from a country of granite hills down through sandy and gravelly intervals.

The Songo itself affords no spawning-ground for the salmon, almost its whole sluggish course being through a low-lying country, and the entire fall, except at the lock, being but a few inches. The Crooked River, however, is rapid through its whole course, except where here and there interrupted by dams and mill-ponds. In old times, doubtless, the whole length of this stream formed the breeding-grounds of the salmon. At present only that portion is accessible which lies below the village of Edes Falls, not exceeding, probably, six miles in length. There are, however, in this short distance, many gravelly rapids where the salmon spawn.

Besides Songo River there is but one other stream known to be a breeding-ground for the salmon inhabiting Sebago Lake, namely, Northwest River. Mr. Buck visited this stream in November and found it accessible to fish for only about a mile from the lake, a mill-dam intercepting further progress. At that time the stream was about 20 feet wide and 18 inches deep, with a moderate current.

It is also currently reported that the salmon spawn on gravelly bars and beaches in the lake itself. This is not improbable, though Mr. Buck explored Sandy Beach, which is singled out by report as the special place for this sort of work, without finding any indications of fish having resorted to it in 1878.

Besides Sebago Lake itself, the same variety of salmon inhabit Long Pond, the most considerable body of water drained by Songo River, eleven miles long but quite narrow, having an area of nine or ten square miles. The principal breeding-ground of the Long Pond salmon is Bear Brook, which comes in from the north near the village of Harrison. Doubtless other streams were once frequented by them, but not in recent years.

2.—CHARACTERISTICS OF SEBAGO SALMON.

First of all the Sebago salmon are distinguished from the sea-going salmon on the one hand, and from the Schoodic and Sebec salmon on the other, by their size. As exhibiting the result of my own observation in 1867 and such researches as I was able to make at that time, I extract the following from the Maine Fisheries Report for 1867: "The average of those taken in the fall is, for the males, 5 pounds; for the females, a little more than three pounds. A female 25 inches long weighs 5 pounds, a male of the same length weighs 7 pounds. Of two males 29 inches long, one weighed 9 pounds 14 ounces, the other 11 pounds 4 ounces. Some extreme weights may be given. One was taken the past season (1867) at Edes Falls that dressed 14½ pounds. The largest on record was caught by Mr. Sawyer, of Raymond. Its weight was 17½ pounds, and is vouched for by Franklin Sawyer, esq., of Portland. These old fish are seldom caught with the hook, and of those taken in the spring and summer, when they are in season, the average weight would be less than indicated by the above." I have been told of still larger specimens having been taken, but am unable now to give any authority. Thus it will be seen that the Sebago salmon average about one-third the size of the sea-going Atlantic salmon and twice the size of the Schoodic salmon. I am aware that from the naturalist's standpoint the matter of size is not important, yet with the fish-culturist it is of the very first moment. It is not, perhaps, a very reliable characteristic, being so much influenced often by the character of the range and feeding-ground, but in the present case there are reasons for thinking that the Sebago salmon have inherited a tendency to rapid growth and the attainment of a large size not possessed by those of the Schoodic Lakes; for not alone in the Sebago Lake and the Sango and Crooked Rivers are fish of such large size found. Those of Long Pond are little, if any, inferior in this respect to those of the Sebago, though Long Pond is a much smaller body of water than several of the Schoodic Lakes, and is not known to offer in depth, in the character of the water, or in food, any special advantages.

In form and color the Sebago salmon approach more nearly to the sea salmon than do the Schoodic or the Sebec fish. In the breeding season the males are much brighter colored and the hook on the lower jaw is more developed. The males, at least, judging from the few specimens measured, are stouter in proportion to their length than any other salmon I have ever examined. The single specimen mentioned above as weighing 11¼ pounds was 29 inches long. An average Penobscot male salmon of an equal weight would have been 32 inches long.

The habits of the Sebago salmon are identical, so far as observed, with those of other fresh-water salmon. They dwell and feed in the lakes, occasionally running into the larger streams after food, and at spawning time, which begins the last of October, they seek the gravelly rapids of the streams and there excavate nests, in which they deposit their eggs. The old fish abstain from food at spawning time, but young males are taken with eggs in their mouths and stomachs. The males are found frequenting the spawning-beds when only 6 inches long, retaining still the dark bars and red spots on the sides, and these little fish yield milt abundantly. The females, however, are not found till well grown up.

At the feeding season both sexes take bait and rise to the fly, and are taken in Songo and Crooked Rivers and in Sebago Lake. In Long Pond they are never taken except at the spawning season, while ascending the stream or near its mouth.

3.—FORMER EFFORTS AT CULTIVATION.

But very little has been done in this direction. I myself visited Bear Brook in 1867 and secured about 8,000 eggs, but, being an utter novice in the art, succeeded in impregnating but a very small percentage, and nothing practical ever came of them. Shortly after that Mr. A. B. Crockett, of Norway, and a Mr. Holmes, associated with him, secured small quantities of spawn several seasons in succession, but with what result is unknown. In 1870 Mr. Brackett, of the Massachusetts commission, visited Songo Lock and obtained a number of large fish, which he transported alive to Winchester, Mass., and from these were obtained several thousand eggs.

Several years later Mr. Joseph R. Dillingham, of Songo Lock, began to take spawn of these fish for the Maine commission. He followed it up for several years, but never with any great degree of success.

4.—ORGANIZATION OF OPERATIONS IN 1878.

It was evidently very desirable to cultivate on a large scale a variety of salmon of such superior character. Previous attempts had been on a small scale, and had not demonstrated the existence of great numbers of breeding fish, but there were not wanting reasons for believing that only efficient means of capture were wanting to develop an ample supply. It was finally arranged between the Commissioners of Fisheries of the United States and of the State of Maine that at their joint expense a new attempt should be made by a party well fitted out with all the appliances deemed necessary to a thorough trial of the locality. The management of the affair was placed in my hands. I selected Mr. Harry H. Buck, of Orland, to conduct the experiment, my own presence during the spawning season being impracticable.

On the 14th of August I visited the locality with Mr. Buck for the purpose of selecting sites for fishing and for developing the eggs, and deciding other general questions. There seemed to be no doubt, taking all the testimony at our command, that the most promising site for fishing operations was at Songo Lock, and it was decided to construct here, at the junction of the Songo and Crooked Rivers, a set of pounds, on the principle of an ordinary fish-weir, of fine-meshed nets suspended on stakes and weighted at the bottom by chains. The main net was to cross the mouth of the Crooked River and intercept the ascent of fish and lead them into the pounds, which were built immediately below the Songo dam, aside from the current of Crooked River, and supplied with Songo water. The best evidence we could collect assured us that there was no probability of such a rise of either river as would endanger our work. Should they stand and prove as efficient as we hoped, we should be in position to take almost every fish that entered the river, for all the spawning ground lay above our nets.

No little difficulty was experienced in fixing upon a convenient site for a hatching-house. Mr. Joseph R. Dillingham, whose premises were occupied, had a very good hatching-house of small size fed by a small spring brook, but our anticipations were so great that his supply of water appeared insufficient. After a deal of searching we finally, a few weeks later, found an admirable site at the outlet of Trickey's Pond, a short distance to the westward from the lock.

5.—THE SEASON'S WORK.

Mr. Buck returned to the scene of operations on the 22d of August with a supply of apparatus, and immediately set about the construction of the works. The main net was sufficiently advanced to prevent fish passing up by us on September 12, the date when, we had been assured, the fish invariably made their appearance here. We were ardently expecting to see great numbers of them in the lock, where they can always be seen if present, and where many of them, it is said, always turn aside from Crooked River; but neither on the 12th nor for many days afterwards did any salmon make their appearance. Mr. Buck's diary shows that the first one was taken in the pounds September 20. From this date they continued to straggle in, one or two at a time, at intervals, until the *large number of* 15 were secured. Of these, nine were males and six were females. This was the entire catch.

But meanwhile disasters had occurred. On October 24 a freshet occurred which bore down our net until the top line was three feet under water. Some salmon undoubtedly passed by at that time. The net was again in complete order on the 27th, and so remained until November 24, when the river had again risen to such a height and brought down such an accumulation of leaves, brush, trees, and logs as to completely wreck the net. It was again repaired and kept in position until December 1, when it became evident that it was a hopeless case, and the enterprise was brought to a close.

Among the reasons for our failure, I place, first, an absolute dearth of fish; second, the inability of our fixtures to withstand the freshets. The result of Mr. Buck's observations and other testimony collected satisfies me that there was really a very small number of fish in the river that season. The net was in place and efficient until October 24, nearly six weeks after the date when we were warned to expect the advent of the salmon, and during that time neither did they come into our inclosures, nor did they enter Songo Lock, nor did they accumulate in any considerable numbers below our barrier. Had there been many fish in the river they surely could have been seen. The fatal gap of two or three days after October 24 doubtless allowed some salmon to pass, but I think not a very great number. It cannot be supposed that all the breeding salmon passed up in that brief space so early in the season. Yet during a whole month thereafter the net continued in place and still no great number of fish to be seen anywhere; and during the whole season but two fish were seen in the lock, where they were wont to be taken plentifully. Evidently this was a year of scarcity.

The freshets demonstrated the insufficiency of our fixtures. Had there been no greater rise of water than testimony led us to expect, our barrier would have remained secure to the season's close. But the season's experience has given us new light on this point, and in future it would be unwise to risk the result of a season's work on the chance of such fixtures being able to stand in the current of Crooked River.

I do not doubt that some efficient means of taking Sebago salmon in Songo or Crooked River could be devised after possibly some more unsuccessful experimenting; but unless there were some better reason than now exists with myself to expect a good run of fish, the prospect of success would hardly justify the risk.

In conclusion, I will merely add that I made several visits to the scene of operations early in the season, and myself fixed upon the main points in the schedule for operations. The plans formed were well carried out by Mr. Buck and his assistants, and such matters as were left to his discretion were judiciously managed.

I present Mr. Buck's diary and weather record, which will be found to contain many interesting details.

6.—H. H. BUCK'S DIARY AT SONGO LOCK, 1878.

August 22, 1878.—Commenced working on behalf of Sebago salmon-breeding establishment. Took from Penobscot establishment about 726 feet of chain, 570 pounds of netting, corks, 1 car for transportation of fish alive, 1 punt, 1 pair oars, trays for eggs, 1 shovel, 1 hoe, and 2 net-bows.

August 23.—Proceeded to Portland on steamer City of Richmond.

August 24.—Through courtesy of J. Hamilton, superintendent of the Portland and Ogdensburgh Railroad, was enabled to get everything to foot of Sebago Lake. As the steamer could not delay, left the freight, and arrived at Songo Lock at 3 p. m.

August 25.—Think the water below the lock is more than a foot lower than upon the 14th. It is reported to have fallen $\frac{3}{4}$ inch per day lately. Above the lock it is apparently at the same height as upon the 14th. Selected as a permanent mark to which to refer the height of water above the lock the lowest block of granite in the upper end of the wing at the north end of the dam. Selected as a water-mark below the dam the top of the largest of a group of stones on the east side of Songo River, below the junction.

August 26.—Made partial survey of the premises, and sent sketch to Mr. Atkins. Freight came to hand, with exception of one tent.

August 27.—Stowed away the car; went to Naples village for sundries, and caulked and puttied punt.

August 28.—Commenced clearing bottom of the stream.

August 29.—Continued clearing bottom of the stream.

August 30.—Continued clearing and cut stakes.

August 31.—Continued cutting stakes.

September 2.—Continued clearing stream.

September 3.—Continued clearing stream. Mr. Atkins came, and we visited brook 1½ miles to westward; found no water.

September 4.—Went to Mr. Dillingham's hatching-house; found but very little water running; commenced setting stakes.

September 5.—Continued setting stakes.

September 6, 7, 8.—Absent on trip to Boston.

September 9.—Returned from Boston; find water still falling. Mr. Mitchell reports having seen two salmon up in Crooked River.

September 10, 11, 12.—Worked getting net across Crooked River, assisted by Dillingham and Mitchell. Afternoon of 12th got so far arranged that I think no fish can pass.

September 17.—Completed arrangement of trap on lower side of main pound.

September 19.—This morning found in the trap two brook-trout, weighing about 2 pounds and ¼ pound, respectively, four or five suckers, one bream.

September 20.—Found in the trap this morning one land-locked salmon 20 inches long, apparently a female, not in very good condition; one brook-trout, about 1¼ pounds, apparently a female, as was also the one taken yesterday.

September 21.—Steamer Mount Pleasant stopped running to-day; water is so low that she cannot pass the lock. Went to Andrew Gray's brook; found no good site for a hatching-house.

September 22.—Took from trap this morning five brook-trout weighing 2 pounds and less; saved two of them, think one of each sex; returned three to the stream, one above net, two below.

September 27.—Two brook-trout this morning, one of each sex.

September 28.—One brook-trout this morning.

September 29.—Took from trap six brook-trout, two fine ones weighing 4 pounds each, I should think. Called four of the fish males, two females. One of the latter got meshed in the dip-net, and was hurt considerably, so killed her. Found she was very full of eggs, and there was apparently nothing in her stomach.

September 30.—Took four brook-trout this morning; saved all.

October 1.—Took four brook-trout; saved three of them. Perry Harriman came this afternoon.

October 2.—One small brook-trout this morning. Having heard several times that Crooked River was full of salmon above our net, to-day got J. B. Mitchell to go up and explore. He reports having seen two brook-trout and large numbers of suckers, but no salmon.

October 2.—Sam. Shane reported thousands of salmon in a deep place below the lock; went down there and saw five large fish, four of which I think were salmon. Took two brook-trout from trap at 9 p. m.

October 3.—Took two male brook-trout this morning. One of them had the peculiar formation of lower jaw indicative of male fish, well developed; the first instance I have noticed this season. Large numbers of fish being reported in the river below the nets, went down this afternoon and explored. Looked very carefully the entire length of the river and saw six salmon. For the first two miles had a favorable chance to see them, as the weather was calm and bright. Took one male brook-trout at 9.30 p. m.; think he would weigh nearly 4 pounds.

October 4.—Got the outside pound completed to-day.

October 8.—Took one small brook-trout this evening.

October 9.—Took one male brook-trout this evening.

October 11.—Took two male salmon this morning, length 19 inches; also one fine brook-trout.

October 12.—One female salmon in morning; one male and one female salmon at 9 p. m.

October 13.—Went to the trap about 2 a. m., and took out two male and one female brook-trout. At 9 p. m. got one female salmon 17 inches long. Heavy shower last night; did not raise the water any; continued to fall to-day.

October 14.—Two female *Salmo fontinalis* this morning, and two of the same at 10 p. m.

October 15.—One male *Salmo sebago* this morning.

October 18.—This morning found that some one had been trying to destroy the nets. The new net across Crooked River was cut or torn in several places, and the poles and stakes which supported it disarranged. The net above it used for stopping leaves was dragged out and very badly torn, then thrown back into the water. An attempt had also been made to let the fish out of the inclosure.

October 19.—One eel about 24 inches long.

October 20.—At 10 a. m. one female *Salmo sebago*. Perry thought it would weigh 10 pounds.

October 22.—Three female and one male *Salmo fontinalis*.

October 23.—One male salmon; three female brook-trout.

October 24.—One male brook-trout in evening. Last night we had a very heavy storm of wind and rain, and this morning Perry made the usual round and thought everything was in proper condition. Found that Crooked River had risen 4 inches. During the forenoon it continued rising and was very thick with brown earthy matter; probably immense numbers of leaves came with it below the surface. Our plans with regard to the direction of the current were found wrong. Instead of rushing on and expending its force in the cove (on the north of the lock), turned and ran down throughout the entire length of our net. The net for leaves was not, therefore, in position to get more than half that came, and they went into the main net in large quantities. We also supposed there would be an eddy at the east end of the main net, and so had not braced it very securely upon the lower side. About noon we noticed the leaf-net had partly discharged its contents into the main net, and that the braces upon the lower side of the latter were beginning to give way. Immediately got all the spare line to be had and stayed the hedge to the shore as thoroughly as possible, but could not save it, and by 10 o'clock p. m. it was pretty thoroughly wrecked. To prevent a recurrence of the accident, think it will be necessary to have two strong nets for leaves and a windlass upon the bank for drawing them alternately. Songo River, above the dam, has risen 4 inches, and its flow through the lock and our inclosures east of lock amounts to nothing in checking the force of Crooked River at the main net.

October 25.—Water continued to rise in Crooked River and reached its height at evening. Did not rise any more above the dam. Our main net seems to be whole and in position, except that the top line is about 3 feet under water—its whole length nearly. There has been no outward current (or, at most, very little) through our traps since the first of the rise.

October 26.—Repairs progressing rapidly as possible under the direction of Perry Harriman. I have been unwell and not able to work for several days.

October 27.—Got the main net in place again to-night; found one or two small holes, but it was not much damaged.

November 1.—Swept the main pound to-day for the purpose of turning the brook-trout up into Crooked River. We should have had forty-five on hand, but only found six, and could not account for their disappearance otherwise than by supposing that they went out when the attempt was made to release the fish. The record showed that ten salmon had been taken, but we found thirteen in the pound.

November 3.—Found one salmon in the trap this morning. Have not been able to see any fish from the pier for several days. Reports have come in of Crooked River being "full of them," and Perry Harriman and Dillingham went up to-day, but did not see any.

November 9.—Have seen no fish for a week. The three females in the pound have begun to spawn at its lower side, or at least the fish are doing a good deal of work there. There has been ice in Crooked River all of the past week, and more or less on all of our nets.

November 12.—One male salmon last night.

November 14.—To-day went up by land to Edes's Falls and examined Crooked River pretty thoroughly for fish and their work; saw fifteen salmon and forty-two nests or ridds. I think most of them were made this year; many of the nests seemed to be in an unfinished state, as though the fish had been speared before completing them; found two boats arranged for spearing.

November 15.—Went to Sandy Beach and Northwest River; saw no fish; saw no sign of fish having spawned at Sandy Beach. There is a small brook at the north extremity which is said to have been dry previons to the recent rains. A native said he had not seen any fish there this fall; subsequently, Sam. Nason said only brook-trout were speared there. Explored Northwest River from the mill to the lake; saw 10 nests apparently made by land-locked salmon. At this time, in places where there is a moderate current, the stream is about 20 feet wide and has an average depth of 18 inches. When the mill is running, the stream is raised about 6 inches; in time of freshet the volume is more than double; judge the distance from the mill to the lake by stream to be about one mile; direct line, one-half that distance.

November 16.—Visited the old mouth of Songo River; saw no signs of fish having been there. There was no current coming from it, and the water thereabouts is very shoal.

November 19.—A man on the canal-boat said the net had been cut near the bottom, and we pulled it up and examined it to-day; found it in good condition. Took one small brook-trout this afternoon—a male with milt.

November 20.—Took one male salmon to-night; was in bad condition; evidently had been struck with spear; gave some milt.

November 24.—Wet and rainy weather has kept Crooked River gradnally rising for several days past. We have kept the nets in good condition, but last night leaves, pine foliage, and drift and brush of all kinds began to accumulate in the upper net. We went out about midnight and drew it up, cleared and returned it, but to no purpose. We left it about 3 a. m., and by daylight it was full again and badly wrecked. We cleared it the best we could and secured the main hedge as thoroughly as possible.

November 25.—Water continued rising, and this morning the whole of our works except the main pound and upper trap were completely wrecked. Do not think any arrangement of nets of ordinary strength could have been kept in place. Passed the day clearing the wreck; could not get the main net, but cleared away stakes, braces, &c., so that we hope it has gone nearly to bottom.

June 25, 1880.—After the wrecking of the works above described, we cleared out two or three boat-loads of brush, drift-wood, &c., and got everything in place again about the 27th of November, and kept on exploring the river for fish and watching the traps until December 1, when word came to abandon the enterprise, and we stored everything with Dillingham. (The apparatus was afterwards transported to Bucksport.)

At Trickey's Pond we left a small house, 11 by 15 feet, on the land of L. L. Crockett, and he says it may stand there without paying rent.

Before we began to get any fish, I feared that we should not find it worth while to open the hatching-house. Wrote to Mr. Atkins and received orders to take eggs when we had caught fifteen female fish.

As the season advanced found we were not likely to get that number, and the few on hand were allowed to deposit their eggs in our inclosure just east of Songo Lock.

The following table shows the length and weight of the spent fish released December 1:

Males.		Females.	
Length.	Weight.	Length.	Weight.
Inches.	*Pounds.*	*Inches.*	*Pounds.*
22	$5\frac{1}{2}$	$18\frac{1}{2}$	$2\frac{3}{4}$
15	$1\frac{1}{2}$	$16\frac{1}{2}$	2
21	$3\frac{3}{4}$	$18\frac{1}{2}$	$2\frac{1}{2}$
20	$3\frac{3}{4}$	$18\frac{1}{2}$	$2\frac{3}{4}$
$14\frac{1}{2}$	$1\frac{1}{4}$	$20\frac{1}{2}$	$3\frac{1}{2}$
$17\frac{1}{2}$	$2\frac{3}{4}$	$29\frac{1}{2}$	$8\frac{5}{8}$
17	$2\frac{1}{3}$	....	....
16	2	....	....
19	....	....	....

7.—OBSERVATIONS ON TEMPERATURE AND WEATHER AT SONGO LOCK, MAINE, 1878.

Date.	Temperature							Wind.	Other phenomena.
	Air.			Songo River water.		Crooked River water.			
	7 a. m.	1 p. m.	6 p. m.	7 a. m.	1 p. m.	7 a. m.	1 p. m.		
Aug. 26	59	72	70	68	70	67	70	Southerly, light	Clear
27	58	77	73	69	71	68	70	Southerly, light	Clear.
28	57	...	65	70	....	68	....	Southerly, light	Clear; cloudy towards evening.
29	63	69	69	70	71	69	71	Southerly, light	Cloudy morning, then clear.
30	63	81	79	70	72	70	72	Gentle northerly, p. m	Clear.
31	61	79	74	73	74	72	73	Gentle northerly	Clear
Sept. 1	66	73	79	70	71	72	72	Light southerly	Rainy a m, clear p. m.; thunder shower in evening.
2	70	79	74	72	74	72	74	Light southerly	Partly cloudy, thunder shower at 6 p m.
3	68	....	..	73	...	74	....	Southerly	Partly cloudy.
4	72	....	61	71	....	73	....	Southerly and easterly	Rainy and foggy.
5	61	64	66	71	71	70	70	Calm	Cloudy, foggy, and rainy
6	65	....	....	71	....	69	....	....	
10	58	70	67	68	70	66	67	....	
11	65	76	71	69	70	67	69	....	Cloudy.
12	65	69	71	70	71	....	70	Southerly and easterly	Very wet, with occasional showers
13	67	74	71	71	71	70	70	Southerly and easterly	Very wet, with showers
14	62	71	65	69	70	68	69	Northwest, fresh	Clear
15	58	69	59	68	69	67	68	Northwest, fresh	Clear
16	45	64	57	65	67	63	64	Southerly	Clear
17	51	67	65	65	67	62	63	....	Clear.
18	57	78	73	66	68	63	65	Southerly	Clear.
19	56	....	65	68	....	64	....	Northerly a. m, southerly p m.	Clear
20	59	80	73	66	68	65	66	....	Clear.
21	66	....	72	67	...	67	....	....	Clear.
22	56	70	57	67	67	67	67	Northwest, fresh	Clear.
23	40	65	56	63	65	61	63	Southerly, light	Clear.
24	40	....	60	63	....	60	..	Southerly, light	Clear.
25	58	70	65	64	66	60	63	....	Clear.
26	56	65	64	65	66	63	65	....	Cloudy all day; shower at 5 p. m.
27	48	63	51	63	63	59	60	Strong northwest	Clear.
28	39	62	56	59	60	56	57	....	Clear.
29	....	61	...	..	61	....	58	Southerly	Clear
30	40	66	61	59	62	53	57	Gentle northerly, a m	Clear.

7.—OBSERVATIONS ON TEMPERATURE AND WEATHER AT SONGO LOCK, &c.—Cont'd.

Date		Temperature.							Wind.	Other phenomena.
		Air.			Songo River water.		Crooked River water.			
		7 a. m.	1 p. m.	6 p. m.	7 a. m.	1 p. m.	7 a. m.	1 p. m.		
Oct.	1	48	70	67	61	62	56	60	Light southwest	Clear
	2	46	68	65	61	62	58	60	Fresh southwest	Clear, foggy morning.
	3	57	72	64	62	63	59	62	Light northerly	Clear.
	4	51	....	65	63	64	61	62	....	Clear
	5	40	65	....	62	63	59	61	Southerly and westerly, light	Clear
	6	50	53	50	61	61	56	56	Northeast veering to west	Light rain, clearing at night.
	7	39	58	52	59	60	55	58	Southerly	Clear, cloudy evening.
	8	51	62	57	60	60	56	58	....	Clear, cloudy morning
	9	49	59	67	59	59	56	57	Westerly	Clouding up towards night; thunder shower in evening.
	10	49	55	51	58	58	53	54	Strong northerly	Clear
	11	39	....	....	55	....	....	51	Light southerly	Clear
	12	42	46	48	55	55	52	50	Northeast	Cloudy a. m., rainy p. m.
	13	48	60	....	53	54	48	50	Strong northwest	Clear
	14	42	61	53	53	54	50	52	Westerly a. m.; southerly p. m.	Clear
	15	42	66	64	54	55	50	52	....	Clear in a. m., clouds and rain in p. m.
	16	50	66	61	55	56	52	54	Fresh southwest	Clear weather
	17	49	66	65	55	56	54	56	Southerly	Clear, foggy forenoon.
	18	58	63	60	57	57	56	57	Southerly	Cloudy
	19	56	58	50	57	57	57	58	Southwest	Cloudy
	20	....	....	....	....	....	....	....	....	Clear
	21	50	67	59	53	56	52	55	....	Clear.
	22	46	....	....	53	....	52	....	....	Clear
	23	46	....	....	54	....	53	....	Easterly	Cloudy, commenced raining at 6 p. m.
	24	52	50	50	54	54	53	53	Strong northeast	Cloudy.
	25	45	....	....	53	....	52	....	....	Clear.
	26	43	....	....	52	....	50	....	....	Clear.
	27	48	....	51	54	....	50	....	....	Cloudy.
	28	46	53	41	52	54	50	....	Strong northerly	Clear.
	29	28	50	40	50	....	48	....	....	Clear.
	30	37	....	42	49	....	47	....	Southeasterly	Rainy.
	31	40	....	44	48	....	46	....	....	Clear.
Nov.	1	39	40	39	48	49	46	45	....	
	2	37	57	46	48	49	45	45	....	Clear
	3	39	45	33	48	48	45	43	Westerly	Clear
	4	20	....	26	45	....	40	....	Northerly	Cloudy, with spits of snow.
	5	20	....	27	43	....	34	....	Northerly	Clear.
	6	14	35	....	40	....	32	....	Northerly	½-inch snow nearly last night.
	7	14	....	....	40	41	32	33	Northerly	Clear a. m., cloudy p. m.
	8	28	....	30	40	....	32	....	Strong northerly	Snow in a. m.
	9	30	....	....	40	....	32	....	....	
	10	32	....	....	39	....	32	....	....	
	11	32	....	....	39	....	33	....	Light southerly	Clear.
	12	39	....	48	39	....	34	....	....	Clear.
	13	35	....	48	39	....	36	....	....	
	14	39	....	....	39	....	33	....	Northerly	Clear.
	15	20	....	....	38	....	32	....	Light southerly	Clear.
	16	26	....	....	38	....	32	....	Light southerly	Clear.
	17	....	40	....	....	39	....	36	....	Clear.
	18	....	....	....	....	....	....	....	....	
	19	35	....	....	39	....	36	....	Easterly	Rainy.
	20	35	....	....	39	....	36	....	Northerly	Clear.
	21	32	....	....	39	....	34	....	Northwest	Cloudy.
	22	38	....	....	39	....	36	....	Easterly	Rainy.
	23	44	....	....	39	....	37	....	....	Cloudy, with showers; $1\frac{7}{16}$-inch rain last night and to day.
	24	....	....	....	....	....	....	....	Northerly	Cloudy.
	25	26	....	....	38	....	36	....	Northerly	Commenced snowing and raining middle of p. m.
	28	36	....	....	38	....	36	....	Southerly	Cloudy and wet.

General summary of observations on temperature at Songo Lock, Me., from August 26 to November 28, 1878, inclusive.

Date.	Air.									
	7 a. m.		1 p. m.		6 p. m.		Max.		Min.	
	Number of observations.	Average degrees.	Number of observations.	Average degrees.	Number of observations.	Average degrees.	Date of observation	Number of degrees.	Date of observation.	Number of degrees.
1878.										
August	6	60.16	5	75.60	6	71.66	30	81	28	57
September	26	57.23	21	69.33	24	64.96	20	80	28	39
October	30	46.23	21	60.38	23	55.04	3	72	29	28
November	23	30.87	5	43.40	8	37.12	2	57	6	14

Date.	Songo River water.							
	7 a. m.		1 p. m.		Max.		Min.	
	Number of observations.	Average degrees.	Number of observations.	Average degrees.	Date of observation.	Number of degrees.	Date of observation.	Number of degrees.
1878.								
August	6	70	5	71.60	31	74	26	68
September	26	57.04	21	67.48	2	74	30	59
October	30	55.73	21	58.09	4	64	31	48
November	23	40.60	5	45.20	1	49	15	38

Date.	Crooked River water.							
	7 a. m.		1 p. m.		Max.		Min.	
	Number of observations.	Average degrees.	Number of observations.	Average degrees.	Date of observation.	Number of degrees.	Date of observation.	Number of degrees.
1878.								
August	6	69	5	71.20	31	73	26	67
September	25	65.24	21	65.57	2	74	30	53
October	29	53.13	21	56	3	62	31	46
November	23	35.69	5	40.40	1	45	17	36

XXXVII.—REPORT ON THE COLLECTION AND DISTRIBUTION OF SCHOODIC SALMON EGGS IN 1878-'79.

By Charles G. Atkins.

1.—Preparations.

The preparations for the capture of the breeding fish this year were almost identical with those of the year before, and the work proceeded on the old basis until late in the season, when we were compelled to resort to new expedients, which will be detailed below.

For the incubation of the eggs more extensive preparations were made. The ill-success of many of the lots of eggs sent out the previous season warned me not to depend on the old hatching-house, which evidently did not command water enough in a dry winter to forward two millions of eggs and nourish them properly. It was not easy to find a satisfactory remedy. The water of the old hatching-house was spring water. There were numerous other small springs in the neighborhood, but none of those yet discovered could be led into the old hatching-house, and no one of them afforded alone water enough to supply a half million of eggs. Grand Lake Stream affords water of the very best quality, but unfortunately the facilities for using it are very poor. At the dam which commands the outlet of Grand Lake there is, in the spring of the year, a head of perhaps 6 feet, but in the fall, sometimes less than 2 feet, and any hatching-house located low enough to take this water in without artificial raising, at a low stage of the stream, would infallibly be flooded at time of freshet. Nearly equal and generally similar disadvantages attached to every site along the stream. It was, however, finally decided to put in a temporary hatching-house on the west bank of the stream at the first fall below the dam. Even here we had a fall of but little more than 10 feet, and liability to flooding by spring freshets, but the facilities for taking our supply of water from the stream were better than at the dam; and it was hoped that every year we should have the distribution of the eggs completed and the old hatching-house free for the reception of the 25 per cent., reserved for the stream, before the spring freshets should come.

The new hatching-house was a very humble structure, only 20 feet by 10; but there were placed in it three troughs, each 17 inches deep, which had an aggregate capacity of nearly a million of eggs. Wire trays were employed about 12 inches square, nested in frames carrying 20 trays per frame—the identical apparatus used at this establishment in 1875 and yearly since. The water was taken from the open stream through a covered plank conduit, with the expectation that very pure water would thus be secured. It was afterwards found that the little brook that flows from the old hatching-house through a swampy piece of land discharged its waters into the stream above the new house in such a way that, instead of mingling at once with the water from the lake, they crept down along the shore almost by themselves, as far as the hatching-house. It thus came about that whenever the brook was in flood its turbid waters crowded the pure water of the lake away from the conduit, and took entire possession of the hatching-troughs, making a very dirty piece of work of it. It is not known that any harm resulted beyond the extra work involved in cleaning up and the unpleasant appearance of the fixtures. But steps have been taken to avoid such an occurrence in future by continuing the conduit out under the water of the stream far enough from shore to avoid receiving any part of the brook water. It will be seen that this new hatching-house, though of the greatest service as supplementary to the old one, could not wholly take the place of the latter, which alone afforded facilities for hatching out the reserve for the stream. I therefore turned my attention to the improvement of the old house. In the first place, it appeared advisable to secure, if possible, better aeration of the water; for this end the situation was a very unfavorable one. The spring issued from a gravelly bank, at an elevation scarcely above the level of a large swamp, through which the overflow oozed away. We had the year before cut a wide and deep ditch, nearly half a mile long, for the outflow, so that there was no longer danger of the house being flooded by freshets, but this did not enable us to lower the troughs from their original elevation. We could not curb the spring and thus raise a head, because of the danger that the water would then find a new outlet through the loose gravel and be lost to us altogether. The available head was thus scarcely a foot. The best that could be done was to construct above the hatching-house a narrow, circuitous drain or canal, about 10 feet wide and nearly 70 feet long, through which the water should flow with a surface air-exposure of about 140 square feet, before entering the hatching-house; to have all the overflows and passages, from canal to feeder and from feeder to hatching-trough, at the surface rather than submerged; and to introduce in all the troughs occasional dams which should bring all the water repeatedly to the surface and expose it to the air in wide and shallow currents.

Careful search also revealed a very considerable leak around one end of the hatching-house dam. This was finally, though not without some difficulty, completely stopped. No other change of importance was made in the general hatching arrangements.

The summer, and more especially the early autumn, were rather dry seasons, and both the lake and the springs were at a low stage. A careful measurement of the amount flowing through the troughs at the old hatching-house indicated but a trifle over four gallons per minute on the 23d of August, 4.164 gallons per minute on the 28th of September; on the 9th of November, after the leaks were stopped, this had risen to 8.4-7 gallons per minute; on the 16th of November, to 12.86 gallons per minute.

2.—FISHING AND SPAWNING.

The progress of the season's work is sufficiently portrayed by the following extracts from my note book, and interpolations.

August 23, 1878.—All reports agree in representing the fishing last spring and early summer to have been uncommonly fine. Fish were plenty, and of unwonted size and fatness. At the dam all the gates are now open, and a lot of logs fill the large pools below our works and partially obstructs its outlet. Consequently the water at our catching and spawning ground is unwontedly high (not high in lake, however). At the dam, in the upper end of the sluice-gate-way, that is, on the upper edge of the rollway, the water is 10 inches deep. At our gauge, on the pier above the dam, the water stands at 2 feet 2 inches exactly, with calm air and still water.

September 28.—Arrived yesterday from Bucksport (second visit). The nets were put across the stream and the head of the tannery canal about the middle of this month.

September 29.—The water being very clear and air still, I looked carefully all along the lower side of the dam, but not a single Schoodic salmon was in sight. I went up to the pier where the water gauge is, but saw nothing there. I think that evidently the fish are coming in slower than last year. On October 6, 1877, I found them very plenty below the dam, but then several hundred had been put in from the canal by Mr. Munson. The water is some lower than last year (nearly 3 inches lower than October 8, 1877), and only one gate is now open instead of three at that time. Water has fallen nearly 5 inches since August 23.

Verified the elevation of the water gauge on the pier above the dam, and found it to agree exactly with the position laid down in Mr. Buck's notes last fall; that is, the 4-feet mark is on a level with the lower side of a drill hole on the south face of a large bowlder, lying in the water near the east shore, "about 225 feet above the dam."

October 1.—Again carefully looked about the dam, but saw no Schoodic salmon. Have not seen one since I came here.

October 31.—Fish have begun to descend below the dam, and nightly come down to our net. Munson thinks them very plenty in lake, from what he has seen above the dam and been told by boatmen. He saw some work in the gravel by fish at the head of our main lead on the 28th, but not much done yet.

November 1.—I see three nests begun at head of our main lead, but none elsewhere. Got all ready for the capture of fish. Shall not begin sweeping yet. Fish captured every night after this date. See appended "Statement of Fishing."

November 4.—Don't see any new nests above the dam yet; but the old ones, three, are enlarged every night. No new work of consequence in the main lead. I saw yesterday four or five nests in three feet of water, along the west side of the pool below the dam, in precisely the position where I saw them last year.

November 6.—At 8 p. m. I explored the main lead and two first pounds, and found not a single salmon; never knew such a thing before; yet quite a number are just above the gate. Varnum counted 10 there. I think they are mostly males.

November 7.—Scarcely any more spawning operations in our main lead. Two partially-made nests in pound C, our principal inclosure. Above the dam I can count, close together, 13 nests, most of them pretty complete; these are in the swift water above the sluice-gate.

Began taking spawn this afternoon, and find very few ripe females, only 8 out of 54. Total catch of fish to this date, 246, of which 54 are females.

We found among the salmon one gravid female togue—the first time such an event has occurred at this establishment. Her spawn was milted with salmon milt, but all afterwards perished without giving any certain indication as to the cause.

To prevent the fish stealing their nests in the swift water above the gate, I propose to surround this spot on the upper side by a drop-net, and put in some small pounds to entrap any fish that may venture upon the forbidden grounds. This was put in operation the next day, and from November 10 to 17 over 500 salmon were captured on this spot and placed below the dam. Though sharply followed up, the fish succeeded in doing a great deal of nest-digging there. The first nests dug were completely obscured by new ones. Further, quite a number of fish are spawning above all our nets, especially at a gravelly shallow on the site of an old coffer-dam, about 300 feet above the main dam. In former years there has always been some spawning above the dam, but never to an extent approaching their present operations. I attribute their behavior this year to the low stage of the water, which has never been equalled at this date in any year within my experience. To get below the dam, the fish must pass through the sluice-gate, adown an inclined "rollway" about 40 feet long. At the head of this rollway the water is about 10 inches deep, but it flattens out to less than 3 inches at the lower end, so that a fish of ordinary size cannot go down without rubbing upon the plank flooring of the rollway and being pushed partly out of water, to which they appear to be much averse.

November 16.—So many fish are beginning to spawn above the dam, that to-day we put in a set of pounds at the old coffer-dam, entirely closing the stream at that point. This evening fish are entering our new pounds in great numbers.

The fish are remarkably backward in development. Of 385 females examined up to this date, only 139 (36 per cent.) have yielded spawn. To-day we found among the fish caught last night at our lower works 7 ripe females and 50 unripe.

November 17.—Very good success with our new trap at the coffer-dam. We find in it this morning 113 female fish and 36 males. Ninety-seven females and 54 males were also taken last night at the lower pounds, making a total of 300 fish last night—all Schoodic salmon. The weather was clear and calm.

I think, however, that the new trap is somewhat defective in form, especially at the entrance, and that many fish, after once entering the first pound, find their way out again. Some also broke through the net last night, and so escaped. All the fish captured here this morning taken in a salmon-car to the lower pounds, by dragging the car through the sluice at the dam.

At one o'clock to-day, the sun shining, I saw two female fish in the act of spawning close above the dam, turning on their sides and flapping after their usual manner, with no male in sight. Two hours later I find the same two and one other female spawning near, and still no males near. I made these observations very carefully, and have no doubt of their entire accuracy. There is no sort of difficulty in distinguishing the sexes. I have often watched them on the same spot.

November 18.—Another clear and calm night has given us 246 fish, of which 187 were taken in the new trap. The majority of the females handled to-day are pronounced unripe. All the eggs taken previous to this date have been placed in the old hatching-house, and 155,000 eggs are now there. To-day we place 67,000 in the new house.

November 19.—Last night was stormy, with sleet and snow, and much westerly and northerly wind, which continued all night. Total catch of Schoodic salmon, 201; three-quarters females.

November 20.—Last night the wind was very light, and varied from northeast to northwest; cloudy and clear by turns through the night; 288 fish caught, of which 271 came from the upper trap.

November 21.—A rainy evening and fresh northerly wind, clearing at midnight, brought us in last night 187 fish.

November 22, *a. m.*—Clear and calm last night, and we took 306 fish; the best catch of the season.

In the appended tabular record of fishing will be found notes on the weather of each night during the fishing season. So far as the indications go, they seem to be in favor of the conclusion that stormy weather deters the fish from running.

Among the fish handled to-day was one small one, 12 inches long and weighing 11 ounces. This is an unusual size, of which a few are taken each year. Only two of them have been seen this fall. One of the largest of the females handled to-day, a full and ripe fish, measured 22 inches in length, weighed before spawning 4 pounds 1 ounce, and yielded

15 ounces spawn. I think this a small yield of spawn for so large a fish. Another female to-day, 19 inches long, weighing 2 pounds 1 ounce after spawning, gave 14 ounces spawn, which counted out 2,068 eggs. The average yield is much less than this. Up to this date we have taken 450,000 eggs from 386 female fish, an average of 1,166 per fish. This is a much higher average than any preceding year, the fish averaging larger than usual. There is always a slight uncertainty in the estimates of the number of eggs, but the error from this source is probably not over 5 per cent.

The backwardness of the fish this season is more evident now than ever. Less than half of our female fish have yielded their eggs, and the number of eggs taken is less than ever before at this date. We have, however, over 500 gravid fish on hand, besides the catch of the last two nights.

November 22, *p. m.*—On examining the fish taken during the last two nights, we find 241 ripe out of 406 females. They add 243,000 to our stock of eggs, and will add 25,000 more on second handling.

November 25.—The catch of fish has fallen off rapidly since the 22d. Only 38 taken last night. Evidently the season is drawing to a close.

To-day's work adds 445,000 to our stock of eggs, and brings the total thus far up to 1,170,000, with some hundreds of females yet on hand; 948,000 are now deposited in the new hatching-house, and the remainder will be placed in the old house.

To-day we have taken 445,000 eggs. This unusual feat was accomplished by six men working all day, without weighing or measuring any of the fish handled. That gives an average of 74,000 as a day's work for a man, or, say, 7,400 per hour. This would be accounted very slow progress with sea-going salmon, either of the Atlantic or Pacific. But the Schoodic salmon are among the least prolific of fishes, and to get the eggs taken to-day we had to handle over a thousand salmon. The work of putting the eggs in the trays took over four hours in addition. I find that the best working party at the spawning-shed consists of five or six persons—one to dip the fish and pass them to the spawn-takers, one to keep the tally, three to take spawn, and, perhaps, one more to wash and care for the eggs. In addition to these, it will require one man to carry the eggs to the hatching-house, and another to place them in the troughs, and, if the fish are weighed and measured, another man will be required for this. The weighing and measuring have been done this year, as usual, every day except when the work pressed too much. The general results are tabulated below.

December 3.—This afternoon we began sending off the parent fish. Up to this date all caught have been kept in our inclosures. All unripe, and all awaiting manipulation, were kept in the inclosure below the spawning-shed. After the final manipulation they were placed in a roomy inclosure above the spawning-shed. Having now, as is supposed, taken all the fish possible for this season, there is no objection to setting the old fish free in the lake. With the purpose of securing them as far as possible from destruction, and affording the best facilities for feeding and recovering their condition, we take them in salmon-cars from a mile and a half to two miles up the lake and there set them free. The cars used are fishing "dories" sunk in the water, with grated apertures at either end, the same used on the Penobscot. From one to two hundred of these Schoodic salmon can be transported in one of them at each trip. The weather, fortunately, has been mild for several weeks; otherwise we might be unable to force our way far up the lake by reason of ice.

December 4.—To-day we finished taking spawn by giving the final manipulation to the last fish. We have taken in all 1,723,000 eggs.

December 6.—All the nets and chains were taken out of the water to-day, except a net to guard the passage of the dam by any returning fish.

Grand Lake is not yet frozen over. Big Lake and the other lower lakes are also open, and the steamer continues to run between this stream and Princeton. Ordinarily the lower lakes close about the 18th of November; and this year there have been two interruptions to navigation. On the night of November 15 those lakes froze over and so remained for a week, and once since then they have been frozen over for a single day.

December 9.—Last night the thermometer dropped to 12° F., and ice formed on Grand Lake as far as we can see from the outlet. Within six days the temperature of the water in the new hatching-house has fallen from 40° to 32° F., and we do not expect to see it often above 34° until spring opens.

A good many of the fish that we carried up the lake have come back and got into the canal, and others are hanging about near the dam, just above the net set to intercept their descent. I think they are nearly all males; I looked carefully at about 50 of them and do not see one that I think is a female.

December 11.—A warm rain, with a southeast wind, breaks up the ice in Grand Lake and carries off the snow. There is quite a flood in the brook at the old hatching-house and the water rises to within 3 inches of the tops of the hatching-troughs. A good deal of sediment is deposited in the troughs and on the eggs in both houses, but they can be easily cleansed by the careful hands of Mr. Munson. Grand Lake has risen to 3 feet 6½ inches on our gauge; it has been steadily rising since November 16, when it stood at 1 foot 9 inches.

December 13.—The last of our nets removed from the water to-day. No fish to be seen about the dam, and I think most of them have returned to the lake.

December 14.—The wind blew strongly from the northeast throughout last night, and the lake being open, and the temperature low (15° F.), the rapidly-forming ice-crystals were driven down to the dam, where they were piled up in a mass that clogged the gates and came near being the cause of a serious calamity. At 6.30 a. m. it was discovered that the water in the stream was very low; that it had actually ceased to flow into the hatching-house, and that from the hatching-troughs themselves it had leaked away, till they were only half full and half the eggs were in the air. The air of the hatching-house was far below the freezing point. Ice coated the frames, and the upper layers of eggs in several of the frames appeared to be frozen. The situation was alarming. The greatest exertions were made to free the gates of the dam from the ice blockade, and in half an hour we had the satisfaction of seeing the water flowing through the hatching-troughs as usual. An examination into the condition of the eggs a few hours later dispelled our fears. Not more than a hundred were killed. I suppose that those which appeared to be frozen on the first examination were merely encased in ice, or possibly the outer shell was frozen, the interior of the egg remaining untouched. At any rate, they all thawed out without loss, except the few mentioned. These losses were confined to the outer edges of the upper and more exposed trays. It would probably have taken some hours longer for the freezing temperature to have penetrated to the interior of the frames of eggs (each frame held 20 shallow trays piled one above another). Had the stoppage of water continued for six or eight hours, I question whether the loss from freezing would have been much more serious than the injury that would have resulted from confining the eggs for the same length of time in a trough filled with stagnant water. It is therefore doubtful whether, even in this cold house, there would be any advantage in a tight trough. In a house kept always above 32° F. in temperature, I think a leaky trough would, without doubt, be the safest, and I would advise that pains be taken to provide some small leak, so that in case of stoppage of the water the troughs may be drained and the eggs left exposed to the air. In a covered trough drying would proceed very slowly, and the eggs would be in no way harmed by contact with air, so long as they neither froze nor died. Such an occurrence as the clogging of the gates of the dam by ice can only occur when the lake is open, and at the same time a very strong wind accompanies extreme cold, a conjunction of circumstances not likely to come about very often.

3.—THE DEVELOPMENT OF THE EGGS.

As soon as the lots of eggs successively reached a stage of development, suitable for the determination between impregnated and non-impregnated, they were subjected to a close scrutiny to determine their condition in this respect. In working upon smaller lots, a sample, numbering 20, taken at random from the eggs as they lay upon the trays, was placed in a shallow testing-box, perforated with 20 holes. Holding this up so that a strong light shone up through the bottom, we could see the condition of the eggs with great distinctness. Several trials were made with each lot, not less than 100 eggs being examined in any lot, except in case of some experiments embracing less than 100 eggs in all. In the larger lots, however, this was found to be too tedious a process, and resort was had to another. A whole tray of eggs was held up to a window so as to throw a strong light upon the under side, and all the eggs were so lighted up as to make their condition plainly discernible. A whole row, or two or three whole rows, next to the frame, were critically examined and the unimpregnated counted. Each row being known to contain on an average 40 eggs, the ratio of impregnation was speedily arrived at. In this way, in the largest lots, 2,000 and more were examined, and it is believed that the result obtained must be very nearly correct. The result of the examination was fairly satisfactory, but did not indicate so good a ratio of impregnation as I had hoped. Exclusive of the experimental lots, it ranged from 74 to 93 per cent., and averaged, by careful computation, 90.1 per cent. At Bucksport, in former seasons, with sea-going salmon, an average of 96 to 98 per cent. was attained. The reason for the inferior result at Grand Lake Stream is not apparent, but is most likely connected with these too-well attested phenomena: first, that the Schoodic salmon, far more than the Penobscot salmon, are subject to diseases which affect the eggs before they are laid, so that often a large percentage, and sometimes the entire litter, is damaged past all remedy when laid; second, that unlike the Penobscot salmon, the Schoodic fish often came into our hands while yet unripe, and therefore liable to be prematurely pressed.

The percentage of non-impregnation would account for a loss of 172,300 eggs. The shrinkage up to the time of dividing and shipping the eggs was 253,000, as deduced from the number shipped and turned out to hatch. Probably the actual deaths were not far from the latter number. Aside from the ordinary pickings of white eggs, there were some entire lots which perished. These were all experimental lots except in one instance. A lot taken on the 28th of November, numbering 103,000, was a total loss. There was nothing unusual in the appearance of these eggs until December 14, when it was observed that there were small circular white spots in a great many of the eggs; these were near the embryonic disk in all cases, and in most cases directly over it, moving about with the disk as if attached to the yolk. At this date, the eggs of the embryonic disk were just beginning to expand. In a few days it was observed that the lines showing the progress of the growth of the embryo were less regular in this than in other lots. If allowed to remain in the water, the white spots enlarged and the eggs soon decayed. They were, therefore, picked out as fast as they appeared, the examinations being made every three days. From 1,000 to 8,000 were taken out at a picking, and on one day the number rose to 15,000. By the 9th day of February 80,000 had been taken out, one by one. They were then removed to the new hatching-house, when, in the month of March, about 5,000 of them hatched into weak fish. Unfortunately, the record does not throw any light upon the causes that led to this failure, but I have a suspicion that the eggs were kept too long in contact with the milt, which I have found is sometimes fatal. The eggs and milt were this season kept in contact much longer than I have ever practiced before, with the hope that a more complete impregnation would thereby be secured. The latter object was not attained, and I now believe that a minute is as good as an hour of contact. With the above-named exceptions the eggs were to all appearance a healthy lot, and the shipments were made in a very hopeful spirit.

The water in the new hatching-house fell, during the month of November, from 48° F. to 38°. December 14 it stood at 32°, and thenceforward through the winter between 32° and 33°. The development of the eggs placed therein was, therefore, very slow. The water of the old hatching-house, however, was about as warm as usual, ranging from 44° to 38° in December, from 41° to 39° in January, 39° to 37° in February, 39° to 38° in March, 38° to 34° in April. The eggs placed in this house early in the spawning season came forward rapidly, and were ready for shipment early in February. It was decided to take the 25 per cent. reserve out of these, and, therefore, but 150,000 of them were shipped. These were sent to Connecticut, Massachusetts, and New Hampshire, February 10 and 11. The remainder of the shipments were made from the eggs that passed the winter in the cold water of the new hatching-house, and so much colder was the water and so much more backward the development of these eggs than I had anticipated, that not until April did they reach the state at which I thought it safe to pack them up.

4.—DIVISION AND SHIPMENT.

There were left at time of shipment, as nearly as could be estimated, 1,470,000 eggs; 370,000 of these were retained for hatching, and the balance, 1,100,000 were sent away—616,000 on account of the United States Fish Commission, and 484,000 on account of the States of Massachusetts, Connecticut, and New Hampshire. The distribution is given more in detail in Table IV.

It has been the practice, rarely omitted at this establishment, to remove the unfertilized eggs from each lot before packing it up. At the stage suitable for packing, the unfertilized eggs are very tender, and a moderate agitation suffices to turn most of them white, when they can be easily removed. This was generally done at the time of packing, but in some cases a few days before. The packing-boxes used this year were from 1¾ to 3½ inches deep. The eggs were, as usual, enveloped in mosquito-netting, between layers of wet sphagnous moss. Between the packing-boxes and the outer cases was a space of about 3 inches, filled with chaff, dry moss, or sawdust. The cases generally left Grand Lake Stream in the afternoon, and were carried on sleds to Princeton, where they passed the first night. The second day they went to Calais, and there passed another night. The third day (two days from starting) they took the railroad train at St. Stephen, N. B., and from this point the journey was generally uninterrupted.

The eggs endured the journey in most excellent condition, and very encouraging reports were received from most of the consignees. Out of thirty lots shipped, sixteen were reported as arriving in "good" condition; three, "very good"; one, "splendid"; one, "excellent, and very superior"; only four were reported in inferior condition. The promise of an unusually successful issue was very flattering, but it was only partially fulfilled.

5.—HATCHING AND PLANTING.

The eggs that were reserved at Grand Lake Stream suffered least loss in hatching. Out of 370,000 but 22,000 were lost, and had the unfertilized been all removed, as in the case of those shipped, the loss would have been much less. The young fish were apparently healthy and vigorous. They were, as usual, set free the last of April and early in May, in the shallow waters around the shores of Grand Lake, and a few in Grand Lake Stream.

Next best were the eggs that were hatched at other points in New England. The poorest results were obtained from the eggs that were shipped to the farthest points. The total number of fish planted is reported at 1,145,665; this indicates a total loss of 324,335 after shipment, which, considering the careful removal of the unimpregnated eggs, is not quite satisfactory. I am disposed to locate in the Grand Lake establishment the causes which led to the greater part of this loss. Though unable to point them out with certainty, I think I am on their track and shall be able to ferret them out. The details of the distribution of the young fish are given in Table V. In the other tables, not yet referred to, will be found records of temperature, of fishing, of spawning operations, and of measurement of the parent fish.

TABLE I.—*Record of fishing at Grand Lake Stream, Maine, November, 1878, and October and November, 1879.*

Date.	Weather.	Height of Grand Lake.	Temperature.		Schoodic salmon caught.									Notes.
			Air.	Water.	Nightly catch.					Nightly summary.				
		7 a. m.	7 a. m.	7 a. m.	Hour.	Males.	Females.	Unknown.	Total.	Males.	Females.	Unknown.	Total	
1878.		*Ft. In.*	°	°										
Nov. 1–2	[illegible], clear	1 9½	38	48½	7 a. m	65	5	0	70	65	5	0	70	1 small togue.
2–3	Mostly cloudy; shower in night	1 8¾	34	48	7 a. m	40	11	0	51	40	11	0	51	
3–4	Clear; [illegible]e wind from N	1 9¼	39	48	7 a. m	38	12	0	50	38	12	0	50	
4–5	Evening clear, moonlight; in morning cloudy, beginning to snow.	1 8¾	21	45	7 a. m	31	13	0	44	31	13	0	44	
5–6	Evening [illegible]r, [illegible]g clear, still	1 8¾	18	42½	7 a. m	14	11	0	25	14	11	0	25	
6–7	Perfectly calm and clear; moon nearly full	1 9	20	41	7 a. m	5	1	0	6	5	1	0	6	
7–8	Cloudy all night, nearly calm	1 8¼	14	40	7 a. m	7	3	0	10	7	3	0	10	
8–9	[illegible]dy, with N. wind fresh all night	1 8½	28	40½	7 a. m	25	25	0	50	25	25	0	50	
9–10	Cloudy and light snow all night; full moon 9th; wind NW.		28	43½	7 a. m	24	12	0	36	24	12	0	36	
10–11	Wind slightly strong at nightfall, dying away entirely by midnight; perfectly clear		27	39	7 a. m	11	6	0	17	11	6	0	17	
11–12	Rain	1 9¼	28	38	7, 9, 12, 3, 6	48	17	0	65	48	17	0	65	
12–13	Calm and clear	1 8¾	40	40	7, 9, 12, 3, 6	69	31	0	100	69	31	0	100	
13–14	Rain, snow, and wind until 12.30; after that clear and calm.	1 9½	24	40		118	102	0	220	118	102	0	220	
14–15	Clear all night and calm		27	39		121	89	0	210	121	89	0	210	
15–16	Calm and mostly clear	1 9¼	23	38	7, 9, 12 p. m.; 3, 6, 7 a. m.	49	54	0	103	49	54	0	103	
16–17	Clear and calm	1 9	20	37½	7, 9, 12 p. m.; 3, 6 a. m.	90	210	0	300	90	210	0	300	
17–18	do		30	38½	7, 9, 12, 3, 6	69	177	0	246	69	177	0	246	
18–19	Stormy; sleet and snow, with much westerly or northerly wind all night.	1 9¼	25	38	7 p. m.; 12, 3	48	153	0	201	48	153	0	201	2 whitefish in upper trap.
19–20	Wind very light, variable NE. to NW.; starlight beginning of evening, mostly cloudy till 12, cloudy and half cloudy by turns rest of night.		31	37		38	250	0	288	38	250	0	288	
20–21	Wind fresh northerly, falling to light NW.; rain in evening; clear at 12.		34	37½		35	152	0	187	35	152	0	187	1 togue in upper trap 26½ inches long, 6¾ pounds; 3 whitefish.
21–22	Clear and calm, or with very little westerly wind till 4 a. m., when clouded up; calm, foggy morning.		33	38	2 p. m.; 1, 5½, 7 a. m.	51	254	1	306	51	254	1	306	
22–23	Rained hard all night, with NE. wind	1 11	30	38½	7 a. m	21	63	0	84	21	63	0	84	

23–24		1	11½	47	39½	7 a. m	14	46	0	60	14	46	0	60	
24–25		2	0½	37	40	7 a. m	10	28	0	38	10	28	0	38	
25–26	Snow light nearly all night			30	39		17	12	0	29	17	12	0	29	
26–27	Clear and calm			32	39	7 a. m	12	17	0	29	12	17	0	29	
27–28	Clear evening; rain all night after 10.30	2	1¾	21	37	7 a. m	9	5	0	14	9	5	0	14	
28–29	Clear and calm			31	39½	7 a. m	12	13	0	25	12	13	0	25	
29–30	do			30	38	7 a. m	17	8	0	25	17	8	0	25	
Nov. 30 / Dec. 1	Clear and some NW. wind			30	38½		5	1	0	6	5	1	0	6	
1–2				16	35		9	4	0	13	9	4	0	13	
1879. Oct. 29	Full moon														1 togue; 1 whitefish; 47 suckers.
30–31	Flying clouds; strong NW. wind all night	2	4	36	46½		117	35	1	153	117	35	1	153	All in upper pounds.
Oct. 31 / Nov. 1	Strong NW. wind all night; mostly clear	2	4½	36	47		113	51	0	164	113	51	0	164	1 togue; 1 whitefish; about 40 or 50 suckers; all in upper pounds.
1–2	Clear all night; moon rose at 6 8 p. m.; wind fresh NW. most of night, dying away toward morning.			26	44		63	56	0	119	63	56		119	2 or 3 suckers.
2–3	Cloudy	2	3¾	20	42		33	20	0	53	33	20	0	53	Suckers.
3–4	[illegible]dy; [illegible]w at 11.40 p. m., wind NE.; gentle in evening, changing to N; strong or very strong in early morning; moon rose at 7.52 (Boston).			23	42½		62	25	0	87	62	25	0	87	Do.
4–5	Wind [illegible]ed at 10 p. m.; d[illegible]k evening; clearer t[illegible]d morning.			25	40½		56	47	0	103	56	47	0	103	1 whitefish; suckers.
5–6	Wind light NW., calming completely away before midnight; clear all night; moon rose at 9 56.	2	3½	22	40		23	17	0	40	23	17	0	40	1 whitefish; 1 small male Schoodic salmon, 9½ inches, 7 oz., mature.
6–7	Wind calm in evening, and nearly so all night; clear in early [illegible]ng, but cloudy from 8 or 9 till morning, snow [illegible]ery light, by spells from 10 till morning.			10	38½		55	40	0	95	55	40	0	95	
7–8		2	3	28	40		41	41	0	82	41	41	0	82	
8–9	Mostly overcast; d[illegible], with a slight squall of hail at 2 a. m.			35	40½		23	47	0	70	23	47	0	70	1 whitefish.
9–10	Evening [illegible]r, but v[illegible]y dark; slight snow-squall at 11.30 and rain [illegible]s; [illegible] about 1 o'clock; light after; t[illegible]al fall 0.15 [illegible]ch; moon rose at 1.20 a. m., [illegible].	2	3	38	41		56	99	0	155	56	99	0	155	Do.
10–11	Evening [illegible]lear [illegible]t dark, with fresh NW. wind; wind increased [illegible] midnight, with flying clouds.			54	42½		120	150	0	270	120	150	0	270	Several whitefish.
11–12	Evening dark and [illegible]m; [illegible]w with light northerly wind began at midnight, and [illegible]ed light till morning.			34	41½		38	23	0	61	38	23	0	61	4 whitefish.
12–13	Evening [illegible]loudy, with northerly [illegible]nd gentle; wind incre[illegible]sed to fresh at 10, and so continued all night, with cloudy sky.			34	42		12	52	0	64	12	52	0	64	
											1,934	2,488	2	4,424	

TABLE 11.—*Record of spawning operations at Grand Lake Stream*, 1878.

Date.	Remarks.	Fish at first handling.							Females.				Eggs.					Remarks.
				Females.					Spawned.									
		Total.	Males.	Unripe.	Ripe.	Spent.	Total.	Barren.	First handling.	Afterwards.	Total spawned.	Respawned.	Weight.		Number.	Lots.	Impregnation.	
													Lbs.	*oz.*			*Pr. ct.*	
1878. Nov. 7	Pound C, comprising all the females and part of the males in it; rest of males put right back, not used.	68	14	46	8	0	54	0	8	....	5	......	2	2	4,000	1	87	Two fish called ripe were only partially so.
	Female togue	....	....	....	....	....	....	....	....	....	....	....				31	82	
12	Respawning	....	....	....	....	....	....	....	....	....	....	7	1	1	2,000	2	93	8 females yield —.
	Respawning togue	....	....	....	....	....	....	....	....	....	....	....				32	87	
	Part of last night's catch	55	46	2	7	0	9	0	8	....	8	....	3	2	5,500	3	88	
14	Respawning	....	....	....	....	....	....	....	....	....	....	8	0	11	1,500	4	94	8 females yield —; 16 yield —.
	Last night's catch	217	118	69	29	1	99	0	29	....	29	....	14	13	30,000	5	90	
	Pound C (main stock entire)	398	321	50	27	0	77	0	27	26	53	....	28	15	57,000	6, 28, 29, 37	....	
	Summary to date	738	499	167	71	1	239	0	72	26	98	15	50	12	100,000	....	....	
15	Respawning of yesterday's catch	....	....	....	....	....	....	....	....	....	....	82	5	9	10,000	7	82	98 females, 110,000; 82 females, 97,000.
	Last night's catch	210	121	55	34	0	89	0	34	0	34	....	16	14	33,000	8 & expr.	93⅓	132 females yield —; 34 females yield —.
16	Respawning, yesterday's catch	....	....	....	....	....	....	....	....	....	7	34	2	5	4,000	9	86	Lot 9 washed as usual.
	Last night's fish	112	55	50	7	0	57	0	7	....	....	....	4	6	8,000	10 & expr.	74	
18	Respawning (from 16th)	....	....	....	....	....	....	....	....	....	....	7	0	13	....	....	....	Placed in new hatching-house.
	Fish caught last two nights in lower pounds.	233	76	107	49	1	157	0	49	....	49	....	27	0	} 67,000	11 & 12	90 & 91	
	One draught of seine in pound C: 174 males, 15 ripe 76 unripe females.	....	....	....	....	....	....	....	....	15	15	....	7	1				
20	Respawning fish of 18th	....	....	....	....	....	....	....	....	....	....	64	4	12	9,000	14	90	New hatching-house; 203 females yield 231,000; 64 females, 76,000.
	Fish of last two nights	473	98	184	183	8	375	....	183	....	183	....	91	9	198,000	13, 15, 16, 26	....	New hatching-house.
22	Respawning 183 females	....	....	....	....	....	....	....	....	....	....	183	15	3	21,000	17	91	Do.
	Summary to date	1,766	849	563	344	10	917	0	345	41	386	385	226	4	450,000	....		
	Fish of last two nights	489	82	160	241	5	406	1	241	....	241	....	116	8	243,000	18	88	Do.
23	Respawning 241 females	....	....	....	....	....	....	....	....	....	....	241	16	2	25,000	19	....	Do.
	Last night's fish	76	20	46	10	....	56	....	10	....	10	....	4	5	7,000	20 & 33	....	Do.
25	Respawning same	....	....	....	....	....	....	....	....	....	....	10	0	10	....	....	....	

	To night's fish	96	23	32	34	7	73	....	34	....	34	......	15	3	....	....	.....
	Pound C	.....	.....	.....	.....	.....	.....	....	....	431	431	......	233	13	....	....	.....
	Summary to date	2,427	974	801	629	22	1,452	1	630	472	1,102	636	612	13	1,170,000	....	.....
27	Respawning 431 females	.....	.....	.....	.....	.....	.....	....	....	....	.....	431	24	5	40,000	23	93
	Fish of last two nights	57	26	5	16	10	31	0	16	....	16	......	5	5	211,000	24	93
	Part of pound C	.....	.....	.....	.....	.....	.....	....	....	211	211	......	113	10			
28	Rest of pound C	.....	.....	.....	.....	.....	.....	....	....	111	111	......	60	1	103,000	25	89
	Fish of last night	14	10	0	4	0	4	0	4	....	4	......	1	1			
29	Respawning	.....	.....	.....	.....	.....	.....	....	....	....	.....	342	25	10	40,000	30	89
	Fish of last night	25	12	0	3	10	13	0	3	....	3	......	2	2			
30	Fish of last night	25	17	0	2	6	8	0	2	....	2	......	0	14	106,000	34	91
	Pound C entire	.....	.....	.....	.....	.....	.....	....	....	145	145	......	69	7	20,000	34 expr.	.....
Dec. 2	Remaining on hand	.....	.....	.....	.....	.....	.....	....	....	23	23	......	11	12	31,000	35	81
	Respawning of 152 females	.....	.....	.....	.....	.....	.....	....	....	....	.....	152	8	6			
4	Respawning of 23 fish	.....	.....	.....	.....	.....	.....	....	....	....	.....	......	.....	.....	2,000	36	91
	Summary to date	2,548	1,039	806	654	48	1,508	1	655	962	1,617	1,561	935	6	1,723,000	....	.....

TABLE III.—*Measurement of Schoodic salmon at Grand Lake Stream, Maine*, 1878.

Date.	Males.													Females gravid before spawning.													Females spawned.											
	Number weighed and measured.	Weight.						Length.						Number weighed and measured.	Weight.						Length.						Number weighed and measured.	Weight after spawning.						Length.					
		Average.	Heaviest.		Lightest.		Average.	Longest.	Shortest.						Average.	Heaviest.		Lightest.		Average.	Longest.	Shortest.						Average.	Heaviest.		Lightest.		Average.	Longest.	Shortest.		
		Lbs.	Lbs.	oz.	Lbs.	oz.	In.	In.	In.						Lbs.	Lbs.	oz.	Lbs.	oz.	In.	In.	In.						Lbs.	Lbs.	oz.	Lbs.	oz.	In.	In.	In.		
November 7, 1878	14	2.2	2	12	1	8	18.6	20	16					46	2.2	3	0	1	7	17.8	20½	16½					8	2.0	3	4	1	7	17.7	20½	16½		
12, 1878	46	2.4	4	4	1	5	18.3	23	15½					2	1.9	2	1	1	12	17.2	17½	17					7	1.8	2	9	1	4	16.8	19	13½		
14, 1878	122	2.2	3	4	1	6	18.4	21	15					68	2.3	2	12	1	8	17.2	20	16					30	1.8	2	14	1	3	16.9	19½	15		
15, 1878	121	2.4	3	8	1	4	17.6	21½	15					55	2.3	3	0	1	9	17.5	20	15½					33	2.1	3	3	1	1	16.6	21	15		
16, 1878	55	2.2	3	5	0	14	18.0	22	13					50	2.2	4	3	1	6	17.3	21¾	16					7	2.0	2	13	1	2	17.0	19	15		
22, 1878	33	2.4	3	8	1	6	18.5	22	16					59	2.3	4	1	1	6	17.6	20	16					92	1.7	3	2	1	2	17.3	21	15		
Sums	391	2.3	4	4	0	14	15.5	23	13					280	2.2	4	3	1	6	17.5	21¾	15½					177	1.8	3	4	1	1	17.2	21	13½		

TABLE IV.—*Record of shipment of salmon-spawn from Grand Lake Stream, February and April,* 1879.

Date.	Consignee.	Address.	On whose account.	Number of cases.	For what State.	Number of eggs. Belonging to State.	Number of eggs. Donated by United States.	Total.
1879. Feb. 10	George Jelliffe	Westport, Conn	Fish Commission	1	Connecticut	50,000		50,000
11	E. A. Brackett	Winchester, Mass	do	1	Mass	50,000		50,000
	A. H. Powers	[illegible]h, N. H	do	1	New Hampshire	50,000		50,000
Apr. 7	F. N. Clark	Northville, Mich		2	[illegible]an, Iowa, and Minnesota		60,000	60,000
	T. B. Ferguson	Baltimore, Md	T. B. Ferguson	1	Maryland		50,000	50,000
	Prof. M. McDonald	Lexington, Va	Prof. M. McDonald	1	Virginia		30,000	30,000
	George Jelliffe	Westport, Conn	Fishery Commission	1	Connecticut	70,000		70,000
	C. S. White	Romney, W Va	do	1	West Virginia		30,000	30,000
8	Dr. R. O. Sweeny	Saint Paul, Minn	do	1	Minnesota		45,000	45,000
	B. F. [illegible]	Anamosa, Iowa	do	1	Iowa		35,000	35,000
	William Welch	[illegible], Wis	do	1	Wisconsin		20,000	20,000
	D. B. Long	Ellsworth, Kans	do	1	Kansas		10,000	10,000
	E A. Brackett	Winchester, Mass	do	2	Massachusetts	140,000		140,000
9	G. H. [illegible]	Niles, Mich	do	1	Michigan		40,000	40,000
	F. N. Clark	Northville, Mich		1	do		20,000	20,000
	State Hatching-House	[illegible], N. Y	Fish Commission	1	New York		25,000	25,000
	L. C. Starkel	Belleville, Ill	Belleville Fishing Club	1	Illinois		10,000	10,000
	E. A. Brackett	Winchester, Mass	Fish Commission	1	Massachusetts	30,000	10,000	40,000
10	Alfred Swazey	Bucksport, Me	do	1	Maine		50,000	50,000
14	Mrs J. H. [illegible]	Bloomsbury, N. J	do Hon. A. B. Hewett	1	New Jersey		35,000	35,000
	P. H. Christie	Clove, Dutchess County, N. Y	P. H. [illegible] and others	1	New York		20,000	20,000
	[illegible] Duffy	Marietta, Pa	Fish Commission	1	Pennsylvania		25,000	25,000
	Seth Weeks	[illegible]y, Pa	do	1	do		12,500	12,500
	E. D. Potter	Toledo, Ohio	do	1	Ohio		10,000	10,000
	A. H. [illegible]	Plymouth, N. H	do	1	New Hampshire	38,000	30,000	68,000
15	E. A. [illegible]	Winchester, Mass	do	1	Mass		10,000	10,000
	Alfred Swazey	Bucksport, Me	do	1	Maine		9,000	9,000
	George Jelliffe	Westport, Conn	do	1		56,000	19,000	75,000
	A. E. Neill	Grand Lake Stream			Maine		10,500	10,500
	Total			30		484,000	616,000	1,100,000
	Retained at Grand Lake Stream, 25 per cent. reserve							370,000
	Total divided							1,470,000

TABLE V.—*Statement of the distribution of young Schoodic salmon, 1879.*

State.	Place of hatching.	In charge of hatching.	Number of fish sent out.	Waters stocked.	Tributary to what other water.	Locality.	Number of fish set free.
Maine	Bucksport	Alfred Swazey	55,000	Moosehead Lake	Kennebec River	Moosehead	20,000
				Pond		Damariscotta	5,000
				do		Unity	5,000
				do		[illegible]	5,000
				do		[illegible]	5,000
				Pushaw Pond		Glenburn	10,000
				Cobosseecontee Lake		[illegible]	5,000
	Grand Lake Stream	Wm. H. Munson	348,000	Grand [illegible]	Saint Croix River	Hinkley	348,000
			10,500	Nash's Lake	do	[illegible]	10,500
New Hampshire	Plymouth	A. H. Powers	104,500	Jones Pond		Raymond	5,000
				[illegible] [illegible] Pond		New Durham	5,000
				East Pond		Wak[illegible]	5,000
				Ossipee Lake		[illegible]	5,000
				[illegible] Lake		[illegible]	5,000
				Tarlton Pond		Piermont	5,000
				[illegible]		Newbury	10,000
				Star [illegible]		Springfield	1,500
				Echo Lake		Franconia	2,500
				[illegible] Lake		Bridgewater	5,000
				Squam Lake		Holderness	10,000
				Mascoma Lake		Enfield	5,000
				[illegible] Lake		Pittsburg	5,332
				Winnipiseogee Lake		Centre Harbor	5,000
				[illegible] (no [illegible])		Sandwich	5,000
				Nutt's [illegible]		Manchester	5,000
				Bradford's Pond		Bradford	5,000
				Chestnut Pond		Nor[illegible]	2,000
				Sundry Ponds		[illegible] County	5,000
				do		[illegible]	5,000
Massachusetts	Winchester	E. A. Brackett	221,000	Pond		Harvard	6,000
				do		East Brookfield	10,000
				do		[illegible]	4,000
				do		Natick	3,000
				do		[illegible]	6,000
				do		Wenham	10,000
				do		Framingham	8,000
				do		Lynnfield	8,000
				do		[illegible] burnham	20,000
				do		[illegible]	10,000
				do		East Brookfield	8,000

TABLE V.—*Statement of the distribution of young Schoodic salmon, 1879*—Continued.

State.	Place of hatching.	In charge of hatching.	Number of fish sent out.	Waters stocked.	Tributary to what other water.	Locality.	Number of fish set free.
				Pond		[illegible]	9,000
				do		Wakefield	3,000
				do		Middleborough	20,000
				do		Great Barrington	6,000
				do		Stockbridge	20,000
				do		[illegible]	3,000
				do		West Scituate	3,000
				do		Paxt[illegible]	6,000
				do		[illegible]	20,000
				do		Rochester	4,000
				do		Westborough	4,000
				do		Essex	10,000
				do		Holyoke	6,000
				do		Milford	3,000
				do		Falmouth	3,000
				do		Marshall	6,000
				do		Waltham	2,000
Connecticut	Westport	George Jelliffe	178,715	Round Pond	Housatonic	Ridg field	10,000
				Lake Kenosha	do	Danbury	10,000
				[illegible]	do	Chapinsville	5,000
				Green Pond	do	Sherman	5,000
				Wanouscoponous Lake	do	Lakeville	5,000
				L S Reservoir	Thames	Stafford Springs	10,000
				Long Lake	Naugatuck	West Winsted	10,000
				Saltonstall Lake		East Haven	10,000
				Hog Lake	Connecticut River	Lyme	10,000
				Gardiner's Lake	Thames	Salem	10,000
				Tributary	Housatonic	Newtown	10,000
				Waa-am-ang Lake	do	[illegible] Preston	10,000
				Bantam Lake	do	Litchfield	10,000
				Lake Pocatapang	Connecticut River	East Hampton	10,000
				Tributaries	Thames	[illegible]	10,000
				Wangambomg Lake	do	South Coventry	10,000
				Reservoir	Connecticut River	West Hartford	3,715
				Melrose Pond	do	[illegible]	10,000
				[illegible]	do	Rockville	10,000
				Southport Pond	Mill River	Southport	10,000
New York	Caledonia	Seth Green	15,000	Woodhull Lake		[illegible]	5,000
				Allen Creek	Lake Ontario	Monroe County	10,000
	Clove, Dutchess Co	P. H. Christie	24,000	Silvan Lake		Dutchess [illegible]	3,000

				Upton's Pond		do	3,000
				Long Pond		do	3,000
				Furnace Pond		do	10,000
				Lile Whala Pond		do	5,000
New Jersey	Bloomsbury	A. A. Anderson	33,400	Drake's Pond	Paulinskill River	Sussex County	1,000
					Passaic River	Passaic County	2,000
				Shepherd's Lake	do	Ringwood, N. J	9,400
				Green's Pond	Pequest River	Oxford, Warren County	4,000
				Lake Hopatkong	South branch Raritan	D[illegible]ville, Morris County	16,994
Pennsylvania	Marietta	J. P. Creveling	23,500	Harvies Lake	Susquehanna River	W[illegible]e	6,000
				Tobyhanna	Lehigh River	Scranton	3,000
				Paupock Lake	do	do	11,500
				Pig Pond	do	White Haven	1,500
				Moses Wood Pond	do	do	1,500
	Corry	Seth Weeks	12,000	Bay	Lake Erie	Erie	10,000
						Potter County, Pennsylvania	2,000
Virginia	Lexington	H. W. Williamson	27,350	[illegible] River	James River	Farmville	6,000
				South River	do	Lexington	2,500
				Middle River	Shenandoah River	Staunton	6,500
				Buffalo Creek	James River	Lexington	250
				McKee's Spring	do	do	300
				Jackson River	do	Covington	6,500
				Craig's Creek	do	Fincastle	2,800
				New River	do		2,500
Maryland	Baltimore	Frank Behler	26,500	Stream	Beaverdam Creek	Baltimore	750
				do	Susquehanna River		1,000
				Pond	Charles River	Baltimore	500
				Lake		Druid Hill	8,000
				do		do	2,000
				do		do	2,000
				Pond	Little Pipe Creek	Westminster	1,000
				Cobb's Branch	do	do	2,000
				Pond	Gunpowder River	Hampton	1,000
				Lake S.	Stony Run	Charles Street avenue	500
				Pond	Big Pipe Creek	Union Bridge	500
				do	do	do	1,000
				Lake	Patapsco Falls	Reisterstown	500
				do		Druid Hill	1,000
				Green Spring Run	Jones Falls	Burnsides	1,500
				Lake	do	Baltimore	1,000
				Pond	Miles Creek	Easton	250
				do		do	250
				do	Miles Creek	do	500
				do	do	do	250
						Druid Hill	500
						do	500
West Virginia	Romney	Z. J. Graham	27,200	Pond at Institution for Deaf, &c.	Potomac River	Romney	400
				Dillon's Run	do	Hampshire County	600
				Fountain at Institution for Deaf, &c.	do	Romney	50

TABLE V.—*Statement of the distribution of young Schoodic salmon*, 1879—Concluded.

State.	Place of hatching.	In charge of hatching.	Number of fish sent out.	Waters stocked.	Tributary to what other water.	Locality.	Number of fish set free.
West Virginia	Romney	Z. J. Graham	27,200	West Fork	Monongahela	Clarksburg	2,000
				do	do	Weston	1,500
				Wheeling Creek	Ohio River	Cold Spring	5,000
				Fish Creek	do	Littleton	5,000
				Dry Fork	Greenbrier River	White Sulphur Springs	3,000
				Williams Spring	Patterson's Creek	Williamsport	1,000
				Mill Creek	Potomac River	Hamot's Mill	500
				do	do	do	1,500
				South branch of Potomac River	do		650
				Greenbrier River	New River	Hinton	5,669
Ohio	Toledo	E. D. Potter	1,000	Maumee River	Lake Erie		1,000
Kansas	Ellsworth	D. B. Long	5,000	Smoky Hill River	Kansas River	Ellsworth County	3,000
				Inman Lake		McPherson County	2,000
Minnesota	Willowbrook and Red Wing.	R. O. Sweeny	48,500	Sandy Lake		Cass County	5,000
				Spring Lake		Dakota County	5,000
				Lake Como		Houston County	3,000
				Barnum		do	500
				Canosia Lake		Duluth County	5,000
				Roberts Lake		Rice County	4,000
				Cedar Lake		do	4,000
				Long Lake		Watonwan County	3,000
				Cedar Lake		Martin County	3,000
				Lake Alley		Renville County	5,000
				Lake Preston		do	5,000
				Skellman's Pond		Wabasha County	5,000
				Cannon River	Cannon Falls		1,000

Table VI.—*Observations on temperature and weather at Grand Lake Stream, 1878 and 1879.*

Date.	Temperature.					Wind.	Other phenomena.
	Air.		Stream at dam.		Old hatching-house water.		
	7 a. m.	1 p. m	7 a. m	1 p. m			
1878. Oct. 1	39	70	59½	61½	47	Southerly, light in morning, gentle through day.	Foggy till 9 a m., some thin clouds, but sun four-fifths of time.
2	54	71	60	62½	48	A. M, light NW.; p. m, light S. by E., E. in evening.	Clear all day, cloudy evening.
3	55	70	61	63½	48½	Light N. by W., dying away at night	Foggy till 8½ a m., clear rest of day.
4	51	67	61½	63½	48½	Fresh N. by W. or NW.; light morning, and calm evening.	Clear all day.
5	48	61	61	62	48½	NW, fresh in morning, strong through day.	Do
6	45	57	60	60½	48½	A. M., very light NE., in p. m hauled to light N. and W, dying entirely away at night.	Cloudy till middle of p m, perfectly clear rest of day.
7	44	59	58	60	48	West	Clear.
8	42	56	58	59	48	South	Cloudy.
9	47	58	58	59	48	......do	Do.
10	48	52	57	58	48	West	Do.
11	40	53	54	56	48	North	Clear.
12	38	52	54	56	48	East	Cloudy.
13	40	50	54	54	48	North	Do.
14	41	60	51	54	48	West	Clear.
15	38	68	53	56	48	......do	Do.
16	52	67	54	55	48	......do	Cloudy.
17	50	54	55	55	48	Southwest	Clear
18	55	69	55	57	48	South	Do.
19	60	55	57	57	48	......do	Rainy.
20	44	46	55	55	48	West	Cloudy.
21	42	56	53	54	48	......do	Clear.
22	40	59	53	54	48	......do	Do.
23	37	46	53	53	47	East	Cloudy.
24	49	50	52	53	47	......do	Rainy.
25	49	52	52	53	48	West	Cloudy.
26	46	55	52	52	47	North	Do
27	45	50	52	52	47	South	Do.
28	48	46	52	52	47	North	Do.
29	28	36	47	49	47	......do	Clear
30	22	37	46	47	46	......do	Cloudy.
31	46	48	49	50	47	West	Do.
Nov. 1	38	44	48½	49	....	NW, fresh	Half cloudy
2	34	53	48	49	46½	South, very light	Do
3	39	42	48	49	46½	Northwest	Clear mostly.
4	21	39	45	46½	46	SE, light till 3 p. m.; then NW, fresh.	Clear a m, cloudy p m, snow began at 3 p m., cleared at 5½; 3 inches.
5	18	29	42½	42½	...	NW., strong all day	A little snow in a m, cloudy rest of day
6	20	35	41	42	....	NW, strong, dying away at night	Clear all day
7	14	38	40	41½	...	Southerly, gentle	Clear a. m, cloudy p m
8	28	30	40½	39½	46	NW., fresh in a. m, strong in p m.	Snow began to fall at 8 a. m., continued all day, 5 or 6 inches
9	28	31	43½	43½	45½	NW, fresh or gentle	Snowing very lightly all day
10	27	33	39	39	..	NE to NW, strong	Partly clear, partly cloudy, fully clear at sunset and after
11	28	38	38	39½	....	Northerly in a m, calm in p m	Clear a. m, cloudy p. m.
12	40	46	40	41	...	S E, gentle, late in p. m. changed to SW	Light rain in a m., cloudy in p. m.; cleared in evening
13	24	42	40	41	...	Southerly, light	Light rain from 10 a m to 5 p. m; clear morning and evening.
14	27	35	39	40	....	NW., fresh	Mostly clear.
15	23	33	38	39	...	NW., light, calm at evening	Clear.
16	20	37	37½	39	...	S, light all day	Cloudy mostly.
17	30	41	38½	39½	....	Variable, westerly, easterly, very light.	Clear.
18	25	34	38	39	....	Southeast, gentle	Cloudy, snow and sleet in evening and all night
19	31	35	37	37½	....	N to NE, fresh	Light snow and some sleet all day.
20	34	38	37½	38	..	N., gentle to light	Cloudy, with very light rain.

TABLE VI.—*Observations on temperature and weather, &c.*—Continued.

Date	Temperature.					Wind.	Other phenomena
	Air.		Stream at dam.		Old hatching-house water.		
	7 a. m.	1 p. m.	7 a. m.	1 p. m.			
1878.							
Nov. 21	33	36	38	39	....	N. to NW., light; dying away in p. m.	Clear in morning, but soon cloudy, and cloudy rest of day till sun-set, after that clear
22	30	35	38½	39½	....	NE, light; increasing at evening	Cloudy all day· slight rain in evening
23	47	49	39½	40½	...	Southerly, light to gentle	Cloudy, with occasional light rain.
24	37	41	40	41	....	NW. to W., light................	Clear
25	30	40	39	40	....	Westerly a. m, southerly p m....	Clear a m, cloudy p. m, snow in evening and nearly all night.
26	32	36	39	40	....	NW, fresh	Clear
27	21	54	37	39	....	Westerly	Do
28	31	32	39½	40½	....	Southeast and southwest	Rain since 10½ last evening and nearly all to-day
29	30	98	38	39	....	NW., strong	Clear
30	30	36	38½	39	....	NW., light a. m., strong p. m	Mostly clear.

Date	Temperature.			Wind.	Other phenomena
	Air.	Water at new hatching-house.	Water at old hatching-house.		
	7 a. m.	7 a. m.			
1878.					
Dec. 1	16	35	44	Westerly	A m., clear, p. m., cloudy.
2	.....	34½	43½	Southerly................................	Cloudy.
3	.....	40	43	Southeasterly...........................	Cloudy, rain, heavy rain last night
4	.....	37	44	Southerly, light.........................	Cloudy, some rain.
5	36	38	44	Northerly, light.........................	Cloudy.
6	25	37	43½	Westerly, gentle	A. m, cloudy; p. m, clear.
7	.....	34	42½	Northwest, gentle........................	
8	18	33	42	Northwest, fresh.........................	
9	12	31½	42	Westerly, light to gentle................	Cloudy
10	23	32½	42	SE. and SW., gentle to strong............	A. m, snow
11	45	36	42	SE. and S., strong	Rained heavily all last night, and moderate rain through day.
12	29	36	43	Northwest, fresh.........................	Cloudy.
13	21	33	43	Northwest, strong	Cloudy, mostly
14	15	32	42	Strong, NW, through night and a m; p. m changed to easterly.	A m, clear, p m, cloudy.
15	27	33	41½	Light, southerly	Cloudy, snow early a. m.
16	28	33	41½	Strong, northerly	Cloudy till evenng.
17	13	33	38	Gentle, westerly	Clear
18	10	32	38	....	
19	17	32	38½	Northerly	Clear.
20	10	32	39	Westerly	Do
21	2	32	40	Snow, calm a. m..........................	Mostly cloudy, gale in night.
22	38	33	40½	Southeasterly............................	Mostly cloudy
23	16	32½	38½	Westerly	Clear.
24	9	32½	39	....	Do
25	.....	32	40	Southwesterly	Do
26	.....	31½	41	......do	Do.
27	3	32½	40½	....	Do
28	13	32	41	Northerly................................	Do.
29	5	32	41	....	Do.
30	.....	32	41	....	Cloudy; snow in evening
31	5	32	40½	....	Clear.

TABLE VI.—*Observations on temperature and weather, &c.*—Continued.

Date.	Temperature.			Wind.	Other phenomena.
	Air.	Water at new hatching-house.	Water at old hatching-house.		
	7 a. m.	7 a. m.			
1879.					
Jan. 1	6	....	41	Northerly, light	Clear.
2	0	....	40	Southeasterly, light	Cloudy to snow.
3	8	....	41	Northeast, fresh	Snow.
4	12	....	41	Northerly, light	Clear.
5	18	....	40½	Northerly, squalls	Do.
6	12	....	41		
7	....	....	41		
8	12	....	40	Southerly	Cloudy and snow
9	16	....	41		
10	12	....	40	Northerly	Clear
11	-4	....	40	......do	Do.
12	-7	....	40		Do.
13	8	....	40½		Cloudy
14	18	....	40	Northwest	Clear
15	-10	....	39	......do	Do.
16	-10	....	39	Northeast	Cloudy
17	-16	....	39	Northerly	Clear.
18	20	....	40	Southerly	Snowy to clear
19	-1	....	39		Clear.
20	2	....	39	Easterly	Snow last night; clear.
21	-20	....	39	Easterly and northerly	Clear
22	9	....	39	Northerly and westerly	Do.
23	11	....	39	Northeast	Snow, clear
24	6	....	39	Northerly to southerly	Clear
25	26	....	39½		Snow
26	-7	....	39	Strong northerly	Clear
27	-6	....	39		Cloudy and little snow.
28	15	....	39		Do.
29	23	....	39	Northwest, fresh	Cloudy to clear.
30	5	....	39	......do	Clear
31	6	....	39	......do	Do.
Feb. 1	8	....	39	Northerly	Clear to snow.
2	20	....	39	......do	Snow to clear
3	28	....	39	......do	Mostly cloudy, clear evening.
4	20	....	39		Clear
5	10	....	39		Do
6	18	....	39		Do
7	10	....	39	Northerly or northwest	Do
8	2	....	39	......do	Do
9	-7	....	38	Southerly and southeasterly	Snow
10	16	....	38	Northerly and northwest	Clear.
11	9	....	38⅓	Southerly and easterly	Clear to cloudy.
12	40	....	37	Southerly and southeasterly	Rain.
13	8	....	37	Northerly	Clear.
14	-2	....	37	Northwest	Do.
15	-8	....	37	......do	Do.
16	2	....	39	Northerly and westerly	Do
17	8	....	38½	Variable	Mostly clear.
18	12	....	38½	Northeast	Snow storm.
19	4	....	39½	Northerly	Clear
20	4	....	38½	Easterly	With snow
21	7	....	38	Northerly, gale	Cloudy and snow
22	3	....	38½	Northerly, fresh	Clear
23	....	....	38⅓	Southwesterly and southerly	Clear to cloudy
24	17	....	39	Variable, snow squalls	Partially clear
25	-5	....	39	Northerly	Clear
26	19	....	39	Easterly to southeast	Cloudy to hail, to rain, snow last night.
27	29	....	39	Northeasterly	Snow to clear weather.
28	-4	....	38	Northerly	Clear weather.
Mar. 1	0	....	38	Variable, and very light	Cloudy.
2	19	....	38½	Northerly, light	Clear; a little snow last night.
3	7	....	38½	......do	Clear.
4	27	....	38½	Southerly	Clear to cloudy.
5	29	....	38½	Northerly	Clear.

TABLE VI.—*Observations on temperature and weather, &c.*—Continued.

Date.	Temperature. Air. 7 a. m.	Water at new hatching-house. 7 a. m.	Water at old hatching-house.	Wind.	Other phenomena.
1879.					
Mar. 6	— 2	....	38	Northerly to southeast	Clear, followed by snow.
7	17	....	38½	Northeast	Four or five inches light snow; clear evening.
8	8	....	38	Easterly	Clear.
9	....	....	39	......do	Clear morning, hail and rain at night.
10	27	...	39		Rain.
11	30	....	38	Easterly	Do.
12	22	....	38	Northerly	Clear.
13	21	....	38½	......do	Do
14	....	....	39	Northerly and westerly, light	Clear, a very little snow last night.
15	36	....	39	Northerly	Clear.
16	15	....	38	North and northwesterly	Do.
17	16	....	38	Northeast	Snow storm, six inches at 6 p m.
18	22	....	38	Light, northerly and westerly	Clear.
19	— 4	....	38		Clear weather.
20	14	....	38	Westerly	Clear.
21	2	....	38½	Southerly and easterly	Clear to snow.
22	16	....	38½	Southerly, a. m; northerly, p. m.	Clear to cloudy.
23	22	....	39	Easterly	Snow storm, nine inches.
24	15	....	38		Clear, a. m., cloudy, p m.
25	33	....	39		Cloudy, a. m., clear, p. m.
26	30	....	39		Light snow, a m; clear, p m.
27	27	....	38½	Easterly	Cloudy.
28	28	....	38½	Westerly	Clear
29	27	....	39	Easterly	Cloudy
30	37	....	39	......do	Foggy and wet.
31	37	....	39	Northeast	Cloudy to rain, with thunder and lightning
Apr. 1	36	....	34	Northeast to northwest	Rain to snow, cleared off in evening.
2	22	....	36	North and west	Clear
3	20	....	37	Northerly	Snow, a m., clear, p. m
4	14	....	37	Northeast	Snow storm, five inches
5	24	....	38		Cloudy, snow, p m
6	26	....	38	Northerly	Mostly cloudy.
7	28	....	38	Variable	Cloudy to clear, snow squalls.
8	33	....	38½	Northerly	Clear weather
9	34	....	38	Strong, northerly	Clear
10	25	....	38		Clear weather
11	28	....	37½	Northeast	Snow storm
12	28	....	38	Northerly, with snow squalls	
13	....	....	38	Southwest or westerly	Clear
14	....	....	38		Fair weather
15	26	....	38	Northerly	Clear.
16	36	....	38½	Northerly, gentle	Do.
17	38	....	38½	..... do	Do
18	....	....	38	Northerly to northeast	Clear to snow.
19	29	....	37	Northeast	Cloudy, cleared off at night.
20	28	....	36½	Gentle, northerly	Clear
21	35	....	37	.....do	Do
22	36	....	37½	..... do	Do
23	40	....	37	Northeast to heavy north	
24	34	....	37	Strong, northerly	Clear.
25	34	....	37	Light, southerly	Do.
26	38	....	37½	..... do	Do.
27	40	....	38	Light, southwest	Do.
28	42	....	38	Light, southeast	Cloudy.
29	46	....	38	Easterly air, no wind	Do.
30	47	....	38	Southeast	Rain all day.
May 1	46	....	38½	No wind	Clear, a. m.; thunder shower, p. m.
2	41	....	38½	Northerly	Clear.
3	....	....	39		

TABLE VI.—*Observations on temperature and weather, &c.*—Concluded.

Date.	Temperature.			Wind.	Other phenomena.
	Air.	Water at new hatching-house.	Water at old hatching-house.		
	7 a. m.	7 a. m.			
1879.					
May 4	....	....	39	....	
5	.....	....	39½	....	
6	.....	....	39½	....	
7	44	....	39½	....	
8	.....	....	39	....	
9	.....	....	39	....	
10	.....	...	38½	....	
11	.....	....	39½	....	
12	.....	....	39½	....	
13	....	....	39½	....	
14	.....	....	40	....	
15	.....	....	40	....	
16	.....	....	40	....	
17	.....	....	40	....	
18	.....	....	40	....	
19	.....	....	41	....	
20	.....	....	41½	....	

TABLE VII.—*General summary of observations on temperatures at Grand Lake Stream, from October, 1878, to May, 1879, inclusive.*

Date.	Air.								Water of stream.								Water at old hatching-house.						Water at new hatching-house.					
	7 a. m.		1 p. m.		Max.		Min.		7 a. m.		1 p. m.		Max.		Min.		7 a. m.		Max.		Min.		7 a. m.		Max.		Min.	
	No. of observations.	Degrees.	No. of observations.	Degrees.	Date.	Degrees.	Date.	Degrees.	No. of observations.	Degrees.	No. of observations.	Degrees.	Date.	Degrees.	Date.	Degrees.	No. of observations.	Degrees.	Date.	Degrees.	Date.	Degrees.	No. of observations.	Degrees.	Date.	Degrees.	Date.	Degrees.
1878.																												
October	31	44.6	31	55.8	2	71	30	22	31	54.7	31	55.9	3	63½	30	46	31	47.7	5	48½	30	46	...	...	...	...	...	...
November	30	29	30	38.3	27	54	7	14	30	40.2	30	41.1	1	49	19	37	5	46.1	3	46½	9	45½	...	...	...	...	...	...
December	31	18.2	...	...	11	45	21	2	...	...	...	...	...	...	...	...	31	41.3	1	44	17	38	31	33.5	3	40	9	31½
1879.																												
January	30	5.5	...	...	25	26	21	—20	...	...	...	...	...	...	...	...	31	39.7	1	41	15	39	...	...	...	...	...	...
February	27	9.7	...	...	12	40	15	— 8	...	...	...	...	...	...	...	...	28	38.5	19	39½	12	37	...	...	...	...	...	...
March	29	19.9	...	...	30	37	19	— 4	...	...	...	...	...	...	...	...	31	38.5	9	39	1	38	...	...	...	...	...	...
April	27	32.1	...	...	30	47	4	14	...	...	...	...	...	...	...	...	30	37.5	8	38½	1	34	...	...	...	...	...	...
May	3	43.7	...	...	1	46	2	41	...	...	...	...	...	...	...	...	20	39.5	20	41½	1	38½	...	...	...	...	...	...

XXXVIII.—DO THE SPAWNING SALMON ASCENDING THE SACRAMENTO RIVER ALL DIE WITHOUT RETURNING TO SEA?

BY HORACE D. DUNN.

[With notes by Livingston Stone.]

A.

[Communication from San Francisco, Cal., September 26, 1876.

Prof. SPENCER F. BAIRD,

Assistant Secretary Smithsonian Institution, Washington, D. C.:

SIR: I am the author of the article lately published in the San Francisco Bulletin on fish-culture, which was sent to you by B. S. Redding, esq., fish-commissioner of the State of California. Mr. Redding has kindly loaned me his copy of the United States Fish Commission Report, 1872–'73, and at his suggestion I write you. I have been a resident of California since January, 1850, and during the last ten years have taken great interest in its fisheries. As a consequence I have been much interested in the article of Mr. Livingston Stone on the Sacramento salmon in the report. In some particulars my experience differs from that of Mr. Stone, and on this account Mr. Redding has asked me to write to you.

On page 180, Mr. Stone states that all the Sacramento salmon die after spawning. No doubt great numbers do, but a very large portion of the run return to sea again, as before the close season between August 1, and November 1, was established it was a common occurrence to find spent salmon in market between the dates named. These salmon were very much emaciated, had no scales, and varied in color from a rusty black on the backs to a faded brown on the belly. Some were of a dirty white color all over, as if they had been parboiled.

In regard to the quality of the Sacramento salmon, I think they compare favorably with those caught in Maine. The mode of treatment here of salmon is simply barbarous. The fish are caught in drift-nets in the Sacramento River, near Rio Vista, about eighty miles from this city. They, as a rule, lie in a boat for several hours exposed to the sun before being brought to the steamer's wharf. There they lie in large heaps for several hours more, and are dragged on board and put in large heaps again. At San Francisco the fish are dragged ashore and roughly thrown into wagons, and on arrival at the markets experience the same treatment again; as a consequence, the salmon have been dried and heated and much bruised before they are sold to the consumer, and their flavor and firmness of meat much impaired. Treat the Maine salmon in the same way and its best qualities would be gone.

Our salmon do take bait readily in the salt water at the mouths of rivers and creeks, and in the bays along the coast. At such places fly-fishing is generally not successful.

I have had several conversations with Mr. Redding about his observations of the San Joaquin salmon and of your desire to procure a species of salmon that can be successfully introduced into the waters of the States south of the Potomac. It seems to me that the San Joaquin salmon will not be as good for such purpose as the salmon which frequent the rivers which empty direct into the Pacific along the California coast from Monterey north. This last variety makes its appearance at the mouths of the coast streams from the middle of October to November, awaiting the annual winter rains to swell the streams, up which they go to their spawning-beds. The spawning takes place in December and January, the spent fish returning to the ocean in February and March. These fish, in good condition, have been caught weighing 25 pounds. As salmon spawn at set times, regardless of change of location, the coast species would arrive in the rivers of the Southern States when the streams would be swollen with winter rains, the water of a low temperature, and such enemies as blue-fish, Spanish mackerel, &c., withdrawn to warmer waters near the Florida coast. The rivers, also, being higher than in summer, would, with the colder weather, insure the salmon a better chance for successful spawning than would be the case with the San Joaquin salmon, which would be in the rivers in August and September, and easily caught, which is not desired until the rivers become fully stocked.

I would also call your attention to a fish commonly called salmon-trout, which visits our coast rivers about the same time the salmon do, probably two weeks later. This fish is trout-shaped, being longer and rounder than the salmon, and of proportionately less weight. This fish spawns as late as February, and possibly early in March, after which they return to the mouths of the streams and to the ocean in the early part of April. They have been caught weighing 20 pounds, from 8 to 10 pounds being a common weight. I think this variety would do well if introduced into Virginia and the two Carolinas. They are a very game fish, flesh before spawning a faint cream-color, changing to white when returning to sea. A 7-pound fish in good condition measured 31½ inches long, while the largest run to about 40 inches. This variety of fish stock the coast streams, where the mountain trout have been completely fished out. They are, however, fast disappearing under the effects of netting, giant-powder, and spearing, and promise to become extinct within the next ten years unless protected by law from such dangers.

Apologizing for so long and disjointed a letter, I remain your obedient servant,

HORACE D. DUNN.

B.

[Notes on letter of Mr. Horace D. Dunn to Prof. Spencer F. Baird, dated San Francisco, September 26, 1876. By Livingston Stone.]

In the first part of his letter Mr. Dunn says that on page 180, United States Fish Commission Report, 1872 and 1873, Mr. Stone states that "all the Sacramento salmon die after spawning," and criticises this statement as follows:

"No doubt that great numbers do, but a very large portion of the run return to sea again, as before the close season between August 1, and November 1, was established it was a common occurrence to find spent salmon in market between the dates named."

Now, Mr. Dunn, undoubtedly without designing to, has misquoted what I said. By referring to page 180, of the report referred to, it will be seen that I did not say that *all* the Sacramento salmon die after spawning, but limited my statement to the salmon of the McCloud River. Indeed, in reference to the *Salmo quinnat* dying in the California rivers I took particular care in my report to say that my remarks applied only to the salmon of the McCloud and upper tributaries, because these salmon I was familiar with, while I was not familiar with the habits of the Sacramento salmon of other portions of the river. The exact words of the statement on page 180, of the report are as follows:

"Soon after this they (the salmon) become foul, diseased, and very much emaciated, and in the McCloud River, at least, they die a short time after spawning."

This statement I am prepared to support by evidence collected since that time, and by the testimony of many others who have had opportunities of observing the McCloud River salmon with particular reference to this peculiarity.

The fact further mentioned by Mr. Dunn in confirmation of his statement that spent salmon were frequently seen in San Francisco in the fall is no evidence in regard to the McCloud River salmon, for no salmon in any stage were ever sent from this river to the San Francisco market, while it is a fact within my own knowledge that spent salmon were formerly sent to market from the vicinity of Cloverdale and also from Tehama. My own opinion on the subject, confirmed by five seasons' observation on the McCloud River, is that the Sacramento salmon which spawn near the sea are, many of them, able to return to the salt water, but that the salmon which spawn as far away from the ocean as the McCloud River and upper tributaries of the Sacramento are too much exhausted after spawning to find their way back to the sea alive.

The remarks of Mr. Dunn, farther on in his letter, in regard to the quality of the flesh of the Sacramento salmon, and also in regard to taking them with a hook and line, I entirely agree with and approve of. I do not, however, attach as much importance as Mr. Dunn does to the distinction which he makes between the coast salmon of California and the San Joaquin salmon. He says that the coast variety is better than the San Joaquin fish for introducing into the Atlantic rivers of the South, because they spawn in December and January, which would be more favorable months for the salmon to ascend the rivers in.

Now, I am inclined to think that the coast fish are the same variety as the San Joaquin fish, and that salmon do not spawn at set times, regardless of change of location, as Mr. Dunn asserts.

On the contrary, I think that the spawning season is a mere accident of place, and that if the California coast salmon should be introduced into the Atlantic rivers they would adopt a time for spawning suited to their new location which would be entirely independent of the season at which their progenitors spawned on the Pacific coast, and consequently it would make no difference as far as the favorableness of their spawning season was concerned whether it was the coast salmon or the San Joaquin salmon that were introduced.

The suggestions in regard to the introduction of the California salmon-trout into Eastern waters, with which Mr. Dunn closes his letter, is, I think, a very valuable one. This fish, the local name of which is salmon-trout, is a large and excellent variety, and is certainly fast disappearing. If it is not practicable at present to transport any of them to the Atlantic States, some effort ought to be made at once to, at least, protect and increase them where they are.

In conclusion I take the liberty to say that Mr. Dunn's contributions on fish-culture to the newspapers have been of a high order of merit, and rank among the best specimens of newspaper literature on the subject.

LIVINGSTON STONE.

XXXIX.—PRESENT STAGE OF THE SALMON EXPERIMENT IN TASMANIA.

By Morton Allport, F. L. S., F. Z. S., &c.

(Read November 12, 1877.)

[Papers and proceedings and report of The Royal Society of Tasmania, for 1877, pp. 109 to 114.]

Though grilse weighing from 3 to 7 pounds have, during the last four years, been taken in the Derwent, how is it that no mature salmon—that is, fish weighing from 15 to 30 pounds, have been captured? This is a question frequently asked both here and in the neighboring colonies, but it will be necessary before attempting to answer it to refer to what is known of the early life-history of the salmon in Europe and Tasmania.

It has been calculated by able British authorities that in specially good salmon rivers, such as the Tay in Scotland, not more than one egg in every 1,500 deposited ever becomes a salmon, the diminution in number taking place chiefly during the earlier stages of life, and especially during the journey of the smolt to the sea, and the first few weeks of their residence there, though even the grilse appear liable to have their number considerably decreased by the attacks of marine enemies before their return as veritable salmon.

The limited number of mature salmon we can yet have in the Derwent might, therefore, alone account for their non-capture, but we must add to that disadvantage the want of adequate appliances to ensnare large-sized fish. The chance of taking one with the rod is infinitesimal while the fish are scarce, the fishermen scarcer, food very abundant, and the difficulties with which the angler in the Upper Derwent has to contend great. The one or two fine-meshed seine nets worked down the river, though well adapted to scrape out smolt, are quite unfitted for the capture of salmon, as they are shot so as to leave a considerable space between the net and the shore, and take so long to haul that the wary old salmon would, before the end of the net reached the land, pass round one or the other, and so escape.

The majority of the 18 or 20 grilse caught have been taken in an ordinary grab-all net, having a mesh of such a size that only the fish of from 3 to 5 pounds weight can mesh themselves, and no larger salmon is at all likely to be taken by the same net, as in this method of fishing it is essential that the fish should be able to get the gill covers through the mesh, or by backing it can at once free itself. If a grab-all net, having a mesh of two and a half inches from knot to knot, was used, the chance of catching a mature salmon would be largely increased, provided only such mature salmon are there to be caught, which has yet to be proved.

We know that in our own waters the capture of sea-going salmonoids was at first, that is in the year 1869, confined to a few smolts only, and these were taken in the small meshed seines after strong freshets had come down the Derwent in the end of October and the beginning of November. In subsequent years, and always in the same months, many of these fish came to the hands of the salmon commissioners till the river was very properly closed to the seine-nets above Hobart Town, and many more of the same fish were doubtless taken of which the commissioners knew nothing. The capture of these smolts was in several seasons followed by the taking in December, January, and February, of salmonoids intermediate in size between smolts and grilse—that is, weighing from three-quarters of a pound to one pound and a half—and it was one of these fish taken in December, 1869, of which Dr. Günther wrote that it presented all the characteristics usually found in the true salmon (*Salmo salar*).

Time passed on and one grilse was taken in December, 1873, followed by two or three others in subsequent years, and in the beginning of January, 1876, between two or three hundred of the salmonoids intermediate between smolts and grilse were taken at a few hauls of the seine-nets on the open sea beaches, some distance below Hobart Town, since which the river has been wisely closed still lower down.

Eight of the last-mentioned fish, taken at random, were carefully examined and dissected, and of these eight, six proved to be unmistakably true salmon (*Salmo salar*), while the remaining two exhibited characteristics common to both the true salmon and the salmon trout (*Salmo trutta*), so that their species could not be positively determined. We next come to the comparatively frequent capture of grilse this season in one place, and by one small net ill-suited for the purpose. And so far therefore, the sequence of events has been marvelously regular, and exactly what was to be expected if all went well. Yet it is not absolutely certain that this regular sequence will be followed by the crowning triumph in the shape of the capture of a 30-pound salmon, though the probabilities are greatly in favor of such a capture being soon made if proper means are used to effect it.

It is certain, from the life history of our salmonoids, as already detailed, that the smolts descending the Derwent find ample food and sufficiently salt water in the estuary immediately below Hobart Town to carry them on to the stage referred to as intermediate between smolts and grilse, after which stage we altogether lose sight of these fish for a time, during which they probably go with the floods of autumn and farther out on to the open coast, for when we next see them it is as grilse in early spring (August and September), and they then appear to be working their way up the Derwent estuary and following the myriads of indigenous small fry, which are then constantly hatching out. As the female of these grilse taken in early autumn invariably exhibit the ova considerably more matured than when taken in the spring, there can be no doubt that they are finding their way to the fresh-water spawning-beds, and would reach them in the early floods of winter, though among the wealth of suitable streams running through scores of miles of uninhabited, mountainous, and inhospitable country we have never yet (except, perhaps, in one instance) been able to ascertain the exact locality of such spawning-beds.

It is quite possible that the grilse after spawning, and on its return with the last of the winter floods to salt water, requires some greater change and a longer journey seawards than when it was passing from the smolt to the grilse stage. And, if so, it may have to encounter more formidable marine enemies than on its first journey, or some unfavorable physical features of our coast of which we, as yet, know nothing. Unfortunately, the most scientific ichthyologists and the most practical fishermen are still equally ignorant of the precise habits of the mature salmon when at sea, and experience can alone prove whether the final stage is to be successfully reached, and, if so, when?

During the last few years, and since the commencement of the salmon experiment, large numbers of specimens of our coast fish have been forwarded to and examined by Dr. Albert Günther, of the British Museum, whose determination of the species proves that many of our fish are not merely representatives of but identical with British forms, such, for instance, as the John Dory (*Zeus faber*), the horse-mackerel (*Trachurus trachurus*), the dog-fish (*Acanthias vulgaris*), the sprat (*Clupea sprattus*), and the conger (*Conger vulgaris*). And this fact goes far to show that there can be no vast difference between the physical features of the Tasmanian and British coasts.

If, therefore, there is any truth in the doctrine of natural selection and survival of the fittest, we may rest assured that as the grilse are rapidly increasing in number, some few out of the thousands sent time after time to sea will be able to adapt themselves to their altered circumstances, escape their foes, and find their way back as salmon. After the second migration is accomplished, the increased speed and cunning of the fish will materially improve its chance of successfully overcoming the dangers of all subsequent journeys.

For each of the grilse which have been taken in one minute spot of the wide estuary of the Derwent, by a net ill-suited for the purpose, there must be hundreds, and more probably thousands, passing of which we hear and see nothing; and if this is true of the grilse after the manifold risks to which they have been exposed on our coasts, what must be the number of smolts that have passed down the Derwent, and what the still greater number of fry in the earlier parr stage on the gravelly rapids of some tributary or tributaries of the Upper Derwent? Can we set such numbers down at less than hundreds of thousands?

And yet, marvelous to relate, not one single parr has yet been seen (so far as the salmon commissioners are aware) in the fresh waters of the Derwent or any of its tributaries; and this is more amazing, because these fish take the worm or artificial fly with the greatest readiness, and would have been almost certain to make their presence known to any angler in their immediate neighborhood.

A writer in Queensland, a few months ago, also referred to this extraordinary absence of the parrs, and used it as a powerful argument against there being any salmon in Tasmania; but he went rather too far, and used the same argument to prove that the migratory salmonoids, which he admitted were taken in the Lower Derwent, were only salmon trout, ignoring the fact that the parrs of the salmon trout (identical in appearance and habits with those of the salmon) were equally remarkable for their apparent absence. If amongst the dozens of suitable tributaries of the Upper Derwent we are unable to find a trace of these hundreds of thousands of salmon parrs, which it is impossible to doubt must be there, we need scarcely be surprised at our inability to light upon the mere handful of mature salmon which we are yet likely to have in the wide waters of the deep Derwent estuary. Some day an errant fisherman on one of the small streams about or beyond the lakes, such as the Clarence, the Pine, the Nive, or the Cuvier, where nobody ever thinks of fishing now, will probably drop on such myriads of these parrs as will enlighten us as to the supply of grilse below, and the knowledge so gained may lead to the obtaining fresh supplies of ova for the stocking of our Northern and Western rivers, because the parrs never move far from the original spawning-place before assuming the smolt dress; and their detection would enable us in the following winter to watch for and take the parent fish on their certain return to the same spawning-beds.

Before concluding, it may be as well to refer to the one instance in which it is just possible we have hit upon the spawning-bed of a true salmon. In the early part of the past winter a pair of large fish were observed spawning in the Plenty, and were netted by the bailiff in charge at the ponds after the bulk of the ova were deposited.

The female, after having parted with the greater part of her ova, weighed more than twenty pounds, and the male weighed nearly nineteen pounds. Mr. Read, one of the salmon commissioners, examined these fish carefully, and both he and the bailiff are of opinion, from the external appearance of the fish, that they were true salmon, or at any rate belonged to one of the two migratory species.

After stripping the remaining ova (almost one thousand in number) from the female and applying the artificial process of impregnation, both fish were returned to the river.

Subsequently a few of the naturally-deposited ova were, with judicious forethought, taken from the rid, placed in one of our hatching-boxes mentioned, and then kept carefully separated from other ova.

The artificially-impregnated ova failed, but that so prudently taken from the rid has hatched out; and an examination of both the eggs and the newly-hatched fry has very materially strengthened the impression in my mind that these fish were salmon, for the eggs were not only larger than any we have yet taken in the colony, but had exactly the pink tinge which characterized the salmon eggs received from England. The umbilical sac attached to the newly-hatched fry is longer in proportion to the width than that of the trout, and this was a marked peculiarity in the fry hatched from the imported salmon eggs. It is quite true that there is considerable diversity both in the size and color of the eggs of the brown trout (*Salmo fario*); but the size of the eggs in that species by no means depends upon the size of the fish, as large eggs are often found in small fish; and no cause can yet be assigned for this diversity in size, but the difference in color clearly depends on the quality of the fish, the red-fleshed fish invariably producing red eggs, and the white-fleshed fish the pale straw-colored eggs. As an actual fact, none of the originally imported salmon-trout or trout eggs approached in size either these eggs taken from the rid in the Plenty or the imported salmon eggs; and very great interest will therefore attach to the subsequent stages of the fry now hatched, because, if they are true emigrants, that fact must be made manifest when the deciduous silvery scales which first hide the parr marks are put on, and the young fish assume the smolt stage, though it may even then (as long since pointed out) be difficult, if not impossible, to determine accurately to which of the two migratory species the smolts may belong.

A few days after the foregoing was written, namely, on the 15th day of October last, a strong freshet came down the Plenty, during which a school of about a dozen salmonoids found their way into the water-course which supplies the ponds, being evidently bound seaward. Mr. Read was so much struck with the difference between these fish and trout-fry of the same size that he preserved two of them in spirits and forwarded them to me for examination. Externally, both fish presented the characteristics of true salmon, and upon dissection the number of pyloric appendages was found to be sixty-two in one and sixty-five in the other—numbers which prove these specimens to have been salmon and not salmon-trout. This capture, therefore, lends additional force to the presumption that the 20-pound fish taken in the Plenty was a salmon.

XL.—CORRESPONDENCE CONNECTED WITH THE TRANSMISSION OF EGGS OF THE QUINNAT SALMON AND WHITEFISH TO AUSTRALIA AND NEW ZEALAND, 1877, 1878, AND PRIOR YEARS.*

AUSTRALIA.

Sir Samuel Wilson to the Melbourne Argus, published November 30, 1877.

SIR: The shipment of salmon ova which I received from California by way of Auckland arrived in very good condition. Only about 6 per cent. had perished during the voyage, and as there had been two transshipments, this was a very small proportion of loss. On the arrival of the ova at Sydney, they were taken out of the ice-house on board the San Francisco mail-steamer and transferred to the City of Adelaide.

The ova were packed in layers in a box or ice chest, about 3 feet by 4 feet, and about 2 feet in depth. They were placed between two pieces of thinly-woven cotton stuff, about 7,000 in each layer, and a layer of moss about 2 inches deep between each two layers, and also above and below the eggs. Six inches of ice was placed over the ova, and the bottom was pierced with holes to allow the escape of water from the melting ice. The ice was renewed every 12 hours on the voyage from Sydney to Melbourne. The box had an inner lining, inclosing about 4 inches of sawdust to act as a non-conductor, and which answered the object sufficiently well.

The weak points in the packing were the use of cotton stuff, which rots and gets mouldy, while the moss remains green and fresh, and also that the ova were too closely packed together. For a short voyage this matters little, but in a long distance the difference is great, as when one egg loses vitality it soon decays, and the byssus, or fungus, which quickly forms, attacks all the ova within reach, which adhere to each other, and although little altered in appearance these ova invariably perish in the hatching.

On the arrival of the ova at Sandridge they were removed, and taken by train and wagon, well packed on an elastic cushion of straw, and opened at the spring on Ercildoune estate, where the hatching-boxes were ready for their reception. On the pads covering the ice being taken off, a layer of moss, fresh and green as if newly gathered, was to be seen covering the ova. On this being removed, the eggs were visible

* For the purpose of completing the record, begun in the report for 1875-'76, vol. 4, I give some letters relating to sendings prior to 1877.—S. F. B.

through the thin net-like web which covered them, and at once it was evident, to my great delight, that they were in splendid preservation, and far advanced in hatching, the eyes of the young fish being clearly visible. With little loss of time the ova were transferred to the gravel of the hatching-boxes, which had been prepared with great care, by being screened to insure a uniform size, and by boiling to destroy insect germs which might be injurious to the ova. All were got into the hatching-boxes the same evening, except one layer of about 7,000, which were left under the ice till next morning. About 6 per cent. of dead eggs were taken out at once, but many were adhering in clusters, most of which I knew could not live, but which looked healthy enough at the time.

The next morning over 100 young salmon were hatched, and they were lively little fellows even at that early stage of their existence. When touched with a feather they would start off and swim round in a circle, and settle down again amongst the gravel. On the remainder of the ova being transferred to the hatching-boxes several of the young fish were found to have hatched out during the night, and during the day 400 or 500 made their appearance.

The ovum of the American salmon is larger than that of the British species. It measures almost exactly a quarter of an inch in diameter. It is of a transparent pink color, and is nearly globular, being slightly elongated. The young fish is about an inch long, and it has attached to it the umbilical sac containing the yolk of the egg, which is a clean transparent red color, and seems quite as large as the egg from which it has emerged. This sac contains the food of the young salmon for three or four weeks, and is gradually absorbed, becoming smaller as the young fish grows.

The hatching process is effected simply by placing the ova on a layer of gravel, over which a stream of water is allowed to run. The temperature of the water is a most important point, and I selected a spring from its being of a uniform degree of cold and from its freedom from sediment, which by settling on the egg interferes with the supply of oxygen necessary for its vivification. As the supply from the spring is limited, being only four or five pints per minute, I had a pipe laid down from the creek supplying two filters which are used to increase the supply. The water from the pipe can be shut off entirely when its temperature is too high, but for so far the difference has not been great between the water from it and that of the spring. The permanent temperature of the spring is 55° and the pipe supply has risen on hot days to 62°, but the young fish did not seem to suffer in the slightest, and those in the warmest water are further advanced. The hatching went on favorably, but a large number of the eggs arrived at a certain stage and failed to produce live fish. Sometimes after a struggle the head would appear, and the little creature would perish in the effort to emerge from the shell. In others, after the eyes were plainly visible, the living principle became extinguished, as shown by the ovum becoming white or opaque. The fish which were hatched, however, were strong and healthy. For a time the dead eggs picked out were over 3,000 a day, and prospects were rather gloomy, but circumstances proved that it was more from the conditions to which the ova had been subjected before their arrival that the losses were attributable than to their treatment after landing.

The total shipment was supposed to be 50,000 ova, but from a rough count the number received was estimated at 55,000. When the eggs were opened, one layer of about 7,000 ova was put in each box. The combined stream runs through the boxes from one to eight, the first boxes getting the fresh, cool water, and having the best chance, the water heating 2° in passing through the boxes in hot weather. When the hatching was nearly finished, a very marked difference was observable in the number hatched in each box. No. 1 had only about 1,000 live fish out of 7,000 eggs. These eggs were on the top nearest the ice. The next layer in No. 2 had probably 3,000 fish out of 7,000 ova. No. 3 was the best of all, and there were probably 6,000 live fish out of 7,000 ova. Nos. 4, 5, and 6 were pretty equal, and hatched over 50 per cent. The eggs in No. 7 hatched out much earlier, but the percentage of loss was above the average. No. 8 opened a day later, and the lowest layer of eggs hatched out very quickly, having the warmest water, and produced probably 5,000 fish.

It will thus be seen that the different lots of eggs, when treated exactly in the same way, varied very materially in the number hatched, showing that the causes of this difference were to be looked for in the treatment of the ova when first taken, or in the mode of packing, rather than in their management after their arrival in Australia.

After the young fish were fairly hatched but few losses occurred, probably not 50 in the whole number. Of the ova, only about 500 remained to hatch on the 24th, and that day, although the hottest of the season, did not appear to injure the *alevins*, as the young fish at this stage are called. The number of live fish is now about 28,000, which is a not unsuccessful result. Had the ova arrived a week earlier, probably three-fourths could have been saved. If they had been a week later, probably they would have been a mass of putrefaction from the fish hatching out, as happened with the first lot of 20,000 California ova which I had brought over three years ago.

Success in most things is the result of good arrangements, made with a thorough knowledge of the subject, and combined with favorable circumstances, where these are beyond control. The result in this case shows what a narrow line may lie between success and failure.

The question now becomes, What is it best to do with these 28,000 young salmon? While they are in the *alevin* stage the losses are comparatively slight, little feeding is required, and they only require to be kept from enemies and allowed a good supply of cool water. When the sac is absorbed they become perfect salmon in miniature, but with a different coat, and can do well for themselves in the open river, being quick and active in eluding their enemies. In my opinion they cannot, then, be too soon turned out into suitable streams.

The streams really well suited to the salmon in Victoria are few in number, and are only to be found on the slopes running to the southern coast from the Cape Otway or the Gipps Land Ranges. The Snowy River and the Gellibrand River are probably the two best for this purpose, but the rivers running through the Gipps Land Lakes into the sea may probably suit the Californian salmon, which can bear higher temperature in the waters and seas which it frequents than the species found in more northerly latitudes, and I am not unhopeful that even the Yarra, and the headwaters of the streams south of Mount Macedon, may be found not entirely unfitted for the purpose.

But it will be asked, How are the young salmon to get from Ercildoune to the Gellibrand, to the Snowy River, or even to the Yarra? This difficulty has been solved by actual experiment. By the aid of ice to keep down the temperature, and a force-pump to aerate the water and maintain the needful supply of oxygen (without which no life can exist, whether of vegetable or animal nature), the fish can be carried for 24 hours or more with every chance of success. The Yarra, or the creek at Wooling, near Macedon, is within six or seven hours' traveling. The Gellibrand River is within 18 hours, as I have proved by carrying successfully a few young trout and salmon from Tasmanian ova with a very small percentage of loss. The Gipps Land Rivers are further away, but with a steamer like the Victoria, within 25 hours the feat might be accomplished, and the experiment, which has been successful so far, might be carried to a successful result, or, at any rate, as far as human enterprise might conduce to that end. Would it be too much to to ask that the government steamer Victoria should be allowed to convey a portion of these young salmon to the Snowy River and the Gipps Land Lakes?

Since writing the above I have heard from Ercildoune, and the report is—During the last 24 hours six ova have died, and seven *alevins*. The fish are beginning to develop the gregarious instinct, and are collecting in large crowds, and continue to do well.

I remain, &c.,

SAMUEL WILSON.

OAKLEIGH HALL,
East St. Kilda, Melbourne.

Editorial of Melbourne Argus of December 1, 1877.

The suggestion with which Sir Samuel Wilson closed his very interesting letter in our yesterday's issue on the subject of the introduction of the salmon into the colony—viz, that the government should give the services of the Victoria to complete the work already so far advanced—is eminently reasonable. When private liberality and enterprise have done so much towards the conferring of a great benefit upon the country, it would be churlish to a degree if the state were to refuse to assist in the little that still remains to be done. Sir Samuel imported the ova at his own cost, and hatched them out in his own hatching-boxes, and it only now remains to transport the 28,000 infant salmon that form the magnificent result of his labor and outlay to waters suitable for their growth and increase. The Victoria could not be more usefully employed than on this errand. It would only be a matter of 20 or 30 tons of coal, and some extra wages. Probably £100 would cover the whole cost. Further, if there is anything that the railway department can do to further the distribution of the young fish, that also should be done, promptly and cheerfully. When the history of fish acclimatization in Victoria comes to be written, there will be several of our older colonists who will be entitled to warm commendation for their exertions in the cause. About twenty-one years ago Mr. Edward Wilson showed what could be done in the matter by transferring the cod from an affluent of the Murray into an affluent of the Yarra, thus giving this noble fish an entirely new habitat of great extent. A little later Mr. Learmonth introduced the same fish into Lake Burrumbeet.

J. C. Firth to the Melbourne Argus of January 9, 1878:

SIR: My attention has been drawn to a letter appearing in your issue of November 30 from Sir Samuel Wilson, detailing his operations relating to the hatching of some 50,000 salmon ova recently received in Victoria from California. Every one will be pleased at so successful a result, and grateful to Sir S. Wilson for the care he has taken with so valuable a consignment after it reached his hands.

It is, however, I think, to be regretted that Sir S. Wilson, through inadvertence or some other cause, omitted to state to whom the colony of Victoria is indebted for so great a boon as the Californian salmon. As I have had opportunities of becoming acquainted with the matter, I will, with your permission, supply the information so innocently omitted by Sir S. Wilson.

In June last Sir Samuel applied to me to procure 50,000 salmon ova from California. I informed him that, though almost too late, I would write the Hon. Spencer F. Baird, chief of the United States Fish Commission at Washington, by the next mail.

I wrote Mr. Baird, acquainting him with Sir S. Wilson's request, and asked him, if possible, to confer a lasting obligation on the colony of Victoria by sending 50,000 salmon ova. In due course, Mr. Baird courteously replied, acceding to the request, should the lateness of the order permit of its execution. By next mail I conveyed this intimation to Sir S. Wilson. On November 2, 1877, the mail-steamer brought the 50,000 salmon ova for Victoria.

It will be seen that this noble gift of salmon is due to the generous good-will of the people of the United States, directed by the high-minded and courteous chief of their Fish Commission, the Hon. Spencer F. Baird. No charge whatever is made beyond the cost of transport and packing, and even this has not yet been made. When we consider that the United States has constructed expensive fish-breeding establishments on the M'Cloud and other rivers, which are kept up by large annual appropriations by the Legislature of the United States, for the purpose of stocking their own rivers with fish, the noble generosity of their gifts of salmon ova to New Zealand and Victoria, the colonies of a foreign nation, will be fully appreciated, and I trust you will pardon me for thus supplementing Sir S. Wilson's letter.

I regret, also, that Sir S. Wilson should have felt it necessary to complain of the imperfect manner in which the ova had been packed in California. This packing had been done by the officers of the Fish Commission at the M'Cloud River, whose experience ought to have enabled them to pack the ova in such a manner as to secure success. That they have done so is evident from the fact that about 95 per cent. of the ova arrived in good condition. Sir S. Wilson will, I think, regret that in this instance he "looked a gift-horse in the mouth," and found nothing worth a complaint even then.

Regarding Sir S. Wilson's statement that this consignment of ova narrowly escaped destruction, because in about two days after arrival most of them hatched, I may say that during the last three years the United States Fish Commission have presented one million salmon ova to this colony, the whole of which have been distributed throughout the colony by me, one-third of which I have personally placed in the hatching-boxes and shingle-beds of this part of the colony, and in every instance I have noticed that two days after the removal of the ova from the low temperature secured by the ice the retardation ceased, and within forty-eight hours of the increase of temperature from 35° Fahrenheit to 60° or 65° the hatching process was nearly completed.

I regret, also, that Sir S. Wilson has published no acknowledgment to Mr. A. S. Webster, of Sydney, a gentleman whom I had requested to see the ova transferred from the Californian steamer to the first Melbourne steamer in case no person had been sent by Sir S. Wilson to take charge in Sydney. To the admirable manner in which Mr. Webster carried out my instructions the safe arrival of the ova in Melbourne is largely due. To the agents in California, the proprietors, captain, and officers of the California mail-steamer City of Sydney, in my capacity of president of the Auckland Acclimatization Society I have forwarded and published the thanks of the Auckland council. I have on this and all similar occasions taken care that the services of every helper in this good work have been duly recognized and published, not merely on the ground of policy, to secure future co-operation, but because they have well deserved recognition. Of my own humble services I say nothing beyond this, that the cost of transit from San Francisco, cost of ice, ova, and ice-chests, provided in Auckland, have been defrayed by me, which I pray you to permit me through your columns to present to the people of Victoria as my contribution to the great work of introducing salmon into your noble colony.

I am, &c.,

J. C. FIRTH.

AUCKLAND, *Decemb.* 27.

Sir Samuel Wilson to S. F. Baird.

ERCILDOUNE, BURRUMBEET, *March* 16, 1878.

SIR: I have learned from Mr. Firth, the president of the Auckland Acclimatization Society, that 50,000 salmon ova which were received by me from New Zealand have been supplied by the liberality and generosity of the United States Government, and I now, on behalf of myself, who carried out the experiment to a successful result, and on behalf of the colony of Victoria, which will, I hope, benefit greatly by the acclimatization of such a valuable fish as the salmon, beg to offer my best thanks for the very valuable consignment of ova, and for your care and trouble in sending them so safely.

I have to request that you will convey to the Government of the United States the warm appreciation by the people of Victoria of the noble and generous spirit which prompts them to support so liberally an establishment calculated to do so much good to the human race. On behalf of this colony I tender them my most hearty thanks.

It will interest you to know that the ova arrived, after their long voyage and transshipment at various ports, with a loss of only 6 per cent., and out of 50,000 about 28,000 were hatched successfully. The bulk of these were distributed safely in all the rivers and streams likely to suit them in the colony, and, from their capability of resisting high temperatures, they promise to succeed admirably, so far as can be judged at present. The next report of the zoological and acclimatization society will contain a report of the experiment, and I shall do myself the honor of sending you a copy.

I should much like to introduce here some of your *Salmo fontinalis* and *Coregonus albus*, if the ova are procurable. I observe that the latter has not hitherto succeeded in New Zealand.

I have just received a small quantity of English salmon ova (*Salmo, salar*), which, I hope, will succeed as well as your *S. quinnat.*

I have the honor to be your obedient servant,

SAML. WILSON.

Sir Samuel Wilson to S. F. Baird.

OAKLEIGH HALL, EAST ST. KILDA,
Victoria, Australia, July 22, 1878.

DEAR SIR: I have the pleasure of acknowledging the receipt of your letter of 24th May, and have to offer my best thanks for your continued kind offers of salmon ova for this colony. I have been making inquiries as to the temperatures of the waters of the Murray River, which is the largest stream in Australia, and from what I can gather it will, I think, be found suitable to the Californian salmon. To be successful, the thing should be done on a large scale, and not less than 500,000 or 1,000,000 ova obtained for it. As the undertaking would benefit three colonies it should be a joint affair, and would be rather too heavy for any individual to attempt to carry through. I intend to propose the matter to the governments of the three colonies interested, and hope that it may be taken up by them.

Regarding the *Salmo fontinalis* ova, which you so kindly offer to send, the best way will be to place them in a box similar to those in which the salmon ova were sent here, but smaller, with a supply of ice inside the box. The case should be sent in the ice-house of a steamer to London or Liverpool, thence by rail to London, to be placed in an ice-house till forwarded by the Peninsula and Oriental Company's steamer in their ice-house, or by one of the new fast line of steamers to Melbourne. This will obviate the dangers of freezing in crossing the Continent.

Be so good as to consign the eggs to Messrs. Robert Brooks & Co., Cornhill, London, who will follow instructions and forward them to me. They will also attend to their forwarding from Liverpool should there be no steamers direct to London. Should there be no ice-house, a large box of ice would do very well instead as far as England, as the weather is then cold.

You do not mention the *Coregonus albus*, which is a very desirable fish to acclimatize, but which may need lower temperatures than our waters here.

I have the pleasure to inform you that the council of the Zoölogical and Acclimatization Society have, in token of their appreciation of your very valuable services to the cause of acclimatization, awarded to you their silver medal, of which you will have official notice from the secretary.

I shall send you a report of the different attempts that have been made to introduce the salmon here, which will appear in the next volume of the society's proceedings, now in the press.

The shipment of English salmon ova by the Chimborazo was almost a complete failure. I received three boxes containing about 1,700 ova, but two of these only produced one live fish. The third box contained fine, large pink ova, but there were only 320 in it, and of these 200 looked well, and 150 live fish were hatched. From some cause they all died but 32, which are still alive and doing well. The remainder, about 52,000, that went on to New Zealand, were nearly all bad, but a few hundred were hatched. I have not heard how many survived. The English trout ova, by the same shipment, came to New Zealand in good order, but they are the large species *Salmo fario ausonii*, of Günther, and prove too large for many of our streams. We want here your smaller variety—the *S. fontinalis* or the *S. fario gamiardi*—the "burn trout" of the Highlands of Scotland.

I cannot close this letter without again expressing my admiration of the generosity and public spirit, in a cosmopolitan sense, which is displayed in the kind manner in which you, as the representative of the United States, have so freely supplied ova of your valuable fish for the purpose of stocking the waters of Australia and other countries, and I only wish that we had something of equal value to offer in return.

Pray accept my best thanks on behalf of the colony, which is greatly indebted to you,

And believe me to be, yours, very truly,

SAM'L WILSON.

P. S.—Any expenditure incurred I shall be happy to liquidate on hearing the amount.

S. W.

NEW ZEALAND.

J. C. Firth to S. F. Baird.

AUCKLAND, *April* 11, 1877.

DEAR SIR: I have duly received your valued favors of December 21 and 19th January, and have read their contents with a great deal of pleasure.

I thank you for your cordial and generous offer to supply a further quantity of salmon ova, if necessary, under certain conditions detailed in your letter of January 19.

I have now the pleasure to inform you that the council of Auckland Acclimatization Society, of which society the members have done me the honor to elect me president for the current year, are desirous of obtaining a further supply of salmon ova on terms of your letter before cited,* and I have the honor to ask you to be good enough to cause to be forwarded to us at the proper time, through our agents, Messrs. Cross & Co., of San Francisco, 200,000 salmon ova.

R. J. Creighton, esq., New Zealand agent in San Francisco, will also interest himself in seeing that every facility be offered by the mail steamer.

About a month ago I located the last batch of young salmon for the season.

---

* Payment of actual expenses of obtaining and forwarding eggs.—S. F. BAIRD.

As you may perhaps feel a little interest in knowing what sort of a country it is that you are so much benefiting, I inclose you a copy of the account of my last "salmon excursion."

I am, dear sir, yours sincerely,

J. C. FIRTH,
President of the Auckland Acclimatization Society.

Hon. SPENCER F. BAIRD,
Chief Commissioner of United States Fisheries.

Same to the same.

AUCKLAND, NEW ZEALAND, *October* 24, 1877.

DEAR SIR: I am duly in receipt of your favor of August 6, and thank you for your kind accession to my requests on behalf of the Victoria and Canterbury societies so far as circumstances will permit.

I am pleased to have to report that in one of the rivers, the Rapurapu, in the shingle-beds of which I last year placed a large number of the salmon ova you so kindly sent, a great success has been achieved; large numbers of young salmon 5 inches in length being reported as swarming in the river for miles.

I am, dear sir, your obedient servant,

J. C. FIRTH.

Hon. SPENCER F. BAIRD,
Commissioner United States Fish Commission, Washington.

Robert Houghton to S. F. Baird.

SAN FRANCISCO, *November* 1, 1877.

Hon. Professor BAIRD, *Washington:*

DEAR SIR: I have to thank you for the shipment of 500,000 salmon eggs to New Zealand, per Cross & Co., San Francisco, on the 10th ultimo. Fortunately the shipment was successful, Mr. Hooper, of Cross & Co., having attended to the matter personally, as, owing to misdirection, Mr. Stone's letter to me did not reach me until yesterday, when, having been told about it, I applied personally at the post-office. No doubt the delay was occasioned by Cross & Co.'s business troubles—that firm having more pressing matters to attend to.

On the 24th ultimo, Mr. Pratt, secretary to Mr. Stone, called upon me and informed me of the shipment; he also gave me memorandum of account and receipt of payment for $750, being $1.50 per 1,000, as notified by you in previous correspondence.* At Mr. Pratt's suggestion,

* This amount represented the supposed extra expense to the United States of obtaining, developing, and packing the eggs for shipment, the condition of the donation.—S. F. BAIRD.

in the absence of Mr. Stone, I paid the money into Myron Green's public account with the Capital Bank, Sacramento, and notified him of the fact. I did so because I was anxious to save expense to the acclimatization societies necessarily incurred by remitting to Washington, and because Mr. Pratt assured me that it would be equally agreeable to the United States fish commissioners, who otherwise might be called upon to transmit an equal amount from the Capital.

I trust this explanation may be satisfactory.

The shipment was carefully packed in ice in a refrigerator, and as the commander of the City of Sidney had taken two parcels of eggs previously, no doubt every pains will be taken to have it landed in good condition. Besides, he was promised a bonus by Mr. Hooper.

I regret exceedingly that the white fish ova, shipped last season, failed, not through any want of care or attention here, but on the other side. A portion of the first shipment was hatched out at Christchurch, but, through want of proper precaution, they were swept out of the breeding-ponds one night by a heavy rainfall. The second parcel went safe, having been three weeks in ice in San Francisco. They were landed in good condition in Auckland, were sent a tedious coasting voyage of 700 miles, and landed sound at Invercargill, but, the small steamer intended to take the eggs to their ultimate destination having been wrecked, they all died from exposure. This experiment, however, clearly demonstrates that whitefish eggs may be sent to New Zealand under unfavorable conditions, and with care be hatched out. I should be extremely desirous of seeing another attempt made, leaving the eggs at Auckland on the arrival of the steamer, where the experiment would have a better chance of succeeding.

The salmon are thriving admirably.

I remain, yours, very sincerely,

ROBERT HOUGHTON.

[Telegram.]

Livingston Stone to S. F. Baird.

CHARLESTON, N. H., *January* 28, 1878.

Received $750 for New Zealand salmon eggs, and $25 from Stone & Hooper for 10,000 salmon eggs which went to France.

L. STONE.

James Hector to S. F. Baird.

COLONIAL MUSEUM OF NEW ZEALAND,
Wellington, December 13, 1877.

DEAR PROFESSOR BAIRD: You will be glad to learn that the last shipment of salmon ova, which reached us in the beginning of November, has been a most complete success.

The boxes were distributed as suggested in my letter of the 28th of July, and from the eight centers over forty distinct river systems have been successfully stocked. From all quarters the most favorable reports have come in, generally to the effect that 90 per cent. of the ova produced strong fish that survived the early stages; even rivers in the King or rebel Maori country have been stocked, and the natives take a great interest in protecting the fish. Only in one case—Dunedin—is there a report of mortality among the young fish, the reason of which is not stated.

It must be satisfactory to you that your simple and inexpensive arrangements have produced such good results, considering the large sums and the many years (sixteen) that have been spent in the endeavor to introduce *Salmo salar* into this colony and Tasmania, as yet without any certainty of success.

Everything is now ready for the whitefish ova, which I hope will arrive next month. They are all to be placed in Le Dun Lake.

I remain, yours, very sincerely,

JAMES HECTOR.

Professor BAIRD,
Washington.

J. C. Firth to S. F. Baird.

AUCKLAND, *December* 17, 1877.

Prof. S. F. BAIRD:

At request of Government of New Zealand, I inform you I have successfully distributed throughout colony 500,000 salmon eggs. Thanks to care of gentlemen in charge, eggs arrived in splendid condition. Had insulated chests and tons of ice in readiness; by working through night on arrival of steamer had all placed on board; each provided with arrangement for drainage and ventilation; one-half by rail across island by steamers to the south; remainder by steamer to east coast; 50,000 to Sir Samuel Wilson for Victoria. From Nelson, Greymouth, Wellington, Napier, Christchurch, Dunedin, Invercargill, and Victoria 95 per cent. hatched. 100,000 by government apportioned Auckland; 10,000 at our establishment near city; 40,000 on shingle in country of Maori King. At request of most powerful supporters, Punier, where last year placed a large number; 50,000 placed in Mangakahia River—fine river for salmon, through wooded country, cold water from high ranges, through little falls, long rapids, deep pools; ova placed on shingle. Disappointed that order for Auckland and Canterbury (250,000) overlooked.

J. C. FIRTH.

T. F. Cheeseman to S. F. Baird.

AUCKLAND ACCLIMATIZATION SOCIETY,
Auckland, December 17, 1877.

DEAR SIR: I forwarded a short note by last mail acknowledging the safe reception of the consignment of salmon ova so liberally forwarded by the United States Fish Commission, and conveying our best thanks for the same. I have now the pleasure of informing you that the ova have turned out even better than those sent last year, the hatching having been in every way successful. Our share of the consignment was divided into three portions; the first, containing about 40,000 eggs, was taken to the upper part of the Puniu River, an affluent of the Waipa, which is the principal tributary of the Waikato River. It is estimated that 38,000 healthy young fish have been hatched out in this locality. Another lot of 50,000 was placed in the Northern Wairoa River, and has hatched out almost as well as the Puniu portion. The remaining 10,000 have been treated at our fish-house near Auckland, where we have now slightly over 9,000 healthy young fish.

You will have doubtless received particulars from Dr. Hector respecting the ova sent to the southern provinces. I believe that the result there is also very encouraging.

I can now only again express our sense of the deep obligations under which you have placed us, and beg to remain,

Yours, obediently,

T. F. CHEESEMAN,
Secretary.

Professor BAIRD,
Commissioner of Fish and Fisheries, Washington.

J. C. Firth to S. F. Baird.

AUCKLAND, *December* 17, 1877.

Hon. SPENCER F. BAIRD,
Chief of the United States Fish Commission, Washington:

DEAR SIR: I have great pleasure in informing you that, at the request of the Government of New Zealand, I have successfully distributed throughout this colony the 500,000 salmon ova so generously presented by the United States Fish Commission to New Zealand. The boxes of ova arrived at this port, thanks to the care of the gentleman in charge of your establishment at the McCloud River, in splendid condition. The circumstance of the ova boxes being stripped from their crates in San Francisco, to get them into the steamer's ice-house, renders it necessary to provide each ova box with chests properly insulated. All these, together with some tons of ice I had in readiness, and by working through the night on the arrival of the mail steamer, I had all safely inclosed in the insulators, each provided with a simple arrangement for drainage and ventilation. I dispatched one-half by rail across the island, thence by steamer to their respective destinations south. Two days after I dispatched the remainder by steamer down the east coast. I also dispatched the 50,000 to Sir Samuel Wilson for Victoria. I have since received advices that each parcel had arrived in perfect safety and condition at its destination, viz: Nelson, Greymouth, Wellington, Napier, Christ Church, Dunedin, Invercargill, and Victoria; and that about 95 per cent. had hatched into healthy fry. The 100,000 ova apportioned by government to Auckland I then located; hatching out about 10,000 at our fish establishment near this city; 40,000 I placed in a fine shingle river in the country of the Maori King, at the request of his most powerful supporter. This river (Puniu) is a tributary of the Waikato River, where I had last year placed a large number of salmon ova and fry. The remainder (50,000) I placed in the Mangakahia River, which I think a fine river for salmon, flowing, as it does, through a wooded country, fed by streams from high ranges, with clear, bright, cold water rippling over shingle beds, rushing over little falls, now dashing through a long rapid, and anon loitering in deep and placid pools. Backed by a party of strong and willing hands, after a most toilsome, but exciting and pleasurable, ascent up this beautiful river, I placed the ova, securely guarded from all their enemies, in the shingle beds of the river. I was a little disappointed that our order for Auckland and Canterbury (250,000) had been overlooked, but I doubt not you had good reason for doing so.

Yours, truly,

J. C. FIRTH.

P. S.—Excuse this scrawl, as it is written on the desk of a small river steamer by which I am conveying 8,000 fine, healthy salmon fry up river.—J. C. F.

Same to the same.

AUCKLAND, *February* 4, 1878.

DEAR SIR: Since my letter of January, I have been greatly pleased to learn that the whole of the last shipment of ova—500,000 to New Zealand, 50,000 to Victoria—have been located in both colonies with great success.

I am so impressed with the genuine international courtesy displayed by you in sending about a million salmon ova to New Zealand, the generous gift of the United States to the colony of New Zealand—a dependency of a foreign nation—that I have written to the colonial secretary, the Hon. Colonel Whitmore, asking him to bring your goodness under the notice of his excellency the Marquis of Normanby, governor of New Zealand, with the view of asking the imperial secretary for the colonies, the Earl Caernarvon, to convey the thanks of the colony to the Government of the United States for their munificent gift of salmon ova to New Zealand. In due course you will probably hear more of it.

The 50,000 ova you were good enough to forward to Victoria came duly to hand, and were forwarded by me in insulating cases (your own boxes being stripped of all covering to get them into the steamer's ice-house at San Francisco). By means of abundant supplies of ice and the precautions I had taken, and with the assistance of Mr. A. S. Webster, of Sydney, the ova arrived in perfect order at Melbourne, where they were taken charge of by Sir Samuel Wilson, the gentleman who was to bear the whole cost of the experiment.

Judge of my surprise at reading in the Argus, the leading journal of Victoria, a letter from Sir S. Wilson, in which he ignores you, Mr. Webster, and myself. I inclose you this letter and my reply to it.

Sir S. Wilson is, I believe, annoyed at his having overlooked your services, or at my having called attention to it. It does not matter which, as you will probably receive proper acknowledgments by the same mail in which this letter goes in.

Having distributed nearly the whole of the ova you have so kindly sent, at my own personal cost, and with an infinite deal of pleasure to myself, and my third term of office expiring on March 5 next, I shall not have another opportunity of officially communicating with you. You may rest assured of my warm sympathy for you in your great and good enterprise.

I am, dear sir, yours,

J. C. FIRTH.

The Hon. SPENCER F. BAIRD,
Chief of Fish Commission, Washington, D. C.

R. J. Creighton to S. F. Baird.

SAN FRANCISCO, CAL., *February* 5, 1878.

DEAR SIR: In reply to your inquiries regarding the shipment of salmon ova from San Francisco to New Zealand and Australia, I regret that I am unable to give you the details as fully as I could wish, owing to my inability to follow it throughout.

The City of Sydney, with 500,000 eggs from McCloud River, sailed hence October 10, 1877, at noon. The eggs were deposited in an ice-chest made specially for the purpose by the Auckland, New Zealand, Acclimatization Society; but it was found after the ship sailed that the waste of ice was so great it would hardly last the voyage; and Captain Dearborn took the boxes out and deposited them in the ship's ice-house, surrounded by ice, where they remained the residue of the voyage. I attribute the success of the shipment in great measure to Captain Dearborn's care. It was impossible to pack the eggs in the ship's ice-house at the outset as they arrived too late; the ice-house was full before they were put on board.

Arrived at Honolulu October 18, where there was several hours' detention; arrived at Auckland November 3. Here the eggs were transshipped, save the parcel of 50,000 for Victoria to the order of Sir Samuel Wilson, which remained on board. The City of Sydney arrived at Sydney on the 8th of November, and here a transshipment took place, the eggs for Melbourne being transferred by a steamer sailing for that port. I am not aware what detention took place in Sydney, but the voyage would be about thirty hours or two days. At all events the eggs were properly cared for, and arrived at their ultimate destination in sound condition, for I observed from a Victoria newspaper that they had been hatched out successfully. This, I should say, is the longest voyage yet made by California salmon eggs which preserved their vitality.

Reverting to the New Zealand shipment, I inclose extract from Auckland Weekly News, November 20, 1877, which fully explains the preparations made for the transportation of the eggs in ice to the various parts of the colony. Captain Dearborn informed me that the transshipment and repacking were accomplished inside of two hours, and the eggs sent across the isthmus, seven miles by the railroad, to the steamer on the western harbor sailing for southern ports. The ordinary time of the coasting steamers (writing from memory) is: To Nelson from Ouehunga, the port of departure, 2 days; Nelson to Wellington, 24 hours; Nelson to Greymouth, 1 to 2 days; Wellington to Napier, about 30 hours; Wellington to Canterbury, 1 day; Canterbury to Dunedin, 1 day; Dunedin to Southland, about 24 hours.

I give the time approximately; it certainly was not under this. It should be borne in mind that the eggs passed through the tropics and arrived in New Zealand and Australia at midsummer, when the heat is great. The distribution in New Zealand was specially trying, owing to the frequent transshipment and handling. For example, the consignment of 50,000 eggs for Freymouth would be landed at Nelson and transshipped to another steamer going down the coast. As the English mail was forward, little or no detention would occur at Nelson, but there is always a weather risk on the west coast of the middle island. How that matter was I am unable to say.

Again, the consignment of 50,000 for Napier would be landed at Wellington, put on board a government steamer, and carried north along the east coast, the steamboat traffic being chiefly on the west coast of the north island. No time would be lost, however, on this line. A further transshipment for Southland would take place at Dunedin, the larger class of boats not going so far south as Foveaux Strait. It was on this section that the whitefish eggs perished.

As I remarked at the outset, I am unable to follow this distribution closely. I infer from what Captain Dearborn tells me, and a note from Mr. Firth, that the original programme was carried out, and as I have not heard any reference to failure or miscarriage I have confidence in stating that the distribution was successful. The eggs have hatched out beautifully, and every promise is given that the noble streams of New Zealand will be stocked with California salmon, thanks to your kind co-operation and the liberality of the United States Fish Commission.

Several years ago Scotch salmon were hatched out in Otago, but the experiment was considered a failure as none of the fish returned, whereas salmon-trout became plentiful, having been introduced at the same time. By last mail I observe a statement to the effect that two of the Scotch colonists of Otago had seen a salmon in the shallows of a Southland river recently, from which it is presumed that the fish are returning. Should any further evidence be given on this point I shall apprise you of it, as it is one of very great interest in the practical work of acclimatizing food fish.

Relating to the shipment of whitefish last year, I need hardly particularize. They arrived in New Zealand in good order, and a portion of the eggs were hatched out in Canterbury, having been transshipped at Auckland, but were lost owing to ignorance regarding their habits, as the little fellows get up and swim the moment they are hatched out. The remainder were lost through a series of misadventures, and generally from want of preparedness. Great care was taken at this side, the second shipment having been nearly three weeks in the Pacific Company's ice-house here, and besides being packed in ice were frequently drenched with water of the temperature of the eggs. The boxes were occasionally turned while in the ice-house. This relieved the pressure on the lower layers. I received most valuable assistance from Mr. Woodbury, foreman of the State fish-hatching establishment, San Leandro, to whom the credit of preserving the eggs is wholly due.

The shipment of 500,000 whitefish eggs for New Zealand this season fortunately went by the City of Sidney, Captain Dearborn. The crates were opened and the eggs examined in the ice-house by Mr. Woodbury and myself, and were in good condition; whereas a shipment from Mr. Clarke, packed in the same way, for the California and Nevada State fish commission, was baked. The cause was overcare in the express car. The eggs were kept near the stove, whereas it should be generally known that they cannot be kept too cool in transit across the continent.

We unscrewed the lids, ascertained the temperature, and gave them a good drenching; then replaced the lids and packed them in ice. Two days afterwards I had the boxes surrounded by ice, and saw them placed in the ship's ice-house, to prevent the possibility of an accident. I put two tons of ice on board for the ship's use, packed in the Auckland Society's ice-chest, to avoid disturbing the eggs as much as possible. I can only hope that they will arrive safe.

I informed the New Zealand Government by cable of the consignment, and wrote the colonial secretary and Mr. Firth fully on the subject, giving the latter valuable hints regarding the mode of hatching and feeding the young fish from Mr. Woodbury. I also arranged for a telegram to Canterbury and Otago, informing them that 50,000 eggs were at each of their disposal if prepared to receive them; but requested Mr. Firth to hatch all out in Auckland if there was the slightest risk of losing any. The steamer sailed on January 21, at 11 p. m., and will probably make a short run. I shall advise you of the result. Apologizing for this rambling letter,

I am yours, very truly,

ROBT. J. CREIGHTON.

Prof. SPENCER F. BAIRD,
Washington, D. C.

Extracts from the Fifteenth Annual Report of the Canterbury Acclimatization Society for 1878.*

"In February the society received 20,000 ova of the whitefish (*Coregonus albus*) through the New Zealand Government from America, but we regret to state that only 12 were hatched, out of which only 8 survived. These were taken to Lake Coleridge and liberated in a small tributary by the chairman, Sir J. Cracroft Wilson.

"The 240 young salmon remaining in the fish-house from last season were liberated in the river Heathcote by the chairman, and during the year reports have been received of salmon having been seen and caught, and little doubt remains about their success, some having been taken 12 and 14 inches in length, and sold among some trout."

"Correspondence with agents and others in America have been instituted, whereby the society may anticipate the arrival of a variety of suitable game and fish from California and the Eastern States, as also seeds of the sugar maple and other useful and ornamental trees. The hearty thanks of the society are due to Prof. S. Baird, Commissioner of American Fisheries, and also to Livingstone Stone, esq., for the great interest taken in furthering the wishes of the society. R. Creighton, esq., of San Francisco, had also taken a great interest in procuring 'prairie chickens' for the society, but owing to circumstances his attempts have proved unsuccessful. But steps have been taken which may prove a success next season; our thanks, therefore, are due to him for what he has done."

From the Report of the Auckland Acclimatization Society for 1877-'78.

SALMON.—At the last annual meeting the council were instructed to make arrangements for a further supply of Californian salmon ova, and

*The | Fifteenth Annual Report | of | Canterbury | Acclimatization Society, | as adopted at the annual meeting of the society, | held at | the Commercial Hotel, Cathedral Square, | Christchurch, N. Z., | together with | the rules and list of subscribers. | — | Christchurch: | Printed at the "Press" office, Cashel street. | 1879. (8vo pamphlet, pp. 16.)

accordingly one of their first acts was to forward to the United States Fish Commission an order for 200,000 ova, 50,000 of which were afterwards reserved for the Canterbury Society. Subsequently Sir Samuel Wilson, of Melbourne, asked that 50,000 should be obtained for Victoria; and the order was consequently increased by that number. A most courteous reply was received from Professor Baird, promising that the ova should be forwarded, if the usual appropriations were made by Congress; but in the mean time the New Zealand Government determined to introduce a consignment of 500,000 to be divided among the various acclimatization societies in the colony. The result was that the Fish Commission forwarded the government order, together with the 50,000 for Victoria, the share of the Auckland Society of the half-million ova being fixed at 100,000. As in previous years, the consignment was supplied by the McCloud River establishment. Excellently packed there, it was conveyed to San Francisco, and shipped by the Pacific Mail Company's steamer City of Sydney, arriving in Auckland in the best possible condition.

In anticipation of the arrival of the ova, the government had made arrangements with the president—Mr. J. C. Firth—to superintend the transshipment of the portions intended for the Southern Provinces. Under his instructions, double boxes, with the interspaces filled with sawdust, and furnished with ice-racks at the top, were prepared; and on arrival of the City of Sydney, the ova boxes were rapidly transferred from the steamer's ice-house to these cases, and then shipped south by the steamers Rotorua and Wanaka—arriving at their final destination in superb order.

The disposition of the 100,000 ova retained in Auckland was as follows: 10,000 were placed in the hatching-boxes in the Domain; 40,000 were then taken by Mr. Firth to the Puniu River, at the special invitation of the Ngatimaniapoto chieftain Rewi, and safely deposited there. Returning from thence, Mr. Firth then conveyed the balance (50,000) to the Mangakahia River, the chief affluent of the Northern Wairoa. In this stream the ova were also very successfully placed. In all the localities the hatching was very satisfactory. In the Domain 9,000 fish were hatched out of the 10,000 ova deposited; and although it was impossible to ascertain the exact proportion of fish produced at the Puniu and Mangakahia, the result was evidently not far different. In every respect the consignment must be considered as being most successful.

The council have now to tender the warmest thanks of the society—or, to speak more correctly, of the whole community—to the United States Fish Commission for their liberality in presenting the ova. During this season and the two preceding ones, nearly a million of salmon eggs have been forwarded to New Zealand, for no portion of which has any charge been made, save the actual expenses of packing and transit to San Francisco, &c. This alone would be a gift of no inconsiderable magnitude; but when, in addition, it is considered how great are the benefits in an economic point of view that must result from the successful establishment of so valuable a food-fish as the salmon in our rivers, it is difficult to estimate the extent of our obligations to Professor Baird and his coadjutors on the Fish Commission, or to place too high a value on their active and zealous co-operation.

In the next place, special acknowledgments are due to Mr. J. C. Firth, the president of the society. Upon this gentleman devolved the whole of the arrangements for the reception of the ova and the transshipment of the portions intended for the Southern Provinces and Australia. Special expeditions were also made by him to the Puniu and Mangakahia Rivers, for the purpose of depositing the ova; and he has personally attended to the distribution of the fry hatched in the Domain fish-house. The entire cost of the transit of the ova from San Francisco, and its distribution throughout the colony, together with that of the journeys alluded to and all other expenses connected with the Auckland portion, have been also defrayed by Mr. Firth, so that the consignment has been absolutely without cost to the society. The council are convinced that but for Mr. Firth's energetic labors and careful oversight the enterprise could not have resulted in so satisfactory a manner.

Thanks are also due to the following gentlemen, many of whom have afforded valuable assistance: To Messrs. Cross & Co., Mr. Edwin Hooper, and Mr. R. J. Creighton, who attended to the shipment of the ova at San Francisco; to the Pacific Mail and Union Steamship Companies, who very liberally made no charge for freight; to Captain Dearborn and the officers of the City of Sydney, for the care bestowed on the ova during the voyage to Auckland; to Captains Kennedy and MacGillivray, with their chief officers, Messrs. Cromarty and Gerrard, for similar attentions on board the Rotorua and Wanaka; to Mr. G. S. Cooper, under colonial secretary, who afforded valuable assistance in many ways; to Mr. W. Seed, the secretary to the customs, who kindly granted the use of a steam-launch to convey the ova for the Mangakahia River; to Mr. A. V. Macdonald, the railway officials, and the Waikato Steam Navigation Company, who gave every assistance in their power towards the conveyance of the Waikato portion of the consignment, making no charge for transit; to the proprietors of the steamers Durham and Ruby, for the free conveyance of the young salmon to the Thames River; to Messrs. J. H. Smith, Tremain; A. Kay, E. Mitchelson, Uloth; Major Jackson, Cowan; D. McGregor, H. Wilson, J. Wilson, Waymouth; Captain Lowrie, and many others, for their hearty assistance in contributing to the success of the enterprise.

A pleasing feature, and one worthy of record, is the great interest taken in salmon importation by many of the Maories. It has already been mentioned that a share of the ova was forwarded to the Puniu River at the special invitation of the eminent Ngatimaniapoto chieftain, Rewi, not very long ago engaged in open warfare against the European settlers. Not only did Rewi, together with his kinsman Te Puke, afford every assistance during the work of depositing the ova, and most hospitably entertain the party at his settlement, but he has also taken the young fish under his special protection, giving orders that if any should be caught in the Maorie eel-weirs or fishing-nets they shall be immediately restored to the water. Similarly the well-known northern chief, Tirarau, rendered considerable assistance to Mr. Firth while conveying the Mangakahia portion of the consignment.

With reference to the salmon importation of 1876, it is satisfactory to report that young fish have been repeatedly seen. In August last, numbers of fry, about five inches in length, were noticed in the Rapurapu stream, an affluent of the Upper Thames; and only a short time ago comparatively large fish, undoubtedly salmon, were observed at Omahu, on the Thames River itself.

WHITEFISH (*Coregonus albus*).—A box of ova of this valuable lake fish, taken from a large consignment received by the government from the United States Fish Commission, was placed in the hands of the society for treatment, but unfortunately proved a complete failure, only nine fish hatching, and of these all but two died shortly afterwards. The council trust that a future attempt will be more successful, as it is a fish that would probably do well in Lake Taupo, and possibly also in Tarawera and other of our lakes.

BROOK TROUT (*Salmo fontinalis*).—Mr. T. Russell, who has done, and is doing, so much for acclimatization in New Zealand, has, through his agent, Mr. Hugh Craig, of San Francisco, forwarded to the society a box of 5,000 ova of this little trout, said to be one of the best of the Western American species, both as an article of food and as affording capital sport to the fly-fisher. The box did not arrive in as good condition as could have beeen desired; but nevertheless 400 young fish were successfully hatched. They have since been liberated, half the number in a tributary of the Waikato near Cambridge, and the remainder in the upper part of the Kaukapakapa stream, Kaipara district.

CATFISH (*Pimelodus Catus*).—Two consignments of this well-known fish have also been introduced from America by Mr. T. Russell. In all, 140 living fish arrived, which have been liberated in St. John's Lake. Of late years considerable attention has been paid to the distribution of this species in the United States. It is said to do well in small lakes, ponds, mill-dams, and even swamps ; to be good eating, easily caught by hook and line, and to be not destructive to the young of other fish.

It should here be mentioned that the entire cost of these importations is borne by Mr. Russell, who has certainly earned the warmest thanks of the society for the services he has so unostentatiously rendered to the colony.

From a New Zealand paper, published toward the close of the year 1877.

ACCLIMATIZING SALMON.

Our readers are already aware that some months ago the Auckland Acclimatization Society requested the United States Fish Commissioners to be good enough to forward 200,000 salmon ova from their establishment on the McCloud River, California, to aid in stocking the rivers in this province. Subsequently the Canterbury Acclimatization Society, and Sir Samuel Wilson on behalf of the Victorian Acclimatization Society, requested the Auckland society to obtain 50,000 ova for each applicant. By the August mail, Mr. J. C. Firth, president of the Auckland society, received a letter from the Hon. Spencer F. Baird, chief commissioner, in which he very courteously offered to supply the number of ova wanted should the supply of eggs be sufficient to warrant it. After that the New Zealand Government requested Mr. Baird to dispatch 500,000 salmon ova for New Zealand. These are expected by the mail steamer to-day. By Parliamentary papers we observe they are intended to be distributed as follows: Auckland, 100,000; Napier, 50,000; Nelson, 50,000; Greymouth, 50,000; Wellington, 50,000; Christchurch, 50,000; Dunedin, 50,000; and Makarewa (Southland), 100,000. In reference to the distribution of the coming ova, Mr. Firth has received the following letter from the colonial secretary's office:

"COLONIAL SECRETARY'S OFFICE,
"*Wellington*, 11*th October*, 1877.

"SIR: I have the honor, by direction of the colonial secretary, to inform you that Professor Baird was, on the 28th July last, requested to be good enough to have the next shipment of salmon ova packed, if possible, in cases containing 50,000 each, of which two cases are intended for your society. This shipment may be expected to arrive by the next, or at latest the following, San Francisco mail, and I am to request that you will be prepared to receive it immediately on arrival of the steamer at Auckland. I inclose a copy of the papers which have been laid before Parliament on the subject generally, for the information of your society. Should it appear to you that any additional expenditure for supply of ice, or on any other account, to insure the chance of success for the shipments to southern societies, I am directed to request that you will kindly make such arrangements and incur such expenditure on behalf of the government as may, in your judgment, appear necessary to attain the object in view.

"I have, &c.,

"G. S. COOPER.

"J. C. FIRTH, Esq.,
"*President of the Acclimatization Society, Auckland.*"

Mr. Firth, who was busily engaged in making preparations for the reception, preservation, and safe distribution of the 300,000 ova for

Auckland, Canterbury, and Victoria, when the above letter came to hand, at once heartily acceded to the request of the colonial secretary. His great experience enabled him to make the necessary preparations which, under ordinary circumstances, secure success. It must be understood that the ova boxes are transmitted from San Francisco in the ice-house of the mail steamer, by which means the hatching of the ova is retarded. If the ice were to run short during the voyage, or the boxes to be exposed to the sun for even a short time after leaving the mail steamer, the retarding effects of the cold would be destroyed, and under the influence of the high temperature here premature hatching would take place in the boxes, and the whole experiment prove a disastrous failure. The ova boxes, as we have stated, are simply placed in the ice-house of the steamer on being sent from San Francisco, and on arrival here it is necessary that a separate case be provided for each box of ova to safely convey the ova to their destination. These cases were made in accordance with the plan which Mr. Firth's experience has shown him to obtain the largest measure of success. A large number of boxes have been prepared for the immediate reception of the ova on its arrival here. Each of these boxes is provided with an internal division, which admits of packing three inches of sawdust between the outside of the case and the internal division. In the open inner space is placed the ova box, which is protected from the heat on the sides by the sawdust, and on the top of the ova box is placed about five inches of ice to keep the ova cool and moist, and in the bottom are perforations for carrying away the water from the melting ice. The ice thus placed on the top of the ova is protected from the heat by pads containing a thick packing of sawdust. The ova box having been placed in position, the ice on top and the sawdust pads on the top of the ice, the lid of the outer case is then closed and securely fastened, which makes the affair complete. By a very simple arrangement the two very necessary requisites—ventilation and drainage—are provided. It is perhaps necessary to say that not only has increase of temperature to be provided against, but also any risk of concussion must as far as possible be avoided. To secure this latter, a simple and effective means of carrying the boxes has been designed by Mr. Firth, which will prevent any concussion during the transference from ship to wharf, in their final transport down the sides of the ravines or up the shingle-beds of the rivers to their final destination. Mr. Firth also provided a number of boxes to contain a reserve supply of ice to guard against the exhaustion of ice in the ova-boxes. These boxes are constructed so that there is a three-inch space between the inner and outer boxes packed with sawdust. The inner box is then filled with ice, and covered with a sawdust pad, as in the case of the ova-boxes. The box-lid is then closed and fastened, and there is very little doubt that the ice so stored will be available in case of need.

Mr. Firth's experience warrants him in thinking that the most successful way of hatching out ova is the natural process on the shingle-beds of rivers. In confirmation of this, Mr. Firth has received intelligence of the salmon having been seen in the various rivers in which ova was placed, but the experiment made in the Rapurapu River, one of the upper branches of the Thames, is the most successful and interesting. In that river, for miles above and below the point where Mr. Firth deposited the ova upon the shingle-bed last November, swarms of young salmon have been seen five inches long. There can be very little doubt that the great experiment now being undertaken will, with previous efforts, successfully establish the king of fish in the rivers of this colony.

After transmitting each box to Victoria and the south, Mr. Firth will take 40,000 ova to the Upper Puniu. Rewi, the Ngatimaniapoto chief, having some time ago requested Mr. Firth to meet him in the King Country to see if any of the rivers were suitable for salmon, Mr. Firth accordingly went up and selected the Puniu as one of the most suitable of the Waikato system of rivers. A parcel will also be taken to the Upper Thames. A box will also be hatched in the society's hatching-boxes in the domain, and the remainder will be placed by Mr. Firth in the Mangakahia River and another stream which fall into the Northern Wairoa. These northern rivers have been selected by Mr. Firth during his recent visit to that part of the country. It is hardly necessary to add that the society are satisfied that the only proper way to secure success in the important experiments is to concentrate their operations upon the three most important river systems in this province—Waikato, the Thames, and the Northern Wairoa—rather than by placing small quantities in the innumerable creeks all over the country. If success is achieved, every suitable stream in the country can be stocked at leisure without difficulty from one or other of the rivers named. The future importance of the salmon-fishing industry, the foundation of which Mr. Firth and the Acclimatization Society are now laying, can hardly be estimated.

When the steamer arrived it was found that a less quantity of ova was on board than was expected. There were eleven boxes of salmon ova, containing, it was estimated, about 550,000, which are distributed as follows: For Auckland, 100,000; for Napier, 50,000; for Nelson, 50,000; for Greymouth, 50,000; for Wellington, 50,000; for Canterbury, 50,000; for Dunedin, 50,000; for Southland, 100,000; for Victoria, 50,000. No time will be lost in distributing through the Auckland streams and rivers the proportion assigned to this district. Mr. J. C. Firth, who continues to be so energetic in this kind of public usefulness, left on Wednesday morning for Te Awamutu, in which neighborhood some of the eggs are to be deposited. It is intended to deposit 30,000 in the Puniu. There will be deposited some 50,000 at various points in the upper waters of the Waikato, and the remainder, about 20,000, in the Rapurapu, one of the tributaries of the Upper Thames. It is satisfactory to be able to state that the ova have arrived in excellent condition, and for this result thanks are due to Professor Spencer Baird, chief of the Fish Commission of the United States, and to Mr. Livingstone Stone, the commissioner for the Pacific States, who bestowed the greatest pains in packing the consignment for shipment.

Robert Houghton to S. F. Baird..

SAN FRANCISCO, *September* 11, 1878.

DEAR SIR: I have just received a letter from the New Zealand Government, in reply to a communication from me, in which they state that they will take one million whitefish ova on the terms stated by Mr. Clark, namely, 60 cents per 1,000 f. o. b. at San Francisco, and that Mr. Clark undertakes personally to superintend the shipment at that port, provided his actual expenses across the continent are defrayed jointly by the State fish commissioners of California and Nevada and New Zealand. In my letter to the government, however, I inclosed express charges which they appear to have overlooked in their letter to me, but this item should be included in the bill by Mr. Clark, to whom I send a copy of the letter.

I have forwarded your note of acknowledgment of remittance from the colony.

I am, dear sir, very truly yours,

ROBT. HOUGHTON.

Prof. SPENCER F. BAIRD, &c.

S. C. Farr to S. F. Baird.

CANTERBURY ACCLIMATIZATION SOCIETY,
Christchurch, September 13, 1878.

The Hon. SPENCER F. BAIRD,
United States Commissioner Fish and Fisheries, Washington:

DEAR SIR: We are very anxious to introduce into Canterbury some of the most useful game and insectivorous birds from America, and it was resolved at the last meeting of council that I should communicate with you upon the subject, and feeling assured that you will help us in the matter I take the liberty of asking you to kindly inform us which are the best, with something of their habits, best season for procuring them, and probable cost delivered on steamer at San Francisco. I am convinced this will be an intrusion upon your valuable and much occupied time, which I trust you will pardon.

I am also directed to inquire if you could secure for us in the season 100,000 ova of the silver trout? If so, at what cost delivered on board steamer at 'Frisco.

You will, I have no doubt, be pleased to hear that the salmon are doing well with us, especially so in one of our rivers, Waimakiriri, some having been taken 9 and 10 inches in length, and, of course, returned to their natural element to mature.

Here permit me to offer a suggestion in repacking of ova. It occurred to me, when unpacking the ova received from the Fish Commission, that an undue pressure presented itself in the center of each box, at which place the greatest loss was experienced, the ova being compressed thus [drawing], and void of any appearance of vitality, while those protected somewhat by the sides of the box from like pressure were all right. To prevent such a disaster, I thought if small twigs or laths, about the same substance as the ova is, in diameter, were laid crosswise, so as to divide the box into compartments, and thus support the screen and moss, might probably prevent it. [Drawing.]

I have taken part in unpacking the ova received by us, and have noticed the same thing in each case; therefore venture the suggestion.

In reference to the packing, I consider (with the exception of the above mentioned) nothing could have been more systematic or precisely executed, hence the success.

Apologizing for thus imposing upon you, I am, dear sir, yours faithfully,

S. C. FARR,
Honorable Secretary.

CHRISTCHURCH ACCLIMATIZATION SOCIETY.

An adjourned meeting of this society was held yesterday afternoon at the Gardens. Present, Hon J. T. Peacock, chairman; Drs. Nedwill and Poweli, Messrs. Hill, Boys, Carrick, Jameson, honorable treasurer; Farr, honorable secretary; Johnstone, Foreday, Haumer, and Blackiston.

The secretary said since last meeting he had received £20 from the Auckland society, balance of the £70 refund on account of the California salmon ova.

A telegram was read from Dr. Hector, requesting that the majority of the whitefish might be sent to Lake Coleridge as soon as they were fit for carriage, a few to be kept by the society for experimental purposes.

The curator, who was present, said that only about half a dozen of the fish were now alive. He had put some muslin in the boxes to retain the food; this had caused the boxes to overflow, and the fish had been thrown on to the floor of the breeding-house.

A very general regret was expressed that such a mishap should have occurred, and which had all the appearance of having resulted from very great carelessness.

The secretary was instructed to telegraph the fact to Dr. Hector.

In reply to Mr. Boys, the secretary said about 200 of these fish had been hatched out.

Robert Creighton to S. F. Baird.

SAN FRANCISCO, CAL., *January* 15, 1878.

S. F. BAIRD:

Thanks for the contribution of food-fish for New Zealand. I have forwarded your letters and telegrams to government, at New Zealand. Clark shipped ova on the 11th, and to-night they have arrived and are on the City of Sydney, and will sail to-morrow, 21st of January. I have sent a cablegram to the Government of New Zealand. Shipment of salmon arrived safely and have proved a great success. I think New Zealand is now fully stocked with salmon, at least to such an extent as to render further shipments of ova unnecessary for some time to come. Small parcels of eastern trout have been sent from time to time and have been successful. I attach greater importance to whitefish than any other, because of delicacy of flesh and commercial value. New Zealand is a country of lakes and rivers peculiarly adapted for white-fish. I hope that this consignment will survive better than last year.

ROBT. CREIGHTON.

Robert Creighton to S. F. Baird.

[Telegram.]

SAN FRANCISCO, CAL., *January* 19, 1878.

S. F. BAIRD:

Whitefish eggs arrived in good order. Shipped per steamer City of Sydney. Sails 21st instant.

CREIGHTON.

James Hector to S. F. Baird.

COLONIAL MUSEUM OF NEW ZEALAND,
Wellington, April 27, 1878.

MY DEAR PROFESSOR BAIRD: I have been away for the last two months and find that you have not been informed of the result of the whitefish shipment of January last, which reached Auckland on the 15th February. I inclose a copy of my report to government, of 8th March, which you should have received by last mail. You will see that the experiment has been so far successful as to prove that these fish *can* be introduced with proper care into the most distant part of the colony. The partial failure must be attributed to some error during the transit. If due to overpacking with moss, as suggested by some, I don't see how any could have survived. On looking through the papers I find that Mr. Creighton states, as follows:

"The entire shipment of whitefish ova for California and Nevada, from Northville, Michigan, packed precisely as those for New Zealand by Mr. Clark, and coming in the same car, were spoiled in transit; on being opened by Mr. Woodbury and myself they stunk and were putrid. They had been placed near the stove by the express agent to prevent them freezing. Ours had been less considerately treated and arrived (in 'Frisco) *sound and lively*, as I had proof, every box having been opened and *examined* by Mr. Woodbury in my presence. We then ascertained their temperature and *gave them a drenching with water at a similar heat*, screwed them up, reversed their position, and placed them upon and under ice in the Pacific Mail Company's ice-house."

It is evident, therefore, that the ova were all right so far; whether the treatment I have underlined was judicious you will be able to judge. My own impression is that the mischief commenced toward this end of the journey. Don't you think it would be better to pack them in tin boxes inside the wood? The wood boxes were quite sodden and rotting, and four of them had the lids loose. The holes, top and bottom, seem also a mistake, as they promote drainage of the melting ice-water through the ova and may cause them to hatch. Holes on side and bottom would be better. Also, I would suggest that each piece of screen carrying ova should be stitched on a light frame resting on corner-pieces, so as to take the weight off the bottom layers and to prevent sagging in the central part. But I hope to get authority to ask you to repeat the experiment, when I will write all my suggestions at length.

The shipment of *S. salar* from Great Britain has been again a failure. Fifty thousand ova were packed in fifty-six boxes! Most of them seem not to have been impregnated, and at most only a few hundred hatched out. This is a great contrast to the success of the California salmon.

Yours, very truly,

JAMES HECTOR.

Prof. SPENCER F. BAIRD,
Washington.

J. C. Firth to S. F. Baird.

AUCKLAND, N. Z., *May* 2, 1878.

DEAR SIR: I regret to have to inform you that the half million white-fish ova which you were good enough to transmit to this colony, and the transshipment of which at this port the New Zealand Government intrusted to me, have turned out badly so far as yet known. I think, probably, that the ova and moss were too much compressed—the moss being very hard and the netting adhering, the ova presenting the appearance of having been crushed. In the box left at Auckland all but 30 ova were dead, and these only appear to have escaped by reason of there being less pressure at the sides than elsewhere. Mr. Creighton, our secretary, appears to have taken every precaution to secure success. Captain Dearborn, of the City of Sydney, Halifax mail line, and his officers, did all in their power to secure success. The government of this colony will doubtless furnish you with full particulars.

Of the 30 only 9 hatched; 6 of these died immediately; 2 died yesterday—only one remaining alive.

You will be glad to learn those fine healthy salmon from your ova have been seen a week ago, about 15 to 18 inches long, in the river Thames, not far from the point where I placed them two years ago.

I am, dear sir, yours truly,

J. C. FIRTH,
President A. A. Society.

I send newspaper with account of whitefish.

W. M. Evarts to S. F. Baird.

DEPARTMENT OF STATE,
Washington, D. C., April 20, 1878.

SPENCER F. BAIRD, Esq.,
Commissioner, &c., Washington, D. C.:

SIR: I inclose herewith for your information copy of a note of the 18th instant, from the British minister at this capital, and of its inclosures, relating to the manner of the shipment under your direction of salmon ova to New Zealand.

I am, sir, your obedient servant,

WM. M. EVARTS.

Sir Edward Thornton to W. M. Evarts.

WASHINGTON, D. C., *April* 18, 1878.

SIR: In compliance with an instruction which I have received from the Earl of Derby, I have the honor to inform you that the governor of New Zealand, at the instance of his ministers, has requested that the thanks of the colony may be conveyed to the Government of the United States for the very handsome and effective manner in which salmon ova have been shipped to New Zealand by the Fishery Commission of the United States, under the direction of the chief Commissioner, the honorable Spencer F. Baird.

I have the honor to transmit herewith copy of the dispatch and of its inclosure upon this subject from the governor of New Zealand to the secretary of state for the colonies.

I have, &c.,

EDWD. THORNTON.

The Marquis of Normanby to the Earl of Carnarvon.

WELLINGTON, *February* 1, 1878.

MY LORD: I have the honor to inclose a memorandum which I have received from my government, by which you will see that they are anxious to convey the thanks of this colony to the Government of the United States for the very handsome and effective manner in which salmon ova has been shipped to this colony by the Fishery Commission of the United States, under the direction of the chief Commissioner, the honorable Spencer F. Baird.

I venture also to express a hope on my own part that your lordship will see no objection to adopt the course proposed by my government, as I think that the action of the American Government has evinced such a feeling of friendship and generosity towards New Zealand in a matter in which deep interest is taken as to demand a special mark of acknowledgment and thanks on the part of this colony.

I have, &c.,

NORMANBY.

C. S. Whitmore to the Governor of New Zealand.

MEMORANDUM FOR HIS EXCELLENCY.

Ministers desire respectfully to inform his excellency the governor that the half million salmon ova which arrived by the mail steamer from San Francisco in November last have been successfully hatched and distributed to the various rivers in the colony, and that, by information which has reached the government from various directions, it has been demonstrated that owing to the extreme care with which the ova was packed in America the very satisfactory result of about 95 per cent. of live fish has been obtained.

In addition to the half million sent at the request of the government an equal quantity has been sent to the various acclimatization societies in the colony, and this handsome gift of salmon ova has been made to the colony without charge, except cost of packing and transit, by the Fish Commission of the United States, under the direction of the Hon. Spencer F. Baird, as chief commissioner.

Ministers venture to think that so generous an action on the part of a foreign nation is worthy of being acknowledged in a special manner; they would therefore respectfully ask his excellency to bring the matter under the notice of Her Majesty's Government, through the secretary of state for the colonies, in the hope that Her Majesty's Government will permit a communication to be made to the Government of the United States of the thanks of the colony of New Zealand for the generous and valuable gift of a million salmon ova to the colony.

C. S. WHITMORE.

WELLINGTON, *February* 1, 1878.

S. F. Baird to Wm. M. Evarts.

UNITED STATES COMMISSION, FISH AND FISHERIES,
Washington, April 23, 1878.

SIR: I have the honor to acknowledge the receipt of your letter of the 20th of April, with the inclosures, and, of course, feel much gratified at the appreciation manifested by the Government of New Zealand and the Foreign Office in London of the efforts made by the United States Fish Commission to supply desirable food fishes to a sister country.

I have the honor to be, very respectfully, your obedient servant,

SPENCER F. BAIRD,
Commissioner.

Hon. WILLIAM M. EVARTS,
Secretary of State.

The following is the substance of an official document relative to the introduction of Quinnat salmon, published by the New Zealand Government in 1878. Although its substance is contained in the preceding correspondence, it embraces many facts relative to the California salmon of much interest, and worthy of reproduction.

H.—11.

CALIFORNIA SALMON AND WHITEFISH OVA, (PAPERS RELATIVE TO THE INTRODUCTION OF).

Presented to both Houses of the General Assembly by command of His Excellency.

No. 1.

The Under-Secretary to the Hon. Spencer Baird.

WELLINGTON, 31*st May*, 1877.

SIR: With reference to the offer which you kindly made in your letter of the 7th ultimo, addressed to Dr. Hector, I have the honor to request that arrangements may be made for the transmission to this colony during next season of 500,000 of the ova of the Californian salmon, and 250,000 of the ova of lake whitefish (*Coregonus albus*).

I have, &c.,

G. S. COOPER.

Professor BAIRD,
Commissioner United States Fisheries Commission, Washington.

No. 2.

Dr. Hector to Professor Baird.

WELLINGTON, 28*th July*, 1877.

DEAR SIR: I am directed by government to ask you to be good enough to have the next shipments of salmon ova packed, if possible, in cases containing 50,000 ova each, in order to facilitate their transit to the different districts throughout the colony.

The government proposes to distribute the ova as follows:

Auckland	2
Napier	1
Nelson	1
Greymouth	1
Wellington	1
Christchurch	1
Dunedin	1
Makarewa	2
	10 = 500,000

I have, &c.,

JAMES HECTOR.

The Hon. SPENCER F. BAIRD.

No. 3.

Professor S. F. Baird to the Hon. the Colonial Secretary.

WASHINGTON, 10*th July*, 1877.

SIR: I have the honor to acknowledge the receipt of your letter of the 31st of May, asking for 500,000 eggs of the California salmon and 250,000 of the whitefish, to be sent to New Zealand during the present year.

This request I shall take pleasure in supplying, and in the mean time beg to be advised of the proper address of the packages, and whether they shall be subdivided into smaller quantities. Of course I can only promise them conditionally—in the event of nothing untoward happening to the fisheries.

I have, &c.,

SPENCER F. BAIRD.

The Hon. the COLONIAL SECRETARY.

No. 4.

Mr. W. Arthur to the Hon. the Colonial Secretary.

DUNEDIN, *25th September*, 1877.

SIR: The Acclimatization Society of Otago had intended procuring a supply of American whitefish ova this season from the States.

In the course of our inquiries, however, we were informed that the Colonial Government of New Zealand had already taken up the matter, and were going to import a variety of the ova of *Salmonidæ*, and that a portion was to be forwarded to Otago.

Under these circumstances, the society will gladly await the government experiment, and give any assistance in its power to secure success, and

I have, &c.,

W. ARTHUR,

Acting Secretary Otago Acclimatization Society.

The Hon. the COLONIAL SECRETARY.

No. 5.

Circular.

To the SECRETARY OF ——— ACCLIMATIZATION SOCIETY:

SIR: I have the honor, by direction of the colonial secretary, to inform you that Professor Baird was, on the 28th of July last, requested to be good enough to have the next shipment of salmon ova packed, if possible, in cases containing 50,000 ova each, of which ——— case is intended for your society.

This shipment may be expected to arrive by the next, or, at latest, the following San Francisco mail, and I am to request that you will be prepared to receive it immediately on arrival of the steamer at ———.

I inclose a copy of the papers which have been laid before Parliament on the subject generally, for the information of your society.

To Mr. Firth:

Should it appear to you that any additional expenditure for supplying of ice, or any other account, is required to insure the chance of success for the shipments to southern societies, I am directed to request that you will kindly make such arrangements, and incur such expenditure on behalf of the government as may in your judgment appear necessary to attain the object in view.

I have, &c.,

G. S. COOPER.

No. 6.

Mr. J. C. Firth to the Under Secretary.

[Telegram.]

AUCKLAND, 2*d November.*

Preparations for safe distribution of salmon ova completed. Shall send ova-boxes and ice-chests for Nelson or Greymouth, Wellington, and Christchurch per Wanaka; those for Napier, Dunedin, and Invercargill per Rotorua. Make prior arrangements for forwarding ova for Invercargill from Dunedin if mail arrives to-morrow. I wish to convey to King country and Upper Thames. Pray ask minister to authorize of running of locomotive to Newcastle on Sunday morning.

No. 7.

Mr. Frederick Huddlestone to the Under Secretary.

NELSON, 16*th October*, 1877.

SIR: I have the honor to acknowledge the receipt of your letter of the 11th instant, wherein you inform me that a shipment of 50,000 salmon ova may be expected by the Nelson Acclimatization Society by the next or, at the latest, the following San Francisco mail.

In reply, I have to request that you will be good enough to convey the thanks of this society to the government, and inform the Hon. the Colonial Secretary that the ponds will be ready for the reception of the ova before the arrival of the next mail, and every care will be taken to hatch the fish.

I have, &c.,

FREDERICK HUDDLESTONE,
Hon. Secretary Nelson Acclimatization Society.

The UNDER SECRETARY,
Colonial Secretary's Office, Wellington.

No. 8.

Mr. James Payne to the Hon. the Colonial Secretary.

GREYMOUTH, 25*th October*, 1877.

SIR: I have the honor to acknowledge receipt of your letter of date and number as per margin, and to inform you that this society will have all its hatching-boxes and ponds in perfect readiness to receive the salmon ova on its arrival.

The boxes have been cleaned from all trout, and are available at any moment.

I am further directed by this society to request that its claims for portion of the whitefish to arrive be recognized, and that you will be good enough to put such upon record.

This society acknowledge with deep gratitude the attention of the government in securing salmon ova for it.

I have, &c.,

JAMES PAYNE,

Hon. Secretary Grey District Acclimatization Society.

The Hon. the COLONIAL SECRETARY.

No. 9.

Mr. S. C. Farr to the Hon. the Colonial Secretary.

CHRISTCHURCH, 18*th October*, 1877.

DEAR SIR: I beg leave to acknowledge receipt of yours of the 12th instant, covering papers for our information, for which receive our most sincere thanks.

I have, &c.,

S. C. FARR.

The Hon. the COLONIAL SECRETARY.

No. 10.

Mr. W. Arthur to the Hon. the Colonial Secretary.

DUNEDIN, 1*st November*, 1877.

SIR: I have the honor to acknowledge your letter of 11th ultimo, regarding a box of American salmon ova to arrive soon. In reply, I have to state that the Otago Society has given the necessary instructions to Mr. Deans, the curator, to make his preparations for accommodating 50,000 ova in our hatching-boxes.

I have, &c.,

W. ARTHUR,

Acting Secretary Otago Acclimatization Society.

The Hon. the COLONIAL SECRETARY.

No. 11.

Mr. Henry Howard to the Under Secretary.

SALMON PONDS, *Wallacetown*, 20*th October*, 1877.

SIR: I have the honor to acknowledge the receipt of your letter of the 11th instant, informing me of the expected arrival of salmon ova, and to inform you that everything is ready for its reception.

I should feel thankful if the government could give such directions to the railway authorities at Invercargill as would prevent any unnecessary delay in its transit from Bluff to the Makarewa station.

I have, &c.,

HENRY HOWARD.

The UNDER SECRETARY, *Wellington.*

No. 12.

The Under Secretary to Mr. Henry Howard.

SIR: With reference to your letter of the 20th instant, relative to preparations being made for the reception of salmon ova at the Makarewa ponds, I am directed to inform you that the railway authorities have been instructed to give you every facility in the transit of the ova from the bluff to its destination.

You had better place yourself in communication with the station-master at the bluff on the subject.

I have, &c.,

G. S. COOPER.

No. 13.

The Hon. Mathew Holmes, M. L. C., to the Hon. the Colonial Secretary.

WELLINGTON, *November* 2, 1877.

SIR: On behalf of the Oamaru Acclimatization Society I beg to thank you for the manner in which you were prepared to meet their application for salmon ova for that district, and am sorry to find that all the shipment now on its way from San Francisco was promised before my application was made.

As further shipments are to follow, I now beg to apply for two cases salmon and two cases whitefish ova, out of the first shipment to arrive from America, for the Oamaru Acclimatization Society.

I may state that suitable provision has been made to receive and hatch the ova, and that Mr. Young (one of the most successful in this line) has undertaken to conduct the experiment.

I have, &c.,

MATHEW HOLMES.

The Hon. the COLONIAL SECRETARY.

No. 14.

Mr. J. C. Firth to the Under Secretary.

[Telegram.]

AUCKLAND, *November* 6, 1877.

I have shipped per Rotorua, sailing this evening, one case of fifty thousand salmon ova to Williams, Napier, with one chest of ice in reserve. Same quantity ova and two chests reserve ice each to Travers, Wellington, Farr, Christchurch, Perkins, Invercargill, and have wired advice of shipment to each party. Did Greymouth ova arrive?

J. C. FIRTH.

No. 15.

Mr. J. C. Firth to the Under Secretary.

AUCKLAND, *November* 20, 1877.

SIR: I perceive by your telegram of yesterday that some misapprehension exists as to the quantity and distribution of the salmon ova received by the November mail-steamer.

By way of putting the matter fully before you, I may state that in answer to my letter of 11th of April to the Hon. S. F. Baird that gentleman arranged to send 200,000 ova for the Auckland Acclimatization Society, and, in answer to a subsequent request of mine, a further shipment of 50,000 for the Canterbury Society, and 50,000 for the Victorian Society. On receiving your letter of the 11th October, asking me to receive and provide for the safe distribution of the 500,000 salmon ova the New Zealand Government were expecting to arrive by steamer on November 3, or at latest by next mail-steamer, and, knowing that the ova-boxes are shipped from their crates in San Francisco so that they may be placed in the steamer's ice-house, I immediately set to work to provide a double chest (the interspace packed with sawdust) for each ova-box expected (16 in number), with the necessary ice-boxes for a reserve of ice. I had provided also 2 tons of ice as a first installment, if the whole 800,000 ova arrived. These preparations were fully completed on November 2, when the mail-steamer arrived at Auckland. On her arrival I found that 11 boxes only had arrived, consigned on ship's manifest to Auckland Acclimatization Society. I could learn nothing of any for the New Zealand Government.

I had a staff of 8 men on the wharf, but the difficulty of getting the ova-boxes out of the ice-house, where they lay imbedded in tons of ice, was so great that I had not completed the packing of the 11 boxes till 5 o'clock on the morning of the 3d November, though I and my men had been hard at work all through the night.

Not wishing to disappoint the more suitable localities in the south, I arranged to ship some of the Auckland ova to Christchurch (in addition to their own parcel), to Dunedin, to Invercargill, and Napier, to be returned to us on receipt by government of the ova ordered by them. I therefore placed on board the Wanaka steamship, before 7 o'clock a. m., November 3, 4 boxes with reserves of ice for the three places first named, intending to ship to Napier by the Rotorua on the 6th. When on my return from Onehunga, the secretary of our society, having obtained his advices, waited upon me with a letter from Messrs. Cross & Co., our San

Francisco agents, advising shipping 11 boxes salmon ova for the Auckland Acclimatization Society, and inclosing press copy of a letter from Professor Baird's deputy at Redding, in which there fortunately happened to be a copy of the names of places to which the 10 boxes were to be sent—identical with Dr. Hector's list of 28th July, 1877—embodied in the Parliamentary papers you sent to me (with one for the Victorian Society). I then found that for some reason or other the United States Fish Commissioners had not forwarded the Auckland and Can bury orders. I at once telegraphed Captain McGillivray, of the Wanaki steamship, to deliver the two boxes marked "Christchurch" to Nelson and Greymouth. On the 6th I dispatched per Rotorua:

1 box to Napier,
1 box to Wellington,
1 box to Christchurch,
1 box to Invercargill,
} With 7 ice-boxes in reserve.

4

Per Wanaka—

1 box to Nelson (as above),
1 box to Greymouth (as above),
1 box to Dunedin,
1 box to Invercargill,
} With 5 boxes ice in reserve.

8

Leaving for Auckland $\frac{2}{10}$ and $\frac{1}{11}$ for Victoria Society (not included in government order.)

Having made every arrangement at great expense and much personal inconvenience for the safe reception and proper dispersion of the full quantity of 800,000 ova, I must confess to a little disappointment at being therefore rendered unable to stock the Auckland rivers to the number and extent I had intended.

Since the arrival of the mail steamer on November 2, I have been actively engaged in carrying out the work you intrusted to me, of packing and transshipping the ova to southern ports, and in placing the Auckland portions in the King country to the south, and in the Wairoa River and its tributaries to the north.

From telegrams I have received, I am pleased to think that the work, arduous though it has been, has not been in vain.

Pray pardon the length of this letter, as I could not permit any misapprehension as to the proper disposal of the ova to exist in your mind without endeavouring to remove it.

I have, &c.,

J. C. FIRTH,
President of the Auckland Acclimatization Society.

G. S. COOPER, Esq.,
Under Secretary, Wellington.

No. 16.

Mr. J. C. Firth to the Under Secretary.

AUCKLAND, *21st January*, 1878.

SIR: I beg to inform you that I have successfully deposited the 100,000 salmon ova placed at the disposal of the Auckland Acclimatization Society, as follows:

40,000 in the Puniu River, in the King country, the chief Rewi Maniapoto co-operating with me and assisting me.

8,000 in the river Thames.

7,000 in a small stream near the chief Tirarau's settlement, Wairoa North.

7,000 in the Mangakahia River, near the Hikurangi stream.

36,000 in the Mangakahia River, near Te Wero's settlement.

About 95 per cent. of these hatched out, and, though the occurrence of a fresh in the Mangakahia River interfered somewhat with the success of the enterprise, I have no doubt that a very fair measure of success has been attained.

I inclose (1) duplicate receipt from Mr. Myron Green for $750, paid by Mr. Creighton to United States Fish Commission, for package and transit charges of 500,000 salmon ova, and (2) letter from Prof. Spencer F. Baird confirming same. For this sum Mr. Creighton drew upon me, which I honored, and was subsequently refunded a like amount by the Treasury at Wellington, £164 1*s.* 3*d.*

I have to thank you for the very efficient aid you have rendered me in the distribution of the half million ova.

I have, &c., J. C. FIRTH.

G. S. COOPER, Esq.,
Under Secretary, Wellington.

No. 17.

Frederick Huddleston, Esq., to the Hon. the Colonial Secretary.

NELSON, *7th January*, 1878.

SIR: I have the honor to report for the information of the government the success that has so far attended the introduction of American salmon ova into the rivers of this district.

The ova arrived from San Francisco on the evening of Sunday, the 4th of November. On Monday morning I opened the box said to contain 50,000. I found eight layers, each about a quart, and packed between a thin material like scrim, and each layer separated by moss. I caused all the dead eggs to be picked out (about 1,500). The sound ones were then put into the hatching ponds, and the ponds covered with boards to protect the eggs from the sun. On Friday, the 9th November, the first fish made its appearance, and by Monday, the 19th, all were hatched out, with the exception of about 1,000 bad eggs. They were thus left undisturbed until the 8th December, when, finding they had begun to feed, I caught about half of them and turned them into the Wairoa River, close by the railway bridge. On the following Saturday, 15th December, the remainder were caught and placed into two large *tin-lined* cases and sent by rail to Fox Hill, from which place they were taken by spring conveyances over Spooner's Range, a distance of about fourteen miles, and placed into the Motueka River, with a loss only of about fifteen on the road.

I estimate the total number turned out at about 25,000, and the bad ova at about 2,500. It will thus be seen that the box contained little more than half the estimated quantity, viz, 50,000.

The ova was certainly most carefully and beautifully packed, and the arrangements for supplying ice were exceedingly good. Great credit is due to the shippers, and it would be well if Dr. Buckland and others in England interested in the acclimatization of fish would take a lesson in packing ova from our American friends.

In conclusion, I hope the government will continue the good work so well commenced until salmon is established in New Zealand waters beyond a doubt and our rivers well stocked.

I have, &c., FREDERICK HUDDLESTONE,
Hon. Secretary Nelson Acclimatization Society.

The Hon. the COLONIAL SECRETARY.

No. 18.

Mr. W. Arthur to the Hon. the Colonial Secretary.

DUNEDIN, 16*th December*, 1877.

SIR: I have the honor to inform you that the box of American salmon ova (supposed 50,000) arrived here safely by the Taupo on the 7th, and contents transferred to the breeding boxes of the Otago Acclimatization Society with as little delay as possible. The supply of ice was not exhausted, and the ova were in very good condition, only four or five per cent. having gone bad. I am sorry, however, to say that after being four days in the hatching-boxes many of them died, but others are healthy, and some are hatching out.

The society will be glad to hear soon as to when the supply of white-fish ova may be expected for our lakes. Our accommodation is limited, and besides the salmon ova we have a great number of young trout recently hatched out still in the hatching-boxes.

I have, &c., W. ARTHUR,
Acting Secretary Otago Acclimatization Society.

The Hon. the COLONIAL SECRETARY.

No. 19.

The Hon. J. A. R. Menzies to the Hon. the Colonial Secretary.

WYNDHAM, 25*th January*, 1878.

SIR: I have the honor to inform you that Mr. Howard reports that he has placed the following numbers of California salmon fry in the rivers named:

In the Oreti	35,000
In the Waipahi	10,000
In the Makarewa	18,000
Total	63,000

He retains for the present about 800 fry in the ponds. Mr. Howard remarks that only 25,000 fry were available from the ova contained in the second box he received, that box, as you may remember, having been transshipped in Auckland, by mistake, to the Rotorua, whereby it reached the ponds above a week later than the other box, the hatching of the ova of which seems to have produced 80 per cent. of fry.

Mr. Howard also says "the young fish are exceedingly healthy and strong, and the arrangements for the transport of the ova from America, though simple, were almost perfect."

Have you any intelligence of the dispatch of the English salmon ova ordered?

I have, &c.,

J. A. R. MENZIES,
Chairman of Commissioners of Salmon Ponds.

The Hon. the COLONIAL SECRETARY.

No. 20.

The Hon. the Colonial Secretary to His Excellency the Governor.

WELLINGTON, 1*st February*, 1878.

Ministers desire respectfully to inform his excellency the governor that the half million salmon ova which arrived by the mail steamer from San Francisco in November last have been successfully hatched and distributed to the different rivers of the colony, and that, by information that has reached the government from various directions, it has been demonstrated that owing to the extreme care with which the ova were packed in America the very satisfactory result of about 95 per cent. of the fish has been obtained.

In addition to the half million sent at the request of the government, an equal quantity has been sent to the various acclimatization societies in the colony, and this handsome gift of salmon ova has been made to the colony without charge, except cost of package and transit, by the

Fish Commission of the United States, under the direction of the Hon. Spencer F. Baird, as Chief Commissioner.

Ministers venture to think that so generous an action on the part of a foreign nation is worthy of being acknowledged in a special manner. They would, therefore, respectfully ask his excellency to bring the matter under the notice of Her Majesty's Government, through the secretary of state for the colonies, in the hope that Her Majesty's Government will permit a communication to be made to the Government of the United States of the thanks of the colony of New Zealand for the generous and valuable gift of a million salmon ova to the colony.

I have, &c.,

G. S. WHITMORE,
Colonial Secretary.

His Excellency the GOVERNOR.

No. 21.

The Hon. the Colonial Secretary to Mr. J. C. Firth.

WELLINGTON, *6th December*, 1877.

SIR: Referring to the correspondence which has taken place on the subject of the salmon ova supplied by the American Fish Commissioner, and which reached New Zealand by the November mail, I have the honor to inform you that communications have been received from all the acclimatization societies to which consignments were sent, stating that the importation seems likely to turn out perfectly successful.

It gives me great pleasure to offer you the thanks of the government for the readiness with which you undertook the arduous task of attending to the shipment on its arrival and for the judicious arrangements you made for the distribution of the portions assigned to southern societies. There can be no doubt that to those arrangements is largely attributable the success which has attended the experiment.

I have, &c.,

G. S. WHITMORE.

J. C. FIRTH, Esq., *Auckland.*

No. 22.

Mr. J. C. Firth to the Hon. the Colonial Secretary.

AUCKLAND, *11th February*, 1878.

SIR: I have the honor to thank you for your letter of 6th December last, conveying the thanks of the government to me for my services in distributing the salmon ova recently presented to this colony by the United States Government.

I have also to thank you for bringing under the notice of his excellency the governor the act of genuine international courtesy displayed by the Government of the United States in the noble gift of one million salmon ova to the colony of New Zealand, and for the information that his excellency has communicated with the secretary of state for the colonies, requesting that the Government of the United States may be thanked on behalf of this colony.

I have, &c.,

J. C. FIRTH.

The Hon. the COLONIAL SECRETARY.

No. 23.

Mr. R. J. Creighton to the Hon. the Colonial Secretary.

SAN FRANCISCO, CAL., 19*th January*, 1878.

SIR: I have the honor to inform you that I have consigned to your government from the United States Fish Commission, per favor of Professor Baird, 500,000 whitefish eggs, which I hope will arrive in good condition and hatch out. I inclose Professor Baird's letters and telegrams to me on this subject; also, telegrams from and to Mr. Clark, deputy fish commissioner at Northville, Mich. In further explanation, however, I may state that I wrote to Professor Baird on this subject several months ago, and expressed a desire of obtaining, if possible, another supply of whitefish eggs for the colony, in consequence of the failure of previous shipments. I explained to him the geographical position of the leading settlements, and the risk of failure in distributing the ova on arrival about midsummer along such an extended seaboard, and he promised that the next consignment would be left to my discretion in that regard.

Accordingly I have written to J. C. Firth, esq., president of the Auckland Acclimatization Society, requesting him to take charge of at least 250,000 eggs, and hatch out the same in the breeding ponds at Auckland, from which stock the North Island lakes should be supplied. It is necessary that there should be running water. I should be gratified if, in addition to Lake Taupo and other lakes on the line of the Waikato, the Wairarapa could be speedily stocked with this valuable fish. The lesser lakes could be attended to subsequently.

I have likewise telegraphed to the Christchurch and Dunedin Acclimatization Societies, requesting them to put themselves in communication with you; but I am of the opinion that only these leading societies, and perhaps Nelson, should be supplied with eggs, and these only if, upon examination in Auckland, the eggs could fairly stand the journey. In any contingency, or if there should be a doubt of the eggs spoiling, I should recommend that the entire consignment should be hatched at Auckland, and the young fish thence distributed over the colony. But as there is always a reasonable feeling of pride in such matters, the societies named are entitled to the utmost consideration consistent with the preservation of this valuable contribution to the food fish of the colony. One hundred thousand eggs might be shipped to Canterbury, 100,000 to Dunedin, and 50,000 to Nelson. This would dispose of the entire shipment, which is in ten (50,000) boxes.

In this connection I have consulted several gentlemen experienced in the American fisheries, and they unhesitatingly place whitefish as the most valuable of all fresh-water fish, ranking as a food fish above all other varieties. They are prolific, grow to a large size, and are equally good for food fresh or salted. Should they be successfully acclimatized in New Zealand, the colony will derive an immense return for the small outlay incident to introducing them.

I have taken advantage of the refrigerator-box of the Auckland Society, in which the late consignment of salmon ova were shipped, and filled it with ice in lieu of the ship's ice-house, which Captain Dearbour has placed at my disposal for the whitefish. This will economize ice and give a more reasonable certainty of the consignment arriving safely. I may here state that Captain Dearbour, of the City of Sydney, takes a very deep interest in this work of acclimatization, and, I think, deserves some recognition by the government.

I have also consulted Mr. Redding, Fish Commissioner for California, from whom, and his Deputy, Mr. Woodbury, I have received every possible aid. I need not, however, encumber this communication by inclosing my correspondence with these gentlemen.

The fact that I received intimation of this shipment by telegram on the 5th instant compelled me to wire a message through by cable to prevent the possibility of the consignment failing for want of preparedness on arrival. It was addressed to the premier. As I was not in funds to meet this and other disbursements on account of the colony, I have drawn for the amount, as per vouchers and statement of account annexed, which please honor. I also inclose statement of account from Mr. Clark, to whom you will be good enough to remit the amount by return mail, apprising me of the fact. You will observe what Professor Baird states upon this subject—and I would respectfully suggest that the government convey to him an expression of their appreciation of the interest he has taken in the acclimatizing of food fish in New Zealand.

I have acted in this matter without instructions, but in the belief that my conduct will meet with your approval.

I would suggest, in conclusion, that the government in future would prevent risk of loss by apprizing me when they order fish eggs from the United States Commission. I had no knowledge of the last order for salmon until after the ship sailed, and it was by a mere accident that the entire consignment was not left behind.

I have, &c., ROBT. J. CREIGHTON.

The Hon. the COLONIAL SECRETARY,
Wellington.

[Inclosure 1 in No. 23.]

Professor Baird to Robert J. Creighton, Esq.

WASHINGTON, *5th January*, 1878.

SIR: Mindful of the desire of New Zealand to obtain an additional supply of whitefish eggs, I arranged with Mr. F. N. Clark, of Northville, Mich., for half a million, and to bring them forward to a proper stage for shipment. I am informed that the eggs are now ready, and he has been instructed to forward them to you at once. They are to be put up in ten packets of 50,000 each, so as to be more conveniently divided.

It may be well for you to confer with Mr. B. B. Redding, Commissioner for California, in regard to the proper treatment of these eggs. They are not quite so far advanced as those of last year.

Mr. Clark's charge for these eggs is $1 per thousand, or $500 for the lot, exclusive, I presume, of packing and expressage. If you have not this amount on hand you can collect it at your earliest convenience from the colony and send it direct to Mr. Clark. My own appropriation did not permit me to incur so large an expense during the present season.

It is possible that for greater security the eggs may be shipped in two lots at intervals of two or three days, so that if one is lost the other may not be.

Presuming that you have ample instructions from New Zealand as to the distribution of these eggs, and leaving it to you to attend to their specific assignment,

I have, &c.,

SPENCER F. BAIRD,
Commissioner.

R. J. CREIGHTON, Esq.,
Agent for New Zealand, San Francisco, Cal.

[Inclosure 2 in No. 23.]

Mr. R. J. Creighton to Professor Baird.

SAN FRANCISCO, *15th January*, 1878.

SIR: Accept my best thanks for your letters and telegrams, and the valuable contributions of food-fish for New Zealand which you have been good enough to make on behalf of the United States Fish Commission.

I have forwarded your letters and telegrams to the New Zealand Government, which will not fail to appreciate your kindness. Mr. Clark telegraphed me of the departure of the ova from Northville on the 11th, and I expect their arrival to-night or to-morrow. I have made arrangement for their shipment per City of Sydney, which sails for New Zealand and Australia on the 21st instant, and have apprised the government by cablegram of the consignment. Mr. Clark's bill for the eggs will be forwarded to the colony, and a remittance direct made by the government. I shall write to him to that effect. If I had been in funds, I should have had pleasure in paying the amount at once.

I am happy to say that the shipment of salmon ova arrived at its destination safely, and has proved a great success. I think New Zealand is now fully stocked with salmon, at least to such an extent as to render further shipment for some time to come unnecessary. Small parcels of eastern trout have been sent and are successful; but I attach greater importance to the acclimatization of whitefish than to all the others, as well from the delicacy of the flesh as from its commercial value. New Zealand is a country of lakes and rivers peculiarly adapted for it. I can only express a hope that this consignment may fare better than the consignment of last year.

I have, &c.,

ROBERT J. CREIGHTON.

Prof. S. F. BAIRD, *Washington.*

[Inclosure 3 in No. 23.]

Mr. Frank N. Clark to Mr. R. J. Creighton.

NORTHVILLE, MICH., *11th January,* 1878.

SIR: I have this day shipped you two crates (500,000) of whitefish eggs for your government, and telegraphed you to that effect. Please have your government report condition upon opening of the same, to me.

I have, &c.,

FRANK N. CLARK.

[Inclosure 4 in No. 23.]

Mr. R. J. Creighton to Mr. F. N. Clark.

SAN FRANCISCO, CAL., *20th January,* 1878.

SIR: I have pleasure in acknowledging the safe arrival of ten boxes whitefish eggs for New Zealand from your fish-hatching house. They arrived early Friday, and were opened and examined by Mr. Woodbury, foreman of the State hatching house, San Leandro. They are in good condition, and promise to arrive safely at their destination. I have had them packed in ice in the ice-chest of the mail steamship City of Sydney, which sails on the 21st. They will remain in ice all the voyage, and be hatched out prompt on arrival.

I regret that the consignment to the State Fish Commissioners of California and Nevada was valueless, as on opening them they were all found to be dead and stinking. They had been placed near the stove in transit; hence the total failure of the shipment. As it is impossible to freeze fish-eggs in the express car, owing to the fact that a stove is always kept alight, I should suggest that in future consignments instructions be given that they be kept as cool as possible. The instructions on the commissioner's crate not to let the eggs get below zero appears to have been literally followed. The sawdust packing was at blood-heat when opened by Mr. Woodbury. To the absense of this special instruction I attribute the safe arrival of the New Zealand consignment, and a parcel of trout from Wisconsin.

I forward Professor Baird's letter to the New Zealand Government, in which he intimates that your charges for the eggs would be $1 per thousand and packing. You did not send me an account, but I presume this to be correct. The communication with the colony is monthly. I have requested the New Zealand Government to transmit the amount direct to you, and inform me of the fact. I likewise forward your letter to me with a request that the government should report the condition of the eggs upon opening the same.

I can only express the hope that the consignment may àrrive at its destination in as prime condition as it leaves San Francisco.

I have, &c.,

ROBT. J. CREIGHTON.

F. N. CLARK, Esq.,
United States Fish Commissioner, Northville, Mich.

No. 24.

Mr. R. J. Creighton to the Hön. the Colonial Secretary.

SAN FRANCISCO, *January* 20, 1878.

SIR: I have the honor to state, in reference to my previous letter, that I have had a conversation with the members of the State Fish Commission, and learned several facts of great practical value in reference to the propagation of whitefish, which I have embodied in a letter to Mr. Firth, of Auckland, in the belief that the Auckland Acclimatization Society will have the task of hatching out the bulk, if not all, the whitefish eggs. I am unable to copy the letter in time for this mail. Should the suggestions given therein be acted upon, I have no doubt of the success of the experiment.

I have further to request that you will cause the request in Mr. Clark's letter to be attended to. As Mr. Clark did not send any statement of account, I infer that the $1 per 1,000 mentioned by Professor Baird covers cost of package. It may not be the case, however. If so, $500 is due the Fish Commission at Northville, and should be remitted. I have sent two tons of ice—not three, as I originally intended. I think two tons will be ample. I may mention that I received very great assistance from Mr. Woodbury, who came a long distance twice in very inclement weather, to examine and repack, after drenching the eggs with water at proper temperature. I should be pleased if the government would authorize me to thank him for his gratuitous help.

The accompanying telegrams and correspondence give the history of the transaction. It will be observed from my reply to Mr. Clark that the New Zealand shipment was fortunate in not sharing the same fate as those consigned to the State Fish Commissioners of California at Nevada, which perished by the way.

I have, &c.,

ROBT. J. CREIGHTON.

The Hon. the COLONIAL SECRETARY,
Wellington, N. Z.

No. 25.

Mr. J. C. Firth to the Under Secretary.

[Telegram.]

AUCKLAND, *February* 15, 1878.

Mail steamer arrived last night at seven o'clock. I shipped on board Hawea eight boxes containing your hundred thousand whitefish ova—packed ice in two insulating chests with hundred weight ice in reserve. Hawea cleared wharf at half-past eight. Owing to having no information of dimensions of ova boxes, I could not pack the remaining two boxes containing one hundred thousand ova. These I forward per Rotorua. Creighton sends full instructions, which I will wire you to-day for information of Southern Society. Creighton's exertions well deserve the thanks of the government.

J. C. FIRTH.

G. S. COOPER, Esq.,
Under Secretary.

No. 26.

Mr. J. C. Firth to the Under-Secretary.

AUCKLAND, *April* 19, 1878.

SIR: I have this day forwarded one box whitefish ova said to contain 50,000 ova, properly packed in ice in insulating box, and one box containing ice in reserve. I inclose Mr. Creighton's instructions. Having fully acquainted you of all matters relating to this shipment of whitefish ova, it is not necessary for me to enter into any recapitulations. My account for cost incurred will be forwarded to you shortly.

I have, &c.,

J. C. FIRTH.

G. S. COOPER, Esq.,
Under-Secretary, Wellington.

(NOTE.—This box was forwarded from Wellington to A. M. Johnson, Christchurch, on 22d April.—J. H.)

[Inclosure in No. 26.]

Mr. Creighton to Mr. J. C. Firth.

SAN FRANCISCO, 20*th January*, 1878.

MY DEAR SIR: Since I wrote to you *re* whitefish, as per inclosure, I have learned some facts which are of interest relative to the artificial hatching of them, from the State Fish Commissioner (Mr. Redding), and the foreman (Mr. Woodbury), which you should know.

1st. Mr. Redding declares that it is almost essential that they should be hatched out at the first point of landing, owing to their delicacy. They will thrive anywhere if the water is deep enough, their food being small crustacea adhering to rocks in fresh water lakes, having a current running through them. They should have a sandy and gravelly bottom.

2d. They are much more difficult to manage than salmon, and, until recently, little was known of their habits. They lose their sacks in ten days at a temperature of 35°, and earlier at a higher temperature. It will be necessary to feed them three days afterwards, or perhaps earlier, if they are to be transported any distance. The Fish Commissioners of Wisconsin discovered this year that whitefish could be fed with blood for an indefinite period, and in the San Leandro hatching establishment, and at Lake Chabot in this State, the same experiment has been tried with success. Mr. Woodbury, therefore, suggests that you keep twenty of the fish in the hatching trough and feed them with blood, which can be squirted into the water with a syringe and thoroughly mixed. This would serve a double purpose. It would establish as a fact what is now experiment, that whitefish may be fed upon coagulated blood, and also give you a permanent stock for purposes of spawning, by which your society might derive no little profit. The Fish Commissioners here are very anxious in regard to this matter, and I would be glad if you could give it a fair trial and report the result. As fish culture is now becoming a leading industry, the economic side of the question will readily suggest itself to your mind.

3d. Whitefish, as soon as hatched out, rise and swim, unlike trout and salmon, which lie dormant. The little fellows are, therefore, carried down the trough with the current, and, unless fine wire screens are placed across it to intercept them, they are almost certain to be lost. It was in this way, I suspect, the Christchurch Society lost their whitefish, and not by a fresh during the night, as reported. No. 18 mesh (eighteen) will keep them in. They should have as much back-water as possible to swim in. In ten days, as I have said, they lose their sack, at a temperature of 35°, but, as they may lose it earlier, it is necessary that a register of the daily temperature of the water be kept, and food be furnished as above described.

4th. In the interest of science and acclimatization, should any portion of these eggs be sent south, I have to request that you communicate these facts to the persons in charge of them for their guidance. One way and other I have written a decent volume in this connection, and cannot possibly duplicate or quadruple these notes, which are in the rough. I have not written on this subject to the government, which must depend upon your society and similar bodies for the propagation and distribution of the whitefish.

I may remark here that the acclimatization of whitefish is in its infancy, and much has yet to be learned regarding it. It was thought, less than four years ago, that the eggs could not be sent across this continent. Several parcels failed, but at length a few were hatched and placed in Lake Tahoe, in the north. This was less than three years ago, and now the fishes which come to the sandy, pebbly banks on the Californian side of the lake are being netted and sent to Virginia City Market. They spawn, it is believed, the third year. Last year ten men and two teams were employed by the lake commissioners to cut a road several miles through the snow to place whitefish in another Northern California lake, and Lake Tulare in the south, warmer than Taupo, and about as large, has been stocked. The entire shipment of whitefish ova for California and Nevada, from Northville, Mich., packed precisely as those for New Zealand by Mr. Clark, and coming in the same car, were spoiled in transit. On being opened by Mr. Woodbury and myself they stank and were putrid. They had been placed near the stove by the express agents to prevent their freezing. Ours had been less considerately treated, and arrived sound and lively, as I had proof, every box having been opened and examined by Mr. Woodbury in my presence. We then ascertained their temperature, and gave them a drenching with water at a similar heat; screwed them up, reversed their position, placed them upon and surrounded them with ice in the Pacific Company's ice-houses. I telegraphed to Mr. Woodbury, and brought him twice from a considerable distance, by road and rail, in extremely wet weather, to assist me, and as it was a labor of love, I am anxious that he should, at least, have honorable mention. I should also remark that Woodbury has invented a hatching basket, in which 30,000 salmon eggs may be hatched with certainty. It occupies about two feet square, and would, I think, be a great assistance to you. I don't know the price, but it is trifling, and I thought I would mention it to you. If I can get one by next steamer, I will send it down.

Perhaps it would not be trespassing too much upon your kindness to ask the secretary of your society to make copies of this letter, or so much of it as may be necessary for their guidance, and forward one to the Christchurch, Dunedin, and Nelson societies; or send one to the government requesting them to communicate the same to those bodies.

I dare say I have nearly wearied you, but I know your enthusiastic love for acclimatizing such natural products as animals and fishes as may be useful to man, and therefore presume upon your time and patience. I forgot to say that whitefish take bait. They should be closely protected for, *at least, four years.* The wire screen referred to in paragraph 3 should be higher than the water to prevent loss of fish by overflow.

I have, &c.,

ROBT. J. CREIGHTON.

No. 27.

Mr. J. C. Firth to the Under-Secretary.

[Telegram.]

AUCKLAND, 16*th February*, 1878.

Since writing last I find I can push on preparations at the hatching-house, and will therefore take charge of one box. The other goes on by Rotorua.

J. C. FIRTH.

G. S. COOPER,
Under-Secretary.

No. 28.

Mr. J. C. Firth to the Under-Secretary.

AUCKLAND, 18*th February*, 1878.

SIR: Whitefish ova turned out very badly in the box you wished me to take charge of. All destroyed but thirty. Some of these died in hatching, others died soon after. Two fish living; eight ova yet to hatch. Cause of destruction, too many in one box and too much compression. Shall I forward the second box or open it here?

J. C. FIRTH.

G. S. COOPER, Esq.,
Under-Secretary.

No. 29.

The Hon. the Colonial Secretary to James Hector, M. D.

WELLINGTON, 15*th February*, 1878.

SIR: As you are already aware, a shipment of 250,000 whitefish ova sent from San Francisco by the United States Fishery Commission has arrived by the City of Sydney at Auckland, and has been transshipped with the mail on board the Hawea.

I should be much obliged if you would hold yourself in readiness to take charge of the ova on arrival here, and to proceed with them to the Bluff, and superintend their deposition in Lake Te Anau, taking with you, if necessary, an assistant from the staff of the museum.

The necessary instructions have been sent to the railway officers at Invercargill to co-operate with you, and rendering every assistance in the transport of the ova.

I have, &c.,

G. S. WHITMORE.

JAMES HECTOR, M. D., F. R. S., C. M. G.,
&c., &c., &c.

No. 30.

James Hector, Esq., M. D., to the Hon. the Colonial Secretary.

WELLINGTON, *March* 5, 1878.

SIR: I have the honor to report that, in accordance with your instructions, I have distributed the cases of whitefish ova received by the last San Francisco mail in the following manner:

Eight boxes, each containing 50,000 ova, were received in Wellington by the steamship Hawea on the 19th ultimo, packed in two large ice-chests, two boxes having been left in Auckland. The four ova-boxes half filled each chest, the space above being filled with broken ice and non-conducting pads. The chests stood on the fore-hatch, which is a convenient and safe position, but liable to the objection that the ova-boxes have to be moved at every port, and that they might be influenced by the vibration of the steam-winch.

At Lyttelton one chest was opened, and two of the small ova-boxes were left with Mr. G. S. Farr, honorary secretary to the Christchurch Acclimatization Society. I should state that one of these boxes had the cover loose. The space in that chest was filled up with ice and blanketing, and at Port Chalmers it was delivered, with the two remaining ova-boxes, to Mr. Arthur, of the Otago Acclimatization Society, with instructions to hand one of them to Mr. Connell, or his agent, for the Oamaru Acclimatization Society, on application.

The other chest and the spare ice, of which I got a fresh supply at Dunedin, were then transshipped to the steamship Wanganui, the sailing of which had been delayed twenty-four hours through the liberality of the owners—Messrs. Houghton & Co. Notice having been previously given, a special train was awaiting my arrival at the Bluff, but the steamer being later than was expected there was a little delay at Invercargill, so that it was not until 1 o'clock p. m. that we reached the Elbow.

The two chests, one containing the spare ice, and the other the ova, weighing about 600 pounds, were transferred to an American wagon with leather braces, and, having covered them with blankets and our tent, a start was made at 2.30 p. m.

The arrangements for the conveyance of the ova from the Elbow to Lake Te Anau, upon which the success of the experiment so much depended, had been made by Captain Hankinson with great judgment.

Traveling at about 4 miles an hour, by sundown we reached Centre Hill Station, and halted to rest two hours, until the moon rose. At 11 p. m. we again started, guided by Mr. Connor—the road, and especially the fords, being difficult to find in the dark. By daylight the first ford of the Mararoa River was reached, and we again halted for an hour, and repacked the chest containing the ova, filling it up with all the ice that was left, and leaving the spare ice-chest, and so lightening the load. At 11 a. m., on the 23d, we arrived at Messrs. Hankinson's Station and obtained fresh horses, and by 3 p. m. the most difficult part of the road, which is that crossing the mountains bounding the east side of the lake, had been overcome, and the journey safely accomplished. By previous arrangement the hatching-troughs had been prepared by Mr. F. Hankinson, so that with his assistance no time was lost in unpacking the ova, and by 6 p. m. the operation was completed and the result of the experiment ascertained. I regret to say that this was not very satisfactory, as out of the four boxes of ova three were almost completely destroyed by the growth of white fungus, and the young fish, which had evidently been hatched out for some time, were reduced to a pulpy jelly. In the fourth box, in which there was only a slight growth of fungus, a considerable number of the ova were found in sound condition, and hatched out rapidly as they were transferred to the trough. The trough was not placed actually in the lake, but in a small stream fed by a spring close to the shore, the temperature of the water being a little below 50° Fahr. After completing the arrangements I returned to Messrs. Hankinson's Station, leaving Mr. Burton, taxidermist to the Colonial Museum, in charge of the young fish, with instructions to camp beside them, and tend them until they were sufficiently advanced to turn out in the lake.

I should state that the supply of ice proved to be quite sufficient, more than 50 pounds being left in the ice-chest at the end of the journey.

The reason of the failure of the ova was evidently defective treatment during some part of the long journey from Lake Michigan. Each box contained four layers of eggs placed between layers of gauze-net and moss. The ova-boxes, which were 11 inches square by 5 inches deep, had several holes bored in both top and bottom, and the only sound ova were in the top layers, and out of reach of these holes. I may state that this was also found to be the case in one out of the two boxes left at Dunedin, the other being a total failure.

At Christchurch, also, a few sound ova were found in a similar position in one of the boxes.

I am inclined to think that the ova-boxes, when placed in the ice-chests, should have been surrounded with ice instead of having it only on the top, as, if great care was not taken to cool the ice-chests thoroughly before the ova-boxes were placed in them, it is obvious that the temperature of the ova-boxes would be at first considerably raised, while at the same time the water of the melting ice would drip through the holes and saturate the contents, and so cause the ova to hatch.

The white fungus growth which was found so abundantly in most of the boxes seemed to spring from that portion of the moss in contact with the layers of dead fish; but one of the boxes was nearly free from it, except in the bottom layers, and in that the moss was green and springy. It is probable, therefore, that the decay of the moss and the growth of the fungus commenced after the hatching out and death of the young fish, and was not the cause of the failure. From the circumstances that the other boxes which were opened at Christchurch and Dunedin were in the same condition, it is to be concluded that the failure of those taken to the Te Anau Lake was not due to the long and rough land journey to which they were subjected, so that with the experience now gained, and with some modification of the method adopted in packing the ova-boxes, so that they may be thoroughly *surrounded* with ice, I feel confident that future consignments can be safely conveyed to our large Alpine lakes, where they have the best chance of thriving. The ova that escaped destruction were those which were protected from the drip of the melting ice, and were therefore comparatively dry, and in such a position that they were at the same time kept at a low temperature by the ice resting immediately above them. It did not appear to me that too much moss had been placed in the boxes, which has been suggested as a reason for the failure, but, when the fungus had grown, the moss was necessarily crushed into less space and formed into a sodden mass. At the same time I would recommend that in future experiments the gauze on which the eggs are spread should be stretched on light frames supported at proper intervals by intermediate corner pieces; but these and other suggestions I will defer for another report upon the subject, after conferring with Mr. Firth at Auckland.

The experiment on this occasion has been so far successful that a few hundred fish, at least, will be turned out in Te Anau Lake, and I have recommended that the fish hatched in Dunedin, of which there are about a thousand, should be sent to the Wanaka Lake, and the small number (about a dozen) obtained at Christchurch to Lake Coleridge.

I have, &c.,

JAMES HECTOR.

The Hon. the COLONIAL SECRETARY,
Wellington.

[Inclosure 1 in No. 30.]

Mr. S. Herbert Cox to Dr. Hector.

TE ANAU, *February* 20, 1878.

SIR: You will be pleased to hear that the whitefish are doing very well. They are all hatched out and are feeding well on the blood which they are having given them.

But very few have died, and, if cold be an essential to their existence, it has been cold enough to-day for almost anything. Burton says he would be afraid to turn the fish into the lagoon now, as the distance is rather far, so they will, I presume, be let loose in the lake about Saturday, if it is calm enough.

I have, &c., S. HERBERT COX.

[Inclosure 2 in No. 30.]

Mr. W. Arthur to Dr. Hector.

ACCLIMATIZATION SOCIETY,
Dunedin, 10th July, 1878.

SIR: You will be sorry to hear that our American whitefish experiment has failed. I suppose we had about 1,000 young fish which throve very well at the breeding-ponds. The last I know of them is that Deans started with the whole lot for the Wanaka before they had reached that age and size which, in conversation with you, we all agreed to be most prudent before turning them out. He got as far as the Teviot, but they had nearly all died or escaped during the night into a creek where the cans were put. Both Maitland and I knew nothing about it until Deans returned, or we should certainly never have sanctioned so rash a step. I hope those in the Te Anau will get on better, and be the means of stocking our deep lakes.

I have, &c., W. ARTHUR, *Secretary.*

No. 31.

Mr. A. M. Johnson to the Hon. the Minister for Public Works.

AMERICAN WHITEFISH.

TROUTDALE FARM, OPAWA,
Christchurch, 6th February, 1878.

SIR: If you should receive any whitefish ova, will you kindly consider my application for a portion.

I have every facility for fish culture, and have this season hatched out about 70,000 ova (English trout and American salmon).

My establishment being a private one, I am not in receipt of public moneys in the shape of subscriptions, licenses, and fines, like the various acclimatization societies, although I have to compete with them in the sale of young fish for stocking purposes; therefore, I trust you will see that I have an equal, if not a greater, claim on your consideration.

I may also add that the English brown trout, English perch, and the American brook-trout (*Salmo fontinalis*) were first introduced into New Zealand at my expense.

I have, &c.,

A. M. JOHNSON.

The Hon. the MINISTER FOR PUBLIC WORKS.

No. 32.

Mr. A. M. Johnson to the Hon. the Colonial Secretary.

TROUTDALE FARM, OPAWA,
Christchurch, 23*d April*, 1878.

SIR: The whitefish ova received to-day by the Rotorua I regret to report as all hopelessly bad, with the exception of three.

From the appearance of the ova the failure most probably arises from the eggs having been obtained too long, or kept without ice before the starting of the steamer.

It is quite possible that a further supply might be obtained this season, if instructions are sent by the outgoing mail so that the order could reach the collector direct from San Francisco. The actual cost of eggs in America is not much. I have had out many lots of trout ova, and seldom paid more than $4 per thousand.

With a view to increasing the chances of success in future similar shipments, I would suggest that the lids of the ova boxes be screwed down instead of nailed, a larger number of holes made in the lid, and the inside of the boxes slightly burnt.

Again thanking you for your kindness in forwarding me the ova,

I have, &c.,

A. M. JOHNSON.

The Hon. the COLONIAL SECRETARY, WELLINGTON.

No. 33.

Sir J. Cracroft Wilson to Mr. S. C. Farr.

CASHMERE, 2*d April*, 1878.

SIR: I have the honor to report the following circumstances in connection with the fry of the whitefish:

On the 26th of February you reported that you had opened the two boxes supposed to contain 20,000 whitefish ova, a present from the United States Fish Commission to the Government of New Zealand; that there were a few of the ova hatching out, but that the majority of them had hatched on the voyage from San Francisco or Auckland, the fry from which were dead. Finally, about 20 eggs produced fry in the hatching-boxes of the society. Two of these died previous to Sunday, the 17th of March. On that day, in consequence of a hot wind from the northwest raising the temperature of the water to 62° Fahr., six more died, and it was evident the remaining twelve would not survive such hot weather.

On Wednesday afternoon, the 20th of March, I started, according to promise, by the 4.20 p. m. train for Coalgate station, taking with me an American vehicle, a pair of horses, two servants, one small fish-can, with an aerating ball and tube, containing the twelve surviving fish, two large fish-cans filled with fresh artesian-well water, a four-gallon block-tin bucket, and 6 packets, each containing ten ounces muriate of ammonia, and 6 packets, each containing ten ounces of niter, prepared for the trip by Dr. Macdonald, of Lyttelton. The whole party was franked by the general government, and the thanks of the society are due to all the railway authorities in Christchurch.

Having taken up my position in the guard's van with one servant, nine parts of water were placed in the four-gallon bucket, and three packets of muriate of ammonia and three packets of niter being added, the mixture was well stirred. The thermometer was then placed in it, and it fell, in a short space of time, to 34° Fahr. The thermometer having been withdrawn, the can containing the fish was placed in the bucket. The servant kept continually aerating the water in the fish-can, and thus, without changing the water or interfering with the mixture, we arrived at Coalgate station at 7.15 p. m., the temperature of the mixture during the journey never having exceeded 38°.

After giving the fry fresh water and preparing the freezing mixture as before, four of us started in the American trap, Mr. James McIlraith having kindly volunteered to accompany and show me the new road, which skirts the swamp known by the name of Dr. Turnbull. We, however, lost our way, and nearly two hours of our valuable time. As we were approaching the hotel at Windwhistle, we were joined, according to appointment, by Mr. F. E. Upton, who, on horseback, piloted us to Snowden, the residence of Mr. W. Gerard, where we arrived between eleven and twelve o'clock, midnight.

Having partaken of some refreshments and given the fry fresh water, we were supplied with another pair of horses by Mr. Gerard. Mr. Upton having taken Mr. McIlraith's place in the vehicle, we continued our journey towards Mr. Cotton's house, on the border of Lake Coleridge, which we reached about 3 o'clock a. m.

Owing to a cold northwesterly wind which had prevailed all night, there was a considerable surf rolling onto the shores of the lake. We therefore thought it advisable to liberate the fry in a small rivulet about two hundred yards from the lake. Previously to liberating them we took the can into a stable, lighted a candle, and satisfied ourselves that not one of them was dead or injured. We then retraced our steps to Snowden, and took possession of our beds about 5 a. m. Thursday, 21st March.

Lamentable as is the outcome of this handsome present from the American Fish Commission, I congratulate the Canterbury Acclitimasation Society on the fact that nothing was left undone to insure success. Ice was prepared according to Dr. Hector's instructions, and taken by you on board the steamer which conveyed the boxes of ova to Canterbury, but it is evident that the ice, *en route* from San Francisco or Auckland, must have failed, and the ova hatched out only to die.

In conclusion, I cannot help making a few observations. The fry of the American whitefish are evidently more delicate than the fry of any other fish known to me, and I am persuaded that not a fry would have reached Lake Coleridge alive had it not been for the freezing mixtures and the great cold we experienced after reaching Windwhistle.

It is greatly to be desired that all the parties to whom the boxes of this consignment of ova were trusted should write detailed reports as to results, which reports, if printed and circulated, might help us to discover some means of rearing to maturity these far-famed fish.

One thing is very certain, that they cannot succeed in any place in New Zealand not situated in the mountains.

Trusting that the council will admit that I have faithfully fulfilled the promises which I made to them respecting these fish, and that the Government of New Zealand will be satisfied with the endeavors of our society,

I have, &c.,

J. CRACROFT WILSON,
Chairman.

S. C. FARR, Esq.,
Secretary Canterbury Acclimatization Society, Christchurch.

No. 34.

Mr. R. J. Creighton to the Hon. the Colonial Secretary.

SAN FRANCISCO, CAL., 18*th February*, 1878.

SIR: I inclose herewith letter from Mr. Clark, of Michigan, relative to the shipment of whitefish eggs per City of Sydney, for New Zealand. From it I gather that the charge for the eggs and packing, as per Professor Baird's letter, will be $500, at $1 per thousand, which amount you will be good enough to cause to be forwarded to Mr. Clark.

I hope the consignment arrived in good order, and has been hatched out and distributed successfully. Whitefish is more highly esteemed than salmon where it is known. It is difficult to acclimatize it, but should the colony succeed, it will add a valuable food fish to its other attractions for settlement, and solve a difficult problem for scientists.

Professor Baird has written to me for the history of salmon acclimatization in New Zealand, so far as the California salmon is concerned, and I was only able to make a very fragmentary report in relation to the last shipment. He is solicitous of obtaining full information for his annual report to Congress, and lays great stress upon the New Zealand experiments, similar shipments to Germany having entirely failed. I have, therefore, to request that you will cause a report to be forwarded to me, supplemental to that made by me, showing the date of arrival of the eggs at the several ports of the colony; by what conveyance, and the time occupied in transshipping and handling them; how packed during the coasting voyage; proportion of eggs hatched out in each province, and how the young fish were distributed. I approximated the time on the coasting voyage, but I was ignorant of the success, if any, except in Auckland, the newspapers of which contained a general statement that the eggs distributed by Mr. Firth had hatched out. I trust this information will be supplied by return mail. It may not be too late for Professor Baird's report, and will complete my otherwise imperfect one.

I observe by the London Times, that Sir Julius Vogel sent out a consignment of salmon ova from England, per steamer Chimborazo, via Melbourne.

The result of this experiment will be of great interest to the United States Fish Commission, and to the California State Fish Commission, to both of which New Zealand is under great and lasting obligations.

I would, therefore, esteem it a favor if you would advise me, in due course, of the success had in introducing British salmon, and the relative cost of the two sources of acclimatization.

As the Sacramento salmon may now be said to be introduced permanently into New Zealand, details regarding its habits, &c., will be interesting and of value to the colony. I therefore append extracts from the biennial report of the California fisheries commissioners, presented to the State legislature recently, bearing upon the point. It will be seen that it possesses many special advantages over the British salmon, and for commercial and food purposes is decidedly superior. On economic grounds alone, the acclimatization of this excellent food fish is an event of very great importance. I likewise extract the passages relating to whitefish and catfish (the latter introduced, I understand, by Mr. Thomas Russel, C. M. G).

I have, &c.,

R. J. CREIGHTON.

[Inclosure in No. 34.]

Extract from Biennial Report of the California State Fisheries Commissioners, 1876–'77.

SALMON (*Salmo quinnat*).

1. Before the discovery of the gold mines in California, nearly all of the tributaries of the Sacramento and San Joaquin Rivers were the spawning beds of the salmon. Soon after mining commenced the sediment deposited by gold washing covered the gravel bottoms of the streams. The fish found no proper place on which to deposit its eggs, and after three or four years became extinct in those tributaries. The instinct of the fish leads it to return from the ocean to the stream in which it was born for the purposes of reproduction. If this place, for any reason, is rendered unfit, it will not seek a new and appropriate place. In 1850

the salmon resorted in vast numbers to the Feather, Yuba, American, Mokolumne, and Tuolumne Rivers for purposes of spawning, and many places, such as Salmon Falls, on the American, were named from the abundance of these fish. On the Yuba River, as late as 1853, the miners obtained a large supply of food from this source. At the present time no salmon enter these streams. It would be safe to estimate that one-half the streams in this State to which salmon formerly resorted for spawning, have, for this purpose, been destroyed by mining. As mining is the more important industry, of course, for this evil there is no remedy other than by artificial means to increase the supply in those tributaries that are still the resort of these fish. The principal spawning grounds remaining are the McCloud, Klamath, Little Sacramento, and Pit Rivers in the northern part of the State, and the San Joaquin and Merced in the southern. The short streams entering into the ocean from the coast range of mountains from Point Conception, in latitude 34° 20′ north to the boundary of Oregon, are also spawning grounds for salmon. The fish of the coast streams deposit their eggs in January and February, during the winter rains, when the streams are full, while the salmon of the tributaries of the Sacramento and San Joaquin spawn in August and September, when the water is at its lowest stage. The salmon of the short coast rivers do not average as large as the Sacramento salmon, but they are probably the same fish, with habits modified to suit the streams to which they resort.

2. The *Salmo quinnat* readily adapts itself to a life in fresh water, and reproduces its kind where it has no opportunity to go to the ocean. When the dams were constructed on the small streams that go to make the reservoirs of San Andreas and Pillarcitos—which supply the city of San Francisco with water—as also when the dam was constructed on the San Leandro, to supply the city of Oakland, the young of the salmon that had spawned the year previous to the erection of these dams remained in the reservoirs and grew to weigh, frequently, as much as ten pounds; these reproduced until the reservoirs have been stocked. As the supply of fish increased the quantities of food lessened, so that the salmon have gradually decreased in weight until now, after nine years, they do not average more than two pounds. From the fact that, when food was in abundance, they grew to weigh from eight to twelve pounds, and that, as they increased in numbers, they averaged less in size, but still continued to spawn and produce young fish, it would seem that the Sacramento salmon may be successfully introduced into large lakes in the interior of the continent, where, in consequence of dams or other obstructions, they would be prevented from reaching the ocean. The history of this fish in these small reservoirs shows that all that is requisite for their successful increase is the abundant supply of food, to be found in large bodies of fresh water. Salmon, fully mature, weighing two pounds, and filled with ripe eggs, were taken, in September, 1877, in the waters of San Leandro reservoir. These fish were hatched in the

stream which supplies the reservoir, and by no possibility have ever been to the ocean. The San Leandro is a coast stream, not exceeding fifteen miles in length, and empties into the Bay of San Francisco. It contains water in the winter and spring, at which time, before the reservoir was constructed, the salmon sought its sources for the purpose of spawning. There was never sufficient water in the months of August or September to permit the fish to reach their spawning grounds. After the construction of the reservoir, large numbers of the salmon that came in from the ocean in January and February were caught at the foot of the dam and transported alive and placed in the reservoir above. The descendents of these fish thus detained in fresh water and not permitted to go to the ocean, have so far modified the habits of their ancestors that they now spawn in September, instead of in January and February. Inasmuch as these fish spawn in the McCloud, in the headwaters of the Sacramento, and at the sources of the San Joaquin, in the Sierra Nevada, in September, and in short coast range rivers in January and February, and as, when changed to other waters, their eggs ripen at a time when the conditions of their new homes are most favorable for reproduction, they show a plastic adaptability, looking to their future distribution, of much practical, as well as scientific, importance.

3. The statistics hereafter given of the temperature of the water through which the Sacramento and San Joaquin salmon pass to reach their spawning grounds, show that they swim for hundreds of miles through the second hottest valley in the United States, during the hottest portion of the year, where the mean temperature of the air is 92° Fahr., and of the water 75°. These statistics have been obtained from the record kept by the Central Pacific Railroad Company, and are for the months of August and September of the years 1875–'76–'77. They are of importance as showing that the Sacramento salmon will enter rivers for spawning purposes where the water is so warm that the eastern salmon (*Salmo salar*), if it were to meet it, would turn back to the ocean. They are also of importance as illustrating the probability that there are many streams on the Atlantic coast, from the Potomac to the Rio Grande, into which this fish could be successfully introduced.

4. Mr. Livingston Stone, deputy United States fish commissioner, in charge of the government hatching establishment on the McCloud River, reports officially that in his opinion, all of the salmon of that river die after depositing their spawn. This is possibly true; but it does not account for the fact that in the spawning season the McCloud contains grilse and fish evidently three, four, and five years old, unless we are to imagine that some salmon, after being hatched and going to the ocean, remain there two, three, or more years without returning to the parent stream for purposes of spawning. Beyond doubt the salmon that spawn in the coast streams go back to the ocean, as they are frequently taken in the lagoons at the mouths of these rivers on their return. Somewhere on the tributaries of the Sacramento or San Joaquin,

there are salmon that do not die after the act of spawning, for they are frequently taken in the nets of the fishermen in the brackish waters at Collinsville and Rio Vista on their return from their spawning grounds. If it were a fact that the Sacramento salmon so widely differed from other fish that it spawned but once and then died, it would detract from its value. This subject is one of importance, but at present the facts are so obscure that we have made considerable effort to obtain the opinions and the result of the observations of the men who are practically engaged in the taking of salmon in the Sacramento River.

5. The following, from the letter of a fisherman who has pursued the business of taking salmon for the San Francisco market during more than fifteen years, gives some facts and his theory, based on his observations. In reply to an inquiry on the subject, he says: "As to the return of the seed salmon to the sea after depositing the spawn, I am inclined to the opinion of Mr. Stone, so far as the greater part of the female fish is concerned. I think very few of these, but many, though not all of the males, return. I should judge that 5 per cent. of females and 20 per cent. of males might be an approximation. I express this opinion diffidently. It is based on the style of fish caught in the lower part of the river (from Sacramento to Collinsville). After about the 20th of September, of the fish then dropping down, the nets catch but few, for the reason that the net is drifting with the current and the fish are doing the same thing, and in consequence, as a rule, the two do not come together, and the greater part of the return fish escape. When the run is upward, the net drifts with the current, and the fish swim against it, and the rule is reversed. The percentage named above is not that of return fish caught, but of fish that I estimate may have returned, judging by the very few return fish that are caught. It is a very cloudy subject to all fishermen. I have heard, perhaps, a thousand discussions on the river, at all times of day and night, at the head of the 'drift,' among men of the largest experience—men right in the teeth of the business—men born to a boat and net, and grown gray and grizzled in their use—upon the point you raise, and the average conclusion always was that nobody quite knew how it was. Of one thing I am convinced, to wit, that return fish need no protection from the drifting gill net. Not one fish in ten could be caught in that way. No such thing as a run of salmon down the river ever occurs. The normal position of salmon is head to the current. Though drifting with the current, his head is toward it. In the light (or darkness) of these facts, you see how difficult it is to say, positively, what proportion of these fish that have delivered seed return to the ocean. No man can say positively that the mass do not return. That some return is beyond doubt of a reasonable nature. If they all perish, it is certain that many survive long enough to reach the fishing grounds lying in the bays nearest the ocean. But I fail to see why the value of the California salmon is affected by the fact (if it is a fact) that the

fish never spawn but once. I have a theory of the salmon of this river. It may not be scientific, but it is mine, and I can give reasons for it. It is this: The female salmon seldom or never spawns but once. The exceptions to the rule, if any, are few, and the second product of these exceptions is found in a salmon differing slightly from the mass of fish found in the river. A goodly, though not the larger, part of the male salmon that have assisted in reproduction return to the ocean, and 'live long and grow broad,' and return to the river many times. On their return these fish constitute that class far above the average size. They reach 30, 40, 50, and even a greater number of pounds in weight, while the average weight for which our meshes are sized is from 16 to 20 pounds. The female spawn is not ripe for delivery, nor the male fish sufficiently mature for milting, until they have made repeated trips between the ocean and the river. The yearly broods return periodically and in regular cycles; the youngest fishes arrive earliest in the season, which begins about the 1st of November, and do not penetrate far the first time. In the order of their birth, the other broods arrive and return to the sea until in August and September, the great seed run, consisting of mature fish, always on time, always urgent in their movements and purposes, passes up to the headwaters. Salmon of different ages are always coming in and going out to sea. The older the fish the longer his stay in fresh water. The younger the fish (after he once leaves for the ocean) the more of flirting about the bays and brackish water near the mouths of the river, with short excursions up the river. The foregoing is the outline of a theory, though it is derived from, and apparently justified by, known truths in the history of the Sacramento salmon during the last twenty years. I believe it to be correct; that is to say, that in any year representatives of the brood of any other year not yet extinct enter the river, and that not one-fifth of the fish that enter the river in any given year go to the headwaters that year, but that more than four-fifths return to the ocean, and, consequently, that of all the fish that come into the river each year, but one-fifth go to the headwaters for purposes of reproduction."

6. The habits of the Sacramento salmon, while on their spawning grounds in the McCloud River, have been closely observed by Deputy United States Fish Commissioner Livingston Stone, and the result of his investigations has been published by Congress in the report of the United States Fish Commissioner, Spencer F. Baird. But little is known of their habits while in the ocean. They probably feed on shoals not many miles from the shore. They are occasionally taken in the nets of fishermen in the ocean not far from Golden Gate. Many grilse, and a few mature fish, make their appearance in the Bay of San Francisco in December, and remain several weeks feeding upon smelts and other small fish. During this period thousands are taken with hook and bait on lines from the Oakland pier and other wharves. Many more are also taken in the nets of fishermen. After leaving the salt water

of the bay, they go to the brackish waters, where the currents of the Sacramento and San Joaquin meet the tide from the ocean. After entering the fresh water of the river they cease to feed. No food has ever been found in all the tens of thousands caught in the Sacramento. As it is of importance to obtain a knowledge of the habits of the salmon while it remains at the mouths of the rivers, playing back and forth between brackish and fresh water, before it makes its long and perilous journey to the head of the stream, we select from our correspondence extracts from a letter from Mr. Samuel N. Norton, of Rio Vista. Mr. Norton is a practical fisherman of many years' experience, and the record of his close observation is of much value. He says: "I will give you a synopsis of one year's trip with the salmon, showing the general habits of the fish in all years while remaining in or passing through that part of the Sacramento River lying between its mouths and the point where the Feather River empties into it. For this purpose the Georgian Slough, the Three-mile Slough around the head of Sherman Island, the San Joaquin River between these sloughs and the bay, and the Montezuma Slough leading into the northern arm of Suisun Bay from the Sacramento River, are considered as mouths of the river, with like functions and processes as the main trunk of the river. Indeed, some of the best fishing ground, at certain seasons, is found in the Montezuma, Three-mile, and San Joaquin. To commence with an anachronism, the spring run begins in the fall! In November and December a very few small (as fishermen use the word—say twelve or fourteen pounds each) bright salmon appear in the river, and if no rains occur, or only slight rains, an increase in their numbers is noticed, yet they are always very scarce in those months. There are never enough to half supply the local demand of the San Francisco and other home markets. At first, in November, we pick up occasionally on their return, the last dregs of the old seed run which occurred during August and September. These are usually male fish, very dark, ill-conditioned, lank-jawed, disconsolate looking fellows, who through misfortune, incompetency, or other cause—to me not more than presumable—seemed to have failed in their mission up the river, or to have fallen into disgrace. The last of these soon disappear. The bright ones are the *avant couriers* of the great spring run, which thus, as I said, begins in the fall. With the first heavy rains the fish that have penetrated the river recede, or, as we say, back down before the thick muddy stream, retreat to tide-water in the bays, and remain there reconnoitering and waiting a steady river current. Now is the time for good fishing in the bay and just in the mouths of the river. The fish are not very plentiful, but none being caught within the river proper, there is a great demand and great price against a small area of fishing ground, where all that had before penetrated the river are now concentrated. When the river becomes steady, that is, neither rising nor falling, the fish start up again, no matter how high the water may be, and by the varying moods of the river in sudden rise or fall, is the spring run mainly governed. Sudden rise or fall alike will check them. Thus it often happens that for many weeks the fish will be taken in numbers at Benicia and Collinsville, in smaller numbers at Rio Vista, and none at all farther up. Again, there have been seasons when a steady run commenced in the early part of January, and by an almost uniform rate of increase reached its culmination in May. But this is exceptional. The spring run may be stated as commencing in November and ending in July, and having its greatest strength in May. Under the most favorable conditions the months of November and December might be classed 'very scarce;' January and February, 'scarce;' March, 'not scarce;' April 'plenty;' May, 'very plenty;' June, 'not scarce;' July, 'scarce.' Under unfavorable conditions, November, December, January, and February would have almost none at all; March, 'scarce;' April, 'not scarce;' May, 'plenty;' June, 'scarce;' July, 'almost none at all.' In defining the terms here adopted, let them be applied to the product of the labor of two men with their boat and net per day: 'Almost none at all, would mean two fish per week; 'very scarce,' two fish per day; 'scarce,' six fish per day; 'not scarce,' eighteen per day; 'plenty,' thirty-six per day; 'very plenty,' seventy-two per day.' There are times in the height of the run when a greater number than is here named might be caught with ease; but these are exceptional. In the great run three years ago, three hundred salmon per day might be caught with ease; but in no other year, since the Anglo-American occupation, has there been such a run. It must not be understood that salmon can be caught at all times by fishing for them, even in the most limited numbers above stated. There are times when one could not be caught in a month, if life were at stake upon it. I only intend to give a fair idea of the average business. You will readily deduce from it that there are not more than two months, during the spring run, when fish can be caught in excess of the demand for home consumption. After the subsidence of the spring run, in July, they are often found in great numbers near the confluence of the Feather River with the Sacramento. They have a taste for variety, it would seem, and the marked difference between the cool, muddy water of the former, and the warmer, limpid, and clear stream of the latter, affords them great satisfaction. During the first half of August, the mature seed fish start for the spawning grounds. All along the line, from the ocean to the most advanced posts along the river, the word (if fishes have words—if not, then wag) is onward and upward. They are on business, and on time; they do not shy much, nor stop for trifles; they rush at a drifting gill-net determined to do or die, and, of course, generally die if the net is sound. The run of August and September, I have before described. As for the few belated fellows that are about in October, they might as well be caught as not—and so, my year is out."

7. At the time our last report was made, Mr. Charles Crocker had

requested us to cause to be hatched, at his expense, and placed in streams that do not reach the ocean, a half million of Sacramento salmon. One-half of these we determined to put in Kern River, which empties into Buena Vista and Tulare Lakes, and the other half in the Truckee River, which empties into Pyramid Lake, in the State of Nevada. The quarter of a million of eggs sent to Kern River, where their hatching was to be completed, unfortunately were lost. At the point of the river selected for hatching the water contains too much alkali, it is supposed, and all the eggs died within twenty-four hours from the time they were placed in the hatching troughs. The other quarter of a million, sent to the Truckee, were successfully hatched out and turned into that stream. They will go to Pyramid Lake the present season. They should return during the summer of 1878, and we are confident they will be taken in the Truckee weighing five or six pounds. Pyramid Lake is a body of water forty miles long, and averaging ten miles in width, and has no outlet. It contains an abundance of food. This experiment will demonstrate how large the Sacramento salmon will grow, with plenty of food, when confined entirely to fresh water.

8. Since the organization of the commission, we have caused to be hatched and placed in the streams of this State 8,350,000 young salmon. These include 1,000,000 paid for in 1875, and presented by Ex-Governor Leland Stanford. As the salmon is our most important food-fish, we deemed it of the most importance to keep up the supply. The numbers of fishermen are yearly increasing, as are also the numbers of persons who are consuming the fish. As railroad facilities are increased, and reach new points, the market becomes extended. The sea-lions and seals at the outlet of the bay, being preserved and protected by law, are also increasing. They now number thousands, and as each requires from ten to thirty pounds of fish daily, it was a serious question whether we could keep up the supply by the addition of 2½ millions artificially hatched each year. Since our last report, a salmon "cannery" has been established on the Sacramento, at Collinsville, and another opposite the city of Sacramento. This Collinsville canning establishment reports as having canned this year 8,542 cases, of four dozen cans in a case, equivalent to 34,168 fish, weighing 546,688 pounds.

Under the enlightened superintendence of Prof. Spencer F. Baird, United States Fish Commissioner, the Sacramento salmon is being widely distributed to streams throughout the United States. The government establishment on the McCloud River annually hatches from six to ten million eggs. These are distributed to all States having appropriate waters, whose legislatures have appointed fish commissioners. From this source the State of California has received, as a donation, a half million fish each year since 1874. In addition, we have expended a large part of our appropriation annually, in payment for the hatching of one or two million young fish, which, through the kindness of Professor Baird, have been furnished at the actual cost of hatching. The introduction of more than 8,000,000 young salmon into the headwaters of the Sacramento, since the organization of the commission, in addition to the natural increase, has had the effect to keep up the supply, and to reduce the local market price of these fish. It is reported that the "cannery" at Collinsville has purchased all the salmon it could consume during the past season at from 25 to 40 cents each.

9. Over-fishing, the absence of any close season, and no effort at artificial increase, has at last had an effect on the salmon of the Columbia River, in Oregon, and complaint is made that this river, once thought inexhaustible, has begun to fail in its accustomed supply. This decrease has been so marked during the season, that the "canners" have been compelled to pay from 30 to 50 cents each for salmon. In the absence of legislation, the canning companies on this river have subscribed $20,000, which has been placed under the control of Mr. Livingston Stone, deputy United States Fish Commissioner, to be expended in artificial hatching, and restocking that stream. Fortunately, intelligent legislation in California made provision for continuing the supply of fish in the Sacramento before there was any marked decrease by over-fishing. It is not disputed that the salmon were more numerous in the Sacramento before their spawning grounds on the American, Yuba, Feather, and other rivers had been destroyed by mining. After the fish were destroyed in these tributaries, the supply of the State had to come from the other tributaries of the Sacramento and San Joaquin, on which there was no mining, and these latter streams furnished the normal supply. Before these became exhausted, the natural increase was supplemented by artificial hatching.

10. In this connection a fact, of much practical as well as scientific importance, may be stated as showing the advantages in numbers to be obtained by artificial hatching in comparison with the increase, by natural methods. In 1876, Mr. Myron Green, foreman for Mr. Livingston Stone, United States deputy fish commissioner, at the McCloud River, having observed in the river a favorite gravel bed where many salmon were depositing their eggs, carefully dug up the gravel and several thousand eggs. He separated the eggs from the gravel, and placed the former, after counting them, in the hatching-boxes. After twenty-four hours, he found large numbers of these eggs turning white, showing that the milt had failed to come in contact with the eggs. After throwing out all the eggs not found to be fecund, there were left 8 per cent. of the whole number gathered, which were found to be fertile. When the eggs and milt are artificially brought in contact out of the water, it would be carelessness or inexperience that would prevent 95 per cent. of the eggs from being fertilized.

11. The following tables will show the number and weight of salmon transported on the railroads and steamboats from the Sacramento and San Joaquin Rivers to the cities of San Francisco and Stockton, from points on the river below the cities of Sacramento and Stockton, from

1st November, 1874, to 1st August, 1876; and from 1st November, 1876, to 1st August, 1877. They do not include the catch of the fisheries at Tehama or near the mouth of the Feather River, nor do they include the fish taken on the upper waters of the Sacramento and San Joaquin, nor the salmon brought to market by fishermen in their own boats; therefore, to the totals should be added, at least, 25 per cent., to show an approximation of the actual catch.

12. In our last report, after adding 25 per cent. to the statements of the catch which we obtained, we showed the total weight as transported from the same places, from 1st November, 1874, to 1st August, 1875, to be 5,098,781 pounds. Adding the same percentage to the totals in the above tables, and they show the catch from 1st November, 1875, to 1st August, 1876, to be 5,311,423 pounds; and from 1st November, 1876, to 1st August, 1877, 6,493,563 pounds.

13. This shows a gain of more than a million pounds in the legal catch over any year since the organization of the commission, and may be ascribed to the fact that our waters are now beginning to feel the beneficial effects of the millions of salmon hatched artificially and turned into the headwaters. We have no means of ascertaining the weight of fish taken out of season, but estimate that between 1st August and 1st November of this year, not less than 2,000,000 pounds were taken in defiance of law.

CLOSE SEASON FOR SALMON.

14. We are informed that a determined effot will be made to induce the legislature to alter the time of the close season, so that fishing for salmon may be permitted in August and September, and that the close season may be changed from these months to July. With this object in view, it is reported that the proprietors of the present "canneries," and capitalists, who have in contemplation the construction of other "canneries," have been obtaining the evidence of fishermen, to present to the legislature to show that July is the proper month when fishing should not be permitted.

15. As we have shown, in July the spring run of fish has about ceased and the fall run but commencing. It is one of the months when fish are most scarce. To permit unlimited fishing during all the months in the year except July, would have the effect of exhausting our rivers of salmon within ten years. It is a simple proposition that if some of the ripe fish are not permitted to reach their spawning-grounds, they cannot reproduce naturally, neither can the United States nor the State obtain eggs from which to restock the river by artificial hatching. One of the fishermen who was approached with the object of obtaining his testimony in favor of a change to July, wrote to the commissioners, 30th September, as follows: "The close season should never, on any possible pretense or persuasion, be pressed outside the months of August and September to give opportunity for fishing in those months. Right there

is the life of the matter. The regularity, multitudes, and urgency of the seed run, the consequent ease and certainty of the catch, the fine weather for work, all present a weighty temptation to both catcher and canner." The object of a close season is, that some of the fish may be permitted to reach the headwaters to spawn. If they are not allowed to do so the race will soon be extinct. Cupidity and desire for immediate profit should not be permitted to influence legislation with the ultimate result of the extinction of the last fish. The interest of the public is that the fish be continued in the river. A change in the law that will omit August and September from the close season cannot but result in material and permanent injury.

TEMPERATURE OF AIR AND WATER.

16. The following statistics will be found of much importance. They exhibit the temperature of the water and air at two stations, each on the Sacramento and San Joaquin Rivers, taken for three years during the months the great army of salmon are passing up to their spawning-grounds. They will show conclusively that the Sacramento salmon lives for weeks, if not months, in water much warmer than any other fish of the same family. They also show the strong probability that these fish may be successfully introduced into rivers in still lower latitudes than those of which they are native—without doubt into the waters that flow into the Gulf of Mexico, and with many prospects of success into the rivers of Europe emptying into the Mediterranean.

TEMPERATURE (Fahrenheit).

Railroad crossing at Sacramento, Sacramento River, latitude 38° 35′ N., longitude 121° 30′ W.

	August.									September.								
	1875.			1876.			1877.			1875.			1876.			1877.		
	Air.	Water at surface.	Water at bottom.	Air.	Water at surface.	Water at bottom.	Air.	Water at surface.	Water at bottom.	Air.	Water at surface.	Water at bottom.	Air.	Water at surface.	Water at bottom.	Air.	Water at surface.	Water at bottom.
	°	°	°	°	°	°	°	°	°	°	°	°	°	°	°	°	°	°
Maximum	106	81	81	98	80	79	99	80	80	96	75	75	97	75	75	105	77	77
Minimum	71	75	75	75	72	71	80	73	73	72	71	71	73	70	69.50	76	70	70
Mean	92.96	78.83	78.83	87.93	76.40	75.37	91.54	77.22	77.22	88.93	73	73	85.53	72.13	71.30	90.56	73.76	73.76
Railroad crossing at Sacramento River, latitude 40° 01′ 30″ N., longitude 122° 06′ W.																		
Maximum	104	78	78	100	74	74	100	76	76	99	70	70	98	70	70	104	72	72
Minimum	78	70	70	69	68	68	84	69	69	80	66	66	70	66	66	80	66	66
Mean	95.64	75.51	75.51	91.38	70.61	70.61	93.74	72.90	72.90	90.63	69.60	69.60	87.26	67.66	67.66	91.23	68.73	68.73
Lower railroad crossing, San Joaquin River, latitude 37° 50′ N., longitude 121° 22′ W.																		
Maximum	98	82	81	97	79	78	95	81	81	94	78	78	93	75	75	102	78	78
Minimum	73	72	71	75	75	74	78	71	71	73	72	72	73	70	69	70	71	71
Mean	88.16	78.67	78.3	86.16	76.93	76.09	89.58	77.87	77.87	85.63	74.08	74.43	83.43	72.56	72.06	87	73.80	73.80
Upper railway crossing, San Joaquin River, latitude 36° 52′ N., longitude 119° 54′ W.																		
Maximum	107	84	83	111	77	76	112	77	76	104	82	83	108	78	77	105	78	77
Minimum	82	74	73	81	73	72	90	73	72	82	74	73	80	74	73	76	75	74
Mean	100.61	80.67	79.67	101.09	74.96	73.96	99.64	75.80	74.80	95.53	78.83	77.83	94.00	76.76	75.76	92.96	76.63	75.63

ILLEGAL FISHING.

17. There is a prevalent opinion throughout the States that it is the especial duty of the fish commissioners to act as local police in each neighborhood and prevent violations of the law in relation to fishing during the close season. Much time is consumed in answering questions on this subject and informing correspondents by letter that it is the duty of every citizen to see that the law is obeyed. We believe the law which prohibits the catching or having in possession salmon from 1st August to 1st November has been more extensively violated during the present year than ever before. It is true the fish are not sold openly in the city markets, but we are informed that the fishermen have erected salting establishments and smoke-houses in various by-places in the sloughs between the Sacramento and San Joaquin, where the work of salting and smoking has been prosecuted more extensively than in any previous year. We learned that the canning establishment of Messrs. Emersen Corville & Co., at Collinsville, only made a pretense of ceasing work on the 1st of August, and that they secretly persisted in violating the law. We caused them to be arrested and fined, upon which they quit work and promised hereafter to obey the law. The canning establishment near Sacramento was also reported as at work during the close season. The proprietors have been indicted by the grand jury of Sacramento, and will be fined, if found guilty, during the next term of court. It is well known that salmon, during the spawning season, are unfit for food. The fish canned, salted, or smoked at this period, if consumed or sold, will have the effect of giving the Sacramento salmon a bad reputation in the market. For this reason the "canners" on the Columbia River cease work on the 1st of August in their own interest and without any requirement of law. It is useless for the State to hatch fish and turn them into the river if there is no time in the year when they are permitted to reach their spawning grounds for purposes of reproduction. It would seem that when the State expends money in filling the river with valuable fish for the benefit of the public, and especially for the benefit of fishermen, that there should be sufficient intelligence and public spirit among local officers and the fishermen themselves to see the law obeyed and give the fish an opportunity to keep up the supply. If the commissioners are to expend the appropriation in prosecuting violations of the law, there will be no money to pay for the hatching of additional fish. Many of the fishermen acknowledge the justice and ultimate benefit of an observance of the law and obey it, but very properly complain that their work ceases, while those who violate it reap a greater benefit.

18. The following extracts from a letter received by the commissioners from a fisherman who has followed the business of catching salmon on the Sacramento and San Joaquin for the San Francisco market during twenty years will illustrate that, at least, the more intelligent and thoughtful of these men acknowledge the necessity of an observance of

the law. His letter also gives facts of importance as to the habits of the Sacramento salmon. Writing from Rio Vista, August 17, 1877, he says: "I understand the 'cannery' has shut down, but the greed for salmon is so great I would not trust them without watching. As to the fishermen, they will be salting them all along the banks of the Sacramento and Lower San Joaquin (as far up as the mouth of the Mokelumne) unless special means are taken to prevent it. The Three-mile Slough, leading from one river to the other, around the head of Sherman Island, is also fine fishing ground, and more retired from public observation than any other. Many of the fishermen started off with their tanks, &c., the very day the 'cannery' was reported to have stopped. Many of them are energetic, restless men, and the idea of doing something sly or contrary to law gives zest to their labor. Right here, where I write, a few boards have been thrown up shed-fashion by a party I need not now name. You may well believe salted salmon will be under it if some stranger does not prevent it. You may rest assured that the people who reside here will not be known as the initial instruments in punishing any one for the violation of the salmon laws, although there are many who feel it ought to be respected. No doubt public feeling and practice will occupy about the same status at Collinsville and wherever salmon fishing is a business. As I wrote to you the other day, now (August) is the time to protect the salmon. In review of long experience and observation I opine that of all the salmon passing in the months of August, September, and October, more than 90 per cent. pass between August 10 and October 1. The seed run is always on time, not being like the spring run, accelerated or retarded by the different moods of the river, caused by the winter and spring rains. If, during the last-named period (August 10 to October 1) the law was rigidly enforced, you would find seed enough for home use and a good part of all creation besides. Indeed, I think that one month out of the thickest of them, say August 20 to September 20, would be quite sufficient, and therein I differ with you in opinion, no doubt. But you have not, perhaps, observed in person, as I have, the multitudes and urgency of the run at that time; and this is almost uniform—it has not varied in time ten days in twenty years. Now, during the period of four or six weeks, the State, in view of the magnitude of the producing interest involved, ought surely to provide, beyond peradventure, for the enforcement of the law. The statute names the taking or possession of salmon a crime, but in the public mind this crime is only an illegal act. You cannot force sentiment by act of the legislature. The absence of sentiment excuses the citizens' apathy, and between ignorance and cupidity the salmon will suffer unless special agents of the State do for the public what the public have not yet quite learned they ought to do for themselves. Strangers are the best agents for this business. Citizens living in a fishing neighborhood do not feel like subjecting themselves to the enmity and revenge of a rough class by complaint. And, again, in this salting business, the criminal acts are beyond observation, except by express intention, as the fish are caught chiefly in the night, and the salteries are usually situated away from public highways and thoroughfares."

19. We have expended a part of the appropriation in prosecuting offenders against the law, but the field is so large and the profit so great that but little good has been accomplished. The more fish hatched and placed in the river the more numerous the fishermen, and the greater, apparently, the desire to make a profit from a violation of the law. As has been stated, unless the fish are allowed, in their season, to reach their spawning grounds, the rivers will be exhausted. Until the fishermen realize that the object of the law in creating a close season is the perpetuation and increase of the numbers of fish the law will continue to be violated. We see no remedy at present except, hereafter, to devote a larger portion of the appropriation in preventing illegal fishing and in prosecuting offenders against the law. This will require the use of a part of the appropriation which should be devoted to increasing the number of fish placed in the river. If it is expected that the commission shall employ special means to enforce an observance of the law, and also employ attorneys to prosecute offenders, it is necessary that the appropriation should be increased. It is not now sufficient for these purposes, and also for the hatching of any large quantity of salmon with which to keep pace with the increased fishing and the increasing numbers of sea-lions. We have consulted with many of the fishermen, and they admit that the law creating a close season should be obeyed, provided all be made to obey it. It is but proper to say, however, that they at the same time urge that the close season for salmon (August 1 to November 1) is too long a period. In correspondence with one of these men, who has made a business of fishing for salmon on the Sacramento and San Joaquin for many years past, as to the necessity for an observance of the law, he says: "I do not wish to be known as urging the enforcement of the law, or as a special informer against any party who has violated it. My reasons for this reservation affect alike my own peace and safety and that of many persons whom I know have no worse intention than to earn a living and obey the law, provided that others less honest are prevented from violating it with impunity. Your idea of a patrol boat, or boats with officers, is the correct one, and I firmly believe that if by this or other means the prohibition were strictly maintained from Benicia upward, wherever there are practical fishing grounds, during the period of one month at the right time, that the perpetuation of salmon in our rivers would be abundantly secured. Between the 10th of August and 1st of October more than 90 per cent. of the seed run passes, and has not failed to pass during twenty years of my observation. If the whole of the seed run is not wanted for seed, they ought not to be so used, for the fish is just as good food then as at any other time, only the wastage is something more, the spawn being larger. On the Columbia River I understand that the fall run is almost or quite

worthless. Not so on the Sacramento. Well, we may be proud of our river; it is the paradise of the salmon, and they seem determined to resist the devils—who also seem determined to drive them out—better than could be expected; but they will need help in the future. The nets for taking them are being multiplied and improved. The fishing grounds are better known than formerly. Such obstructions as snags in the river bottom are less common—many of them having been broken off or taken up by the nets and put out of the way, or covered by sediment, so that a wider and longer sweep may be taken by the drifting net. Altogether, the salmon is sure to be exterminated, fight he ever so persistently, unless we help him. Surely the State can afford to guard him effectually one month in the year. The cupidity of the fish speculator, who only cares for the greatest number of cases he can pack and ship, should not be allowed to influence the statement of that time. Let it be somewhere between the 10th of August and the 1st of October. By the way, it seems to me that at the extreme upper waters, on the spawning grounds, the fish should be protected during their entire stay, excepting as needed solely for the purpose of artificial hatching. But of this you are a better judge than I can be."

20. While not agreeing with this intelligent fisherman as to the propriety of shortening the close season, we fully concur as to the absolute necessity of a patrol to prevent unlawful fishing while the salmon are passing up to their spawning grounds. We also concur in his suggestion that the salmon should be protected on their breeding beds. The most important spawning ground left in this State is the McCloud River, in Shasta County. Its banks are mainly composed of lava and limestone, and, so far as known, they contain no mines. By some inadvertence or intentional manipulation, this county was exempted from the law creating a close season for salmon, and the fish are persistently taken in this county for market while in the act of reproduction on their spawning beds. We respectfully urge that Shasta County be reincorporated in the law, and that no salmon be allowed to be taken there during the close season except for purposes of artificial propagation.

21. The Chinese and others continue to use nets of a mesh much finer than is allowed by law, and the young of all kinds of salt-water fish that spawn in the bays and estuaries are persistently caught, dried, and shipped to China. The records of the custom-house show that there were shipped to China, from San Francisco, during the year ending 1st July, 1877, dried fish and dried shell-fish valued at $293,971.

22. We have caused several arrests to be made for violations of this law, but it is impossible for the commissioners to act as local police on all parts of the bay and rivers, and we see no remedy except in increasing the penalties for violations of the law, involving even, if necessary, the destruction of the nets, when used out of season. Unless in some way the wise provisions of the statute are compelled to be observed, we can see no reason why our present abundance of fish will not decrease,

as they have decreased in other States, in consequence of the disregard of wise enactments made for their preservation and increase. Ordinarily, salmon should reach their spawning grounds on the McCloud and Little Sacramento by the 20th of August. As will be seen by the statistics heretofore stated, the catch was never so great as during the past fishing season. At the commencement of the close season, 1st August, the river was filled with fish, yet they were not permitted to reach their spawning places. Mr. Myron Green, the deputy in charge of the United States fish-hatching establishment on the McCloud, reported, 15th September, that there were ten salmon in the McCloud in 1876 to one in 1877. Up to that time but 5,000,000 eggs had been taken, while nearly 10,000,000 had been taken in a corresponding period in 1876. The fish were, in the Lower Sacramento, more numerous than ever before, but they were caught, canned, salted, and smoked, in defiance of the law. It is estimated that the "canneries" took 50,000 after the 1st of August, and that there were salted and smoked on the banks of the sloughs and other by-places at least 100,000 more. If this is to continue, the government hatching-works will have to be removed to the Columbia, and we will be compelled to import eggs from some other State, even to keep up a partial supply of salmon in the Sacramento River.

23. In addition to making the penalties more severe for violations of the law, we would recommend that the law be so amended that it shall be made a misdemeanor to fish for salmon with nets or traps between sunset on Saturday and sunrise on Monday of each week. This would give the salmon the freedom of the river one day in the week, do no injury to the fisherman, and go far towards continuing the supply in our rivers.

WHITEFISH (*Coregonus alba*).

24. In January last we received from the United States Fish Commissioner a donation of 300,000 eggs of the whitefish. These were successfully hatched under the superintendence of Mr. J. G. Woodbury, at the State hatching-house at Berkeley, and the young fish were distributed as follows: 75,000 in Donner Lake; 50,000 in Sereno and other lakes near the Summit, in Placer County; and 175,000 in Lake Tahoe. Including 25,000 placed in Clear Lake in 1873, and 25,000 in Tulare Lake in 1875, there have been planted in the waters of this State 350,000 of these valuable food-fish. We believe they have lived in Clear Lake, also in Tulare. It was reported in a Lake County paper that a whitefish was taken in Clear Lake on 10th April, 1876, which measured a foot in length. We have no positive information that they have found a congenial home in Tulare Lake, but have heard reports that a few have been seen. As these fish can only be taken with a net, and as these are rarely used on these lakes, their waters will have an opportunity to become fully stocked before they are extensively fished. There can hardly be any doubt but they will succeed in Tahoe and other lakes near the summit of the Sierra—the climate, water, and food being not dissimilar

to those of Lakes Michigan, Huron, and Superior, in which they are indigenous. These fish live upon small crustacea, found on the rocky and gravel bottoms of lakes. They grow to weigh an average of one and a half pounds, and constitute the most important food-fish of the people living near the great lakes. Professor Baird, in his report to Congress, says: "Few fishes of North America will better repay efforts for their multiplication." We are promised a further supply of eggs during the present winter, and shall continue receiving eggs, and hatching and distributing these fish to all the mountain lakes that are accessible during the winter months.

CATFISH (*Pimelodus cattus*).

25. The seventy-four Schuylkill catfish imported in 1874, and placed in lakes near Sacramento, have increased to a vast extent. They already furnish an important addition to the fish food supply of the city of Sacramento and vicinity. From the increase we have distributed 8,400 to appropriate waters, in the counties of Napa, Monterey, Los Angeles, Fresno, Tulare, Santa Cruz, Shasta, Solano, Alameda, San Diego, Yolo, Santa Barbara, and Siskiyou. These, should they thrive and increase as they have in Sacramento, will furnish an abundance of valuable food in the warm waters of the lakes and sloughs of the interior, and replace the bony and worthless chubs and suckers that now inhabit these places. It may be proper to call attention to the fact that these fish have become so numerous in the lakes near Sacramento that they can now be obtained in any quantity for stocking other appropriate waters in any part of the State.

No. 35.

Dr. Hector to Professor Baird (April 27, 1878).

[See page 851.]

No. 36.

Mr. R. J. Creighton to the Hon. the Colonial Secretary.

SAN FRANCISCO, CAL., *April* 15, 1878.

SIR: I have received the inclosed letter from Mr. Clark, and in reply explained that the Government of New Zealand had not put me in funds to meet the payment of $500, but that I had forwarded his claim and a reply could not possibly be expected before the incoming mail arrived. I trust this matter will have been attended to. I regret to learn through the newspapers that the last shipment of whitefish failed. I think it was unfortunate that an attempt was not made to hatch them out in Auckland. In all probability sufficient would have been saved to stock the lakes. All experts here declare that the eggs should be hatched out where the ship first touches. The young fish can be fed on blood and taken anywhere over the country. It is a mistake to suppose that whitefish will not thrive in Taupo or Waikare. They are thriving in Lake Tulare, Southern California, the water of which is at least of as high a temperature as either of the Auckland lakes; and they thrive at San Leandro, Alameda County, which is quite as warm as the central heat of the North Island. There should be no local jealousy or feeling in a great national enterprise like acclimatizing food-fish, and I cannot divest myself of the idea that the order of the government for the distribution of eggs after such a perilous journey was given with the view of conciliating local opinions. Doubtless it would be very agreeable for gentlemen in every important section of the country to have an opportunity of hatching out these fish, and watching over them until their waters had been fairly stocked, but the risk of failure is too great. The acclimatization of whitefish is still a difficult problem. Their acclimzatiation in New Zealand would be a feat, apart from its economic results, of which the country might well be proud.

Should the government resolve upon testing the experiment next year, I will take precautions against failure such as the fish-packing establishments of the Union suggest.

I have, &c., ROBT. J. CREIGHTON.

The Hon. the COLONIAL SECRETARY,
Wellington, N. Z.

No. 37.

The Hon. the Colonial Secretary to Prof S. F. Baird.

WELLINGTON, *20th June*, 1878.

SIR: Mr. Creighton informs government that the sum of $500 is due on account of the transshipment of whitefish ova, and the matter is also referred to in your letter to Mr. Creighton, of January 5th, but no account has been sent for the amount. Mr. Creighton, in his letter to the government, states: "Mr. Clark did not send any accounts; so I infer that the $1 per thousand mentioned by Professor Baird covers the cost of package. This may not be the case, however, and if so, $500 is due to the fish commission in Northville."

Under the circumstances it is desirable that the payment should be made through you, and I beg, therefore, to inclose bill of exchange for the amount, $500, with a voucher form, and request that you will be good enough to pay the money and procure a receipt from the person to whom the money is due, as it is not clear if "Mr. Clark" and the fish commission in Northville are one and the same.

Apologizing for having to trouble you in this matter,

I have, &c., G. S. WHITMORE.

Prof. SPENCER F. BAIRD,
Washington, D. C., U. S.

No. 38.

The Under Colonial Secretary to R. J. Creighton, esq., San Francisco.

WELLINGTON, *20th June*, 1878.

SIR: I have the honor, by direction of the colonial secretary, to acknowledge the receipt of your letter of the 15th April, in which you inclose one from Mr. F. N. Clark, of Northville, and ask that a sum of $500 may be remitted to that gentleman.

As you name no one in your letter, and as it is not clear on what account and for what service the $500 is claimed, the government have, in order to avoid any possible mistake, remitted the money to Professor Baird in a letter, a copy of which is inclosed for your information.

I have, &c.,

G. S. COOPER.

No. 39.

Mr. J. C. Firth to the Hon. the Colonial Secretary.

AUCKLAND, *7th June*, 1878.

SIR: Referring to your letter of 11th October, 1877, asking me to undertake the transshipment of half a million salmon ova, expected by the next San Francisco mail steamer from Professor Baird, of the United States Fish Commission, and authorizing me to incur the necessary expenditure to insure the success of the importation, and having now received final accounts, I have the honor to inform you that the total expenditure incurred in this behalf has amounted to the sum of £195 17*s*., minus £22 5*s*., cost of sending ova to Sir Samuel Wilson = £173 12*s*.

These charges are heavy, but I am happy to learn from various sources that perfect success has been obtained, which would not have been secured under a less liberal expenditure.

Having taken the keenest possible interest in the great work of establishing American salmon in this colony from the first, my personal services have been most cheerfully rendered, and I beg you will permit me to present the above sum of £173 12*s*. as my contribution to the good work of introducing so valuable a food-fish in New Zealand.

I have further the honor to inform you that I have frequent reports of the success of the experiment, young salmon in various stages being reported to me as seen in nearly all the rivers in this provincial district, in which I placed the ova or fry.

I have, &c.,

J. C. FIRTH.

The Hon. the COLONIAL SECRETARY, *Wellington*.

No. 40.

The Hon. the Colonial Secretary to Mr. J. C. Firth.

COLONIAL SECRETARY'S OFFICE,
Wellington, 20th June, 1878.

SIR: I have the honor to acknowledge the receipt of your letter of the 7th instant, reporting the perfect success which has so far attended the last importation of salmon ova from America, and informing me that your total expenses in connection with the ova amounted to £173 12*s*., which sum you desire to present as your contribution to the good work of introducing so valuable a food-fish into New Zealand.

I can only again tender you my thanks on behalf of the government for your very successful exertions in this cause, and, at the same time, say that while they regret your refusal to allow them to reimburse you for the expenditure you have incurred, the government feel that the colony is deeply indebted to you for your generous aid in the introduction of American salmon.

I have, &c.,
G. S. WHITMORE.

J. C. FIRTH, Esq., *Auckland.*

No. 41.

Professor Baird to Dr. Hector.

UNITED STATES COMMISSION FISH AND FISHERIES,
Washington, 12th June, 1878.

DEAR Dr. HECTOR. Yours of the 27th April is to hand. I had been prepared for the account of the failure of the whitefish eggs, having been previously advised to that effect.

If you wish to renew the experiment this year, I will send Mr. Clark through to San Francisco in charge. He can then see that they are properly packed in the vessel.

If you want any more salmon eggs, let me know in time.

I have, &c.,

SPENCER F. BAIRD,
Commissioner.

Dr. JAMES HECTOR,
Wellington, N. Z.

No. 42.

Extract from private letter from Mr. R. J. Creighton to the Hon. James Macandrew.

Mr. Clark made a proposal to me, which I consider highly favorable to the colony, and I promised to submit it, which I do through you. It is this: "He is willing, if an order be received by him for several million whitefish eggs, jointly from the New Zealand Government and the State fish commissioners of California and Nevada, to furnish the eggs, carefully packed, at·65 cents per 1,000, and further, to insure their safe delivery at San Francisco for shipment to the colony and deposit in our lakes and rivers here; he would come across the continent in charge, on receiving his traveling expenses to and fro, asking nothing whatever for his time—this extra to be borne proportionately by the colony and California and Nevada." I think the proposal is an extremely liberal one. New Zealand might procure 1,000,000 whitefish eggs in this way for a trifling sum, under conditions which would insure the absolute success of the experiment. I have no doubt I could arrange matters with the fish commissioners of these States. I should state that Mr. Clark explained that to insure success the order for the eggs should be in by October, or early in November. The order passing through Professor Baird, came at a time when the eggs were in a too advanced state. The ova should have been packed at least a month earlier. As it was, the Pacific-coast shipment all went back *en route*, and I saved our lot by the best of good luck.

No. 43.

The Hon. the Minister for Public Works to Mr. R. J. Creighton.

PUBLIC WORKS OFFICE,
Wellington, New Zealand, 17th August, 1878.

DEAR SIR: Referring to your private letter to me of 8th July, in which you inform me that Mr. Clark has offered to supply whitefish ova at 65 cents per 1,000 f. o. b. at San Francisco, and that Mr. Clark undertakes personally to superintend the shipment at that port provided his actual expenses across the continent are defrayed jointly by the State fish commissioners of California, Nevada, and this colony.

The Government of New Zealand will be glad to be a party to this arrangement, and will take 1,000,000 ova on these terms.

If Mr. Clark will forward his account along with the ova, the amount will be remitted to him in due course.

Thanking you for the interest and trouble which you have taken in this matter,

I have, &c.,

J. MACANDREW.

R. J. CREIGHTON, Esq.,
Evening Post Office, San Francisco.

No. 44.

Professor Baird to the Hon. the Colonial Secretary.

UNITED STATES COMMISSION FISH AND FISHERIES,
Gloucester, Mass., *29th July*, 1878.

DEAR SIR: I have the honor to acknowledge the receipt of your letter of the 20th of June with the accompanying check for £104 3*s.* 4*d.*, being the amount of indebtedness to Mr. N. W. Clark for eggs of whitefish furnished by him at my request for the use of the New Zealand Government. The charge was for the cost of collecting and keeping in the hatching-house one month, so as to bring forward the embryo, and for packing and shipping; and of course the price of one dollar per thousand was merely nominal.

I greatly regret that, after all, the eggs arrived in an unsatisfactory condition; but, if you desire to renew the order, I think I can promise better results.

I have sent the account to Mr. Frank N. Clark, a son of the deceased N. W. Clark, for his signature, and on receiving it will forward it promptly to you.

I have, &c.,

SPENCER F. BAIRD,
Commissioner.

The Hon. the COLONIAL SECRETARY,
Wellington, N. Z.

No. 45.

The Hon. the Colonial Secretary to Prof. Spencer F. Baird.

COLONIAL SECRETARY'S OFFICE,
Wellington, *21st October*, 1878.

SIR: I have the honor to acknowledge the receipt of your letter of the 29th July, and to thank you for your offer to endeavor to procure a better result than was before obtained should this government think fit to renew the order for whitefish ova.

An order for 1,000,000 ova had been sent to Mr. Clark through Mr. Creighton, an old New Zealand colonist who is settled in San Francisco, before your letter arrived, and it will, therefore, be unnecessary that the government of this colony should avail itself of your kind offer to send a shipment of ova this season.

I have, &c.,

G. S. WHITMORE.

XLI.—CORRESPONDENCE CONNECTED WITH THE TRANSMISSION OF EGGS OF THE QUINNAT SALMON AND OTHER SALMONIDÆ TO EUROPEAN COUNTRIES IN 1878 AND PRIOR YEARS.

GERMANY.

H. Bartels to S. F. Baird.

IMPERIAL GERMAN COMMISSION FOR THE
CENTENNIAL EXHIBITION,
GERMAN PAVILION, CENTENNIAL GROUNDS,
Philadelphia, October 19, 1876.

Professor BAIRD,
United States Building:

SIR: I have the honor to inform you that the Prussian minister of the agricultural department has notified me by a cable telegram that he accepts, with his greatest thanks, your very kind offer for sending eggs of the Californian salmon to Prussia, and he begs you to forward these eggs to the address of the "Fischzucht-Anstalt, Hameln, Province Hannover."

I am, yours, very respectfully,

H. BARTELS.

H. Bartels to S. F. Baird.

IMPERIAL GERMAN COMMISSION FOR THE
CENTENNIAL EXHIBITION,
GERMAN PAVILION, CENTENNIAL GROUNDS,
Philadelphia, December 23, 1876.

In reply to your valued favor of the 22d instant, I beg leave to inform you that, about the forwarding of salmon eggs to Germany, it will be the best you write officially to his excellency the Prussian minister of the agricultural department, Dr. Friedenthal, in Berlin, who will give the necessary information to Hameln.

For losing no time any more I have already communicated to the minister your very kind offer, so that the necessary information can be given to the agency in Bremen for accepting and forwarding the eggs immediately on their arrival in Bremen; therefore it will be sufficient if you will be kind enough as to inform the minister when you send the eggs and by what steamer they will be shipped.

I shall leave here next week and shall give me the honor to call on your office in Washington, where I have business for some days.

I am, yours very respectfully,

H. BARTELS.

Prof. SPENCER F. BAIRD,
Washington, D. C.

Closing this letter I have received a letter from the minister, Excellenz Friedenthal, who informs me that he has communicated to the magistrate of Hameln, Hannover, your kind offer. Will you please inform the magristat at Hameln when the salmon eggs will be shipped and by what steamer.

Yours, H. B.

The German Minister to S. F. Baird.

IMPERIAL GERMAN LEGATION,
Washington, 21st March, 1877.

SIR: In the letter you directed on the 6th of February last to the minister of the agricultural department of Prussia, you were kind enough to offer to Dr. Friedenthal some eggs of the California salmon, provided Congress would continue the appropriations, and that the Prussian department would pay the cost of packing and shipment.

Dr. Friedenthal has now requested me to express to you his warmest thanks for this kind offer, and to beg you that, if possible, 50,000 embryonated eggs of the California salmon may at the proper time of this year be forwarded to the "Fischzucht-Anstalt in Hameln, Provinz Hannover."

I beg you, sir, to be kind enough to inform me if, and when, such transportation would be possible, and to inform me also of the expenses aforesaid, in order to be restituted by this legation.

Accept, sir, the renewed assurances of my high consideration.

SCHLOEZER,
German Minister.

Prof. SPENCER F. BAIRD.

H. E. Rockwell to the German Minister.

WASHINGTON, D. C., *March* 22, 1877.

SIR: I have the honor to acknowledge, in behalf of Professor Baird, who is temporarily absent from the city, the receipt of your letter of the 21st, and to say that a memorandum has been made of the request contained therein for 50,000 eggs of the Californian salmon.

On the return of Professor Baird you will doubtless receive a suitable response to your communication.

Respectfully, H. E. ROCKWELL,
Secretary.

Baron SCHLOEZER,
German Minister, Washington, D. C.

Fred Mather to S. F. Baird.

STEAMER ODER, IN NORTH SEA,
October 23, 1878.

MY DEAR PROFESSOR: Last evening at seven I delivered three boxes of eggs to the agent of the N. G. Lloyds at Southampton for Paris. They were in perfect condition, but I regret that I did not put a caution in each box concerning the temperature.

The eggs were down to 42°, and if they plunge them into water from 15° to 20° higher it may be fatal.

I wrote a caution to Mr. Wattel by mail but fear that the eggs may reach him first. The other eggs are in equally good order.

I telegraphed Dr. Finsch from Southampton, as per request by letter from him received there, "all good."

Package for Southport sent with request to divide with Mr. Moore. I repacked those in moss and don't know condition. Also for the aquarium, two *Menopoma*, present from Blackford; five horse-feet (all I could get), and three tortoises, *Emys picta* and *Pseudemys rugosa*, all alive.

I find such good accommodations for soles in the Oder that I had a plan to leave Bremen on the 27th and return in her to Southampton November 5, but Dr. Finsch says that Mr. von Behr wants me to go to Hameln on Weser. Still I *may* do it if time and soles permit.

Very truly,

FRED MATHER.

Prof. S. F. BAIRD,
Washington, D. C.

Fred Mather to S. F. Baird.

BREMEN, GERMANY, *October* 24, 1878.

MY DEAR PROFESSOR: I arrived at Bremerhaven at 9 a. m. and delivered 100,000 eggs for Holland to Mr. Garrell on telegram order from Mr. Heck, whom Mr. Bottemanne informed me by letter, received at Southampton, would meet me. He will arrive at night.

Dr. Finsch and Director Haack met me. The eggs were in splendid condition, and people have flocked to see them. To-morrow I go to Hameln with eggs, and hope to sail for England on Saturday or Sunday next. I have not get given up the hope of getting back in the Oder, but will go to Cunard office, London, to see Mr. Franklyn or learn if he is in Liverpool.

Will keep you advised of all movements, and if soles are not ready will come back in Oder without waiting, as last year's experience in that line was bad.

Very truly, yours,

FRED MATHER.

Prof. S. F. BAIRD,
Washington, D. C.

Extract from the Weser Zeitung of October 25, 1878.

[Translation.]

When we made a detailed report relative to the first transport of California Salmon eggs to Europe last year (No. 11062 and 11063, 1st and 2d November, 1877), we unfortunately could record but a partial success: of 300,000 eggs, only 25,000 arrived in a perfect state. This was the lot brought over in the chest constructed by Mr. Mather, and described in last year's report. The rest of the eggs, which had been packed in moss and gauze, which Mr. Mather was not authorized to repack after his plan, were all spoiled, as, in consequence of the heat produced by the decay of the moss, all the eggs had been hatched. In spite of this failure, the problem of transporting salmon eggs over great distances has been brilliantly solved. The second transport of 250,000 California salmon eggs, which arrived to-day, in the Lloyd steamer Oder, Captain Leist, has confirmed the views formerly expressed. The eggs, just arrived, like last year's, come from the United States breeding establishment on the McCloud River, in California, about 200 miles from Sacramento. They left the latter city in an ice-car of the Pacific Railroad, September 28, and on reaching Chicago were taken in charge by Mr. Mather, who was commissioned to escort the sending to Europe. Mr. Mather repacked the eggs in his chests, which had been improved since last year, and embarked on board the Oder, which, after a quick and pleasant passage, arrived in the Weser this morning at about nine o'clock. Mr. Mather was received by Mr. Finsch, who has repeatedly represented the German Society of Fish-breeders, also by Mr. Haack, director of the Imperial Fishbreeding Establishment, in Hüningen, both congratulating Mr. Mather most sincerely and heartily upon his brilliant success, which was immediately announced by telegraph to the highly-deserving president of the society, Herr von Behr-Schmoldow. According to the orders of the managers of the society, 45,000 eggs will be transferred to Hüningen, to go eventually to the Rhine; 115,000 to the renowned breeding establishment of Mayor Schuster, near Freiburg (Baden), for the Rhine and Danube; 2,000 to Bonn and Münden (Hannover); 30,000 to the establishment in Hameln, for the Weser, and 58,000 have been sent off to Berlin this evening, where they are to be subdivided among various smaller establishments, under the management of the well-known fish-breeder, Max von dem Borne; the Mark, part of Silesia, Saxony, and Mecklenburg being the recipients. Besides the eggs intended for Germany, Mr. Mather brought over 100,000 for Holland, 100,000 for France, and 15,000 for England; all of which arrived in an equally good condition, the loss amounting to scarcely ½ per cent.; that is no more than would be the case in breeding establishments. After this success there is no longer a doubt that salmon eggs can be carried just as well to Australia. The California salmon (*Salmo quinnat*) is a different species from ours, and shows its peculiarity in its development. It has been previously remarked that the young fish, though only just hatched, possesses much more vitality than our salmon. It is more vigorous, lively, and voracious than ours. It is then not astonishing that the Californians developed themselves very well in Hüningen, and grew with surprising rapidity. The little one-year-old fishes which Director Haack keeps in a pond for the sake of observation and study are already a span long, quick and lively, whose well-being in our rivers is not to be doubted. A considerable number have been transferred to the Rhine, Danube, and Weser. The new sending will now supply our rivers with the stranger in much greater quantities, and we owe this to the society of fish-breeders, whose beneficent efforts deserve a much more lively sympathy, as also to Prof. Spencer F. Baird, in Washington, the Commissioner of Fish and Fisheries of the United States, who presented to the society, and consequently to Germany, this valuable sending.

Deutsche Fischerei-Verein to S. F. Baird.

BERLIN, *December* 14, 1878.

DEAR SIR: We cannot allow our seventh circular of 1878 to cross the Atlantic without offering our special tribute of thanks to that kind friend in America who has enabled us to proclaim in Mr. Haack's report (No. 11) that the introduction of *Salmo quinnat* into German waters and its domestication may henceforth be considered as in a manner accomplished. Be pleased, therefore, to accept our renewed assurance that we are fully alive to a sense of your unvarying and helpful courtesy.

Mr. Mather's skill has again obtained a signal triumph. Very few losses occurred on the road. We may confidently hope that a few weeks hence nearly a quarter of a million young Californian salmon will be lustily permeating the various river highways of this country. The Danube and its tributaries have claimed our special attention, inasmuch as they possess no migratory salmon and seemed to wait for the arrival of one so constituted as the quinnat.

We should hail the day, dear sir, when we might be permitted to offer you, for the benefit of American rivers or lakes, any inhabitants of our waters unknown beyond the ocean.

You will receive copies of a prospectus lately published for the international fishery exhibition of 1880, to which we beg to draw your very especial attention. As we said in our November circular, when forming the scheme of that exhibition, we reckoned chiefly upon the willingness of America to send specimens of that gigantic progress which pisciculture and other cognate matters have there achieved.

Thanking you again and again, we remain, dear sir, yours very sincerely,

The committee of the Deutsche Fischerei-Verein:

VON BEHR.	O. HERMES.
G. VON BUNSEN.	P. MAGNUS.
W. PETERS.	L. WITTMACK.
MARCARD.	FASTENAU.
V. BAUMBACH.	FINNOR.
F. JAGOR.	GREIFF.
E. FRIEDEL.	

FRANCE.

Society d'Acclimatation to S. F. Baird.

(Cable message received at Halifax, October 9, 1877.)

COMMISSIONER FISHERIES,
Halifax, N. S.

Envoyez œufs.

SOCIÉTÉ D'ACCLIMATATION.

11 Paris.

Drouyn de Lhuys to S. F. Baird.

SOCIÉTÉ D'ACCLIMATATION,
Paris, le 19 *octobre*, 1877.

MONSIEUR: La société d'acclimatation a reçu avec la plus vive reconnaissance l'offre si généreuse contenue dans votre lettre du 20 septembre dernier et je m'empresse d'être auprès de vous, en cette circonstance, l'interprète de tous ses remerciements.

L'introduction du saumon de Californie dans beaucoup de nos rivières aurait une trop haute importance économique pour que la société d'acclimatation ne soit pas fort heureuse de pouvoir en tenter l'utile essai.

Un de nos agents sera chargé d'aller recevoir les œufs, dont vous avez bien voulu nous annoncer, par le télégraphe, la prochaine arrivée à Southampton. Ces œufs seront confiés aux soins des personnes les plus compétentes, et rien ne sera négligé en vue de mener à bonne fin la très-intéressante expérience qu'il va nous être permis d'entreprendre, grâce à la libéralité du gouvernement des États-Unis.

Veuillez agréer, Monsieur, l'assurance de mes sentiments de haute considération.

Le Président,

DROUYN DE LHUYS.

Monsieur SPENCER F. BAIRD,
Halifax, Nova Scotia.

Raveret-Wattel to S. F. Baird.

SOCIÉTÉ D'ACCLIMATATION,
Paris, october 29, 1878.

MONSIEUR: J'ai la satisfaction de vous informer que nous verrons de recevoir, dans les meilleures conditions possibles, le généreux envoi que vous avez bien voulu nous faire encore, d'œufs de saumon de Californie.

Ces œufs ont immédiatement été confiés à ceux de nos sociétaires la plus en situation de leur donner de bons soines, et une large part a été réservée au ministère des travaux publics (service des péches); le bon état dans lequel ils nous sont parvenus nous donne tout lieu d'espérer une réussite complète.

En vous réitérant l'expression de la reconnaissance de la société pour votre concours si bienveillant, je vous prie, monsieur, d'agréer la nouvelle assurance de mes sentiments les plus respectueux et dévoués.

Le sécrétaire des séances,

RAVERET-WATTEL.

Hon. SPENCER F. BAIRD,
Président de la commission des pêcheries nationales
à Washington, États-Unis.

---

*Louis de Bebian to Fred Mather.**

COMPAGNIE GENERALE TRANSATLANTIQUE,
55 *Broadway, New York, February* 13, 1878.

DEAR SIR: Your favor of 11th at hand. I shall be pleased to take the case of salmon eggs free of charge to Havre. The steamer will sail early in the morning of the 20th, therefore case must be here on the 19th. Mark case to the consigner's address in Havre, and send it to me with *all charges prepaid.* There is an ice-box on our steamer belonging to party who ship "fresh beef" and I will endeavor to get him to allow me to put the case in there, where no doubt it will be in good condition on its arrival in Havre. Mark case distinctly and notify me of its shipment to me, and I will send receipt to you and consignee in Havre.

Yours truly,

LOUIS DE BEBIAN.

Mr. FRED MATHER,
271 *High Street, Newark, N. J.*

P. S.—If case is destined for Paris we can notify our agent in Havre to forward, you guaranteeing charges between Havre and Paris in case receiver refuses.

---

*In connection with a proposed sending of eggs of land-locked salmon.

THE NETHERLANDS.

H. Cazaux to S. F. Baird.

NETHERLANDS AMERICAN STEAM NAVIGATION COMPANY,
27 *South William Street, New York, August* 31, 1877.

SPENCER F. BAIRD, Esq.,
United States Commission Fish and Fisheries,
Halifax, Nova Scotia:

DEAR SIR: In reply to your esteemed favor of the 27th instant, I beg to inform you that the steamers of our line are appointed to leave this port for Rotterdam as follows: Steamship Rotterdam, September 13; steamship W. A. Scholten, September 27; steamship Maas, October 11; steamship P. Caland, October 25; steamship Schiedam, November 1.

There is no direct steamship communication between New York and Amsterdam.

Thanking you for your courtesy to our government, I remain, with great respect, yours, very truly,

H. CAZAUX.

C. J. Bottemanne to S. F. Baird.

BERGEN-OP-ZOOM, 7 *November*, 1877.

DEAR SIR: Inclosed you will receive the report of the assistant director of the zoölogical garden at Amsterdam to me about the *S. quinnat* eggs you presented to our government. I am very sorry that it is such a total failure. The chief of the department (finance), whom I hold my commission under, has taken the necessary steps already, by his colleague of the interior, to make a proper acknowledgment to the United States Commission for the truly magnificent donation.

Mr. Mather had telegraphed from New York the starting; this enabled me to prepare in time. The 23d I wired to Southampton, in order to know the quantum Mr. Mather was bringing (your letter of the 9th of October only reached me the 26th), and received an answer on the 24th. Having no time to spare to run over to Bremerhaven, I had arranged with Dr. Westerman, the managing director of the zoölogical garden, that the asst. director should go to take charge of the eggs at Bremerhaven. In the gardens, I may say, all were in high glee. Mr. Noordhoch Hegt telegraphed the time of his arrival, and so the whole board of directors was present in the breeding-house, to be driven off in less than no time as soon as the first box was opened, by the terrible smell; it was really a pity. As the report is pretty exhaustive, I will say nothing more here.

When going to Amsterdam, to make my arrangements for the eggs, I touched at Leiden, and got from Professor Buys (secretary of the board of commissioners of the sea fisheries) all the yearly reports since 1870, which I forwarded to your address in Washington by bookpost in the latter days of October last, and I will take care to send the same every year as soon as published, which is commonly not before September; rather late. At the same time I forwarded two pamphlets of Mr. de Bont, one French, one Dutch. This gentleman, who belongs to the board of directors of the zoölogical garden, is amateur pisciculturist, has always superintended the fish-breeding establishment, and is lately very successful.

In our rivers we find in the fall, September and November, the salmon almost and entirely ready to spawn, but at the same time there are in September commonly a few, later on more, and now nothing else but heavy salmon, averaging about 25 pounds, which we call "winter salmon;" in England, I believe, "fresh run salmon." The ovary in those salmon are so minute that they were formerly entirely overlooked, and so this salmon was declared to be sterile. When I investigated our rivers in 1869 and 1870 with Mr. Pollen, we found this to be not so. We gathered the roe and milt for a period of more than a year, and so we got the successive development from the winter up to the spawning salmon; and this proved that the winter salmon was not sterile. We proved also that the *Salmo hamatus* was nothing else but the male *Salmo salar* in breeding time, and that the gray and dark colored red-spotted one was nothing but the female in the same condition. After spawning she became just as silvery as a spring salmon and the male lost his hook before he died. Up till now it is still an open question "what makes the winter salmon come into the river the same time the others come into spawn?" She is not ready to do so before next fall, and in the spring there is caught once in a while a salmon with worms hanging out of his head, emaciated, and in terribly poor condition: a fish that in good condition would weigh 25 pounds does not weigh more than 11 or 12 pounds; the body, always broader than the gills, is so much shrunk that the gills protrude considerably. Almost invariably are those that have been so-called winter salmon, drifting seaward, which is proven by the fact that the first caught are always high up in the river and later on in the tideway, but we never catch many, and in later years but few. Now, as to what winter salmon is, I gave the following explanation:

The salmon is bound to come into the rivers for reproduction, but if they all came at the same time, viz, in the fall, being all in nearly the same state of development, they would altogether reach to nearly the same height in the river and be compelled to spawn there and then. Of course there was not place enough for the whole lot, so the one would root up the nest of the other, and it would be a wonder if the whole progeny was not destroyed. No, says mother nature, not so. *You* go so much earlier in the river; and *you* so much; that leaves you time to go so much farther, and in this way you will all find a good place to spawn, and so you will find spawning-beds all along the Rhine up to

Schaffhausen. Now this looked to me a very good solution of the question, but what in the case? At Basel we find just as well and at the same time the salmon in prime spawning condition and the winter salmon; this upsets my whole theory, unless I am able to prove that the salmon caught in spawning condition at Basel has been a round year in the river, which is not probable and which I don't believe, as being in too good condition for that. If you have an opinion about this I will be very glad to know it.

Oyster culture is taking here such a swing that it is becoming a national interest. The board has proposed to exhibit next year in Paris. As soon as the decision is known I will let you know. I think to go there anyhow. Whenever you come across, either to France or England, try to stop here too. When arriving in England, by leaving London via Queensborough-Flushing for the Continent, it takes about eight hours to cross the North Sea, and from Flushing one hour per rail to get here, and from here you are within 25 minutes on the route Amsterdam, Brussels, or Paris.

Offering you my sincere thanks for your great liberality in sending the eggs,

I remain, dear sir, yours respectfully,

C. J. BOTTEMANNE, M. D.,
Superintendent of Fisheries, Netherlands.

Prof. SPENCER F. BAIRD,
United States Commissioner Fish and Fisheries,
Washington, D. C.

---

*Report on the salmon eggs sent by the United States Government to the Netherlands.**

On the 24th October, Mr. Bottemanne, superintendent of fisheries in the Netherlands, wired to us that the steamer Mosel, with destination for Bremerhaven, has arrived at Southampton, and that Mr. Fred Mather had under his care four crates, with one hundred thousand eggs.

At the same time Mr. Bottemanne sent us a letter to Mr. Fred Mather, to authorize the undersigned to receive the above salmon eggs.

Immediately we made here all preparations to place the eggs on arrival, and I went to Bremerhaven to receive them. I waited there for the steamer bringing down the passengers from the Mosel, and the 26th October, about eleven o'clock in the forenoon, I met Mr. Mather and received the four crates destined for the Netherlands. I found there Messrs. Dr. Haack, director of the Imperial German Institution at Hünningen; Dr. Finch, director of the museum in Bremen; Mr. Schiever, superintendent of the breeding station at Hameln. They all, came for the same purpose. We agreed to move the crates unopened,

*Made by the Koninglijk Zoölogisch Genootschap Natura Artis Magistra, Amsterdam.

and, not to lose a moment, to start immediately. We had them landed and brought to the railway station at Geestemunde.

While waiting there for the custom-house officers and the starting of the train, we perceived that a milky stuff of a nasty smell was running out of the crates. We took the precaution to buy a hundred weight of ice to keep them cool, but it proved useless.

On arriving at Bremen, we agreed, by general consult, and as there was some time left before any of us could start for our different destinations, to open one of the crates and the interior boxes. We did so at the station, and found the first crate for the Netherlands all spoiled, except in a corner where there were a few eggs in apparently good condition. All the interior, however, was heated up to about 70° or 80° Fah. There were a few *eclosions* with some of the fish still alive. The whole, however, gave the impression that the enterprise had failed, the fault being the packing, which, perhaps quite sufficient for a short journey of a few days, was not adequate to a transport of three or four weeks.

As the only direct train to go to Amsterdam started at one o'clock at night, I resolved to let my four crates quiet and proceed with them immediately in order not to lose 24 hours, as I could reach Amsterdam next morning. Mr. Schiever did the same and left the same night to his destination close to Hanover. Dr. Haack, who had to go further, resolved to remain in Bremen for the night and we proceeded together to a hotel, where he opened his crates. The contents proved to be in much the same condition as those of the crate opened at the railway station, the heat of the boxes' interior being 80° Fah. The best eggs were taken out of the box and put separate, and it took about four hours to do one crate.

It seems that Mr. Fred Mather had received orders to leave the salmon eggs quiet and not open them at all, as the gentleman who sent them off from California expected they would arrive safe in Europe. It is to be regretted that this was the case. Mr. Mather had opened one case in New York, and took the trouble to bring over the contents in a case of his own construction. The result was truly magnificent—out of 25,000 eggs, shipped in New York, only about 400 were lost. Through the arrangement with slides, the eggs lay in single layers and could be taken out very easily and cleaned, there being some space left at the top and the bottom. Ice could be brought close to the eggs without touching them, and the mean temperature was kept without much trouble at from 41° to 45° Fah. As I have said, the success was wonderful and the arrangement was perfect, without incurring much extra cost. As these eggs belonged to the lot for Hünningen, Director Haack had the means of putting the eggs saved from the crates in a few trays, which were empty in the above named box. What will be the result of these I cannot say here, but I have no doubt Director Haack will report on them later

With my four crates I proceeded further to Amsterdam where I arrived 27th October, just 24 hours after receiving them in Bremerhaven.

What I expected proved true; the contents were all spoiled through heat generated in the interior. The thin pieces of white cloth where the eggs lay between were so rotten that they could not be taken up, but broke on being touched. A few of the eggs in some of the corners were still in what seemed a sound condition; the great mass was a white rotten mess with a very offensive smell. We tried to save the best and let them cool off slowly and put them afterward in water. Some *eclosions* followed, even at the moment the eggs were uncovered, but most of the apparently good eggs burst on being touched; a few hundred only came in the water, but those were all dead next day, except two or three *eclosions* which died also shortly afterwards.

It is very much to be regretted that this most liberal trial of the United States Government to enrich our European rivers with some of the valuable products of their waters has this time failed. For many reasons it is to be hoped that the Government will repeat this trial once more next year, and that they may have occasion to make use of this intelligent case of Mr. Fred Mathers. I have not the least doubt that if he can make use of the boxes with slides, as used by him this time for a small number of eggs trusted to his care, success must follow for any number, and a great boon will be bestowed on Europe for the introduction of a new population on many of its rivers.

Amsterdam, 28th October, 1877.

J. NOORDHOCH HEGT,
Adj. Director of the Royal Zoological Society.

The Minister of the Netherlands to S. F. Baird.

NETHERLANDS LEGATION,
Washington, January 28, 1878.

SIR: I have the honor to transmit to you the thanks of the Government of the Netherlands for the valuable gift of 100,000 eggs of the California salmon, made last year by you to Mr. Bottemanne, commissioner of fisheries on the Scheldt and the waters of Iceland, and to express the hope that, though the hatching of these eggs in Holland has not succeeded, through various causes relating, as it is believed, to the packing and handling of this collection on the sea voyage, this circumstance may not prevent you from having the experiment made again this year under more favorable auspices.

I have the honor to be, sir, your obedient servant,

VON PESTEL,
Minister Resident of the Netherlands.

Prof. SPENCER F. BAIRD,
Commissioner Fish and Fisheries,
Smithsonian Institution, Washington.

C. J. Bottemanne to S. F. Baird.

BERGEN-OP-ZOOM, *September* 4, 1878.

Prof. SPENCER F. BAIRD,
Washington, D. C., U. S. A.:

DEAR SIR: Your favor came to hand in due time, and beg to inform you that the sailing days of the Rotterdam steamers where the eggs may go by are, September 28, Schiedam; October 19, Scholten; November 9, Caland. Between each of those is a smaller steamer, but neither of those have a proper ice-house, so they are of no use.

The captains of the Scholten and Caland I'll give instructions to before they leave. I missed the captain of the Schiedam at Rotterdam, but I'll write to him. I have not seen the accommodation on board the Schiedam, but the directors say she has the best of the three; still the place on board the other two is in my opinion well suited, too.

The captains will be instructed that in case you give them instructions, they have to obey yours punctually without taking any regard of mine.

In case neither of those dates of departure suit, and there is a steamer for Bremerhaven leaving at better time, having proper accommodation for the eggs, you are free to ship thereby on my account. In this case it will be necessary that you provide them with instructions. If you can manage to ship by the Rotterdam line I think we have better chance of success, as the captains of those steamers, if on no other account, will take an interest in the affair on my account. The directors, too, are interested, so it seems, as they are willing to deliver the eggs at Rotterdam free of cost. One of the directors, Mr. Plate, is in New York now; a very nice gentleman he is.

I am convinced that the whole trouble of last year was nothing else but that the eggs have not been kept cool enough and so come out too soon.

If you can do me the favor please advise them a couple of days before the departure of the steamer where the eggs are to go by, as it will give me a chance to await the arrival of the ships in port.

The New York address of the company is Funch, Edye & Co., 27 South William street.

Oyster culture is quite a success this year. The catch of spat on the tiles and shells is very plentiful; the season for it is ended just now. It is really worth the while seeing, and whenever you come to this side of the water be sure not to miss it. Of course as long as I am here, I am at your disposal.

With kind regards, I remain, dear sir, your obedient servant,

C. J. BOTTEMANNE,
Superintendent of Fisheries.

C. J. Bottemanne to S. F. Baird.

BERGEN-OP-ZOOM, *October* 30, 1878.

DEAR SIR: October 20 I dispatched a letter to Southampton, addressed to Mr. Mather, care of the agents of the Bremen Lloyd steamer Oder, begging to wire me his arrival there. For what reason I don't know, but I did not get a message. As you know in country places, newspapers are behind, and so I knew only the 23d the Oder arrived the day before in England. Immediately I wired to Amsterdam, not being able to go myself, and Mr. N. Hegt, the same gentleman who went for the eggs last year, started in the afternoon for Bremen, but failed to meet Mr. Mather. He got the eggs and started with them with the first train, and arrived in Amsterdam at noon October 26, where I met him in the Zoo, and unpacked the eggs. They looked very healthy.

The number in the three boxes we made 85,000, out of which we picked that day and the next morning, 3,000 bad ones, and so we had Sunday morning 82,000 on the trays, splendid looking. Towards evening, when I had to leave for home, I looked them over and found about 70 bad ones more. I did not hear since from Amsterdam, but if anything was wrong they would have written or telegraphed, so I conclude all is going right.

As things stand now we may consider the transport as a splendid success, and I am very much obliged for your kindness and the trouble taken. Of course in due time you will receive an official acknowledgment.

Mr. Hegt not meeting Mr. Mather, leaves me in the dark about the costs, and not having any address of Mr. Mather in Germany I am obliged to address you about this. Please let me know at your earliest convenience and I'll remit by post-office order.

I remain, dear sir, yours, very truly,

C. J. BOTTEMANNE,
Superintendent of Fisheries.

C. J. Bottemanne to S. F. Baird.

BERGEN-OP-ZOOM, 27 *Janvier*, 1879.

DEAR SIR: Your favor of November 12, 1878, duly received. I am sorry I had to delay the answer so long, but different reasons prompted me thereto.

On the arrival of the eggs, which I found to be irregular in size, their temperature was 8°, the water 12° Celsius; slowly we rose the temperature of the eggs up to the water.

They came out very irregular, the first about the middle of November and the last had not absorbed their sack a week ago.

In the first time we lost many by smothering, not being able to get their head out of the shell. When the head came first it was all right, but if the sack burst through the little fellow seemed to have not power enough to clear the shell. The first impression was the shell was hardened somehow so they did not open far enough to admit an easy exit, but afterwards I felt inclined to believe it was weakness of the fry caused by the comparatively high temperature of the water, the weather being very mild at the time.

The eggs that came out later, when the temperature of the water had gone down considerable, we had less trouble with, till it gradually left off entirely. After all were out, the loss in fish became small, going down to ten and less per day. By sharp watching many were saved by helping them out of the shell when this burst.

I had a good deal hunting round to find a suitable place to deposit the fry. December 26, I fixed the place for the first lot of 10,000. January 4, when I came with them, the river had only risen about 15 feet; still the place held good and the water was pretty clear; had gone down already about 3 feet.

The transport took place in tin cans of one-half meter diameter and about one meter high. In the middle of the slanting top is a large round hole fitted with an inverted cover. The bottom of this cover and the slanting top of the can are perforated with small holes. In the top are fastened three tubes, 1 inch diameter, reaching about half way down the can. In one of them, reaching down farther and by netting preventing the fry of getting in, is a tin pump screwed in. Once in a while this pump is applied; the water runs in the middle cover and on the top of the can, is prevented from running off by the sides of the can being run up as high as the cover, and so the water runs through the little openings back in the can, which is only filled half with water and is aerated in this style.

In the front of the can is, a little under half the height, a top-screw; this serves to change the temperature of the water in the can, if necessary. The water out of the river is put on the top, where it leaks through, while at the same time the water in the tin can runs off by taking the screw out.

The 10,000 fish were put in five cans at 6 a. m., and we had them in the river in a shoal, gravelly place at 1 p. m., with the loss of not a single one. Temperature of the cans 4½°, river 6° Celsius. January 11, 20,000 made the same trip in 6 cans under a sharp frost. The cans were covered with straw and a basket and had hot-water stoves between them in the cars; and by transporting them to the river, temperature 1½°, river 1½° Cels. Had to chop a hole in the ice to get the fry in. Loss, a dozen. January 18, about 20,000 went the same way, also in 6 cans; the weather thawing, no stoves used; river full of floating ice; temperature cans and river alike, 1½°; loss a couple. River gone down considerable—about 5 feet. The place of deposit of these 50,000 was in the river Maas, opposite the city of Venlo, in Limburg, Netherlands, near the railway bridge. After the first lot was put in, they have been noticed many days.

I selected that river because salmon is almost extinct and very bad for seining, having a clear gravel bottom, with very shoal water almost the year round, only in open winter or the spring a sudden rise two or three, and is even then not half as muddy as the Rhine.

In Gelderland, near Apeldoorn, I found a splendid brook with a fall at the end of three or four meters. No mills on, nor fish in it, only some sticklebacks, and teeming with food.

A few ponds were dug in the lower ground aside of the brook, which were fed by a screen and pipe through the dike. On the 7th of January, 5,600 were put in the ponds; sharp frost; northeast gale; two and one-half hours by rail, one-half by cart. In the cart only with a couple of hot-water stoves with them. Loss, ten in the two cans; pumps frozen. The fish cannot get away. There is plenty of pure spring-water and plenty of food. I put them there to see how they will thrive in captivity. In the Zoö at Amsterdam are left now 1,000 to be kept there and fed, and 4,000 which I intended to place in the ponds of the Loo, near Apeldoorn, the King's summer residence and his favorite place. He is principally there. The fry would have been put in long since, but first his marriage, and now the death of our very much lamented Prince Henri, prevented me to ask his permission.

You see thus we have not been doing so bad after all; out of the sound eggs packed out (82,000 we put on the trays) I have got about 62,000 fry deposited. Of *unimpregnated* eggs we found only a *couple;* of twin fish and crooked-backs about 300, and a few with water bellies. Size of the fry 26–30 days after birth, 4 and 4½ centimeters, quite a lot of them.

I suppose it is settled now that eels are oviparous and not viviparous. Can you give me anything about Professor Packard finding the spermatozoa? Was the eel caught in fresh or salt water? What success you have got with your cod-fish hatching? Somebody has told me the *S. quinnat* does not go to sea, but stays always in the rivers. What is the temperature of the water on the breeding places in McCloud River, California?

Did Mr. Livingston Stone find out something about the so-called *winter salmon* I wrote about to you last year?

The American oyster is taking in England so much that our merchants had to come down in price, although our oyster is much superior to the American; ours sell for £4 or £5 sterling per 1,000.

I am afraid a great many will have been killed by the frost we have had. Theromometer now at 0 since yesterday. Barometer stationary at 775, noon. From yesterday has slowly risen. Wind light, east; expect more frost yet.

If the English demand for American oysters keeps up, you had better look out, otherwise you'll be out before long. The French is a bad, poor thing, and the Portuguese is no oyster at all. I prefer our mussel to either of them.

I hope you will have this in hand before the meeting in February of the Fishcultural Society.

With kind regards, yours truly,

C. J. BOTTEMANNE,
Superintendent of Fisheries.

Official report on California salmon bred in the zoölogical gardens at Amsterdam for the Netherlands Government.

EGGS LOST.

Date.	Temperature of water.	Quantity.	Date.	Temperature of water.	Quantity.
1878.			1878.		
October 26	On arrival..	3,000	November 13	Celsius 9°	600
28	Celsius 12°	428	14	9	490
29	12	453	15	9	490
30	11½	520	16	9	390
31	11	1,850	17	9	420
November 1	11½	985	18	9	380
2	11½	672	19	9	360
3	11	820	20	9	370
4	11	800	21	8	350
5	11	1,100	22	8	170
6	10½	1,300	23	8	250
7	10½	1,450	24	8	87
8	10	1,480	25	8	280
9	10	1,006	26	8	126
10	10	840			
11	9½	620			22,497
12	9	610			

FISH LOST.

Date.	Temperature of water.	Quantity.	Date.	Temperature of water.	Quantity.
1878.			1878.		
November 27	Celsius 8°	113	December 25	Celsius 4½°	8
28	8	49	26	4½	5
29	8	47	27	4½	8
30	8	45	28	4½	8
December 1	7½	16	29	4½	7
2	7½	55	30	5	10
3	7½	75	31	5	38
4	7½	51	1879.	5	31
5	7½	24	January 1		
6	7½	24	2	5	24
7	7½	45	3	5	14
8	7	47	4	5	24
9	7	14	5	5	20
10	7	31	6	5½	17
11	7	23	7	5	16
12	6½	14	8	5	23
13	6½	11	9	4½	16
14	6	2	10	4½	23
15	6	8	11	4½	17
16	6	2	12	4	9
17	5½	8	13	4	13
18	5½	9	14	4½	7
19	5½	2	15	4	10
20	5	8	16	4	12
21	5	8	17	4	*290
22	5	8			
23	5	15			1,410
24	5	6			

* Twins, hunchbacks, &c.

FRY DEPOSITED.

1879—January 4. Maas, near Venlo, Limburg	10,000
January 8. Zwaanspring, Gelderland	6,000
January 11. Maas, near Venlo, Limburg	20,000
January 18. Maas, near Venlo, Limburg	20,000
	56,000

RECAPITULATION.

Eggs received	85,000
Eggs lost	22,497
Fry lost	1,410
Fry deposited	56,000
Fry kept for the Loo	4,000
Fry to be kept in the Zoölogical Garden	1,093
Total	85,000

November 10. The first fish broke shell.

November 26. Last fish out shell.

The few fry that died between November 10 and November 26 are put in with those of November 27.

Bergenop-Zoom, January 28, 1879.

C. J. BOTTEMANNE,
Superintendent of Fisheries.

XLII.—SUMMARY OF REPORTS FOR 1878, BY STATE FISH COMMISSIONERS RESPECTING THE INCREASE OF FOOD FISHES BY ARTIFICIAL PROPAGATION.*

MAINE.

Atlantic salmon.—The results of our salmon planting have been most satisfactory, as exhibited on the Penobscot, Androscoggin, St. Croix, and Medomac. On the Penobscot the yield has been very large. Perhaps no better summary can be given than the paragraph we here quote from the Bangor Commercial:

"The salmon fishers must now cease their fishing, as the close time for this season commenced to-day. The run has, in the main, been quite large. There was a time early in the season when they reached the remarkably low price of 8 cents a pound at Bucksport, but, as a rule, the prices have been good. A good deal of money is brought to the Penobscot Valley by this important industry."—July 15.

Since July the number of salmon on the east and west branches has been reported to us as very large by the river drivers. Parties of excursionists have likewise represented young salmon as being very numerous, and annoying much by their numbers, and rising to their flies when fishing for trout. On the St. Croix the yield was large for that river. It may be remembered that in 1873 we transported 10,000 salmon fry to Vanceboro, which we turned into the St. Croix at that place. In 1874, 50,000 salmon ova were hatched for us in the hatching-house of the Dobsis club, and turned into the St. Croix tributaries. We think the inference is fairly deduced that these contributions have materially added to the stock of the river. On the Androscoggin a good many salmon have been taken the last two or three years. Mr. Ambrose T. Storer, the fish warden at Brunswick, writes under date of August 25, 1878: "I have tried to ascertain the number of salmon caught on the Kennebec, but was unable to learn the exact number, but think it larger than usual. Mr. Trott caught seven in one day. I don't know how many have been caught in the small rivers tributary to the Androscoggin, but on this river the number caught by our fishermen was fourteen, which is more than has been caught before for some years. I

* The report of the commissioners of New Hampshire for 1878 contains an excellent summary of general results of fish culture by the States, from which numerous extracts have been made and inserted under their respective headings. The selections for this article have been made by Mr. C. W. Smiley.

have been hoping to see the fishway completed." Another gentleman writes: "Brunswick, July 6, 1878. I have the pleasure of informing you that a fine large salmon has been the admiration of many of our citizens, playing around above the falls near the short bridge. He was so tame that some one undertook to catch him by a spear or hook, and by that means wounded him, so that he was this morning found dead. Of course no one knows who did it, but it was learned with manifold regret though his existence establishes the fact that we have young salmon in our river. Now, if we had good fishways in good condition on our falls, there is no doubt but that we would have a plenty of these beautiful fish in our river." Still another, under date of July, 1878, says: "Can anything be done by us to enable you to have our fishways made more practicable?"

On the Medomac, "laige salmon have been seen jumping in the basin, above the dam, where such a sight has not been witnessed before for forty years." (Twelfth report of the commissioners of fisheries of the State of Maine, for the year 1878, p. 8–9.)

Alewives.—We transported seventy alewives in cans from Bucksport to Enfield, part way by wagon and the rest by railroad, on the 17th of May. The 10th of September the first school of young fry were seen on their way down to the Penobscot; two other schools followed at intervals of a few days. These fish, it is estimated, will make their first return from the ocean in two years. (Twelfth report of the commissioners of fisheries of the State of Maine, 1878, p. 17.)

We quote from several of their recent reports, as follows:

Maine says: "The salmon fisheries of the State have been largely productive, that of the Penobscot being reported as greater than for the last twenty-five years. The take of alewives in those parts of the State where fishways have been provided and the fish protected was likewise very large and remunerative. The most gratifying feature of this year's experience is the wide interest awakened in the State in fish culture among all classes, as evidenced in the extensive demand for brook-trout, land-locked salmon, and black bass to stock waters for private enterprises, as well as for towns and counties. The black bass we apply in all cases as an antidote to the worthless pickerel. It costs more to feed a pickerel than any other fish; it costs more to make a pound of pickerel than a pound of any other fish; the pickerel consumes everything that swims or that it can swallow; it is very destructive to young water-fowl.

"For the last four or five years large numbers of young salmon have made their appearance in the Penobscot River below Bangor. Even the Kenduskeag River, below Morse & Co.'s mill, has been full of them. Large numbers have been taken this year below the dam of the Holly Water-Works, at Treat's Falls, and in Barr's Brook, by both men and boys. In dipping for smelts in Brewer, sixty young salmon were picked from among the captured smelts in the course of two hours and returned to the water. They were recognized by an intelligent bystander and their distinctive marks pointed out, when all parties immediately took a deep interest in protecting them. One man, in fishing for suckers in the Kenduskeag, with coarse line and baited hook sunk on the bottom, caught sixteen young salmon in two hours, and carefully returned them to the water again. The Bangor Commercial says: 'On visiting his weir yesterday in Marsh River, Mr. Reuben Hopkins found one hundred and forty young salmon in it, varying from 8 inches to 1 foot. He turned them all loose in the river. We learn that these young salmon are found in all the weirs in the river in large numbers.'

"The salmon fishery of the Penobscot is estimated to be the largest for many years, so much beyond the product of years past as to leave no doubt in the minds of the most incredulous that the work of restoration by planting and protection is an entire and unmistakable success. Many of the salmon were of very large size.

"Of one of the large fishes the following paragraph, cut from the Belfast Journal, will be read with interest, as conveying some important facts in relation to their growth and habits: 'In our issue of May 3, we made mention of a very large salmon caught at Cape Jellison, Stockton, by Josiah Parsons, and purchased by Frank Collins, of this city. The fish measured 50 inches in length and weighed thirty-three and one-half pounds. Attached to the fish was a metallic tag, numbered 1019, indicating that it was one liberated from the Bucksport breeding works. The tag was forwarded to Mr. Atkins, the superintendent of the works, who keeps a record of all fish used for spawning purposes and then liberated. We now chronicle the record of the fish as learned from a letter from Mr. Atkins to Mr. Collins. He writes that the salmon was liberated at Bucksport, November 10, 1875. It was a female fish, 39½ inches in length, and yielded five pounds and six ounces of spawn, or about 16,000 eggs. After spawning it weighed sixteen pounds. He judges that in the preceding May (1875) the fish weighed twenty-five pounds; thus the fish in two years had grown nearly an additional foot in length and eight and a half pounds in weight. One important fact in the habits of the salmon has been demonstrated by the use of these tags, and that is that the fish, after it becomes large, does not visit the river every year, as was formerly supposed, but only every second year. Those liberated in the Penobscot in 1873 were caught again in 1875, and those let loose in 1875 are now being caught. One dollar premium is paid for every tag thus found. Among others of the large fish, one was taken at Veazie, by Mr. Albert Spencer, weighing 38 pounds. The salmon presented by our worthy mayor, Dr. A. C. Hamlin, to Mayor Prince, of Boston, and which was captured at Sandy Point on the Penobscot, was said to have weighed forty pounds when first taken. A very good run of salmon has visited the St. Croix the last year.' (New Hampshire fish commission report for 1878, pp. 25, 26, 27.)

"In 1873 some thousands of young salmon were turned loose by us in the St. Croix at Vanceborough. In 1874, 50,000 fry were hatched and turned into Dobsis stream for us by the courtesy of the Hon. Harvey Jewell, of the Dobsis club. The inference is but fair that these contributions to the stock of the river had a marked influence in adding to the number that constituted the good run of this year." (New Hampshire fish commission report, 1878, p. 27.)

NEW HAMPSHIRE.

Atlantic salmon.—More salmon were seen during the summer at Amoskeag Falls than were noted in the fish-way at Lawrence; and a pair of very large ones, estimated by Mr. Kidder at sixteen or seventeen pounds each, were seen about September 1, and a similar pair were seen by Mr. Powers, jumping the falls above the hatching-house, September 15. (Report of the fish commissioners of New Hampshire, 1879, p. 4.)

"Atlantic salmon, 7 inches long, of the planting of 1876, were so plentiful up to about the middle of August that it was impossible to fish without frequently hooking them. Mr. R. R. Holmes actually hooked three at one cast, and remarked that the river was alive with them. In August they began to disappear, and at this date very few are seen. On the 6th of November I dipped up a small Atlantic salmon, about 3 inches long, at the outlet of the hatching-house brook, which must have resulted from last year's run of salmon in this river, as there has been no plant since 1876, which, as above stated, have grown to the length of 7 inches." (Report of the fish commissioners, New Hampshire, 1879, p. 11.)

Quinnat salmon.—"The California salmon fry turned into the river in 1878 were very numerous up to the last of July, and had grown to the length of about 3 inches. On the 20th of June they were so plenty as to be seen in numbers in any locality near the hatching-house." Report, p. 11.)

Salmonidæ.—"L. D. Butler, of Woodbine, writes March 23, 1877: 'The California salmon, planted in our streams last February a year ago, are now from 7 to 9 inches long. One of the former plant was caught that weighed one and a half pounds.'

"A. A. Mosher, of Spirit Lake, writes March 13, 1877: 'The fish you sent us last year are doing wonderfully well. They are now about 7 inches long and take to these waters.'

"Large numbers of letters and newspaper paragraphs of this kind are in the possession of the commissioners, and these are given only as samples, while great numbers of people have given testimony as to having seen and caught the young fish.

"Mr. E. Bush, station agent, reports the catching of a dozen salmon, weighing two and a half pounds each, in the North Fork of the Maquoketa.

"The principal of the high school at Marion reports catching a half dozen, weighing from one and a half to two pounds each.

"Dr. French reports having seen one at Davenport that would weigh two and three-fourths pounds.

"George Brown caught two in Wapsie that would weigh one and a half pounds each.

"Mrs. H. Ruble has in her pond at North McGregor a number of Penobscot salmon, three years old, some individuals of which will, it is estimated, weigh ten pounds. They have never been out of the pond they are now in, and, notwithstanding their confinement in fresh water, are perfectly healthy and hearty, and as fine a sight as it is possible to conceive of." (New Hampshire fish commissioners' report, 1878, p. 31.)

"As the salmon did not loiter, but passed quickly over, it is fair to conclude that hundreds passed up unnoticed; and this conclusion is confirmed by well-authenticated reports of the large number seen at Manchester as well as all along the Pemigewasset.

"Mr. Tomkinson, of Livermore Falls, counted twenty ascending the rapids in about two hours. Indeed, so common a thing was it to see them scaling the falls, that the White Mountain stage frequently stopped on the bridge to allow the passengers to see them. Mr. White, of Boston, who spent the summer at the Profile House, reports having seen, in one pool, thirteen large salmon from 2½ to 3 feet long.

"The report shows that forty-seven salmon were found in the fish-way during an examination of thirty minutes a day for twenty-eight days. If we assume the running time at twelve hours a day, the total number that passed over would be in this proportion, 47 by 24=1,128 salmon, to which must be added a certain number that passed over in October. Taking the weights as roughly estimated, we may say that about one in seven were rather small fish, of about eight pounds; one in seven were large fish, of fifteen pounds or more; and the great majority, or five in seven, were medium salmon, of ten or twelve pounds.

"The following table will show the dates at which the batches of parrs were put in the river and their respective ages up to the spring of 1877:

Put in the river.	1873, spring.	1874, spring.	1875, spring	1876, spring.	1877, spring.
Spring 1872, 16,000 parrs	1 year old	2 years old	3 years old	4 years old	5 years old.
Spring 1873, 185,000 parrs		1 year old	2 years old	3 years old	4 years old.
Spring 1875, 230,000 parrs				1 year old	2 years old.
Spring 1876, 400,000 parrs					1 year old.

"The few salmon of fifteen to eighteen pounds that ran up may have been of the batch of 1872; the smallest, of six and eight pounds (including those of the October run) may have been late or under-fed fish. Evidently the bulk of the salmon were of the plant of 1873, because the sixteen thousand parrs put in the year previous could not by any calculation have furnished one-fifth of the adult salmon that returned in 1877."

"PLYMOUTH, *February* 22, 1878.

"SAMUEL WEBBER, Esq.:

"DEAR SIR: It is with pleasure that I answer your inquiries in regard to the salmon that came up the Pemigewasset River this season. And at the same time allow me to congratulate you upon the complete success that has attended the labors of the Massachusetts and New Hampshire commissioners in their attempts to restock our beautiful river with the king of fish, the *Salmo salar*.

"The work is no longer an experiment, but an assured success, as not a single salmon has been seen in the Pemigewasset until this year since the erection of the dam at Lawrence some thirty years ago. That they have returned this year in large numbers is beyond a doubt. And this fact must be a source of congratulation to the gentlemen who had charge of the work of transferring the young fry from the hatching house at Winchester to the headwaters of the Pemigewasset and Baker's Rivers, as they no doubt will remember the discouragement and even ridicule they met with from the time they would leave Winchester until the cans were emptied into the river. It is impossible to say at what time the first salmon made their appearance at Livermore's Falls, as no one was looking for them.

"On my return from New Brunswick the last of June, I learned that the salmon were passing the fish-way at Lawrence. I immediately requested the Messrs. Tompkinson, at the falls, to watch for them. The first one was seen about the 1st of July. It was a full-grown fish about 3 feet in length, and for several weeks following there was hardly a day but what they could be seen in their endeavors to pass over the falls. I have counted from eight to ten in an hour, but do not think they were different fish, for it was very seldom that one would pass the rapids at the first attempt. Many of them would make leaps of 10 to 15 feet and pass up, but if they fell short of that they would be carried back into the pool below. It is impossible to say how many passed the falls, but there must have been some hundreds. They were seen all along the river as far up as the Woodstock dam, but as far as I can learn, and I have made careful inquiries, none have been sent above that point. The young salmon went many miles above there this season.

"Full-grown fish, that is, from 2 to 3 feet in length, were seen by many from the bridge in this village during July, and at the falls as late as November.

"There must be more stringent measures taken to prevent the destruction of the small salmon by fishermen, as they were taken by hundreds this season.

"I remain, respectfully, yours,

"E. B. HODGE."

We have given Mr. Hodge's letter exactly as received, but other advices lead us to believe that the heavy rains of the first week of September, 1877, carried away so large a part of the dam at Woodstock as to give a free passage to the large salmon, thus allowing them to ascend the river nearly to the Profile House, as stated in the report of the Massachusetts commissioners.

Mr. Tompkinson's letter is interesting as giving positive details of his observations, and from it we quote as follows:

"The first of our seeing the salmon go up through the Livermore Falls was in the early part of July, 1877, when our attention was called by Mr. Hodge to see if we could see any salmon going up the falls. The first day we saw seven, at four different times during the day, stopping only about ten to twenty minutes each time. This was the first day we began to look for them. We reported the same to every one that came along. Almost every day afterward, for about six or seven weeks, there were salmon seen. The largest number in one day (seen by my brother) was twenty. I myself saw five go up in forty-five minutes. We never lost much time in watching for them, as we could not afford to lose any time, for we have so much work on hand. I saw eleven on another day in about two hours. On another day my brother saw seventeen in about two hours. We never stopped a whole day to watch at one time, but state what we have seen. The above were seen about the 20th of July. The largest one my brother saw was nearly 3 feet long, and he was within 8 feet of it when he saw it. * * *

"Yours, most truly,

"J. R. TOMKINSON."

We must now go back a little in our dates to connect the thread of our story. Early in June we were notified by Mr. Brackett, of the Massachusetts fish commission, that salmon were passing up the fish-way at Lawrence.

The dam at Lowell had recently been rebuilt, discarding the old fish-way, but running the north end of the dam on to a gently-sloping ledge in such a manner that it only needed to take off one flash-board to leave an easy passage 10 feet wide, with 12 or 16 inches depth of water, over a fall of about the same height, and the fish found no difficulty in passing it; so that on the 13th of June we were notified by Mr. Kidder, the keeper of the gates and locks of the Amoskeag Company at Manchester, that he had secured for us the first salmon seen at Amoskeag Falls for thirty years. We had requested Mr. Kidder to look out for the first fish that came up and let us know, and he had done so literally. The fish was a male, apparently of four years of age, two feet four inches in length and a half inch in depth, and weighed eight pounds and five ounces.

Mr. Kidder unintentionally transgressed the law in his anxiety to please the commissioners, but his fine was settled by his many friends in Manchester, and the salmon that followed were allowed to pass "free of toll." Within a week from the capture of this first one a report was brought us by Conductor Colby, of the Concord Railroad, of a large salmon having been seen at the mouth of Martin's Brook, four miles above Manchester; and almost daily after that date we heard of them farther and farther up the Merrimack River. (New Hampshire fish commission, 1878, pp. 6, 7, 8, 9, 10.)

"In addition to the above record there was a full run of salmon, which commenced October 11 and ended October 30. These fish, so far as seen in the way, were from six to ten pounds in weight. Much larger ones may have passed over, as Mr. R. R. Holmes saw one 3 feet long near the hatching-house, at Plymouth, the 1st of November." (New Hampshire fish commission report, p. 6, 1878.)

MASSACHUSETTS.

Schoodic salmon.—Some of the land-locked salmon received from Maine were turned into Halfway Pond in Plymouth. The returns received from many of those who had charge of these fish are very favorable. It is quite certain that they are well established in Halfway Pond. And in Mystic Pond, situated in Medford and Winchester, where they were first introduced, they are appearing in considerable numbers. On the 11th of September a land-locked salmon, 22½ and a half inches long and weighing three and one-quarter pounds, was caught in Lower Mystic Pond by a boy while fishing for perch. The boy, not knowing what it was, sold it to J. P. Richardson, of Medford, who forwarded it to the commissioners for identification. A careful inspection of the pond, made in October, showed quite a large school of them, weighing from two to eight pounds each, at the mouth of one of the streams entering the pond. The large fish are probably the Sebago salmon, put in about six years ago. One of the persons making the inspection hooked one of them; but, being in a small cloth canoe, barely large enough to carry one person, and having the fish on a light fly-rod, he found it impossible to get him into the boat; and, in attempting to reach the shore, the salmon recovered himself, and with a sudden leap left hook, line, boat, and fisherman behind him. (Thirteenth annual report of the commissioners of inland fisheries for the year ending September 30, 1878. 8vo. pamph., Boston, 1879, paper, p. 13.)

Atlantic salmon.—Our experience with young salmon in the Merrimack shows pretty conclusively that they do not go down to the sea until the third year. The salmon put in the river in 1876 have been carefully watched, and were found to be very numerous all along the river, especially near the mouths of trout brooks, showing no disposition to change their quarters until about the middle of last August, when they began slowly to move downstream. (p. 18.)

Atlantic salmon, 7 inches long, of the planting of 1876, were so plentiful up to about the middle of August, that it was impossible to fish without frequently hooking them. Mr. R. R. Holmes actually hooked three at one cast, and remarked that the river was alive with them. In August they began to disappear, and at this date very few are seen. On the 6th of November I dipped up a small Atlantic salmon, about 3 inches long, at the outlet of the hatching-house brook, which must have resulted from last year's run of salmon in this river, as there has been no plant since 1876, which, as before stated, have grown to the length of 7 inches. (p. 19.)

Quinnat salmon.—The California salmon-fry turned into the Merrimack River in 1878 were very numerous up to the last of July, and had grown to the length of about 3 inches. On the 20th of June they were so plenty as to be seen in numbers in any locality near the hatching-house. (p. 19.)

Condensed report of Thomas S. Holmes, of fish found in the Lawrence fish-way from May 1 to August 1, 1877.

May 31. Two salmon, 12 to 18 pounds each.
June 2. Two large shad.
3. Three large shad.
4. One salmon, 12 to 18 pounds.
10. Two 12-pound salmon.
11. One 8-pound salmon.
12. Two 6 to 8 pound salmon.
13. One 10-pound salmon.
14. One 8-pound salmon.
15. One 8-pound salmon.
16. One 10-pound salmon.
19. One 18-pound salmon.
20. One salmon.
22. One 8-pound salmon.
23. Three 12-pound salmon.
25. One 10-pound salmon.
26. One 12-pound salmon.
28. Two 8-pound salmon.
29. One 10-pound salmon.
July 1. One 12-pound salmon.
2. Two 10 or 12 pound salmon.
3. Two 10 or 12 pound salmon.
4. Four 10 to 15 pound salmon.
6. Five 8 to 18 pound salmon.
7. One salmon.
9. One 12-pound salmon.
12. One 8-pound salmon.
From this to the 23d no salmon.
23. Three 8 to 12 pound salmon.
30. Two large salmon.

(Extract from the twelfth report of the fish commissioners of the State of Connecticut, 1878, pp. 10, 11.)

STATE OF MASSACHUSETTS,
DEPARTMENT OF INLAND FISHERIES,
Winchester, Mass., October 12, 1877.

MY DEAR HUDSON: The rise in the river has brought another run of salmon, which are now passing over the fishway at Lawrence. There seems to be no end to our success on the Merrimack.

Yours, with sincere regard,

E. A. BRACKETT.

RHODE ISLAND.

Atlantic salmon.—We have to report that a good many salmon have been taken in the past two years that we have record of, and no doubt many more unreported.

The largest weighed ten pounds, and was taken at the foot of the falls, at Pawtucket, last June. Smaller ones were taken in the Pawtuxet, between the first dam and Pontiac, and a number near Westerly, below the first dam on the Pawcatuck; none larger than two and one-half pounds. (Eighth report of the fish commissioners of the State of Rhode Island, 1878, p. 4.)

CONNECTICUT.

Atlantic salmon.—Under date of June 18, 1878, Mr. D. W. Clark writes from Saybrook, Conn.:

"The first salmon caught this season was taken in a gill-net, April 30, and weighed 12 pounds. From that time to May 25 salmon were cought more or less nearly every day. Since May 25 they have been more scattering, so that from that date to June 18 but three have been taken. The above number does not include any that the pounds have liberated when caught, but those only which have gone to market.

"The average weight of those caught has been about fourteen pounds. The whole number taken up to this date in the towns of Saybrook and Westbrook is forty-five, of which three-quarters were caught by gill-nets on the river and one-quarter in seines on the river and pounds on Long Island Sound."

Under date of July 12, 1878, Mr. Clark again writes:

"I give you full results of the season of 1878. The salmon caught by the pounds and put back into the water may be given as about twelve, and the whole number caught by pounds as about thirty. In the river the salmon caught by gill-nets were almost all taken while the water was thick with mud in freshet. Experience this season proves that the gill-nets are not sufficiently strong for taking salmon, nor are they of the right-sized mesh. The fishermen found many torn places in the nets, which had the appearance of being caused by salmon. When these fish are caught the nets are hanging slack in the water and the fish are caught by many folds of twine. But when the current is strong and the meshes are all drawn the salmon easily break through.

"I have been unable to obtain the number of salmon caught by gill-nets in Lyme, but the dealers estimate them from seventy-five to a hundred."

The commissioners have authentic evidence that the greater portion of the salmon caught were sent out of the State. Not less than three hundred and twenty-two were sent to the New York markets, and they are reported as weighing about twelve pounds apiece on an average, and to be superior to every other salmon in the market. From all the facts which the commissioners have been able to gather, they feel no hesitation in asserting that over *five hundred* full-grown marketable fish were caught in and near the river during the past season, and with the exception of the few reported as returned to the water from the pounds, every one of them was destroyed; a most lamentable example of reckless improvidence and wastefulness. (Thirteenth report of the commissioners on fisheries of the State of Connecticut, 1879, pp. 5, 6.)

About a dozen salmon, weighing each from nine to eighteen pounds, have been taken in the Connecticut River or the pounds west of its mouth during the past season, but no information has been given your commissioners of even one having succeeded in passing above Portland. Great numbers of the young, from one to three years old, in good condition, have been seen in different parts of the river and some have been taken, specimens of which have been sent to your commissioners. (Page 10.)

NEW YORK.

Trout, &c.

DEPOSIT, *October* 26, 1877.

SETH GREEN, Esq.:

DEAR SIR: Yours of the 15th instant received. I have not had an opportunity to observe the condition of the brook trout placed near the head of the Oquago Creek, but those we placed in a little tributary near this place are doing well, and there are no reasons to doubt that the others are doing equally well. They were about 3 or 4 inches long when I saw them. The trout placed in the lake two years ago and last spring have not been heard from. I do not think there has been any fishing specially for them. There is no reason why they may not do well, as the water, depth, and bottom are adapted to that kind of fish. The black and the rock bass put in the lake six years ago last spring have increased wonderfully. A great many fine bass have been caught this fall, ranging from one-half to three pounds six ounces, the largest that has been taken. There will be fine fishing next year. A few have been taken in the Delaware; they probably came from the lake, as they were caught below the mouth of the outlet. We have succeeded in having a law passed removing the eel-weirs, which will make it an object to stock the river. I think it would be advisable to place a quantity of young bass in the river at this point this coming winter and spring; it would be better to place them in after the spring ice-freshet, if possible, as they would not then be liable to be driven down the stream and killed by the ice-jams. The west branch of our river is equally as good as the waters in the main river at Port Jervis, where large quantities have been taken in the last three months. I should be pleased to hear from you, if you have any advice or suggestion. I shall take the first opportunity to examine into the condition of the trout in the creek, and will inform you if I find anything new.

Yours, respectfully,

F. STURDEVANT, M. D.

(Tenth annual report of the New York commissioners of fisheries, 1877, p. 45.)

COOPERSTOWN, *October* 20, 1877.

Friend GREEN:

DEAR SIR: Mr. Jarvis informed me that you would like to know about the fish we put into the streams and lakes. The salmon-trout are increasing very fast. One man took in one day, a-trolling, seven trout, the smallest weighing two pounds, the largest six pounds. Another man caught eight, and had a number of more near the boat, all in the same day. This was in June; and hundreds of smaller size were seen. The brook trout are all right and are doing well; also black bass. One man took four at once catch, and lost a number of more. We shall want some more whitefish and trout this winter. Hoping this will find you well, I remain as ever,

Your friend,

A. W. THAYER.

(New York fishery commissioners' report, 1877, p. 47.)

CORNING, N. Y., *October* 20, 1877.

SETH GREEN, Esq.:

DEAR SIR: Yours received, and in reply I would say the trout received from you are doing nicely in all the streams, and we expect to have fine trouting in this section again. The black bass are multiplying very fast in the Chemung River, and fine strings are taken below the dams here. The State dam is 8 feet high, and proves an obstacle that a fishway would overcome. The canal will probably be abandoned in another year; if not, the canal commissioners will be required to comply with the law.

Yours, respectfully,

J. H. WAY.

(New York fishery commissioners' report, 1877, p. 46.)

Black Bass.

OSWEGO, *October* 16, 1877.

SETH GREEN, Esq.:

DEAR SIR: Your postal card of the 15th instant, making inquiries as to what has been heard from the fish shipped us from the New York State hatching-house received, and I reply that the Susquehanna River stocked with black bass seems to be well stocked. There were a few caught last year, but this year, I presume, there has been taken tons of them in the river within eight miles each way of this place. It does not seem possible that the fish put in here could have produced as many fish as there seems to be in the river. There seems to be no end to the bass. I have only been out once this year, and then a gentleman and myself caught 35, several of which would weigh at least two pounds each.

There have several been caught that weighed as high as three pounds and over.

As to the trout in the small streams, I cannot tell what they will come to yet, as it has not been long enough yet.

Yours, respectfully,

BARNEY M. STEBBINS.

(Tenth annual report New York fishery commissioners, 1877, p. 47.)

Atlantic salmon.

"PEEKSKILL, N. Y., *March* 11, 1878.

"I wish to mention the capture of a salmon, a true *Salmo salar*, in the Hudson, about two miles north of our village. It was taken on the flats this morning, near the mouth of Snake Hole Creek, just below Iona Island, in an ordinary seine, while its captors were hauling for perch and other small fish. It measured 33 inches in length and weighed but 8½ pounds, being in very poor condition, and presenting the appearance of having recently spawned. Small fish of this species have been taken through the ice during the past winter in T-nets, but nothing approached this in size. I regret my inability to forward you the fish, but it was disposed of before I saw it. Am I justified in supposing it to be one of the fry introduced into the upper part of our rivers a few years since?"

This confirms the observations of Mr. Atkins, that the salmon which spawn in the fall and winter of each year return to the salt water the year following, and again return to the fresh water the next year; so that while one stock of spawners will ascend the rivers in the even years, as in 1874, 1876, 1878, &c., another body of fish comes up in 1875, 1877, 1879, &c. (New Hampshire fish commission report, 1878, p. 29.)

Shad.—Forest and Stream says: "Syracuse papers of the 10th instant are congratulating Mr. Seth Green upon accumulating evidence of his success in cultivating shad in Lake Ontario. Very recently a fine male shad, weighing five and a half pounds, was caught in a gill-net, six or seven miles out in Lake Ontario, off Port Ontario, at the mouth of Salmon River. The fish is the largest of its kind yet caught in the lakes, and is one of those placed in its waters by Mr. Green in the year 1872. The attempt to introduce this fish in fresh water was an experiment. It is now no longer in the list of experiments, but a matter of certainty. The fish have been caught at various points on the lake ever since the fry were put in, and appear to grow as rapidly and possess all the qualities of the shad that are caught from salt water. (New Hampshire fish commission report, 1878, p. 29.)

NEW JERSEY.

Atlantic Salmon.—No attempt has been made to capture any adult salmon which may have returned to our river, and it was not expected that they would make their reappearance until four or five years after they were placed in the stream. In the spring and summer of 1877, however, six or seven fish were taken in shad nets at different points on the river. They were medium-sized fish, averaging about ten pounds, but had evidently been to the sea and had returned to the river to deposit their eggs. This was deemed highly encouraging, and the next season was looked forward to with much anxiety by those who were interested in fish culture and who appreciated the immense importance of the success of the efforts to establish this valuable fish in the rivers of the State. On the 5th of April, in the present year, a magnificent salmon was taken in the Delaware River, within two miles of Trenton. This fish, which was three feet five inches in length and weighed twenty-three and a quarter pounds, came into the hands of the commissioners and was by them forwarded to Prof. Spencer F. Baird, at Washington, who addressed the following letter to one of the commissioners:

"UNITED STATES COMMISSION FISH AND FISHERIES,
"*Washington, D. C., April* 11, 1878.

"DEAR SIR: You have rendered the United States Fish Commission a very great service by sending on the specimen of Delaware salmon as advised in yours of the 6th of April. It reached me in good condition Tuesday, and I have already had the pleasure of exhibiting it to the President, and the greater part of his Cabinet, and a number of members of Congress who are interested in such matters, and who came to witness the realization of the efforts made toward stocking the Delaware with this noble fish. I shall have a plaster cast made, colored from nature, and the specimen itself will be prepared and kept in alcohol in a jar of suitable size. I am waiting the result of a conference of some experienced salmon fishermen as to whether this is to considered as a fresh-run fish from the sea, or a fish that has been in the river all winter, as is quite frequently the habit of salmon. The slight development of the hook of the jaw is rather an indication of the former supposition.

"From the size of the fish, I incline to refer it to the lot of Rhine salmon of which about 500,000 eggs were imported in 1873, but which, owing to the unprecedented heat of the weather in Germany and on board the vessel, arrived in poor condition, only about 5,000 surviving, and being hatched out at Dr. Slack's place at Bloomsbury. These were introduced into the Musconetcong, and doubtless made their way to the sea. A fish of this weight would require five years for its growth.

"I hope you will continue to gather all the data possible in regard to the occurrence of salmon in the Delaware, and that you may be able to detect among them some of the California salmon, which should be making their appearance.

"I am happy thus to open a communication with yourself as one of the commissioners of New Jersey, and shall take pleasure in acting with you in the promotion of the common work of stocking our rivers with useful food-fish.

Yours, truly,

"SPENCER F. BAIRD,

"*United States Commissioner of Fish and Fisheries.*

"To E. J. ANDERSON,

"*Commissioner of Fisheries, State of New Jersey.*"

During the shad season, which closed below Trenton June 10, and above Trenton June 15, 1878, a number of salmon were taken by shad fishermen at different points on the Delaware. It has been impossible to procure information of all that were taken, but a sufficient number were reported to warrant the assertion that from fifty to one hundred were taken before June 10. All of those reported to the commissioners were larger fish than any of those taken in the preceding year, and ranged in weight from 12 to 29 pounds, only two or three weighing less than 15 pounds. After the shad season closed and the nets were taken from the water, there was nothing to interrupt the progress of the salmon from the sea to the headwaters of the stream, and doubtless many passed up and deposited their eggs, since the commissioners are informed of a number of large ones having been seen at different points in the river between Trenton and Port Jervis.

In the Raritan River, one large fish was taken near New Brunswick in the summer of 1878; but none have been reported as yet from the Passaic and Hackensack Rivers.

The facts above stated concerning the presence of salmon in the Delaware were deemed to go far toward demonstrating the success of the efforts to convert that river into a salmon stream. (Report of the commissioners of fisheries of the State of New Jersey, 1878, pp. 15, 16, 17.)

New Jersey makes a very favorable report of the general progress of fish culture. Shad are increasing in numbers, and very greatly in size and quality; and salmon have made their appearance in the Delaware, as mentioned by the Maryland commissioners, nine having been taken this year, though their report does not say whether they were Penobscot or California salmon, both of which have been planted. Two were taken at Newcastle in May; two at Riverton in May; one between Bordentown and Trenton in May; two at the Delaware Water Gap in October; one in October at Carpenter's Point, the extreme northwest corner of New Jersey, and one in the Bushkill in November. The fisherman who took the two at the Gap was ignorant of the species till informed by Mr. A. A. Anderson. The taking of the five last mentioned, in the fall, and so far up stream, some sixty or one hundred miles above tide, shows that they were seeking spawning-grounds at the headwaters of the river, and, if of the California variety, except the last, at the usual season of their spawning. Whether others have been taken by persons ignorant of their kind, we know not. It is fair to suppose, however, that not all those that returned from the sea were taken. Many, measuring from 6 to 12 inches, have also been caught the past season with the hook.

The commissioners also report an enormous increase from the black bass that they have previously distributed in various waters, and excellent fishing obtained from this source. They have distributed nearly 10,000 of these fish this year, besides 4,230,000 shad-fry, 400,000 smelts, and 250,000 California salmon, and are now earnestly at work on fishways. (New Hampshire fish commission report, 1878, pp. 29, 30.)

PENNSYLVANIA.

Atlantic salmon.—The Free Press, of Easton, Pa., under date of November 10, 1877, says:

"We referred briefly yesterday to a salmon being captured in the Bushkill, and have since verified the report.

"The fish was discovered in Groetzinger's mill-race, on the Bushkill, at the foot of Fourth street, and its unusual size immediately attracted the attention of a number of people, who resorted to various devices for its capture. Hooks and lines were used, and it was hooked but broke loose. It was also shot with bird-shot. This did not kill it. It was finally shot with a rifle by a young man named James Young, the bullet passing into its body and stomach just at the junction of the head and body, and the strange fish was secured. Mr. Young presented his prize to his uncle, Mr. J. E. Stair, and it was very generally believed to be a salmon. Mr. Stair appreciated its important bearing on certain mooted points of the history and habits of this fish, and, in the interest of fish culture thoughtfully placed it at the disposal of Fish Commissioner Howard J. Reeder.

"The point at issue with scientific men, referring to salmon, is whether this fish placed in rivers as far south as the Delaware and Susquehanna will, with the instinct of their class, return to the grounds where they were hatched, and as nothing but experiment will prove this, the importance of all evidence bearing upon the controversy will be realized. At different times during the past four years a great many thousand salmon eggs and salmon fry have been deposited in the Bushkill and Delaware Rivers, under the supervision of Commissioner Reeder, and at various times reports have been circulated of salmon of considerable size being caught at Bordentown, Trenton, Carpenter's Point, and other points on the Delaware, ranging from five to eight pounds weight; but, unfortunately, these have fallen into the hands that did not perceive anything in the fact beyond the table, and their evidence was lost to the scientific world. But this fish is a fact, and in official hands will be irrefutable evidence that the stocking of our rivers with the most valuable fish in the world is not visionary, but practicable." (Report fish commissioners Pennsylvania, 1878, p. 9.)

MARYLAND.

Atlantic salmon.—We were not hopeful of any results from the introduction of the salmon of Maine, as it has been known only in the coldest waters. We, therefore, devoted our attention rather to the salmon of the Pacific slope, which, on the contrary, were known to ascend rivers in which the water at times reached a very high temperature. Contrary to our expectations, the true salmon have returned to the Delaware River in some abundance, a great many adults having been taken during the last two years in this river. On the night of the 11th of May Mr. Frank Parr, one of the gillers of Havre de Grace, who had been in the habit of furnishing us with the ripe shad taken in his gill-net, secured the first adult Atlantic salmon of which we have any record taken in Maryland waters. This fish was a female, measuring 3 feet 4½ inches, and weighing about seventeen pounds, fresh run from the sea.

The fish was captured off Spesutie Island, having been entangled in the gill-net, which was much torn; and Mr. Parr, who captured it, is confident that at least one other fish accompanied the one taken, but made its escape.

The gillers are in the habit of having their nets much torn by sturgeon, and no doubt have attributed to them many casualties which may have been occasioned by salmon. (Report Fish Commissioners Maryland, 1879, p. xiv.)

VIRGINIA.

Salmon.—"To us in Maryland of more importance is the reappearance in the Delaware of salmon of both varieties, the *Salmo salar*, of the North Atlantic, and the *Salmo quinnat*, of the Pacific. I have received authenticated accounts of the capture of one weighing eight and a half pounds, at Newcastle; one weighing eight and a quarter pounds, at Riverton; and one weighing nine pounds, taken between Bordentown and Trenton; and have myself seen a large female Penobscot salmon, with the mature eggs running from her, which was taken at Easton in the act of spawning; and there have been several others reported, even weighing as high as twenty pounds. These indications of the successful introduction of salmon into the Delaware, commenced two years prior to the establishment of a fish commission in Maryland, strengthen our hopes and confidence in the result of our efforts. More important to us still than the accumulated evidences of the laws which govern the migrations of the salmon are the proofs which have been added during the year that these laws are as surely applicable to the migrations of the shad." We cannot spare space to quote further from the very full and interesting report of Major Ferguson, and can only say that the hatching-house at Druid Hill Park is working very successfully on salmon and trout, while outside the commissioners are devoting their chief attention to shad and smelts, with every prospect of success, which another year will manifest in all probability.

Commissioner Moseley, of Virginia, says: "In the fall of 1876, our limited means being devoted to trout and land-locked salmon, we turned over the State's quota of California eggs to the Maryland commission. In return, that commission hatched and deposited, of young salmon, during the winter of 1876–'77, in the Shenandoah, 78,400; in Occoquan, 16,000; and in Goose Creek, Loudoun, 32,000. Besides, a very large portion of the above hatch was deposited in other tributaries of the Potomac, in the fish of which stream the people of this State have a common interest. It is no longer deemed a problem that this salmon will flourish in our waters and return by instinct to the stream in which it spent its infancy. Several have been caught in the Delaware and Susquehanna Rivers, the first streams in which they were placed, weighing from ten to fifteen pounds. In May last a fish weighing four pounds was caught in James River, at Bosher's Dam, nine miles above Richmond. The fisherman, never having seen such a fish before, brought it to Manchester, where gentlemen familiar with the *Salmonidæ* recognized it as one of that family. No doubt it was a California grilse, one of the lot put in James River in 1874–'75, at Lynchburg, by Dr. Robertson, which had straggled back before its time. It is probable that more of them may make their appearance next spring. Of the large number of these fish placed in James River by the commission, in the winter of 1875–'76, we hear that early in April last several were caught, from nine to ten inches long, twelve miles above Norfolk. They were said to be moving in solid column and with great rapidity oceanward, and only the few that fell out of line were captured in fyke-nets." (New Hampshire Report 1878, pp. 27, 28, 29.)

MISSISSIPPI VALLEY.

Shad.—The report of Kentucky has not yet come to hand, but the report of Iowa states that "shad were caught at several places on the Ohio River, the most notable case being at Louisville, Ky., where the catch during the run was reported at from forty to one hundred per day." This was in May and June, 1877, and a letter from Prof. Spencer F. Baird, United States Commissioner, to Forest and Stream, vouches for the fact that "a specimen sent him was the genuine white shad." Other letters to Forest and Stream state that 600 genuine Atlantic shad were caught at Louisville during the season, and trace them to the young fry planted, in behalf of the United States Fish Commission, by Seth Green in 1872, viz: 30,000 in the Alleghany River, at Salamanca, N. Y., and 25,000 in the Mississippi River, near Saint Paul; and 200,000 planted by the United States Commissioner in July, 1872, also at Salamanca. "In 1873, 100,000 shad-fry were placed in Greenbrier and New Rivers, in Virginia, and about 55,000 in the Monongahela, in Pennsylvania, and the Wabash, in Indiana; and these may or may not have contributed toward the supply met with at Louisville. The latter is possible, if the assumption of a four years' period is correct. If five years are required, then we must look to the stock of 200,000 in 1872 exclusively." (New Hampshire Fish Commission Report, 1878, p. 33.)

CALIFORNIA.

"*Shad*, in their season, are becoming quite numerous in the Sacramento River. The experiment of their importation to this coast has resulted satisfactorily. The river is of proper temperature, and furnishes an abundance of food for the young fish before they go to the ocean. There can be no doubt that the first shad brought from the Hudson River in 1871 have been to the ocean, returned and spawned. No shad were placed in this river during the years 1874 and 1875; yet shad two years old were quite numerous this year, and they must have been the product of the first importation.

"It may be safely asserted that we now have shad born in the Sacramento. As it is illegal to take this fish prior to December of this year, probably there has been no systematic fishing for them, yet numbers have been accidentally caught in traps and nets; probably not less than 1,000 were thus taken during the winter and spring of 1877." (New Hampshire Fish Commissioner Report, 1878, p. 35.)

XLIII.—CHEAP FIXTURES FOR THE HATCHING OF SALMON.

By Chas. G. Atkins.

1.—SCOPE OF THE PAPER.

It is proposed to limit this paper to the consideration of the construction, fitting, and management of the simplest houses and apparatus suitable for the hatching and rearing of salmon up to the complete absorption of the yolk sack, that being the time when it is customary to turn the young fish out to shift for themselves; and it is hoped that the instructions given will be so plain and yet so complete that a person previously entirely ignorant of the whole business can without further direction set up an efficient establishment. No attempt will be made to explain the construction of the more elaborate devices that have lately come into so general use, since these devices have for their main purpose the saving of space in establishments where large quantities of eggs are to be developed up to the shipping point, and few or none to be hatched out; though some of these are also available for hatching, and to a certain extent for the rearing of the young fish.

For the most part the same apparatus and management are applicable to Atlantic and land-locked salmon, Pacific salmon, and brook trout. The Atlantic and land-locked salmon, both in the egg and in the sack stage, are so closely alike as to be practically indistinguishable. Indeed, the latest conclusion of special students of the *Salmonidæ* is that they all belong to the same species, *Salmo salar*. Be this as it may, their habits and requirements during these early stages are, so far as known, identical. On these fish my personal observations have been mostly made, and to them, therefore, the instructions of this paper may be considered as more especially applicable. Yet the difference between the treatment they require and that applicable to Pacific salmon is so slight that all the rules laid down may, it is believed, with perfect safety be followed in the management of the latter, except in certain minutiæ, which depend mainly on the greater size and hardihood of the Pacific salmon, partly on their adaptation to warmer water, and when not specially mentioned will readily suggest themselves to the common sense of the operator. Similar observations may be made with reference to the brook trout. The same apparatus, with some possible change in management, will answer also the very best purpose in the hatching of lake trout.

2.—WATER.

The first thing to be sought is an ample supply of wholesome water, on a site where it can be brought completely under control and the requisite fall secured. In this matter there is quite a range of choice. The very best is the water from a stream fed by a clean lake taken a short distance below the outlet of the lake, with an intervening rapid. Such water is commonly quite even in volume and temperature, and comparatively free from sediment and harmful impregnations. It is cold in winter and warms up slowly in spring, giving assurance of a slow and normal development, which is more conducive to health and vigor than a very rapid development. The passage down a rapid, though by no means an essential point, will further improve this water by charging it highly with air. After this, I would choose the water of a brook that is fed largely by springs, so as to insure constancy in the supply and some moderation of the temperature on warm days; but it is better to have the water flow a long distance in an open channel before using, and, if possible, over a rough and descending bed, that it may be well aerated, and in cold weather somewhat cooled down from the temperature with which it springs from the ground. Thirdly, choose pure spring water; but in all cases where this is necessary provide a cooling and aerating pond, that you may have the original warmth of the water subdued by the cold of the air before it reaches the hatching troughs, and that it may absorb more or less air by its wide surface. Lastly, choose ordinary river or brook water, as clean as possible. These kinds are considered inferior to spring water by reason of their liability to floods, drought, muddiness and foulness of other sorts, and in cold climates to anchor ice. The water of a stream that has its source in a not very distant lake or spring is not considered *ordinary* river or brook water, but is advanced thereby into the first or second rank. Between these different sorts there is of course an infinite number of gradations. If lake water cannot be obtained, it would be of some advantage to have a supply of both spring water and brook water, depending for ordinary use on the brook water or a mixture of the two, and on the spring water for emergencies, such as the freezing, drying, or excessive heating of the brook, floods with accompanying muddiness, etc. Avoid water that comes from boggy and stagnant ponds and marshes; for though excellent water, capable of bringing out the most vigorous of fish, may sometimes be had in such places, yet when not supplied by springs it is dependent for its freshness and good qualities upon sufficiently copious rains, and if these fail, as they are liable to, the water may become foul and unfit. The best time to select a site for a hatching establishment is in time of extreme drought. If the site in question has at that time an ample supply of pure, sweet water, the first requisites are fulfilled. But if such an examination discloses any lack in this respect, the site must be rejected. It would be well, also, to visit the place in time of flood and, if in a very cold climate, in severe winter weather, to know what dangers are to be guarded against on those scores.

The volume of water necessary will depend on several circumstances, mainly on the following: 1st, the proposed capacity of the establishment; 2d, the temperature of the water; 3d, its character as to aeration; 4th, the facilities existing in the house for the aeration and repeated use of the water. With water of the highest quality and low temperature and with unlimited facilities for aeration, possibly a gallon a minute or even less can be made to answer for the incubation of 100,000 eggs of salmon. As the temperature rises or the facilities for aeration are curtailed, a larger volume becomes necessary. In case of spring water, cooled only to 40°, and aerated only by exposure to air in a pool of about a square rod surface, with no facilities in the house for aeration, and with the eggs and fry crowded in the troughs at the rate of 4,000 per square foot, 4 gallons a minute is the least that can be trusted to support that number, (100,000,) while 6, 8, or 10 gallons per minute would be much better. While the minimum is, as stated above, possibly less than a gallon a minute, no novice can be advised to trust to less than 3 gallons per minute for each hundred thousand eggs or fish under the most favorable circumstances. These statements are about as definite as can be made. The question of volume must be decided for each case according to the peculiar circumstances existing, and the novice must first acquaint himself with the mode of arranging the fixtures in the house, and especially with the means and facilities for aeration, for which directions will be given below, and then study the possibilities of the proposed site. It should be borne in mind that the volumes of water stated above are strictly minimum quantities, meant to apply to the very lowest stage of water that can possibly occur during the hatching season.

If the water supply is to be drawn from a small brook or a spring, it will be necessary to measure the volume carefully. The following is an easy and accurate mode, applicable to most cases. Take a wide board one inch thick, (or two or three of them carefully jointed or matched,) and bore a smooth inch hole through the middle of it. With this make a tight dam across the stream so that all the water will have to flow through the hole. If the water on the upper side rises just to the top of the hole, it indicates a volume of 2.3 gallons per minute; a rise of half an inch above the top of the hole indicates a volume of $3\frac{1}{2}$ gallons per minute; 2 inches rise, 5 gallons per minute; 3 inches, 6 gallons per minute; 6 inches, 8 gallons per minute; 12 inches, 12 gallons per minute. If two one-inch holes are bored, the same rise will of course indicate twice the volume. The volume vented by holes of different sizes is in proportion to the squares of their diameters; thus a two-inch hole vents four times as much as a one-inch hole. A cylindrical tube whose length is three times its diameter will vent 29 per cent. more water than a hole of same diameter through a thin plate or board.

3.—SITE.

A satisfactory supply of water having been found, it is next necessary to select a site for the hatching-house that combines in as great a degree as possible the various desiderata, of which the most important are, first, facilities for creating a head of water to provide for the requisite fall into and through the troughs; second, security against inundation; third, if in a cold climate, security against too much freezing; fourth, general safety and accessibility.

The fall required in the hatching-house cannot be stated very definitely, but it can hardly be too great. The minimum for the most favorable cases is as low as three inches, but only under the most favorable circumstances in other respects will this answer, and even then it is subject to several very serious disadvantages. It is only admissible where there is an ample supply of aerated water, and the troughs are very short, and there is absolutely no danger of inundation; and the disadvantages are the impracticability of introducing any aerating apparatus and the necessity of having the troughs sunk below the floor of the hatching-house, which makes the work of attending the eggs and fish very laborious.

A fall of one foot will do pretty well if there is entire safety from inundation. This will permit the troughs to be placed *on* the floor instead of below it, (a better position, though still an inconvenient one,) and some of the simpler aerating devices can be introduced. Better is a fall of three feet, and far better a fall of six feet. The latter will allow the lowest hatching-troughs to be placed two feet above the floor, to the great relief of the backs of the attendants, and leave ample room for complete aeration. Of course the necessities of the case are dependent largely upon the volume and character of the water. If there is plenty of it, and if it is well aerated before reaching the hatching-house, there will be no occasion in a small establishment of additional aeration in the house, and, therefore, no need of more than three feet fall, and, except for convenience in working and for guarding against inundation, one foot fall is enough.

As to liability to inundation, actual inspection of the premises at time of floods will generally suggest what safeguards are needed. If located by a brook-side, the hatching-house should not obtrude too much on the channel, and below the house there should be an ample outlet for everything that may come. By clearing out and enlarging a natural water-course much can often be done to improve an originally bad site.

In a cold climate it is an excellent plan to have the hatching-house partly under ground, which will protect it wonderfully against outside cold. When spring water is used there is rarely any trouble, even in a cool house, from the formation of ice in the troughs; but lake, river, or brook water is, in the latitude of the northern tier of States, so cold in winter that if the air of the hatching-house is allowed to remain much below the freezing point, ice will form in the troughs and on the floor, if there is any leakage, to such an extent as to be a serious annoyance, and sometimes, if not watched, will form in the hatching-troughs and extend so deep as to freeze the eggs and destroy them. Stoves are needed in such climates to warm the air enough for the comfort of the attendants; but the house should be so warmly located and constructed that it may be left without a fire for weeks without any dangerous accumulation of ice. The easiest way to effect this is to have the house partly under ground; but if the site does not permit this, the same result can be brought about by thorough construction of the walls and by banking well with earth, sawdust, or other material. In warmer climates no trouble will be experienced from this source.

4.—DAMS AND CONDUITS.

In some cases the best way to get the requisite head is to throw a dam across the stream and locate the hatching-house close to it. The dam will form a small pond which will serve the triple purpose of cooling, aerating, and cleansing the water. But unless the character of the bed and banks of the stream be such as to warrant against undermining or washing out at the ends of the dam, it is best not to undertake to raise a great head in this way. With any bottom except one of solid ledge there is always great danger, and to guard against it when the dam is more than two feet high may be very troublesome. If there is any scarcity of water, or if it be desirable for any other reason, for aerating or other purposes, to secure a considerable fall, it is better to construct the dam at some distance above the hatching-house, on higher ground, where a very low dam will suffice to turn the water into a conduit which will lead it into the hatching-house at the desired height.

The conduit is best made of wood. A square one of boards or planks, carefully jointed and nailed, is in nearly all cases perfectly satisfactory. For an ordinary establishment a very small conduit will suffice. The volume of water that will flow through a pipe of given form depends first, upon the the size of the pipe, and second, upon the inclination at which it is laid. A straight cylindrical pipe, one inch in diameter, inclined one foot in ten, will convey about eleven gallons of water per minute. The same pipe, with an inclination of one in twenty, will convey eight gallons per minute; with an inclination of one in fifty, five gallons per minute; with an inclination of one in one hundred, three gallons and a half per minute; with an inclination of one in one thousand, one gallon per minute. A two-inch pipe will convey about 5½ times as much water as an inch pipe; a three-inch pipe nearly fifteen times as much. A one-inch pipe with an inclination of one in 1,000 will convey water enough for hatching 25,000 eggs; with an inclination of one in fifty, enough for 100,000 eggs; with an inclination of one in twenty, enough for nearly 200,000 eggs. A square conduit will convey one-quarter more water than a cylindrical pipe of same diameter. If there are any angles or abrupt bends in the pipe its capacity will be considerably reduced. It should be remembered that if the water completely fills the aqueduct it is thereby entirely shut out from contact with air during its passage, whereas if the pipe be larger than the water can fill the remainder of the space will be occupied by air, of which the water, rushing down the incline, will absorb a considerable volume and be thereby greatly improved. It will therefore be much better, when practicable, (and this includes nearly all cases,) to make the conduit twice or thrice the size demanded by the required volume of water. If the bottom and sides be rough, so as to break up the water, so much the better; and the wider the conduit is of course the more surface does the water present to the air. It is not at all necessary to cover the conduit, unless from its position it is exposed to inundation or to pollution by the visits of mischievous animals or other agencies, or unless, as may sometimes, but rarely, occur, the water would be in danger of freezing up. If the water comes from springs or a spring brook, or a lake or pond, there is no danger on that side, unless the aqueduct is a very long one; on the contrary, the spring water will only receive a wholesome cooling down.

5.—AERATION.

This is perhaps the most important branch of the whole subject. The water which fishes breathe is but the medium for the conveyance of air, which is the real vivifying agent. Without air every fish and every egg must surely die, and with a scanty supply the proper development of the growing embryo becomes impossible. Water readily absorbs air whenever it comes in contact with it, and the more intimate and long continued the contact the greater the volume it will absorb. The ample aeration of the water to be used in the hatching-house has already been mentioned as a desideratum of the first importance, and some of the devices by which it is to be secured have been incidentally alluded to. But a little more remains to be said.

Water from either a brook or river that has been torn into froth by dashing down a steep bed has absorbed all the air that will be needed in ten or twenty feet of hatching-trough, and demands no further attention on this score. But if the water must be taken from a lake, a spring, or a quiet brook, its burden of air is much less and is liable to become so reduced before it gets through the hatching-house as to be unable to do its proper work. It is therefore desirable to adopt all practicable means of re-inforcing it. If the site of the hatching-house commands a fall of five feet or more, the thing is easily done. Either in the conduit, outside the house, or in the hatching-troughs themselves, a series of miniature cascades may be contrived. The broader and thinner the sheet of water, the more thoroughly it is exposed to the air, and if, instead of allowing it to trickle down the face of a perpendicular board, we carry it off so that it must fall free through the air, as in Figure 1, both surfaces of the sheet are exposed

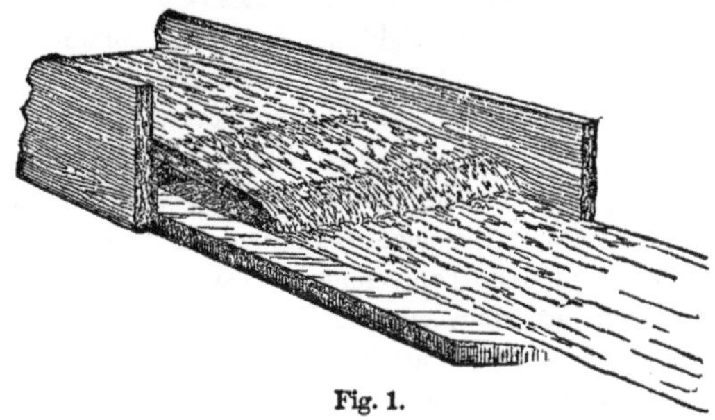

Fig. 1.

and the effect doubled. When the circumstances permit, it is best to introduce these in the conduit, which, as already suggested, may be made wide and open for that purpose. If the aeration cannot be effected outside the house there is still opportunity inside. Two long troughs may be placed side by side, leveled carefully, and the water be received in one of them and pour over into the other in a sheet the whole length of the trough, which, of course, would be a very thin sheet, and very effective. In the hatching-troughs themselves, also, there is an opportunity for aeration, either by making short troughs with a fall from one to another, or by inclining the troughs and creating falls at regular distances by partitions or dams, each with its cascade, after the fashion already described.

The only serious difficulty is encountered where the ground is very flat, so that the requisite fall cannot be obtained. In this case the best that can be done is to make a very large pool, several square rods at least, outside the house, and make all the conduits as wide as possible, so that the water shall flow in a wide and shallow stream.

It will of course be borne in mind that the better the aeration the smaller the volume required to do a given work; and on the other hand it is equally true that the greater the volume the less aeration is necessary. When so large a volume as six gallons per minute for every hundred thousand eggs is at command a comparatively low degree of aeration will answer. But so far as known the higher the degree of aeration the better the result, without limit, other things being equal, and it is therefore advised to make use of *all* the facilities existing for this purpose.

6.—FILTERING.

Before the introduction of wire or glass trays for hatching fish-eggs it was customary to lay them on gravel, and under these circumstances it was absolutely necessary to filter all but the purest water. Even ordinary spring water deposits a very considerable sediment, which might accumulate upon the eggs to such an extent as to deprive them of a change of water and thereby smother and destroy them. When the eggs are deposited on trays, however, even though their upper sides be covered with sediment, underneath they are clean and bright, and remain in communication with the water beneath the tray, though of course the circulation of water through the tray is not perfect. The trays, moreover, offer the best facilities for cleansing the eggs as often as may be necessary, and establishments for the hatching of eggs of the *salmonidæ* do not commonly receive them until they have arrived at the stage when they can be safely subjected to whatever washing and disturbance may be desired. It is not, therefore, deemed necessary to introduce any considerable devices for filtering water which is naturally very pure, as are lake and spring water commonly when not subject to intermixture with surface water during rains. There are, however, so many cases in which it is necessary to use water subject to constant or occasional turbidness that some directions for filtering are indispensable.

In the first place, let the water from the conductor be led into a deep tank, which may be termed the "settling tank," where the coarser and heavier dirt will sink to the bottom. This may as well be located outside the hatching-house, and for a small establishment a hogshead sunk in the ground will answer. From the settling tank the water should be led into a filtering trough inside the house, as shown in Figure 2, which exhibits

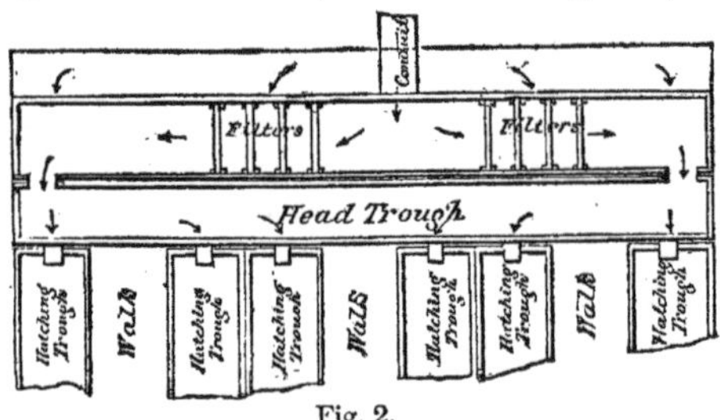

Fig. 2.

one out of many convenient arrangements. This trough may be just the length of the head or distributing trough alongside which it lies, or may be much shorter, four feet answering well where little work is demanded of it. For depth and width 15 to 18 inches are convenient dimensions. If the water is introduced near the middle of the filtering trough the current may be subdivided, part going to the right and part to the left, each part through its own set of filters, as shown by the arrows. This makes the single long trough equivalent to two shorter troughs, and since the shorter trough would be amply long to receive the requisite screens, the filtering capacity of the trough is thus doubled. When either the volume or excessive turbidness of the water demands an extraordinary capacity in the filter, the water may be introduced at several points by means of an additional long distributing trough placed alongside the filtering trough, as shown by the dotted outline in Figure 2, and each of the separate currents be subdivided as already described. In this way six separate sets of filters may be introduced into a single trough 12 feet long.

The filters to be used with the foregoing arrangement are made by stretching woolen flannel on wooden frames. The best device consists of two separate frames, one fitting inside the other, (without nails,) as in Fig. 3, and holding stretched between them a piece of flannel considerably larger than the frame, to allow for shrinkage and for a margin to close the interstices on either side and at the bottom between the frame and the trough. This filter slides down into the trough obliquely, between two pairs of cleats on opposite sides, as shown in Fig. 4. Strips of wood half an inch thick are suitable for the construction of these frames, giving a total thickness of one inch to each filter, and if it is desired to save room, the space intervening between the frames may be as narrow as half an inch, so that it is possible to get eight filters into a single foot of the length of the trough. They should slide easily into place, so that they may be removed whenever necessary to clean them. The cloth can be removed from the frames and washed or dried and brushed. There should be a large surplus of them on hand, so that a clean one for immediate use should always be ready. The filters should not come quite to the top of the trough, so that if they become completely clogged with dirt the water may flow over their tops to the hatching-troughs; for dirty water is much better than stagnation. It is better to have flannel (or baize) of several grades of fineness, and pass the water through the coarser ones first. If leaves and other coarse rubbish are liable to enter the filtering-trough they must be arrested by a coarse grating of wood or metal above each set of filters; it is better to stop all such coarse material outside the house.

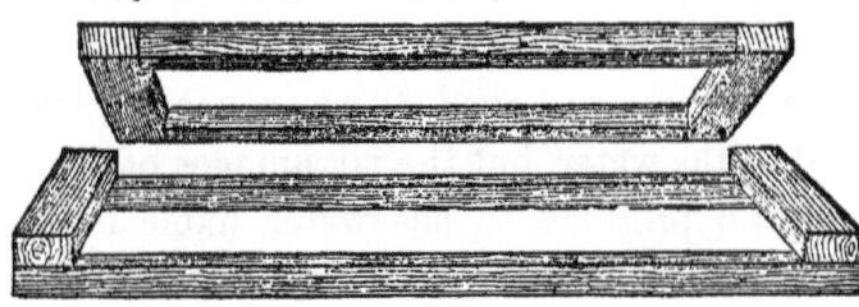
Fig. 3.

Fig. 4.

The filters will of necessity obstruct the water somewhat, and a slight head be created by each one,—perhaps an inch each will be a rough approximation to the truth. Allowance must be made for this by having the filtering-trough several inches higher than the hatching-trough. But do not draw the water away from the lower sides of the filters so as to expose them to the air, for the water will pass through much freer when it is backed up nearly as high on the lower side as on the upper. The number of filters to be used depends upon the amount of foreign matter in suspension in the water, and can only be determined by observation and experiment in each case.

Another mode of filtering sometimes resorted to, either alone or in connection with the flannel screens, consists in passing the water through a bed of gravel; but the method already described will answer every purpose and is much easier of application.

As already remarked, there are many places where it is a waste of effort to filter the water, but the advantages of cleanliness are so great that every one who proposes to use water liable to become at any time muddy is advised to put in the necessary troughs, or at any rate to leave space for them. If, however, a hatching-house has been already fitted up without any provision of this sort, a set of filters can be fitted into the upper part of each hatching-trough and be just as effective as if in a trough by themselves.

7.—HATCHING-TROUGHS AND FITTINGS.

We come now to the hatching apparatus proper, the troughs and trays. Whatever may be the advantages derived from the use of very compact apparatus, some forms of which allow us to mature 30,000 eggs to every square foot of trough room, they do not pertain to the hatching out and rearing of the fry. For this work nothing has yet been found better than a long, straight, shallow trough. Ten feet is the length I would recommend as most desirable. In no case have them longer than fifteen feet. In passing down a well populated trough fifteen feet any ordinary volume of water will be deprived of so much of its air and oxygen that a new supply is needed, and if necessary to make further use of this water it is best to let it fall in a thin sheet into another trough set a few inches lower. In some cases, where the water as introduced into the house is deficient in aeration, it is best to make troughs as short as five feet, or, what will amount to the same thing, (though a less convenient and less satisfactory mode,) incline the trough from one to two or three inches for every five feet in length, and check the water and keep it up to the proper height in different parts of the trough by a series of transverse partitions or dams. Under ordinary circumstances, with well aerated water at the start, a trough ten or fifteen feet long may be set perfectly level.

Figure 5 shows the interior of a hatching-house supposed to have a

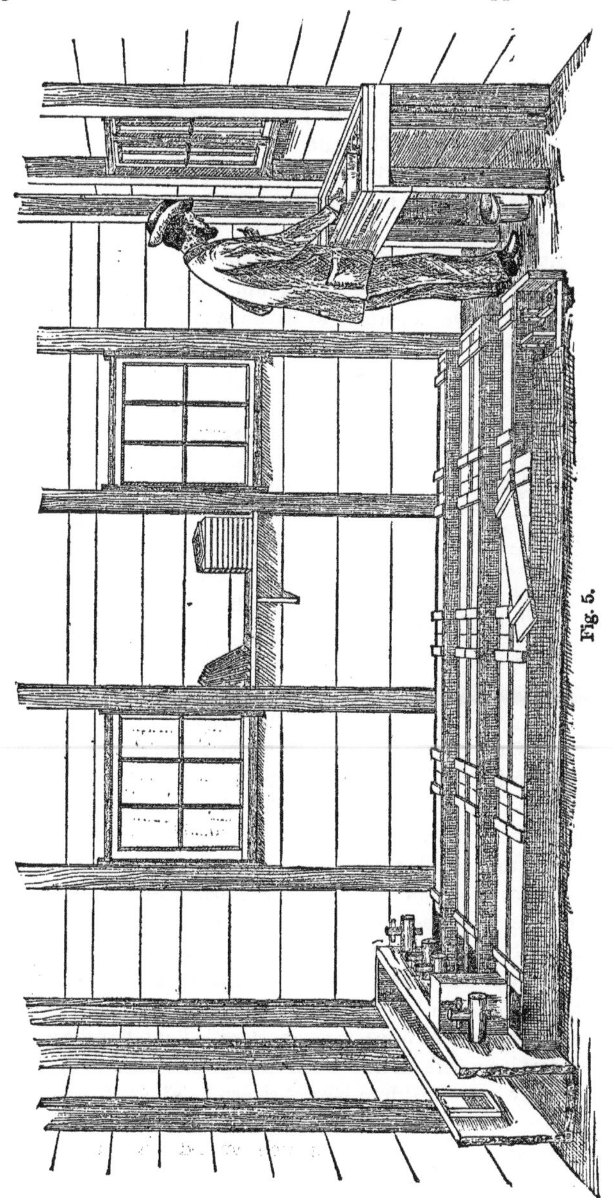

Fig. 5.

capacity of 150,000 Atlantic salmon, or say 100,000 to 125,000 Pacific salmon. The troughs are about ten feet long and six inches deep, arranged in pairs (except the one next the wall) with walks between. These troughs are placed upon the floor, but when circumstances permit well aerated water to be brought into the house high enough, it is better to place them two or three feet above the floor. This is, however, entirely a question of convenience for the attendant. The water used is supposed to be unfiltered, and is therefore received in a deep and wide head trough, which will serve as a settling tank. From the head trough the water is delivered by wooden faucets to the hatching-troughs, the fall at this point affording an opportunity for aeration, which can be improved by letting the water fall on a slanting board, from the edge of which it will fall in a thin sheet into the trough. It is important to have the faucets all exactly on the same level; otherwise those which are lowest will, unless carefully regulated, rob the others of their share of the water. The style of faucet represented is very convenient and safe, but a plain spout of lead or wood, three or four inches long, and closed by a slide on the upper side, as shown in Figure 6, is just as good and easier made. Avoid any kind of a faucet that is liable to be accidentally closed, like a molasses faucet, an occurrence that I have known to be followed by very serious results. The bore of the faucet should not be less than one inch for a trough a foot wide.

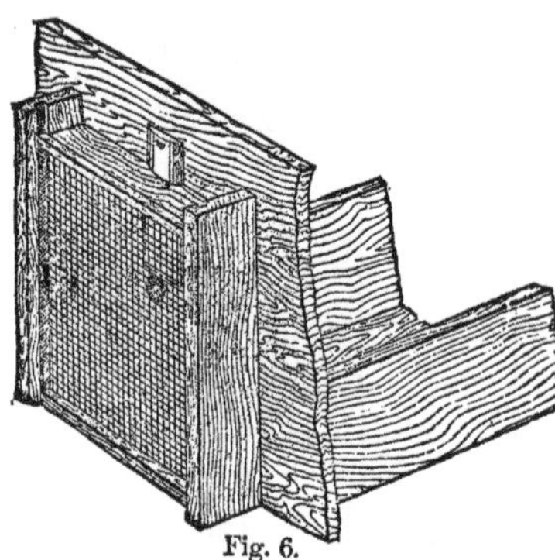
Fig. 6.

A very convenient outlet for a hatching-trough is formed by a two-inch lead pipe set into the bottom of the trough and running down through the floor. The water is maintained at the proper height by a movable partition, or dam of thin boards sliding down between cleats nailed to the sides of the trough, as shown in Figure 5. The height of the water depends upon the number of pieces brought into use at any time. These boards must be carefully jointed and fit nicely between the cleats, that there be no waste of water. A dam of the same sort should be used to hold the water at several points in an inclined trough.

The troughs should be fitted throughout with light board covers from two to four feet long, with cleats or other fittings convenient to lift them by. The faucets may be covered by a box, as shown in Figure 5, on the second trough. Screens fine enough to shut out all vermin should be placed at both ends of the trough.

Almost any kind of easily worked wood may be used for building the troughs. White pine is the favorite wood in northern sections. Arborvitæ, (*Thuja occidentalis*,) known in the north as white cedar, is unfit; water in which shavings of this wood have been soaked is deadly to grown trout. Caution should also be used in employing the southern white cedar, or cypress, (*Cupressus thyoides*,) red cedar, or savin, (*Juniperus virginianá*,) or any other odorous woods.

Inch boards are heavy enough for troughs not more than six inches deep, whatever their length or width. For deep distributing or filtering-troughs use plank an inch and a half or two inches thick.

All the wood-work about the troughs should be varnished with several heavy coats of asphaltum varnish, thoroughly dried in before the wood is wet. This makes a smooth, shining black surface, very easy to clean.

8.—WIRE TRAYS.

The practice of covering the bottom of the hatching-trough with gravel and depositing the fish eggs directly upon that has deservedly become nearly obsolete. Its principal disadvantages are, that it is impossible to spread the eggs evenly on such a bed; that there is great danger of suffocation by sediment because of the absence of any circulation of water beneath the eggs; that the operation of cleaning them is tedious in the extreme, and that the gravel seriously interferes with moving the fish about in the trough or even dipping them out.

The receptacle for the eggs which in one form or another has come into general use is a shallow tray, made by attaching wire-cloth to a narrow wooden frame. In its original form this was known as the "Brackett tray," and that name properly applies to the sort recommended below. The prominent advantages of this piece of apparatus are: first, the more perfect circulation of water amongst the eggs, insuring a better supply of the air demanded for their healthy development; second, almost entire safety from suffocation by sediment; third, the facility with which the eggs can be cleaned and moved about in the trough or be taken out for cleaning and examination. These advantages are so great and save so much labor that the wire tray is almost indispensable.

Trays of the following construction will be found most serviceable: Make the frame of any easily worked wood, ("white wood," the product of the tulip-tree, *Liriodendron*, is firstrate.) Half an inch in width and thickness are the best dimensions of material. Stouter frames would be likely to float the wire, whereas it is better that they should sink. The completed frame should be 12½ inches wide. This precise width is chosen because it is best fitted to receive wire-cloth one foot wide,—the size found to be most eligible. If the cloth were cut of the full width of the frame there would be many projecting rough edges, which would be an annoyance by scratching up everything they came in contact with and would be constantly rusting. Trays of this width fit well in troughs $12\frac{3}{4}$ inches wide. Their length may be equal to their width, as I prefer, or greater.

The wire-cloth heretofore commonly used is woven of annealed iron wire, in square meshes. This answers admirably when the wires are from $\frac{1}{8}$ to $\frac{1}{5}$ inch apart, (not wider than $\frac{1}{7}$ if brook-trout eggs are in hand,) so long as the fishes remain in the egg. But as soon as hatched they begin to poke their heads and tails down through the meshes, or sometimes their sacks are drawn through, and being unable to extricate themselves, they perish miserably. If, therefore, square meshes are to be used they should be very small,—not over $\frac{1}{12}$ inch wide. This sort of wire-cloth has, however, still this slight drawback,—that while the eggs are hatching the picking must be done in the trough, or if the trays are taken out the young alevins must come out into the air also. There is not, to be sure, much danger of injuring the fry by exposure for a moment to the air, but a good deal of extra care is involved, and it is much better not to have to take them from the water at all. These little troubles are all avoided by using cloth with a long mesh, (see Fig. 7,)—for Atlantic and land-locked salmon a mesh $\frac{1}{5}$ inch wide and $\frac{5}{8}$ to $\frac{3}{4}$ inch long,—through which the soft bodies of the fishes easily slide

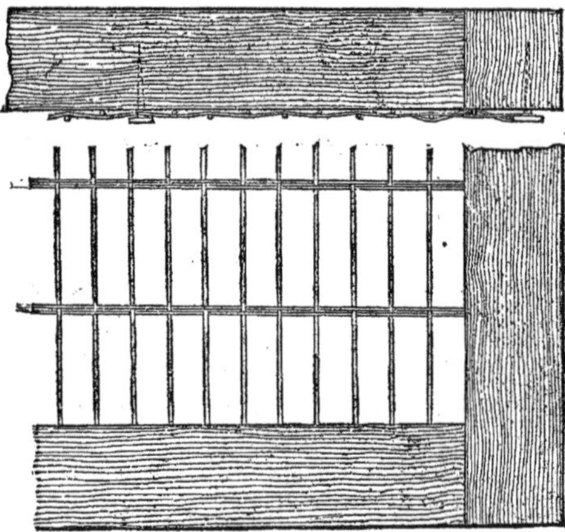

Fig. 7.

as soon as they have broken the shell, while the whole eggs are retained upon the trays and can at any time be lifted out without lifting the fish. Any one who is so situated as to get wire woven to order had better adopt the long meshes, woven of wire as small as can be well worked, which may be left to the judgment of the weaver. If, however, this cannot be had, then choose common wire-cloth, 12 wires to the inch or finer. The article sold at all the hardware shops for window screens is very suitable; being already painted thoroughly it requires but a single coat of asphaltum varnish to fit it for use.

All iron wire must be protected from rust by painting or varnishing in a most thorough manner. The commonly used material for this purpose is asphaltum varnish. The so-called paraffine varnish, a coal-tar product, is much inferior. It is very uneven in quality, but generally dries very slowly and has a penetrating and disagreeable odor. It is best to have the wire-cloth cut of the proper size, rolled perfectly flat, and then varnished with two or three coats on the edges that are to lie against the wooden frame. The rest of the varnishing can be done after the wire is attached to the frames. Two good coats, very carefully laid on, is the least that will answer for iron not previously painted, and three coats are much to be preferred. For nailing to the frames use tinned tacks.

There is, after all, a good deal of trouble in securing a thorough spreading and adhesion of the varnish, and it is much to be hoped that some better material will soon be discovered. I have tried iron wire, tinned after cutting up, and for a single season it has worked well; but I fear the tin will not be permanent enough. Brass wire, nickel-plated, is admirable but expensive,—costing about 60 or 70 cents per square foot. For the present, therefore, iron wire is recommended. There should always be a surplus of trays, so that if any of those in use are found to rust badly they can be exchanged for newly varnished ones.

9.—ARRANGING THE TRAYS FOR WORK.

The trays must not be placed on the bottom of the trough, but on a support raised a little distance above the bottom. As it is very desirable to have the trough as free as possible from obstructions, it is best to provide a temporary support for the trays, like that shown in Figure 8. Take a long, narrow strip of wood a quarter of an inch thick and drive

Fig. 8.

through it, at proper distances, nails one inch long. Set the points of the nails a quarter of an inch into the floor of the trough and the top of the strip will then be three-quarters of an inch above the floor. On two of these supports, placed at a distance of a quarter or half inch from the sides and running the whole length of the trough, rest the hatching-trays. Supports touching the sides of the trough will not answer, because they form, with the trays and the sides, narrow crevices into which the young fish may wriggle, to the great danger of being crushed to death. After all are hatched the trays are no longer of service, and the support can then be taken out without injuring the fish, leaving an unobstructed floor.

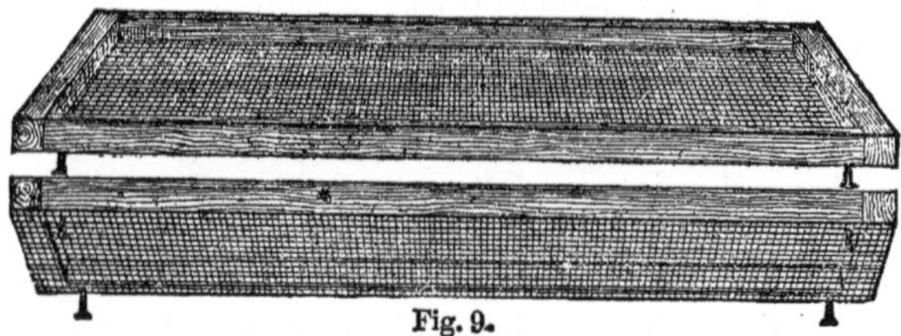

Fig. 9.

The trays which rest on the supports just described need no legs. To use a trough to its full capacity, however, another series of trays, resting on the first series, is necessary, and sometimes a third series resting on the

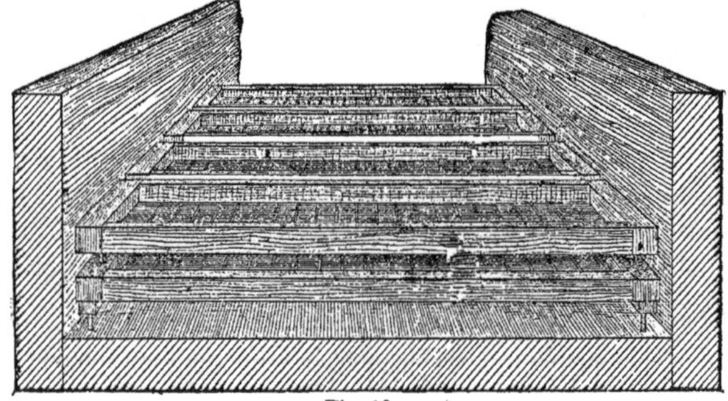

Fig. 10.

second. The trays used in these upper series must be provided with legs half an inch long, obtained by driving four nails into the under side of the frame. (See Figures 9 and 10.) This keeps the trays half an inch apart, the proper distance when there is a space of three-quarters or an inch under the lower trays. It is, however, recommended to partially close the lower space at first by a few movable cleats, which can be removed when the fish begin to come out of the shell and accumulate on the floor. These precautions are to guard against a too free flow of water underneath the trays, where it would at that time be wasted, and perhaps leave a scanty supply for the eggs above. As a further precaution, with the same end in view, if the trays do not fit the troughs pretty closely they may be placed obliquely, so that two opposite corners will prevent a draft of water down the side. (See Fig. 11.)

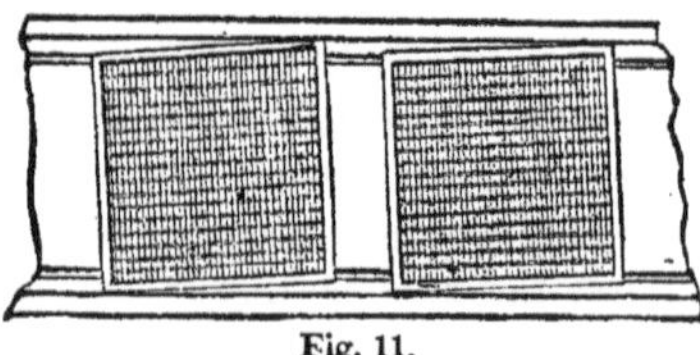

Fig. 11.

10.—CAPACITY OF THE TROUGHS.

The trays may be placed close together, allowing merely space enough to admit the fingers when handling them. Each tray should receive a single layer of eggs. They will count, of Atlantic salmon, about 2,000 per square foot; of Schoodic salmon, about 1,800, and of California salmon, about 1,200 per square foot. Allowing for all the waste space, a trough ten feet long with a single series of trays will hold about 13,000 eggs,* a very light stock. On two series of trays there would be 26,000 eggs—a fair stock—and on three series of trays, 39,000 eggs. The latter number would give us, after hatching, about 4,300 alevins for every square foot of trough-floor. With plenty of well-aerated water, a person with some experience will have no difficulty in bringing as heavy a stock as this through in safety. Indeed I have known a stock of over 5,000 per square foot to be brought through without serious loss. If the fish would lie evenly distributed over the floor there would be no difficulty, but at certain times they are seized with a perverse inclination to collect together in heaps, and, if they remain so a long time, those underneath are suffocated. Therefore, though it is wonderful how much crowding they will endure, the novice is advised not to attempt more than two series of trays, or 3,000 fish per square foot of trough.

11.—SCREENS.

If the trough is level there will be no occasion for any dams or barriers until the eggs are hatched, but, as something of the sort is needed to keep the alevins well distributed, it is better to provide for it in the beginning. At regular distances, not more than five nor less than two feet apart, attach to the opposite sides of the trough pairs of cleats, as if for a dam, such as has already been described for the outlet. Connect these opposite pairs of cleats by a low cross-piece or sill about half an inch high. As soon as the fish begin to move about a fine wire screen can be slipped down between the cleats until it rests upon the cross-piece; this is shown near the lower end of the front trough in Figure 5. The screen should not be coarser than twelve wires to the inch, and finer still will be better. Wherever dams occur in the trough or at its outlet the fish must be kept away from them by similar screens placed a few inches above the dams, or by one of another pattern, shown in Figure 12, which may be termed a safety screen. This form is worthy of special recommendation.

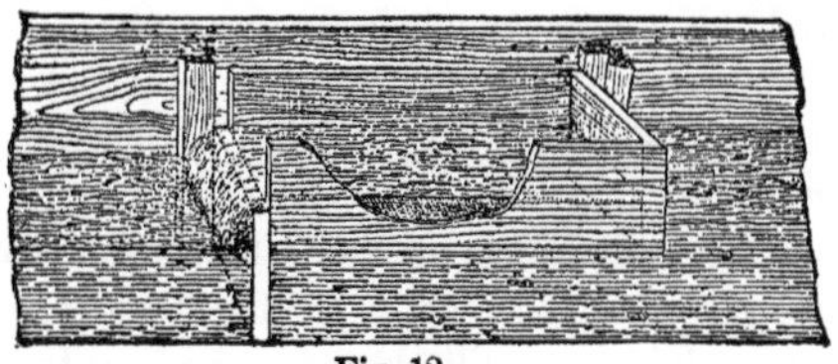

Fig. 12.

* Eggs of Schoodic salmon referred to when not otherwise specified.

The water passes through from below upward, and the weight of the fishes constantly tends to keep them away from it and assists them to clear themselves if once drawn against it. If there is a very strong current this is the only safe screen. It is nothing unusual for young fish to get against an upright screen and, the current being pretty strong, be unable to get away from it, and if the screen be too coarse their sacks are often drawn through, to their almost certain destruction. The safety screen should be sunk an inch or two below the top of the dam.

12.—TREATMENT OF THE EGGS.

If the foregoing instructions for the erection and fitting of the hatching-house have been judiciously followed, the task of caring for the eggs and young fry will not be a very difficult one, but will nevertheless demand constant alertness.

When eggs are received from other stations, it is important to lose no time in opening the package and ascertaining their condition. If the eggs are packed in moss, plunge the bulb of a thermometer into the moss under several layers of eggs, taking care to admit the least possible amount of outside air; cover it up and wait fifteen or twenty minutes, when it can be examined and the general temperature of the package ascertained pretty nearly. If it is within six degrees of the temperature of the hatching-water the eggs may be immediately placed on the hatching-trays. If, however, the temperature of the moss is six degrees higher or lower than the hatching-water, it is better to drench the boxes with water of intermediate temperature, several times if the difference be very great, to bring the eggs gradually to the temperature of the water. After this the sooner the eggs are placed on the trays the better. If it is impossible to avoid waiting, (over night, for instance,) let the packages stand in a room of safe and uniform temperature, (hatching-house or cellar,) but *never let packages of eggs stand in water.* If the eggs are packed in the mode now commonly adopted, between folds of mosquito net and layers of moss, first remove the upper moss carefully and then lift them out, a whole layer at a time, on the cloth on which they lie, and turn them into a pan of water, from which bits of moss, &c., can be picked out or rinsed off. An even distribution of them on the trays will be facilitated by measuring them out in a measure holding just enough to cover a tray.

Once deposited on the trays the necessary work is comprised in a simple routine. The dead eggs and fish turn white and must be removed before they taint the water. It is better, but not essential, to have a table or sink to do this work on, and a broad shallow square or oblong pan to set the tray of eggs in while picking, that they may not remain long out of water. This pan will also be convenient to rinse the eggs in, should they become very dirty. At any time after the eyes of the embryo become black a good deal of rough handling can be practiced without the slightest harm, and they can be safely shaken about upon the tray until thoroughly washed. A pair of tweezers will be needed to pick out the white eggs, and I would recommend a home-made article, shown in Figure 13, consisting of two pieces of wood tacked together and tipped with

Fig. 13.

wire loops. They are much easier to the hand and altogether better than metallic tweezers. In water of 46° F. the dead eggs should be removed daily; at 45°, every two days will answer; at 40°, every three days; at 33°, once a week; but these are maximum periods and should never be overstepped.

If the eggs are neglected, the first result is that the dead ones begin to decay and taint the water, rendering it unfit for the healthy eggs. In the next place, if left long enough in the water, the decaying egg is attacked by a fungoid growth, of which the technical name is *Achlya prolifera.* This is what is commonly termed "fungus," though some writers have applied the term "fungus" to a totally different plant, a kind of *Conferva* or slime, which is either colorless or green, grows in long fine threads, and where too much light is admitted to the trough multiplies often to such an extent as to prove a nuisance, but never is troublesome in a darkened trough, and never, so far as known, feeds on animal matter. The *Achlya,* on the contrary, feeds on animal matter, and, so far as my own observations go, always on *dead and decaying animal matter,* never attacking a living egg. It grows in long white threads which radiate from the object upon which it is feeding, giving it a woolly appearance. It grows rapidly, spreads over all surrounding objects, and may do harm to good eggs by shutting off the circulation of water from them and thus exposing them to the poisonous exudations from the decaying substance. The presence of this growth in a hatching-trough is a sure sign of neglect; for, if the dead matter is removed before decay sets in, *Achlya* will never make its appearance.

The screens and filters must be daily or oftener examined to see that they are not choked up, for a few hours' stoppage of the flow might have disastrous results. If any emergency arises requiring a stoppage of the water for several hours, before the fish have broken the shell, it can be safely done if, at the same time, the water be drawn off from the trough, for which purpose a movable plug should be put in the bottom of every trough. Eggs are not injured by exposure to the air for however long time, provided they do not freeze nor get too warm nor dry up. But after the fish are hatched of course this cannot be done.

The height of the water in the hatching-troughs should be carefully attended to, so that it be high enough to have a current over the upper trays but not high enough to let the bulk of the water flow over the tops, depriving the lower layers of their share. If through neglect this robbery takes place, a lot of eggs with white stripes across them will be found some day, and close examination will show that the trunk of the embryo in each one is white, opaque, and dead—sure symptoms of suffocation.

The trays must be carefully watched, and those that rust be exchanged for newly varnished ones. The change is easily made by turning the new tray bottom up over the eggs, when, by a dexterous movement of the hands, the two are inverted and the eggs fall upon the new tray. This should be done over the broad pan, but the knack of doing it with very little spilling is soon acquired.

Strong light should not be allowed to shine for any great length of time on the eggs. Total darkness is as good as anything. But if covers are provided for the troughs, the house may be kept well lighted, and no harm will come from leaving the covers of a single trough off long enough to do any necessary work. In examining and picking the eggs, too, they may be brought into a strong light. But sunshine should never touch them.

13.—TREATMENT OF THE FISH.

After the eggs are all hatched the trays may be removed from the troughs. The principal thing to be looked after now is that the fish do not crowd up in heaps and smother each other. As soon as they begin to move about a great deal the screens described above should be put in place to prevent their congregating too much. If it becomes necessary to move them about in the troughs, to disperse improper gatherings, or to get them away from a spot that it is desirable to clean, it can be easily done by means of a sweeping board, (Fig. 14.) This effective implement is simply a thin board, a little shorter than the width of the trough, with the lower corners cut away as shown, so that they cannot touch the sides of the trough and perchance catch and crush the young fry. It depends for its efficiency on the fact that if a surface current is created in the trough in any direction there will be a corresponding bottom current in the opposite direction, and if this bottom current be moderately strong it will sweep along the young fish with it. To move the fish down the trough the sweeping board is placed in about the position shown in the cut and moved *up* the trough.

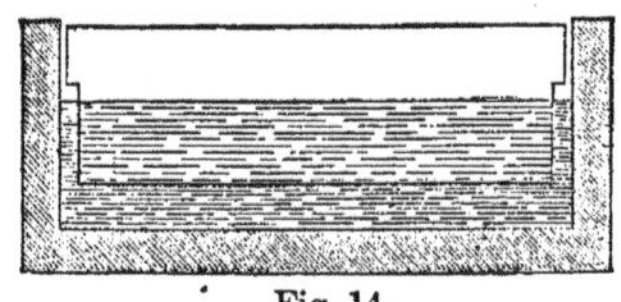
Fig. 14.

If the young fish are to be set free this must be done as soon as the yolk sack is absorbed, which will be from three weeks to three months after they are hatched, according to the temperature of the water. It is better to be too early than too late in this matter. For the young fish is well able to take care of himself, and in fact will sometimes begin to feed some days before the sack has entirely disappeared, while we know not how serious may be the result of two or three days' hunger. To remove them from the trough a scoop nearly as wide as the trough, made of a wooden frame with a shallow bag of mosquito net attached, after the fashion of Figure 15, will do good service. If the troughs are raised above the floor of the hatching-house the fish can also be drawn out from the outlets with water into a pail.

Fig. 15.

It is sometimes desirable to keep fish over night ready for an early morning start on a journey. This can be accomplished by taking a long box that nearly fits a hatching-trough, knocking out the ends and supplying their places with wire-cloth fine enough to hold the fish. When the time comes to put them into the cans they can be poured in from the box. When several cans are to be filled the fish for each may be put into a separate box.

14.—CONCLUSION.

In conclusion it is urged upon every person attempting the management of spawn and young fish that, however careful the construction of the houses and fixtures, the necessity for constant watchfulness is not to be escaped. There is no insurance so good as frequent and careful inspection. Especially in case of a severe storm or uncommonly cold weather, the attendant should be on the alert early enough to watch for the coming of danger and avert it. Nothing must be taken for granted until the establishment has demonstrated its security. Experience will show how far vigilance can afterward be safely relaxed.

Another matter that cannot be too strongly urged upon the attention of fish-culturists is the importance of complete records of all occurrences at the hatching-house. Not only the receipts of spawn and its condition, the losses occurring from day to day, and the shipment of young fish should be promptly, fully, and carefully entered upon the record-book, but the temperature of the air, the temperature, volume, and condition of the water should be regularly observed and recorded, and occasional notes made regarding the hatching and behavior of the fish, the presence and progress of maladies, if any occur, and any other phenomena of importance or interest. In no other way can the results of experience be so well preserved and made available, and it is much to be regretted that it has not been the practice of all fish-culturists to keep such records.

TABLE OF CONTENTS.

APPENDIX H.

MISCELLANEOUS.

XLIV.—ON THE NATURE OF THE PECULIAR REDDENING OF SALTED CODFISH DURING THE SUMMER SEASON.*

By W. G. Farlow, M. D.

Prof. S. F. Baird:

Dear Sir: At your request, I have made an examination of codfish for the purpose of ascertaining the cause of the peculiar redness which is found on the dried fish during the hot and damp weather of summer. The red fish, as is well known, putrefy comparatively quickly, and this fact, taken in connection with the disagreeable, and, in fishes, unusual color, renders them unfit for the market, so that, in seasons when the redness prevails, dealers suffer a loss which is certainly considerable, although exact statistics with regard to the amount are wanting.

For the purpose of examining fresh material, and in order to make a personal inspection of the drying apparatus and storehouses, I went to Gloucester in the beginning of September, 1878, at which date the weather was hot and damp, and the codfish then being prepared for market were largely affected by the redness, the cause of which it was my object to discover. With the assistance of Captain Martin, of the United States Fish Commission, I was able not only to procure an abundance of the red fish for study, but also to examine several different buildings used in salting and packing fish, as well as a schooner which had just returned from a voyage to the banks.

Before speaking of the immediate cause of the redness, I may say that all persons of whom I made inquiry agreed in stating that the redness makes its appearance to such an extent as to be troublesome only during the hot weather, and that it disappears with the return of cool weather. I ascertained farther that the redness in most cases does not appear until the fish have been landed from the vessel. In some cases, however, the fish become red while in the vessel, but this happens only when the weather has been unusually hot at the time of catching.

A microscopic examination shows that the redness is owing to a very minute plant, known to botanists by the name of *Clathrocystis roseo-persicina*. The plant consists simply of very minute cells filled with red coloring-matter and imbedded in a mass of slime. The cells, as usually seen, are arranged without order, but under the most favorable conditions of observation they are found to be grouped in spheroidal masses. In relation to the botanical characteristics of the plant nothing more need be said in the present connection. Its development has been studied by several well-known botanists, who agree in considering

* As observed more particularly at Gloucester, Mass., during the summer of 1878.

it closely related to *Clathrocystis æruginosa*, a common species growing in fresh-water ponds, which has lately come into public notice in consequence of the so-called *pig-pen* odor which it exhales when decaying.

The *Clathrocystis* in question belongs to the lowest group of plants, the *Schizophytæ*, many of which are the cause of decomposition or putrefaction of different animal and vegetable substances. *Clathrocystis roseo-persicina* is very widely diffused, being known both in Europe and America. It is found in summer along our shores, and at times is so abundant as to cover the ground with a purplish tinge, as one may see in the marshes near Lynn. It is also known in dissecting-rooms, where it grows in tubs in which bones are macerating. Wherever found it does not flourish nor increase rapidly at a temperature below 65° Fahr.

The next point to be considered is the manner in which the *Clathrocystis* is communicated to the fish. An examination of several different packing-houses and the wharves on which the fish are landed showed that the *Clathrocystis* was present in large quantities on the wood-work of all kinds; on walls, floors, and the flakes on which the fish are laid. How it might have been originally introduced into the buildings is a question easily answered when we consider how abundant the plant is on the marshes in the vicinity of Gloucester. It might have been brought in on the boots of fishermen, on sea-weed, on grass, or in other evident ways. Once in the buildings it would grow and increase on the damp wood-work, which contains usually more or less animal matter coming from the fish in process of drying. Why the plant is found at times on board the fishing-vessels themselves admits of explanation in two ways. It will easily be seen that, when it is common in and around the buildings on the wharves, it would be carried on the feet of fishermen on board the vessels. But there is also another reason why it should be found on the vessels. Large quantities of salt are of course used in packing the fish in the hold of the vessels. The two kinds of salt most commonly used by the fishermen of Gloucester are the Cadiz and the Trapani. I procured specimens of both kinds and submitted them to microscopic analysis. The Cadiz salt has a slight rose-colored tinge; the Trapani is nearly a pure white. The microscope shows that the reddish color of the Cadiz salt is owing to the presence in considerable quantities of precisely the same minute plant which is found in the red fish. The Trapani is a much purer salt, and the *Clathrocystis*, if it is found in it at all, exists in very small quantities. What must happen then is plain. The Cadiz salt, as it comes into the hands of fishermen, is already impregnated with a considerable quantity of the *Clathrocystis*. It is sprinkled in large quantities upon the fish as they are packed in the hold of the vessel, and if the weather is warm enough for the favorable growth of the plant, which, fortunately for the fishermen, is not the case in this latitude except for a short period, the fish must inevitably be affected during the voyage. As soon as the fish are landed, the circumstances are much more favorable for the rapid growth of the *Clathrocystis*. The temperature is higher, more salt is added, and the fish are exposed either in buildings or on flakes which are themselves more or less covered by the red plant.

I have endeavored to ascertain whether a similar trouble arising from the growth of *Clathrocystis* has been observed in the fisheries of other countries, but I have not been able to obtain any information on the subject from the botanists who are best informed in these matters. Such questions, however, are not often discussed in scientific journals, and the trouble may perhaps be known to fishermen, although it has not yet, as far as I know, been called to the attention of scientific men. In Norway, where the cod-fisheries are of great extent, we might expect the redness to occur, but we must remember that in the region of Bergen and northwards the temperature is rarely high enough to favor the rapid growth of *Clathrocystis*.

Having ascertained the cause of the redness, let us consider the means of preventing or diminishing the evil. Nature herself, in bestowing upon the New England coast a cold climate, has practically set a limit to the trouble, and has enabled the inhabitants of our coast to carry on the business of curing fish with a degree of success which would be quite impossible in a more southern latitude, no matter how abundant the fish might be. In attempting to diminish the trouble in New England, we must bear in mind that the disease, if we may call it so, is transmitted to the fish from the wood-work and drying-apparatus, and, in some cases at least, from the salt used. The question, in short, is how to get rid of the pest already established in our fish-houses.

To speak, in the first place, of the treatment to be pursued in purifying the drying-establishments on shore: The conditions of life of the *Clathrocystis* are such that it could be killed by a temperature equal to that of boiling water, by applications of strong solutions of carbolic acid, of the mineral acids, &c. As a matter of fact, however, it is very doubtful whether the application of boiling water or of steam, if possible, would be serviceable. It is difficult so to saturate the different parts of a drying-house with boiling water as to be sure that the different parts have really been raised to the boiling point. Generally a great part of the wood-work fails to reach anything like a temperature of 212° F. The application of carbolic acid, or the mineral acids, is expensive and troublesome, and, unless judiciously managed, the remedy might prove worse than the disease. What is wanted is some means so simple that it can be applied without trouble and without much expense. It is useless to try to eradicate the trouble completely; one can only expect to diminish it perceptibly; and for the purpose I can think of no practical way better than scraping, painting, and frequent washing with hot water. In midsummer the houses used for curing fish are not always kept as clean as they should be. Unpainted wood is generally used, and every one knows how difficult it is, by washing, really to clean wood which has been softened by the action of salt substances. The wood-work of all kinds, floors, walls, &c., should be thoroughly scraped several times a season. At present this is not the case, for one sees at Gloucester many gratings on which wood-mosses (lichens) have begun to grow, and even attained considerable size, proof positive, to any one who knows how slowly such plants grow, that no thorough scraping nor cleansing has been attempted for a long time. I should recommend that everything made of wood used in the curing should be painted at least once a year with white paint, and that it should be washed at frequent intervals with hot water. It is easier to paint than to scrape wood, and wood-work which has been painted white can be cleansed by washing with hot water with a thoroughness which is never the case with unpainted wood. Rough, unplaned wood should never be used, as the roughnesses are sure to be filled with a growth of *Clathrocystis* in course of time. Everything should be smooth and painted, so as to give as little possible chance for the lodgment of foreign matter, and so that washing can be surely and quickly accomplished. Iron or metallic instruments, of course, should be frequently washed and scoured, but I am inclined to think that the cleanliness of these is better cared for than in the case of wood-work.

With regard to the fishing-vessels themselves, apart from the salt which is used (which will be considered presently), not very much can be said. They are not generally exposed to as high a temperature as the wharves and buildings, and in them the *Clathrocystis* does not often develop to a marked extent. What has been said about the painting and scraping of wood-work applies, however, with practical modifications, to vessels, but, of course, at sea one cannot be as neat as on land.

There remains the important question with relation to the salt used. As I have before said, I have carefully examined specimens of Cadiz and Trepani salt, and I have no reason to suppose that the specimens examined were other than fair samples of what are in general use by fishermen. Microscopic examination shows conclusively to my mind that the Trepani is more free from impurities, and that the Cadiz salt contains a decided amount of the *Clathrocystis*, which, when communicated to the fish, is so detrimental to its sale. Judging from the examination which I have made, I should certainly advise the use of Trepani salt as less likely to produce the redness in the fish themselves. I have no means of ascertaining how the amount annually saved by using Cadiz salt instead of Trepani compares with the amount annually lost by the "red fish." If it is the case that more is saved by the use of Cadiz salt than is lost by the unmarketableness of "red fish," then, of course, it will be useless to advise the use of Trepani salt.

I have delayed transmitting to you my report in the hope that I might learn something concerning the prevalence of "red fish" in Europe, but having made numerous inquiries without obtaining any information having any economical bearing, I present the results at which I have arrived from my own examination of the subject, without being able to add to it the results of the experience of others. The question, after all, is one of dollars and cents, but looking at it abstractedly, as I have been obliged to do, I think that my statement of the cause of the trouble and of the examination of the two kinds of salt most generally used should furnish useful hints to those who, from their occupation, are most directly interested in the matter.

Yours, respectfully,

W. G. FARLOW.

CAMBRIDGE, MASS.,
June 22, 1879.

NOTE.—With regard to the presence of *Clathrocystis roseo-persicina* in salt coming from the Mediterranean, perhaps the following may have some significance: In the Annales des Sciences Naturelles, series 2, vol. 9, p. 112, is an article entitled "Extrait d'un Mémoire de M. F. Dunal, sur les algues qui colorent en rouge certaines eaux des marais salins Méditerranéens." In this article an attempt is made to explain the presence of a red substance in the salt works at Villa Franca. M. Dunal denies that the redness is owing to the remains of the crustacean *Artemia salina*, and maintains that the redness is due to a minute plant, *Protococcus salinus* Dunal, found in the bottom of the tanks. It is not impossible that the *P. salinus* of Dunal may be what is now known as *Clathrocystis roseo-persicinia.* The development of the last-named species has occupied the attention of several botanists and zoölogists, and the reader interested in such matters is referred to Cohn's Beiträge zur Biolagie der Pflanzen, vol. 1, part 3, p. 157, and to an article on "A peach-colored Bacterium," by Prof. E. Ray Lankaster, in the Quarterly Journal of Microscopical Science, vol. 13, new series, p. 408, and to articles by the same writer in subsequent numbers of the same journal.

Besides the *Clathrocystis* which was found on the red codfish at Gloucester, another form of microscopic plant was observed, which deserves at least a passing notice. Small colonies of cells, destitute of coloring matter and arranged in fours, were not unfrequent on the infected codfish. The absence of color and the arrangement of cells in fours at once suggests the genus *Sarcina*, of which *S. ventriculi* is found in the fluids vomited in certain diseases of the stomach, in the lungs, and occasionally in other tissues. The species in question, however, differs materially from *S. ventriculi.* The individual cells are larger and the colonies are irregular in outline and not arranged in regular cubes as in *S. ventriculi*, nor does the membrane inclosing the cells contain any silicate, as is said to be the case in that species. Treated with strong acids, as nitric acid, the cells at once expand and soon disintegrate. On seeing the species on codfish, the first thing that struck me was the strong resemblance which it bore to *Glœocopsa crepidinum* Thuret, except in the absence of coloring matter. The *Glœocopsa* is common on the wood-work of wharves at Gloucester near high-water mark, and it might easily have been communicated to the fish. When growing and in good condition, however, it always has a brownish or yellowish-brown color. The species on codfish was always colorless, and yet it seemed to be alive and in good condition, and I am inclined to reject my first belief that the form was a discolored *Glœocopsa crepidinum*, but think it rather an undescribed species of *Sarcina.* My stay at Gloucester being short, and having other things which demanded my attention, I was unable to make any continued observation on this curious form, which may be described as follows:

SARCINA? MORRHUÆ n. sp. *Cells colorless, cuboidal,* 5-8ᵐ *in diameter, united in fours and surrounded by a thin hyaline envelope. Colonies* 10-20ᵐ *in diameter, formed by division of the cells in three dimensions. Colonies heaped together in irregularly-shaped, lobulated masses.*

HAB.—On putrifying codfish, in company with *Clathrocystis roseo-persicina,* Gloucester, Mass., September, 1878.

ALPHABETICAL INDEX.*

*For a special index to the report on New England Isopoda by Harger, see page 459.

Q.

R.

T.

○